13th edition

MEMMLER'S The HUMAN BODY in Health and Disease

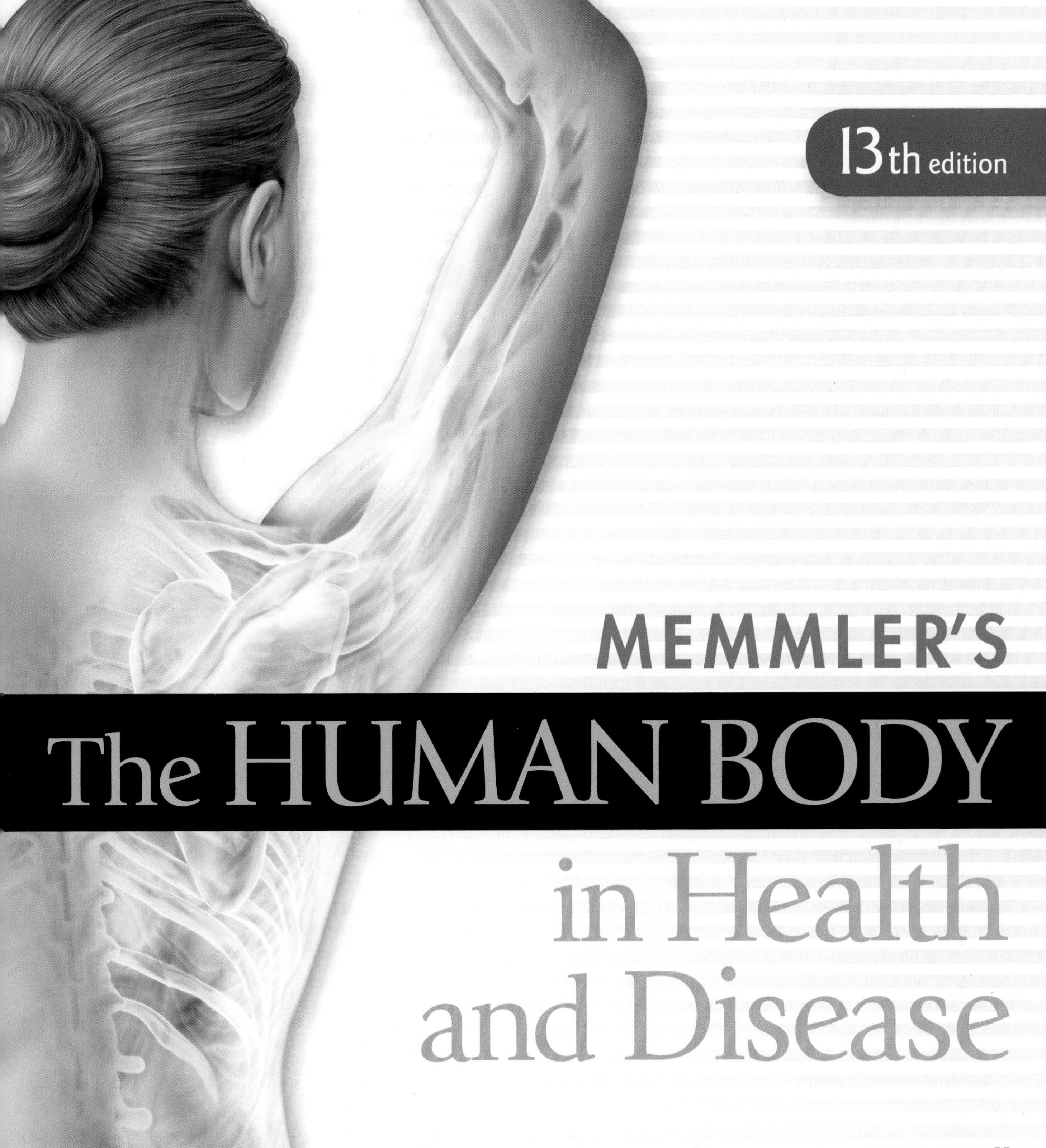

13th edition

MEMMLER'S The HUMAN BODY in Health and Disease

Barbara Janson Cohen • Kerry L. Hull

Wolters Kluwer

Philadelphia • Baltimore • New York • London
Buenos Aires • Hong Kong • Sydney • Tokyo

Senior Acquisitions Editor: Michael Nobel
Product Development Editor: Staci Wolfson
Editorial Assistant: Tish Rogers
Production Project Manager: Marian Bellus
Design Coordinator: Terry Mallon
Illustration Coordinator: Jennifer Clements
Artist: Dragonfly Media Group
Manufacturing Coordinator: Margie Orzech
Prepress Vendor: SPi Global

13th edition

9 8 7 6 5 4 3 2

Printed in China

Library of Congress Cataloging-in-Publication Data

Cohen, Barbara J., author.
Memmler's the human body in health and disease. — 13th edition / Barbara Janson Cohen, Kerry L. Hull.
p. ; cm.
Human body in health and disease
Includes bibliographical references and index.
ISBN 978-1-4511-9280-3 (alk. paper)
I. Hull, Kerry L., author. II. Title. III. Title: Human body in health and disease.
[DNLM: 1. Physiological Phenomena. 2. Anatomy. 3. Pathology. QT 104]
QP34.5
612—dc23

2014021555

LWW.com

Reviewers

We gratefully acknowledge the generous contributions of the reviewers whose names appear in the list that follows.

Beatrice Avila, BS
Instructor
Palo Alto College
San Antonio, Texas

Christi Blair, MSN, RN
Faculty
Holmes Community College
Goodman, Mississippi

Ernema Boettner, RN, BSN
Instructor
Northwest Technical School
Maryville, Missouri

Beth Gagnon, MSN
Nursing Instructor
Wayne County Schools Career Center
Barberton, Ohio

Michael Hawkes, MEd, RT (R) (ARRT)
Program Director
Pima Medical Institute – Mesa Campus
Mesa, Arizona

Camille Humphreys, BS
Program Director
AmeriTech College
Provo, Utah

Joshua Kramer, MBA
Program Director
Arizona College
Mesa, Arizona

Constance Lieseke, AAS
Faculty and Program Coordinator
Olympic College
Port Orchard, Washington

Maryagnes Luczak, BS, MA
Training Director
Career Training Academy
Pittsburgh, Pennsylvania

Patty Bostwick Taylor, Masters
Instructor
Florence Darlington Technical College
Florence, South Carolina

Robyn Wilhelm, PT, DPT
Assistant Professor
A.T. Still University – Mesa Campus
Mesa, Arizona

Memmler's The Human Body in Health and Disease is a textbook for introductory-level health professions and nursing students who need a basic understanding of anatomy and physiology, the interrelationships between structure and function, and the effects of disease on body systems.

Like preceding editions, the 13th edition remains true to Ruth Memmler's original vision. The features and content specifically meet the needs of those who may be starting their health career preparation with little or no science background. This book's primary goals are

- To provide the essential knowledge of human anatomy, physiology, and the effects of disease at an ideal level of detail and in language that is clear and understandable
- To illustrate the concepts discussed with anatomic art of appropriate detail with accuracy, simplicity, and style that is integrated seamlessly with the narrative
- To incorporate the most recent scientific findings into the fundamental material on which Ruth Memmler's classic text is based
- To include pedagogy designed to enhance interest in and understanding of the concepts presented
- To teach the basic anatomic and medical terminology used in healthcare settings, preparing students to function efficiently in their chosen health careers
- To present an integrated teaching–learning package that includes all of the elements necessary for a successful learning experience

This revision is the direct result of in-depth market feedback solicited to tell us what instructors and students at this level most need. We listened carefully to the feedback, and the results we obtained are integrated into many features of this book and into the ancillary package accompanying it. The text itself has been revised and updated where needed to improve organization of the material and to reflect current scientific thought.

Because visual learning devices are so important to students, this new edition continues to include "The Body Visible," a series of illustrations with labeled transparent overlays of the major body systems described in the text. In addition to being a learning and testing tool, these illustrations provide enrichment and are a valuable general reference.

The 13th edition retains its extensive art program with updated versions of figures from previous editions and many new figures. These features appear in a modified design that makes the content more user-friendly and accessible than ever. Our innovative ancillary package on *thePoint* helps students match their learning styles to a wealth of resources, while the comprehensive package of instructor resources provides instructors with maximum flexibility and efficiency. The online Instructor's Manual describes all of the updates in this new edition and presents teaching and learning strategies for traditional classrooms, flipped classrooms, and online courses.

Organization and Structure

Like previous editions, this 13th edition uses a body systems approach to the study of the normal human body and the effects of disease. The book is divided into seven units, grouping related information and body systems together as follows:

- Unit I, The Body as a Whole (Chapters 1–4), focuses on the body's organization, basic chemistry needed to understand body functions, cells and their functions, and tissues, glands, and membranes.
- Unit II, Disease and the First Line of Defense (Chapters 5 and 6), presents information on disease, organisms that produce disease, and the integumentary system, which is the body's first line of defense against injury and disease.
- Unit III, Movement and Support (Chapters 7 and 8), includes the skeletal and muscular systems.
- Unit IV, Coordination and Control (Chapters 9–12), focuses on the nervous system, the sensory system, and the endocrine system.
- Unit V, Circulation and Body Defense (Chapters 13–17), includes the blood, the heart and heart disease, blood vessels and circulation, the lymphatic system, and the immune system.
- Unit VI, Energy: Supply and Use (Chapters 18–22), includes the respiratory system; the digestive system; metabolism, nutrition, and temperature control; body fluids; and the urinary system.
- Unit VII, Perpetuation of Life (Chapters 23–25), covers the male and female reproductive systems, development and birth, and heredity and hereditary diseases.

The main Glossary defines all the chapters' key terms and many additional terms emphasized in the text. An additional Glossary of Word Parts is a reference tool that not only teaches basic medical and anatomic terminology but also helps students learn to recognize unfamiliar terms. Appendices include a variety of supplementary information that students will find useful as they work with the text, including a photographic Dissection Atlas (Appendix 5) and answers to the Chapter Checkpoint questions and Zooming In illustration questions (Appendix 4) that are found in every chapter.

Pedagogic Features

Every chapter contains pedagogy that has been designed with the health professions and nursing student in mind.

- **Learning Objectives:** Chapter objectives at the start of every chapter help the student organize and prioritize learning.
- **Ancillaries At-A-Glance:** Learning Tools, Learning Resources, and Learning Activities are highlighted in a one-stop overview of the supplemental materials available for the chapter.
- **Disease in Context:** Familiar scenarios transport chapter content into a real-life setting, bringing the information to life for students and showing how disease may affect the body's state of internal balance.
- **A Look Back:** With the exception of Chapter 1, each chapter starts with a brief review of how its content relates to prior chapters.
- **Chapter Checkpoints:** Brief questions at the end of main sections test and reinforce the student's recall of key information in that section. Answers are in Appendix 4.
- **Key Points:** Critical information in figure legends spotlights essential aspects of the illustrations.
- **"Zooming In" questions:** Questions in the figure legends test and reinforce student understanding of concepts depicted in the illustration. Answers are in Appendix 4.
- **Phonetic pronunciations:** Easy-to-learn phonetic pronunciations are spelled out in the narrative, appearing in parentheses directly following many terms—no need for students to understand dictionary-style diacritical marks (see the "Guide to Pronunciation").
- **Special interest boxes:** Each chapter contains special interest boxes focusing on topics that augment chapter content. The book includes five kinds of boxes altogether:
  - **Disease in Context Revisited:** Traces the outcome of the medical story that opens each chapter and shows how the cases relate to material in the chapter and to others in the book.
  - **A Closer Look:** Provides additional in-depth scientific detail on topics in or related to the text.
  - **Clinical Perspectives:** Focuses on diseases and disorders relevant to the chapter, exploring what happens to the body when the normal structure–function relationship breaks down.
  - **Hot Topics:** Focuses on current trends and research, reinforcing the link between anatomy and physiology and related news coverage that students may have seen.
  - **Health Maintenance:** Offers supplementary information on health and wellness issues.
- **Figures:** The art program includes full-color anatomic line art, many new or revised, with a level of detail that matches that of the narrative. Photomicrographs, radiographs, and other scans give students a preview of what they might see in real-world healthcare settings. Supplementary figures are available on the companion website on *thePoint.*
- **Tables:** The numerous tables in this edition summarize key concepts and information in an easy-to-review form. Additional summary tables are available on the companion website on *thePoint.*
- **Color figure and table callouts:** Figure and table numbers appear in color in the narrative, helping students quickly find their place after stopping to look at an illustration or table.
- **Word Anatomy:** This chart defines and illustrates the various word parts that appear in terms within the chapter. The prefixes, roots, and suffixes presented are grouped according to chapter headings so that students can find the relevant text. This learning tool helps students build vocabulary and promotes understanding even of unfamiliar terms based on a knowledge of common word parts.
- **Chapter Wrap-Up:** A graphic outline at the end of each chapter provides a concise overview of chapter content, aiding in study and test preparation.
- **Key Terms:** Selected boldface terms throughout the text are listed at each chapter's end and defined in the book's glossary.
- **Questions for Study and Review:** Study questions are organized hierarchically into three levels. New in this edition, the section includes questions that direct students to "The Body Visible" and the various appendices to promote use of these resources. Question levels include the following:
  - **Building Understanding:** Includes fill-in-the-blank, matching, and multiple choice questions that test factual recall
  - **Understanding Concepts:** Includes short-answer questions (define, describe, compare/contrast) that test and reinforce understanding of concepts
  - **Conceptual Thinking:** Includes short-essay questions that promote critical thinking skills. Included are thought questions related to the Disease in Context case stories.

For Students

Look for callouts throughout the chapters for pertinent supplementary material on the companion website on *thePoint.*

The companion website on *thePoint* includes a practical system that lets students learn faster, remember more, and achieve success. Students consider their unique learning styles then choose from a wealth of resources for each learning style, including animations, a pre-quiz, and 10 different types of online learning activities; an audio glossary; and other supplemental materials such as health professions career information, additional charts and images, and study and test-taking tips and resources. Throughout the textbook, the graphic icon shown above alerts students to pertinent supplementary material.

See the inside front cover of this text for the passcode you will need to gain access to the companion website, and see pages xv–xvii for details about the website and a complete listing of student resources.

Instructor Ancillary Package

All instructor resources are available to approved adopting instructors and can be accessed online at http://thepoint.lww.com/MemmlerHBHD13e

- Instructor's Manual is available online as a PDF.
- Brownstone Test Generator allows you to create customized exams from a bank of questions.
- PowerPoint Presentations use visuals to emphasize the key concepts of each chapter.
- Image Bank includes labels-on and labels-off options.
- Supplemental Image Bank with additional images can be used to enhance class presentations.
- Lesson Plans are organized around the learning objectives and include lecture notes, in-class activities, and assignments, including student activities from the student companion website.
- Answers to "Questions for Study and Review" provide responses to the quiz material found at the end of each chapter in the textbook.
- Strategies for Effective Teaching provide sound, tried-and-true advice for successful instruction in traditional, flipped, and online learning environments.
- WebCT/Blackboard/Angel Cartridge allows easy integration of the ancillary materials into learning management systems.

Instructors also have access to all student ancillary assets, via *thePoint* website.

Guide to Pronunciation

The stressed syllable in each word is shown with capital letters. The vowel pronunciations are as follows:

Any vowel at the end of a syllable is given a long sound, as follows:

a as in say
e as in be
i as in nice
o as in go
u as in true

A vowel followed by a consonant and the letter e (as in rate) also is given a long pronunciation.

Any vowel followed by a consonant receives a short pronunciation, as follows:

a as in absent
e as in end
i as in bin
o as in not
u as in up

The letter *h* may be added to a syllable to make vowel pronunciation short, as in *vanilla* (vah-NIL-ah).

Summary

The 13th edition of *Memmler's The Human Body in Health and Disease* builds on the successes of the previous 12 editions by offering clear, concise narrative into which accurate, aesthetically pleasing anatomic art has been woven. We have made every effort to respond thoughtfully and thoroughly to reviewers' and instructors' comments, offering the ideal level of detail for students preparing for careers in the health professions and nursing and the pedagogic features that best support them. With the online resources, we have provided students with an integrated system for understanding and using their unique learning styles—and ultimately succeeding in the course. We hope you will agree that the 13th edition of *Memmler's* suits your educational needs.

User's Guide

For today's health careers, a thorough understanding of human anatomy and physiology is more important than ever. Memmler's *The Human Body in Health and Disease,* 13th edition not only provides the conceptual knowledge you'll need but also teaches you how to apply it. This User's Guide introduces you to the features and tools that will help you succeed as you work through the materials.

Your journey begins with your textbook, ***Memmler's The Human Body in Health and Disease***. Newly updated and fully illustrated, this easy-to-use textbook is filled with resources and activities that will enhance your personal learning style.

- **Disease in Context** provides an interesting case story that uses a familiar, real-life scenario to illustrate key concepts in anatomy and physiology. Later in the chapter, the case story is revisited in more detail—improving your understanding and helping you remember the information.
- **Ancillaries At-A-Glance** highlights the Learning Resources and Learning Activities available for the chapter.
- **Learning Objectives** help you identify learning goals and familiarize yourself with the materials covered in the chapter. These objectives are referenced to page numbers in the text.

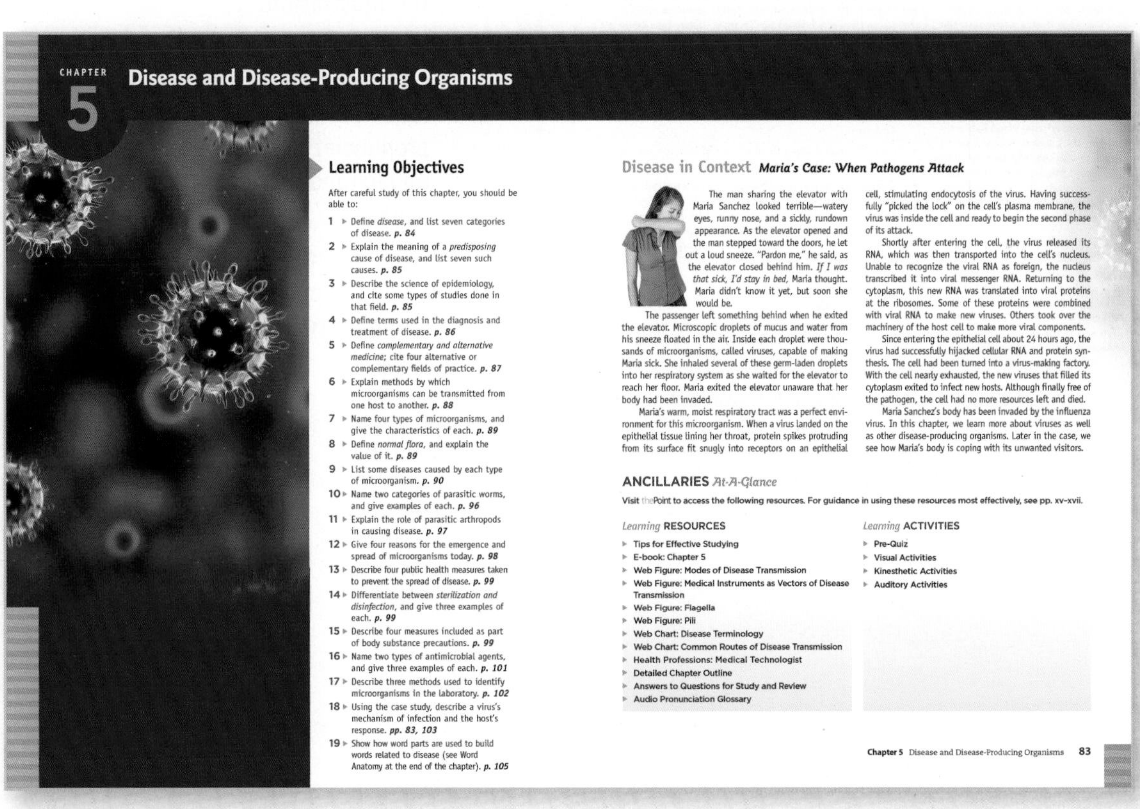

CHAPTER 5 **Disease and Disease-Producing Organisms**

Learning Objectives

After careful study of this chapter, you should be able to:

1 ▸ Define *disease*, and list seven categories of disease. ***p. 84***
2 ▸ Explain the meaning of a *predisposing* cause of disease, and list seven such causes. ***p. 85***
3 ▸ Describe the science of epidemiology, and cite some types of studies done in that field. ***p. 85***
4 ▸ Define terms used in the diagnosis and treatment of disease. ***p. 86***
5 ▸ Define *complementary and alternative medicine*; cite four alternative or complementary fields of practice. ***p. 87***
6 ▸ Explain methods by which microorganisms can be transmitted from one host to another. ***p. 88***
7 ▸ Name four types of microorganisms, and give the characteristics of each. ***p. 89***
8 ▸ Define *normal flora*, and explain the value of it. ***p. 89***
9 ▸ List some diseases caused by each type of microorganism. ***p. 90***
10 ▸ Name two categories of parasitic worms, and give examples of each. ***p. 96***
11 ▸ Explain the role of parasitic arthropods in causing disease. ***p. 97***
12 ▸ Give four reasons for the emergence and spread of microorganisms today. ***p. 98***
13 ▸ Describe four public health measures taken to prevent the spread of disease. ***p. 99***
14 ▸ Differentiate between *sterilization and disinfection*, and give three examples of each. ***p. 99***
15 ▸ Describe four measures included as part of body substance precautions. ***p. 99***
16 ▸ Name two types of antimicrobial agents, and give three examples of each. ***p. 101***
17 ▸ Describe three methods used to identify microorganisms in the laboratory. ***p. 102***
18 ▸ Using the case study, describe a virus's mechanism of infection and the host's response. ***pp. 83, 103***
19 ▸ Show how word parts are used to build words related to disease (see Word Anatomy at the end of the chapter). ***p. 105***

Disease in Context ***Maria's Case: When Pathogens Attack***

The man sharing the elevator with Maria Sanchez looked terrible—watery eyes, runny nose, and a sickly, rundown appearance. As the elevator opened and the man stepped toward the doors, he let out a loud sneeze. "Pardon me," he said, as the elevator closed behind him. *If I was that sick, I'd stay in bed,* Maria thought. Maria didn't know it yet, but soon she would be.

The passenger left something behind when he exited the elevator. Microscopic droplets of mucus and water from his sneeze floated in the air. Inside each droplet were thousands of microorganisms, called viruses, capable of making Maria sick. She inhaled several of these germ-laden droplets into her respiratory system as she waited for the elevator to reach her floor. Maria exited the elevator unaware that her body had been invaded.

Maria's warm, moist respiratory tract was a perfect environment for this microorganism. When a virus landed on the epithelial tissue lining her throat, protein spikes protruding from its surface fit snugly into receptors on an epithelial cell, stimulating endocytosis of the virus. Having successfully "picked the lock" on the cell's plasma membrane, the virus was inside the cell and ready to begin the second phase of its attack.

Shortly after entering the cell, the virus released its RNA, which was then transported into the cell's nucleus. Unable to recognize the viral RNA as foreign, the nucleus transcribed it into viral messenger RNA. Returning to the cytoplasm, this new RNA was translated into viral proteins at the ribosomes. Some of these proteins were combined with viral RNA to make new viruses. Others took over the machinery of the host cell to make more viral components.

Since entering the epithelial cell about 24 hours ago, the virus had successfully hijacked cellular RNA and protein synthesis. The cell had been turned into a virus-making factory. With the cell nearly exhausted, the new viruses that filled its cytoplasm exited to infect new hosts. Although finally free of the pathogen, the cell had no more resources left and died.

Maria Sanchez's body has been invaded by the influenza virus. In this chapter, we learn more about viruses as well as other disease-producing organisms. Later in the case, we see how Maria's body is coping with its unwanted visitors.

ANCILLARIES *At-A-Glance*

Visit thePoint to access the following resources. For guidance in using these resources most effectively, see pp. xv–xvii.

Learning **RESOURCES**

- Tips for Effective Studying
- E-book: Chapter 5
- Web Figure: Modes of Disease Transmission
- Web Figure: Medical Instruments as Vectors of Disease Transmission
- Web Figure: Flagella
- Web Figure: Pili
- Web Chart: Disease Terminology
- Web Chart: Common Routes of Disease Transmission
- Health Professions: Medical Technologist
- Detailed Chapter Outline
- Answers to Questions for Study and Review
- Audio Pronunciation Glossary

Learning **ACTIVITIES**

- Pre-Quiz
- Visual Activities
- Kinesthetic Activities
- Auditory Activities

Chapter 5 Disease and Disease-Producing Organisms 83

A LOOK BACK

The skin is introduced in Chapter 4 as one of the epithelial membranes, the cutaneous (ku-TA-ne-us) *membrane, overlying a connective tissue membrane, the superficial fascia. In this chapter, we describe the skin in much greater detail as it forms the major portion of the integumentary system. This system provides a first line of defense against infectious microorganisms, described in Chapter 5, as well as other harmful agents.*

- **A Look Back** relates each chapter's content to concepts in the preceding chapters.

116 **Unit 2** Disease and the First Line of Defense

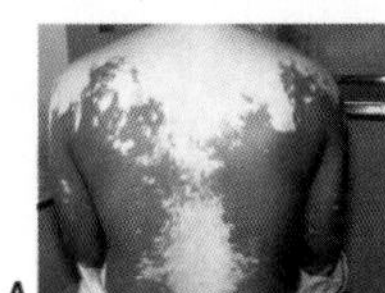

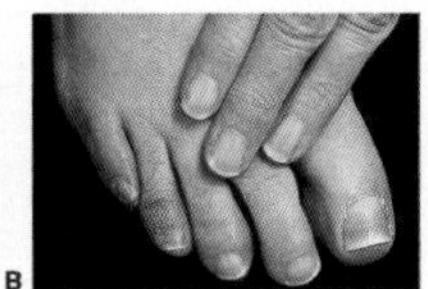

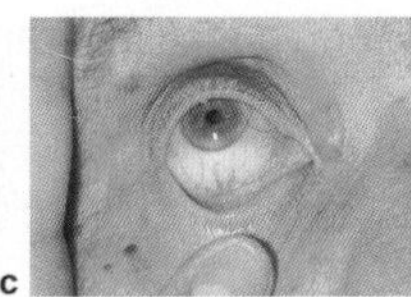

Figure 6-6 **Discoloration of the skin.** **KEY POINT** Changes in skin color can reveal illness. **A.** Vitiligo results from regional defects in melanocyte action. **B.** Cyanosis is a bluish discoloration caused by lack of oxygen. It is seen here in the toes as compared to normal fingertips. **C.** Jaundice is a yellowish discoloration caused by bile pigments in the blood. **ZOOMING IN** What color is associated with cyanosis? What color is associated with jaundice?

is found in the hair, the middle coat of the eyeball, the iris of the eye, and certain tumors. It is common to all races, but darker people have a much larger quantity in their tissues because their melanocytes are more active. The melanin in the skin helps to protect against sunlight's damaging UV radiation. Thus, skin that is exposed to the sun shows a normal increase in this pigment, a response we call tanning.

Sometimes, there are abnormal increases in the quantity of melanin, which may occur either in localized areas or over the entire body surface. For example, diffuse spots of pigmentation may be characteristic of some endocrine disorders. In Addison disease, malfunction of the adrenal gland indirectly stimulates melanocytes, giving an unusual bronze cast to the skin from excess melanin. In contrast, **albinism** (AL-bih-nizm) is a hereditary disorder that impairs melanin production, resulting in a lack of pigment in the skin, hair, and eyes. **Vitiligo** is patchy local blanching of skin to near whiteness, reflecting a regional defect in melanocyte action (Fig. 6-6A).

Hemoglobin Hemoglobin (he-mo-GLO-bin) is the pigment that carries oxygen in red blood cells (further described in Chapters 13 and 18). It gives blood its color and is visible in the skin through vessels in the dermis. **Pallor** (PAL-or) is paleness of the skin, often caused by reduced blood flow or by reduction in hemoglobin, as occurs in cases of anemia. Pallor is most easily noted in the lips, nail beds, and mucous membranes. **Flushing** is diffuse redness caused by increased blood flow to the skin. It is often related to fever and is most noticeable in the face and neck.

When there is not enough oxygen in circulating blood, the skin may take on a bluish discoloration termed **cyanosis** (si-ah-NO-sis) (Fig. 6-6B). This is a symptom of heart failure, breathing problems, such as emphysema, or respiratory obstruction.

Carotene is a skin pigment obtained from carrots and other orange and yellow vegetables. Excessive intake of these vegetables can result in carotene accumulation in blood, a condition known as **carotenemia** (kar-o-te-NE-me-ah) (the suffix *-emia* refers to blood). The excess carotene is deposited in the stratum corneum, resulting in a yellowish red skin discoloration known as carotenoderma.

Bile Pigments A yellowish skin discoloration may be caused by excessive amounts of bile pigments, mainly bilirubin (BIL-ih-ru-bin), in the blood (see Fig. 6-6C). (Bile is a substance produced by the liver that aids in fat digestion; see Chapter 19.) This condition, called **jaundice** (JAWN-dis) (from the French word for "yellow"), may be a symptom of certain disorders, including

- A tumor pressing on the common bile duct or a stone within the duct, either of which would obstruct bile flow into the small intestine
- Inflammation of the liver (hepatitis), commonly caused by a virus
- Certain blood diseases in which red blood cells are rapidly destroyed (hemolyzed)
- Immaturity of the liver. Neonatal (newborn) jaundice occurs when the liver is not yet capable of processing bilirubin. Most such cases correct themselves without treatment in about a week, but this form of jaundice may be treated by exposure to special fluorescent light that helps the body eliminate bilirubin.

CHECKPOINTS

- [ ] 6-10 Name some pigments that give color to the skin.
- [ ] 6-11 What is the term for a bluish skin discoloration caused by insufficient oxygen?

Repair of the Integument

Repair of the integument after injury can occur only in areas that have actively dividing stem cells or cells that can be triggered to divide by injury. These cells are found in the skin's epithelial tissues, and to a lesser extent, connective tissues. Mainly, they are located in the stratum basale of the epidermis and in the hair follicles of the dermis. If both layers of the skin are destroyed along with their stem cells, healing may require skin grafts.

Repair of a skin wound or lesion begins after blood has clotted and an inflammatory response occurs. Blood brings

- **Chapter Checkpoints** pose brief questions at the end of main sections that test and reinforce student recall.
- **Key Points** in the figure captions spotlight essential aspects of the illustrations.
- **"Zooming In" questions** in the figure captions test and reinforce student understanding of concepts depicted in the illustration.
- **Phonetic pronunciations** spelled out in the narrative directly following many terms make learning pronunciation easy—no need to understand dictionary-style diacritical marks.
- **Color figure and table callouts** help students quickly find their place after stopping to look at an illustration or table.

Special interest boxes focus on topics that augment chapter content.

Disease in Context Revisited

Regina's Healing Process

Dr. Stanford told Regina that minor burns covering less than 10% of the total body surface area usually are managed as outpatient visits in a medical office. Regina's burns were assessed at 10%, so she fell into this category according to the American Burn Association Grading System. Regina was given instructions on washing the burned area using sterile technique. She also received a prescription to continue with Silvadene, the topical cream used for preventing wound infection following burns.

Wearing sterile gloves, Regina applied the cream thinly to the burned area twice a day. Strict compliance with the treatment plan was helping to prevent infection and promote healing. Dr. Stanford evaluated Regina's progress when she returned for follow-up appointments. He noted positive results after about two weeks. Scar tissue was beginning to form, and the area remained clean and free from infection.

In this case, we saw how burns are classified and treated. Regina's burns were serious, but careful management prevented the breach in the skin's b... from causing a life-threatening infection.

- **Disease in Context Revisited** boxes provide the outcome of the clinical case story that opens each chapter.

A CLOSER LOOK Box 5-1

The CDC: Making People Safer and Healthier

The CDC in Atlanta, Georgia, is responsible for protecting and improving the health of the American public—at home and abroad. Established in 1946, the CDC has become a world leader in the fight against infectious disease, with an expanded role that now includes control and prevention of chronic diseases such as cancer, heart disease, and stroke. The CDC also works to protect the public from environmental hazards such as waterborne illnesses, weather emergencies, biologic and chemical terrorism, and dangers in the home and workplace. In addition, the CDC provides education to guide informed health and lifestyle decisions. The CDC's stated goal is "healthy people in a healthy world—through prevention." Some of its past accomplishments include the following:

- In the 1950s, the CDC participated in the fight against polio, ... has now been eliminated in industrialized countries ...ost other areas of the world.
- In the 1960s, the CDC joined the WHO in efforts to eradicate smallpox worldwide.
- In the 1970s, it identified the pathogen responsible for Legionnaires disease.
- In the 1980s, it reported the first cases of AIDS and began intensive research on the disease, which continues today.
- In the 1990s, it investigated an outbreak of deadly Ebola virus in Zaire.

Currently, the CDC is working on hundreds of public health issues, including control of food- and waterborne illnesses; tracking emerging diseases, such as new types of influenza; and reduction of the obesity epidemic in America. These professionals are focusing on supporting state and local health departments, improving global health, strengthening surveillance and epidemiology, and reforming health policies. The CDC employs about 8,500 people in state, federal, and foreign locations. They work in more than 170 occupations, including health information, laboratory science, and microbiology.

- **A Closer Look** boxes give in-depth scientific detail on topics in or related to the text.

CLINICAL PERSPECTIVES Box 6-2

Medication Patches: No Bitter Pill to Swallow

For most people, pills are a convenient way to take medication, but for others, they have drawbacks. Pills must be taken at regular intervals to ensure consistent dosing, and they must be digested and absorbed into the bloodstream before they can begin to work. For those who have difficulty swallowing or digesting pills, **transdermal (TD) patches** offer an effective alternative to some oral medications.

TD patches deliver a consistent dose of medication that diffuses at a constant rate through the skin into the bloodstream. There is no daily schedule to follow, nothing to swallow, and no stomach upset. TD patches can also deliver medication to unconscious patients, who would otherwise require intravenous drug delivery. TD patches are used in hormone replacement therapy, to treat heart disease, to manage pain, and to suppress motion sickness. Nicotine patches are also used as part of programs to quit smoking.

TD patches must be used carefully. Drug diffusion through the skin takes time, so it is important to know how long the patch must be in place before it is effective. It is also important to know how long the medication's effects will persist after the patch is removed. Because the body continues to absorb what has already diffused into the skin, removing the patch does not entirely remove the medicine. Also, increased heat may elevate drug absorption to dangerous levels.

A recent advance in TD drug delivery is **iontophoresis**. Based on the principle that like charges repel each other, this method uses a mild electric current to move ionic drugs through the skin. A small electrical device attached to the patch uses positive current to "push" positively charged drug molecules through the skin and a negative current to push negatively charged ones. Even though very low levels of electricity are used, people with pacemakers should not use iontophoretic patches. Anot... vantage is that they can move only ionic drug... the skin.

- **Clinical Perspectives** boxes focus on diseases and disorders relevant to the chapter, exploring what happens to the body when the normal structure–function relationship breaks down.

HOT TOPICS Box 11-1

Eye Surgery: A Glimpse of the Cutting Edge

Cataracts, glaucoma, and refractive errors are the most common eye disorders affecting Americans. In the past, cataract and glaucoma treatments concentrated on managing the diseases. Refractive errors were corrected using eyeglasses and, more recently, contact lenses. Today, laser and microsurgical techniques can remove cataracts, reduce glaucoma, and allow people with refractive errors to put their eyeglasses and contacts away. These cutting-edge procedures include the following:

- *Laser in situ keratomileusis (LASIK)* to correct refractive errors. During this procedure, a surgeon uses a laser to reshape the cornea so that it refracts light directly onto the retina, rather than in front of or behind it. A microkeratome (surgical knife) ...d to cut a flap in the cornea's outer layer. A computer-...lled laser sculpts the middle layer of the cornea, and ...he flap is replaced. The procedure takes only a few minutes, and patients recover their vision quickly and usually with little postoperative pain.
- *Laser trabeculoplasty* to treat glaucoma. This procedure uses a laser to help drain fluid from the eye and lower intraocular pressure. The laser is aimed at drainage canals located between the cornea and iris and makes several burns that are believed to open the canals and allow fluid to drain better. The procedure is typically painless and takes only a few minutes.
- *Phacoemulsification* to remove cataracts. During this surgical procedure, a very small incision (approximately 3 mm long) is made through the sclera near the cornea's outer edge. An ultrasonic probe is inserted through this opening and into the center of the lens. The probe uses sound waves to emulsify the lens's central core, which is then suctioned out. Then, an artificial lens is permanently implanted in the lens capsule. The procedure is typically painless, although the patient may feel some discomfort for one to two days afterward.

- **Hot Topics** boxes examine current trends and research.

HEALTH MAINTENANCE Box 5-2

The Cold Facts about the Common Cold

Every year, an estimated 1 billion Americans suffer from the symptoms of the common cold—runny nose, sneezing, coughing, and headache. Although most cases are mild and usually last about a week, colds are the leading cause of doctor visits and missed days at work and school.

Colds are caused by a viral infection of the upper respiratory mucous membranes. More than 200 different viruses are known to cause cold symptoms. Most belong to the group known as rhinoviruses (the word root *rhino* means "nose"). Whereas there is some evidence that cold viruses live longer at low temperatures, the incidence of colds is probably higher in winter because people spend more time indoors, increasing the chances that the virus will spread from person to person.

Colds spread primarily from contact with a contaminated surface. When an infected person coughs or sneezes, small droplets of water filled with viral particles are propelled through the air. One unshielded sneeze may spread hundreds of thousands of viral particles several feet. The dry indoor air of winter allows greater airborne spread of cold virus droplets. Depending upon temperature and humidity, these particles may live as long as three to six hours, and others who touch the contaminated surface may pick up the particles on their hands.

To help prevent the transmission of cold viruses:

- Avoid close contact with someone who is sneezing or coughing.
- Wash hands frequently to remove any viral particles you may have picked up.
- Avoid touching or rubbing your eyes, nose, or mouth with contaminated hands.
- Clean contaminated surfaces with disinfectant.

Because of the large number of virus types involved in causing colds, and because these germs mutate so rapidly into new forms, medical scientists have not been able to develop an effective cold vaccine. One antiviral drug has been effective against colds, but most "treatments" only ease their symptoms. Antibiotics are of no benefit against these viruses. While treating symptoms with over-the-counter medications may help relieve some discomfort, getting plenty of rest and drinking lots of fluids are the best ways to speed up recovery.

- **Health Maintenance** boxes offer supplementary information on health and wellness issues.

- **Figures:** The art program includes full-color anatomic line art, many new or revised, with a level of detail that matches that of the narrative. Photomicrographs, radiographs, and other scans give students a preview of what they might see in real-world healthcare settings. Supplementary figures are available on the companion website on *thePoint*.

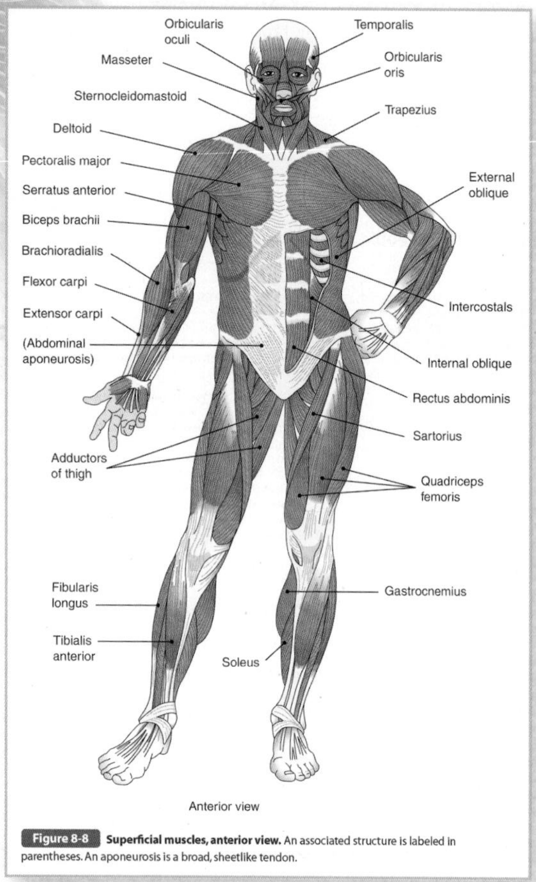

Figure 8-8 **Superficial muscles, anterior view.** An associated structure is labeled in parentheses. An aponeurosis is a broad, sheetlike tendon.

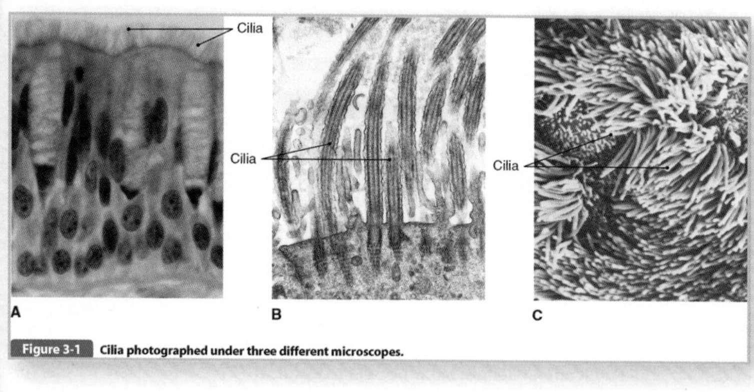

Figure 3-1 **Cilia photographed under three different microscopes.**

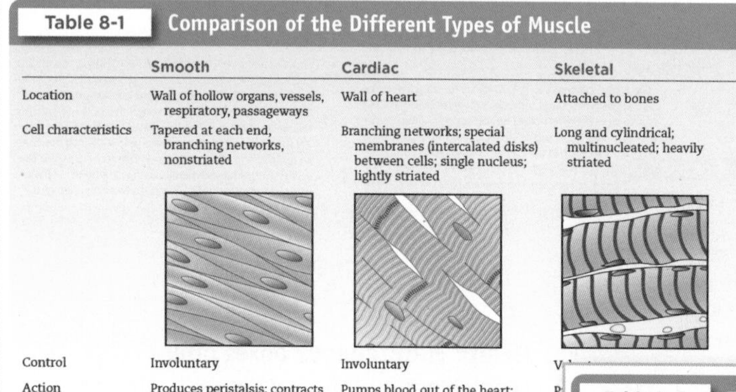

Table 8-1 Comparison of the Different Types of Muscle

	Smooth	Cardiac	Skeletal
Location	Wall of hollow organs, vessels, respiratory, passageways	Wall of heart	Attached to bones
Cell characteristics	Tapered at each end, branching networks, nonstriated	Branching networks; special membranes (intercalated disks) between cells; single nucleus; lightly striated	Long and cylindrical; multinucleated; heavily striated
Control	Involuntary	Involuntary	[illegible]
Action	Produces peristalsis; contracts and relaxes slowly; may sustain contraction	Pumps blood out of the heart; self-excitatory but influenced by nervous system and hormones	[illegible]

- **Tables:** The numerous tables in this edition summarize key concepts and information in an easy-to-review form. Additional summary tables are available on the companion website on *thePoint*.

Table 3-6 The Genetic Code

Amino Acid	Transcribed DNA Triplet	mRNA	tRNA
Glycine	CCC	GGG	CCC
Proline	GGG	CCC	GGG
Valine	CAC	GUG	CAC
Phenylalanine	AAA	UUU	AAA

The nucleotide triplet code in DNA and RNA is shown for four amino acids.

- **Chapter Wrap-Up** at the end of each chapter outlines the chapter content.
- **Key Terms** sections provide a concise list of selected boldface terms used in the chapter and defined in the book's glossary.

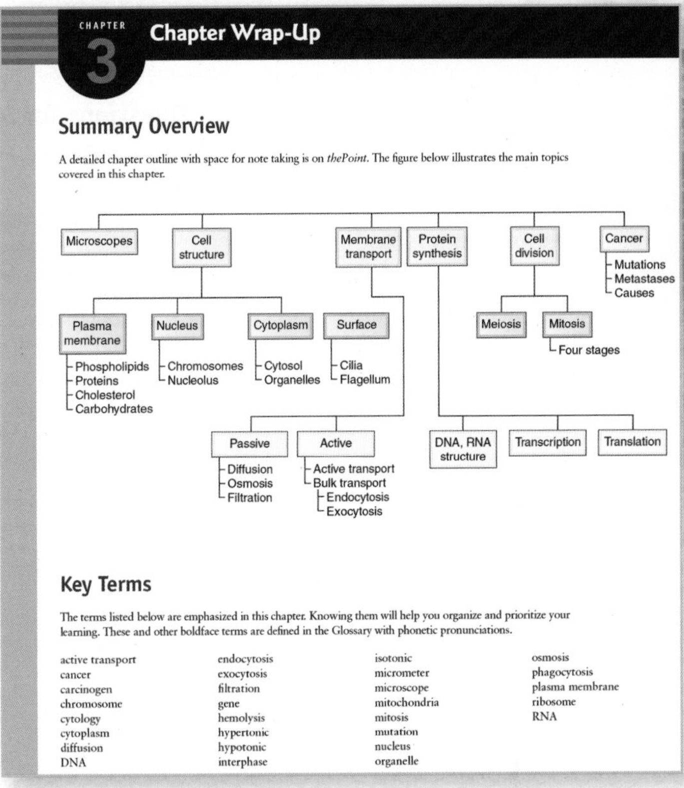

CHAPTER 3 **Chapter Wrap-Up**

Summary Overview

A detailed chapter outline with space for note taking is on *thePoint*. The figure below illustrates the main topics covered in this chapter.

Key Terms

The terms listed below are emphasized in this chapter. Knowing them will help you organize and prioritize your learning. These and other boldface terms are defined in the Glossary with phonetic pronunciations.

active transport
cancer
carcinogen
chromosome
cytology
cytoplasm
diffusion
DNA
endocytosis
exocytosis
filtration
gene
hemolysis
hypertonic
hypotonic
interphase
isotonic
micrometer
microscope
mitochondria
mitosis
mutation
nucleus
organelle
osmosis
phagocytosis
plasma membrane
ribosome
RNA

Word Anatomy

Medical terms are built from standardized word parts (prefixes, roots, and suffixes). Learning the meanings of these parts can help you remember words and interpret unfamiliar terms.

WORD PART	MEANING	EXAMPLE
The Muscular System		
aer/o	air, gas	An *aerobic* organism can grow in the presence of air (oxygen).
an-	not, without	*Anaerobic* metabolism does not require oxygen.
iso-	same, equal	In an *isotonic* contraction, muscle tone remains the same, but the muscle shortens.
-lysis	separation, dissolving	*Glycolysis* is the breakdown of glucose.
metr/o	measure	In an *isometric* contraction, muscle length remains the same, but muscle tension increases.

- **Word Anatomy** defines and illustrates the various word parts that constitute the chapter's specialized terminology, helping to build vocabulary and promote understanding of unfamiliar terms.

- **Questions for Study and Review** sections organize study questions hierarchically into three levels.

Questions for Study and Review

BUILDING UNDERSTANDING

Fill in the Blanks

1. Chemical messengers secreted by the endocrine glands are called ______.
2. The part of the brain that regulates pituitary gland activity is the ______.
3. Red blood cell production in the bone marrow is stimulated by the hormone ______.
4. The main androgen produced by the testes is ______.
5. A hormone produced by the heart is ______.

Matching > Match each numbered item with the most closely related lettered item.

___ **6.** A disorder caused by overproduction of growth hormone in the adult

___ **7.** A disorder caused by underproduction of parathyroid hormone

___ **8.** A disorder caused by overproduction of insulin

___ **9.** A disorder caused by overproduction of growth hormone in a child

___ **10.** A disorder caused by underproduction of antidiuretic hormone

a. hypoglycemia

b. gigantism

c. tetany

d. diabetes insipidus

e. acromegaly

- **Building Understanding:** Includes fill-in-the-blank, matching, and multiple choice questions that test factual recall.

UNDERSTANDING CONCEPTS

16. With regard to regulation, what are the main differences between the nervous system and the endocrine system?
17. Explain how the hypothalamus and pituitary gland regulate certain endocrine glands. Use the thyroid as an example.
18. Name the two divisions of the pituitary gland. List the hormones released from each division, and describe the effects of each.
21. Compare and contrast the following diseases:
    a. Hashimoto thyroiditis and Graves disease
    b. type 1 diabetes and type 2 diabetes
    c. Addison disease and Cushing syndrome
22. Name the hormone released by the kidneys and by the pineal body. What are the effects of each?
23. List several hormones released during stress. What is the relationship between prolonged stress and disease?

- **Understanding Concepts:** Includes short-answer questions (define, describe, compare/contrast) that test and reinforce understanding of ideas. This section now includes questions pertaining to "The Body Visible" and the diverse information in the appendices.

CONCEPTUAL THINKING

26. In the case study Dr. Carter noted that Becky presented with the three cardinal signs of type 1 diabetes mellitus. What are they? What causes them?
27. How is type 1 diabetes mellitus similar to starvation?
28. Mr. Jefferson has rheumatoid arthritis, which is being treated with glucocorticoids. During a recent checkup, his doctor notices that Mr. Jefferson's face is "puffy" and his arms are bruised. Why does the doctor decide to lower his patient's glucocorticoid dosage?

- **Conceptual Thinking:** Includes short-essay questions that promote critical thinking skills. This section includes thought questions related to the Disease in Context case stories.

GETTING STARTED WITH THE STUDENT RESOURCES

Your journey begins with your textbook, *Memmler's Human Body in Health and Disease*, 13th edition. The textbook has icons that guide you to resources and activities designed for your personal learning style.

Look for this icon throughout the book for pertinent supplementary material on the companion website.

thePoint

Here's how to begin:

1. Scratch off the personal access code inside the front cover of your textbook.
2. Log on to **http://thepoint.lww.com/MemmlerHBHD13e**, the companion website for *Memmler's The Human Body in Health and Disease*, 13th edition, on *thePoint*.
3. Click on "Student Resources," and explore the wide variety of auditory, visual, and kinesthetic activities to fit your learning style.

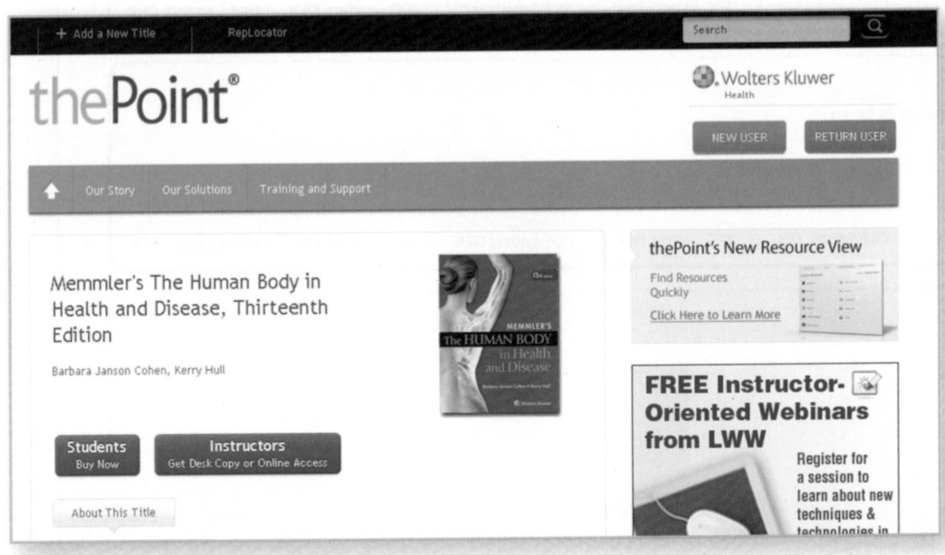

Resources and activities available to instructors include the following:

PowerPoints
Image Bank
Answer Key
Customizable Test Generator
WebCT, Angel, and Blackboard-Ready Cartridges

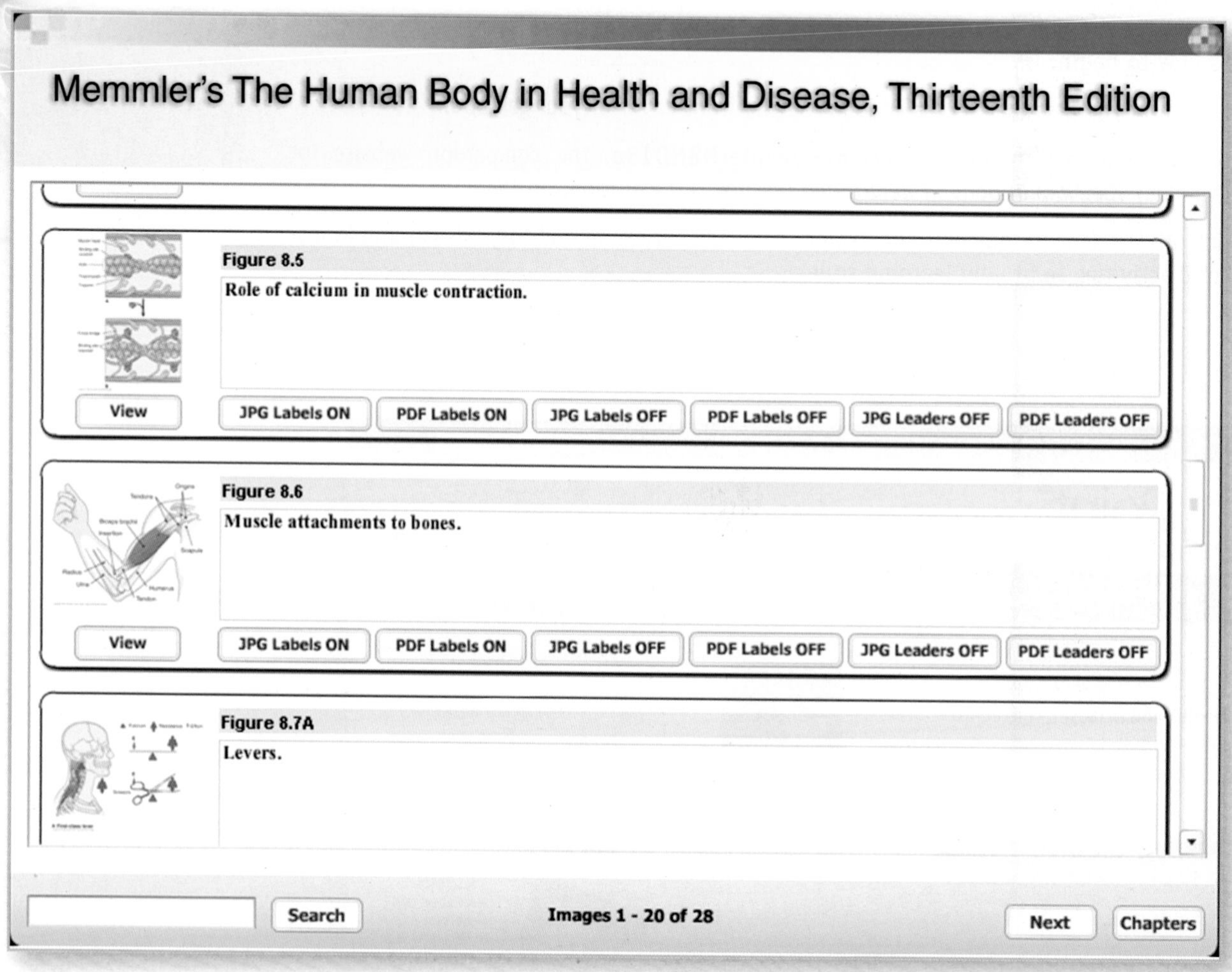

Resources and activities available to students include the following:

- Pre-Quiz
- True or False?
- Key Terms Categories
- Fill-in-the-Blank
- Crossword Puzzle
- Audio Flash Cards
- Word Anatomy
- Look & Label
- Listen & Label
- Zooming In
- Body Building
- Animations
- Supplemental Images
- Audio Pronunciation Glossary
- Health Professions Career Information
- Tips for Effective Studying
- Chapter Outlines and Student Note-Taking Guides

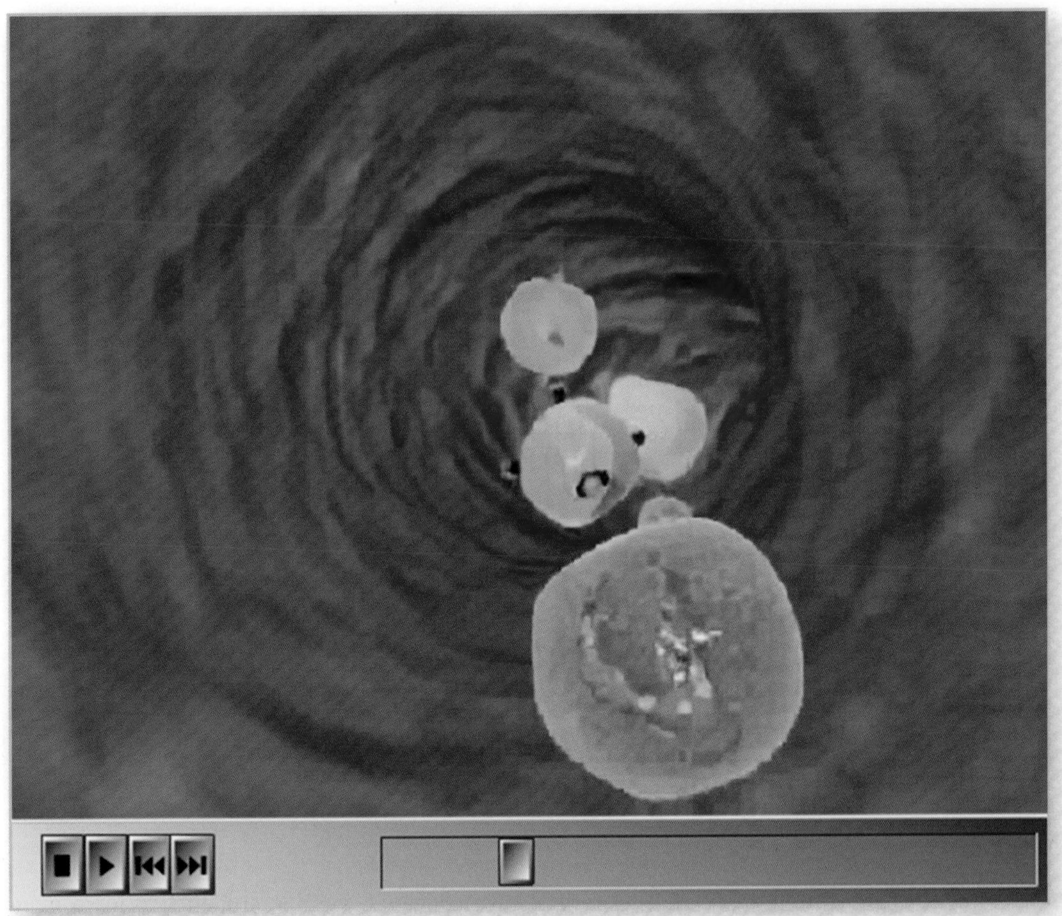

PrepU: AN INTEGRATED ADAPTIVE LEARNING SOLUTION

PrepU, Lippincott's adaptive learning system, is an integral component of *Memmler's The Human Body in Health and Disease*.

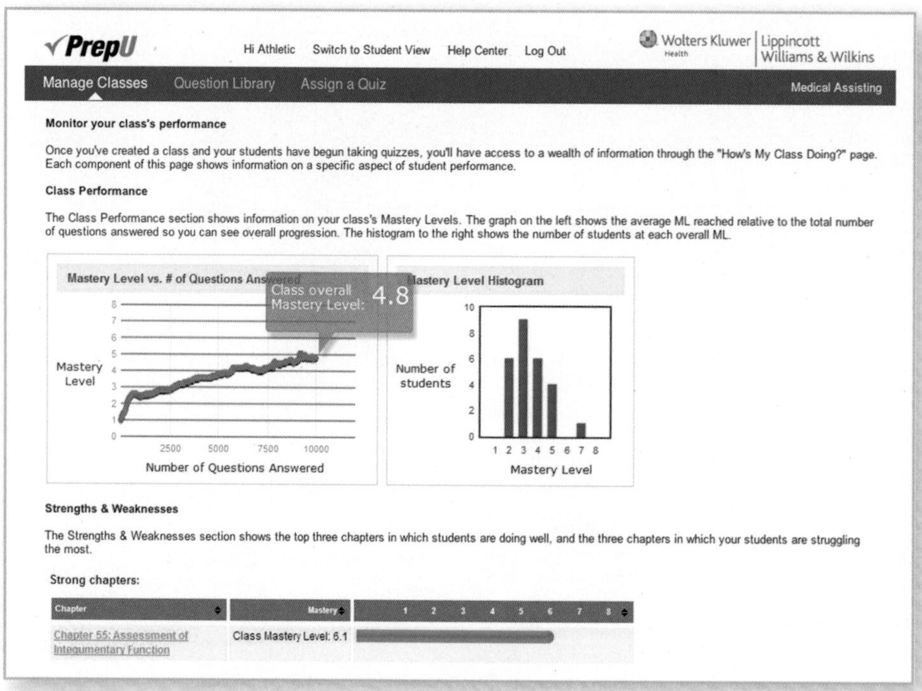

PrepU uses repetitive and adaptive quizzing to build mastery of A&P concepts, helping students to learn more while giving instructors the data they need to monitor each student's progress, strengths, and weaknesses. The hundreds of questions in PrepU offer students the chance to drill themselves on A&P and support their review and retention of the information they have learned. Each question provides not only an explanation for the correct answer but also references of the text page for the student to review the source material. PrepU for *The Human Body in Health and Disease* challenges students with questions and activities that coincide with the materials they have learned in the text and gives students a proven tool to learn A&P more effectively. For instructors, PrepU provides tools to identify areas and topics of student misconception; instructors can use this rich course data to assess students' learning and better target their in-class activities and discussions, while collecting data that are useful for accreditation.

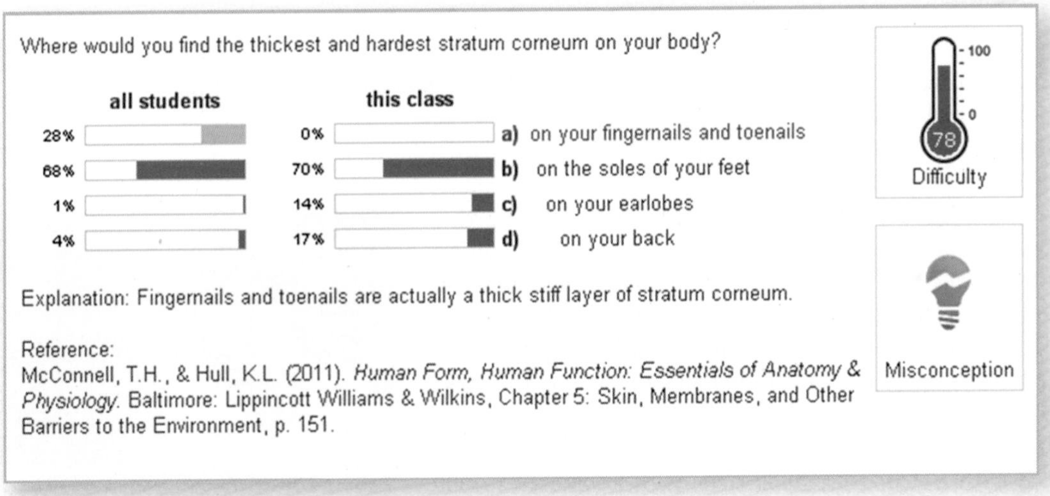

Where would you find the thickest and hardest stratum corneum on your body?

all students	this class	
28%	0%	**a)** on your fingernails and toenails
68%	70%	**b)** on the soles of your feet
1%	14%	**c)** on your earlobes
4%	17%	**d)** on your back

78 Difficulty

Misconception

Explanation: Fingernails and toenails are actually a thick stiff layer of stratum corneum.

Reference:
McConnell, T.H., & Hull, K.L. (2011). *Human Form, Human Function: Essentials of Anatomy & Physiology.* Baltimore: Lippincott Williams & Wilkins, Chapter 5: Skin, Membranes, and Other Barriers to the Environment, p. 151.

A learning experience individualized to each student. Being an adaptive learning engine, PrepU offers questions customized for each student's level of understanding, challenging students at an appropriate pace and difficulty level, while dispelling common misconceptions. As students review and master PrepU's questions, the system automatically increases the difficulty of questions, effectively driving student understanding of A&P to a mastery level. PrepU not only helps students to improve their knowledge but also helps foster their test-taking confidence.

PrepU works! PrepU works, and not just because we say so. PrepU efficacy is *backed by data*:

1. In an introductory nursing course at Central Carolina Technical College, student course outcomes were positively associated with PrepU usage. The students who answered the most PrepU questions in the class also had the best overall course grades.
2. In a randomized, controlled study at UCLA, students using PrepU (for biology) achieved 62% higher learning gains than those who did not.

To see a video explanation of PrepU, go to http://download.lww.com/wolterskluwer_vitalstream_com/mktg/prepuvid/prepupromo01.html

Study Guide for Memmler's The Human Body in Health and Disease, 13th edition
Kerry L. Hull, BSC, PhD
Barbara Janson Cohen, BA, MSEd

Along with the companion website on *thePoint*, this *Study Guide* is the ideal companion to the 13th edition of *Memmler's The Human Body in Health and Disease*. Following the text's organization chapter by chapter, the *Study Guide* provides a full range of self-study aids that actively engage you in learning and enable you to assess and build your knowledge as you advance through the text. Most importantly, the *Study Guide* allows you to get the most out of your study time, with a variety of exercises that meet the needs of all types of learners.

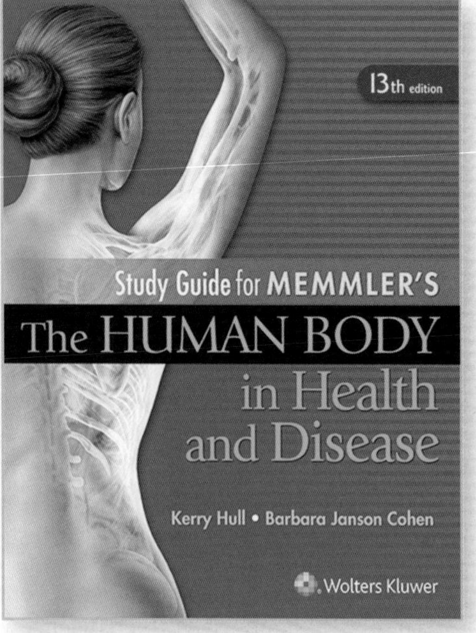

Inside the Study Guide you'll find the following:

- Chapter Overview summarizes the chapter's critical concepts.
- Addressing the Learning Objectives includes labeling, coloring, matching, and short-answer questions, all designed to foster active learning.
- Making the Connections integrates information from each chapter's learning objectives into concept mapping exercises.

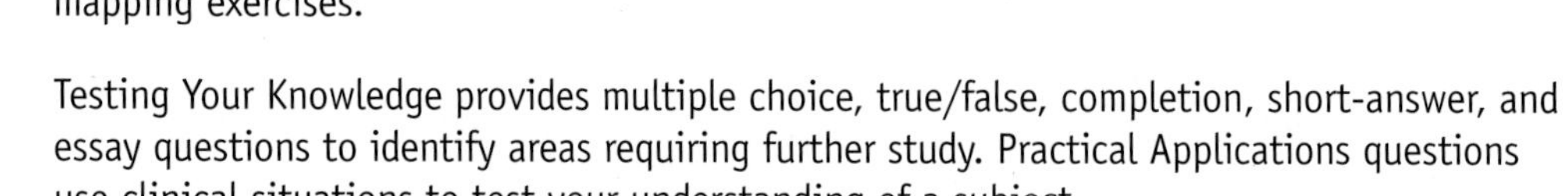

- Testing Your Knowledge provides multiple choice, true/false, completion, short-answer, and essay questions to identify areas requiring further study. Practical Applications questions use clinical situations to test your understanding of a subject.
- Expanding Your Horizons helps you learn from the world around you and highlights emerging issues and discoveries in the health professions.

Visit **www.lww.com**, and reference
ISBN 978-1-4511-9348-0 to order your copy
of this important resource.

Acknowledgments

It is with great satisfaction that we welcome Kerry Hull as coauthor for this 13th edition of *Memmler's The Human Body in Health and Disease*. Kerry has extensive experience with these books, as she has worked with diligence and dedication as a contributing editor for several past editions of the text and author of the ancillary materials. In addition to her knowledge and skills, Kerry has brought talent, patience, and great attention to detail to the preparation of this learning package. I could not have a better coworker or someone more trusted to carry on the traditions of this fine text. I also thank Ann DePetris for reviewing all aspects of the books, for her clinical expertise, and for her contributions to the case studies and ancillary materials.

The skilled staff at Lippincott Williams & Wilkins, as always, has been instrumental in the development of these texts. Consistently striving for improvements and high quality, they have helped achieve the great success of these books over their long history. Specifically, I'd like to acknowledge Michael Nobel, Senior Acquisitions Editor; Staci Wolfson, Product Development Editor, who showed great insights and was always ready with support and suggestions; and Jennifer Clements, Art Director, who assisted ably with art and illustrations.

Ongoing words of appreciation to Dragonfly Media Group, the artists whose skills, knowledge, and imagination have contributed so much to these books.

Thanks to the reviewers, listed separately, who made valuable comments on the text. Their suggestions and insights formed the basis for this revision.

As always, thanks to my husband, Matthew, an instructor in anatomy and physiology, who not only gives consistent support but also contributes advice and suggestions for the text.

—Barbara Janson Cohen

My greatest thanks go to Barbara Cohen, who welcomed me into "Team Memmler" and has served as a mentor for the past 12 years. Thanks to Barbara, I have discovered an entirely new area of scholarship that has proven enormously rewarding. I echo Barbara in her thanks to Michael Nobel, Jennifer Clements, and Staci Wolfson, whose creativity, flexibility, and gentle whip-cracking enabled us to produce the best book and learning package possible. My thanks also go out to the reviewers for their expertise and careful proofreading.

And, finally, my heartfelt thanks to my husband Norm for his support (and cappuccino-making skills), and to my children, Evan and Lauren, for their patience and hugs.

—Kerry L. Hull

Brief Contents

Contents

The Body Visible

The Body Visible is a unique study tool designed to enhance your learning of the body's systems in this course and in your future work.

The Body Visible illustrates the systems discussed in the text in the same sequence in which they appear in the text. Each full-color detailed illustration also contains numbers and lines for identifying the structures in the illustration. A transparent overlay with labels for all of the numbered structures in the art accompanies each image.

With the labels in place, *The Body Visible* allows you to study each illustration and helps you learn the body's structures. When you view each system without the overlay in place, *The Body Visible* becomes a self-testing resource. As you test your knowledge and identify each numbered part, you can easily check your answers with the overlay.

Many of the images in *The Body Visible* have somewhat more detail than is covered in the text. We encourage you to keep *The Body Visible* available as a general reference and as a useful study tool as you progress to more advanced levels in your chosen healthcare career.

*The Body Visible** begins on the next page.

*The images in *The Body Visible* are adapted with permission from Anatomical Chart Company, *Rapid Review: A Guide for Self-Testing and Memorization*, 3rd ed. Philadelphia, PA: Lippincott Williams & Wilkins, 2010.

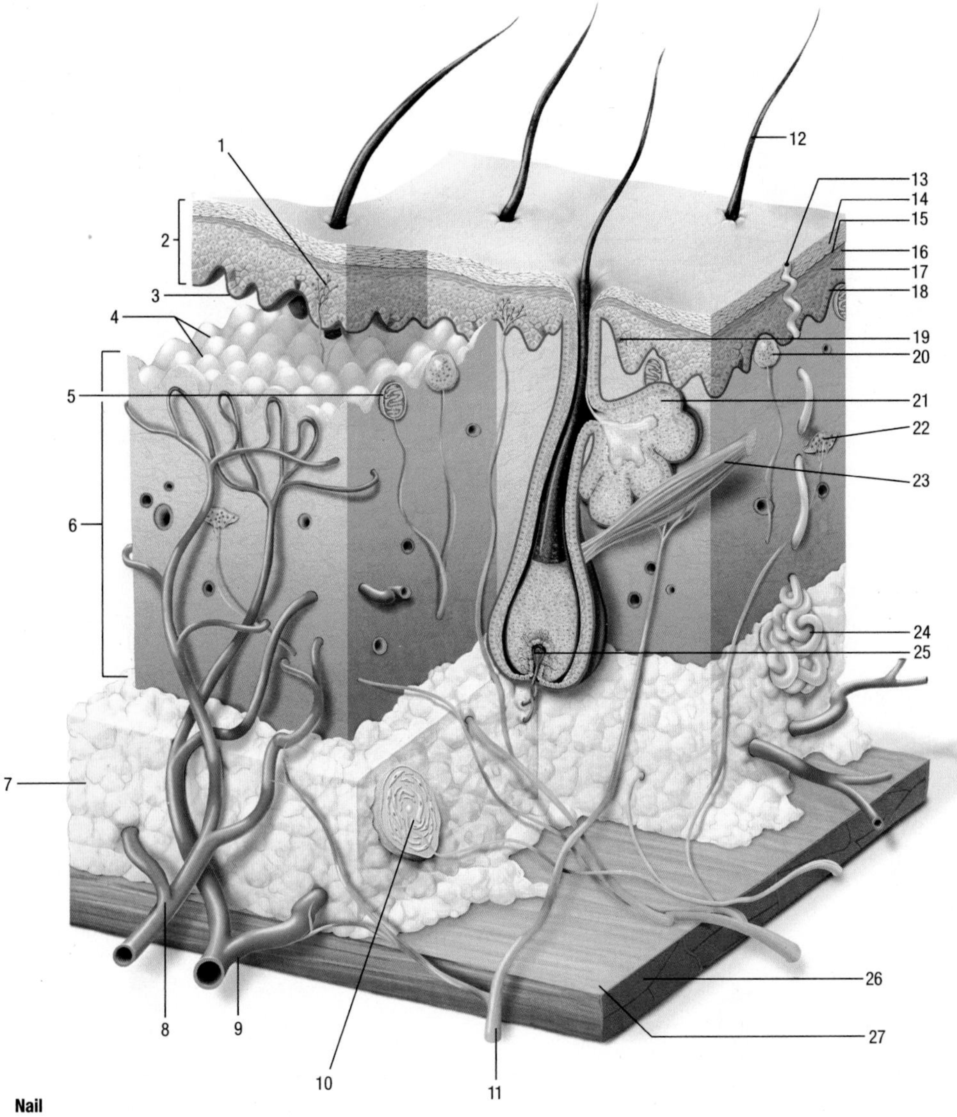

1
2
3
4
5
6
7
8
9
10
11
12
13
14
15
16
17
18
19
20
21
22
23
24
25
26
27

Nail

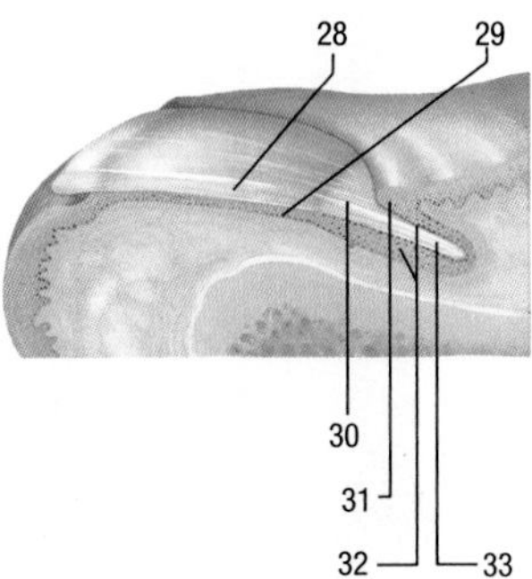

The Skeletal System

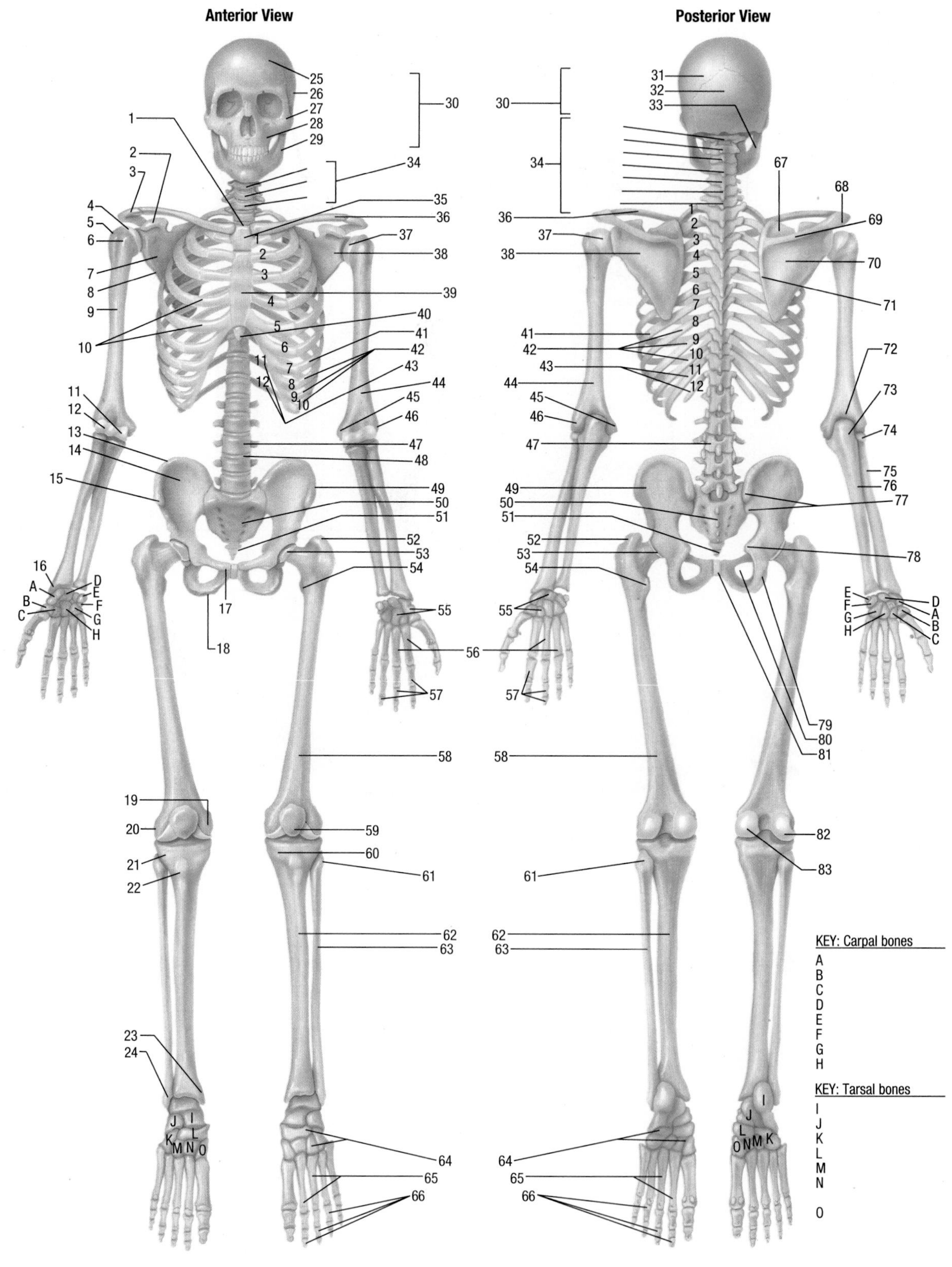

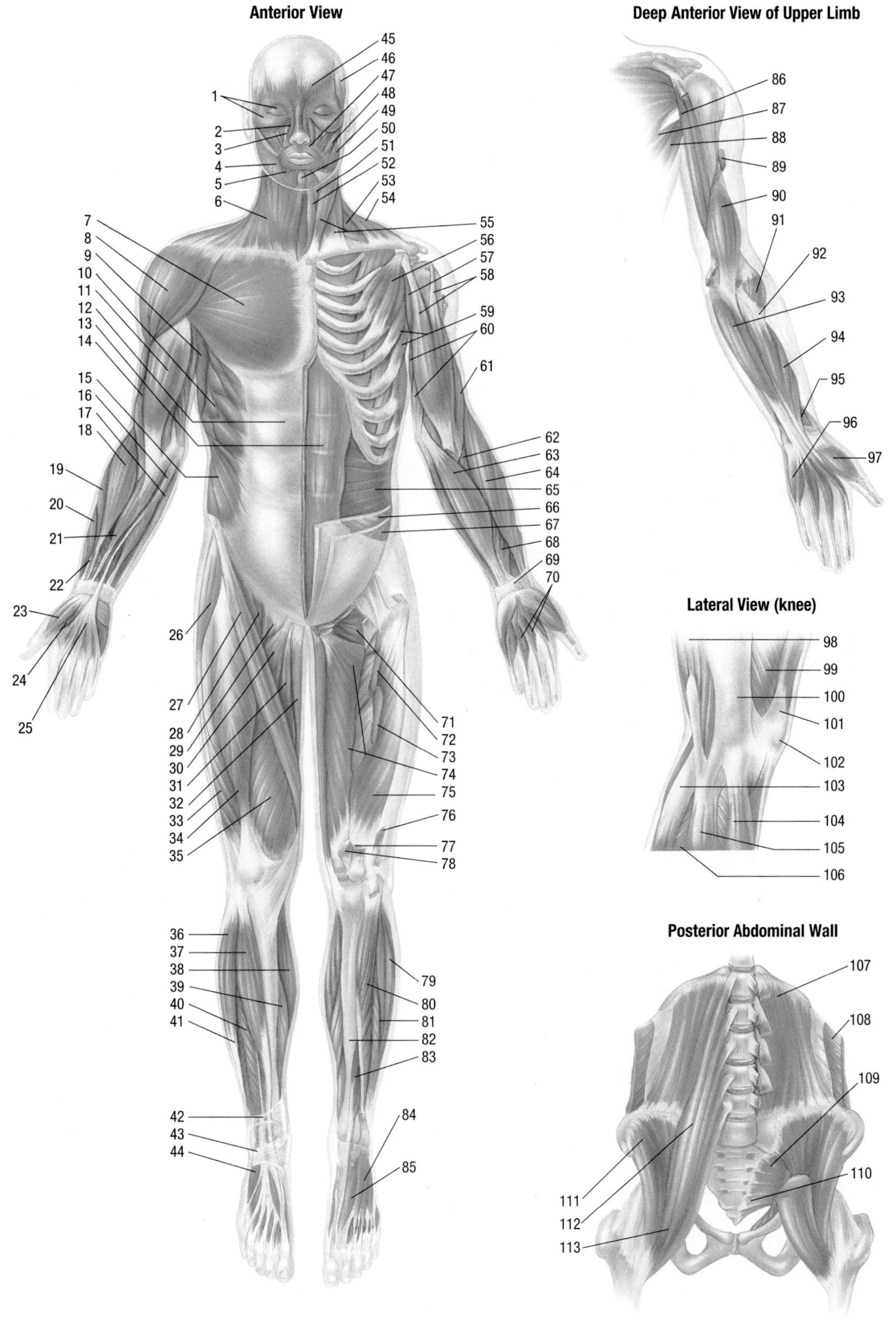
Anterior View
Deep Anterior View of Upper Limb
Lateral View (knee)
Posterior Abdominal Wall

The Muscular System—Posterior View

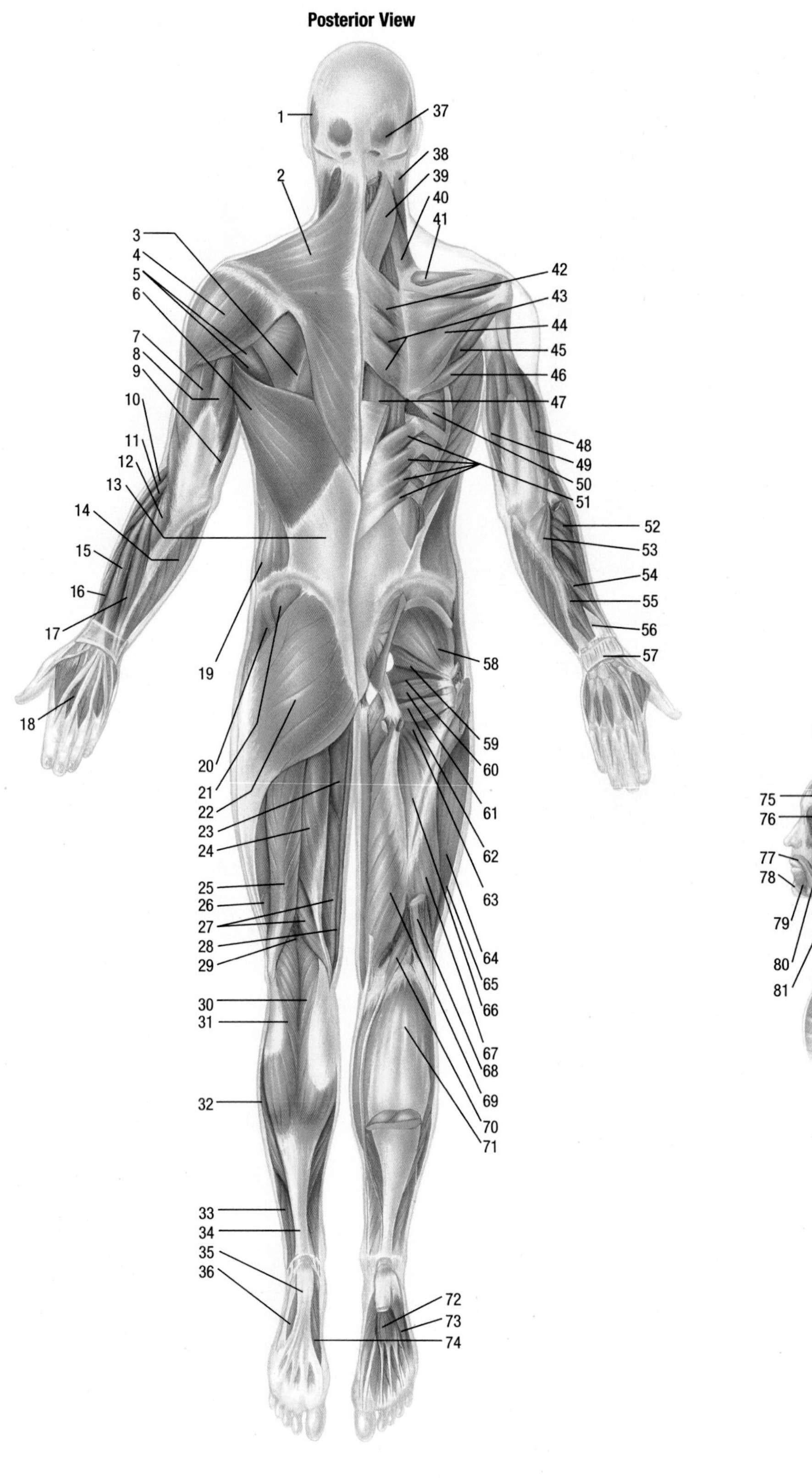

Lateral View
75
76
77
78
79
80
81
82
83
84
85
86
87
88
89
90
91
92
93
94
95
96

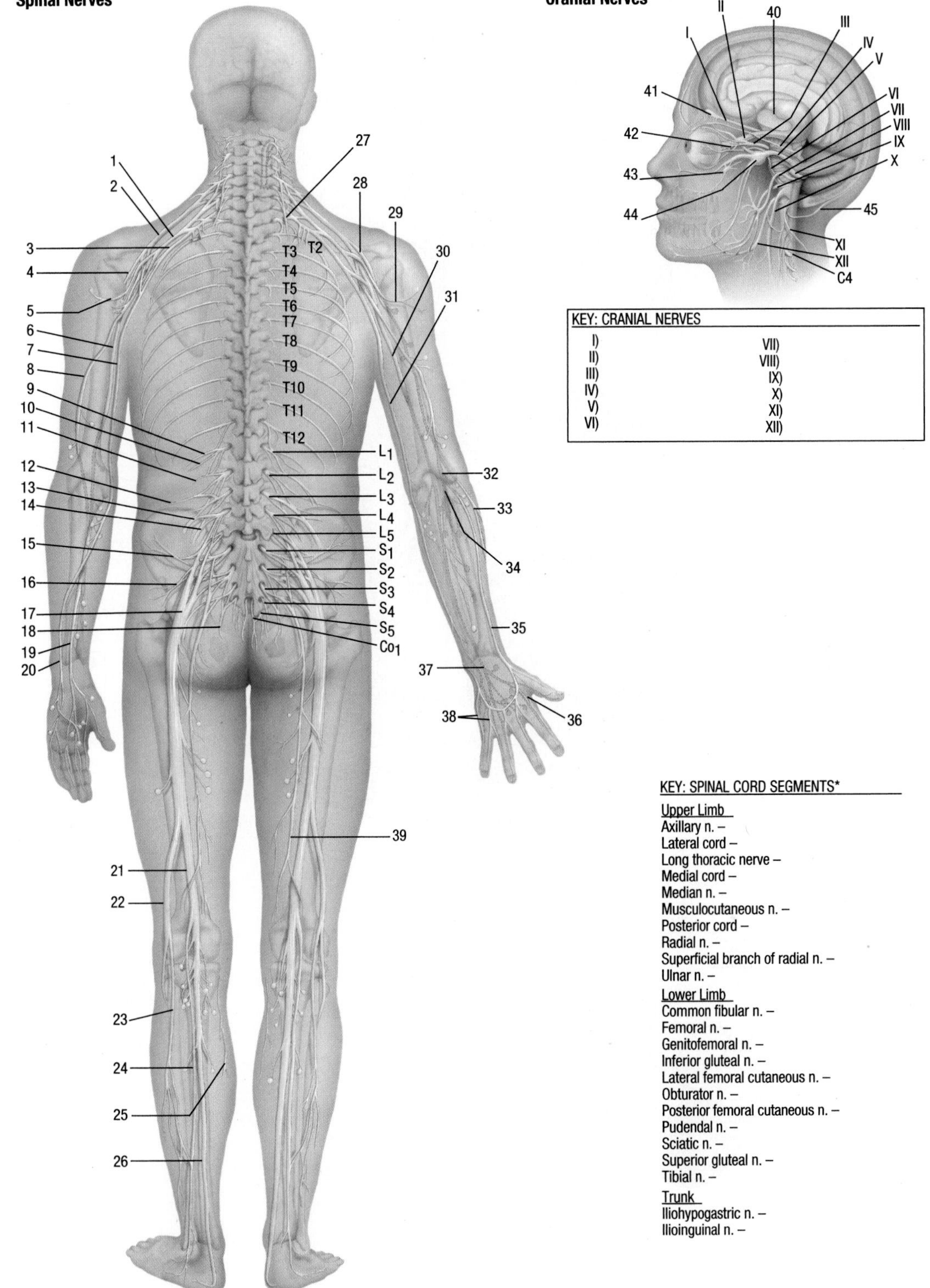

KEY: CRANIAL NERVES

I)
II)
III)
IV)
V)
VI)
VII)
VIII)
IX)
X)
XI)
XII)

KEY: SPINAL CORD SEGMENTS*

Upper Limb
Axillary n. –
Lateral cord –
Long thoracic nerve –
Medial cord –
Median n. –
Musculocutaneous n. –
Posterior cord –
Radial n. –
Superficial branch of radial n. –
Ulnar n. –

Lower Limb
Common fibular n. –
Femoral n. –
Genitofemoral n. –
Inferior gluteal n. –
Lateral femoral cutaneous n. –
Obturator n. –
Posterior femoral cutaneous n. –
Pudendal n. –
Sciatic n. –
Superior gluteal n. –
Tibial n. –

Trunk
Iliohypogastric n. –
Ilioinguinal n. –

The Brain

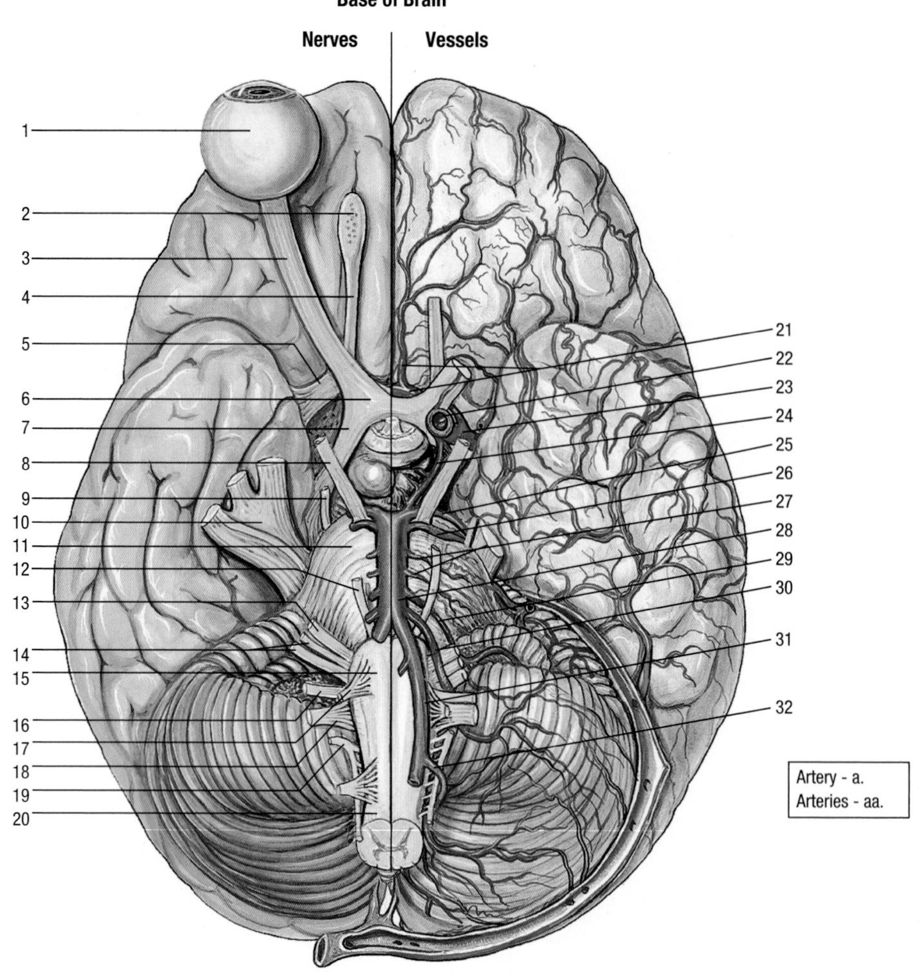
Base of Brain
Nerves
Vessels
1
2
3
4
5
6
7
8
9
10
11
12
13
14
15
16
17
18
19
20
21
22
23
24
25
26
27
28
29
30
31
32
Artery - a.
Arteries - aa.

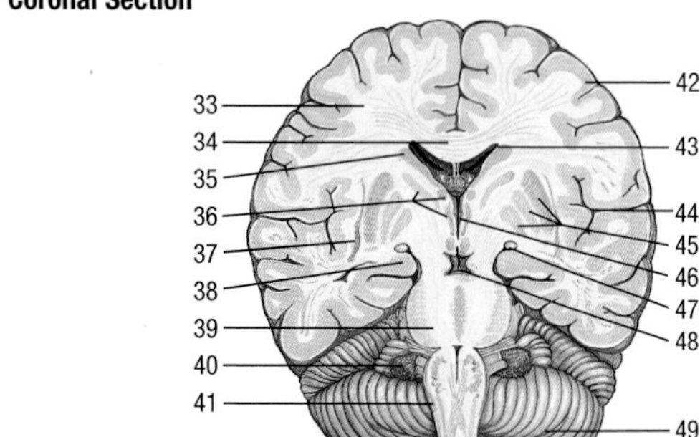
Coronal Section
33
34
35
36
37
38
39
40
41
42
43
44
45
46
47
48
49

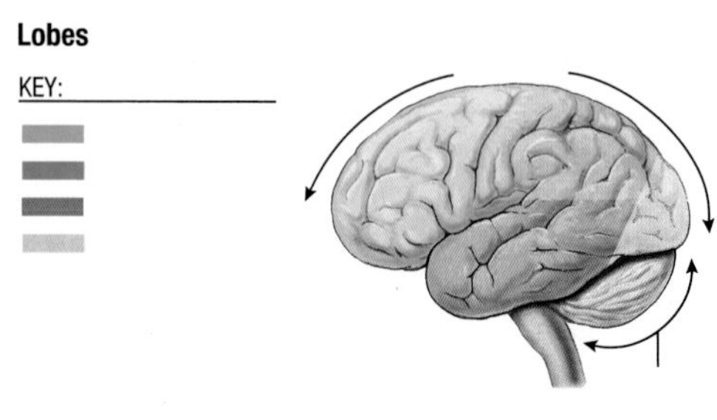
Lobes
KEY:

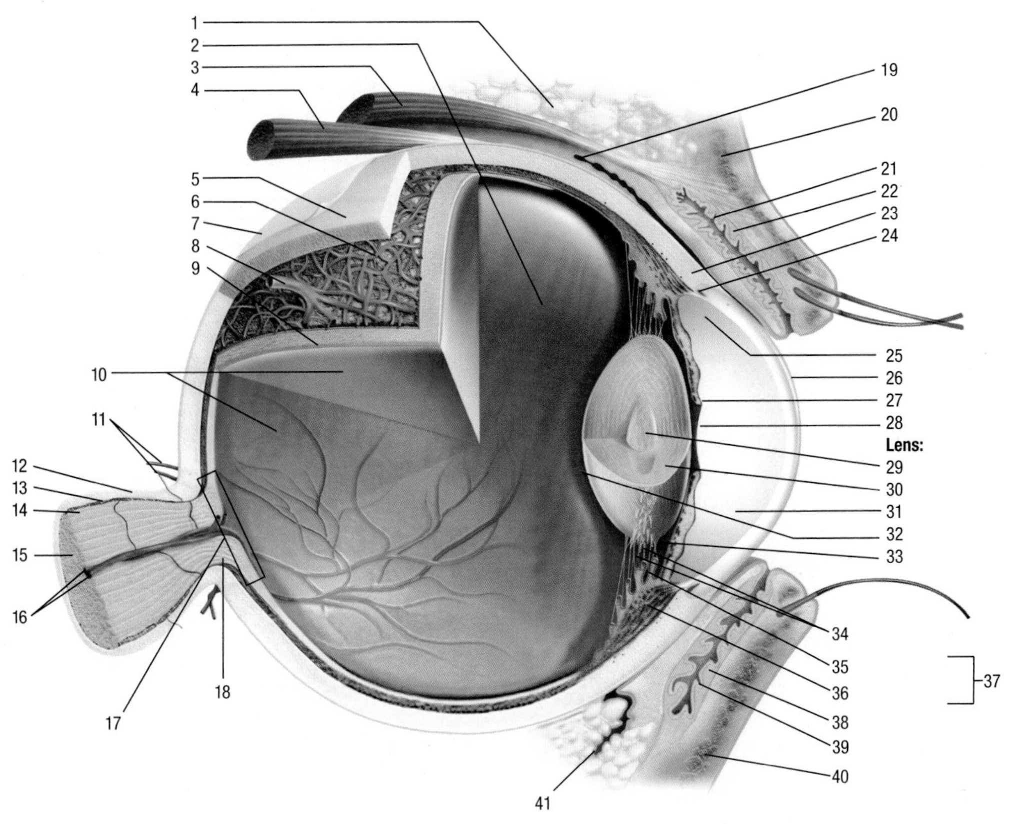

Lacrimal Gland:

42
43
44
45
46
47
48
49
50
51

Eye Muscles Superior View

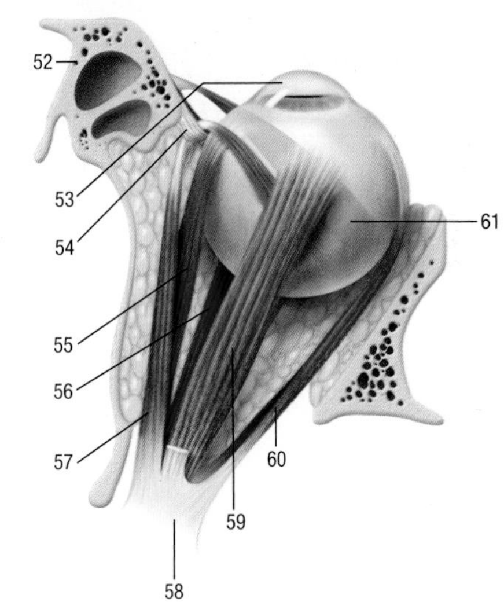

The Ear

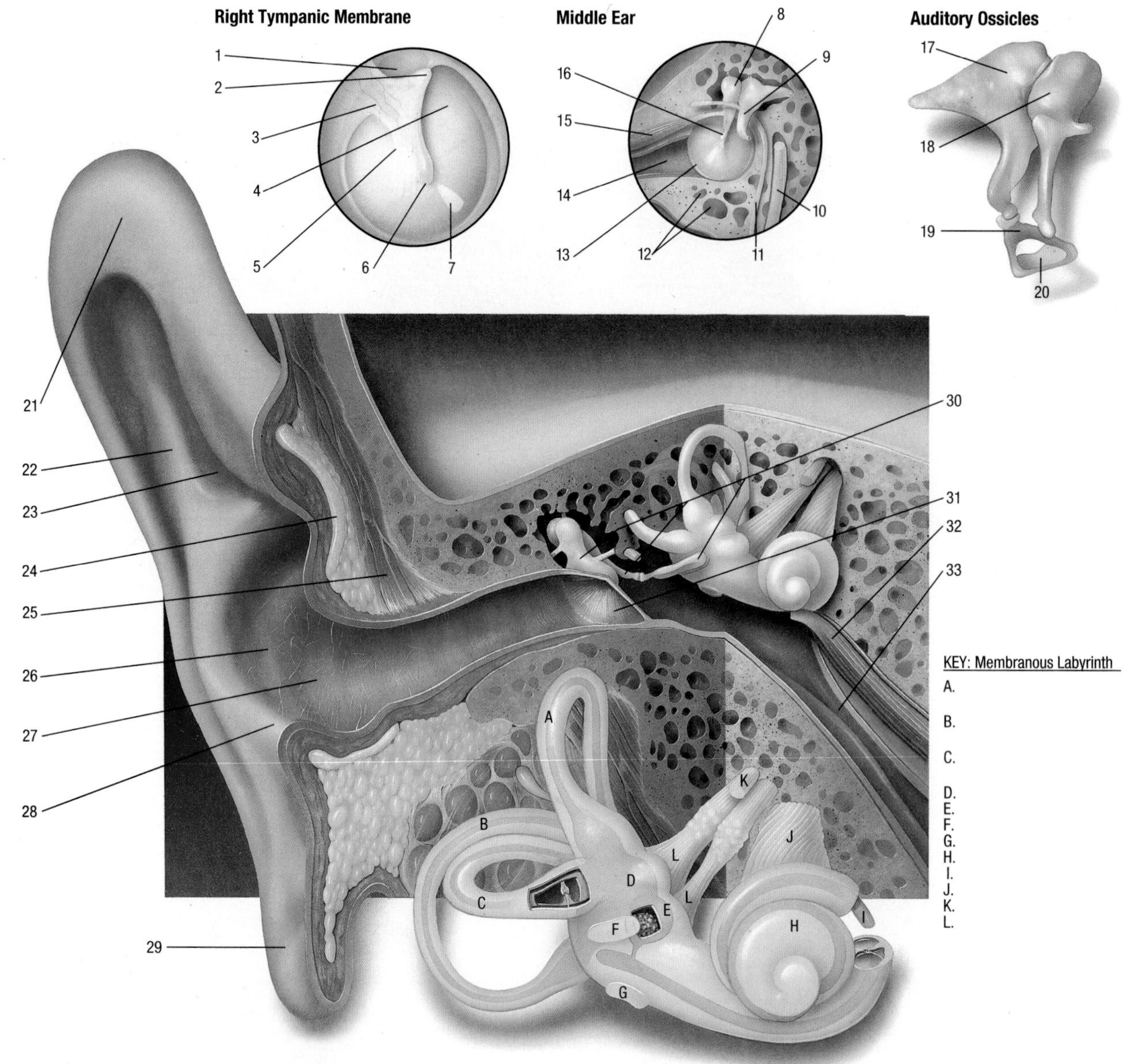

KEY: Membranous Labyrinth

A.

B.

C.

D.

E.

F.

G.

H.

I.

J.

K.

L.

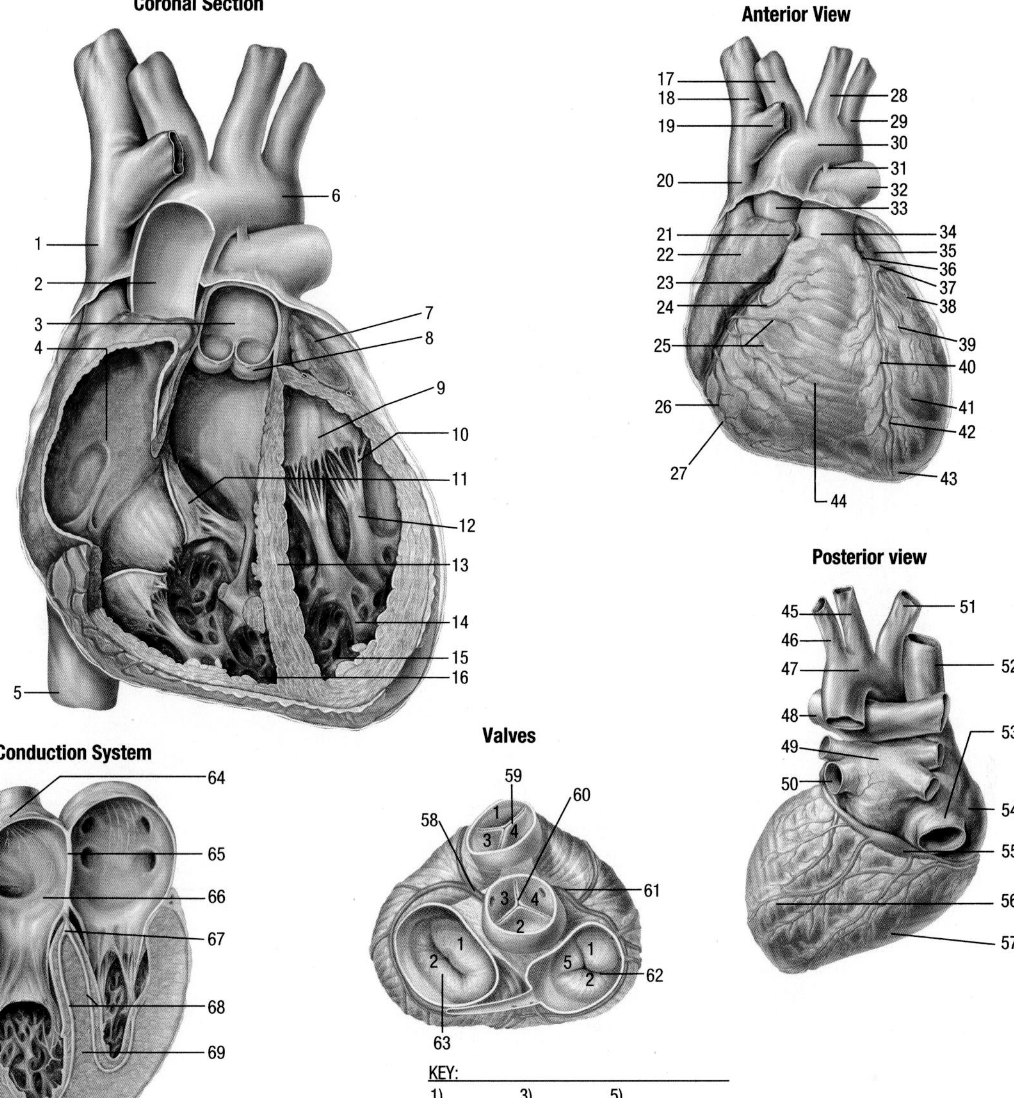

Coronal Section
1
2
3
4
5
6
7
8
9
10
11
12
13
14
15
16
Anterior View
17
18
19
20
21
22
23
24
25
26
27
28
29
30
31
32
33
34
35
36
37
38
39
40
41
42
43
44
Posterior view
45
46
47
48
49
50
51
52
53
54
55
56
57
Conduction System
64
65
66
67
68
69
Valves
58
59
60
61
62
63
KEY:
1)
2)
3)
4)
5)

The Arteries

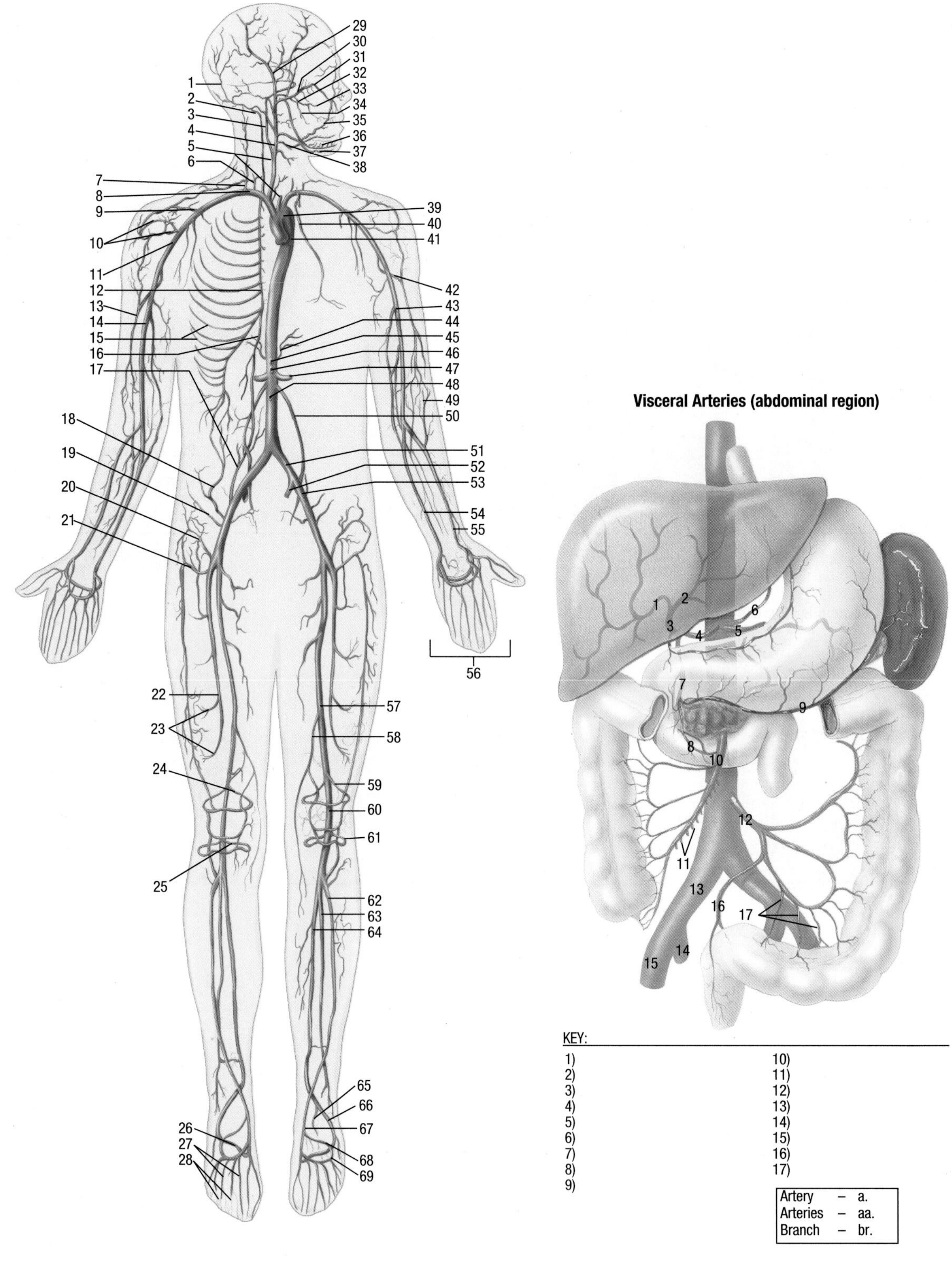

KEY:

1)
2)
3)
4)
5)
6)
7)
8)
9)
10)
11)
12)
13)
14)
15)
16)
17)

Artery	–	a.
Arteries	–	aa.
Branch	–	br.

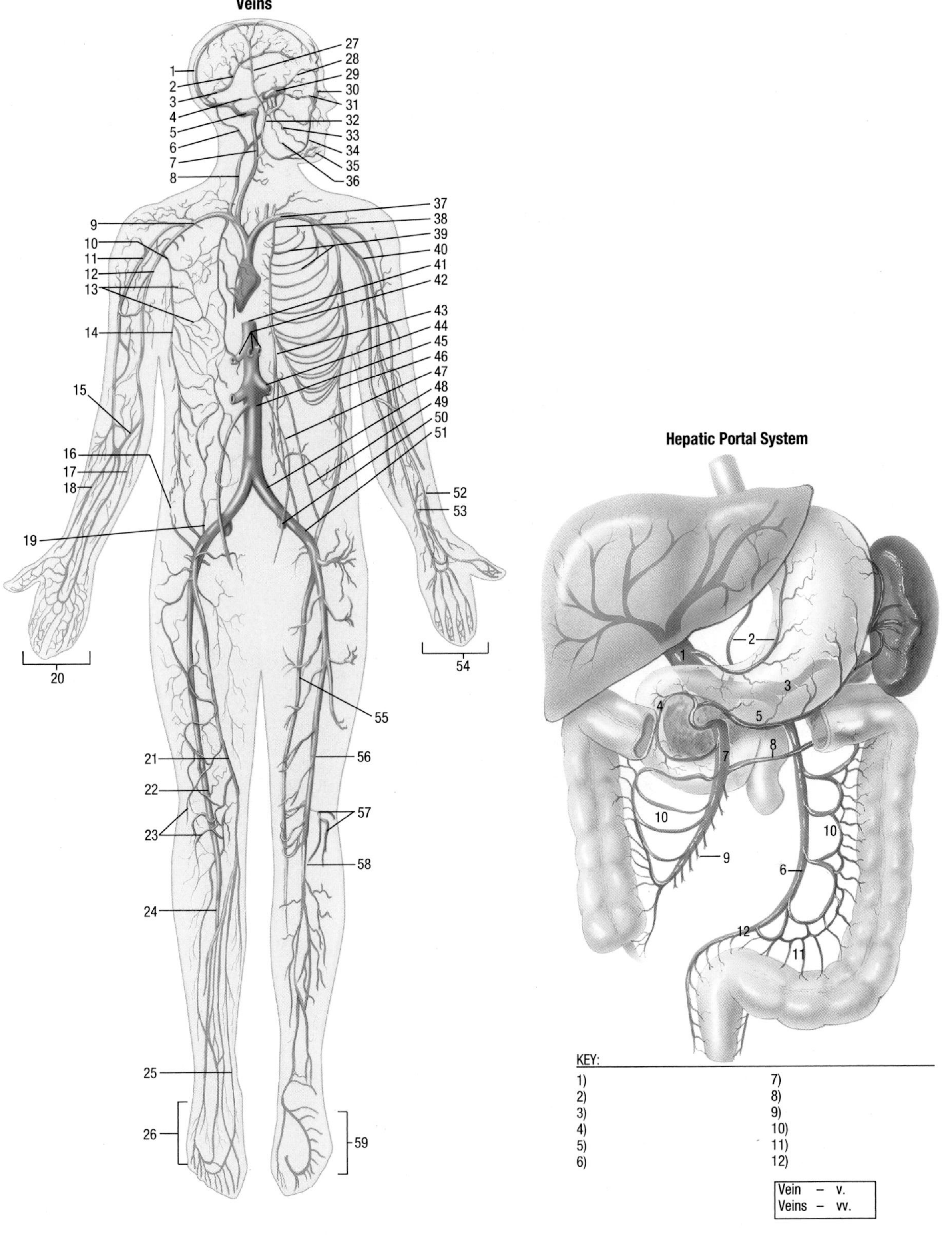
Veins
1
2
3
4
5
6
7
8
9
10
11
12
13
14
15
16
17
18
19
20
21
22
23
24
25
26
27
28
29
30
31
32
33
34
35
36
37
38
39
40
41
42
43
44
45
46
47
48
49
50
51
52
53
54
55
56
57
58
59
Hepatic Portal System
1
2
3
4
5
6
7
8
9
10
11
12
KEY:
1)
2)
3)
4)
5)
6)
7)
8)
9)
10)
11)
12)
Vein – v.
Veins – vv.

The Respiratory System

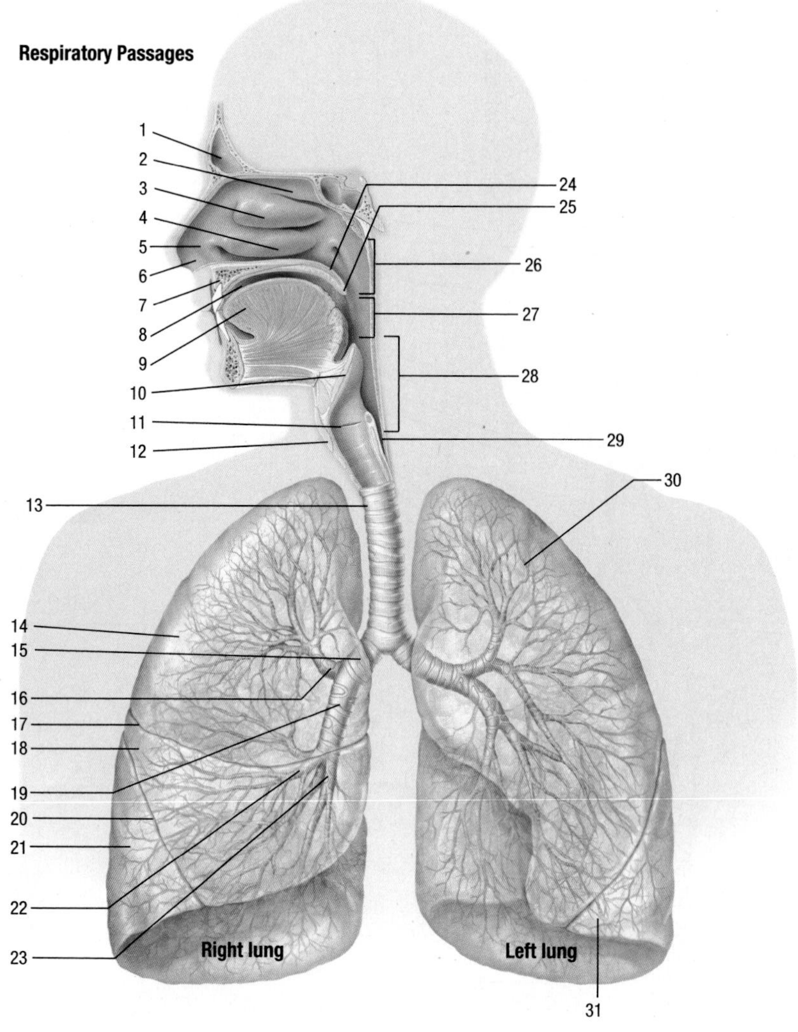

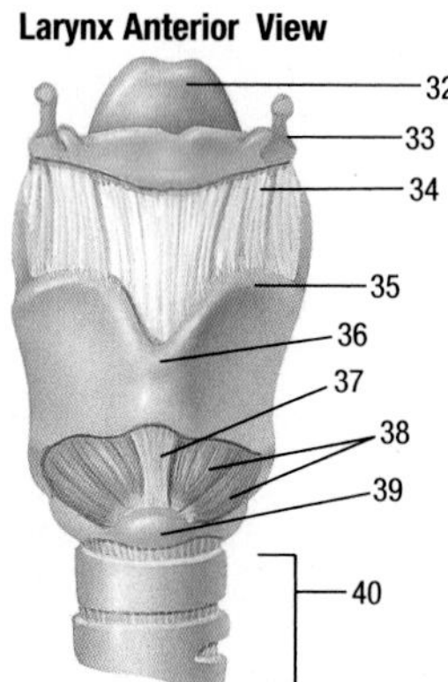

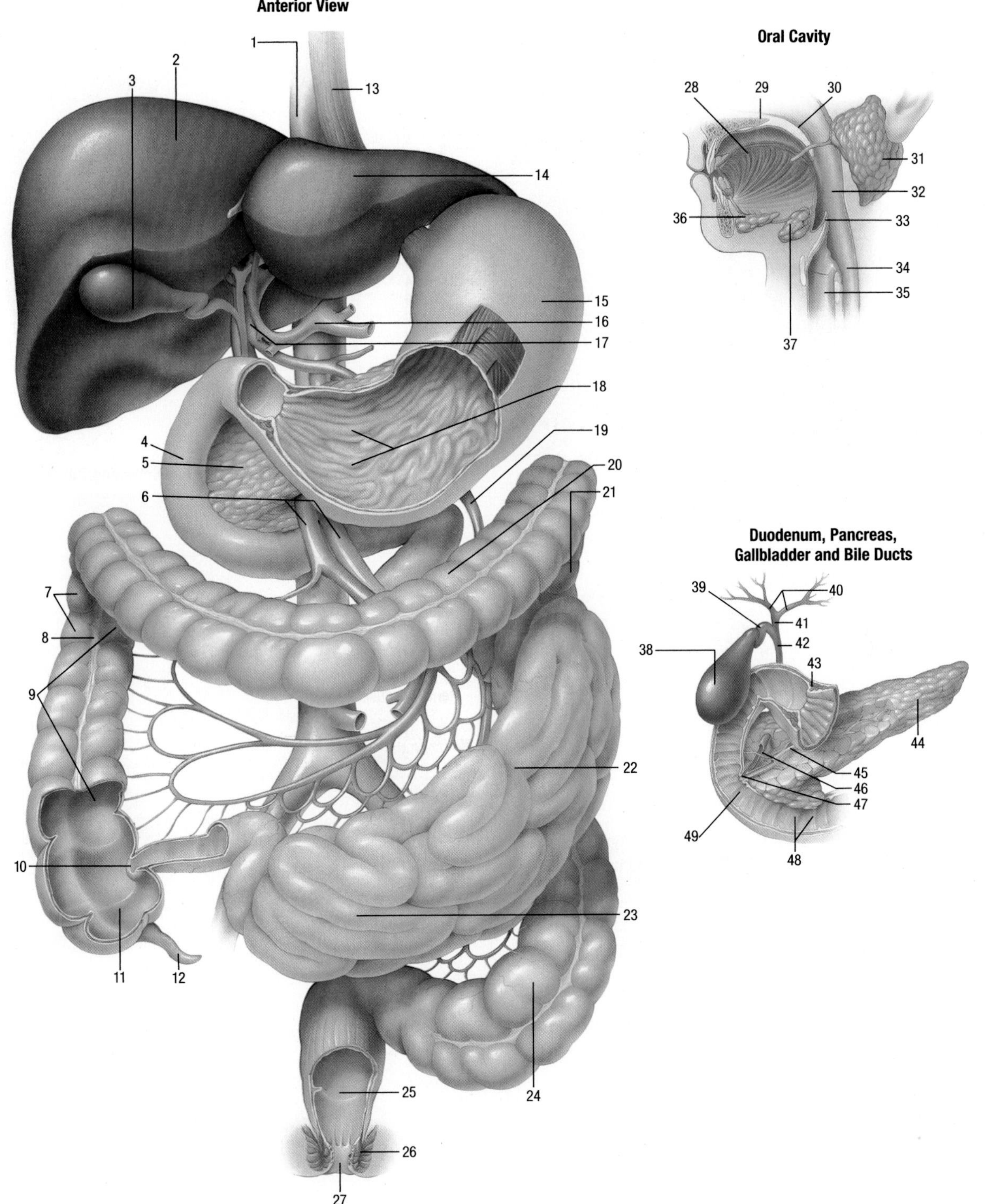
Anterior View
1
2
3
4
5
6
7
8
9
10
11
12
13
14
15
16
17
18
19
20
21
22
23
24
25
26
27
Oral Cavity
28
29
30
31
32
33
34
35
36
37
Duodenum, Pancreas, Gallbladder and Bile Ducts
38
39
40
41
42
43
44
45
46
47
48
49

The Urinary System

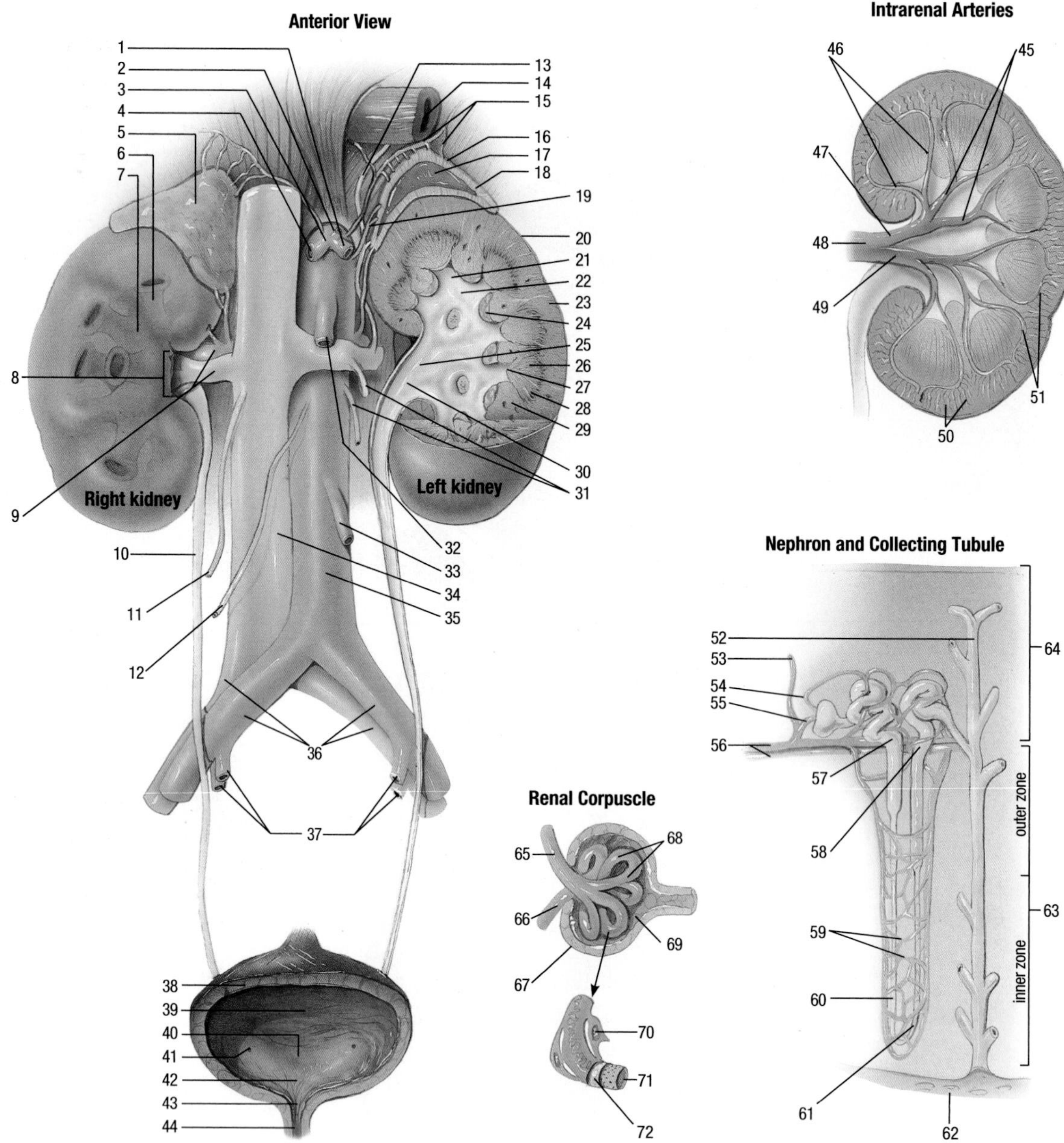

Anterior View
Right kidney
Left kidney
Intrarenal Arteries
Nephron and Collecting Tubule
outer zone
inner zone
Renal Corpuscle
1
2
3
4
5
6
7
8
9
10
11
12
13
14
15
16
17
18
19
20
21
22
23
24
25
26
27
28
29
30
31
32
33
34
35
36
37
38
39
40
41
42
43
44
45
46
47
48
49
50
51
52
53
54
55
56
57
58
59
60
61
62
63
64
65
66
67
68
69
70
71
72

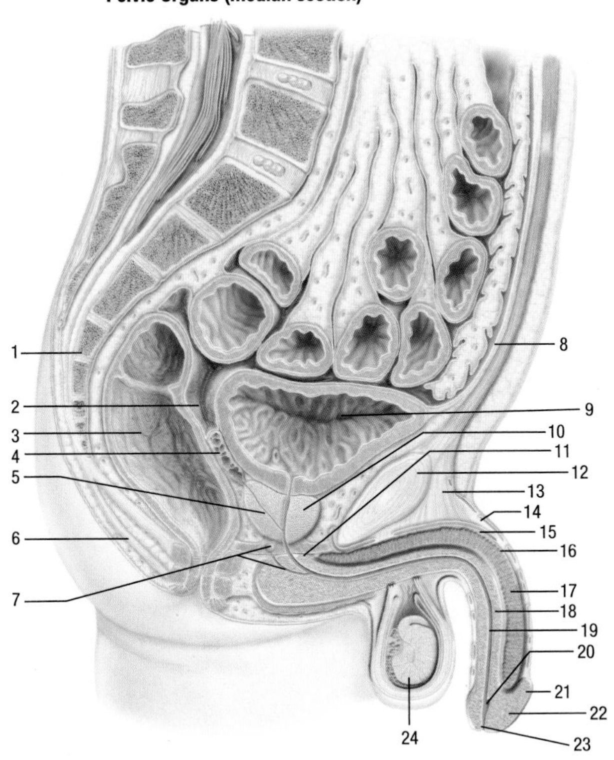
Pelvic Organs (median section)
1
2
3
4
5
6
7
8
9
10
11
12
13
14
15
16
17
18
19
20
21
22
23
24

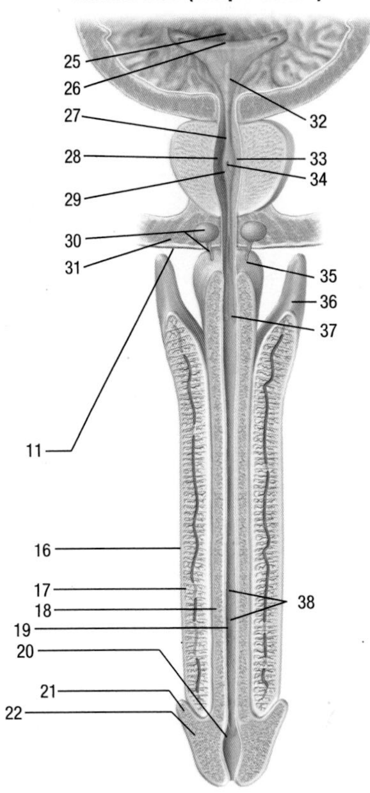
Anterior View (oblique section)
25
26
27
28
29
30
31
32
33
34
35
36
37
11
16
17
18
19
20
21
22
38

The Female Reproductive System

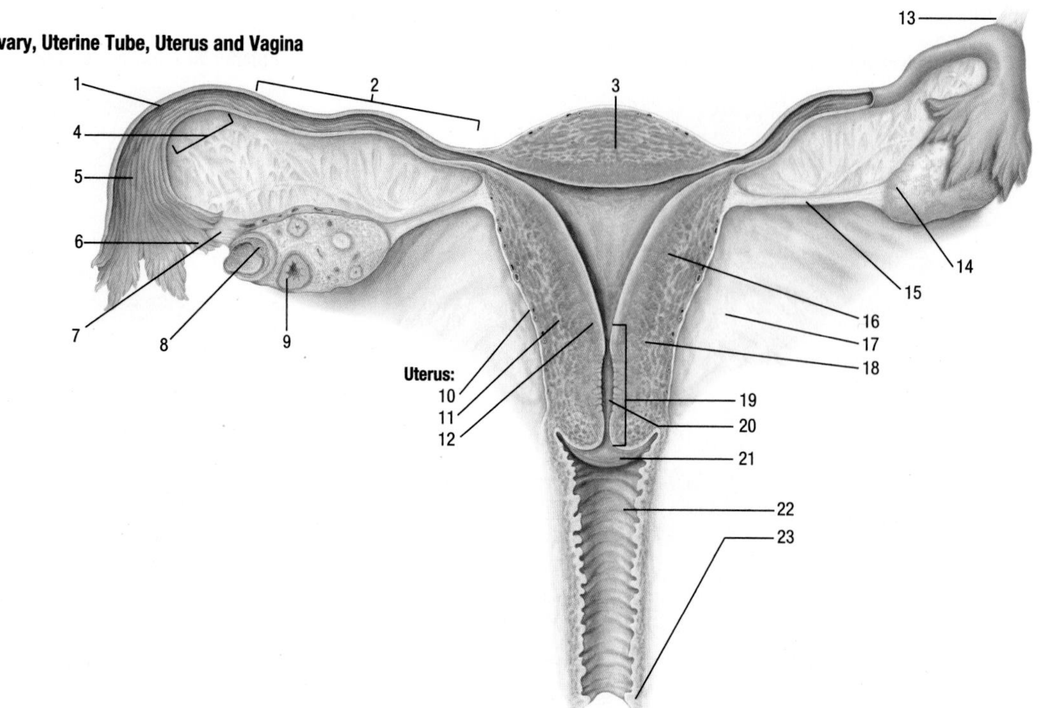

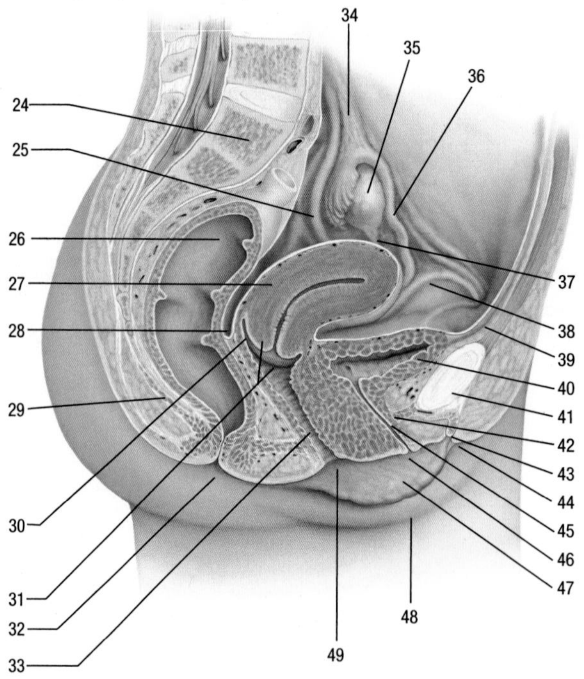

UNIT I

The Body as a Whole

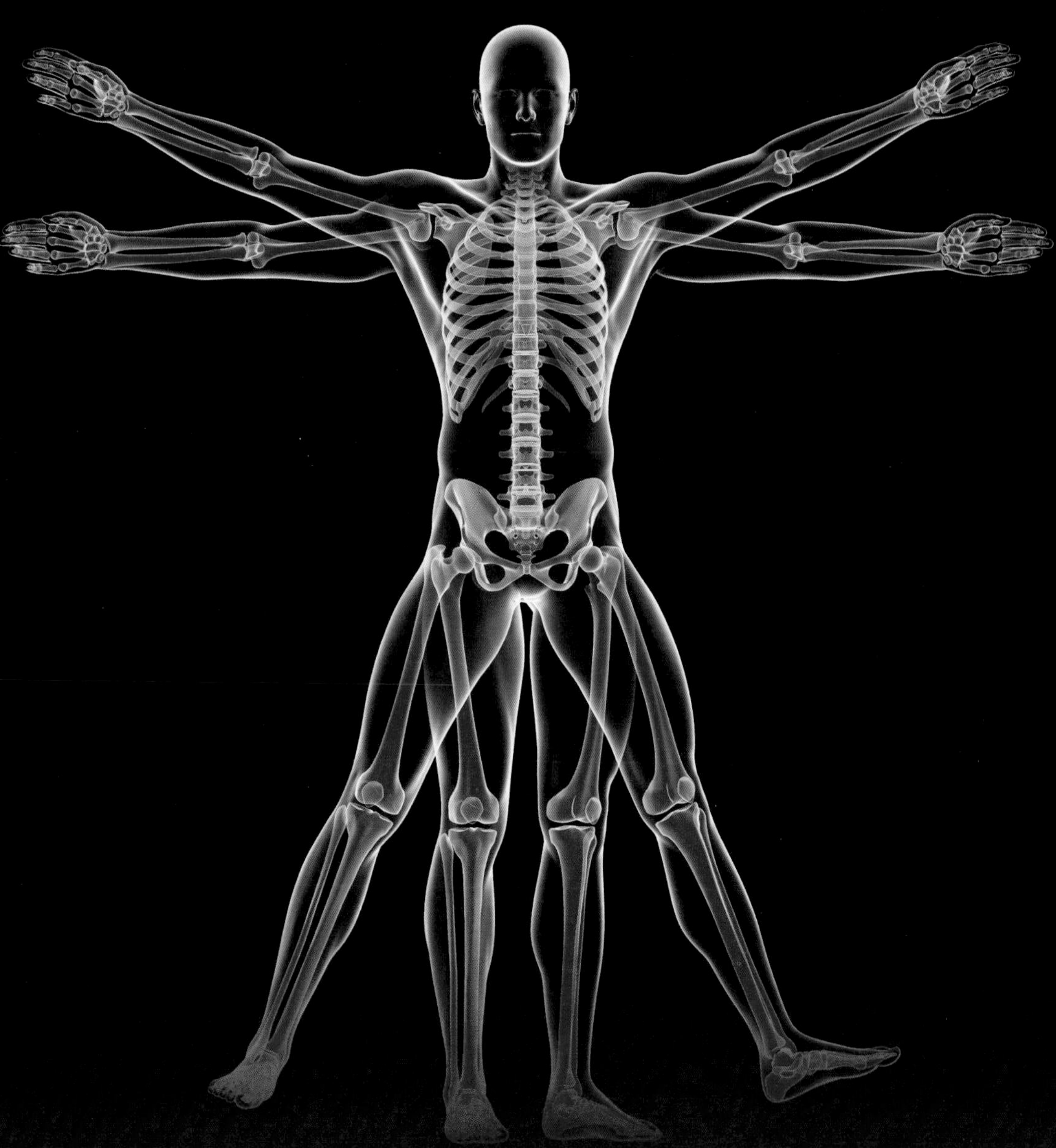

CHAPTER 1 ▸ **Organization of the Human Body**

CHAPTER 2 ▸ **Chemistry, Matter, and Life**

CHAPTER 3 ▸ **Cells and Their Functions**

CHAPTER 4 ▸ **Tissues, Glands, and Membranes**

The chapters in this unit provide the foundation for further studies of the human body. The unit begins with a broad overview of concepts in human anatomy and physiology and then zooms in to discuss the smallest units of matter—atoms and molecules. We then widen our view to discuss the smallest units of life, called cells, and continue to enlarge our scope even further to discuss groupings of similar cells, known as tissues. These chapters will prepare you for the more detailed study of individual body systems

CHAPTER 1

Organization of the Human Body

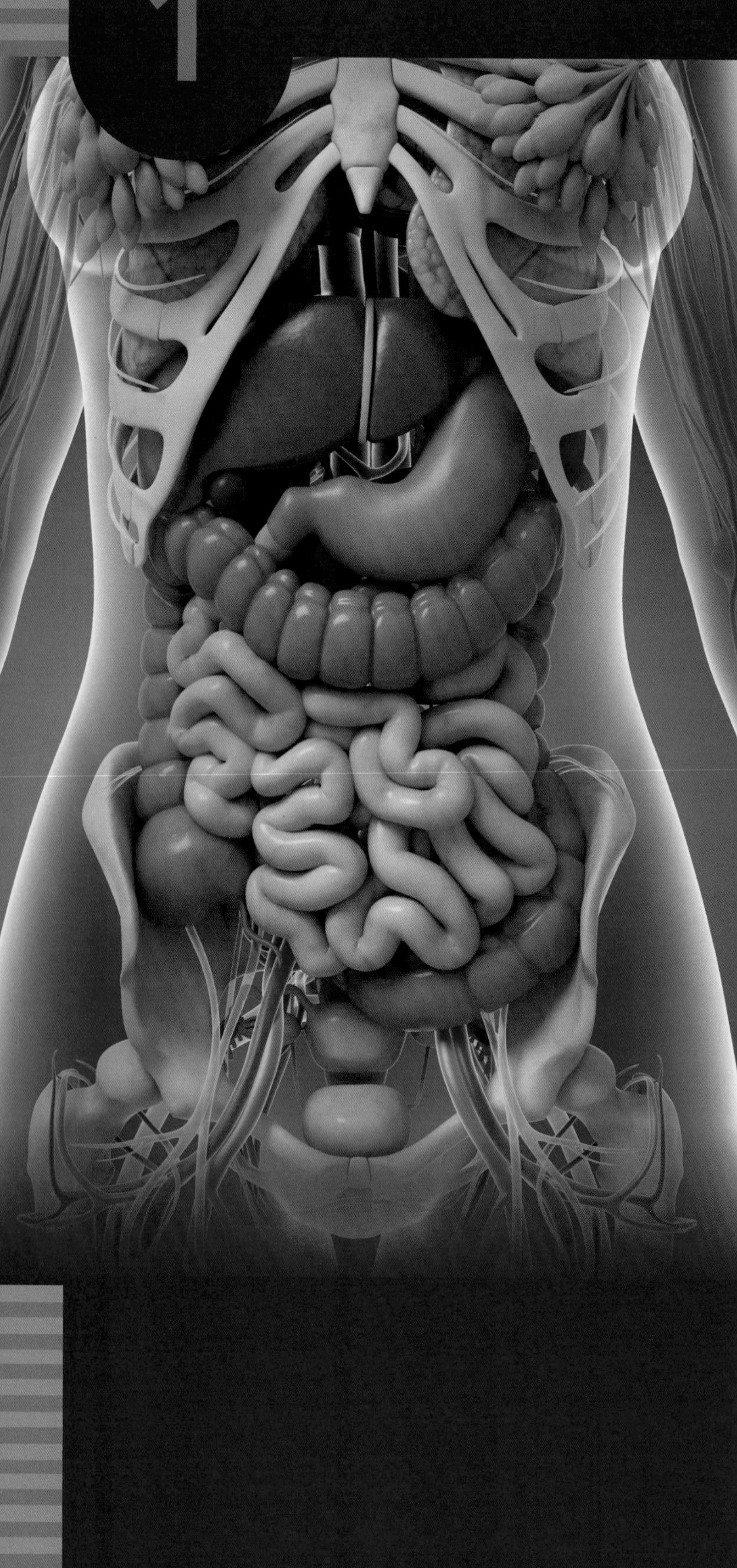

Learning Objectives

After careful study of this chapter, you should be able to:

1 ▶ Define the terms *anatomy*, *physiology*, and *pathology*. ***p. 4***

2 ▶ Describe the organization of the body from chemicals to the whole organism. ***p. 4***

3 ▶ List 11 body systems, and give the general function of each. ***p. 4***

4 ▶ Define and give examples of homeostasis. ***p. 5***

5 ▶ Using examples, discuss the components of a negative feedback loop. ***p. 6***

6 ▶ Define *metabolism*, and name the two types of metabolic reactions. ***p. 8***

7 ▶ List and define the main directional terms for the body. ***p. 8***

8 ▶ List and define the three planes of division of the body. ***p. 9***

9 ▶ Name the subdivisions of the dorsal and ventral cavities. ***p. 10***

10 ▶ Name and locate the subdivisions of the abdomen. ***p. 12***

11 ▶ Cite some anterior and posterior body regions along with their common names. ***p. 13***

12 ▶ Find examples of anatomic and physiologic terms in the case study. ***pp. 3, 14***

13 ▶ Show how word parts are used to build words related to the body's organization (see Word Anatomy at the end of the chapter). ***p. 16***

Disease in Context *Mike's Case: Emergency Care and Possible System Failure*

"Location—Belle Grove Road. Single MVA. Male. Early 20s. Fire and police on scene," crackled the radio. "Medic 12. Respond channel 2."

"Medic 12 responding. En route to Belle Grove Road," Ed radioed back, while his partner, Samantha, flipped the switch for the lights and siren and hit the accelerator. When they arrived at the scene, police officers were directing traffic, and a fire crew was at work on the vehicle. Samantha parked the ambulance just as the crew breached the door of the crumpled minivan. Samantha and Ed grabbed their trauma bags and approached the wreck.

Ed bent down toward the injured man. "I hear your name is Mike. Mine is Ed. I'm a paramedic. My partner and I are going to take a quick look at you and then get you out of here."

Samantha inspected the vehicle. "Looks like the impact sent him up and over the steering wheel. Guessing from the cracked windshield, he may have a head injury. The steering column is bent, so I wouldn't rule out thoracic or abdominal injuries either."

Ed agreed. "He's got forehead lacerations, and he's disoriented. Chest seems fine, but his abdominal cavity could be a problem. There is significant bruising across the left lumbar and umbilical regions—probably from the steering wheel. When I palpated his left upper quadrant, it caused him considerable pain."

Samantha and Ed carefully immobilized Mike's cervical spine and with the help of the fire crew, transferred him to a stretcher. Samantha started IV fluid while Ed performed a detailed physical examination beginning at the head and working inferiorly. Mike's blood pressure was very low, and his heart rate was very high—both signs of a cardiovascular emergency. In addition, he had become unresponsive to questions.

Ed shared his findings with Samantha while she placed an oxygen mask over Mike's nose and mouth. "He's hypotensive and tachycardic. With the pain he reported earlier, signs are pointing to intra-abdominal hemorrhage. We've got to get him to the trauma center right now."

Ed depends on his understanding of anatomy and physiology to help his patient and communicate with his partner. He suspects that Mike is bleeding internally and that his heart is working hard to compensate for the drastic decrease in blood pressure. As we will see later, Mike's state of internal balance, known as homeostasis, must be restored, or his body systems will fail.

ANCILLARIES *At-A-Glance*

Visit thePoint to access the following resources. For guidance in using these resources most effectively, see pp. xv–xvii.

Learning RESOURCES

- Tips for Effective Studying
- Web Figure: Abdominal Regions
- Web Figure: Abdominal Quadrants
- Web Chart: Body Systems and Their Functions
- Web Chart: Directional Terms
- Web Chart: The Metric System
- Web Chart: Abdominal Quadrants (Details)
- Animation: Negative Feedback
- Health Professions: Health Information Technician
- Detailed Chapter Outline
- Answers to Questions for Study and Review
- Audio Pronunciation Glossary

Learning ACTIVITIES

- Pre-Quiz
- Visual Activities
- Kinesthetic Activities
- Auditory Activities

Studies of the body's normal structure and functions are the basis for all medical sciences. It is only from understanding the normal that we can analyze what is going wrong in cases of disease. These studies give us an appreciation for the design and balance of the human body and for living organisms in general.

Studies of the Human Body

The scientific term for the study of body structure is **anatomy** (ah-NAT-o-me). The *-tomy* part of this word in Latin means "cutting," because a fundamental way to learn about the human body is to cut it apart, or **dissect** (dis-sekt) it. **Physiology** (fiz-e-OL-o-je) is the term for the study of how the body functions; *physio* is based on a Latin term meaning "nature" and *logy* means "study of." Anatomy and physiology are closely related—that is, structure and function are intertwined. The stomach, for example, has a pouch-like shape for storing food during digestion. The cells in the lining of the stomach are tightly packed to prevent strong digestive juices from harming underlying tissue. Anything that upsets the normal structure or function of the body is considered a **disease** and is studied as the science of **pathology** (pah-THOL-o-je).

LEVELS OF ORGANIZATION

All living things are organized from very simple levels to more complex levels (**Fig. 1-1**). Living matter is derived from chemicals, including simple substances, such as water and salts, and more complex materials, such as sugars, fats, and proteins. These chemicals assemble into living **cells**—the basic units of all life. Specialized groups of cells form **tissues**, such as muscle tissue and connective tissue. Tissues function together as **organs**, for example, a muscle contains both muscle tissue and connective tissue. Organs working together for the same general purpose make up the body **systems**, discussed below. All of the systems work together to maintain the body as a whole organism.

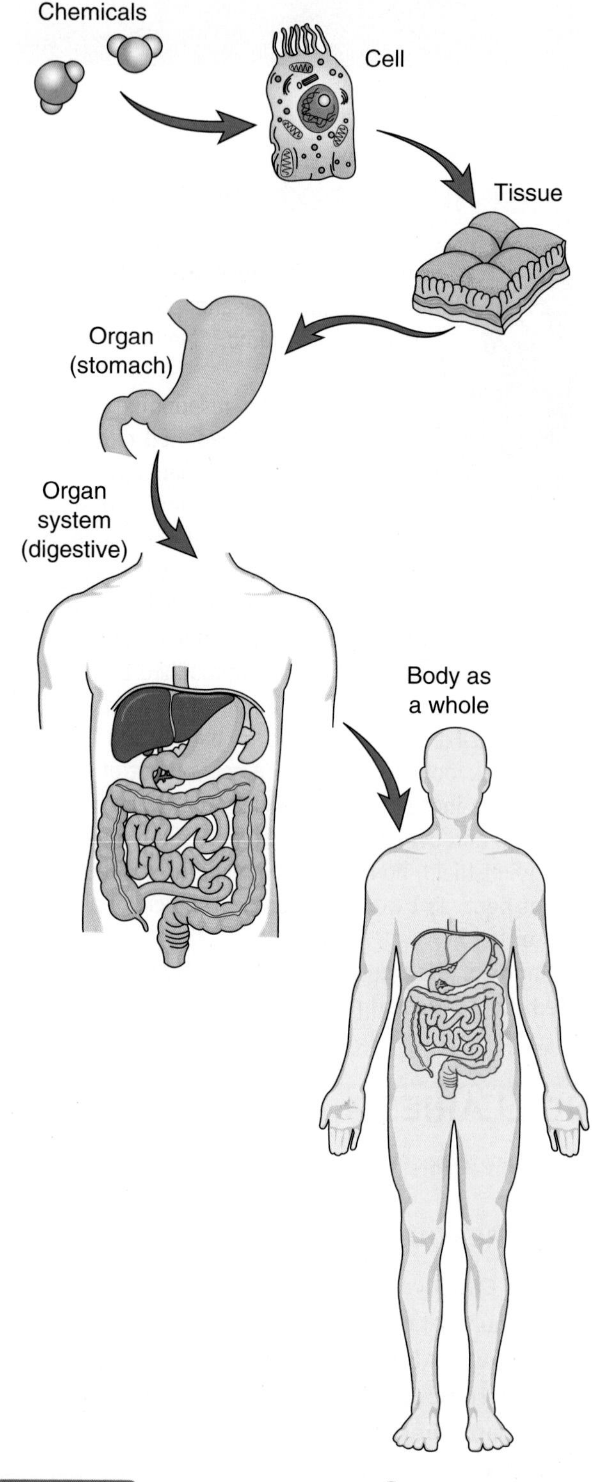

Figure 1-1 **Levels of organization.** **KEY POINT** The body is organized from the level of simple chemicals by increasing levels of complexity to the whole organism. The organ shown here is the stomach, which is part of the digestive system.

BODY SYSTEMS

We can think of the human body as organized according to the individual systems, as listed below, grouped according to their general functions.

- Protection, support, and movement
    - The **integumentary** (in-teg-u-MEN-tar-e) **system.** The word ***integument*** (in-TEG-u-ment) means "skin." The skin with its associated structures is our outermost body system. The skin's associated structures include the hair, nails, sweat glands, and oil glands.
    - The **skeletal system.** The body's basic framework is a system of 206 bones and the joints between them, collectively known as the **skeleton.**
    - The **muscular system.** The muscles in this system are attached to the bones and produce movement of the skeleton. These skeletal muscles also give the body structure, protect organs, and maintain posture.

      Two additional types of muscle contribute to other body systems. Smooth muscle is present in the walls of blood vessels and many organs; cardiac muscle constitutes the bulk of the heart wall.

- Communication and control
  - The **nervous system.** The brain, spinal cord, and nerves make up this complex system by which the body is controlled and coordinated. The nervous system also includes the special sense organs (the eyes, ears, taste buds, and organs of smell) and the receptors of the general senses, such as pain and touch. When sense organs or receptors detect changes in the external and internal environments, electrical signals are transmitted along nerves to the brain, which directs responses.
  - The **endocrine** (EN-do-krin) **system.** The scattered organs known as endocrine glands are grouped together because they share a similar function. All produce special substances called hormones, which regulate such body activities as growth, nutrient utilization, and reproduction. Examples of endocrine glands are the thyroid, pituitary, and adrenal glands.
- Circulation and immunity
  - The **cardiovascular system.** The heart and blood vessels make up the system that pumps blood to all body tissues, bringing with it nutrients, oxygen, and other needed substances. This system then carries waste materials away from the tissues to points where they can be eliminated.
  - The **lymphatic system.** Lymphatic vessels assist in circulation by returning fluids from the tissues to the blood. Lymphatic organs, such as the tonsils, thymus, and spleen, play a role in immunity, protecting against disease. The lymphatic system also aids in the absorption of dietary fats. The fluid that circulates in the lymphatic system is called *lymph*.
- Energy supply and fluid balance
  - The **respiratory system.** This system includes the lungs and the passages leading to and from the lungs. This system takes in air and conducts it to the areas in the lungs designed for gas exchange. Oxygen passes from the air into the blood and is carried to all tissues by the cardiovascular system. In like manner, carbon dioxide, a gaseous waste product, is taken by the circulation from the tissues back to the lungs to be expelled through the respiratory passages.
  - The **digestive system.** This system is composed of all the organs that are involved with taking in nutrients (foods), converting them into a form that body cells can use, and absorbing them into the circulation. Organs of the digestive system include the mouth, esophagus, stomach, small and large intestines, liver, gallbladder, and pancreas.
  - The **urinary system.** The chief purpose of the urinary system is to rid the body of waste products and excess water. This system's main components are the kidneys, the ureters, the bladder, and the urethra. (Note that some waste products are also eliminated by the digestive and respiratory systems and by the skin.)
  - The **reproductive system.** This system includes the external sex organs and all related internal structures that are concerned with the production of offspring.

References may vary in the number of body systems cited. For example, some separate the sensory system from the nervous system. Others have a separate entry for the immune system, which protects the body from foreign matter and invading organisms. The immune system is identified by its function rather than its structure and includes elements of both the cardiovascular and lymphatic systems. Bear in mind that, even though you will study the systems as separate units, they are interrelated and must cooperate to maintain health.

THE EFFECTS OF AGING

With age, changes occur gradually in all body systems. Some of these changes, such as wrinkles and gray hair, are obvious. Others, such as decreased kidney function, loss of bone mass, and formation of deposits within blood vessels, are not visible. However, they may make a person more subject to injury and disease. Changes due to aging are described in chapters on the body systems.

CHECKPOINTS

- [ ] 1-1 What are the studies of body structure and body function called?
- [ ] 1-2 What do organs working together combine to form?

See the Student Resources on thePoint for a chart summarizing the body systems and their functions.

Homeostasis and Metabolism

Despite changing environmental conditions, normal body function maintains a state of internal balance. Conditions such as body temperature, the volume and composition of body fluids, blood gas concentrations, and blood pressure must remain within a somewhat narrow range, known as a *set point*, if we are to stay healthy. This overall steady state within the organism is called **homeostasis** (ho-me-o-STA-sis), which literally means "staying (stasis) the same (homeo)."

Body fluids play such an important role in homeostasis that they deserve a special mention. One type of body fluid bathes the cells, carries nutrients to and from the cells, and transports some substances into and out of the cells. This type is called **extracellular fluid** because it includes all body fluids outside the cells (the prefix *extra-* means "outside"). Examples of extracellular fluids are the blood plasma (the fluid portion of blood), lymph, and the fluid between the cells in tissues. A second type of fluid, **intracellular fluid**, is contained inside the cells (the prefix *intra-* means "within"). The volume and composition of these body fluids must be kept in homeostatic balance at all times. Body fluids are discussed in more detail in Chapter 21.

NEGATIVE FEEDBACK

The main method for maintaining homeostasis is **negative feedback**, a control system based on information returning to a source. We are all accustomed to getting feedback about the results of our actions and using that information to regulate our behavior. Poor marks on tests and assignments, for example, may inspire us to work harder to reverse the downward slide of our grades.

Negative feedback systems keep body conditions within a set normal range by reversing any upward or downward shift. Note that negative feedback doesn't always mean less response; it just means an opposite response to a stimulus. Any negative feedback loop must contain three components:

- A *sensor* gathers information about a given body condition.
- A *control center* compares the sensor inputs with the set point and sends a corrective signal if necessary.
- An *effector* responds to the signal.

A familiar example of negative feedback is the regulation of room temperature by means of a thermostat (**Fig. 1-2**). The user determines the desired room temperature (the set point). Within the thermostat, a thermometer (sensor) measures the actual room temperature, and other components (forming the control center) compare the measured temperature with the set point. If the measured temperature is too low, the control center signals the furnace (effector) to produce heat and increase room temperature. When the room temperature reaches an upper limit (as detected by the sensor), the control center shuts off the furnace. The control center regulates the effector by means of electrical signals traveling through wires.

In the body, sensors in the brain and other organs constantly monitor body temperature and send signals to a specific brain region (the control center). This center activates effectors (such as sweat glands) to cool or warm the body if body temperature deviates too far above or below the average set point of approximately 37°C (98.6°F) (**Fig. 1-3**). As with our thermostat example, electrical impulses transmitted through the nervous system act as signals between the components of the feedback loop.

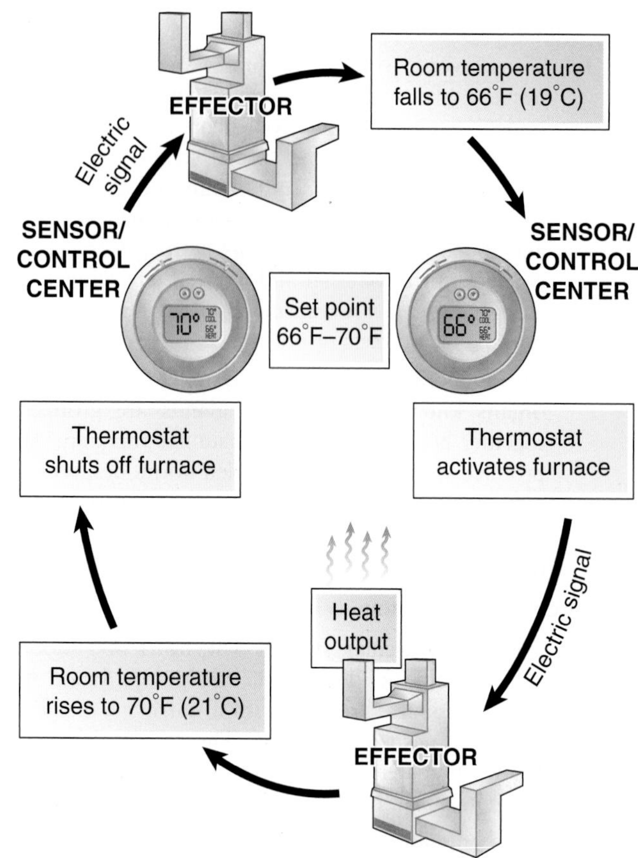

Figure 1-2 Negative feedback. **KEY POINT** A home thermostat illustrates how negative feedback keeps temperature within a set range. This thermostat is set to keep the average temperature at 68°F.

As another example, let's say you've just finished eating breakfast—a bowl of cereal and a glass of orange juice. As a result, the level of glucose (a simple sugar) increases in your bloodstream. Blood glucose levels must be tightly regulated

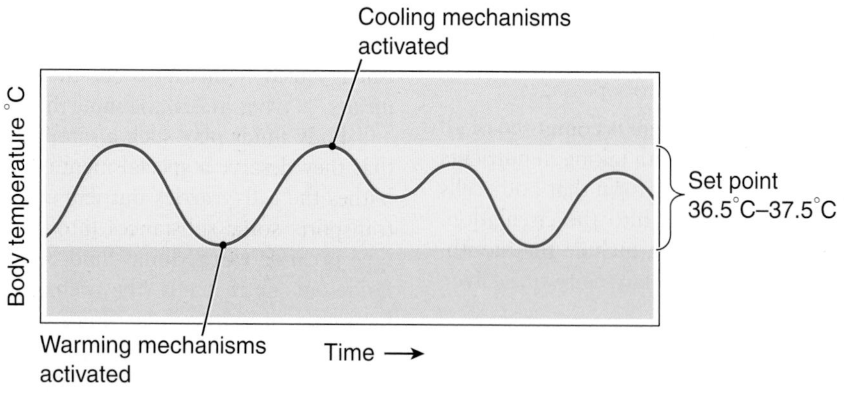

Figure 1-3 Negative feedback and body temperature. **KEY POINT** Body temperature is kept within a narrow range by negative feedback acting on a center in the brain. The set point is 37°C.

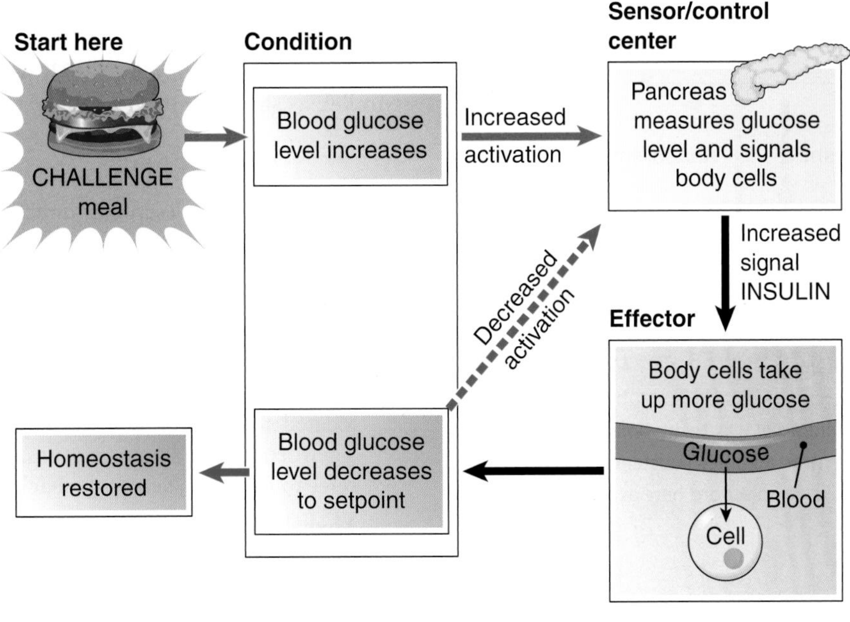

Figure 1-4 **Negative feedback in the endocrine system.** **KEY POINT** The pancreas regulates blood glucose concentration using insulin as the signal.

to prevent disease, so this increase constitutes a challenge to homeostasis. Endocrine cells in your pancreas act as both sensors and the control center; they sense blood glucose levels and compare them with the set point. When blood glucose levels rise above the set point, pancreatic cells send a signal to muscle and fat cells (the effectors) to increase their glucose uptake. A hormone called *insulin* acts as the signal, traveling through the bloodstream from the pancreas to the effector cells. Increased glucose uptake and the subsequent drop in the blood glucose level cause the pancreas to reduce insulin secretion, and homeostasis is restored (**Fig. 1-4**).

Many more examples of internal regulation by negative feedback will appear throughout this book, so study **Figures 1-2 through 1-4** closely. Clearly, negative feedback systems are critical for maintaining our health. If negative feedback mechanisms cannot restore homeostasis in response to a change, disease can result (**see Box 1-1**). Mike's case illustrates the dangers of homeostatic imbalance.

See the Student Resources on thePoint to view an animation on negative feedback.

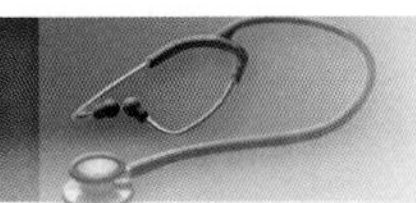

Box 1-1

CLINICAL PERSPECTIVES

Homeostatic Imbalance: When Feedback Fails

Each body structure contributes in some way to homeostasis, often through negative feedback mechanisms. The nervous and endocrine systems are particularly important in feedback. The nervous system's electrical signals act quickly in response to deviations from normal, whereas the endocrine system's chemical signals (hormones) act more slowly but over a longer time. Often both systems work together to maintain homeostasis.

As long as feedback keeps conditions within normal limits, the body remains healthy, but if feedback cannot maintain these conditions, the body enters a state of homeostatic imbalance. Moderate imbalance causes illness and disease, while severe imbalance causes death. At some level, all illnesses and diseases can be linked to homeostatic imbalance.

For example, feedback mechanisms closely monitor and maintain normal blood pressure. When blood pressure rises, negative feedback mechanisms lower it to normal limits. If these mechanisms fail, hypertension (high blood pressure) develops. Hypertension further damages the cardiovascular system and, if untreated, may lead to death. With mild hypertension, lifestyle changes in diet, exercise, and stress management may lower blood pressure sufficiently, whereas severe hypertension often requires drug therapy. The various types of antihypertensive medication all help negative feedback mechanisms lower blood pressure.

Feedback mechanisms also regulate body temperature. When body temperature falls, negative feedback mechanisms raise it back to normal limits, but if these mechanisms fail and body temperature continues to drop, hypothermia develops. The main effects of hypothermia are uncontrolled shivering, lack of coordination, decreased heart and respiratory rates, and, if left untreated, death. Cardiac surgeons use hypothermia to their advantage during open heart surgery. Cooling the body lets surgeons stop the heart during an operation without damaging its tissues.

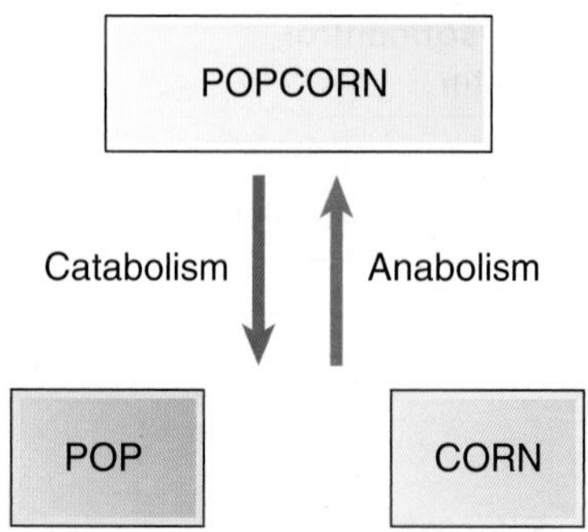

Figure 1-5 Metabolism. **KEY POINT** Metabolism includes two types of reactions. In catabolism, substances are broken down into their building blocks. In anabolism, simple components are built into more complex substances. We use the breakdown and building of a simple word here as an example of these reactions.

METABOLISM

All the life-sustaining reactions that occur within the body systems together make up **metabolism** (meh-TAB-o-lizm). Metabolism can be divided into two types of activities:

- In **catabolism** (kah-TAB-o-lizm), complex substances are broken down into simpler compounds (**Fig. 1-5**). The breakdown of food, for example, yields simple chemical building blocks and energy to power cellular activities. The energy obtained from the catabolism of nutrients is used to form a compound often described as the cell's "energy currency." It has the long name of **adenosine triphosphate** (ah-DEN-o-sene tri-FOS-fate) but is commonly abbreviated ATP. Chapters 2 (**see Figure 2-12**) and 20 have more information on metabolism and ATP.
- In **anabolism** (ah-NAB-o-lizm), simple compounds are used to manufacture materials needed for growth, function, and tissue repair. Anabolism consists of building, or synthesis, reactions. These synthesis reactions are fueled by ATP, as discussed in Chapter 2.

CHECKPOINTS

- [ ] 1-3 Where are intracellular fluids located? Extracellular fluids?
- [ ] 1-4 What is the definition of homeostasis?
- [ ] 1-5 What are the three components of a negative feedback loop?
- [ ] 1-6 What are the two types of metabolic activities, and what happens during each?

Body Directions

Because it would be awkward and inaccurate to speak of bandaging the "southwest part" of the chest, for example, healthcare professionals use standardized terms to designate body positions and directions. For consistency, all descriptions assume that the body is in the **anatomic position.** In this posture, the subject is standing upright with face front, arms at the sides with palms forward, and feet parallel, as shown in **Figure 1-6**.

DIRECTIONAL TERMS

The main terms for describing directions in the body are as follows (**see Fig. 1-6**):

- **Superior** is a term meaning above, or in a higher position. Its opposite, **inferior**, means below, or lower. The heart, for example, is superior to the intestine.
- **Anterior** and **ventral** have the same meaning in humans: located toward the belly surface or front of the body. Their corresponding opposites, **posterior** and **dorsal**, refer to locations nearer the back.
- **Medial** means nearer to an imaginary plane that passes through the midline of the body, dividing it into left and right portions. **Lateral**, its opposite, means farther away

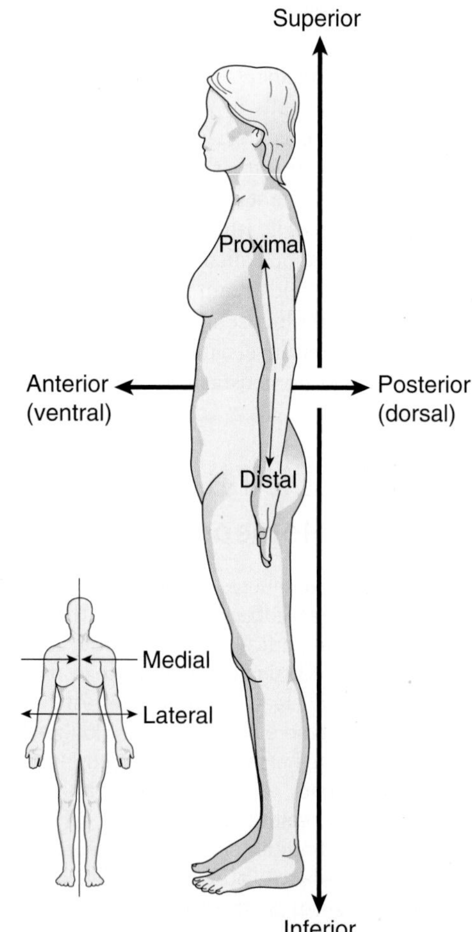

Figure 1-6 Directional terms. **KEY POINT** Healthcare professionals use standardized terms to describe body directions. **ZOOMING IN** What is the scientific name for the position in which the figures are standing?

from the midline, toward the side. For example, your nose is medial to your ears.

- **Proximal** means nearer to the origin or attachment point of a structure; **distal** means farther from that point. For example, the part of your thumb where it attaches to your hand is its proximal region; the tip of the thumb is its distal region. Considering the mouth as the beginning (origin) of the digestive tract, the small intestine is distal to the stomach.

See the Student Resources on thePoint for a chart of directional terms with definitions and examples.

PLANES OF DIVISION

To visualize the various internal structures in relation to each other, anatomists can divide the body along three planes, each of which is a cut through the body in a different direction (**Fig. 1-7**), as follows:

- **Frontal plane.** If the cut were made in line with the ears and then down the middle of the body, you would see an anterior, or ventral (front), section and a posterior, or dorsal (back), section. Another name for this plane is *coronal plane*.
- **Sagittal** (SAJ-ih-tal) **plane.** If you were to cut the body in two from front to back, separating it into right and left portions, the sections you would see would be sagittal sections. A cut exactly down the midline of the body, separating it into equal right and left halves, is a **midsagittal plane.**
- **Transverse plane.** If the cut were made horizontally, across the other two planes, it would divide the body into a superior (upper) part and an inferior (lower) part. A transverse plane is also called a *horizontal plane*.

Some additional terms are used to describe sections (cuts) of tissues, as used to prepare them for study under the microscope. A cross-section (**Fig. 1-8**) is a cut made perpendicular to the long axis of an organ, such as a cut made across a banana to give a small round slice. A longitudinal section is made parallel to the long axis, as in cutting a banana from tip to tip to make a slice for a banana split. An oblique section is made at an angle. The type of section used will determine what is seen under the microscope, as shown with a blood vessel in **Figure 1-8**.

These same terms are used for images taken by techniques such as computed tomography (CT) or magnetic resonance imaging (MRI) (**see Box 1-2**). In imaging studies, the term *cross-section* is used more generally to mean any two-dimensional view of an internal structure obtained by imaging, as shown in **Figure 1-9**.

CHECKPOINTS

- ☐ 1-7 What term describes a location farther from an origin, such as the wrist in comparison to the elbow?
- ☐ 1-8 What are the three planes in which the body can be cut?

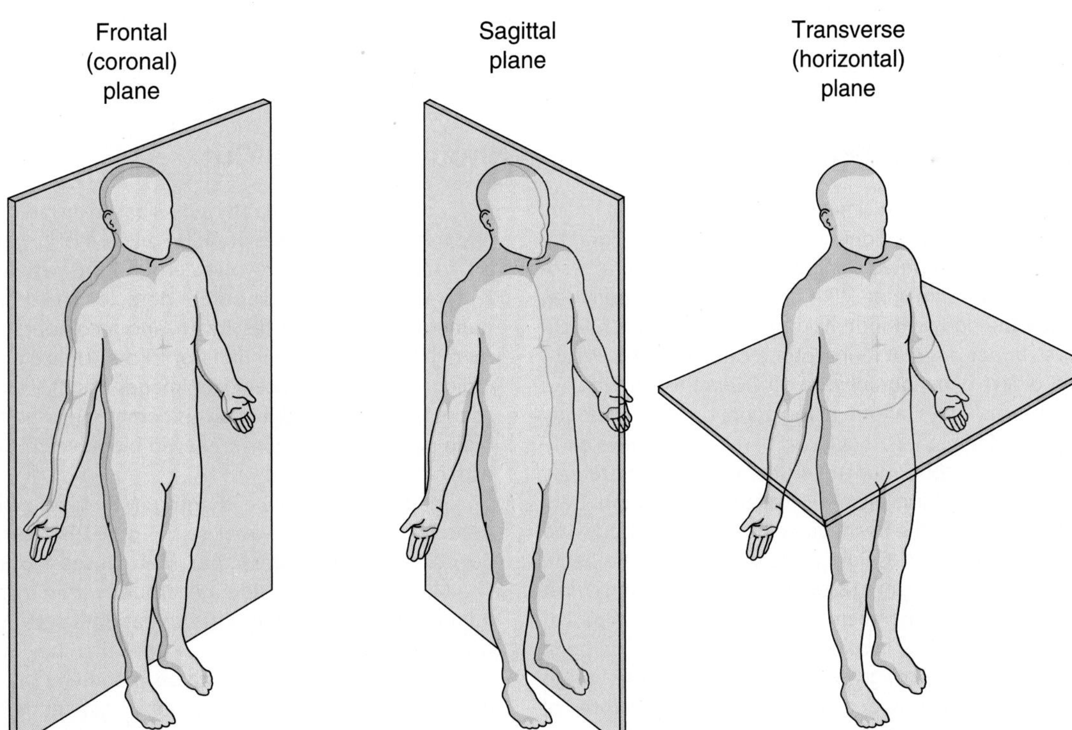

Figure 1-7 **Planes of division.** **KEY POINT** The body can be divided along three different planes. **ZOOMING IN** Which plane divides the body into superior and inferior parts? Which plane divides the body into anterior and posterior parts?

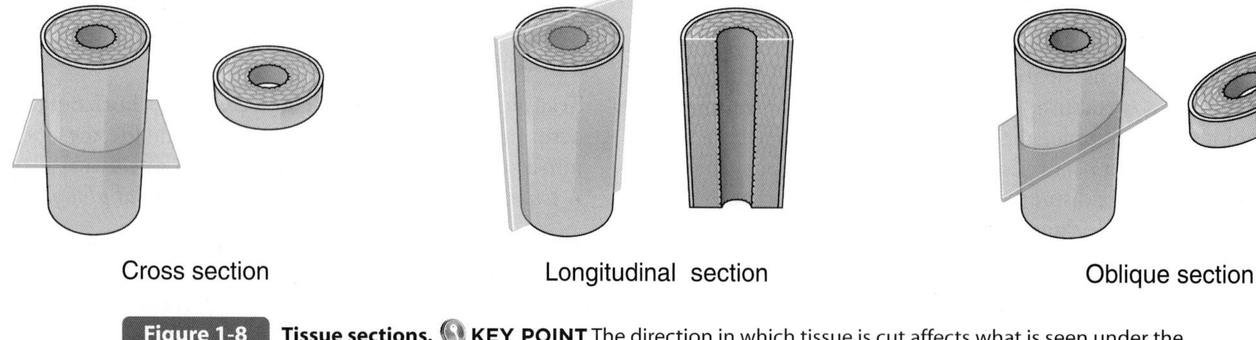

Figure 1-8 **Tissue sections.** **KEY POINT** The direction in which tissue is cut affects what is seen under the microscope.

Body Cavities

Internally, the body is divided into a few large spaces, or cavities, which contain the organs. The two main cavities are the **dorsal cavity** and **ventral cavity** (Fig. 1-10).

DORSAL CAVITY

The dorsal body cavity has two subdivisions: the **cranial cavity**, containing the brain, and the **spinal cavity (canal)**, enclosing the spinal cord. These two areas form one continuous space.

VENTRAL CAVITY

The ventral cavity is much larger than the dorsal cavity. It has two main subdivisions, which are separated by the **diaphragm** (DI-ah-fram), a muscle used in breathing. The **thoracic** (tho-RAS-ik) **cavity** is superior to (above) the diaphragm. Its contents include the heart, the lungs, and the large blood vessels that join the heart. The heart is contained in the pericardial cavity, formed by the pericardial sac, the tissue that surrounds the heart; the lungs are in the pleural cavity, formed by the pleurae, the membranes that enclose the lungs (Fig. 1-11). The **mediastinum** (me-de-as-TI-num) is the space between the lungs, including the organs and vessels contained in that space.

The **abdominopelvic** (ab-dom-ih-no-PEL-vik) **cavity** (see Fig. 1-10) is inferior to (below) the diaphragm. This space is further subdivided into two regions. The superior portion, the **abdominal cavity**, contains the stomach, most of the intestine, the liver, the gallbladder, the pancreas, and the spleen. The inferior portion, set off by an imaginary line across the top of the hip bones, is the **pelvic cavity**. This cavity contains the urinary bladder, the rectum, and the internal parts of the reproductive system. See **Figures A5-5 and A5-7** in Appendix 5, Dissection Atlas, for dissection photographs showing organs of the ventral cavity.

HOT TOPICS

Box 1-2

Medical Imaging: Seeing without Making a Cut

Three imaging techniques that have revolutionized medicine are radiography, computed tomography, and magnetic resonance imaging. With them, physicians today can "see" inside the body without making a single cut. Each technique is so important that its inventor received a Nobel Prize.

The oldest is radiography (ra-de-OG-rah-fe), in which a machine beams x-rays (a form of radiation) through the body onto a piece of film. Like other forms of radiation, x-rays damage body tissues, but modern equipment uses extremely low doses. The resulting picture is called a radiograph. Dark areas indicate where the beam passed through the body and exposed the film, whereas light areas show where the beam did not pass through. Dense tissues (bone, teeth) absorb most of the x-rays, preventing them from exposing the film. For this reason, radiography is commonly used to visualize bone fractures and tooth decay as well as abnormally dense tissues like tumors. Radiography does not provide clear pictures of soft tissues because most of the beam passes through and exposes the film, but contrast media can help make structures like blood vessels and hollow organs more visible. For example, radiologists use ingested barium sulfate (which absorbs x-rays) to coat the digestive tract for imaging.

Computed tomography (CT) is based on radiography and also uses very low doses of radiation (see Fig. 1-9A). During a CT scan, a machine revolves around the patient, beaming x-rays through the body onto a detector. The detector takes numerous pictures of the beam, and a computer assembles them into transverse sections, or "slices." Unlike conventional radiography, CT produces clear images of soft structures such as the brain, liver, and lungs. It is commonly used to visualize brain injuries and tumors and even blood vessels when used with contrast media.

Magnetic resonance imaging (MRI) uses a strong magnetic field and radio wave (see Fig. 1-9B). So far, there is no evidence to suggest that MRI causes tissue damage. The MRI patient lies inside a chamber within a very powerful magnet. The molecules in the patient's soft tissues align with the magnetic field inside the chamber. When radio waves beamed at the region to be imaged hit the soft tissue, the aligned molecules emit energy that the MRI machine detects, and a computer converts these signals into a picture. MRI produces even clearer images of soft tissue than does CT and can create detailed pictures of blood vessels without contrast media. MRI can visualize brain injuries and tumors that might be missed using CT.

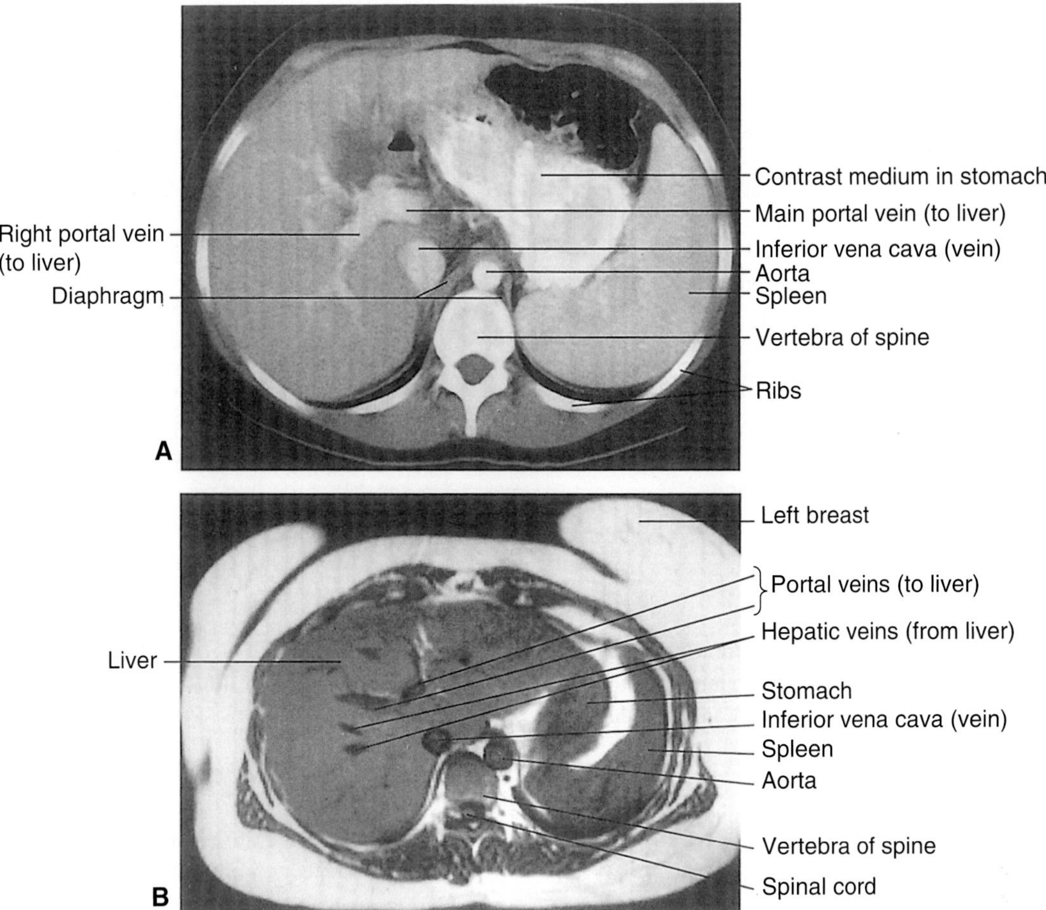

Figure 1-9 **Cross-sections in imaging.** Images taken across the body through the liver and spleen by **(A)** computed tomography and **(B)** magnetic resonance imaging.

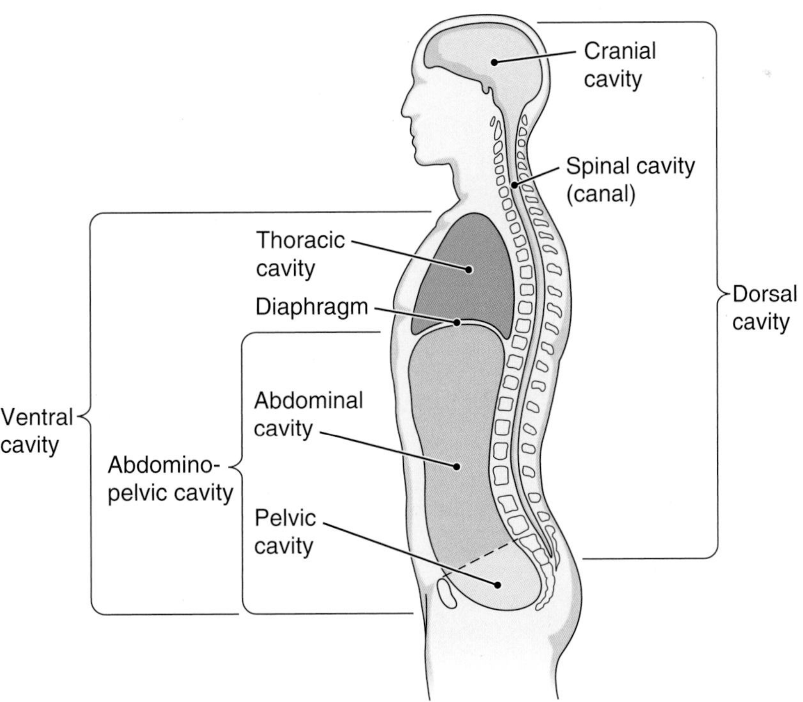

Figure 1-10 **Body cavities, lateral view.** Shown are the dorsal and ventral cavities with their subdivisions. **ZOOMING IN** What cavity contains the diaphragm?

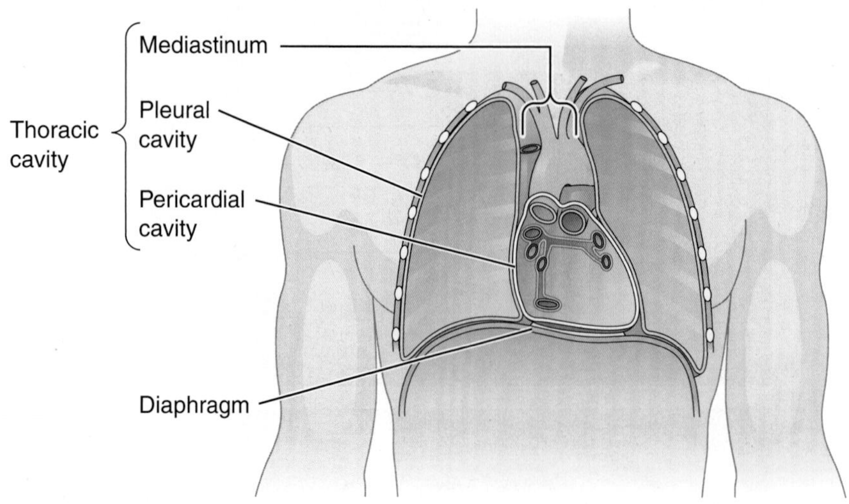

Figure 1-11 The thoracic cavity.
KEY POINT Among other structures, the thoracic cavity encloses the pericardial cavity, which contains the heart, and the pleural cavity, which contains the lungs.

DIVISIONS OF THE ABDOMEN

It is helpful to divide the abdomen for examination and reference into nine regions (**Fig. 1-12**).

The three central regions, from superior to inferior, are the following:

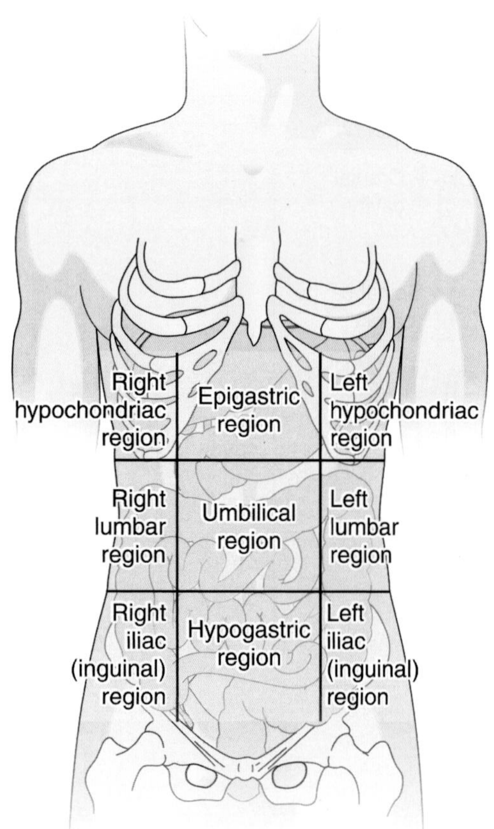

Figure 1-12 The nine regions of the abdomen. **KEY POINT** Internal structures can be localized within nine regions of the abdomen.

- **Epigastric** (ep-ih-GAS-trik) **region,** located just inferior to the breastbone
- **Umbilical** (um-BIL-ih-kal) **region,** around the umbilicus (um-BIL-ih-kus), commonly called the *navel*
- **Hypogastric** (hi-po-GAS-trik) **region,** the most inferior of all the midline regions

The regions on the right and left, from superior to inferior, are the following:

- **Hypochondriac** (hi-po-KON-dre-ak) **regions,** just inferior to the ribs
- **Lumbar regions,** which are on a level with the lumbar regions of the spine
- **Iliac,** or **inguinal** (IN-gwih-nal), **regions,** named for the upper crest of the hip bone and the groin region, respectively

A simpler but less precise division into four quadrants is sometimes used. These regions are the right upper quadrant, left upper quadrant, right lower quadrant, and left lower quadrant (**Fig. 1-13**).

For your reference, **Figures 1-14 and 1-15** give anatomic adjectives for some other body regions along with their common names.

CHECKPOINTS

- ☐ 1-9 Name the two main body cavities.
- ☐ 1-10 Name the three central regions and the three left and right lateral regions of the abdomen.

See the Student Resources on thePoint for photographic versions of Figures 1-12 and 1-13 and a list of the organs in each quadrant. You can also find information on the metric system, which is used for all scientific measurements.

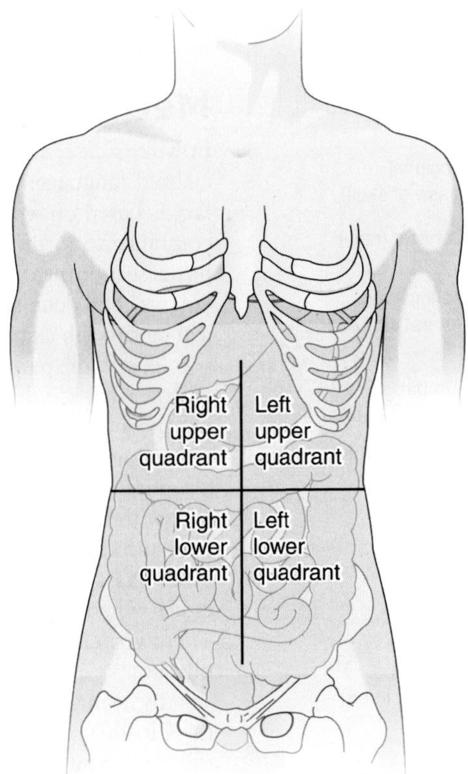

Figure 1-13 **Quadrants of the abdomen.** The organs within each quadrant are shown. **ZOOMING IN** Which four abdominal regions are represented in the left lower quadrant?

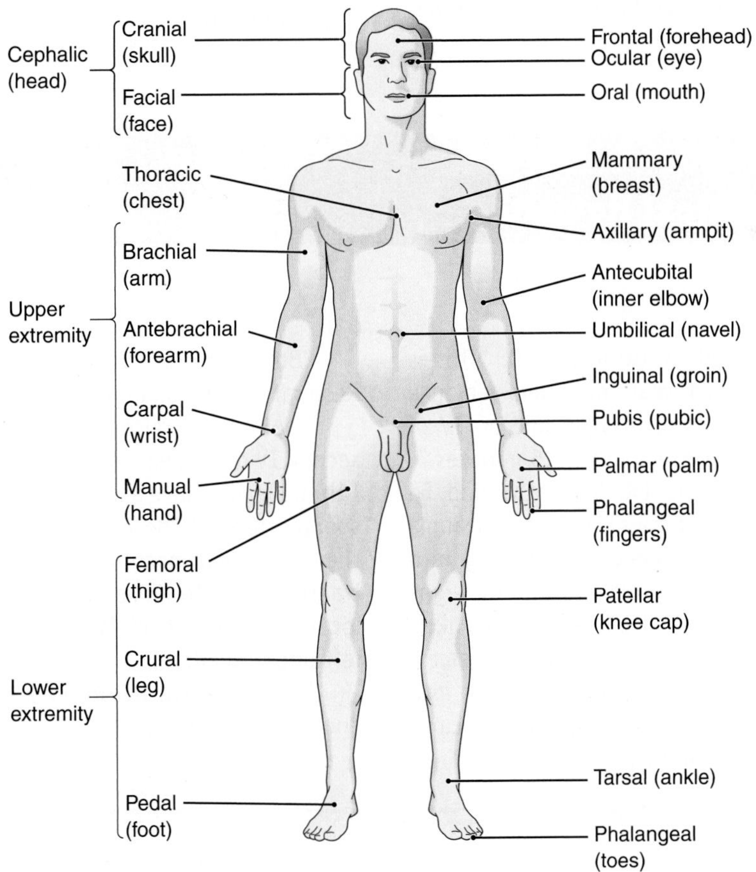

Figure 1-14 **Adjectives for some anterior body regions.** The names of the regions are in parentheses.

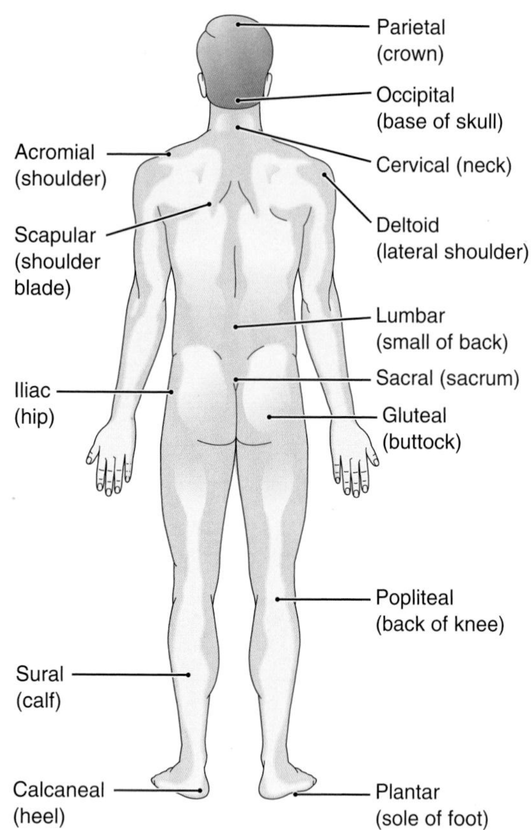

Figure 1-15 **Adjectives for some posterior body regions.** The names of the regions are in parentheses.

Medical Terminology

In Mike's case, we saw that health professionals share a specialized language: medical terminology. This special vocabulary is based on word parts with consistent meanings that are combined to form different words. Each chapter in this book has a section near the end entitled "Word Anatomy." Here, you will find definitions of word parts commonly used in medical terms with examples of their usage.

The main part of a word is the **root**. Some compound words, such as wheelchair, gastrointestinal, and lymphocyte, use more than one root. A **prefix** is a short part that starts a word and modifies the root. A **suffix** follows the root and also modifies it. In the Word Anatomy charts, the word parts follow the chapter sequence. Prefixes are followed by a dash, and suffixes are preceded by a dash. A root has no dash but often has a combining vowel added to make pronunciation easier when it is combined with another root or a suffix. These vowels are separated from the root with a slash, as in physi/o.

By using the Word Anatomy charts, the Glossary, and the Glossary of Word Parts (the last two found at the back of this text), you too can learn to speak this language.

See the box Health Information Technicians in the Student Resources on thePoint for a description of a profession that requires knowledge of medical terminology.

Disease in Context Revisited

Mike's Homeostatic Emergency

The dispatch radio crackled to life in the ER. "This is Medic 12. We have Mike, 21 years old. Involved in a head-on collision. Patient is on oxygen and an IV of normal saline running wide open. ETA is 15 minutes."

When they arrived at the ER, Samantha and Ed wheeled their unconscious patient into the trauma room. Immediately, the emergency team sprang into action. The trauma nurse measured Mike's vital signs while a technician drew blood from a vein in Mike's antecubital region for testing in the lab. The emergency physician inserted an endotracheal tube into Mike's pharynx to keep his airway open and then carefully examined his abdominopelvic cavity.

"Blood pressure is 80 over 40. Heart rate is 146. Respirations are shallow and rapid," said the nurse.

"We need to raise his blood pressure—let's start a second IV of plasma. His abdomen is as hard as a board. I think he may have a bleed in there—we need an ultrasound," replied the doctor. The sonographer wheeled the ultrasound machine into position and placed the transducer onto Mike's abdomen. Immediately, she located the cause of Mike's symptoms—blood in the left upper quadrant.

"OK. We have a ruptured spleen here," said the doctor. "Call surgery—they need to operate right now."

CHAPTER 1

Chapter Wrap-Up

Summary Overview

A detailed chapter outline with space for note taking is on *thePoint*. The figure below illustrates the main topics covered in this chapter.

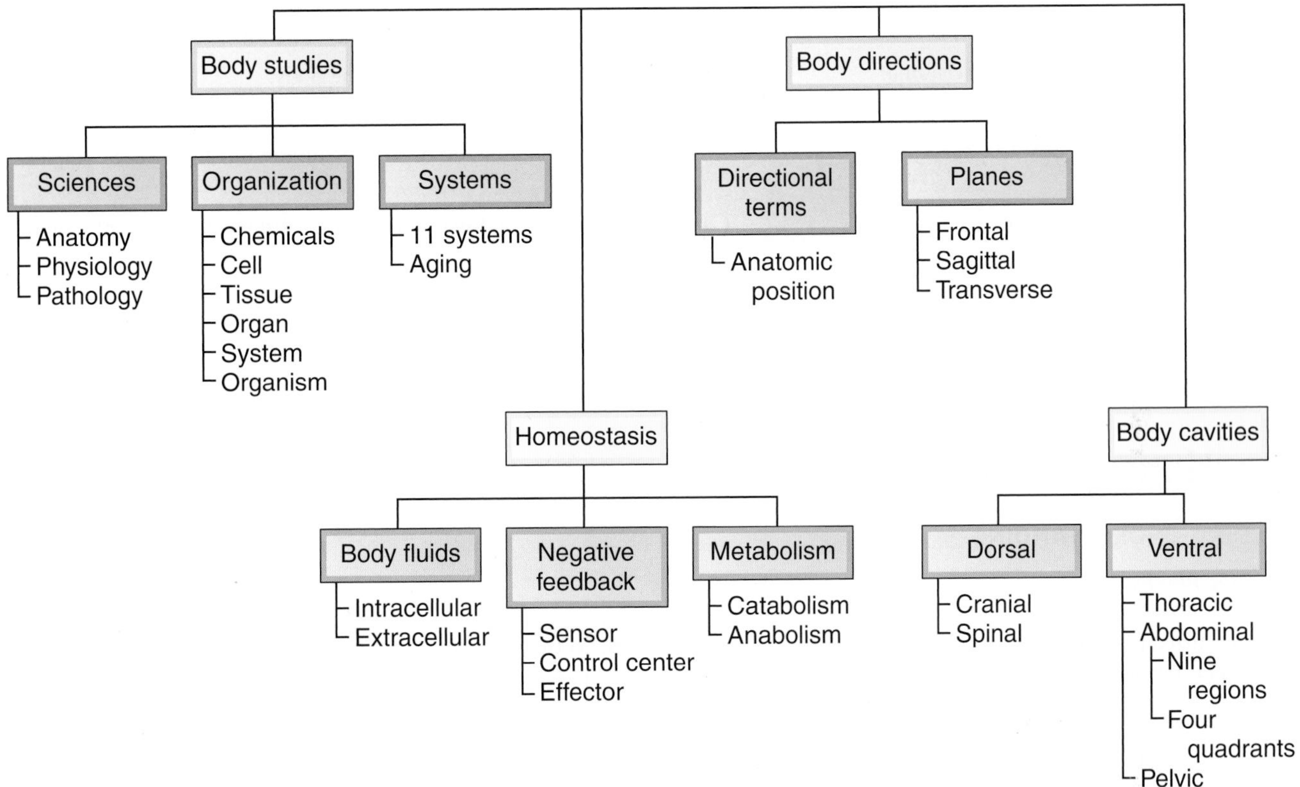

Key Terms

The terms listed below are emphasized in this chapter. Knowing them will help you organize and prioritize your learning. These and other boldface terms are defined in the Glossary with phonetic pronunciations.

anabolism
anatomic position
anatomy
catabolism
cell
disease
extracellular fluid
homeostasis
intracellular fluid
metabolism
negative feedback
organ
pathology
physiology
system
tissue

Word Anatomy

Medical terms are built from standardized word parts (prefixes, roots, and suffixes). Learning the meanings of these parts can help you remember words and interpret unfamiliar terms.

WORD PART	MEANING	EXAMPLE
Studies of the Human Body		
dis-	apart, away from	To *dissect* is to cut apart.
-logy	study of	*Radiology* is the study and use of radioactive substances.
path/o	disease	*Pathology* is the study of disease.
physi/o	nature, physical	*Physiology* is the study of how the body functions.
-tomy	cutting, incision of	*Anatomy* can be revealed by cutting the body.
Metabolism		
ana-	upward, again, back	*Anabolism* is the building up of simple compounds into more complex substances.
cata-	down	*Catabolism* is the breakdown of complex substances into simpler ones.
extra-	outside of, beyond	*Extracellular* fluid is outside the cells.
home/o-	same	*Homeostasis* is the steady state (sameness) within an organism.
intra-	within	*Intracellular* fluid is within a cell.
stat, -stasis	stand, stoppage, constancy	In *homeostasis*, "-stasis" refers to constancy.

Questions for Study and Review

BUILDING UNDERSTANDING

Fill in the Blanks

1. Specialized groups of cells working together for the same general purpose form _____.

2. In location, the nose is _____ to the eyes.

3. Normal body function maintains a state of internal balance called _____.

4. In the word *physiology*, *-logy* is an example of a word part called a(n) _____.

5. In the opening case study, Mike's intra-abdominal hemorrhage is in the _____ body cavity.

Matching < Match each numbered item with the most closely related lettered item.

___ **6.** One of two systems that control and coordinate other systems

___ **7.** The system that brings needed substances to the body tissues

___ **8.** The system that converts foods into a form that body cells can use

___ **9.** The outermost body system

___ **10.** The system of glands that produce hormones

a. nervous system

b. integumentary system

c. cardiovascular system

d. endocrine system

e. digestive system

Multiple Choice

___ **11.** Which science studies normal body structure?

a. homeostasis
b. anatomy
c. physiology
d. pathology

___ **12.** Where is intracellular fluid located?

a. between body cells
b. in blood plasma
c. in lymph
d. inside body cells

___ **13.** What is the main way of regulating homeostasis?

a. anabolism
b. biofeedback
c. catabolism
d. negative feedback

___ **14.** Which cavity contains the mediastinum?

a. abdominal
b. dorsal
c. thoracic
d. pelvic

___ **15.** In location, the ankle is _____ to the knee.

a. distal
b. inferior
c. proximal
d. superior

___ **16.** A plane that divides the body into right and left portions is a

a. frontal plane
b. transverse plane
c. sagittal plane
d. horizontal plane

___ **17.** The most inferior midline region of the abdomen is the

a. superior region
b. hypogastric region
c. umbilical region
d. epigastric region

UNDERSTANDING CONCEPTS

18. What do you study in anatomy? In physiology? Would it be wise to study one without the other?

19. List in sequence the levels of organization in the body from simplest to most complex. Give an example for each level.

20. Compare and contrast the anatomy and physiology of the nervous system with that of the endocrine system.

21. What is the difference between catabolism and anabolism? Give an example of each type of activity.

22. Name in order of action the components of a negative feedback loop.

23. Use The Body Visible overlays at the beginning of this book to name the lateral bone of the lower leg. Name the proximal bone of the arm.

24. List the subdivisions of the dorsal and ventral cavities. Name some organs found in each subdivision. Use the Dissection Atlas in Appendix 5 to locate organs in the ventral cavity.

CONCEPTUAL THINKING

25. The human body is organized from very simple levels to more complex levels. With this in mind, describe why a disease at the chemical level can have an effect on organ system function.

26. Use a car operating under cruise control as an example of a negative feedback loop, identifying the set point and the components of the system.

27. In Mike's case, the paramedics discovered bruising of the skin over Mike's left lumbar region and umbilical region. Mike also reported considerable pain in his upper left quadrant. Locate these regions on your own body. Why it is important for health professionals to use medical terminology when describing the human body?

For more questions, see the Learning Activities on thePoint.

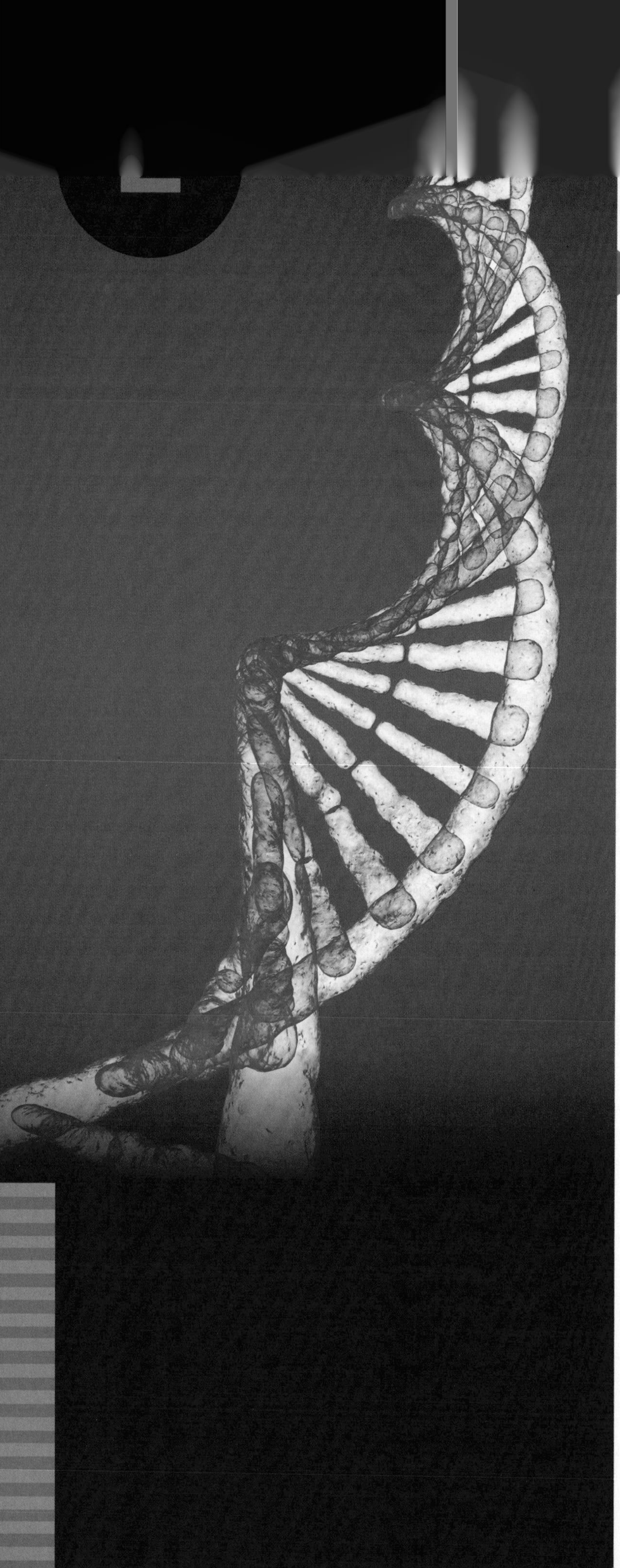

Learning Objectives

After careful study of this chapter, you should be able to:

1 ▶ Define a chemical element. *p. 20*
2 ▶ Describe the structure of an atom. *p. 21*
3 ▶ Differentiate between ionic and covalent bonds. *p. 22*
4 ▶ Define an electrolyte. *p. 23*
5 ▶ Differentiate between molecules and compounds. *p. 23*
6 ▶ Define *mixture*; list the three types of mixtures, and give two examples of each. *p. 25*
7 ▶ Explain why water is so important in metabolism. *p. 25*
8 ▶ Compare acids, bases, and salts. *p. 26*
9 ▶ Explain how the numbers on the pH scale relate to acidity and alkalinity. *p. 26*
10 ▶ Explain why buffers are important in the body. *p. 27*
11 ▶ Define *radioactivity*, and cite several examples of how radioactive substances are used in medicine. *p. 27*
12 ▶ Name the three main types of organic compounds and the building blocks of each. *p. 28*
13 ▶ Define *enzyme*; describe how enzymes work. *p. 29*
14 ▶ List the components of nucleotides, and give some examples of nucleotides. *p. 31*
15 ▶ Use the case study to discuss the importance of regulating body fluid quantity and composition. *pp. 19, 32*
16 ▶ Show how word parts are used to build words related to chemistry, matter, and life. *p. 34*

Disease in Context *Margaret's Case: Chemistry's Role in Health Science*

"Ugh," sighed Angela as she pulled into her hospital parking spot. The heat wave was into its second week, and she was getting tired of it. It was beginning to take its toll on the city too, especially on its infants and older residents. As Angela walked toward the hospital, she thought back to yesterday's ICU shift. One elderly patient stood out in her mind, probably because she reminded Angela of her own grandmother.

The patient, Margaret Ringland, a 78-year-old widow, lived alone in her apartment on New York's Upper East Side. Yesterday, her niece found Margaret collapsed on the floor, weak and confused. She called 911, and Margaret was rushed to the emergency room. According to her medical chart, Margaret presented with flushed dry skin, a sticky oral cavity, and a furrowed tongue. She was confused and disoriented. She also had hypotension (low blood pressure) and tachycardia (an elevated heart rate). All were classic signs of dehydration, a severe deficiency of water. Without adequate water, Margaret's body was unable to perform essential metabolic processes, and her tissues and organs were not in homeostatic balance.

Her neurologic symptoms were caused by changes in water volume. Although it was difficult to get a blood sample from Margaret's flattened veins, her blood work confirmed the initial diagnosis. Margaret's electrolyte levels were out of balance; specifically, she had a high blood sodium ion concentration, a condition called hypernatremia. Her hematocrit was also high, indicating low blood volume. This decrease was seriously affecting her cardiovascular system. Margaret's blood pressure had dropped, which forced her heart to beat faster to ensure proper delivery of blood to her tissues.

The emergency team started an IV line in Margaret's antebrachium. An aqueous solution of 5% dextrose (a sugar) was delivered through the IV at a rate of 500 mL/hour. A catheter was inserted into Margaret's urethra to allow for urinary drainage. Once stabilized, Margaret was moved to ICU for recovery.

Angela depends on her knowledge of chemistry to make sense of the signs and symptoms she observes in her patients. As you read this chapter, keep in mind that a firm understanding of the chemistry presented in this chapter will help you understand the anatomy and physiology of the cells, tissues, and organ systems discussed in subsequent chapters.

ANCILLARIES *At-A-Glance*

Visit thePoint to access the following resources. For guidance in using these resources most effectively, see pp. xv–xvii.

Learning RESOURCES

- Tips for Effective Studying
- Animation: Enzymes
- Health Professions: Pharmacist and Pharmacy Technician
- Detailed Chapter Outline
- Answers to Questions for Study and Review
- Audio Pronunciation Glossary

Learning ACTIVITIES

- Pre-Quiz
- Visual Activities
- Kinesthetic Activities
- Auditory Activities

A LOOK BACK

In Chapter 1, we learned that chemicals are the fundamental components of living organisms. In this chapter, we explore chemicals—some of their properties and how they react.

Greater understanding of living organisms has come to us through **chemistry**, the science that deals with the composition and properties of matter. Knowledge of chemistry and chemical changes helps us understand the body's normal and abnormal functioning. Food digestion in the intestinal tract, urine production by the kidneys, the regulation of breathing, and all other body activities involve the principles of chemistry. The many drugs used to treat diseases are also chemicals. Chemistry is used for their development and for understanding their actions in the body.

To provide some insights into the importance of chemistry in the life sciences, this chapter briefly describes elements, atoms, molecules, compounds, and mixtures, which are fundamental forms of matter. We also describe the chemicals that characterize organisms—organic chemicals.

Elements

Matter is anything that takes up space, that is, the materials from which the entire universe is made. **Elements** are the unique substances that make up all matter. The food we eat, the atmosphere, and water—everything around us and everything we can see and touch—are made from just 92 naturally occurring elements. (Twenty additional elements have been created in the laboratory.) Examples of elements include various gases, such as hydrogen, oxygen, and nitrogen; liquids, such as mercury used in barometers and other scientific instruments; and many solids, such as iron, aluminum, gold, silver, and zinc. Graphite (the so-called lead in a pencil), coal, charcoal, and diamonds are different forms of the element carbon.

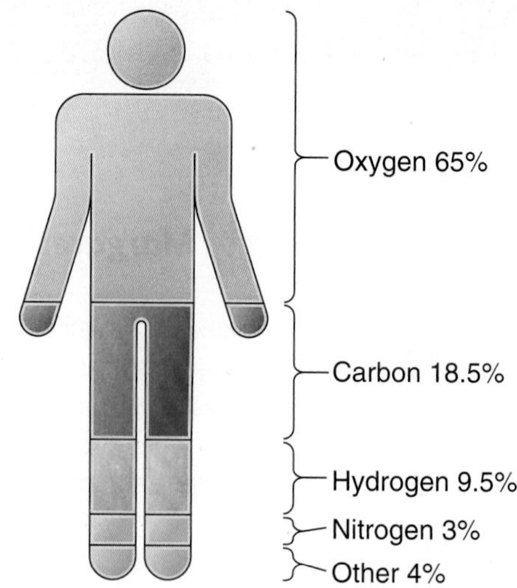

Figure 2-1 The body's chemical composition by weight. **KEY POINT** Oxygen, carbon, hydrogen, and nitrogen make up about 96% of body weight.

Elements can be identified by their names or their chemical symbols, which are abbreviations of their modern or Latin names. Each element is also identified by its own number, which is based on its atomic structure, discussed shortly. The periodic table is a chart used by chemists to organize and describe the elements. Appendix 1 shows the periodic table and gives some information about how it is used.

Of the 92 elements that exist in nature, only 26 have been found in living organisms. Hydrogen, oxygen, carbon, and nitrogen make up about 96% of body weight (**Fig. 2-1**). Nine additional elements—calcium, sodium, potassium, phosphorus, sulfur, chlorine, magnesium, iron, and iodine—make up most of the remaining 4% of the body's elements. The remaining 13, including zinc, selenium, copper, cobalt, chromium, and others, are present in extremely small (trace) amounts totaling about 0.1% of body weight. **Table 2-1** lists some of these elements along with their functions.

Table 2-1 Some Common Elements

Name	Symbol	Function
Oxygen	O	Part of water; needed to metabolize nutrients for energy
Carbon	C	Basis of all organic compounds; component of carbon dioxide, the gaseous byproduct of metabolism
Hydrogen	H	Part of water; participates in energy metabolism; determines the acidity of body fluids
Nitrogen	N	Present in all proteins, ATP (the energy-storing compound), and nucleic acids (DNA and RNA)
Calcium	Ca	Builds bones and teeth; needed for muscle contraction, nerve impulse conduction, and blood clotting
Phosphorus	P	Active ingredient in ATP; builds bones and teeth; component of cell membranes and nucleic acids
Potassium	K	Active in nerve impulse conduction; muscle contraction
Sulfur	S	Part of many proteins
Sodium	Na	Active in water balance, nerve impulse conduction, and muscle contraction
Iron	Fe	Part of hemoglobin, the compound that carries oxygen in red blood cells

The elements are listed in decreasing order by weight in the body.

ATOMIC STRUCTURE

The smallest units of elements are **atoms**. As such, atoms are the smallest complete units of matter. They cannot be broken down or changed into another form by ordinary chemical and physical means. Atoms are so small that millions of them could fit on the sharpened end of a pencil.

Despite the fact that the atom is so tiny, chemists have studied it extensively and have found that it has a definite structure composed of even smaller, or subatomic, particles. These particles differ as to their electric charge. Two types of electric charges exist in nature: positive (+) and negative (−). Particles with the same type of charge repel each other, but particles with different charges attract each other. So, a negatively charged particle would be repelled by another negatively charged particle but attracted by a positively charged particle. At the center of each atom is a nucleus composed of positively charged particles called **protons** (PRO-tonz) and noncharged particles called **neutrons** (NU-tronz) (**Fig. 2-2**). Together, the protons and neutrons contribute nearly all of the atom's weight.

In orbit around the nucleus are **electrons** (e-LEK-tronz). These nearly weightless particles are negatively charged. It is the electrons that determine how (or if) the atom will react chemically. The protons and electrons of an atom are equal in number so that the atom as a whole is electrically neutral (**see Fig. 2-2**). However, as we will see later, most atoms gain or lose electrons by interacting with other atoms and thus become electrically charged.

The **atomic number** of an element is equal to the number of protons that are present in the nucleus of its atoms. Because the number of protons is equal to the number of electrons, the atomic number also represents the number of electrons orbiting the nucleus. As you can see in **Figure 2-2**, oxygen has an atomic number of 8. No two elements share the same atomic number. Oxygen is the only element with the atomic number of 8. As another example, a carbon atom has six protons in the nucleus and six electrons orbiting the nucleus, so the atomic number of carbon is 6. In the Periodic Table of the Elements (see Appendix 1), the atomic number is located at the top of the box for each element. The atomic weight (mass), the sum of the protons and neutrons, is the number at the bottom of each box. It takes about 1,850 electrons to equal the weight of a single neutron or proton, so electrons are not counted in the determination of atomic weight.

The positively charged protons keep the negatively charged electrons in orbit around the nucleus by means of the opposite charges on the particles. Positively charged protons attract negatively charged electrons.

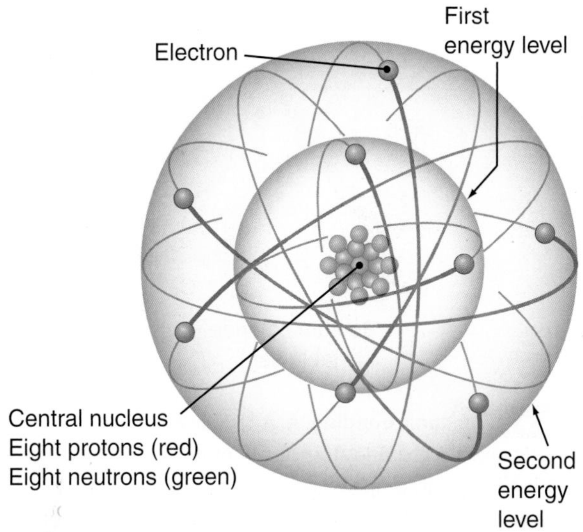

Figure 2-2 Representation of the oxygen atom. **KEY POINT** Eight protons and eight neutrons are tightly bound in the central nucleus. The eight electrons are in orbit around the nucleus, two in the first energy level, and six in the second. **ZOOMING IN** How does the number of protons in this atom compare with the number of electrons?

Energy Levels An atom's electrons orbit at specific distances from the nucleus in regions called energy levels. The first energy level, the one closest to the nucleus, can hold only two electrons. The second energy level, the next in distance away from the nucleus, can hold eight electrons. More distant energy levels can hold more than eight electrons, but they are stable (nonreactive) when they have eight.

The electrons in the energy level farthest away from the nucleus determine how the atom will react chemically. Atoms can donate, accept, or share electrons with other atoms to make the outermost level complete. In so doing, they form chemical bonds, as described shortly.

If the outermost energy level has more than four electrons but less than its capacity of eight, the atom typically completes this level by sharing or gaining electrons from one or more other atoms. The oxygen atom in **Figure 2-3**, illustrated with only the protons in the nucleus and the electrons in fixed position in their energy levels, has six electrons in its second, or outermost, level. When oxygen enters into chemical reactions, it must gain or share two electrons to achieve a complete outermost level.

In contrast, if an atom's outermost energy level has fewer than four electrons, the atom typically loses those electrons to empty the level. Magnesium (**see Fig. 2-3B**) has two electrons in the outermost energy level. In chemical reactions, it gives up those electrons, leaving the second level, complete with eight electrons, as the outermost level. Hydrogen (**see Fig. 2-3C**), having just one electron, can lose or share that one electron.

Carbon, which has four electrons in its outermost energy level, usually shares its electrons with multiple atoms in order to complete its outer energy level (**see Fig. 2-3D**). Atoms with a stable number of electrons in the outermost energy level are not reactive. Examples are the inert or "noble" gases, including helium, neon, and argon.

CHECKPOINTS

- ☐ **2-1** What are atoms?
- ☐ **2-2** What are three types of particles found in atoms?
- ☐ **2-3** Which of these atoms would be more likely to participate in a chemical reaction—an atom with eight electrons in its outermost energy level or an atom with six electrons in its outermost energy level?

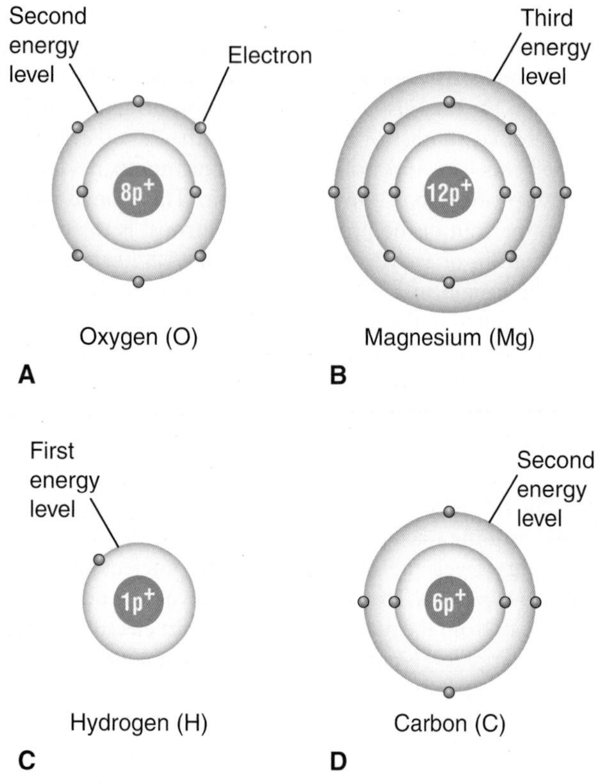

Figure 2-3 **Examples of atoms.** **KEY POINT** The first energy level can hold two electrons, and the second and third can hold eight. The outermost energy level determines chemical reactivity. **ZOOMING IN** How many electrons does oxygen need to complete its outermost energy level? How does magnesium achieve a stable outermost energy level?

Chemical Bonds

When an atom interacts with other atoms to stabilize its outermost energy level, a bond is formed between the atoms. In these chemical reactions, electrons may be transferred from one atom to another or may be shared between atoms. The number of bonds an atom needs to form in order to stabilize its outermost energy level is called its **valence** (from a Latin word that means "strength"). An atom needs to form one bond for every electron it donates, accepts, or shares, so valence can also be defined as the number of electrons lost, gained, or shared by atoms of an element in chemical reactions. Referring back to our examples in the previous section, oxygen has six electrons in the outer energy level and must form two bonds to reach a stable outer energy level of eight. Therefore, oxygen's valence is two. Magnesium's valence is also two, because it must transfer two electrons to achieve a stable outermost energy level.

IONIC BONDS

When electrons are transferred from one atom to another, the type of bond formed is called an **ionic** (i-ON-ik) **bond**. The sodium atom, for example, tends to lose the single electron in its outermost shell leaving the now outermost shell with a stable number of electrons (eight) (**Fig. 2-4**). Removal of a single electron from the sodium atom leaves one more proton than electrons, and the sodium then has a single net positive charge. The sodium in this form is symbolized as Na^{+}. Calcium loses two electrons when it participates in ionic bonds, so the calcium ion has two positive charges and is abbreviated Ca^{2+}.

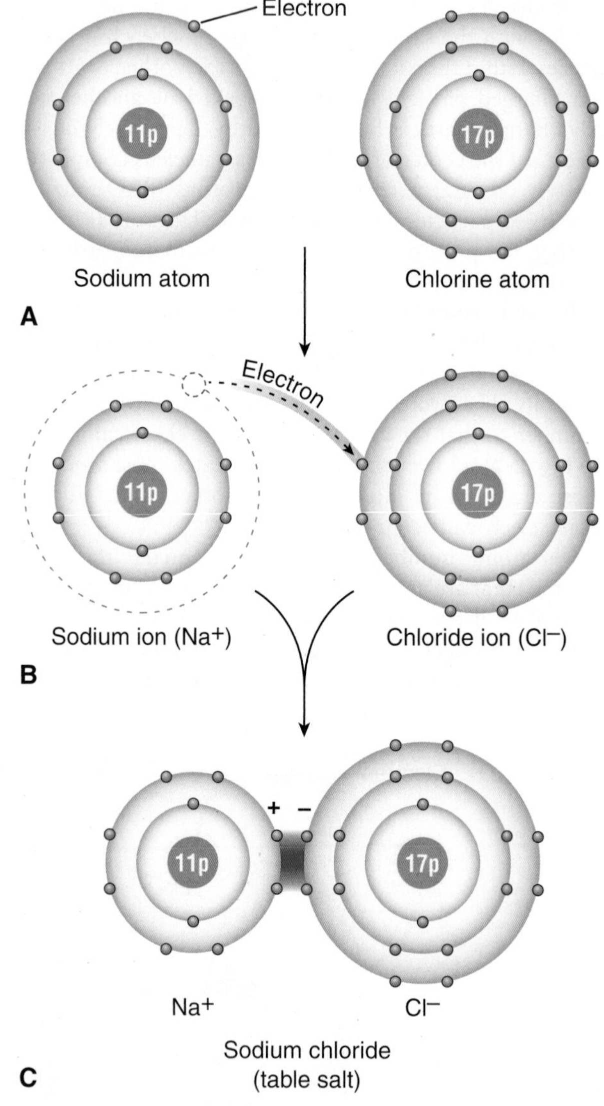

Figure 2-4 **Ionic bonding. A.** A sodium atom has 11 protons and 11 electrons. A chlorine atom has 17 protons and 17 electrons. **B.** A sodium atom gives up one electron to a chlorine atom in forming an ionic bond. The sodium atom now has 11 protons and 10 electrons, resulting in a positive charge of one. The chlorine becomes negatively charged by one, with 17 protons and 18 electrons. **C.** The ionic bond between the sodium ion (Na^{+}) and the chloride ion (Cl^{-}) forms the compound sodium chloride (table salt). **ZOOMING IN** How many electrons are in the outermost energy level of a sodium atom? Of a sodium ion?

Alternately, atoms can gain electrons so that there are more electrons than protons. Chlorine, which has seven electrons in its outermost energy level, tends to gain one electron to fill the level to its capacity. The resultant chlorine is negatively charged (Cl^-) (**see Fig. 2-4**). (Chemists refer to this charged form of chlorine as *chloride*.) An atom or group of atoms that has acquired a positive or negative charge is called an **ion** (I-on). Any ion that is positively charged is a **cation** (CAT-i-on). Any negatively charged ion is an **anion** (AN-i-on).

Imagine a sodium atom coming in contact with a chlorine atom. The sodium atom gives up its outermost electron to the chlorine and becomes positively charged; the chlorine atom gains the electron and becomes negatively charged. The two newly formed ions (Na^+ and Cl^-), because of their opposite charges, attract each other to produce sodium chloride, ordinary table salt (**see Fig. 2-4**). The attraction between the oppositely charged ions forms an ionic bond. Sodium chloride and other ionically bonded substances tend to form crystals when solid and to dissolve easily in water.

Electrolytes When ionically bonded substances dissolve in water, the atoms separate as ions. Compounds that release ions when they dissolve in water are called **electrolytes** (e-LEK-tro-lites). Note that in practice, the term *electrolytes* is also used to refer to the ions themselves in body fluids. Electrolytes include a variety of salts, such as sodium chloride and potassium chloride. They also include acids and bases, which are responsible for the acidity or alkalinity of body fluids, as described shortly. Electrolytes must be present in the proper concentrations in the intracellular and extracellular fluids, or damaging effects will result, as seen in Margaret's case study, which opens this chapter.

Ions in the Body Body fluids contain many different ions. Indeed, many of the elements listed in **Table 2-1** are only active in their ionic forms. Sodium (Na^+) and potassium (K^+) ions, for instance, play critical roles in the transmission of electric signals by virtue of their positive charges. The concentration of many different ions in body fluids must be kept within narrow limits in order to maintain homeostasis.

Because ions are charged particles, electrolytic solutions can conduct an electric current. Records of electric currents in tissues are valuable indications of the functioning or malfunctioning of tissues and organs. The **electrocardiogram** (e-lek-tro-KAR-de-o-gram) and the **electroencephalogram** (e-lek-tro-en-SEF-ah-lo-gram) are graphic tracings of the electric currents generated by the heart muscle and the brain, respectively (see Chapters 10 and 14).

COVALENT BONDS

Although ionic bonds form some chemicals, many more are formed by another type of chemical bond. This bond involves not the exchange of electrons but a sharing of electrons between the atoms and is called a **covalent bond**. This name comes from the prefix *co-*, meaning "together," and *valence*, referring to the electrons involved in chemical reactions between atoms. In a covalently bonded substance, the shared electrons orbit around both of the atoms, making both of them stable. Covalent bonds may involve the sharing of one, two, or three pairs of electrons between atoms.

In some covalent bonds, the electrons are equally shared, as in the combination of two identical atoms of hydrogen, oxygen, or nitrogen (**Fig. 2-5**). Electrons may also be shared equally in some bonds involving different atoms—methane (CH_4), for example. If electrons are equally shared in forming a bond, the electric charges are evenly distributed around the atoms and the bond is described as a *nonpolar covalent bond*. That is, no part of the combined particle is more negative or positive than any other part. More commonly, the electrons are held closer to one atom than the other, as in the case of water (H_2O), shown in **Figure 2-6**. In water, the shared electrons are actually closer to the oxygen atom than the hydrogen atoms at any one time, making the oxygen region more negative. Such bonds are called *polar covalent bonds*, because one region of the combination is more negative and one part is more positive at any one time.

MOLECULES AND COMPOUNDS

When two or more atoms unite covalently, they form a **molecule** (MOL-eh-kule). A molecule is thus the smallest unit of a covalently bonded substance that retains all the properties of that substance. A molecule can be made of like atoms—the oxygen molecule is made of two identical atoms, for example—but more often a molecule is made of atoms of two or more different elements. For example, a water molecule (H_2O) contains one atom of oxygen (O) and two atoms of hydrogen (H) (**see Fig. 2-6**). Any particle resulting from a polar covalent bond is known as a polar molecule. Some polar molecules can gain or lose electrons and thus become ions. Bicarbonate ions (HCO_3^-), for instance, participate in the regulation of body fluid acidity. Polar molecules, whether they are charged or uncharged, can interact using weak bonds called hydrogen bonds, as discussed in **Box 2-1**.

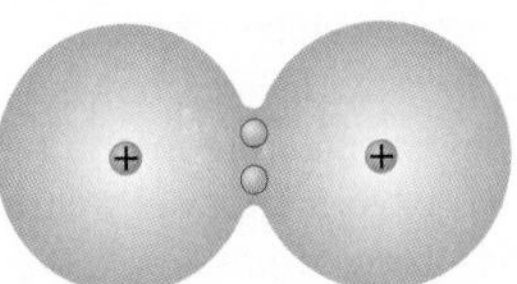

Hydrogen molecule (H_2)

Figure 2-5 **A nonpolar covalent bond.** **KEY POINT** The electrons involved in the bonding of two hydrogen atoms are equally shared between the two atoms. The electrons orbit evenly around the two. **ZOOMING IN** How many electrons are needed to complete the energy level of each hydrogen atom?

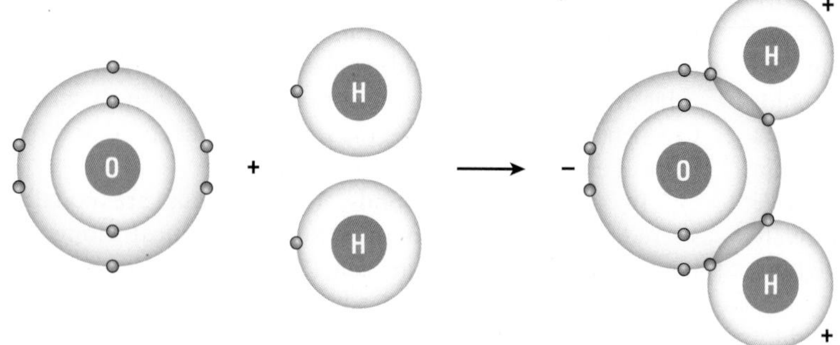

Figure 2-6 **Formation of water.** **KEY POINT** Polar covalent bonds form water. The unequal sharing of electrons makes the region near the oxygen nucleus more negative and the region near the hydrogen nucleus more positive. **ZOOMING IN** How many hydrogen atoms bond with an oxygen atom to form water?

Note that chemists do not consider ionically bonded substances to be composed of molecules, as their atoms are held together by electrical attraction only. The bonds that hold these atoms together are weak, and the components separate easily in solution into ions, as already described. Thus, unlike ionic bonds and hydrogen bonds, only covalent bonds form molecules.

Any substance composed of two or more different elements is called a **compound**. This definition includes both ionically and covalently bonded substances. The formula for a compound shows all the elements that make up that compound in their proper ratio, such as NaCl, H_2O, and CO_2. Some compounds are made of a few elements in a simple combination. For example, molecules of the gas carbon monoxide (CO) contain one atom of carbon (C) and one atom of oxygen (O). Other compounds have very large and complex molecules. Such complexity characterizes many of the compounds found in living organisms. Some protein molecules, for example, have thousands of atoms.

It is interesting to observe how different a compound is from any of its constituents. For example, a molecule of liquid water is formed from oxygen and hydrogen, both of which are gases. Another example is the sugar glucose ($C_6H_{12}O_6$). Its constituents include 12 atoms of the gas hydrogen, six atoms of the gas oxygen, and six atoms of the solid element carbon. The component gases and the solid carbon do not in any way resemble the glucose.

CHECKPOINTS

- 2-4 Which type of chemical bond is formed by an exchange of electrons? Which type is formed by a sharing of electrons?
- 2-5 What happens when an electrolyte goes into solution?
- 2-6 What are molecules, and what are compounds?

Box 2-1

A CLOSER LOOK

Hydrogen Bonds: Strength in Numbers

In contrast to ionic and covalent bonds, which hold atoms together, hydrogen bonds hold molecules together. Hydrogen bonds are much weaker than ionic or covalent bonds—in fact, they are more like "attractions" between molecules. While ionic and covalent bonds rely on electron transfer or sharing, hydrogen bonds form bridges between two molecules. A hydrogen bond forms when a slightly positive hydrogen atom in one molecule is attracted to a slightly negative atom in another molecule. Even though a single hydrogen bond is weak, many hydrogen bonds between two molecules can be strong.

Hydrogen bonds hold water molecules together, with the slightly positive hydrogen atom in one molecule attracted to a slightly negative oxygen atom in another. Many of water's unique properties come from its ability to form hydrogen bonds. For example, hydrogen bonds keep water liquid over a wide range of temperatures, which provides a constant environment for body cells.

Hydrogen bonds form not only between molecules but also within large molecules. Hydrogen bonds between regions of the same molecule cause it to fold and coil into a specific shape, as in the process that creates the precise three-dimensional structure of proteins. Because a protein's structure determines its function in the body, hydrogen bonds are essential to protein activity.

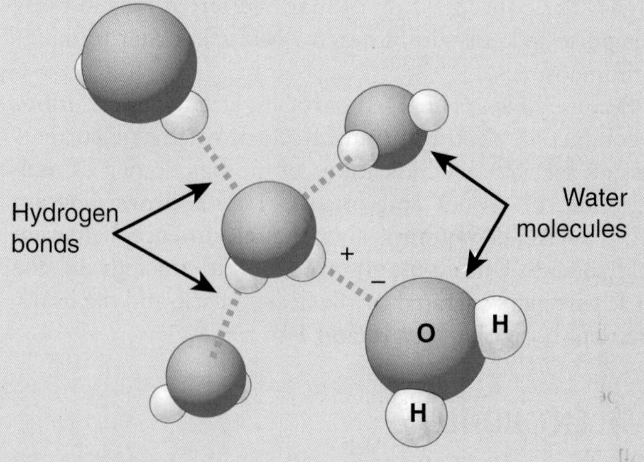

Hydrogen bonds. The bonds shown here are holding water molecules together.

Mixtures

Not all elements or compounds react chemically when brought together. The air we breathe every day is a combination of gases, largely nitrogen, oxygen, and carbon dioxide, along with smaller percentages of other substances. The constituents in the air maintain their identity, although the proportions of each may vary. Blood plasma—the fluid portion of blood—is also a combination in which the various components maintain their identity. The many valuable compounds in the plasma remain separate entities with their own properties. Such combinations are called **mixtures**—blends of two or more substances (**Table 2-2**).

SOLUTIONS AND SUSPENSIONS

A mixture formed when one substance dissolves in another is called a **solution**. One example is salt water. In a solution, the component substances cannot be distinguished from each other and remain evenly distributed throughout; that is, the mixture is homogeneous (ho-mo-JE-ne-us). The dissolving substance, which in the body is water, is the **solvent**. The substance dissolved, table salt in the case of salt water, is the **solute**. An **aqueous** (A-kwe-us) **solution** is one in which water is the solvent. Aqueous solutions of glucose, salts, or both of these together are used for intravenous fluid treatments.

In some mixtures, the substance distributed in the background material is not dissolved and will settle out unless the mixture is constantly shaken. This type of nonuniform, or heterogeneous (het-er-o-JE-ne-us), mixture is called a **suspension**. The particles in a suspension are separate from the material in which they are dispersed, and they settle out because they are large and heavy. Examples of suspensions are milk of magnesia, finger paints, and in the body, red blood cells suspended in blood plasma.

One other type of mixture is important in body function. Some organic compounds form **colloids**, in which the molecules do not dissolve yet remain evenly distributed in the suspending material. The particles have electric charges that repel each other, and the molecules are small enough to stay in suspension. The fluid that fills the cells (cytosol) is a colloid, as is blood plasma.

Many mixtures are complex, with properties of solutions, suspensions, and colloids. For instance, blood plasma has dissolved compounds, making it a solution. The red blood cells and other formed elements give blood the property of a suspension. The proteins in the plasma give it the property of a colloid. Chocolate milk also has all three properties.

THE IMPORTANCE OF WATER

Water is the most abundant compound in the body. No plant or animal can live very long without it. Water is of critical importance in all physiological processes in body tissues. A deficiency of water, or dehydration (de-hi-DRA-shun), can be a serious threat to health, as illustrated by Margaret's case study. Water carries substances to and from the cells and makes possible the essential processes of absorption, exchange, secretion, and excretion. What are some of the properties of water that make it such an ideal medium for living cells?

- Water can dissolve many different substances in large amounts. For this reason, it is called the *universal solvent*. Many of the body's necessary materials, such as gases and nutrients, dissolve in water to be carried from place to place. Substances, such as salt, that mix with or dissolve in water are described as *hydrophilic* ("water-loving"); substances, such as fats, that do not dissolve in water are described as *hydrophobic* ("water-fearing").
- Water is stable as a liquid at ordinary temperatures. It does not freeze until the temperature drops to 0°C (32°F) and does not boil until the temperature reaches 100°C (212°F). This stability provides a consistent environment for living cells. Water can also be used to distribute heat throughout the body and to cool the body by evaporation of sweat from the body surface.
- Water participates in the body's chemical reactions. It is needed directly in the digestive process and in many of the metabolic reactions that occur in the cells.

Table 2-2 Mixtures

Type	Definition	Example
Solution	Homogeneous mixture formed when one substance (solute) dissolves in another (solvent)	Table salt (NaCl) dissolved in water; table sugar (sucrose) dissolved in water
Suspension	Heterogeneous mixture in which one substance is dispersed in another but will settle out unless constantly mixed	Red blood cells in blood plasma; milk of magnesia
Colloid	Heterogeneous mixture in which the suspended particles remain evenly distributed based on the small size and opposing charges of the particles	Blood plasma; cytosol

CHECKPOINTS

- 2-7 What is the difference between solutions and suspensions?
- 2-8 What is the most abundant compound in the body?

Acids, Bases, and Salts

An **acid** (AH-sid) is a chemical substance capable of releasing a hydrogen ion (H^+) when dissolved in water. A common example is hydrochloric acid (HCl), the acid found in stomach juices. HCl releases hydrogen ions in solution as follows:

$$\underset{\text{(hydrochloric acid)}}{HCl} \rightarrow \underset{\text{(hydrogen ion)}}{H^+} + \underset{\text{(chloride ion)}}{Cl^-}$$

A **base** is a chemical substance that can accept (react with) a hydrogen ion. A base is also called an **alkali** (AL-kah-li), and bases are described as alkaline. Most bases release a hydroxide ion (OH^-) in solution, and the hydroxide ion subsequently accepts a hydrogen ion to form water. Sodium hydroxide is an example of a base:

$$\underset{\text{(sodium hydroxide)}}{NaOH} \rightarrow \underset{\text{(sodium ion)}}{Na^+} + \underset{\text{(hydroxide ion)}}{OH^-}$$

$$\underset{\text{(hydroxide ion)}}{OH^-} \rightarrow \underset{\text{(hydrogen ion)}}{H^+} + \underset{\text{(water)}}{H_2O}$$

A reaction between an acid and a base produces a **salt** and also water. In the reaction, the hydrogen of the acid is replaced by the positive ion of the base. A common example of a salt is sodium chloride (NaCl), or table salt, produced by the reaction:

$$HCl + NaOH \rightarrow NaCl + H_2O$$

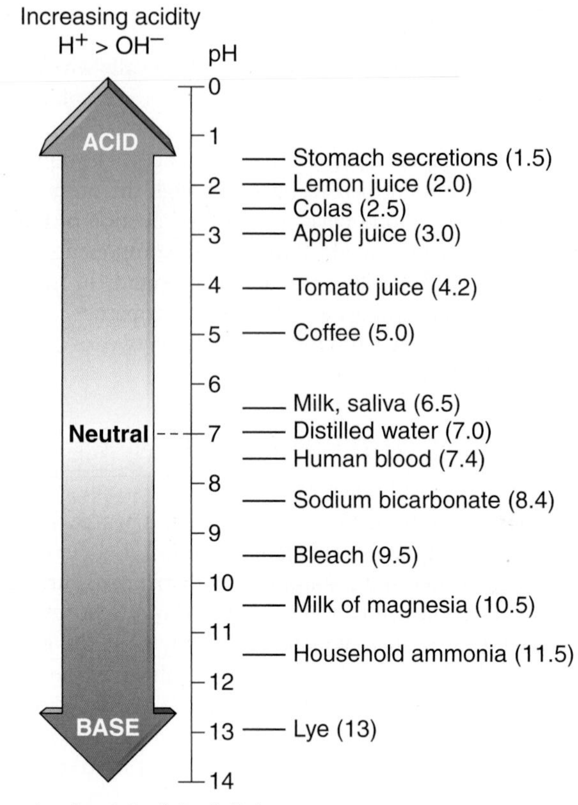

Figure 2-7 The pH scale. Degree of acidity or alkalinity is shown in pH units. This scale also shows the pH of some common substances. **ZOOMING IN** What happens to the amount of hydroxide ion (OH^-) present in a solution when the amount of hydrogen ion (H^+) increases?

THE pH SCALE

The greater the concentration of hydrogen ions in a solution, the greater the acidity of that solution. The greater the concentration of hydroxide ion (OH^-), the greater the alkalinity of the solution. The concentrations of H^+ and OH^- in a solution are inversely related; as the concentration of hydrogen ions increases, the concentration of hydroxide ions decreases. Conversely, as the concentration of hydroxide ions increases, the concentration of hydrogen ions decreases.

Acidity and alkalinity are indicated by **pH** units, which represent the relative concentrations of hydrogen and hydroxide ions in a solution. The pH units are listed on a scale from 0 to 14, with 0 being the most acidic and 14 being the most basic (**Fig. 2-7**). A pH of 7.0 is neutral. At pH 7.0, the solution has an equal number of hydrogen and hydroxide ions. Pure water has a pH of 7.0. Solutions that measure less than 7.0 are acidic; those that measure above 7.0 are alkaline (basic).

Because the pH scale is based on multiples of 10, each pH unit on the scale represents a 10-fold change in the number of hydrogen and hydroxide ions present. A solution registering 5.0 on the scale has 10 times the number of hydrogen ions as a solution that registers 6.0. The pH 5.0 solution also has one-tenth the number of hydroxide ions as the solution of pH 6.0. A solution registering 9.0 has one-tenth the number of hydrogen ions and 10 times the number of hydroxide ions as one registering 8.0. Thus, the lower the pH reading, the greater is the acidity, and the higher the pH, the greater is the alkalinity.

Blood and other body fluids are close to neutral but are slightly on the alkaline side, with a pH range of 7.35 to 7.45. Urine averages pH 6.0 but may range from 4.6 to 8.0 depending on body conditions and diet. **Figure 2-7** shows the pH of some other common substances.

Because body fluids are on the alkaline side of neutral, the body may be in a relatively acidic state even if the pH does not drop below 7.0. For example, if a patient's pH falls below 7.35 but is still greater than 7.0, the patient is described as being in an acidic state known as *acidosis*. Thus, within this narrow range, physiologic acidity differs from acidity as defined by the pH scale.

An increase in pH to readings greater than 7.45 is termed *alkalosis*. Any shifts in pH to readings above or below the normal range can be dangerous, even fatal.

BUFFERS

In a healthy person, body fluids are delicately balanced within narrow limits of acidity and alkalinity. This balanced chemical state is maintained in large part by **buffers**. Chemical buffers form a system that prevents sharp changes in hydrogen ion concentration and thus maintains a relatively constant pH. Buffers are important in maintaining stability in the pH of body fluids. More information about body fluids, pH, and buffers can be found in Chapter 21.

CHECKPOINTS

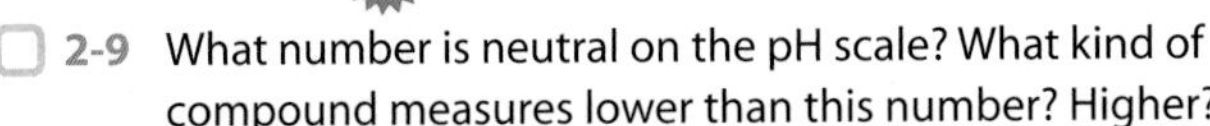

☐ 2-9 What number is neutral on the pH scale? What kind of compound measures lower than this number? Higher?

☐ 2-10 What is a buffer?

Isotopes and Radioactivity

Elements may exist in several forms, each of which is called an **isotope** (I-so-tope). These forms are alike in their numbers of protons and electrons but differ in their atomic weights because of differing numbers of neutrons in the nucleus. The most common form of oxygen, for example, has eight protons and eight neutrons in the nucleus, giving the atom an atomic weight of 16 atomic mass units (amu). But there are some isotopes of oxygen with only six or seven neutrons in the nucleus and others with nine to 11 neutrons. The isotopes of oxygen thus range in atomic weight from 14 to 19 amu.

Some isotopes are stable and maintain constant characteristics. Others fall apart and radiate (give off) subatomic particles and/or electromagnetic (energy) waves called *gamma rays*. (Other types of electromagnetic waves are visual light, ultraviolet light, and x-rays.) Isotopes that fall apart easily are said to be **radioactive**. Radioactive elements, also called *radioisotopes*, may occur naturally, as is the case with isotopes of the very heavy elements radium and uranium. Others may be produced artificially by placing the atoms of lighter, nonradioactive elements in accelerators that smash their nuclei together.

The radiation given off by some radioisotopes is used in the treatment of cancer because it can penetrate and destroy tumor cells. A growing tumor contains immature, dividing cancer cells, which are more sensitive to the effects of radiation than are mature body cells. The greater sensitivity of these younger cells allows radiation therapy to selectively destroy them with minimal damage to normal tissues. Modern radiation instruments produce tremendous amounts of energy (in the multimillion electron-volt range) that can destroy deep-seated cancers without causing serious skin reactions.

In radiation treatment, a radioisotope, such as cobalt 60, is sealed in a stainless steel cylinder and mounted on an arm or crane. Beams of radioactivity are then directed through a porthole to the area to be treated. Implants containing radioisotopes in the form of needles, seeds, or tubes also are widely used in the treatment of different types of cancer.

In addition to its therapeutic values, radiation is extensively used in diagnosis. Radioactive elements that can be administered and then detected internally to identify abnormalities are called *tracers*. Radioactive iodine, for instance, can diagnose problems of the thyroid gland (see Box 2-2). X-rays are electromagnetic waves produced using high voltage instead of radioisotopes. X-rays penetrate tissues and produce an image of internal structures on a photographic plate.

When using radiation in diagnosis or therapy, healthcare personnel must follow strict precautions to protect themselves and the patient, because the rays can destroy healthy as well as diseased tissues.

CHECKPOINT

☐ 2-11 What word is used to describe isotopes that give off radiation?

Box 2-2

HOT TOPICS

Radioactive Tracers: Medicine Goes Nuclear

Like radiography, computed tomography (CT), and magnetic resonance imaging, **nuclear medicine imaging** (NMI) offers a noninvasive way to look inside the body. An excellent diagnostic tool, NMI not only shows structural details but also provides information about body function. NMI can help diagnose cancer, stroke, and heart disease earlier than do techniques that provide only structural information.

NMI uses **radiotracers**, radioactive substances that specific organs absorb. For example, radioactive iodine is used to image the thyroid gland, which absorbs more iodine than does any other organ. After a patient ingests, inhales, or is injected with a radiotracer, a device called a gamma camera detects the radiotracer in the organ under study and produces a picture, which is used in making a diagnosis. Radiotracers are broken down and eliminated through urine or feces, so they leave the body quickly. A patient's exposure to radiation in NMI is usually considerably lower than with x-ray or CT scan.

Three NMI techniques are **positron emission tomography** (PET), **bone scanning**, and the **myocardial perfusion imaging** (MPI) stress test. PET is often used to evaluate brain activity by measuring the brain's use of radioactive glucose. PET scans can reveal brain tumors because tumor cells are often more metabolically active than are normal cells and thus absorb more radiotracer. Bone scanning detects radiation from a radiotracer absorbed by bone tissue with an abnormally high metabolic rate, such as a bone tumor. The MPI test is used to diagnose heart disease. A nuclear medicine technologist injects the patient with a radionuclide (e.g., thallium, technetium), and a gamma camera images the heart during exercise and later rest. When compared, the two sets of images help evaluate blood flow to the working, or "stressed," heart.

Organic Compounds

The complex molecules that characterize living things are called **organic compounds.** All of these are built on the element **carbon.** Because carbon atoms can form covalent bonds with a variety of different elements and can even covalently bond to other carbon atoms to form long chains, most organic compounds consist of large, complex molecules. The starch found in potatoes, the fat and protein in tissues, hormones, and many drugs are examples of organic compounds. These large molecules are often formed from simpler molecules called *building blocks*, or *monomers (mono-* means "one"), which bond together in long chains.

The main types of organic compounds are carbohydrates, lipids, and proteins. All of these organic compounds contain carbon, hydrogen, and oxygen as their main ingredients. Carbohydrates, lipids, and proteins (in addition to minerals, vitamins, and water) must be taken in as part of a normal diet. These nutrients are discussed further in Chapters 19 and 20.

CARBOHYDRATES

The building blocks of **carbohydrates** (kar-bo-HI-drates) are simple sugars, or **monosaccharides** (mon-o-SAK-ah-rides) (**Fig. 2-8**). (The word root *sacchar/o* means "sugar.") **Glucose** (GLU-kose), a simple sugar that circulates in the blood as a cellular nutrient, is an example of a monosaccharide. Two simple sugars may be linked together to form a **disaccharide** (**see Fig. 2-8B**), as represented by sucrose, or table sugar. More complex carbohydrates, or **polysaccharides**, consist of many simple sugars linked together (**see Fig. 2-8C**). (The prefix *di-* means "two," and *poly-* means "many.") Examples of polysaccharides are starch, which is manufactured in plant cells, and **glycogen** (GLI-ko-jen), a storage form of glucose found in liver cells and skeletal muscle cells. Carbohydrates in the form of sugars and starches are important dietary sources of energy.

LIPIDS

Lipids are a class of organic compounds that are not soluble in water. They are mainly found in the body as **fat.** Simple fats are made from a substance called glycerol (GLIS-er-ol), commonly known as glycerin, in combination with three fatty acids (**Fig. 2-9**). One fatty acid is attached to each of the three carbon atoms in glycerol, so simple fats are described as **triglycerides** (tri-GLIS-er-ides) (the prefix *tri-* means "three"). Fats insulate the body and protect internal organs. In addition, fats are the main form in which energy is stored, and most cells use fatty acids for energy.

Glucose (dextrose)

HOCH₂ O OH HO OH OH

A Monosaccharide

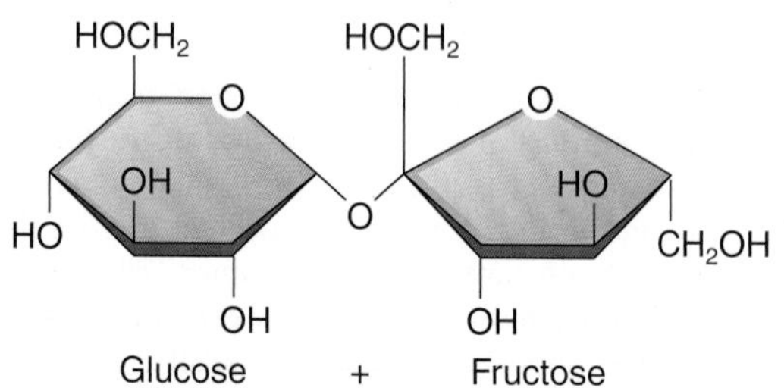

B Disaccharide

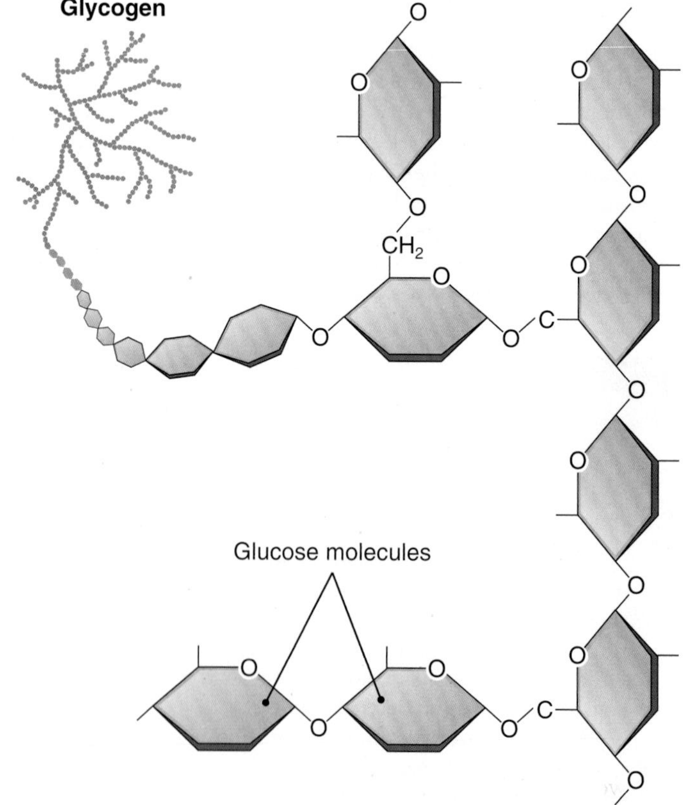

C Polysaccharide

Figure 2-8 **Examples of carbohydrates.** **KEY POINT** A monosaccharide **(A)** is a simple sugar. A disaccharide **(B)** consists of two simple sugars linked together, whereas a polysaccharide **(C)** consists of many simple sugars linked together in chains. **ZOOMING IN** What are the building blocks (monomers) of disaccharides and polysaccharides?

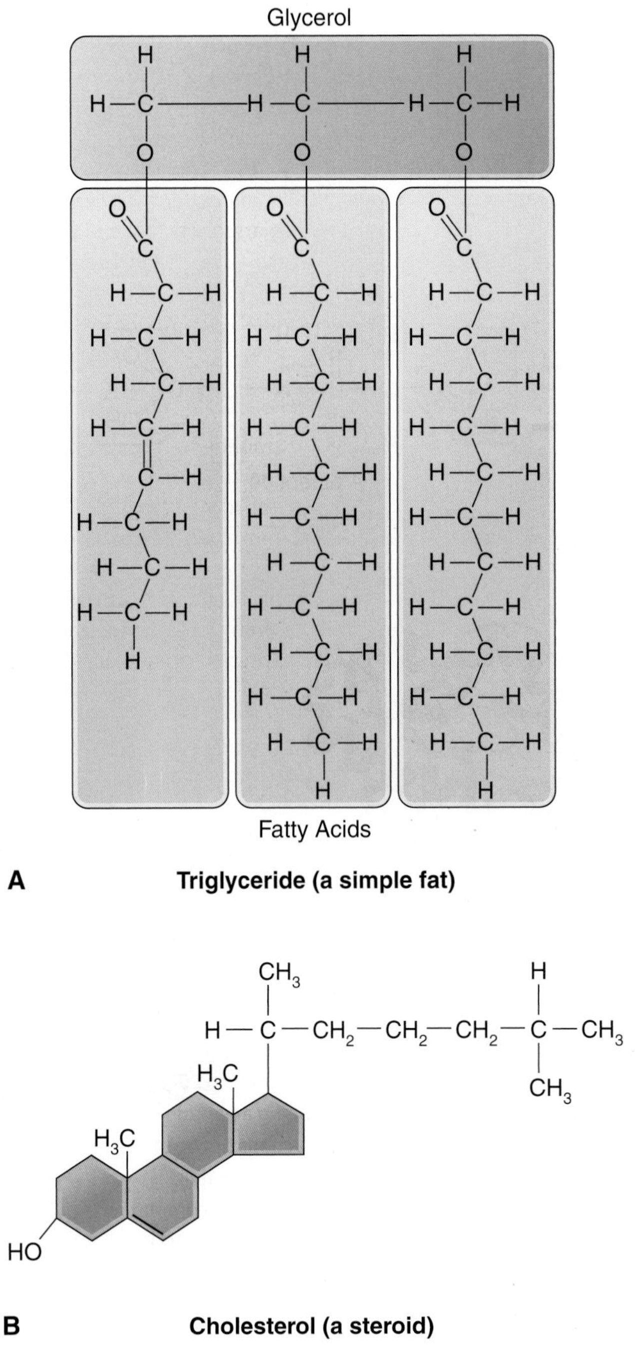

Glycerol

Fatty Acids

A **Triglyceride (a simple fat)**

B **Cholesterol (a steroid)**

Figure 2-9 **Lipids. A.** A triglyceride, a simple fat, contains glycerol combined with three fatty acids. **B.** Cholesterol is a type of steroid, a lipid that contains rings of carbon atoms. **ZOOMING IN** How many carbon atoms are in glycerol?

Two other types of lipids are important in the body. Phospholipids (fos-fo-LIP-ids) are complex lipids containing the element phosphorus. Among other functions, phospholipids make up a major part of the membrane around living cells. **Steroids** are lipids that contain rings of carbon atoms. The most important sterol is **cholesterol** (ko-LES-ter-ol), another component of cellular membranes (**see Fig. 2-9B**). Cholesterol is also used to make steroid hormones, including cortisol, testosterone, and estrogen.

PROTEINS

All **proteins** (PRO-tenes) contain, in addition to carbon, hydrogen, and oxygen, the element **nitrogen** (NI-tro-jen). They may also contain sulfur or phosphorus. Proteins are the body's structural materials, found in muscle, bone, and connective tissue. They also make up the pigments that give hair, eyes, and skin their colors. It is proteins that make each individual physically distinct from others. Proteins also serve functional roles. For instance, some act as transporters, moving substances across cell membranes. Other proteins, known as *enzymes*, promote metabolic reactions. Enzymes are discussed further below.

Proteins are composed of monomers called **amino** (ah-ME-no) **acids** (**Fig. 2-10**). Although only about 20 different amino acids exist in the body, a vast number of proteins can be made by linking them together in different combinations.

Each amino acid contains an acid group (COOH) and an amino group (NH_2), the part of the molecule that has the nitrogen. These groups are attached to either side of a carbon atom linked to a hydrogen atom. The remainder of the molecule, symbolized by R in **Figure 2-10A**, is different in each amino acid, ranging from a single hydrogen atom to a complex chain or ring of carbon and other elements. These variations in the R region of the molecule account for the differences in the amino acids.

In forming proteins, the acid group of one amino acid covalently bonds with the amino group of another amino acid (**Fig. 2-10B**). This bond is called a *peptide bond*. Many amino acids linked together in this way form a protein, which is essentially a long chain of amino acids. (Shorter chains of amino acids are sometimes called polypeptides). The linear chain of amino acids can fold into specific shapes because of chemical attractions between nonadjacent amino acids. The most common of these simple shapes is a helix (spiral) (**Fig. 2-10C**). The final, functional form of a protein depends on further interactions between these simple shapes (**Fig. 2-10D**). Some proteins consist of multiple protein chains, each folded into a helix, coiled together into ropelike structures (**Fig. 2-10D, left side**). These proteins are known as fibrous proteins, and they play important roles in body structure. Collagen, for instance, provides structure to bones and cartilage. Other proteins, known as globular proteins, consist of helices (or other simple shapes) folded back on themselves into complex three-dimensional structures (**Fig. 2-10D, right side**). Myoglobin, for example, is a globular protein similar to hemoglobin that stores oxygen in muscle cells. Other globular proteins include hormones, antibodies needed for immunity, enzymes (described below), and many other metabolically active compounds. The overall three-dimensional shape of a protein is important to its function, as can be seen in the activity of enzymes.

Enzymes **Enzymes** (EN-zimes) are proteins that are essential for metabolism. They are **catalysts** (KAT-ah-lists) in the hundreds of reactions that take place within cells. Without these catalysts, which increase the speed of

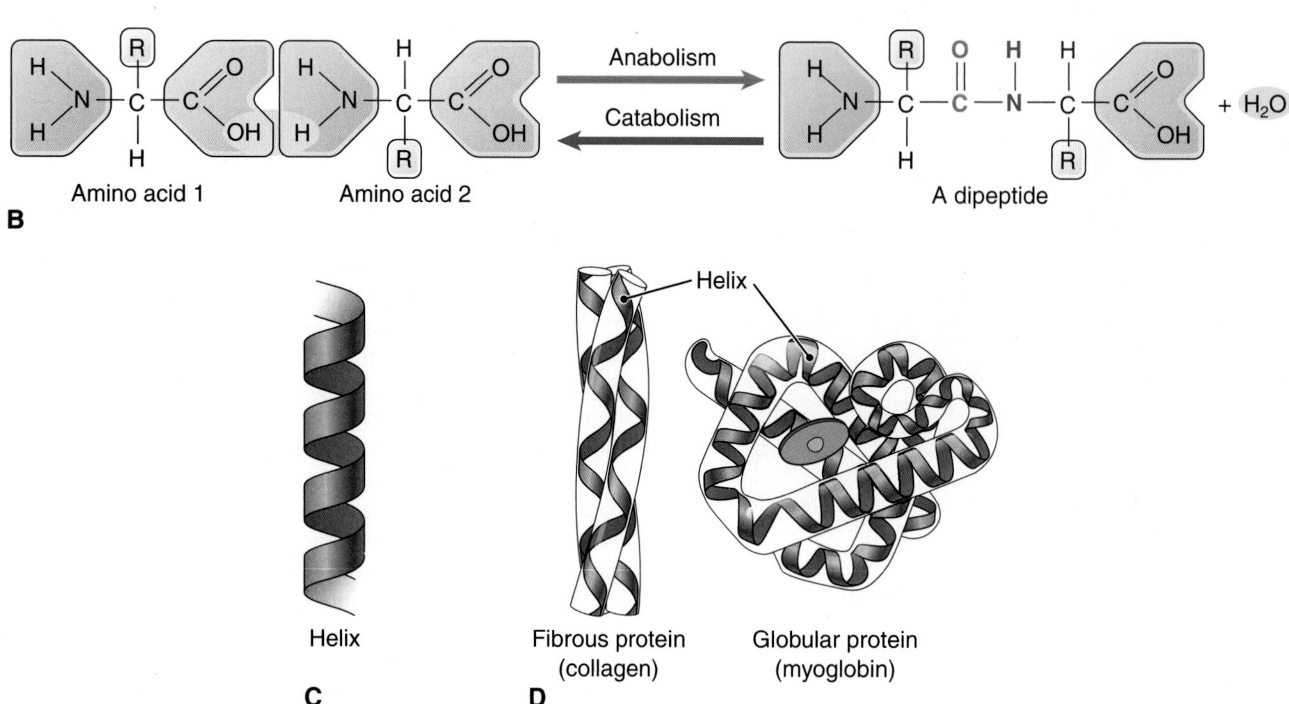

Figure 2-10 **Proteins. A.** Amino acids are the building blocks of proteins. Each amino acid contains an amino group and an acid group attached to a carbon atom. The remainder of the molecule (shown by R) can vary in 20 different ways. **B.** The acid group of one amino acid can react with the amino group of another forming a peptide bond. Further additions of amino acids result in formation of a polypeptide chain. **C.** Chemical attractions between nonadjacent amino acids form simple shapes, such as a helix. **D.** Fibrous proteins consist of multiple protein helices coiled together. Globular proteins consist of helices (or other simple shapes) folded back on themselves into complex three-dimensional structures. The characteristic shape of each protein is critical to its function. **ZOOMING IN** What part of an amino acid contains nitrogen?

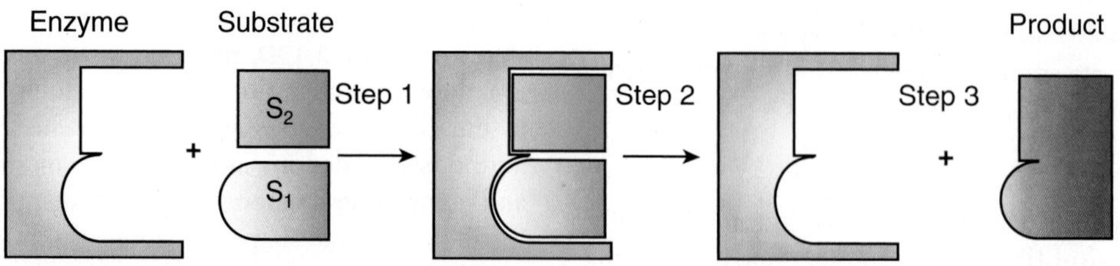

Figure 2-11 **Diagram of enzyme action.** **KEY POINT** An enzyme joins with substrate 1 (S_1) and substrate 2 (S_2) and speeds up the chemical reaction in which the two substrates bond. Once a new product is formed from the substrates, the enzyme is released unchanged. **ZOOMING IN** How does the shape of the enzyme before the reaction compare with its shape after the reaction?

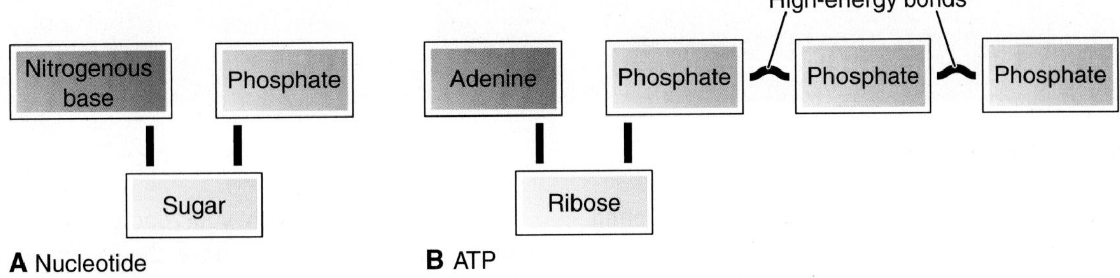

Figure 2-12 **Nucleotides. A.** A nucleotide consists of a nitrogenous base, a sugar, and one or more phosphate groups. **B.** ATP has high-energy bonds between the phosphates. When these bonds are broken, energy is released. **ZOOMING IN** What does the prefix *tri-* in adenosine triphosphate mean?

chemical reactions, metabolism would not occur at a fast enough rate to sustain life. Because each enzyme works only on a specific substance, or **substrate**, and does only one specific chemical job, many different enzymes are needed. Like all catalysts, enzymes take part in reactions only temporarily; they are not used up or changed by the reaction. Therefore, they are needed in small amounts. Many of the vitamins and minerals required in the diet are parts of enzymes.

An enzyme's shape is important in its action. Just as the shape of a key must fit that of its lock, an enzyme's shape must match the shape of the substrate it acts on. This so-called "lock-and-key" mechanism is illustrated in **Figure 2-11**. Harsh conditions, such as extremes of temperature or pH, can alter the shape of any protein, such as an enzyme, and destroy its ability to function. The alteration of a protein's shape so that it can no longer function is termed **denaturation**. Such an event is always harmful to the cells.

You can usually recognize the names of enzymes because, with few exceptions, they end with the suffix *-ase*. Examples are lipase, protease, and oxidase. The first part of the name usually refers to the substance acted on or the type of reaction in which the enzyme is involved.

NUCLEOTIDES

One additional class of organic compounds is composed of building blocks called **nucleotides** (NU-kle-o-tides) (**Fig. 2-12**). A nucleotide contains

- A nitrogenous (nitrogen-containing) subunit called a base (not to be confused with an alkali).
- A sugar, usually a sugar called ribose or a related sugar called deoxyribose.
- A phosphate group, which contains phosphorus. There may be more than one phosphate group in the nucleotide.

The nucleic acids deoxyribonucleic acid (DNA) and ribonucleic acid (RNA), involved in the transmission of genetic traits and their expression in the cell, are composed of nucleotides. These are discussed in further detail in Chapter 3. Adenosine triphosphate (ATP), the cell's high-energy compound, is also a nucleotide. The extra energy in ATP is stored in special bonds between the nucleotide's three phosphates (**see Fig. 2-12B**). When these bonds are broken catabolically, energy is released for cellular activities. Learn more about ATP and its role in metabolism in Chapter 20.

CHECKPOINTS

- ☐ **2-12** What element is the basis of organic chemistry?
- ☐ **2-13** What are the three main categories of organic compounds?
- ☐ **2-14** What is an enzyme?
- ☐ **2-15** What is in a nucleotide, and what compounds are made of nucleotides?

See the Student Resources on thePoint to view an animation on enzymes. In addition, the Health Professions topic, "Pharmacists and Pharmacy Technicians," describes some professions that require knowledge of chemistry.

Disease in Context Revisited

Margaret: Back in Balance

"Good morning, Mrs. Ringland. How are you feeling today?" asked Angela.

"Much better, thank you," replied Margaret. "I'm so grateful that my niece found me when she did."

"I'm glad, too," said Angela. "With the heat wave we're having, dehydration can become a serious problem. Older adults are particularly at risk of dehydration because with age there is usually a decrease in muscle tissue, which contains a lot of water, and a relative increase in body fat, which does not. So, older adults don't have as much water reserve as do younger adults. But," Angela continued as she flipped through Margaret's chart, "it looks like you're well on your way to a full recovery. Your electrolytes are back in balance. Your blood pressure is back to normal, and your heart rate is good too. Your increased urine output tells me that your other organs are recovering as well."

"Does that mean I can get rid of this IV?" asked Margaret.

"Well, I'll check with your doctor first," replied Angela. "But when you do have the IV removed, you will need to make sure that you drink plenty of fluids."

It was the end of another long shift, and Angela was at her locker, changing into a pair of shorts and a T-shirt. As she closed her locker, she thought of Margaret once again. It always amazed her that chemistry could have such a huge impact on the body as a whole. She grabbed her water bottle, took a long drink, and headed out into the scorching heat.

In this case, we see that health professionals require a background in chemistry to understand how the body works—when healthy and when not. As you learn more about the human body, consider referring back to this chapter when necessary. For more information about the elements that make up every single substance within the body, see Appendix 1: Periodic Table of the Elements at the back of this book.

CHAPTER 2

Chapter Wrap-Up

Summary Overview

A detailed chapter outline with space for note taking is on *thePoint*. The figure below illustrates the main topics covered in this chapter.

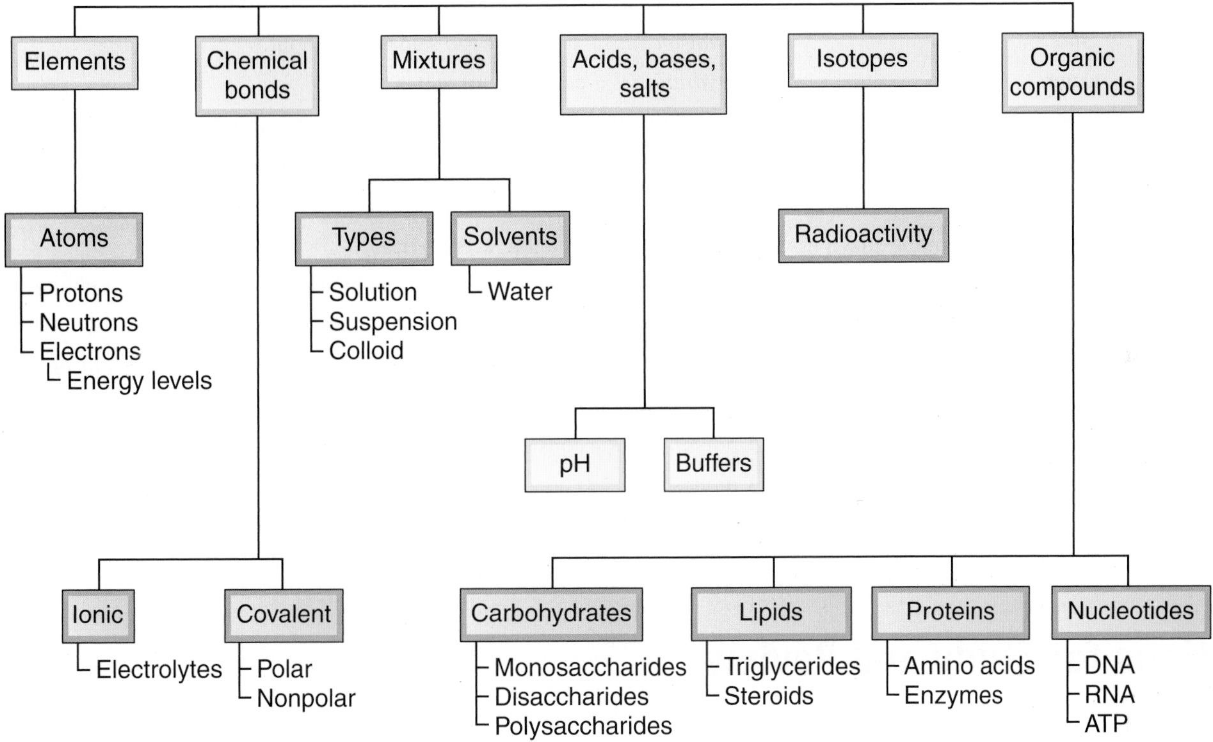

Key Terms

The terms listed below are emphasized in this chapter. Knowing them will help you organize and prioritize your learning. These and other boldface terms are defined in the Glossary with phonetic pronunciations.

acid
amino acid
anion
aqueous
atom
base
buffer
carbohydrate
catalyst
cation
chemistry
colloid
compound
denaturation
electrolyte
electron
element
enzyme
glucose
glycogen
ion
isotope
lipid
molecule
nucleotide
neutron
pH
protein
proton
radioactive
salt
solute
solution
solvent
steroid
substrate
suspension
valence

Word Anatomy

Medical terms are built from standardized word parts (prefixes, roots, and suffixes). Learning the meanings of these parts can help you remember words and interpret unfamiliar terms.

WORD PART	MEANING	EXAMPLE
Chemical Bonds		
co-	together	*Covalent* bonds form when atoms share electrons.
Solutions and Suspensions		
aqu/e	water	In an *aqueous* solution, water is the solvent.
heter/o-	different	*Heterogeneous* solutions are different (not uniform) throughout.
hom/o-	same	*Homogeneous* mixtures are the same throughout.
hydr/o-	water	*Dehydration* is a deficiency of water.
-phil	to like	*Hydrophilic* substances "like" water—they mix with or dissolve in it.
-phobia	fear	*Hydrophobic* substances "fear" water—they repel and do not dissolve in it.
Organic Compounds		
-ase	suffix used in naming enzymes	A *lipase* is an enzyme that acts on lipids.
de-	remove	*Denaturation* of a protein removes its ability to function (changes its nature).
di-	twice, double	A *disaccharide* consists of two simple sugars.
glyc/o-	sugar, glucose, sweet	*Glycogen* is a storage form of glucose. It breaks down to release glucose.
mon/o-	one	In a *monosaccharide*, "mono-" refers to one.
poly-	many	A *polysaccharide* consists of many simple sugars.
sacchar/o-	sugar	A *monosaccharide* consists of one simple sugar.
tri-	three	*Triglycerides* have one fatty acid attached to each of three carbon atoms.

Questions for Study and Review

BUILDING UNDERSTANDING

Fill in the Blanks

1. The subunits of elements are _____.
2. The atomic number is the number of _____ in an atom's nucleus.
3. A mixture of solute dissolved in a solvent is called a(n) _____.
4. Blood has a pH of 7.35 to 7.45. Gastric juice has a pH of about 2.0. The more alkaline fluid is _____.
5. Proteins that catalyze metabolic reactions are called _____.

Matching > Match each numbered item with the most closely related lettered item.

___ 6. A simple carbohydrate such as glucose

___ 7. A complex carbohydrate such as glycogen

___ 8. An important component of cell membranes

___ 9. Examples include DNA, RNA, and ATP

___ 10. The basic building block of protein

a. polysaccharide

b. phospholipid

c. nucleotide

d. amino acid

e. monosaccharide

Multiple Choice

___ **11.** What type of mixture is red blood cells "floating" in plasma?

a. compound
b. suspension
c. colloid
d. solution

___ **12.** What is the most abundant compound in the body?

a. carbohydrate
b. protein
c. lipid
d. water

___ **13.** Which compound releases ions when in solution?

a. solvent
b. electrolyte
c. anion
d. colloid

___ **14.** Which substance releases a hydrogen ion when dissolved in water?

a. acid
b. base
c. salt
d. catalyst

___ **15.** Which element is found in all organic compounds?

a. oxygen
b. carbon
c. nitrogen
d. phosphorus

UNDERSTANDING CONCEPTS

16. Compare and contrast the following terms:

a. proton, neutron, and electron
b. ionic bond and covalent bond
c. anion and cation
d. polar and nonpolar covalent bonds
e. acid and base

17. What are some of the properties of water that make it an ideal medium for living cells?

18. What is pH? Discuss the role of buffers in maintaining a steady pH in the body.

19. Describe some uses of radioactive isotopes in medicine.

20. What are the characteristics of organic compounds?

21. Compare and contrast carbohydrates, lipids, and proteins, and give examples of each.

22. List the components of nucleotides, and give three examples of nucleotides.

23. Define the term *enzyme*, and discuss the relationship between enzyme structure and enzyme function.

CONCEPTUAL THINKING

24. Explain the statement, "All compounds are composed of molecules, but not all molecules are compounds."

25. Based on your understanding of strong acids and bases, why does the body have to be kept at a close-to-neutral pH?

26. In Margaret's opening case study, her hematocrit reading was high. The hematocrit measures the percentage of red cells in blood. Using Appendix 2 at the back of this book, give the normal range of the hematocrit for women, and explain how the high reading relates to Margaret's condition.

27. In Margaret's case, an aqueous solution of 5% dextrose was used to rehydrate her. Name the solution's solute and solvent.

For more questions, see the Learning Activities on thePoint.

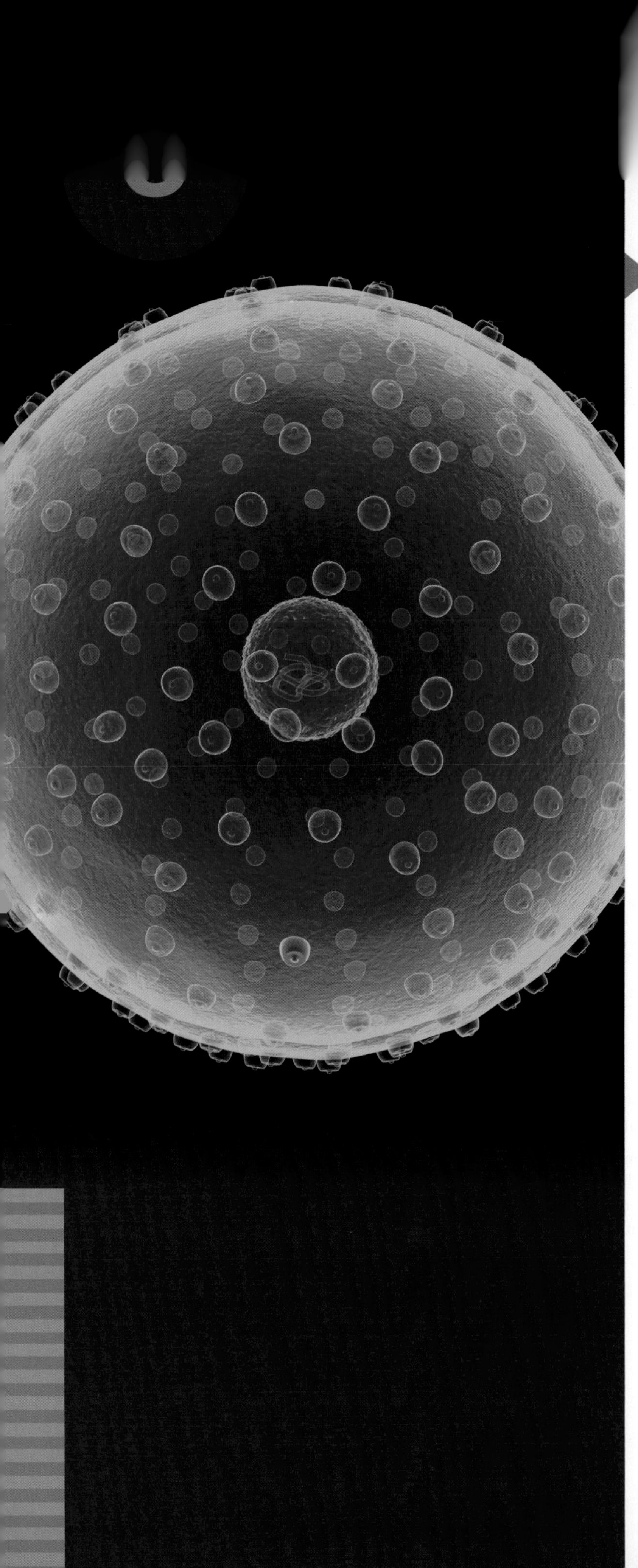

Learning Objectives

After careful study of this chapter, you should be able to:

1 ▶ List three types of microscopes used to study cells. ***p. 38***

2 ▶ Describe the composition and functions of the plasma membrane. ***p. 39***

3 ▶ Describe the cytoplasm of the cell, and cite the names and functions of the main organelles. ***p. 40***

4 ▶ Describe methods by which substances enter and leave cells that do not require cellular energy. ***p. 44***

5 ▶ Explain what will happen if cells are placed in solutions with concentrations the same as or different from those of the cytoplasm. ***p. 46***

6 ▶ Describe methods by which substances enter and leave cells that require cellular energy. ***p. 47***

7 ▶ Describe the composition, location, and function of the DNA in a cell. ***p. 49***

8 ▶ Compare the functions of three types of RNA in cells. ***p. 51***

9 ▶ Explain briefly how cells make proteins. ***p. 51***

10 ▶ Name and briefly describe the stages in mitosis. ***p. 53***

11 ▶ Discuss the cellular changes that may lead to cancer, and list several cancer risk factors. ***p. 54***

12 ▶ Use the case study to discuss the importance of the plasma membrane to the functioning of the body as a whole. ***pp. 37, 55***

13 ▶ Show how word parts are used to build words related to cells and their functions (see Word Anatomy at the end of the chapter). ***p. 57***

Disease in Context *Ben's Case: How a Tissue Failure Affects the Entire Body*

"Cough. Cough. Cough." Alison awoke with a start. "Not again," she thought as she stumbled out of bed toward the baby's room. For the last few days, Alison's 1-year-old was sick with what appeared to be a nasty chest infection. This wasn't unusual for Ben—he had come down with several lung infections in the past year and often seemed congested, but Alison had chalked this up to normal childhood illnesses. Lately though, Alison had become more worried, especially after taking Ben to their community center, where she noticed that he seemed smaller than the other children of his age and was not as active. "I'll take him in to see the doctor tomorrow," Alison thought as she sat down in the rocking chair beside Ben's crib and began patting his back.

At the medical center, Ben's doctor examined him carefully. Ben was smaller and weighed less than did most boys of his age, despite his mom's observation that he had a good appetite. His recurrent respiratory infections were also cause for worry. In addition, Alison reported that Ben had frequent bowel movements with stools that were often foul smelling and greasy. The doctor's next question caught Alison off guard. "When you kiss your son, does he taste saltier than what you might expect?" The doctor wasn't surprised when Alison answered yes. "I need to run a few more tests before I can make a diagnosis," he said. "In the meantime, let's start Ben on some oral antibiotics for his chest infection."

A few days later, Ben's doctor reviewed his chart and the lab test results. Chest and sinus radiography showed evidence of bacterial infection and thickening of the membrane lining Ben's respiratory passages. The blood test indicated that Ben had elevated levels of the pancreatic enzyme immunoreactive trypsinogen. Genetic testing revealed mutations in a specific gene called CFTR. The sweat test revealed that Ben's sweat glands excreted abnormally high concentrations of sodium chloride. With the evidence he had, the doctor was ready to make his diagnosis. Ben had cystic fibrosis (CF).

CF is caused by a mutation in a gene that codes for a channel protein in the plasma membrane of only certain types of cells. Its consequences, however, are seen in many different organs and systems—especially the respiratory and digestive systems. We will learn more about the implications of this disease later in the chapter.

ANCILLARIES *At-A-Glance*

Visit thePoint to access the following resources. For guidance in using these resources most effectively, see pp. xv–xvii.

Learning RESOURCES

- Tips for Effective Studying
- Web Figure: Electron Micrograph of an Animal Cell Magnified over 20,000 Times
- Web Figure: Electron Micrograph of an Animal Cell Magnified over 48,000 Times
- Web Figure: Electron Micrograph of a Replicated Chromosome
- Animation: Osmosis
- Animation: Function of Proteins in the Plasma Membrane
- Animation: The Cell Cycle and Mitosis
- Health Professions: Cytotechnologist
- Detailed Chapter Outline
- Answers to Questions for Study and Review
- Audio Pronunciation Glossary

Learning ACTIVITIES

- Pre-Quiz
- Visual Activities
- Kinesthetic Activities
- Auditory Activities

A LOOK BACK

The chemicals we learned about in Chapter 2 are the building blocks of cells, the fundamental structures of all organisms. In this chapter, we also learn more about the nucleotides, first introduced in Chapter 2.

The cell is the basic unit of all life. It is the simplest structure that shows all the characteristics of life, including organization, metabolism, responsiveness, homeostasis, growth, and reproduction. In fact, it is possible for a single cell to live independently of other cells. Examples of some free-living cells are microscopic organisms such as protozoa and bacteria, some of which produce disease. As we saw in Chapter 1, cells make up all tissues in a multicellular organism; the human body, for example, contains trillions of cells. All of the body's abilities, including thinking, running, and generating energy from food reflect activities occurring in individual cells. So, understanding how the body accomplishes these complex actions requires first that we understand the structures and abilities of individual cells.

Microscopes

The study of cells is **cytology** (si-TOL-o-je). Scientists first saw the outlines of cells in dried plant tissue almost 350 years ago. They were using a **microscope**, a magnifying instrument that allowed them for the first time to examine structures not visible to the naked eye. Study of a cell's internal structure, however, depended on improvements in the design of the single-lens microscope used in the late 17th century. The following three microscopes, among others, are used today:

- The **compound light microscope** is the microscope most commonly used in laboratories. This instrument, which can magnify an object up to 1,000 times, usually has two lenses and uses visible light for illumination, although some may use other light sources (such as ultraviolet light).
- The **transmission electron microscope (TEM)** uses an electron beam in place of visible light and can magnify an image up to 1 million times.
- The **scanning electron microscope (SEM)** does not magnify as much as does the TEM (100,000 times) and shows only surface features; however, it provides a three-dimensional view of an object.

These microscopes are commonly linked to cameras and computers to record and digitally analyze images. **Figure 3-1** shows some cell structures viewed with each of these types of microscopes. The structures are cilia—short, hairlike projections from the cell that move nearby fluids. The metric unit used for microscopic measurements is the **micrometer** (MI-kro-me-ter). This unit is 1/1,000 of a millimeter and is abbreviated as mcm.

Before scientists can examine cells and tissues under a light microscope, they must usually color them with special dyes called *stains* to aid in viewing. These stains produce the variety of colors seen in photographs (micrographs) of cells and tissues taken under a microscope.

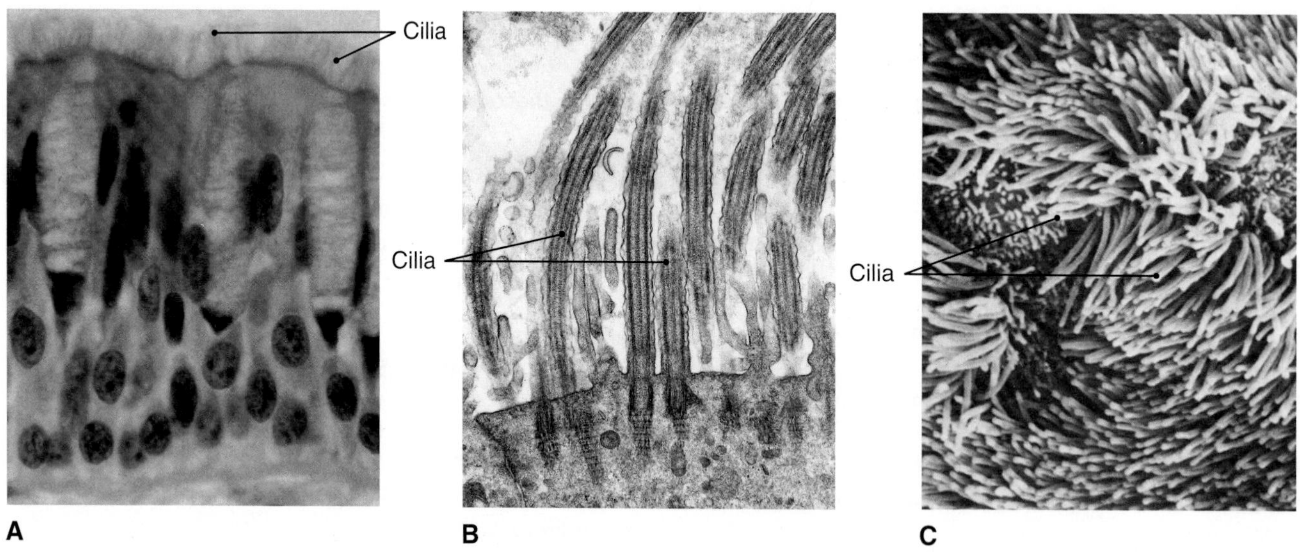

Figure 3-1 Cilia photographed under three different microscopes. **KEY POINT** Each type of microscope produces a different type of image that reveals different aspects of structure. **A.** Cilia (hairlike projections) in cells lining the trachea under the highest magnification of a compound light microscope (1,000 times). **B.** Cilia in the bronchial lining viewed with a TEM. Internal components are visible at this much higher magnification. **C.** Cilia on cells lining an oviduct as seen with a SEM (7,000 times). A three-dimensional view is visible. **ZOOMING IN** Which microscope shows the most internal structure of the cilia? Which shows the cilia in three dimensions?

CHECKPOINTS

- 3-1 List six characteristics of life shown by cells.
- 3-2 Name three types of microscopes.

See the Student Resources on thePoint for information on careers in cytotechnology, the clinical laboratory study of cells, as well as to view electron micrographs of the cell.

Cell Structure

Just as people may look different but still have certain features in common—two eyes, a nose, and a mouth, for example—all cells share certain characteristics. Refer to **Figure 3-2** as we describe some of the structures that are common to most animal cells. A summary table follows the descriptions.

PLASMA MEMBRANE

The outer layer of the cell is the **plasma membrane** (**Fig. 3-3**). (This cell part is still often called the *cell membrane*, although this older term fails to distinguish between the cell's outer membrane and other internal cellular membranes.) The plasma membrane not only encloses the cell contents but also participates in many cellular activities, such as growth, reproduction, and communication between cells, and it is especially important in regulating what can enter and leave the cell.

Some cells specialize in the uptake, or *absorption*, of materials from the extracellular fluid. The plasma membrane of these cells is often folded into multiple small projections called *microvilli* (mi-kro-VIL-li; **see Fig. 3-2**). These projections increase the membrane's surface area, allowing for greater absorption, much as a sponge's many holes provide increased surface for absorption. Microvilli are found on cells that line the small intestine, where they promote absorption of digested foods into the circulation. They are also found on kidney cells, where they reabsorb materials that have been filtered out of the blood.

Components of the Plasma Membrane The main substance of the plasma membrane is a double layer—or bilayer—of lipid molecules. Because these lipids contain the element phosphorus, they are called *phospholipids*. We introduced these lipids in Chapter 2, along with cholesterol, another type of lipid found in the plasma membrane. Molecules of cholesterol are located between the phospholipids, and they make the membrane stronger and more flexible.

Carbohydrates are present in small amounts on the outer surface of the membrane, combined either with proteins (glycoproteins) or with lipids (glycolipids). These carbohydrates help cells recognize each other and stick together.

A variety of different proteins float within the lipid bilayer. Some of these proteins extend all the way through the membrane, and some are located near the membrane's inner or outer surface. The importance of these proteins is revealed

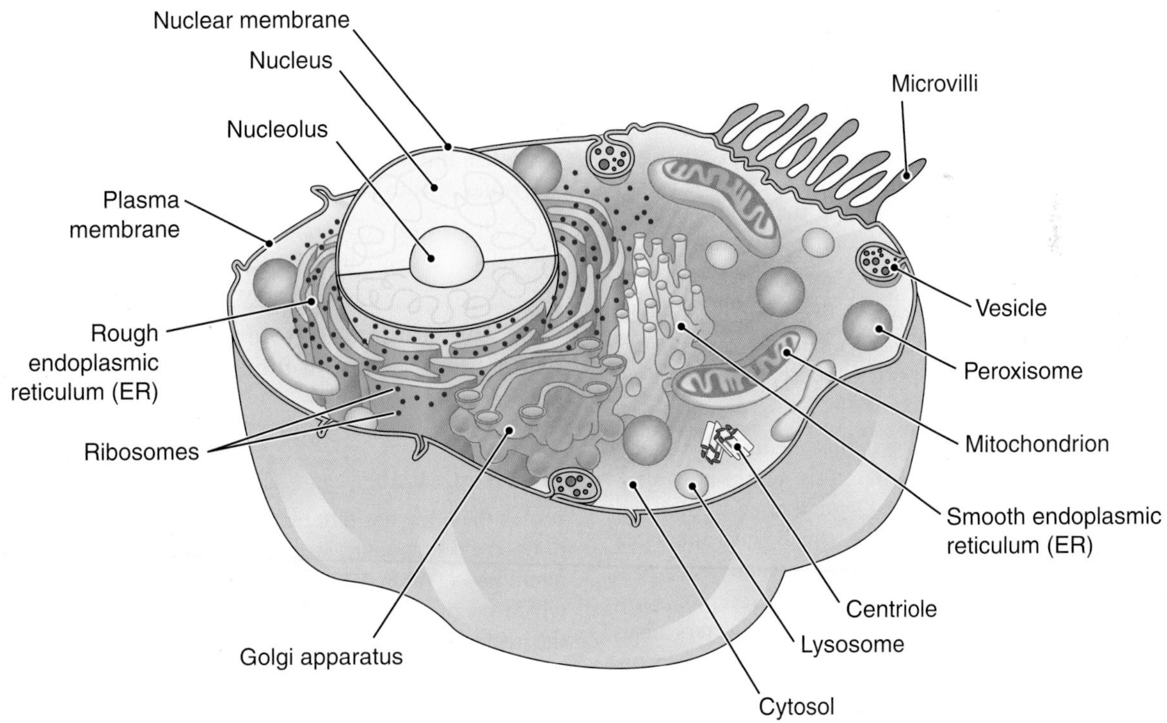

Figure 3-2 **A generalized animal cell, sectional view.** **ZOOMING IN** What is attached to the ER to make it look rough? What is the liquid part of the cytoplasm called?

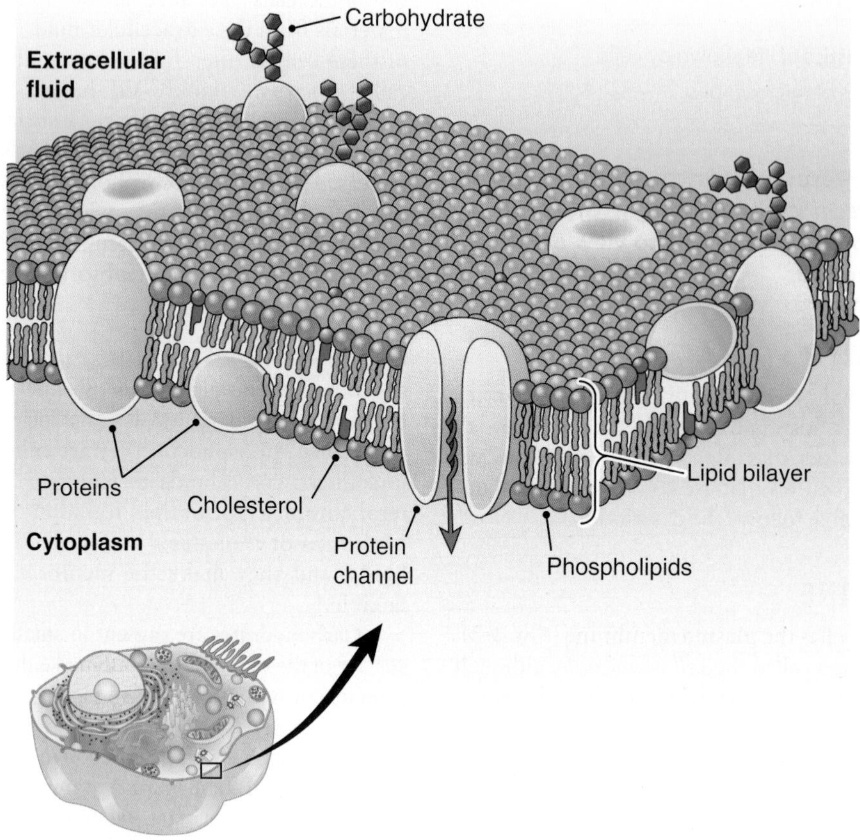

Figure 3-3 **The plasma membrane.** This drawing shows the current concept of its structure. **KEY POINT** The membrane is composed of a double layer of phospholipids with proteins and other materials embedded in it. **ZOOMING IN** Why is the plasma membrane described as a bilayer?

in later chapters, but they are listed here along with their functions, as well as summarized and illustrated in **Table 3-1**.

- Channels—pores in the membrane that allow specific substances to enter or leave. Certain ions travel through channels in the membrane.
- Transporters—shuttle substances from one side of the membrane to the other. Unlike channels, transporters change shape during transport. Glucose, for example, is carried into cells using transporters.
- Receptors—specialized proteins that mediate the effects of chemical signals on cells. Chemicals such as hormones or neurotransmitters (chemical signals used by the nervous system) bind to a receptor, which then alters cell function. For example, the hormone insulin binds to a receptor on muscle cells, and the bound receptor stimulates the production of the glucose transporters mentioned above. Some neurotransmitters bind to receptors that open or close membrane channels. Chapter 9 discusses neurotransmitter receptors in greater detail, and Chapter 12 discusses hormone receptors.
- Enzymes—participate in reactions occurring at the plasma membrane.
- Linkers—link to other proteins within the cell to stabilize the membrane and link to membrane proteins of other cells to attach cells together.
- Cell identity markers—proteins unique to an individual's cells. These are important in the immune system.

See the Student Resources on thePoint to view an animation on the functions of proteins in the plasma membrane.

THE NUCLEUS

Just as the body has different organs to carry out special functions, the cell contains specialized structures that perform different tasks (**Table 3-2**). These structures are called **organelles**, which means "little organs." The largest of the organelles is the **nucleus** (NU-kle-us), which is surrounded by a membrane, the *nuclear membrane*, that encloses its contents.

The nucleus is often called the *control center* of the cell because it contains the **chromosomes** (KRO-mo-somes), the threadlike structures of heredity that are passed on from parents to their children. It is information contained in the chromosomes that governs all cellular activities, as described

Table 3-1 Proteins in the Plasma Membrane and Their Functions

Type of Protein	Function	Illustration
Channels	Pores in the membrane that allow passage of specific substances, such as ions	
Transporters	Proteins that change shape as they shuttle substances, such as glucose, across the membrane	
Receptors	Allow for attachment of substances, such as hormones, to the membrane	
Enzymes	Participate in reactions at the membrane surface	
Linkers	Help stabilize the plasma membrane and attach cells together	
Cell identity markers	Proteins unique to a person's cells; important in the immune system and in transplantation of tissue from one person to another	

later in this chapter. Most of the time, the chromosomes are loosely distributed throughout the nucleus, giving it a uniform, dark appearance when stained and examined under a microscope (**see Fig. 3-2**). When the cell is dividing, however, the chromosomes tighten into their visible threadlike forms.

Within the nucleus is a darker stained region called the **nucleolus** (nu-KLE-o-lus), which means "little nucleus." The job of the nucleolus is to assemble **ribosomes** (RI-bo-somz), small bodies outside the nucleus that are involved in the manufacture of proteins.

THE CYTOPLASM

The remaining organelles are part of the **cytoplasm** (SI-to-plazm), the material that fills the cell from the nuclear membrane to the plasma membrane. The liquid part of the cytoplasm is the **cytosol**, a suspension of nutrients, electrolytes, enzymes, and other specialized materials in water. The main organelles are described here (**see Table 3-2**).

Recall that ribosomes are small organelles that assemble proteins. Ribosomes begin the process of protein synthesis while floating freely in the cytoplasm. Then, they usually migrate to the surface of a different organelle, the endoplasmic reticulum. The **endoplasmic reticulum** (en-do-PLAS-mik re-TIK-u-lum) is a membranous network located between the nuclear membrane and the plasma membrane. Its name literally means "network" (reticulum) "within the cytoplasm" (endoplasmic), but for ease, it is almost always called simply the **ER**. Sections of the ER studded with ribosomes have a gritty, uneven surface, causing them to be described as *rough ER*. An attached ribosome feeds the protein into the rough ER, where enzymes add sugar chains and help the protein fold into the correct shape. The part of the ER that is not

Table 3-2 Cell Parts

Name	Description	Function
Plasma membrane	Outer layer of the cell; composed mainly of lipids and proteins	Encloses the cell contents; regulates what enters and leaves the cell; participates in many activities, such as growth, reproduction, and interactions between cells
Microvilli	Short extensions of the plasma membrane	Absorb materials into the cell
Nucleus	Large, membrane-bound, dark-staining organelle near the center of the cell	Contains the chromosomes, the hereditary structures that direct all cellular activities
Nucleolus	Small body in the nucleus	Makes ribosomes
Cytoplasm	Colloid that fills the cell from the nuclear membrane to the plasma membrane	Site of many cellular activities; consists of cytosol and organelles
Cytosol	The fluid portion of the cytoplasm; contains water, enzymes, nutrients, and other substances	Surrounds the organelles; site of many chemical reactions and nutrient storage
Endoplasmic reticulum (ER)	Network of membranes within the cytoplasm. Rough ER has ribosomes attached to it; smooth ER does not	Rough ER modifies, folds, and sorts proteins; smooth ER participates in lipid synthesis
Ribosomes	Small bodies free in the cytoplasm or attached to the ER; composed of RNA and protein	Manufacture proteins
Golgi apparatus	Layers of membranes	Further modifies proteins; sorts and prepares proteins for transport to other parts of the cell or out of the cell
Mitochondria	Large organelles with internal folded membranes	Convert energy from nutrients into ATP
Lysosomes	Small sacs of digestive enzymes	Digest substances within the cell
Peroxisomes	Membrane-enclosed organelles containing enzymes	Break down harmful substances
Proteasomes	Barrel-shaped organelles	Destroy improperly synthesized proteins
Vesicles	Small membrane-bound sacs in the cytoplasm	Store materials and move materials into or out of the cell in bulk
Centrioles	Rod-shaped bodies (usually two) near the nucleus	Help separate the chromosomes during cell division
Surface projections	Structures that extend from the cell	Move the cell or the fluids around the cell
Cilia	Short, hairlike projections from the cell	Move the fluids around the cell
Flagellum	Long, whiplike extension from the cell	Moves the cell

covered with ribosomes appears to have an even surface and is described as *smooth ER*. This type of ER is involved with the synthesis of lipids.

The rough ER sends proteins to the nearby **Golgi** (GOL-je) **apparatus** (also called the Golgi complex), a large organelle consisting of a stack of membranous sacs. As the proteins pass through this organelle, they are further modified, sorted, and packaged for export from the cell.

The **mitochondria** (mi-to-KON-dre-ah) are large, round, or bean-shaped organelles with folded membranes on the inside. Enzymes within the mitochondria convert the energy from nutrients into cellular energy in the form of adenosine triphosphate (ATP). Mitochondria are the cell's "power plants." Active cells, such as muscle cells or sperm cells, need lots of energy and thus have large numbers of mitochondria.

Several types of organelles appear as small sacs in the cytoplasm. These include **lysosomes** (LI-so-somz), which contain digestive enzymes. (The root *lys/o* means "dissolving" or "separating.") Lysosomes remove waste and foreign materials from the cell. They are also involved in destroying old and damaged cells as needed for repair and remodeling of tissue. **Peroxisomes** (per-OK-sih-somz) have enzymes that destroy harmful substances produced in metabolism. Read Box 3-1 to learn about the importance of lysosomes and peroxisomes in health and disease. **Vesicles** (VES-ih-klz) are small, membrane-bound storage sacs. They can be used to move materials into or out of the cell, as described later.

Very small, barrel-shaped protein complexes called **proteasomes** (not shown in Fig. 3-2) also participate in waste removal. They specialize in the destruction of any proteins produced by the ribosomes and ER that do not meet quality control specifications. Sometimes this quality control system is too sensitive, destroying relatively functional proteins. CF,

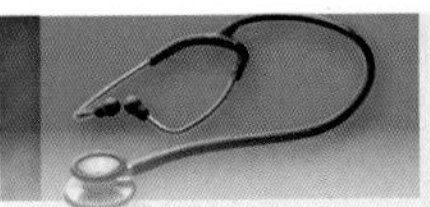

Box 3-1

CLINICAL PERSPECTIVES

Lysosomes and Peroxisomes: Cellular Recycling

Two organelles that play a vital role in cellular disposal and recycling are lysosomes and peroxisomes. **Lysosomes** contain enzymes that break down carbohydrates, lipids, proteins, and nucleic acids. These powerful enzymes must be kept within the lysosome because they would digest the cell if they escaped. In a process called **autophagy** (aw-TOF-ah-je), the cell uses lysosomes to safely recycle cellular structures, fusing with and digesting worn-out organelles. The digested components then return to the cytoplasm for reuse. Lysosomes also break down foreign material, as when cells known as **phagocytes** (FAG-o-sites) engulf bacteria and then use lysosomes to destroy them. The cell may also use lysosomes to digest itself during **autolysis** (aw-TOL-ih-sis), a normal part of development. *Auto-* means "self," and cells that are no longer needed "self-destruct" by releasing lysosomal enzymes into their own cytoplasm.

Peroxisomes are small membranous sacs that resemble lysosomes but contain different kinds of enzymes. They break down toxic substances that may enter the cell, such as drugs and alcohol, but their most important function is to break down free radicals. These substances are byproducts of normal metabolic reactions but can kill the cell if not neutralized by peroxisomes.

Disease may result if either lysosomes or peroxisomes are unable to function. In Tay-Sachs disease, nerve cells' lysosomes lack an enzyme that breaks down certain kinds of lipids. These lipids build up inside the cells, causing malfunction that leads to brain injury, blindness, and death.

for instance, results when proteasomes destroy a mutated but relatively functional version of an ion channel. Without this channel, mucus accumulates in the respiratory and digestive systems, eventually causing death. CF is the subject of Ben's opening case study.

Centrioles (SEN-tre-olz) are rod-shaped bodies near the nucleus that function in cell division. They help to organize the cell and divide the cell contents during this process.

SURFACE ORGANELLES

Some cells have structures projecting from their surfaces that are used for motion. **Cilia** (SIL-e-ah) are small, hairlike projections that wave, creating movement of the fluids around the cell (see Fig. 3-1). For example, cells that line the passageways of the respiratory tract have cilia that move impurities out of the system. Ciliated cells move the egg cell from the ovary to the uterus in the female reproductive tract.

A long, whiplike extension from a cell is a **flagellum** (flah-JEL-lum). The only type of cell in the human body that has a flagellum is the male sperm cell. Each human sperm cell has a flagellum that propels it toward the egg in the female reproductive tract (see Fig. 3-4E).

CELLULAR DIVERSITY

Although all cells have some fundamental similarities, individual cells may vary widely in size, shape, and composition according to their functions. The average cell size is 10 to 15 mcm, but cells may range in size from the 7 mcm of a red blood cell to the 200 mcm or more in the length of a muscle cell.

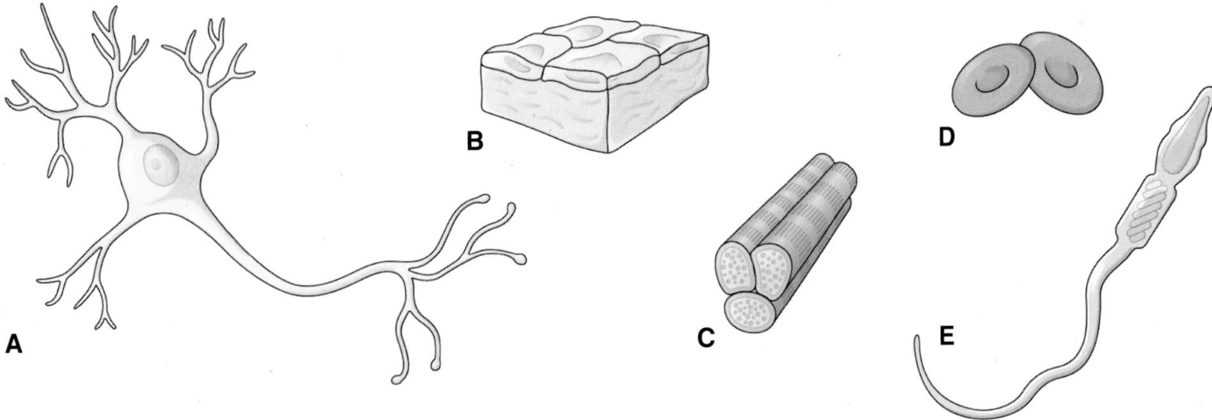

Figure 3-4 **Cellular diversity.** **KEY POINT** Cells vary in structure according to their functions. **A.** A neuron has long extensions that pick up and transmit electric impulses. **B.** Epithelial cells cover and protect underlying tissue. **C.** Muscle cells have fibers that produce contraction. **D.** Red blood cells lose most organelles and have a round, indented shape to facilitate blood flow through large vessels. **E.** A sperm cell is small and light and swims with a flagellum. **ZOOMING IN** Which of the cells shown would best cover a large surface area?

Cell shape is related to cell function (**Fig. 3-4**). A neuron (nerve cell) has long fibers that transmit electric signals over distances up to 1 m (3 feet). Red blood cells are small and flexible, assuming an indented shape in large blood vessels to facilitate blood flow but a cigar-like shape in small vessels to maximize gas exchange. As a red blood cell matures, it loses its nucleus and most other organelles, freeing up space to carry oxygen.

Aside from cilia and flagella, all the organelles described above are present in most human cells. They may vary in number, however. For example, cells producing lipids have lots of smooth ER. Cells that secrete proteins have lots of ribosomes and a prominent Golgi apparatus. All active cells have lots of mitochondria to manufacture the ATP needed for energy.

CHECKPOINTS

- 3-3 List four substances found within the plasma membrane.
- 3-4 What are cell organelles?
- 3-5 Why is the nucleus called the cell's control center?
- 3-6 What are the two types of organelles used for movement, and what do they look like?

Movement of Substances across the Plasma Membrane

The plasma membrane serves as a barrier between the cell and its environment. Nevertheless, nutrients, oxygen, and many other substances needed by the cell must be taken in, and waste products must be eliminated. Clearly, some substances can be exchanged between the cell and its environment through the plasma membrane. For this reason, the plasma membrane is described at a simple level as **semipermeable** (sem-e-PER-me-ah-bl). It is permeable, or passable, to some molecules but impassable to others.

The plasma membrane is freely permeable to lipid-soluble (fat-soluble) substances because they can dissolve in and pass through the lipid bilayer. Steroid hormones and gases (O_2, CO_2, N_2) are examples of lipid-soluble substances. Small, uncharged molecules, such as urea, can squeeze between the phospholipids and also pass through the membrane. Nutrients, ions, and other water-soluble substances cannot pass through the lipid bilayer, so they must use transporters or ion channels to cross. The permeability of the membrane to these substances thus depends on the presence of specific membrane proteins. For instance, we ingest starch, but intestinal cells only possess transporters for the product of digested starch (glucose). So, the intestinal cell membrane is permeable to glucose but not starch.

Because the permeability of the plasma membrane varies among substances and over time, the membrane is most accurately described not as simply semipermeable but as **selectively permeable**. It alters what can enter and leave based on the cell's needs. Various physical processes are involved in exchanges through the plasma membrane. One way of grouping these processes is according to whether they do or do not require cellular energy.

MOVEMENT THAT DOES NOT REQUIRE CELLULAR ENERGY

The adjective *passive* describes movement through the plasma membrane that does not directly require energy output by the cell. Passive mechanisms depend on **gradients**, which are differences in a particular quality between two regions. For instance, a sled moves freely down an altitude gradient from a higher altitude to a lower altitude. In the body, many substances move because of differences in solute concentrations, but other types of gradients (such as pressure gradients) can also drive transport.

Diffusion **Diffusion** is the net movement of particles from a region of relatively higher concentration to one of lower concentration. Just as couples on a crowded dance floor spread out into all the available space to avoid hitting other dancers, diffusing substances spread throughout their available space until their concentration everywhere is the same—that is, they reach equilibrium (**Fig. 3-5**). Diffusion uses the particles' internal energy and does not directly require cellular ATP. The particles are said to follow, or move down, their *concentration gradient* from higher concentration to lower concentration.

Particles can only enter or exit the cell by diffusion if they can cross the plasma membrane. A particle cannot diffuse through, regardless of the gradient strength, if the plasma membrane is impermeable to it. So lipid-soluble substances such as gases and steroid hormones diffuse freely in and out of cells whenever a concentration gradient exists. Water-soluble substances, on the other hand, will only diffuse across the plasma membrane if a suitable membrane protein (ion channel or transporter) is available to permit passage through the inhospitable lipid bilayer. **Figure 3-6** illustrates how glucose uses a transporter to diffuse across the plasma membrane.

Osmosis **Osmosis** (os-MO-sis) is a special type of diffusion. The term applies specifically to the diffusion of water through a semipermeable membrane. Water moves rapidly through the plasma membrane of most cells with the help of channels called *aquaporins* (a-kwa-POR-ins). The water molecules move, as expected, from an area where there are more of them to an area where there are fewer of them. That is, the solvent (the water molecules) moves from an area of lower *solute* concentration to an area of higher *solute* concentration, as demonstrated in **Figure 3-7**.

Figure 3-5 **Diffusion of a solid in a liquid.** **KEY POINT** The molecules of the solid tend to spread evenly throughout the liquid as they dissolve.

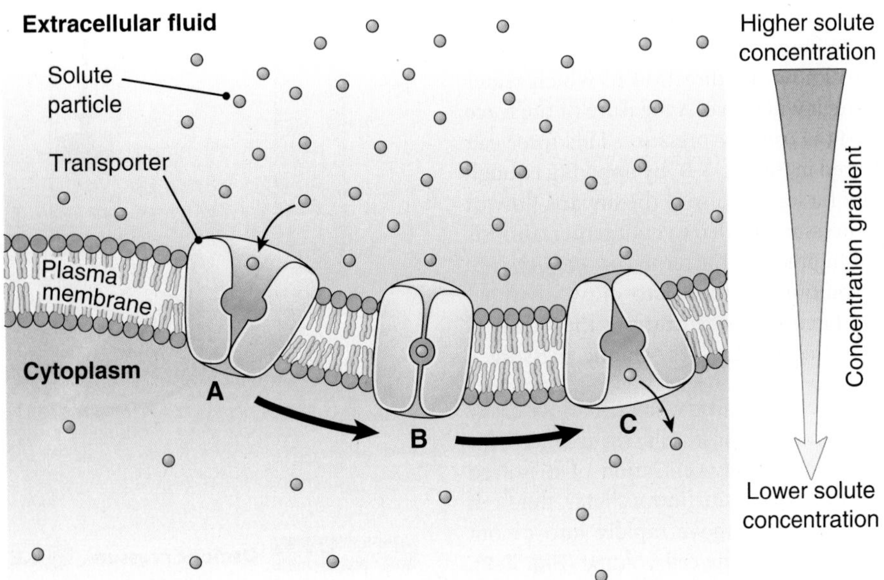

Figure 3-6 **Diffusion using transporters.** **KEY POINT** Protein transporters in the plasma membrane move solute particles through a membrane from an area of higher concentration to an area of lower concentration. **A.** A solute particle enters the transporter. **B.** The transporter changes shape. **C.** The transporter releases the solute particle on the other side of the membrane. **ZOOMING IN** How would a decrease in the number of transporters affect this solute's movement?

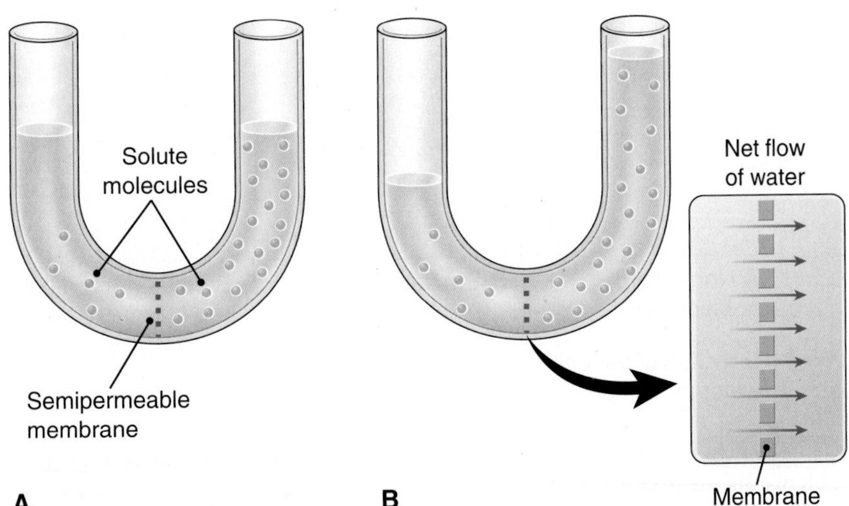

Figure 3-7 **A simple demonstration of osmosis.** **KEY POINT** The direction of water flow tends to equalize concentrations of solutions. Solute molecules are shown in *yellow*. All of the solvent (*blue*) is composed of water molecules. **A.** Two solutions with different concentrations of solute are separated by a semipermeable membrane. Water can flow through the membrane, but the solute cannot. **B.** Water flows into the more concentrated solution, raising the level of the liquid in that side. **ZOOMING IN** What would happen in this system if the solute could pass through the membrane?

For a physiologist studying water's flow across membranes, as in exchange of fluids through capillaries in the circulation, it is helpful to know the direction in which water will flow and at what rate it will move. A measure of the force driving osmosis is called the *osmotic pressure*. This force can be measured, as illustrated in **Figure 3-8**, by applying enough pressure to the surface of a liquid to stop the inward flow of water by osmosis. The pressure needed to counteract osmosis is the osmotic pressure. In practice, the term *osmotic pressure* is used to describe a solution's tendency to draw in water. This force is directly related to concentration; the higher a solution's concentration, the greater is its osmotic pressure.

How Osmosis Affects Cells Because water can move easily through the plasma membrane of most cells, the extracellular fluid must have the same overall concentration of dissolved substances (solutes) as the cytoplasm (intracellular fluid). If this balance is altered, water will move rapidly into or out of the cell by osmosis and change the cell volume (**Fig. 3-9**). Solutions with concentrations equal to the concentration of the cytoplasm are described as isotonic (i-so-TON-ik). Tissue fluids and blood plasma are isotonic for body cells. Manufactured solutions that are isotonic for the cells and can thus be used intravenously to replace body fluids include 0.9% salt, or normal saline, and 5% dextrose (glucose).

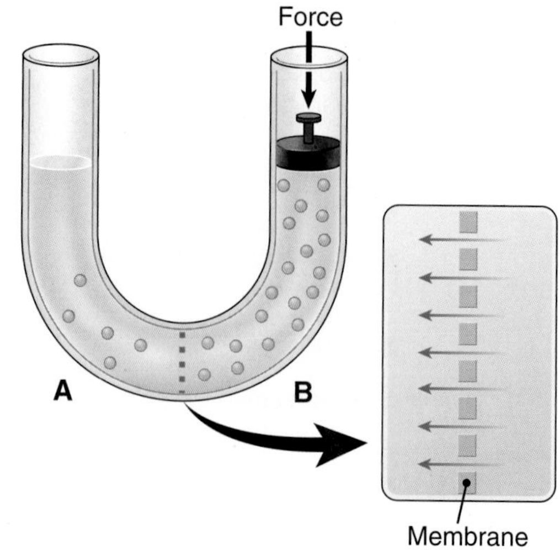

Figure 3-8 **Osmotic pressure.** **KEY POINT** Osmotic pressure is the force needed to stop the flow of water by osmosis. Pressure on the surface of the fluid in side B counteracts the osmotic flow of water from side A to side B. **ZOOMING IN** What would happen to osmotic pressure if the concentration of solute were increased on side B of this system?

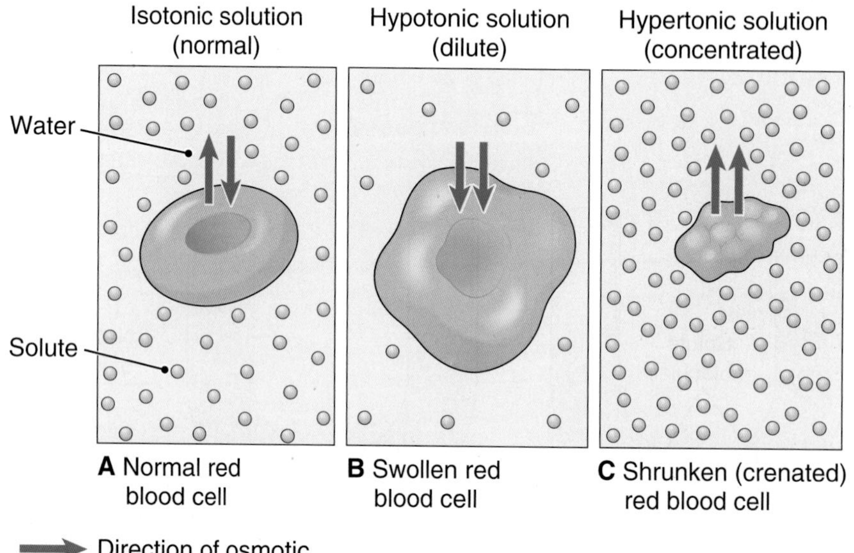

Figure 3-9 **The effect of osmosis on cells.** **KEY POINT** Cells must be kept in fluids that are compatible with the concentration of their cytoplasms. This figure shows how water moves through a red blood cell membrane in solutions with three different concentrations of solute. **A.** The isotonic (normal) solution has the same concentration as the cytoplasm, and water moves into and out of the cell at the same rate. **B.** A cell placed in a hypotonic (more dilute) solution draws water in, causing the cell to swell and perhaps undergo hemolysis (bursting). **C.** The hypertonic (more concentrated) solution draws water out of the cell, causing it to shrink, an effect known as crenation. **ZOOMING IN** What would happen to red blood cells in the body if blood lost through injury were replaced with pure water?

A solution that is less concentrated than the cytoplasm is described as **hypotonic**. Based on the principles of osmosis already explained, a cell placed in a hypotonic solution draws water in, swells, and may burst. When a red blood cell draws in water and bursts in this way, the cell is said to undergo **hemolysis** (he-MOL-ih-sis). If a cell is placed in a **hypertonic** solution, which is more concentrated than the cellular fluid, it loses water to the surrounding fluids and shrinks, a process termed **crenation** (kre-NA-shun) (**see Fig. 3-9**).

Fluid balance is an important facet of homeostasis and must be properly regulated for health. You can figure out in which direction water will move through the plasma membrane if you remember the saying "water follows salt," salt meaning any dissolved material (solute). The total amount and distribution of body fluids is discussed in Chapter 21. **Table 3-3** summarizes the effects of different solution concentrations on cells.

Filtration Filtration is the passage of water and dissolved materials through a membrane down a pressure gradient from an area of higher pressure to an area of lower pressure. A mechanical ("pushing") force is usually responsible for the high pressure. The membrane acts as a filter, preventing larger substances from crossing. An everyday example of filtration is an espresso machine, which uses steam to increase the pressure in the machine above atmospheric pressure. This gradient forces water and dissolved chemicals (such as caffeine) into the cup, but the filter retains the grounds. In the body, heart contractions increase the pressure in capillaries (i.e., the blood pressure) above the pressure of the surrounding fluid. The gradient pushes water and electrolytes out of the capillary, but the capillary wall retains the larger proteins and blood cells (see Chapter 15). In the same way, water and dissolved substances are filtered out of blood in the first step of urine formation in the kidney (see Chapter 22).

See the Student Resources on thePoint to view an animation on osmosis and osmotic pressure.

MOVEMENT THAT REQUIRES CELLULAR ENERGY

Some materials move across the plasma membrane without depending on a gradient. For instance, intestinal cells import glucose when glucose is more concentrated inside the cell than outside it. Other substances, such as bacteria or complex solutions, are too large or too heterogeneous for channels or transporters to handle. Cellular energy must drive transport in both of these situations. *Active transport* uses transporters and ATP to move ions and nutrients, and *bulk transport* uses vesicles and ATP to move large amounts of substances at once.

Active Transport While any method that uses cellular energy can be defined as "active," the term **active transport** usually refers to the movement of solutes against their concentration gradients using membrane transporters. These transporter proteins move specific solute particles from an area where they are in relatively lower concentration to an area where they are in higher concentration. This movement requires energy just as getting a sled to the top of a hill requires energy. Instead of the physical energy needed to push a sled, this process uses the chemical energy of ATP. The nervous system and muscular system, for example, depend on the active transport of sodium, potassium, and calcium ions for proper function. The kidneys also carry out active transport in regulating the composition of urine, and the digestive system uses active transport to absorb virtually all of the nutrients in our ingested food. By means of active transport, the cell can take in what it needs from the surrounding fluids and remove materials from the cell.

Bulk Transport There are several active methods for moving large quantities of material into or out of the cell. These methods are grouped together as **bulk transport**, because of the amounts of material moved. They are also referred to as **vesicular transport**, because small sacs, or vesicles, are needed for the processes. These processes are grouped according to whether materials are moved into or out of the cells, as follows:

Table 3-3 Solutions and Their Effects on Cells

Type of Solution	Description	Examples	Effect on Cells
Isotonic	Has the same concentration of dissolved substances as the fluid in the cell	0.9% salt (normal saline); 5% glucose (dextrose)	None; cell in equilibrium with its environment
Hypotonic	Has a lower concentration of dissolved substances than the fluid in the cell	Less than 0.9% salt or 5%	Cell takes in water, swells, and may burst; red blood cell undergoes hemolysis
Hypertonic	Has a higher concentration of dissolved substances than the fluid in the cell	Higher than 0.9% salt or 5% dextrose	Cell will lose water and shrink; cell undergoes crenation

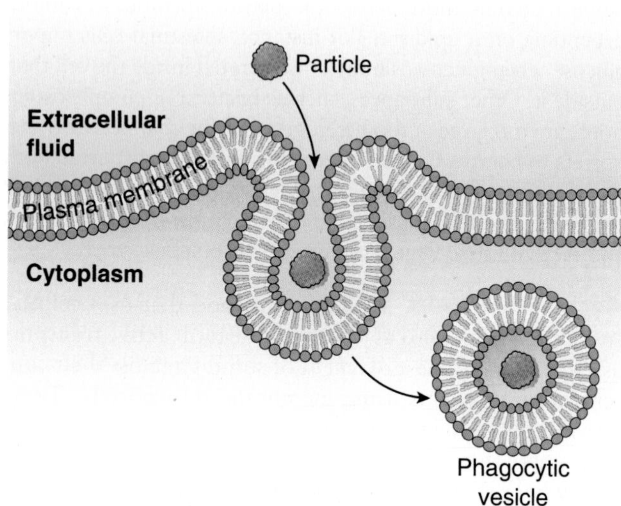

Figure 3-10 **Phagocytosis.** **KEY POINT** The plasma membrane encloses a particle from the extracellular fluid. The membrane then pinches off, forming a vesicle that carries the particle into the cytoplasm. **ZOOMING IN** What organelle would likely help to destroy a particle taken in by phagocytosis?

- **Endocytosis** (en-do-si-TO-sis) is a term that describes the bulk movement of materials into the cell. Some examples are
  - **Phagocytosis** (fag-o-si-TO-sis), in which relatively large particles are engulfed by the plasma membrane and moved into the cell (**Fig. 3-10**). (The root *phag/o* means "to eat.") Certain white blood cells carry out phagocytosis to rid the body of foreign material and dead cells. Material taken into a cell by phagocytosis is first enclosed in a vesicle made from the plasma membrane and is later destroyed by lysosomes.
  - **Pinocytosis** (pi-no-si-TO-sis), in which the plasma membrane engulfs droplets of fluid. This is a way for large protein molecules in suspension to travel into the cell. The word *pinocytosis* means "cell drinking."
  - **Receptor-mediated endocytosis,** which involves the intake of substances using specific binding sites, or receptors, in the plasma membrane. The bound material, or *ligand* (LIG-and), is then drawn into the cell by endocytosis. Some examples of ligands are lipoproteins (complexes of cholesterol, other lipids, and proteins) and certain vitamins.
- In **exocytosis,** the cell moves materials out in vesicles (**Fig. 3-11**). One example of exocytosis is the export of neurotransmitters from neurons (neurotransmitters are chemicals that control the activity of the nervous system).

All the transport methods described above are summarized in **Table 3-4**.

CHECKPOINTS

☐ 3-7 What types of movement through the plasma membrane do not directly require cellular energy, and what types of movement do require cellular energy?

☐ 3-8 What term describes a fluid that is the same concentration as the cytoplasm? What type of fluid is less concentrated? More concentrated?

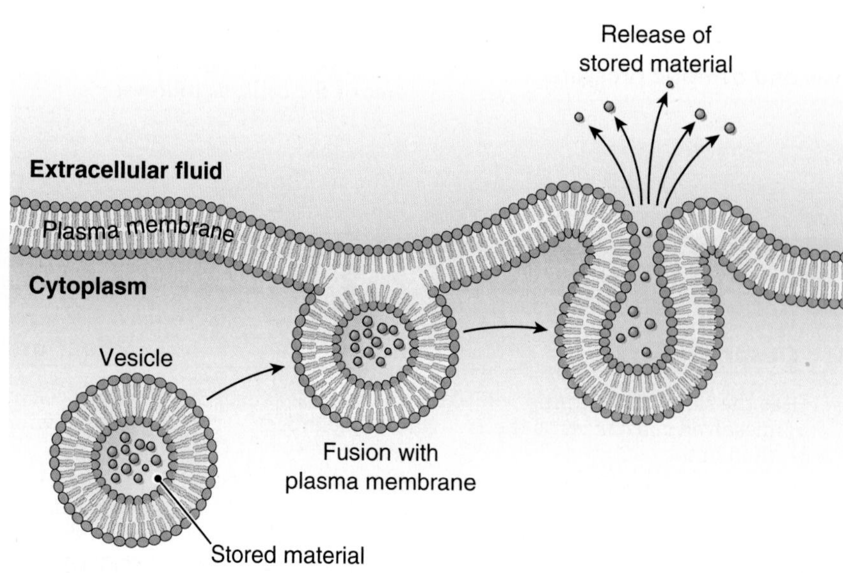

Figure 3-11 **Exocytosis.** **KEY POINT** A vesicle fuses with the plasma membrane and then ruptures and releases its contents.

Table 3-4 Membrane Transport

Process	Definition	Example
Do not require cellular energy (passive)		
Diffusion	Random movement of particles down the concentration gradient (from higher concentration to lower concentration)	Movement of gases through the membrane, ions through an ion channel, or nutrients via transporters
Osmosis	Diffusion of water through a semipermeable membrane	Movement of water across the plasma membrane through aquaporins
Filtration	Movement of materials through a membrane down a pressure gradient	Movement of materials out of the blood under the force of blood pressure
Require cellular energy		
Active transport (pumps)	Movement of materials through the plasma membrane against the concentration gradient using transporters	Transport of ions (e.g., Na^+, K^+, and Ca^{2+}) in neurons
Bulk transport	Movement of large amounts of material through the plasma membrane using vesicles; also called vesicular transport	
Endocytosis	Transport of bulk amounts of materials into the cell using vesicles	Phagocytosis—intake of large particles, as when white blood cells take in waste materials; also pinocytosis (intake of fluid), and receptor-mediated endocytosis, requiring binding sites in the plasma membrane
Exocytosis	Transport of bulk materials out of the cell using vesicles	Release of neurotransmitters from neurons

Protein Synthesis

Because proteins play an indispensable part in the body's structure and function, we need to identify the cellular substances that direct protein production. As noted earlier, the hereditary structures that govern the cell are the chromosomes in the nucleus. Each chromosome in turn is divided into multiple units, called **genes** (Fig. 3-12). It is the genes that carry the messages for the development of particular inherited characteristics, such as brown eyes, curly hair, or blood type, and they do so by directing protein manufacture in the cell.

STRUCTURE OF DNA AND RNA

Genes are distinct segments of the complex organic chemical that makes up the chromosomes, a substance called **deoxyribonucleic** (de-ok-se-RI-bo-nu-kle-ik) **acid**, or **DNA**. DNA is composed of subunits called nucleotides, introduced in Chapter 2 (see Fig. 3-12). A related compound, **ribonucleic** (RI-bo-nu-kle-ik) **acid**, or **RNA**, which participates in protein synthesis but is not part of the chromosomes, is also composed of nucleotides. As noted, a nucleotide contains a sugar, a phosphate, and a nitrogen-containing base. The sugar and phosphate are constant in each nucleotide, although DNA has the sugar deoxyribose and RNA has the sugar ribose. The sugars and phosphates alternate to form a long chain to which the nitrogen bases are attached. The five different nucleotides that appear in DNA and RNA thus differ in the nature of their nitrogen base. Three of the five nucleotides are common to both DNA and RNA. These are the nucleotides containing the nitrogen bases adenine (A), guanine (G), and cytosine (C). However, DNA has one nucleotide containing thymine (T), whereas RNA has one containing uracil (U). Table 3-5 compares the structure and function of DNA and RNA.

DNA AND PROTEIN SYNTHESIS

Most of the DNA in the cell is organized into chromosomes within the nucleus (a small amount of DNA is in the mitochondria located in the cytoplasm). Figure 3-12A and B show a section of a chromosome and illustrate that the DNA exists as a double strand. Visualizing the complete molecule as a ladder, the sugar and phosphate units of the nucleotides make up the "side rails" of the ladder, and the nitrogen bases project from the side rails to make up the ladder's "steps" (Fig. 3-12C and D). The two DNA strands are paired very specifically according to the identity of the nitrogen bases in the nucleotides. Adenine (A) always pairs with thymine (T); guanine (G) always pairs with cytosine (C). The two strands of DNA are held together by weak bonds (hydrogen bonds; see Box 2-1). The doubled strands then coil into a spiral, giving DNA the descriptive name *double helix*.

The message of the DNA that makes up the individual genes is actually contained in the varying pattern of the four nucleotides along the strand. Consider the four nucleotides

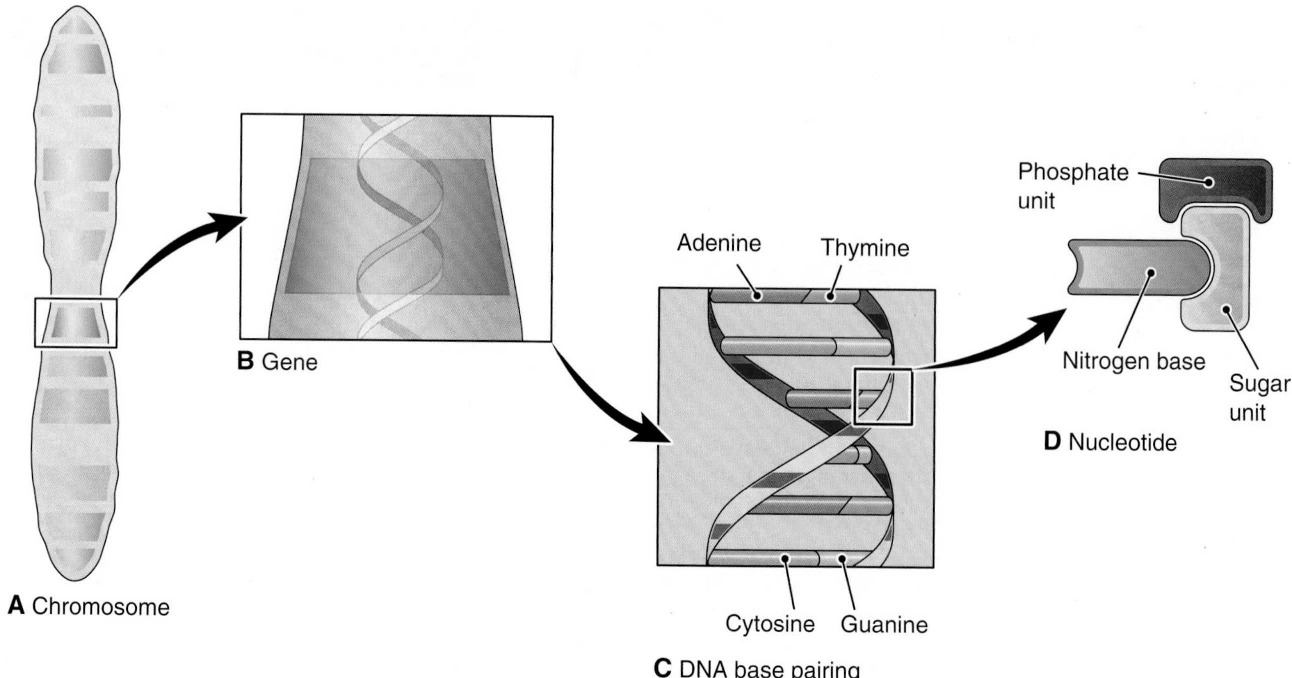

Figure 3-12 **Chromosomes and DNA. A.** A gene is a distinct region of a chromosome. **B.** The DNA making up genes consists of paired nucleic acid strands twisted into a double helix. **C.** The two DNA strands are held together by bonds between the nitrogen bases of complementary nucleotides. **D.** Each structural unit, or nucleotide, consists of a phosphate unit and a sugar unit attached to a nitrogen base. The sugar unit in DNA is deoxyribose. **KEY POINT** There are four different nucleotides in DNA. Their arrangement "spells out" the genetic instructions that control all activities of the cell. **ZOOMING IN** Two of the DNA nucleotides (A and G) are larger in size than the other two (T and C). How do the nucleotides pair up with regard to size?

as a small alphabet consisting of four different letters. These "letters" are combined to make different three-letter "words," or triplets, and each word is the code for a specific amino acid (remember that amino acids are the building blocks of proteins). For instance, the sequence CCC is the code for the amino acid glycine (**Table 3-6**, first and second columns). Each gene thus consists of a string of three-letter words that codes for a string of amino acids—in other words, an entire protein. Remember that all enzymes are proteins, and enzymes are essential for all cellular reactions. DNA is thus the cell's master blueprint.

In light of observations on cellular diversity, you may wonder how different cells in the body can vary in appearance and function if they all have the same amount and same kind of DNA. The answer to this question is that only portions of the DNA in a given cell are active at any one time. In some cells, regions of the DNA can be switched on and off, under the influence of hormones, for example. However,

Table 3-5 **Comparison of DNA and RNA**

	DNA	RNA
Location	Almost entirely in the nucleus	Almost entirely in the cytoplasm
Composition	Nucleotides contain adenine (A), guanine (G), cytosine (C), or thymine (T) Sugar: deoxyribose	Nucleotides contain adenine (A), guanine (G), cytosine (C), or uracil (U) Sugar: ribose
Structure	Double-stranded helix formed by nucleotide pairing A–T; G–C	Single strand
Function	Makes up the chromosomes, hereditary units that control all cellular activities; divided into genes that carry the nucleotide codes for the manufacture of proteins	Manufacture proteins according to the codes carried in the DNA; three main types: mRNA, rRNA, and tRNA

3

Table 3-6 The Genetic Code

Amino Acid	Transcribed DNA Triplet	mRNA	tRNA
Glycine	CCC	GGG	CCC
Proline	GGG	CCC	GGG
Valine	CAC	GUG	CAC
Phenylalanine	AAA	UUU	AAA

The nucleotide triplet code in DNA and RNA Is shown for four amino acids.

as cells differentiate during development and become more specialized, regions of the DNA are permanently shut down, leading to the variations in the different cell types. Scientists now realize that the control of DNA action throughout a cell's life span is a very complex matter involving not only the DNA itself but proteins as well.

ROLE OF RNA IN PROTEIN SYNTHESIS

A blueprint is only a guide. The information it contains must be interpreted and acted upon, and RNA is the substance needed for these steps. RNA is much like DNA except that it exists as a single strand of nucleotides and has uracil (U) instead of thymine (T). Thus, when RNA pairs up with another molecule of nucleic acid to manufacture proteins, as explained below, adenine (A) bonds with uracil (U) instead of thymine (T).

A detailed account of protein synthesis is beyond the scope of this book, but a highly simplified description and illustrations of the process are presented. The process begins with the copying of information from DNA to RNA in the nucleus, a process known as *transcription* (**Fig. 3-13**). The RNA copy is called messenger RNA (mRNA) because it carries the DNA message from the nucleus to the cytoplasm. Before transcription begins, the DNA separates into single strands. Then, enzymes assemble a matching strand of RNA along one of the DNA strands by the process of nucleotide pairing. Information on which strand will be used for transcription is contained in the chromosomes themselves. For

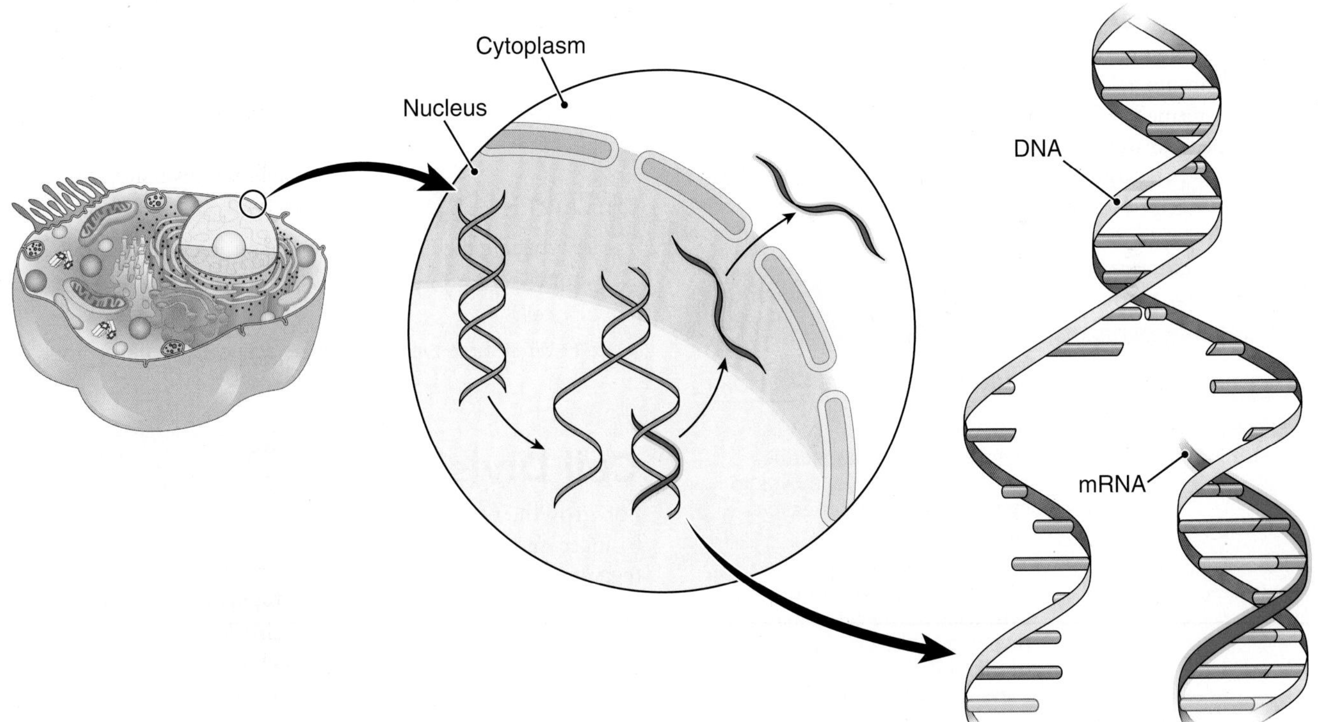

Figure 3-13 Transcription. **KEY POINT** In the first step of protein synthesis, the DNA code is transcribed into messenger RNA (mRNA) by nucleotide base pairing. An enlarged view of the nucleic acids during transcription shows how mRNA forms according to the nucleotide pattern of the DNA. Note that adenine (A, *red*) in DNA bonds with uracil (U, *brown*) in RNA.

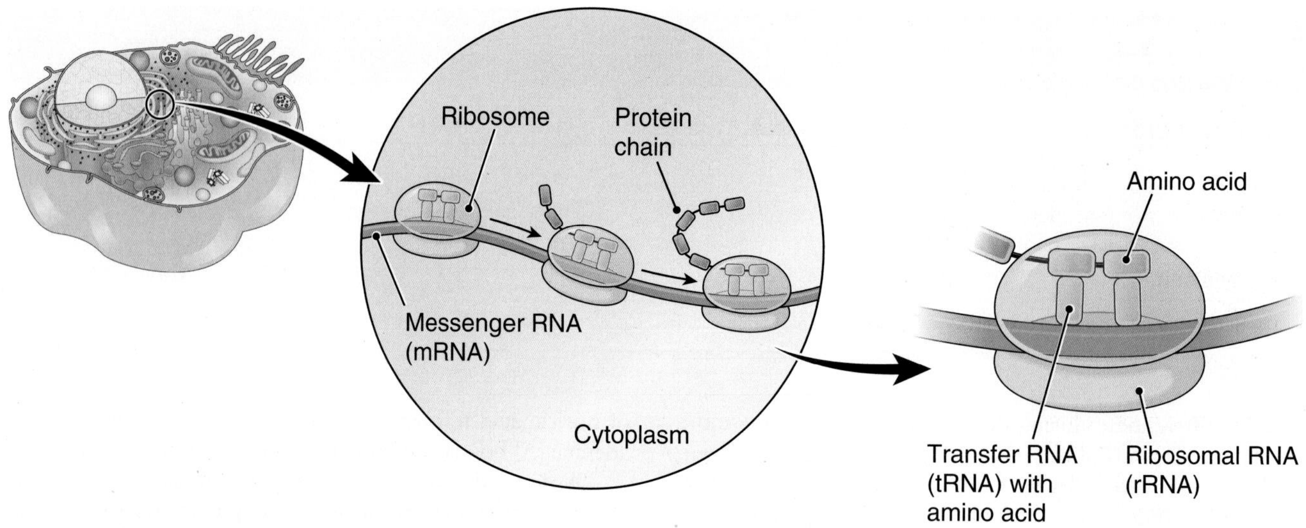

Figure 3-14 Translation. **KEY POINT** In protein synthesis, messenger RNA (mRNA) travels to the ribosomes in the cytoplasm. The information in the mRNA codes for the building of proteins from amino acids. Transfer RNA (tRNA) molecules bring amino acids to the ribosomes to build each protein.

example, if the DNA strand reads CAC, the corresponding mRNA will read GUG (remember that RNA has U instead of T to bond with A) (**Table 3-6**, third column). When complete, this mRNA leaves the nucleus and travels to a ribosome in the cytoplasm (**Fig. 3-14**). Recall that ribosomes are the site of protein synthesis in the cell.

Ribosomes are composed of an RNA type called ribosomal RNA (rRNA) and also protein. At the ribosomes, the genetic message now contained within mRNA is decoded to assemble amino acids into the long chains that form proteins, a process termed *translation*. This final step requires a third RNA type, transfer RNA (tRNA), present in the cytoplasm (**see Fig. 3-14**). Note that both rRNA and tRNA are formed by the transcription process illustrated in **Figure 3-13**.

Table 3-7 RNA

Types	Function
Messenger RNA (mRNA)	Is built on a strand of DNA in the nucleus and transcribes the nucleotide code; moves to cytoplasm and attaches to a ribosome
Ribosomal RNA (rRNA)	With protein makes up the ribosomes, the sites of protein synthesis in the cytoplasm; involved in the process of translating the genetic message into a protein
Transfer RNA (tRNA)	Works with other forms of RNA to translate the genetic code into protein; each molecule of tRNA carries an amino acid that can be used to build a protein at the ribosome

Remember that each amino acid is coded by a nucleotide triplet. Every tRNA contains the complementary nucleotides to one of these sequences and carries the corresponding amino acid (**Table 3-6**, fourth column). When the matching triplet is present in the mRNA, the tRNA binds to the mRNA, and the ribosome adds its amino acid to the growing protein chain. After the amino acid chain is formed, it must be coiled and folded into the proper shape for that protein by the endoplasmic reticulum, as discussed above. **Table 3-7** summarizes information on the different types of RNA. Also see **Box 3-2**, "Proteomics: So Many Proteins, So Few Genes."

CHECKPOINTS

- ☐ 3-9 What are the building blocks of nucleic acids?
- ☐ 3-10 What category of compounds does DNA code for in the cell?
- ☐ 3-11 What three types of RNA are active in protein synthesis?

Cell Division

For growth, repair, and reproduction, cells must multiply to increase their numbers. The cells that form the sex cells (egg and sperm) divide by the process of *meiosis* (mi-O-sis), which cuts the chromosome number in half to prepare for union of the egg and sperm in fertilization. If not for this preliminary reduction, the number of chromosomes in the offspring would constantly double. The process of meiosis is discussed in Chapters 23 and 25. All other body cells, known as *somatic cells*, are formed by a process called **mitosis** (mi-TO-sis). In this process, each original parent cell becomes two identical daughter cells. Somatic cells develop from actively dividing cells called *stem cells*, which we will discuss in more detail in Chapter 4.

Box 3-2

HOT TOPICS

Proteomics: So Many Proteins, So Few Genes

To build the many different proteins that make up the body, cells rely on instructions encoded in the genes. Collectively, all the different genes on all the chromosomes make up the **genome**. Genes contain the instructions for making proteins, and proteins perform the body's functions.

Scientists are now studying the human **proteome**—all the proteins that can be expressed in a cell—to help them understand protein structure and function. Unlike the genome, the proteome changes as the cell's activities and needs change. In 2003, after a decade of intense scientific activity, investigators mapped the entire human genome. We now realize that it probably contains no more than 25,000 genes, far fewer than initially expected. How could this relatively small number of genes code for several million proteins? They concluded that genes were not the whole story.

Gene transcription is only the beginning of protein synthesis. In response to cellular conditions, enzymes can snip newly transcribed mRNA into several pieces, each of which a ribosome can use to build a different protein. After each protein is built, enzymes can further modify the amino acid strands to produce several more different proteins. Other molecules help the newly formed proteins to fold into precise shapes and interact with each other, resulting in even more variations. Thus, while a gene may code for a specific protein, modifications after gene transcription can produce many more unique proteins. There is much left to discover about the proteome, but scientists hope that future research will lead to new techniques for detecting and treating disease.

PREPARATION FOR MITOSIS

Before mitosis can occur, the genetic information (DNA) in the parent cell must be replicated (doubled), so that each of the two new daughter cells will receive a complete set of chromosomes. For example, a human cell that divides by mitosis must produce two cells with 46 chromosomes each, the same number of chromosomes that are present in the original parent cell. DNA replicates during **interphase**, the stage in the cell's life cycle between one mitosis and the next. During this phase, DNA uncoils from its double-stranded form, and enzymes assemble a matching strand of nucleotides for each old strand according to the pattern of A–T, G–C pairing. There are now two double-stranded DNA molecules, each identical to the original double helix. The two double helices are held together at a region called the *centromere* (SEN-tro-mere) until they separate toward the end of mitosis. A typical stem cell lives in interphase for most of its life cycle and spends only a relatively short period in mitosis. For example, a cell reproducing every 20 hours spends only about one hour in mitosis and the remaining time in interphase. Most mature body cells spend their entire lives in interphase and never enter mitosis.

See the Student Resources on thePoint for a photomicrograph of a replicated chromosome.

STAGES OF MITOSIS

Although mitosis is a continuous process, distinct changes can be seen in the dividing cell at four stages (**Fig. 3-15**).

- In **prophase** (PRO-faze), each replicated chromosome winds up tightly and separates from the other replicated chromosomes. The nucleolus and the nuclear membrane begin to disappear. In the cytoplasm, the two centrioles move toward opposite ends of the cell, and a spindle-shaped structure made of thin fibers begins to form between them.
- In **metaphase** (MET-ah-faze), the chromosomes line up across the center (equator) of the cell attached to the spindle fibers.
- In **anaphase** (AN-ah-faze), the centromere splits, and the replicated chromosomes separate and begin to move toward opposite ends of the cell.
- As mitosis continues into **telophase** (TEL-o-faze), a membrane appears around each group of separated chromosomes, forming two new nuclei.

Also during telophase, the plasma membrane pinches off to divide the cell. The midsection between the two areas becomes progressively smaller until finally the cell splits into two. There are now two new cells, or daughter cells, each with exactly the same kind and amount of DNA as was present in the parent cell. In just a few types of cells, skeletal muscle cells, for example, the cell itself does not divide following nuclear division. The result, after multiple mitoses, is a giant single cell with multiple nuclei. This pattern is extremely rare in human cells.

CHECKPOINTS 

- ☐ 3-12 What must happen to the DNA in a cell before mitosis can occur? During what stage in the cell life cycle does this occur?
- ☐ 3-13 What are the four stages of mitosis?

See the Student Resources on thePoint to view the animation "The Cell Cycle and Mitosis."

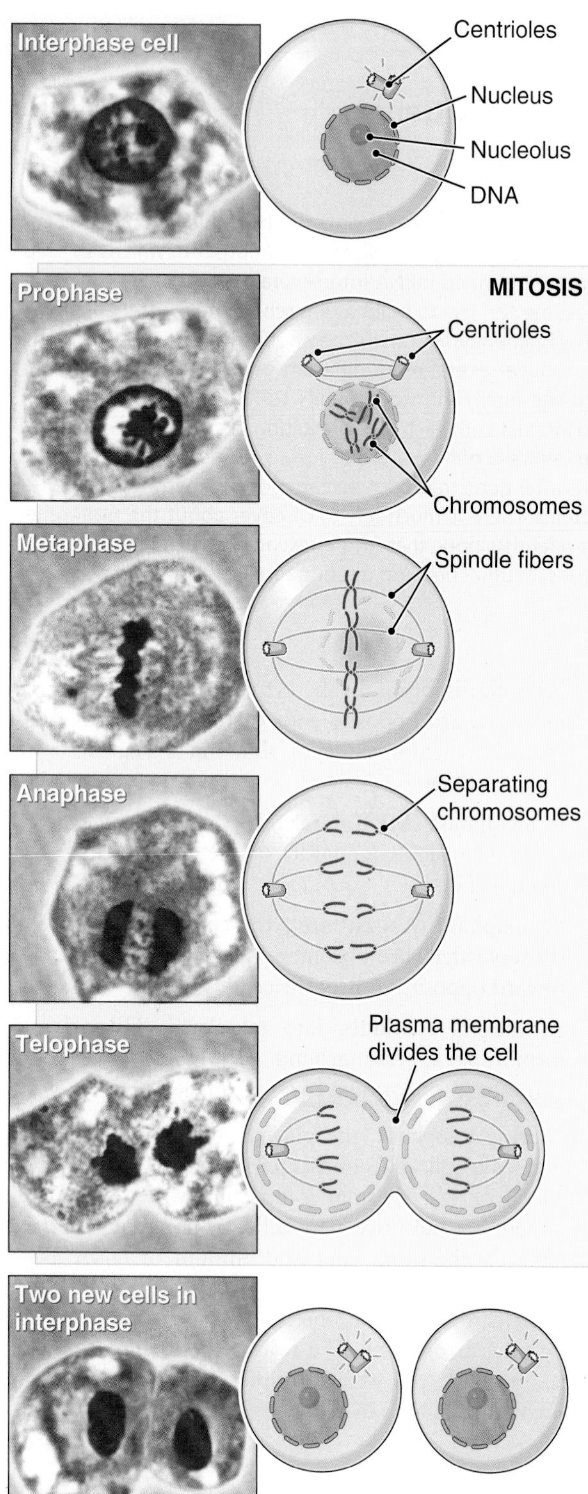

Figure 3-15 **The stages of mitosis.** **KEY POINT** Although it is a continuous process, mitosis can be seen in four stages. When it is not dividing, the cell is in interphase. The cell shown is for illustration only. It is not a human cell, which has 46 chromosomes. **ZOOMING IN** If the original cell shown has 46 chromosomes, how many chromosomes will each new daughter cell have?

Cell Aging

As cells multiply throughout life, changes occur that may lead to their damage and death. Harmful substances known as *free radicals*, produced in the course of normal metabolism, can injure cells unless they are destroyed. Chapter 20 covers free radicals in more detail. Lysosomes may deteriorate as they age, releasing enzymes that can harm the cell. Alteration of the genes, or **mutation**, is a natural occurrence in the process of cell division and is increased by exposure to harmful substances and radiation in the environment. Mutations sometimes harm cells and may lead to cancer.

As a person ages, stem cells divide less frequently, and mature body cells become less active. These changes slow down repair processes, which rely on the production of new cells and the production of substances from existing cells. A bone fracture, for example, takes considerably longer to heal in an old person than in a young person.

One theory on aging holds that cells are preprogrammed to divide only a certain number of times before they die. Support for this idea comes from the fact that cells taken from a young person divide more times when grown in the laboratory than do similar cells taken from an older individual. This programmed cell death, known as *apoptosis* (ah-pop-TO-sis), is a natural part of growth and remodeling before birth in the developing embryo. For example, apoptosis removes cells from the embryonic limb buds in the development of fingers and toes. Apoptosis also is needed in repair and remodeling of tissue throughout life. Cells subject to wear and tear regularly undergo apoptosis and are replaced. For example, the cells lining the digestive tract are removed and replaced every two to three days. This "cellular suicide" is an orderly, genetically programmed process. The "suicide" genes code for enzymes that destroy the cell quickly without damaging nearby cells. Phagocytes then eliminate the dead cells.

Cells and Cancer

Certain mutations (changes) in a cell's genetic material may cause that cell to reproduce without control. Cells that normally multiply at a fast rate, such as epithelial stem cells, are more likely than are slower-growing cells to undergo such transformations. If these altered cells do not die naturally or get destroyed by the immune system, they will continue to multiply and may spread (metastasize) to other tissues, producing **cancer**. Cancer cells form tumors, which interfere with normal functions, crowding out normal cells and robbing them of nutrients. There is more information on the various types of tumors in Chapter 4.

The causes of cancer are complex, involving interactions between cellular factors and the environment. Because cancer may take a long time to develop, it is often difficult to identify its cause or causes. *Risk factors* are aspects of an individual's heredity, lifestyle, or environment that increase his or her chances of developing a disease. For cancer, risk factors include the following:

- **Heredity.** Certain types of cancer occur more frequently in some families than in others, indicating that genes influence cancer development. For example, an inherited mutation in the BRCA1 gene is associated with increased risk of breast or ovarian cancer.
- **Chemicals.** Certain industrial, dietary, and environmental chemicals are known to increase the risk of cancer. Any chemical that causes cancer is called a **carcinogen** (kar-SIN-o-jen). The most common carcinogens in our society are those present in cigarette smoke. Carcinogens are also present in substances we consume; for example, alcohol is a carcinogen, and certain drugs also may be carcinogenic.
- **Radiation.** Certain types of radiation can produce damage to cellular DNA that may lead to cancer. These include x-rays, gamma rays from radioactive substances, and ultraviolet rays. For example, the ultraviolet rays received from sun exposure and tanning beds increase the risk of skin cancer.
- **Obesity.** Excessive fat tissue is a risk factor for certain cancers. Fat cells produce substances that directly or indirectly promote tumor development.
- **Physical inactivity.** A lack of physical activity increases the risk of colon cancer and breast cancer (and possibly others), even in people who are not overweight.
- **Poor nutrition.** A poor-quality diet is a cancer risk factor, even in people of healthy body weight. While further studies are needed, it appears that diets high in animal fats and/or lacking in fiber, fruits, and vegetables are associated with an increased occurrence of certain cancers.
- **Infectious agents.** These have been implicated in specific cancers. For instance, the human papillomavirus can cause cervical cancer, and the bacteria that causes stomach ulcers (*Helicobacter pylori*) promote the development of stomach cancer.

Disease in Context Revisited

Ben's parents were shocked when the doctor diagnosed their 1-year-old with CF. Their immediate concern was, of course, for their son. The doctor reassured them that with proper treatment, their son could lead a relatively normal life for the present and that, in the future, new therapies might extend the life span of those with CF and even offer a cure. He asserted that they were not to blame for Ben's condition. CF is an inherited disease—Ben's parents each carried a defective gene in their DNA and both had, by chance, passed copies to Ben. As a result, Ben was unable to synthesize a channel protein found in the plasma membranes of certain cells. Normally, this channel regulates the movement of chloride into the cell. Because the channels did not work in Ben's case, chloride was trapped outside the cells. The negatively charged chloride ions attract positively charged sodium ions normally found in extracellular fluid. These two ions form the salt, sodium chloride, which is lost in high amounts in the sweat of individuals with CF.

Abnormal chloride channel function causes cells in many organs to produce thick, sticky mucus. In the lungs, this mucus causes difficulty breathing, inflammation, and frequent bacterial infections. The thick mucus also decreases the ability of the large and small intestines to absorb nutrients, resulting in low weight gain, poor growth, and vitamin deficiencies. This problem is compounded by damage to the pancreas, preventing production of essential digestive enzymes.

In this case, we saw that a defective plasma membrane channel in some of Ben's cells had widespread effects on his whole body. In later chapters, as you learn about the body's organs, remember that their structure and function are closely related to the condition of their cells and tissues.

CHAPTER 3

Chapter Wrap-Up

Summary Overview

A detailed chapter outline with space for note taking is on *thePoint*. The figure below illustrates the main topics covered in this chapter.

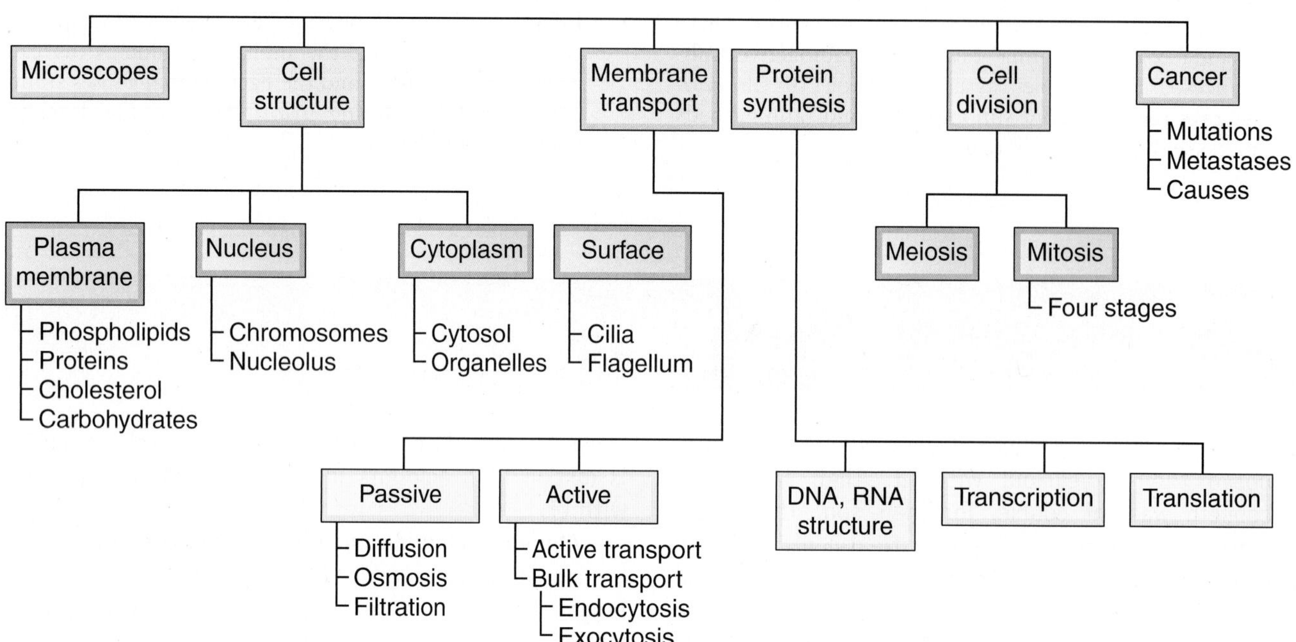

Key Terms

The terms listed below are emphasized in this chapter. Knowing them will help you organize and prioritize your learning. These and other boldface terms are defined in the Glossary with phonetic pronunciations.

active transport
cancer
carcinogen
chromosome
cytology
cytoplasm
diffusion
DNA
endocytosis
exocytosis
filtration
gene
hemolysis
hypertonic
hypotonic
interphase
isotonic
micrometer
microscope
mitochondria
mitosis
mutation
nucleus
organelle
osmosis
phagocytosis
plasma membrane
ribosome
RNA

Word Anatomy

Medical terms are built from standardized word parts (prefixes, roots, and suffixes). Learning the meanings of these parts can help you remember words and interpret unfamiliar terms.

WORD PART	MEANING	EXAMPLE
Microscopes		
cyt/o	cell	*Cytology* is the study of cells.
micr/o	small	*Microscopes* are used to view structures too small to see with the naked eye.
Cell Structure		
bi-	two	The lipid *bilayer* is a double layer of lipid molecules.
chrom/o-	color	*Chromosomes* are small, threadlike bodies that stain darkly with basic dyes.
end/o-	in, within	The *endoplasmic* reticulum is a membranous network within the cytoplasm.
lys/o	loosening, dissolving, separating	*Lysosomes* are small bodies (organelles) with enzymes that dissolve materials (*see also hemolysis*).
-some	body	*Ribosomes* are small bodies in the cytoplasm that help make proteins.
Movement across the Plasma Membrane		
ex/o-	outside, out of, away	In *exocytosis*, the cell moves material out from vesicles.
hem/o	blood	*Hemolysis* is the destruction of red blood cells.
hyper-	above, over, excessive	A *hypertonic* solution's concentration is higher than that of the cytoplasm.
hypo-	deficient, below, beneath	A *hypotonic* solution's concentration is lower than that of the cytoplasm.
iso-	same, equal	An *isotonic* solution has the same concentration as that of the cytoplasm.
phag/o	to eat, ingest	In *phagocytosis*, the plasma membrane engulfs large particles and moves them into the cell.
pin/o	to drink	In *pinocytosis*, the plasma membrane "drinks" (engulfs) droplets of fluid.
semi-	partial, half	A *semipermeable* membrane lets some molecules pass through but not others.
Cell Division		
ana-	upward, back, again	In the *anaphase* stage of mitosis, chromosomes move to opposite sides of the cell.
inter-	between	*Interphase* is the stage between one cell division (mitosis) and the next
meta-	change	*Metaphase* is the second stage of mitosis when the chromosomes change position and line up across the equator.
pro-	before, in front of	*Prophase* is the first stage of mitosis.
tel/o-	end	*Telophase* is the last stage of mitosis.
Cells and Cancer		
carcin/o	cancer, carcinoma	A *carcinogen* is a chemical that causes cancer.
-gen	agent that produces or originates	See preceding example.

Questions for Study and Review

BUILDING UNDERSTANDING

Fill in the Blanks

1. The part of the cell that regulates what can enter or leave is the _____.
2. The cytosol and organelles make up the _____.
3. If Solution A has more solute and less water than does Solution B, then Solution A is _____ to Solution B.
4. Mechanisms that require cellular energy to move substances across the plasma membrane are called _____ transport mechanisms.
5. Distinct segments of DNA that code for specific proteins are called _____.

Matching > Match each numbered item with the most closely related lettered item.

___ 6. DNA replication occurs

___ 7. DNA is tightly wound into chromosomes

___ 8. Chromosomes line up along the cell's equator

___ 9. Chromosomes separate and move toward opposite ends of the cell

___ 10. Cell membrane pinches off, dividing the cell into two new daughter cells

a. metaphase
b. anaphase
c. telophase
d. interphase
e. prophase

Multiple Choice

___ 11. The nucleus is called the cell's control center because it contains the
- a. nucleolus
- b. chromosomes
- c. cytosol
- d. cilia

___ 12. Where in the cell does ATP synthesis occur?
- a. endoplasmic reticulum
- b. Golgi apparatus
- c. mitochondria
- d. nucleus

___ 13. The movement of solute from a region of high concentration to one of lower concentration is called_____.
- a. diffusion
- b. endocytosis
- c. exocytosis
- d. osmosis

___ 14. A DNA sequence reads: TGAAC. What is its mRNA sequence?
- a. ACTTG
- b. ACUUG
- c. CAGGT
- d. CAGGU

___ 15. Rupture of red blood cells placed in a hypotonic solution is called ______.
- a. crenation
- b. hemolysis
- c. permeability
- d. mitosis

UNDERSTANDING CONCEPTS

16. List the components of the plasma membrane, and state a function for each.

17. Compare and contrast the following cellular components and processes:

a. microvilli and cilia
b. rough ER and smooth ER
c. lysosome and peroxisome
d. endocytosis and exocytosis
e. DNA and RNA
f. chromosome and gene

18. List and define five methods by which materials cross the plasma membrane. Which of these requires cellular energy?

19. Why is the plasma membrane described as selectively permeable?

20. What will happen to a body cell placed in a 5.0% salt solution? In distilled water?

21. Describe the role of each of the following in protein synthesis: DNA, nucleotide, RNA, ribosomes, rough ER, and Golgi apparatus.

22. Discuss the link between genetic mutation and cancer. List seven risk factors associated with cancer.

CONCEPTUAL THINKING

23. Kidney failure causes a buildup of waste and water in the blood. A procedure called hemodialysis removes these substances from the blood. During this procedure, the patient's blood passes over a semipermeable membrane within the dialysis machine. Waste and water from the blood diffuse across the membrane into dialysis fluid on the other side. Based on this information, compare the osmotic concentration of the blood with that of the dialysis fluid.

24. CF can result from a change in the CFTR gene on chromosome number 7. This gene codes for a membrane protein that transports chloride. Describe the process that produces an abnormal transporter beginning with a change in the DNA of the CFTR gene.

25. Changes at the cellular level can ultimately affect the entire organism. Using Ben's case, explain why this is so.

For more questions, see the Learning Activities on thePoint.

CHAPTER 4

Tissues, Glands, and Membranes

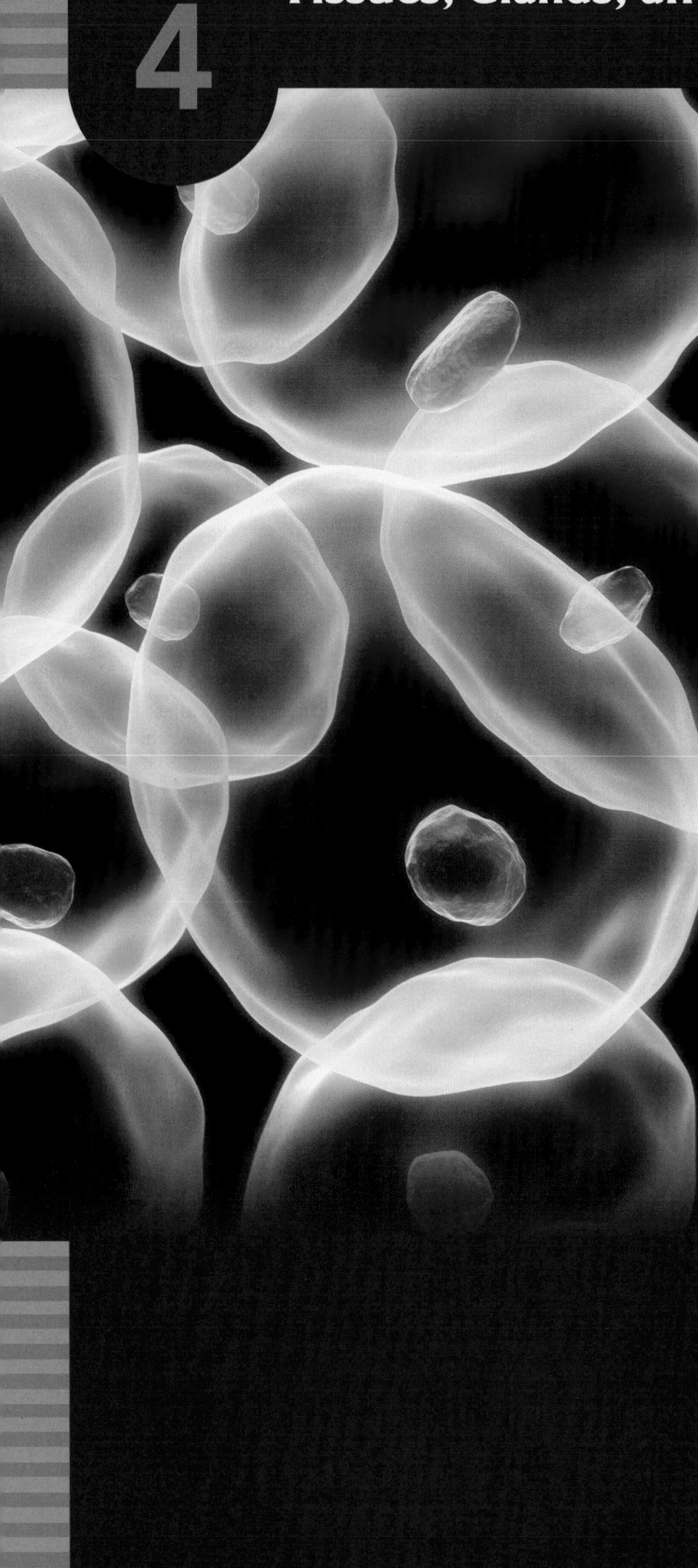

Learning Objectives

After careful study of this chapter, you should be able to:

1 ▶ Define stem cells and describe their role in development and repair of tissue. ***p. 62***

2 ▶ Name the four main groups of tissues and give the location and general characteristics of each. ***p. 62***

3 ▶ Describe the difference between exocrine and endocrine glands, and give examples of each. ***p. 63***

4 ▶ Classify the different types of connective tissue. ***p. 65***

5 ▶ Describe three types of epithelial membranes. ***p. 69***

6 ▶ List six types of connective tissue membranes. ***p. 70***

7 ▶ Explain the difference between benign and malignant tumors, and give several examples of each type. ***p. 72***

8 ▶ Identify the most common methods of diagnosing and treating cancer. ***p. 73***

9 ▶ Using the case study and information in the text, describe the warning signs of cancer. ***pp. 61, 76***

10 ▶ Show how word parts are used to build words related to tissues, glands, and membranes (see Word Anatomy at the end of the chapter). ***p. 78***

Disease in Context *Paul's Case: Sun-Damaged Skin*

"Wait a minute," Paul said to his reflection in the mirror as he examined his face after shaving. He had noticed a small nodule to the side of his left nostril. The lump was mostly pink with a pearly white border and painless to the touch. *I haven't seen that before. Probably just a pimple, or maybe a small cyst,* Paul thought, although he couldn't help thinking back to the many hours he had spent as a kid sailing competitively at the seashore. *I know sun exposure isn't great for your skin, even dangerous, and I wasn't real careful about wearing sunscreen. Even if I did, it would have washed off anyway while I was sailing,* he thought. Paul finished his trimming and decided the lump was probably nothing.

Despite his attempts to forget about the lump, Paul was concerned. Over the next several days, he showed the small, rounded mass to several people to get their opinions. No one had an answer when he asked, "What do you think this is?" When several weeks produced no change, except maybe a little depression in the center of the mass, worry led him to make an appointment with a dermatologist.

"Well Paul, I'm not sure. It could be nothing, but we'd better look a little closer," said Dr. Nielsen. "It could be benign, but we have to be sure that it's not a small skin cancer. This is a very common site for such a lesion. Basal cell and squamous cell carcinomas arise from the top layer of skin cells, especially in sun-exposed areas. UV rays from the sun can damage DNA, causing the cells to divide more rapidly than normal, resulting in an abnormal growth. Basal cell and squamous cell carcinomas are the most common forms of cancer but are usually completely treatable. We'll remove this and send it to the pathology lab to see what's going on." Paul left Dr. Nielsen's office with a small bandage over the site of excision, some ointment to apply, and instructions to call the office in three days.

Paul's dermatologist suspects that he may have skin cancer—a disease affecting the cutaneous membrane. Later in the chapter, we revisit Paul and learn the final diagnosis of that lump on his nose.

ANCILLARIES *At-A-Glance*

Visit thePoint to access the following resources. For guidance in using these resources most effectively, see pp. xv–xvii.

Learning RESOURCES

- Tips for Effective Studying
- Web Chart: Epithelial Tissue
- Web Chart: Connective Tissue
- Health Professions: Histotechnologist
- Detailed Chapter Outline
- Answers to Questions for Study and Review
- Audio Pronunciation Glossary

Learning ACTIVITIES

- Pre-Quiz
- Kinesthetic Activities
- Auditory Activities

In Chapter 3, we learned about cells and their structures and functions. Cells work together to form tissues, which are the subject of this chapter. We'll also look further into when and why cells multiply and discuss cancer and its treatment.

Tissues are groups of cells similar in structure, arranged in a characteristic pattern, and specialized for the performance of specific tasks. The tissues in our bodies might be compared with the different materials used to construct a building. Think for a moment of the great variety of building materials used according to need—wood, stone, steel, plaster, insulation, and others. All the functions of the building depend upon the properties and organization of the individual building materials. Similarly, an organ's ability to accomplish its functions depends on the organization, structure, and abilities of its tissues. The study of tissues is known as **histology** (his-TOL-o-je).

Tissue Origins

During development, all tissues derive from young, actively dividing cells known as **stem cells** (Box 4-1). Most stem cells gradually differentiate into the mature, functioning cells that make up different body tissues. These mature body cells no longer undergo mitosis, but remain in interphase, as described in Chapter 3.

A variable number of stem cells persist in each tissue, producing new cells that can differentiate into mature cells. Tissues subject to wear and tear, such as skin and the lining of the digestive and respiratory tracts, maintain a large population of stem cells that continually divide in order to replace lost or damaged cells. These tissues can repair themselves relatively easily. Other tissues, especially nervous tissue and muscle tissue, maintain few stem cells that divide infrequently, so these tissues repair themselves slowly, if at all. Brain tissue injured by a stroke or heart muscle tissue injured by a heart attack has limited regenerative ability. Between these two extremes are organs, such as the liver, that maintain enough stem cells to replace the entire organ within months or years. So, for example, a portion of the liver can be transplanted from one person to another, and the donor's organ will be restored.

Stem cells give rise to four main tissue groups, as follows:

- **Epithelial** (ep-ih-THE-le-al) **tissue** covers surfaces, lines cavities, and forms glands.
- **Connective tissue** supports and forms the framework of all parts of the body.
- **Muscle tissue** contracts and produces movement.
- **Nervous tissue** conducts nerve impulses.

This chapter concentrates mainly on epithelial and connective tissues, the less specialized of the four types. As discussed later in the chapter, epithelial and connective tissues often form thin sheets of tissues called **membranes**. Muscle and nervous tissues receive more attention in later chapters.

Epithelial Tissue

Epithelial tissue, or **epithelium** (ep-ih-THE-le-um), forms a protective covering for the body. It is the main tissue of the skin's outer layer. It also forms membranes, ducts, and the lining of body cavities and hollow organs, such as the organs of the digestive, respiratory, and urinary tracts.

HOT TOPICS — Box 4-1

Stem Cells: So Much Potential

At least 200 different types of cells are found in the human body, each with its own unique structure and function. All originate from unspecialized precursors called stem cells, which exhibit two important characteristics: they can divide repeatedly and have the potential to become specialized cells.

Stem cells come in two types. **Embryonic stem cells**, found in early embryos, are the source of all body cells and can potentially differentiate into any cell type. **Adult stem cells**, found in babies and children as well as adults, are stem cells that remain in the body after birth and can differentiate into different cell types. They assist with tissue growth and repair. For example, in red bone marrow, these cells differentiate into blood cells, whereas in the skin, they differentiate into new skin cells to replace cells in surface layers that are shed continually or cells that are damaged by a cut, scrape, or other injury.

The potential healthcare applications of stem cell research are numerous. In the near future, stem cell transplants may be used to repair damaged tissues in treating illnesses such as diabetes, cancer, heart disease, Parkinson disease, and spinal cord injury. This research may also help explain how cells develop and why some cells develop abnormally, causing birth defects and cancer. Scientists may also use stem cells to test drugs before trying them on animals and humans.

But stem cell research is controversial. Some argue that it is unethical to use embryonic stem cells because they are obtained from aborted fetuses or fertilized eggs left over from in vitro fertilization. Others argue that these cells would be discarded anyway and have the potential to improve lives. A possible solution is the use of adult stem cells. However, adult stem cells are less abundant than are embryonic stem cells and lack their potential to differentiate, so more research is needed to make this a viable option.

Epithelium repairs itself quickly after it is injured. In areas of the body subject to normal wear and tear, such as the skin, the inside of the mouth, and the lining of the intestinal tract, epithelial stem cells reproduce frequently, replacing dead or damaged cells. Certain areas of the epithelium that form the outer layer of the skin are capable of modifying themselves for greater strength whenever they are subjected to unusual wear and tear; the growth of calluses is a good example of this response.

STRUCTURE OF EPITHELIAL TISSUE

Epithelial cells are tightly packed to better protect underlying tissue or form barriers between systems. The cells vary in shape and arrangement according to their functions. In shape, the cells may be described as follows:

- **Squamous** (SKWA-mus)—flat and irregular
- **Cuboidal**—square
- **Columnar**—long and narrow

The cells may be arranged in a single layer, in which case the epithelium is described as **simple** (**Fig. 4-1**). Simple epithelium functions as a thin barrier through which materials can pass fairly easily. For example, simple epithelium allows for absorption of materials from the lining of the digestive tract into the blood and allows for passage of oxygen from the blood to body tissues. Areas subject to wear and tear that require protection are covered with epithelial cells in multiple layers, an arrangement described as **stratified** (**Fig. 4-2**). If the cells are staggered so that they appear to be in multiple layers but really are not, they are termed *pseudostratified*.

Terms for both shape and arrangement are used to describe epithelial tissue. Thus, a single layer of flat, irregular cells would be described as *simple squamous epithelium*, whereas tissue with many layers of these same cells would be described as *stratified squamous epithelium*. This is the type of tissue involved in Paul's opening case study.

Some organs, such as the urinary bladder, must vary a great deal in size as they work. These organs are lined with **transitional epithelium**, which is capable of great expansion but returns to its original form once tension is relaxed—as when, in this case, the urinary bladder is emptied.

See the Student Resources on thePoint for a summary chart on epithelial tissue.

GLANDS AND SECRETIONS

Most epithelial tissues produce secretions, such as **mucus** (MU-kus) (a clear, sticky fluid), digestive juices, sweat, hormones, and/or other substances. These secretions are produced by specialized epithelial structures known as *glands*. A **gland** is an organ or cell specialized to produce a substance that is sent out to other parts of the body. The gland manufactures these secretions from materials removed from the blood. Glands are divided into two categories based on how they release their secretions: exocrine glands and endocrine glands.

Exocrine Glands **Exocrine** (EK-so-krin) **glands** produce secretions that are carried out of the body (recall that *ex/o* means "outside" or "away from"). The exocrine glands usually have ducts or tubes to carry their secretions away from the glands. Their secretions are delivered into an organ, a cavity, or to the body surface and act in a limited area near their sources. Examples of exocrine glands include the glands in the stomach and intestine that secrete

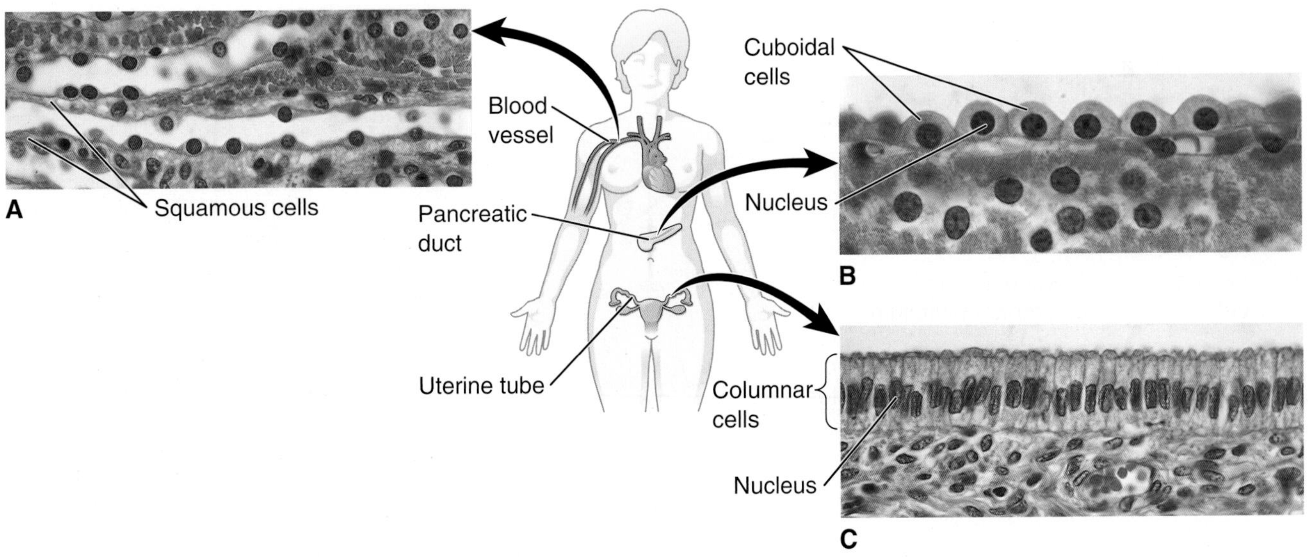

Figure 4-1 **Simple epithelial tissues.** **KEY POINT** Epithelial tissue can be described by the shape of its cells. **A.** Simple squamous epithelium has flat, irregular cells with flat nuclei. **B.** Cuboidal epithelial cells are square in shape with round nuclei. **C.** Columnar epithelial cells are long and narrow with ovoid nuclei. **ZOOMING IN** In how many layers are these epithelial cells?

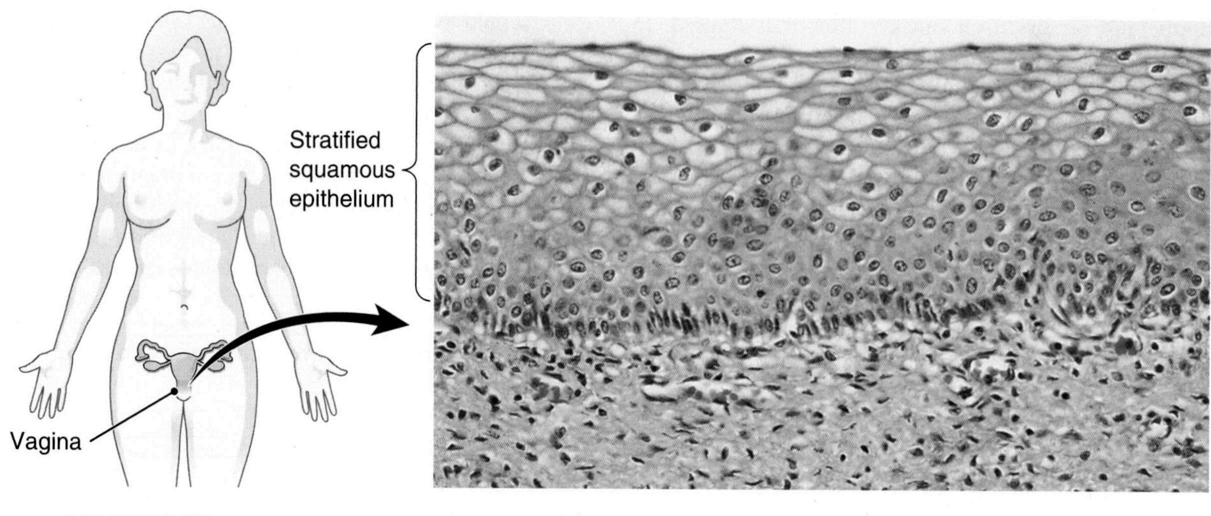

Figure 4-2 **Stratified squamous epithelium.** **KEY POINT** Stratified epithelium has multiple layers of cells. **ZOOMING IN** What is the function of stratified epithelium?

digestive juices, the salivary glands, the sweat and sebaceous (oil) glands of the skin, and the lacrimal glands that produce tears.

Most exocrine glands are composed of multiple cells in various arrangements, including tubular, coiled, or saclike formations. **Goblet cells**, in contrast, are single-celled exocrine glands that secrete mucus. Goblet cells are scattered among the epithelial cells lining the respiratory and digestive passageways (**Fig. 4-3**). As discussed further below, the mucus lubricates the passageways and protects the underlying tissue.

Endocrine Glands **Endocrine** (EN-do-krin) **glands** secrete not through ducts but directly into surrounding tissue fluid. Most often the secretions are then absorbed into the bloodstream, which distributes them internally, as indicated by the prefix *end/o,* meaning "within." These secretions, called **hormones**, have effects on specific tissues known as the *target tissues*. Endocrine glands have an extensive network of blood vessels. These so-called ductless glands include the pituitary, thyroid, adrenal glands, and others described in greater detail in Chapter 12.

CHECKPOINTS

- [ ] 4-1 What are the three basic shapes of epithelial cells?
- [ ] 4-2 What are the two categories of glands based on their methods of secretion?

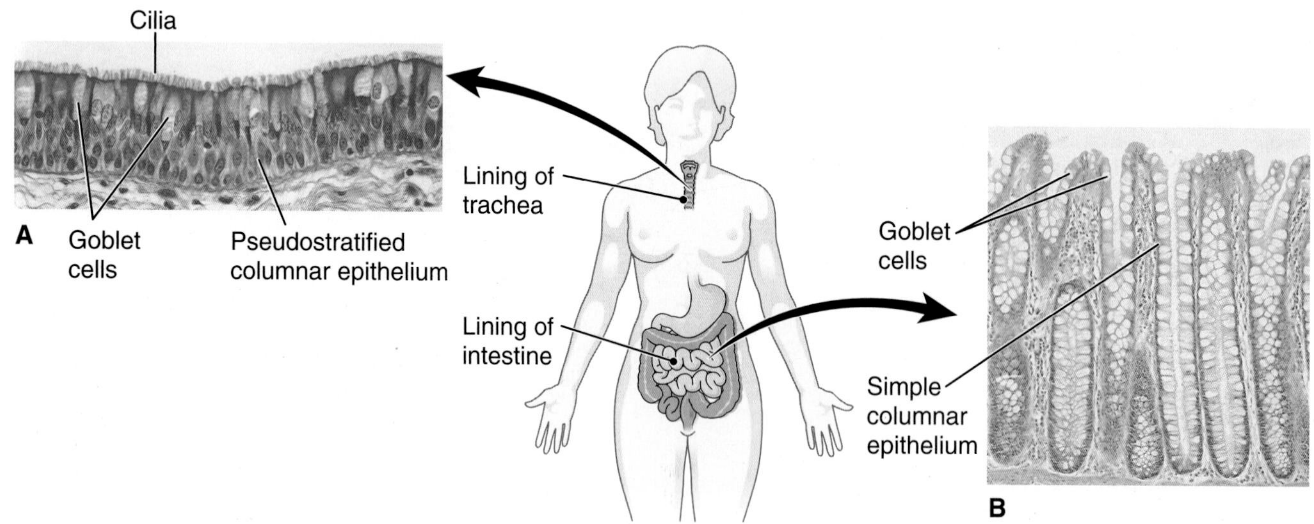

Figure 4-3 **Goblet cells.** **KEY POINT** Goblet cells in epithelium secrete mucus. **A.** The lining of the trachea showing cilia and goblet cells that secrete mucus. **B.** The lining of the intestine showing goblet cells.

Connective Tissue

The supporting fabric everywhere in the body is connective tissue. This is so extensive and widely distributed that if we were able to dissolve all the tissues except connective tissue, we would still be able to recognize the entire body. Connective tissue has large amounts of nonliving material between the cells. This intercellular background material or **matrix** (MA-trix) contains varying amounts of water, protein fibers, and hard minerals.

Histologists, specialists in the study of tissues, have numerous ways of classifying connective tissues based on their structure or function. Here, we describe the different types in order of increasing hardness.

- **Circulating connective tissue** has a fluid consistency; its cells are suspended in a liquid matrix. The two types are blood, which circulates in blood vessels (**Fig. 4-4**), and lymph, a fluid derived from blood that circulates in lymphatic vessels.
- **Loose connective tissue** has a soft consistency, similar to jelly.
- **Dense connective tissue** contains many fibers and is quite strong, similar to a rope or canvas fabric.
- **Cartilage** has a very firm consistency. The gristle at the end of a chicken bone is an example of this tissue type.
- **Bone tissue**, the hardest type of connective tissue, is solidified by minerals in the matrix.

Chapters 13 and 16 have more information on circulating connective tissue, which is highly specialized in both composition and function. The other types of connective tissue are discussed in greater detail below.

See the Student Resources on thePoint for a summary chart of the connective tissue types.

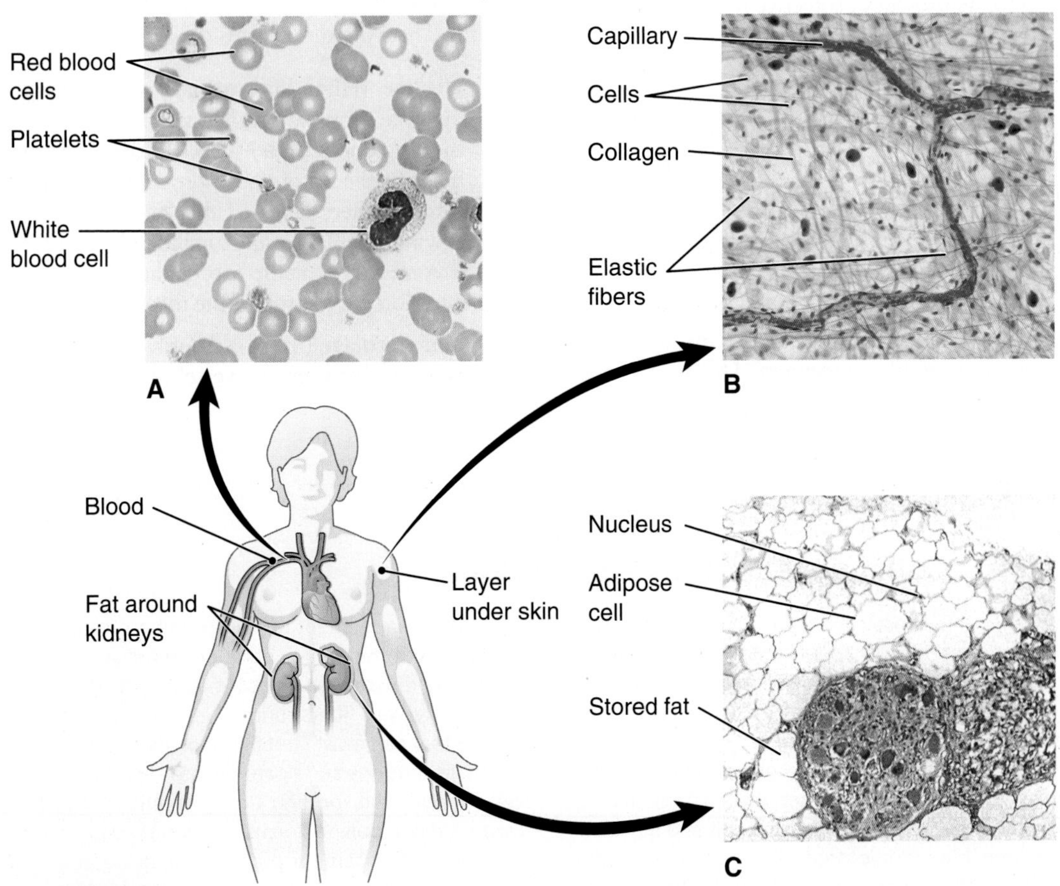

Figure 4-4 Circulating and loose connective tissue. **KEY POINT** Connective tissue is classified according to its distribution and the consistency of its matrix. **A.** Blood smear showing various blood cells in a liquid matrix. **B.** Areolar connective tissue, a mixture of cells and fibers in a jelly-like matrix. **C.** Adipose tissue shown here surrounding dark-staining glandular tissue. The micrograph shows areas where fat is stored and nuclei at the edge of the cells. **ZOOMING IN** Which of these tissues has the most fibers? Which of these tissues is modified for storage?

LOOSE CONNECTIVE TISSUE

As the name implies, loose connective tissue has a soft or semiliquid consistency. There are two types:

- **Areolar** (ah-RE-o-lar) **tissue** is named from a word that means "space" because of its open composition (**see Fig. 4-4B**). It contains cells and fibers in a soft, jellylike matrix. The main cell type is the **fibroblast**, which produces the protein fibers and other components of the matrix (the word ending *-blast* refers to a young and active cell). Fibroblasts produce **collagen** (KOL-ah-jen), a flexible white protein (**see Box 4-2**), as well as elastic fibers. Areolar tissue forms an important component of many tissue membranes (discussed later) and is the most common type of connective tissue.
- **Adipose** (AD-ih-pose) **tissue** is primarily composed of fat cells (adipocytes) with minimal intercellular matrix. Adipocytes are able to store large amounts of fat that serves as a reserve energy supply for the body (**see Fig. 4-4C**). Adipose tissue underlying the skin acts as a heat insulator, and adipose tissue surrounding organs and joints provides protective padding.

DENSE CONNECTIVE TISSUE

Dense connective tissue, like areolar tissue, contains fibroblasts that synthesize a collagen-rich matrix. However, dense connective tissue contains significantly more protein fibers, so it is stronger, firmer, and more flexible than areolar tissue. The different types of dense connective tissue vary in the arrangement of the collagen fibers:

- **Irregular dense connective tissue** has mostly collagenous fibers in random arrangement. This tissue makes up the strong membranes that cover joints and various organs, such as the kidney and liver, and strengthen the skin.
- **Regular dense connective tissue** also has mostly collagenous fibers, but they are in a regular, parallel alignment like the strands of a cable. This tissue can pull in one direction. Examples are the cordlike **tendons**, which connect muscles to bones, and the **ligaments**, which connect bones to other bones (**Fig. 4-5A**). The regular dense connective tissue in the walls of blood vessels, the respiratory passageways, and the vocal cords contains large amounts of elastic fibers. As a result, these structures can stretch and return to their original dimensions.

CARTILAGE

Because of its strength and flexibility, cartilage is a structural material and provides reinforcement. It is also a shock absorber and a bearing surface that reduces friction between moving parts, as at joints. The cells that produce cartilage are **chondrocytes** (KON-dro-sites), a name derived from the word root *chondro*, meaning "cartilage" and the root *cyto*, meaning "cell." There are three forms of cartilage:

- **Hyaline** (HI-ah-lin) **cartilage** is the tough translucent material, popularly called gristle, that covers the ends of the long bones (**see Fig. 4-5B**). You can feel hyaline cartilage at the tip of your nose and along the front of your throat, where rings of this tissue reinforce the trachea ("windpipe"). Hyaline cartilage also reinforces the larynx ("voice box") at the top of the trachea, and can be felt anteriorly at the top of the throat as the "Adam's apple."
- **Fibrocartilage** (fi-bro-KAR-tih-laj) is firm and rigid and is found between the vertebrae (segments) of the spine,

Box 4-2

A CLOSER LOOK

Collagen: The Body's Scaffolding

The most abundant protein in the body, making up about 25% of total protein, is collagen. Its name, derived from a Greek word meaning "glue," reveals its role as the main structural protein in connective tissue.

Fibroblasts secrete collagen molecules into the surrounding matrix, where the molecules are then assembled into fibers. These fibers give the matrix its strength and its flexibility. Collagen fibers' high tensile strength makes them stronger than steel fibers of the same size, and their flexibility confers resilience on the tissues that contain them. For example, collagen in skin, bone, tendons, and ligaments resists pulling forces, whereas collagen found in joint cartilage and between vertebrae resists compression. Based on amino acid structure, there are at least 19 types of collagen, each of which imparts a different property to the connective tissue containing it.

The arrangement of collagen fibers in the matrix reveals much about the tissue's function. In the skin and membranes covering muscles and organs, collagen fibers are arranged irregularly, with fibers running in all directions. The result is a tissue that can resist stretching forces in many different directions. In tendons and ligaments, collagen fibers have a parallel arrangement, forming strong ropelike cords that can resist longitudinal pulling forces. In bone tissue, collagen fibers' meshlike arrangement promotes deposition of calcium salts into the tissue, which gives bone strength while also providing flexibility.

Collagen's varied properties are also evident in the preparation of a gelatin dessert. Gelatin is a collagen extract made by boiling animal bones and other connective tissue. It is a viscous liquid in hot water but forms a semisolid gel on cooling.

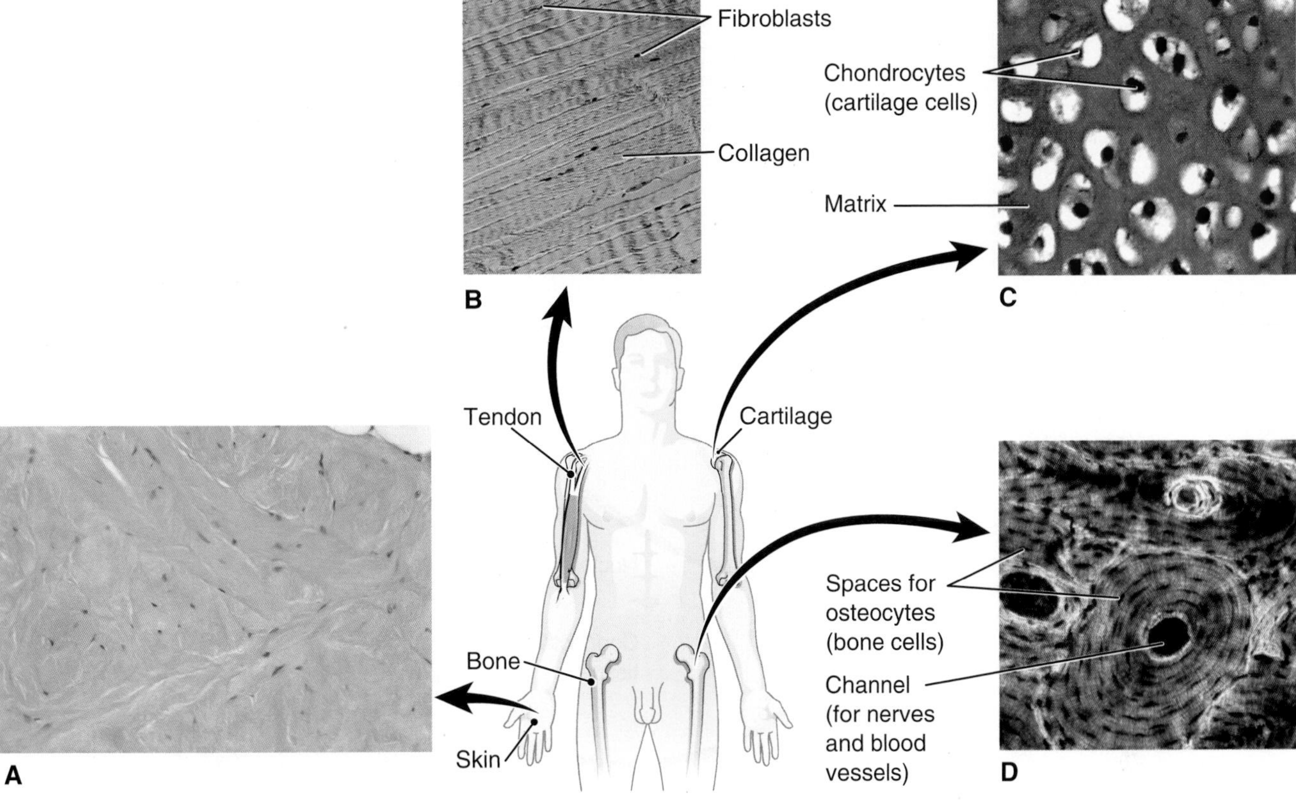

Figure 4-5 **Dense connective tissue, cartilage, and bone.** **KEY POINT** Fibers are a key component of connective tissue. **A.** Dense irregular connective tissue. **B.** Dense regular connective tissue. In tendons and ligaments, collagenous fibers are arranged in the same direction. **C.** In cartilage, the cells (chondrocytes) are enclosed in a firm matrix. **D.** Bone is the hardest connective tissue. The cells (osteocytes) are within the hard matrix.

at the anterior joint between the pubic bones of the hip, and in the knee joint.

- **Elastic cartilage** can spring back into shape after it is bent. An easy place to feel the properties of elastic cartilage is in the outer portion of the ear. It is also located in the larynx.

BONE

The tissue that composes bones, called **osseous** (OS-e-us) **tissue**, is much like cartilage in its cellular structure (see Fig. 4-5C). In fact, the fetal skeleton in the early stages of development is made almost entirely of cartilage. This tissue gradually becomes impregnated with salts of calcium and phosphorus that make bone characteristically solid and hard. The cells that form bone are called **osteoblasts** (OS-te-o-blasts), a name that combines the root for bone (*osteo*) with the ending *blast*. As these cells mature, they are referred to as **osteocytes** (OS-te-o-sites). Within the osseous tissue are nerves and blood vessels. A specialized type of tissue, the bone marrow, is enclosed within bones. The red bone marrow contained in certain regions produces blood cells. Chapter 7 has more information on bones.

CHECKPOINTS

- [ ] 4-3 What is the general name for the intercellular material in connective tissue?
- [ ] 4-4 What protein makes up the most abundant fibers in connective tissue?
- [ ] 4-5 What type of cell characterizes dense connective tissue? Cartilage? Bone tissue?

Muscle Tissue

Muscle tissue is capable of producing movement by contraction of its cells, which are called *muscle fibers* because most of them are long and threadlike. If you were to pull apart a piece of well-cooked meat, you would see small groups of these muscle fibers. Muscle tissue is usually classified as follows:

- **Skeletal muscle**, which works with tendons and bones to move the body (Fig. 4-6A). This type of tissue is described as **voluntary muscle** because we make it contract by conscious thought. The cells in skeletal muscle are very large and are remarkable in having multiple nuclei and a pattern of dark and light banding described

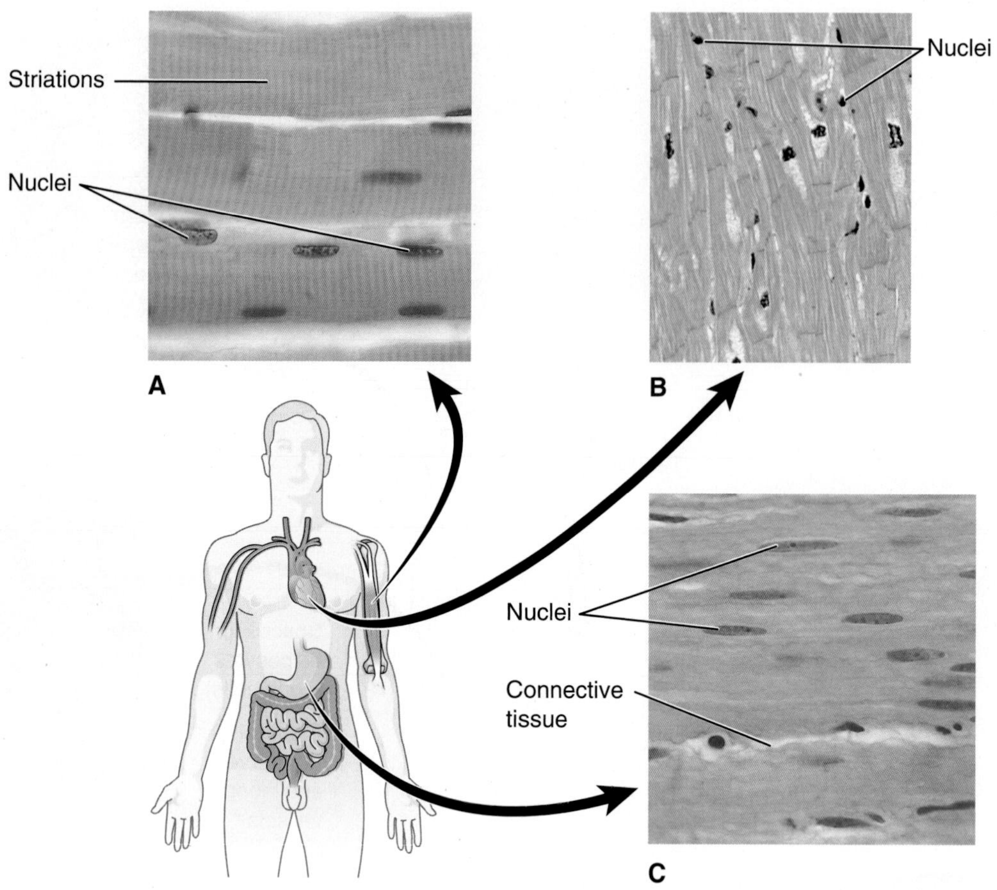

Figure 4-6 **Muscle tissue.** **KEY POINT** There are three types of muscle tissue. **A.** Skeletal muscle cells have bands (striations) and multiple nuclei. **B.** Cardiac muscle makes up the wall of the heart. **C.** Smooth muscle is found in soft body organs and in vessels.

as **striations**. For this reason, skeletal muscle is also called *striated muscle*. Chapter 8 has more details on skeletal muscles.

- **Cardiac muscle**, which forms the bulk of the heart wall and is known also as **myocardium** (mi-o-KAR-de-um) (see Fig. 4-6B). This muscle produces the regular contractions known as *heartbeats*. Cardiac muscle is described as **involuntary muscle** because it typically contracts independently of thought. Most of the time we are not aware of its actions at all. Cardiac muscle has branching cells that form networks. The heart and cardiac muscle are discussed in Chapter 14.
- **Smooth muscle** is also involuntary muscle (see Fig. 4-6C). It forms the walls of the hollow organs in the ventral body cavities, including the stomach, intestines, gallbladder, and urinary bladder. Together these organs are known as viscera (VIS-eh-rah), so smooth muscle is sometimes referred to as *visceral muscle*. Smooth muscle is also found in the walls of many tubular structures, such as the blood vessels and the tubes that carry urine from the kidneys to the bladder. A smooth muscle is also attached to the base of each body hair. Contraction of these muscles causes the condition of the skin that we call *gooseflesh*. Smooth muscle cells are of a typical size and taper at each end. They are not striated and have only one nucleus per cell. Structures containing smooth muscle are discussed in the chapters on the various body systems.

Muscle tissue, like nervous tissue, repairs itself only with difficulty or not at all once a major injury has been sustained. When severely injured, muscle tissue is frequently replaced with connective tissue.

CHECKPOINT

☐ 4-6 What are the three types of muscle tissue?

Nervous Tissue

The human body is made up of countless structures, each of which contributes to the action of the whole organism. This aggregation of structures might be compared to a large corporation. For all the workers in the corporation to coordinate their efforts, there must be some central control, such as the president or CEO. In the body, this central agent is the **brain**

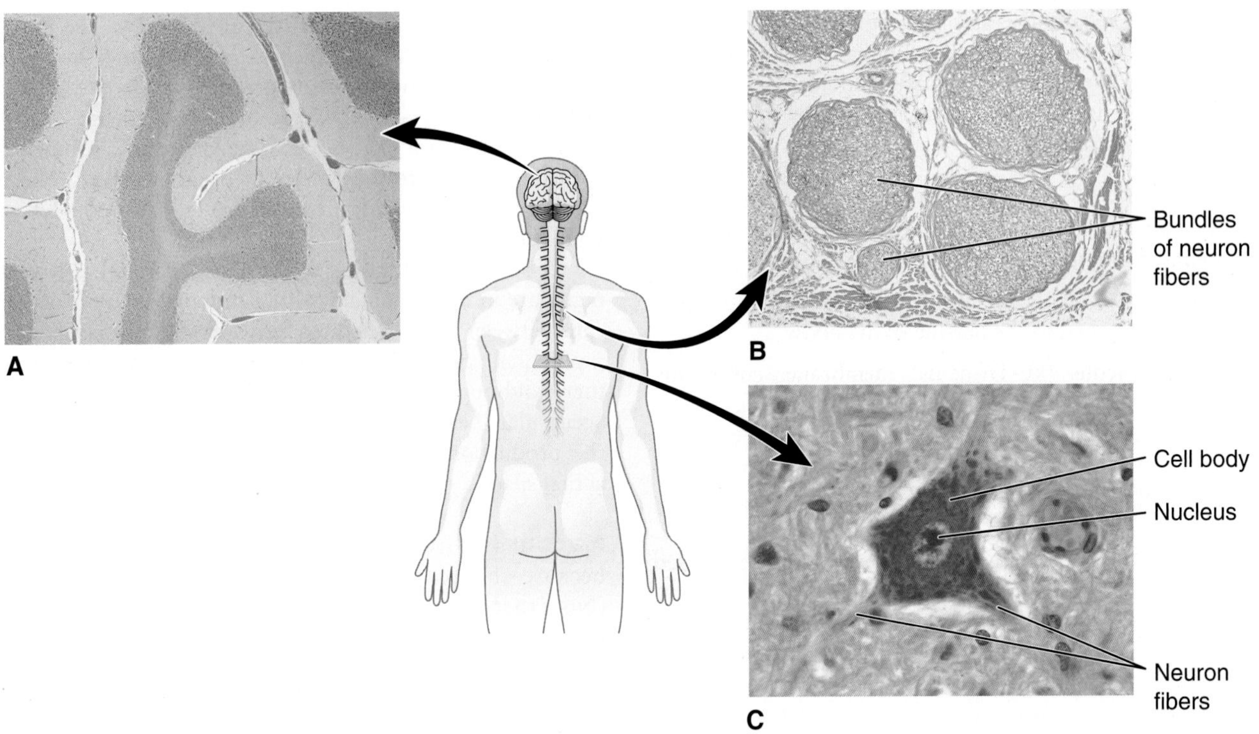

Figure 4-7 **Nervous tissue. A.** Brain tissue. **B.** Cross-section of a nerve. **C.** A neuron, or nerve cell.

(Fig. 4-7A). Each body structure is in direct communication with the brain by means of its own set of "wires," called **nerves** (see Fig. 4-7B). Nerves from even the most remote parts of the body come together and feed into a great trunk cable called the **spinal cord**, which in turn leads into the central switchboard of the brain. Here, messages come in and orders go out 24 hours a day. Some nerves, the cranial nerves, connect directly with the brain and do not communicate with the spinal cord. This entire control system, including the brain, is made of nervous tissue.

THE NEURON

The basic unit of nervous tissue is the **neuron** (NU-ron), or nerve cell (see Fig. 4-7C). A neuron consists of a nerve cell body plus small branches from the cell called *fibers*. These fibers carry nerve impulses to and from the cell body. Neurons may be quite long; their fibers can extend for several feet. A **nerve** is a bundle of such nerve cell fibers held together with connective tissue (see Fig. 4-7B).

NEUROGLIA

Nervous tissue is supported and protected by specialized cells known as **neuroglia** (nu-ROG-le-ah) or *glial* (GLI-al) *cells*, which are named from the Greek word *glia* meaning "glue." Some of these cells protect the brain from harmful substances; others get rid of foreign organisms and cellular debris; still others form the myelin sheath around axons. They do not, however, transmit nerve impulses.

A more detailed discussion of nervous tissue and the nervous system appears in Chapters 9 and 10.

CHECKPOINTS

- ☐ 4-7 What is the basic cell of the nervous system, and what is its function?
- ☐ 4-8 What are the nonconducting support cells of the nervous system called?

Membranes

Recall that membranes are thin sheets of tissue. Their properties vary: some are fragile, and others tough; some are transparent, and others opaque (i.e., you cannot see through them). Membranes may cover a surface, may be a dividing partition, may line a hollow organ or body cavity, or may anchor an organ. They may contain cells that secrete lubricants to ease the movement of organs, such as the heart and lung, and the movement of joints. Membranes are classified based on the tissues they contain. Epithelial membranes contain epithelium and supporting tissues, but connective tissue membranes consist exclusively of connective tissue.

EPITHELIAL MEMBRANES

An **epithelial membrane** is so named because its outer surface is made of epithelium. Underneath, however, there is a layer of areolar and/or dense irregular connective tissue that strengthens the membrane, and in some cases, there is a thin layer of smooth muscle under that. Epithelial membranes are made of closely packed active cells that manufacture lubricants and protect the deeper tissues from invasion by microorganisms. Epithelial membranes are of several types:

- **Serous** (SE-rus) **membranes** line the walls of body cavities and are folded back onto the surface of internal organs, forming their outermost layer.
- **Mucous** (MU-kus) **membranes** line tubes and other spaces that open to the outside of the body.
- The **cutaneous** (ku-TA-ne-us) **membrane**, commonly known as the skin, has an outer layer of stratified squamous epithelium and an inner layer of dense irregular connective tissue. This membrane is complex and is discussed in detail in Chapter 6.

Serous Membranes Serous membranes line the closed ventral body cavities and do not connect with the outside of the body. They secrete a thin, watery lubricant, known as serous fluid, that allows organs to move with a minimum of friction. The thin epithelium of serous membranes is a smooth, glistening kind of tissue called *mesothelium* (mes-o-THE-le-um) overlying areolar tissue. The membrane itself may be referred to as the **serosa** (se-RO-sah).

There are three serous membranes:

- The **pleurae** (PLU-re), or **pleuras** (PLU-rahs), line the thoracic cavity and cover each lung.
- The **serous pericardium** (per-ih-KAR-de-um) forms part of a sac that encloses the heart, which is located in the chest between the lungs.
- The **peritoneum** (per-ih-to-NE-um) is the largest serous membrane. It lines the walls of the abdominal cavity, covers the abdominal organs, and forms supporting and protective structures within the abdomen (**see Fig. 19-2** in Chapter 19).

Serous membranes are arranged so that one portion forms the lining of a closed cavity, while another part folds back to cover the surface of the organ contained in that cavity. The relationship between an organ and the serous membrane around it can be visualized by imagining your fist punching into a large, soft balloon (**Fig. 4-8**). Your fist is the organ and the serous membrane around it is in two layers, one against your fist and one folded back to form an outer layer. Although in two layers, each serous membrane is continuous.

The portion of the serous membrane attached to the wall of a cavity or sac is known as the **parietal** (pah-RI-eh-tal) **layer**; the word *parietal* refers to a wall. In the example above, the parietal layer is represented by the outermost layer of the balloon. Parietal pleura lines the thoracic (chest) cavity, and parietal pericardium lines the fibrous sac (the fibrous pericardium) that encloses the heart (**see Fig. 4-8B**).

Because internal organs are called *viscera*, the portion of the serous membrane attached to an organ is the **visceral layer**. Visceral pericardium is on the surface of the heart, and each lung surface is covered by visceral pleura. Portions of the peritoneum that cover organs in the abdomen are named according to the particular organ involved. The visceral layer in our balloon example is in direct contact with your fist.

A serous membrane's visceral and parietal layers normally are in direct contact with a minimal amount of lubricant between them. The area between the two layers forms a potential space. That is, it is *possible* for a space to exist there, although normally one does not. Only if substances accumulate between the layers, as when inflammation causes the production of excessive amounts of fluid, is there an actual space.

Mucous Membranes Mucous membranes are so named because they contain goblet cells that produce mucus. (Note that the adjective *mucous* contains an "o," whereas the noun *mucus* does not). These membranes form extensive continuous linings in the digestive, respiratory, urinary, and reproductive systems, all of which are connected with the outside of the body. The membranes vary somewhat in both structure and function, but they all have an underlying layer of areolar tissue known as the *lamina propria*. The epithelial cells that line the nasal cavities and the respiratory passageways have cilia. The microscopic cilia move in waves that force secretions outward. In this way, foreign particles, such as bacteria, dust, and other impurities trapped in the sticky mucus, are prevented from entering the lungs and causing harm. Ciliated epithelium is also found in certain tubes of both the male and the female reproductive systems.

The mucous membranes that line the digestive tract have special functions. For example, the stomach's mucous membrane protects its deeper tissues from the action of powerful digestive juices. If for some reason a portion of this membrane is injured, these juices begin to digest a part of the stomach itself—as happens in cases of peptic ulcers. Mucous membranes located farther along in the digestive system are designed to absorb nutrients, which the bloodstream then transports to all cells.

The noun **mucosa** (mu-KO-sah) refers to the mucous membrane of an organ.

CONNECTIVE TISSUE MEMBRANES

The following list is an overview of membranes that consist of connective tissue with no epithelium. These membranes are described in greater detail in later chapters.

- **Synovial** (sin-O-ve-al) **membranes** are thin layers of areolar tissue that line the joint cavities. They secrete a lubricating fluid that reduces friction between the ends of bones, thus permitting free movement of the joints. Synovial membranes also line small cushioning sacs near the joints called **bursae** (BUR-se).

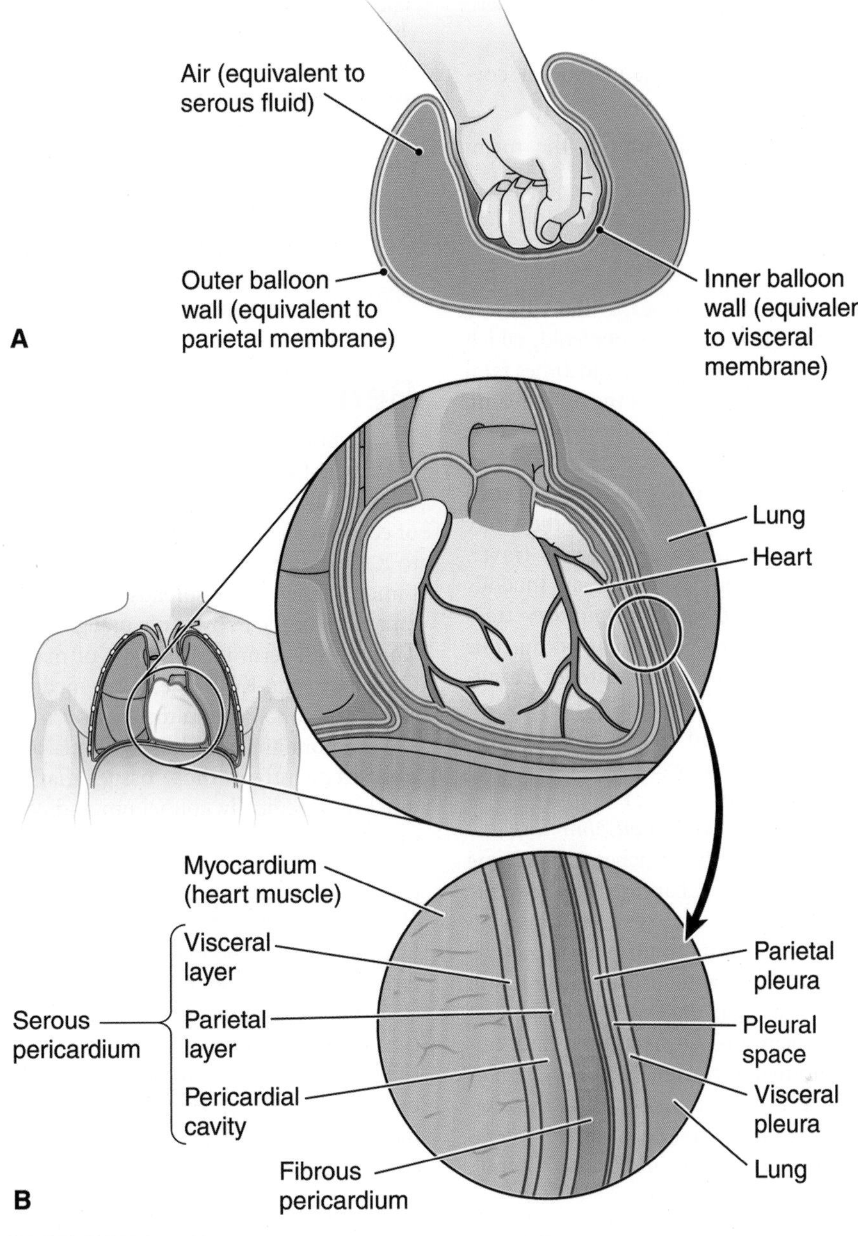

Figure 4-8 **Organization of serous membranes.** **KEY POINT** A serous membrane that encloses an organ has a visceral and a parietal layer. **A.** An organ fits into a serous membrane like a fist punching into a soft balloon. **B.** The outer layer of a serous membrane is the parietal layer. The inner layer is the visceral layer. The fibrous pericardium reinforces the serous pericardium around the heart. The pleura is the serous membrane around the lungs.

- The **meninges** (men-IN-jeze) are several membranous layers covering the brain and the spinal cord.

Fascia (FASH-e-ah) refers to fibrous bands or sheets that support organs and hold them in place. Fascia is found in two regions:

- **Superficial fascia** is the continuous sheet of tissue that underlies the skin. Composed of areolar and adipose tissue, this membrane insulates the body and cushions the skin. This tissue is also called *subcutaneous fascia* because it is located beneath the skin.
- **Deep fascia** covers, separates, and protects skeletal muscles, nerves, and blood vessels. It consists of dense connective tissue.

Finally, there are membranes whose names all start with the prefix *peri* because they are around organs. These tough, protective coverings are made of dense irregular connective tissue.

- The **fibrous pericardium** (per-e-KAR-de-um) forms the cavity that encloses the heart, the pericardial cavity. This fibrous sac and the serous pericardial membranes described above are often described together as the pericardium (see Fig. 4-8B).

- The **periosteum** (per-e-OS-te-um) is the membrane covering a bone.
- The **perichondrium** (per-e-KON-dre-um) is the membrane covering cartilage.

MEMBRANES AND DISEASE

We are all familiar with a number of diseases that directly affect membranes. These range from the common cold, which is an inflammation of the nasal mucosa, to the sometimes fatal condition known as **peritonitis**, an infection of the peritoneum, which can follow rupture of the appendix and other mishaps in the abdominal region.

Although membranes usually help to prevent the spread of infection from one area of the body to another, they may sometimes act as pathways along which disease may travel. For example, throat infections may travel along mucous membranes into other parts of the upper respiratory tract and even into the sinuses and the ears. The inflamed mucous membrane swells and increases its secretory activity, resulting in pain, congestion, and increased mucus. In females, an infection may travel up the tubes and spaces of the reproductive system into the peritoneal cavity (see Fig. 23-15 in Chapter 23), potentially causing pain and scarring.

Membranes are frequently involved in *autoimmune disorders*, in which the body mistakenly launches an immune attack against its own tissues (see Chapter 17). **Systemic sclerosis (scleroderma)**, for instance, occurs when abnormal immune reactions stimulate fibroblasts to produce too much collagen. Tissue membranes thicken and harden, causing skin disfiguration and organ dysfunction. In **systemic lupus erythematosus** (LU-pus er-ih-them-ah-TO-sus) (SLE), autoimmunity can cause inflammation of serous membranes, such as the pleura, pericardium, and peritoneum, as well as the connective tissue membranes supporting blood vessels. Chapter 6 discusses systemic sclerosis and SLE in greater detail. The synovial membrane becomes inflamed and swollen in **rheumatoid** (RU-mah-toyd) **arthritis**, an autoimmune disorder affecting the joints. As the disease progresses, fibrous connective tissue replaces cartilage in the affected joints.

CHECKPOINTS

- [ ] 4-9 What are the three types of epithelial membranes?
- [ ] 4-10 What is the difference between a parietal and a visceral serous membrane?
- [ ] 4-11 What is fascia, and where is it located?

Benign and Malignant Tumors

For a variety of reasons, the normal pattern of cell and tissue development may be broken by an uncontrolled growth of cells having no purpose in the body. Any abnormal growth of cells is called a **tumor**, or **neoplasm.** If the tumor is confined to a local area and does not spread, it is a **benign** (be-NINE) tumor. If the tumor spreads to neighboring tissues or to distant parts of the body, it is a **malignant** (mah-LIG-nant) tumor. The general term for any type of malignant tumor is **cancer. Oncology** (ong-KOL-o-je) is the medical specialty that studies and treats cancer (from the root *onc/o* meaning "tumor").

Tumors are found in all kinds of tissue, but they occur most frequently in those tissues that repair themselves most quickly, specifically epithelium and connective tissue, in that order.

BENIGN TUMORS

Benign tumors, theoretically at least, are not dangerous in themselves; they do not spread. Their cells stick together, and often they are encapsulated, that is, surrounded by a containing membrane. The cells in a benign tumor are very similar in appearance to the normal cells from which they are derived (Fig. 4-9A and B). Benign tumors grow as a single mass within a tissue, which makes them good candidates for complete surgical removal. Medically, they are described as

A
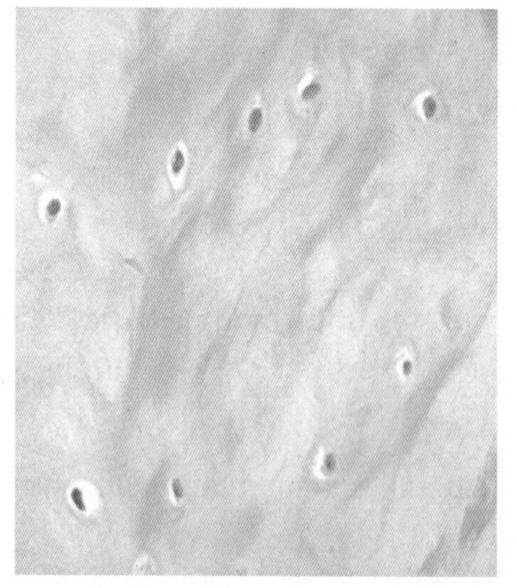
B
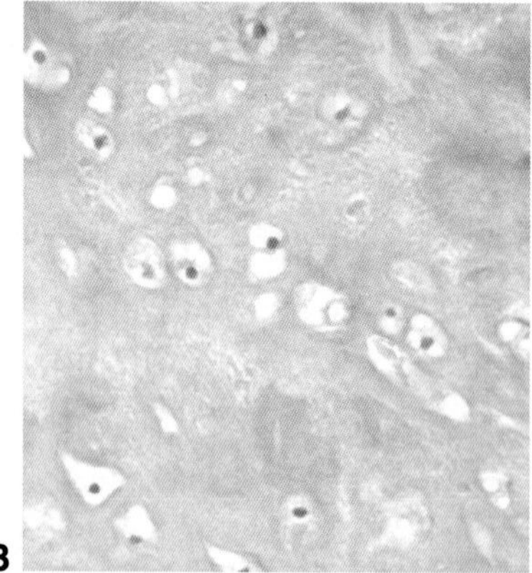
C
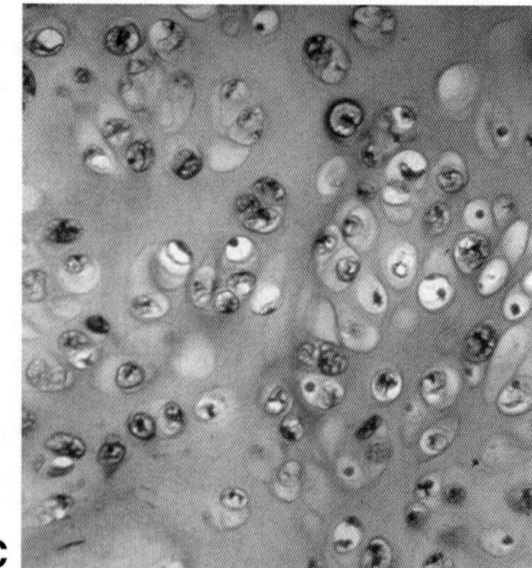

Figure 4-9 Benign and malignant tumors. A. Normal cartilage. **B.** A benign chondroma closely resembles normal cartilage. **C.** Chondrosarcoma of bone. The tumor is composed of malignant chondrocytes, which have bizarre shapes and abnormal nuclei.

growing in situ (SI-tu), meaning that they are confined to their place of origin and do not invade other tissues or spread to other sites. Of course, some benign tumors can be quite harmful; they may grow within an organ, increase in size, and cause considerable mechanical damage. A benign brain tumor, for example, can kill a person just as a malignant one can because it grows in an enclosed area and compresses vital brain tissue. Some examples of benign tumors are given below (note that most of the names end in *oma*, which means "tumor").

- **Papilloma** (pap-ih-LO-mah)—a tumor that grows in epithelium as a projecting mass. One example is a wart.
- **Adenoma** (ad-eh-NO-mah)—an epithelial tumor that grows in and about the glands (*adeno-* means "gland").
- **Lipoma** (lip-O-mah)—a connective tissue tumor originating in fatty (adipose) tissue.
- **Osteoma** (os-te-O-mah)—a connective tissue tumor that originates in the bones.
- **Myoma** (mi-O-mah)—a tumor of muscle tissue. Rare in skeletal muscle, it is common in some types of smooth muscle, particularly in the uterus. When found in the uterus, however, it is ordinarily called a *fibroid*.
- **Angioma** (an-je-O-mah)—a tumor that usually is composed of small blood or lymphatic vessels; an example is a birthmark.
- **Nevus** (NE-vus)—a small skin tumor. Brown or black nevi are better known as moles and can develop into malignant skin tumors called *melanomas*, described and illustrated in Chapter 6. Other nevi develop from blood vessels (angiomas) or epithelial tissue and rarely, if ever, become malignant.
- **Meningioma** (men-in-je-O-mah)—a benign tumor that originates in the linings of the brain and spinal cord.
- **Chondroma** (kon-DRO-mah)—a tumor of cartilage cells that may remain within the cartilage or develop on the surface, as in the joints (**see Fig. 4-9B**).

MALIGNANT TUMORS

Malignant tumors, unlike benign tumors, can cause death no matter where they occur. Malignant cells are very different in appearance from their parent cells and are unable to function normally (**see Fig. 4-9C**). Malignant tumors, moreover, grow much more rapidly than benign tumors. The word *cancer* means "crab," and this is descriptive: a cancer sends out clawlike extensions into neighboring tissue. Cancer cells can also break away from a tumor mass and enter either the blood or the lymph (another circulating fluid). In this process, called **metastasis** (meh-TAS-ta-sis), malignant cells spread to distant body regions. When they arrive at other sites, they form new (secondary) growths, or **metastases** (meh-TAS-tah-seze).

Malignant tumors are classified as to the tissue in which they originate:

- **Carcinoma** (kar-sih-NO-mah). This type of cancer originates in epithelium and is by far the most common form of cancer. Usual sites of carcinoma are the skin, mouth, lung, breast, stomach, colon, prostate, and uterus. Paul's case study involves a carcinoma of the skin. Carcinomas are usually spread by the lymphatic system (see Chapter 16).
- **Sarcoma** (sar-KO-mah) develops from connective tissue and therefore may be found anywhere in the body. Their cells are usually spread by the blood, and they often form metastases in the lungs. A chondrosarcoma, a tumor that arises from cartilage cells, is shown in **Figure 4-9C**.
- **Leukemia** (lu-KE-me-ah) is a cancer of white blood cells.
- **Lymphoma** (lim-FO-mah) is a malignant neoplasm of lymphatic tissue.
- **Glioma** (gli-O-mah) is a cancer of the support (neuroglial) tissue of the brain or spinal cord.

SIGNS OF CANCER

Everyone should be familiar with certain warning signs that may indicate early cancer and should report them to their healthcare providers immediately for further investigation. Early signs may include

- A thickening or lump
- Changes in the color, shape, or size of a mole
- A sore that does not heal
- Unusual bleeding or discharge
- Difficulty in swallowing
- Persistent indigestion or discomfort after eating
- Hoarseness or a persistent cough
- White patches in the mouth
- Changes in bowel or bladder habits
- Unexplained weight gain or loss
- Feeling weak or very tired

CANCER DIAGNOSIS

Many cancers are now diagnosed by routine screening tests that are part of the standard physical examination. Advanced methods of cancer detection lead to earlier and more successful treatment. Some of these methods are discussed here.

Cell and Tissue Studies Microscopic study of cells or tissue samples removed from the body may reveal abnormalities, as shown in **Figure 4-9**. **Biopsy** (BI-op-se) is the removal of living tissue for the purpose of microscopic examination. Specimens can be obtained by needle withdrawal (aspiration) of fluid; by a small punch, as of the skin; by an endoscope (lighted tube) introduced into a body cavity; or by surgical removal (excision). Pleural, peritoneal, and other fluids can be examined for signs of cancerous cells, as these often slough off into their environment. In the Pap (Papanicolaou) test, a physician collects cells from the outer opening of the uterine cervix, the narrow inferior region of the uterus, to be examined for any signs of abnormality.

Diagnostic Imaging Physicians now use a variety of imaging techniques to detect tumors. These include the following:

- **Radiography** is the use of x-rays to obtain images of internal structures. A common application of this method is mammography, x-ray study of the breast (**Fig. 4-10A**). Radiologists can examine other structures such as the lungs, nervous system, and digestive system by this method, although they often need a contrast medium to show changes in soft tissues.
- **Ultrasound** (ultrasonography) is the use of reflected high-frequency sound waves to differentiate various kinds of tissue.
- **Computed tomography** (CT) is the use of x-rays to produce a cross-sectional picture of body parts, such as the brain (**see Fig. 4-10B**).
- **Magnetic resonance imaging** (MRI) is the use of magnetic fields and radio waves to show changes in soft tissues (**see Fig. 4-10C**).
- **Positron emission tomography** shows activity within an organ by computerized interpretation of radiation emitted following administration of a radioactive substance, such as glucose. This method has been used to diagnose lung and brain tumors.

Molecular Methods of Diagnosis All tumors have DNA mutations that alter the type and/or quantity of proteins that they produce. Tumor markers are substances, such as enzymes or other proteins, produced in greater than normal quantity by cancerous cells. Laboratories

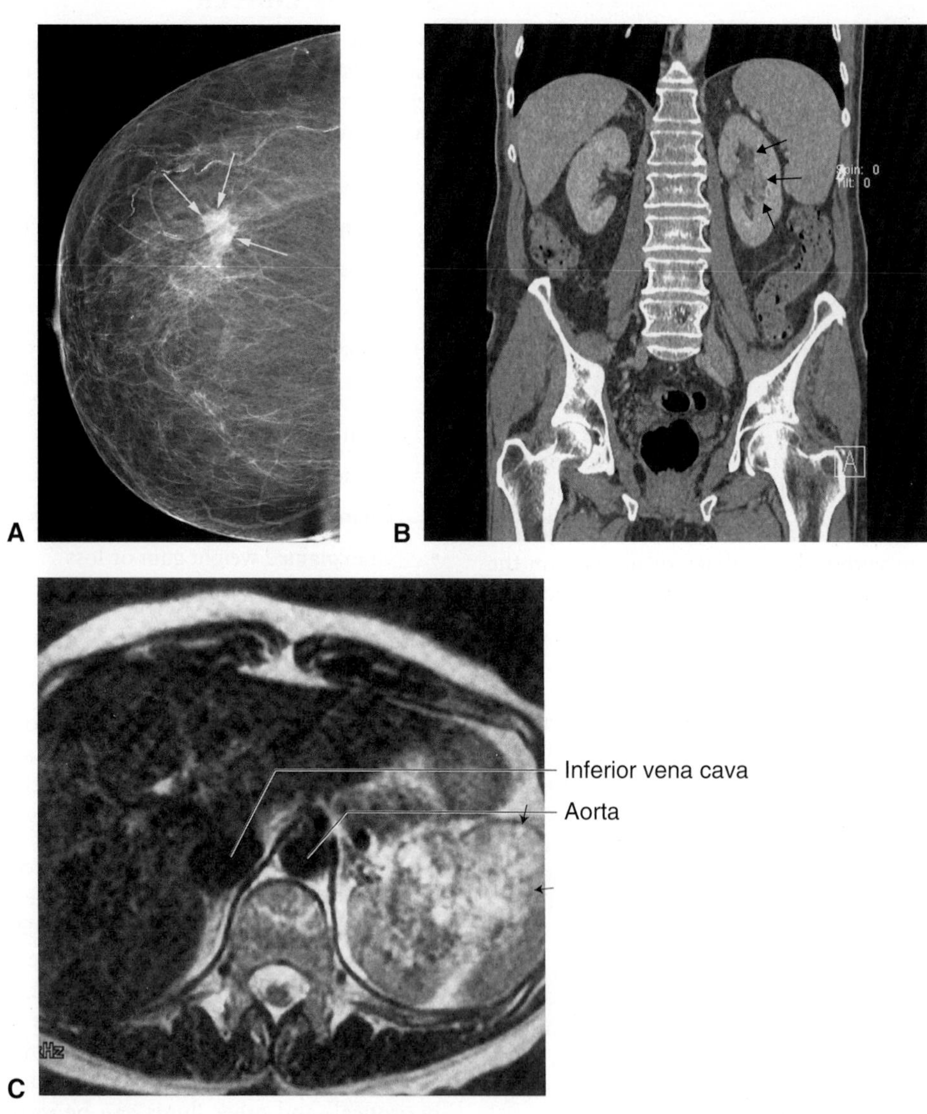

Figure 4-10 Diagnostic imaging for tumors. A. Mammography. A malignant mass (*arrows*) shows poorly defined borders. **B.** A CT scan shows a mass (*arrows*) in the left kidney. **C.** MRI shows a large tumor (*arrows*) in the left kidney.

can detect these markers in cells or in body fluids, such as blood. The most widely used of these screenings is for prostate-specific antigen, a protein produced in large quantity by prostatic tumors. Some others are CA-125 in ovarian cancer, alpha-fetoprotein in liver cancer, carcinoembryonic antigen in digestive system and breast cancers, and human chorionic gonadotropin for testicular cancer. Although tumor markers are often produced in amounts too small to be detected early in the disease process, physicians still can use them to monitor treatment.

Cancer of a specific organ, such as the breast, is not a single disorder. Rather, there are many types of breast cancer, each resulting from a different set of genetic mutations and responding best to a particular treatment regimen. *Genetic profiling* identifies which genes are mutated in a particular tumor. Oncologists can use this information to provide the most effective treatment.

While most of the mutations in a tumor occur over time, some mutations can be inherited. In these cases, a certain type of cancer would "run in families," and people could be genetically tested to determine whether or not they have the gene associated with an increased risk for that cancer. The clearest correlation between genetics and cancer has been established in a small percentage of breast cancers.

Staging Oncologists (cancer specialists) use diagnostic studies for a process called **staging**, which is a procedure for establishing the extent of tumor spread, both at the original site and in other locations (metastases). Staging is important for selecting and evaluating therapy and for predicting the disease's outcome. The method commonly used for staging is the TNM system. The letter T stands for tumor and describes both the size of the primary tumor and whether or not it has invaded nearby tissues. The letter N for nodes identifies any involvement of regional lymph nodes. The letter M for metastases indicates any spread to distant body regions. Based on TNM results, a stage ranging in Roman numerals from I to IV in severity is assigned. Cancers of the blood, lymphatic system, and nervous system are evaluated by different methods.

See the Student Resources on thePoint for information on careers in histotechnology—the laboratory study of tissues.

TREATMENT OF CANCER

Some standard cancer treatments are described first, followed by some newer approaches. Note that various treatment methods may be combined. For example, chemotherapy alone or with radiation therapy may precede surgery, or surgery may be performed before other treatment is given.

Surgery Surgery can usually completely remove benign tumors. Malignant tumors cannot be treated so easily. If cancerous tissue is removed surgically, there is always the possibility that a few hidden cells will be left behind to resume tumor growth. If the cells have metastasized to distant locations, the predicted outcome is usually much less favorable.

The **laser** (LA-zer) is a device that produces a highly concentrated and intense beam of light. It is used to destroy tumors or as a cutting device for removing a tumorous growth. The laser's important advantages are its ability to coagulate blood and prevent bleeding and its capacity to direct a narrow beam of light accurately and attack harmful cells while avoiding normal cells.

Other methods for destroying cancerous tissue include electrosurgery, which uses high-frequency current; cryosurgery, which uses extreme cold generated by liquid nitrogen; and chemosurgery, which employs chemicals.

Radiation Sometimes, surgery is preceded or followed by radiation. Radiation therapy is administered by x-ray machines or by the placement of small amounts of radioactive material within the involved organ. These materials can be in the form of needles, beads, seeds, or other devices. Some radioactive chemicals can be injected or taken orally. Radiation destroys the more rapidly dividing cancer cells while causing less damage to the more slowly dividing normal cells. A goal in radiation therapy is accurate focus of the radiation beam to reduce damage to normal body structures.

Chemotherapy **Chemotherapy** (ke-mo-THER-ah-pe) used to be a general term for treatment with drugs, but now the term is understood to mean the treatment of cancer with **antineoplastic** (an-ti-ne-o-PLAS-tik) **agents.** These agents are drugs that act to greatest effect on actively growing cells, and they are most effective when used in combination. Oncologists often effectively treat certain types of leukemia, various cancers of the lymphatic system, and many other forms of cancer by this means. Researchers continue to develop new drugs and more effective drug combinations, and genetic tumor profiling can help oncologists match each tumor with the most effective treatment. Although antineoplastic agents are more toxic to tumor cells than to normal cells, they do also damage normal dividing cells, such as stem cells in the lining of the digestive tract, bone marrow, and skin. This unwanted activity explains some of the adverse side effects of chemotherapy—hair loss, nausea, and anemia. For this reason, chemotherapeutic agents are typically administered in special oncology clinics under careful supervision by healthcare professionals who understand the complications they may cause. Patients who receive these drugs, because of their generally weakened state and the destruction of white blood cells important in body defenses, are at increased risk for *opportunistic infections*; that is, infections that develop in a person who has been weakened by disease.

Newer Approaches to Cancer Treatment The immune system constantly works to protect the body against abnormal and unwanted cells, a category that includes tumor cells. **Biological therapy**, sometimes called immunotherapy, involves the administration of substances that stimulate the immune system as a whole or promote an immune reaction against a type of cancer cell. For example, an agent called *rituximab* coats a type of cell that is overproduced in lymphoid cancers. The immune system can easily recognize and destroy the coated malignant cells.

Some cancers, such as those of the breast, testis, and prostate, are stimulated to grow more rapidly by hormones. **Hormone therapy** suppresses the production of these hormones or prevents tumors from responding to them by blocking receptor sites.

For a tumor to establish itself, new blood vessels must develop and supply the rapidly growing cells with nutrients and oxygen. **Angiogenesis inhibitors** attempt to block tumor growth by preventing this process of angiogenesis (growth of blood vessels). These drugs have significant side effects and are still in clinical trials.

CHECKPOINTS

- ☐ 4-12 What is the difference between a benign and a malignant tumor?
- ☐ 4-13 What is a biopsy?
- ☐ 4-14 What are the three standard approaches to the treatment of cancer?

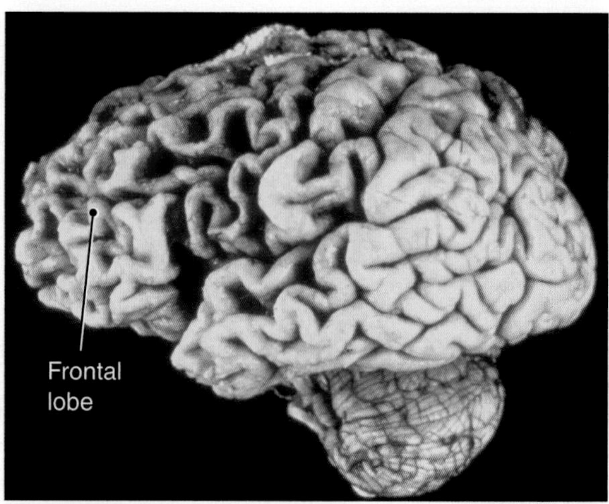

Figure 4-11 **Atrophy of the brain.** **KEY POINT** Brain tissue has thinned, and large spaces appear between sections of tissue, especially in the frontal lobe.

Tissues and Aging

With aging, connective tissues lose elasticity, and collagen becomes less flexible. These changes affect the skin most noticeably, but internal changes occur as well. The blood vessels, for example, have a reduced capacity to expand. Less blood supply, lower metabolism, and decline in hormone levels slow the healing process. Tendons and ligaments stretch, causing a stooped posture and joint instability. Bones may lose calcium salts, becoming brittle and prone to fracture. With age, muscles and other tissues waste from loss of cells, a process termed *atrophy* (AT-ro-fe) (**Fig. 4-11**). Changes that apply to specific organs and systems are described in later chapters.

Disease in Context Revisited

Squamous Cell Carcinoma

Paul was edgy during the three days before he telephoned the dermatologist's office. *What if I have skin cancer? Even if it's treatable, I may have a scar—and right smack in the center of my face!* Finally, he made the call and learned from Dr. Nielsen that he did indeed have a small squamous cell carcinoma.

"I recommend that you consult Dr. Morris, a local surgeon who specializes in a procedure that guarantees the removal of all abnormal cells," the dermatologist advised. "MOHS surgery is done in stages, with the surgeon first removing just the visible lesion and then checking microscopically to be sure that the margins of the excised tissue are free of cancerous cells. If not, additional tissue is removed by degrees until the margins are clean."

Fortunately, Dr. Morris had to repeat the procedure only once after the first pathology examination to be sure of success. Paul left reassured after several hours. Dr. Morris was confident that Paul was safe from the cancer and that scarring would be minimal. "Let's make an appointment for a follow-up visit when I'll remove the stitches and you can see for yourself," he said.

That evening Paul described his day to his wife and told her about his additional instructions. "I need to see my regular dermatologist every six months now, as I may be prone to these types of carcinomas." Paul can't undo earlier damage, but he can prevent further insult to his skin by wearing sunscreen outdoors and reapplying it often. The doctor also advised him to cover up in the sun and avoid times of high sun intensity. "Come to think of it," Paul said to his wife, "that's good advice for you too!"

In this case, we saw the carcinogenic effect of sun damage on skin tissue. Refer back to Chapter 3 to review information on the causes of cancer and its risk factors. Chapter 6 has more specific information on skin cancer.

CHAPTER 4

Chapter Wrap-Up

Summary Overview

A detailed chapter outline with space for note taking is available on thePoint. The figure below illustrates the main topics covered in this chapter.

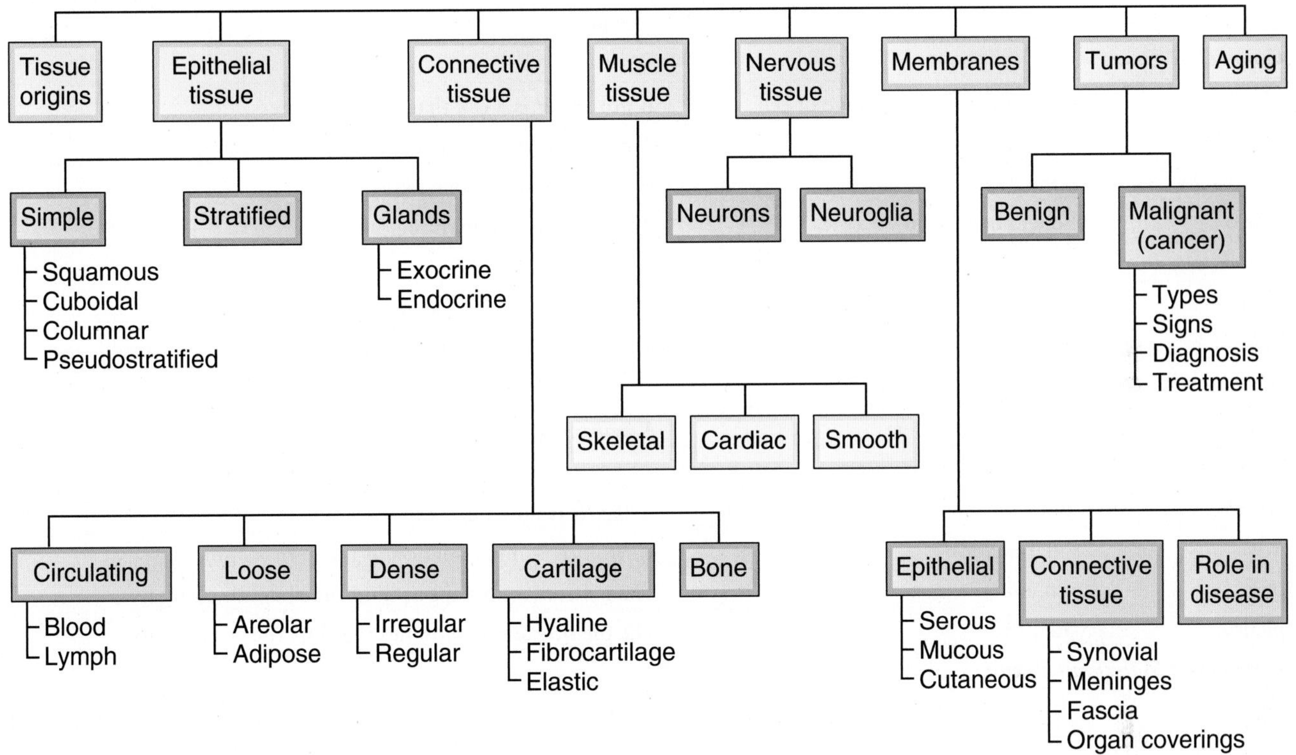

Key Terms

The terms listed below are emphasized in this chapter. Knowing them will help you organize and prioritize your learning. These and other boldface terms are defined in the Glossary with phonetic pronunciations.

adipose
areolar
benign
biopsy
cancer
cartilage
chemotherapy
chondrocyte
collagen
endocrine
epithelium
exocrine
fascia
fibroblast
histology
malignant
matrix
membrane
metastasis
mucosa
mucus
neoplasm
neuroglia
neuron
osteocyte
parietal
serosa
staging
stem cell
visceral

Word Anatomy

Medical terms are built from standardized word parts (prefixes, roots, and suffixes). Learning the meanings of these parts can help you remember words and interpret unfamiliar terms.

WORD PART	MEANING	EXAMPLE
hist/o	tissue	*Histology* is the study of tissues.
Epithelial Tissue		
epi-	on, upon	*Epithelial* tissue covers body surfaces.
pseud/o-	false	*Pseudostratified* epithelium appears to be in multiple layers but is not.
Connective Tissue		
blast/o	immature cell, early stage of cell	A *fibroblast* is a cell that produces fibers.
chondr/o	cartilage	A *chondrocyte* is a cartilage cell.
oss, osse/o	bone, bone tissue	*Osseous* tissue is bone tissue.
oste/o	bone, bone tissue	An *osteocyte* is a mature bone cell.
Muscle Tissue		
cardi/o	heart	The *myocardium* is the heart muscle.
my/o	muscle	See preceding example.
Nervous Tissue		
neur/o	nerve, nerve system	A *neuron* is a nerve cell.
Membranes		
arthr/o	joint	*Arthritis* is inflammation of a joint.
-itis	inflammation	*Peritonitis* is inflammation of the peritoneum.
peri-	around	The *peritoneum* wraps around the abdominal organs.
pleur/o	side, rib	The *pleurae* are membranes that line the chest cavity.
Benign and Malignant Tumors		
aden/o	gland	An *adenoma* is a tumor of a gland.
angi/o	vessel	An *angioma* is a tumor composed of small vessels.
ant/i-	against	An *antineoplastic* agent is a drug active against cancer.
graph/o	writing, record	*Mammography* is x-ray imaging (radiography) of the breast (mamm/o).
leuk/o-	white, colorless	*Leukemia* is a cancer of white blood cells.
mal-	bad, disordered, diseased, abnormal	A *malignant* tumor spreads to other parts of the body.
neo-	new	A *neoplasm* is an abnormal growth of new cells, a tumor.
-oma	tumor, swelling	A *lipoma* is a tumor that originates in adipose (fatty) tissue.
onc/o	tumor	An *oncologist* is a specialist in cancer treatment.
papill/o	nipple	A *papilloma* is a projecting (nipple-like) tumor, such as a wart.
ultra-	beyond	*Ultrasound* is high-frequency sound waves.

Questions for Study and Review

BUILDING UNDERSTANDING

Fill in the Blanks

1. A group of similar cells arranged in a characteristic pattern is called a(n) _____.

2. Glands that secrete their products directly into the blood are called _____ glands.

3. Tissue that supports and forms the framework of the body is called _____ tissue.

4. A tumor that is confined to a local area and does not spread is a(n) _____ tumor.

5. The removal of living tissue for the purpose of microscopic examination is called _____.

Matching > Match each numbered item with the most closely related lettered item.

___ **6.** Membrane around the heart

___ **7.** Membrane around each lung

___ **8.** Membrane around bone

___ **9.** Membrane around cartilage

___ **10.** Membrane around abdominal organs

a. perichondrium
b. pericardium
c. peritoneum
d. periosteum
e. pleura

Multiple Choice

___ **11.** You look under the microscope and see tissue composed of a single layer of long and narrow cells. What is it?

a. simple cuboidal epithelium
b. simple columnar epithelium
c. stratified cuboidal epithelium
d. stratified columnar epithelium

___ **12.** What tissue forms tendons and ligaments?

a. areolar connective tissue
b. loose connective tissue
c. regular, dense connective tissue
d. cartilage

___ **13.** Which tissue is composed of long striated cells with multiple nuclei?

a. smooth muscle
b. cardiac muscle
c. skeletal muscle
d. nervous tissue

___ **14.** A bundle of nerve cell fibers held together with connective tissue is a(n)

a. dendrite
b. axon
c. nerve
d. neuroglia

___ **15.** Which membrane is formed from connective tissue?

a. cutaneous
b. mucous
c. serous
d. fascia

UNDERSTANDING CONCEPTS

16. Define stem cells, and explain how stem cells function in healing.

17. Explain how epithelium is classified, and discuss at least three functions of this tissue type.

18. Compare the structure and function of exocrine and endocrine glands, and give two examples of each type.

19. Put the following terms in a logical order: cartilage—dense connective tissue—circulating connective tissue—bone—loose connective tissue.

20. Compare and contrast the three different types of muscle tissue.

21. Compare serous and mucous membranes.

22. Describe the difference between a benign tumor and a malignant tumor.

23. Define the term *cancer,* and name some of its early symptoms. How is cancer diagnosed? How is it treated?

CONCEPTUAL THINKING

24. Prolonged exposure to cigarette smoke causes damage to ciliated epithelium that lines portions of the respiratory tract. Discuss the implications of this damage.

25. The hereditary disease, osteogenesis imperfecta, is characterized by abnormal collagen fiber synthesis. Which tissue type would be most affected by this disorder? List some possible symptoms of this disease.

26. In Paul's case, sun damage caused skin cancer. Name the type of membrane that was damaged. Why is this tissue more likely to become cancerous than, for example, muscle tissue?

For more questions, see the Learning Activities on thePoint.

UNIT II

Disease and the First Line of Defense

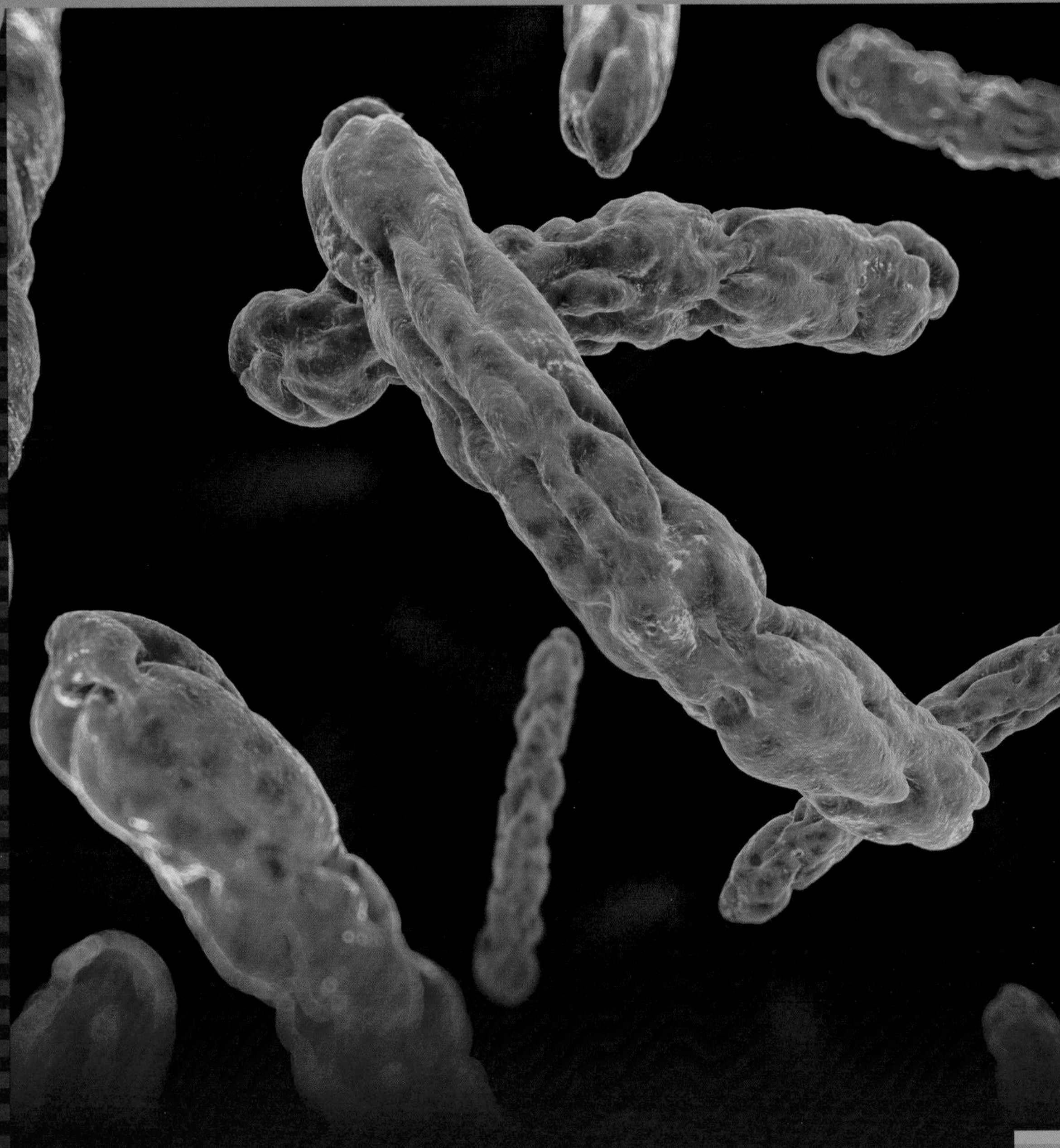

The first chapter in this unit discusses deviation from the normal, which is the basis for disease. After a general discussion of different types of diseases, the chapter concentrates on infectious diseases and the organisms that cause them, including bacteria, viruses, fungi, protozoa, and worms. Other forms of disease are discussed in chapters on the individual body systems. The skin is the first defense against infections, harmful substances, and other sources of injury. The properties and functions of the skin and its associated structures are discussed in the second chapter of this unit.

CHAPTER 5

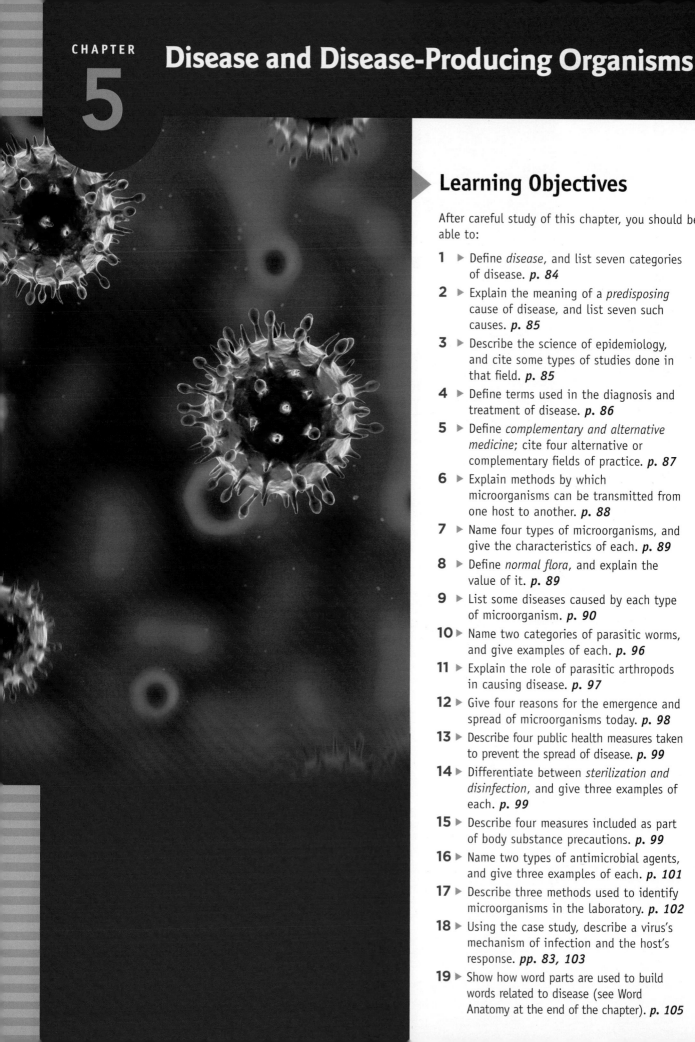

Disease and Disease-Producing Organisms

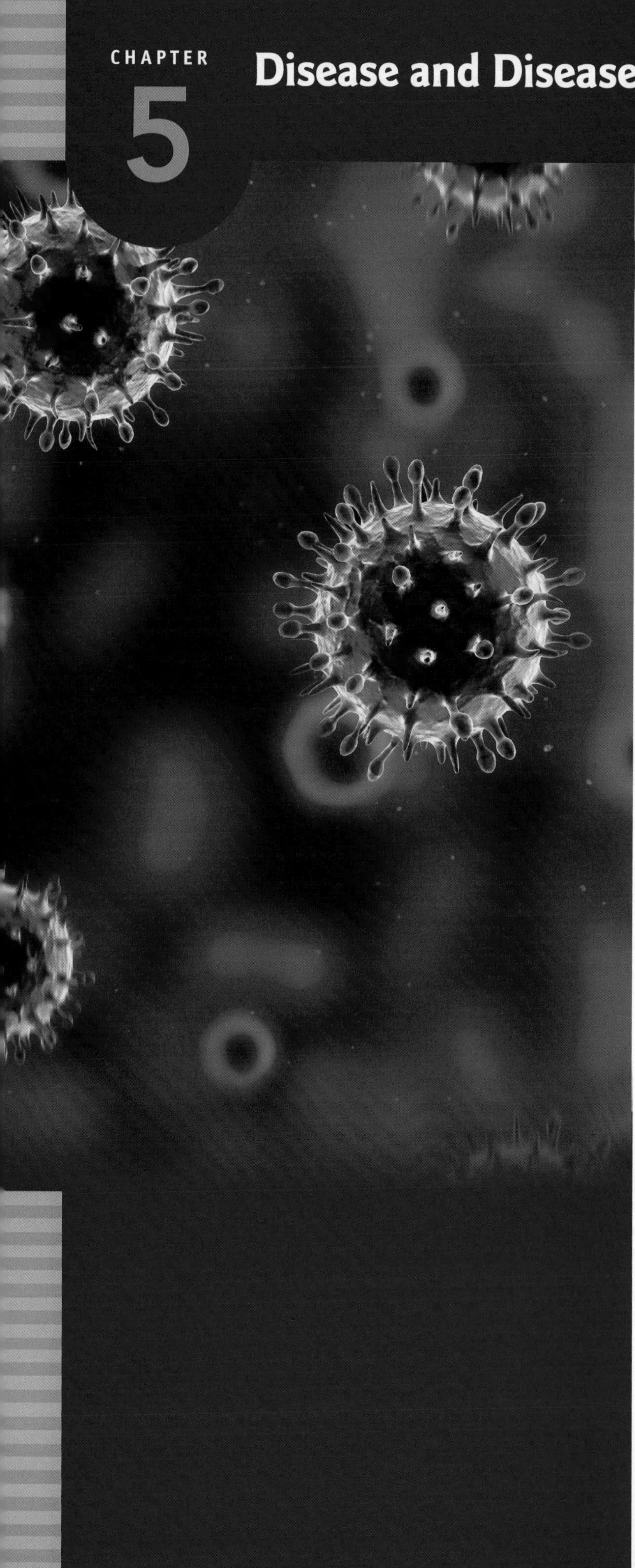

Learning Objectives

After careful study of this chapter, you should be able to:

1 ▶ Define *disease*, and list seven categories of disease. ***p. 84***

2 ▶ Explain the meaning of a *predisposing* cause of disease, and list seven such causes. ***p. 85***

3 ▶ Describe the science of epidemiology, and cite some types of studies done in that field. ***p. 85***

4 ▶ Define terms used in the diagnosis and treatment of disease. ***p. 86***

5 ▶ Define *complementary and alternative medicine*; cite four alternative or complementary fields of practice. ***p. 87***

6 ▶ Explain methods by which microorganisms can be transmitted from one host to another. ***p. 88***

7 ▶ Name four types of microorganisms, and give the characteristics of each. ***p. 89***

8 ▶ Define *normal flora*, and explain the value of it. ***p. 89***

9 ▶ List some diseases caused by each type of microorganism. ***p. 90***

10 ▶ Name two categories of parasitic worms, and give examples of each. ***p. 96***

11 ▶ Explain the role of parasitic arthropods in causing disease. ***p. 97***

12 ▶ Give four reasons for the emergence and spread of microorganisms today. ***p. 98***

13 ▶ Describe four public health measures taken to prevent the spread of disease. ***p. 99***

14 ▶ Differentiate between *sterilization and disinfection*, and give three examples of each. ***p. 99***

15 ▶ Describe four measures included as part of body substance precautions. ***p. 99***

16 ▶ Name two types of antimicrobial agents, and give three examples of each. ***p. 101***

17 ▶ Describe three methods used to identify microorganisms in the laboratory. ***p. 102***

18 ▶ Using the case study, describe a virus's mechanism of infection and the host's response. ***pp. 83, 103***

19 ▶ Show how word parts are used to build words related to disease (see Word Anatomy at the end of the chapter). ***p. 105***

Disease in Context *Maria's Case: When Pathogens Attack*

The man sharing the elevator with Maria Sanchez looked terrible—watery eyes, runny nose, and a sickly, rundown appearance. As the elevator opened and the man stepped toward the doors, he let out a loud sneeze. "Pardon me," he said, as the elevator closed behind him. *If I was that sick, I'd stay in bed,* Maria thought. Maria didn't know it yet, but soon she would be.

The passenger left something behind when he exited the elevator. Microscopic droplets of mucus and water from his sneeze floated in the air. Inside each droplet were thousands of microorganisms, called viruses, capable of making Maria sick. She inhaled several of these germ-laden droplets into her respiratory system as she waited for the elevator to reach her floor. Maria exited the elevator unaware that her body had been invaded.

Maria's warm, moist respiratory tract was a perfect environment for this microorganism. When a virus landed on the epithelial tissue lining her throat, protein spikes protruding from its surface fit snugly into receptors on an epithelial cell, stimulating endocytosis of the virus. Having successfully "picked the lock" on the cell's plasma membrane, the virus was inside the cell and ready to begin the second phase of its attack.

Shortly after entering the cell, the virus released its RNA, which was then transported into the cell's nucleus. Unable to recognize the viral RNA as foreign, the nucleus transcribed it into viral messenger RNA. Returning to the cytoplasm, this new RNA was translated into viral proteins at the ribosomes. Some of these proteins were combined with viral RNA to make new viruses. Others took over the machinery of the host cell to make more viral components.

Since entering the epithelial cell about 24 hours ago, the virus had successfully hijacked cellular RNA and protein synthesis. The cell had been turned into a virus-making factory. With the cell nearly exhausted, the new viruses that filled its cytoplasm exited to infect new hosts. Although finally free of the pathogen, the cell had no more resources left and died.

Maria Sanchez's body has been invaded by the influenza virus. In this chapter, we learn more about viruses as well as other disease-producing organisms. Later in the case, we see how Maria's body is coping with its unwanted visitors.

ANCILLARIES *At-A-Glance*

Visit thePoint to access the following resources. For guidance in using these resources most effectively, see pp. xv–xvii.

Learning RESOURCES

- Tips for Effective Studying
- Web Figure: Modes of Disease Transmission
- Web Figure: Medical Instruments as Vectors of Disease Transmission
- Web Figure: Flagella
- Web Figure: Pili
- Web Chart: Disease Terminology
- Web Chart: Common Routes of Disease Transmission
- Health Professions: Medical Technologist
- Detailed Chapter Outline
- Answers to Questions for Study and Review
- Audio Pronunciation Glossary

Learning ACTIVITIES

- Pre-Quiz
- Visual Activities
- Kinesthetic Activities
- Auditory Activities

A LOOK BACK

So far, we've looked primarily at how the body is constructed and how it functions under normal circumstances. When, for a variety of reasons, a person's proper homeostatic state is altered, disease results. This chapter takes a general look at disease to prepare for discussion of diseases that affect specific systems.

Disease may be defined as abnormality of the structure or function of a part, organ, or system. The effects of a disease may be felt by a person or observed by others. Diseases may be of known or unknown causes and may show marked variation in severity and effects on an individual.

Studies of Disease

Individuals have been treating disease for millennia. In the past, treatments were sometimes based on a philosophical system, such as Chinese medicine or Indian Ayurvedic medicine. Other treatments were simply based on what had worked in the past. The modern, Western approach to the study of disease emphasizes understanding how the normal (physiologic) function has become abnormal (pathologic) at the systems, cellular, chemical, or even molecular level. The term used for this combined study in medical science is **pathophysiology**. Medical professionals study diseases as they affect an individual or groups.

DISEASE CATEGORIES

Diseases fall into a number of different, but often overlapping, categories. These include the following:

- **Infectious diseases.** Infectious organisms are believed to play a part in at least half of all human illnesses. Examples of diseases caused by infectious organisms are colds, acquired immunodeficiency syndrome (AIDS), "strep" throat, tuberculosis, and food poisoning. Microorganisms may also contribute to more complex disorders, for example, stomach ulcers, some forms of heart disease, and even certain types of cancer. This chapter focuses on infectious diseases. Other types of diseases listed below are discussed in other chapters.
- **Degenerative diseases.** These are disorders that involve tissue degeneration (breaking down) in any body system. Examples are muscular dystrophy, cirrhosis of the liver, Alzheimer disease, osteoporosis, and arthritis. Some of these disorders are hereditary; that is, they are passed on by parents through their reproductive cells. Others are due to infection, injury, substance abuse, or normal wear and tear. For some, such as multiple sclerosis (MS), there is no known cause at present.
- **Nutritional disorders.** Most of us are familiar with diseases caused by a dietary lack of essential vitamins, minerals, proteins, or other substances required for health: scurvy due to a lack of vitamin C; beriberi due to a lack of thiamine; rickets due to a lack of vitamin D for bone development; iron deficiency anemia that affects the blood; or kwashiorkor, a disease of children in underdeveloped countries caused by protein deficiency. This category also includes problems caused by excess intake of substances, such as alcohol, vitamins, minerals, animal fats, proteins, and the intake of too many calories leading to obesity (see Chapters 19 and 20).
- **Metabolic disorders.** This general category describes any disruption of cellular metabolism. Hormones regulate many metabolic reactions. Hormone-producing glands and the diseases caused by hormonal excess or deficiency, such as diabetes mellitus, are the subject of Chapter 12. Hereditary errors of metabolism result from genetic changes that affect enzymes. The basics of heredity are described in Chapter 25.
- **Immune disorders.** These relate to the system that protects us against infectious diseases (see Chapter 17). Some deficiencies in the immune system are inherited; some, such as AIDS, are the result of infection. This category also includes allergies, in which the immune system reacts against normally harmless substances, and autoimmune diseases, which occur when the immune system becomes active against a person's own tissues. Examples of diseases that involve autoimmunity are rheumatoid arthritis (**Fig. 5-1**), multiple sclerosis, and systemic lupus erythematosus.
- **Neoplasms.** The word *neoplasm* means "new growth" and refers to cancer and other types of tumors. These were described in Chapter 4.
- **Psychiatric disorders.** Psychiatry is the medical field that specializes in the treatment of mental disorders. The brain and the nervous system as a whole are discussed in Chapters 9 and 10. Note, however, that it is often impossible to separate mental from physical factors in any discussion of disease.

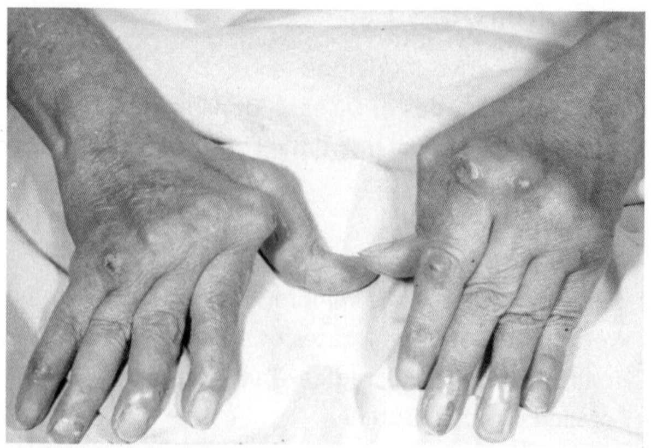

Figure 5-1 Rheumatoid arthritis. The hands of a patient with this disease show joint swelling and deviation of the fingers. **KEY POINT** A disease may fall into more than one category. Rheumatoid arthritis is an immune disorder resulting in a degenerative disease.

CAUSES OF DISEASE

The study of a disease's cause or the theory of its origin is its **etiology** (e-te-OL-o-je). The causes of disease are complex, involving each person's individual characteristics as influenced by behaviors and the environment. Factors known to increase one's susceptibility to a specific disease are known as *predisposing causes*. Although a predisposing cause may not in itself give rise to a disease, it increases the probability of a person's becoming ill. Examples of predisposing causes include the following:

- **Age**. Tissues degenerate with age, becoming less active and less capable of performing normal functions. The normal wear and tear of life, continuous infection, or repeated minor injuries all may speed up decline. Age may also be a factor in the incidence of specific diseases. For example, measles is more common in children than in adults. Other diseases may appear most commonly in middle-aged people or older adults.
- **Gender**. Certain diseases are more characteristic of one gender than the other. Men are more susceptible to early heart disease; women are more likely to develop autoimmune diseases.
- **Heredity**. Our genetic makeup affects our susceptibility to diseases from every category listed above, including cancer, obesity, Alzheimer disease, multiple sclerosis, allergies, and a tendency to catch colds.
- **Living conditions and lifestyle**. Poor lifestyle habits, including inadequate sleep or exercise and poor dietary choices, predispose individuals to all categories of disease. The abuse of drugs, including alcohol and tobacco, also decreases disease resistance. Overcrowding and poor sanitation invite the spread of infectious diseases among groups of people.
- **Emotional disturbance**. Some physical disturbances have their basis in emotional upsets, stress, and anxiety in daily living. Some headaches and so-called nervous indigestion are examples.
- **Physical and chemical damage**. Injuries that cause burns, cuts, fractures, or crushing damage to tissues predispose people to infection and degeneration. Some chemicals that may be poisonous, carcinogenic, or otherwise injurious if present in excess are lead compounds (in old paint), pesticides, solvents, carbon monoxide and other air pollutants, and a wide variety of other environmental toxins. Exposure to radiation, such as the ultraviolet radiation in sunlight, is associated with an increased incidence of cancer. Many of the so-called occupational diseases are caused by exposure to harmful agents on the job. For example, inhalation of coal dust and other types of dusts or fibers has caused lung damage among miners. Metals or toxins may cause skin reactions, and exposure to pesticides and other agricultural chemicals has been associated with neurologic disorders and cancer.
- **Preexisting illness**. Any preexisting illness, especially a chronic disease such as high blood pressure or diabetes, increases one's chances of contracting another disease.

A term used in describing a disease without a known cause is **idiopathic** (id-e-o-PATH-ik), a word based on the Greek root *idio* meaning "self-originating." These diseases are of unknown origin and as yet have no explanation. An **iatrogenic** (i-at-ro-JEN-ik) **disease** results from the adverse effects of treatment, including drug treatment and surgery. The Greek root *iatro* relates to a physician or to medicine.

5

EPIDEMIOLOGY

Some health specialists study diseases in populations, a science known as **epidemiology** (ep-ih-de-me-OL-o-je). They collect information on a disease's geographical distribution and its tendency to appear in one gender, age group, or race more or less frequently than another. Some statistics they collect are as follows:

- **Incidence rate**, the number of new disease cases appearing in a particular population during a specific time period divided by the size of the population
- **Prevalence rate**, the overall frequency of a disease in a given group, that is, the number of cases of a disease present in a given population during a specific period or at a particular time. **Morbidity rate** can refer to either the incidence rate or the prevalence rate.
- **Mortality rate**, the percentage of the population that dies from a given disease within a given time period

If many people in a given region acquire a certain disease at the same time, that disease is said to be **epidemic**. Epidemics of influenza occur at times within populations, as illustrated in Marie's case study. Epidemics of measles and polio still occur in groups that do not receive vaccinations. If a given disease is found to a lesser extent but continuously in a particular region, the disease is considered **endemic** to that area. The common cold is endemic in human populations. Occasionally, an epidemic can spread throughout an entire country, continent, or planet, resulting in a **pandemic** (*pan-* means "all"). Pandemics of influenza have appeared periodically throughout history (**Fig. 5-2**), and a pandemic of bubonic plague (the "black death") wiped out up to one-half of Europe's population in the Middle Ages. AIDS is now considered to be pandemic in certain areas of the globe.

See the Student Resources on thePoint to view a chart summarizing terms related to disease.

CHECKPOINTS

- [ ] 5-1 What is disease?
- [ ] 5-2 Name several categories of disease.
- [ ] 5-3 What is the definition of a predisposing cause of disease?
- [ ] 5-4 List several predisposing causes of disease.
- [ ] 5-5 Identify three types of statistics typically collected by epidemiologists.

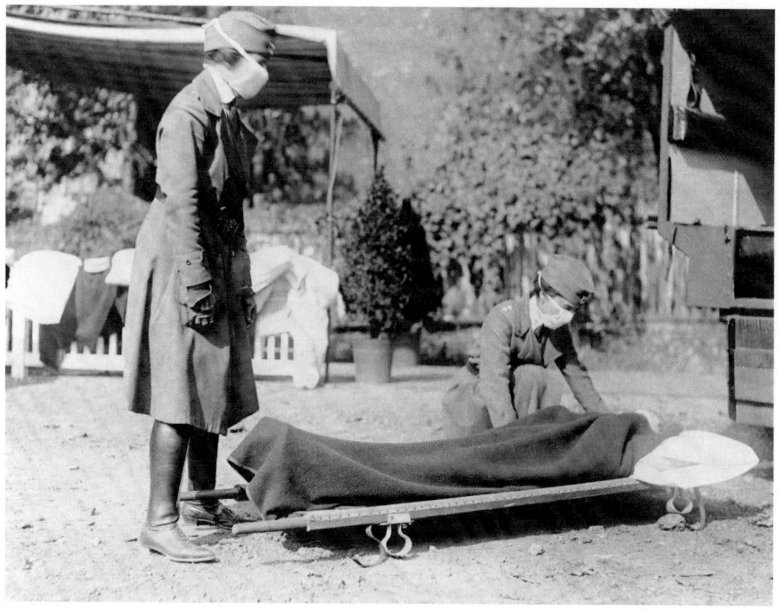

Figure 5-2 **Influenza pandemic.** Nurses attend to a patient in the 1918 flu pandemic. **KEY POINT** Epidemics have appeared throughout history.

Disease Diagnosis, Treatment, and Prevention

In this section, we discuss the conventional medical approach to diagnosis, treatment, and prevention of disease. We also take a brief look at complementary and alternative medicine (CAM).

DIAGNOSIS

Before treating a patient, a physician must make a **diagnosis** (di-ag-NO-sis), that is, reach a conclusion as to the nature or identity of the illness. To do this, the physician relies on the patient's report of his or her **symptoms** (SIMP-toms), disease conditions experienced by the patient, such as pain, fatigue, join stiffness, or nausea. In addition, the physician looks for **signs** of disease, objective manifestations such as a rash, a rapid pulse, or wheezing that the physician or other healthcare providers can observe. Frequently, the physician orders radiology or laboratory tests to gather further signs that can help establish or rule out a diagnosis. Common tests include imaging studies; blood, urine, and stool tests; and study of tissues removed in biopsy. Note that many diseases prompt a variety of symptoms and signs involving more than one body system.

A **syndrome** (SIN-drome) is a complex disorder characterized by a cluster of typical symptoms and signs. For example, patients are diagnosed with metabolic syndrome if they have at least three of the following signs: elevated blood pressure, high plasma triglycerides (fats), low levels of "healthy" high-density lipoproteins, increased blood glucose, and obesity. Individuals with this syndrome are at high risk of developing diabetes mellitus and cardiovascular diseases.

Diseases are often classified on the basis of severity and duration as follows:

- **Acute**. These diseases are relatively severe but usually last a short time.
- **Chronic**. These diseases are often less severe but are likely to be continuous or recurring for long periods.
- **Subacute**. These diseases are intermediate between acute and chronic, not as severe as acute disorders nor as long lasting as chronic ones.

The word *latent* is often used to describe an infection or chronic disorder that is not currently causing symptoms but will in the future. Along with the diagnosis, a physician typically gives the patient a **prognosis** (prog-NO-sis), a prediction of the probable outcome of the disease (*diagnosis* and *prognosis* are from the Greek word *gnosis* meaning "knowledge"). A prognosis is based on the patient's condition and the physician's knowledge about the typical course of the disease. Nurses and other healthcare professionals play an extremely valuable role in the diagnostic process by observing closely for signs, collecting and organizing information from the patient about his or her symptoms, and then reporting this information to the physician.

TREATMENT

Once a patient's disorder is known, the physician prescribes a course of treatment, known as **therapy**. In conventional Western medicine, therapy typically includes medications or surgery. Other commonly prescribed forms of therapy include the following:

- Physical therapy focuses on developing, maintaining, or restoring movement (**Fig. 5-3**).
- Occupational therapy is helpful for restoring functional ability related to specific tasks.

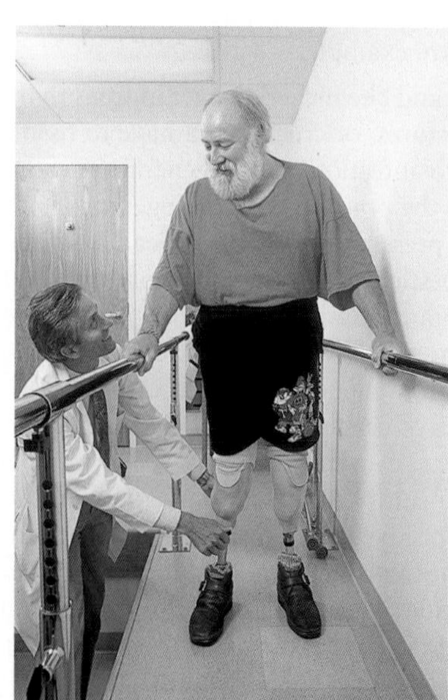

Figure 5-3 **Physical therapy. KEY POINT** Medical treatment often involves several approaches and healthcare teams, including physicians, nurses, and other therapists.

- Respiratory therapy aims to improve the patient's respiratory status.
- Nutritional therapy promotes a healthful intake of nutrients and total energy, including nutritional supplements.
- Psychotherapy, counseling, and support groups may be helpful for patients with psychiatric illness.

Healthcare professionals who specialize in providing different forms of therapy often work together in multidisciplinary teams. Bear in mind that many diseases, including many infectious diseases, can be cured, but others, such as high blood pressure and diabetes, are chronic and can only be managed.

COMPLEMENTARY AND ALTERNATIVE MEDICINE

The term *complementary and alternative medicine* refers to methods of disease prevention or treatment that can be used along with or instead of conventional medical practices. Many of these approaches have a long history in other countries. Some examples of complementary and alternative practices are

- **Naturopathy** (na-chur-OP-a-the), a philosophy of helping people to heal themselves by developing healthy lifestyles
- **Chiropractic** (ki-ro-PRAK-tik), a field that stresses manipulation to correct misalignment for treatment of musculoskeletal disorders
- **Acupuncture** (AK-u-punk-chur), an ancient Chinese method of inserting thin needles into the body at specific points to relieve pain or promote healing
- **Biofeedback** (bi-o-FEED-bak), which teaches people to control involuntary responses, such as heart rate and blood pressure, by means of electronic devices that monitor changes and feed information back to a person

Exercise, massage, yoga, meditation, and other health-promoting practices are also included under this heading. The U.S. National Institutes of Health has established the National Center for Complementary and Alternative Medicine to study the value of these therapies. Another area within the umbrella of CAM is the use of botanicals, plant-derived remedies sometimes described as herbal remedies.

Some pharmaceutical (conventional) drugs are derived from plants, but they are highly purified, extensively tested, and often more potent than are botanicals. Pharmaceutical drugs (both prescription and over-the-counter) are tightly regulated by the U.S. Food and Drug Administration (FDA). In contrast, botanicals are classified as dietary supplements. The FDA does not test them prior to marketing, and there are no requirements to report their adverse effects. Not surprisingly, questions of purity, safety, dosage, and effectiveness have arisen in their use. The FDA does restrict the health claims that supplement manufacturers can make, and the agency can withdraw products from the market that cause unreasonable risk of harm at the recommended doses. The U.S. Office of Dietary Supplements supports and coordinates research on botanicals.

DISEASE PREVENTION

Throughout most of medical history, the physician's aim has been to cure patients of existing diseases or at least to manage the symptoms. The modern concept of prevention, however, seeks to stop disease before it actually happens—to keep people well through the promotion of health. The most significant preventive behaviors include cessation of tobacco use, limiting alcohol use, consumption of a nourishing diet, maintenance of a healthful weight, and participation in regular exercise. In recent years, physicians, nurses, and other healthcare workers have taken on increasing responsibilities in disease prevention. In addition, a vast number of organizations exist for the purpose of promoting health, ranging from the World Health Organization (WHO) on an international level to local private, governmental, and community health programs. In the United States, the Centers for Disease Control and Prevention (CDC) plays an important role in preventing disease (see Box 5-1).

Box 5-1

A CLOSER LOOK

The CDC: Making People Safer and Healthier

The CDC in Atlanta, Georgia, is responsible for protecting and improving the health of the American public—at home and abroad. Established in 1946, the CDC has become a world leader in the fight against infectious disease, with an expanded role that now includes control and prevention of chronic diseases such as cancer, heart disease, and stroke. The CDC also works to protect the public from environmental hazards such as waterborne illnesses, weather emergencies, biologic and chemical terrorism, and dangers in the home and workplace. In addition, the CDC provides education to guide informed health and lifestyle decisions. The CDC's stated goal is "healthy people in a healthy world—through prevention." Some of its past accomplishments include the following:

- In the 1950s, the CDC participated in the fight against polio, which has now been eliminated in industrialized countries and most other areas of the world.
- In the 1960s, the CDC joined the WHO in efforts to eradicate smallpox worldwide.
- In the 1970s, it identified the pathogen responsible for Legionnaires disease.
- In the 1980s, it reported the first cases of AIDS and began intensive research on the disease, which continues today.
- In the 1990s, it investigated an outbreak of deadly Ebola virus in Zaire.

Currently, the CDC is working on hundreds of public health issues, including control of food- and waterborne illnesses; tracking emerging diseases, such as new types of influenza; and reduction of the obesity epidemic in America. These professionals are focusing on supporting state and local health departments, improving global health, strengthening surveillance and epidemiology, and reforming health policies. The CDC employs about 8,500 people in state, federal, and foreign locations. They work in more than 170 occupations, including health information, laboratory science, and microbiology.

Educating patients on the maintenance of total health, both physical and mental, is an increasingly important responsibility of all healthcare practitioners.

CHECKPOINTS

- 5-6 What is the identification of an illness called?
- 5-7 What is a symptom? A sign?

Infectious Disease

A predominant cause of human disease is the invasion of the body by disease-producing **microorganisms** (mi-kro-OR-gan-izms). The word *organism* means "anything having life;" *micro* means "very small." Thus, a microorganism is a tiny living thing, too small to be seen by the naked eye. Other terms for microorganism are *microbe* and, more popularly, *germ*. Although most microorganisms are harmless to humans and many are beneficial, some are **parasites**; that is, they live on or within a living **host** and at the host's expense.

Any disease-causing microorganism is a **pathogen** (PATH-o-jen) and is described as **pathogenic** (path-o-JEN-ic). Invasion of the body by microorganisms, which may result in illness, is called **infection.** The **virulence** (VIR-u-lens) of a pathogen describes its ability to cause disease. Avirulent microbes do not cause any signs or symptoms, whereas virulent microbes may have serious impacts on health. If the infection is restricted to a relatively small area of the body, it is local; a generalized, or **systemic** (sis-TEM-ik), infection is one that affects the whole body. Systemic infections are usually spread by the blood.

An infection that takes hold because the host has been compromised (weakened) by disease is described as an **opportunistic infection.** For example, people with depressed immune systems, such as those with AIDS, can become ill from infection with organisms that are ordinarily harmless.

MODES OF TRANSMISSION

A **communicable** disease is one that can be transmitted from one person to another; it is contagious or "catching." Organisms may be transmitted from an infected host to a new host by direct or indirect contact. For example, infected human hosts may transfer their microorganisms directly to other individuals by touching, shaking hands, kissing, or having sexual intercourse. Microorganisms may be transferred indirectly through touched objects, such as bedding, toys, food, and dishes.

The atmosphere is a carrier of microorganisms. Although microbes cannot fly, the dust in the air is alive with them. In close quarters, germ-laden droplets discharged by sneezing, coughing, and even normal conversation contaminate the atmosphere. Pathogens are also spread by such pests as rats, mice, fleas, lice, flies, and mosquitoes. Pets may be an indirect source of some infections, depositing infectious material on food, skin, or clothing.

An insect bite may introduce infectious organisms into the body. An insect or other animal that transmits a disease-causing organism from one host to another is termed a **vector** (VEK-tor), and the disease can be described as zoonotic. Crowded conditions and poor sanitation increase the spread of disease organisms by all of these mechanisms (see Box 5-2).

HEALTH MAINTENANCE **Box 5-2**

The Cold Facts about the Common Cold

Every year, an estimated 1 billion Americans suffer from the symptoms of the common cold—runny nose, sneezing, coughing, and headache. Although most cases are mild and usually last about a week, colds are the leading cause of doctor visits and missed days at work and school.

Colds are caused by a viral infection of the upper respiratory mucous membranes. More than 200 different viruses are known to cause cold symptoms. Most belong to the group known as rhinoviruses (the word root *rhino* means "nose"). Whereas there is some evidence that cold viruses live longer at low temperatures, the incidence of colds is probably higher in winter because people spend more time indoors, increasing the chances that the virus will spread from person to person.

Colds spread primarily from contact with a contaminated surface. When an infected person coughs or sneezes, small droplets of water filled with viral particles are propelled through the air. One unshielded sneeze may spread hundreds of thousands of viral particles several feet. The dry indoor air of winter allows greater airborne spread of cold virus droplets. Depending upon temperature and humidity, these particles may live as long as three to six hours, and others who touch the contaminated surface may pick up the particles on their hands.

To help prevent the transmission of cold viruses:

- Avoid close contact with someone who is sneezing or coughing.
- Wash hands frequently to remove any viral particles you may have picked up.
- Avoid touching or rubbing your eyes, nose, or mouth with contaminated hands.
- Clean contaminated surfaces with disinfectant.

Because of the large number of virus types involved in causing colds, and because these germs mutate so rapidly into new forms, medical scientists have not been able to develop an effective cold vaccine. One antiviral drug has been effective against colds, but most "treatments" only ease their symptoms. Antibiotics are of no benefit against these viruses. While treating symptoms with over-the-counter medications may help relieve some discomfort, getting plenty of rest and drinking lots of fluids are the best ways to speed up recovery.

PORTALS OF ENTRY AND EXIT

While undamaged skin may be the site of local infections, systemic infections usually access body tissues via the mucous membranes of the respiratory, digestive, urinary, and reproductive tracts, known as *portals of entry*. Any physical damage to skin or mucous membranes increases their susceptibility to microbes. As discussed in Chapter 17, the "soldiers" of the immune system are especially abundant at these sites. Microbes must penetrate the epithelial membranes and the immune defenses before they can establish an infection.

Portals of entry may also serve as exit routes, leading to the spread of infection. For example, discharges from the respiratory and intestinal tracts may spread infection through air, by contamination of hands, and by contamination of food and water supplies.

Control of infectious disease involves breaking the "chain of infection" by which microorganisms spread through a population. Microbial control is discussed later in this chapter.

See the Student Resources on thePoint for a chart and illustrations pertaining to disease transmission.

CHECKPOINTS

- ☐ 5-8 What is the relationship between a parasite and a host?
- ☐ 5-9 What is a communicable disease?
- ☐ 5-10 What term describes any disease-causing organism?
- ☐ 5-11 What are some portals of entry and exit for microorganisms?

Microbiology—The Study of Microorganisms

Microorganisms are simple forms of life, usually a single cell. The group includes bacteria, viruses, fungi, protozoa, and algae. The study of these microscopic organisms is **microbiology** (mi-kro-bi-OL-o-je). The organisms included in the study of microbiology, along with their scientific specialties, are listed here and summarized in Table 5-1. (Algae are not included here, as they do not infect humans.)

- **Bacteria** (bak-TE-re-ah) are primitive, single-cell organisms that grow in a wide variety of environments. The study of bacteria, both beneficial and disease producing, is **bacteriology** (bak-te-re-OL-o-je). The group includes rickettsiae and chlamydiae, which are extremely small bacteria that multiply within living cells.
- **Viruses** (VI-rus-es) are extremely small infectious agents that can multiply only within living cells. **Virology** (vi-ROL-o-je) is the study of viruses.
- **Fungi** (FUN-ji) are a group that includes single-celled yeasts and multicellular molds. **Mycology** (mi-KOL-o-je) is the study of fungi (the root *myc/o* refers to a fungus).
- **Protozoa** (pro-to-ZO-ah) are single-cell animals. Their study is **protozoology** (pro-to-zo-OL-o-je). Although the term *parasitology* (par-ah-si-TOL-o-je) is the study of parasites in general, in practice, it usually refers to the study of protozoa and worms (helminths).

Despite the fact that this discussion centers on pathogens, most microorganisms are harmless to humans and are actually essential to the continuation of all life on earth. Bacteria and other microorganisms decompose dead animals and plants and transform them into substances that enrich the soil. Certain bacteria and fungi produce the antibiotics that make our lives safer. Others produce the fermented products that make our lives more enjoyable, such as beer, wine, cheese, chocolate, and yogurt.

NORMAL FLORA

We have a population of microorganisms that normally grows on body surfaces and cavities open to the environment, such as the mouth, throat, nasal cavities, and large

Table 5-1 Organisms Studied in Microbiology

Type of Organisms	Name of Study	Characteristics of Organism
Bacteria	Bacteriology	Simple, single-cell organisms. Grow in many environments. Lack a true nucleus and most organelles.
Viruses	Virology	Composed of nucleic acid (DNA or RNA) and protein. Can reproduce only within living cells—obligate intracellular parasites.
Fungi	Mycology	Very simple, nongreen, plantlike organisms. Single-cell, globular forms are yeasts; multicellular filamentous forms are molds.
Protozoa	Protozoology	Single-cell, animal-like organisms.

intestine. We live in balance with these organisms, which make up the **normal flora.** These populations are beneficial in helping to break down food residues and in producing vitamin K and other nutrients. The normal flora also provide important training opportunities for the immune system during childhood. Children pick up microbes that subsequently colonize skin and the digestive system. The immune system "learns" to tolerate these harmless residents, and, by extension, other harmless environmental agents such as pollens. Recent studies suggest that a more diverse intestinal flora decreases the risk of autoimmune disorders such as allergies and asthma.

The normal flora protects against infectious diseases by preventing pathogens from attaching to body tissues and by using up available space and nutrients. Resident bacteria also produce antibiotics and other chemicals that kill other microbes or prevent them from dividing.

Elements of the normal flora can, on occasion, cause disease if they are present in excessive numbers or in hosts with compromised immunity. For instance, the common skin disorder acne vulgaris results from normal resident bacteria. When the hormones of adolescence increase the bacterial food supply (oil in the hair follicles), bacteria multiply out of control and cause pimples. Any defect in the host's immune system can also result in infection from normally harmless microbes. The fungus *Candida,* for instance, is part of the normal flora, but it can establish infections in individuals afflicted with HIV.

The normal flora is a finely tuned ecosystem, and any disruption can cause disease. Some antibiotics, for instance, kill the normal bacterial flora but spare other microorganisms. Without competing bacteria, *Candida,* for example, can thrive and reach levels sufficient to cause an infection in the mouth or vaginal tract. To help maintain a healthy flora, many healthcare workers recommend the consumption of probiotics, formulations of harmless bacteria, when taking antibiotics. In another experimental approach, physicians are using fecal transplants from healthy donors to repopulate the gut with a normal flora when there are inadequate numbers or disease-producing bacteria present. The current procedure is as unpleasant as it sounds, but there's hope that biologists can isolate the beneficial intestinal bacteria and administer them in pure cultures.

BACTERIA

Bacteria are single-cell organisms that are among the most primitive forms of life on earth. They are unique in that their genetic material is not enclosed in a membrane; that is, they do not have a true nucleus. They also lack most of the organelles found in plant and animal cells. They can be seen only with a microscope; from 10 to 1,000 bacteria (depending on the species) would, if lined up, span a pinhead. Staining of the cells with dyes helps make their structures more visible and reveals information about their properties.

Bacteria are found everywhere: in soil, in hot springs, in polar ice, and on and within plants and animals. Their requirements for water, nutrients, oxygen, temperature, and other factors vary widely according to species. Some are capable of carrying out photosynthesis, as do green plants; others must take in organic nutrients, as do animals. Some, described as **anaerobic** (an-air-O-bik), can grow only in the absence of oxygen; others, called **aerobic** (air-O-bik), require oxygen. Other groups of bacteria are described as **facultative anaerobes**. These cells will use oxygen if it is present but are able to grow without oxygen if it is not available. *Escherichia coli*, an intestinal organism, is an example of a facultative anaerobe.

Some bacteria produce resistant forms, called **endospores**, that can tolerate long periods of dryness or other adverse conditions (**Fig. 5-4**). Because these endospores become airborne easily and are resistant to ordinary methods of disinfection, pathogenic organisms that form endospores are particularly dangerous. Note that it is common to shorten the name endospore to just "spore," but these structures are totally different in structure and purpose from the reproductive spores produced by fungi and plants. The organisms that cause tetanus, botulism (a deadly form of food poisoning), and anthrax are examples of spore-forming species.

Many types of bacteria are capable of swimming rapidly by means of flagella (**Fig. 5-5**). These may be located all around the cell, at one end, or at both ends. Short flagella-like structures called **pili** (PI-li) help bacteria to glide along solid surfaces. Pili also help to anchor bacteria to surfaces, such as to the surface of a liquid to get oxygen.

See the Student Resources on thePoint for electron micrographs showing flagella and pili.

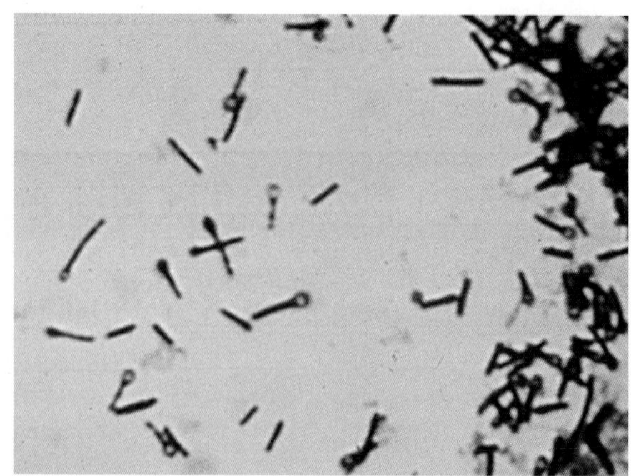

Figure 5-4 **Endospores.** **KEY POINT** These bacteria have endospores that are visible as clear enlargements at the ends of the cells.

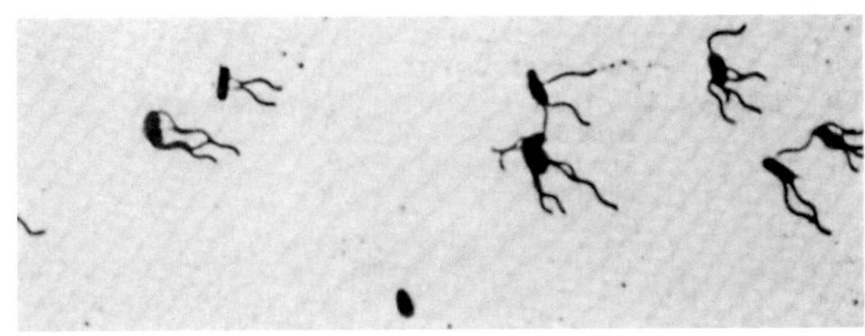

Figure 5-5 **Flagella.** **KEY POINT** Bacteria use these threadlike appendages to swim.

Bacteria comprise the largest group of human pathogens. When conditions are suitable, the organisms reproduce by binary fission (simple cell division). Depending on the species and the growth conditions, cells can divide as rapidly as once every 20 minutes or as slowly as just once every 24 hours. When growing rapidly, populations can increase with unbelievable speed. Just 10 cells dividing at a rate of once every 20 minutes becomes a population of over 40,000 within four hours. Imagine this activity occurring in a wound or in a bowl of food left out at a picnic without refrigeration.

Bacteria can directly cause damage in two ways: by producing poisons, or **toxins**; or by growing within body cells and impairing their function. Indirectly, bacteria (and other pathogens) cause damage when they activate the immune system. A pneumonia infection, for instance, can kill when an immune response promotes fluid accumulation in the lungs. **Table A3-1** in Appendix 3 lists some typical pathogenic bacteria and the diseases they cause.

Shape and Arrangement of Bacteria There are so many different types of bacteria that their classification is complicated. For our purposes, a convenient and simple grouping is based on their shape and arrangement as seen with a microscope:

- **Cocci** (KOK-si). These cells are round and are seen in characteristic arrangements (**Fig. 5-6**). Those that are in pairs are called **diplococci**. Those that are arranged in chains, like a string of beads, are called **streptococci**. A third group, seen in large clusters, is known as **staphylococci** (staf-ih-lo-KOK-si). Among the diseases caused by diplococci are gonorrhea and meningitis; streptococci and staphylococci are responsible for a wide variety of infections, including pneumonia, rheumatic fever, and scarlet fever.
- **Bacilli** (bah-SIL-i). These cells are straight, slender rods (**Fig. 5-7**), although some are cigar shaped, with tapering ends. All endospore-forming bacteria are bacilli. Typical diseases caused by bacilli include tetanus, diphtheria, tuberculosis, typhoid fever, and Legionnaires disease.
- **Curved rods**, which includes several categories:
    - **Vibrios** (VIB-re-oze) are short rods with a slight curvature, like a comma (**Fig. 5-8A and B**). Cholera is caused by a vibrio.
    - **Spirilla** (spi-RIL-a) are long and wavelike cells, resembling a corkscrew. The singular is *spirillum*. *Helicobacter pylori* is a spirillum that colonizes mucus in the digestive tract. It initiates an inflammatory reaction in the stomach or intestine that promotes ulcer formation.

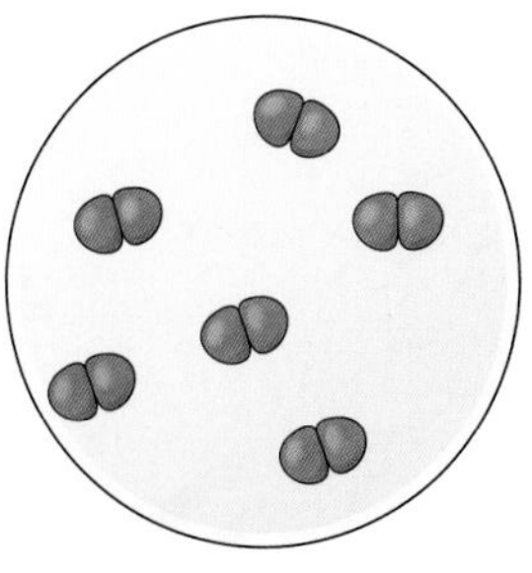

A Diplococci

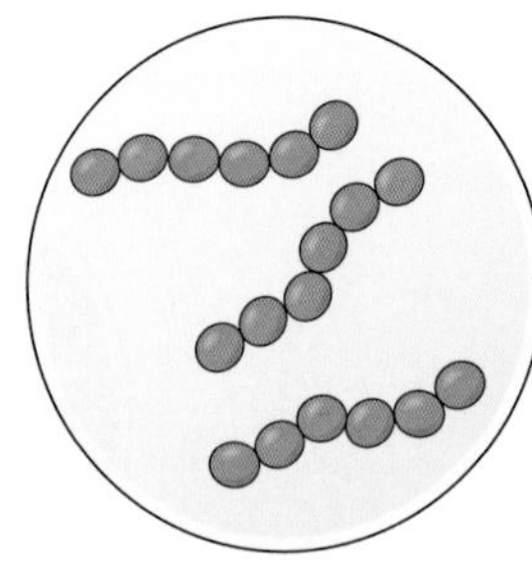

B Streptococci

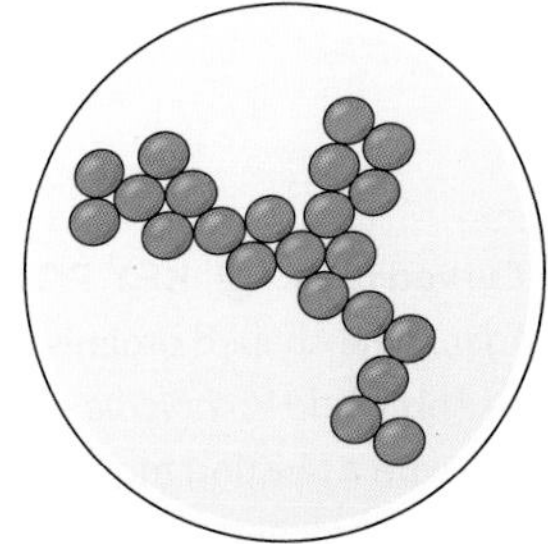

C Staphylococci

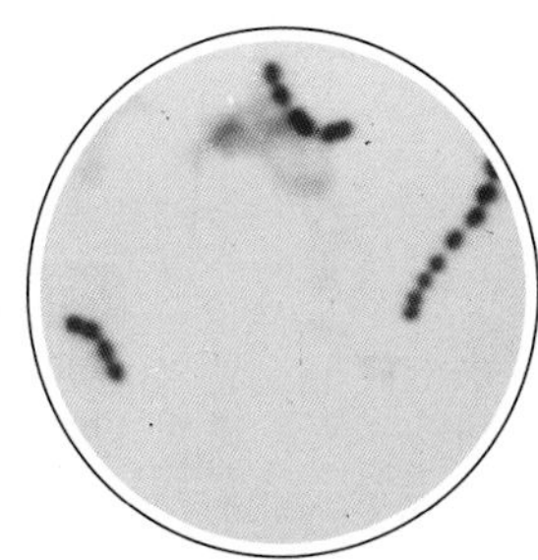

D Streptococci, photomicrograph

Figure 5-6 **Cocci (Gram stained).** **KEY POINT** Cocci are round bacteria that appear in characteristic arrangements. **A.** Diplococci, cocci in pairs. **B.** Streptococci, cocci in chains. **C.** Staphylococci, cocci in clusters. **D.** Streptococci stained and viewed under a microscope.

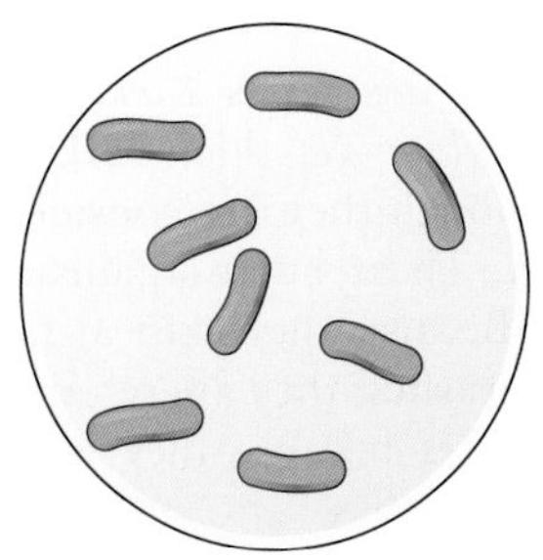

A Bacilli

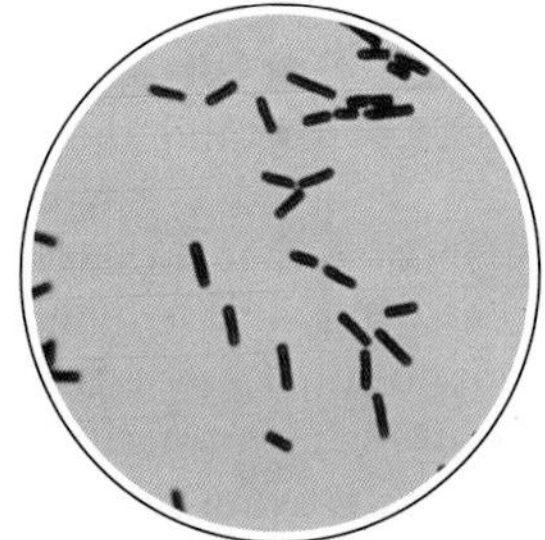

B Bacilli, photomicrograph

Figure 5-7 **Bacilli.** **KEY POINT** Bacilli are rod-shaped bacteria. **A.** Bacilli. **B.** Bacilli stained and viewed under a microscope.

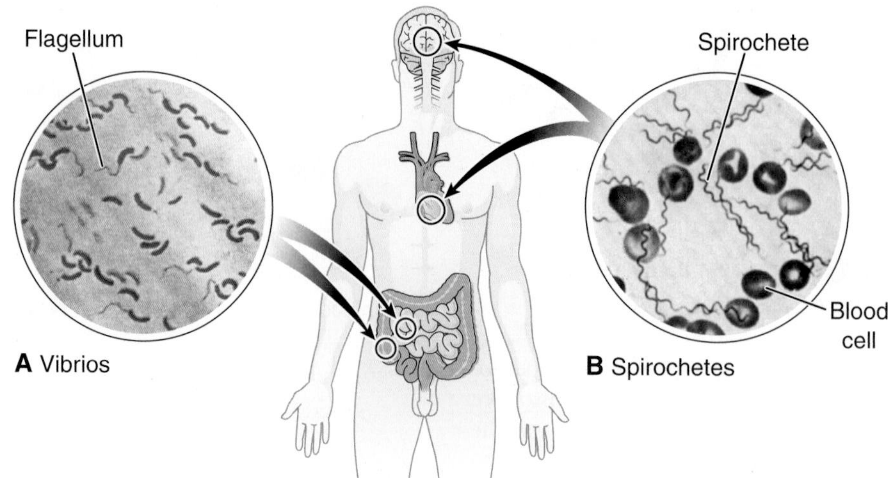

Figure 5-8 **Curved rods.** **KEY POINT** Curved rods include short, slightly curved organisms and longer, spiral-shaped organisms. **A.** Vibrios, comma-shaped organisms, as seen under a microscope. This vibrio *V. cholerae* causes Asiatic cholera. **B.** Spirochetes, spiral-shaped organisms that move with a twisting motion, as seen under the microscope. The large round cells are blood cells. This spirochete causes relapsing fever. **ZOOMING IN** What feature indicates that the cells in **(A)** are capable of movement?

- **Spirochetes** (SPI-ro-ketes) are similar to the spirilla, but are capable of waving and twisting motions (**see Fig. 5-8B**). One infection caused by a spirochete is syphilis. The syphilis spirochetes enter the body at a contact point, usually through the genital skin or mucous membranes. They then travel into the bloodstream and set up a systemic infection (see **Table A3-1** in Appendix 3 for a summary of the three stages of syphilis).

A spirochete carried by deer ticks is also responsible for Lyme disease, which has increased in the United States since it first appeared in the early 1960s. People who walk in or near woods and fields are advised to wear light-colored clothing (so that ticks can be more easily spotted) and to tuck the bottoms of shirts into pants, and pants into socks. They should examine their bodies for the freckle-sized ticks that carry the disease. **Figure 5-8B** illustrates another tick-borne spirochete that causes a similar disease, relapsing fever.

Atypical Bacteria Members of the genus *Rickettsia* (rih-KET-se-ah) and the genus *Chlamydia* (klah-MID-e-ah) are classified as bacteria, although they are considerably smaller than other bacteria. These microorganisms can exist only inside living cells. Because they exist at the expense of their hosts, they are parasites; they are referred to as *obligate intracellular parasites* because they must grow within living cells.

The rickettsiae are the cause of several serious diseases in humans, such as typhus and Rocky Mountain spotted fever. In almost every instance, these organisms are transmitted through the bites of insects, such as lice, ticks, and fleas. A few common diseases caused by rickettsiae are listed in **Table A3-1** in Appendix 3.

The chlamydiae are smaller than the rickettsiae. They are the causative organisms in trachoma (a serious eye infection that ultimately causes blindness), parrot fever or psittacosis, the sexually transmitted infection lymphogranuloma venereum, and some respiratory diseases (see **Table A3-1** in Appendix 3).

Naming Bacteria As is common in naming higher plants and animals, the names of bacteria include a genus name, written with a capital letter, and a species name, written with a small letter, both names italicized. The genus or species names may be taken from the name of an organism's discoverer, as in *Escherichia*, named for Theodor Escherich, or *Rickettsia*, named for Howard T. Ricketts. Some other criteria are shape (e.g., genus *Staphylococcus, Bacillus*), the disease caused (e.g., *Streptococcus pneumoniae*, which causes pneumonia; *Neisseria meningitidis*, the cause of meningitis), habitat (*Staphylococcus epidermidis*, which grows on the skin), or growth characteristics. *Streptococcus pyogenes* produces pus, and colonies of *Staphylococcus aureus*, based on the Latin word for gold, have a golden yellow color. More specific information is conveyed by adding names for subgroups such as type, subtype, strain, and variety.

CHECKPOINTS

- ☐ 5-12 What categories of organisms are studied in microbiology?
- ☐ 5-13 What term refers to microorganisms that normally live in or on the body?
- ☐ 5-14 What are resistant forms of bacteria called?
- ☐ 5-15 What are the three basic shapes of bacteria?

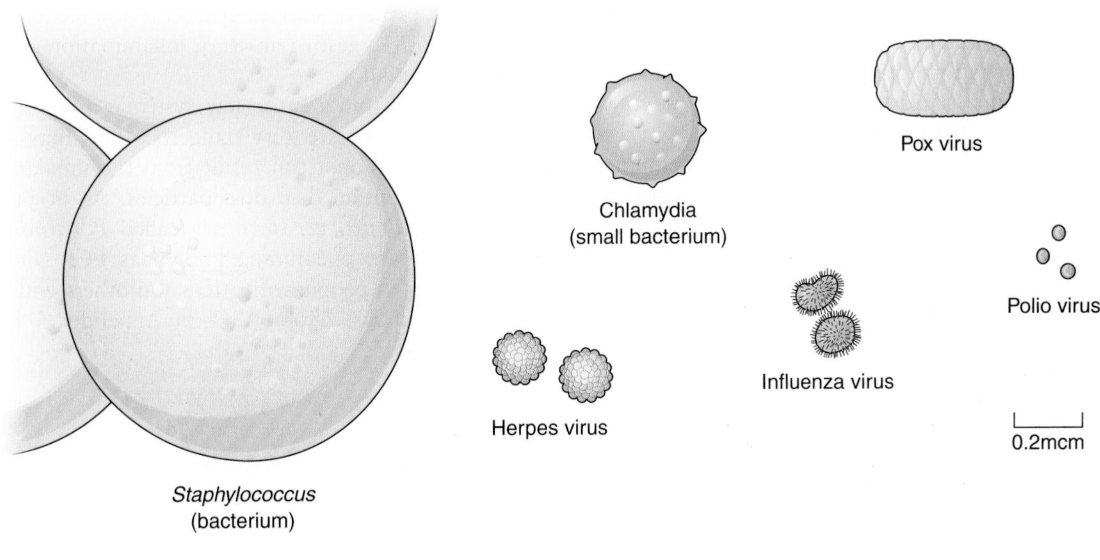

Figure 5-9 **Virus size comparison.** **KEY POINT** Viruses are much smaller than bacteria. A chlamydia and a *Staphylococcus* are shown for reference.

VIRUSES

Although bacteria seem small, they are enormous in comparison with viruses (**Fig. 5-9**). Viruses are comparable in size to large molecules, but unlike other molecules, they contain genetic material and are able to reproduce. Viruses are so tiny that they are invisible with a light microscope; they can be seen only with an electron microscope. Because of their small size and the difficulties associated with growing them in the laboratory, viruses were not studied with much success until the middle of the 20th century.

Viruses have some of the fundamental properties of living matter, but they are not cellular, and they have no organelles. They consist of a nucleic acid core, either DNA or RNA, surrounded by a protein coat (**Fig. 5-10**). Like the rickettsiae and the chlamydiae, they can grow only within living cells—they are obligate intracellular parasites. Unlike these organisms, however, viruses are not susceptible to antibacterial agents (antibiotics) and must be treated with antiviral drugs.

Viruses are classified according to the type of nucleic acid they contain—DNA or RNA—and whether that nucleic acid is single stranded (ss) or double stranded (ds). They are further grouped according to the diseases they cause, of which there are a considerable number—measles, poliomyelitis, hepatitis, chickenpox, and the common cold, to name just a few. Maria's disease in the case study is caused by a virus. AIDS is a life-threatening viral disease discussed in Chapter 17. The virus that causes AIDS and other representative viruses are listed in **Table A3-2** in Appendix 3.

Viruses are named according to where they were isolated (Hanta, Ebola, West Nile), the symptoms they cause (yellow fever virus, which causes jaundice; hepatitis virus, which causes inflammation of the liver), the host (HIV, swine influenza), or the vector that carries them (Colorado tick fever).

INFECTIOUS PROTEINS

We take a moment at this point to describe unusual infectious agents even smaller and simpler than viruses. **Prions** (PRI-ons) are infectious particles composed solely of protein (the name

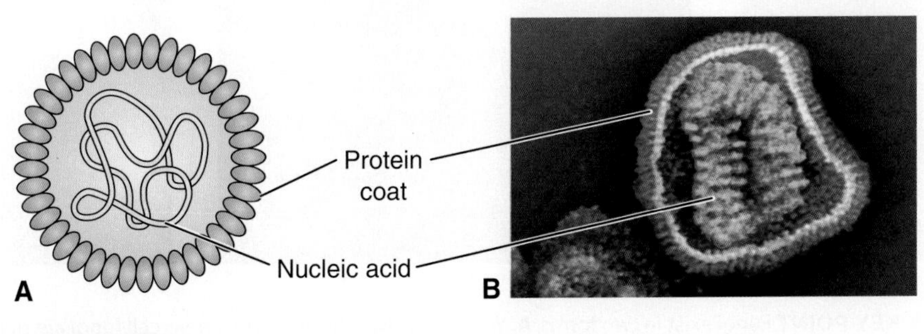

Figure 5-10 **Virus structure.** **KEY POINT** A virus consists of a nucleic acid core enclosed in a protein coat. **A.** Diagram of a virus. **B.** Photomicrograph of the influenza virus.

comes from proteinaceous infectious agent). Researchers have linked prions to several fatal diseases in humans and animals. They are slowly growing and hard to destroy, producing spongy degeneration of brain tissue, described as *spongiform encephalopathy* (en-sef-ah-LOP-a-the). Some examples of diseases caused by prions are Creutzfeldt-Jakob disease in humans; bovine spongiform encephalopathy, the so-called "mad cow disease" in cows, a variant of which affects humans; and scrapie in sheep. Some diseases caused by prions are described in Table A3-3 of Appendix 3.

FUNGI

The true fungi are a large group of simple plantlike organisms. Only a few types are pathogenic. Although fungi are much larger and more complicated than bacteria, they are a simple life form. They differ from higher plants in that they lack the green pigment chlorophyll, which enables most plants to use sunlight's energy to manufacture food. Like bacteria, fungi grow best in dark, damp places. Round, single-cell fungi are generally referred to as **yeasts**; the fuzzy, filamentous, multicellular forms are called **molds** (Fig. 5-11). Molds reproduce in several ways, including by simple cell division and by production of multiple reproductive spores. Yeasts reproduce by simple cell division and can also form buds that pinch off as new cells. Familiar examples of fungi are mushrooms, puffballs, bread molds, and the yeasts used in baking and brewing.

Diseases caused by fungi are called **mycotic** (mi-KOT-ik) infections. Tinea (TIN-e-ah) refers to fungal infections of the skin's outer layer, the epidermis. The rash is often ring shaped, so the infection is also known as ringworm (but worms are not involved). The infection is named according to the site of infection; tinea capitis (KAP-ih-tis) involves the scalp, tinea pedis (athlete's foot) involves the feet, and tinea corporis (kor-PO-ris) involves nonhairy body surfaces.

One yeastlike fungus that commonly infects humans is *Candida* (KAN-did-ah) (see Fig. 5-11C). As previously mentioned, this normal inhabitant of the mouth and digestive tract may produce skin lesions, an oral infection commonly called *thrush*, digestive upset, or inflammation of the vaginal tract (vaginitis).

Although fungi cause few systemic diseases, some diseases they cause are very dangerous and may be difficult to cure. Pneumonia can result from the inhalation of fungal spores contained in dust particles. An atypical fungus, *Pneumocystis jiroveci* (formerly called *P. carinii*), causes a previously rare pneumonia known as PCP (*Pneumocystis* pneumonia) in people with AIDS and others with weakened immune systems. Table A3-4 in Appendix 3 lists typical fungal diseases.

PROTOZOA

With the protozoa, we come to the only group of microbes that can be described as animal-like. Although protozoa are also single-cell organisms, they are much larger than bacteria (Fig. 5-12).

Protozoa are found all over the world in the soil and in almost any body of water from moist grass to mud puddles to the sea. There are four main divisions of protozoa:

- **Amebas** (ah-ME-bas). An ameba (also spelled *amoeba*) is an irregularly shaped organism that propels itself by extending part of its cell (a pseudopod, or "false foot") and then flowing into the extension. Amebic dysentery is caused by a pathogen of this group (see Fig. 5-12A and B).
- **Ciliates** (SIL-e-ates). This type of protozoon is covered with cilia that wave to propel the organism. The only human pathogen in this group is *Balantidium coli*, which is transmitted by infected pig feces and can cause diarrhea or constipation.
- **Flagellates** (FLAJ-eh-lates). Flagella propel these organisms. One of these groups, a **trypanosome** (tri-PAN-o-some), causes African sleeping sickness, which is spread by the tsetse fly (see Fig. 5-12C and D). *Giardia* is a flagellated protozoon that contaminates water supplies throughout the world. It infects the intestinal tract,

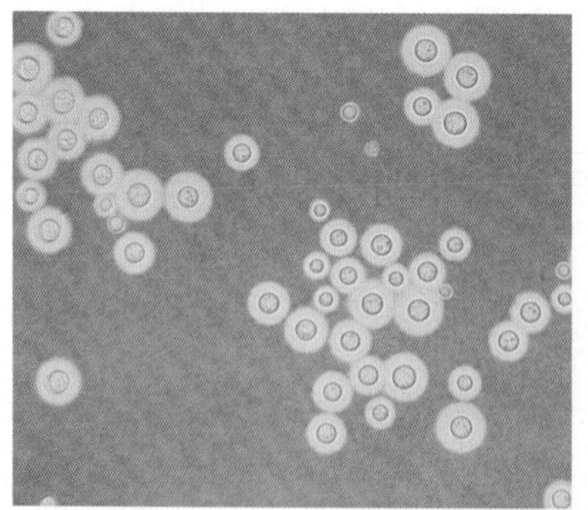

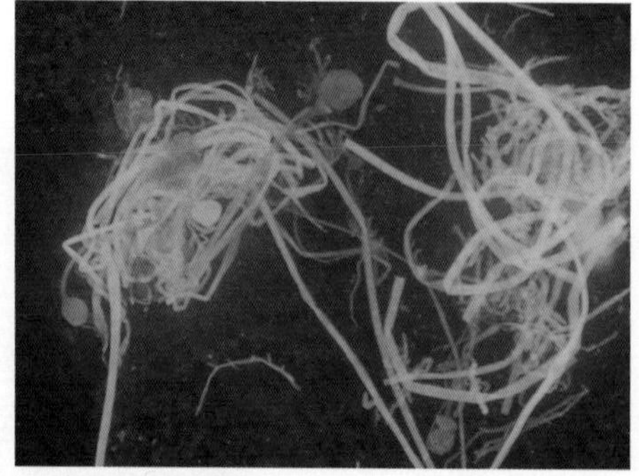

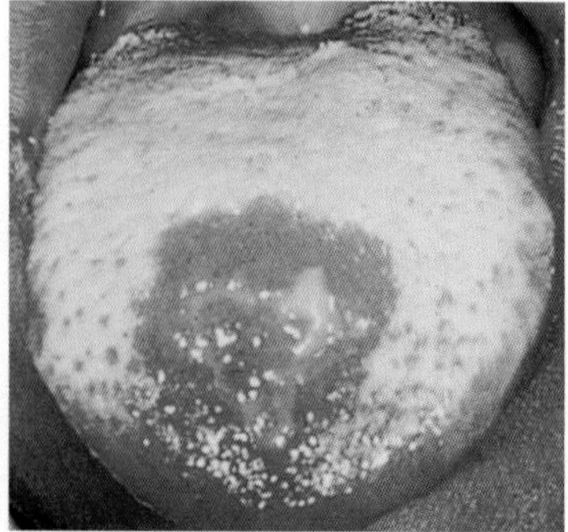

Figure 5-11 **Fungi.** **KEY POINT** Fungi exist in two forms. **A.** Yeasts. Some of these round, single-cell fungi are budding to form new cells. **B.** Molds: multicellular, filamentous fungi. **C.** Oral infection with *Candida*, a yeastlike fungus.

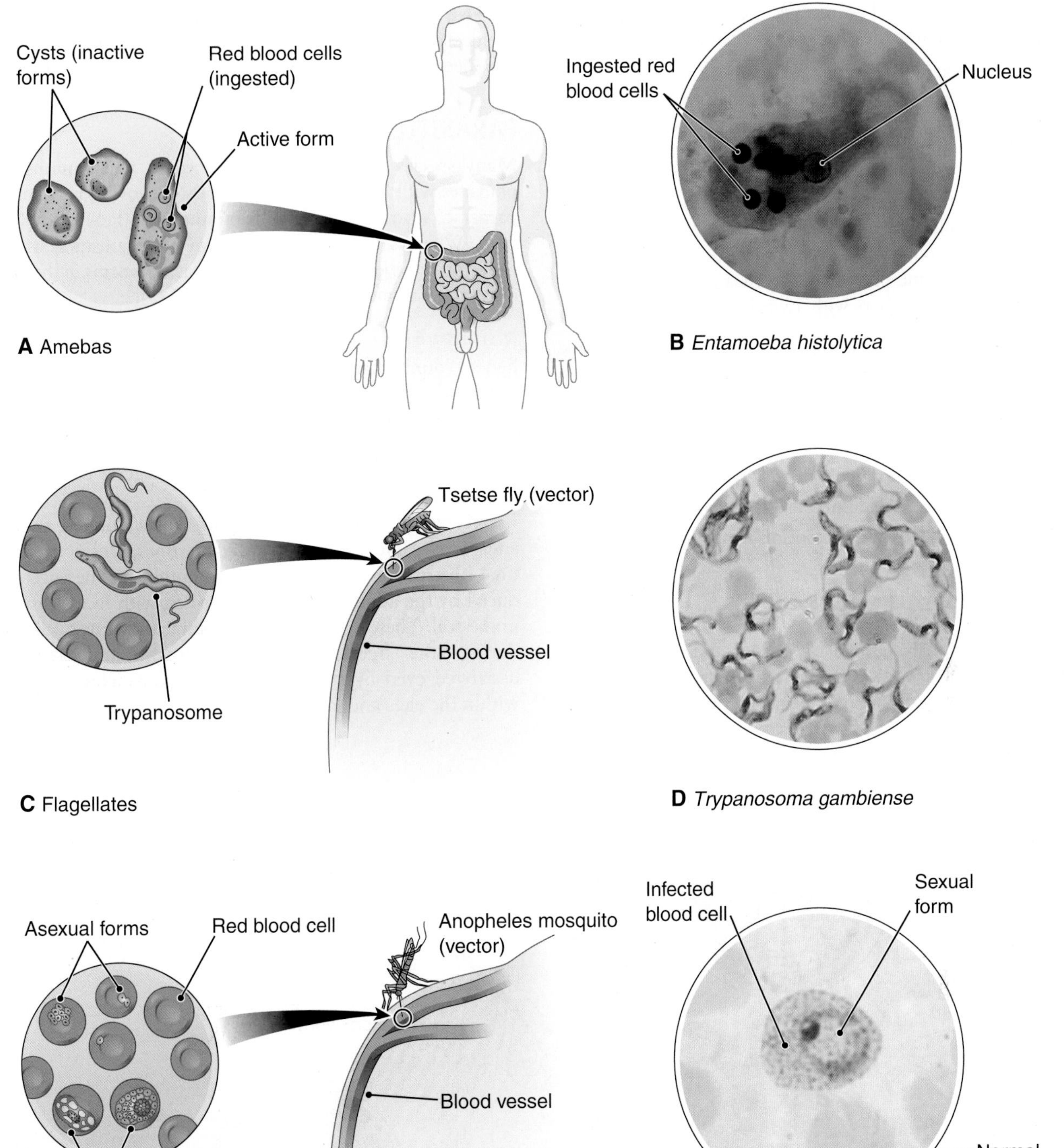

Figure 5-12 **Some parasitic protozoa. A.** Amebas. **B.** *Entamoeba histolytica*, cause of amebic dysentery, seen under the microscope. **C.** Flagellates. Trypanosomes cause African sleeping sickness. **D.** Trypanosomes in a blood sample seen under the microscope. **E.** Apicomplexans. *Plasmodium vivax* causes malaria. **F.** An enlarged red blood cell with a single parasite seen under the microscope.

ZOOMING IN Why are the parasites in **(E)** described as intracellular? What is the role of the vectors shown in **(C)** and **(E)**?

causing diarrhea. The disease, giardiasis, is the most common waterborne disease in the United States.

- **Apicomplexans** (ap-i-kom-PLEK-sans), formerly called Sporozoa (spor-o-ZO-ah), unlike other protozoa, cannot propel themselves. They are obligate parasites, unable to grow outside a host. Members of the genus *Plasmodium* (plaz-MO-de-um) cause malaria (**see Fig. 5-12E and F**). These protozoa, carried by a mosquito vector, cause over 1 million deaths each year. The apicomplexan *Cryptosporidium* is an opportunistic pathogen that causes severe and prolonged diarrhea in people suffering from AIDS and others with a weakened immune system.

Table A3-5 in Appendix 3 presents a list of typical pathogenic protozoa with the diseases they cause.

CHECKPOINTS

- ☐ 5-16 How do viruses differ from bacteria?
- ☐ 5-17 What type of organism causes a mycotic infection?
- ☐ 5-18 What group of microorganisms is most animal-like?

Parasitic Animals

Multicellular animals, particularly worms and insects, can act as human parasites. These organisms most often colonize body surfaces and cavities, but some forms can travel through the bloodstream. Unlike microorganisms, parasitic organisms are multicellular organisms that can be seen with the naked eye. One needs a microscope only to see their eggs or larval forms.

PARASITIC WORMS

Many species of worms, also referred to as **helminths**, are parasites with human hosts. The study of worms, particularly parasitic worms, is called **helminthology** (hel-min-THOL-o-je). Whereas invasion by any form of organism is usually called an *infection*, the presence of parasitic worms in the body also can be termed an *infestation* (**Fig. 5-13**).

Roundworms Many human parasitic worms are classified as roundworms, and the large worm **Ascaris** (AS-kah-ris) is one of the most common (**see Fig. 5-13A and D**). This worm is prevalent in many parts of Asia, where it is found mostly in larval form. In the United States, it is found most frequently among children (ages 4 to 12 years) in rural areas with warm climates.

Ascaris is a long, white-yellow worm pointed at both ends. It may infest the lungs or the intestines, producing intestinal obstruction if present in large numbers. The eggs produced by the adult worms are deposited with feces (excreta) in the soil. These eggs are very resistant; they can live in soil during either freezing or hot, dry weather and cannot be destroyed even by strong antiseptics. New worms develop within the eggs and later reach the digestive system of a host

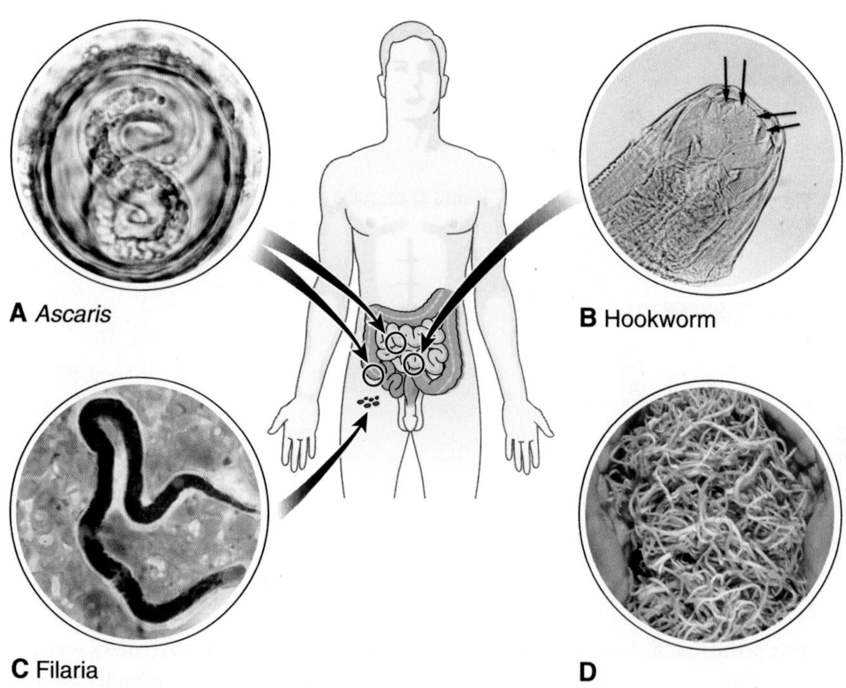

Figure 5-13 Common parasitic roundworms. **KEY POINT A.** A young ascaris preparing to emerge from a cyst. **B.** Hookworms use their mouth parts (indicated by *arrows*) to suck blood from the cells lining the digestive tract. **C.** Filaria may cause elephantiasis. **D.** A population of *Ascaris* extracted from a child's intestine.

by means of contaminated food. Ascaris infestation may be diagnosed by a routine stool examination.

Another fairly common infestation, particularly in children, is the seat worm or **pinworm** (*Enterobius vermicularis*), which is also hard to control and eliminate. The worms average 12 mm (slightly less than 1/2 in) in length and live in the large intestine. The adult female moves outside the vicinity of the anus to lay its thousands of eggs. A child's fingers often transfer these eggs from the itching anal area to his or her mouth. In the digestive system of the host, the eggs develop to form new adult worms, and thus, a new infestation is begun. A child also may infect others by this means. In addition, pinworm eggs that are expelled from the body constitute a hazard because they may live in the external environment for several months. Patience and every precaution, with careful attention to medical instructions, are necessary to get rid of the worms. It is essential to wash the hands, keep the fingernails clean, and avoid finger sucking.

Hookworms are a type of parasitic roundworm that live in the small intestine (**Fig. 5-13B**). They are dangerous because they suck blood from the host, causing such a severe blood deficiency that the victim becomes sluggish, both physically and mentally. Most victims become susceptible to various chronic infections because of extremely reduced resistance following continuous blood loss.

Hookworms lay thousands of eggs, which are distributed in the soil by contaminated feces. The eggs develop into small larvae, which are able to penetrate the intact skin of bare feet. They enter the blood and are carried to the lungs and the upper respiratory tract, finally reaching the digestive system. Proper disposal of body wastes, attention to sanitation, and wearing shoes in areas where the soil is contaminated best prevent this infestation.

Biting insects, such as flies and mosquitoes, transmit the tiny, threadlike filarial worm that causes **filariasis** (fil-ah-RI-ah-sis) (**see Fig. 5-13C**). The worms grow in large numbers, causing various disturbances. If they clog the lymphatic vessels, a condition called **elephantiasis** (el-eh-fan-TI-ah-sis) results, in which the lower extremities, the scrotum, the breasts, and other areas may become tremendously enlarged (**Fig. 5-14**). Filariasis is most common in tropical and subtropical lands, such as southern Asia and many of the South Pacific islands.

Flatworms Some flatworms resemble long ribbons, whereas others have the shape of a leaf. Tapeworms, ribbon-like flatworms, grow in the intestinal tract, reaching a length of 1.5 to 15 m (5 to 50 ft) (**Fig. 5-15**). They are spread by infected, improperly cooked meats, including beef, pork, and fish. Leaf-shaped flatworms, known as *flukes*, may invade various parts of the body, including the blood, lungs, liver, and intestine (**see Fig. 5-15C**). Like that of most parasitic intestinal worms, the flatworm's reproductive system is highly developed, so that each worm produces an enormous number of eggs, which may then contaminate food, water, and soil.

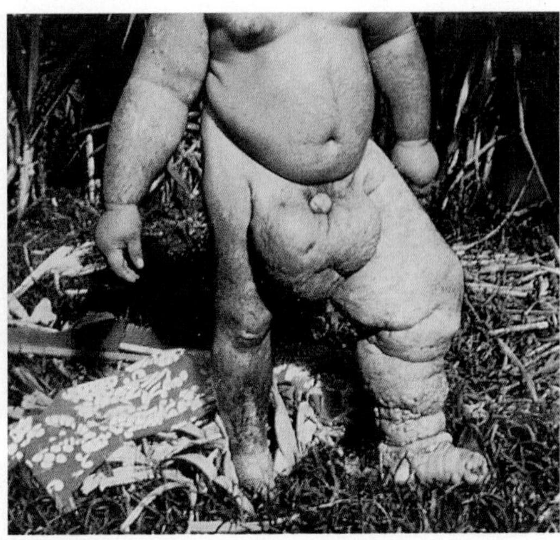

Figure 5-14 Filariasis. **KEY POINT** Filaria worms can block lymphatic vessels and cause tissue swelling. The photo shows massive enlargement (elephantiasis) of the scrotum and left leg.

PARASITIC ARTHROPODS

Arthropods (AR-thro-pods) are a diverse group of organisms, including insects and arachnids (spiders and mites). As discussed earlier, many arthropods act as disease vectors as they feed on blood. Mosquitoes, for instance, can spread malaria, dengue fever, and viral encephalitis.

Some arthropods establish infestations that, while not fatal, can cause significant discomfort. Head lice (pediculosis capitis) are common in school-age children and their families (**Fig. 5-16A and B**). Lice use their sharp mouth parts to obtain blood from the scalp, and their saliva can induce the itching and redness of an allergic reaction. An adult louse can lay up to nine eggs per day, anchoring them tightly to the hair shaft. Lice infesting pubic hair (pediculosis pubis) are often called crabs, but they are insects, not shellfish. Body lice, which infest clothing, are quite rare in North America but are still observed in populations without access to laundry facilities. Unlike other lice forms, body lice are vectors for many diseases, including typhus.

A tiny arachnid known as *Sarcoptes scabiei* causes scabies, which spreads through close contact or via contaminated clothes or bedding. The mites form tunnels in the epidermis, where they deposit eggs and feces (**Fig. 5-16C**). The body's allergic reaction to the scabies' byproducts can cause intense itching; infested individuals often scratch to the point of bleeding. Both lice and scabies can be treated with insecticidal creams.

5-19 What is the study of worms called?

5-20 Name two types of arthropods.

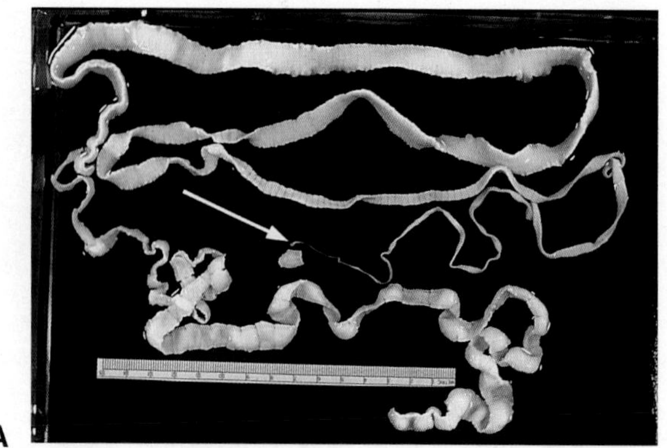

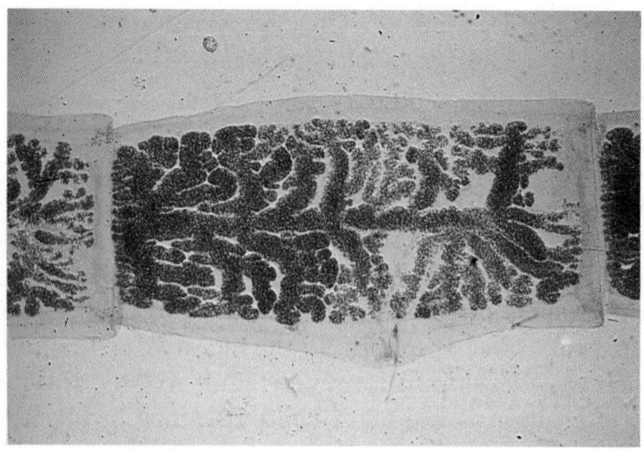

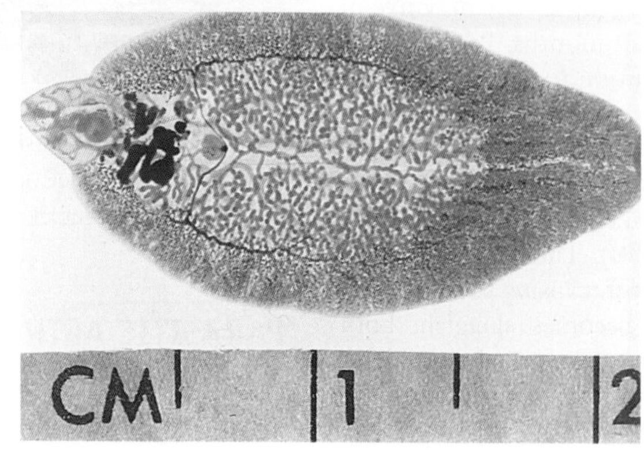

Figure 5-15 **Flatworms.** **KEY POINT** Flatworms are long and ribbon-like or leaf shaped. **A.** Tapeworms can measure more than 30 ft in length. The *arrow* shows the headlike attachment point. **B.** Segments of a tapeworm. **C.** A fluke.

Microbial Control

As a result of immunization programs and the development of antibiotics in the early 20th century, public health authorities believed that infectious diseases would be conquered by the end of the century. Indeed, smallpox was eradicated from the world by 1980. However, more than a decade into the 21st century, we are far from reaching that ambitious goal. Infectious diseases are actually increasing because of factors that include

- Increase in the world population, with more crowding of people into cities and poor sanitation. These conditions increase the spread of microorganisms by direct contact, through the air, and by pests.
- Disruption of animal habitats, with more contact between humans and animals, allowing the spread of animal pathogens to human hosts. Some organisms that have made this shift include HIV, the cause of AIDS, which originated in chimpanzees; West Nile virus from birds; severe acute respiratory syndrome (SARS), which probably came from a small wild mammal used for food; and various strains of influenza.
- Increased travel, especially air travel, which can spread an infectious organism throughout the globe in a day.

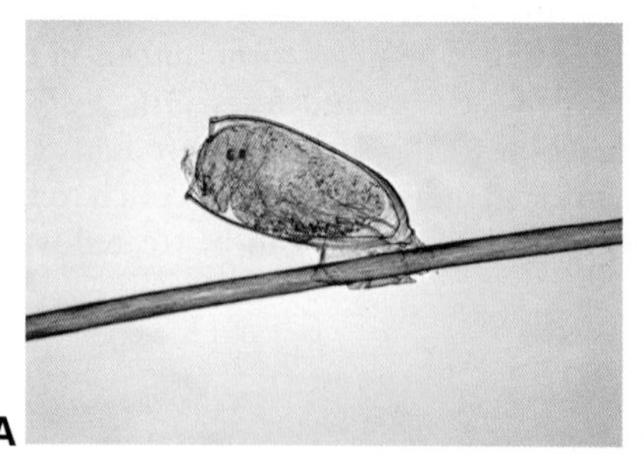

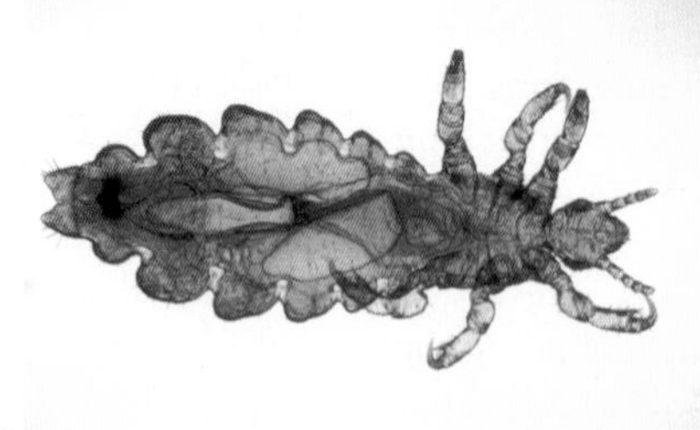

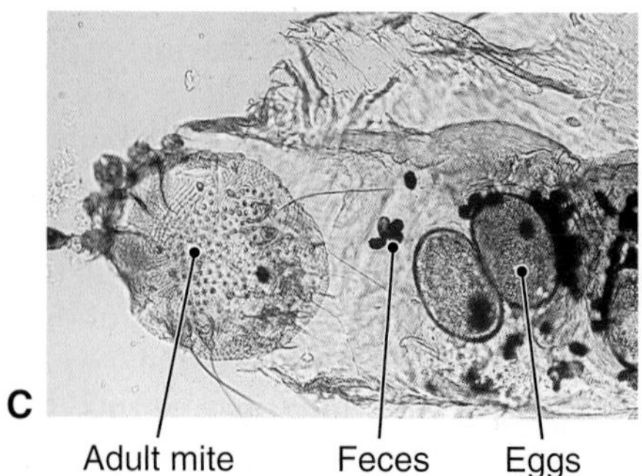

Figure 5-16 **Arthropods. A.** A young louse (nymph) preparing to emerge from the egg (nit). **B.** An adult hair louse. **C.** A scabies mite in its burrow.

SARS spread rapidly from China to other countries in the spring of 2003.

- Medical advances that keep people alive longer, but in a debilitated state, subject to opportunistic infections.

MICROBES AND PUBLIC HEALTH

All societies establish and enforce measures designed to protect the health of their populations. Most of these practices are concerned with preventing the spread of infectious organisms. A few examples of fundamental public health considerations are listed below:

- **Sewage and garbage disposal.** In times past, when people disposed of the household "slops" by simply throwing them out the window, great epidemics were inevitable. Modern practice is to divert sewage into a processing plant in which harmless microbes are put to work destroying the pathogens. The resulting noninfectious "sludge" makes excellent fertilizer.
- **Water purification.** Drinking water that has become polluted with untreated sewage may be contaminated with such dangerous pathogens as typhoid bacilli, polio and hepatitis viruses, and dysentery amebas. A filtering process usually purifies the municipal water supply, and a close and constant watch is kept on its microbial population.
- **Prevention of food contamination.** Various national, state, and local laws seek to prevent disease outbreaks through contaminated food. Certain animal diseases (e.g., tuberculosis and tularemia) can be passed to humans through food, and food is also a natural breeding place for many dangerous pathogens. Some of the organisms that cause food poisoning are the botulism bacillus (*Clostridium botulinum*) that grows in improperly canned foods, the so-called staph (*S. aureus*), and species of *Salmonella* transmitted in eggs, poultry, dairy products, and a wide variety of other foods. In recent years, a virulent variety of the normally harmless intestinal bacillus *E. coli* (*E. coli* 0157:H7) has caused outbreaks of severe and even fatal food poisoning from undercooked meat and from produce. For further information, see **Table A3-1** in Appendix 3. Most cities have sanitary regulations requiring, among other things, compulsory periodic inspection of food-handling establishments.
- **Pasteurization.** The number of microbes in milk and juice can be significantly reduced by pasteurization, a process by which the liquid is heated to 72°C (145°F) for 15 seconds and then cooled rapidly before being packaged. In ultra-high temperature pasteurization, higher temperatures (140°C) are used for several seconds in order to kill all microbes, a procedure more accurately described as sterilization (see below). The entire pasteurization process, including the cooling and packing, is accomplished in a closed system, without any exposure to air. Pasteurization is also used to preserve wine, vinegar, and some foods.

ASEPTIC METHODS

In the practice of medicine, surgery, nursing, and other health fields, specialized procedures are followed to reduce or eliminate the growth of pathogenic organisms. The term *sepsis* refers to the presence of pathogenic organisms in the blood or tissues; **asepsis** (a-SEP-sis) is its opposite—a condition in which no pathogens are present. Procedures that are designed to kill, remove, or prevent the growth of microbes are called *aseptic methods*.

There are several terms designating aseptic practices. These are often confused with one another. Some of the more commonly used terms and their definitions are as follows:

- **Sterilization.** To sterilize an object or liquid means to kill *every* living microorganism on or within it (with the exception of prions). In operating rooms and delivery rooms especially, staff members keep as much of the environment as possible sterile, including the clothing worn by personnel and the instruments used. The most common sterilization method is by means of steam under pressure in an **autoclave** (**Fig. 5-17A**). Dry heat is also used. The gas ethylene oxide is used to sterilize supplies and equipment that are not able to withstand high temperatures. Most pathogens can be killed by exposure to boiling water for four minutes. However, the time and temperature required to ensure the destruction of all spore-forming organisms in sterilization are much greater than are those required to kill most pathogens.
- **Disinfection. Disinfectants** (disinfecting agents, antimicrobials) are chemicals that kill most microorganisms (**Fig. 5-17B**). Pathogens vary in their sensitivity to different disinfectants, but generally speaking, prions and bacterial endospores are the most resistant, viruses and Gram-positive bacteria are the least resistant, and fungi, other bacteria, and protozoa show intermediate resistance. Examples of disinfectants are chlorine compounds, such as household bleach, phenol compounds, and mercury compounds. Commercial disinfectant products contain more than one chemical agent in order to kill a variety of organisms. The term **antiseptic** is often used for disinfectants applied to skin and other living surfaces. Examples are alcohol, organic iodine solutions, and hydrogen peroxide. Regular soap is also a simple but effective antiseptic. (**Fig. 5-17C**)

INFECTION CONTROL TECHNIQUES

In the 1980s, concern about the spread of blood-borne infections, such as hepatitis B and HIV, led to the development of a variety of infection control techniques.

Body Substance Precautions Isolation and barrier techniques for handling blood and all other potentially infective body substances are referred to as *body substance precautions* or *body substance isolation*. According to guidelines established by the CDC, healthcare personnel must use barriers for any contact with moist body

A Sterilization

Autoclave

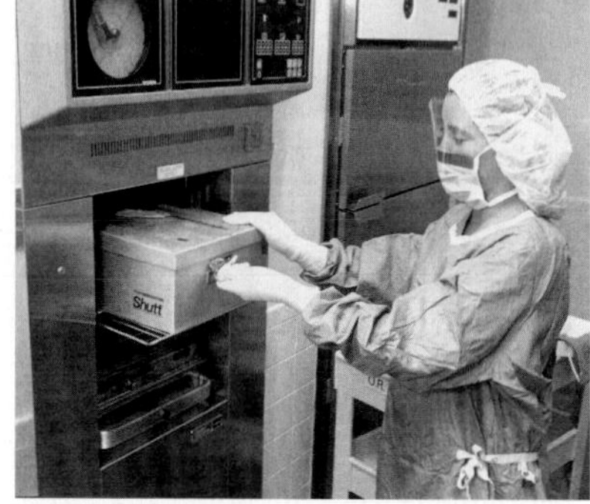

B Disinfection

Examples:
Chlorine bleach
Ammonia
Phenol

C Antisepsis

Examples:
Alcohol
Hydrogen peroxide
Antibacterial soap

Figure 5-17 **Aseptic methods. A.** Sterilization. A large, built-in autoclave. **B.** Disinfection. During an outbreak of foot-and-mouth disease, travelers were required to disinfect their shoes before traveling to the United States. **C.** Antisepsis. A sign created by the Minnesota Department of Health in the 1930s, promoting good hygiene among food services workers. **KEY POINT** Methods differ in how effectively they kill or inhibit microorganisms.

substances, mucous membranes, or nonintact skin whether or not blood is visible and regardless of the patient's diagnosis. The following measures are mandated:

- Gloves must be worn for each patient contact and changed if necessary during care.
- Protective coverings such as a mask, eye protection, face shield, or fluid-repellent gown should be worn during procedures that may generate sprays of blood or body fluids.
- Soiled linen, trash, and other waste must be treated as if contaminated and disposed of properly.
- Needles and other sharp instruments must be handled safely and disposed of in puncture-proof containers. To avoid the risk of needlestick injuries, *needles are not recapped*. In circumstances when it is necessary to recap, the needle should be slipped into the cap with one hand or by using a recapping device.

Additional isolation precautions are instituted for infections that are spread by airborne routes, such as tuberculosis; measles (rubeola); and influenza for those spread by droplets or direct contact, and for infections involving antibiotic-resistant organisms. These measures may include keeping a patient in a private room and limiting visitor contact, filtering circulating air, and having health personnel wear protective clothing.

Handwashing Handwashing is the single most important measure for preventing the spread of infection in all settings. Thorough washing promptly after patient contact and after contact with any body secretions is of utmost importance in infection control. Standard precautions include handwashing after glove removal due to the rapid multiplication of normal flora inside the gloves. Gloves are not considered a substitute for handwashing because they may have small defects, they may be torn, or hands may become contaminated when the gloves are removed.

Occupational Safety and Health Administration The Occupational Safety and Health Administration is a U.S. government agency that establishes minimum health and safety standards for workers. The agency has issued regulations for protection against infectious materials based on the CDC guidelines and has enforced these regulations. Employers at healthcare facilities must provide workers with the equipment and supplies needed to prevent their exposure to infectious materials.

ANTIMICROBIAL AGENTS

Antimicrobial (anti-infective) agents are drugs that act to kill or inhibit infectious microorganisms. These include antibacterial, antiviral, antifungal, and antiparasitic substances, which interfere with vital metabolic processes that the infecting agents need to survive and reproduce. A drug that acts on intestinal worms is an anthelmintic (ant-hel-MIN-tik) agent or vermifuge (VER-mih-fuj). The term *antibiotic*, in its most general sense, refers to any substance that acts against a living organism, but the term has come to be used only for drugs that act against bacteria.

Antibiotics (Antibacterial Agents) An **antibiotic** is a substance produced by living cells that has the power to kill or arrest the growth of bacteria. Most antibiotics are derived from fungi (molds) and soil bacteria. Penicillin, the first widely used antibiotic, is made from a common blue mold, *Penicillium*. Often, the drugs derived from penicillin can be recognized by the ending *-cillin* in the name. Other fungi that produce a large number of antibiotics are members of the group *Cephalosporium*. The soil bacteria *Streptomyces* produce some frequently used antibiotics. These drug names often end in *-mycetin*.

The development of antibiotics has been of incalculable benefit to humanity. Since the time that penicillin saved many lives on the battlefields of World War II in the 1940s, antibiotics have been considered miracle drugs. Enthusiasm for their use, however, has given rise to some undesirable effects.

One danger of antibiotic use is the development of opportunistic infections. As noted, there is a normal flora of microorganisms in the body that competes with pathogens. Antibiotics, especially those that kill a wide variety of bacteria (broad-spectrum antibiotics), eliminate these competitors and allow pathogens to thrive. For example, antibiotics often destroy the normal flora of the vaginal tract and allow a troublesome yeast infection to develop.

The widespread use of antibiotics has resulted in the natural selection of pathogens that are resistant to these medications. Under some circumstances, bacteria can even transfer genes for resistance directly from one cell to another. Some strains of common organisms, such as streptococci, staphylococci, pneumococci, *E. coli*, and tuberculosis are now resistant to most antibiotics (see Appendix 3 for diseases caused by these microorganisms).

The prevalence of antibiotic-resistant pathogens is a serious problem in hospitals today. In the United States, about 5% of acute care hospital patients contract one or more such infections. Patients who are elderly or severely debilitated are most susceptible to these **nosocomial** (nos-o-KO-me-al) (hospital-acquired) infections (the word *nosocomial* comes from a Greek word for hospital). Some strains currently causing nosocomial infections are methicillin-resistant *Staphylococcus aureus* and vancomycin-resistant enterococci.

A variety of measures can help reverse the trend toward increasing antibiotic resistance. Public health authorities advise that, when taking antibiotics, it is important to complete the entire course of treatment to guarantee the destruction of all pathogens. If not, the more resistant cells will be able to survive treatment and grow out in great numbers, leading to the development of strains that do not respond to that antibiotic. Also, patients should not press physicians to prescribe antibiotics when they will do no good, as for the treatment of viral infections such as colds and flu. Combination products containing two or more antibacterials are now available and can help eliminate all the bacteria causing an infection. In addition, pharmaceutical companies are working to develop new antibiotics.

The problem of antibiotic resistance involves not only healthcare but also industry. Large quantities of antibiotics are now used on animal feedlots to control disease and increase productivity. As a result, resistant infections—both of animals and of feedlot employees and their family members—are a significant concern. We may be able to reverse the trend toward antibiotic resistance among animals by regulating the use of these drugs more tightly. In addition, consumers can shop for meats that are free of antibiotics. There is some evidence that susceptibility to a given drug will reappear when a bacterial population is no longer exposed to it.

Antiviral Agents Effective antiviral drugs are difficult to develop. Thus, only a few are available, and each one has a limited range of action. These agents function to block

- Removal of the protein coat of the virus after it enters a cell, as in treatment of influenza A virus (rimantadine) and rhinovirus (pleconaril)
- Production or function of viral nucleic acid, as does the drug AZT (zidovudine) used to treat HIV infections or acyclovir used for herpesvirus
- Enzymes that are needed to assemble and release new virus particles, as does the drug indinavir, a so-called protease inhibitor, used to treat HIV infection
- Release of viral particles from infected cells, as does Tamiflu (oseltamivir) that is used to treat influenza

Note that viruses mutate rapidly to become resistant to these drugs, and none can eliminate latent (temporarily inactive) infections. In treating AIDS, the drugs are commonly used in combinations.

CHECKPOINTS

- [ ] 5-21 What are the two levels of asepsis?
- [ ] 5-22 What is the single most important measure for preventing the spread of infection?
- [ ] 5-23 What is an antibiotic?

Laboratory Identification of Pathogens

The nurse, physician, or laboratory worker often obtains specimens from patients to identify bacteria and other organisms. Specimens most frequently studied are blood, spinal fluid, feces, urine, and sputum, along with swabbings from other areas. Swabs are used to collect specimens from the nose, throat, eyes, and cervix as well as from ulcers or other infected areas. Correct labeling and timely delivery of the specimen to the laboratory are of the essence. Parasitic worms, insects, protozoa, and some fungi can be identified by sight, either by the naked eye or with the assistance of magnifying glasses or microscopes. Bacteria and viruses, however, require special techniques.

BACTERIAL ISOLATIONS AND TESTS

Bacterial identification involves analyzing a large population of identical bacteria, known as a pure **culture**. Bacterial cells are cultured (grown) using nutrient-rich substances called **media**. Media can be solidified using agar, a gel derived from seaweed. Individual bacteria are isolated, that is, separated from other cells in the specimen, by streaking (scraping) the sample over the surface of a solid medium in a particular pattern (**Fig. 5-18**). Each bacterium then begins to divide by binary fission, producing a large population of identical cells (a colony) within a few days. Laboratory technicians then perform a variety of tests to identify the organism or organisms obtained from the specimen. At this time, the organism is usually tested for sensitivity to various antibiotics that might be used to treat the infection. This whole procedure is described as a culture and sensitivity (C&S).

Fungi can also be cultured and identified using a similar procedure.

STAINING TECHNIQUES

One of the most frequently used methods for beginning the identification process involves examining the cells under the microscope. To make the cells visible under the microscope, they must first be stained with colored dyes (stains), which are applied to a thin smear of the culture on a glass slide. The most commonly used staining procedure is known as the **Gram stain** (**Fig. 5-19**). A bluish purple dye (such as crystal violet) is applied, and then a weak iodine solution is added. This causes a colorfast combination within certain organisms, so that washing with alcohol does not remove the dye. These bacteria are said to be *Gram-positive* (GM^+) and appear bluish purple under the microscope (**see Fig. 5-19A**). Examples are the pathogenic staphylococci and streptococci; the cocci that cause certain types of pneumonia; and the bacilli that produce diphtheria, tetanus, and anthrax. Other organisms are said to be *Gram-negative* (GM^-) because the coloring can be removed from them with a solvent. These cells are then stained for visibility, usually with a red dye (**see Fig. 5-19B**). Examples of Gram-negative organisms are the diplococci that cause gonorrhea and epidemic meningitis and the bacilli that produce typhoid fever, pneumonia, and one type of dysentery. The colon bacillus (*E. coli*) normally found in the intestine is also Gram-negative, as is the cholera vibrio. Can you tell which of the organisms in **Figure 5-6** are Gram-positive and which are Gram-negative?

Another stain used to identify organisms is the **acid-fast stain**. After being stained with a reddish dye (carbol fuchsin), the smear is treated with acid. Most bacteria quickly lose their stain upon application of the acid, but the organisms that cause tuberculosis and other acid-fast cells retain the red stain. The negative cells are then stained with a different dye, usually a blue one.

A few organisms, such as the spirochetes of syphilis and the rickettsiae, do not stain with any of the commonly used dyes. Special techniques must be used to identify these organisms.

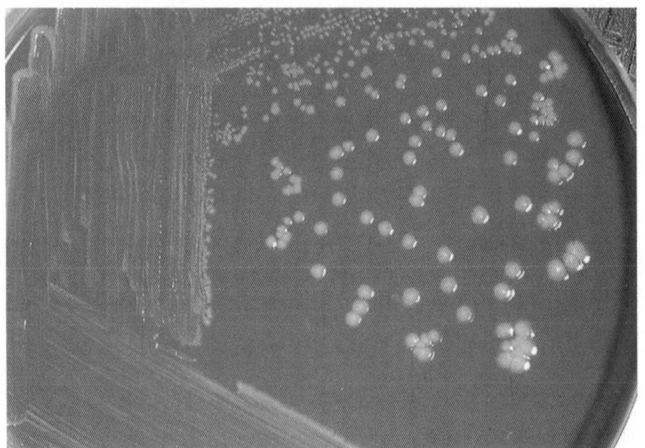

Figure 5-18 **Isolated colonies of bacteria growing on a solid medium.** **KEY POINT** Each colony contains all the offspring of a single cell.

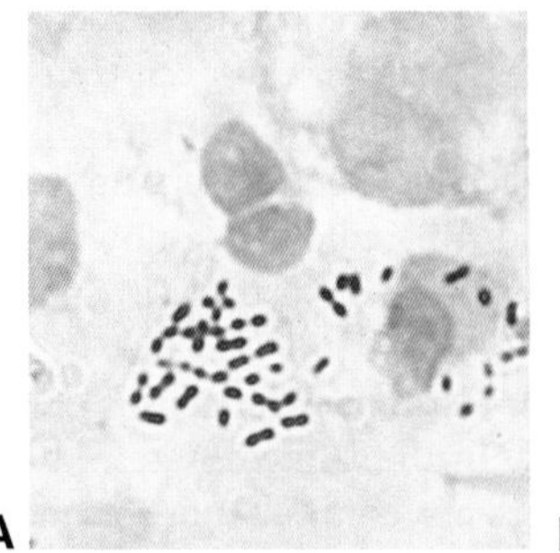

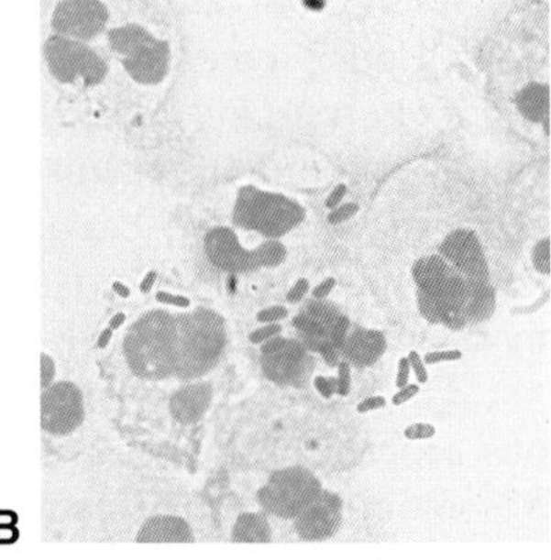

Figure 5-19 **Gram-stained bacteria.** **KEY POINT** Gram-positive cells stain bluish purple; Gram-negative cells usually are stained red. **A.** Gram-positive diplococci. **B.** Gram-negative bacilli among white blood cells in a urine sediment.

OTHER METHODS OF BACTERIAL IDENTIFICATION

In addition to the various staining procedures, laboratory techniques for identifying bacteria include

- Observing the cultures' growth characteristics in liquid and solid medium
- Studying the cells' oxygen requirements
- Observing the ability of the bacteria to utilize (ferment) various carbohydrates (sugars)
- Analyzing reactions to various test chemicals
- Studying the bacteria by serologic (immunologic) tests based on the antigen–antibody reaction (see Chapter 17)

MOLECULAR METHODS OF MICROBE IDENTIFICATION

Traditional methods of bacterial identification can take up to a week, by which time the patient has usually started treatment with a broad-acting antibiotic that might be useless (if the infection was viral) or inappropriate. Moreover, viruses cannot be identified microscopically because they are too small. Newer techniques can identify common bacterial and viral agents within hours or even minutes, resulting in less antibiotic use, shorter hospitalizations, and reduced healthcare costs. For instance, rapid detection tests analyze samples for the presence of proteins (antigens) characteristic of certain microbes. Tests have been developed for streptococcus (strep throat), HIV, malaria, and others that do not require large-scale equipment and provide results in minutes.

The polymerase chain reaction (PCR) uses enzymes that promote replication of nucleic acids to make multiple copies of an organism's unique DNA or RNA sequences. These sequences are then compared with known sequences in a database. Scientists have used PCR to identify the infectious agent for diseases such as AIDS, hepatitis, and Lyme disease, and for identifying the pathogens that cause emerging diseases such as West Nile disease and SARS. Epidemiologists use PCR to identify the source and monitor the spread of infections in a population, because it can identify individuals infected with the same strain of an organism.

CHECKPOINT

- [ ] 5-24 What are the dyes used to color microorganisms called?

See the Student Resources on thePoint for information on careers in medical technology.

Disease in Context Revisited

Maria Battles the Influenza Virus

Maria Sanchez looked terrible—watery eyes, runny nose, and a sickly, rundown appearance. Just yesterday, she felt fine. But today she was so fatigued that she could hardly get out of bed. She had a fever, chills, and a cough that left her throat aching. Maria was certainly feeling the effects of the influenza virus that had invaded the epithelial lining of her respiratory tract about three days ago.

Many of Maria's local symptoms were due to the virus. Her watery eyes and runny nose were the result of extracellular fluid leaking from holes in the mucous membrane lining her respiratory tract—holes left from the death and shedding of millions of epithelial cells. Without these ciliated cells, Maria wasn't able to move the extra fluid up and out of her respiratory tract, so she periodically coughed in an effort to do so. Maria's systemic symptoms were due, in large part, to her body's counterattack against the virus. Since viruses work most efficiently at near-normal body temperature, her fever actually slowed viral replication. Her fatigue helped beat the virus too. By staying in bed, her body was able to conserve energy that her immune system could use to fight the virus. In a few days, with rest, fluids, and aspirin, that system would beat the virus, and she would make a complete recovery.

During this case, we saw that the influenza virus that invaded Maria's body is designed to hijack her epithelial cells and force them to make new viruses. Maria's body is well designed to combat the virus, but she planned to take precautions against contracting the flu again by receiving an annual flu vaccine shot from her doctor. We'll learn more about the immune system and vaccines in Chapter 17.

CHAPTER 5

Chapter Wrap-Up

Summary Overview

A detailed chapter outline with space for note taking is on *thePoint*. The figure below illustrates the main topics covered in this chapter.

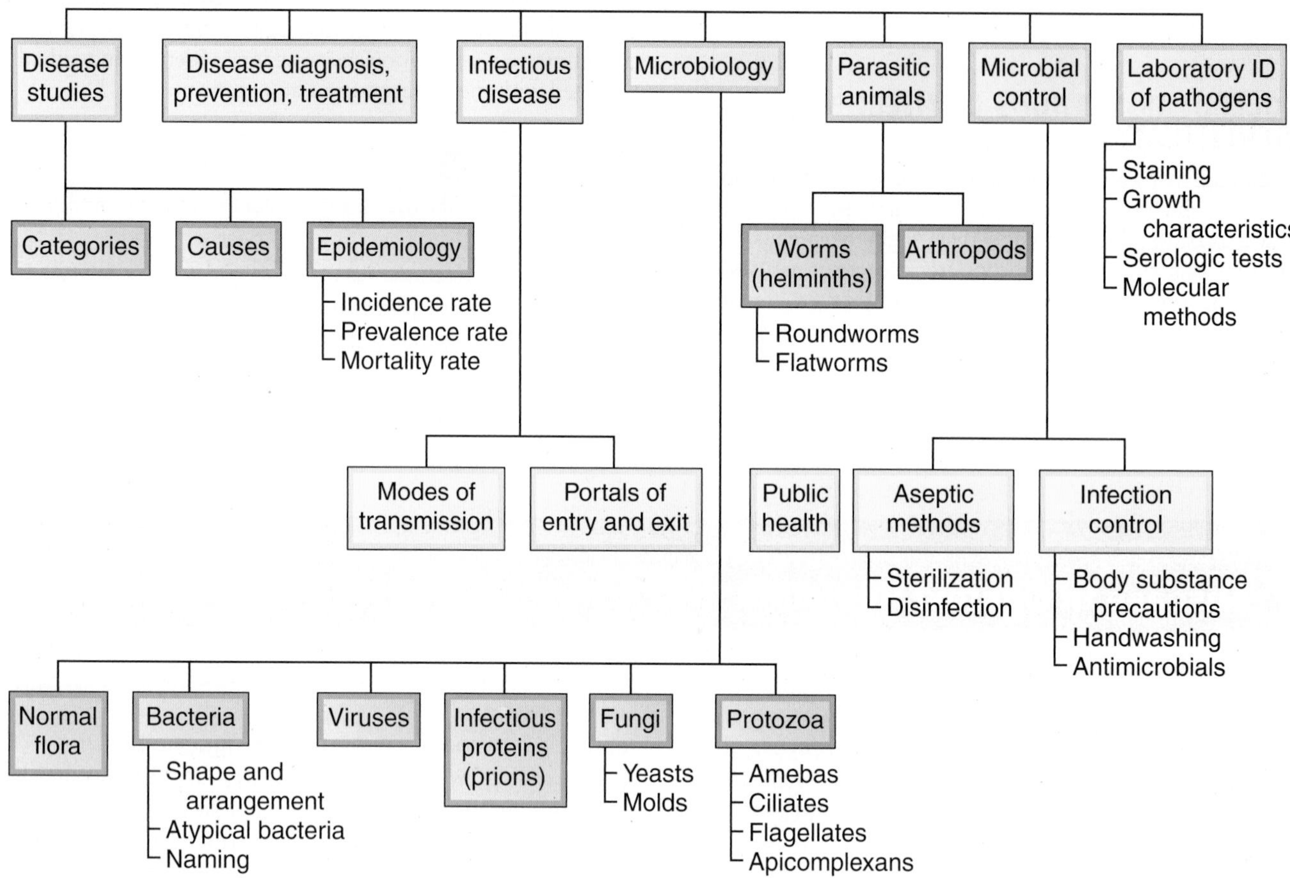

Key Terms

The terms listed below are emphasized in this chapter. Knowing them will help you organize and prioritize your learning. These and other boldface terms are defined in the Glossary with phonetic pronunciations.

acute
antibiotic
antiseptic
arthropod
asepsis
bacteria
chronic
diagnosis
disease
disinfectant
endemic
endospore
epidemic
etiology
fungus
helminth
host
infection
microorganism
nosocomial infection
opportunistic infection
pandemic
parasite
pathogen
pathophysiology
prion
prognosis
protozoa
sign
sterilization
symptom
syndrome
systemic
therapy
toxin
vector
virulence
virus

Word Anatomy

Medical terms are built from standardized word parts (prefixes, roots, and suffixes). Learning the meanings of these parts can help you remember words and interpret unfamiliar terms.

WORD PART	MEANING	EXAMPLE
Studies of Disease		
iatro	physician, medicine	An *iatrogenic* disease results from the adverse effects of treatment.
idio	self, separate, distinct	An *idiopathic* disease has no known cause; it is "self-originating."
pan-	all	A *pandemic* disease is prevalent throughout an entire country or continent, or the world.
pre-	before	A *predisposing* cause enters into production of disease.
psych/o	mind	The medical field of *psychiatry* specializes in treatment of mental disorders.
Disease Diagnosis, Treatment, and Prevention		
chir/o	hand	*Chiropractic* treatment involves use of the hands for manipulation to correct misalignment of the body.
syn-	together	A *syndrome* is a group of symptoms and signs that together characterize a disease.
Infectious Disease		
aer/o	air, gas	An *aerobic* organism requires air (oxygen) to grow.
an-	absent, deficient, lack of	An *anaerobic* organism does not require air (oxygen) to grow.
diplo-	double	*Diplococci* are round bacteria arranged in pairs.
myc/o	fungus	*Mycology* is the study of fungi.
py/o	pus	The species name *pyogenes* indicates that an organism produces pus.
staphylo-	grapelike cluster	*Staphylococci* are bacteria in clusters.
strepto-	chain	*Streptococci* are round bacteria arranged in chains.
tox/o	poison	Bacteria can harm the body by producing *toxins*.
Microbial Control		
-cide	kill or destroy	A *bactericide* is an agent that kills bacteria.
septic	poison, rot, decay	*Aseptic* methods are used to kill or prevent the growth of microorganisms.

Questions for Study and Review

BUILDING UNDERSTANDING

Fill in the Blanks

1. An inadequate diet may result in _____ disorders such as scurvy or rickets.

2. The study of the cause of any disease or the theory of its origin is _____.

3. A(n) _____ infection attacks an individual already weakened by disease.

4. The term for a complex disorder characterized by a cluster of typical symptoms and signs is a(n) _____.

5. Certain molds and soil bacteria produce bacteria-killing substances called _____.

Matching > Match each numbered item with the most closely related lettered item.

___ **6.** Organisms that cause Lyme disease, tuberculosis, and gonorrhea

___ **7.** Organisms that cause AIDS, hepatitis, and influenza

___ **8.** Organisms that cause athlete's foot, ringworm, and thrush

___ **9.** Organisms that cause amebic dysentery, giardiasis, and malaria

___ **10.** Organisms that cause pinworm, hookworm, and filariasis

a. protozoa
b. helminths
c. viruses
d. fungi
e. bacteria

Multiple Choice

___ **11.** The incidence of early heart disease is higher in men than in women. What is the predisposing cause of the disease in this example?

a. age
b. gender
c. heredity
d. living conditions and habits

___ **12.** What is a prognosis?

a. a form of therapy
b. a prediction of the probable outcome of a disease
c. an alternative medical practice
d. a communicable disease

___ **13.** Which of the following statements best describes an iatrogenic disease?

a. It occurs with high prevalence in a population.
b. It is of unknown origin.
c. It is resistant to antibiotics.
d. It results from adverse effects of treatment.

___ **14.** What enables some bacteria to swim rapidly?

a. vibrios
b. pili
c. flagella
d. spirilla

___ **15.** Which protozoa are all obligate parasites?

a. amebas
b. ciliates
c. flagellates
d. apicomplexans

___ **16.** Which of the following are examples of arthropods?

a. insects, spiders, and mites
b. filaria, tinea, and ascaris
c. candida, trypanosome, and apicomplexans
d. pinworm, fluke, and pneumocystis

___ **17.** What is meant by the virulence of an organism?

a. the location in which it thrives
b. its ability to cause disease
c. its host range
d. the type of nucleic acid it has

UNDERSTANDING CONCEPTS

18. Explain the difference between the terms in each of the following pairs:

- **a.** acute and chronic
- **b.** epidemic and endemic
- **c.** symptom and sign
- **d.** aerobic and anaerobic
- **e.** incidence rate and prevalence rate

19. Define a portal that pathogens may use to enter or exit the body. List five examples of portals.

20. What is a bacterial endospore? What is the significance of endospores in the spread of disease?

21. What is meant by the term "normal flora?" How does the normal flora influence health and disease?

22. What is a vector? Give some examples of vectors and diseases involving vectors.

23. Why are body substance precautions followed? What measures are included in the use of standard precautions?

24. Refer to Appendix 3 to answer the following questions:

- **a.** How many types of hepatitis virus are there?
- **b.** What does human papillomavirus (HPV) cause?
- **c.** What are some diseases caused by prions?
- **d.** What are the symptoms of malaria?

CONCEPTUAL THINKING

25. While you are on a work exchange program in the tropics, a cholera epidemic sweeps through the area where you are staying. Of the 1,000 people living in the area, 100 people contract cholera, and 10 die of it in one month. What are the incidence rate and mortality rate of cholera during that time? What precautions could you take to lessen your risk of contracting this bacterial disease?

26. While working in the lab, you are given the job of identifying bacteria cultured from a patient's sputum sample. You smear a sample of the bacteria onto a glass slide and Gram stain it. Microscopic examination reveals bluish purple cells arranged in chains. What do you think the bacteria could be? What disease could the patient have? What kind of drug therapy may be prescribed?

27. In Maria's case, what was the virus's portal of entry? To prevent Maria from spreading the virus to other people, what infection control technique would you suggest?

For more questions, see the Learning Activities on thePoint.

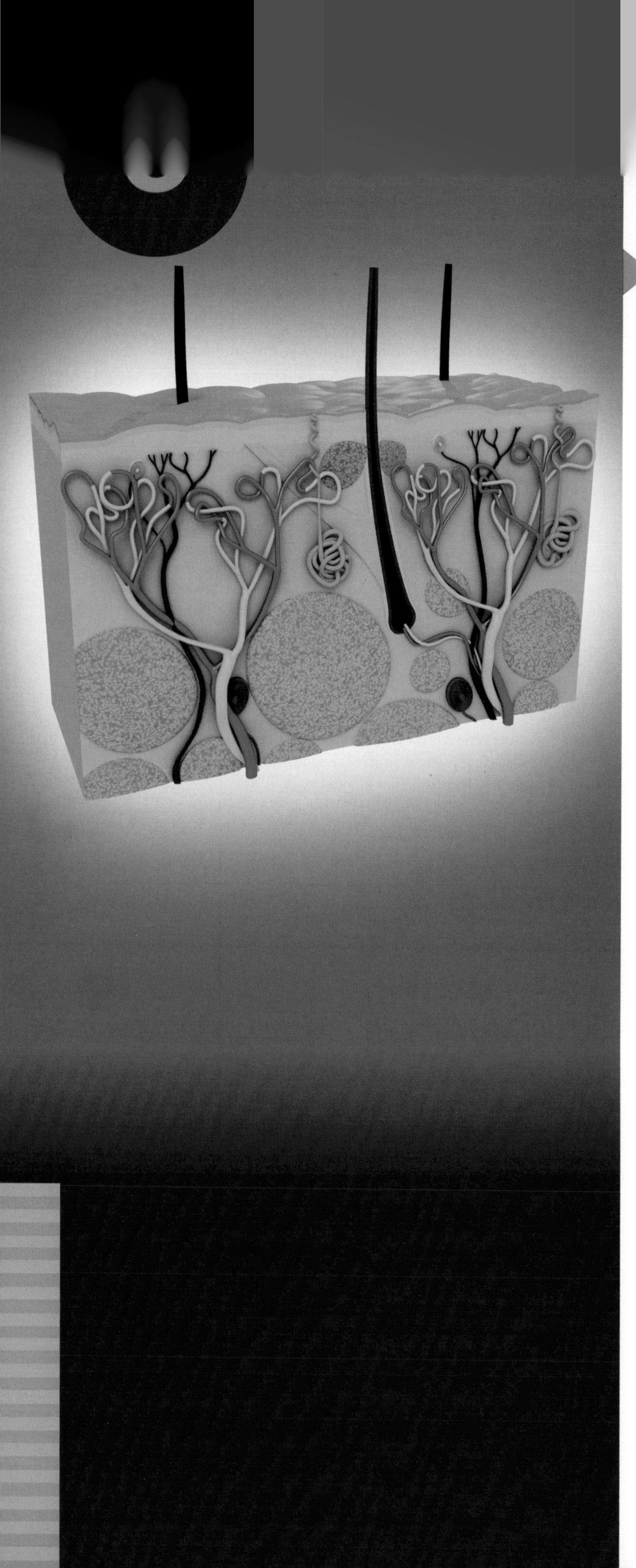

Learning Objectives

After careful study of this chapter, you should be able to:

1 ▶ Name and describe the layers of the skin. ***p. 110***

2 ▶ Describe the subcutaneous layer. ***p. 111***

3 ▶ Give the locations and functions of the accessory structures of the integumentary system. ***p. 112***

4 ▶ List the main functions of the integumentary system. ***p. 114***

5 ▶ Summarize the information to be gained by observation of the skin. ***p. 115***

6 ▶ Cite the steps in repair of skin wounds and the factors that affect healing. ***p. 116***

7 ▶ Describe how the skin changes with age. ***p. 117***

8 ▶ List the main disorders of the integumentary system. ***p. 117***

9 ▶ Using information in the case study and the text, describe the causes, classification, and dangers of a burn. ***pp. 109, 123***

10 ▶ Show how word parts are used to build words related to the integumentary system (see Word Anatomy at the end of the chapter). ***p. 125***

Disease in Context: *Regina's Accidental Burn*

It was end of the summer, and Regina's tomato plants had produced a bountiful yield. It was time to plan for the all-day canning event that she and her sister looked forward to every year.

The two gathered all of the supplies and began the annual canning process. While her sister was preparing the tomatoes, Regina started sterilizing the jars by immersing them into the hot water bath in the canner following the recommended procedure. She used the special tongs to lift out the jars, but one slipped and splashed the steaming water directly onto her upper chest, shoulders, face, and neck. The scalding water and steam caused the tissue to become reddened and blistered. Regina was experiencing considerable pain but thought it best to apply an antibiotic cream she had on hand to the burned skin.

The next morning Regina saw that the blisters had broken and were weeping a liquid. The top layers of the burned skin were peeling off, and the underlying tissue was bright red and swollen. The pain had worsened, and Regina was worried about infection. She called her physician and scheduled an appointment for that afternoon.

Dr. Stanford took a history of the event and examined the damaged tissue on Regina's upper body. Using the "rule of nines" to assess the extent of the damage, he gauged that approximately 5% of the chest was burned, along with 1% of the neck and face, 3% of the upper right arm, and 1% of the left shoulder. "You were smart to come in now," Dr. Stanford told Regina. "You've experienced a pretty serious burn to about 10% of your body." He debrided the damaged skin and applied Silvadene, an antimicrobial cream. He then applied a sterile dressing and explained, "This type of burn is called a deep partial-thickness burn. It involves the top layer of skin, called the epidermis, and portions of the next layer, the dermis. Infections are the main complication in this situation. You will have to be meticulous in keeping the area clean and following the medication regime."

The body's first line of defense, intact skin, was compromised when Regina suffered these burns. In this chapter, we will learn the functions of the skin and the importance of this protective barrier.

ANCILLARIES *At-A-Glance*

Visit thePoint to access the following resources. For guidance in using these resources most effectively, see pp. xv–xvii.

Learning RESOURCES

- Tips for Effective Studying
- Web Figure: Nail Abnormalities
- Web Figure: Clinical Findings in Systemic Lupus Erythematosus (SLE)
- Web Chart: Skin Structure
- Web Chart: Accessory Skin Structures
- Animation: Wound Healing
- Health Professions: Registered Nurse
- Detailed Chapter Outline
- Answers to Questions for Study and Review
- Audio Pronunciation Glossary (*if available*)

Learning ACTIVITIES

- Pre-Quiz
- Visual Activities
- Kinesthetic Activities
- Auditory Activities

A LOOK BACK

The skin is introduced in Chapter 4 as one of the epithelial membranes, the cutaneous (ku-TA-ne-us) *membrane, overlying a connective tissue membrane, the superficial fascia. In this chapter, we describe the skin in much greater detail as it forms the major portion of the integumentary system. This system provides a first line of defense against infectious microorganisms, described in Chapter 5, as well as other harmful agents.*

Although the skin may be viewed simply as a membrane enveloping the body, it is far more complex than are the other epithelial membranes previously described. The skin is associated with accessory structures, also known as appendages, which include glands, hair, and nails. Together with blood vessels, nerves, and sensory organs, the skin and its associated structures form the **integumentary** (in-teg-u-MEN-tar-e) **system.** This name is from the word integument (in-TEG-u-ment), which means "covering," but the skin has many functions and reflects an individual's health and emotional state.

See the Student Resources on thePoint for a summary chart of skin structure.

Structure of the Skin

Recall from Chapter 4 that the skin, or cutaneous membrane, is an epithelial membrane. Like all epithelial membranes, it contains an outer layer of epithelium and an inner layer of connective tissue (**Fig. 6-1**):

- The **epidermis** (ep-ih-DER-mis), the outermost portion, which itself is subdivided into thin layers called **strata** (STRA-tah) (sing. stratum). The epidermis is composed entirely of epithelial cells and contains no blood vessels.

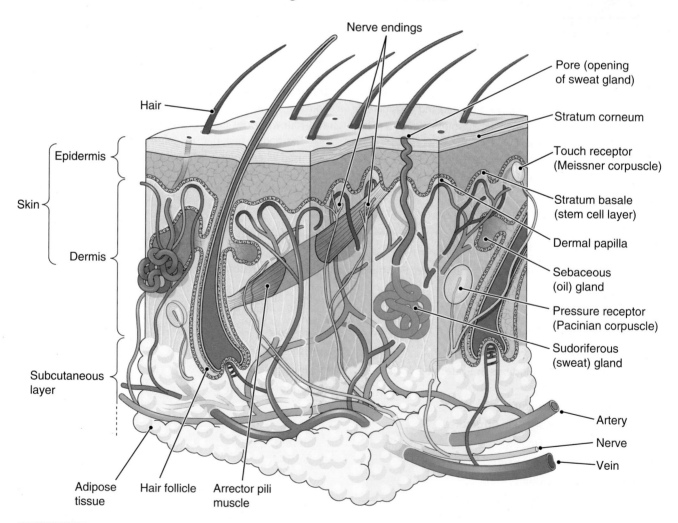

Figure 6-1 **Cross-section of the skin.** **KEY POINT** The dermis and epidermis make up the skin. The skin and its associated structures make up the integumentary system. **ZOOMING IN** How is the epidermis supplied with oxygen and nutrients? What tissue is located beneath the skin?

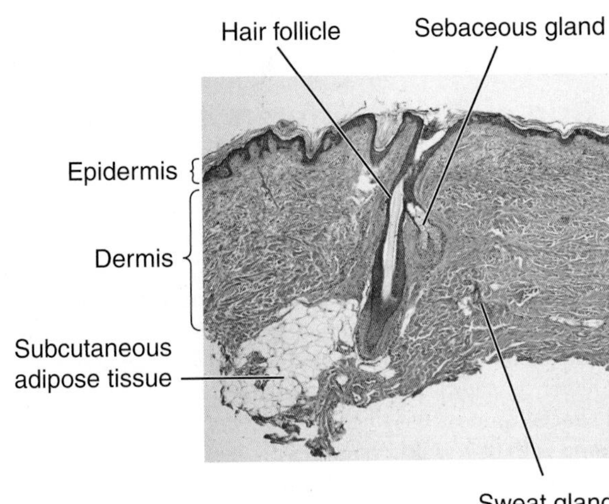

Figure 6-2 **Microscopic view of thin skin.** Tissue layers and some accessory structures are labeled.

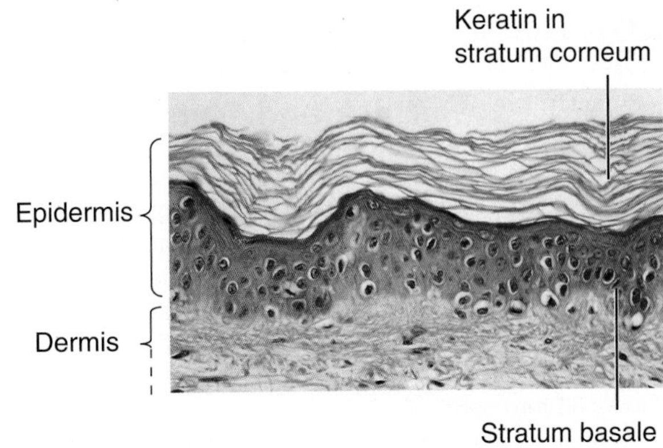

Figure 6-3 **Upper portion of the skin.** KEY POINT Layers of keratin in the stratum corneum are visible at the surface. Below are layers of stratified squamous epithelium making up the remainder of the epidermis.

- The **dermis**, which is the connective tissue portion of the cutaneous membrane, contains many blood vessels, nerve endings, and glands.

Figure 6-2 is a photograph of the skin as seen through a microscope showing the layers and some accessory structures.

EPIDERMIS

The epidermis is the skin's surface portion, the outermost cells of which are constantly lost through wear and tear. Because there are no blood vessels in the epidermis, the cells must be nourished by capillaries in the underlying dermis. New epidermal cells are produced from stem cells in the deepest layer, which is closest to the dermis. The cells in this layer, the **stratum basale** (bas-A-le), or *stratum germinativum* (jer-min-a-TI-vum), are constantly dividing and producing new cells, which are then pushed upward toward the skin surface. As the epidermal cells die from the gradual loss of nourishment, they undergo changes. Mainly, their cytoplasm is replaced by large amounts of a protein called **keratin** (KER-ah-tin), which thickens and protects the skin (**Fig. 6-3**).

By the time epidermal cells approach the surface, they have become flat and keratinized, or cornified, forming the uppermost layer of the epidermis, the **stratum corneum** (KOR-ne-um). The stratum corneum is a protective layer and is more prominent in thick skin than in thin skin. Cells at the surface are constantly being lost and replaced from below, a process that gets rid of accumulated environmental toxins and microbes. Although this process of **exfoliation** (eks-fo-le-A-shun) occurs naturally at all times, many cosmetics companies sell products to promote exfoliation, presumably to "enliven" and "refresh" the skin.

Between the stratum basale and the stratum corneum are additional layers of stratified epithelium that vary in number and quantity depending on the skin's thickness. Some places, such as the soles of the feet and the palms of the hands, are covered with very thick, hairless skin, such as illustrated in **Figure 6-3**. Other regions, such as the eyelids, are covered with very thin and delicate layers. Box 6-1 describes the structural and functional differences between thick and thin skin. Like the epidermis, the dermis varies in thickness in different areas.

Some cells in the deepest layer of the epidermis produce **melanin** (MEL-ah-nin), a dark pigment that colors the skin and protects it from sunlight's harmful rays. The cells that produce this pigment are the **melanocytes** (MEL-ah-no-sites). Irregular patches of melanin are called freckles.

DERMIS

The **dermis** has a framework of dense irregular connective tissue and is well supplied with blood vessels and nerves. It is called the "true skin" because it carries out the skin's vital functions. Because of dermal elasticity, the skin can stretch, even dramatically as in pregnancy, with little damage. Most of the skin's accessory structures, including the sweat glands, the oil glands, and the hair, are located in the dermis and may extend into the subcutaneous layer under the skin.

Portions of the dermis extend upward into the epidermis, allowing blood vessels to get closer to the superficial cells (**see Fig. 6-1**). These extensions, or **dermal papillae**, can be seen on the surface of thick skin, such as at the tips of the fingers and toes. Here they form a distinct pattern of ridges that help to prevent slipping, as when grasping an object. The unchanging patterns of the ridges are determined by heredity. Because they are unique to each person, fingerprints and footprints can be used for identification.

SUBCUTANEOUS LAYER

The skin rests on a connective tissue membrane, the **subcutaneous** (sub-ku-TA-ne-us) **layer**, sometimes referred to as the hypodermis or the superficial fascia (**see Fig. 6-1**). This layer connects the skin to the deep fascia covering the underlying

Box 6-1

A CLOSER LOOK

Thick and Thin Skin: Getting a Grip on Their Differences

The skin is the largest organ in the body, weighing about 4 kg. Though it appears uniform in structure and function, its thickness in fact varies. Many of the functional differences between skin regions reflect the thickness of the epidermis and not the skin's overall thickness. Based on epidermal thickness, skin can be categorized as **thick** (about 1 mm deep) or **thin** (about 0.1 mm deep).

Areas of the body exposed to significant friction (the palms, fingertips, and bottoms of the feet and toes) are covered with thick skin. It is composed of a thick stratum corneum and an extra layer not found in thin skin, the stratum lucidum, both of which make thick skin resistant to abrasion. Thick skin is also characterized by epidermal ridges (e.g., fingerprints) and numerous sweat glands, but lacks hair and sebaceous (oil) glands. These adaptations make the thick skin covering the hands and feet effective for grasping or gripping. The dermis of thick skin also contains many sensory receptors, giving the hands and feet a superior sense of touch.

Thin skin covers body areas not exposed to much friction. It has a very thin stratum corneum and lacks a distinct stratum lucidum. Though thin skin lacks epidermal ridges and has fewer sensory receptors than does thick skin, it has several specializations that thick skin does not. Most thin skin is covered with hair, which may help prevent heat loss from the body. In fact, hair is most densely distributed in skin that covers regions of great heat loss—the head, axillae (armpits), and groin. Thin skin also contains numerous sebaceous glands, making it supple and free of cracks that might let infectious organisms enter.

muscles. It consists of areolar connective tissue and variable amounts of adipose (fat) tissue. The fat serves as insulation and as a reserve energy supply. Continuous bundles of elastic fibers connect the subcutaneous tissue with the dermis, so there is no clear boundary between the two.

The blood vessels that supply the skin with nutrients and oxygen and help to regulate body temperature run through the subcutaneous layer and send branches into the dermis. This tissue is also rich in nerves and nerve endings, including those that supply nerve impulses to and from the dermis and epidermis. The thickness of the subcutaneous layer varies in different parts of the body; it is thinnest on the eyelids and thickest on the abdomen.

CHECKPOINTS

- ☐ 6-1 What is the name of the system that comprises the skin and all its associated structures?
- ☐ 6-2 Moving from the superficial to the deeper layer, what are the names of the two layers of the skin?
- ☐ 6-3 What is the composition of the subcutaneous layer?

Accessory Structures of the Skin

The integumentary system includes some structures associated with the skin—glands, hair, and nails—that protect the skin and serve other functions.

See the Student Resources on thePoint for a chart summarizing the skin's accessory structures.

SEBACEOUS (OIL) GLANDS

The **sebaceous** (se-BA-shus) **glands** are saclike in structure, and their oily secretion, **sebum** (SE-bum), lubricates the skin and hair and prevents drying. The ducts of the sebaceous glands open into the hair follicles (Fig. 6-4A).

Babies are born with a covering produced by these glands that resembles cream cheese; this secretion is called the **vernix caseosa** (VER-niks ka-se-O-sah), which literally means "cheesy varnish." Modified sebaceous glands, **meibomian** (mi-BO-me-an) **glands**, are associated with the eyelashes and produce a secretion that lubricates the eyes.

SUDORIFEROUS (SWEAT) GLANDS

The **sudoriferous** (su-do-RIF-er-us) **glands**, or sweat glands, are coiled, tubelike structures located in the dermis and the subcutaneous tissue (see Fig. 6-4B).

Most of the sudoriferous glands function to cool the body. They release sweat, or perspiration, that draws heat from the skin as the moisture evaporates at the surface. These **eccrine**-type (EK-rin) sweat glands are distributed throughout the skin. Each gland has a secretory portion and an excretory tube that extends directly to the surface and opens at a pore (see also Fig. 6-1). Because sweat contains small amounts of dissolved salts and other wastes in addition to water, these glands also serve a minor excretory function.

Present in smaller number, the **apocrine** (AP-o-krin) sweat glands are located mainly in the armpits (axillae) and groin area. These glands become active at puberty and release their secretions through the hair follicles in response to emotional stress and sexual stimulation. The apocrine glands release some cellular material in their secretions. Body odor develops from the action of bacteria in breaking down these organic cellular materials.

Several types of glands associated with the skin are modified sudoriferous glands. These are the **ceruminous** (seh-RU-min-us) **glands** in the ear canal that produce ear

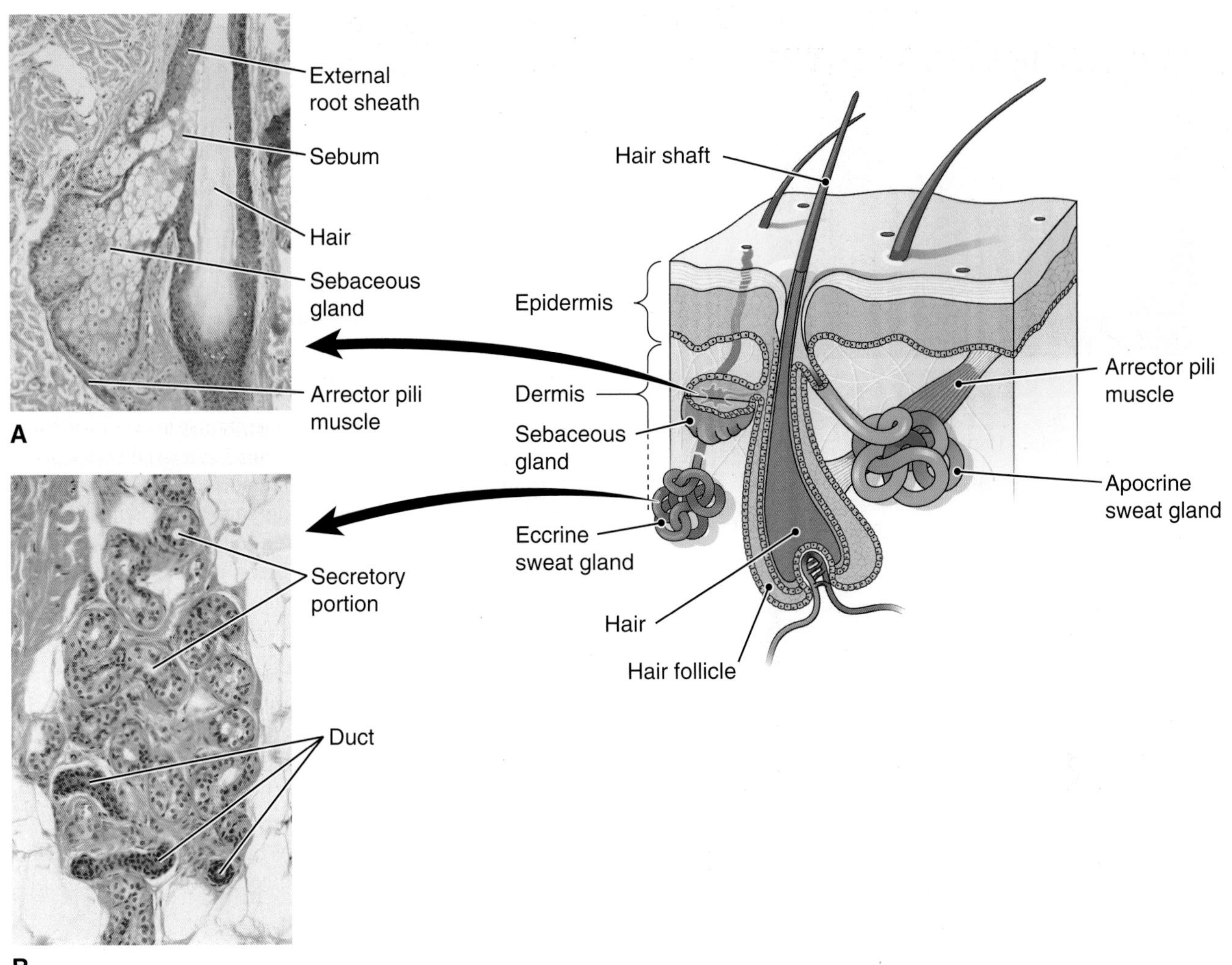

Figure 6-4 Portion of skin showing associated glands and hair. **KEY POINT A.** A sebaceous (oil) gland and its associated hair follicle. **B.** An eccrine (temperature-regulating) sweat gland. **ZOOMING IN** How do the sebaceous glands and apocrine sweat glands secrete to the outside? What kind of epithelium makes up the sweat glands?

wax, or **cerumen**; the **ciliary** (SIL-e-er-e) **glands** at the edges of the eyelids; and the **mammary glands** in the breasts.

HAIR

Almost all of the body is covered with hair, which in most areas is soft and fine. Hairless regions are the palms of the hands, soles of the feet, lips, nipples, and parts of the external genitalia. Hair is composed mainly of keratin and is not living. Each hair develops, however, from stem cells located in a bulb at the base of the **hair follicle**, a sheath of epithelial and connective tissue that encloses the hair (**see Fig. 6-4**). Melanocytes in this growth region add pigment to the developing hair. Different shades of melanin produce the various hair colors we see in the population. The part of the hair that projects above the skin is the **shaft**; the portion below the skin is the hair's **root**.

Attached to most hair follicles is a thin band of involuntary muscle (**see Fig. 6-4**). When a person is frightened or cold, this muscle contracts, raising the hair and forming "goose bumps" or "chicken skin." The name of this muscle is **arrector pili** (ah-REK-tor PI-li), which literally means "hair raiser." This response is not important in humans but is a warning sign in animals and helps animals with furry coats to conserve heat. As the arrector pili contracts, it presses on the sebaceous gland associated with the hair follicle, causing the release of sebum to lubricate the skin.

NAILS

Nails protect the fingers and toes and also help in grasping small objects with the hands. They are made of hardened keratin formed by the epidermis (**Fig. 6-5**). New cells develop continuously in a growth region (nail matrix) located under the nail's proximal end, a portion called the **nail root**. The remainder of the **nail plate** rests on a **nail bed** of epithelial tissue. The color of the dermis below the nail bed can be seen through the clear nail. The pale **lunula** (LU-nu-lah), literally "little moon,"

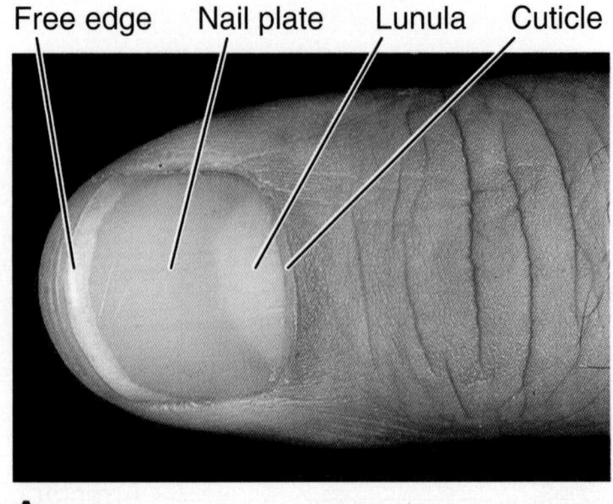

A

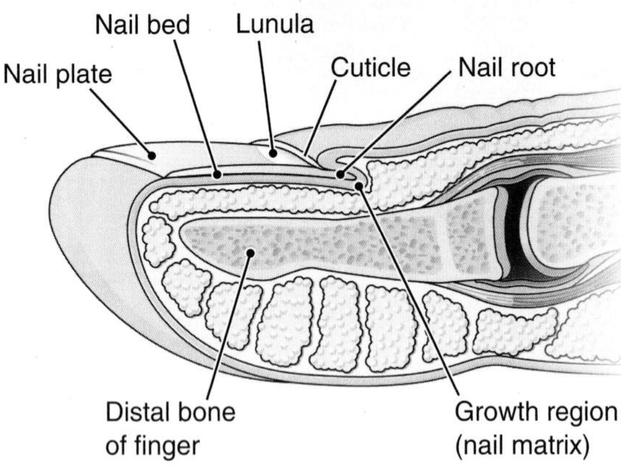

B

Figure 6-5 **Nail structure.** **KEY POINT** Nails protect the fingertips and toes. They form from epidermal cells at the nail root. **A.** Photograph of a nail, superior view. **B.** Midsagittal section of a fingertip.

at the nail's proximal end appears lighter because it lies over the nail's thicker growing region. The **cuticle**, an extension of the stratum corneum, seals the space between the nail plate and the skin above the root.

Nails of both the toes and the fingers are affected by general health. Changes in nails, including abnormal color, thickness, shape, or texture (e.g., grooves or splitting), occur in chronic diseases such as heart disease, peripheral vascular disease, malnutrition, and anemia.

See the Student Resources on thePoint to view some of these conditions affecting nails.

CHECKPOINTS

- 6-4 What is the name of the skin glands that produce an oily secretion?
- 6-5 What is the scientific name for the sweat glands?
- 6-6 What is the name of the sheath in which a hair develops?
- 6-7 Where are the active cells that produce a nail located?

Functions of the Integumentary System

Among the main functions of the integumentary system are the following:

- protection against infection
- protection against dehydration (drying)
- regulation of body temperature
- collection of sensory information

PROTECTION AGAINST INFECTION

Intact skin forms a primary barrier against invasion of pathogens. The cells of the stratum corneum form a tight interlocking pattern that resists penetration. The surface cells are constantly shed, thus mechanically removing pathogens. Rupture of this barrier, as in cases of wounds or burns, invites infection of deeper tissues. The skin also protects against bacterial toxins (poisons) and some harmful chemicals in the environment.

PROTECTION AGAINST DEHYDRATION

Both keratin in the epidermis and the oily sebum released to the skin's surface from the sebaceous glands help to waterproof the skin and keep it moist and supple, even in dry environments. These substances also prevent excessive water loss by evaporation. The skin itself forms a boundary that encloses body fluids, limiting water loss. When the skin is burned, fluid losses are significant, and burn patients commonly complain of intense thirst.

REGULATION OF BODY TEMPERATURE

Both the loss of excess heat and protection from cold are important functions of the integumentary system. Indeed, most of the blood flow to the skin is concerned with temperature regulation. In cold conditions, vessels in the skin constrict (become narrower) to reduce blood flow to the surface and diminish heat loss. The skin may become visibly pale under these conditions.

To cool the body, the skin forms a large surface area for radiating body heat to the surrounding air. When the blood vessels dilate (widen), more blood is brought to the surface so that heat can dissipate. The other mechanism for cooling the body involves the sweat glands, as noted above. The evaporation of perspiration draws heat from the skin. A person feels uncomfortable on a hot and humid day because water does not evaporate as readily from the skin into the surrounding air. A dehumidifier makes one more comfortable even when the temperature remains high.

As is the case with so many body functions, temperature regulation is complex and involves other areas, including certain centers in the brain. See Chapter 20 for more information.

COLLECTION OF SENSORY INFORMATION

Because of its many nerve endings and other special receptors, the integumentary system may be regarded as one of the body's chief sensory organs. Free nerve endings detect pain and moderate changes in temperature. Other types of sensory receptors in the skin respond to light touch and deep pressure. **Figure 6-1** shows some free nerve endings, a touch receptor (Meissner corpuscle), and a deep pressure receptor (Pacinian corpuscle) in a section of skin.

Many of the reflexes that make it possible for humans to adjust themselves to the environment begin as sensory impulses from the skin. For example, touching a hot stove activates skin receptors, which initiate a reflex that ends with withdrawing your finger. As elsewhere in the body, the skin receptors work with the brain and the spinal cord to accomplish these important functions.

OTHER ACTIVITIES OF THE INTEGUMENTARY SYSTEM

Substances can be absorbed through the skin in limited amounts. Some drugs—for example, estrogens, other steroids, anesthetics, and medications to control motion sickness—can be absorbed from patches placed on the skin (**see Box 6-2**). Most medicated ointments used on the skin, however, are for the treatment of local conditions only. Even medication injected into the subcutaneous tissue is absorbed very slowly.

There is also a minimal amount of excretion through the skin. Water and electrolytes are excreted in sweat (perspiration). Some nitrogen-containing wastes are eliminated through the skin, but even in disease, the amount of waste products excreted by the skin is small. Therefore, claims that saunas and hot yoga "detoxify" the body are not substantiated by science.

Vitamin D needed for the development and maintenance of bone tissue is manufactured in the skin under the effects of ultraviolet (UV) radiation in sunlight.

Note that the human skin does not "breathe." The pores of the epidermis serve only as outlets for perspiration from the sweat glands and sebum (oil) from the sebaceous glands. They are not used for exchange of gases.

CHECKPOINTS

- [ ] **6-8** What two substances produced in the skin help to prevent dehydration?
- [ ] **6-9** What two mechanisms involving the skin are used to regulate temperature?

Observation of the Skin

The skin is the one organ that can be inspected in its entirety without requiring surgery or special equipment. The skin not only gives clues to its own health but also reflects the health of other body systems. What can the skin tell you? What do its color, texture, and other attributes indicate? Much can be learned by an astute observer. In fact, the first indication of a serious systemic disease may be a skin disorder.

COLOR

Skin color is determined by pigments present in the skin itself and in blood circulating through the skin. The main pigments that impart color to the skin are melanin, hemoglobin, and carotene, but pigments present in bile, a digestive secretion produced by the liver, and certain poisons may affect skin color as well. Any distinct change in skin color from the normal is termed *discoloration*.

Melanin Melanin is the skin's main pigment. It is produced by melanocytes in the stratum germinativum of the epithelium. In addition to its presence in the skin, it

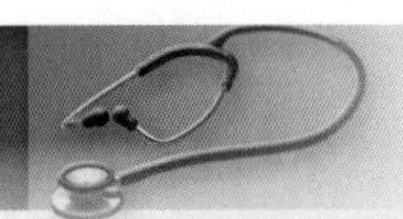

Box 6-2

CLINICAL PERSPECTIVES

Medication Patches: No Bitter Pill to Swallow

For most people, pills are a convenient way to take medication, but for others, they have drawbacks. Pills must be taken at regular intervals to ensure consistent dosing, and they must be digested and absorbed into the bloodstream before they can begin to work. For those who have difficulty swallowing or digesting pills, **transdermal (TD) patches** offer an effective alternative to some oral medications.

TD patches deliver a consistent dose of medication that diffuses at a constant rate through the skin into the bloodstream. There is no daily schedule to follow, nothing to swallow, and no stomach upset. TD patches can also deliver medication to unconscious patients, who would otherwise require intravenous drug delivery. TD patches are used in hormone replacement therapy, to treat heart disease, to manage pain, and to suppress motion sickness. Nicotine patches are also used as part of programs to quit smoking.

TD patches must be used carefully. Drug diffusion through the skin takes time, so it is important to know how long the patch must be in place before it is effective. It is also important to know how long the medication's effects will persist after the patch is removed. Because the body continues to absorb what has already diffused into the skin, removing the patch does not entirely remove the medicine. Also, increased heat may elevate drug absorption to dangerous levels.

A recent advance in TD drug delivery is **iontophoresis**. Based on the principle that like charges repel each other, this method uses a mild electric current to move ionic drugs through the skin. A small electrical device attached to the patch uses positive current to "push" positively charged drug molecules through the skin and a negative current to push negatively charged ones. Even though very low levels of electricity are used, people with pacemakers should not use iontophoretic patches. Another disadvantage is that they can move only ionic drugs through the skin.

A 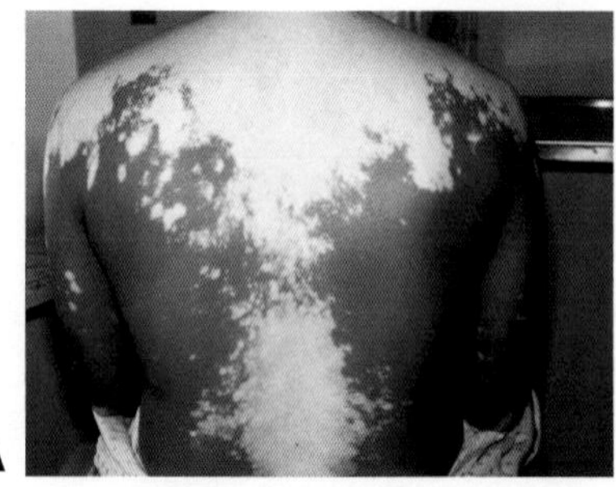

B 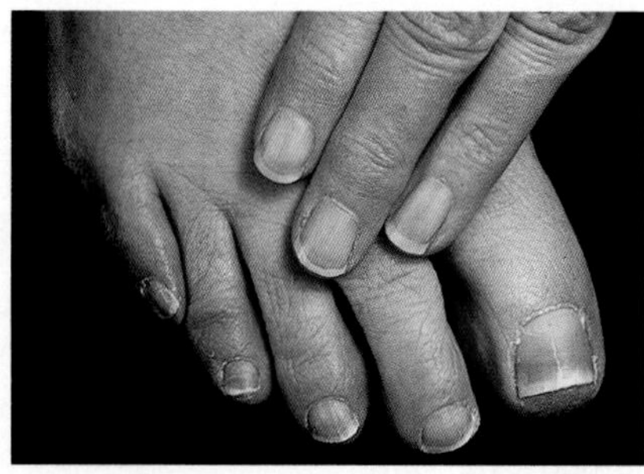

C 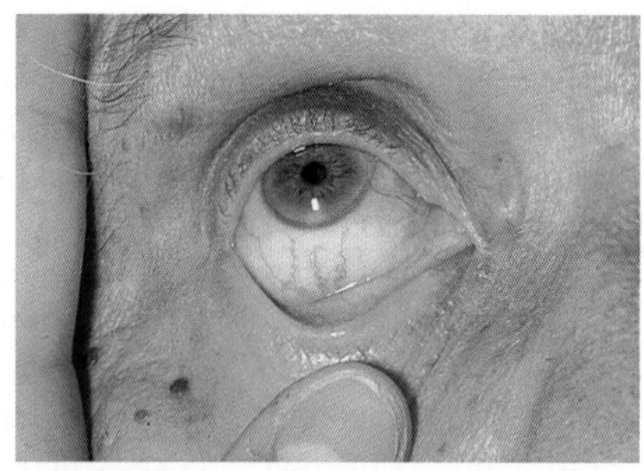

Figure 6-6 **Discoloration of the skin.** **KEY POINT** Changes in skin color can reveal illness. **A.** Vitiligo results from regional defects in melanocyte action. **B.** Cyanosis is a bluish discoloration caused by lack of oxygen. It is seen here in the toes as compared to normal fingertips. **C.** Jaundice is a yellowish discoloration caused by bile pigments in the blood. **ZOOMING IN** What color is associated with cyanosis? What color is associated with jaundice?

is found in the hair, the middle coat of the eyeball, the iris of the eye, and certain tumors. It is common to all races, but darker people have a much larger quantity in their tissues because their melanocytes are more active. The melanin in the skin helps to protect against sunlight's damaging UV radiation. Thus, skin that is exposed to the sun shows a normal increase in this pigment, a response we call tanning.

Sometimes, there are abnormal increases in the quantity of melanin, which may occur either in localized areas or over the entire body surface. For example, diffuse spots of pigmentation may be characteristic of some endocrine disorders. In Addison disease, malfunction of the adrenal gland indirectly stimulates melanocytes, giving an unusual bronze cast to the skin from excess melanin. In contrast, **albinism** (AL-bih-nizm) is a hereditary disorder that impairs melanin production, resulting in a lack of pigment in the skin, hair, and eyes. **Vitiligo** is patchy local blanching of skin to near whiteness, reflecting a regional defect in melanocyte action (**Fig. 6-6A**).

Hemoglobin Hemoglobin (he-mo-GLO-bin) is the pigment that carries oxygen in red blood cells (further described in Chapters 13 and 18). It gives blood its color and is visible in the skin through vessels in the dermis. **Pallor** (PAL-or) is paleness of the skin, often caused by reduced blood flow or by reduction in hemoglobin, as occurs in cases of anemia. Pallor is most easily noted in the lips, nail beds, and mucous membranes. **Flushing** is diffuse redness caused by increased blood flow to the skin. It is often related to fever and is most noticeable in the face and neck.

When there is not enough oxygen in circulating blood, the skin may take on a bluish discoloration termed **cyanosis** (si-ah-NO-sis) (**Fig. 6-6B**). This is a symptom of heart failure, breathing problems, such as emphysema, or respiratory obstruction.

Carotene is a skin pigment obtained from carrots and other orange and yellow vegetables. Excessive intake of these vegetables can result in carotene accumulation in blood, a condition known as **carotenemia** (kar-o-te-NE-me-ah) (the suffix *-emia* refers to blood). The excess carotene is deposited in the stratum corneum, resulting in a yellowish red skin discoloration known as carotenoderma.

Bile Pigments A yellowish skin discoloration may be caused by excessive amounts of bile pigments, mainly bilirubin (BIL-ih-ru-bin), in the blood (**see Fig. 6-6C**). (Bile is a substance produced by the liver that aids in fat digestion; see Chapter 19.) This condition, called **jaundice** (JAWN-dis) (from the French word for "yellow"), may be a symptom of certain disorders, including

- A tumor pressing on the common bile duct or a stone within the duct, either of which would obstruct bile flow into the small intestine
- Inflammation of the liver (hepatitis), commonly caused by a virus
- Certain blood diseases in which red blood cells are rapidly destroyed (hemolyzed)
- Immaturity of the liver. Neonatal (newborn) jaundice occurs when the liver is not yet capable of processing bilirubin. Most such cases correct themselves without treatment in about a week, but this form of jaundice may be treated by exposure to special fluorescent light that helps the body eliminate bilirubin.

CHECKPOINTS

- ☐ 6-10 Name some pigments that give color to the skin.
- ☐ 6-11 What is the term for a bluish skin discoloration caused by insufficient oxygen?

Repair of the Integument

Repair of the integument after injury can occur only in areas that have actively dividing stem cells or cells that can be triggered to divide by injury. These cells are found in the skin's epithelial tissues, and to a lesser extent, connective tissues. Mainly, they are located in the stratum basale of the epidermis and in the hair follicles of the dermis. If both layers of the skin are destroyed along with their stem cells, healing may require skin grafts.

Repair of a skin wound or lesion begins after blood has clotted and an inflammatory response occurs. Blood brings growth factors that promote the activity of restorative cells and agents that break down tissue debris and fight infection. New vessels branch from damaged capillaries and grow into the injured tissue. Fibroblasts (connective tissue cells) manufacture collagen to close the gap made by the wound. If the wound is large, underlying muscle tissue may contract to bring the edges of the wound closer together.

A large injury requires extensive growth of new connective tissue, which develops from within the wound. This new tissue sometimes forms a **scar**, also called a **cicatrix** (SIK-ah-triks), which may continue to show at the surface as a white line. Scar tissue is strong but is not as flexible as normal tissue and does not function like the tissue it replaces. Suturing (sewing) the edges of a clean wound together, as is done for operative wounds, decreases the amount of connective tissue needed for repair and thus minimizes scarring. Excess collagen production in the formation of a scar may result in the development of **keloids** (KE-loyds), tumor-like masses or sharply raised areas on the skin surface. These are not dangerous but may be removed for the sake of appearance.

WOUND CARE

A caregiver should protect a wound from any further injury, keep the wound clean, and control infection. Water or saline solution is best for cleansing, as harsher antiseptics may further damage tissue. Antibacterial ointments may be used if needed. Keeping a wound dry will promote formation of a scab, or crust, over the wound. This may protect the injured tissue and keep out germs, but it does slow healing and encourage scar formation. Health professionals now recommend maintaining a proper moisture balance to prevent scabbing. A moist environment stimulates the activity of healing cells, retains growth factors, and speeds the removal of damaged tissue. If needed, a variety of wound dressings to maintain proper moisture balance are available.

FACTORS THAT AFFECT HEALING

Wound healing is a complex process involving multiple body systems. It is affected by the following:

- Nutrition—A complete and balanced diet will provide the nutrients needed for cell regeneration. All required vitamins and minerals are important, especially vitamins A and C, which are needed for collagen production.
- Blood supply—The blood brings oxygen and nutrients to the tissues and also carries away waste materials and toxins (poisons) that might form during the healing process. White blood cells attack invading bacteria at the site of the injury. Poor circulation, as occurs in cases of diabetes, for example, will delay wound healing.
- Infection—Contamination prolongs inflammation and interferes with the formation of materials needed for wound repair.
- Age—Healing is generally slower among the elderly reflecting their slower rate of cell replacement. The elderly also may have lowered immune responses to infection.

CHECKPOINTS 

- ☐ 6-12 What two categories of tissues repair themselves most easily?
- ☐ 6-13 Name four factors that affect skin healing.

6

See the Student Resources on thePoint to view an animation on wound healing.

Effects of Aging on the Integumentary System

As people age, wrinkles, or crow's feet, develop around the eyes and mouth owing to the loss of fat, elastic fibers, and collagen in the underlying tissues. The dermis becomes thinner, and the skin may become transparent and lose its elasticity, an effect sometimes called "parchment skin." Exposure to the UV radiation in sunlight degrades collagen and elastic fibers, thereby accelerating these changes. Pigment formation decreases with age. However, there may be localized areas of extra pigmentation in the skin with the formation of brown spots ("liver spots"), especially on areas exposed to the sun (e.g., the backs of the hands). Circulation to the dermis decreases, so white skin looks paler. Wounds heal more slowly and are more susceptible to infection.

The hair does not replace itself as rapidly as before and thus becomes thinner on the scalp and elsewhere on the body. Decreased melanin production leads to gray or white hair. Hair texture changes as the hair shaft becomes less dense, and hair, like the skin, becomes drier as sebum production decreases.

The sweat glands decrease in number, so there is less output of perspiration and lowered ability to withstand heat. The elderly are also more sensitive to cold because of having less fat in the skin and poor circulation. The fingernails may flake, become brittle, or develop ridges, and toenails may become discolored or abnormally thickened.

Like all cancers, skin cancer occurs more frequently in older individuals. Repeated and prolonged exposure to UV radiation causes genetic mutations in skin cells that interfere with repair mechanisms and may lead to cancer. Note that radiation exposure in tanning booths is no safer than sun tanning. Appropriate application of sunscreen before and during time spent in the sun, especially after swimming, can prevent UV-related skin damage. Limiting sun exposure during midday and covering up with protective clothing are also important.

Disorders of the Integumentary System

Skin disorders can result from physical and chemical trauma or from disease. Skin diseases range from simple superficial nuisances, such as acne and rashes, to more deep-seated problems that may lead to systemic disease. **Dermatosis** (der-mah-TO-sis) is a general term referring to any skin disease.

LESIONS

A **lesion** (LE-zhun) is any wound or local damage to tissue. Many skin diseases are diagnosed based on the lesions they cause. In examining the skin for lesions, it is important to make note of their type, arrangement, and location. Lesions may be flat or raised or may extend below the skin surface.

Surface Lesions A surface lesion is often called a **rash** or, if raised, an **eruption** (e-RUP-shun). Skin rashes may be localized, as in diaper rash, or generalized, as in measles and other systemic infections. Often, these lesions are accompanied by **erythema** (er-eh-THE-mah), or redness of the skin. The following are some terms used to describe surface skin lesions:

- **Macule** (MAK-ule)—a spot that is neither raised nor depressed. Macules are typical of measles and descriptive of freckles (Fig. 6-7A).
- **Papule** (PAP-ule)—a firm, raised area, as in some stages of chickenpox and in the second stage of syphilis (see Fig. 6-7B). A pimple is a papule. A large firm papule is called a **nodule** (NOD-ule).
- **Vesicle** (VES-ih-kl)—a blister or small fluid-filled sac as seen in some stages of chickenpox or shingles eruptions (see Fig. 6-7C). Another term for a vesicle is a **bulla** (BUL-ah).
- **Pustule** (PUS-tule)—a vesicle filled with pus. Pustules may develop if vesicles become infected (see Fig. 6-7D).

Deeper Lesions A deeper skin lesion may develop from a surface lesion or may be caused by **trauma** (TRAW-mah), that is, a wound or injury. Because such breaks may be followed by infection, wounds should be cared for to prevent the entrance of pathogens and toxins into deeper tissues and body fluids. Deeper injuries to the skin include the following:

- **Excoriation** (eks-ko-re-A-shun)—a scratch into the skin
- **Laceration** (las-er-A-shun)—a rough, jagged wound made by tearing of the skin
- **Ulcer** (UL-ser)—a sore associated with disintegration and death of tissue (see Fig. 6-7E)
- **Fissure** (FISH-ure)—a crack in the skin. Athlete's foot and other skin disorders can produce fissures (see Fig. 6-7F).

Pressure Ulcers **Pressure ulcers** are skin lesions that appear where the body rests on skin that covers bony projections, such as the spine, heel, elbow, or hip. The pressure interrupts circulation leading to ulceration and death of tissue. Poor general health, malnutrition, age, obesity, diabetes, and infection contribute to the development of pressure ulcers.

Lesions first appear as skin redness. If ignored, they may penetrate the skin and underlying muscle, extending even to bone and requiring months to heal.

Pads or mattresses to relieve pressure, regular cleansing and drying of the skin, frequent change in position, and good nutrition help prevent pressure ulcers. Prevention of pressure ulcers by these methods is far easier than treatment of an established ulcer.

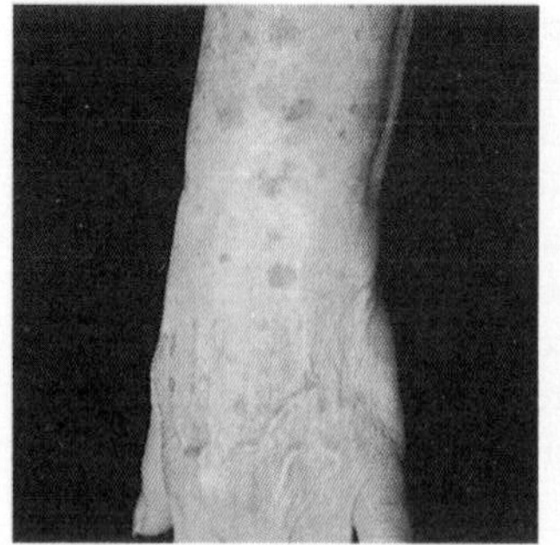

A Macule

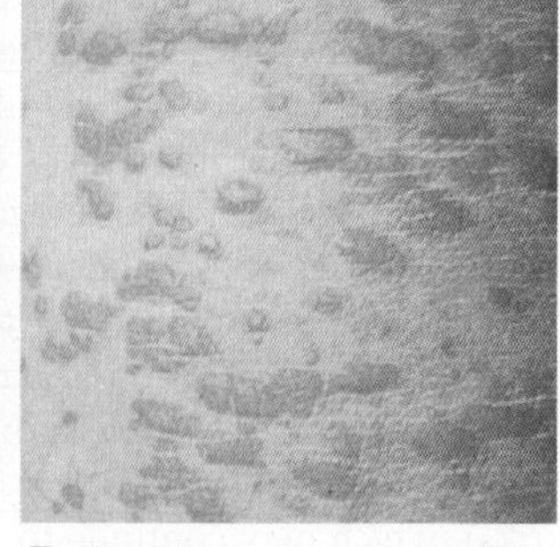

B Papule

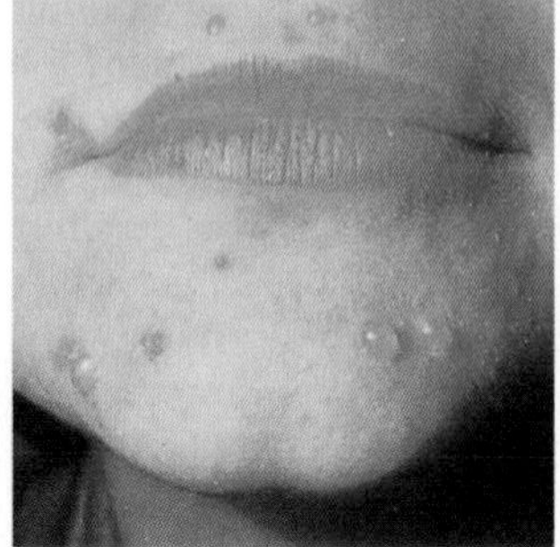

C Vesicle

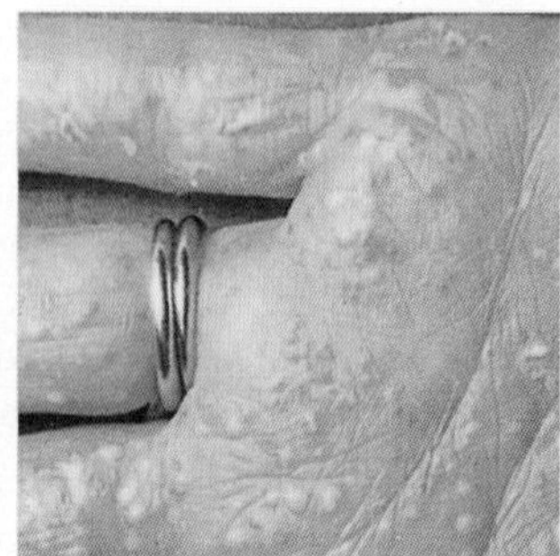

D Pustule

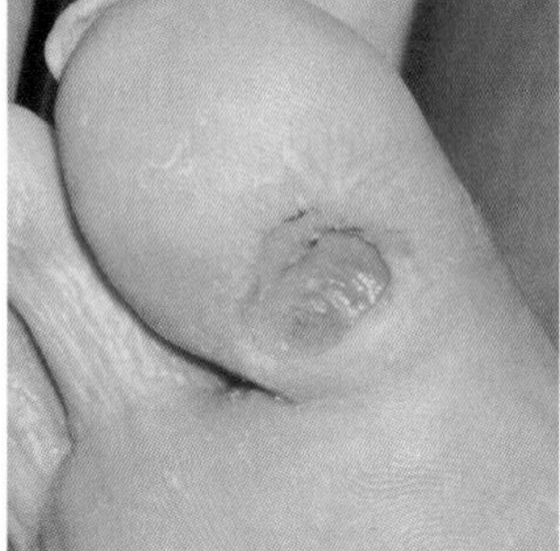

E Ulcer

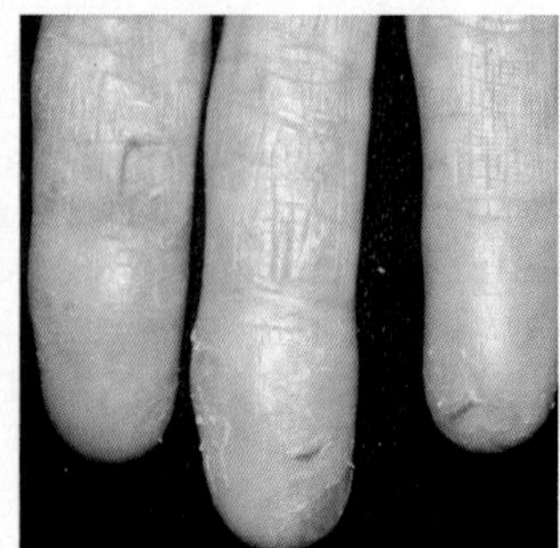

F Fissures

Figure 6-7 **Skin lesions.** **KEY POINT** A lesion is a wound or local damage to tissue. **A.** Macules on the dorsal surface of the hand, wrist, and forearm. **B.** Papules on the knee. **C.** Vesicles on the chin. **D.** Pustules on the palm. **E.** Ulcer. Early stage of a pressure ulcer with breakdown of the dermis. **F.** Fissures on the distal fingers in atopic dermatitis (eczema).

Other terms for pressure ulcers are *decubitus ulcer* and *bedsore*. Both of these terms refer to lying down, although pressure ulcers may appear in anyone with limited movement, not only those who are confined to bed.

BURNS

Most burns are caused by contact with hot objects, explosions, or scalding with hot liquids. They may also be caused by electrical injuries, contact with harmful chemicals, or abrasion. Sunlight can also cause severe burns that result in serious illness. On exposure, the skin first becomes reddened (erythematous) and then may become swollen and blistered.

Burns are assessed by the depth of damage and the percentage of body surface area (BSA) involved. Depth of tissue destruction is categorized as follows:

- **Superficial**—involves the epidermis only (**Fig. 6-8A**). The skin is red and dry; there is minimal pain. Typical causes are mild sunburn and very short heat exposure. This type of burn is also called a first-degree burn.
- **Superficial partial thickness**—involves the epidermis and a portion of the dermis (**Fig. 6-8B**). The tissue reddens and blisters and is painful, as in cases of severe sunburn or scalding.
- **Deep partial thickness**—involves the epidermis and the dermis (**Fig. 6-8C**). The tissue may be blistered with a weeping surface or dry because of sweat gland damage. These burns may be less painful than superficial burns because of nerve damage. Causes include scalding and exposure to flame or hot grease. Superficial and deep partial-thickness burns are also classified as second-degree burns. Regina's case study involved this type of burn.
- **Full thickness**—involves the full skin and sometimes subcutaneous tissue and underlying tissues as well (**Fig. 6-8D**). The tissue is broken, dry and pale, or charred. These injuries may require skin grafting and may result in loss of digits or limbs. Full-thickness burns are also classified as third-degree burns.

The amount of BSA involved in a burn may be estimated by using the **rule of nines**, in which surface areas are assigned percentages in multiples of nine (**Fig. 6-9**). The more accurate Lund and Browder method divides the body into small areas and estimates the proportion of BSA that each contributes.

Infection is a common complication of burns, because the skin, a major defense against invasion of microorganisms, is damaged. Respiratory complications may be caused by inhalation of smoke and toxic chemicals, and circulatory problems may result from loss of fluids and electrolytes. Treatment of burns includes respiratory care, administration of fluids, wound care, and pain control. Patients must be monitored for circulatory complications, infections, and signs of posttraumatic stress.

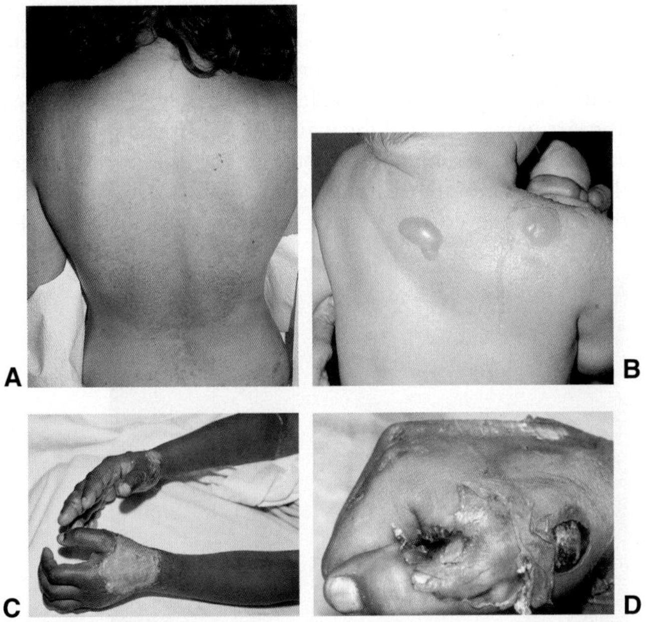

Figure 6-8 **Burns.** **KEY POINT** Burns are classified based on the depth of the damage. **A.** Superficial. **B.** Superficial partial thickness. **C.** Deep partial thickness. **D.** Full thickness.

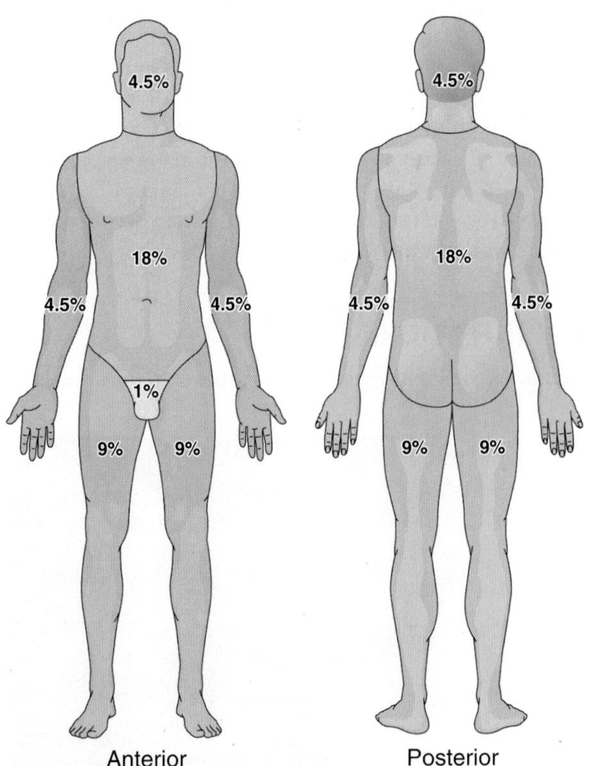

Figure 6-9 **The rule of nines.** **KEY POINT** This method is used to estimate percentages of body surface area (BSA) in treatment of burns.

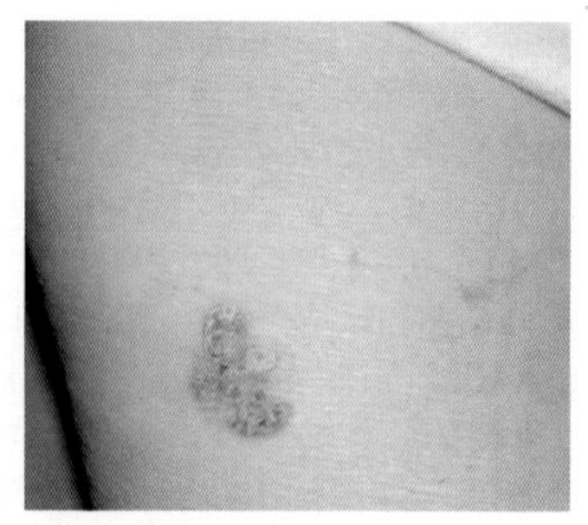
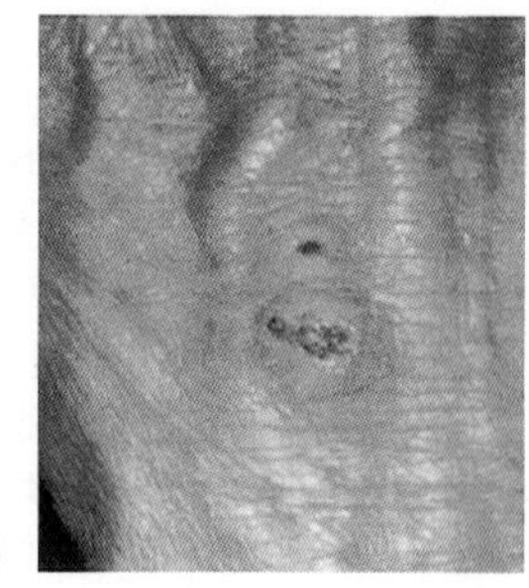
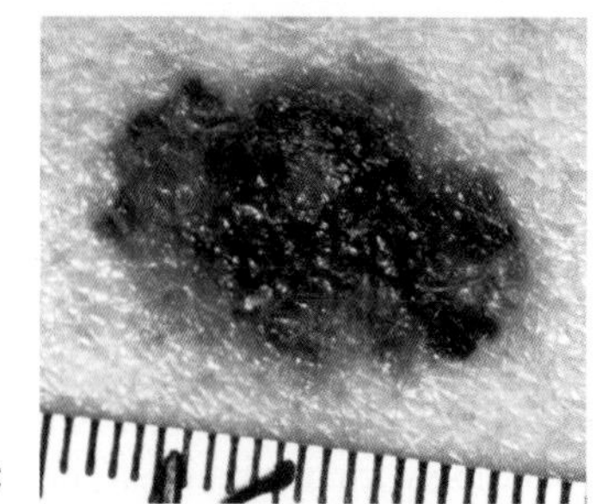

Figure 6-10 **Skin cancer.** **KEY POINT** Skin cancers are the most common form of cancer. **A.** Basal cell carcinoma. **B.** Squamous cell carcinoma. **C.** Malignant melanoma.

SKIN CANCER

Skin cancer is the most common form of cancer in the United States. Exposure to sunlight predisposes one to development of skin cancer, which, in the United States, is most common among people who have fair skin and who live in the Southwest, where exposure to the sun is consistent and may be intense.

Basal cell and squamous cell carcinomas arise in the epidermis and generally appear on the face, neck, and hands (**Fig. 6-10A and B**). Early detection and treatment in these cases usually results in a cure, although squamous cell carcinoma is the more likely to metastasize.

Melanoma (mel-ah-NO-mah) is a malignant tumor of melanocytes. This type of cancer originates in a **nevus** (NE-vus), a mole or birthmark, anywhere in the body (**see Fig. 6-10C**). Unlike a normal mole, which has an evenly round shape and well-defined border, a melanoma may show irregularity in shape. Other signs of melanoma are a change in color or uneven color and increase in size of a mole. A predisposing factor for melanoma is severe, blistering sunburn, although these cancers can appear in areas not sun exposed, such as the soles of the feet, between fingers and toes, and in mucous membranes.

SKIN INFECTIONS

Bacteria, viruses, and fungi may all cause skin infections.

Bacterial Infections The skin is the site of many bacterial infections, many of which are secondary to scratches or other skin lesions. If the infection spreads to the dermis, it is described as cellulitis.

Impetigo (im-peh-TI-go) is an acute contagious disease of staphylococcal origin that may be serious enough to cause death in newborn infants (**Fig. 6-11A**). It takes the form of blister-like lesions that become filled with pus and contain millions of virulent bacteria. It is found most frequently among poor and undernourished children. Affected people may reinfect themselves or infect others.

Viral Infections One virus that involves the integument is **herpes** (HER-peze) **simplex virus**, which causes the formation of watery vesicles (cold sores, fever blisters) on the skin and mucous membranes (**see Fig. 6-7C**). Type I herpes causes lesions around the nose and mouth; type II is responsible for genital infections (**see Table A3-2** in Appendix 3).

Shingles (herpes zoster) is seen in adults and is caused by the same virus that causes chickenpox (varicella). Infection follows nerve pathways, producing small, vesicular skin lesions along the course of a nerve. The disease's name comes from the Latin word for "belt," as the shingles rash often appears near or around the waist. Pain, increased sensitivity, and itching are common symptoms that usually last longer than a year. Prompt treatment with antiviral drugs decreases the disease's severity. A vaccine for shingles is now available.

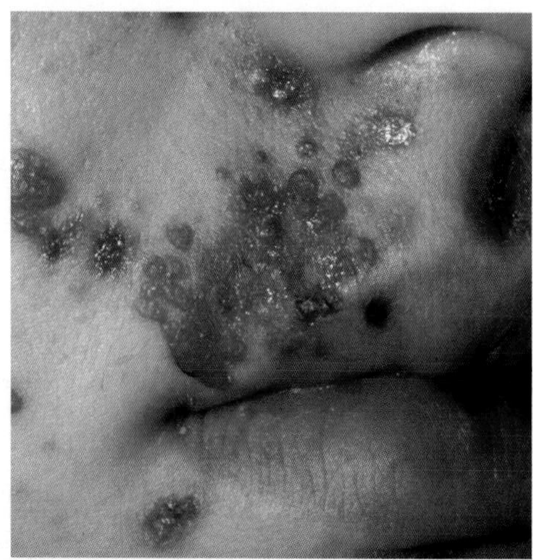
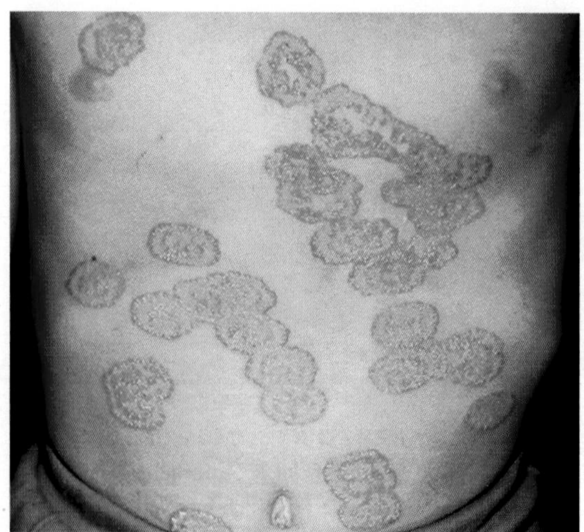

Figure 6-11 **Infectious skin diseases. A.** Impetigo. **B.** Tinea (ringworm) of the body.

A **wart**, or **verruca** (veh-RU-kah), is a small tumor caused by a virus of the human papillomavirus group. Warts may appear anywhere on the body, including the genital region and the soles of the feet (plantar wart). A dermatologist can remove them by chemical treatment or surgery. Usually benign, warts have been associated with cancer, especially in the case of genital warts and cancer of the cervix (neck of the uterus).

Fungal Infections Fungi may cause surface infections of the skin. These superficial mycotic infections, commonly known as **tinea**, may appear on the face, body, scalp, hands, or feet (**Fig. 6-11B**). The infection often creates a red, ring-shaped lesion, so tinea is also known as ringworm (even though worms are not involved). When on the foot, the condition is usually called athlete's foot, as fungal growth is promoted by the dampness of perspiration, for example, within an athletic shoe. Fungal nail infections commonly result from wearing false nails or acrylic nails, as fungal growth is promoted by the moisture that accumulates under the artificial nails.

Fungal infections are difficult to treat. Topical antifungal agents may be effective, but often the patient must take an oral antifungal drug.

INFLAMMATORY DISORDERS AFFECTING THE SKIN

If you've ever had a sunburn or a mosquito bite, you've experienced the redness, heat, and itching of skin inflammation, known as **dermatitis** (der-mah-TI-tis). While uncomfortable, acute skin inflammation is rarely dangerous. Chronic skin inflammation, on the other hand, frequently results from autoimmune diseases, some of which are life threatening.

Acute Inflammatory Disorders **Urticaria** (ur-tih-KA-re-ah), or hives, is an allergic reaction characterized by the temporary appearance of elevated red patches known as *wheals* (**Fig. 6-12A**). An allergic reaction is an unfavorable immune response to a substance that is normally harmless to most people (see Chapter 17). The inducing agent can be internal, such as an antibiotic, aspirin, inhaled pollens, food, or venoms, or it can be external, such as contact with cat dander.

Acute eczematous dermatitis or **eczema** (EK-ze-mah) is characterized by intense itching and skin inflammation (**Fig. 6-12B**). The affected areas show redness (erythema), blisters (vesicles), pimple-like lesions (papules), and scaling and crusting of the skin surface. Scratching (excoriation) of the skin can lead to a secondary bacterial infection. Oral or topical corticosteroids or antihistamines can reduce the inflammation and treat the unpleasant symptoms.

Eczema can result from contact with external irritants, such as the oil of poison oak or poison ivy plants, detergents, and strong acids, alkalis, or other chemicals. Prompt removal of the irritant is the most effective method of prevention and treatment. A thorough cleansing as soon as possible after contact with plant oils may prevent the development of itching eruptions. In some individuals, food or medications can result in eczema.

Atopic dermatitis (ah-TOP-ik der-mah-TI-tis) (also known as atopic eczema) describes recurrent bouts of eczema beginning in childhood. The skin inflammation can be initiated by irritants such as detergents, rough fabrics, soaps, or even perspiration, by environmental allergens, or by food allergens, so it can be difficult to prevent outbreaks. The person with atopic dermatitis may also be subject to other allergic disorders, such as hay fever or asthma.

Chronic Inflammatory Disorders An autoimmune disease results from an immune reaction to one's own tissues. The following chronic inflammatory diseases that involve the skin are believed to be caused, at least in part, by autoimmune reactions.

Pemphigus (PEM-fi-gus) is characterized by the formation of blisters, or bullae (BUL-e) in the skin and mucous membranes caused by a separation of epidermal cells from underlying tissue layers (**Fig. 6-13A**). Rupture of these lesions leaves deeper areas of the skin unprotected from infection and fluid loss, much as in cases of burns. Pemphigus is fatal unless treated by methods to suppress the immune system.

Lupus erythematosus (LU-pus er-ih-the-mah-TO-sus) is a chronic, inflammatory, autoimmune disease of connective tissue (**Fig. 6-13B**; also see Chapter 4). The more widespread form of the disease, systemic lupus erythematosus (SLE), involves the skin and other organs. The discoid form (DLE) involves only the skin. It is seen as rough, raised, violet-tinted papules, usually limited to the face and scalp. There may also be a butterfly-shaped rash across the nose and cheeks,

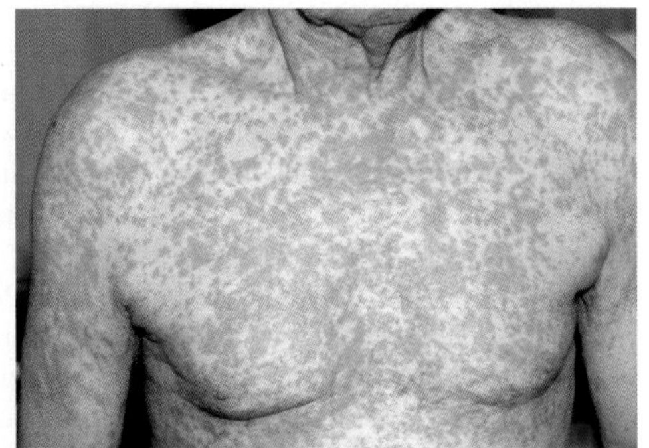
A

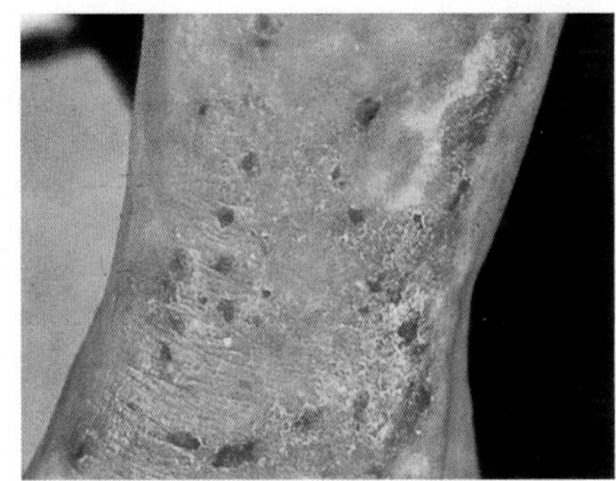
B

Figure 6-12 **Acute inflammatory skin disorders. A.** Urticaria (hives), resulting from a drug allergy. **B.** Atopic dermatitis (atopic eczema). Scratches (excoriation) are visible in the photo.

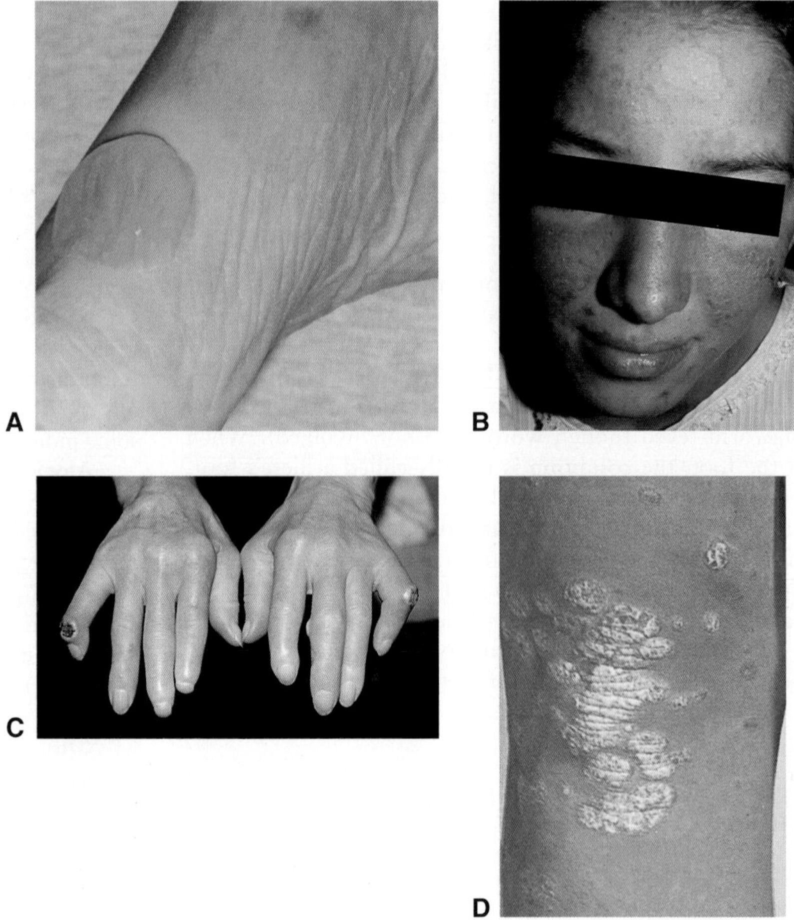

Figure 6-13 Chronic inflammatory skin disorders. A. Pemphigus. **B.** Lupus. **C.** Scleroderma. **D.** Psoriasis. Silvery surface scales are visible.

described as a malar (cheekbone) rash. The skin lesions of lupus are worsened by exposure to the UV radiation in sunlight. SLE is more prevalent in women than in men and has a higher incidence among Asians and African Americans than in other populations.

> **See the Student Resources on thePoint for a figure on SLE.**

Systemic sclerosis (scleroderma) is an autoimmune disease of unknown cause that involves overproduction of collagen with thickening and tightening of the skin (the root *scler/o* means "hard") (also see Chapter 4). In severe cases, the skin hardening makes facial movements impossible, resulting in a masklike appearance, and it also renders breathing and hand movements difficult (**Fig. 6-13C**). Sweat glands and hair follicles are also involved. A very early sign of scleroderma is numbness, pain, and tingling on exposure to cold caused by constriction of blood vessels in the fingers and toes (known as Raynaud phenomenon). Skin symptoms first appear on the forearms and around the mouth. Eventually, visceral connective tissue is also affected.

Psoriasis (so-RI-ah-sis) is a chronic overgrowth of keratinocytes within the epidermis leading to large, sharply outlined, red (erythematous), flat areas (plaques) covered with silvery scales (**Fig. 6-13D**). Scientists have identified gene mutations that may contribute to the development of psoriasis, but not everyone with the gene mutation develops the disorder. In some cases, physical trauma can trigger its development in susceptible individuals. Psoriasis is treated with topical corticosteroids and exposure to UV light.

DISORDERS OF THE ACCESSORY ORGANS

Acne **Acne** (AK-ne) is a disorder caused by overactivity of the sebaceous (oil) glands connected with the hair follicles. The common type, called **acne vulgaris** (vul-GA-ris), is seen most often in people between the ages of 14 and 25 years, when hormones that control sebaceous secretions are particularly active (**Fig. 6-14A**).

Acne results from excessive sebum production and reduced keratinocyte shedding. Sebum combined with keratinocytes can form a plug within the hair follicle. Air causes the plug to darken, resulting in a blackhead, or open comedo (KOM-e-do). If the follicle becomes completely blocked, backed up sebum causes the follicle to bulge and form a whitehead (closed comedo). Certain normally occurring bacteria thrive in this closed environment, feasting on the abundant sebum and reproducing rapidly. The resulting infection activates the immune response, resulting in a swollen, inflamed pimple.

Acne medications include oral or topical antibiotics to reduce bacterial numbers, and topical agents to reduce sebum

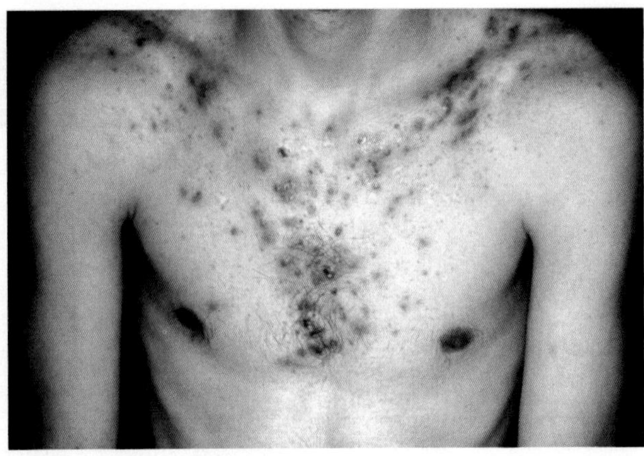

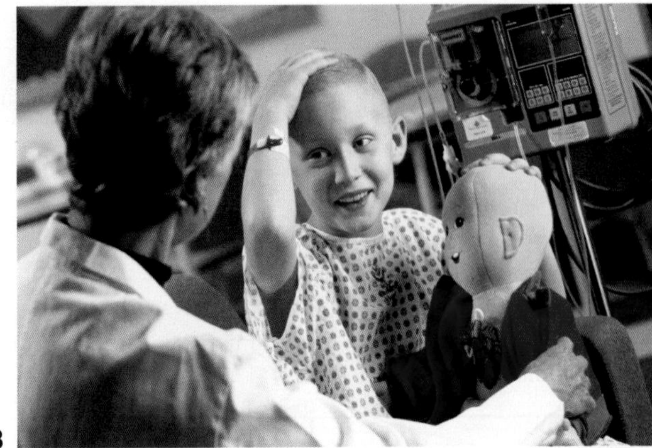

Figure 6-14 **Accessory organ disorders. A.** Acne vulgaris. **B.** Alopecia resulting from chemotherapy.

production and promote keratinocyte shedding. Laser, light, and cosmetic procedures that reduce sebum and correct scarring are also available.

Alopecia (Baldness) **Alopecia** (al-o-PE-she-ah), or baldness, may be caused by a number of factors. The most common type, known as male pattern baldness, is an expression of heredity and aging; it is influenced by male sex hormones. Topical applications of the drug minoxidil (used as an oral medication to control blood pressure) have produced hair growth in this type of baldness. Alopecia also may result from a systemic disease, such as uncontrolled diabetes, thyroid disease, or malnutrition. In such cases, control of the disease results in hair regrowth. An expanding list of drugs has been linked with baldness, including the chemotherapeutic drugs used in treating cancer (**Fig. 6-14B**). Sometimes, hair falls out in patches for unknown reasons. This condition, called *alopecia areata*, can occur in children as well as adults and affects both males and females.

CHECKPOINTS

- 6-14 What is a lesion?
- 6-15 What two factors are used to assess the severity of burns?
- 6-16 What are some microorganisms that affect the skin?
- 6-17 What is the general term for inflammation of the skin?
- 6-18 What is the probable cause of chronic inflammatory skin diseases?
- 6-19 What skin disorder results from overactivity of the sebaceous glands?
- 6-20 What is the technical term for baldness?

Observation and care of the skin are important in nursing as well as other healthcare professions. The Student Resources on thePoint have information on nursing careers.

Disease in Context Revisited

Regina's Healing Process

Dr. Stanford told Regina that minor burns covering less than 10% of the total body surface area usually are managed as outpatient visits in a medical office. Regina's burns were assessed at 10%, so she fell into this category according to the American Burn Association Grading System. Regina was given instructions on washing the burned area using sterile technique. She also received a prescription to continue with Silvadene, the topical cream used for preventing wound infection following burns.

Wearing sterile gloves, Regina applied the cream thinly to the burned area twice a day. Strict compliance with the treatment plan was helping to prevent infection and promote healing. Dr. Stanford evaluated Regina's progress when she returned for follow-up appointments. He noted positive results after about two weeks. Scar tissue was beginning to form, and the area remained clean and free from infection.

In this case, we saw how burns are classified and treated. Regina's burns were serious, but careful management prevented the breach in the skin's barrier from causing a life-threatening infection.

CHAPTER 6

Chapter Wrap-Up

Summary Overview

A detailed chapter outline with space for note taking is on *thePoint*. The figure below illustrates the main topics covered in this chapter.

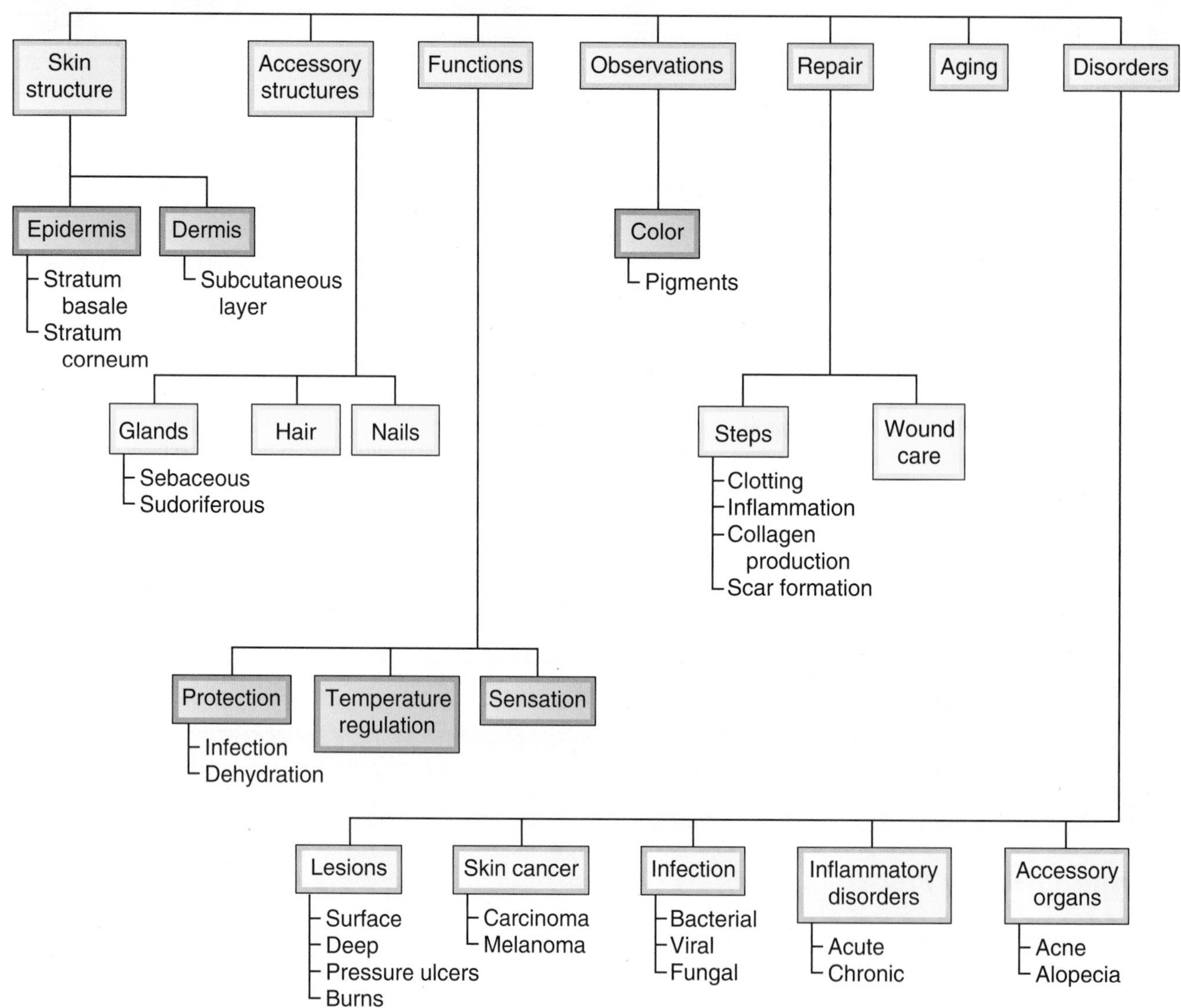

Key Terms

The terms listed below are emphasized in this chapter. Knowing them will help you organize and prioritize your learning. These and other boldface terms are defined in the Glossary with phonetic pronunciations.

alopecia
apocrine
arrector pili
cerumen
cicatrix
cyanosis
dermatitis
dermis
eccrine
epidermis
erythema
exfoliation
hair follicle
integumentary system
jaundice
keloid
keratin
lesion
melanin
melanocyte
scar
sebaceous gland
sebum
stratum basale
subcutaneous layer
sudoriferous gland

Word Anatomy

Medical terms are built from standardized word parts (prefixes, roots, and suffixes). Learning the meanings of these parts can help you remember words and interpret unfamiliar terms.

WORD PART	MEANING	EXAMPLE
Structure of the Skin		
corne/o	cornified, keratinized	The stratum *corneum* is the outermost thickened, keratinized layer of the skin.
derm/o	skin	The *epidermis* is the outermost layer of the skin.
melan/o	dark, black	A *melanocyte* is a cell that produces the dark pigment melanin.
sub-	under, below	The *subcutaneous* layer is under the skin.
Accessory Structures of the Skin		
ap/o-	separation from, derivation from	The *apocrine* sweat glands release some cellular material in their secretions.
pil/o	hair	The *arrector pili* muscle raises the hair to produce "goose bumps."
Observation of the Skin		
alb/i	white	*Albinism* is a condition associated with a lack of pigment, so the skin appears white.
bili	bile	*Bilirubin* is a pigment found in bile.
cyan/o	blue	*Cyanosis* is a bluish discoloration of the skin caused by lack of oxygen.
-emia	condition of blood	In *carotenemia*, vegetable pigments, as from carrots, appear in the blood and give color to the skin.
eryth	red	*Erythema* is redness of the skin.
-ism	state of	See *alb/i* example.
-sis	condition, process	See *cyan/o* example.
Disorders of the Integumentary System		
dermat/o	skin	*Dermatosis* is any skin disease.
scler/o	hard	*Scleroderma* is associated with a hardening of the skin.

Questions for Study and Review

BUILDING UNDERSTANDING

Fill in the Blanks

1. Cells of the stratum corneum contain large amounts of a protein called _____.

2. Sweat glands located in the axillae and groin are classified as _____ glands.

3. The name of the muscle that raises the hair is the _____.

4. A dark-colored pigment that protects the skin from the rays in sunlight is called _____.

5. Refer to the integumentary system in "The Body Visible" overlays to find that Meissner and Ruffini corpuscles are receptors for the sense of _____.

Matching > Match each numbered item with the most closely related lettered item.

___ **6.** Skin sensitivity characterized by intense itching and inflammation

___ **7.** A viral infection that follows nerve pathways, producing small lesions on the overlying skin

___ **8.** A fungal infection of the skin

___ **9.** Allergic reaction characterized by the appearance of wheals

___ **10.** Chronic skin disease characterized by red flat areas covered with silvery scales

a. urticaria

b. tinea

c. shingles

d. psoriasis

e. eczema

Multiple Choice

___ **11.** The dermis is _____ to the epidermis.

a. superficial
b. deep
c. lateral
d. medial

___ **12.** What type of gland is involved in acne?

a. sudoriferous
b. sebaceous
c. ceruminous
d. meibomian

___ **13.** Which of the following is not a type of skin lesion?

a. pallor
b. vesicle
c. papule
d. laceration

___ **14.** Which skin discoloration is caused by an accumulation of bile pigment in the blood?

a. pallor
b. cyanosis
c. jaundice
d. carotenemia

___ **15.** Which type of cells are involved in basal cell and squamous cell carcinomas?

a. epidermal cells
b. dermal cells
c. melanocytes
d. adipocytes

UNDERSTANDING CONCEPTS

16. Compare and contrast the epidermis, dermis, and hypodermis. How are the outermost cells of the epidermis replaced?

17. Referring to the integumentary system in "The Body Visible" overlays and information in Chapter 4, name the type of tissue that comprises number 27.

18. Describe the location and function of the two types of skin glands.

19. What are the four most important functions of the skin?

20. Describe the events associated with skin wound healing.

21. What changes may occur in the skin with age?

22. List some chronic inflammatory disorders of the skin and explain their probable causes.

23. What is a pressure ulcer? List measures for preventing pressure ulcers. Cite two other names for this type of lesion.

24. Define alopecia and list some possible causes of alopecia.

CONCEPTUAL THINKING

25. Why is the skin described as a membrane? An organ? A system?

26. Explain the correlation between the functions of the skin and the possible dangers of burns, including the events in Regina's case study.

For more questions, see the Learning Activities on thePoint.

UNIT

III

Movement and Support

This unit deals with the skeletal and muscular systems, which work together to execute movement and to support and protect vital organs. The chapter on the skeleton identifies the bones and joints and discusses bone formation, growth, and repair. It also covers bone and joint disorders. The chapter on muscles describes the characteristics of all types of muscles and then concentrates on the muscles that are attached to the skeleton and how they function. We name and locate the main skeletal muscles and describe their actions.

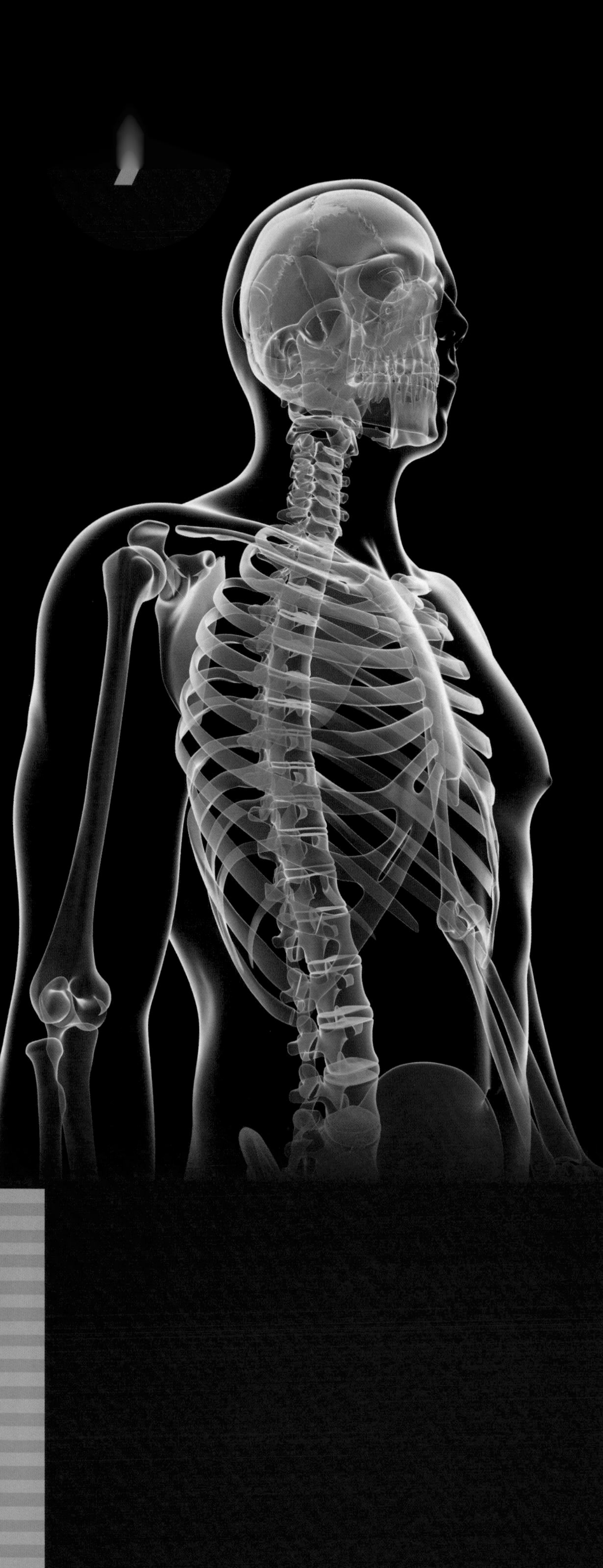

Learning Objectives

After careful study of this chapter, you should be able to:

1 ▶ List the functions of bones. ***p. 130***

2 ▶ Describe the structure of a long bone. ***p. 130***

3 ▶ Name the three different types of cells in bone, and describe the functions of each. ***p. 130***

4 ▶ Differentiate between compact bone and spongy bone with respect to structure and location. ***p. 131***

5 ▶ Explain how a long bone grows. ***p. 132***

6 ▶ Name and describe nine markings found on bones. ***p. 133***

7 ▶ Name, locate, and describe the bones in the axial skeleton. ***p. 134***

8 ▶ Describe the normal curves of the spine, and explain their purpose. ***p. 135***

9 ▶ Name, locate, and describe the bones in the appendicular skeleton. ***p. 140***

10 ▶ Describe five types of bone disorders. ***p. 145***

11 ▶ Name and describe eight types of fractures. ***p. 148***

12 ▶ Describe three categories of joints based on degree of movement, and give examples of each. ***p. 149***

13 ▶ Name six types of synovial joints, and demonstrate the movements that occur at each. ***p. 150***

14 ▶ Describe three types of joint disorders. ***p. 153***

15 ▶ Describe methods used to correct diseased joints. ***p. 154***

16 ▶ Describe how the skeletal system changes with age. ***p. 155***

17 ▶ Using the case study, discuss how fractures heal. ***pp. 129, 155***

18 ▶ Show how word parts are used to build words related to the skeleton (see Word Anatomy at the end of the chapter). ***p. 157***

Disease in Context *Reggie's Case: A Footballer's Fractured Femur*

"Donnelly throws deep for a touchdown. Wilson makes a beautiful catch! Ooh, a nasty hit from number 26." The crowd roared their approval for the wide receiver. On the ground, Reggie Wilson knew that something was wrong with his hip. In fact, he thought he had actually heard the bone break. It didn't take long for the coaches and medical staff to realize that Reggie needed help. And it didn't take long for the ambulance to get him to the trauma center closest to the stadium.

At the hospital, the emergency team examined Reggie. His injured leg appeared shorter than the other and was adducted and laterally rotated—all signs of a hip fracture. An x-ray confirmed the team's suspicions; Reggie had sustained an intertrochanteric fracture of his right femur. He would need surgery, but luckily for Reggie, the fracture line extended from the greater trochanter to the lesser trochanter and didn't involve the femoral neck. This meant that the blood supply to the femoral head was not in danger, so the surgery would be more straightforward.

In the operating room, the surgical team applied traction to Reggie's right leg, pulling on it to reposition the broken ends of his proximal femur back into anatomic position (verified with another x-ray). Then, the orthopedic surgeon made an incision beginning at the tip of the greater trochanter and continuing distally along the lateral thigh through the skin, subcutaneous fat, and vastus lateralis muscle. After exposing the proximal femur, the surgeon drilled a hole and installed a titanium screw through the greater trochanter and neck and into the femoral head. He then positioned a titanium plate over the screw and fastened it to the femoral shaft with four more screws. Confident that the broken ends of the femur were firmly held together, the surgeon closed the wound with sutures and skin staples. Reggie was then wheeled into the recovery room.

The surgical team successfully realigned the fractured ends of Reggie's femur. Now Reggie's body will begin the healing process. In this chapter, we learn more about bones and joints. Later in the chapter, we see how Reggie's skeletal system is repairing itself.

ANCILLARIES *At-A-Glance*

Visit thePoint to access the following resources. For guidance in using these resources most effectively, see pp. xv–xvii.

Learning RESOURCES

- Tips for Effective Studying
- Web Figure: Bone Markings and Formations
- Web Figure: Skeletal Features of the Head and Neck
- Web Figure: Skeletal Features of the Shoulder
- Web Figure: Skeletal Features of the Arm
- Web Figure: Skeletal Features of the Thigh
- Web Figure: Skeletal Features of the Leg and Foot
- Web Figure: Spine Curvatures
- Web Figure: Cleft Palate
- Web Chart: Bones of the Skull
- Animation: Bone Growth
- Health Professions: Radiologic Technologist
- Detailed Chapter Outline
- Answers to Questions for Study and Review
- Audio Pronunciation Glossary

Learning ACTIVITIES

- Pre-Quiz
- Visual Activities
- Kinesthetic Activities
- Auditory Activities

A LOOK BACK

Bone tissue is the densest form of the connective tissues introduced in Chapter 4. We now describe its characteristics in greater detail and show how it is built into the skeletal system.

The skeleton is the strong framework on which the body is constructed. Much like the frame of a building, the skeleton must be strong enough to support and protect all the body structures. Bones work with muscles to produce movement at the joints. The bones and joints together form the skeletal system.

Bones

Bones have a number of functions, several of which are not evident in looking at the skeleton. They

- form a sturdy framework for the entire body
- protect delicate structures, such as the brain and the spinal cord
- work as levers with attached muscles to produce movement
- store calcium salts, which may be resorbed into the blood if calcium is needed
- produce blood cells (in the red marrow)

BONE STRUCTURE

The complete bony framework of the body, known as the **skeleton** (**Fig. 7-1**), consists of approximately 206 bones (the precise number can vary somewhat among individuals). The axial skeleton includes the bones of the head and torso, and the appendicular skeleton includes the bones of the extremities. The individual bones in these two divisions are described in detail later in this chapter. The bones of the skeleton can be of several different shapes. They may be flat (ribs, cranium), short (carpals of wrist, tarsals of ankle), or irregular (vertebrae, facial bones). The most familiar shape, however, is the **long bone**, the type of bone that makes up most of the appendicular skeleton. The long narrow shaft of this type of bone is called the **diaphysis** (di-AF-ih-sis). At the center of the diaphysis is a **medullary** (MED-u-lar-e) **cavity**, which contains bone marrow. The long bone also has two irregular ends, a proximal and a distal **epiphysis** (eh-PIF-ih-sis) (**Fig. 7-2**).

Bone Tissue Bones are living organs with their own systems of blood vessels and nerves. Bone tissue, also known as **osseous** (OS-e-us) **tissue**, is the hardest form of connective tissue.

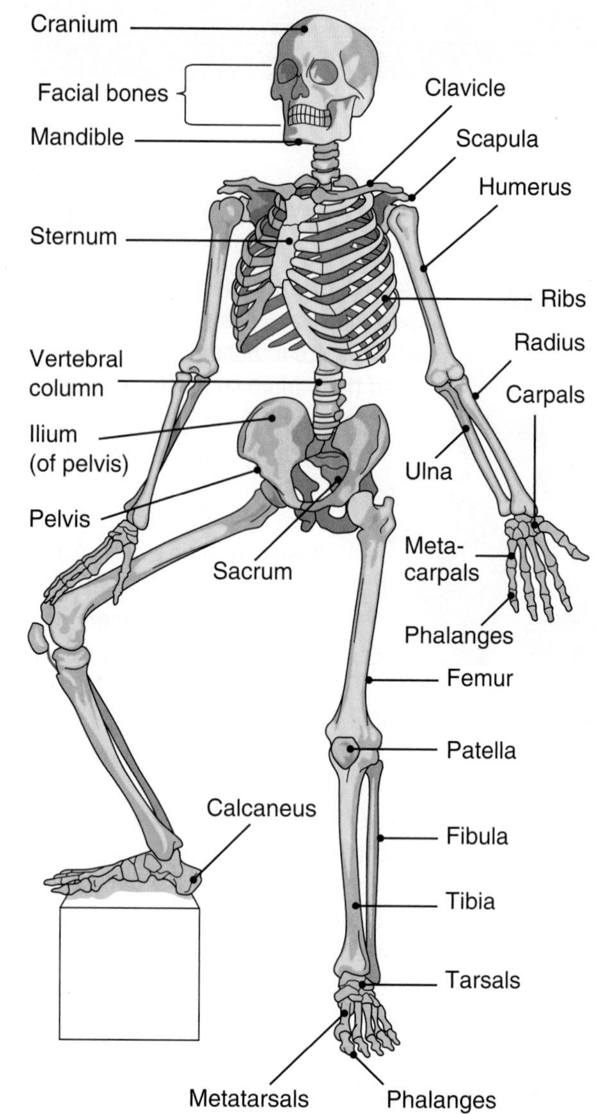

Figure 7-1 The skeleton **KEY POINT: The skeleton is divided into two portions.** The axial skeleton is shown here in *yellow*; the appendicular in *blue*.

Bone's hardness and strength reflect the components of the **matrix**, the material between the living bone cells. This material is rich in the protein **collagen** (KOL-ah-jen) and in calcium salts. Both substances are necessary for healthy bones; without minerals, bones would bend easily, but without collagen, the bones would shatter easily, like sticks of chalk. See the "Diseases of Bone" section later for bone diseases resulting from the lack of minerals or collagen.

Bone tissue contains three types of cells:

- **Osteoblasts** (OS-te-o-blasts) build bone tissue. (You can use the mnemonic "Blasts Build" to remember the role of these cells.)

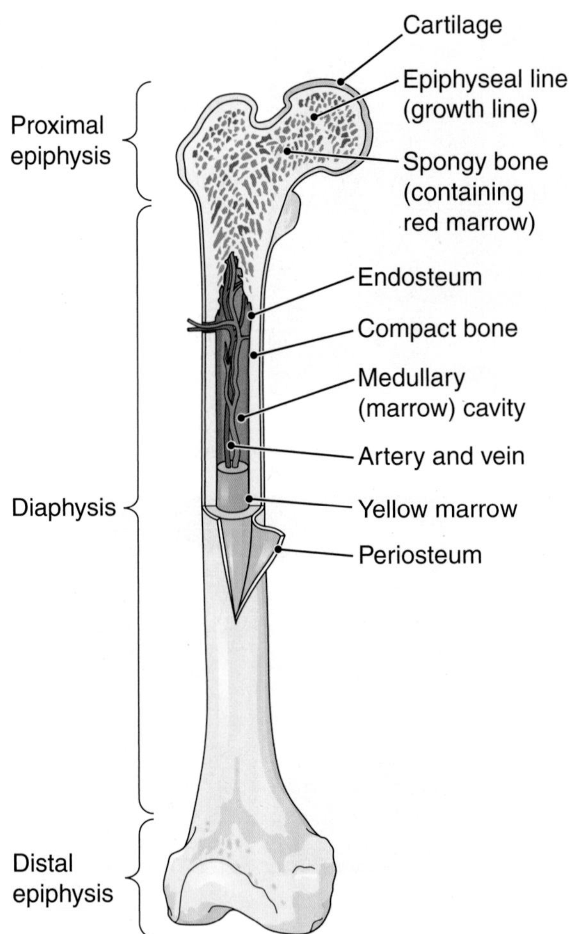

Figure 7-2 The structure of a long bone KEY POINT A long bone has a long, narrow shaft, the diaphysis, and two irregular ends, the epiphyses. The medullary cavity has yellow marrow. Red marrow is located in spongy bone. ZOOMING IN What are the membranes on the outside and the inside of a long bone called?

- **Osteocytes** (OS-te-o-sites) are mature osteoblasts that become trapped in the bone matrix. They maintain bone tissue.
- **Osteoclasts** (OS-te-o-klasts) are large, multinucleated cells responsible for the process of **resorption**, which is the breakdown of bone tissue. Osteoclasts develop from a type of white blood cell (monocyte). (You can use the mnemonic "Clasts Cleave.")

Types of Osseous Tissue There are two types of osseous tissue: compact and spongy. **Compact bone** is hard and dense (**Fig. 7-3**). This tissue makes up the main shaft of a long bone and the outer layer of other bones. The osteocytes (mature bone cells) in this type of bone are located in rings of bone tissue around a **central canal**, also called a *haversian* (ha-VER-shan) *canal*, containing nerves and blood vessels. Osteocytes live in spaces (lacunae) between the rings and extend out into many small radiating channels so that they can be in contact with nearby cells. Each ringlike unit with its central canal makes up an **osteon** (OS-te-on) or haversian system (**see Fig. 7-3B**). **Perforating** (Volkmann) **canals** form channels across the bone, from one side of the shaft to the other, permitting the passage of blood vessels and nerves.

The second type of bone tissue, called **spongy bone**, or cancellous bone, has more spaces than does compact bone. It is made of a meshwork of small, bony plates filled with red marrow. Spongy bone is found at the epiphyses (ends) of the long bones and at the center of other bones. It also lines the medullary cavity of long bones. **Figure 7-3C** shows a photograph of both compact and spongy bone tissue.

Bone Marrow Bones contain two kinds of marrow. In adults, **red marrow** is found in the spongy bone at the ends of the long bones and at the center of other bones (**see Fig. 7-2**). Red bone marrow manufactures blood cells. **Yellow marrow** is found chiefly in the central cavities of the long bones. Yellow marrow is composed largely of fat. The long bones of babies and children contain mostly red marrow, reflecting their greater need for new blood cells.

Bone Membranes Bones are covered (except at the joint region) by a membrane called the **periosteum** (per-e-OS-te-um) (**see Fig. 7-2**). This membrane's inner layer contains osteoblasts that build bone tissue and osteoclasts that break down bone tissue. The coordinated actions of these cells build, repair, and maintain bone throughout life. Blood vessels in the periosteum play an important role in the nourishment of bone tissue. Nerve fibers in the periosteum make their presence known when a person suffers a fracture or receives a blow, such as on the shinbone. A thinner membrane, the **endosteum** (en-DOS-te-um), lines the bone's marrow cavity; it too contains osteoblasts and osteoclasts.

CHECKPOINTS

- ☐ 7-1 What are the scientific names for the shaft and the ends of a long bone?
- ☐ 7-2 What compounds are deposited in the intercellular matrix of the embryonic skeleton to harden it?
- ☐ 7-3 What are the three types of cells found in bone, and what is the role of each?
- ☐ 7-4 What are the two types of osseous (bone) tissue, and where is each type found?

BONE GROWTH, MAINTENANCE, AND REPAIR

The process of bone formation begins in the earliest weeks of embryonic life and continues until young adulthood.

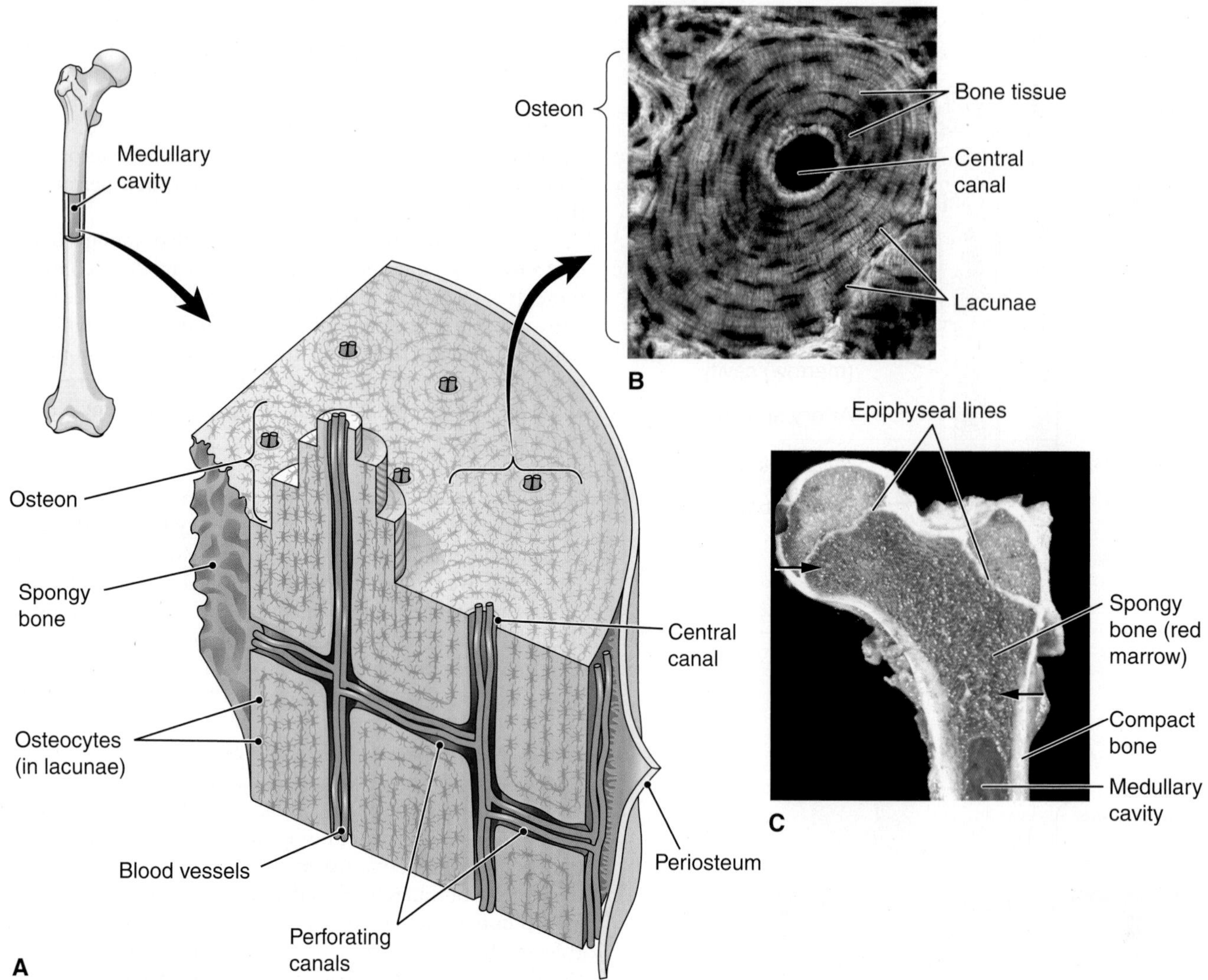

Figure 7-3 **Bone tissue.** **KEY POINT** There are two types of bone tissue—compact and spongy. **A.** This section shows osteocytes (bone cells) within osteons (haversian systems) in compact bone. It also shows the canals that penetrate the tissue. **B.** Microscopic view of compact bone in cross-section (×300) showing a complete osteon. In living tissue, osteocytes (bone cells) reside in spaces (lacunae) and extend out into channels that radiate from these spaces. **C.** Longitudinal section of a long bone showing both types of bone tissue. There is an outer layer of compact bone; the remainder of the tissue is spongy bone, shown by the *arrows*. Transverse growth lines are also visible. **ZOOMING IN** What cells are located in the spaces of compact bone?

Fetal Ossification During early development, the long bones of the embryonic skeleton are composed primarily of cartilage. The conversion of cartilage to bone, a process known as **ossification**, begins during the second and third months of embryonic life. At this time, osteoclast-like cells remove the cartilage, and osteoblasts deposit bone tissue in place of the cartilage.

Once this intercellular material has hardened, the cells remain enclosed within the lacunae (small spaces) in the matrix. These cells, now known as osteocytes, are still living and continue to maintain the existing bone matrix, but they do not produce new bone tissue. Osteoblasts and osteoclasts in the bone membranes are responsible for bone growth, repair, and remodeling later in life. You will see the importance of these cells in Reggie's case study.

The flat bones of the skull and other regions develop from fibrous connective tissue membranes instead of from cartilage. Osteoblasts deposit bone tissue within these fibrous membranes.

Formation of a Long Bone In a long bone, the transformation of cartilage into bone begins at the center of the shaft during fetal development. As bone synthesis continues, osteoclasts (the bone removers) degrade the bone tissue at the center of the bone, producing the medullary cavity. Around the time of birth, secondary bone-forming centers, or **epiphyseal** (ep-ih-FIZ-e-al) **plates**, develop across the ends of the bones. The long bones continue to grow in length at these centers by the production of new cartilage within the plate and calcification of older cartilage. The large amount of cartilage in a child's bones renders them more pliable and tougher to break.

Finally, by the late teens or early 20s, the bones stop growing in length. Bone tissue replaces all of the cartilage in the epiphyseal plate, which can be seen in x-ray films as a thin line, known as the epiphyseal line (**see Figs. 7-2 and 7-3C**). Physicians can use the presence of the epiphyseal plate or line on x-rays to evaluate a patient's age.

As a bone grows in length, it also grows in width. To prevent bones from becoming too heavy, osteoclasts remove bone tissue from the shaft to enlarge the central marrow cavity as osteoblasts add bone tissue to the outside.

Bone Tissue Regulation Even after skeletal growth is complete, osteoblasts and osteoclasts actively maintain and repair bone tissue and remodel it according to need. For instance, a right-handed person uses the right arm more than the left arm and the right arm bones adapt by becoming larger and stronger. Astronauts (in the absence of gravity) lose bone mass because their bones are no longer subjected to stress. Resorption is also necessary for repair of bone injury. In addition, bone tissue is resorbed when the body needs its stored minerals.

Both the formation and resorption of bone tissue are regulated by hormones. Vitamin D, consumed in the diet and produced by the skin, promotes calcium absorption from the intestine. Parathyroid hormone is produced by the parathyroid glands in the neck (posterior to the thyroid gland). Parathyroid hormone stimulates osteoclast activity, resulting in bone resorption and release of calcium into the blood. The sex hormones, estrogen and testosterone, also contribute to bone growth and maintenance. Hormones are discussed more fully in Chapter 12.

The balance between osteoblast and osteoclast activity dictates changes in bone mass. Bones increase in density until the early 20s in females and the late 20s in males, at which point bones are at peak density and strength. Most people maintain peak bone density until about age 40. As people age, there is a slowing of bone tissue renewal. As a result, the bones become weaker and damage heals more slowly.

See the Student Resources on thePoint to view the animation "Bone Growth," showing the growth process in a long bone.

BONE MARKINGS

In addition to their general shape, bones have other distinguishing features, or **bone markings**. These markings include raised areas and depressions, which help form joints or serve as points for muscle attachments, and various holes, which allow the passage of nerves and blood vessels. Some of these identifying features are described next.

Projections

- **Head**—a rounded, knoblike end separated from the rest of the bone by a slender region, the neck
- **Process**—a large projection of a bone
- **Condyle** (KON-dile)—a rounded projection; a small projection above a condyle is an epicondyle
- **Crest**—a distinct border or ridge, often rough, such as over the top of the hip bone
- **Spine**—a sharp projection from the surface of a bone, such as the spine of the scapula (shoulder blade)

Depressions or Holes

- **Foramen** (fo-RA-men)—a hole that allows a vessel or a nerve to pass through or between bones. The plural is foramina (fo-RAM-ih-nah).
- **Sinus** (SI-nus)—A cavity or hollow space. Most commonly, an air-filled chamber found in some skull bones (**Fig. 7-4**).
- **Fossa** (FOS-sah)—a depression on a bone surface. The plural is fossae (FOS-se).
- **Meatus** (me-A-tus)—a short channel or passageway, usually the external opening of a canal. An example is the channel in the skull that leads to the inner ear.

Examples of these and other markings can be seen on the bones illustrated in this chapter. To find out how these markings can be used in healthcare, see **Box 7-1**, "Landmarking: Seeing with Your Fingers."

CHECKPOINTS

- 7-5 What are the centers for secondary growth of a long bone called?
- 7-6 What are some functions of bone markings?

See the Student Resources on thePoint to view bone markings on an illustration of a whole skeleton.

7

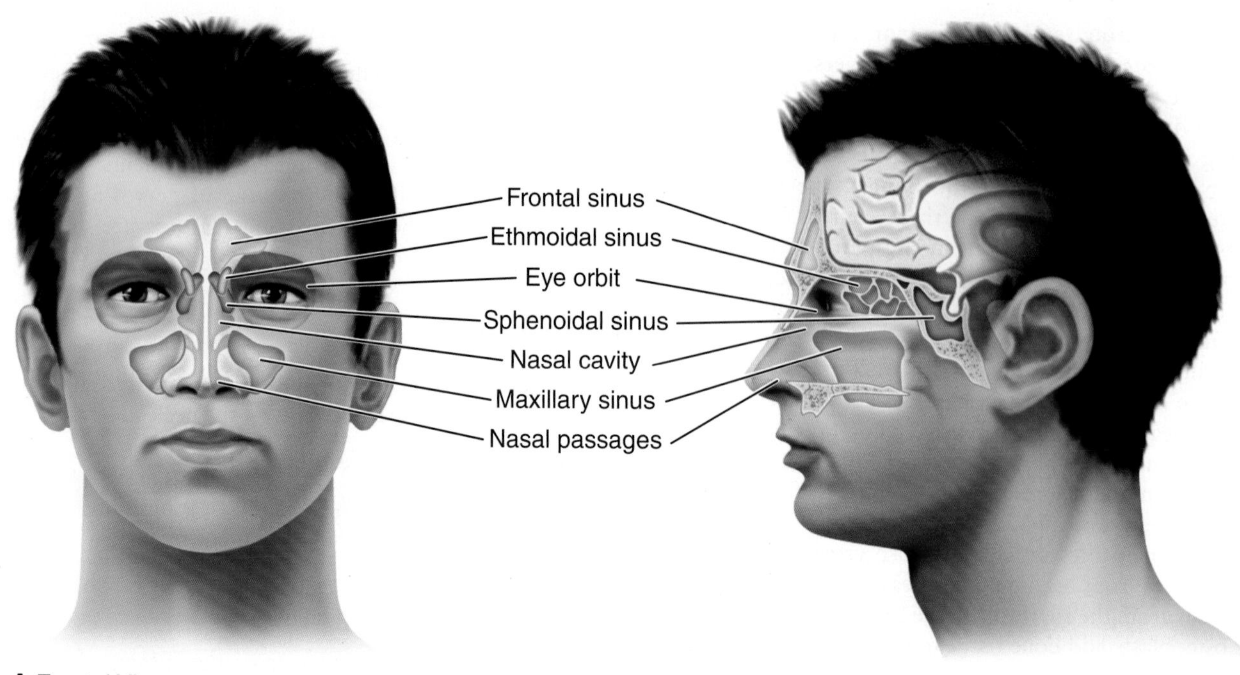

Figure 7-4 **Sinuses.** **KEY POINT** A sinus is a cavity or hollow space, such as the air-filled chambers in certain skull bones. View of the skull showing the sinuses from frontal **(A)** and lateral **(B)** aspects.

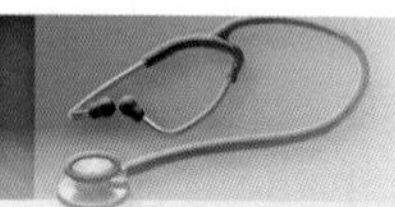

CLINICAL PERSPECTIVES

Box 7-1

Landmarking: Seeing with Your Fingers

Most body structures lie beneath the skin, hidden from direct view except in dissection. A technique called **landmarking** allows healthcare providers to locate hidden structures simply and easily. Bony prominences, or landmarks, can be palpated (felt) beneath the skin to serve as reference points for locating other internal structures. Landmarking is used during physical examinations and surgeries, when giving injections, and for many other clinical procedures. The lower tip of the sternum, the xiphoid process, is a reference point in the administration of cardiopulmonary resuscitation (CPR).

Practice landmarking by feeling for some of the other bony prominences. You can feel the joint between the mandible and the temporal bone of the skull (the temporomandibular joint, or TMJ) anterior to the ear canal as you move your lower jaw up and down. Feel for the notch in the sternum (breast bone) between the clavicles (collar bones). Approximately 4 cm below this notch you will feel a bump called the sternal angle. This prominence is an important landmark because its location marks where the trachea splits to deliver air to both lungs. Move your fingers lateral to the sternal angle to palpate the second ribs, important landmarks for locating the heart and lungs. Feel for the most lateral bony prominence of the shoulder, the acromion process of the scapula (shoulder blade). Two to three fingerbreadths down from this point is the correct injection site into the deltoid muscle of the shoulder. Place your hands on your hips and palpate the iliac crest of the hip bone. Move your hands forward until you reach the anterior end of the crest, the anterior superior iliac spine (ASIS). Feel for the part of the bony pelvis that you sit on. This is the ischial tuberosity. This and the ASIS are important landmarks for locating safe injection sites in the gluteal region.

Bones of the Axial Skeleton

As noted earlier, the skeleton may be divided into two main groups of bones (see Fig. 7-1):

- The **axial** (AK-se-al) **skeleton** consists of 80 bones and includes the bony framework of the head and the trunk. Think of the axial skeleton as the body's "axis."
- The **appendicular** (ap-en-DIK-u-lar) **skeleton** consists of 126 bones and forms the framework for the **extremities** (limbs) and for the shoulders and hips. Think of the appendicular skeleton as the body's "appendages."

We describe the axial skeleton first and then proceed to the appendicular skeleton. A table at the end of this section summarizes all of the bones described. Also see Appendix 5: Dissection Atlas, **Figures A5-1 and A5-2**, for pictures of a complete human skeleton, both anterior and posterior views. You can refer to these pictures as you study this chapter.

FRAMEWORK OF THE SKULL

The bony framework of the head, called the **skull**, is subdivided into two parts: the cranium and the facial portion. Refer to **Figure 7-5**, which shows different views of the skull, as you study the following descriptions. The individual bones are color coded to help you identify them as you study the skull in different views.

Note the many features of the skull bones as you examine these illustrations. For example, openings in the base of the skull provide spaces for the entrance and exit of many blood vessels, nerves, and other structures. Bone projections and fossae (depressions) provide for muscle attachment. Some portions protect delicate structures, for example, the eye orbit (socket) and the part of the temporal bone at the lateral skull that encloses the inner ear. The sinuses provide lightness and serve as resonating chambers for the voice (which is why your voice sounds better to you as you are speaking than it sounds when you hear it played back as a recording).

Cranium This rounded chamber that encloses the brain is composed of eight distinct cranial bones.

- The **frontal bone** forms the forehead, the anterior of the skull's roof, and the roof of the eye orbit. The **frontal sinuses** communicate with the nasal cavities (**see Fig. 7-4**). These sinuses and others near the nose are described as paranasal sinuses.
- The two **parietal** (pah-RI-eh-tal) bones form most of the top and the side walls of the cranium.
- The two **temporal bones** contribute to the sides and the base of the skull. Each contains one ear canal, eardrum, and the ear's entire middle and inner portions. The **mastoid process** of the temporal bone projects downward immediately behind the outer ear (**see Fig. 7-5B**). It is a place for muscle attachments and contains air cells (spaces) that make up the **mastoid sinus** (not illustrated).
- The **ethmoid** (ETH-moyd) **bone** is a light, fragile bone located between the eyes (**see Fig. 7-5A and C**). It forms a part of the medial wall of the eye orbit, a small portion of the cranial floor, and most of the nasal cavity roof. It also forms the superior and middle nasal conchae (KON-ke), bony plates that extend into the nasal cavity (the name *concha* means "shell"). The mucous membranes covering the conchae help filter, warm, and moisten air as it passes through the nose. The ethmoid houses several air cells, comprising some of the paranasal sinuses. A thin, platelike, downward extension of this bone (the perpendicular plate) forms much of the **nasal septum**, the midline partition in the nose (**see Fig. 7-5A**).
- The **sphenoid** (SFE-noyd) **bone**, when seen from a superior view, resembles a bat with its wings extended (**see Fig. 7-5D**). It lies at the base of the skull anterior to the temporal bones and forms part of the eye orbit. It contains the sphenoid sinuses. It also contains a depression called the **sella turcica** (SEL-ah TUR-sih-ka), literally "Turkish saddle," that holds and protects the pituitary gland like a saddle.
- The **occipital** (ok-SIP-ih-tal) **bone** forms the skull's posterior portion and a part of its base. The **foramen magnum**, located at the base of the occipital bone, is a large opening through which the spinal cord attaches to the brain (**see Fig. 7-5C and D**).

Uniting the skull bones is a type of flat, immovable joint known as a **suture** (SU-chur) (**see Fig. 7-5B**). Some of the most prominent cranial sutures are as follows:

- The coronal (ko-RO-nal) suture joins the frontal bone with the two parietal bones along the coronal plane.
- The squamous (SKWA-mus) suture joins the temporal bone to the parietal bone on the cranium's lateral surface (named because it is in a flat portion of the skull).
- The lambdoid (LAM-doyd) suture joins the occipital bone with the parietal bones in the posterior cranium (named because it resembles the Greek letter lambda).
- The sagittal (SAJ-ih-tal) suture joins the two parietal bones along the superior midline of the cranium, along the sagittal plane. Although this suture is not visible in **Figure 7-5B**, you can feel it if you press your fingertips along the top center of your skull.

Facial Bones The facial portion of the skull is composed of 14 bones (**see Fig. 7-5A**):

- The **mandible** (MAN-dih-bl), or lower jaw bone, is the skull's only movable bone.
- The two **maxillae** (mak-SIL-e) fuse in the midline to form the upper jaw bone, including the anterior part of the hard palate (roof of the mouth). Each maxilla contains a large air space, called the **maxillary sinus**, that communicates with the nasal cavity.
- The two **zygomatic** (zi-go-MAT-ik) **bones**, one on each side, form the prominences of the cheeks. The zygomatic forms an arch over the cheek with a process of the temporal bone (**see Fig. 7-5B**).
- Two slender **nasal bones** lie side by side, forming the bridge of the nose.
- The two **lacrimal** (LAK-rih-mal) **bones**, each about the size of a fingernail, form the anterior medial wall of each orbital cavity.
- The **vomer** (VO-mer), shaped like the blade of a plow, forms the inferior part of the nasal septum (**see Fig. 7-5A**).
- The paired **palatine** (PAL-ah-tine) **bones** form the posterior part of the hard palate (**see Fig. 7-5C**).
- The two **inferior nasal conchae** (KON-ke) extend horizontally along the lateral wall (side) of the nasal cavities. (As noted, the paired superior and middle conchae are part of the ethmoid bone, as shown in **Fig. 7-5C**.)

In addition to the cranial and facial bones, there are three tiny bones, or **ossicles** (OS-sik-ls), in each middle ear

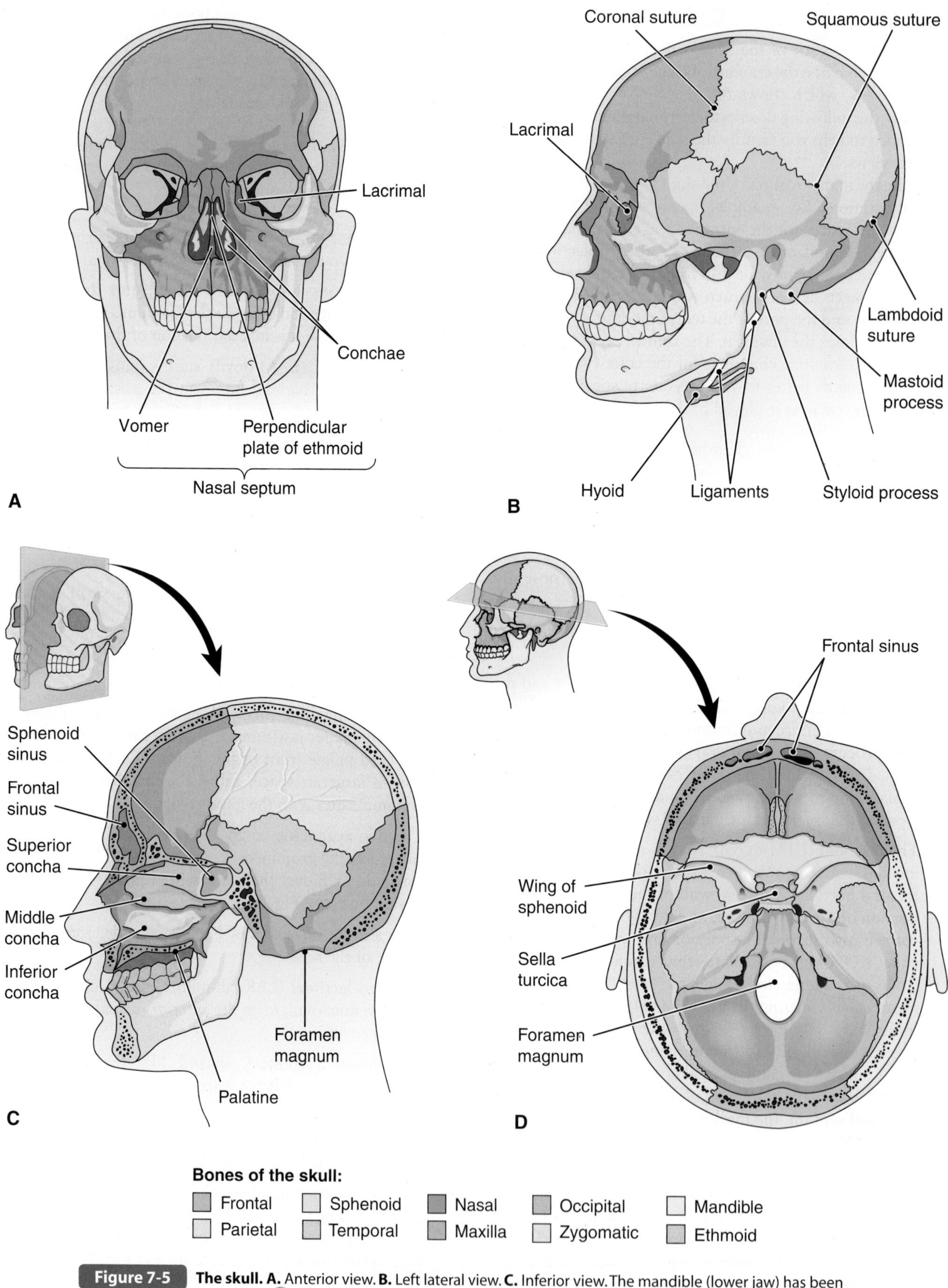

Figure 7-5 **The skull. A.** Anterior view. **B.** Left lateral view. **C.** Inferior view. The mandible (lower jaw) has been removed. **D.** Sagittal section. **ZOOMING IN** What two bones make up each side of the hard palate? What is a foramen? What bone makes up the superior and middle conchae?

(see Chapter 11), and just below the mandible (lower jaw), a single horseshoe, or U-shaped, bone called the **hyoid** (HI-oyd) **bone,** to which the tongue and other muscles are attached (**see Fig. 7-5B**).

Infant Skull The infant's skull has areas in which the bone formation is incomplete, leaving membranous "soft spots," properly called **fontanels** (fon-tah-NELS) (also spelled *fontanelles*) (**Fig. 7-6**). These flexible regions allow the skull to compress and change shape during the birth process. They also allow for rapid brain growth during infancy. Although there are a number of fontanels, named for their location or the bones they border, the largest and most recognizable is near the front of the skull at the junction of the two parietal bones and the frontal bone. This anterior fontanel usually does not close until the child is about 18 months old.

See the Student Resources on thePoint for a summary table of the cranial and facial bones and figures on the skeletal features of the head and neck.

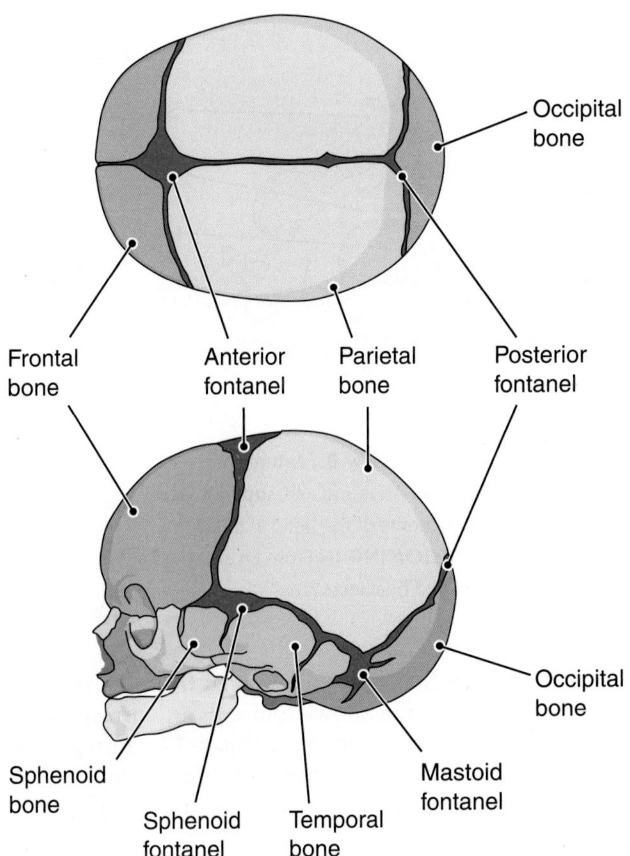

Figure 7-6 **Infant skull, showing fontanels.** **KEY POINT** Fibrous membranes between the skull bones allow the skull to compress during childbirth. Sutures later form in these areas. **ZOOMING IN** Which is the largest fontanel?

FRAMEWORK OF THE TRUNK

The bones of the trunk include the spine, or **vertebral** (VER-teh-bral) **column**, and the bones of the chest, or **thorax** (THO-raks).

Vertebral Column This bony sheath for the spinal cord is made of a series of irregularly shaped bones. These number 33 or 34 in the child, but because of fusions that occur later in the lower part of the spine, there usually are just 26 separate bones in the adult spinal column. **Figure 7-7A** shows a lateral view of the vertebral column.

Each **vertebra** (VER-teh-brah) (aside from the first two) has a drum-shaped **body** located anteriorly (toward the front) that serves as the weight-bearing part; disks of cartilage between the vertebral bodies absorb shock and provide flexibility (**Fig. 7-7A and B**). In the center of each vertebra is a large hole, the vertebral foramen. When all the vertebrae are linked in series by strong connective tissue bands (ligaments), these spaces form the spinal canal, a bony cylinder that protects the spinal cord. Projecting posteriorly (toward the back) from the bony arch that encircles the spinal cord is the **spinous process**, which usually can be felt just under the skin of the back. Projecting laterally is a **transverse process** on each side. These processes are attachment points for muscles. Other processes form joints with adjacent vertebrae. A lateral view of the vertebral column shows a series of **intervertebral foramina**, formed between the vertebrae as they join together. Spinal nerves emerge from the spinal cord through these openings (**see Fig. 7-7A**).

The bones of the vertebral column are named and numbered from superior to inferior and according to location. There are five groups:

- The **cervical** (SER-vih-kal) **vertebrae**, seven in number (C1 to C7), are located in the neck. The first vertebra, called the **atlas**, supports the head (**Fig. 7-7C**). (This vertebra is named for the mythologic character who was able to support the world in his hands.) When you nod your head, the skull rocks on the atlas at the occipital bone. The second cervical vertebra, the **axis** (**Fig. 7-7C**), serves as a pivot when you turn your head from side to side. It has an upright toothlike part, the dens, that projects into the atlas as a pivot point. The absence of a body in these vertebrae allows for the extra movement. Only the cervical vertebrae have a hole in the transverse process on each side (**see Fig. 7-7B and C**). These **transverse foramina** accommodate blood vessels and nerves that supply the neck and head.
- The **thoracic vertebrae**, 12 in number (T1 to T12), are located in the chest. They are larger and stronger than the cervical vertebrae and have a longer spinous process that points downward (**see Fig. 7-7B**). The posterior ends of the 12 pairs of ribs are attached to the transverse processes of these vertebrae.
- The **lumbar vertebrae**, five in number (L1 to L5), are located in the small of the back. They are larger and heavier than the vertebrae superior to them and can

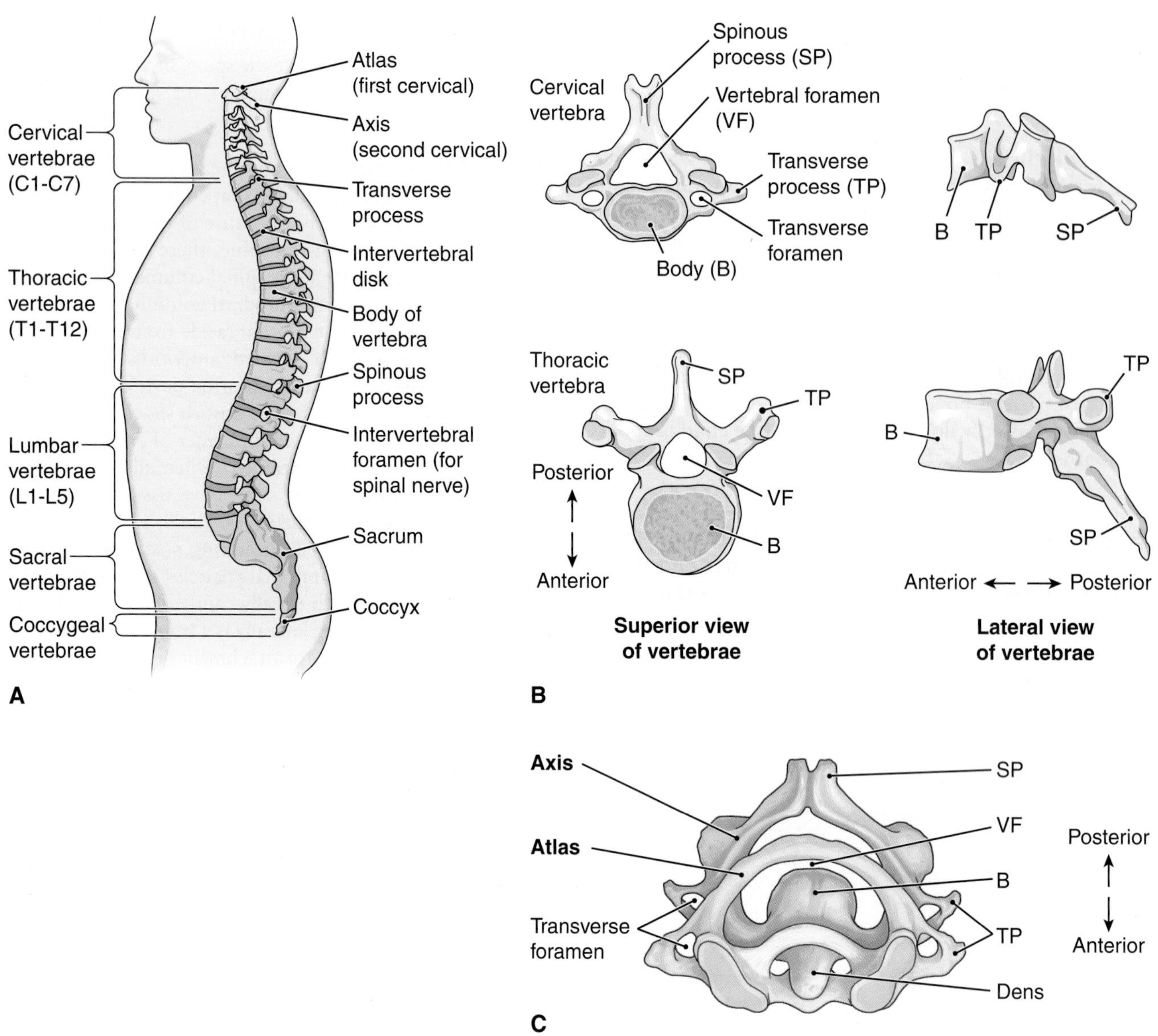

Figure 7-7 **The vertebral column and vertebrae. A.** Vertebral column, left lateral view. **B.** Features of the vertebrae. The blue areas on the vertebrae show points of contact with other bones. **C.** Atlas and axis, superior view. The first two cervical vertebrae are adapted to support the skull and allow for movements of the head in different directions. **KEY POINT** The adult spine has five regions and four curves. **ZOOMING IN** From an anterior view, which group(s) of vertebrae form a convex curve? Which vertebrae are the largest and heaviest? Why?

support more weight (**see Fig. 7-7A**). All of their processes are shorter and thicker.

- The **sacral** (SA-kral) **vertebrae** are five separate bones in the child. They eventually fuse to form a single bone, called the **sacrum** (SA-krum), in the adult. Wedged between the two hip bones, the sacrum completes the posterior part of the bony pelvis.
- The **coccygeal** (kok-SIJ-e-al) **vertebrae** consist of four or five tiny bones in the child. These later fuse to form a single bone, the **coccyx** (KOK-siks), or tail bone, in the adult.

Curves of the Spine When viewed from the side, the adult vertebral column shows four curves, corresponding to the four vertebral groups (**see Fig. 7-7A**). In the fetus, the entire column is concave forward (like a letter "C" and your spine when you assume a "fetal position"). This is the primary curve.

When an infant begins to assume an erect posture, secondary curves develop. The cervical curve is convex and appears as the baby holds its head up at about 3 months of age. The lumbar curve is also convex and appears when the child begins to walk. The thoracic and sacral curves remain the two primary concave curves. These curves of the vertebral

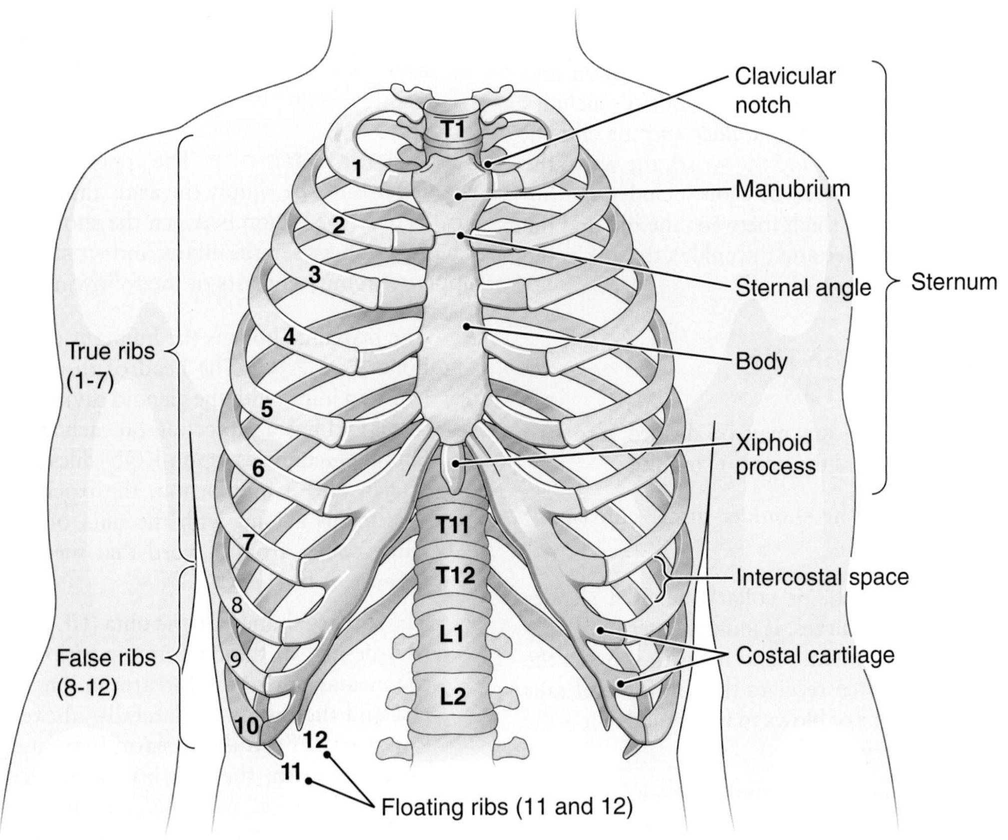

Figure 7-8 **Bones of the thorax, anterior view.** **KEY POINT** The first seven pairs of ribs are the true ribs; pairs 8 through 12 are the false ribs, of which the last two pairs are also called floating ribs. **ZOOMING IN** To what bones do the costal cartilages attach?

column provide some of the resilience and spring so essential in balance and movement.

Thorax The bones of the thorax form a cone-shaped cage (**Fig. 7-8**). Twelve pairs of **ribs** form the bars of this cage, completed anteriorly by the **sternum** (STER-num), or breastbone. These bones enclose and protect the heart, lungs, and other organs contained in the thorax.

The superior portion of the sternum is a roughly triangular **manubrium** (mah-NU-bre-um) that joins laterally on the right and left with a clavicle (collarbone). (The name manubrium comes from a Latin word meaning "handle.") The point on the manubrium where the clavicle joins can be seen on **Figure 7-8** is the clavicular notch. Laterally and inferiorly, the manubrium joins with the anterior ends of the first pair of ribs. The sternum's body is long and bladelike. It joins along each side with ribs two through seven. Where the manubrium joins the body of the sternum, there is a slight elevation, the **sternal angle**, which easily can be felt as a surface landmark.

The inferior end of the sternum consists of a small tip that is made of cartilage in youth but becomes bone in the adult. This is the **xiphoid** (ZIF-oyd) **process**. It is used as a landmark for cardiopulmonary resuscitation (CPR) to locate the region for chest compression.

All 12 ribs on each side are attached to the vertebral column posteriorly. However, variations in the anterior attachment of these slender, curved bones have led to the following classification:

- **True ribs**, the first seven pairs, are those that attach directly to the sternum by means of individual extensions called *costal* (KOS-tal) *cartilages*.
- **False ribs** are the remaining five pairs. Of these, the eighth, ninth, and 10th pairs attach to the cartilage of the rib above. The last two pairs have no anterior attachment at all and are known as **floating ribs**.

The spaces between the ribs, called *intercostal spaces*, contain muscles, blood vessels, and nerves.

CHECKPOINTS

- 7-7 What bones make up the skeleton of the trunk?
- 7-8 What are the five regions of the vertebral column?

Bones of the Appendicular Skeleton

The appendicular skeleton includes an upper division and a lower division. The upper division on each side includes the shoulder, the arm (between the shoulder and the elbow), the forearm (between the elbow and the wrist), the wrist, the hand, and the fingers. The lower division includes the hip (part of the pelvic girdle), the thigh (between the hip and the knee), the leg (between the knee and the ankle), the ankle, the foot, and the toes.

THE UPPER DIVISION OF THE APPENDICULAR SKELETON

The bones of the upper division may be divided into two groups, the shoulder girdle and the upper extremity.

The Shoulder Girdle The shoulder girdle consists of two bones (**Fig. 7-9**).

- The **clavicle** (KLAV-ih-kl), or collarbone, is a slender bone with two shallow curves. It joins the sternum anteriorly and the scapula laterally and helps to support the shoulder. Because it often receives the full force of falls on outstretched arms or of blows to the shoulder, it is the most frequently broken bone.
- The **scapula** (SKAP-u-lah), or shoulder blade, is shown from anterior and posterior views in **Figure 7-9**. The spine of the scapula is the posterior raised ridge that can be felt behind the shoulder in the upper portion of the back. Muscles that move the arm attach to fossae (depressions), known as the **supraspinous fossa** and the **infraspinous fossa,** superior and inferior to the scapular spine. The **acromion** (ah-KRO-me-on) is the process that joins the clavicle. You can feel this as the highest point of your shoulder. Below the acromion, there is a shallow socket, the **glenoid cavity**, that forms a ball-and-socket joint with the arm bone (humerus). Medial to the glenoid cavity is the **coracoid** (KOR-ah-koyd) **process**, to which arm and chest muscles and ligaments attach.

The Upper Extremity The upper extremity is also called the upper limb, or simply the arm, although technically, the arm is only the region between the shoulder and the elbow. The region between the elbow and wrist is the forearm. The upper extremity consists of the following bones:

- The proximal bone is the **humerus** (HU-mer-us), or arm bone (**Fig. 7-10**). The head of the humerus articulates (forms a joint) with the glenoid cavity of the scapula. The distal end has a projection on each side, the medial and lateral **epicondyles** (ep-ih-KON-diles), to which tendons attach, and a midportion, the **trochlea** (TROK-le-ah), that forms a joint with the ulna of the forearm. (The name comes from a word that means "pulley wheel" because of its shape.)
- The forearm bones are the **ulna** (UL-nah) and the **radius** (RA-de-us). In the anatomic position, the ulna lies on the medial side of the forearm in line with the little finger, and the radius lies laterally, above the thumb (**see Fig. 7-10**). When the forearm is supine, with the palm up or forward, the two bones are parallel; when the forearm is prone, with the palm down or back, the distal end of the radius rotates around the ulna so that the shafts of the two bones are crossed (**Fig. 7-11**). In this position, a distal projection (styloid process) of the ulna shows at the outside of the wrist.

The proximal end of the ulna has the large **olecranon** (o-LEK-rah-non), a process that forms the point of the elbow (**Fig. 7-12**). At the posterior elbow joint, the olecranon fits into a depression

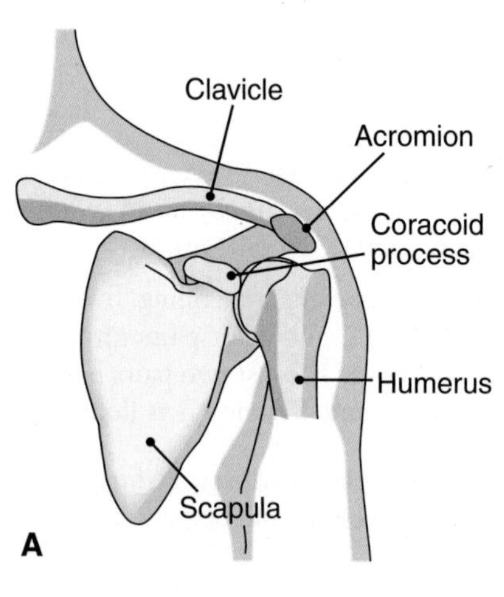

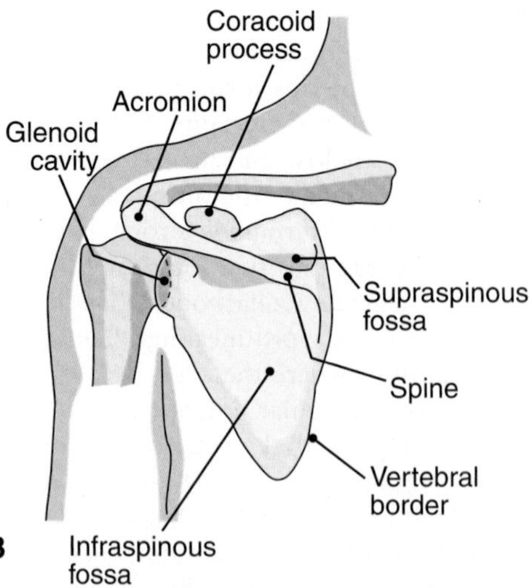

Figure 7-9 **The shoulder girdle.** **KEY POINT** The shoulder girdle consists of the clavicle and scapula. **A.** Bones of the left shoulder girdle, anterior view. **B.** Bones of the left shoulder girdle, posterior view. **ZOOMING IN** What does the prefix *supra-* mean? What does the prefix *infra-* mean?

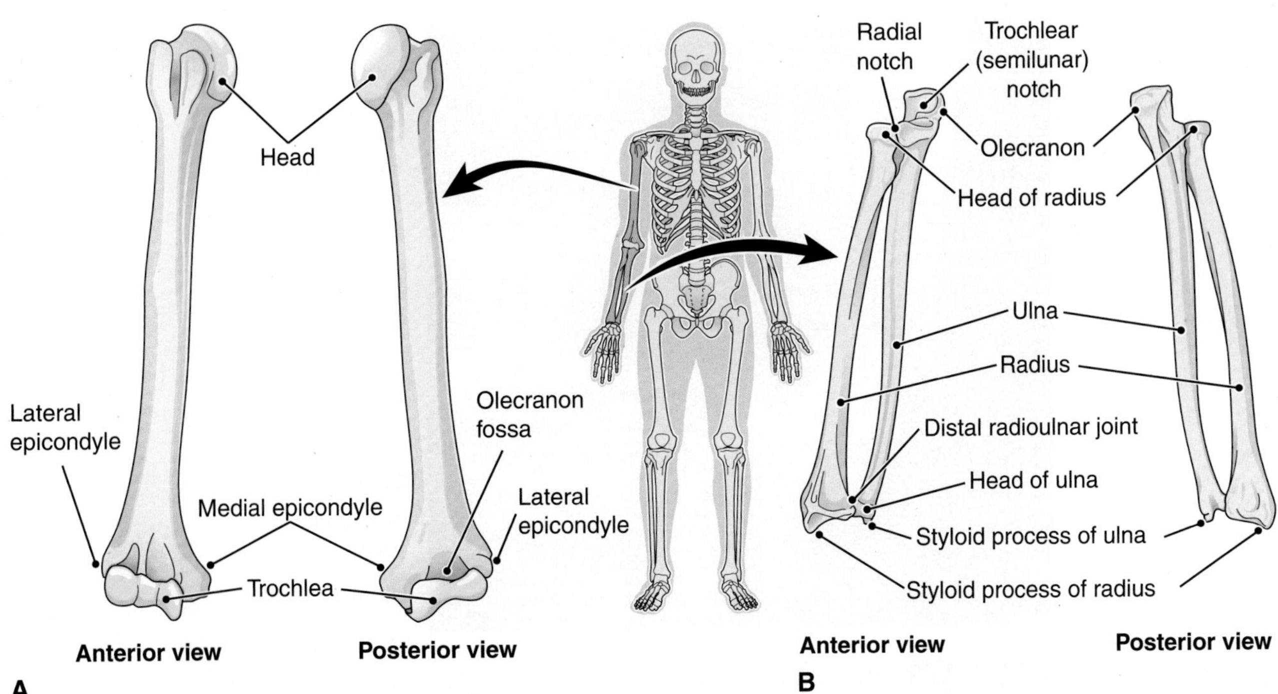

Figure 7-10 **Bones of the upper extremity.** **KEY POINT** The upper extremity consists of the arm and forearm. **A.** The humerus of the right arm in anterior and posterior view. **B.** The radius and ulna of the right forearm in anterior and posterior view. **ZOOMING IN** What is the medial bone of the forearm?

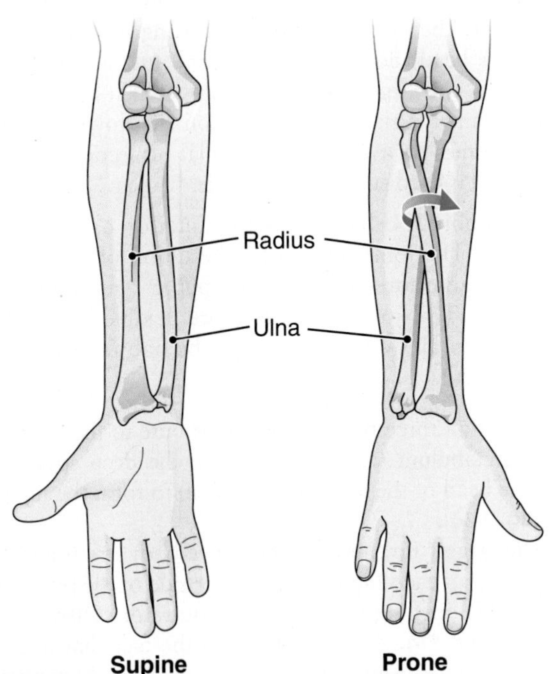

Figure 7-11 **Movements of the forearm.** **KEY POINT** When the palm is supine (facing up or forward), the radius and ulna are parallel. When the palm is prone (facing down or to the rear), the radius crosses over the ulna.

of the distal humerus, the **olecranon fossa**. The trochlea of the distal humerus fits into the ulna's deep **trochlear notch**, allowing a hinge action at the elbow joint. This ulnar depression, because of its deep half-moon shape, is also known as the semilunar notch (**see Fig. 7-12**).

- The wrist contains eight small **carpal** (KAR-pal) **bones** arranged in two rows of four each. The names of these eight different bones are given in **Figure 7-13**. Note that the anatomic wrist, composed of the carpal bones, is actually the heel of the hand. We wear a "wristwatch" over the distal ends of the radius and ulna.
- Five **metacarpal bones** are the framework for the palm of each hand. Their rounded distal ends form the knuckles.
- There are 14 **phalanges** (fah-LAN-jeze), or finger bones, in each hand, two for the thumb and three for each finger. Each of these bones is called a **phalanx** (FA-lanx). They are identified as the proximal, which is attached to a metacarpal; the middle; and the distal. Note that the thumb has only two phalanges, a proximal and a distal (**see Fig. 7-13**).

See the Student Resources on thePoint for figures on the skeletal features of the shoulder and arm.

THE LOWER DIVISION OF THE APPENDICULAR SKELETON

The bones of the lower division also fall into two groups, the pelvis and the lower extremity.

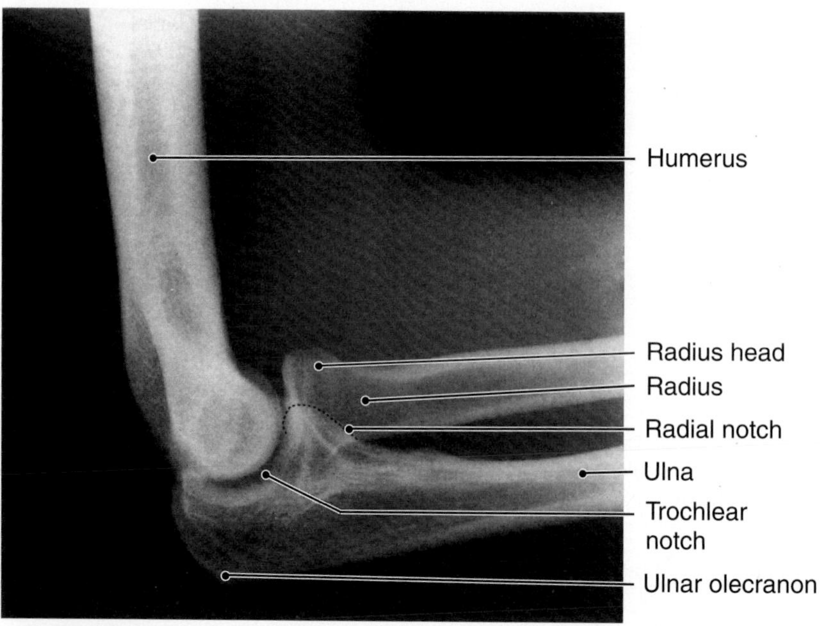

Figure 7-12 **Left elbow, lateral view.** **ZOOMING IN** What part of what bone forms the bony prominence of the elbow?

The Pelvic Bones The hip bone, or **os coxae**, begins its development as three separate bones that later fuse (**Fig. 7-14**). These individual bones are the following:

- The **ilium** (IL-e-um) forms the upper, flared portion. The **iliac** (IL-e-ak) **crest** is the curved rim along the ilium's superior border. It can be felt just below the waist. At either end of the crest are two bony projections. The most prominent of these is the **anterior superior iliac spine**, which is often used as a surface landmark in diagnosis and treatment.
- The **ischium** (IS-ke-um) is the lowest and strongest part. The **ischial** (IS-ke-al) **spine** at the posterior of the pelvic outlet is used as a reference point during childbirth to indicate the progress of the presenting part (usually the baby's head) down the birth canal. Just inferior to this spine is the large **ischial tuberosity**, which helps support the trunk's weight when a person sits down. You may sometimes be aware of this ischial projection when sitting on a hard surface for a while.
- The **pubis** (PU-bis) forms the anterior part of the os coxae. The joint formed by the union of the two hip bones anteriorly is called the **pubic symphysis** (SIM-fih-sis). This joint becomes more flexible late in pregnancy to allow for passage of the baby's head during childbirth.

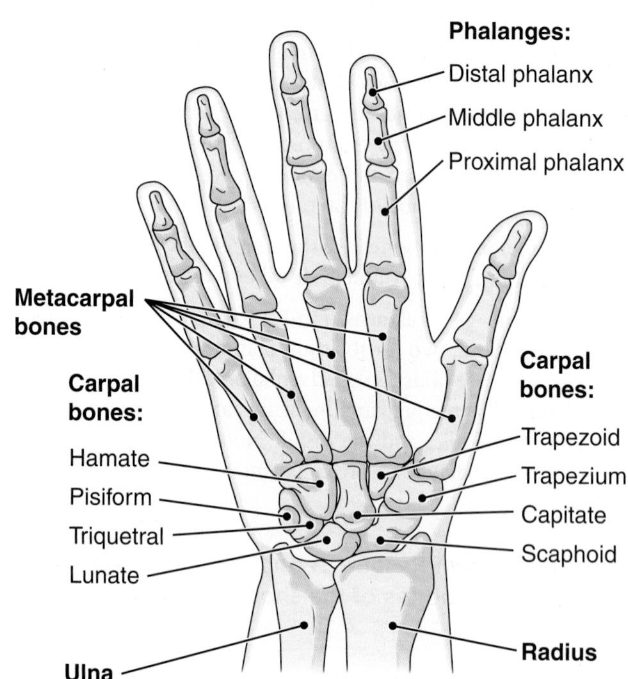

Figure 7-13 **Bones of the right hand, anterior view.** **ZOOMING IN** How many phalanges are there on each hand?

Portions of all three pelvic bones contribute to the formation of the **acetabulum** (as-eh-TAB-u-lum), the deep socket that holds the head of the femur (thigh bone) to form the hip joint (**see Fig. 7-14**).

The largest foramina in the entire body are found near the anterior of each hip bone on either side of the pubic symphysis. This opening is named the **obturator** (OB-tu-ra-tor) **foramen** (**see Fig. 7-14**), referring to the fact that it is partially closed by a membrane and has only a small opening for passage of blood vessels and a nerve.

The two ossa coxae join in forming the pelvis, a strong bony girdle completed posteriorly by the sacrum and coccyx of the spine. The pelvis supports the trunk and surrounds the

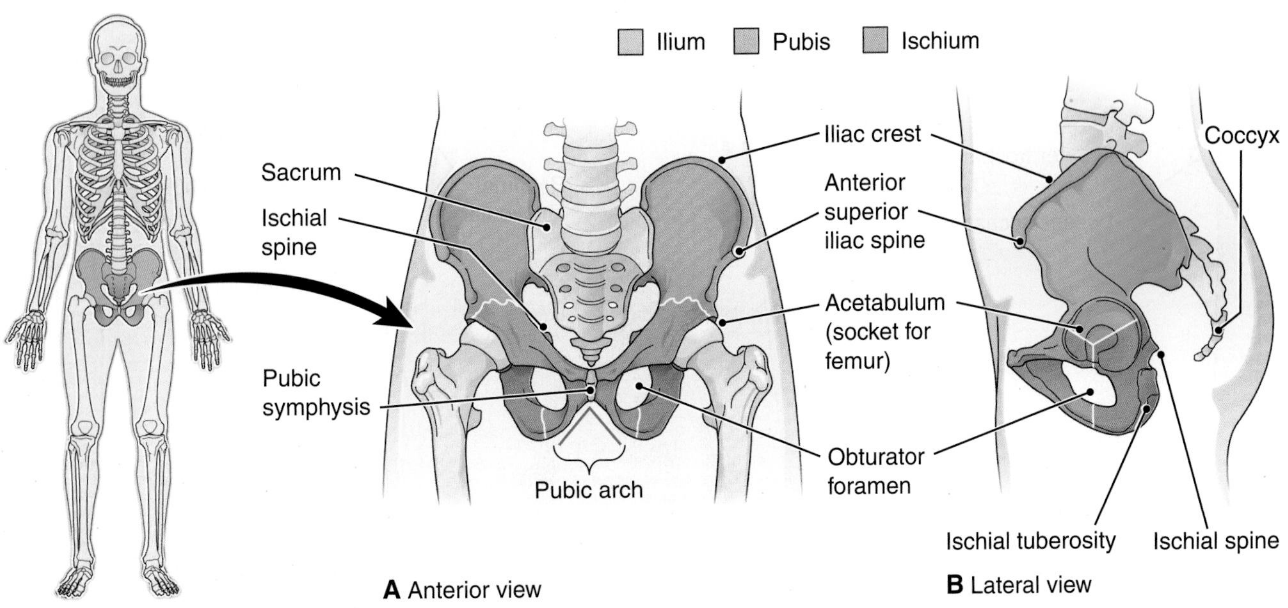

Figure 7-14 **The pelvic bones.** **KEY POINT** The hip bone, or os coxae, is formed of three fused bones. **A.** Anterior view. **B.** Lateral view showing the joining of the three pelvic bones to form the acetabulum. **ZOOMING IN** What bone is nicknamed the "sit bone?"

organs in the pelvic cavity, including the urinary bladder, the internal reproductive organs, and parts of the intestine.

The female pelvis is adapted for pregnancy and childbirth (**Fig. 7-15**). Some ways in which the female pelvis differs from that of the male are as follows:

- It is lighter in weight.
- The ilia are wider and more flared.
- The pubic arch, the anterior angle between the pubic bones, is wider.
- The pelvic inlet, the upper opening, bordered by the pubic joint and sacrum, is wider and more rounded.
- The pelvic outlet, the lower opening, bordered by the pubic joint and coccyx, is larger.
- The sacrum and coccyx are shorter and less curved.

The Lower Extremity The lower extremity is also called the lower limb, or simply the leg, although technically the leg is only the region between the knee and the ankle. The portion of the extremity between the hip and

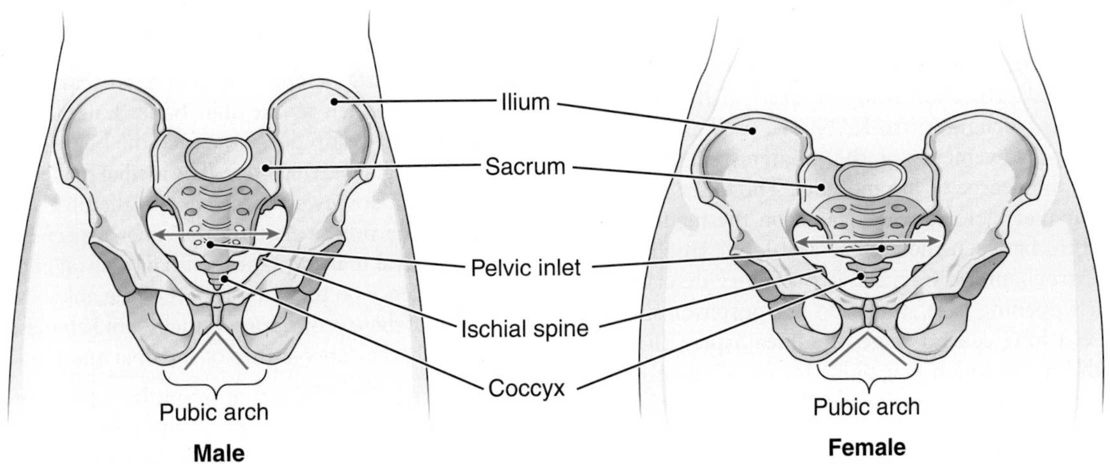

Figure 7-15 **Comparison of male and female pelvis, anterior view.** **KEY POINT** The female pelvis is adapted for pregnancy and childbirth. Note the broader angle of the pubic arch and the wider pelvic outlet in the female. Also, the ilia are wider and more flared; the sacrum and coccyx are shorter and less curved.

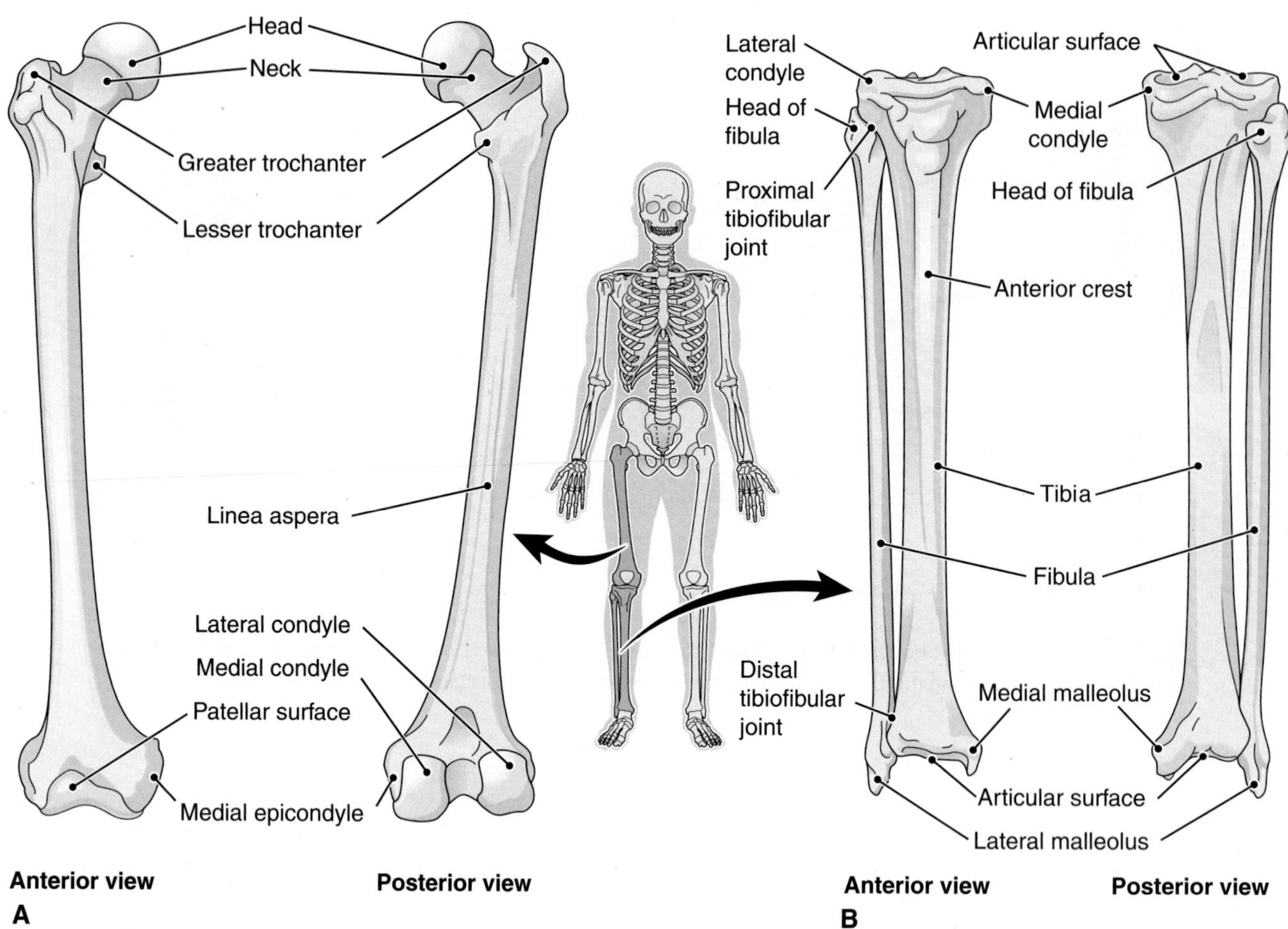

Figure 7-16 **Bones of the lower extremity.** **KEY POINT** The lower extremity consists of the thigh and leg. **A.** The femur of the right thigh. **B.** The tibia and fibula of the right leg. **ZOOMING IN** What is the lateral bone of the leg? Which bone of the leg is weight bearing?

the knee is the thigh. The lower extremity consists of the following bones:

- The **femur** (FE-mer), the thigh bone, is the longest and strongest bone in the body. Proximally, it has a large ball-shaped head that joins the os coxae (**Fig. 7-16**). The large lateral projection near the head of the femur is the **greater trochanter** (tro-KAN-ter), used as a surface landmark. Movements of the greater trochanter can indicate the degree of hip mobility. The **lesser trochanter**, a smaller elevation, is located on the medial side. A fracture between the greater and lesser trochanters, or intertrochanteric fracture, is the injury described in Reggie's opening case study. On the posterior surface, there is a long central ridge, the **linea aspera** (literally "rough line"), which is a point for attachment of hip muscles. The distal anterior **patellar surface** articulates with the kneecap.
- The **patella** (pah-TEL-lah), or kneecap (**see Fig. 7-1**), is embedded in the tendon of the large anterior thigh muscle, the quadriceps femoris, where it crosses the knee joint. It is an example of a **sesamoid** (SES-ah-moyd) **bone**, a type of bone that develops within a tendon or a joint capsule.
- There are two bones in the leg (**see Fig. 7-16**). Medially (on the great toe side), the **tibia**, or shin bone, is the longer, weight-bearing bone. Its proximal surface articulates with the distal femur. The tibia has a sharp anterior crest that can be felt as the shin bone. Laterally, the slender **fibula** (FIB-u-lah) does not reach the knee joint; thus, it is not a weight-bearing bone. The **medial malleolus** (mal-LE-o-lus) is a downward projection at the tibia's distal end; it forms the prominence on the inner aspect of the ankle. The **lateral malleolus**, at the fibula's distal end, forms the prominence on the outer aspect of the ankle. Most people think of these projections as their "ankle bones," whereas, in truth, they are features of the tibia and fibula.
- The structure of the foot is similar to that of the hand. However, the foot supports the body's weight, so it is stronger and less mobile than the hand. There are seven **tarsal bones** associated with the ankle and foot. These are named and illustrated in **Figure 7-17**. The largest of these is the **calcaneus** (kal-KA-ne-us), or heel bone. The **talus** above it forms the ankle joint with the tibia.

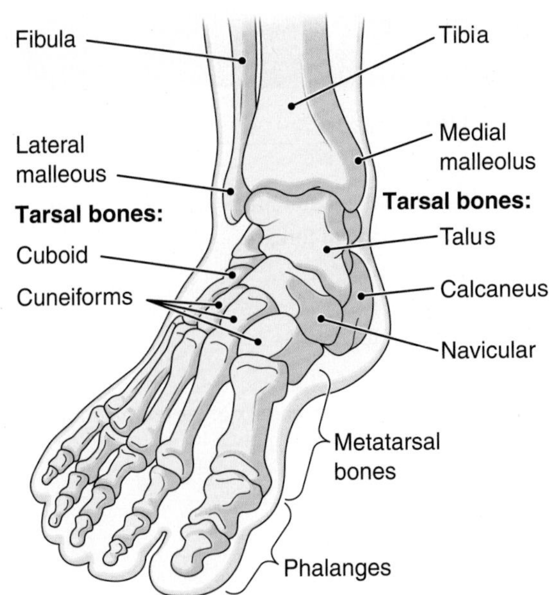

Figure 7-17 **Bones of the right foot.** **ZOOMING IN** Which tarsal bone is the heel bone? Which tarsal bone forms a joint with the tibia?

- Five **metatarsal bones** form the framework of the instep, and the heads of these bones form the ball of the foot (**see Fig. 7-17**).
- The phalanges of the toes are counterparts of those in the fingers. There are three of these in each toe except for the great toe, which has only two.

See **Table 7-1** for a summary outline of all the bones of the skeleton.

CHECKPOINT

☐ **7-9** What are the four regions of the appendicular skeleton?

See the Student Resources on thePoint for figures on the skeletal features of the leg and foot.

Disorders of Bone

Bone disorders include metabolic diseases, in which there is a lack of normal bone formation or excess loss of bone tissue; tumors; infections; structural problems; and fractures. Many bone disorders can be diagnosed by x-ray studies.

See the Student Resources on thePoint for information on careers in radiology.

METABOLIC DISORDERS

Recall that osteoblasts build bone tissue, and osteoclasts break down bone tissue. Metabolic disorders of bone can result from excess or insufficient activity of these two cell types, or from a deficiency in the minerals or proteins making up the bone matrix.

Osteopenia and Osteoporosis **Osteoporosis** (os-te-o-po-RO-sis) is a disorder of the skeletal system associated with porous, fragile bones (**Fig. 7-18**) and a significant loss of bone mass. Osteoporotic bones are prone to fracture, particularly weight-bearing bones such as the spine, pelvic girdle, and long bones. A mild to moderate reduction in bone mass below average levels is described as **osteopenia** (os-te-o-PE-ne-ah).

Table 7-1 Bones of the Skeleton

Region	Bones	Description
Axial Skeleton		
Skull		
Cranium	Cranial bones (8)	Chamber enclosing the brain; houses the ear and forms part of the eye socket
Facial portion	Facial bones (14)	Form the face and chambers for sensory organs
Hyoid		U-shaped bone under lower jaw; used for muscle attachments
Ossicles	Ear bones (3)	Transmit sound waves through middle ear
Trunk		
Vertebral column	Vertebrae (26)	Enclose the spinal cord
Thorax	Sternum	Anterior bone of the thorax
	Ribs (12 pairs)	Enclose the organs of the thorax
Appendicular Skeleton		
Upper division		
Shoulder girdle	Clavicle	Anterior; between sternum and scapula
	Scapula	Posterior; anchors muscles that move arm

(continued)

Table 7-1 Bones of the Skeleton (*continued*)

Region	Bones	Description
Upper extremity	Humerus	Arm bone
	Ulna	Medial bone of forearm
	Radius	Lateral bone of forearm
	Carpals (8)	Wrist bones
	Metacarpals (5)	Bones of palm
	Phalanges (14)	Bones of fingers
Lower division		
Pelvis	Os coxae (2)	Join sacrum and coccyx of vertebral column to form the bony pelvis
Lower extremity	Femur	Thigh bone
	Patella	Kneecap
	Tibia	Medial bone of leg
	Fibula	Lateral bone of leg
	Tarsal bones (7)	Ankle bones
	Metatarsals (5)	Bones of instep
	Phalanges (14)	Bones of toes

The most common type of osteoporosis is postmenopausal osteoporosis, reflecting the decreased production of the female reproductive hormone estrogen. This decline in estrogen levels leads to increased osteoclast activity and thus increased bone breakdown. To make things worse, osteoblast activity declines with age, so the increased bone breakdown is accompanied by decreased bone production.

The treatment of postmenopausal osteopenia and osteoporosis remains an active and controversial area of research. The most obvious approach is to replace the missing estrogen, a treatment known as hormone replacement therapy (HRT). However, long-term HRT increases the risk of breast cancer and cardiovascular problems. Current practice favors the use of short-term (five years) HRT only in younger menopausal women. Researchers are seeking drugs that can selectively activate the estrogen receptors in bones but not in other tissues. Nonhormonal medications are available to reduce bone resorption and even promote the development of new bone tissue.

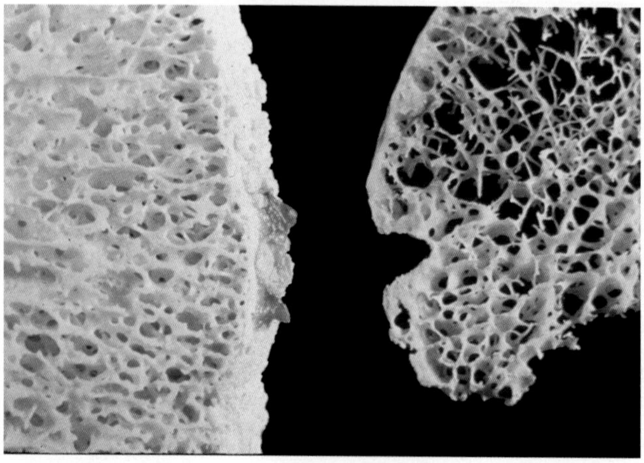

Figure 7-18 Osteoporosis. KEY POINT A section of the vertebral column shows loss of bone tissue and a compression fracture of a vertebral body **(top)**.

While it is impossible to completely prevent age-related osteoporosis, adequate intake of calcium and vitamin D throughout life, especially during childhood and adolescence, can maximize bone density and reduce the risk. Peak bone mineral density is achieved in early adulthood, and every subsequent cycle of bone remodeling results in a small net loss of bone tissue. Weight-bearing exercises, such as weight lifting and brisk walking, are also important to stimulate growth of bone tissue (see Box 7-2). Changes in bone can be followed with radiographic bone mineral density tests to determine possible loss of bone mass.

Whereas postmenopausal osteoporosis is the most common form, other conditions can also lead to a loss of bone mass. Osteoporosis in elderly men can result from the age-related decrease in osteoblast activity. Other causes include disuse, as in paralysis or immobilization in a cast; nutritional deficiencies; and excess corticosteroids (hormones) from the adrenal gland or from medical treatment. Treatments address the cause of the osteoporosis (such as nutritional deficiency) and, if the cause cannot be corrected, stimulate osteoblast activity and inhibit osteoclast activity.

Other Metabolic Diseases of Bone Paget disease of bone, or **osteitis deformans** (os-te-I-tis de-FOR-mans), results from excessive and abnormal osteoclast activity followed by excess osteoblast activity (Fig. 7-19A). The net result is increased bone mass, but the bones are weak, deformed, and easily fractured. The cause of Paget disease remains uncertain, but mutations in genes that regulate osteoclast activity have been detected in some patients.

Box 7-2

HEALTH MAINTENANCE

Three Steps toward a Strong and Healthy Skeleton

The skeleton is the body's framework. It supports and protects internal organs, helps to produce movement, and manufactures blood cells. Bone also stores nearly all of the body's calcium, releasing it into the blood when needed for processes such as nerve transmission, muscle contraction, and blood clotting. Proper nutrition, exercise, and a healthy lifestyle can help the skeleton perform all these essential roles.

A well-balanced diet supplies the nutrients and energy needed for strong, healthy bones. Calcium, phosphorus, and magnesium make up the mineral crystals of bone and confer strength and rigidity. Foods rich in both calcium and phosphorus include dairy products, fish, beans, and leafy green vegetables. When body fluids become too acidic, bone tissue releases calcium and phosphate, and the bone becomes weakened. Both magnesium and potassium help regulate the pH of body fluids, with magnesium also helping bone absorb calcium. Foods rich in magnesium and potassium include beans, potatoes, and leafy green vegetables. Bananas and dairy products are high in potassium.

Protein supplies the amino acids needed to make collagen, which gives bone tissue flexibility. Meat, poultry, fish, eggs, dairy, soy, and nuts are excellent sources of protein.

Vitamin C helps stimulate collagen synthesis, and vitamin D helps the digestive system absorb calcium into the bloodstream, making it available for bone. Most fruits and vegetables are rich in vitamin C. Few foods supply vitamin D. Reliable sources include fatty fish and fortified milk. Liver, butter, and eggs also contain very small amounts.

Like muscle, bone becomes weakened with disuse. Consistent exercise promotes a stronger, denser skeleton by stimulating bone to absorb more calcium and phosphate from the blood, reducing the risk of osteoporosis. A healthy lifestyle also includes avoiding smoking and excessive alcohol consumption, both of which decrease bone calcium and inhibit bone growth. High levels of caffeine in the diet may also rob the skeleton of calcium.

Bones can become decalcified owing to the effect of a parathyroid gland tumor. As mentioned earlier, parathyroid hormone causes the release of calcium from bone.

In **osteomalacia** (os-te-o-mah-LA-she-ah), bone tissue softens due to lack of calcium salt deposition, and osteopenia or osteoporosis can result. Possible causes include vitamin D deficiency, renal disorders, liver disease, and certain intestinal disorders. When osteomalacia occurs in children, the disease is known as **rickets** (Fig. 7-19B). The disorder is usually caused by a deficiency of vitamin D, which is found in only a few foods but can be synthesized by the body using sunlight's UV energy. Rickets was common among children in past centuries who had poor diets and inadequate exposure to sunlight. Rickets affects the bones and their growth plates, causing the skeleton to remain soft and become distorted.

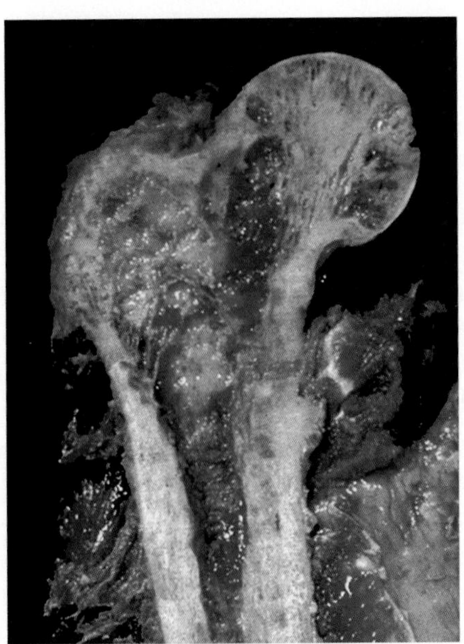

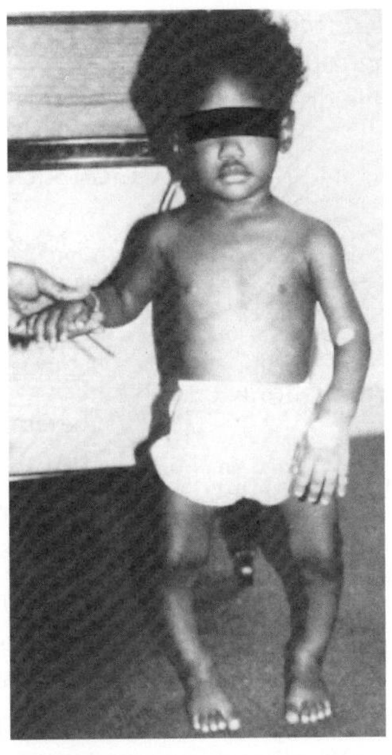

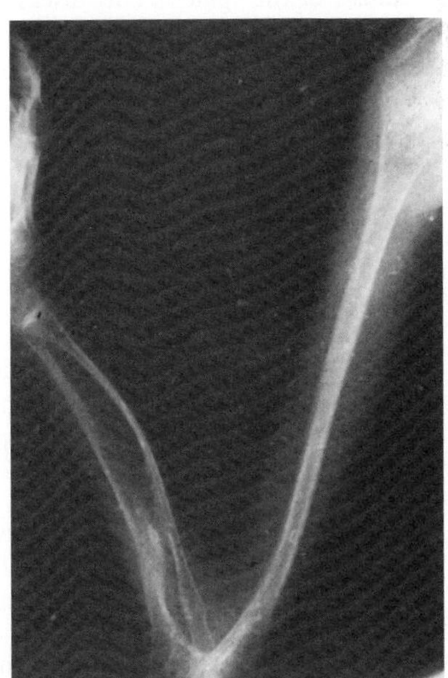

Figure 7-19 **Metabolic bone diseases. A. Paget disease.** A section of the femur shows bone overgrowth in the diaphysis. **B. Rickets. C. Osteogenesis imperfecta.** This x-ray of the upper extremity shows the thin bones and fractures that result from defective collagen production.

7

Osteogenesis imperfecta is a genetic disorder resulting in defective collagen production (**Fig. 7-19C**). Mild cases result in brittle and easily fractured bones, because collagen gives bones resilience. Severe cases are fatal, because the bones are so brittle that they fracture before birth.

TUMORS

Tumors, or neoplasms, that develop in bone tissue may be benign, as is the case with certain cysts, or they may be malignant, as are **osteosarcomas** and **chondrosarcomas**. Osteosarcoma most commonly occurs in a young person in a bone's growing region, especially around the knee. Chondrosarcoma arises in cartilage and usually appears in midlife. In older people, tumors at other sites often metastasize (spread) to bones, most commonly to the spine, causing intense pain.

INFECTION

Osteomyelitis (os-te-o-mi-eh-LI-tis) is an inflammation of bone caused by pyogenic (pi-o-JEN-ik) (pus-producing) bacteria. These organisms may reach the bone through the bloodstream or by way of an injury in which the skin has been broken. The infection may remain localized, or it may spread through the bone to involve the marrow and the periosteum.

Before the advent of antibiotics, bone infections were resistant to treatment, and the prognosis for people with such infections was poor. Now, there are fewer cases because many blood-borne infections are prevented or treated early and do not progress to affect the bones. If those bone infections that do appear are treated promptly, the chance of a cure is usually excellent.

Tuberculosis may spread to bones, especially the long bones of the extremities and the wrist and ankle bones. Tuberculosis of the spine is **Pott disease**. Infected vertebrae are weakened and may collapse, causing pain, deformity, and pressure on the spinal cord. Antibiotics can control the disease if the strains involved are not resistant to the drugs and the host is not weakened by other diseases.

STRUCTURAL DISORDERS

Abnormalities of the spinal curves are known as **curvatures of the spine** (**Fig. 7-20**) and include

- **Kyphosis** (ki-FO-sis), an exaggeration of the thoracic curve, commonly referred to as "hunchback"
- **Lordosis** (lor-DO-sis), an excessive lumbar curve, commonly known as "swayback"
- **Scoliosis** (sko-le-O-sis), a lateral curvature of the vertebral column

Scoliosis is the most common of these disorders. In extreme cases, it may cause compression of internal organs. Scoliosis occurs in the rapid growth period of the teens, more often in girls than in boys. Early discovery and treatment produce good results.

Cleft palate occurs when the maxillary bones do not fuse during embryonic development, resulting in an opening in the roof of the mouth. An infant born with this defect has difficulty nursing because the mouth communicates with the nasal cavity above, and the baby therefore sucks in air rather than milk. Surgery is usually performed to correct the condition.

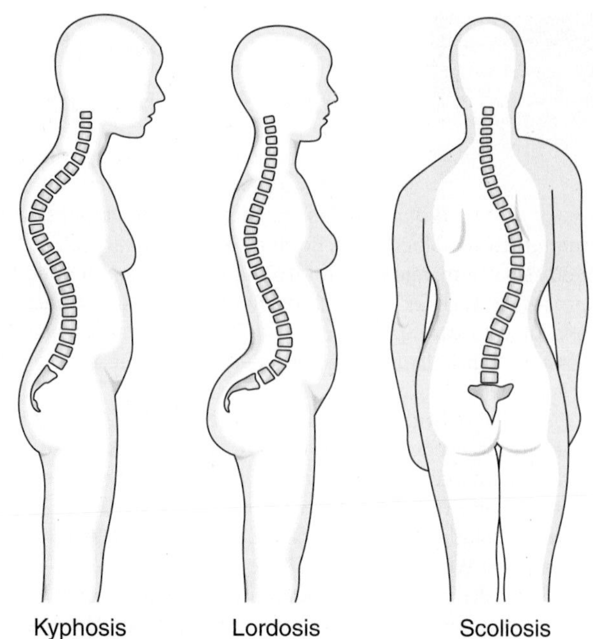

Figure 7-20 Abnormalities of the spinal curves.
ZOOMING IN Which abnormal curve is an exaggerated concave curve? Which is an exaggerated convex curve?

See the Student Resources on thePoint for pictures of scoliosis and other spinal curvatures and a photograph of cleft palate.

FRACTURES

A **fracture** is a break in a bone, usually caused by trauma (**Fig. 7-21**). Almost any bone can be fractured with sufficient force. Such injuries may be classified as follows:

- **Closed fracture**—a simple bone fracture with no open wound
- **Open fracture**—a broken bone protrudes through the skin or an external wound leads to a broken bone
- **Greenstick fracture**—one side of the bone is broken and the other is bent; most common in children
- **Impacted fracture**—the broken ends of the bone are jammed into each other
- **Comminuted** (KOM-ih-nu-ted) **fracture**—there is more than one fracture line, and the bone is splintered or crushed
- **Spiral fracture**—the bone has been twisted apart; relatively common in skiing accidents
- **Transverse fracture**—the fracture goes straight across the bone
- **Oblique fracture**—the break occurs at an angle across the bone

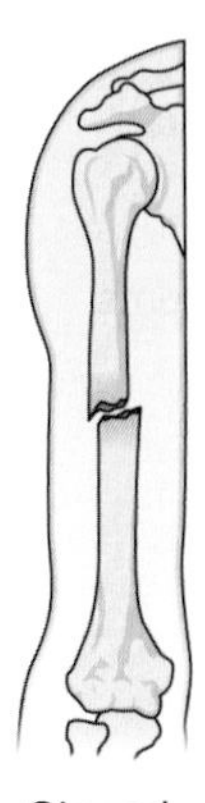

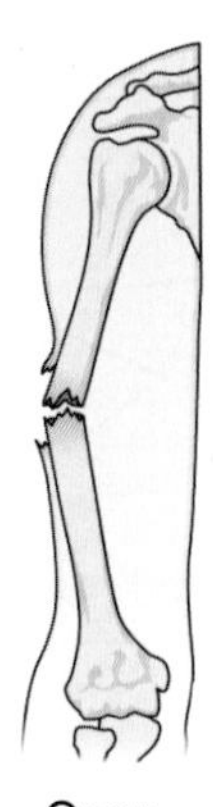

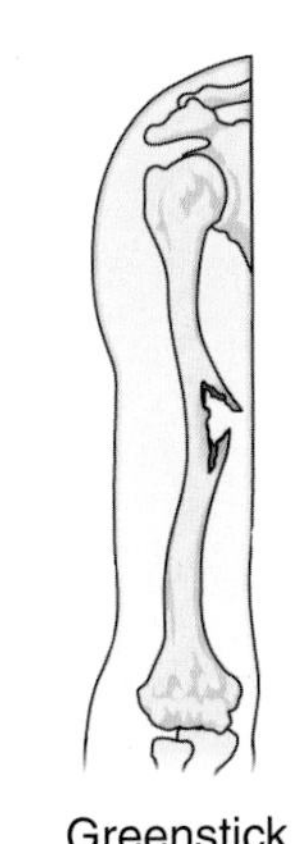

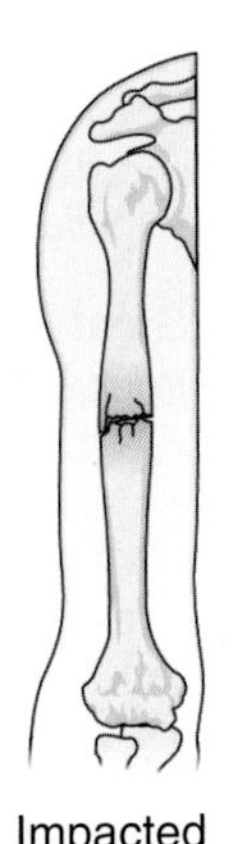

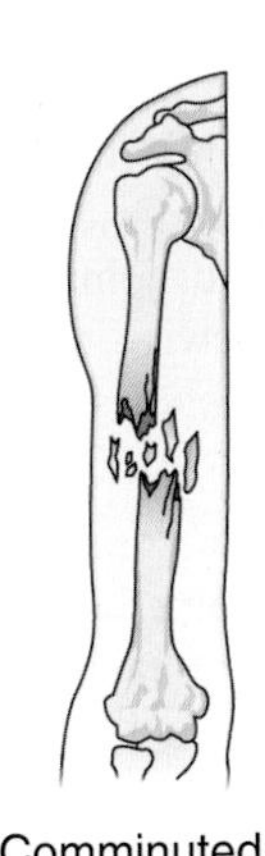

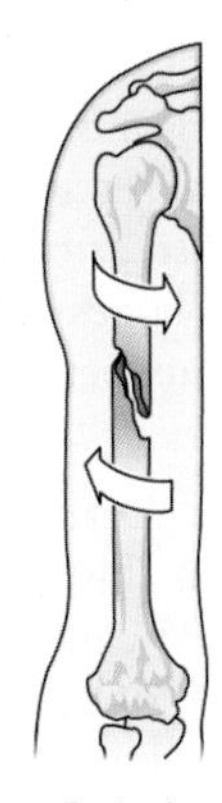

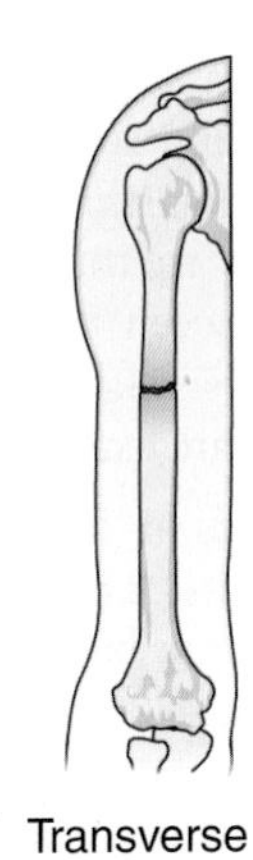

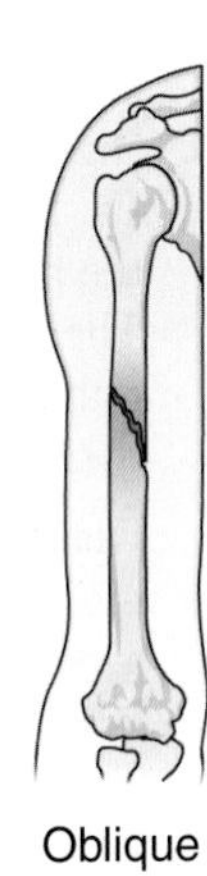

Figure 7-21 **Types of fractures.**

The most important step in first aid care of fractures is to prevent movement of the affected parts. Protection by simple splinting after careful evaluation of the situation, leaving as much as possible "as is," and a call for expert help are usually the safest measures. People who have back injuries may be spared serious spinal cord damage if they are carefully and correctly moved on a firm board or door. If trained paramedics or rescue personnel can reach the scene, untrained people should follow a "hands-off" rule. If there is no external bleeding, covering the victim with blankets may help combat shock. First aid should always be directed immediately toward the control of hemorrhage.

CHECKPOINT

☐ 7-10 What are five categories of bone disorders?

The Joints

An **articulation**, or **joint**, is an area of junction or union between two or more bones. Joints are classified into three main types according to the degree of movement permitted. The joints also differ in the type of material between the adjoining bones (Table 7-2):

- **Synarthrosis** (sin-ar-THRO-sis). The bones in this type of joint are held together so tightly that they cannot move in relation to one another. Most synarthroses use fibrous tissue to join the bones, so they are often described as **fibrous joints**. An example is a suture between bones of the skull.
- **Amphiarthrosis** (am-fe-ar-THRO-sis). This type of joint is slightly moveable. For example, the radius and ulna

Table 7-2 Joints

Type	Material between the Bones	Examples
Immovable (synarthrosis)	Fibrous: No joint cavity; fibrous connective tissue between bones	Sutures between skull bones
Slightly movable (amphiarthrosis)	No joint cavity; cartilage (or sometimes fibrous tissue) between bones	Pubic symphysis; joints between vertebral bodies
Freely movable (diarthrosis)	Joint cavity containing synovial fluid	Gliding, hinge, pivot, condyloid, saddle, ball-and-socket joints

are joined by a large band of fibrous tissue that permits slight movement, so this joint is a fibrous amphiarthrosis. Most amphiarthroses, however, use cartilage to join the bones and are thus described as **cartilaginous joints.** The joint between the pubic bones of the pelvis—the pubic symphysis—and the joints between the bodies of the vertebrae are examples.

- **Diarthrosis** (di-ar-THRO-sis). Diarthroses are freely moveable joints. The bones in this type of joint have a potential space between them called the **joint cavity,** which contains a small amount of thick, colorless fluid. This lubricant, **synovial fluid,** resembles uncooked egg white (*ov* is the root, meaning "egg") and is secreted by the membrane that lines the joint cavity. For this reason, diarthroses are also called **synovial** (sin-O-ve-al) **joints.** Most of the body's joints are synovial joints; they are described in more detail next.

MORE ABOUT SYNOVIAL JOINTS

The bones in freely movable joints are held together by **ligaments,** bands of dense regular connective tissue. Additional ligaments reinforce and help stabilize the joints at various points (**Fig. 7-22**). Also, for strength and protection, there is a **joint capsule** of connective tissue that encloses each joint and is continuous with the periosteum of the bones (**see Fig. 7-22B**). A smooth layer of hyaline cartilage called the **articular** (ar-TIK-u-lar) **cartilage** protects the bone surfaces in synovial joints. Some complex joints may have additional cushioning cartilage between the bones, such as the crescent-shaped medial meniscus (meh-NIS-kus) and lateral meniscus in the knee joint (**Fig. 7-23**). Fat may also appear as padding around a joint.

Near some joints are small sacs called **bursae** (BER-se), which are filled with synovial fluid (**see Fig. 7-23**). These lie in areas subject to stress and help ease movement over and around the joints.

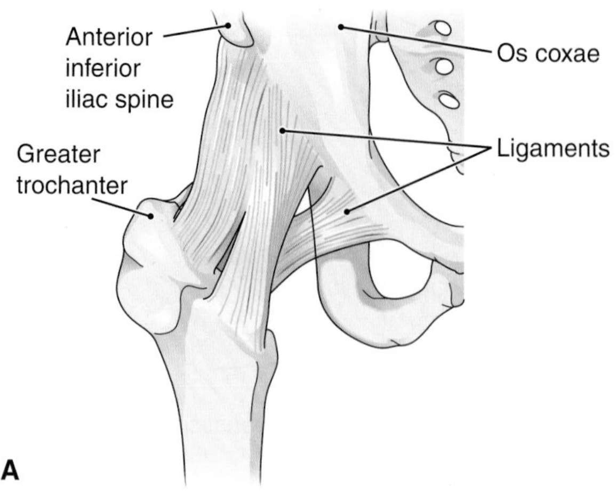

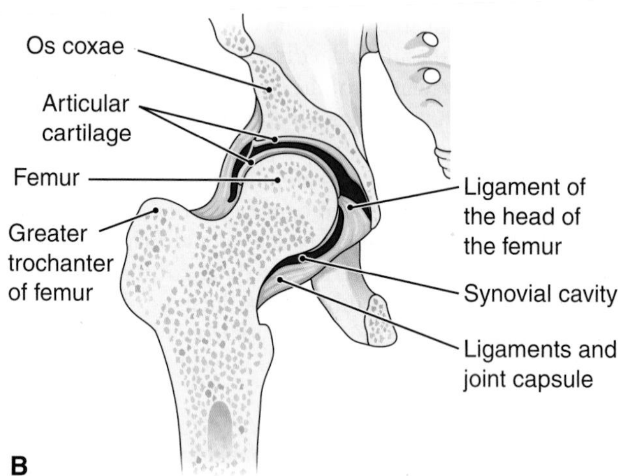

Figure 7-22 Structure of a synovial joint. **KEY POINT** Connective tissue structures stabilize and protect synovial joints. **A.** Anterior view of the hip joint showing ligaments that reinforce and stabilize the joint. **B.** Frontal section through right hip joint showing protective structures. **ZOOMING IN** What is the purpose of the greater trochanter of the femur? What type of tissue covers and protects the ends of the bones?

Movement at Synovial Joints Freely movable joints allow the articulating bones to move in relation to each other. For instance, bending the knee joint moves the leg in relation to the thigh. The specific terms describing movements assume the anatomic position introduced in Chapter 1 (**see Fig. 1-6**). Experiment with moving your body in these different directions as you read these descriptions and examine the illustrations. There are four kinds of angular movement, or movement that changes the angle between bones (**Fig. 7-24**):

- **Flexion** (FLEK-shun) is a bending motion that decreases the angle between bones away from the anatomic position, as in bending the fingers to close the hand. Arm flexion at the shoulder involves raising the upper limb in front of the body, as in raising your hand to ask a question. Specialized terms describe flexion at the ankle:
  - **Dorsiflexion** (dor-sih-FLEK-shun) bends the foot upward at the ankle, narrowing the angle between the leg and the top of the foot.
  - **Plantar flexion** bends the foot so that the toes point downward, as in toe dancing.
- **Extension** is a straightening motion that increases the angle between bones and returns the joint toward the anatomic position, as in straightening the fingers to open the hand. Arm extension lowers the arm from the flexed position. In **hyperextension,** a part is extended beyond its anatomic position, as in opening the hand to its maximum by hyperextending the fingers or hyperextending the thigh at the hip in preparation for kicking a ball from a standing position.
- **Abduction** (ab-DUK-shun) is movement away from the midline of the body, as in moving the arm straight out to the side.
- **Adduction** is movement toward the midline of the body, as in bringing the arm back to its original position beside the body.

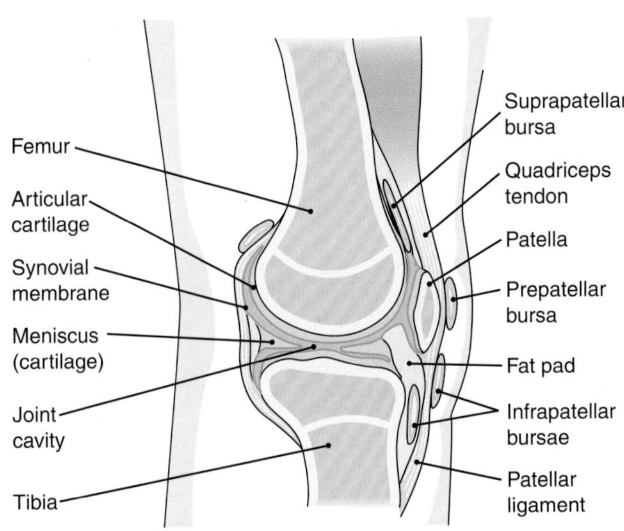

Figure 7-23 The knee joint, sagittal section. Protective structures are also shown.

Specialized terms describe movements of the foot in the lateral plane:

- **Inversion** (in-VER-zhun) is the act of turning the sole inward, so that it faces the opposite foot.
- **Eversion** (e-VER-zhun) turns the sole outward, away from the body.

A combination of angular movements enables one to execute a movement referred to as **circumduction** (ser-kum-DUK-shun). To perform this movement, stand with your arm outstretched and draw a large imaginary circle in the air. Note the smooth combination of flexion, abduction, extension, and adduction that makes circumduction possible.

Rotation refers to a twisting or turning of a bone on its own axis, as in turning the head from side to side to say no. Specialized terms describe rotation of the forearm:

- **Supination** (su-pin-A-shun) is the act of turning the palm up or forward.
- **Pronation** (pro-NA-shun) turns the palm down or backward.

Types of Synovial Joints Synovial joints are classified according to the types of movement they allow, as described and illustrated in **Table 7-3**. Locate these types of joints on your body and demonstrate the different movements they allow. Listed in order of increasing range of motion, they are

- Gliding joint—two relatively flat bone surfaces slide over each other with little change in the joint angle. Examples are the joints between the tarsal and carpal bones.
- Hinge joint—a convex (curving outward) surface of one bone fits into the concave (curving inward) surface of another bone, allowing movement in one direction. Hinge joints allow flexion and extension only. Examples are the elbow joint and the joints between the phalanges.
- Pivot joint—a rounded or pointed portion of one bone fits into a ring in another bone. This joint allows rotation only, as in the joint between the atlas and axis of the

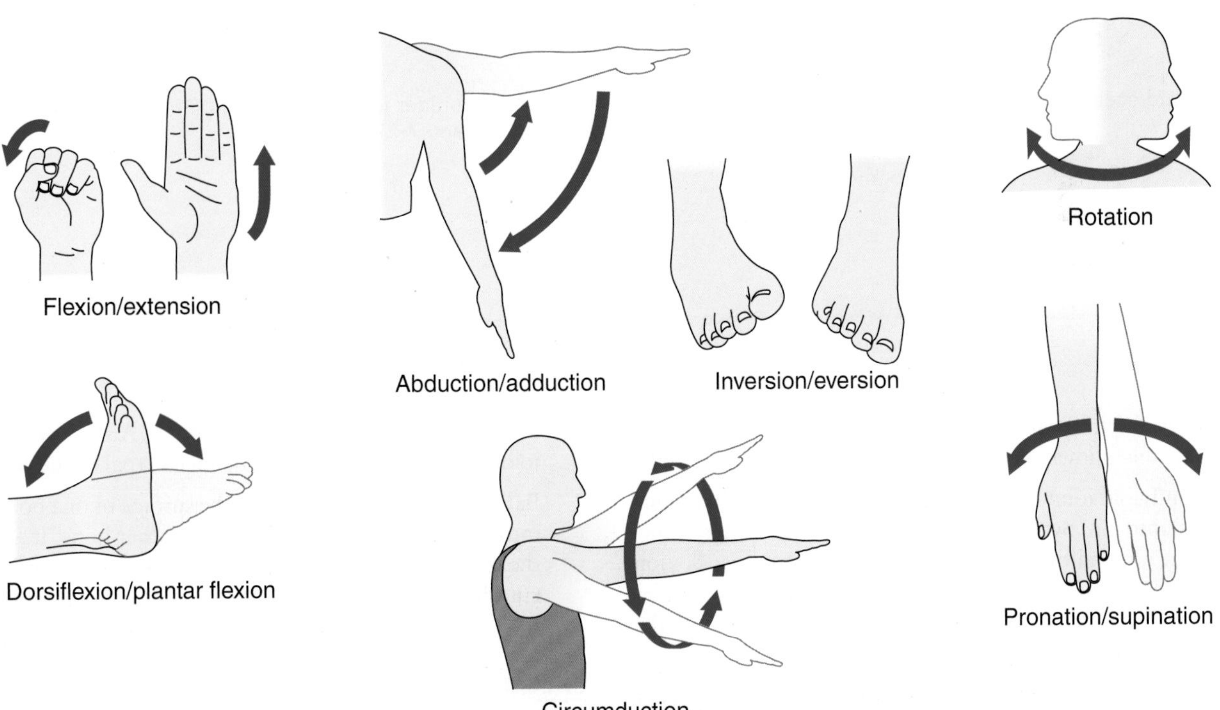

Figure 7-24 Movements at synovial joints. **KEY POINT** Synovial joints allow the greatest range of motion. All movements are in reference to the anatomic position.

Table 7-3 Synovial Joints

Type of Joint	Type of Movement	Examples
Gliding joint	Flat bone surfaces slide over one another with little change in the joint angle	Joints in the wrist and ankles (**Figs. 7-13 and 7-17**)
Hinge joint	Allows movement in one direction, changing the angle of the bones at the joint, as in flexion and extension	Elbow joint; joints between phalanges of fingers and toes (**Figs. 7-12, 7-13, and 7-17**)
Pivot joint	Allows rotation around the length of the bone	Joint between the first and second cervical vertebrae; joint at proximal ends of the radius and ulna (**Figs. 7-7 and 7-11**)
Condyloid joint	Allows movement in two directions: flexion and extension, abduction and adduction	Joint between the occipital bone of the skull and the first cervical vertebra (atlas) (**Fig. 7-7**); joint between the metacarpal and the first phalanx of the finger (knuckle) (**Fig. 7-13**)
Saddle joint	Like a condyloid joint, but with deeper articulating surfaces and movement in three directions, rotation in addition to flexion and extension, abduction and adduction	Joint between the wrist and the metacarpal bone of the thumb (**Fig. 7-13**)
Ball-and-socket joint	Allows the greatest range of motion. Permits movement in three directions around a central point, as in circumduction.	Shoulder joint and hip joint (**Figs. 7-9, 7-14, and 7-22**)

cervical spine or the proximal joint between the radius and ulna that allows supination and pronation of the forearm.

- Condyloid joint—an oval-shaped projection of one bone fits into an oval-shaped depression on another bone. This joint allows movement in two directions: flexion and extension, and abduction and adduction. Examples are the joints between the metacarpal bones and the proximal phalanges of the fingers.
- Saddle joint—similar to the condyloid joint, but deeper and allowing greater range of motion. One bone fits into a saddle-like depression on another bone. It allows movement in three directions: flexion and extension, abduction and adduction, and rotation. An example is the joint between the wrist and the metacarpal of the thumb.
- Ball-and-socket joint—a ball-like surface of one bone fits into a deep cuplike depression in another bone. It allows the greatest range of motion in three directions, as in circumduction. Examples are the shoulder and hip joints.

CHECKPOINTS

- ☐ 7-11 What are three types of joints based on the degree of movement they allow?
- ☐ 7-12 What is the most freely moveable type of joint?

DISORDERS OF JOINTS

Joints are complex structures, composed of multiple tissue types and subject to considerable mechanical stress. They are thus susceptible to both injuries and inflammatory diseases.

Stress (Mechanical) Injuries A **dislocation** is a derangement of joint parts. Ball-and-socket joints, which have the widest range of motion, also have the greatest tendency to dislocate. The shoulder joint is the most frequently dislocated joint in the body.

Sprains A **sprain** is the wrenching of a joint with rupture or tearing of the ligaments, usually as a result of abnormal or excessive joint movement. There may also be injuries to the cartilage within the joint, most commonly in the knee joint. In severe cases, the ligament can actually tear away from its bone. The ankle is a common site of sprains, as when the ankle is "turned;" that is, the foot is turned inward and body weight forces excessive inversion at the joint. Improper lifting or poor posture can sprain the ligaments of the spine, causing back pain. The pain and swelling accompanying a sprain can be reduced by the immediate application of ice packs, which constrict some of the smaller blood vessels and reduce internal bleeding.

Bursitis **Bursitis** is inflammation of a bursa. Bursitis can be very painful with swelling and limitation of motion. Some examples of bursitis are listed below:

- **Olecranon bursitis**—inflammation of the bursa over the point of the elbow (olecranon). Another name for the disorder is student's elbow, as it can be caused by long hours of leaning on the elbows while studying.
- **Ischial bursitis**—common among people who must sit a great deal, such as taxicab drivers and truckers.
- **Prepatellar bursitis**—inflammation of the bursa anterior to the patella. This form of bursitis is found in people who must often kneel, hence the name from an earlier time, *housemaid's knee.*
- **Subdeltoid bursitis** and **subacromial bursitis** appear in the shoulder region. They are fairly common occurrences in baseball, swimming, and other sports requiring repeated overhead movements.

In some cases, a local anesthetic, corticosteroids, or both may be injected to relieve bursitis pain.

Bunions are enlargements commonly found at the base and medial side of the great toe. Usually, prolonged pressure has caused the development of a bursa, which has then become inflamed. Special shoes may be necessary if surgery is not performed.

Vertebral Disorders The 24 bones of the spine are held together by the cartilaginous joints between the vertebral bodies (the intervertebral disks) and synovial joints between the articulating surfaces of adjacent vertebra. A host of ligaments and muscles stabilize the entire structure. With all of this complexity, it's no wonder that spinal problems are a common cause of back pain (other causes include muscle strains and disorders of the abdominal and pelvic organs). In addition to arthritis, which is discussed shortly, lower back pain can reflect:

- Diseases of the vertebrae, such as infections or tumors, osteoarthritis (see below) or bone atrophy (wasting away) following long illnesses and lack of exercise
- Abnormalities of the lower vertebrae or of the ligaments and other supporting structures
- Strains on the lumbosacral joint (where the lumbar region joins the sacrum) or strains on the sacroiliac joint (where the sacrum joins the ilium of the pelvis)

7

Herniated Disk The disks between the vertebrae consist of an outer ring of fibrocartilage and a central mass known as the nucleus pulposus. In the case of a **herniated disk,** this central mass protrudes through a weakened outer cartilaginous ring into the spinal canal (**Fig. 7-25**). The herniated or "slipped" disk puts pressure on the spinal cord or spinal nerves, often causing back spasms or pain along the sciatic nerve that travels through the leg, a pain known as *sciatica* (see Chapter 9). It is sometimes necessary to remove the disk and fuse the vertebrae involved. Newer techniques make it possible for surgeons to remove only a specific portion of the disk.

Arthritis The most common type of joint disorder is termed **arthritis,** which means "inflammation of the joints." There are different kinds of arthritis, including the following:

- **Osteoarthritis** (os-te-o-arth-RI-tis) (OA), also known as degenerative joint disease (DJD), usually occurs in elderly people as a result of normal wear and tear and

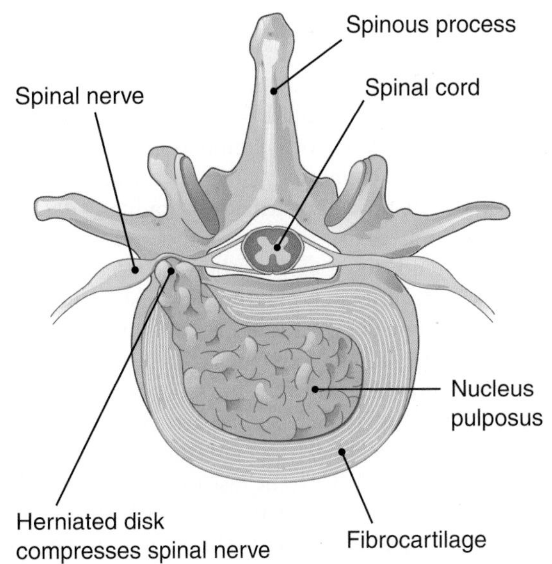

Figure 7-25 **Herniated disk.** **KEY POINT** The central portion (nucleus pulposus) of an intervertebral disk protrudes through the outer rim of cartilage to put pressure on a spinal nerve.

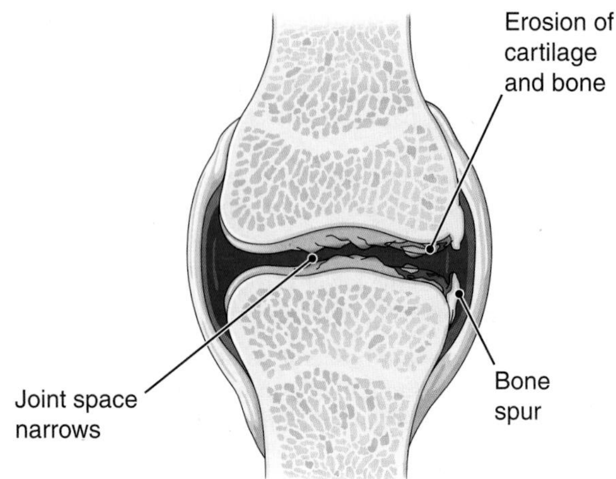

Figure 7-26 Joint changes in osteoarthritis (OA).
KEY POINT The left side of the joint shows early changes with breakdown of cartilage and narrowing of the joint space. The right side shows progression of the disease with loss of cartilage and formation of bone spurs.

low-level inflammation. Genetic and environmental factors, such as injuries and obesity, play a role in the severity and onset of the disease. OA occurs mostly in the hips in men and in the knees and hands in women. The cartilage cushioning the bone ends atrophies and dies so the uncovered bones now rub against each other. This friction causes bone remodeling with bone spurs often developing at the edges of the articulating bones (**Fig. 7-26**).The cartilaginous joints of the spine can also degrade, resulting in thinned intervertebral disks. Osteoarthritis can also cause **lumbar stenosis**, a narrowing of the spinal canal or vertebral foramina, a common cause of lower back pain.

- **Rheumatoid arthritis** (RA) is a crippling condition characterized by joint swelling in the hands, the feet, and elsewhere as a result of inflammation and overgrowth of the synovial membranes and other joint tissues (**see Fig. 5-1**). The articular cartilage is gradually destroyed, and the joint cavity develops adhesions—that is, the surfaces tend to stick together—so that the joints stiffen and ultimately become useless. Rheumatoid arthritis is considered to be an autoimmune disorder, in which the immune system attacks the body's own tissues. Individuals with a genetic susceptibility to RA develop the disease when they are exposed to a specific *antigen* (an antigen is a substance that induces an immune reaction). Current research attempts to identify common antigens that trigger RA, but different individuals appear to react to different ones. Successful treatment includes rest, appropriate exercise, medications to reduce pain and swelling, and suppression of the immune response.
- **Septic (infectious) arthritis** arises when bacteria in the blood or in a nearby local infection colonize a joint. Bacteria introduced during invasive medical procedures, illegal drug injections, or by other means can also settle in joints. A variety of organisms are commonly involved, including *Streptococcus*, *Staphylococcus*, and *Neisseria* species. The bacteria that cause Lyme disease and tuberculosis as well as various viruses can also cause septic arthritis.
- **Gout** is a kind of arthritis caused by a metabolic disturbance. One waste product of metabolism is uric acid, which normally is excreted in the urine. If there happens to be an overproduction of uric acid, or for some reason not enough is excreted, the accumulated uric acid forms crystals that are deposited as masses around the joints and in other areas. As a result, the joints become inflamed and extremely painful. Any joint can be involved, but the one most commonly affected is the big toe. Genetics, increased age, and heavy alcohol consumption all contribute to the development of gout.

JOINT REPAIR

Physicians can examine injured joints and even repair them surgically with a lighted instrument known as an **arthroscope** (AR-thro-skope), a type of endoscope (**Fig. 7-27**). Using arthroscopic techniques, orthopedic surgeons (specialists in the treatment of musculoskeletal disorders) can repair or replace ligaments and remove or reshape cartilage with minimal invasion. Abnormal amounts of fluid that accumulate in a joint cavity as a result of injury can be drained by a tapping procedure called **arthrocentesis** (ar-thro-sen-TE-sis).

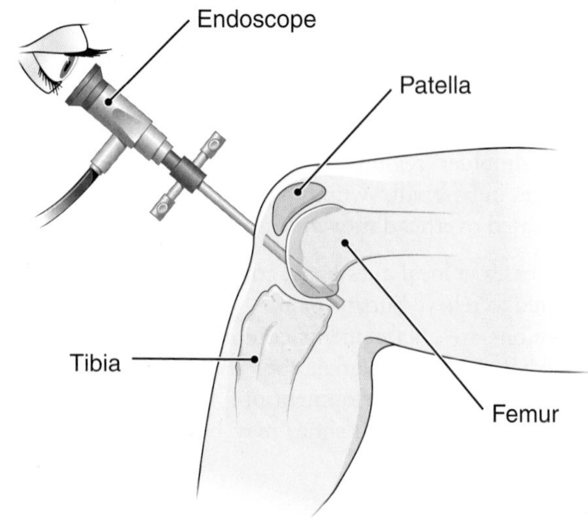

Figure 7-27 Arthroscopic examination of the knee.
KEY POINT An endoscope is inserted between projections at the end of the femur to view the posterior of the knee.

If joint degeneration is severe, and conservative therapies, such as medication, physical therapy, and weight loss, have failed, a joint replacement or **arthroplasty** (AR-thro-plas-te) may be required. Millions of such surgeries have been performed successfully to restore function in cases of injury, arthritis, and other chronic degenerative bone diseases. Hips and knees are the joints most commonly replaced, but surgeons can also replace shoulder, elbow, wrist, hand, ankle, and foot joints.

The strong, nontoxic, and corrosion-resistant prosthetic (artificial) joints are made of plastic, ceramic, or metal alloys and are bonded in place with screws or glue. Some are designed to promote bone growth into them to aid in attachment. Computer-controlled machines now produce individualized joints. Until recently, arthroplasty was rarely performed on young people because prosthetic joints had a life span of only about 10 years. Current methods have increased this range to 20 years or more, so arthroplasties will require fewer replacements.

CHECKPOINTS 

- ☐ 7-13 What is the most common type of joint disorder?
- ☐ 7-14 What type of endoscope is used to examine and repair joints?

Effects of Aging on the Skeletal System

The aging process includes significant changes in all connective tissues, including bone. There is a loss of calcium salts and a decreased ability to form the protein framework on which calcium salts are deposited. Cellular metabolism slows, so bones are weaker, less dense, and more fragile; fractures and other bone injuries heal more slowly. Muscle tissue is also lost throughout adult life. Loss of balance and diminished reflexes may lead to falls. Thus, there is a tendency to decrease the exercise that is so important to the maintenance of bone tissue.

Changes in the vertebral column with age lead to a loss in height. Approximately 1.2 cm (about 0.5 in) are lost each 20 years beginning at 40 years of age, owing primarily to a thinning of the intervertebral disks (between the bodies of the vertebrae). Even the vertebral bodies themselves may lose height in later years. The costal (rib) cartilages become calcified and less flexible, and the chest may decrease in diameter by 2 to 3 cm (about 1 in), mostly in the lower part.

At the joints, reduction of collagen in bone, tendons, and ligaments contributes to the diminished flexibility so often experienced by older people. Thinning of articular cartilage and loss of synovial fluid may contribute to joint damage. By the process of calcification, minerals may be deposited in and around the joints, especially at the shoulder, causing pain and limiting mobility.

Disease in Context Revisited

Reggie's Fracture Begins to Heal Itself

"So, Doc, what's the chance my leg's going to heal up enough to play football again?" asked Reggie.

"Well," replied the doctor, "It's going to take some time before you're catching footballs again, but once your hip heals, it will be better than new."

The surgeon knew that even before the surgery to realign its broken ends, Reggie's femur had already begun to heal itself. Immediately after the injury occurred on the football field, a blood clot formed around the fracture. A day or two later, chemical messengers within the clot would stimulate blood vessels from the periosteum and endosteum to invade the clot, bringing connective tissue cells with them. Over the next several weeks, fibroblasts and chondroblasts in the clot would secrete collagen and cartilage, converting it into a soft callus. Meanwhile, macrophages would remove the remains of the blood clot and osteoclasts would digest dead bone tissue. Soon after, osteoblasts in the callus would convert it into spongy bone called a hard callus. Months after the injury, osteoclasts and osteoblasts would work together to remodel the outer layers of the hard callus into compact bone, resulting in a repair even stronger than the original bone tissue in Reggie's femur.

During this case, we saw how fractured bones are repaired using screws and plates. We also saw that the body has its own "orthopedic surgeons"—cells like osteoblasts and osteoclasts, which can engineer a bone repair that is even stronger than the original.

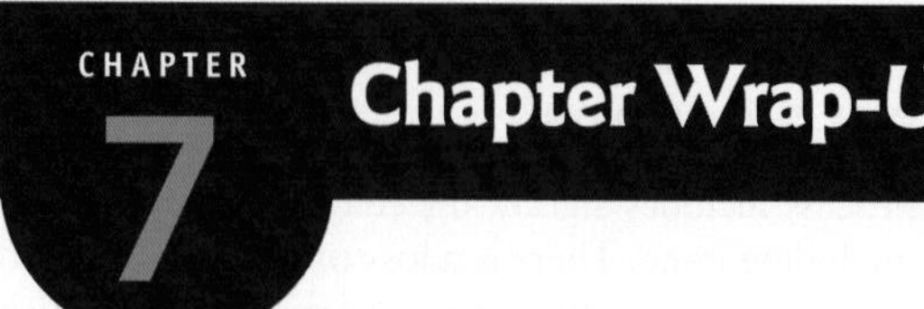

Summary Overview

A detailed chapter outline with space for note taking is on *thePoint*. The figure below illustrates the main topics covered in this chapter.

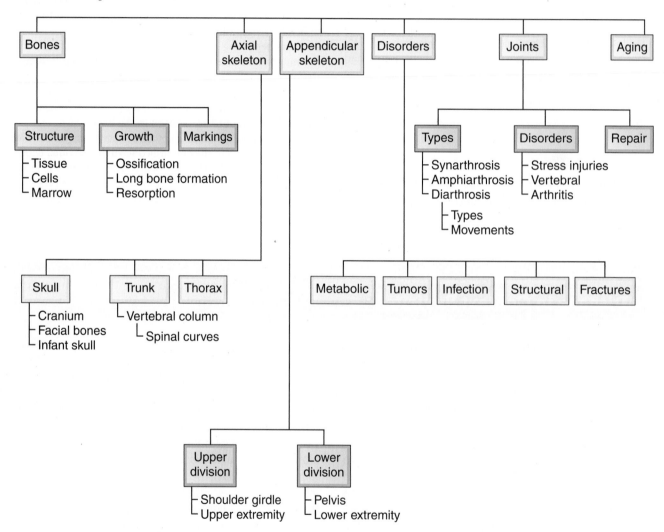

- Bones
  - Structure
    - Tissue
    - Cells
    - Marrow
  - Growth
    - Ossification
    - Long bone formation
    - Resorption
  - Markings
- Axial skeleton
  - Skull
    - Cranium
    - Facial bones
    - Infant skull
  - Trunk
    - Vertebral column
      - Spinal curves
  - Thorax
- Appendicular skeleton
  - Upper division
    - Shoulder girdle
    - Upper extremity
  - Lower division
    - Pelvis
    - Lower extremity
- Disorders
  - Metabolic
  - Tumors
  - Infection
  - Structural
  - Fractures
- Joints
  - Types
    - Synarthrosis
    - Amphiarthrosis
    - Diarthrosis
      - Types
      - Movements
  - Disorders
    - Stress injuries
    - Vertebral
    - Arthritis
  - Repair
- Aging

Key Terms

The terms listed below are emphasized in this chapter. Knowing them will help you organize and prioritize your learning. These and other boldface terms are defined in the Glossary with phonetic pronunciations.

amphiarthrosis
arthritis
arthroscope
arthroplasty
articulation
bone marrow
bursa
circumduction
diaphysis
diarthrosis
endosteum
epiphysis
extremity
fontanel
joint
osteoblast
osteoclast
osteocyte
osteon
osteopenia
osteoporosis
periosteum
resorption
skeleton
synarthrosis
synovial

Word Anatomy

Medical terms are built from standardized word parts (prefixes, roots, and suffixes). Learning the meanings of these parts can help you remember words and interpret unfamiliar terms.

WORD PART	MEANING	EXAMPLE
Bones		
-clast	break	An *osteoclast* breaks down bone in the process of resorption.
dia-	through, between	The *diaphysis*, or shaft, of a long bone is between the two ends, or epiphyses.
oss, osse/o	bone, bone tissue	*Osseous* tissue is another name for bone tissue.
oste/o	bone, bone tissue	The *periosteum* is the fibrous membrane around a bone.
Bones of the Axial Skeleton		
cost/o	rib	*Intercostal* spaces are located between the ribs.
para-	near	The *paranasal* sinuses are near the nose.
pariet/o	wall	The *parietal* bones are the side walls of the skull.
Bones of the Appendicular Skeleton		
infra-	below, inferior	The *infraspinous* fossa is a depression inferior to the spine of the scapula.
meta-	near, beyond	The *metacarpal* bones of the palm are near and distal to the carpal bones of the wrist.
supra-	above, superior	The *supraspinous* fossa is a depression superior to the spine of the scapula.
Disorders of Bone		
-malacia	softening	*Osteomalacia* is a softening of bone tissue.
-penia	lack of	In *osteopenia*, there is a lack of bone tissue.
The Joints		
ab-	away from	*Abduction* is movement away from the midline of the body.
ad-	toward, added to	*Adduction* is movement toward the midline of the body.
amphi-	on both sides, around, double	An *amphiarthrosis* is a slightly movable joint.
arthr/o	joint, articulation	A *synarthrosis* is an immovable joint, such as a suture.
circum-	around	*Circumduction* is movement around a joint in a circle.

Questions for Study and Review

BUILDING UNDERSTANDING

Fill in the Blanks

1. The shaft of a long bone is called the _____.

2. The structural unit of compact bone is the _____.

3. Red bone marrow manufactures _____.

4. Bones are covered by a connective tissue membrane called _____.

5. Bone cells active in resorption are _____.

Matching > Match each numbered item with the most closely related lettered item.

___ **6.** A rounded bony projection

___ **7.** A sharp bony prominence

___ **8.** An air-filled space in bone

___ **9.** A bony depression

___ **10.** A short channel or passageway in bone

a. condyle
b. sinus
c. fossa
d. meatus
e. spine

Multiple Choice

___ **11.** Where is the growth plate of a long bone located?
a. epiphysis
b. articular cartilage
c. marrow cavity
d. endosteum

___ **12.** Which bone contains the foramen magnum?
a. temporal
b. hyoid
c. occipital
d. fibula

___ **13.** What is an abnormal exaggeration of the lumbar spinal curve called?
a. kyphosis
b. lordosis
c. osteomyelitis
d. Pott disease

___ **14.** Which type of bone fracture involves one broken side and one bent side?
a. open
b. impacted
c. greenstick
d. comminuted

___ **15.** What kind of synovial joint is between the atlas and axis?
a. gliding
b. hinge
c. saddle
d. pivot

UNDERSTANDING CONCEPTS

16. List five functions of bone, and describe how a long bone's structure enables it to carry out each of these functions.

17. Explain the differences between the terms in each of the following pairs:
a. osteoblast and osteocyte
b. red marrow and yellow marrow
c. compact bone and spongy bone
d. synarthrosis and amphiarthrosis
e. osteoporosis and osteopenia

18. Discuss the process of long bone formation during fetal development and childhood. What role does resorption play in bone formation?

19. Name the five groups of vertebrae. Explain how the different structures of the different vertebrae correspond to their functions.

20. Referring to the "The Body Visible" at the beginning of the book give the numbers of the following:
a. greater trochanter of the femur
b. maxilla
c. manubrium
d. iliac crest
e. xiphoid process
f. medial malleolus

21. Referring to Appendices 2 and 3, cite the normal range for calcium in the blood. Give two possible reasons for excess blood calcium levels. What is the meaning of the abbreviation dL? (The Student Resources for Chapter 1 on *thePoint* has this information.) How many mL are in a dL?

22. Describe the structure of a synovial joint. Explain how the structure of synovial joints relates to their function.

23. What is circumduction? What type of joint allows for circumduction, and where are such joints located?

24. Name three effects of aging on the skeletal system.

25. What are the similarities and differences between osteoarthritis, rheumatoid arthritis, and gout?

26. Differentiate between the terms in each of the following pairs:
    a. flexion and extension
    b. abduction and adduction
    c. supination and pronation
    d. dorsiflexion and plantar flexion

CONCEPTUAL THINKING

27. Nine-year-old Alek is admitted to the emergency room with a closed fracture of the right femur. Radiography reveals that the fracture crosses the distal epiphyseal plate. What concerns should Alek's healthcare team have about the location of his injury?

28. In Reggie's case, he fractured the proximal end of his right femur, an integral component of his hip. Name the joint disorders that Reggie is more at risk of in (a) the short term and (b) the long term.

For more questions, see the Learning Activities on thePoint.

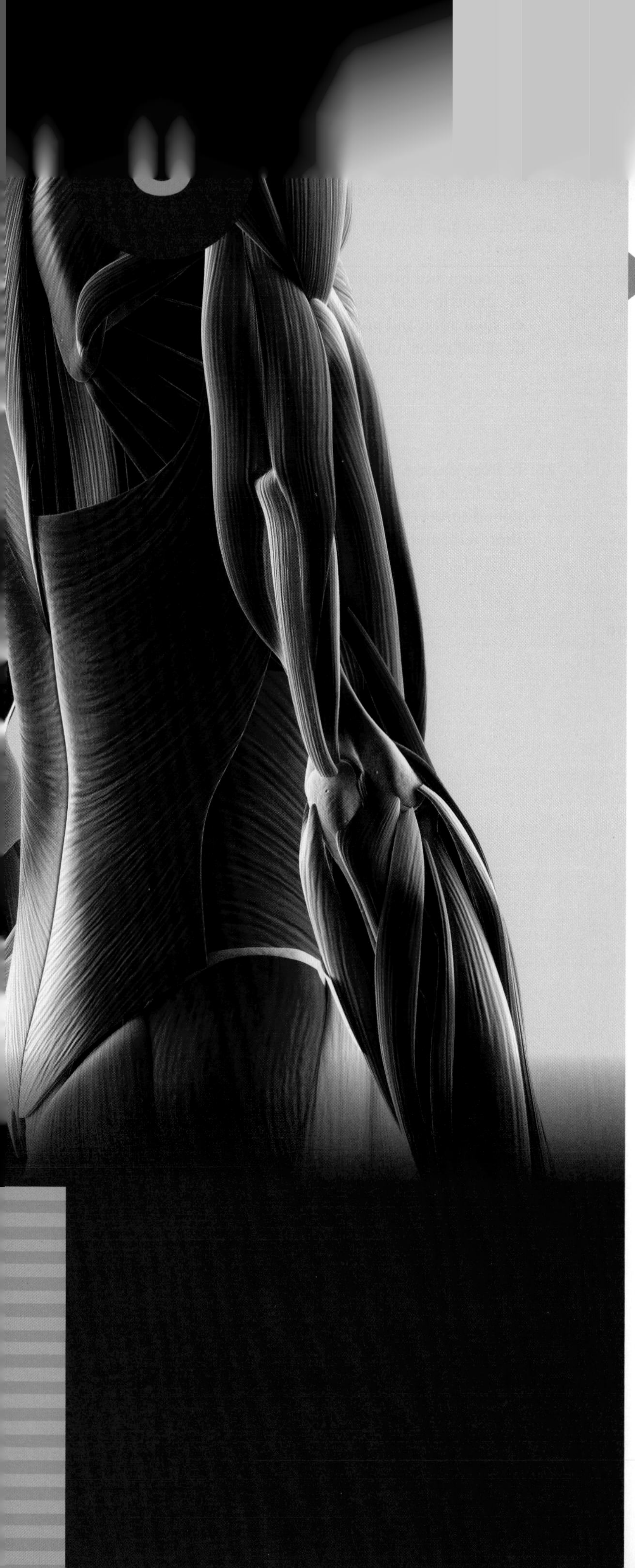

Learning Objectives

After careful study of this chapter, you should be able to:

1 ▶ Compare the three types of muscle tissue. ***p. 162***

2 ▶ Describe three functions of skeletal muscle. ***p. 163***

3 ▶ Describe the structure of a skeletal muscle to the level of individual cells. ***p. 163***

4 ▶ Outline the steps in skeletal muscle contraction. ***p. 163***

5 ▶ List compounds stored in muscle cells that are used to generate energy. ***p. 167***

6 ▶ Explain what happens in muscle cells contracting anaerobically. ***p. 168***

7 ▶ Cite the effects of exercise on muscles. ***p. 168***

8 ▶ Compare isotonic and isometric contractions. ***p. 169***

9 ▶ Explain how muscles work together to produce movement. ***p. 170***

10 ▶ Compare the workings of muscles and bones to lever systems. ***p. 171***

11 ▶ Explain how muscles are named. ***p. 172***

12 ▶ Name some of the major muscles in each muscle group, and describe the locations and functions of each. ***p. 172***

13 ▶ Describe how muscles change with age. ***p. 181***

14 ▶ List the major disorders of muscles and their associated structures. ***p. 182***

15 ▶ Describe some of the diagnostic signs of muscular dystrophy based on the case study. ***pp. 161, 183***

16 ▶ Show how word parts are used to build words related to the muscular system (see Word Anatomy at the end of the chapter). ***p. 185***

Disease in Context *Shane's Daycare Incident*

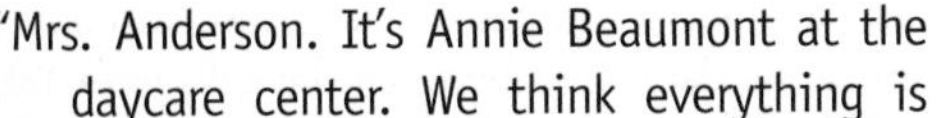

"Mrs. Anderson. It's Annie Beaumont at the daycare center. We think everything is OK, but Shane fell off a small plastic slide this morning, and we think you should take a look at him and maybe have him checked by his doctor."

Two-year-old Shane had started daycare a few days earlier. His mother Kathy, a single working parent, had been using babysitters to watch him since he was 2 months old. This year, she felt it was time to enroll him in a daycare program and was able to register him at a reputable center near the city, not too far from where she worked. She was excited that he would be able to play with other children, as there weren't many social opportunities for him where they lived and with her work schedule. She was hoping the teachers and a more structured daytime environment would benefit Shane until she got home from work and could care for him herself.

"Everything is probably fine," Ms. Beaumont told Kathy when she arrived at the center. "We're just concerned that Shane had a difficult time standing up when he fell, and just to be safe, we think he should be evaluated by his physician."

Later that afternoon, Shane's pediatrician, Dr. Schroeder, listened to Kathy as she described the daycare center incident as it had been related to her. Then he examined Shane for any injuries that may have occurred from the fall. During the evaluation, he observed Shane's voluntary movements. He noticed that the boy's calf muscles were enlarged (pseudohypertrophic), and his thighs were thin. He asked Shane to sit on the floor and then stand up. Shane had to use his hands and arms to "walk" up his own body (Gowers sign), reflecting weak thigh muscles. Dr. Schroeder took Kathy into his office and talked to her privately.

"I haven't seen Shane in some time for his well check-ups," the doctor said.

"He has seemed healthy so I didn't think it was necessary," Kathy replied.

"The fall didn't result in any injury, but after evaluating Shane I am concerned that his muscles are underdeveloped," Dr. Schroeder said. "This may have led to the fall. I want to do some more tests to figure out why Shane's muscles are weak. Let's start with a blood test and go from there."

Dr. Schroeder suspected that Shane had a condition called Duchenne muscular dystrophy or DMD, a hereditary disease that causes damage to muscle cells. In this chapter, we will learn about muscle tissue and how it interacts with the nervous system to produce movement. Later in the chapter, we will find out more on the progression of Shane's disease.

ANCILLARIES *At-A-Glance*

Visit thePoint to access the following resources. For guidance in using these resources most effectively, see pp. xv–xvii.

Learning RESOURCES

- Tips for Effective Studying
- Web Figure: Muscles of the Lateral Head and Neck
- Web Figure: Muscles of the Anterior Head and Neck
- Web Figure: Muscles of the Chest and Shoulder
- Web Figure: Muscles of the Shoulder and Upper Back
- Web Figure: Anatomy of the Rotator Cuff
- Web Figure: Muscles of the Anterior Forearm and Arm
- Web Figure: Muscles of the Posterior Arm
- Web Figure: Muscles of the Posterior Forearm
- Web Figure: Muscles of the Thigh, Anterior View
- Web Figure: Muscles of the Leg, Anterior View
- Web Figure: Muscles of the Leg, Posterior View
- Animation: The Neuromuscular Junction
- Health Professions: Physical Therapist
- Detailed Chapter Outline
- Answers to Questions for Study and Review
- Audio Pronunciation Glossary

Learning ACTIVITIES

- Pre-Quiz
- Visual Activities
- Kinesthetic Activities
- Auditory Activities

A LOOK BACK

The voluntary muscles discussed in this chapter attach to the skeleton to create the movements described in Chapter 7. At this time, we learn much more about the structure and function of skeletal muscle, introduced in Chapter 4, and we will see how neurotransmitters and membrane receptors function in muscle contraction.

Muscle Tissue

There are three kinds of muscle tissue: smooth, cardiac, and skeletal muscle, as introduced in Chapter 4. After a brief description of all three types (**Table 8-1**), this chapter concentrates on skeletal muscle.

SMOOTH MUSCLE

Smooth muscle makes up the walls of the hollow body organs as well as those of the blood vessels and respiratory passageways. It contracts involuntarily and produces the wavelike motions of peristalsis that move substances through a system. Smooth muscle can also regulate the diameter of an opening, such as the central opening of blood vessels, or produce contractions of hollow organs, such as the uterus.

Smooth muscle fibers (cells) are tapered at each end and have a single, central nucleus. The cells appear smooth under the microscope because they do not contain the visible bands, or **striations**, that are seen in the other types of muscle cells. Smooth muscle may contract in response to a nerve impulse, hormonal stimulation, stretching, and other stimuli. The muscle contracts and relaxes slowly and can remain contracted for a long time.

CARDIAC MUSCLE

Cardiac muscle, also involuntary, makes up the heart's wall and creates the pulsing action of that organ. The cells of cardiac muscle are striated, like those of skeletal muscle. They differ in having one nucleus per cell and branching interconnections. The membranes between the cells are specialized to allow electric impulses to travel rapidly through them, so that contractions can be better coordinated. These specialized membrane regions appear as dark lines between the cells (**see Table 8-1**) and are called intercalated (in-TER-kah-la-ted) disks, because they are "inserted between" the cells. The electric impulses that produce cardiac muscle contractions are generated within the muscle itself but can be modified by nervous stimuli and hormones.

SKELETAL MUSCLE

When viewed under the microscope, skeletal muscle cells appear heavily striated. The arrangement of protein threads within the cell that produces these striations is described later. The cells are very long and cylindrical, and because of their great length compared to other cells, they are often described as muscle *fibers*. They have multiple nuclei per cell because, during development, groups of precursor cells (myoblasts) fuse to form large multinucleated cells. Each mature muscle fiber contracts as a single unit when stimulated.

Skeletal muscle is under the control of the nervous system division known as the voluntary, or somatic, nervous system. Because it is under conscious control, skeletal muscle is described as voluntary. This muscle tissue usually contracts and relaxes rapidly.

Skeletal muscle is so named because most of these muscles are attached to bones and produce movement at the joints. There are a few exceptions. The muscles of the abdominal

Table 8-1 Comparison of the Different Types of Muscle

	Smooth	Cardiac	Skeletal
Location	Wall of hollow organs, vessels, respiratory, passageways	Wall of heart	Attached to bones
Cell characteristics	Tapered at each end, branching networks, nonstriated	Branching networks; special membranes (intercalated disks) between cells; single nucleus; lightly striated	Long and cylindrical; multinucleated; heavily striated
Control	Involuntary	Involuntary	Voluntary
Action	Produces peristalsis; contracts and relaxes slowly; may sustain contraction	Pumps blood out of the heart; self-excitatory but influenced by nervous system and hormones	Produces movement at joints; stimulated by nervous system; contracts and relaxes rapidly

wall, for example, are partly attached to other muscles, and the muscles of facial expression are attached to the skin. Skeletal muscles constitute the largest amount of the body's muscle tissue, making up about 40% of the total body weight. This muscular system is composed of more than 600 individual skeletal muscles. Although each one is a distinct structure, muscles usually act in groups to execute body movements.

8-1 What are the three types of muscle?

The Muscular System

The three primary functions of skeletal muscles are as follows:

- Movement of the skeleton. Muscles are attached to bones and contract to change the position of the bones at a joint.
- Maintenance of posture. A steady partial contraction of muscle, known as **muscle tone**, keeps the body in position. Some of the muscles involved in maintaining posture are the large muscles of the thighs, back, neck, and shoulders as well as the abdominal muscles.
- Generation of heat. Muscles generate most of the heat needed to keep the body at 37°C (98.6°F). Heat is a natural byproduct of muscle cell metabolism. When we are cold, muscles can boost their heat output by the rapid small contractions we know of as shivering.

MUSCLE STRUCTURE

In forming whole muscles, individual muscle fibers (cells) are arranged in bundles, or **fascicles** (FAS-ih-kls), held together by fibrous connective tissue (**Fig. 8-1**). These layers are as follows:

- The **endomysium** (en-do-MIS-e-um) is the deepest layer of this connective tissue and surrounds the individual fibers within fascicles.
- The **perimysium** (per-ih-MIS-e-um) is a connective tissue layer around each fascicle.
- The **epimysium** (ep-ih-MIS-e-um) is a connective tissue sheath that encases the entire muscle. The epimysium forms the innermost layer of the **deep fascia**, the tough, fibrous connective tissue membrane that encloses and defines a muscle.

Note that all these layers are named with prefixes that describe their position: *endo-* meaning "within," *peri-* meaning "around," and *epi-* meaning "above." (These prefixes are added to the root *my/o*, meaning "muscle.") All of these supporting tissues merge to form the **tendon**, a band of dense regular connective tissue that attaches a muscle to a bone (**see Fig. 8-1**).

MUSCLE CELLS IN ACTION

Nerve impulses coming from the brain and the spinal cord stimulate skeletal muscle fiber contractions (see Chapter 9). Because these impulses often stimulate movement, they are

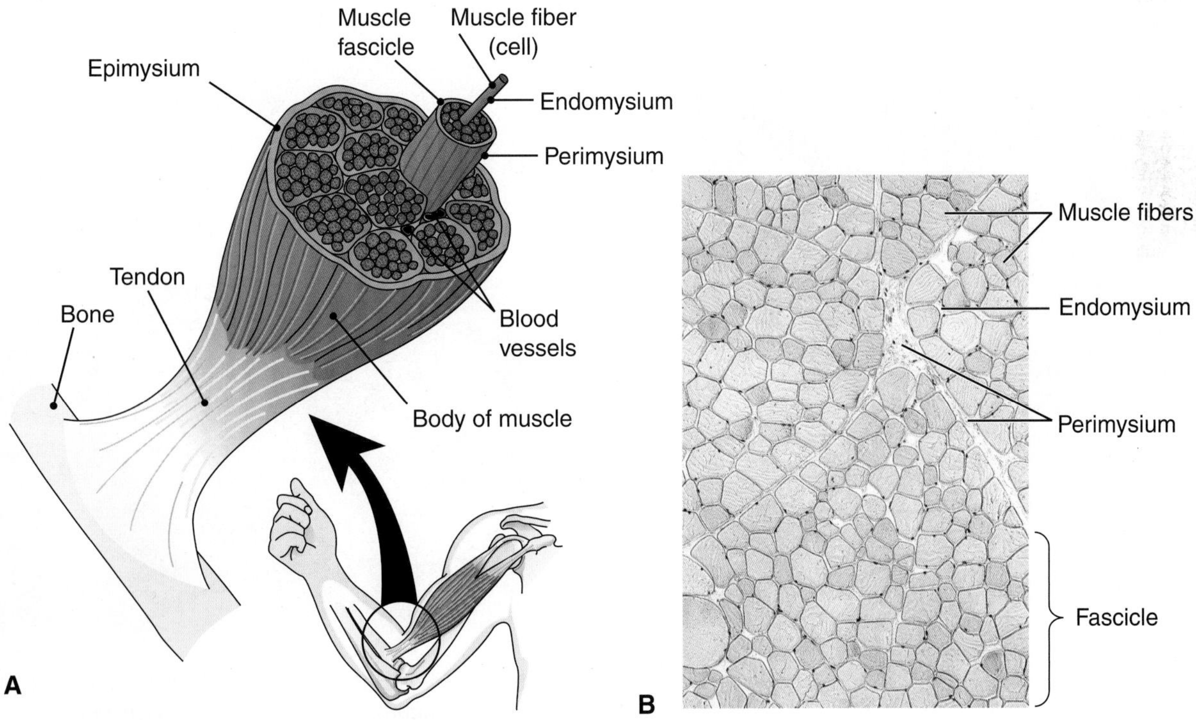

Figure 8-1 **Structure of a skeletal muscle.** **KEY POINT** Muscles are held together by layers of connective tissue. These layers merge to form the tendon that attaches the muscle to a bone. **A.** Muscle structure. **B.** Muscle tissue seen under a microscope. Portions of several fascicles are shown with connective tissue coverings. **ZOOMING IN** What is the innermost layer of connective tissue in a muscle? What layer of connective tissue surrounds a fascicle of muscle fibers?

described as *motor impulses*, and the neurons (nerve cells) that carry these impulses are described as motor neurons. In contrast, sensory impulses travel in sensory neurons from the periphery toward the central nervous system. Each neuron has a long extension called an axon. The axons of motor neurons branch to supply from a few to hundreds of individual muscle cells, or in some cases more than 1,000 (**Fig. 8-2A**).

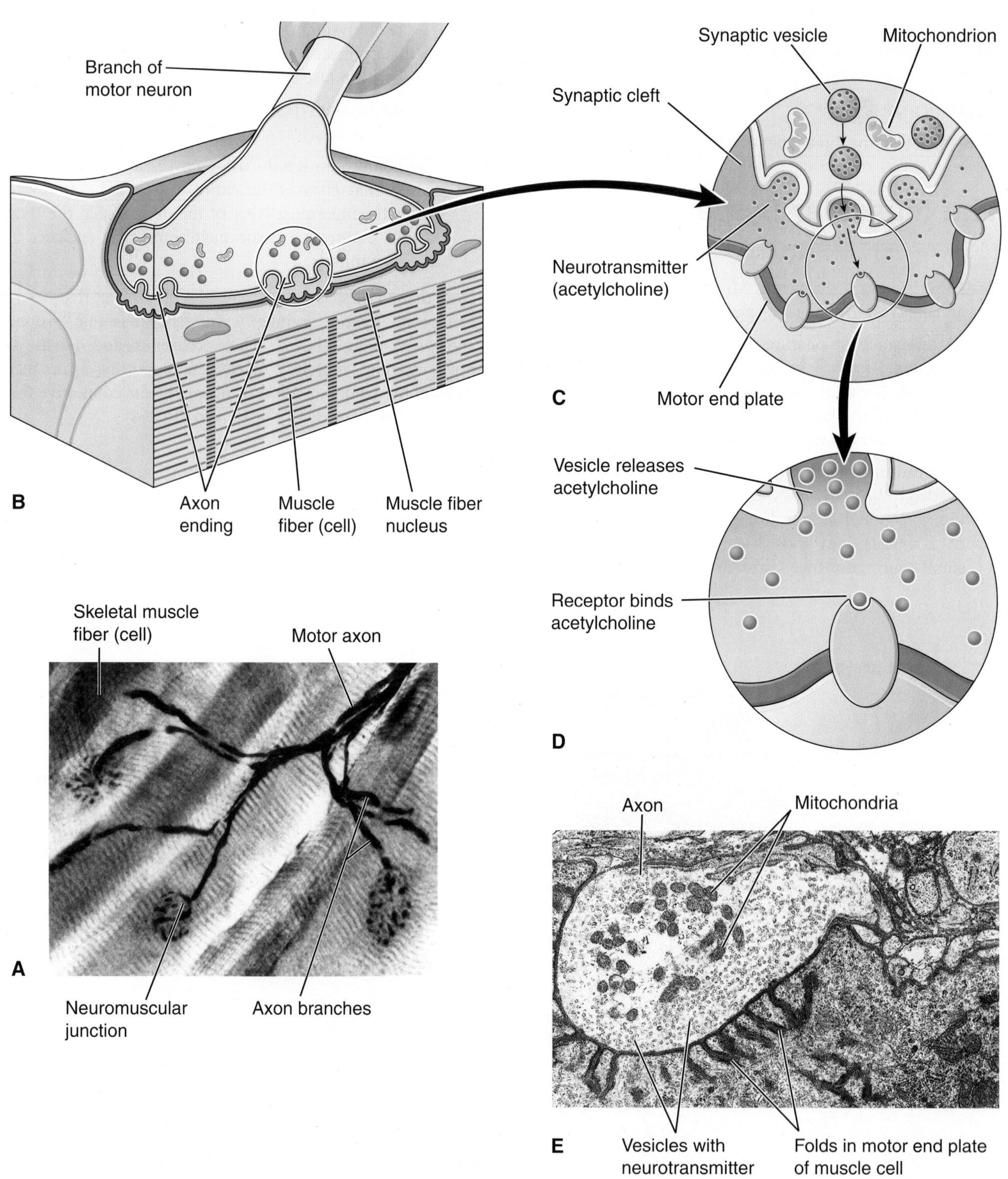

Figure 8-2 **The neuromuscular junction (NMJ).** **KEY POINT** Motor neurons stimulate skeletal muscle cells at the NMJ. **A.** A motor axon branches to stimulate multiple muscle fibers (cells). **B.** An axon branch makes contact with the membrane of a muscle fiber (cell) at the NMJ. **C.** Enlarged view of the NMJ showing release of neurotransmitter (acetylcholine) into the synaptic cleft. **D.** Acetylcholine attaches to receptors in the motor end plate, whose folds increase surface area. **E.** Electron microscope photograph of the NMJ.

A single neuron and all the muscle fibers it stimulates constitute a **motor unit.** Stimulation of the neuron activates all of the associated muscle fibers, so stronger or weaker contractions use more or fewer motor units (respectively). Muscles containing small motor units (with few muscle fibers) provide more control because they can change contraction strength in small increments. Muscles controlling the hand and the eye, for example, contain small motor units and perform very precise movements. Muscles containing larger motor units are used for maintaining posture or for broad movements, such as walking or swinging a tennis racquet.

The Membrane Potential Understanding how signals travel through neurons and muscle cells requires us to revisit the chemical properties of ions. Both the extracellular and intracellular fluids contain large numbers of positive and negative ions. For the most part, each negative ion can pair with a positive ion. Just as when we add "+1" to "−1" to get zero, equal numbers of positive and negative ions cancel each other out. But in a resting cell, the intracellular fluid contains a small excess of negative ions, and the extracellular fluid contains a small excess of positive ions. As a result of these unpaired charges, the plasma membrane of a living cell carries a difference in electric charge (voltage) on either side that is known as a **membrane** (or transmembrane) **potential** (po-TEN-shal). Membrane potential is measured inside the cell, so in resting cells, it is negative (about −70 millivolts, or mV).

Muscle cells and neurons show the property of **excitability,** because their membrane potential can change. For instance, the membrane potential becomes more negative if negative ions enter the cell. It becomes less negative (more positive) if positive ions enter the cell to neutralize the unpaired negative ions. These changes create electric signals, because they spread along the membrane, much like an electric current spreads along a wire. This spreading wave of electric current is called the **action potential** because it calls the cell into action.

The Neuromuscular Junction The point at which a nerve fiber contacts a muscle cell is called the **neuromuscular junction** (NMJ) (**see Fig. 8-2**). It is here that a chemical classified as a **neurotransmitter** is released from the neuron to stimulate the muscle fiber. The specific neurotransmitter released here is **acetylcholine** (as-e-til-KO-lene), abbreviated ACh, which is found elsewhere in the body as well. A great deal is known about the events that occur at this junction, and this information is important in understanding muscle action.

The NMJ is an example of a **synapse** (SIN-aps), a point of communication between a neuron and another cell (the term comes from a Greek word meaning "to clasp"). At every synapse, there is a tiny space, the **synaptic cleft,** across which the neurotransmitter must travel. Until its release, the neurotransmitter is stored in tiny membranous sacs, called vesicles, in the nerve fiber's endings. Once released, the neurotransmitter crosses the synaptic cleft and attaches to a receptor, which is a protein embedded in the muscle cell membrane. The muscle cell membrane forms multiple folds at this point, and these serve to increase surface area and hold a maximum number of receptors. The muscle cell's receiving membrane is known as the **motor end plate.**

Once acetylcholine binds the receptor in the motor end plate, the bound receptor initiates an action potential in the muscle cell that spreads rapidly along the muscle cell membrane. We will see later how this action potential results in muscle contraction. Chapter 9 provides more information on synapses and the action potential.

See the Student Resources on thePoint to view the animation "The Neuromuscular Junction."

8

Contraction Another important property of muscle tissue is **contractility.** This is a muscle fiber's capacity to undergo shortening, becoming thicker. Studies of muscle chemistry and observation of cells under the powerful electron microscope have increased our understanding of how muscle cells work.

These studies reveal that each skeletal muscle fiber contains many threads, or filaments, made primarily of two kinds of proteins, called **actin** (AK-tin) and **myosin** (MI-o-sin). Filaments made of actin are thin and light; those made of myosin are thick and dark. The filaments are present in alternating bundles within the muscle cell (**Fig. 8-3**).

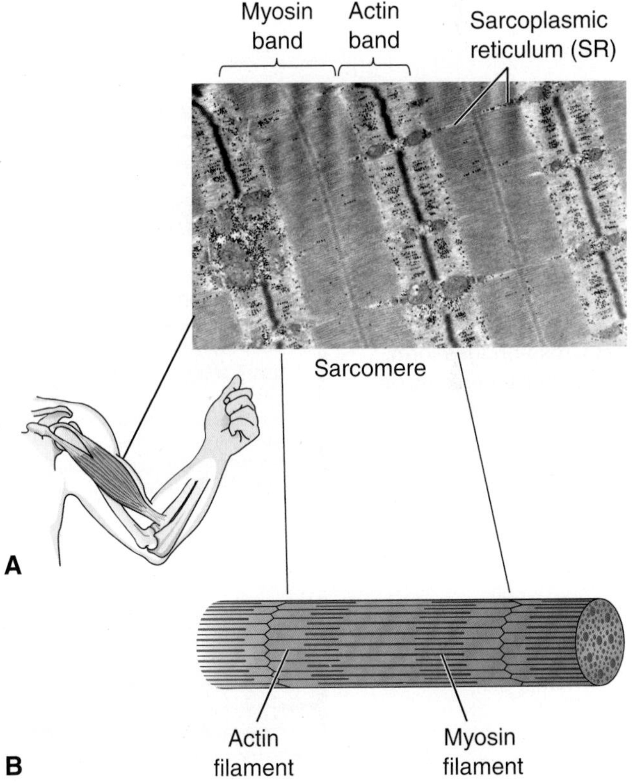

Figure 8-3 Detailed structure of a skeletal muscle cell. **KEY POINT A.** Photomicrograph of skeletal muscle cell (×6,500). Actin makes up the light band and myosin makes up the dark band. The dark line in the actin band marks points where actin filaments are held together. A sarcomere is a contracting subunit of skeletal muscle. The sarcoplasmic reticulum is the ER of muscle cells. **B.** Diagram of the photographic image.

It is the alternating bands of light actin and heavy myosin filaments that give skeletal muscle its striated appearance. They also give a view of what occurs when muscles contract.

Note that the actin and myosin filaments overlap where they meet, just as your fingers overlap when you fold your hands together. A contracting subunit of skeletal muscle is called a **sarcomere** (SAR-ko-mere). It consists of a band of myosin filaments and the actin filaments on each side (**see Fig. 8-3**). The myosin molecules are shaped like two golf clubs twisted together with their paddle-like heads projecting away from the sarcomere's center. The actin molecules are twisted together like two strands of beads, each bead having a myosin-binding site (**Fig. 8-4**).

Figure 8-4 shows a section of muscle as it contracts. In movement, the myosin heads "latch on" firmly to the actin filaments in their overlapping region, forming attachments between the filaments that are described as *cross-bridges*. Using the energy of stored adenosine triphosphate (ATP), the myosin heads, like the oars of a boat moving water, pull all the actin strands closer together within each sarcomere. New ATP molecules trigger the release of the myosin heads and move them back to position for another "power stroke." With repeated movements, the overlapping filaments slide together, and the muscle fiber contracts, becoming shorter and thicker. This action is aptly named the *sliding filament mechanism* of muscle contraction. Note that the filaments overlap increasingly as the cell contracts. (In reality, not all the myosin heads are moving at the same time. About one-half are forward at any time, and the rest are preparing for another swing.) During contraction, each sarcomere becomes shorter, but the individual filaments do not change in length. As in shuffling a deck of cards, as you push the cards together, the deck becomes smaller, but the cards do not change in length.

The Role of Calcium In addition to actin, myosin, and ATP, calcium is needed for muscle contraction. It enables cross-bridges to form between actin and myosin so the sliding filament action can begin. When muscles are at rest, two additional proteins called **troponin** (tro-PO-nin) and **tropomyosin** (tro-po-MI-o-sin) block the sites on actin filaments where cross-bridges can form (**Fig. 8-5**). When calcium attaches to the troponin, these proteins move aside, uncovering the binding sites. In resting muscles, the calcium is not available because it is stored within the cell's endoplasmic reticulum, which, in muscle cells, is called the

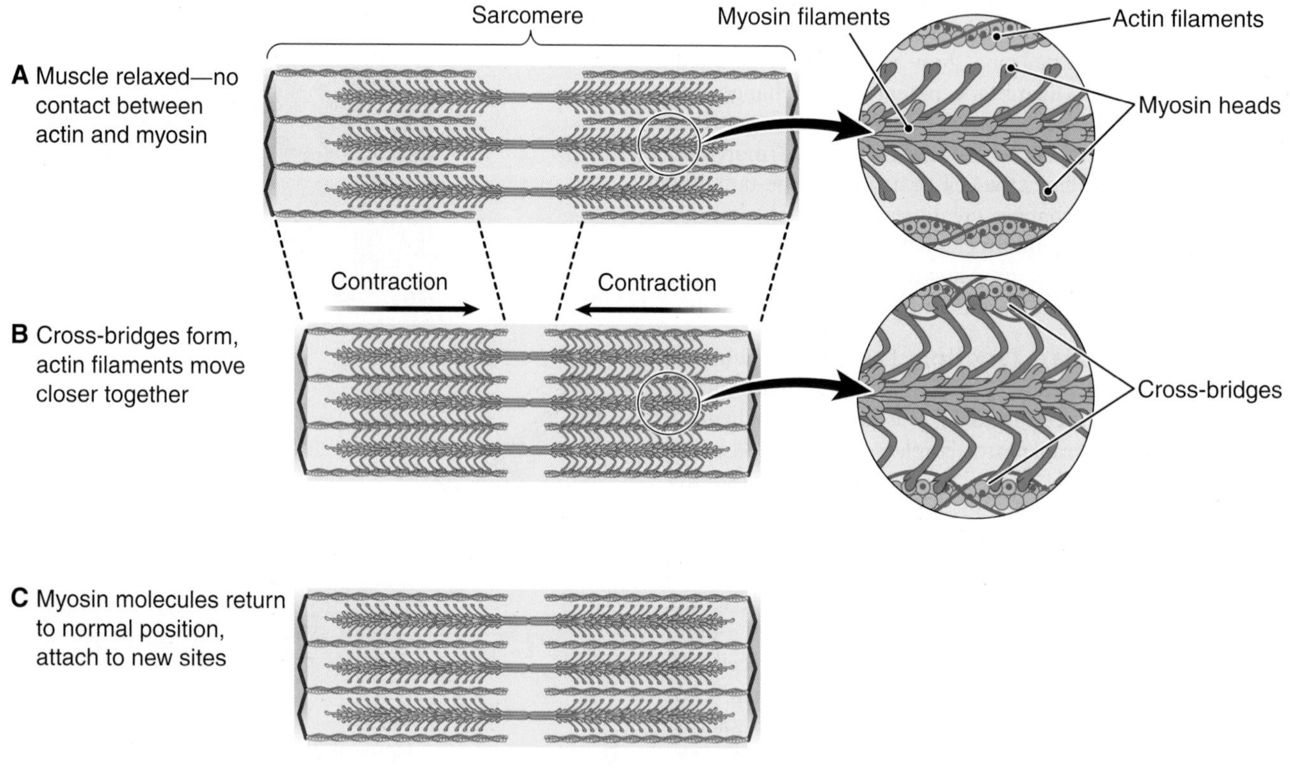

Figure 8-4 Sliding filament mechanism of skeletal muscle contraction. **KEY POINT** Muscle contraction depends on the interaction of actin and myosin filaments within the cell. **A.** Muscle is relaxed, and there is no contact between the actin and myosin filaments. **B.** Cross-bridges form, and the actin filaments are moved closer together as the muscle fiber contracts. **C.** The cross-bridges break, and the myosin heads attach to new sites to prepare for another pull on the actin filaments and further contraction. **ZOOMING IN** Do the actin or myosin filaments change in length as contraction proceeds?

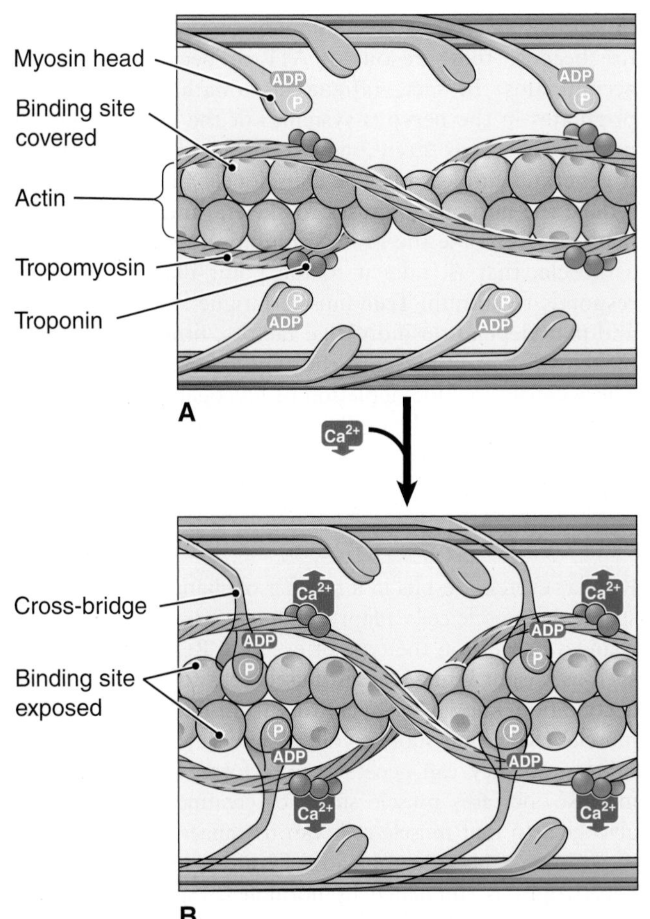

Figure 8-5 Role of calcium in muscle contraction.

KEY POINT Calcium unblocks sites where cross-bridges can form between actin and myosin filaments to begin muscle contraction. **A.** Troponin and tropomyosin cover the binding sites where cross-bridges can form between actin and myosin. **B.** Calcium shifts troponin and tropomyosin away from binding sites so cross-bridges can form.

sarcoplasmic reticulum (SR). Calcium is released into the cytoplasm in response to the action potential in the muscle cell that we discussed earlier. Muscles relax when nervous stimulation stops, and the calcium is pumped back into the SR, ready for the next contraction.

A summary of the events in a muscle contraction is as follows:

1. ACh is released from a neuron ending into the synaptic cleft at the NMJ.
2. ACh binds to the muscle's motor end plate and produces an action potential.
3. The action potential travels to the SR.
4. The SR releases calcium into the cytoplasm.
5. Calcium shifts troponin and tropomyosin so that binding sites on actin are exposed.
6. Myosin heads bind to actin, forming cross-bridges.
7. Using stored energy, myosin heads pull actin filaments together within the sarcomeres, and the cell shortens.
8. New ATP is used to detach myosin heads and move them back to position for another "power stroke."
9. Muscle relaxes when stimulation ends, and the calcium is pumped back into the SR.

The phenomenon of rigor mortis, a state of rigidity that occurs after death, illustrates ATP's crucial role in muscle contraction. Shortly after death, muscle cells begin to degrade. Calcium escapes into the cytoplasm, stimulating cross-bridge formation and muscle contraction. Metabolism has ceased, however, and there is no ATP to disengage the filaments or pump calcium out of the cytoplasm. The cross-bridges cannot detach, so the muscle remains locked in a contracted rigor mortis that lasts about 24 hours, gradually fading as enzymes break down the muscle filaments.

8

CHECKPOINTS

- ☐ 8-2 What are the three main functions of skeletal muscle?
- ☐ 8-3 What are bundles of muscle fibers called?
- ☐ 8-4 What is the term for the difference in electrical charge on the two sides of a plasma membrane?
- ☐ 8-5 What is the name of the special synapse where a nerve cell makes contact with a muscle cell?
- ☐ 8-6 What neurotransmitter is involved in the stimulation of skeletal muscle cells?
- ☐ 8-7 What filaments interact to produce muscle contraction?
- ☐ 8-8 What mineral is needed for interaction of the contractile filaments?

ENERGY SOURCES

As noted earlier, all muscle contraction requires energy in the form of ATP. Most of this energy is produced by the oxidation (commonly called "burning") of nutrients within the cell, especially the oxidation of glucose and fatty acids. Metabolism that requires oxygen is described as *aerobic* (the root *aer/o* means "air" or "gas," but in this case refers to oxygen).

Storage Compounds The circulating blood constantly brings nutrients and oxygen to the cells, but muscle cells also store a small supply of each for rapid ATP generation, such as during vigorous exercise. For example:

- **Myoglobin** (mi-o-GLO-bin) stores oxygen. This compound is similar to the hemoglobin in blood, but it is located specifically in muscle cells as indicated by the root *my/o* in its name.
- **Glycogen** (GLI-ko-jen) is the storage form of glucose. It is a polysaccharide made of multiple glucose molecules, and it can be broken down into glucose when needed by the muscle cells.
- Fatty acids are stored as triglycerides formed into fat droplets. These droplets can be broken down into fatty acids when needed by the muscle cells.

Anaerobic Metabolism Oxidation is very efficient and yields a large amount of ATP per nutrient molecule, but it has some limitations. First, it takes a while to start generating ATP, so oxidation cannot supply enough energy for the first few seconds of muscle contraction. Second, it requires an abundant oxygen supply. During strenuous activity, oxygen delivery to the tissues cannot keep up with the demands of hard-working muscles. These two situations rely on alternate, rapid mechanisms of ATP production that do not require oxygen. They are described as *anaerobic* (*an-* means "not" or "without"). Note that these anaerobic processes are always occurring, but they are particularly important at the beginning of exercise or during very strenuous exercise.

1. Breakdown of **creatine** (KRE-ah-tin) **phosphate**. Creatine phosphate is a compound similar to ATP in that it has a high-energy bond that breaks down to release energy. This energy is used to make ATP for muscle contraction. It generates ATP very rapidly, but its supply is limited (see Box 8-1).
2. **Anaerobic glycolysis**. This process breaks glucose down incompletely without using oxygen (*glyc/o* means "glucose" and *-lysis* means "separation"). A few ATPs are generated in these reactions, as is a byproduct called lactic acid, which is later oxidized for energy when oxygen is available.

When a person stops exercising, the body must generate enough ATP to reestablish a resting state by replenishing stored materials. ATP is also needed to restore normal body temperature. The person must take in extra oxygen by continued rapid breathing, known as *excess postexercise oxygen consumption*. Chapter 20 has more details on metabolism.

Muscle Fatigue It is commonly thought that muscles tire because they are out of ATP or because lactic acid accumulates. In fact, fatigue in nonathletes frequently originates in the nervous system, not the muscles. People unaccustomed to strenuous exercise find the sensations it generates unpleasant and consciously or unconsciously reduce the nervous impulses to skeletal muscles. It is difficult to overcome the brain's inhibition and truly fatigue a muscle, that is, take it to the point that it no longer responds to stimuli. True muscle fatigue has many causes and may depend on individual factors, fitness and genetic makeup for example, and the type of exercise involved. These causes include depletion of glycogen reserves, inadequate oxygen supply, or the accumulation of phosphates from ATP breakdown.

EFFECTS OF EXERCISE

Regular exercise results in a number of changes in muscle tissue as the muscle cells adapt to the increased workload. The changes depend on the type of exercise. Resistance training, such as weight lifting, causes muscle cells to increase in size, a condition known as hypertrophy (hi-PER-tro-fe). Larger muscle cells contain more myofibrils and can form more cross-bridges, so they can generate more force. Resistance training also increases muscle stores of creatine phosphate and glycogen, so that muscle cells can use anaerobic metabolism to generate a large amount of ATP in a short time. Muscle hypertrophy is stimulated by hormones, especially the male sex steroids. Box 8-2 has information on how some athletes abuse these steroids to increase muscle size and strength at the expense of their health.

A CLOSER LOOK — Box 8-1

Creatine Kinase: Muscle's Backup Energy Enzyme

At rest, muscle cells store some of the ATP they produce in the sarcoplasm. But the amount stored is sufficient for only a few seconds of contraction, and it takes several more seconds before anaerobic glycolysis and oxidation can replenish it. So how do muscle fibers power their contractions in the meantime? They have a backup energy source called creatine phosphate to tide them over.

When muscle cells are resting, they manufacture creatine phosphate by transferring energy from ATP to creatine, a substance produced by the liver, pancreas, and kidneys. When muscle cells are exercising actively, they transfer that energy to ADP to create ATP. There is enough creatine phosphate in the sarcoplasm to produce four or five times the original amount of stored ATP—enough to power the cell until the other ATP-producing reactions take over.

Creatine kinase (CK) catalyzes the transfer of energy from creatine phosphate to ADP. CK is found in all muscle cells and other metabolically active cells, such as neurons. It is composed of two subunits, which can be either *B* (brain type) or *M* (muscle type). Therefore, there are three forms of CK: CK-BB, CK-MM, and CK-MB, which are present at different levels in various tissues. CK-BB is found mainly in nervous and smooth muscle tissue. CK-MM is found predominantly in skeletal muscle. Cardiac muscle contains both CK-MM and CK-MB.

Normally, the blood level of CK is low, but damage to CK-containing tissues increases it. Thus, clinicians can use blood CK levels in diagnosis. Elevated blood CK-BB may indicate a nervous system disorder, such as stroke or amyotrophic lateral sclerosis (Lou Gehrig disease). Elevated blood levels of CK-MM may indicate a muscular disorder, such as myositis or muscular dystrophy, as seen in Shane's opening case. Elevated blood levels of both CK-MM and CK-MB may indicate cardiac muscle damage following a myocardial infarction (heart attack).

Box 8-2

HOT TOPICS

Anabolic Steroids: Winning at All Costs?

Anabolic steroids mimic the effects of the male sex hormone testosterone by promoting metabolism and stimulating growth. These drugs are legally prescribed to promote muscle regeneration and prevent atrophy from disuse after surgery. However, some athletes also purchase them illegally, using them to increase muscle size and strength and improve endurance.

When steroids are used illegally to enhance athletic performance, the doses needed are large enough to cause serious side effects. They increase blood cholesterol levels, which may lead to atherosclerosis, heart disease, kidney failure, and stroke. They damage the liver, making it more susceptible to cancer and other diseases, and suppress the immune system, increasing the risk of infection and cancer. In men, steroids cause impotence, testicular atrophy, low sperm count, infertility, and the development of female sex characteristics, such as breasts (gynecomastia). In women, steroids disrupt ovulation and menstruation and produce male sex characteristics, such as breast atrophy, enlargement of the clitoris, increased body hair, and deepening of the voice. In both sexes, steroids increase the risk for baldness and, especially in men, cause mood swings, depression, and violence.

Aerobic exercise, that is, exercise that increases oxygen consumption, such as running, biking, or swimming, leads to improved muscular endurance. Endurance training increases the muscle cells' blood supply and number of mitochondria, improving their ability to generate ATP aerobically and to get rid of waste products. Endurance-trained muscles can contract more frequently and for longer periods without fatiguing.

Cardiovascular changes are perhaps the most important physical benefits of endurance exercise. You can think of endurance exercise as "strength training for the heart." That organ has to pump up to five times as much blood during endurance exercise as it does at rest. The heart muscle (especially the left ventricle) adapts to its increased workload by growing larger and stronger. Its increased pumping efficiency means that the heart doesn't have to work very hard at rest, so the resting heart rate declines. Endurance exercise also benefits the body by decreasing the amount of less healthy (LDL) cholesterol in the blood, reducing blood pressure, and improving blood glucose control.

These beneficial changes may derive in part from the reduced body fat and improved psychological well-being that result from regular physical activity. Whatever the cause, it's well established that aside from not smoking, exercise is the most important thing you can do to improve your health. Recent studies show that a few short bouts of high-intensity exercise are equally or more effective than longer, low-intensity workouts. Even people such as the elderly and type II diabetics show significant improvements when they exercise for 10 intervals of 60 seconds, each at maximum intensity.

The benefits of one form of exercise do not significantly carry over to the other—endurance exercise does not significantly increase muscle strength, and resistance training does not significantly increase muscle endurance. An exercise program thus should include both methods with periods of warm-up and cool-down before and after working out. Stretching generally improves the range of motion at the joints and improves balance. However, studies show that static stretching (holding an extended position for 30 seconds to two minutes) just before a strenuous workout actually decreases muscle strength and increases the risk of injury.

TYPES OF MUSCLE CONTRACTIONS

Muscle tone refers to a muscle's partially contracted state that is normal even when the muscle is not in use. The maintenance of this tone, or **tonus** (TO-nus), is due to the action of the nervous system in keeping the muscles in a constant state of readiness for action. Muscles that are little used soon become flabby, weak, and lacking in tone.

In addition to the partial contractions that are responsible for muscle tone, there are two other types of contractions on which the body depends:

- In **isotonic** (i-so-TON-ik) **contractions**, the tone or tension within the muscle remains the same, but muscle length changes, and the muscle bulges as it accomplishes work (*iso-* means "same" or "equal" and *ton* means "tension"). Within this category, there are two forms of contractions:
  - **Concentric contractions.** These contractions are more familiar, as they produce more obvious changes in position. In concentric contractions, a muscle as a whole shortens to produce movement. Try flexing your arm at the elbow to pick up a dumbbell or heavy can. The anterior arm flexors, the biceps brachii and brachialis, move the forearm at the elbow, lifting the weight, and you can see that the muscles change shape and bulge outward.
  - **Eccentric contractions.** In these contractions, the muscle lengthens as it exerts force. Think of gradually lowering that dumbbell or heavy can. The arm flexors tense as they lengthen. These contractions strengthen muscles considerably but are more likely to cause soreness later, perhaps because of microscopic tears in the muscle fibers.
- In **isometric** (i-so-MET-rik) **contractions**, there is no change in muscle length, but there is a great increase in muscle tension (*metr/o* means "measure"). Pushing against an immovable force produces an isometric contraction, as in trying to lift a weight that is too heavy to move. Try pushing the palms of your hands hard against

each other. There is no movement or change in muscle length, but you can feel the increased tension in your arm muscles.

Most body movements involve a combination of both isotonic and isometric contractions. When walking, for example, some muscles contract isotonically to propel the body forward, but at the same time, other muscles are contracting isometrically to keep your body in position.

CHECKPOINTS 

- 8-9 What compound is formed in oxidation of nutrients that supplies the energy for muscle contraction?
- 8-10 What compound stores reserves of oxygen in muscle cells?
- 8-11 What are the two main types of muscle contraction?

The Mechanics of Muscle Movement

Most muscles have two or more points of attachment to the skeleton. The muscle is attached to a bone at each end by means of a cordlike extension called a tendon (**Fig. 8-6**). All of the connective tissue within and around the muscle merges to form the tendon, which then attaches directly to the bone's periosteum (**see Fig. 8-1**). In some instances, a broad sheet called an **aponeurosis** (ap-o-nu-RO-sis) may attach muscles to bones or to other muscles.

In moving the bones, one end of a muscle is attached to a more freely movable part of the skeleton, and the other end is attached to a relatively stable part. The less movable (more fixed) attachment is called the **origin**; the attachment to the body part that moves is called the **insertion** (**Fig. 8-6**). When a muscle contracts, it pulls on both attachment points, bringing the more movable insertion closer to the origin and thereby causing movement of the body part. **Figure 8-6** shows the action of the brachialis (in the arm) in flexing the arm at the elbow. The muscle's insertion on the radius of the forearm is brought toward the origin at the scapula of the shoulder girdle.

MUSCLES WORK TOGETHER

Many of the skeletal muscles function in pairs. The main muscle that performs a given movement is the **prime mover**. For instance, the brachialis is the prime mover for flexion of the arm at the elbow (**see Fig. 8-6**). Because any muscle that performs a given action is technically called an

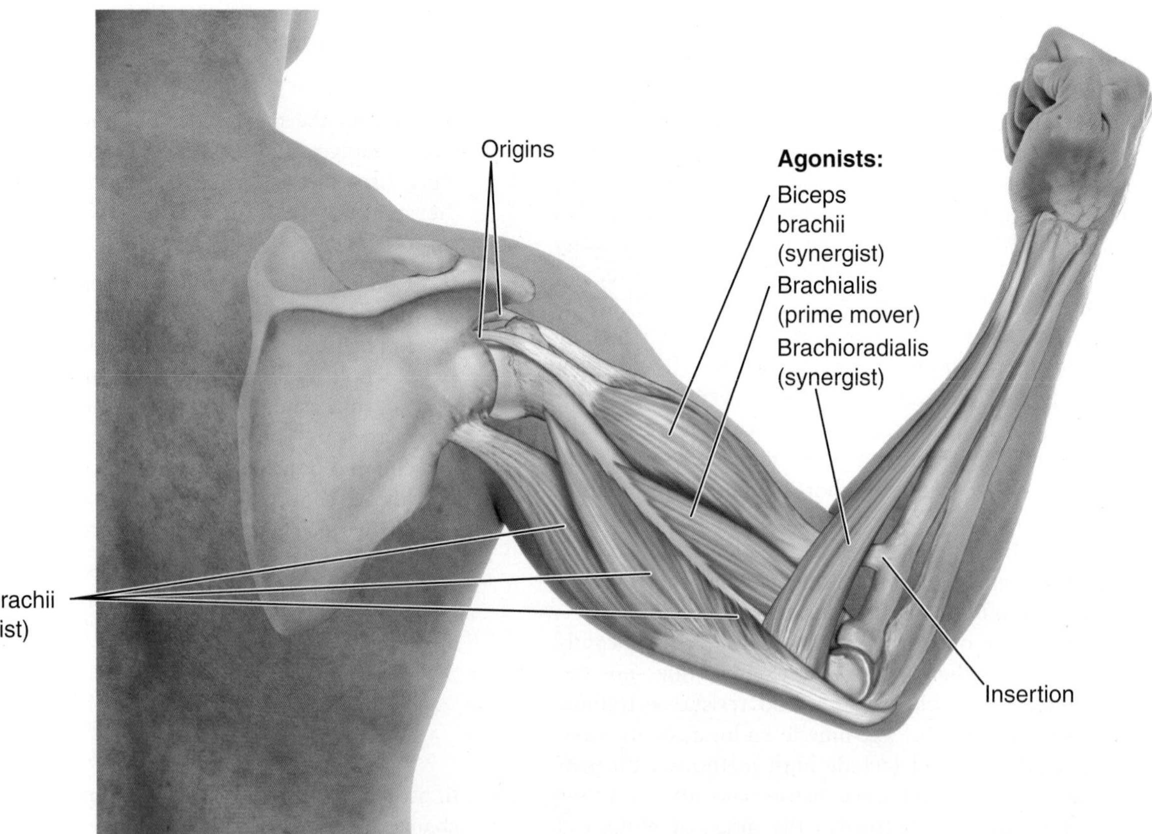

Figure 8-6 Muscle attachments to bones. **KEY POINT** Tendons attach muscles to bones. The stable point is the origin; the movable point is the insertion. In this diagram, three attachments are shown—two origins and one insertion. **ZOOMING IN** Does contraction of the brachialis produce flexion or extension at the elbow?

agonist (AG-on-ist), the muscle that produces an opposite action is termed the **antagonist** (an-TAG-on-ist) (the prefix *anti-* means "against"). Clearly, for any given movement, the antagonist must relax when the agonist contracts. For example, when the brachialis at the anterior arm contracts to flex the arm, the triceps brachii at the back must relax; when the triceps brachii contracts to extend the arm, the brachialis must relax. In addition to prime movers and antagonists, there are also muscles that steady body parts or assist in an action. These "helping" muscles are called **synergists** (SIN-er-jists), because they work with the prime mover to accomplish a movement (*syn-* means "together" and *erg/o* means "work"). For example, the biceps brachii and the brachioradialis are synergists to the brachialis in flexing the arm (**see Fig 8-6**).

As the muscles work together, actions are coordinated to accomplish many complex movements. Note that during development, the nervous system must gradually begin to coordinate our movements. A child learning to walk or to write, for example, may use muscles unnecessarily at first or fail to use appropriate muscles when needed.

LEVERS AND BODY MECHANICS

Proper body mechanics help conserve energy and ensure freedom from strain and fatigue; conversely, such ailments as lower back pain—a common complaint—can be traced to poor body mechanics. Body mechanics have special significance to healthcare workers, who frequently must move patients and handle cumbersome equipment. Maintaining the body segments in correct alignment also affects the vital organs that are supported by the skeleton.

If you have had a course in physics, recall your study of levers. A lever is simply a rigid bar that moves about a fixed pivot point, the fulcrum. There are three classes of levers, which differ only in the location of the fulcrum (F), the effort (E), or force, and the resistance (R), the weight or load:

- In a first-class lever, the fulcrum is located between the resistance and the effort; a seesaw or a pair of scissors is an example of this class (**Fig. 8-7A**).
- The second-class lever has the resistance located between the fulcrum and the effort; a wheelbarrow or a mattress lifted at one end is an illustration of this class. However, there are no significant examples of second-class levers in the body.
- In the third-class lever, the effort is between the resistance and the fulcrum. A forceps or a tweezers is an example of this type of lever. The effort is applied in the tool's center, between the fulcrum, where the pieces join, and the resistance at the tip (**see Fig. 8-7B**).

The musculoskeletal system can be considered a system of levers, in which the bone is the lever, the joint is the fulcrum, and the force is applied by a muscle. An example of a first-class lever in the body is using the muscles at the back of the neck to lift the head at the joint between the skull's occipital bone and the first cervical vertebra (atlas) (**see Fig. 8-7**). However, there are very few examples of first-class levers in the body. Most lever systems in the body are of the third-class type. A muscle usually inserts past a joint and exerts force between the fulcrum and the resistance. That is, the fulcrum is behind both the point of effort and the weight. As shown in **Figure 8-7B**, when the biceps brachii helps flex the forearm

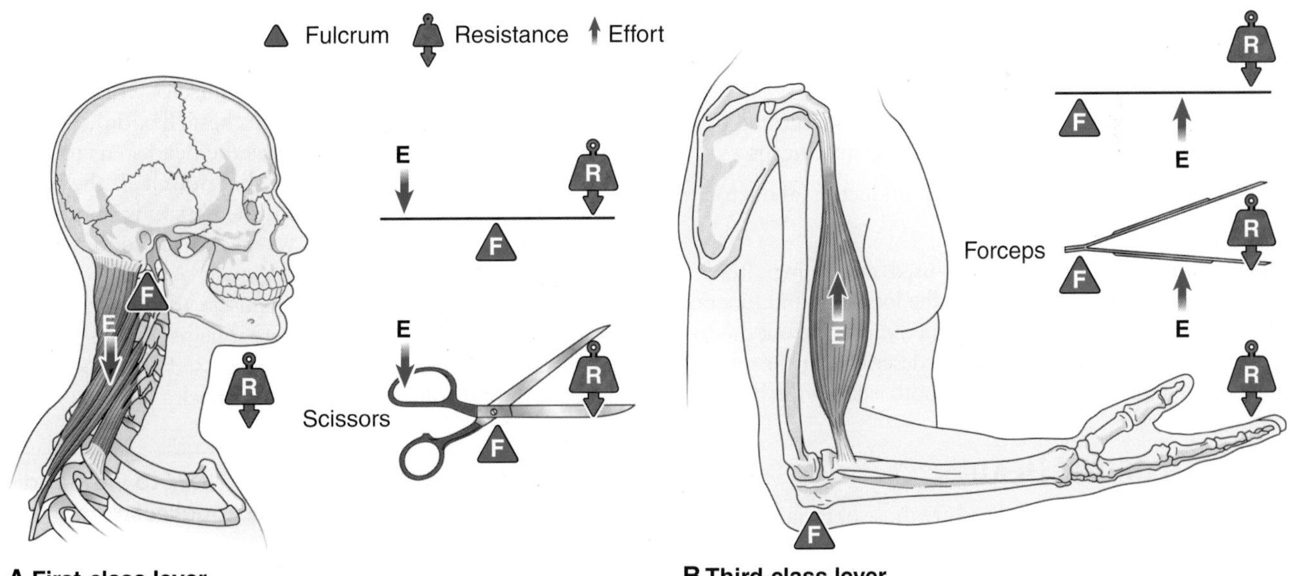

Figure 8-7 **Levers.** **KEY POINT** Muscles work with bones as lever systems to produce movement. Two classes of levers are shown along with tools and anatomic examples that illustrate each type. R, resistance (weight); E, effort (force); F, fulcrum (pivot point). A first-class lever **(A)** and third-class lever **(B)** are shown. **ZOOMING IN** In a third-class lever system, where is the fulcrum with regard to the effort and the resistance?

at the elbow, the muscle exerts its force at its insertion on the radius. The weight of the hand and forearm creates the resistance, and the fulcrum is the elbow joint, which is behind the point of effort.

By understanding and applying knowledge of levers to body mechanics, healthcare workers can reduce the risk of musculoskeletal injury while carrying out their numerous clinical tasks.

CHECKPOINTS 

- ☐ 8-12 What are the names of the two attachment points of a muscle, and how do they function?
- ☐ 8-13 What is the name of the muscle that produces a movement as compared with the muscle that produces an opposite movement?
- ☐ 8-14 Of the three classes of levers, which one represents the action of most muscles?

Skeletal Muscle Groups

The study of muscles is made simpler by grouping them according to body regions. Knowing how muscles are named can also help in remembering them. A number of different characteristics are used in naming muscles, including the following:

- location, using the name of a nearby bone, for example, or a position, such as lateral, medial, internal, or external
- size, using terms such as maximus, major, minor, longus, or brevis
- shape, such as circular (orbicularis), triangular (deltoid), or trapezoidal (trapezius)
- direction of fibers, including straight (rectus) or angled (oblique)
- number of heads (attachment points), as indicated by the suffix *-ceps*, as in biceps, triceps, and quadriceps
- action, as in flexor, extensor, adductor, abductor, or levator

Often, more than one feature is used in naming. Refer to **Figures 8-8 and 8-9** as you study the locations and functions of the superficial skeletal muscles and try to figure out the basis for each name. Although they are described in the singular, most of the muscles are present on both sides of the body.

MUSCLES OF THE HEAD

The principal muscles of the head are those of facial expression and of mastication (chewing) (**Fig. 8-10, Table 8-2**).

The muscles of facial expression include ring-shaped ones around the eyes and the lips, called the **orbicularis** (or-bik-u-LAH-ris) **muscles** because of their shape (think of "orbit"). The muscle surrounding the eye is called the **orbicularis oculi** (OK-u-li), whereas the lip muscle is the **orbicularis oris.** These muscles all have antagonists. For example, the **levator palpebrae** (PAL-pe-bre) **superioris,** or lifter of the upper eyelid, is the antagonist for the orbicularis oculi.

One of the largest muscles of expression forms the fleshy part of the cheek and is called the **buccinator** (BUK-se-na-tor). Used in whistling or blowing, it is sometimes referred to as the trumpeter's muscle. You can readily think of other muscles of facial expression: for instance, the zygomaticus muscles produce a smile, while the depressor anguli oris muscles turn a smile into a grimace. There are a number of scalp muscles that lift the eyebrows or draw them together into a frown.

There are four pairs of mastication (chewing) muscles, all of which insert on and move the mandible. The largest are the **temporalis** (TEM-po-ral-is), which is superior to the ear, and the **masseter** (mas-SE-ter) at the angle of the jaw.

The tongue has two muscle groups. The first group, called the *intrinsic muscles,* is located entirely within the tongue. The second group, the *extrinsic muscles*, originates outside the tongue. Note that *intrinsic* is a generic term to describe muscles located within the moving structure, whereas *extrinsic* describes muscles that connect to the moving structure by tendons. It is because of these many muscles that the tongue has such remarkable flexibility and can perform so many different functions. Consider the intricate tongue motions involved in speaking, chewing, and swallowing. **Figure 8-10** shows some additional muscles of the face.

MUSCLES OF THE NECK

The neck muscles tend to be ribbon-like and extend vertically or obliquely in several layers and in a complex manner (**see Fig. 8-10, Table 8-2**). The one you will hear of most frequently is the **sternocleidomastoid** (ster-no-kli-do-MAS-toyd), sometimes referred to simply as the sternomastoid. This strong muscle extends superiorly from the sternum across the lateral neck to the mastoid process of the temporal bone. When the left and right muscles work together, they bring the head forward on the chest (flexion). Working alone, each muscle tilts and rotates the head so as to orient the face toward the side opposite that muscle. If the head is abnormally fixed in this position, the person is said to have **torticollis** (tor-tih-KOL-is), or wryneck; this condition may be caused by muscle injury or spasm.

A portion of the trapezius muscle (described later) is located at the posterior neck, where it helps hold the head up (extension). Other back muscles, discussed later, extend the entire vertebral column, including the neck.

See the Student Resources on thePoint to view additional pictures of head and neck musculature.

MUSCLES OF THE UPPER EXTREMITIES

Muscles of the upper extremities include the muscles that determine the position of the shoulder, the anterior and posterior muscles that move the arm, and the muscles that move the forearm and hand.

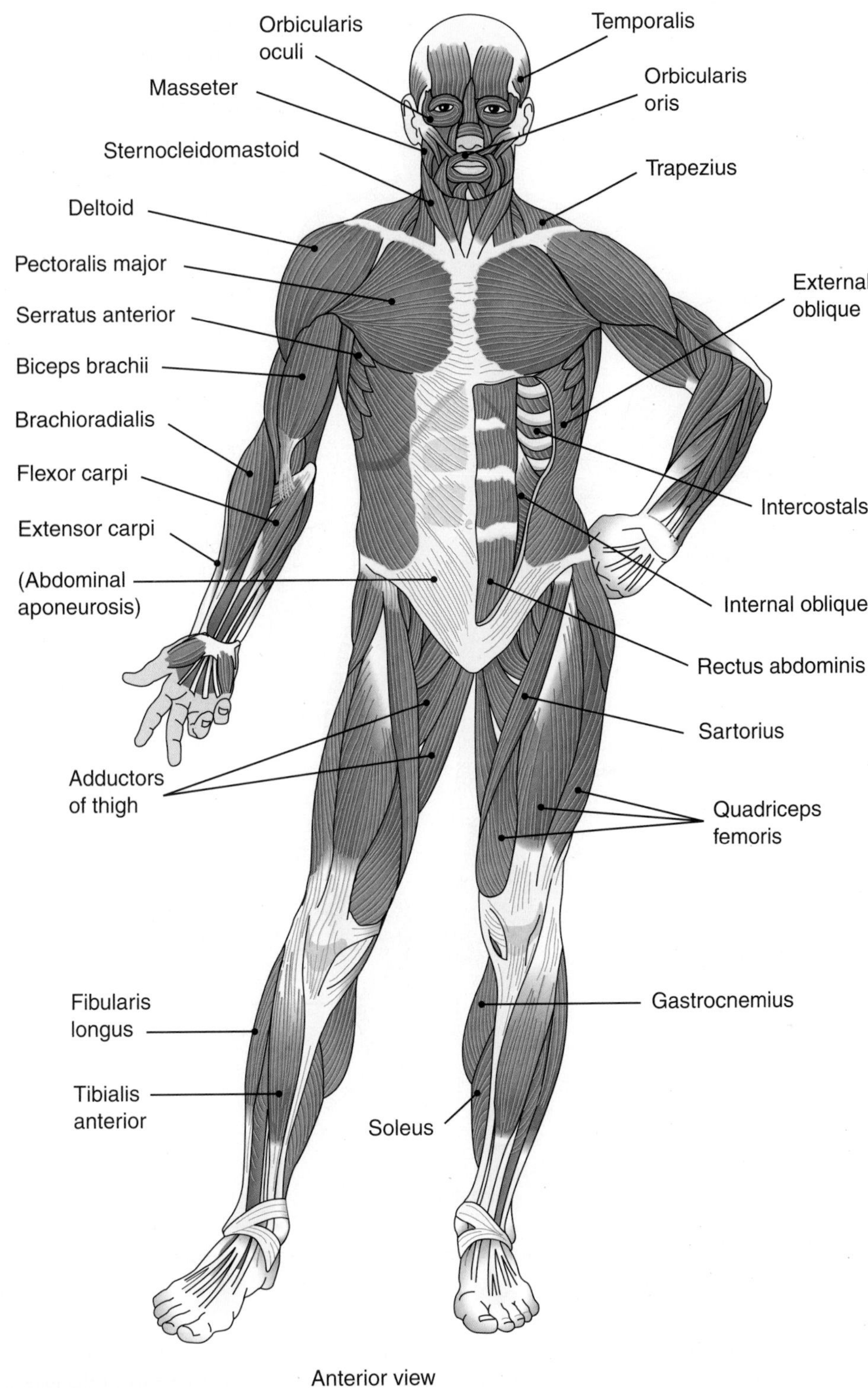

Figure 8-8 **Superficial muscles, anterior view.** An associated structure is labeled in parentheses. An aponeurosis is a broad, sheetlike tendon.

Muscles that Move the Shoulder and Arm The position of the shoulder depends to a large extent on the degree of contraction of the **trapezius** (trah-PE-ze-us), a triangular muscle that covers the posterior neck and extends across the posterior shoulder to insert on the clavicle and scapula (**see Figs. 8-8 and 8-9, Table 8-3**). The trapezius muscles enable one to raise the shoulders and pull them back. The superior portion of each trapezius can also extend the head and turn it from side to side.

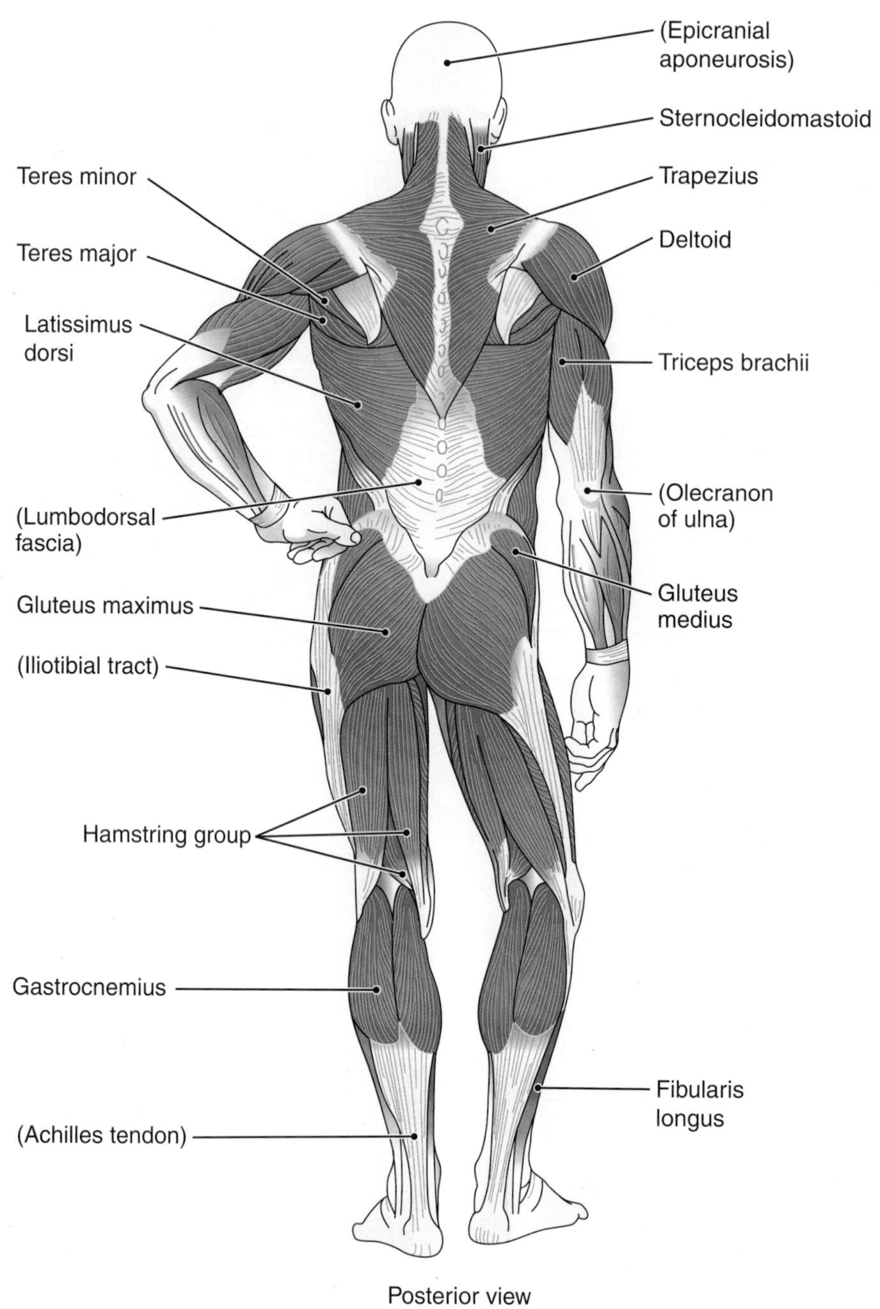

Figure 8-9 **Superficial muscles, posterior view.** Associated structures are labeled in parentheses.

The **latissimus** (lah-TIS-ih-mus) **dorsi** is the wide muscle of the back and lateral trunk (**see Fig. 8-9**). It originates from the vertebral spine in the middle and lower back and covers the inferior half of the thoracic region, forming the posterior portion of the axilla (armpit). The fibers of each muscle converge to a tendon that inserts on the humerus. The latissimus dorsi powerfully extends the arm, bringing it down forcibly as, for example, in swimming. It is also the prime mover in arm adduction.

A large **pectoralis** (pek-to-RAL-is) **major** is located on either side of the superior chest (**see Fig. 8-8**). This muscle arises from the sternum, the upper ribs, and the clavicle and forms the anterior "wall" of the axilla; it inserts on the superior humerus. The superior part of the pectoralis major flexes the arm (raising it overhead). Other portions of the muscle medially rotate the arm, pulling it across the chest, and synergize with the latissimus dorsi to adduct the arm.

The **serratus** (ser-RA-tus) **anterior** is below the axilla, on the lateral chest (**see Fig. 8-8**). It originates on the upper eight or nine ribs on the lateral and anterior thorax and inserts in the scapula on the side toward the vertebrae. The serratus anterior moves the scapula forward and stabilizes it when, for example, one is pushing or punching something. It also aids in flexing and abducting the arm above the horizontal level.

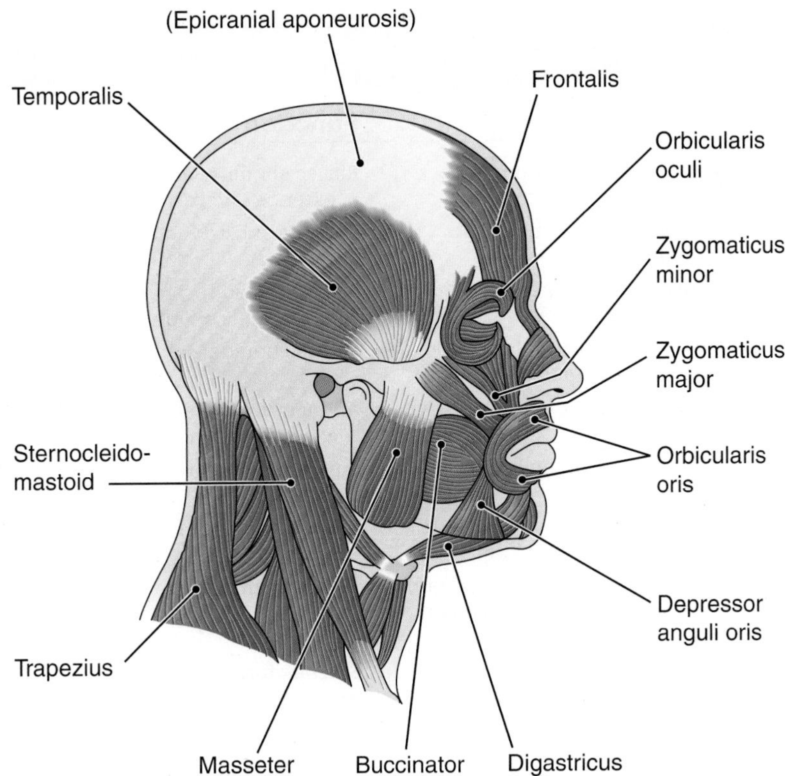

Figure 8-10 **Muscles of the head.** An associated structure is labeled in parentheses. **ZOOMING IN** Which of the muscles in this illustration are named for a nearby bone?

The **deltoid** covers the shoulder joint and is responsible for the roundness of the upper arm just inferior to the shoulder (**see Figs. 8-8 and 8-9**). This muscle is named for its triangular shape, which resembles the Greek letter delta. The deltoid is often used as an injection site. Arising from the shoulder girdle (clavicle and scapula), the deltoid fibers converge to insert on the lateral surface of the humerus. Contraction of the middle portion of this muscle abducts the arm, raising it laterally to the horizontal position. The anterior portion synergizes with the pectoralis major to flex and rotate the arm anteriorly, whereas the posterior portion works with latissimus dorsi to extend and rotate the arm posteriorly.

The shoulder joint allows for a wide range of movements. This freedom of movement is possible because the humerus fits into a shallow scapular socket, the glenoid cavity. This joint requires the support of four deep muscles

Table 8-2 Muscles of the Head and Neck[a]

Name	Location	Function
Orbicularis oculi	Encircles eyelid	Closes eye
Levator palpebrae superioris (deep muscle; not shown)	Posterior orbit to upper eyelid	Opens eye
Orbicularis oris	Encircles mouth	Closes lips
Buccinator	Fleshy part of cheek	Flattens cheek; helps in eating, whistling, and blowing wind instruments
Zygomaticus major and minor	Cheekbone to mouth corners	Raises mouth corners upward and laterally (smile)
Depressor anguli oris	Mandible to mouth corners	Lowers mouth corners (grimace)
Temporalis	Above and near ear	Closes jaw
Masseter	At angle of jaw	Closes jaw
Sternocleidomastoid	Along lateral neck, to mastoid process	Flexes head; rotates head toward opposite side from muscle

[a]These and other muscles of the face are shown in **Figure 8-10**.

Table 8-3 Muscles of the Upper Extremities[a]

Name	Location	Function
Trapezius	Posterior neck and upper back to clavicle and scapula	Raises shoulder and pulls it back; superior portion extends and turns head
Latissimus dorsi	Middle and lower back, to humerus	Extends and adducts arm (prime mover)
Pectoralis major	Superior, anterior chest, to humerus	Flexes and adducts arm; medially rotates arm across chest; pulls shoulder forward and downward
Serratus anterior	Inferior to axilla on lateral chest, to scapula	Moves shoulder forward; synergist in arm flexion and abduction
Deltoid	Covers shoulder joint, to lateral humerus	Abducts arm; synergist in arm flexion, rotation, and extension
Biceps brachii	Anterior arm along humerus, to radius	Supinates the forearm and hand; synergist in forearm flexion
Brachialis	Deep to biceps brachii; inserts at anterior elbow joint	Primary flexor of forearm
Brachioradialis	Lateral forearm from distal end of humerus to distal end of radius	Synergist in forearm flexion
Triceps brachii	Posterior arm, to ulna	Extends forearm to straighten upper extremity
Flexor carpi group	Anterior forearm, to hand	Flexes hand
Extensor carpi group	Posterior forearm, to hand	Extends hand
Flexor digitorum group	Anterior forearm, to fingers	Flexes fingers
Extensor digitorum group	Posterior forearm, to fingers	Extends fingers

[a]These and other muscles of the upper extremities are shown in **Figures 8-8, 8-9, and 8-11**.

and their tendons, which compose the **rotator cuff**. The four muscles are the supraspinatus, infraspinatus, teres minor, and subscapularis, known together as SITS, based on the first letters of their names. In certain activities, such as swinging a golf club, playing tennis, or pitching a baseball, the rotator cuff muscles may be injured, even torn, and may require surgery for repair.

Muscles that Move the Forearm and Hand The **biceps brachii** (BRA-ke-i), located at the anterior arm along the humerus, is the muscle you usually display when you want to "flex your muscles" to show your strength (**Fig. 8-11**). The root *brachi* means "arm" and is found in the names of several arm muscles. The biceps brachii inserts on the radius. It supinates the hand, as in turning a screwdriver, and acts as a synergist in forearm flexion. The **brachialis** (bra-ke-AL-is) lies deep to the biceps brachii and inserts distally over the anterior elbow joint. It flexes the forearm forcefully in all positions, sustains flexion, and steadies the forearm's slow extension.

Another forearm flexor at the elbow is the **brachioradialis** (bra-ke-o-ra-de-A-lis), a prominent forearm muscle that originates at the distal humerus and inserts on the distal radius (**see Fig. 8-11**).

The **triceps brachii**, located on the posterior arm, inserts on the olecranon of the ulna (**see Fig. 8-11B**). It is used to extend the forearm and thus straighten the arm, as in lowering a weight from an arm curl. It is also important in pushing because it converts the arm and forearm into a sturdy rod.

Most of the muscles that move the hand and fingers originate from the distal humerus. Their muscle bellies are usually in the forearm, and they control the hand and fingers by long tendons that are held in place at the wrist by a connective tissue band, the transverse carpal ligament (**see Fig. 9-16**). This arrangement enables precise hand and finger movements unencumbered by bulky muscles.

The action, and sometimes location, of each forearm muscle can be inferred by its name. The **flexors** in the anterior forearm and the **extensors** in the posterior forearm act on the hand (carpi muscles) and fingers (digitorum muscles). The flexor carpi ulnaris is located posteriorly to the other flexors but still inserts on the anterior surface of the carpal bones and accomplishes hand flexion (**see Fig. 8-11**). The flexors carpi and extensors carpi that insert on the radius (thumb side) work together to abduct the hand, and the flexors and extensors carpi that insert on the ulna adduct the hand.

Tiny intrinsic muscle groups in the fleshy parts of the hand fine-tune the intricate movements that can be performed with the thumb and the fingers. The thumb's freedom of movement has been one of humankind's most useful capacities.

See the Student Resources on thePoint for additional pictures of the rotator cuff and the muscles of the upper extremity.

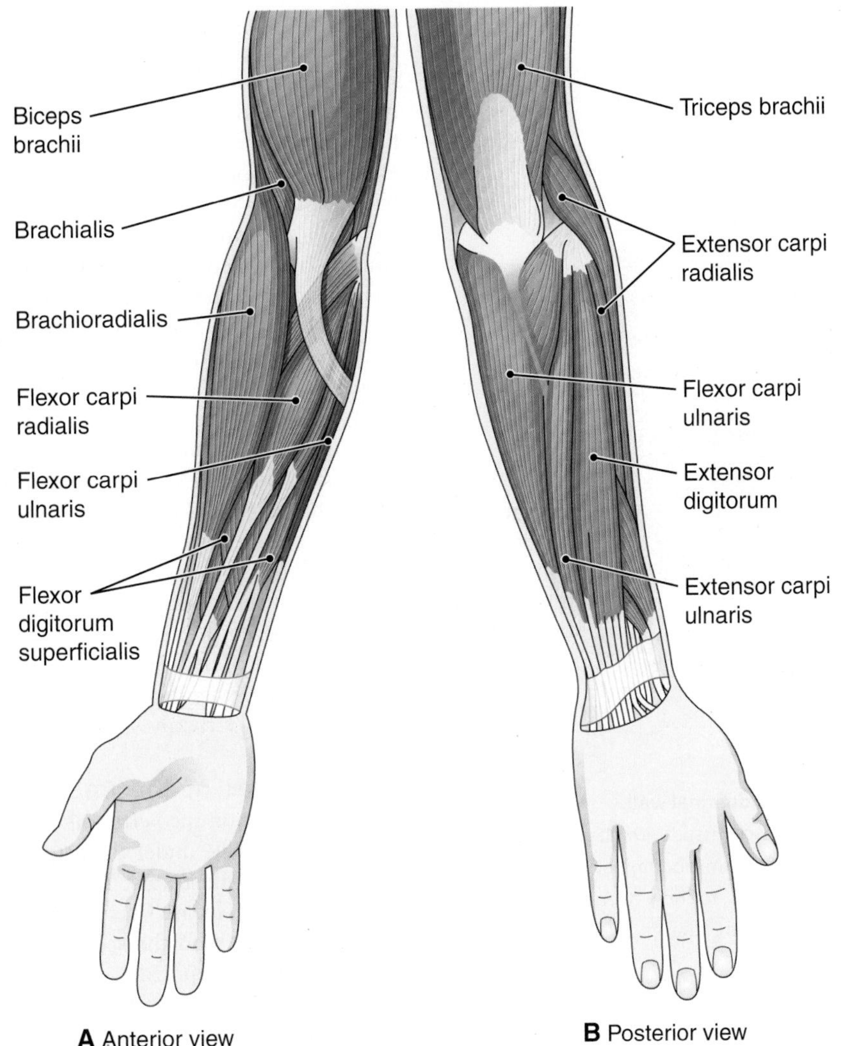

Figure 8-11 **Muscles that move the forearm and hand.** The muscles are shown in anterior **(A)** and posterior **(B)** views. **ZOOMING IN** What does carpi refer to in the names of muscles? Digitorum?

MUSCLES OF THE TRUNK

The trunk muscles include the muscles involved in breathing, the thin muscle layers of the abdomen, and the muscles of the pelvic floor. The following discussion also includes the deep muscles of the back that support and move the vertebral column.

Muscles of Respiration The most important muscle involved in the act of breathing is the **diaphragm.** This dome-shaped muscle forms the partition between the thoracic cavity above and the abdominal cavity below (**Fig. 8-12**). When the diaphragm contracts, the central dome-shaped portion is pulled downward, thus enlarging the thoracic cavity from top to bottom. This action results in inhalation (breathing in).

The **intercostal muscles,** which also act to change thoracic volume, are attached to the ribs and fill the spaces between them. Additional muscles of the neck, chest, and abdomen are employed in forceful breathing. The mechanics of breathing are described in Chapter 18.

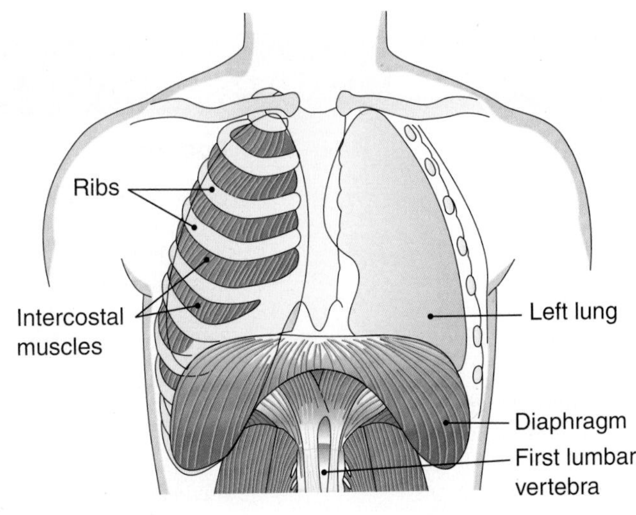

Figure 8-12 **Muscles of respiration.** **KEY POINT** The diaphragm is the main muscle of respiration. The left lung and ribs are also shown.

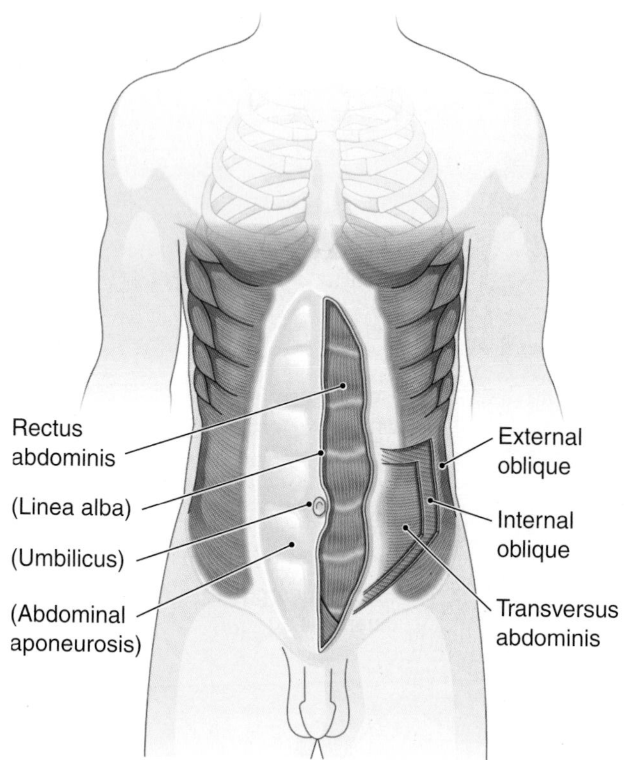

Figure 8-13 **Muscles of the abdominal wall.** **KEY POINT** Thin layers of muscle tissue with fibers running in different directions give strength to the abdominal wall. Surface tissue is removed here on the right side to show deeper muscles. Associated structures are labeled in parentheses. **ZOOMING IN** What does rectus mean? Oblique?

Muscles of the Abdomen and Pelvis The abdominal wall has three muscle layers that extend from the back (dorsally) and around the sides (laterally) to the front (ventrally) (**Fig. 8-13, Table 8-4**). They are the **external oblique** on the exterior, the **internal oblique** in the middle, and the **transversus abdominis,** the innermost. The connective tissue from these muscles extends anteriorly and encloses the vertical **rectus abdominis** of the anterior abdominal wall. The fibers of these muscles, as well as their connective tissue extensions (aponeuroses), run in different directions, resembling the layers in plywood and resulting in a strong abdominal wall. The midline meeting of the aponeuroses forms a whitish area called the **linea alba** (LIN-e-ah AL-ba), which is an important abdominal landmark. It extends from the tip of the sternum to the pubic joint (**see Fig. 8-13**).

These four pairs of abdominal muscles act together to protect the internal organs and compress the abdominal cavity, as in forcefully exhaling, coughing, emptying the bladder (urination) and bowel (defecation), sneezing, vomiting, and childbirth (labor). The two oblique muscles and the rectus abdominis help bend the trunk forward and sideways.

The pelvic floor, or **perineum** (per-ih-NE-um), has its own form of diaphragm, shaped somewhat like a shallow dish. One of the principal muscles of this pelvic diaphragm is the **levator ani** (le-VA-tor A-ni), which acts on the rectum and thus aids in defecation. The superficial and deep muscles of the female perineum are shown in **Figure 8-14** along with some associated structures.

Deep Muscles of the Back The deep muscles of the back, which act on the vertebral column itself, are thick vertical masses that lie under the trapezius and latissimus dorsi and thus are not illustrated. The **erector spinae** muscles make up a large group located between the sacrum and the skull. These muscles extend the spine and maintain the vertebral column in an erect posture. They can be strained in lifting heavy objects if the spine is flexed while lifting. One should bend at the hip and knee instead and use the thigh and buttock muscles to help in lifting.

Even deeper muscles lie beneath the lumbodorsal fascia. These small muscles extend the vertebral column in the lumbar region and are also easily strained.

Table 8-4 Muscles of the Trunk[a]

Name	Location	Function
Diaphragm	Dome-shaped partition between thoracic and abdominal cavities	Dome descends to enlarge thoracic cavity from top to bottom during ventilation
Intercostals	Between ribs	Alter thoracic cavity volume during ventilation
Muscles of abdominal wall: External oblique Internal oblique Transversus abdominis Rectus abdominis	Anterolateral abdominal wall	Compress abdominal cavity and expel substances from body; flex spinal column
Levator ani	Pelvic floor	Aids in defecation
Erector spinae (deep; not shown)	Group of deep vertical muscles between the sacrum and skull	Extends vertebral column to produce erect posture

[a]These and other muscles of the trunk are shown in **Figures 8-12, 8-13, and 8-14**.

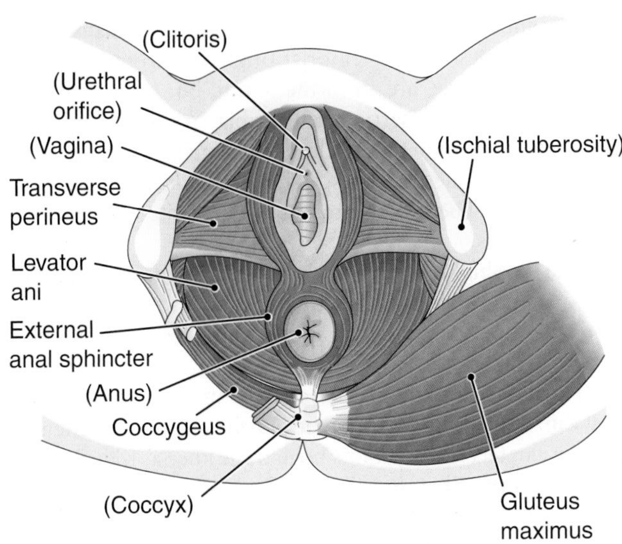

Figure 8-14 **Muscles of the female perineum (pelvic floor).** Associated structures are labeled in parentheses.

MUSCLES OF THE LOWER EXTREMITIES

The muscles in the lower extremities, among the longest and strongest muscles in the body, are specialized for locomotion and balance. They include the muscles that move the thigh and leg and those that control movement of the foot.

Muscles that Move the Thigh and Leg The **gluteus maximus** (GLU-te-us MAK-sim-us), which forms much of the buttock's fleshy part, is relatively large in humans because of its support function when a person is standing erect (**see Figs. 8-9, 8-14, Table 8-5**). This muscle extends the thigh and is important in walking and running by providing a pushing force at the end of a stride. The **gluteus medius**, which is partially covered by the gluteus maximus, abducts the thigh. This movement moves the foot slightly outward and upward with every step. It is one of the sites used for intramuscular injections.

The **iliopsoas** (il-e-o-SO-as) arises from the ilium and the bodies of the lumbar vertebrae; it crosses the anterior hip joint to insert on the femur (**Fig. 8-15A**). It is a powerful

Table 8-5 **Muscles of the Lower Extremities**[a]

Name	Location	Function
Gluteus maximus	Superficial buttock, to femur	Extends thigh
Gluteus medius	Deep buttock, to femur	Abducts thigh
Iliopsoas	Crosses anterior hip joint, to femur	Flexes thigh when trunk is immobilized; flexes trunk when thighs are immobilized
Adductor group (e.g., adductor longus, adductor magnus)	Medial thigh, to femur	Adducts thigh
Sartorius	Crosses anterior thigh; from ilium to medial tibia	Flexes thigh and leg (to sit cross-legged)
Gracilis	Pubic bone to medial surface of tibia	Adducts thigh at hip; flexes leg at knee
Quadriceps femoris:	Anterior thigh, to tibia	Extends leg
Rectus femoris		
Vastus medialis		
Vastus lateralis		
Vastus intermedius (deep; not shown)		
Hamstring group:	Posterior thigh; ischium and femur to tibia and fibula	Flexes leg at knee; extends and rotates thigh at hip
Biceps femoris		
Semimembranosus		
Semitendinosus		
Gastrocnemius	Posterior leg, to calcaneus, inserting by the Achilles tendon	Plantar flexes foot (as in tiptoeing)
Soleus	Posterior leg deep to gastrocnemius	Plantar flexes foot
Tibialis anterior	Anterior and lateral leg, to foot	Dorsiflexes foot at ankle (as in walking on heels); inverts foot (sole inward)
Fibularis (peroneus) longus	Lateral leg, to foot	Everts foot (sole outward)
Flexor digitorum group	Posterior leg and foot to inferior surface of phalanges	Flexes toes
Extensor digitorum group	Anterior surface of leg bones to superior surface of phalanges	Extends toes

[a]These and other muscles of the lower extremities are shown in **Figures 8-15 and 8-16**.

thigh flexor and helps keep the trunk from falling backward when one is standing erect. When the thighs are immobilized, the iliopsoas flexes the trunk, as in doing a sit-up.

The **adductor muscles** are located on the medial thigh (see Fig. 8-15). They arise from the pelvis and insert on the femur. These strong muscles press the thighs together, as in grasping a saddle between the knees when riding a horse. They include the **adductor longus** and **adductor magnus**. The **gracilis** (grah-SIL-is) is also located in the medial thigh, but it crosses both the hip joint and the knee joint. It too adducts the thigh and also flexes the leg at the knee.

The **sartorius** (sar-TO-re-us) is a long, narrow muscle that begins at the iliac spine, winds downward and medially across the anterior thigh, and ends on the tibia's superior medial surface (see Fig. 8-15). It is called the tailor's muscle because it is used in crossing the legs in the manner of tailors, who, in days gone by, sat cross-legged on the floor.

The anterior and lateral femur are covered by the **quadriceps femoris** (KWOD-re-seps FEM-or-is), a large muscle that has four heads of origin (see Fig. 8-15A). The individual parts are as follows: in the center, covering the anterior thigh, the **rectus femoris**; on either side, the **vastus medialis** and **vastus lateralis**; and deeper in the center, the **vastus intermedius**. One of these muscles (rectus femoris) originates from the ilium, and the other three are from the femur, but all four have a common tendon of insertion on the tibia. You may remember that this is the tendon that encloses the patella (kneecap). This muscle extends the leg and straightens the lower limb, as in kicking a ball. The vastus lateralis is also a site for intramuscular injections.

The iliotibial tract is a thickened band of fascia that covers the lateral thigh muscles. It extends from the ilium of the hip to the superior tibia and reinforces the fascia of the thigh (the fascia lata) (see Fig. 8-15).

The **hamstring muscles** are located in the posterior thigh (see Fig. 8-15B). They originate on the ischium and femur, and you can feel their tendons behind the knee as they descend to insert on the tibia and fibula. The hamstrings flex the leg at the knee as in kneeling. They also extend and rotate the thigh at the hip. Individually, moving from lateral to medial position, they are the **biceps femoris**, the **semimembranosus**, and the **semitendinosus**. The name of this muscle group refers to the tendons at the posterior of the knee by which these muscles insert on the leg.

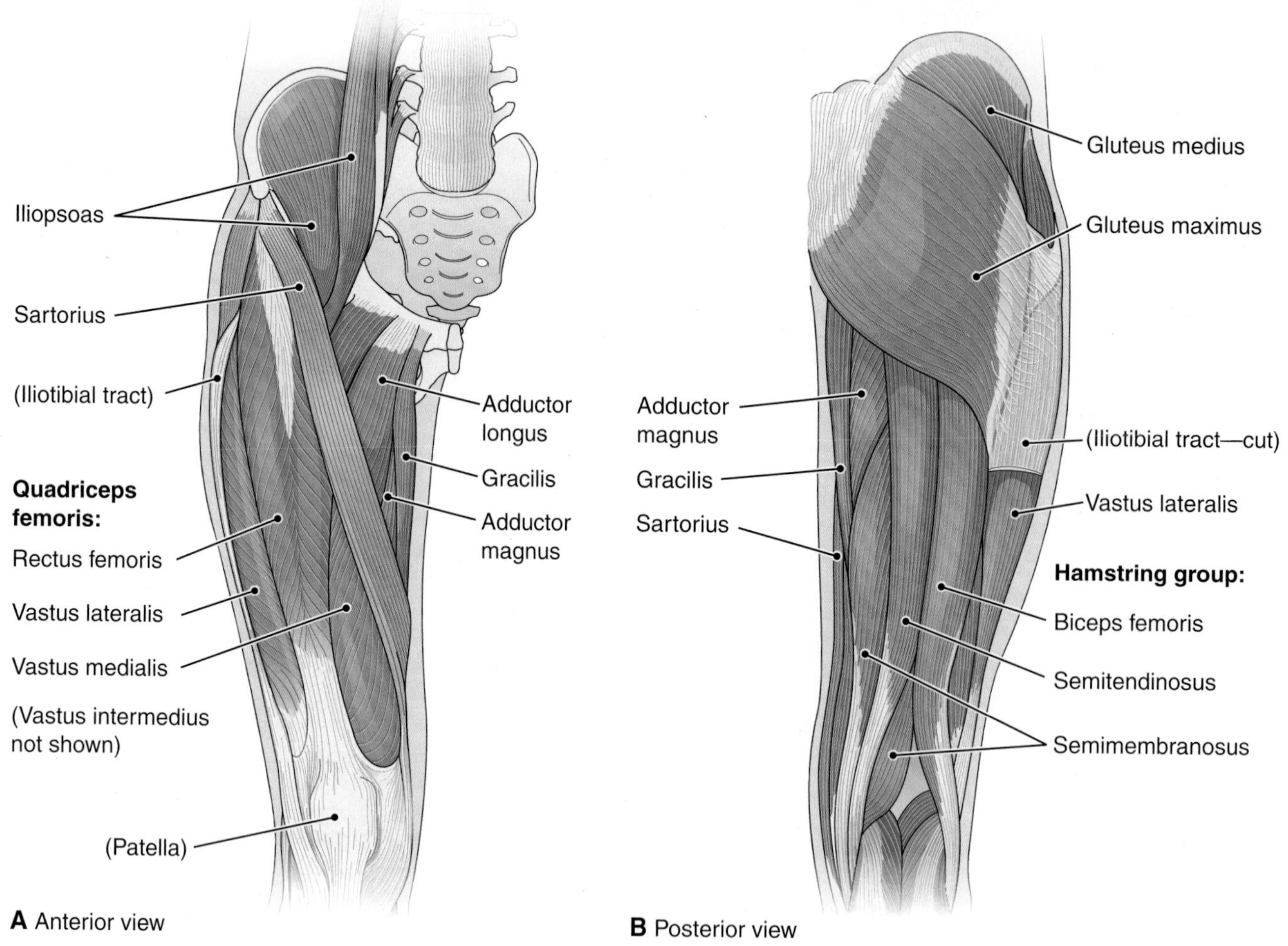

Figure 8-15 **Muscles of the thigh.** Associated structures are labeled in parentheses. Anterior **(A)** and posterior **(B)** views are shown. **ZOOMING IN** How many muscles make up the quadriceps femoris?

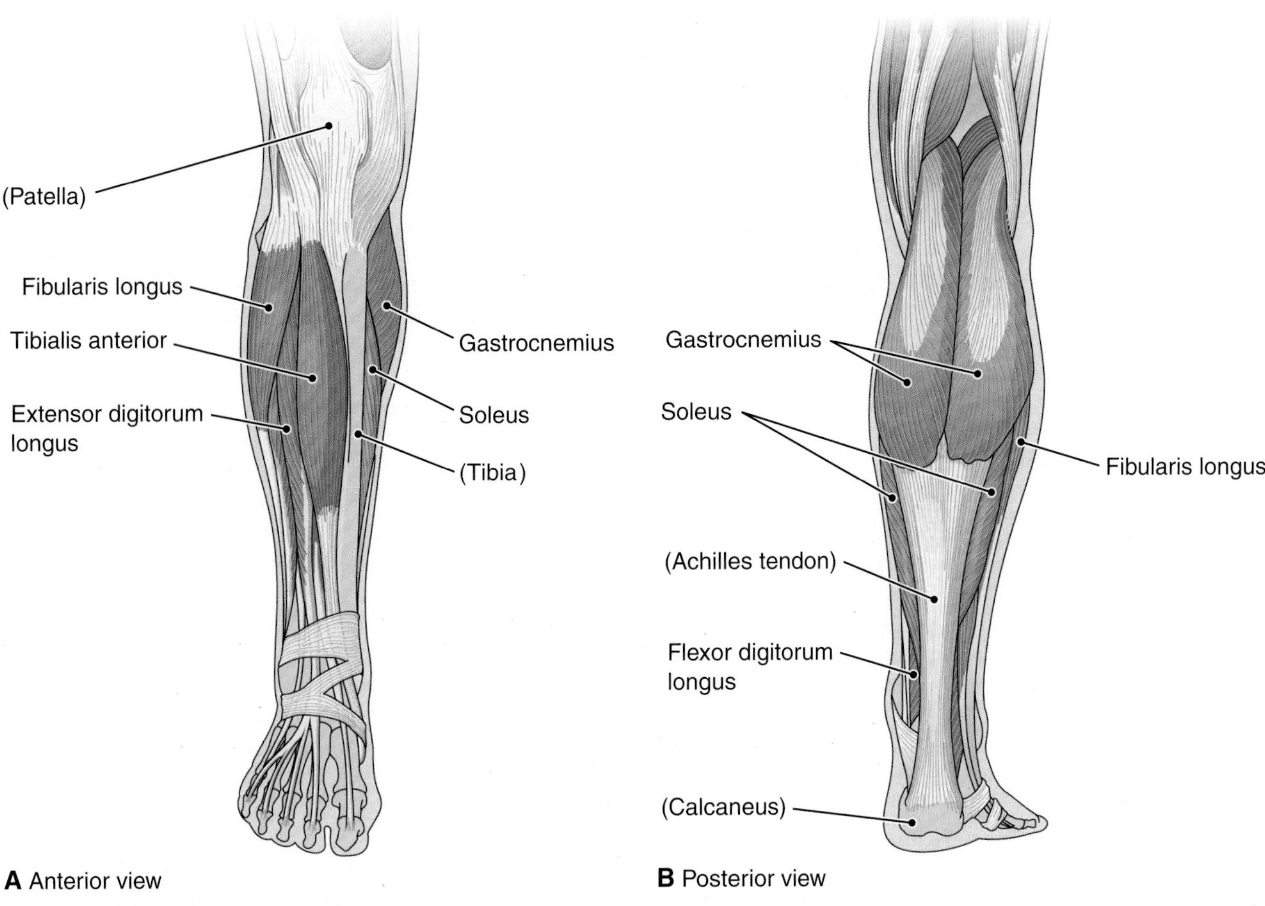

Figure 8-16 **Muscles that move the foot.** Associated structures are labeled in parentheses. Anterior **(A)** and posterior **(B)** views are shown. **ZOOMING IN** On what bone does the Achilles tendon insert?

Muscles that Move the Foot The **gastrocnemius** (gas-trok-NE-me-us) is the chief muscle of the calf of the leg (its name means "belly of the leg") (**Fig. 8-16**). It has been called the toe dancer's muscle because it is used in standing on tiptoe. It ends near the heel in a prominent cord called the **Achilles tendon** (**see Fig. 8-16B**), which attaches to the calcaneus (heel bone). The Achilles tendon is the largest tendon in the body. According to Greek mythology, the region above the heel was the only place on his body where the hero Achilles was vulnerable, and if the Achilles tendon is cut, it is impossible to walk. The **soleus** (SO-le-us) is a flat muscle deep to the gastrocnemius. It also inserts by means of the Achilles tendon and, like the gastrocnemius, flexes the foot at the ankle.

Another leg muscle that acts on the foot is the **tibialis** (tib-e-A-lis) **anterior**, located on the anterior region of the leg (**see Fig. 8-16A**). This muscle performs the opposite function of the gastrocnemius. Walking on the heels uses the tibialis anterior to raise the rest of the foot off the ground (dorsiflexion). This muscle is also responsible for inversion of the foot. The muscle for the foot's eversion is the **fibularis** (fib-u-LA-ris) **longus**, also called the peroneus (per-o-NE-us) longus, located on the lateral leg. This muscle's long tendon crosses under the foot, forming a sling that supports the transverse (metatarsal) arch.

The toes, like the fingers, are provided with flexor and extensor muscles. The tendons of the extensor muscles are located in the superior part of the foot and insert on the superior surface of the phalanges (toe bones). The flexor digitorum tendons cross the sole of the foot and insert on the undersurface of the phalanges (**see Fig. 8-16**).

CHECKPOINTS

- 8-15 What muscle is most important in breathing?
- 8-16 What structural feature gives strength to the muscles of the abdominal wall?

See the Student Resources on thePoint for additional pictures of the muscles that move the lower extremity.

Effects of Aging on Muscles

Beginning at about 40 years of age, there is a gradual loss of muscle cells with a resulting decrease in the size of each individual muscle. The muscle fibers that enable quick, explosive movements die first. The loss of these fibers makes it more difficult for the elderly to recover their balance, leading to more frequent falls. Loss of power in the extensor muscles, such as the large sacrospinalis near the vertebral column, causes the "bent over" appearance of a hunchback (kyphosis; **see Fig. 7-20**). Sometimes, there is a tendency to bend (flex) the hips and knees. In addition to the previously noted changes in the vertebral column (see Chapter 7), these effects on the extensor muscles result in a further decrease in the elderly person's height. Activity and exercise throughout life delay and decrease these undesirable effects of aging. Even among the elderly, resistance exercise, such as weight lifting, increases muscle strength and function.

Muscular Disorders

A **spasm** is a sudden and involuntary muscular contraction, which is always painful. A spasm of the visceral muscles is called **colic**, a good example of which is the intestinal muscle spasm often referred to as a *belly ache*. Spasms also occur in the skeletal muscles. If the spasms occur in a series, the condition may be called a **seizure** or **convulsion**.

Cramps are strong, painful muscle contractions, especially of the leg and foot. They are most likely to follow unusually strenuous activity, but can also indicate nerve compression or dehydration. Cramps that occur during sleep or rest are called *recumbency cramps*. *Charley horse* usually refers to a cramp in a muscle, most commonly in the leg. This term comes from the practice of using Charley as a name for old lame horses that were kept around for family use when they were no longer able to do hard work.

Muscles **atrophy** (AT-ro-fe), or decrease in size, when they are not used, such as when an extremity must be placed in a cast after a fracture or a person is bedridden for a time. Muscle tissue restores itself after such temporary periods of disuse. Atrophy can be significant, however, if muscles are paralyzed, as may occur in a spinal cord injury and other diseases of muscles or the nervous system. With prolonged disuse, a muscle will shrink, and eventually the cells will be irreversibly replaced by fibrous connective tissue.

MUSCLE INJURIES

A muscle **strain** occurs when overuse or overstretching causes tears in the muscle tissue or tears between the muscle and its tendons. The resulting bleeding and inflammation cause pain, stiffness, swelling, and sometimes bruising over the torn regions (**Fig. 8-17A**). Muscle strain in the deep muscles of the back commonly results from poor posture or incorrect lifting techniques and is the most common cause of back pain.

Delayed-onset muscle soreness (**DOMS**) is the stiffness and pain that occurs a day or so after an intense workout. It probably reflects inflammation induced by microtears in muscle tissue, but the underlying cause is not well understood. What is known is that lactic acid is not responsible, and stretching, before or after, does not help prevent or heal DOMS.

Rhabdomyolitis is a potentially life-threatening consequence of serious overtraining. Overworked muscle cells undergo necrosis, releasing their contents (such as creatine kinase) into the bloodstream. These substances can overwhelm the kidney, leading to irreversible kidney damage.

MUSCLE DISEASES

Muscular dystrophy (DIS-tro-fe) is a group of disorders in which there is deterioration of muscles that still have intact nerve function. These disorders all progress at different rates. Duchenne muscular dystrophy, described in Shane's opening case study, is an inherited form of the disease found most frequently in male children. It results from a defect in a protein that links myofibrils to the muscle cell membrane. Without this protein, normal contractions damage the cells, eventually leading to the replacement of muscle tissue with connective tissue, weakness, and paralysis (**Fig. 8-17B**).

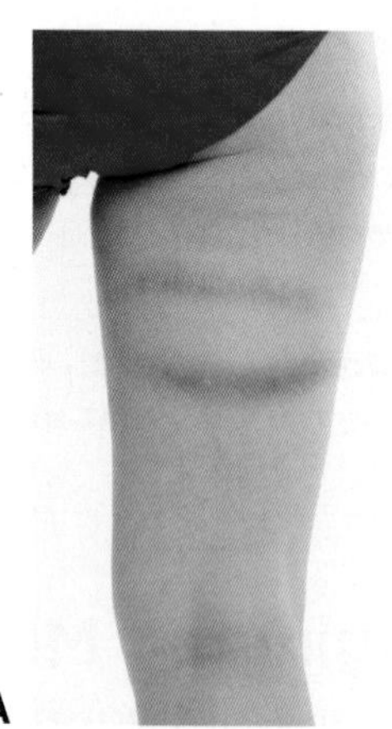

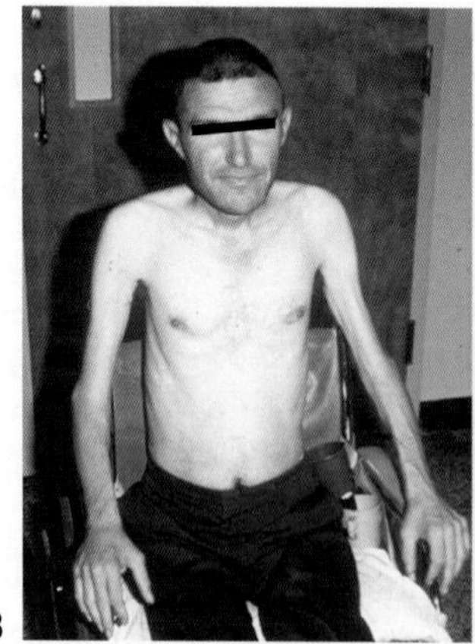

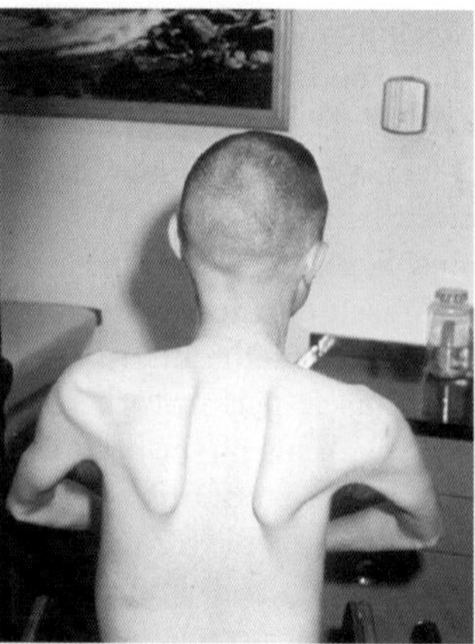

Figure 8-17 Muscle disorders. A. Muscle strain. The muscle tears are deep to the visible bruising. **B.** Muscular dystrophy.

Death results from weakness of the cardiac muscle or paralysis of the respiratory muscles. Life expectancy is about 25 years for the most common type of muscular dystrophy and about 40 years for the other types. Progress toward definitive treatment for inherited forms of the disease may be possible now that scientists have identified the genetic defects that cause them.

Myasthenia gravis (mi-as-THE-ne-ah GRA-vis) is characterized by chronic muscular fatigue brought on by the slightest exertion. It affects adults and begins with the muscles of the head. Drooping of the eyelids (ptosis) is a common early sign. This disease is caused by the loss of acetylcholine receptors in the muscle cell membrane. Without these receptors, the neurons cannot trigger muscle contraction.

Myalgia (mi-AL-je-ah) means "muscular pain;" **myositis** (mi-o-SI-tis) is a term that indicates actual inflammation of muscle tissue. **Fibrositis** (fi-bro-SI-tis) means "inflammation of connective tissues" and refers particularly to those tissues associated with muscles and joints. Usually, these disorders appear in combination as **fibromyositis**, which may be acute with severe pain on motion, or may be chronic. Sometimes, the application of heat together with massage and rest relieves the symptoms.

Fibromyalgia syndrome (FMS) is associated with widespread muscle aches, tenderness, and stiffness along with fatigue and sleep disorders. FMS is difficult to diagnose, and there is no known cause, but it may be an autoimmune disease, in which the immune system reacts to one's own tissues. Treatment may include a controlled exercise program and treatment with pain relievers, muscle relaxants, or antidepressants.

DISORDERS OF ASSOCIATED STRUCTURES

Tendinitis (ten-din-I-tis), an inflammation of muscle tendons and their attachments, occurs most often in athletes who overexert themselves. It frequently involves the shoulder, the hamstring muscle tendons at the knee, and the Achilles tendon near the heel. **Plantar fasciitis** (fash-e-I-tis) is a form of tendinitis that involves the thick band of connective tissue on the bottom of the foot. The plantar fascia originates on the inferior surface of the calcaneus (heel bone) and extends along the sole of the foot toward the toes, helping to maintain the arch. Inflammation results in foot and heel pain. Causes of plantar fasciitis include obesity, prolonged weight bearing, and overuse, as may occur in runners. **Tenosynovitis** (ten-o-sin-o-VI-tis), which involves the synovial sheath that encloses tendons, is found most often in women in their 40s after an injury or surgery. It may involve swelling and severe pain with activity.

Shin-splints is experienced as pain and soreness along the tibia ("shin bone") from stress injury of structures in the leg. Some causes of shin-splints are tendinitis at the insertion of the tibialis anterior muscle, sometimes with inflammation of the tibial periosteum, and even stress fracture of the tibia itself. Shin-splints commonly occurs in runners, especially when they run on hard surfaces without adequate shoe support.

See the Student Resources on thePoint for information on careers in physical therapy and how physical therapists participate in treatment of muscular disorders.

Disease in Context Revisited

Looking at Shane's blood test results, Dr. Schroeder saw that the level of creatine phosphokinase was elevated. Usually, this protein is confined to muscle cells, so this finding indicated that substances were leaking from the weakened walls of his muscle cells. At the next visit, Dr. Schroeder used a small needle to extract a sample of Shane's gastrocnemius (a muscle biopsy). The sample revealed evidence of muscle degeneration and abnormally low levels of dystrophin (the protein that links myofibrils to the cell membrane). A genetic test of Shane's initial blood sample showed that Shane had a gene mutation that causes Duchenne muscular dystrophy.

Dr. Schroeder discussed these results with Kathy. "Kathy, the tests conclusively show that Shane has Duchenne muscular dystrophy. As Shane grows older, his muscles will become weaker, and this disease will considerably shorten his life. With therapy, we can preserve his ability to walk for as long as possible and decrease the development of deformities. I'm going to refer you to a rehabilitation medicine specialist, called a physiatrist, who can further evaluate Shane's condition and formulate a management program. Your family is going to need some help dealing with Shane's condition, so here is the contact information for a DMD support group and some government agencies that can help. While we don't have any treatments for Shane right now, research efforts are ongoing. Don't give up hope."

During this case, we learned that structural components of the muscle cells determine the ability of muscles to contract. Weakness in limb muscles (such as the gastrocnemius) will affect Shane's ability to walk and perform everyday tasks. Eventually, weakness in his respiratory muscles will likely cause his death.

CHAPTER 8 Chapter Wrap-Up

Summary Overview

A detailed chapter outline with space for note taking is on *thePoint*. The figure below illustrates the main topics covered in this chapter.

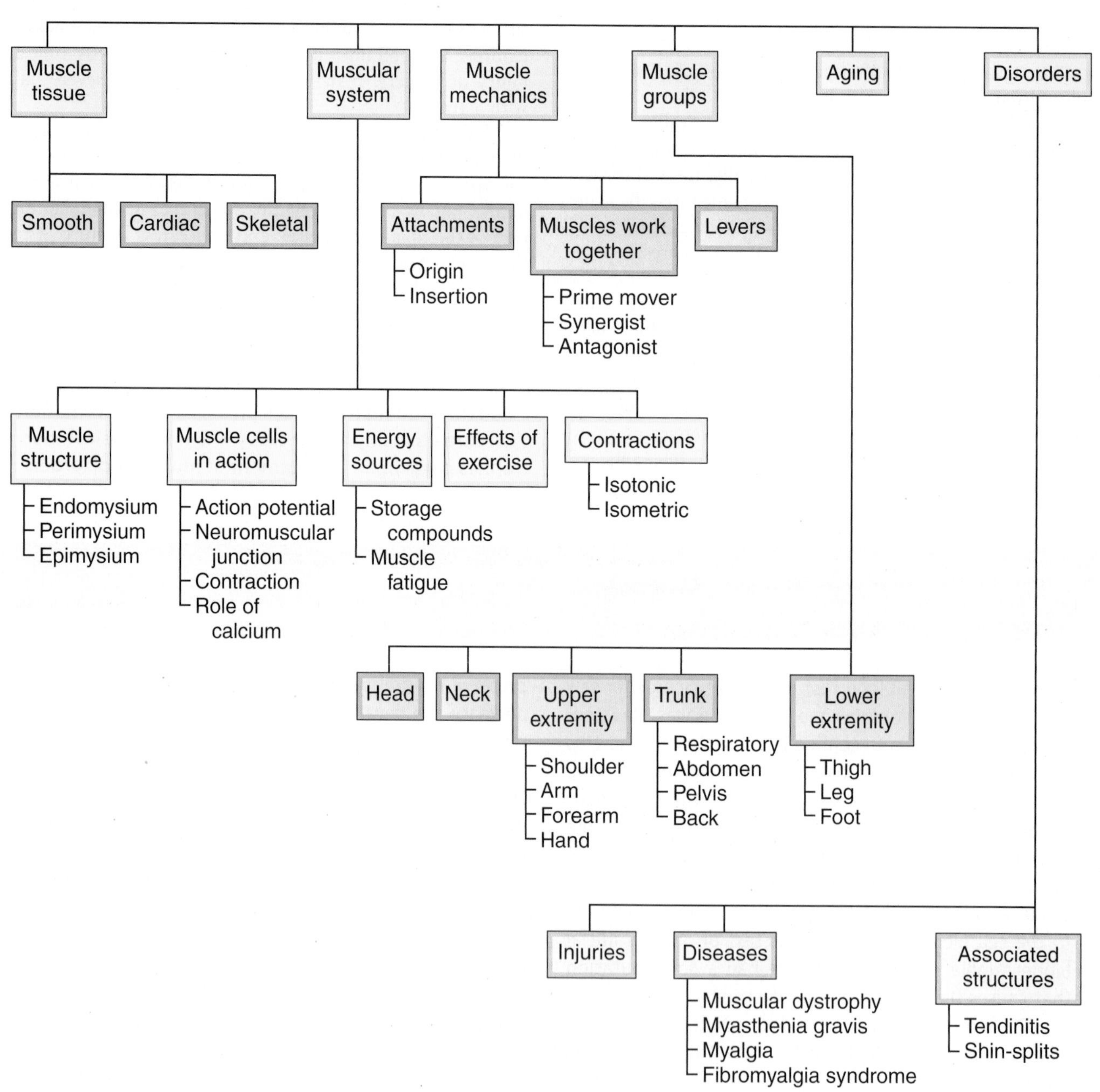

Key Terms

The terms listed below are emphasized in this chapter. Knowing them will help you organize and prioritize your learning. These and other boldface terms are defined in the Glossary with phonetic pronunciations.

acetylcholine
actin
action potential
agonist
antagonist
atrophy
contractility
excitability
fascicle
glycogen
insertion
membrane potential
motor unit
myalgia
myoglobin
myosin
neuromuscular junction
neurotransmitter
origin
prime mover
sarcomere
spasm
synapse
synergist
tendon
tonus
tropomyosin
troponin

Word Anatomy

Medical terms are built from standardized word parts (prefixes, roots, and suffixes). Learning the meanings of these parts can help you remember words and interpret unfamiliar terms.

WORD PART	MEANING	EXAMPLE
The Muscular System		
aer/o	air, gas	An *aerobic* organism can grow in the presence of air (oxygen).
an-	not, without	*Anaerobic* metabolism does not require oxygen.
iso-	same, equal	In an *isotonic* contraction, muscle tone remains the same, but the muscle shortens.
-lysis	separation, dissolving	*Glycolysis* is the breakdown of glucose.
metr/o	measure	In an *isometric* contraction, muscle length remains the same, but muscle tension increases.
my/o	muscle	The *endomysium* is the deepest layer of connective tissue around muscle cells.
sarc/o	flesh	A *sarcomere* is a contracting subunit of skeletal muscle.
ton/o	tone, tension	See "iso-" example.
troph/o	nutrition, nurture	Muscles undergo *hypertrophy*, an increase in size, under the effects of resistance training.
vas/o	vessel	*Vasodilation* (widening) of the blood vessels in muscle tissue during exercise brings more blood into the tissue.
The Mechanics of Muscle Movement		
erg/o	work	*Synergists* are muscles that work together.
syn-	with, together	A *synapse* is a point of communication between a neuron and another cell.
Skeletal Muscle Groups		
brachi/o	arm	The biceps *brachii* and triceps brachii are in the arm.
quadr/i	four	The *quadriceps* muscle group consists of four muscles.
Muscular Disorders		
a-	absent, lack of	*Atrophy* is a wasting of muscle as a result of disuse (lack of nourishment).
-algia	pain	*Myalgia* is muscular pain.
dys-	disordered, difficult	In muscular *dystrophy*, there is deterioration of muscles.
sthen/o	strength	*Myasthenia* gravis is characterized by muscular fatigue (lack of strength).

Questions for Study and Review

BUILDING UNDERSTANDING

Fill in the Blanks

1. The point at which a nerve fiber contacts a muscle cell is called the _____.

2. A single neuron and all the muscle fibers it stimulates make up a(n) ______.

3. A wave of electric current that spreads along a plasma membrane is called a(n) _____.

4. A contraction in which there is no change in muscle length but there is a great increase in muscle tension is _____.

5. A term that means "muscular pain" is _____.

Matching > Match each numbered item with the most closely related lettered item.

___ **6.** Raises and pulls back the shoulder

___ **7.** Active in breathing

___ **8.** Closes jaw

___ **9.** Muscle of the perineum

___ **10.** Closes lips

a. levator ani
b. masseter
c. orbicularis oris
d. trapezius
e. diaphragm

Multiple Choice

___ **11.** From superficial to deep, the correct order of muscle structure is

a. deep fascia, epimysium, perimysium, and endomysium
b. epimysium, perimysium, endomysium, and deep fascia
c. deep fascia, endomysium, perimysium, and epimysium
d. endomysium, perimysium, epimysium, and deep fascia

___ **12.** What is the function of calcium ions in skeletal muscle contraction?

a. bind to receptors on the motor end plate to stimulate muscle contraction
b. cause a pH change in the cytoplasm to trigger muscle contraction
c. block the myosin-binding sites on actin
d. bind to regulatory proteins to expose myosin-binding sites on actin

___ **13.** Which structure is a broad flat extension that attaches muscle to a bone or other muscle?

a. tendon
b. fascicle
c. aponeurosis
d. motor end plate

___ **14.** What is the term for decrease in muscle size, as from disuse?

a. colic
b. strain
c. seizure
d. atrophy

___ **15.** Which disease is characterized by widespread muscle aches, tenderness, and stiffness and has no known cause?

a. shin-splints
b. fibromyalgia
c. myositis
d. tendinitis

UNDERSTANDING CONCEPTS

16. Compare smooth, cardiac, and skeletal muscle with respect to location, structure, and function. Briefly explain how each type of muscle is specialized for its function.

17. Explain what is meant by the statement "The membrane potential in resting muscle cells is about –70 mV."

18. Explain the concept of excitability in muscle cells and how excitability is necessary for muscle function.

19. Describe four substances stored in skeletal muscle cells that are used to manufacture a constant supply of ATP.

20. The first signs of Shane's muscular dystrophy was weakness of his thigh muscles. Name the muscles that are involved in movement of the thigh.

21. Name the following muscle(s):

a. antagonist of the orbicularis oculi
b. prime mover in arm adduction
c. prime mover in hand extension
d. antagonist of zygomaticus major
e. prime mover in dorsiflexion
f. antagonist of the brachialis

22. During a Cesarean section, a transverse incision is made through the abdominal wall. Name the muscles incised, and state their functions.

23. What effect does aging have on muscles? What can be done to resist these effects?

24. What are muscular dystrophies, and what are some of their effects?

CONCEPTUAL THINKING

25. Margo recently began working out and jogs three times a week. After her jog, she is breathless, and her muscles ache. From your understanding of muscle physiology, describe what has happened inside of Margo's skeletal muscle cells. How do Margo's muscles recover from this? If Margo continues to exercise, what changes would you expect to occur in her muscles?

26. Alfred suffered a mild stroke, leaving him partially paralyzed on his left side. Physical therapy was ordered to prevent left-sided atrophy. Prescribe some exercises for Alfred's shoulder and thigh.

For more questions, see the Learning Activities on thePoint.

UNIT

IV

Coordination and Control

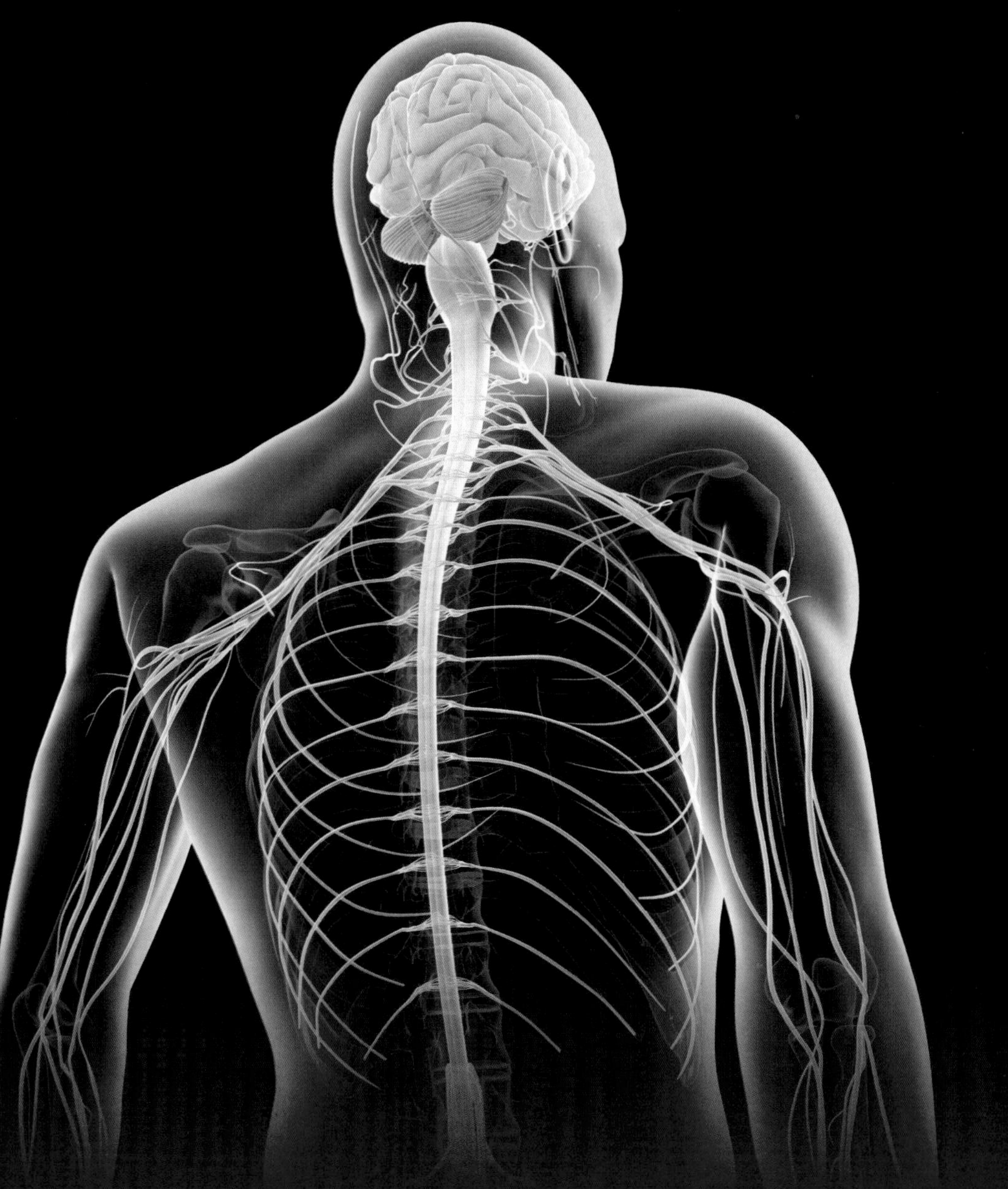

In Chapter 1, we introduced the concept of homeostasis—the maintenance of certain body functions within set limits. This unit discusses how the body detects changes in the external and internal environments and plans the appropriate response. Two body systems focus on communication: the nervous system (which includes the sensory system) and the endocrine system. We begin with the nervous system.

CHAPTER 9

The Nervous System: The Spinal Cord and Spinal Nerves

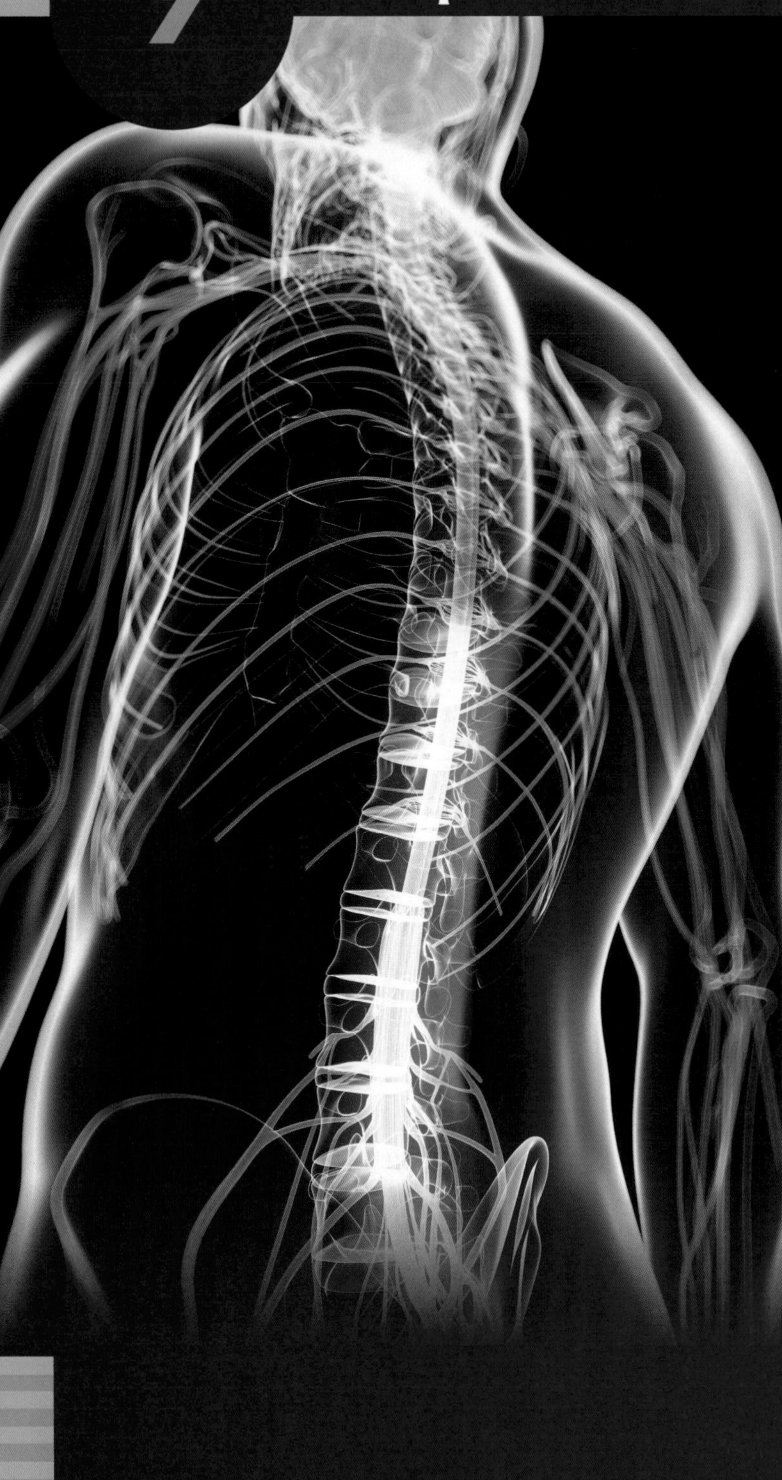

Learning Objectives

After careful study of this chapter, you should be able to:

1 ▶ Outline the organization of the nervous system according to structure and function. ***p. 192***

2 ▶ Describe the structure of a neuron. ***p. 193***

3 ▶ Explain the construction and function of the myelin sheath. ***p. 194***

4 ▶ Describe how neuron fibers are built into a nerve. ***p. 194***

5 ▶ List four types of neuroglia in the central nervous system, and cite the functions of each. ***p. 196***

6 ▶ Diagram and describe the steps in an action potential. ***p. 196***

7 ▶ Explain the role of neurotransmitters in impulse transmission at a synapse. ***p. 198***

8 ▶ Describe the distribution of gray and white matter in the spinal cord. ***p. 200***

9 ▶ Describe and name the spinal nerves and three of their main plexuses. ***p. 200***

10 ▶ List the components of a reflex arc. ***p. 203***

11 ▶ Define a simple reflex, and give several examples of reflexes. ***p. 204***

12 ▶ Compare the locations and functions of the sympathetic and parasympathetic nervous systems. ***p. 204***

13 ▶ Explain the role of cellular receptors in the action of neurotransmitters in the autonomic nervous system. ***p. 207***

14 ▶ Describe eight disorders of the spinal cord and spinal nerves. ***p. 208***

15 ▶ Using the case study, describe the effects of demyelination on motor and sensory function. ***pp. 191, 211***

16 ▶ Show how word parts are used to build words related to the nervous system (see Word Anatomy at the end of the chapter). ***p. 213***

Disease in Context *Sue's Case: The Importance of Myelin*

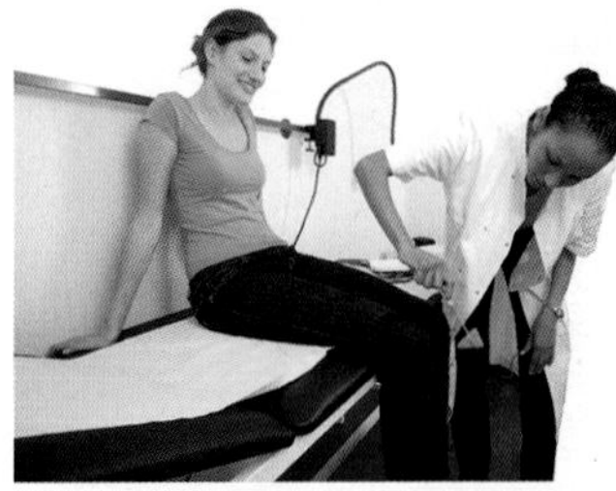

Dr. Jensen glanced at her patient's chart as she stepped into the consulting room. Sue Pritchard, a seemingly healthy 26-year-old Caucasian, had presented to her family doctor with right-hand weakness and difficulty in walking, a feeling that she was off balance. Some preliminary tests of her muscular strength and reflexes had led her to suspect multiple sclerosis (MS), a disease that affects the nerves. In addition to the referral, she had ordered a magnetic resonance image (MRI) of her brain and spinal cord.

"Hi Sue. My name is Dr. Jensen. I'm a neurologist, which means I specialize in the diagnosis and treatment of nervous system disorders. Let's start with a few tests to determine how well your brain and spinal cord communicate with the rest of your body. Then, we'll take a look at your MRI results."

Using a reflex hammer, Dr. Jensen tapped on the tendons of several muscles in Sue's arms and legs to elicit stretch reflexes. Her responses indicated damage to areas of the spinal cord that control reflexes. The doctor also detected muscle weakness in Sue's limbs—an indication of damage to the descending tracts in the spinal cord, which carry motor nerve impulses from the brain to skeletal muscle. In addition, she discovered that Sue's sense of touch was impaired—an indication of damage to the spinal cord's ascending tracts, which carry sensory impulses from receptors in the skin to the brain. Sue was exhibiting several of the most common clinical signs of MS.

Dr. Jensen then showed Sue the results of her MRI scan. "Here's the MRI of your spinal cord. The nervous tissue making up the spinal cord is organized into two regions—this inner region of gray matter and this outer one of white matter. If you look closely at the white matter, you can see several damaged areas, which we call lesions. They are causing many of your symptoms because they prevent your spinal cord from transmitting impulses between your brain and the rest of your body. These lesions, or scleroses, are what give multiple sclerosis its name."

The evidence shows that Sue has MS, a disease of neurons in the central nervous system (CNS). In this chapter, we learn more about neurons and the spinal cord, one part of the CNS.

ANCILLARIES *At-A-Glance*

Visit thePoint to access the following resources. For guidance in using these resources most effectively, see pp. xv–xvii.

Learning RESOURCES

- Tips for Effective Studying
- Web Figure: The Cauda Equina
- Web Chart: Neuroglia
- Animation: The Synapse and the Nerve Impulse
- Animation: The Action Potential
- Animation: The Myelin Sheath
- Animation: The Reflex Arc
- Health Professions: Occupational Therapist
- Detailed Chapter Outline
- Answers to Questions for Study and Review
- Audio Pronunciation Glossary
- Auditory Activities

Learning ACTIVITIES

- Pre-Quiz
- Visual Activities
- Kinesthetic Activities
- Auditory Activities

A LOOK BACK

In Chapter 8, we discussed how action potentials and neurotransmitters are involved in muscle contraction. Now, we broaden our outlook to see how the nervous system uses these same signals to transmit information and coordinate responses to changes in the environment.

Overview of the Nervous System

No body system is capable of functioning alone. All are interdependent and work together as one unit to maintain normal conditions, or homeostasis. The nervous system serves as the chief coordinating agency for most body functions. Conditions both within and outside the body are constantly changing. The nervous system must detect and respond to these changes (known as *stimuli*) so that the body can adapt itself to new conditions.

The nervous system can be compared with a large corporation in which market researchers (sensory receptors) feed information into middle management (the spinal cord), who then transmit information to the chief executive officer or CEO (the brain). The CEO organizes and interprets the information and then sends instructions out to workers (effectors) who carry out appropriate actions for the good of the company. These instructions are communicated by e-mail through a network, which, like the body's nerves, carries information throughout the system.

Although all parts of the nervous system work in coordination, portions may be grouped together on the basis of either structure or function.

DIVISIONS OF THE NERVOUS SYSTEM

The entire nervous system is classified into two divisions (Fig. 9-1):

- The **central nervous system** (CNS) includes the brain and spinal cord.
- The **peripheral** (per-IF-er-al) **nervous system** (PNS) is made up of all the nerves outside the CNS. It includes all the **cranial nerves** that carry impulses to and from the brain and all the **spinal nerves** that carry messages to and from the spinal cord.

The CNS and PNS together include all of the nervous tissue in the body.

Functional Divisions of the PNS The PNS is split by function into two subdivisions. The somatic nervous system controls voluntary functions, whereas the autonomic nervous system (ANS) controls functions we cannot consciously control (Table 9-1). Any tissue or organ that carries out a nervous system command is called an **effector**, all of which are muscles or glands.

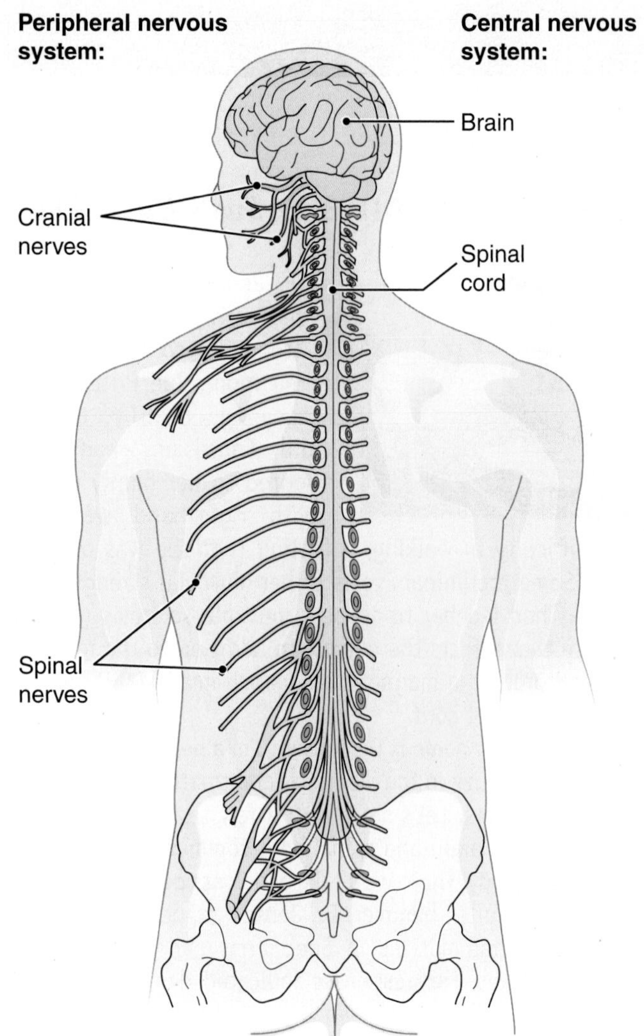

Figure 9-1 Anatomic divisions of the nervous system, posterior view. **KEY POINT** The nervous system is divided structurally into a central nervous system and a peripheral nervous system. **ZOOMING IN** What structures make up the central nervous system? The peripheral nervous system?

The **somatic nervous system** is voluntarily controlled (by conscious will), and all its effectors are skeletal muscles (described in Chapter 8). The nervous system's involuntary division is called the **autonomic nervous system**, making reference to its automatic activity. It is also called the **visceral**

Table 9-1 Functional Divisions of the Peripheral Nervous System

Division	Control	Effectors
Somatic nervous system	Voluntary	Skeletal muscle
Autonomic nervous system	Involuntary	Smooth muscle, cardiac muscle, and glands

nervous system because its effectors are smooth muscle, cardiac muscle, and glands, which are found in the soft body organs, the viscera. The ANS is described in more detail later in this chapter.

Although these divisions are helpful for study purposes, dividing the PNS and its effectors by function can be misleading. Although skeletal muscles *can* be controlled voluntarily, they may function commonly without conscious control. The diaphragm, for example, a skeletal muscle, typically functions in breathing without conscious thought. In addition, we have certain rapid reflex responses involving skeletal muscles—drawing the hand away from a hot stove, for example—that do not involve the brain. In contrast, people can be trained to consciously control involuntary functions, such as blood pressure and heart rate, by training techniques known as *biofeedback*.

CHECKPOINTS

- ☐ 9-1 What are the two divisions of the nervous system based on structure?
- ☐ 9-2 What division of the PNS is voluntary and controls skeletal muscles? What division is involuntary and controls smooth muscle, cardiac muscle, and glands?

Neurons and Their Functions

The functional cells of the nervous system are highly specialized cells called **neurons** (**Fig. 9-2**). These cells have a unique structure related to their function.

STRUCTURE OF A NEURON

The main portion of each neuron, the cell body, contains the nucleus and other organelles typically found in cells. A distinguishing feature of the neurons, however, is the long, threadlike fibers that extend out from the cell body and carry impulses across the cell (**Fig. 9-3**).

Dendrites and Axons Two kinds of fibers extend from the neuron cell body: dendrites and axons.

- **Dendrites** are neuron fibers that conduct impulses *to* the cell body. Most dendrites have a highly branched, treelike appearance (**see Fig. 9-2**). In fact, the name comes from a Greek word meaning "tree." Dendrites function as **receptors** in the nervous system. That is, they receive a

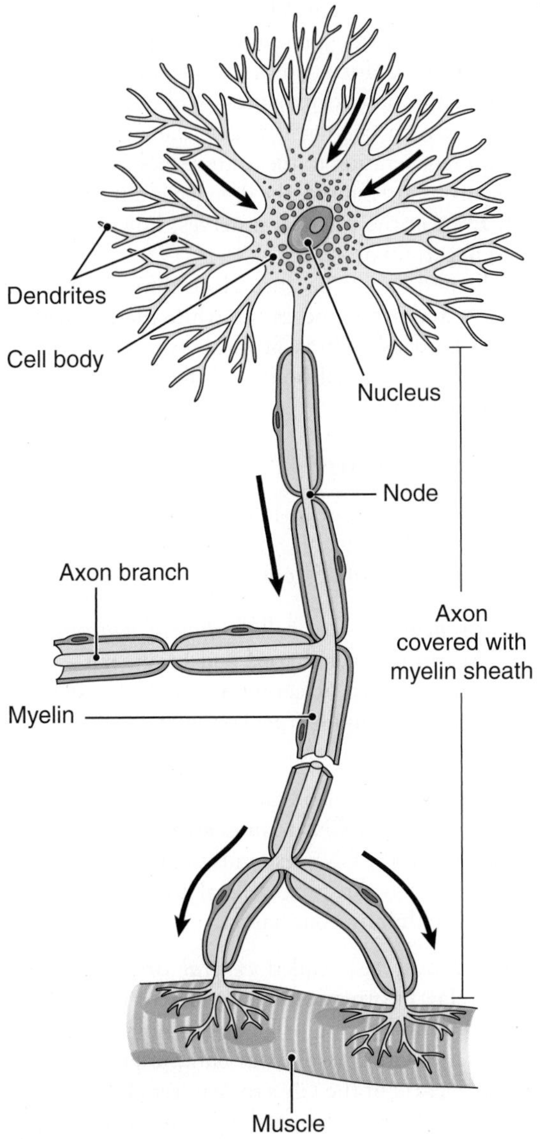

Figure 9-2 Diagram of a motor neuron. **KEY POINT** A neuron has fibers extending from the cell body. Dendrites carry impulses toward the cell body; axons carry impulses away from the cell body. The break in the axon denotes length. The *arrows* show the direction of the nerve impulse. **ZOOMING IN** How do you know the neuron shown here is a motor neuron? Is it part of the somatic or visceral nervous system? Explain.

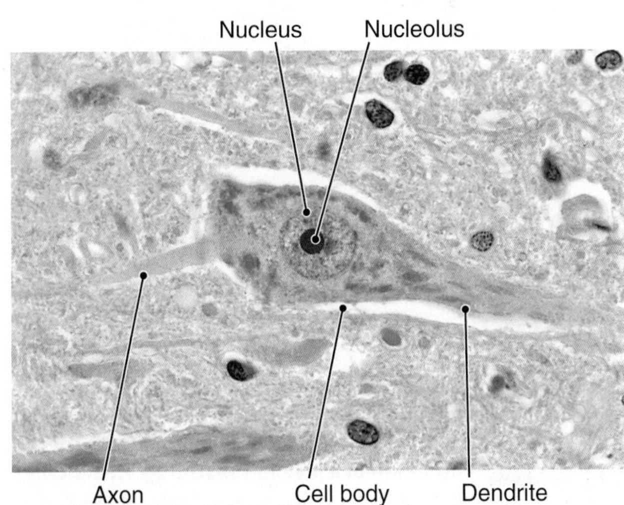

Figure 9-3 Microscopic view of a neuron. Based on staining properties and structure, the fiber on the **left** is identified as an axon; the fiber on the **right** is a dendrite. The clear space around the axon is caused by the staining procedure.

stimulus that begins a neural pathway. In Chapter 11, we describe how the dendrites of the sensory system may be modified to respond to a specific type of stimulus, such as pressure or taste.

- **Axons** (AK-sons) are neuron fibers that conduct impulses *away from* the cell body (**see Fig. 9-2**). These impulses may be delivered to another neuron, to a muscle, or to a gland. A neuron has only one axon, which can extend up to 1 m and give off many branches.

The Myelin Sheath Some axons are covered with a whitish, fatty substance called **myelin** (MI-eh-lin) that insulates and protects the fiber (**see Fig. 9-2**). In the PNS, myelin is actually entire cells, known as **Schwann** (shvahn) **cells**, that are wrapped many times around the axon like a jelly roll (**Fig. 9-4**). The cell bodies and most of the cytoplasm get squeezed into the outermost layer of the Schwann cell wrapping, known as the **neurilemma** (nu-rih-LEM-mah).

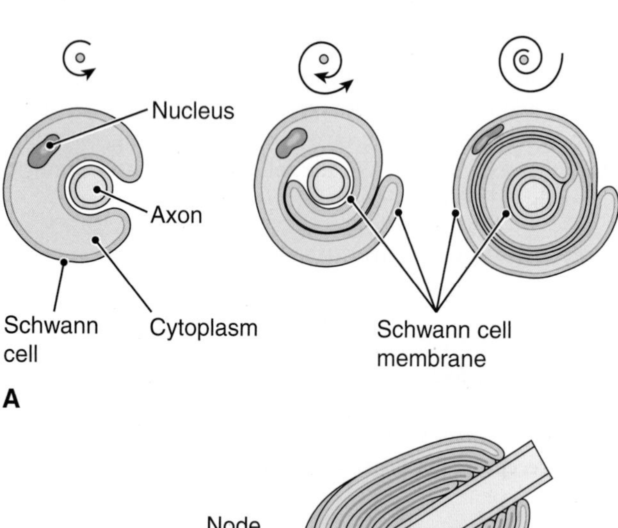

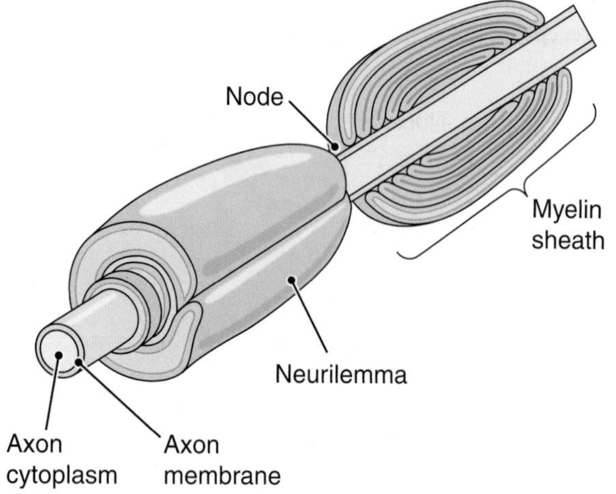

Figure 9-4 **Formation of a myelin sheath.** **KEY POINT** The myelin sheath is formed by Schwann cells in the peripheral nervous system. **A.** Schwann cells wrap around the axon, creating a myelin coating. **B.** The outermost layer of the Schwann cell forms the neurilemma. Spaces between the cells are the nodes (of Ranvier).

When the sheath is complete, small spaces remain between the individual Schwann cells. These tiny gaps, called **nodes** (originally, nodes of Ranvier), are important in speeding nerve impulse conduction. In the CNS, the myelin sheath is formed by another type of cell, the oligodendrocyte (ol-ih-go-DEN-dro-site) (literally meaning "cell with few dendrites"). One oligodendrocyte sends cellular extensions to myelinate several neighboring axons (**see Fig. 9-6** and the later discussion of oligodendrocytes). The cytoplasm and nucleus remain with the oligodendrocyte, so CNS neurons do not have a neurilemma.

Schwann cells help injured neurons regenerate. Undamaged Schwann cells near the injury divide to produce replacement Schwann cells. These new cells remove the damaged tissue and provide a mold (tube) to guide the extension of the new axon. Unfortunately, oligodendrocytes in the CNS do not have the same capabilities. While they can partially restore the myelin coating if it gets damaged, they cannot guide axon regeneration to the same extent as Schwann cells. If CNS neurons are injured, the damage is almost always permanent. Even in the peripheral nerves, however, repair is a slow and uncertain process.

Myelinated axons, because of myelin's color, are called **white fibers** and are found in the **white matter** of the brain and spinal cord as well as in most nerves throughout the body. The fibers and cell bodies of the **gray matter** are not covered with myelin.

TYPES OF NEURONS

The job of neurons is to relay information to or from the CNS or to different places within the CNS itself. There are three functional categories of neurons:

- **Sensory neurons**, also called *afferent neurons,* conduct impulses *to* the spinal cord and brain. For example, if you touch a sharp object with your finger, sensory neurons will carry impulses generated by the stimulus to the CNS for interpretation.
- **Motor neurons**, also called *efferent neurons*, carry impulses *from* the CNS to muscles and glands (effectors). For example, the CNS responds to the pain of touching a sharp object by directing skeletal muscles in your arm to flex and withdraw your hand.
- **Interneurons**, also called *central* or *association neurons*, relay information from place to place within the CNS. Following our original example, in addition to immediate withdrawal from pain, impulses may travel to other parts of the CNS to help retain balance as you withdraw your hand or to help you learn how to avoid sharp objects!

NERVES AND TRACTS

Everywhere in the nervous system, neuron fibers (dendrites and axons) are collected into bundles of varying size (**Fig. 9-5**). A fiber bundle is called a **nerve** in the PNS but a **tract** in the CNS. Tracts are located both in the brain,

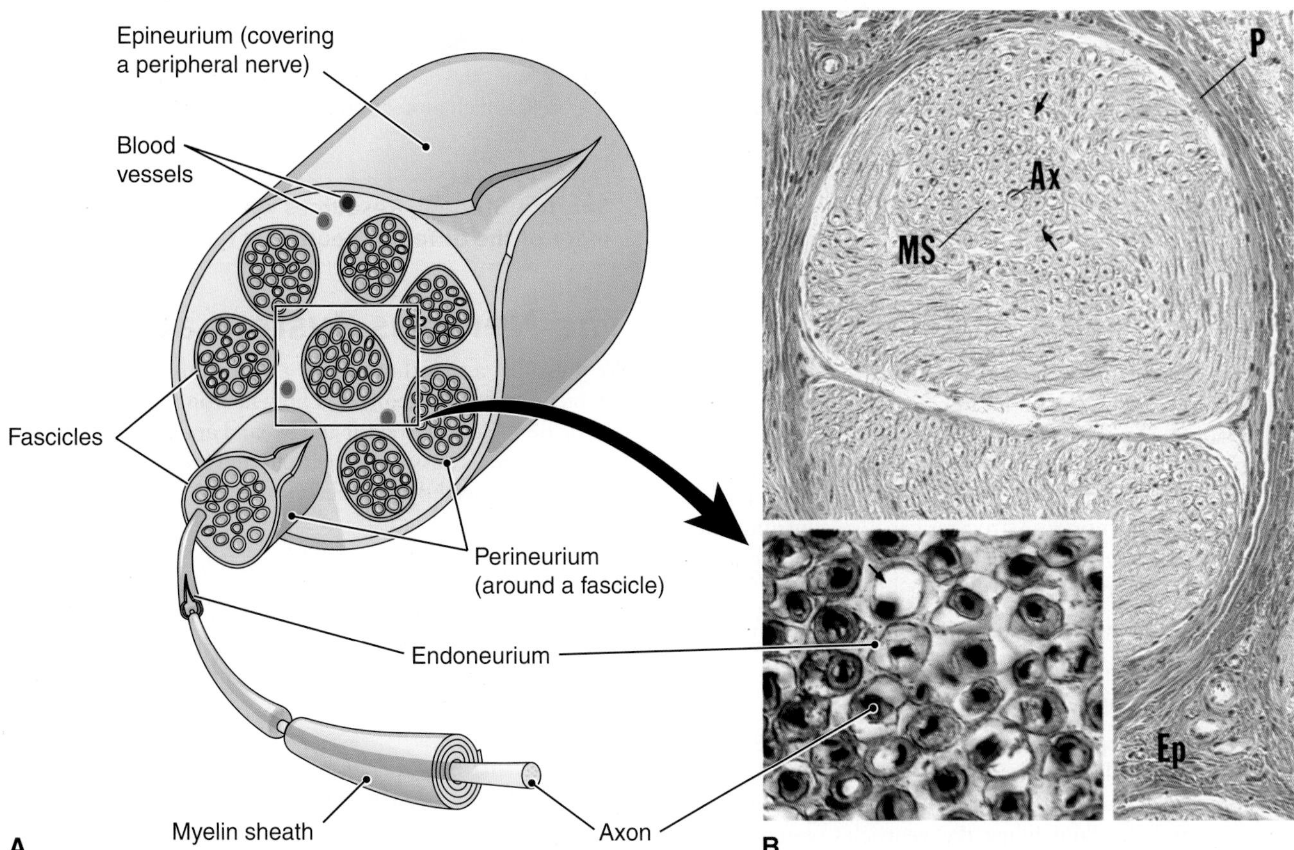

Figure 9-5 **Structure of a nerve.** **KEY POINT** Neuron fibers are collected in bundles called fascicles. Groups of fascicles make up a nerve. Connective tissue holds all components of the nerve together. **A.** Structure of a nerve showing neuron fibers and fascicles. **B.** Micrograph of a nerve (X132). Two fascicles are shown. Perineurium (P) surrounds each fascicle. Epineurium (Ep) is around the entire nerve. Individual axons (Ax) are covered with a myelin sheath (MS). Inset shows myelinated axons surrounded by endoneurium. **ZOOMING IN** What is the deepest layer of connective tissue in a nerve? What is the outermost layer?

where they conduct impulses between regions, and in the spinal cord, where they conduct impulses to and from the brain.

A nerve or tract contains many neuron fibers, just like an electric cable contains many wires. As in muscles, the individual fibers are organized into groups called *fascicles* and are bound together by connective tissue. The names of the connective tissue layers are similar to their names in muscles, but the root *neur/o*, meaning "nerve," is substituted for the muscle root *my/o*, as follows:

- Endoneurium surrounds each individual fiber.
- Perineurium surrounds each fascicle.
- Epineurium surrounds the whole nerve.

A nerve may contain all sensory fibers, all motor fibers, or a combination of both types of fibers. A few of the cranial nerves contain only sensory fibers, so they only conduct impulses toward the brain. These are described as **sensory** (afferent) **nerves.** A few of the cranial nerves, called **motor** (efferent) **nerves,** contain only motor fibers, so they only conduct impulses away from the brain. However, most of the cranial nerves and *all* of the spinal nerves contain both sensory *and* motor fibers and are referred to as **mixed nerves.** Note that in a mixed nerve, impulses may be traveling in two directions (toward or away from the CNS), but each individual fiber in the nerve is carrying impulses in one direction only. Think of the nerve as a large highway. Traffic may be going north and south, for example, but each lane carries cars traveling in only one direction.

CHECKPOINTS

- ☐ 9-3 What is the name of the neuron fiber that carries impulses toward the cell body? What is the name of the fiber that carries impulses away from the cell body?
- ☐ 9-4 What color describes myelinated fibers? What color describes the nervous system's unmyelinated tissue?
- ☐ 9-5 What name is given to nerves that convey impulses toward the CNS? What name is given to nerves that transport away from the CNS?
- ☐ 9-6 What is a nerve? What is a tract?

Neuroglia

Neurons make up only 10% of nervous tissue. The remaining 90% of the CNS and the PNS consists of support cells known as **neuroglia** (nu-ROG-le-ah) or **glial** (GLI-al) **cells**, from a Greek word meaning "glue." The Schwann cell that forms the myelin sheath in the PNS is one type of neuroglia. Other types are located exclusively in the CNS (**Fig. 9-6**). Some of these and their functions are as follows:

- **Astrocytes** (AS-tro-sites) are star-shaped cells serving many functions. They physically support and anchor neurons; they regulate the composition of the extracellular fluid by absorbing and degrading neurotransmitters and excess ions; they form a barrier between blood and brain tissue; and they act as stem cells to make new neurons. These new neurons are important in the generation of memories. Astrocytes can also contribute new neurons to repair damaged areas, but their ability to completely repair brain damage is limited.
- **Oligodendrocytes** form the myelin sheath of CNS neurons.
- **Microglia** (mi-KROG-le-ah) act as phagocytes to remove pathogens, impurities, and dead neurons.
- **Ependymal** (ep-EN-dih-mal) **cells** line ventricles in the brain, fluid-filled cavities discussed in Chapter 10. These cells form a barrier between the nervous tissue of the CNS and the fluid filling the ventricles, cerebrospinal fluid (CSF). In a modified form, these cells also synthesize CSF and promote its movement in the ventricles.

Stem cells produce new neuroglia throughout life. Because neuroglia multiply more frequently than do neurons, most tumors of nervous tissue are glial tumors.

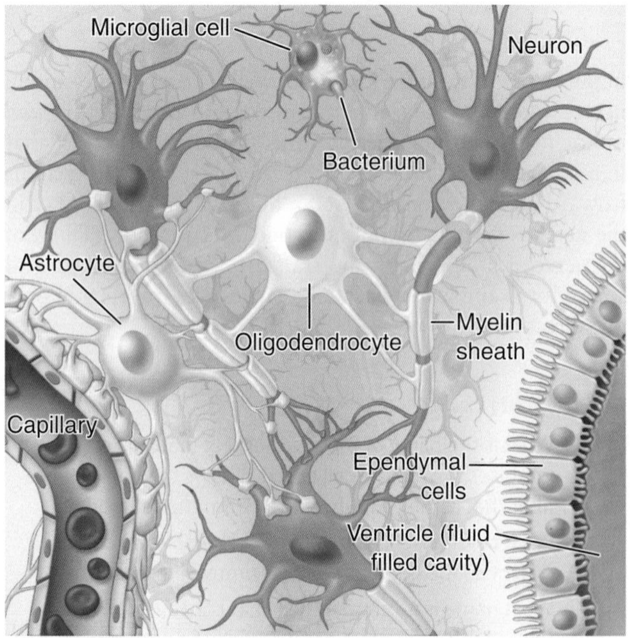

Figure 9-6 Neuroglia in the central nervous system.
KEY POINT Neuroglia serve multiple roles in the nervous system.

CHECKPOINT

☐ 9-7 What is the name of the nervous system's nonconducting cells, which protect, nourish, and support the neurons?

See the Student Resources on thePoint for a summary of the different neuroglial types.

The Nervous System at Work

We presented a snapshot of how the nervous system works in Chapter 8, when we explained how neurons stimulate muscle contraction. Here, we revisit and expand upon the mechanisms by which signals pass down neurons and are transmitted from a neuron to a neighboring cell.

THE NERVE IMPULSE

Recall that neuron fibers can extend over large distances—up to 1 m (3 ft). Specialized electrical signals called **action potentials**, or *nerve impulses*, pass from one end of the neuron to the other, much like an electric current spreads along a wire.

The Resting State Before we discuss the action potential, though, we need to review the basics of membrane potential discussed in Chapter 8. In living cells, positive and negative ions are unequally distributed in the cytoplasm and extracellular fluid. This unequal distribution creates an electric charge across the membrane, known as a membrane (or transmembrane) potential. In this state, the membrane is said to be *polarized*, because oppositely charged particles are separated. In resting cells, the membrane potential is negative (about −70 millivolts, or mV) because of the relative excess of negative ions in the cytoplasm.

Also important for the generation of a nerve impulse are large concentration gradients for sodium and potassium ions across the plasma membrane. Sodium ions are more concentrated along the extracellular side of the plasma membrane than they are along the intracellular side of the membrane. Conversely, potassium ions are in higher concentration on the inside than on the outside of the membrane. The plasma membrane uses active transport to maintain these levels, as ions are constantly diffusing across the membrane in small amounts through channels known as *leak channels* and during nerve impulse transmission. (Remember that substances flow by diffusion from an area where they are in higher concentration to an area where they are in lower concentration.) This transport system requires energy from ATP and is described as the sodium–potassium pump or Na^+–K^+ pump.

Changes in Membrane Potential The movement of ions across the plasma membrane changes the membrane potential (**Fig. 9-7A**). It becomes less negative (more positive) if positive ions enter the cell to neutralize the unpaired negative ions. This change is known as **depolarization**,

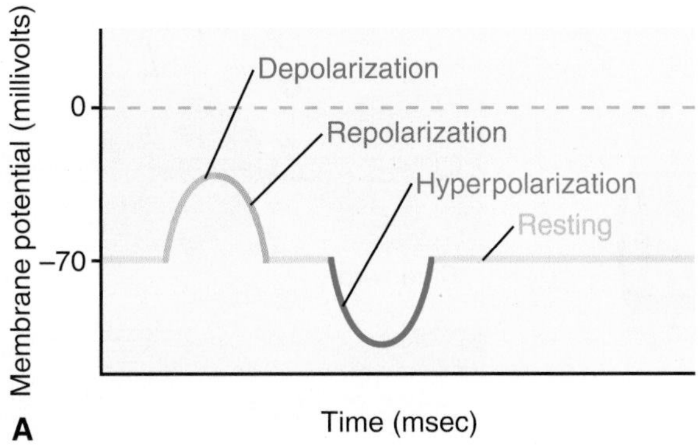

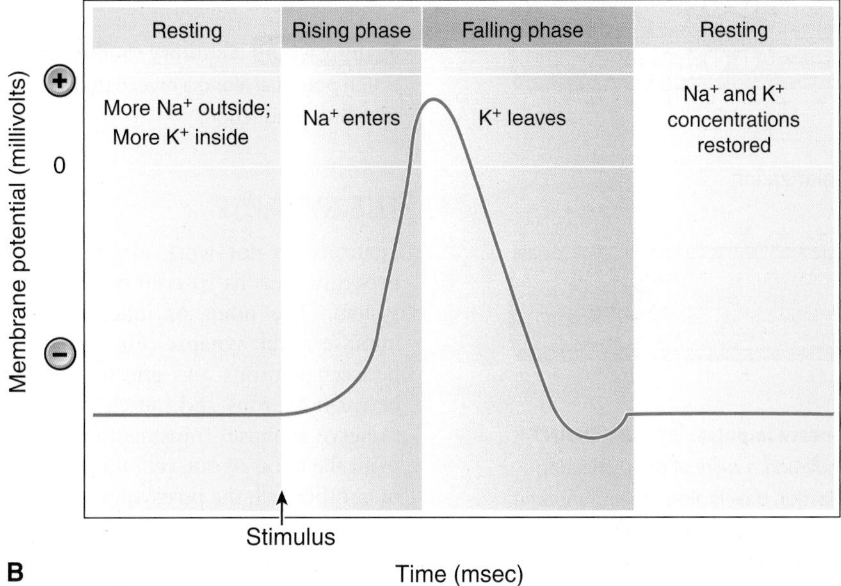

Figure 9-7 **The action potential.** **KEY POINT** In depolarization, Na^+ membrane channels open, and Na^+ enters the cell. In repolarization, K^+ membrane channels open, and K^+ leaves the cell.

because it reduces membrane polarity closer to zero. If positive ions leave the cell (or if negative ions enter the cell), the membrane becomes more negative (or less positive). If this change returns the membrane potential to its resting value, it is known as **repolarization**. If the membrane potential falls below its resting value, the change is known as **hyperpolarization**.

The Action Potential Action potentials result from ion movement across the plasma membrane. A simple description of the events in an action potential is as follows (**Fig. 9-7B**):

- **Rising phase.** A stimulus, such as an electrical, chemical, or mechanical signal of adequate force, causes specific channels in the membrane to open and allow Na^+ ions to flow into the cell. These newly arrived positive ions pair with and eventually outnumber the negative ions, so the membrane potential rises from –70 mV to about +55 mV (**see Fig. 9-7**).
- **Falling phase.** In the next step of the action potential, K^+ channels open to allow K^+ to leave the cell. The departure of positively charged potassium ions causes the membrane potential to fall from about +55 back to –70 mV, so it is described as the falling phase. Since the membrane potential returns to its resting value, it is also known as repolarization. Note that the ions moving in this step are not the same ones that caused the rising phase. During the falling phase, the membrane does not respond to further stimulation. For this reason, the action potential spreads in one direction along the membrane from the point of excitation.

The action potential occurs rapidly, in less than one-thousandth of a second, and is followed by a rapid return to the resting state. However, this local electrical change in the membrane opens the sodium channels in the adjacent membrane region, causing a new action potential (**Fig. 9-8**). And so, the action potential spreads along the membrane as a wave of electric current.

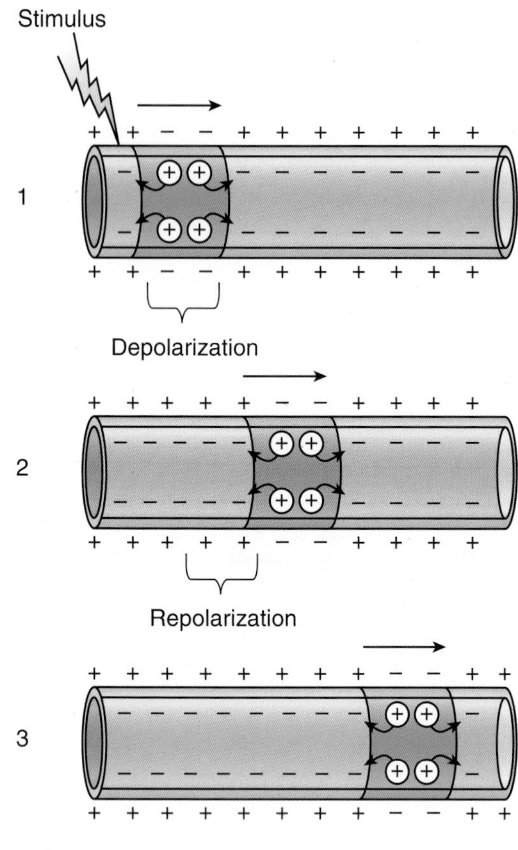

Figure 9-8 **A nerve impulse.** **KEY POINT** From a point of stimulation, a wave of depolarization followed by repolarization travels along the membrane of a neuron. This spreading action potential is a nerve impulse. **ZOOMING IN** What happens to the charge on the membrane at the point of an action potential?

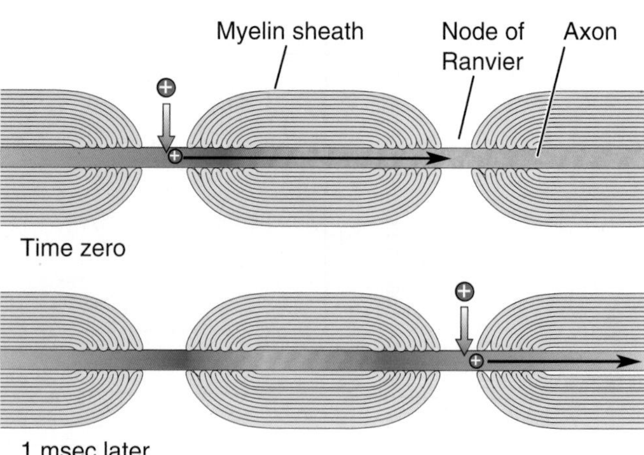

Figure 9-9 **Saltatory conduction.** **KEY POINT** The action potential along a myelinated axon jumps from node to node, speeding conduction.

The Role of Myelin in Conduction As previously noted, some axons are coated with the fatty material myelin (**see Fig. 9-4**). If a fiber is not myelinated, the action potential spreads continuously along the cell's membrane (**see Fig. 9-8**). When myelin is present on an axon, however, it insulates the fiber against the spread of current. This would appear to slow or stop conduction along these fibers, but in fact, the myelin sheath speeds conduction. The reason is that the myelin causes the action potential to "jump" like a spark from node to node along the sheath (**Fig. 9-9**). This type of conduction, called **saltatory** (SAL-tah-to-re) **conduction** (from the Latin verb meaning "to leap"), is actually faster than continuous conduction, because fewer action potentials are needed for an impulse to travel a given distance. It is this type of conduction that is impaired in Sue's case of MS.

> **See the Student Resources on thePoint to view the animations "The Synapse and the Nerve Impulse," "The Action Potential," and "The Myelin Sheath."**

THE SYNAPSE

Neurons do not work alone; impulses must be transferred between neurons to convey information within the nervous system. The point of junction for transmitting the nerve impulse is the **synapse** (**Fig. 9-10**). (There are also synapses between neurons and effector organs. We studied synapses between neurons and muscle cells in Chapter 8.) At a nerve-to-nerve synapse, transmission of an impulse usually occurs from the axon of one cell, the **presynaptic cell**, to the dendrite of another cell, the **postsynaptic cell**.

As described in Chapter 8, information must be passed from one cell to another at the synapse across a tiny gap between the cells, the synaptic cleft. Information usually crosses this gap by means of a chemical signal called a **neurotransmitter**. While the cells at a synapse are at rest, the neurotransmitter is stored in many small vesicles (sacs) within the enlarged axon endings, usually called *end bulbs* or *terminal knobs*, but known by several other names as well.

When an action potential traveling along a neuron membrane reaches the end of the presynaptic axon, some of these vesicles fuse with the membrane and release their neurotransmitter into the synaptic cleft (an example of exocytosis, as described in Chapter 3). The neurotransmitter then acts as a chemical signal to the postsynaptic cell.

On the postsynaptic receiving membrane, usually that of a dendrite, but sometimes another cell part, there are special sites, or receptors, ready to pick up and respond to specific neurotransmitters. Receptors in the postsynaptic cell's membrane influence how or if that cell will respond to a given neurotransmitter.

Neurotransmitters Although there are many known neurotransmitters, some common ones are **norepinephrine** (nor-ep-ih-NEF-rin), **serotonin** (ser-o-TO-nin), **dopamine** (DO-pah-mene), and **acetylcholine** (as-e-til-KO-lene). Acetylcholine (ACh) is the neurotransmitter released at the neuromuscular junction.

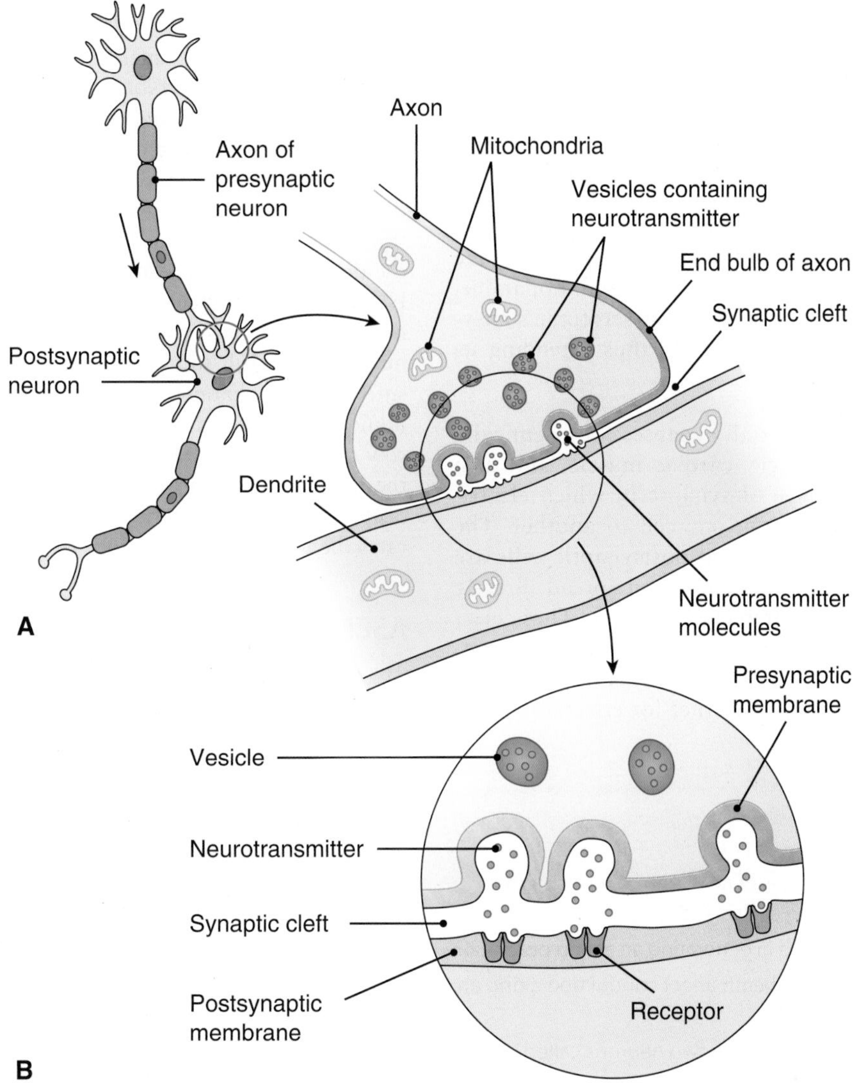

Figure 9-10 **A synapse.** **KEY POINT** Neurotransmitters carry impulses across a synaptic cleft. **A.** The end bulb of the presynaptic (transmitting) axon has vesicles containing neurotransmitter, which is released into the synaptic cleft to the membrane of the postsynaptic (receiving) cell. **B.** Close-up of a synapse showing receptors for neurotransmitter in the postsynaptic cell membrane.

It is common to think of neurotransmitters as stimulating the cells they reach; in fact, they have been described as such in this discussion. Stimulatory neurotransmitters depolarize neurons, increasing the chance that an action potential will occur. Note, however, that some of these chemicals inhibit the postsynaptic cell and keep it from reacting, as will be demonstrated later in discussions of the ANS. These inhibitory neurotransmitters hyperpolarize the cell, making action potentials less likely.

The connections between neurons can be quite complex. One cell can branch to stimulate many receiving cells, or a single cell may be stimulated by a number of different axons. The cell's response is based on the total effects of all the neurotransmitters it receives over a short period of time.

After its release into the synaptic cleft, the neurotransmitter may be removed by several methods:

- It may slowly diffuse away from the synapse.
- It may be rapidly destroyed by enzymes in the synaptic cleft.
- It may be taken back into the presynaptic cell to be used again, a process known as *reuptake*.
- It may be taken up by neuroglial cells, specifically astrocytes.

The method of removal helps determine how long a neurotransmitter will act.

Many drugs that act on the mind, substances known as *psychoactive drugs*, function by affecting neurotransmitter activity in the brain. Prozac, for example, increases the level of the neurotransmitter serotonin by blocking its reuptake into presynaptic cells at synapses. This and other selective serotonin reuptake inhibitors prolong the neurotransmitter's activity and produce a mood-elevating effect. They are used to treat depression, anxiety, and obsessive–compulsive disorder. Similar psychoactive drugs prevent the reuptake of the neurotransmitters norepinephrine and dopamine. Another class of antidepressants prevents serotonin's enzymatic breakdown in the synaptic cleft, thus extending its action.

Electrical Synapses Not all synapses are chemically controlled. In smooth muscle, cardiac muscle, and also in the CNS, there is a type of synapse in which electrical signals travel directly from one cell to another. The membranes of the presynaptic and postsynaptic cells are close together and an electric charge can spread directly between them through an intercellular bridge. These electrical synapses allow more rapid and coordinated communication. In the heart, for example, it is important that large groups of cells contract together for effective pumping action.

CHECKPOINTS

- ☐ 9-8 What are the two stages of an action potential, and what happens during each?
- ☐ 9-9 What ions are involved in generating an action potential?
- ☐ 9-10 How does the myelin sheath affect conduction along an axon?
- ☐ 9-11 What is the junction between two neurons called?
- ☐ 9-12 As a group, what are all the chemicals that carry information across the synaptic cleft called?

The Spinal Cord

The spinal cord is the link between the spinal nerves and the brain. It also helps coordinate some simple actions that do not involve the brain. The spinal cord is contained in and protected by the vertebrae, which fit together to form a continuous tube extending from the occipital bone to the coccyx (**Fig. 9-11**). In the embryo, the spinal cord occupies the entire spinal canal, extending down into the tail portion of the vertebral column. The bony column grows much more rapidly than the nervous tissue of the cord, however, and eventually, the end of the spinal cord no longer reaches the lower part of the spinal canal. This disparity in growth continues to increase so that in adults, the spinal cord ends in the region just below the area to which the last rib attaches (between the first and second lumbar vertebrae). Individual nerves fan out from this point within the vertebral canal, hence its name as the **cauda equina** (KAW-dah eh-KWI-nah), or horse's tail.

STRUCTURE OF THE SPINAL CORD

The spinal cord has a small, irregularly shaped core of gray matter (unmyelinated axons and cell bodies) surrounded by white matter (myelinated axons) (**see Fig. 9-11B and C**). The internal gray matter is arranged so that a column of gray matter extends up and down posteriorly (dorsally), one on each side; another column is found in the anterior (ventral) region on each side. These two pairs of columns, called the **dorsal horns** and **ventral horns**, give the gray matter an H-shaped appearance in cross-section. The bridge of gray matter that connects the right and left horns is the **gray commissure** (KOM-ih-shure). In the center of the gray commissure is a small channel, the **central canal**, containing CSF, the liquid that circulates around the brain and spinal cord. A narrow groove, the **posterior median sulcus** (SUL-kus), divides the right and left portions of the posterior white matter. A deeper groove, the **anterior median fissure** (FISH-ure), separates the right and left portions of the anterior white matter.

ASCENDING AND DESCENDING TRACTS

The spinal cord is the pathway for sensory and motor impulses traveling to and from the brain. Most impulses are carried in the thousands of myelinated axons in the spinal cord's white matter, which are subdivided into tracts (fiber bundles). Sensory impulses entering the spinal cord are transmitted toward the brain in **ascending tracts** of the white matter. Motor impulses traveling from the brain are carried in **descending tracts** toward the PNS.

CHECKPOINTS 

- ☐ 9-13 How are the gray and white matter arranged in the spinal cord?
- ☐ 9-14 What is the purpose of the tracts in the spinal cord's white matter?

See the Student Resources on thePoint to view an illustration of the cauda equina.

The Spinal Nerves

There are 31 pairs of spinal nerves, usually named after the vertebra superior to their point of emergence (**see Fig. 9-11A**). The exception to this rule is the cervical nerves; nerves C1 through C7 emerge above the corresponding vertebra, and C8 arises below vertebra C7. Each nerve passes through an intervertebral foramen linking adjacent vertebrae.

Each nerve is attached to the spinal cord by two roots residing within the spinal canal: the **dorsal root** and the **ventral root** (**see Fig. 9-11B**). The **dorsal root ganglion**, a swelling on each dorsal root, contains the cell bodies of the sensory neurons. A **ganglion** (GANG-le-on) is any collection of nerve cell bodies located outside the CNS.

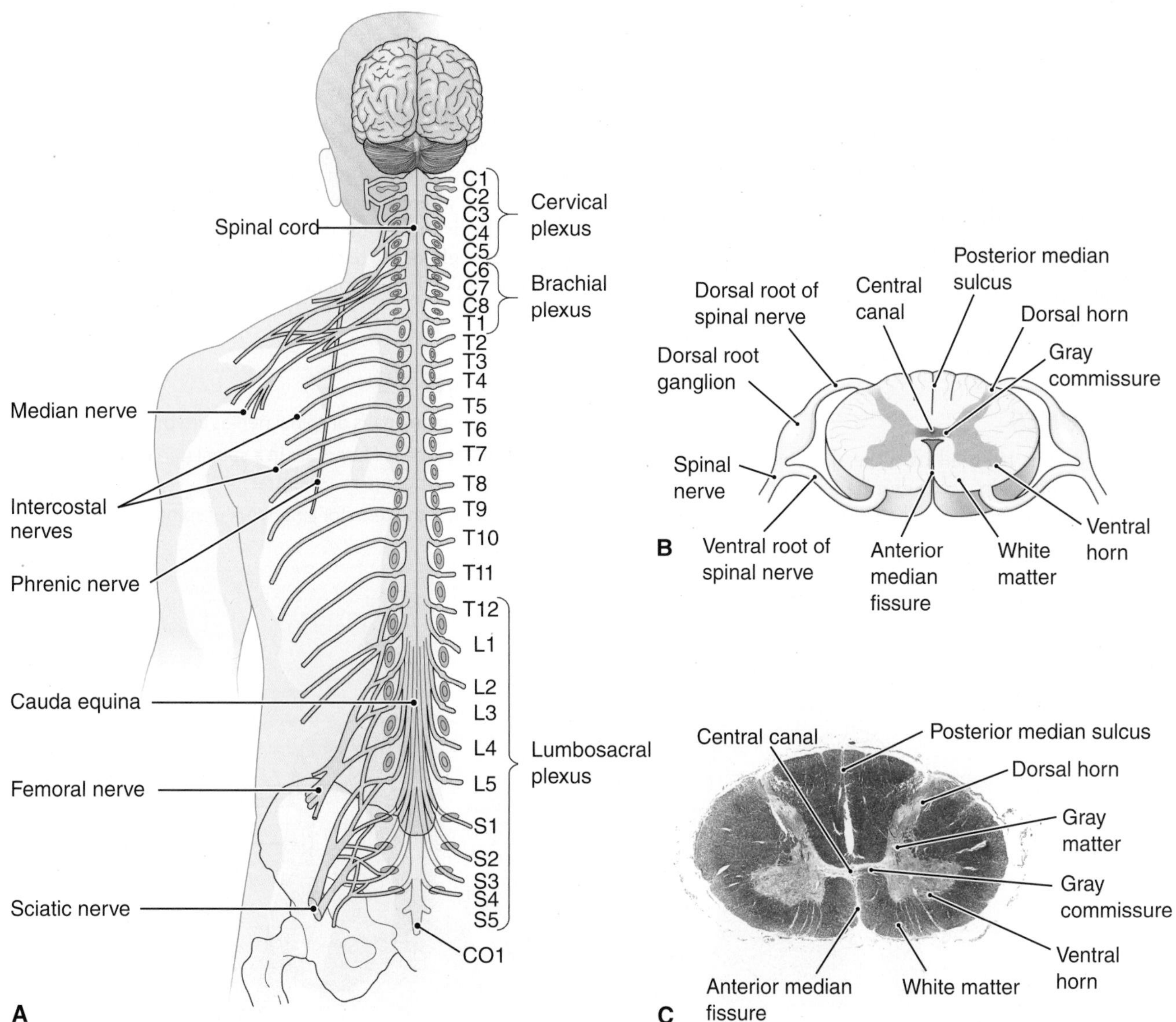

Figure 9-11 **Spinal cord and spinal nerves. A.** Posterior view. Nerve plexuses (networks) are shown. Nerves extending from the distal cord form the cauda equina. **B.** Cross-section of the spinal cord showing the organization of the gray and white matter. The roots of the spinal nerves are also shown. **C.** Microscopic view of the spinal cord in cross-section (×5). **ZOOMING IN** Is the spinal cord the same length as the spinal column? How does the number of cervical vertebrae compare to the number of cervical spinal nerves?

A spinal nerve's ventral root contains motor fibers that supply muscles and glands (effectors). The cell bodies of these neurons are located in the cord's ventral gray matter (ventral horns). Because the dorsal (sensory) and ventral (motor) roots combine to form the spinal nerve, all spinal nerves are mixed nerves.

BRANCHES OF THE SPINAL NERVES

Each spinal nerve continues only a short distance away from the spinal cord and then branches into small posterior divisions and larger anterior divisions. The posterior divisions distribute branches to the back. The anterior branches of the thoracic nerves 2 through 11 become the intercostal nerves supplying the regions between the ribs. The remaining anterior branches interlace to form networks called **plexuses** (PLEK-sus-eze), which then distribute branches to the body parts (**see Fig. 9-11**). The three main plexuses are described as follows:

- The **cervical plexus** supplies motor impulses to the neck muscles and receives sensory impulses from the neck and the back of the head. The phrenic nerve, which activates the diaphragm, arises from this plexus.

- The **brachial** (BRA-ke-al) **plexus** sends numerous branches to the shoulder, arm, forearm, wrist, and hand. For example, the median nerve emerges from the brachial plexus.
- The **lumbosacral** (lum-bo-SA-kral) **plexus** supplies nerves to the pelvis and legs. The femoral nerve to the thigh is part of this plexus. The largest branch in this plexus is the sciatic (si-AT-ik) nerve, which leaves the dorsal part of the pelvis, passes beneath the gluteus maximus muscle, and extends down the posterior thigh. At its beginning, it is nearly 1 in thick, but it soon branches to the thigh muscles. Near the knee, it forms two subdivisions that supply the leg and the foot.

DERMATOMES

Sensory neurons from all over the skin, except for the skin of the face and scalp, feed information into the spinal cord through the spinal nerves. The skin surface can be mapped into distinct regions that are supplied by a single spinal nerve. Each of these regions is called a **dermatome** (DER-mah-tome) (**Fig. 9-12**).

Sensation from a given dermatome is carried over its corresponding spinal nerve. This information can be used to identify the spinal nerve or spinal segment that is involved in an injury, as sensation from its corresponding skin surface will be altered. In some areas, the dermatomes are not absolutely distinct. Some dermatomes may share a nerve supply with neighboring regions. For this reason, it is necessary to numb several adjacent dermatomes to achieve successful anesthesia.

CHECKPOINTS 

- 9-15 How many pairs of spinal nerves are there?
- 9-16 What types of fibers are in a spinal nerve's dorsal root? What types are in its ventral root?
- 9-17 What is the term for a network of spinal nerves?

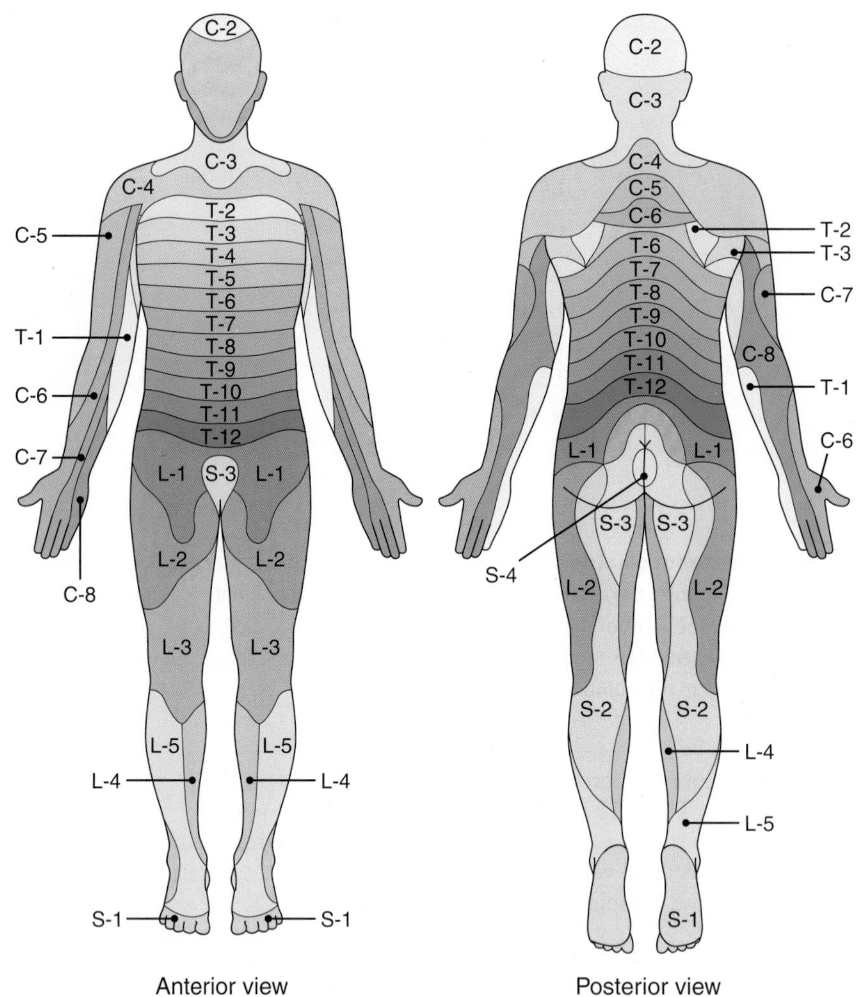

Figure 9-12 Dermatomes. **KEY POINT** A dermatome is a region of the skin supplied by a single spinal nerve. **ZOOMING IN** Which spinal nerves carry impulses from the skin of the toes? From the anterior hand and fingers?

Reflexes

As the nervous system develops, neural pathways are formed to coordinate responses. Most of these pathways are very complex, involving multiple neurons and interactions between different regions of the nervous system. Easier to study are simple pathways involving a minimal number of neurons. Responses controlled by some of these simpler pathways are useful in neurologic studies.

THE REFLEX ARC

A complete pathway through the nervous system from stimulus to response is termed a **reflex arc**. This is the nervous system's basic functional pathway. **Figure 9-13** illustrates the components of a reflex arc, using the example of the pin prick reflex. Try it yourself—if you touch a tack or pin with your finger, your finger will automatically pull away. The fundamental parts of a reflex arc are the following:

1. **Receptor**—the end of a dendrite or some specialized receptor cell, as in a special sense organ, that detects a stimulus. In Step 1 of **Figure 9-13**, a pin prick activates a sensory receptor in the finger.
2. **Sensory neuron**—a cell that transmits impulses toward the CNS. In Step 2, a dendrite carries the signal from the sensory receptor to the cell body in the dorsal root ganglion. An axon carries the signal from the cell body through the dorsal root and into the CNS (in this case, the spinal cord).

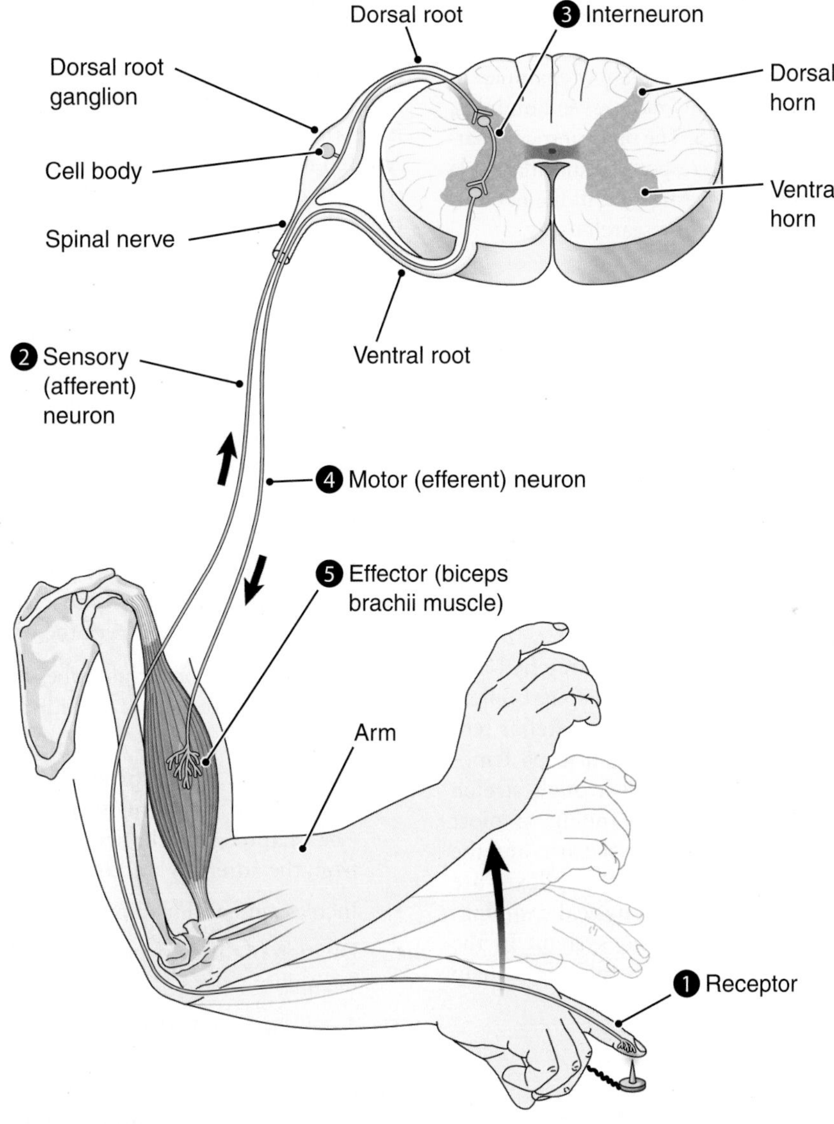

Figure 9-13 Typical reflex arc. **KEY POINT** Numbers show the sequence in the pathway of impulses through the spinal cord (*solid arrows*). Contraction of the biceps brachii results in flexion of the arm at the elbow. **ZOOMING IN** Is this a somatic or an autonomic reflex arc? What type of neuron is located between the sensory and motor neurons in the CNS?

9

3. **Central nervous system**—where impulses are coordinated and a response is organized. In Step 3, the sensory neuron synapses with an interneuron within the CNS. Each interneuron can receive signals from multiple neurons.
4. **Motor neuron**—a cell that carries impulses away from the CNS. The motor neuron in Step 4 conveys a signal from the interneuron to the skeletal muscle. Motor impulses leave the cord through the ventral horn of the spinal cord's gray matter.
5. **Effector**—a muscle or a gland outside the CNS that carries out a response. Step 5 of the figure shows how the biceps brachii responds to the motor signal by contracting, thereby removing the finger.

At its simplest, a reflex arc can involve just two neurons, one sensory and one motor, with a synapse in the CNS. Few reflex arcs require only this minimal number of neurons. (The knee-jerk reflex described below is one of the few examples in humans.) Most reflex arcs involve many more, even hundreds, of connecting neurons within the CNS. The many intricate patterns that make the nervous system so responsive and adaptable also make it difficult to study, and investigation of the nervous system is one of the most active areas of research today.

REFLEX ACTIVITIES

Although reflex pathways may be quite complex, a **simple reflex** is a rapid, uncomplicated, and automatic response involving very few neurons. Reflexes are specific; a given stimulus always produces the same response. When you fling out an arm or leg to catch your balance, withdraw from a painful stimulus, or blink to avoid an object approaching your eyes, you are experiencing reflex behavior. A simple reflex arc that passes through the spinal cord alone and does not involve the brain is termed a **spinal reflex**. Returning to our opening corporation analogy, it's as if middle management makes a decision independently without involving the CEO.

The **stretch reflex**, in which a muscle is stretched and responds by contracting, is one example of a spinal reflex. If you tap the tendon below the kneecap (the patellar tendon), the muscle of the anterior thigh (quadriceps femoris) contracts, eliciting the knee-jerk reflex. Such stretch reflexes may be evoked by appropriate tapping of most large muscles (such as the triceps brachii in the arm and the gastrocnemius in the calf of the leg). Because reflexes are simple and predictable, they are used in physical examinations to test the condition of the nervous system. In the case study, Dr. Jensen tested Sue's stretch reflexes to help in diagnosis.

CHECKPOINT

9-18 What is the name for a pathway through the nervous system from a stimulus to an effector?

See the Student Resources on thePoint to view the animation "The Reflex Arc."

The Autonomic Nervous System

The autonomic (visceral) nervous system regulates the action of the glands, the smooth muscles of hollow organs and vessels, and the heart muscle. These actions are carried out automatically; whenever a change occurs that calls for a regulatory adjustment, it is made without conscious awareness. The ANS consists of the **sympathetic** and **parasympathetic** divisions. These two divisions have distinct functional and structural characteristics (Fig. 9-14), as described below and summarized in Tables 9-2 and 9-3.

FUNCTIONS OF THE AUTONOMIC NERVOUS SYSTEM

Most visceral organs are supplied by both sympathetic and parasympathetic fibers, and the two systems generally have opposite effects (Table 9-2). The sympathetic system tends to act as an accelerator for those organs needed to meet a stressful situation. It promotes what is called the **fight-or-flight response** because in the most primitive terms, the person must decide to stay and "fight it out" with the enemy or to run away from danger. The times when the sympathetic nervous system comes into play can be summarized by the four "Es," that is, times of emergency, excitement, embarrassment, and exercise. If you think of what happens to a person who is in any of these situations, you can easily remember the effects of the sympathetic nervous system:

- Increase in the rate and force of heart contractions
- Increase in blood pressure due partly to the more effective heartbeat and partly to constriction of most small arteries everywhere except the brain
- Dilation of the bronchial tubes to allow more oxygen to enter and more carbon dioxide to leave
- Stimulation of the central portion of the adrenal gland. This gland produces hormones that prepare the body to meet emergency situations in many ways (see Chapter 12). The sympathetic nerves and hormones from the adrenal gland reinforce each other
- Increase in basal metabolic rate
- Dilation of the eye's pupil and increase in distance focusing ability

The sympathetic system also acts as a brake on those systems not directly involved in the stress response, such as the urinary and digestive systems. If you try to eat while you are angry, you may note that your saliva is thick and so small in amount that you can swallow only with difficulty. Under these circumstances, when food does reach the stomach, it seems to stay there longer than usual.

The parasympathetic system normally acts as a balance for the sympathetic system once a crisis has passed. It is the

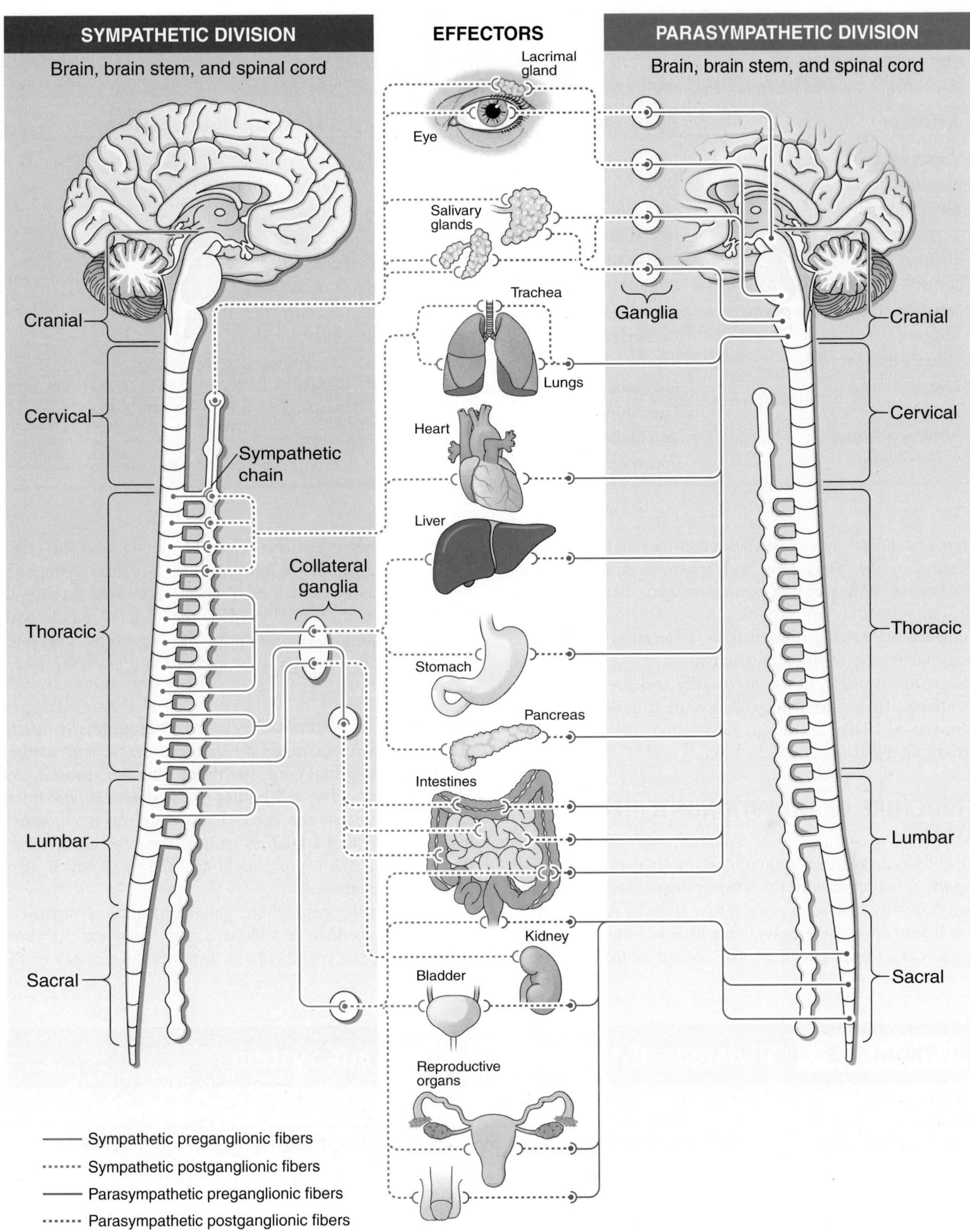

9

Figure 9-14 Autonomic nervous system. **KEY POINT** Most organs have both sympathetic and parasympathetic fibers. The diagram shows only one side of the body for each division. **ZOOMING IN** Which division of the autonomic nervous system has ganglia closer to the effector organ?

Table 9-2 Effects of the Sympathetic and Parasympathetic Systems on Selected Organs

Effector	Sympathetic System	Parasympathetic System
Pupils of eye	Dilation	Constriction
Lacrimal glands	None	Secretion of tears
Sweat glands	Stimulation	None
Digestive glands	Inhibition	Stimulation
Heart	Increased rate and strength of beat	Decreased rate of beat
Bronchi of lungs	Dilation	Constriction
Muscles of digestive system	Decreased contraction	Increased contraction
Kidneys	Decreased activity	None
Urinary bladder	Relaxation	Contraction and emptying
Liver	Increased release of glucose	None
Penis	Ejaculation	Erection
Adrenal medulla	Stimulation	None
Blood vessels	Constriction	Dilation, penis and clitoris only

"rest and digest" system. It causes constriction of the pupils, slowing of the heart rate, and constriction of the bronchial tubes. However, the parasympathetic nervous system also stimulates certain activities needed for maintenance of homeostasis. Among other actions, it promotes the formation and release of urine and activity of the digestive tract. Saliva, for example, flows more easily and profusely under its effects. These stimulatory actions are summarized by the acronym SLUDD: salivation, lacrimation (tear formation), urination, digestion, and defecation.

STRUCTURE OF THE AUTONOMIC NERVOUS SYSTEM

All autonomic pathways contain two motor neurons connecting the spinal cord with the effector organ (Table 9-3). The two neurons synapse in ganglia that serve as relay stations. The first neuron, the preganglionic neuron, extends from the spinal cord to the ganglion. The second neuron, the postganglionic neuron, travels from the ganglion to the effector. This arrangement differs from the voluntary (somatic) nervous system, in which each motor nerve fiber extends all the way from the spinal cord to the skeletal muscle with no intervening synapse. Some of the autonomic fibers are within the spinal nerves; some are within the cranial nerves (see Chapter 10).

Sympathetic Nervous System Pathways Sympathetic motor neurons originate in the thoracic and lumbar regions of the spinal cord, the **thoracolumbar** (tho-rah-ko-LUM-bar) area. The cell bodies of the preganglionic fibers are located within the cord. The axons travel via spinal nerves T1-T5 and L1-L2 to ganglia near the cord, where they synapse with postganglionic neurons, which then extend to the effectors.

Many of the sympathetic ganglia form the **sympathetic chains**, two cordlike strands of ganglia that extend along either side of the spinal column from the lower neck to the

Table 9-3 Divisions of the Autonomic Nervous System

	Divisions	
Characteristics	**Sympathetic Nervous System**	**Parasympathetic Nervous System**
Origin of fibers	Thoracic and lumbar regions of the spinal cord; thoracolumbar	Brain stem and sacral regions of the spinal cord; craniosacral
Location of ganglia	Sympathetic chains and three single collateral ganglia (celiac, superior mesenteric, and inferior mesenteric)	Terminal ganglia in or near the effector organ
Neurotransmitter at effector	Mainly norepinephrine; adrenergic	Acetylcholine; cholinergic
Effects	Response to stress; fight-or-flight response	Reverses fight-or-flight (stress) response (see Table 9-2): stimulates some activities

upper abdominal region. (Note that **Figure 9-14** shows only one side for each division of the ANS.)

In addition, the nerves that supply the abdominal and pelvic organs synapse in three single **collateral ganglia** farther from the spinal cord. These are

- the celiac ganglion, which sends fibers mainly to the digestive organs
- the superior mesenteric ganglion, which sends fibers to the large and small intestines
- the inferior mesenteric ganglion, which sends fibers to the distal large intestine and organs of the urinary and reproductive systems

The postganglionic neurons of the sympathetic system, with few exceptions, act on their effectors by releasing the neurotransmitter norepinephrine (noradrenaline), a compound similar in chemical composition and action to the hormone epinephrine (adrenaline). This system is therefore described as **adrenergic**, which means "activated by adrenaline."

Parasympathetic Nervous System Pathways The parasympathetic motor pathways begin in the **craniosacral** (kra-ne-o-SA-kral) areas, with fibers arising from cell bodies in the brain stem (midbrain and medulla) and the lower (sacral) part of the spinal cord. From these centers, the first fibers extend to autonomic ganglia that are usually located near or within the walls of the effector organs and are called **terminal ganglia**. The pathways then continue along postganglionic neurons that stimulate involuntary muscles and glands.

The neurons of the parasympathetic system release the neurotransmitter ACh, leading to the description of this system as **cholinergic** (activated by ACh).

THE ROLE OF CELLULAR RECEPTORS

An important factor in the actions of neurotransmitters is their "docking sites," that is, their receptors on receiving (postsynaptic) cell membranes. A neurotransmitter fits into its receptor like a key in a lock. Once the neurotransmitter binds, the receptor initiates events that change the postsynaptic cell's activity. Different receptors' responses to the same neurotransmitter may vary, and a cell's response depends on the receptors it contains.

Among the many different classes of identified receptors, two are especially important and well studied. The first is the cholinergic receptors, which bind ACh. Cholinergic receptors are further subdivided into two types, each named for drugs that researchers have discovered bind to them and mimic ACh's effects:

- Nicotinic receptors (which bind nicotine) are found on skeletal muscle cells and stimulate muscle contraction when ACh is present.
- Muscarinic receptors (which bind muscarine, a poison) are found on effector cells of the parasympathetic nervous system. Depending on the type of muscarinic receptor in a given effector organ, ACh can either stimulate or inhibit a response. For example, ACh stimulates digestive organs but inhibits the heart.

The second class of receptors is the adrenergic receptors, which bind norepinephrine. They are found on effector cells of the sympathetic nervous system. They are further subdivided into alpha (α) and beta (β). Depending on the type of adrenergic receptor in a given effector organ, norepinephrine can either stimulate or inhibit a response. For example, norepinephrine stimulates the heart and inhibits the digestive organs.

Some drugs block specific receptors. For example, "betablockers" regulate the heart in cardiac disease by preventing β receptors from binding norepinephrine, the neurotransmitter that increases the rate and strength of heart contractions.

CHECKPOINTS

- [ ] 9-19 How many neurons are there in each motor pathway of the ANS?
- [ ] 9-20 Which division of the ANS stimulates a stress response? Which division reverses the stress response?

Clinical Aspects of the Spinal Cord and Spinal Nerves

In this section, we describe some medical procedures that involve the spinal cord and then discuss various disorders affecting the spinal cord and spinal nerves.

MEDICAL PROCEDURES INVOLVING THE SPINAL CORD

Several types of medical procedures involve the fluid that circulates in and around the brain and spinal cord, known as the *cerebrospinal fluid*, or the membranes that surround these tissues, known as the *meninges* (meh-NIN-jeze). These components of the CNS are discussed in more detail in Chapter 10.

- Lumbar puncture. A lumbar puncture (spinal tap) removes a small amount of CSF to test for inflammation or infection in the nervous system. The ideal location for a spinal tap is between the third and fourth lumbar vertebrae, because this region contains spinal nerves (the cauda equina) instead of the spinal cord (**Fig. 9-15A** syringe 1).
- Drug administration. Pain medication can be delivered into the epidural space, the fat-filled area between the outermost meningeal layer and the vertebra (**Fig. 9-15A** syringe 2). The injected agent diffuses into the spinal cord and interferes with nerve impulse transmission. Low-dose epidurals can achieve analgesia (pain relief) while retaining some sensation and most motor functions, a desirable state for a mother undergoing a vaginal birth. Cesarean sections use a larger dose, achieving anesthesia (loss of sensation) while the patient is still awake. One advantage of an epidural is that it uses a catheter so medication can be administered as required. Epidurals are also used to alleviate back pain and postoperative

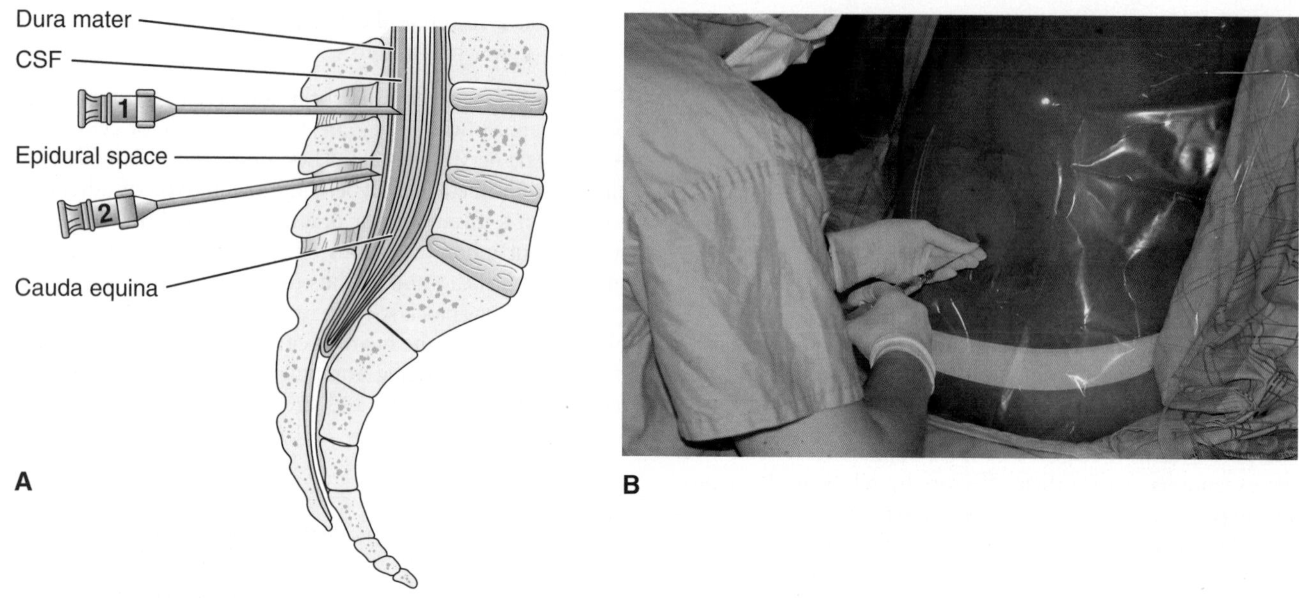

Figure 9-15 **Lumbar puncture and epidural anesthesia. A.** Lumbar puncture withdrawal of a sample of cerebrospinal fluid (CSF) for analysis (syringe 1). This fluid is found just deep to the dura mater, which is the outermost layer of the meninges. Epidural anesthesia delivers painkillers into the layer of fat surrounding the cord (syringe 2). **B.** The position of a patient for an epidural anesthetic.

pain in the abdomen and lower body. Less commonly, medications can be delivered into the CSF surrounding the cauda equina. This procedure is known as spinal or intrathecal anesthesia. It requires smaller doses than an epidural, but is more dangerous because it involves puncturing the outer meninges.

DISORDERS AFFECTING THE SPINAL CORD

Disorders affecting the spinal cord include autoimmune, degenerative, and infectious diseases; tumors; and injuries. Recall from Chapter 5 that autoimmune disorders result when the body's immune cells attack body tissues, causing inflammation and potentially cell death.

Diseases **Multiple sclerosis** is an autoimmune disease resulting in the demyelination (loss of the myelin sheath) of CNS axons and eventually neuronal death. Demyelination slows the speed of nerve impulse conduction and disrupts nervous system communication, as illustrated in Sue's opening case study. Both the spinal cord and the brain can be affected.

MS is the most common chronic CNS disease of young adults in the United States. The disease affects women about twice as often as men, and it is more common in temperate climates and in people of northern European ancestry. Genetics and environment both contribute to the development of MS. Some research suggests that a prior viral or bacterial infection, even one that occurred many years before, may set off the disease, but the study results are not overly convincing. Other researchers have implicated vitamin D deficiency, since individuals in temperature climates have less sun exposure so they make less vitamin D.

MS progresses at different rates depending on the individual, and it may be marked by episodes of relapse and remission. At this point, MS treatments focus on decreasing the immune response, but they are not very effective. Hopefully, some of the new drugs currently under investigation will reduce neuron damage and slow development of the disease.

Amyotrophic (ah-mi-o-TROF-ik) **lateral sclerosis** (also called *Lou Gehrig disease*) is a degenerative disorder in which motor neurons are destroyed. The progressive destruction causes muscle atrophy and loss of motor control until finally the affected person is unable to swallow, talk, or breathe.

Poliomyelitis (po-le-o-mi-eh-LI-tis) ("polio") is a viral disease of the nervous system (*myelo* means "spinal cord") that was intermittently epidemic in children during the first half of the 20th century. Polio is spread by ingestion of water contaminated with feces containing the virus, leading to infection of the gastrointestinal tract. This highly infectious agent usually causes nothing more serious than a minor stomach upset. However, in about 1% of cases, the virus spreads to the motor neurons of the spinal cord and potentially the brain. Paralysis of the limbs and possibly the respiratory muscles can result.

Polio has been virtually eliminated in most countries through the use of vaccines against the disease—first the injected Salk vaccine developed in 1954, followed by the Sabin oral vaccine. A goal of the World Health Organization is the total eradication of polio by worldwide vaccination programs, and several philanthropic organizations have joined in this effort.

Tumors Tumors that affect the spinal cord commonly arise in the support tissue in and around the cord. They are frequently tumors of the meninges or neuroglia. Symptoms are caused by pressure on the cord and the roots of the spinal nerves. These include pain, numbness, weakness, and loss of function. Spinal cord tumors are diagnosed by MRI or other imaging techniques, and treatment is by surgery and radiation.

Injuries Spinal cord injuries may result from wounds, fracture, or dislocation of the vertebrae, herniation of intervertebral disks, or tumors. The most common causes of accidental injury to the cord are motor vehicle accidents; falls; sports injuries, especially diving accidents; and job-related injuries. Spinal cord injuries are more common in the young adult age group, and many are related to alcohol or drug use.

Cord damage may cause paralysis or loss of sensation in structures supplied by nerves below the level of injury. Different degrees of loss are named using the root *plegia*, meaning "paralysis," for example:

- monoplegia (mon-o-PLE-je-ah)—paralysis of one limb
- diplegia (di-PLE-je-ah)—paralysis of both upper or both lower limbs
- paraplegia (par-ah-PLE-je-ah)—paralysis of both lower limbs
- hemiplegia (hem-e-PLE-je-ah)—paralysis of one side of the body
- tetraplegia (tet-rah-PLE-je-ah) or quadriplegia (kwah-drih-PLE-je-ah)—paralysis of all four limbs

Box 9-1 contains information on treatment of these injuries.

DISORDERS OF THE SPINAL NERVES

A **neuropathy** (nu-ROP-ah-the) is any disease of nerves. Peripheral neuropathy involves damage to the nerves of the PNS. **Neuritis** (nu-RI-tis) is a general term for inflammation of a nerve. Since neuropathy usually involves inflammation, these two terms are often used interchangeably. Peripheral neuropathy is not always permanent because peripheral nerves can regenerate. Often, treatment of the underlying cause can also treat the neuropathy, and the patient can recover partial or full functioning.

Mononeuropathy Mononeuropathy affects a single nerve and can result from trauma or tumors. For instance, benign tumors (neuromas) of nerve connective tissue can occur subsequently to severe nerve damage.

Commonly, mononeuropathies result from nerve compression. **Carpal tunnel syndrome** results from compression of the median nerve by the carpal bones of the wrist, causing pain and numbness in the hand (Fig. 9-16). It is commonly thought that carpal tunnel syndrome results from repetitive hand actions, such as typing, video games, or playing an instrument. New studies show that genetics is actually the most important factor in the development of the syndrome. **Morton neuroma,** in which the tarsals and metatarsals of the foot compress the plantar nerve, commonly occurs in women. Wearing pointy or high-heeled shoes contributes to this chronic compression injury. **Sciatica** (si-AT-ih-kah) describes pain, numbness, and tingling along the path of the sciatic nerve. Sciatica results from compression of the sciatic nerve or of the contributing spinal nerves (L3-L5, S1-S3) from a herniated disk, a bone spur, or the weight of a pregnant uterus. Arthritis of the lower spinal column can also cause sciatica.

Polyneuropathy Polyneuropathy (polyneuritis) impacts more than one nerve. Symptoms can include weakness,

9

Box 9-1

HOT TOPICS

Spinal Cord Injury: Crossing the Divide

Approximately 13,000 new cases of traumatic spinal cord injury occur each year in the United States, the majority involving males aged 16 to 30 years. More than 80% of these injuries are due to motor vehicle accidents, acts of violence, and falls. Because neurons show little, if any capacity to repair themselves, spinal cord injuries almost always result in a loss of sensory or motor function (or both), and therapy has focused on injury management rather than cure. However, scientists are investigating four improved treatment approaches:

- *Minimizing spinal cord trauma after injury*. Intravenous injection of the steroid methylprednisolone shortly after injury reduces swelling at the site of injury and improves recovery.
- *Using neurotrophins to induce repair in damaged nerve tissue.* Certain types of neuroglia produce chemicals called neurotrophins (e.g., nerve growth factor) that have promoted nerve regeneration in experiments.
- *Regulation of inhibitory factors that keep neurons from dividing.* "Turning off" these factors (produced by neuroglia) in the damaged nervous system may promote tissue repair. The factor called Nogo is an example.
- *Nervous tissue transplantation.* Successfully transplanted donor tissue may take over the damaged nervous system's functions.

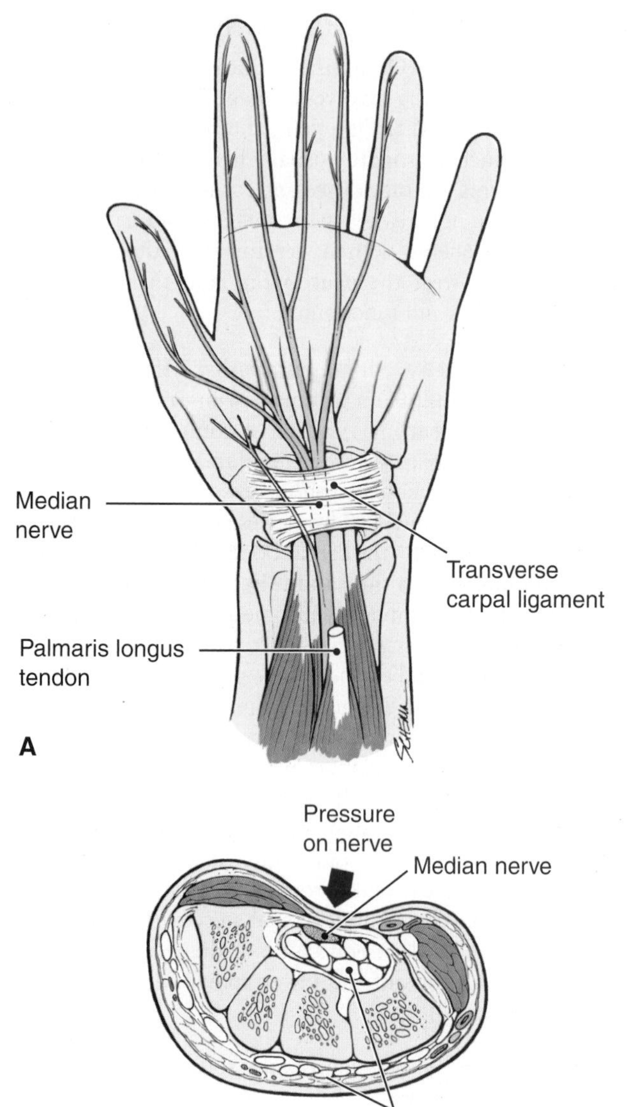

Figure 9-16 Carpal tunnel syndrome. **KEY POINT** Hand symptoms are caused by pressure on the median nerve. **A.** The median nerve as it passes through the carpal bones. **B.** Cross-section of the wrist showing compression of the median nerve.

numbness, paralysis, pain, cramps, and spasms. Balance and coordination may also be affected. If the ANS is involved, involuntary functions, such as control of blood pressure and heart rate, may be impaired. Polyneuropathy can result from tumors, toxins, infection, or autoimmune disorders.

Tumors While tumors can affect single nerves, some genetic diseases result in the appearance of multiple benign tumors. These genetic mutations cause abnormal proliferation of Schwann cells (schwannomas) or fibroblasts (neurofibromas) in the nerve coverings, resulting in nerve compression and sometimes disfiguring skin lesions.

Toxins The toxins that can cause polyneuropathy include alcohol and lead, tumor-produced substances, and medications used to treat cancer and HIV/AIDS. Perhaps the most common toxin is the blood glucose at high levels observed in patients with diabetes mellitus—nearly 80% of these individuals develop polyneuropathy. The resulting loss of sensation (including pain sensation) in the feet can be particularly troublesome, because we depend on this sense to avoid injury. Polyneuropathy in the feet can thus result in frequent foot wounds. Toxins that damage neurons often damage small blood vessels as well, resulting in poor circulation and impaired healing. Wounds can turn into ulcers, and gangrene is not uncommon. This series of events is so common in diabetics that it has a special name (diabetic foot disease). Interdisciplinary medical teams specialize in preventing and treating these foot problems. The autonomic nerves are also affected in diabetic polyneuropathy, potentially resulting in blood pressure problems, difficulties emptying the bladder, and sexual dysfunction.

Infections Infections, particularly HIV/AIDS, hepatitis, and, more rarely, Lyme disease, can cause polyneuropathy. In some cases, it can be difficult to determine if the nerve damage results from the disease agent or from the treatment.

Herpes zoster, commonly known as *shingles*, is an infectious neuropathy characterized by numerous blisters along the course of certain nerves, most commonly the thoracic nerves and their branches. Shingles is caused by a reactivation of a prior chickenpox virus infection. The viruses lurk in the sensory ganglia; when reactivated (often when the immune system is temporarily suppressed), the viruses multiply within the neurons and in the epithelial cells supplied by the neurons. Initial symptoms include fever and pain, followed in two to four weeks by the appearance of vesicles (fluid-filled skin lesions). The drainage from these vesicles contains highly contagious liquid. Neuralgic pains may persist for years and can be distressing. Early treatment of a recurrent attack with antiviral drugs may reduce the neuralgia, and a shingles vaccine is now available.

Autoimmune Disorders Neurons are not immune from self-attack; many chronic autoimmune disorders, including systemic lupus erythematosus and celiac disease, result in nerve damage and polyneuropathy. **Guillain–Barré** (ge-YAN bar-RA) **syndrome** is an acute form of polyneuropathy involving spinal nerves and nerve roots. While MS targets the myelin coating of CNS neurons, Guillain-Barré impacts the myelin coating of PNS neurons. In this disease, there is progressive muscle weakness and paralysis, beginning in the limbs and ascending to the respiratory muscles. Fortunately, PNS neurons can be repaired relatively easily compared to CNS neurons. If the patient can be supported while Schwann cells remyelinate and repair the damaged neurons, a process that can take months or even years, partial or full recovery can occur. Like MS, Guillain-Barré syndrome often follows a bacterial or viral infection.

CHECKPOINTS

- 9-21 What is removed in a lumbar puncture?
- 9-22 What is the meaning of the word root *plegia?*
- 9-23 What term is used for any disorder of the nerves?

Occupational therapists often help care for people with nervous system disorders. See the Student Resources on thePoint for more information about this career.

9

Disease in Context Revisited

Sue Learns about Her MS

"Sue, I can't really answer the question of why you developed multiple sclerosis," Dr. Jensen explained to her patient. "There is evidence that the disease has a genetic component but the environment, and perhaps even a virus, might be involved. We do know that MS affects women more frequently than men and is more prevalent in areas like the northern United States and Canada. We also know that MS is an autoimmune disease. Normally, immune cells travel through the brain and spinal cord looking for pathogens. In MS, the immune cells make a mistake and cause inflammation in healthy nervous tissue. This inflammatory response damages neuroglial cells called oligodendrocytes. These cells form the myelin sheath that covers and insulates the axons of neurons much like the plastic covering on an electrical wire does. When the oligodendrocytes are damaged, they are unable to make this myelin sheath, and the axons can't properly transmit nerve impulses. Right now, it appears that the largest areas of demyelination are in the white matter tracts of your spinal cord."

"Is there a medication I can take to stop the disease?" asked Sue.

"Unfortunately," replied the doctor, "there isn't a cure for MS yet. But we can slow down the disease's progress using antiinflammatory drugs to decrease the inflammation and drugs called interferons that depress the immune response."

During this case, we saw that neurons carrying information to and from the CNS require myelin sheaths. Inflammation and subsequent damage of the myelin sheath in diseases like MS have profound effects on sensory and motor function. For more information about the inflammatory response and interferons, see Chapter 17.

CHAPTER 9

Chapter Wrap-Up

Summary Overview

A detailed chapter outline with space for note taking is on *thepoint*. The figure below illustrates the main topics covered in this chapter.

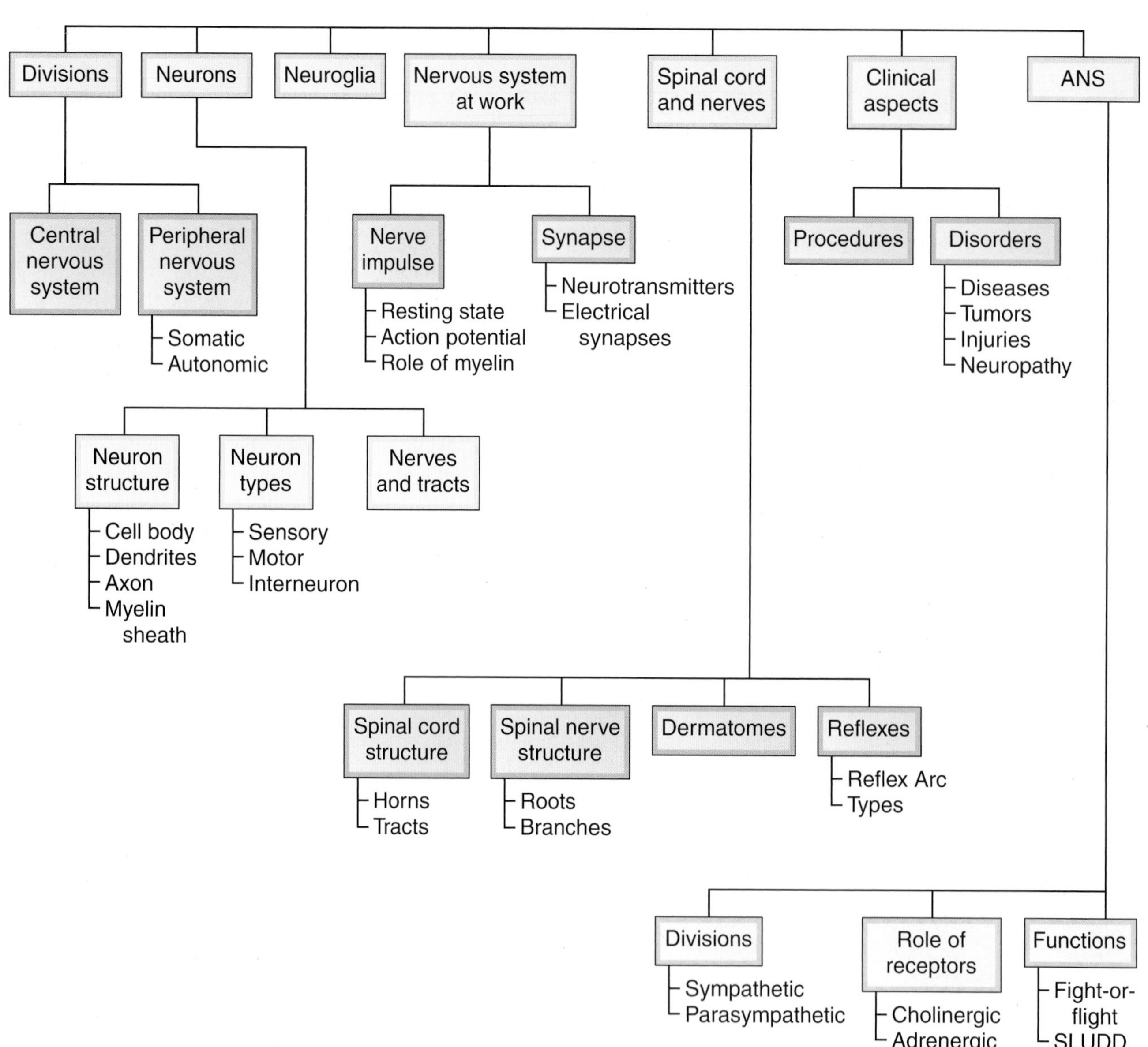

Key Terms

The terms listed below are emphasized in this chapter. Knowing them will help you organize and prioritize your learning. These and other boldface terms are defined in the Glossary with phonetic pronunciations.

acetylcholine
action potential
afferent
autonomic nervous system
axon
dendrite
depolarization
effector
efferent
ganglion
interneuron
motor
nerve
nerve impulse
neuritis
neuroglia
neuron
neurotransmitter
norepinephrine
parasympathetic nervous system
plexus
postsynaptic
presynaptic
receptor
reflex
repolarization
saltatory conduction
sensory
somatic nervous system
sympathetic nervous system
synapse
tract

Word Anatomy

Medical terms are built from standardized word parts (prefixes, roots, and suffixes). Learning the meanings of these parts can help you remember words and interpret unfamiliar terms

WORD PART	MEANING	EXAMPLE
Role of the Nervous System		
aut/o	self	The *autonomic* nervous system is automatically controlled and is involuntary.
-lemma	sheath	See below example.
neur/i	nerve, nervous tissue	The *neurilemma* is the outer membrane of the myelin sheath around an axon.
olig/o	few, deficiency	An *oligodendrocyte* has few dendrites.
soma-	body	The *somatic* nervous system controls skeletal muscles that move the body.
The Nervous System at Work		
de-	remove	*Depolarization* removes the charge on the plasma membrane of a cell.
post-	after	The *postsynaptic* cell is located after the synapse and receives neurotransmitter from the presynaptic cell.
re-	again, back	*Repolarization* restores the charge on the plasma membrane of a cell.
The Spinal Cord		
hemi-	half	*Hemiplegia* is paralysis of one side of the body.
myel/o	spinal cord	*Poliomyelitis* is an infectious disease that involves the spinal cord and other parts of the CNS.
para-	beyond	*Paraplegia* is paralysis of both lower limbs.
plegia	paralysis	*Monoplegia* is paralysis of one limb.
tetra-	four	*Tetraplegia* is paralysis of all four limbs.

Questions for Study and Review

BUILDING UNDERSTANDING

Fill in the Blanks

1. The brain and spinal cord make up the _______ nervous system.
2. The ion that enters a cell to cause depolarization is _______.
3. The term that describes conduction along a myelinated axon is _______.
4. In the spinal cord, sensory information travels in _______ tracts.
5. The common term for herpes zoster is _______.

Matching > Match each numbered item with the most closely related lettered item.

___ **6.** Cells that carry impulses from the CNS
___ **7.** Cells that carry impulses to the CNS
___ **8.** Cells that carry impulses within the CNS
___ **9.** Cells that detect a stimulus
___ **10.** Cells that carry out a response to a stimulus

a. receptors
b. effectors
c. sensory neurons
d. motor neurons
e. interneurons

Multiple Choice

___ **11.** Which system directly innervates skeletal muscles?
a. central nervous system
b. somatic nervous system
c. parasympathetic nervous system
d. sympathetic nervous system

___ **12.** What cells are involved in most nervous system tumors?
a. motor neurons
b. sensory neurons
c. interneurons
d. neuroglia

___ **13.** What is the correct order of synaptic transmission?
a. postsynaptic neuron, synapse, and presynaptic neuron
b. presynaptic neuron, synapse, and postsynaptic neuron
c. presynaptic neuron, postsynaptic neuron, and synapse
d. postsynaptic neuron, presynaptic neuron, and synapse

___ **14.** Where do afferent nerve fibers enter the spinal cord?
a. dorsal horn
b. ventral horn
c. gray commissure
d. central canal

___ **15.** What system promotes the "fight-or-flight" response?
a. sympathetic nervous system
b. parasympathetic nervous system
c. somatic nervous system
d. reflex arc

UNDERSTANDING CONCEPTS

___ **16.** Differentiate between the terms in each of the following pairs:
a. axon and dendrite
b. gray matter and white matter
c. nerve and tract
d. dorsal and ventral spinal nerve root
e. hemiplegia and tetraplegia

17. Describe an action potential. How does conduction along a myelinated fiber differ from conduction along an unmyelinated fiber?

18. What are neuroglia, and what are their functions?

19. Explain the reflex arc using stepping on a tack as an example.

20. Describe the anatomy of a spinal nerve. How many pairs of spinal nerves are there?

21. Define a *plexus*. Name the three main spinal nerve plexuses.

22. What is a dermatome? How are dermatomes used in diagnosis?

23. Differentiate between the functions of the sympathetic and parasympathetic divisions of the ANS.

24. Explain how a single neurotransmitter can stimulate some cells and inhibit others.

25. Define neuropathy, and give several examples.

CONCEPTUAL THINKING

26. Clinical depression is associated with abnormal serotonin levels. Medications that block the removal of this neurotransmitter from the synapse can control the disorder. Based on this information, is clinical depression associated with increased or decreased levels of serotonin? Explain your answer.

27. Mr. Hayward visits his dentist for a root canal and is given Novocaine, a local anesthetic, at the beginning of the procedure. Novocaine reduces membrane permeability to Na^+. What effect does this have on action potential?

28. In Sue's case, her symptoms were caused by demyelination in her CNS. Would her symptoms be the same or different if her spinal nerves were involved? Explain why or why not.

For more questions, see the Learning Activities on thePoint.

CHAPTER 10

The Nervous System: The Brain and Cranial Nerves

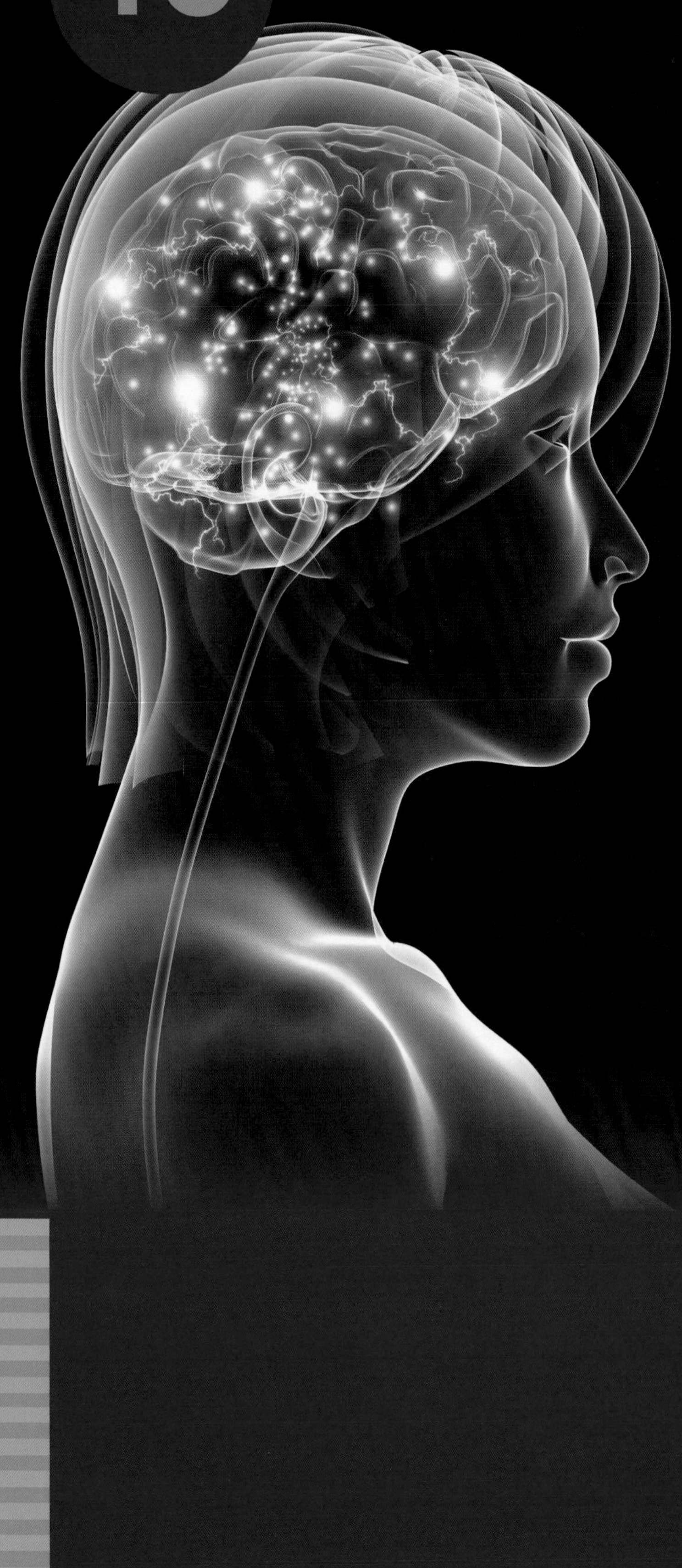

Learning Objectives

After careful study of this chapter, you should be able to:

1 ▶ Give the locations of the four main divisions of the brain. ***p. 218***

2 ▶ Name and describe the three meninges. ***p. 220***

3 ▶ Cite the function of cerebrospinal fluid, and describe where and how this fluid is formed. ***p. 220***

4 ▶ Name and locate the lobes of the cerebral hemispheres. ***p. 222***

5 ▶ Cite one function of the cerebral cortex in each lobe of the cerebrum. ***p. 223***

6 ▶ Name two divisions of the diencephalon, and cite the functions of each. ***p. 225***

7 ▶ Locate the three subdivisions of the brain stem, and give the functions of each. ***p. 225***

8 ▶ Describe the cerebellum, and identify its functions. ***p. 226***

9 ▶ Name three neuronal networks that involve multiple regions of the brain, and describe the function of each. ***p. 226***

10 ▶ Describe four techniques used to study the brain. ***p. 227***

11 ▶ Describe at least six disorders that affect the brain. ***p. 227***

12 ▶ List the names and functions of the 12 cranial nerves. ***p. 232***

13 ▶ Discuss five disorders that involve the cranial nerves. ***p. 234***

14 ▶ Using information in the case study, list the possible effects of mild traumatic brain injury. ***pp. 217, 235***

15 ▶ Show how word parts are used to build words related to the nervous system (see Word Anatomy at the end of the chapter). ***p. 237***

Disease in Context *Natalie's Cerebral Concussion*

Lacey was agitated and a little panicky as she helped her sister into her SUV. Despite Natalie's protests, she insisted on driving her to Mount Desert Island Hospital to be checked. Lacey was an active outdoor person who enjoyed hiking in nearby Acadia National Park. She had been eager to share her favorite trail with her sister, who lived in New York City and rarely strayed from her Wall Street office.

They reached the trail head at dawn and began their hike up Cadillac Mountain. As they climbed, the path became more precarious and lined with boulders. Lacey had experience with climbing and started to scramble up a small boulder. Natalie took a few steps up the rock and fell, striking her head on the ground. Lacey feared her sister had suffered head trauma. Fortunately, they had not climbed very far, and Natalie was conscious, so the two sisters carefully picked their way back down the mountain and drove to the hospital.

Dr. Erickson, the resident on call, performed Natalie's neurologic examination, starting with an evaluation of her mental status and then moving on to a cranial nerves exam. He noted sluggish eye movements during the ocular exam. He proceeded with a motor exam of the extremities, and then assessed her balance and coordination. When questioned, Natalie admitted she felt dizzy and complained of a headache, nausea, and blurred vision. Twice she vomited small amounts. She kept repeating that she was tired and wanted to go to sleep. Dr. Erickson ordered a computed tomography (CT) scan to determine if injury was present and if so, its extent.

Dr. Erickson explained his findings to the sisters.

"As a result of hitting your head, Natalie, you have incurred a cerebral concussion, or mild traumatic brain injury (MTBI). The dizziness, headaches, nausea, and vomiting you are experiencing are caused by injury and swelling in your brain. Thankfully, the CT scan did not show more serious damage, such as bleeding within your brain, but you need to be carefully observed over the next 24 hours."

Natalie's visual symptoms could have resulted from cerebral edema and injury to cranial nerves II, III, IV, or VI. In this chapter, we will learn about the structure and function of the brain and cranial nerves. We will revisit Natalie later in the chapter to see how she is progressing.

ANCILLARIES *At-A-Glance*

Visit thePoint to access the following resources. For guidance in using these resources most effectively, see pp. xv–xvii.

Learning RESOURCES

- Tips for Effective Studying
- Animation: Stroke
- Health Professions: Speech Therapist
- Detailed Chapter Outline
- Answers to Questions for Study and Review
- Audio Pronunciation Glossary

Learning ACTIVITIES

- Pre-Quiz
- Visual Activities
- Kinesthetic Activities
- Auditory Activities

Having discussed the basics of nerve impulse conduction and the reflex arc, we now apply these fundamentals to look at how the various brain regions coordinate information and orchestrate responses. As you might guess, this is a very complex topic, spanning activities from the cellular level to the highest abstract brain function.

Overview of the Brain

The brain is the control center of the nervous system, where sensory information is processed, responses are coordinated, and the higher functions of reasoning, learning, and memory occur. The brain occupies the cranial cavity and is surrounded by membranes, fluid, and the skull bones.

DIVISIONS OF THE BRAIN

For study purposes, the brain can be divided into four regions with specific activities. These divisions are in constant communication as they work together to regulate body functions (**Fig. 10-1**):

- The **cerebrum** (SER-e-brum) is the most superior and largest part of the brain. It consists of left and right hemispheres, each shaped like a small boxer's glove.
- The **diencephalon** (di-en-SEF-ah-lon) sits in the center of the brain between the two hemispheres and superior to the brain stem. It includes the thalamus and the hypothalamus.
- The **brain stem** spans the region between the diencephalon and the spinal cord. The superior portion of the brain stem is the **midbrain.** Inferior to the midbrain is the **pons** (ponz), followed by the **medulla oblongata** (meh-DUL-lah ob-long-GAH-tah). The medulla connects with the spinal cord through a large opening in the base of the skull (foramen magnum).
- The **cerebellum** (ser-eh-BEL-um) is posterior to the brain stem and is connected with the cerebrum, brain stem, and spinal cord by means of the pons. The word *cerebellum* means "little brain."

Each of these divisions is described in greater detail later in this chapter and summarized in **Table 10-1**. Also see Dissection Atlas **Figure A5-3**.

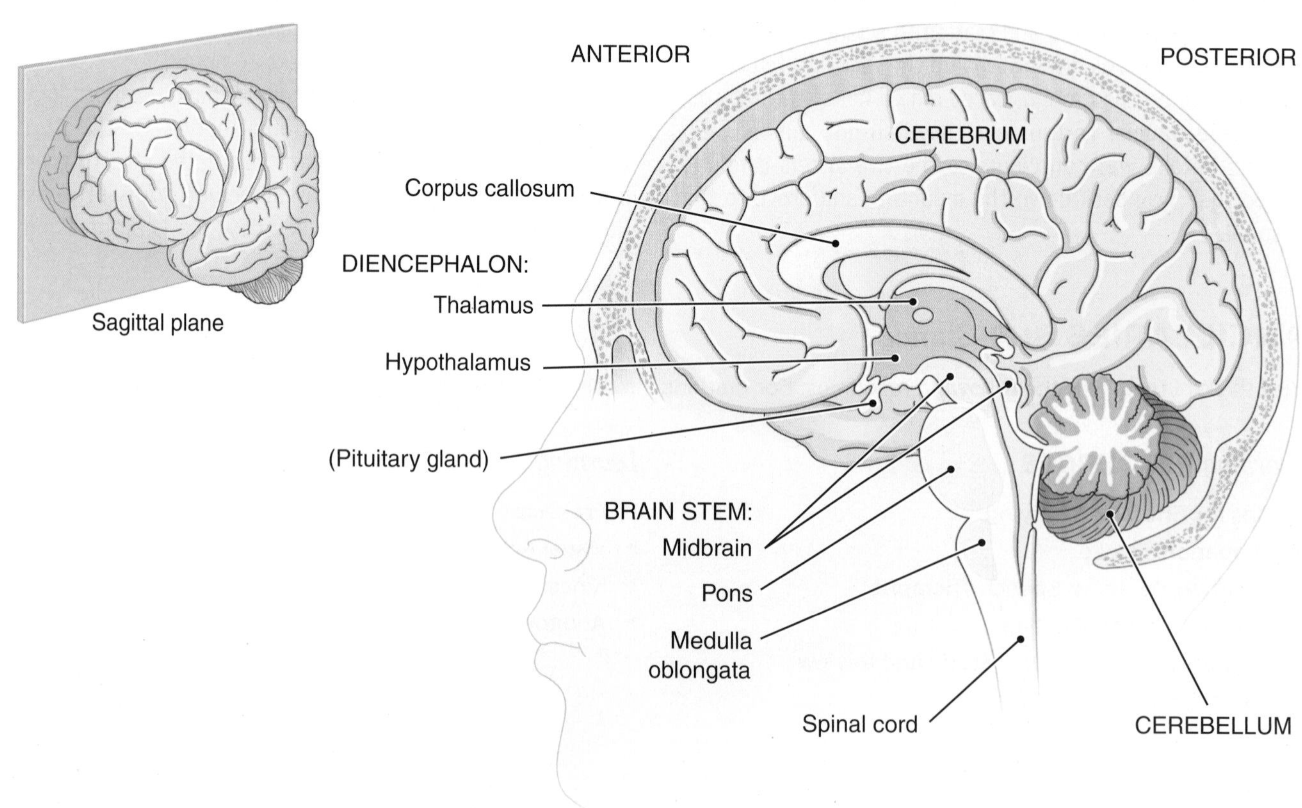

Figure 10-1 Brain, sagittal section. **KEY POINT** The four main divisions of the brain are the cerebrum, diencephalon, brain stem, and cerebellum. The pituitary gland is closely associated with the brain. **ZOOMING IN** What is the largest part of the brain? What part connects with the spinal cord?

Table 10-1 Organization of the Brain

Division	Description	Functions
CEREBRUM 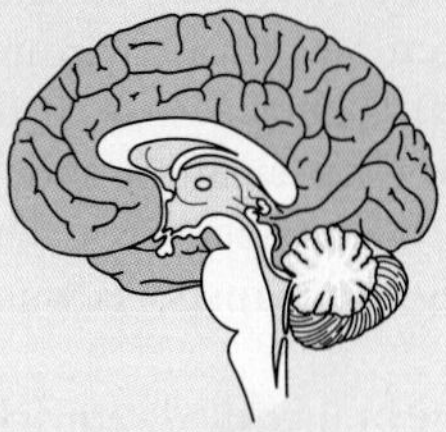	Largest and most superior portion of the brain Divided into two hemispheres, each subdivided into lobes	Cortex (outer layer) is site for conscious thought, memory, reasoning, perception, and abstract mental functions, all localized within specific lobes
DIENCEPHALON 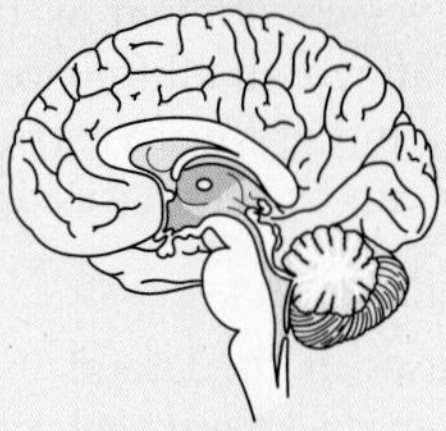	Between the cerebrum and the brain stem Contains the thalamus and hypothalamus	Thalamus sorts and redirects sensory input. Hypothalamus maintains homeostasis, controls autonomic nervous system and pituitary gland
BRAIN STEM	Anterior region below the cerebrum	Connects cerebrum and diencephalon with spinal cord
Midbrain 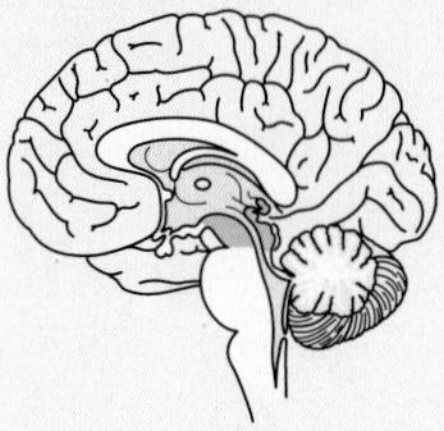	Below the center of the cerebrum	Has reflex centers concerned with vision and hearing. Connects cerebrum with lower portions of the brain
Pons 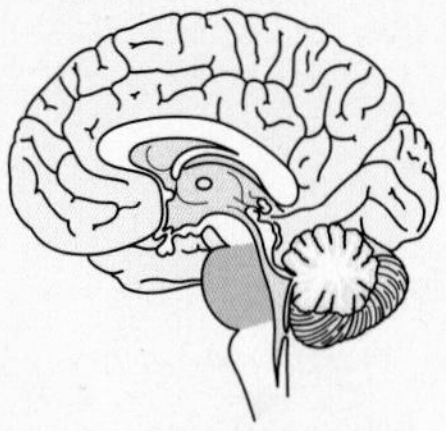	Anterior to the cerebellum	Connects cerebellum with other portions of the brain Helps regulate respiration
Medulla oblongata 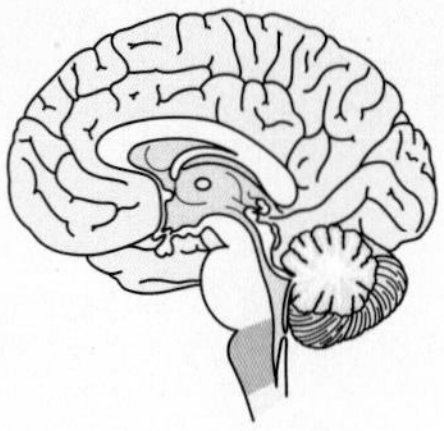	Between the pons and the spinal cord	Links the brain with the spinal cord. Has centers for control of vital functions, such as respiration and the heartbeat
CEREBELLUM 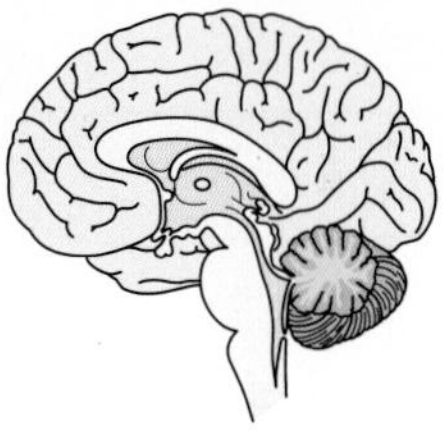	Below the posterior portion of the cerebrum Divided into two hemispheres	Coordinates voluntary muscles Maintains balance and muscle tone

PROTECTIVE STRUCTURES OF THE BRAIN AND SPINAL CORD

The protective structures of the brain include the meninges and the cerebrospinal fluid (CSF). Both the meninges and the CSF also protect the spinal cord.

Meninges The **meninges** (men-IN-jez) are three layers of connective tissue that surround both the brain and spinal cord to form a complete enclosure (**Fig. 10-2**). The innermost layer around the brain, the **pia mater** (PI-ah MA-ter), is attached to the nervous tissue of the brain and spinal cord and follows all the contours of these structures (**see Fig. 10-2**). The pia is made of a delicate connective tissue (*pia* meaning "tender" or "soft"). It holds blood vessels that supply nutrients and oxygen to the brain and spinal cord.

The middle layer of the meninges is the **arachnoid** (ah-RAK-noyd). This membrane is loosely attached to the pia mater by weblike fibers, forming a space (the subarachnoid space) where CSF circulates. Blood vessels also pass through this space. The arachnoid is named from the Latin word for spider because of its weblike appearance.

The outermost **dura mater** (DU-rah MA-ter) is the thickest and toughest of the meninges. (*Mater* is from the Latin meaning "mother," referring to the protective function of the meninges; *dura* means "hard.") Around the brain, the dura mater is in two layers, and the outer layer is fused to the cranial bones. In certain places, these two layers separate to provide venous channels, called **dural sinuses**, for the drainage of blood coming from brain capillaries. Extensions of the arachnoid membrane called *arachnoid villi* project into this space. The dura is in a single layer around the spinal cord.

Cerebrospinal Fluid **Cerebrospinal** (ser-e-bro-SPI-nal) **fluid (CSF)** is a clear liquid that circulates in and around the brain and spinal cord (**Fig. 10-3**). The function of the CSF is to support nervous tissue and to cushion shocks that would otherwise injure these delicate structures. This fluid also carries nutrients to the cells and transports waste products from the cells.

CSF forms in four spaces within the brain called **ventricles** (VEN-trih-klz) (**Fig. 10-3A**). A network of ependymal (specialized neuroglial) cells and blood vessels, known as the **choroid** (KOR-oyd) **plexus**, makes CSF within all four ventricles. You can see a choroid plexus in the third ventricle in **Figure 10-3B**. The journey of CSF can be summarized as follows:

1. CSF formed by the choroid plexuses in the lateral ventricles (ventricles 1 and 2) flows from the lateral ventricles through small openings called **interventricular foramina** (fo-RAM-in-ah) into the third ventricle (**see Fig. 10-3A**). This cavity forms a midline space within the diencephalon (**see Fig. 10-3B**).

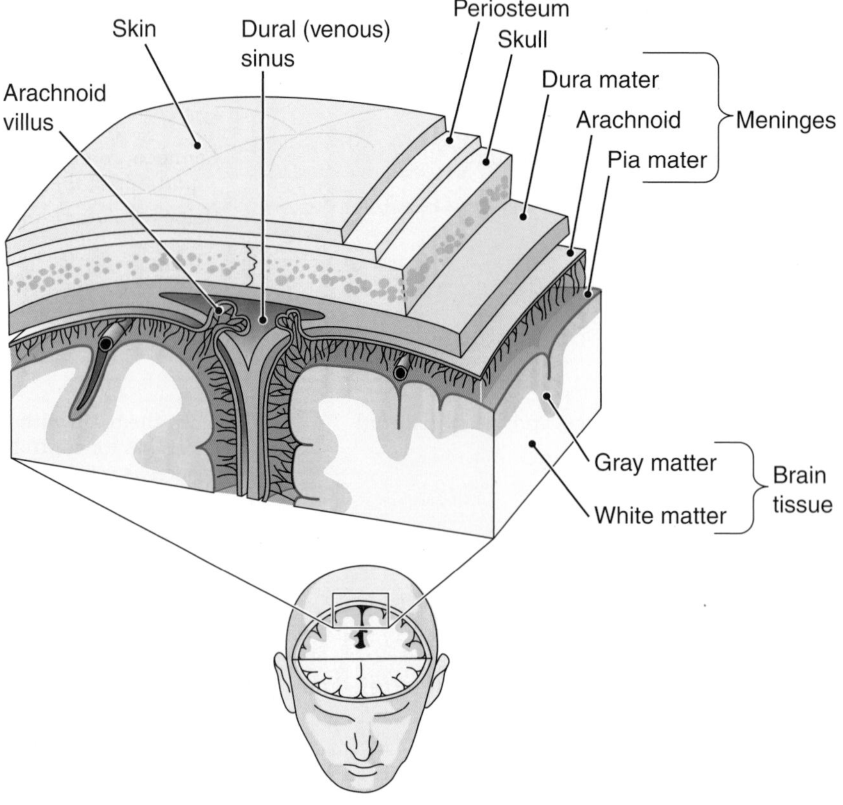

Figure 10-2 **Frontal (coronal) section of the top of the head. The meninges and related parts are shown.** **KEY POINT** The brain has many layers of protective substances. **ZOOMING IN** What are the channels formed where the dura mater divides into two layers? How many layers of meninges are there?

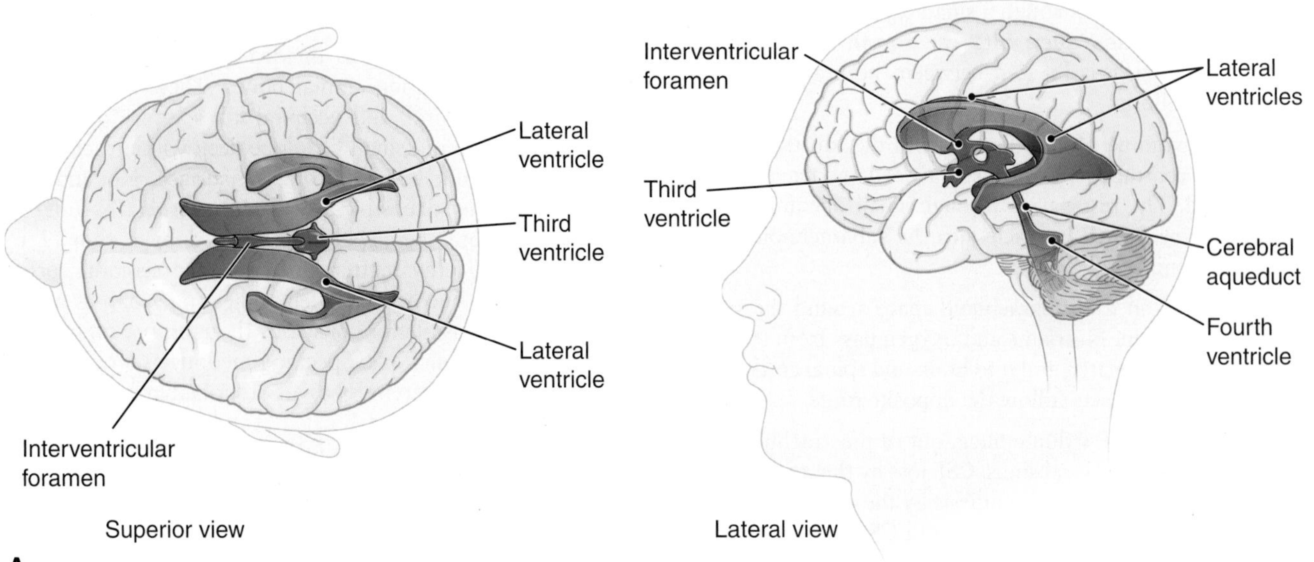

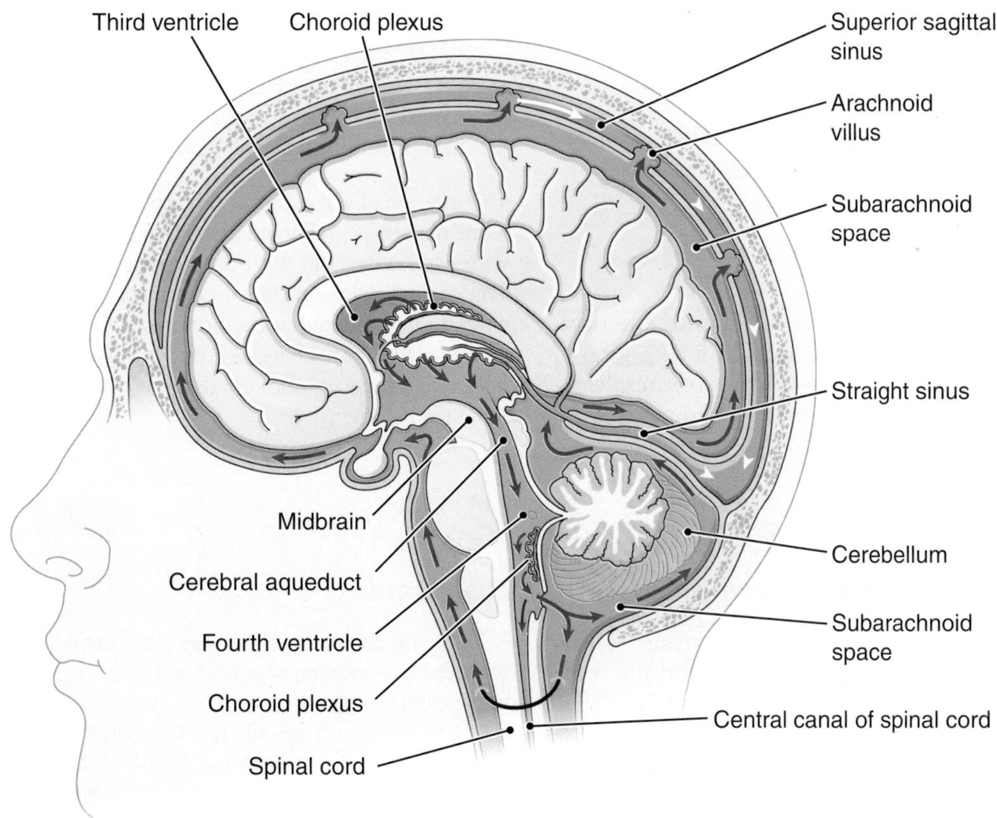

Figure 10-3 **Cerebrospinal fluid (CSF) and the cerebral ventricles. A.** The cerebral ventricles. Note that the first and second ventricles are labeled as the lateral ventricles. **B.** *Black arrows* show the flow of CSF from the choroid plexuses and back to the blood in dural sinuses; *white arrows* show the flow of blood. (The actual passageways through which the CSF flows are narrower than those shown here, which have been enlarged for visibility.) **ZOOMING IN** Which ventricle is continuous with the central canal of the spinal cord?

2. Next, CSF passes through a small canal in the midbrain, the **cerebral aqueduct**, into the fourth ventricle. The fourth ventricle is located between the brain stem and the cerebellum.
3. A small volume of CSF passes from the fourth ventricle into the central canal of the spinal cord and travels down the cord. The rest passes through small openings in the roof of the fourth ventricle into the subarachnoid space of the meninges.
4. CSF travels in the subarachnoid space around the spinal cord and brain. Nutrients and oxygen pass from the CSF, across the pia mater, and into brain and spinal cord tissue, and waste products follow the opposite route.
5. Some of the CSF volume filters out of the arachnoid villi into blood in the dural sinus. CSF loss by this route occurs at the same rate as CSF synthesis by the choroid plexuses of the four ventricles, so the overall CSF volume remains constant.

Box 10-1 presents information on protecting the brain from harmful substances.

CHECKPOINTS 

- 10-1 What are the main divisions of the brain?
- 10-2 What are the names of the three layers of the meninges from the outermost to the innermost?
- 10-3 Where is CSF produced?

The Cerebrum

The cerebrum, the brain's largest portion, is divided into right and left cerebral (SER-e-bral) hemispheres by a deep groove called the longitudinal fissure (Fig. 10-4). The two hemispheres have overlapping functions and are similarly subdivided.

DIVISIONS OF THE CEREBRAL HEMISPHERES

Each cerebral hemisphere is divided into four visible **lobes** named for the overlying cranial bones. These are the frontal, parietal, temporal, and occipital lobes (see Fig. 10-4B). In addition, there is a small fifth lobe deep within each hemisphere that cannot be seen from the surface. Not much is known about this lobe, which is called the **insula** (IN-su-lah).

Each cerebral hemisphere is covered by a thin (2 to 4 mm) layer of gray matter known as the **cerebral cortex** (see Fig. 10-2). The neuronal cell bodies and synapses in this region are responsible for conscious thought, reasoning, and abstract mental functions. Specific functions are localized in the cortex of the different lobes, as described in greater detail later.

The cortex is arranged in folds forming elevated portions known as **gyri** (JI-ri), singular *gyrus*. These raised areas are separated by shallow grooves called **sulci** (SUL-si), singular *sulcus*. Although there are many sulci, the following two are especially important landmarks:

- The **central sulcus** lies between the frontal and parietal lobes of each hemisphere at right angles to the longitudinal fissure (see Fig. 10-4).
- The **lateral sulcus** curves along the side of each hemisphere and separates the temporal lobe from the frontal and parietal lobes (see Fig. 10-4).

The interior of the cerebral hemispheres consists primarily of white matter—myelinated fibers that connect the cortical areas with each other and with other parts of the nervous system. The **corpus callosum** (kah-LO-sum) is an important band of white matter located at the bottom of the longitudinal fissure (see Fig. 10-1). This band is a bridge between the right and left hemispheres, permitting impulses to cross from one side of the brain to the other.

A CLOSER LOOK **Box 10-1**

The Blood–Brain Barrier: Access Denied

Neurons in the central nervous system (CNS) function properly only if the composition of the extracellular fluid bathing them is carefully regulated. The semipermeable blood–brain barrier helps maintain this stable environment by allowing some substances to cross it while blocking others. Whereas it allows glucose, amino acids, and some electrolytes to cross, it prevents passage of hormones, drugs, neurotransmitters, and other substances that might adversely affect the brain.

Structural features of CNS capillaries create this barrier. In most parts of the body, capillaries are lined with simple squamous epithelial cells that are loosely attached to each other. The small spaces between cells let materials move between the bloodstream and the tissues. In CNS capillaries, the simple squamous epithelial cells are joined by tight junctions that limit passage of materials between them.

The blood–brain barrier excludes pathogens, although some viruses, including poliovirus and herpesvirus, can bypass it by traveling along peripheral nerves into the CNS. Some streptococci also can breach the tight junctions. Disease processes, such as hypertension, ischemia (lack of blood supply), and inflammation, can increase the blood–brain barrier's permeability.

The blood–brain barrier is an obstacle to delivering drugs to the brain. Some antibiotics can cross it, whereas others cannot. Neurotransmitters also pose problems. In Parkinson disease, the neurotransmitter dopamine is deficient in the brain. Dopamine itself will not cross the barrier, but a related compound, L-dopa, will. L-Dopa crosses the blood–brain barrier and is then converted to dopamine. Mixing a drug with a concentrated sugar solution and injecting it into the bloodstream is another effective delivery method. The solution's high osmotic pressure causes water to osmose out of capillary cells, shrinking them and opening tight junctions through which the drug can pass.

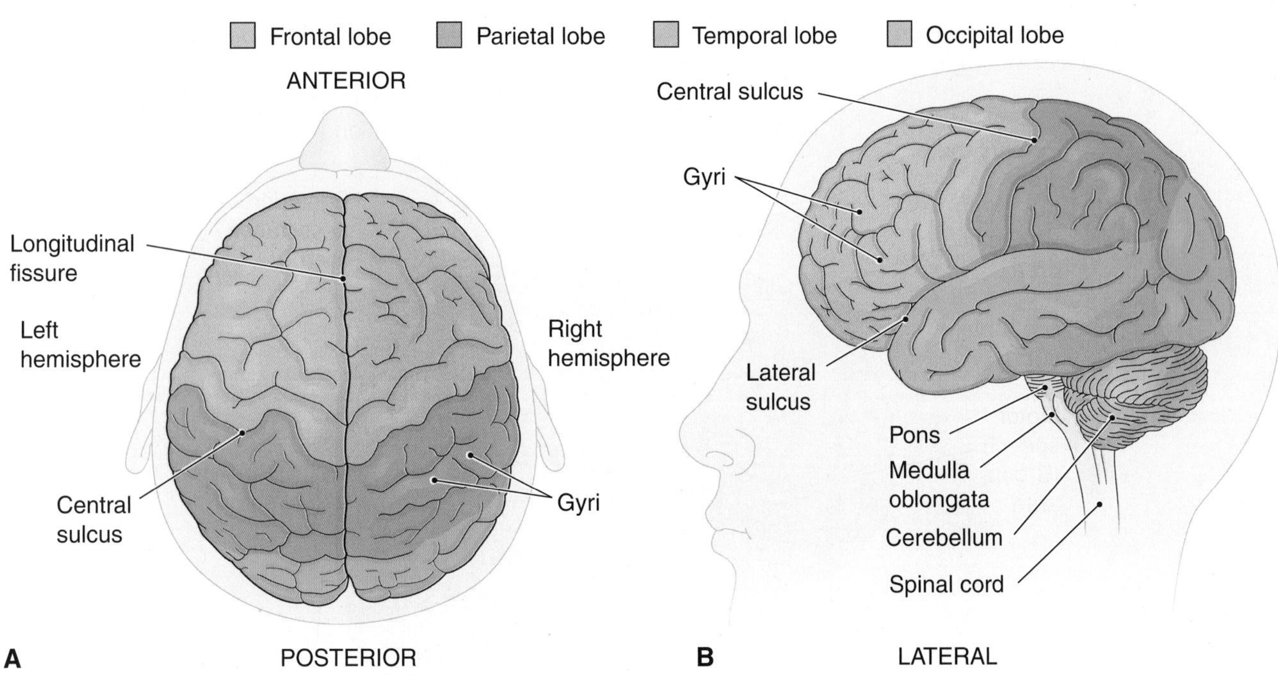

Figure 10-4 **External surface of the brain. A.** Superior view. **B.** Lateral view. **KEY POINT** The brain is divided into two hemispheres by the longitudinal fissure. Each hemisphere is subdivided into lobes. **ZOOMING IN** What structure separates the frontal from the parietal lobe? The temporal lobe from the frontal and parietal lobes?

FUNCTIONS OF THE CEREBRAL CORTEX

The cerebral cortex houses our consciousness—our awareness of the world around us and our ability to voluntarily interact with it. The cerebral cortex "stores" information, much of which can be recalled on demand by means of the phenomenon called *memory*. It is in the cerebral cortex that thought processes such as association, judgment, and discrimination take place.

Although the various brain areas act in coordination to produce behavior, particular functions are localized in the cortex of each lobe (**Fig. 10-5**). Some of these are described below:

Frontal Lobe The **frontal lobe**, which is relatively larger in humans than in any other organism, lies anterior to the central sulcus. The gyrus just anterior to the central sulcus in this lobe contains a **primary motor area**, which provides conscious control of skeletal muscles. Specific segments of the primary motor area control the muscles in different body regions. Relatively larger portions of the cortex are devoted to muscles requiring precise control, such as those of the hand.

Just anterior to the primary motor area is the premotor cortex, which helps plan complex movements. It receives sensory information from other parts of the brain to assist it in this task, and it sends most of its commands to the primary motor area. Anterior to the premotor cortex is the **prefrontal cortex**, involved in memory, problem solving, and conscious thought. Within the prefrontal cortex in one cerebral hemisphere lies the **motor speech area**, or **Broca** (bro-KAH) **area.** This region, which is usually (but not always) found in the left hemisphere, plans the sequences of muscle contractions in the tongue, larynx, and soft palate required to form meaningful sentences. People with damage in this area can understand sentences but have trouble expressing their ideas in words.

Parietal Lobe The **parietal lobe** occupies the superior part of each hemisphere and lies posterior to the central sulcus. The gyrus just posterior to the central sulcus in this lobe contains the **primary somatosensory area**, where impulses from the skin, such as touch, pain, and temperature, are received. As with the motor cortex, the greater the intensity of sensation from a particular area, the tongue or fingers, for example, the more area of the cortex is involved.

Just posterior to this region is the somatosensory association area, which integrates somatosensory input with memories to identify physical sensations. For instance, imagine that you are reaching for a water glass at night, but your hand encounters your pet cat instead. The primary somatosensory area sends information regarding the softness, warmth, and shape of the sensed object. Combined with your memories of what a cat feels like, you are able to identify the object as a cat, not a glass. Damage to this region makes it impossible to identify objects by touch alone, so you would know that the object is fuzzy and warm, but not that it was a cat.

Much of the parietal lobe, as well as portions of the temporal and occipital lobes, form the **posterior association area.** This brain region accepts information from all of the sensory association areas and our memories to construct an integrated view of the world.

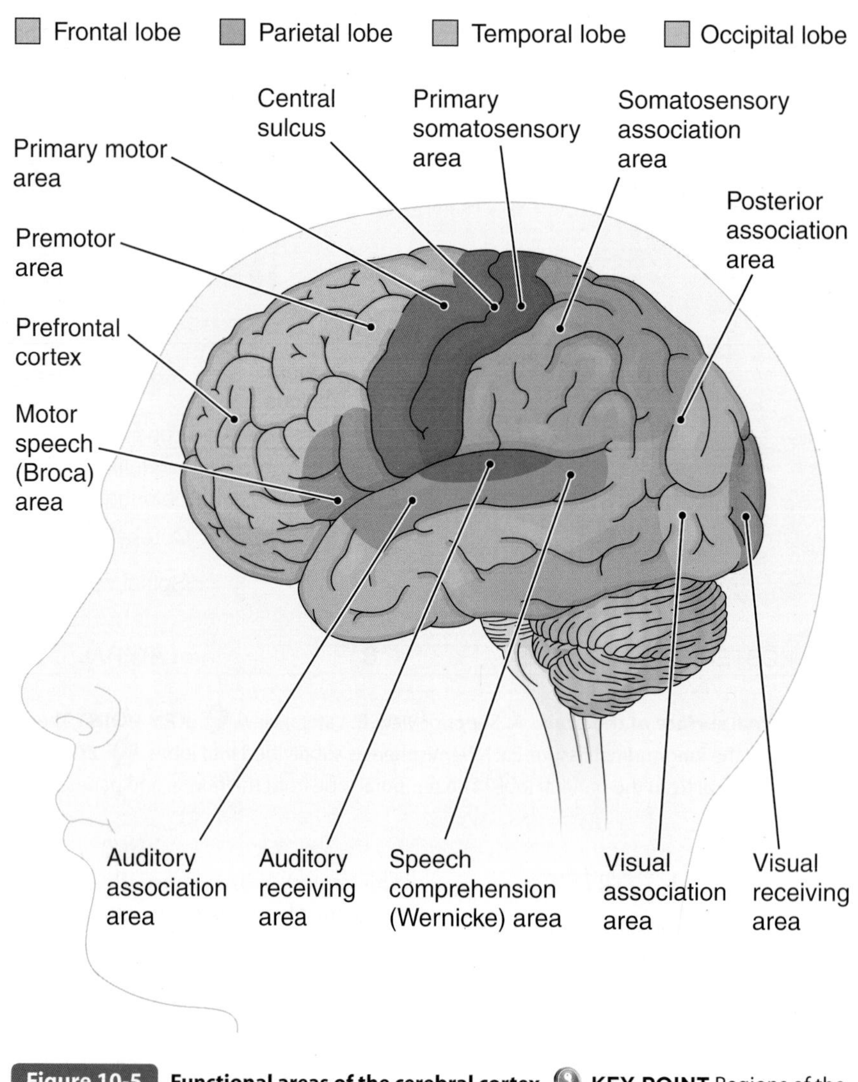

Figure 10-5 **Functional areas of the cerebral cortex.** **KEY POINT** Regions of the cerebral cortex are specialized for specific functions. **ZOOMING IN** What cortical area is posterior to the central sulcus? What area is anterior to the central sulcus?

Temporal Lobe The **temporal lobe** lies inferior to the lateral sulcus and folds under the hemisphere on each side. This lobe processes sounds. The **auditory receiving area** detects sound impulses transmitted from the environment, whereas the surrounding area, the **auditory association area**, interprets the sounds. Another region of the auditory cortex, located on the left side in most people, is the *speech comprehension area*, or **Wernicke** (VER-nih-ke) **area**. This area functions in speech recognition and the meaning of words. Someone who suffers damage in this region of the brain, as by a stroke, will have difficulty in understanding the meaning of speech. The olfactory area, concerned with the sense of smell, is located in the medial part of the temporal lobe and is not visible from the surface; it is stimulated by impulses arising from receptors in the nose.

Occipital Lobe The **occipital lobe** lies posterior to the parietal lobe and extends over the cerebellum. This lobe contains the **visual receiving area**, which collects sensory information from the retina, such as brightness and color, and the **visual association area**, which interprets the impulses into a mental "picture." Additional information processing by the posterior association area is necessary for us to label the mental picture as a flower, an understandable word, or a gesturing friend.

There is a functional relationship among areas of the brain. Many neurons must work together to enable a person to receive, interpret, and respond to verbal and written messages as well as to touch (tactile stimulus) and other sensory stimuli.

MEMORY AND THE LEARNING PROCESS

Memory is the mental faculty for recalling ideas. In the initial stage of the memory process, sensory signals (e.g., visual, auditory) are retained for a very short time, perhaps only fractions of a second. Nevertheless, they can be used for further processing. **Short-term memory** refers to the retention of bits of information for a few seconds or perhaps a few minutes, after which the information is lost unless reinforced. **Long-term memory** refers to the storage of information that can be recalled at a later time. There is a tendency for a memory to become more fixed the more often a person repeats the remembered experience; thus, short-term memory signals can lead to long-term memories. Furthermore, the more often a memory is recalled, the more indelible it becomes; such a memory can be so deeply fixed in the brain that it can be recalled immediately.

Physiologic studies show that rehearsal (repetition) of the same information again and again accelerates and potentiates the degree of short-term memory transfer into long-term memory. It has also been noted that the brain is able to organize information so that new ideas are stored in the same areas in which similar ones had been stored before.

CHECKPOINTS

- [ ] 10-4 Name the four surface lobes of each cerebral hemisphere.
- [ ] 10-5 Name the thin outer layer of gray matter where higher brain functions occur.

The Diencephalon

The diencephalon, or interbrain, is located between the cerebral hemispheres and the brain stem. One can see it by cutting into the central and inferior section of the brain. The diencephalon includes the **thalamus** (THAL-ah-mus) and the **hypothalamus** (**Fig. 10-6**).

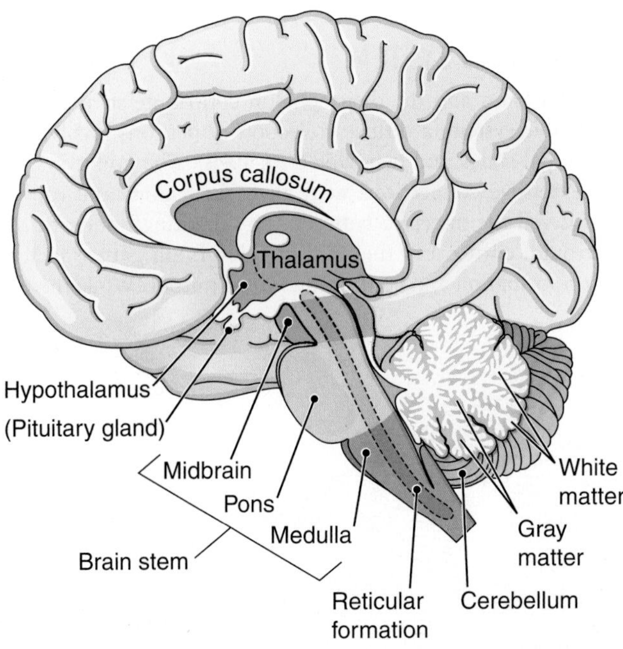

Figure 10-6 The diencephalon, brain stem, and cerebellum. **KEY POINT** The diencephalon consists of the thalamus, hypothalamus, and pituitary gland (hypophysis). The brain stem has three divisions: the midbrain, pons, and medulla oblongata. The white matter of the cerebellum is in a treelike pattern. **ZOOMING IN** To what part of the brain is the pituitary gland attached?

The two parts of the thalamus form the lateral walls of the third ventricle (**see Figs. 10-1 and 10-3**). Nearly all sensory impulses travel through the masses of gray matter that form the thalamus. The role of the thalamus is to sort out the impulses and direct them to particular areas of the cerebral cortex.

The hypothalamus is located in the midline area inferior to the thalamus and forms the floor of the third ventricle. It helps to maintain homeostasis by controlling body temperature, water balance, sleep, appetite, and some emotions, such as fear and pleasure. Both the sympathetic and parasympathetic divisions of the autonomic nervous system are under hypothalamic control, as is the pituitary gland. The hypothalamus thus influences the heartbeat, the contraction and relaxation of blood vessels, hormone secretion, and other vital body functions.

CHECKPOINT

- [ ] 10-6 What are the two main portions of the diencephalon, and what do they do?

The Brain Stem

The brain stem is composed of the midbrain, the pons, and the medulla oblongata (**Fig. 10-6**). These structures connect the cerebrum and diencephalon with the spinal cord.

THE MIDBRAIN

The **midbrain**, inferior to the center of the cerebrum, forms the superior part of the brain stem (**see Fig. 10-6**). Four rounded masses of gray matter that are hidden by the cerebral hemispheres form the superior part of the midbrain. These four bodies act as centers for certain reflexes involving the eye and the ear, for example, moving the eyes in order to track an image or to read. The white matter at the anterior of the midbrain conducts impulses between the higher centers of the cerebrum and the lower centers of the pons, medulla, cerebellum, and spinal cord.

THE PONS

The pons lies between the midbrain and the medulla, anterior to the cerebellum (**see Fig. 10-6**). It is composed largely of myelinated nerve fibers, which connect the two halves of the cerebellum with the brain stem as well as with the cerebrum above and the spinal cord below. Its name means "bridge," and it is an important connecting link between the cerebellum and the rest of the nervous system. It also contains nerve fibers that carry impulses to and from the centers located above and below it. Certain reflex (involuntary) actions, such as some of those regulating respiration, are integrated in the pons.

THE MEDULLA OBLONGATA

The medulla oblongata of the brain stem is located between the pons and the spinal cord (**see Fig. 10-6**). It appears white externally because like the pons, it contains many myelinated nerve fibers. Internally, it contains collections of cell bodies (gray matter) called **nuclei,** or *centers*. Among these are vital centers, such as the following:

- The **respiratory center** controls the muscles of respiration in response to chemical and other stimuli.
- The **cardiac center** helps regulate the rate and force of the heartbeat.
- The **vasomotor** (vas-o-MO-tor) **center** regulates the contraction of smooth muscle in the blood vessel walls and thus controls blood flow and blood pressure.

The ascending sensory fibers that carry messages through the spinal cord up to the brain travel through the medulla, as do descending motor fibers. These groups of fibers form tracts (bundles) and are grouped together according to function.

The motor fibers from the motor cortex of the cerebral hemispheres extend down through the medulla, and most of them cross from one side to the other (decussate) while going through this part of the brain. The crossing of motor fibers in the medulla results in *contralateral* (opposite side) *control*—the right cerebral hemisphere controls muscles in the left side of the body and the left cerebral hemisphere controls muscles in the right side of the body.

The medulla is an important reflex center; here, certain neurons end, and impulses are relayed to other neurons.

CHECKPOINT

☐ 10-7 What are the three subdivisions of the brain stem?

The Cerebellum

The cerebellum is made up of three parts: the middle portion (vermis) and two lateral hemispheres, the left and right (**Fig. 10-7**). Like the cerebral hemispheres, the cerebellum has an outer area of gray matter and an inner portion that is largely white matter (**see Fig. 10-6**). However, the white matter is distributed in a treelike pattern. The functions of the cerebellum are as follows:

- Helps coordinate voluntary muscles to ensure smooth, orderly function. Disease of the cerebellum causes muscular jerkiness and tremors.
- Helps maintain balance in standing, walking, and sitting as well as during more strenuous activities. Messages from the internal ear and from sensory receptors in tendons and muscles aid the cerebellum.
- Helps maintain muscle tone so that all muscle fibers are slightly tensed and ready to produce changes in position as quickly as necessary.

CHECKPOINT

☐ 10-8 What are some functions of the cerebellum?

Widespread Neuronal Networks

Some coordinating networks involve select regions of the diencephalon and brain stem or extend throughout the entire brain. The limbic system, basal nuclei, and reticular formation are three such networks; the limbic system helps control behavior, the basal nuclei participate in motor control, and the reticular formation helps govern awareness.

THE LIMBIC SYSTEM

The cerebrum and the diencephalon contribute structures to the **limbic system,** a diffuse collection of neurons involved in emotional states and behavior (**Fig. 10-8**). The limbic system has extensive connections with all brain regions and can be considered the interface between the "thinking" brain of the prefrontal cortex and the "autonomic" brain gathering sensory information and controlling motor output. While there is

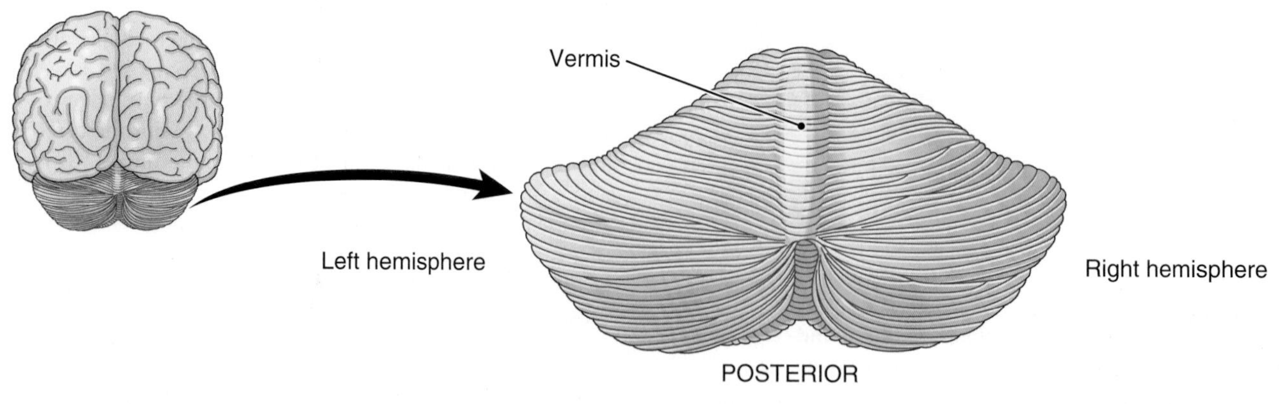

Figure 10-7 The cerebellum. Posterior view.  **KEY POINT** The cerebellum is divided into two hemispheres.

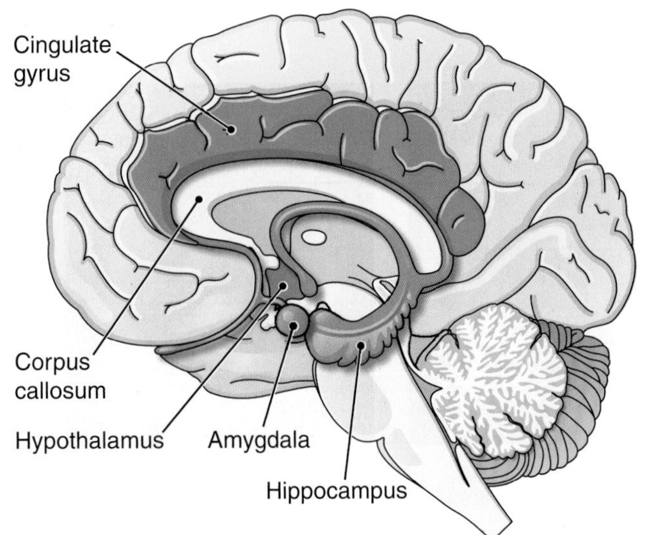

Figure 10-8 **The limbic system.** The system is shown in *red* in this figure. **KEY POINT** The limbic system consists of regions in the cerebrum and diencephalon that are involved in memory and emotion. **ZOOMING IN** Which part of the cerebral cortex contributes to the limbic system?

some disagreement regarding which structures are part of the limbic system, most agree that it includes:

- The **cingulate** (SIN-gu-late) **gyrus**, the portion of the cerebral cortex looping over the corpus callosum. This region of the cortex associates emotions with memories.
- The **hippocampus** (hip-o-KAM-pus), shaped like a sea horse and located under the lateral ventricles. The hippocampus enables us to store new memories—that is, to learn new things. Lesions in the hippocampus impair one's ability to form new memories, but leave old memories intact.
- The **amygdala** (ah-MIG-dah-lah), two clusters of nuclei deep in the temporal lobes. This brain region coordinates our emotional responses to stimuli. It receives extensive input from the olfactory lobe, which is why we often have strong emotional responses to smells.
- Parts of the hypothalamus and nearby nuclei. These regions control our motor responses to emotional stimuli, for instance, activation of the sympathetic nervous system.

BASAL NUCLEI

The **basal nuclei**, also called *basal ganglia*, modulate motor inputs and facilitate practiced, routine motor tasks. The basal nuclei consist of masses of gray matter spread throughout the brain that are extensively interconnected. They include three regions deep within the cerebrum, part of the midbrain called the *substantia nigra*, and a small collection of cell bodies in the diencephalon. As we will see later, death of different basal nuclei cells can cause Huntington or Parkinson disease.

RETICULAR FORMATION

The **reticular** (reh-TIK-u-lar) **formation** is a sausage-shaped network of neuronal cell bodies spanning the length of the brain stem (**see Fig. 10-6**). The reticular activating system (RAS) within this formation sends impulses to the cerebral cortex that keep us awake and attentive. In conjunction with the cortex, the RAS also screens out unnecessary sensory input (such as regular traffic noise) to increase the impact of novel stimuli (such as a car horn). Sleep centers in other brain regions inhibit the RAS, and thus arousal, when we sleep.

CHECKPOINTS

- ☐ **10-9** What are four structures in the limbic system?
- ☐ **10-10** What is the function of the basal nuclei?
- ☐ **10-11** What is the function of the reticular activating system?

10

Brain Studies

Some of the imaging techniques used to study the brain are described in **Box 1-2** in Chapter 1. These techniques include the following:

- CT (computed tomography) scan, which provides photographs of the bone, soft tissue, and cavities of the brain (**Fig. 10-9A**). Anatomic lesions, such as tumors or scar tissue accumulations, are readily seen. (In Natalie's opening case study, a CT scan helped to diagnose her injury.)
- MRI (magnetic resonance imaging), which gives more views of the brain than does CT and may reveal tumors, scar tissue, and hemorrhaging not shown by CT (**see Fig. 10-9B**).
- PET (positron emission tomography), which visualizes the brain in action (**see Fig. 10-9C**).

The interactions of the brain's billions of nerve cells give rise to measurable electric currents. These may be recorded using an instrument called the **electroencephalograph** (e-lek-tro-en-SEF-ah-lo-graf). Electrodes placed on the head pick up the electrical signals produced as the brain functions. These signals are then amplified and recorded to produce the tracings, or brain waves, of an electroencephalogram (EEG).

The electroencephalograph is used to study sleep patterns, to diagnose disease, such as epilepsy, to locate tumors, to study the effects of drugs, and to determine brain death. **Figure 10-10** shows some typical normal and abnormal tracings.

Disorders of the Brain and Associated Structures

Despite its protective structures, many types of disorders can affect the brain. These include seizures, inflammatory conditions, strokes, head injuries, and other disorders. Unlike other organs, the brain cannot increase in volume because of its bony enclosure. So any swelling or added volume tends

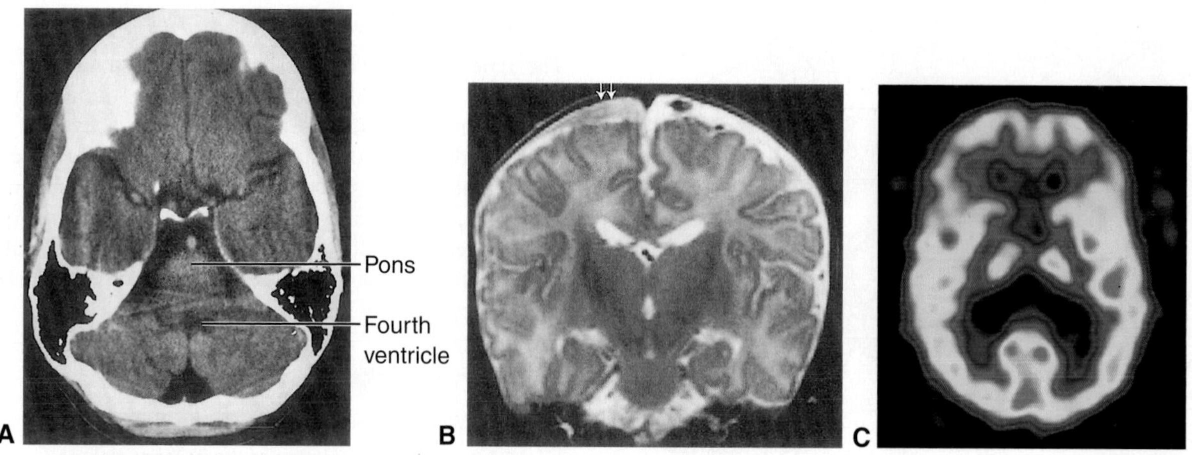

Figure 10-9 **Imaging the brain. A.** CT scan of a normal adult brain at the level of the fourth ventricle. **B.** MRI of the brain showing a point of injury (*arrows*). **C.** PET scan.

to increase pressure within the skull (intracranial pressure). As intracranial pressure rises, brain tissue dies.

SEIZURES AND EPILEPSIES

A **seizure** results from abnormal electrical activity in the brain. Seizures may involve uncontrollable muscle contractions (tonic clonic seizures) but may also consist of loss of awareness without muscle contractions (absence seizures). Seizures can occur in anyone as a result of trauma, infection, electrolyte imbalance or in some cases, fever. The different types of **epilepsy** are syndromes associated with recurrent seizures. In most cases, the cause of epilepsy is idiopathic (not known). An EEG study of brain waves usually shows abnormalities and is helpful in both diagnosis and treatment (**see Fig. 10-10B**). Over 20 medications are now available to prevent seizures, and most (but not all) individuals can lead seizure-free lives. The shortened life expectancy resulting from epilepsy reflects, in part, the damaging impact of the seizures on the brain and the heart, but it also reflects increased preventable, accidental deaths, such as drowning in the bathtub. Epilepsy that appears in young children or teenagers may resolve as they age.

INFLAMMATION

Infection and other factors can cause inflammation of the brain and its protective structures. **Meningitis** (men-in-JI-tis) is an inflammation of the meninges. It is usually caused by bacteria

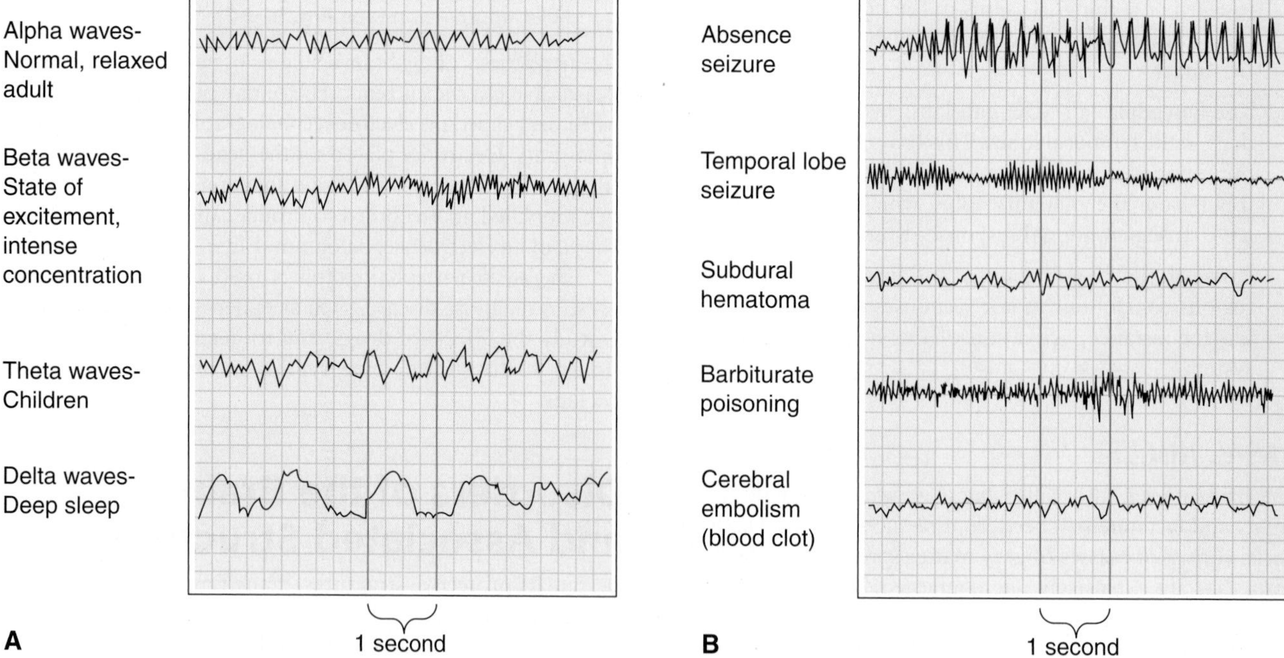

Figure 10-10 **Electroencephalography. A.** Normal brain waves. **B.** Abnormal brain waves.

that enter through the ear, nose, or throat or are carried by the blood. In many cases, an injury, invasive procedure, septicemia (blood infection), or a nearby infection allows the entry of pathogenic organisms. One of these organisms, the meningococcus (*Neisseria meningitidis*), is responsible for epidemics of meningitis among people living in close quarters. Other causative bacteria are *Haemophilus influenzae* (Hib), *Streptococcus pneumoniae*, and *Escherichia coli*. Some viruses, including the mumps virus, can cause meningitis but usually produce mild forms of the disease that require no treatment.

Headache, stiff neck, nausea, and vomiting are common symptoms of meningitis. Diagnosis is by lumbar puncture and examination of the CSF for pathogens and white blood cells (pus). In cases of bacterial meningitis, early treatment with antibiotics can have good results. Untreated cases have a high death rate. Vaccines are available against some of the bacteria that cause meningitis.

Inflammation of the brain is termed **encephalitis** (en-sef-ah-LI-tis), based on the scientific name for the brain, which is *encephalon*. Infectious agents that cause encephalitis include poliovirus, rabies virus, HIV (the cause of AIDS), insect-borne viruses, such as West Nile virus, and, rarely, the viruses that cause chickenpox and measles. Less frequently, exposure to toxic substances or reactions to certain viral vaccines can cause encephalitis. Brain swelling and diffuse nerve cell destruction accompany invasion of the brain by lymphocytes (white blood cells) in cases of encephalitis. Typical symptoms include fever, vomiting, and coma.

HYDROCEPHALUS

An abnormal accumulation of CSF within the brain is termed **hydrocephalus** (hi-dro-SEF-ah-lus) (**Fig. 10-11**). It may result from either overproduction or, more frequently, impaired drainage of the fluid. As CSF accumulates in the ventricles or its transport channels, mounting pressure can squeeze the brain against the skull and destroy brain tissue. Possible causes include congenital malformations present during development, tumors, inflammation, or hemorrhage.

Hydrocephalus is more common in infants than in adults. Because the skull's fontanels have not closed in the developing infant, the cranium itself can become greatly enlarged. In contrast, in the adult, cranial enlargement cannot occur, so that even a slight increase in fluid results in symptoms of increased pressure within the skull and brain damage. Treatment of hydrocephalus involves the creation of a shunt (bypass) to drain excess CSF from the brain.

STROKE

Stroke, or cerebrovascular (ser-e-bro-VAS-ku-lar) accident (CVA), is by far the most common kind of brain disorder. The most common cause is a blood clot that blocks blood flow to an area of brain tissue. Another frequent cause is the rupture of a blood vessel resulting in **cerebral hemorrhage** (HEM-eh-rij) and destruction of brain tissue. Stroke is most common among people older than 40 years of age and those with arterial wall damage, diabetes, or hypertension (high blood pressure). Smoking and excess alcohol consumption also increase the risk of stroke. Restoring blood flow to the affected area can reduce long-term damage. This can be done surgically or by administration of clot-dissolving medication, usually followed by medication to reduce brain swelling and minimize further damage.

A stroke's effect depends on the location of the artery and the extent of the involvement. Damage to the internal capsule's white matter in the inferior part of the cerebrum may cause extensive paralysis of the side opposite the affected area.

A
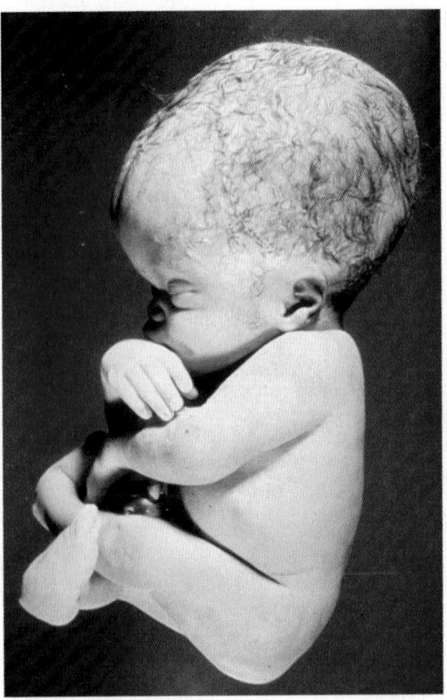

B
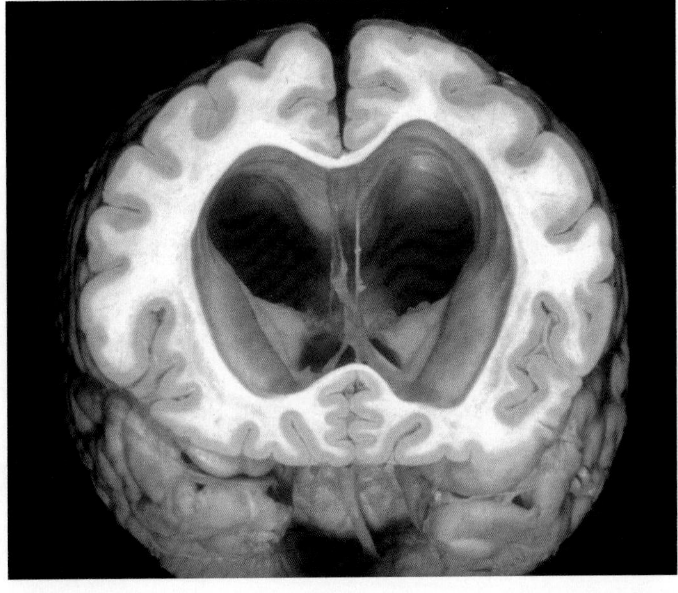

Figure 10-11 **Hydrocephalus. A.** Congenital hydrocephalus causing pronounced enlargement of the head. **B.** Coronal section of the brain showing marked enlargement of the lateral ventricles caused by a tumor that obstructed the flow of CSF.

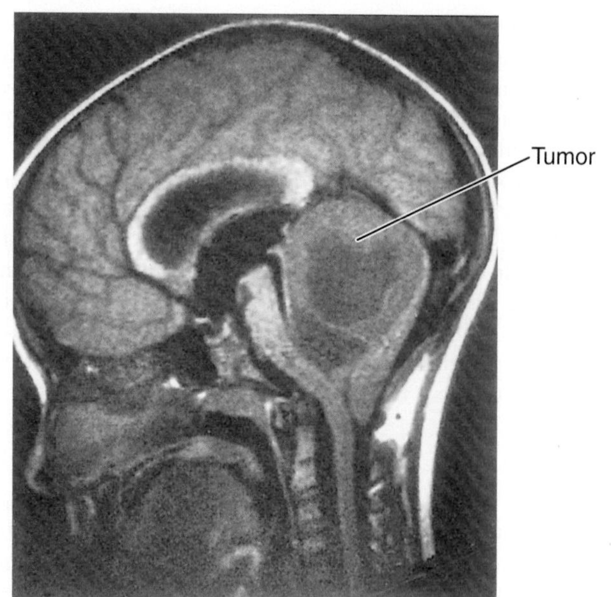

Figure 10-12 Brain tumor. MRI shows a large tumor that arises from the cerebellum and pushes the brain stem forward.

One possible aftereffect of stroke or other brain injury is **aphasia** (ah-FA-ze-ah), a loss or defect in language communication. Losses may involve the ability to speak or write (expressive aphasia) or to understand written or spoken language (receptive aphasia). The type of aphasia present depends on what part of the brain is affected. The lesion that causes aphasia in the right-handed person is likely to be in the left cerebral hemisphere.

Often, much can be done for stroke victims by care and retraining. The brain has tremendous reserves for adapting to different conditions. In many cases, some means of communication can be found even though speech areas are damaged.

See the Student Resources on thePoint to view the animation "Stroke," which illustrates this disorder.

TUMORS

Brain **tumors** may develop in people of any age but are somewhat more common in young and middle-aged adults than in other groups (Fig. 10-12). Most such tumors originate from the neuroglia (support tissue of the brain) and are called **gliomas** (gli-O-mas). These tumors are usually benign and do not metastasize (spread to other areas), but they nevertheless can do harm by compressing brain tissue. Tumors can also develop from the meninges (meningiomas). The symptoms produced depend on the type of tumor, its location, and its degree of destructiveness. Involvement of the cerebrum's frontal portion often causes mental symptoms, such as changes in personality and in levels of consciousness. About half of brain tumors represent metastases from other tissues, especially carcinomas.

Early surgery and radiation therapy offer hope of cure in some cases. The blood–brain barrier, however, limits the effectiveness of injected chemotherapeutic agents (see Box 10-1). A newer approach to chemotherapy for brain tumors is to implant timed-release drugs into a tumor site at the time of surgery.

HEAD INJURY

A common result of head trauma is bleeding into or around the meninges (Fig. 10-13). Damage to an artery from a skull fracture, usually on the side of the head, may result in bleeding between the dura mater and the skull, an **epidural hematoma** (he-mah-TO-mah). The rapidly accumulating blood puts pressure on blood vessels and interrupts blood flow to the brain. Symptoms include headache, partial paralysis, dilated pupils, and coma. If the pressure is not relieved within a day or two, death results.

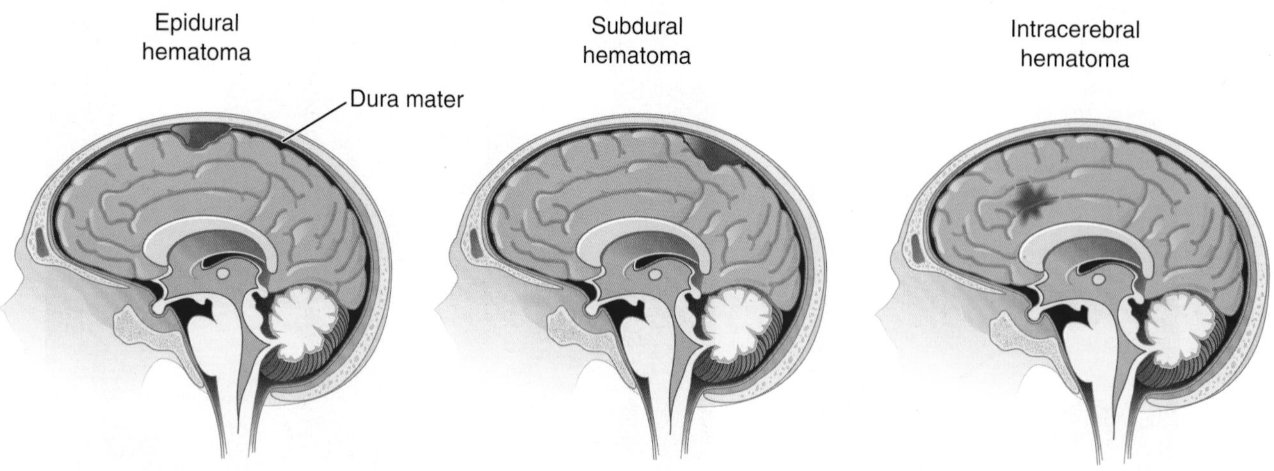

Figure 10-13 Hematomas. **KEY POINT** A hematoma may form outside the dura, below the dura, or within the brain itself. **ZOOMING IN** What type of hematoma forms outside of the dura mater? What type forms below the dura mater?

A tear in the wall of a dural sinus causes a **subdural hematoma.** This often results from a blow to the front or back of the head that separates the dura from the arachnoid, as occurs when the moving head hits a stationary object. Blood gradually accumulates in the subdural space, putting pressure on the brain and causing headache, weakness, and confusion. Death results from continued bleeding.

Bleeding into the brain tissue itself results in an **intracerebral hematoma.** The causes include a penetrating wound, depressive skull fracture, or hemorrhagic stroke. The condition increases intracranial pressure and can lead to coma and death. Symptoms of an intracerebral hematoma depend on the functions of the brain areas involved. Depending on the cause and conditions, treatment may involve surgery and/or medication.

Cerebral **concussion** (kon-CUSH-on), also known as a **mild traumatic brain injury (MTBI),** refers to a transient alteration in brain function resulting from head trauma, as illustrated in Natalie's case. The effects include amnesia, headache, dizziness, vomiting, disorientation, and sometimes loss of consciousness. *Postconcussion syndrome* describes symptoms that persist or develop a month or more after the injury, such as headache, fatigue, mood changes, and cognitive deficits. Individuals who have suffered one concussion are more susceptible to future concussions, and subsequent concussions can cause more severe and long-lasting symptoms. New research has shown that psychological factors play a large role in the severity of the symptoms and the speed of recovery. Concussions are treated with physical and cognitive rest, including restrictions on sports, computer use, reading, and even texting.

Frequent observations of level of consciousness, pupil response, and extremity reflexes are important in the patient with a head injury. (See Box 10-2 for more information on head injuries.)

Cerebral palsy (PAWL-ze) is a disorder caused by brain damage occurring before or during the birth process. Characteristics include diverse muscular disorders that vary in degree from only slight weakness of the lower extremity muscles to paralysis of all four extremities as well as the speech muscles. With muscle and speech training and other therapeutic approaches, children with cerebral palsy can be helped.

DEGENERATIVE DISEASES

Alzheimer (ALZ-hi-mer) **disease** (AD) is a brain disorder resulting from an unexplained degeneration of the cerebral cortex and the hippocampus of the limbic system (**Fig. 10-14**). The disorder develops gradually and eventually causes severe intellectual impairment with mood changes and confusion. Memory loss, especially for recent events, is a common early symptom. Dangers associated with AD are injury, infection, malnutrition, and inhalation of food or fluids into the lungs. Changes in the brain occur many years before noticeable signs of the disease appear. On autopsy, brain cells in AD show accumulation of a protein called *amyloid* (originally mistaken for starch and named with the root *amyl/o*) and protein aggregates called *neurofibrillary tangles* in the neuron fibers. It is not clear whether these materials are a cause or an effect of AD.

At present, there is no cure for AD, but several drugs, if given early, can improve mental function and delay a patient's

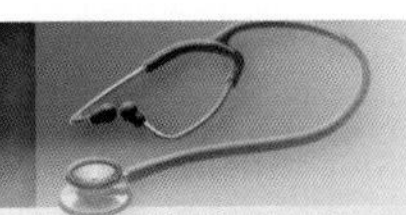

CLINICAL PERSPECTIVES — Box 10-2

Brain Injury: A Heads-Up

Traumatic brain injury is a leading cause of death and disability in the United States. Each year, approximately 1.5 million Americans sustain a brain injury, of whom about 50,000 will die and 80,000 will suffer long-term or permanent disability. The leading causes of traumatic brain injury are motor vehicle accidents, gunshot wounds, sports injuries, and falls. Other causes include shaken baby syndrome (caused by violent shaking of an infant or toddler) and second impact syndrome (when a second head injury occurs before the first has fully healed).

Brain damage occurs either from penetrating head trauma or acceleration–deceleration events where a head in motion suddenly comes to a stop. Nervous tissue, blood vessels, and possibly the meninges may be bruised, torn, lacerated, or ruptured, which may lead to swelling, hemorrhage, and hematoma. The best protection from brain injury is to prevent it. The following is a list of safety tips:

- Always wear a seat belt, and secure children in approved car seats.
- Never drive after using alcohol or drugs or ride with an impaired driver.
- Always wear a helmet during activities such as biking, motorcycling, in-line skating, horseback riding, football, ice hockey, and batting and running bases in baseball and softball.
- Inspect playground equipment, and supervise children using it. Never swing children around to play "airplane" nor vigorously bounce or shake them.
- Allow adequate time for healing after a head injury before resuming potentially dangerous activities.
- Prevent falls by using a nonslip bathtub or shower mat and using a step stool to reach objects on high shelves. Use a safety gate at the bottom and top of stairs to protect young children (and adults with dementia or other disorienting conditions).
- Keep unloaded firearms in a locked cabinet or safe, and store bullets in a separate location.

For more information, contact the Brain Injury Association of America.

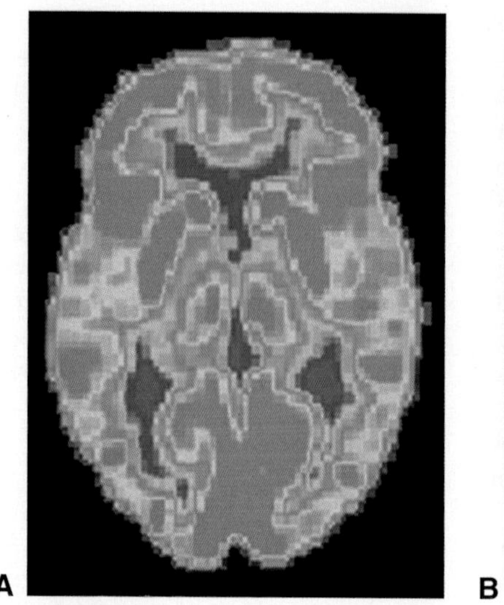

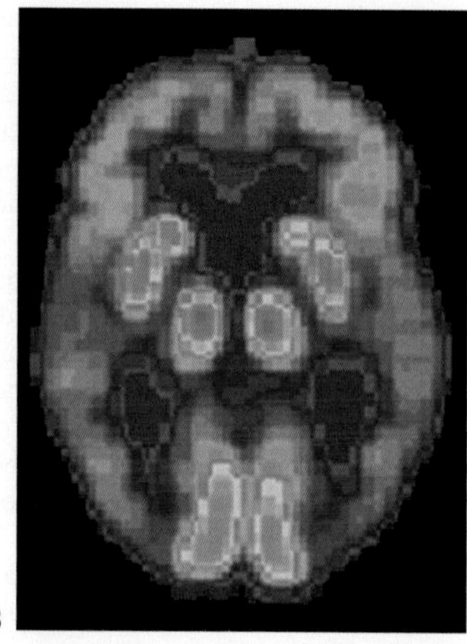

Figure 10-14 Effects of Alzheimer disease. KEY POINT **A.** PET scan of a normal brain. *Red* areas are the most active, followed by *yellow*, *green*, *light blue*, and *dark blue*. **B.** PET scan of the brain of a patient with AD. Notice the dramatically decreased brain activity in virtually every area.

loss of independence. The main drug in use prevents destruction of the neurotransmitter acetylcholine (ACh) to compensate for the loss of neurons that produce ACh. Treatment may also include drugs to manage depression and behavioral disorders, such as agitation and aggression. Also important for the person with AD are exercise, good nutrition, mental and social activities, and a calm, structured environment.

Multi-infarct dementia (de-MEN-she-ah) represents the accumulation of brain damage resulting from chronic ischemia (is-KE-me-ah) (lack of blood supply), such as would be caused by a series of small strokes. There is a stepwise deterioration of function. People with multi-infarct dementia are troubled by progressive loss of memory, judgment, and cognitive function. Many people older than 80 years of age have some evidence of this disorder.

Parkinson disease is a progressive neurologic condition characterized by tremors, rigidity of limbs and joints, and slow movement. These problems reflect the loss of neurons in the substantia nigra of the basal nuclei. These neurons use the neurotransmitter **dopamine** (DO-pah-mene) and usually inhibit basal nuclei output. Their loss results in overactivity of the basal nuclei and thus abnormal signals passing to the motor cortex. The average age of onset is 55 years. As with many diseases, a combination of genetic susceptibility and environmental factors has been implicated. Similar motor changes, together known as *parkinsonism*, may result from encephalitis or other brain diseases, exposure to certain toxins, or repeated head injury, as may occur in boxing.

The main therapy for Parkinson disease is administration of L-dopa, a substance that is capable of entering the brain and converting to dopamine (see Box 10-1). Drugs are now available that mimic the effects of dopamine, prevent its breakdown, or increase the effectiveness of L-dopa. Another approach to treatment is implanting a device that electrically stimulates the brain to control symptoms of Parkinson disease.

Huntington disorder is a hereditary disease resulting in neuron death in the corpus striatum (in the forebrain) of the basal nuclei. These neurons are responsible for modulating motor outputs, so their loss results in uncontrollable, jerky, and writhing movements. Cerebral neurons are eventually affected, resulting in forgetfulness, dementia, and eventually death. There is currently no cure for this disorder, which is also discussed in Chapter 25.

> **Speech therapists treat patients with language or communication problems from any cause. See the Student Resources on thePoint for more information on this career.**

CHECKPOINTS 

- ☐ 10-12 What is the common term for cerebrovascular accident (CVA)?
- ☐ 10-13 What type of cells are commonly involved in brain tumors?

Cranial Nerves

There are 12 pairs of cranial nerves (in this discussion, when a cranial nerve is identified, a pair is meant). They are numbered, usually in Roman numerals, according to their connection with the brain, beginning anteriorly and proceeding posteriorly (Fig. 10-15). Except for the first two pairs, which connect with the cerebrum and diencephalon, respectively, the cranial nerves connect with the brain stem. Also note that anatomists

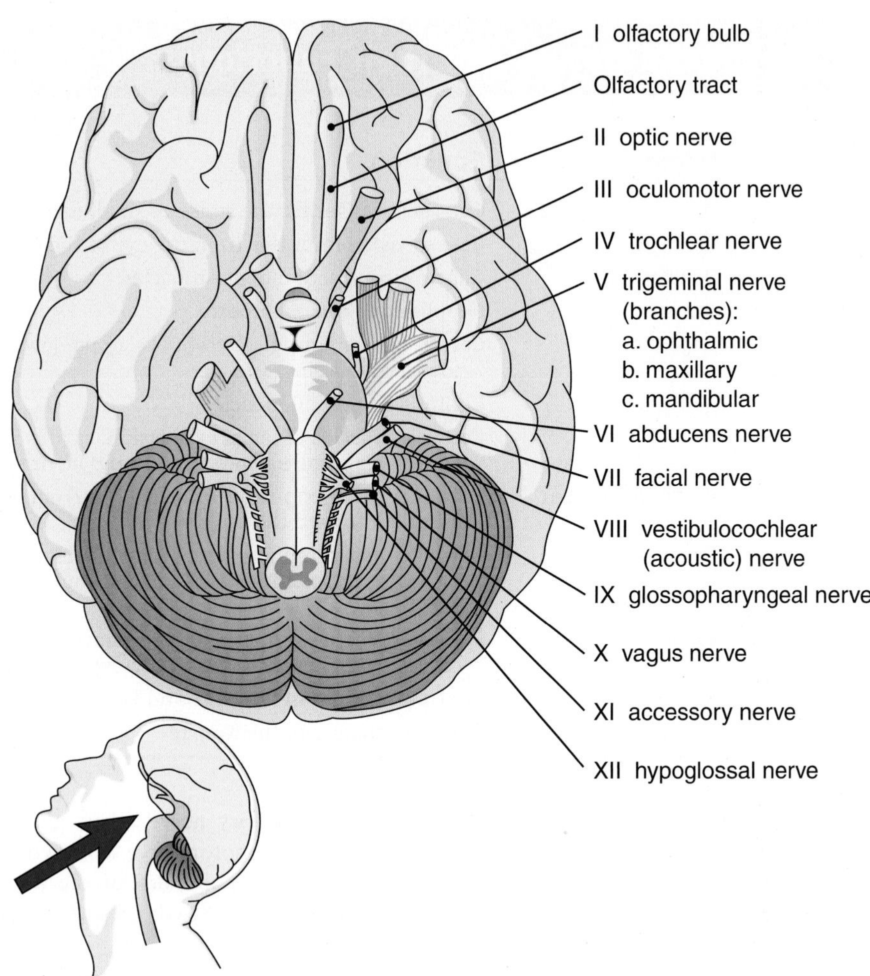

Figure 10-15 **Cranial nerves.** **KEY POINT** There are 12 pairs of cranial nerves, each designated by name and Roman numeral. They are shown here from the base of the brain.

10

now have found that cranial nerve XI connects with the cervical spinal cord. The first nine pairs and the 12th pair supply structures in the head.

From a functional viewpoint, the cranial nerves carry four types of signals. Note that each neuron only carries one signal type, but one cranial nerve can contain neurons carrying different signal types:

- **Special sensory impulses**, such as those for smell, taste, vision, and hearing, originating in special sense organs in the head
- **General sensory impulses**, such as those for pain, touch, temperature, deep muscle sense, pressure, and vibrations. These impulses come from receptors that are widely distributed throughout the body.
- **Somatic motor impulses** that control the skeletal muscles
- **Visceral motor impulses** that control glands, smooth muscle, and cardiac muscle. These motor pathways are part of the autonomic nervous system, parasympathetic division.

NAMES AND FUNCTIONS OF THE CRANIAL NERVES

A few of the cranial nerves (I, II, and VIII) contain only sensory fibers; some (III, IV, VI, XI, and XII) contain all or mostly motor fibers. The remainder (V, VII, IX, and X) contain both sensory and motor fibers; they are known as *mixed nerves*. All 12 nerves are listed below and summarized in Table 10-2:

I. The **olfactory nerve** carries smell impulses from receptors in the nasal mucosa to the brain.

II. The **optic nerve** carries visual impulses from the eye to the brain.

III. The **oculomotor nerve** controls the contraction of most of the eye muscles. These muscles are skeletal muscles and are under voluntary control.

IV. The **trochlear** (TROK-le-ar) **nerve** supplies one eyeball muscle.

V. The **trigeminal** (tri-JEM-in-al) **nerve** is the great sensory nerve of the face and head. It has three branches

Table 10-2 The Cranial Nerves and Their Functions

Nerve (Roman Numeral Designation)	Name	Function
I	Olfactory	Carries impulses for the sense of smell toward the brain
II	Optic	Carries visual impulses from the eye to the brain
III	Oculomotor	Controls contraction of eye muscles
IV	Trochlear	Supplies one eyeball muscle
V	Trigeminal	Carries sensory impulses from eye, upper jaw, and lower jaw toward the brain
VI	Abducens	Controls an eyeball muscle
VII	Facial	Controls muscles of facial expression; carries sensation of taste; stimulates small salivary glands and lacrimal (tear) gland
VIII	Vestibulocochlear	Carries sensory impulses for hearing and equilibrium from the inner ear toward the brain
IX	Glossopharyngeal	Carries sensory impulses from tongue and pharynx (throat); controls swallowing muscles and stimulates the parotid salivary gland
X	Vagus	Supplies most of the organs in the thoracic and abdominal cavities; carries motor impulses to the larynx (voice box) and pharynx
XI	Accessory	Controls muscles in the neck and larynx
XII	Hypoglossal	Controls muscles of the tongue

that carry general sensory impulses (e.g., pain, touch, temperature) from the eye, the upper jaw, and the lower jaw. Motor fibers to the muscles of mastication (chewing) join the third branch. A dentist anesthetizes branches of the trigeminal nerve to work on the teeth without causing pain.

VI. The **abducens** (ab-DU-senz) **nerve** is another nerve sending motor impulses to an eyeball muscle.

VII. The **facial nerve** is largely motor, controlling the muscles of facial expression. This nerve also includes special sensory fibers for taste (anterior two-thirds of the tongue), and it contains secretory fibers to the smaller salivary glands (the submandibular and sublingual) and to the lacrimal (tear) gland.

VIII. The **vestibulocochlear** (ves-tib-u-lo-KOK-le-ar) **nerve** carries sensory impulses for hearing and equilibrium from the inner ear. This nerve was formerly called the auditory or acoustic nerve.

IX. The **glossopharyngeal** (glos-o-fah-RIN-je-al) **nerve** contains general sensory fibers from the posterior tongue and the pharynx (throat). This nerve also contains sensory fibers for taste from the posterior third of the tongue, visceral motor fibers that supply the largest salivary gland (parotid), and motor nerve fibers to control the swallowing muscles in the pharynx.

X. The **vagus** (VA-gus) **nerve** is the longest cranial nerve. (Its name means "wanderer.") It carries autonomic motor impulses to most of the organs in the thoracic and abdominal cavities, including the muscles and glands of the digestive system. This nerve also contains somatic motor fibers supplying the larynx (voice box).

XI. The **accessory nerve** (also called the *spinal accessory nerve*) is a motor nerve with two branches. One branch controls two muscles of the neck, the trapezius and sternocleidomastoid; the other supplies muscles of the larynx.

XII. The **hypoglossal nerve,** the last of the 12 cranial nerves, carries impulses controlling the tongue muscles.

It has been traditional for students in medical fields to use mnemonics (ne-MON-iks), or memory devices, to remember lists of terms. These devices are usually words (real or made-up) or sayings formed from the first letter of each item. We've used the example SLUDD for the actions of the parasympathetic system; for the cranial nerves, students use "On Occasion Our Trusty Truck Acts Funny. Very Good Vehicle Any How." Can you and your classmates make up any other mnemonic phrases for the cranial nerves? You can also check the Internet for sites where medical mnemonics are shared, but be forewarned, students often enjoy making them raunchy!

DISORDERS INVOLVING THE CRANIAL NERVES

Destruction of optic nerve (II) fibers may result from increased pressure of the eye's fluid on the nerves, as occurs in glaucoma; from the effect of poisons; and from some infections. Vision impairment and eventually blindness can result. Certain medications when used in high doses over long periods can damage the branch of the vestibulocochlear nerve involved in hearing and thus cause hearing loss.

Injury to a nerve that contains motor fibers causes paralysis of the muscles supplied by these fibers. The oculomotor nerve (III) may be damaged by certain infections or various poisonous substances. Because this nerve supplies so many muscles connected with the eye, including the levator, which raises the eyelid, injury to it causes a paralysis that usually interferes with eye function.

Bell palsy is a partial facial paralysis caused by inflammation of one of the two facial nerves (VII). This injury temporarily paralyzes the muscles of facial expression on one side of the face. The condition usually resolves within several weeks.

Neuralgia (nu-RAL-je-ah) means "nerve pain." A severe spasmodic pain affecting the fifth cranial nerve is known as **trigeminal neuralgia** or tic douloureux (tik du-lu-RU) (from French, meaning "painful twitch"). At first, the pain comes at relatively long intervals, but as time goes on, intervals between episodes usually shorten while pain durations lengthen. Treatments include microsurgery and high-frequency current.

CHECKPOINTS 

- ☐ 10-14 How many pairs of cranial nerves are there?
- ☐ 10-15 What are the three types of cranial nerves? What is a mixed nerve?

Effects of Aging on the Nervous System

The nervous system is one of the first systems to develop in the embryo. By the beginning of the third week of development, the rudiments of the CNS have appeared. Beginning with maturity, the nervous system begins to undergo degenerative changes. Neurons and glial cells die and are not replaced, decreasing the size and weight of the brain. Neuron loss in the cerebral cortex, accompanied by decreased neurotransmitter production and fewer synapses, results in slower information processing and movements. Memory diminishes, especially for recent events. Narrowing of the cerebral arteries reduces the brain's blood flow. Vascular degeneration increases the likelihood of stroke.

Much individual variation is possible, however, with regard to location and severity of changes. Although age might make it harder to acquire new skills, tests have shown that practice enhances skill retention. As with other body systems, the nervous system has vast reserves, and most elderly people are able to cope with life's demands.

10

Disease in Context Revisited

Natalie's Progress

Dr. Erickson referred Natalie to a neurologist for follow-up after discharge. He also instructed Lacey to watch her sister carefully over the next 24 hours. Natalie was to return to the hospital if her headache worsens, her speech becomes slurred, or she becomes difficult to arouse.

"And," said Dr. Erickson, "take away Natalie's cell phone, laptop computer and any other electronics, as well as all reading material. She needs complete quiet and rest."

At the neurologist appointment later that week, it was obvious that Lacey's patience was wearing thin.

"Natalie is irritable, keeps forgetting where she left her tea, and can't hold a coherent conversation for more than two minutes," Lacey complained.

"Natalie, you are experiencing postconcussion syndrome," said the neurologist. "Although the brain is encased within the skull and cushioned by CSF, it can still be damaged by the force of the brain colliding with the skull. It can take a few weeks for the cerebral swelling to decrease. Until that time, you can expect cognitive, physical, emotional, and behavioral changes."

"I expect you to recover fully," he continued. "But for the time being, you still need rest. Avoid overstimulation and alcohol. Your symptoms should resolve in the next few weeks."

CHAPTER 10

Chapter Wrap-Up

Summary Overview

A detailed chapter outline with space for note taking is on *thePoint*. The figure below illustrates the main topics covered in this chapter.

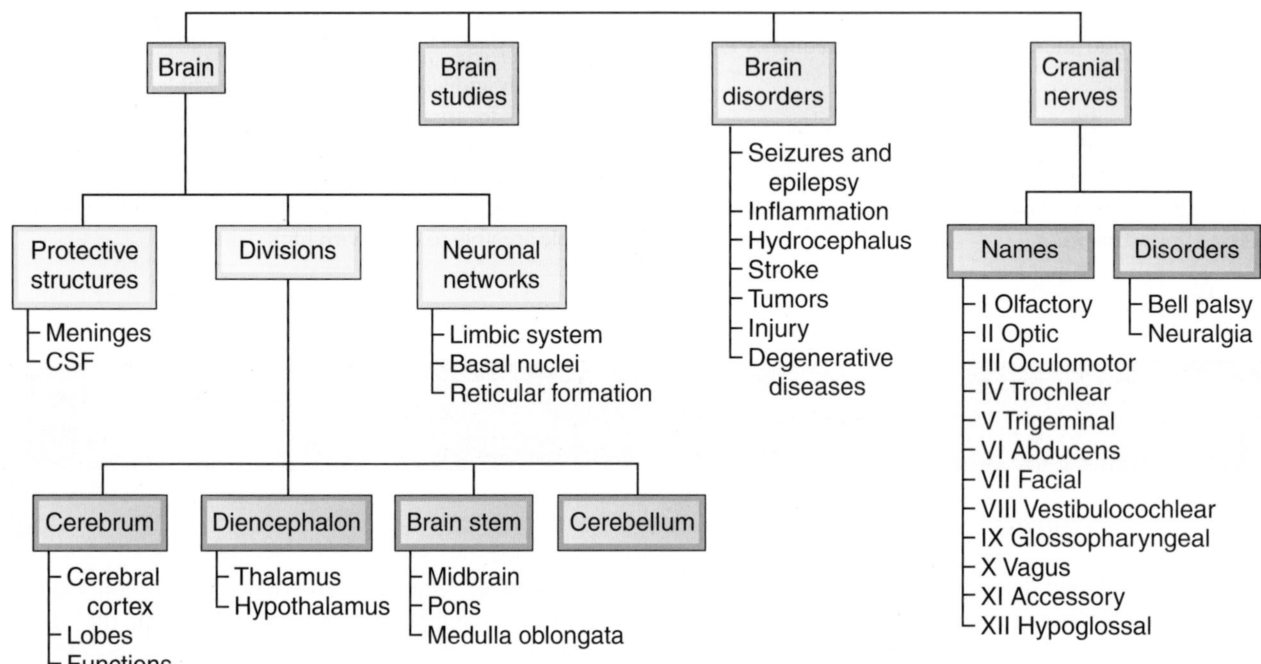

Key Terms

The terms listed below are emphasized in this chapter. Knowing them will help you organize and prioritize your learning. These and other boldface terms are defined in the Glossary with phonetic pronunciations.

aphasia
basal nuclei
brain stem
cerebellum
cerebral cortex
cerebrospinal fluid (CSF)
cerebrum
concussion
corpus callosum
diencephalon
electroencephalograph (EEG)
gyrus (pl. gyri)
hematoma
hypothalamus
limbic system
medulla oblongata
meninges
midbrain
pons
reticular formation
seizure
stroke
sulcus (pl. sulci)
thalamus
ventricle

Word Anatomy

Medical terms are built from standardized word parts (prefixes, roots, and suffixes). Learning the meanings of these parts can help you remember words and interpret unfamiliar terms.

WORD PART	MEANING	EXAMPLE
Protective Structures of the Brain and Spinal Cord		
cerebr/o	brain	*Cerebrospinal* fluid circulates around the brain and spinal cord.
chori/o	membrane	The *choroid* plexus is the vascular membrane in the ventricle that produces CSF.
contra-	opposed, against	The cerebral cortex has *contralateral* control of motor function.
encephal/o	brain	The *diencephalon* is the part of the brain located between the cerebral hemispheres and the brain stem.
gyr/o	circle	A *gyrus* is a circular raised area on the surface of the brain.
later/o	lateral, side	See "contra-" example.
Brain Studies		
tom/o	cut	*Tomography* is a method for viewing sections as if cut through the body.
Disorders of the Brain and Associated Structures		
cephal/o	head	*Hydrocephalus* is the accumulation of fluid within the brain.
-rhage	bursting forth	A cerebral *hemorrhage* is a sudden bursting forth of blood in the brain.
phasia	speech, ability to talk	*Aphasia* is a loss or defect in language communication.
Cranial Nerves		
gloss/o	tongue	The *hypoglossal* nerve controls muscles of the tongue.

Questions for Study and Review

BUILDING UNDERSTANDING

Fill in the Blanks

1. The delicate innermost layer of the meninges is the _____.
2. The large band of white matter that connects the right and left hemispheres is the _____.
3. Sound is processed in the _____ lobe of the brain.
4. Abnormal accumulation of CSF in the brain is termed _____.
5. The cells involved in most brain tumors are _____.

Matching > Match each numbered item with the letter of the most closely related cranial nerve.

___ **6.** The nerve involved with the sense of smell
___ **7.** The large sensory nerve of the face and head
___ **8.** The nerve that controls muscles of the tongue
___ **9.** The sensory nerve for hearing and equilibrium
___ **10.** The long nerve that carries autonomic impulses to the thorax and abdomen

a. olfactory nerve
b. vestibulocochlear nerve
c. trigeminal nerve
d. vagus nerve
e. hypoglossal nerve

Multiple Choice

___ **11.** What divides the cerebrum into left and right hemispheres?

a. central sulcus
b. insula
c. lateral sulcus
d. longitudinal fissure

___ **12.** Which lobe interprets impulses arising from the retina of the eye?

a. frontal
b. occipital
c. parietal
d. temporal

___ **13.** What is a loss or defect in language communication called?

a. stroke
b. hematoma
c. aphasia
d. encephalitis

___ **14.** What disease involves lack of dopamine and overactivity of the basal nuclei?

a. Alzheimer disease
b. Bell palsy
c. Huntington disorder
d. Parkinson disease

___ **15.** What type of impulses are involved in the sense of touch?

a. special sensory
b. general sensory
c. somatic motor
d. visceral motor

UNDERSTANDING CONCEPTS

16. Briefly describe the effects of injury to the following brain areas:

a. Broca area
b. hypothalamus
c. medulla oblongata
d. cerebellum

17. A neurosurgeon has drilled a hole through her patient's skull and is preparing to remove a cerebral glioma. In order, list the membranes she must cut through to reach the cerebral cortex.

18. What is the function of the limbic system? Describe the effect of damage to the hippocampus.

19. Explain the working of an electroencephalograph. What kind of information does the electroencephalograph provide?

20. Compare and contrast the following nervous system disorders:

a. meningitis and encephalitis
b. epidural hematoma and subdural hematoma
c. Alzheimer disease and multi-infarct dementia
d. Bell palsy and trigeminal neuralgia

21. Referring to the 12 cranial nerves and their functions, make a list of the ones that are sensory, motor, and mixed.

22. Explain the function of the four cranial nerves that might have been involved in causing Natalie's visual symptoms discussed in the opening case study.

23. Referring to the brain overlays in The Body Visible at the beginning of the book, give the numbers of the following:

a. a network involved in the manufacture of CSF
b. myelinated fibers
c. a portion of the limbic system
d. a shallow groove in the surface of the cerebral cortex
e. a chamber where CSF is made

CONCEPTUAL THINKING

24. The parents of Molly R., a 2-year-old girl, are informed that their daughter has epilepsy. How would you explain this disorder to her parents? How will her disorder be treated? What can they expect in the future, and what precautions should they take to protect her?

25. A student at a local college has been diagnosed with bacterial meningitis. As the head of health services at the school, what will you tell the students and parents about the causes and symptoms of this disease, and what precautions will you recommend? (See Appendix 3-1.)

For more questions, see the Learning Activities on thePoint.

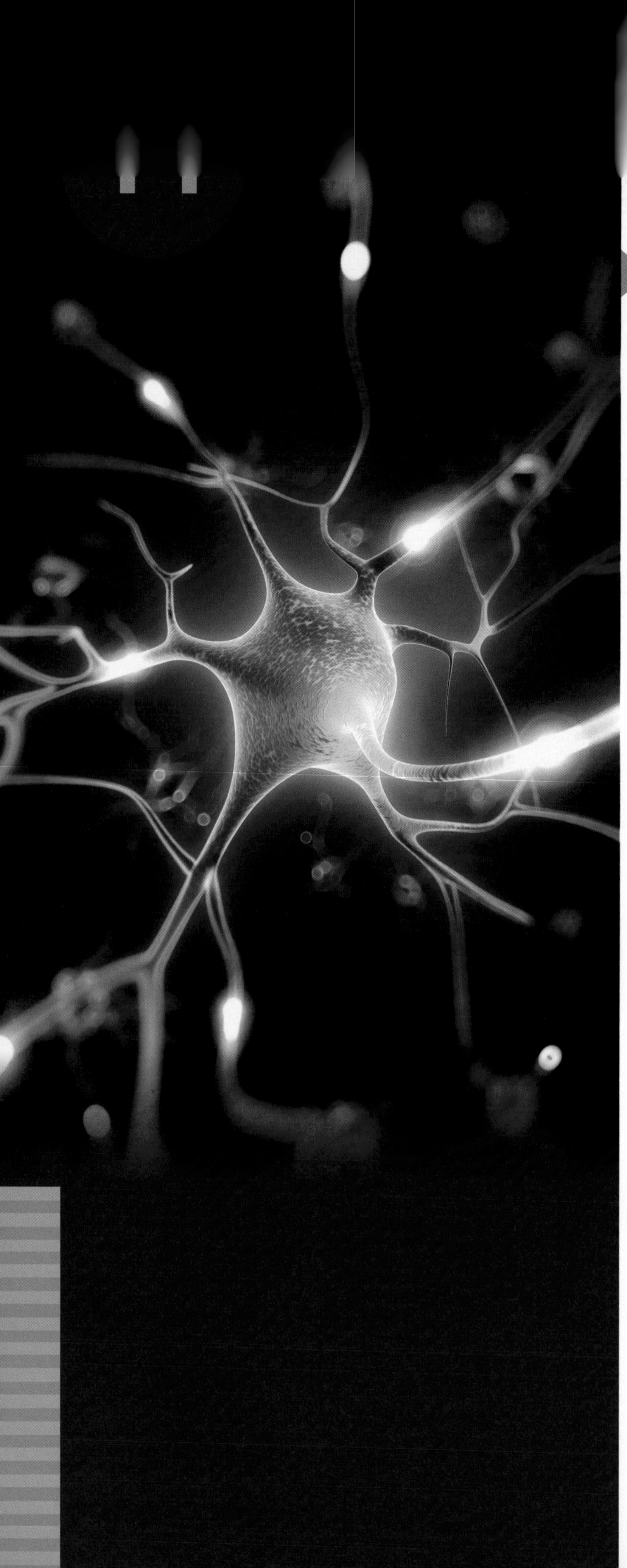

Learning Objectives

After careful study of this chapter, you should be able to:

1 ▶ Describe the functions of the sensory system. ***p. 242***

2 ▶ Differentiate between the different types of sensory receptors, and give examples of each. ***p. 242***

3 ▶ Describe sensory adaptation, and explain its value. ***p. 242***

4 ▶ List and describe the structures that protect the eye. ***p. 243***

5 ▶ Cite the location and the purpose of the extrinsic eye muscles. ***p. 243***

6 ▶ Identify the three tunics of the eye. ***p. 244***

7 ▶ Describe the processes involved in vision. ***p. 244***

8 ▶ Differentiate between the rods and the cones of the eye. ***p. 246***

9 ▶ List seven disorders of the eye and vision. ***p. 248***

10 ▶ Describe the three divisions of the ear. ***p. 252***

11 ▶ Describe the receptor for hearing, and explain how it functions. ***p. 253***

12 ▶ Compare the locations and functions of the equilibrium receptors. ***p. 253***

13 ▶ Describe three disorders that involve the ear. ***p. 256***

14 ▶ Discuss the locations and functions of the sense organs for taste and smell. ***p. 258***

15 ▶ Describe five general senses. ***p. 260***

16 ▶ List five approaches for treatment of pain. ***p. 261***

17 ▶ Referring to the case study, discuss the purpose and mechanism of a cochlear implant. ***pp. 241, 262***

18 ▶ Show how word parts are used to build words related to the sensory system (see Word Anatomy at the end of the chapter). ***p. 264***

Disease in Context: *Evan Needs a Cochlear Implant*

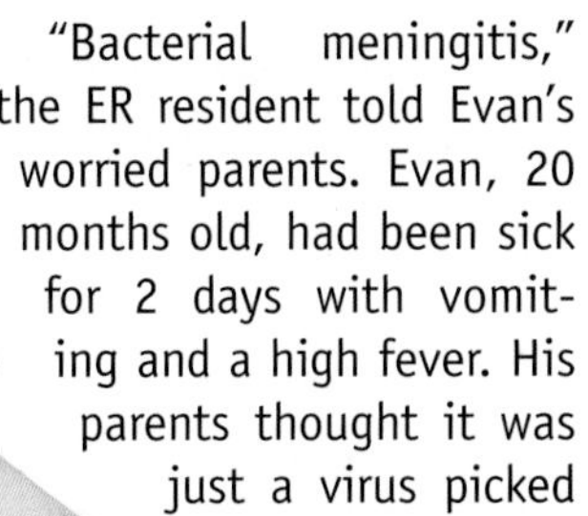

"Bacterial meningitis," the ER resident told Evan's worried parents. Evan, 20 months old, had been sick for 2 days with vomiting and a high fever. His parents thought it was just a virus picked up at daycare, but the sudden development of a rash over his chest and legs sent them to the hospital.

The resident continued, "He needs to be admitted right away for treatment and observation. Strong antibiotics should cure the infection, and steroids should limit the damage caused by Evan's immune system fighting the bacteria. Meningitis is an infection of the central nervous system and can be very dangerous, but we caught it early. Be prepared to stay in the hospital at least a week."

It was 10 days until Evan was home and feeling better. The entire family was so exhausted by the ordeal that it took them another week to notice that Evan wasn't responding to sounds. A follow-up appointment with his pediatrician and subsequent hearing test revealed that Evan had 95% hearing loss in the right ear and 60% in the left.

"I'm referring you to Dr. Sanchez, a specialist in cochlear implants for young children," the pediatrician said. "A cochlear implant can potentially restore some of Evan's hearing."

"A multidisciplinary approach is used for each patient," Dr. Sanchez explained. "A complete audiological, speech, and medical evaluation will be conducted along with a CT scan to determine if Evan is a candidate. If approved, you and Evan will need to be enrolled into the pre-implant program, where you will learn about how a prosthetic implant works. The device will stimulate the cochlear nerve directly, bypassing the receptor cells, and it may restore hearing for medium to loud sounds."

Evan's parents were hopeful and eager to proceed with the evaluation. Dr. Sanchez fully expected Evan to fit the criteria, as his hearing loss was recent. We will check later to see how Evan's case progressed.

ANCILLARIES *At-A-Glance*

Visit thePoint to access the following resources. For guidance in using these resources most effectively, see pp. xv–xvii.

Learning RESOURCES

- Tips for Effective Studying
- Web Figure: Ptosis of the Eyelid
- Web Figure: Strabismus
- Web Figure: Trachoma
- Web Figure: Cataract
- Web Figure: Cataract Surgeries
- Web Figure: Diabetic Retinopathy
- Animation: The Retina
- Health Professions: Audiologist
- Detailed Chapter Outline
- Answers to Questions for Study and Review
- Audio Pronunciation Glossary

Learning ACTIVITIES

- Pre-Quiz
- Visual Activities
- Kinesthetic Activities
- Auditory Activities

A LOOK BACK

In describing the basic organization of the nervous system, we included both sensory and motor functions. Now we concentrate on just the sensory portion of the nervous system and the special receptors that detect environmental changes. These specialized structures initiate the reflex pathways described in the previous chapters.

The Senses

The sensory system provides us with an awareness of our external and internal environments. An environmental change becomes a *stimulus* when it initiates a nerve impulse, which then travels to the central nervous system (CNS) by way of a sensory neuron. A stimulus becomes a sensation—something we experience—only when a specialized area of the cerebral cortex interprets the nerve impulse received. Many stimuli arrive from the external environment and are detected at or near the body surface. Others originate internally and help maintain homeostasis.

SENSORY RECEPTORS

The part of the nervous system that detects a stimulus is the **sensory receptor**. In structure, a sensory receptor may be one of the following:

- The free dendrite of a sensory neuron, such as the receptors for pain and temperature
- A modified ending on the dendrite of a sensory neuron, such as those for touch
- A specialized cell associated with a sensory neuron, such as the rods and cones of the eye's retina

Receptors can be classified according to the type of stimulus to which they respond:

- Chemoreceptors, such as receptors for taste and smell, detect chemicals in solution.
- Photoreceptors, located in the retina of the eye, respond to light.
- Thermoreceptors detect changes in temperature. Many of these receptors are located in the skin.
- Mechanoreceptors respond to movement, such as stretch, pressure, or vibration. These include pressure receptors in the skin, receptors that monitor body position, and the receptors of hearing and equilibrium in the ear, which are activated by the movement of cilia on specialized receptor cells.

Any receptor must receive a stimulus of adequate intensity, that is, at least a **threshold stimulus**, in order to respond and generate a nerve impulse.

SPECIAL AND GENERAL SENSES

Another way of classifying the senses is according to the distribution of their receptors. A **special sense** is localized in a special sense organ; a **general sense** is widely distributed throughout the body.

- Special senses
  - **Vision** from receptors in the eye
  - **Hearing** from receptors in the inner ear
  - **Equilibrium** (balance) from receptors in the inner ear
  - **Taste** from receptors in the tongue
  - **Smell** from receptors in the upper nasal cavities
- General senses
  - **Pressure**, **temperature**, **pain**, and **touch** from receptors in the skin and internal organs
  - Sense of **position** from receptors in the muscles, tendons, and joints

SENSORY ADAPTATION

When sensory receptors are exposed to a continuous and unimportant stimulus, they often adjust so that the sensation becomes less acute. The term for this phenomenon is **sensory adaptation.** For example, when you first put on a watch, you may be aware of its pressure on your wrist. Soon you do not notice it at all. If you are rinsing dishes in very warm water, you may be aware of the temperature at first, but you soon adapt and stop noticing the water's temperature. Similarly, both delicious and horrible odors weaken the longer you smell them. As these examples show, both special and general senses are capable of adaptation. However, different receptors adapt at different rates. Those for warmth, cold, and light pressure adapt rapidly. In contrast, receptors for pain do not adapt. In fact, the sensations from receptors for slow, chronic pain tend to increase over time. This variation in receptors allows us to save energy by not responding to unimportant stimuli while always heeding the warnings of pain.

CHECKPOINTS 

- ☐ 11-1 What is a sensory receptor?
- ☐ 11-2 What are some categories of sensory receptors based on type of stimulus?
- ☐ 11-3 How do the special and general senses differ in location?
- ☐ 11-4 What happens when a sensory receptor adapts to a stimulus?

The Eye and Vision

Vision is arguably the most important of the special senses, contributing more than half of the information we use to perceive the world. Before we discuss the eye itself, we begin with the structures that protect, move, and control the eye.

PROTECTIVE STRUCTURES OF THE EYE

The eye is a delicate organ and is protected by a number of structures:

- The skull bones form the walls of the eye orbit (cavity) and protect the posterior part of the eyeball (**see Fig. 7-5**).
- The upper and lower eyelids aid in protecting the eye's anterior portion (**Figs. 11-1 and 11-2**). The eyelids can be closed to keep harmful materials out of the eye, and blinking helps to lubricate the eye. An eyelid is technically called a palpebra (PAL-peh-brah). A muscle, the levator palpebra, is attached to the upper eyelid (**Fig. 11-2**). When this muscle contracts, it keeps the eye open. If the muscle becomes weaker with age, the eyelids may droop and interfere with vision, a condition called *ptosis*.
- The eyelashes and eyebrow help keep foreign matter out of the eye.
- Tears, produced by the **lacrimal** (LAK-rih-mal) **glands** (**Fig. 11-1**), lubricate the eye and contain an enzyme that protects against infection. As tears flow across the eye from the lacrimal gland located in the orbit's upper lateral part, they carry away small particles that may have entered the eye. The tears then flow into canals near the eye's nasal corner where they drain into the nose by way of the **nasolacrimal** (na-zo-LAK-rih-mal) **duct.** The lacrimal glands, ducts, and canals together make up the **lacrimal apparatus.** An excess of tears causes a "runny nose;" the overproduction causes tears to spill onto the cheeks. With age, the lacrimal glands secrete less, but tears still may overflow if the nasolacrimal ducts become plugged.

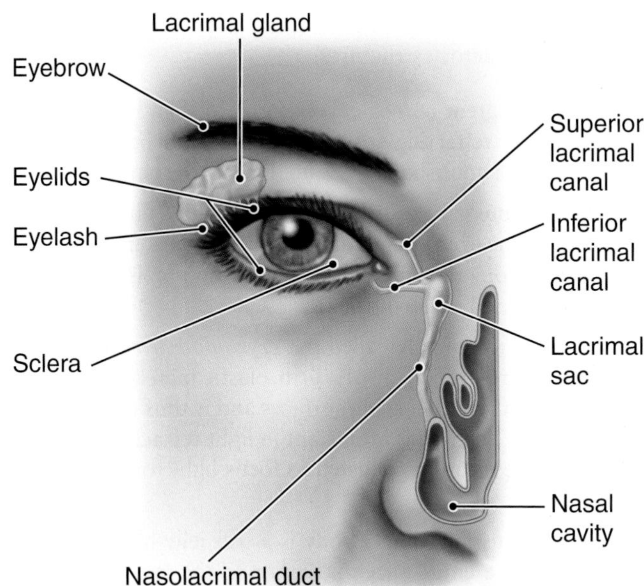

Figure 11-1 The eyelid and lacrimal apparatus. **KEY POINT** Tears are produced in the lacrimal gland, located laterally, and flow across the eye to the lacrimal canals, located medially.

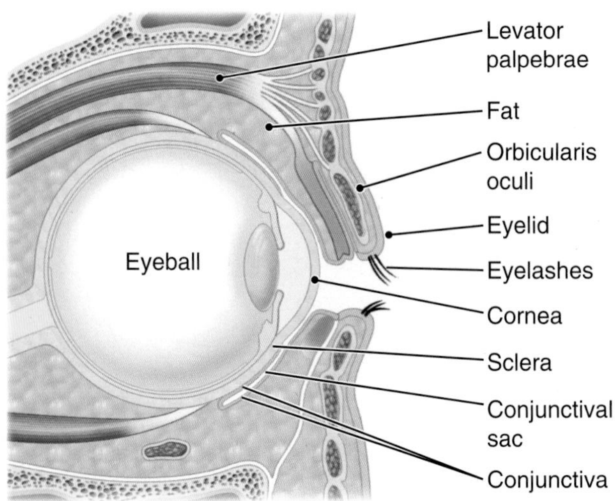

Figure 11-2 A sagittal view of the eye orbit. **KEY POINT** The eye is a delicate organ well guarded by a bony socket and other protective structures.

- A thin membrane, the **conjunctiva** (kon-junk-TI-vah), lines the inner surface of the eyelids and folds back to cover the visible portion of the white of the eye (sclera) (**Fig. 11-1**). Cells within the conjunctiva produce mucus that aids in lubricating the eye. The pocket formed by the folded conjunctiva, known as the conjunctival sac, can be used to instill medication drops (**see Fig. 11-2**). With age, the conjunctiva often thins and dries, resulting in inflammation and dilated blood vessels.

See the Student Resources on thePoint to view an image of ptosis of the eyelid.

THE EXTRINSIC EYE MUSCLES

The **extrinsic muscles** move the eyeball within the eye socket and are attached to the eyeball's outer surface. The six ribbon-like extrinsic muscles connected with each eye originate on the orbital bones and insert on the surface of the sclera (**Fig. 11-3**). They are named for their location and the direction of the muscle fibers. These muscles pull on the eyeball in a coordinated fashion so that both eyes center on one visual field. This process of **convergence** is necessary to the production of a clear retinal image. Having the image come from a slightly different angle from each retina is believed to be important for three-dimensional (stereoscopic) vision, a characteristic of primates.

NERVE SUPPLY TO THE EYE

Two sensory cranial nerves supply the eye (**see Fig. 10-15**):

- The optic nerve (cranial nerve II) carries visual impulses from the eye's photoreceptors to the brain.

11

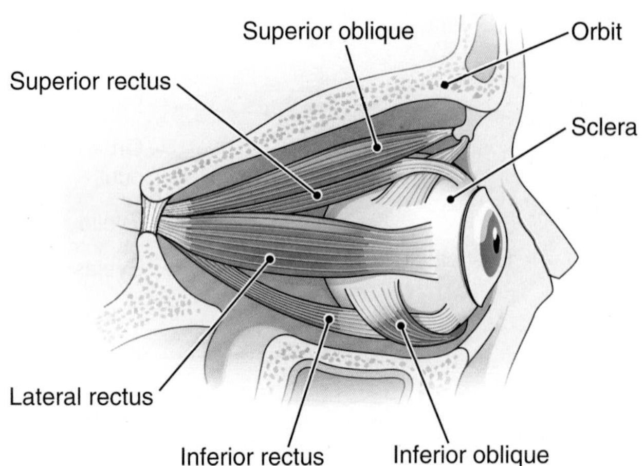

Figure 11-3 **Extrinsic muscles of the eye.** **KEY POINT** The extrinsic muscles coordinate eye movements for proper vision. The medial rectus is not shown. **ZOOMING IN** What characteristics are used in naming the extrinsic eye muscles?

- The ophthalmic (of-THAL-mik) branch of the trigeminal nerve (cranial nerve V) carries impulses of pain, touch, and temperature from the eye and surrounding parts to the brain.

Three cranial nerves carry motor impulses to the eyeball muscles (**see Fig. 10-15**):

- The oculomotor nerve (cranial nerve III) is the largest; it supplies voluntary and involuntary motor impulses to all but two eye muscles.
- The trochlear nerve (cranial nerve IV) supplies the superior oblique extrinsic eye muscle.
- The abducens nerve (cranial nerve VI) supplies the lateral rectus extrinsic eye muscle.

CHECKPOINTS

- [ ] 11-5 What are five structures that protect the eye?
- [ ] 11-6 What is the function of the extrinsic eye muscles?
- [ ] 11-7 Which cranial nerve carries impulses from the retina to the brain?

STRUCTURE OF THE EYEBALL

In the embryo, the eye develops as an outpocketing of the brain, a process that begins at about 22 days of development. The eyeball has three separate coats, or tunics. The components of these tunics are shown in **Figure 11-4**.

1. The outermost tunic is the fibrous tunic. It consists mainly of the **sclera** (SKLE-rah), which is made of tough connective tissue and is commonly referred to as the *white of the eye*. It appears white because of the collagen it contains and because it has no blood vessels to add color. (Reddened or "bloodshot" eyes result from inflammation and swelling of blood vessels in the conjunctiva.) The anterior portion of the fibrous tunic is the forward-curving, transparent, and colorless **cornea** (KOR-ne-ah).
2. The middle tunic is the vascular tunic, consisting mainly of the **choroid** (KO-royd). This layer is composed of a delicate network of connective tissue interlaced with many blood vessels. Its dark coloration reflects the presence of **melanin**, a dark brown pigment. Melanin absorbs light rays and prevents them from reflecting within the eye, much like the black grease athletes use under their eyes to reduce sun glare. At the eye's anterior, the vascular tunic continues as the **ciliary** (SIL-e-ar-e) **muscle** and **suspensory ligaments** (which control the shape of the **lens**, described shortly), and the **iris** (I-ris), the colored, ringlike portion of the eye.
3. The innermost coat is the nervous tunic, consisting of the **retina** (RET-ih-nah), the eye's actual receptor layer. The retina contains light-sensitive cells known as **rods** and **cones**, which generate the nerve impulses associated with vision. The neural tunic covers only the posterior surface of the eye.

PATHWAY OF LIGHT RAYS AND REFRACTION

As light rays pass through the eye toward the retina, they travel through a series of transparent, colorless parts described below and seen in **Figure 11-4**. On the way, they undergo a process known as **refraction**, which is the bending of light rays as they pass from one substance to another substance of different density. (For a simple demonstration of refraction, place a spoon into a glass of water, and observe how the handle appears to bend at the surface of the water.) Because of refraction, light from a very large area can be focused on a very small area of the retina. As light travels from the environment to the retina, it passes through the following refractory structures:

1. The transparent cornea curves forward slightly and is the eye's main refracting structure. The cornea has no blood vessels; it is nourished by the fluids that constantly bathe it.
2. The **aqueous** (A-kwe-us) **humor**, a watery fluid that fills much of the eyeball anterior to the lens, helps maintain the cornea's convex curve. The aqueous humor is constantly produced and drained from the eye.
3. The lens, technically called the *crystalline lens*, is a clear, circular structure made of a firm, elastic material. The lens has two outward-curving surfaces and is thus described as biconvex. The lens is important in light refraction because its thickness can be adjusted to focus light for near or far vision.
4. The **vitreous** (VIT-re-us) **body** is a soft jelly-like substance that fills the entire space posterior to the lens (the adjective *vitreous* means "glasslike"). Like the aqueous humor, it helps maintain the shape of the eyeball.

Accommodation **Accommodation** is the process of adjusting lens thickness to allow for vision at near and far distances. It involves three structures of the vascular tunic:

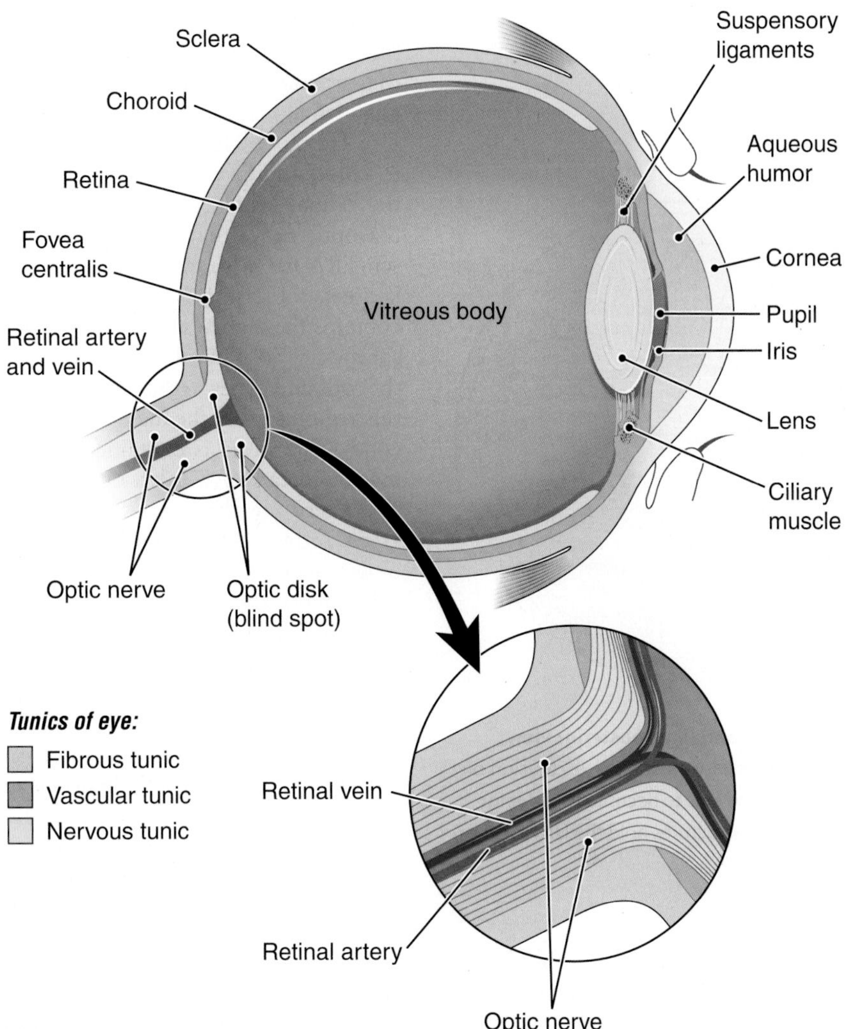

Figure 11-4 **The eye.** **KEY POINT** The eye has three tunics, or coats—the sclera, choroid, and retina. Its refractive parts are the cornea, aqueous humor, lens, and vitreous body. These and other structures involved in vision are shown. **ZOOMING IN** What anterior structure is continuous with the sclera?

the ciliary muscle, the suspensory ligaments, and the lens. The doughnut-shaped ciliary muscle sits posterior to the iris, surrounding the lens. Its central hole is about the same size as the entire iris. Suspensory ligaments extend from the lens periphery to the inner surface of the ciliary muscle, similar to the springs joining a trampoline to its wire frame (**Fig. 11-5**).

Accommodation occurs as follows (**Fig. 11-6**). The ciliary muscle is smooth muscle, so the cells are thin and long when relaxed but short and wide when contracted. For distant vision, the ciliary muscle relaxes and thins, enlarging the central opening. This enlarged opening pulls on the suspensory ligaments, keeping the lens in a more flattened shape. Light rays from a distant object do not require much refraction, so the flattened lens perfectly focuses the light rays on the retina (**Fig. 11-6A**). For close vision, the ciliary muscle contracts and fattens, relaxing tension on the suspensory ligaments. The elastic lens then recoils and becomes thicker, in much the same way that a rubber band thickens when we release the pull on it. The thickened lens refracts the light to a greater extent, focusing the image from a near object on the retina (**Fig. 11-6B**).

Function of the Iris The iris is the pigmented ring that gives an eye its distinctive color. It is composed of two sets of muscle fibers that govern the size of the iris's central opening, the **pupil** (PU-pil) (**Fig. 11-7**). One set of fibers is arranged in a circular fashion, and the other set extends radially like the spokes of a wheel. The iris regulates the amount of light entering the eye. In bright light, the iris's circular muscle fibers contract (and the radial fibers relax), reducing the size of the pupil. This narrowing is termed *constriction*. In contrast, in dim light, the radial fibers contract (and the circular fibers relax), pulling the opening outward and enlarging it. This enlargement of the pupil is known as *dilation*.

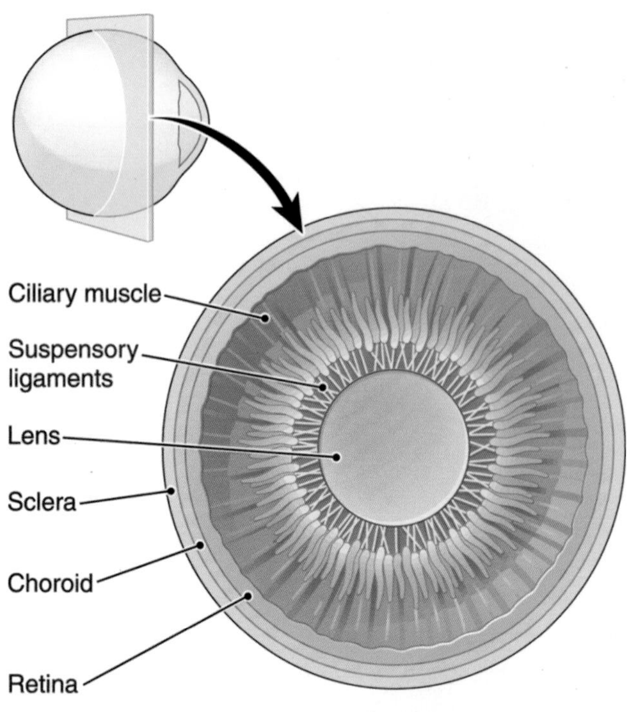

Figure 11-5 The ciliary muscle and lens (posterior view). **KEY POINT** Contraction of the ciliary muscle relaxes tension on the suspensory ligaments, allowing the lens to become rounder for near vision. **ZOOMING IN** What structures hold the lens in place?

FUNCTION OF THE RETINA

The retina has a complex structure with multiple layers of cells (Fig. 11-8). The deepest layer is a pigmented layer just superficial to the choroid. Next are the rods and cones, the eye's receptor cells, named for their respective shapes. Table 11-1 lists the differences between these two cell types. Superficial to the rods and cones are connecting neurons that carry impulses toward the optic nerve. Light rays must pass through these connecting neurons before they can activate the photoreceptors of the retina.

The optic nerve arises from the retina a little toward the medial or nasal side of the eye. There are no photoreceptors in the area of the optic nerve. Consequently, no image can form on the retina at this point, which is known as the blind spot or **optic disk** (see Fig. 11-4).

The optic nerve transmits impulses from the photoreceptors to the thalamus (part of the diencephalon), from which they are directed to the occipital cortex. Note that the light rays passing through the eye are actually overrefracted (overly bent) so that an image falls on the retina upside down and backward (see Fig. 11-6). It is the job of the brain's visual centers to invert the images.

The rods are highly sensitive to light and thus function best in dim light, but they do not provide a sharp image or differentiate colors. They are more numerous than the cones and are distributed more toward the periphery (anterior portion) of the retina. (If you visualize the retina as the inside of a bowl, the rods would be located toward the bowl's lip.) When you enter into dim light, such as a darkened movie theater, you cannot see for a short period. It is during this time that the rods are beginning to function well, a change that is described as **dark adaptation**. When you are able to see again, images are indistinct and appear only in shades of gray.

The cones function best in bright light, are sensitive to color, and give sharp images. The cones are localized at the retinal center, especially in a tiny depressed area near the optic nerve that is called the **fovea centralis** (FO-ve-ah sen-TRA-lis) (Fig. 11-8; see also Fig. 11-4). (Note that *fovea* is a general term for a pit or depression.) Because this area contains the highest concentration of cones, it is the point of sharpest vision. In addition, all of the neurons that connect the rods and cones to the optic nerve are displaced away from this region so that the maximum amount of light reaches the cones. The fovea is contained within a yellowish spot, the **macula lutea** (MAK-u-lah LU-te-ah).

There are three types of cones, each sensitive to red, green, or blue light. Color blindness results from a deficiency of retinal cones. People who completely lack cones are totally color blind; those who lack one type of cone are partially color blind. For instance, individuals without green cones cannot distinguish green from red (red–green color blindness). This disorder, because of its pattern of inheritance, occurs much more commonly in males.

The rods and cones function by means of pigments that are sensitive to light. The light-sensitive pigment in rods is **rhodopsin** (ro-DOP-sin), a complex molecule synthesized from vitamin A. If a person lacks vitamin A, and thus rhodopsin, he or she may have difficulty seeing in dim light, because the rods cannot be activated; this condition is termed **night blindness**. Nerve impulses from the rods and cones flow into sensory neurons that gather to form the optic nerve (cranial nerve II) at the eye's posterior (see Fig. 11-4). The impulses travel to the visual center in the brain's occipital cortex.

When an **ophthalmologist** (of-thal-MOL-o-jist), a physician who specializes in treatment of the eye, examines the retina with an **ophthalmoscope** (of-THAL-mo-skope), he or she can see abnormalities in the retina and in the retinal blood vessels (see Fig. 11-8B). Some of these changes may signal more widespread diseases that affect the eye, such as diabetes and high blood pressure (hypertension).

THE VISUAL PROCESS

To summarize, the events required for proper vision (some of which may be occurring simultaneously) are as follows:

- The extrinsic eye muscles produce convergence.
- Light refracts through the cornea and the aqueous humor.
- The muscles of the iris adjust the pupil.
- The ciliary muscle adjusts the lens (accommodation).
- The light continues to refract through the vitreous body and passes through the superficial layers of the retina.
- Light stimulates retinal receptor cells (rods and cones).
- The optic nerve transmits impulses to the brain.
- The visual areas in the occipital lobe cortex receive and interpret the impulses.

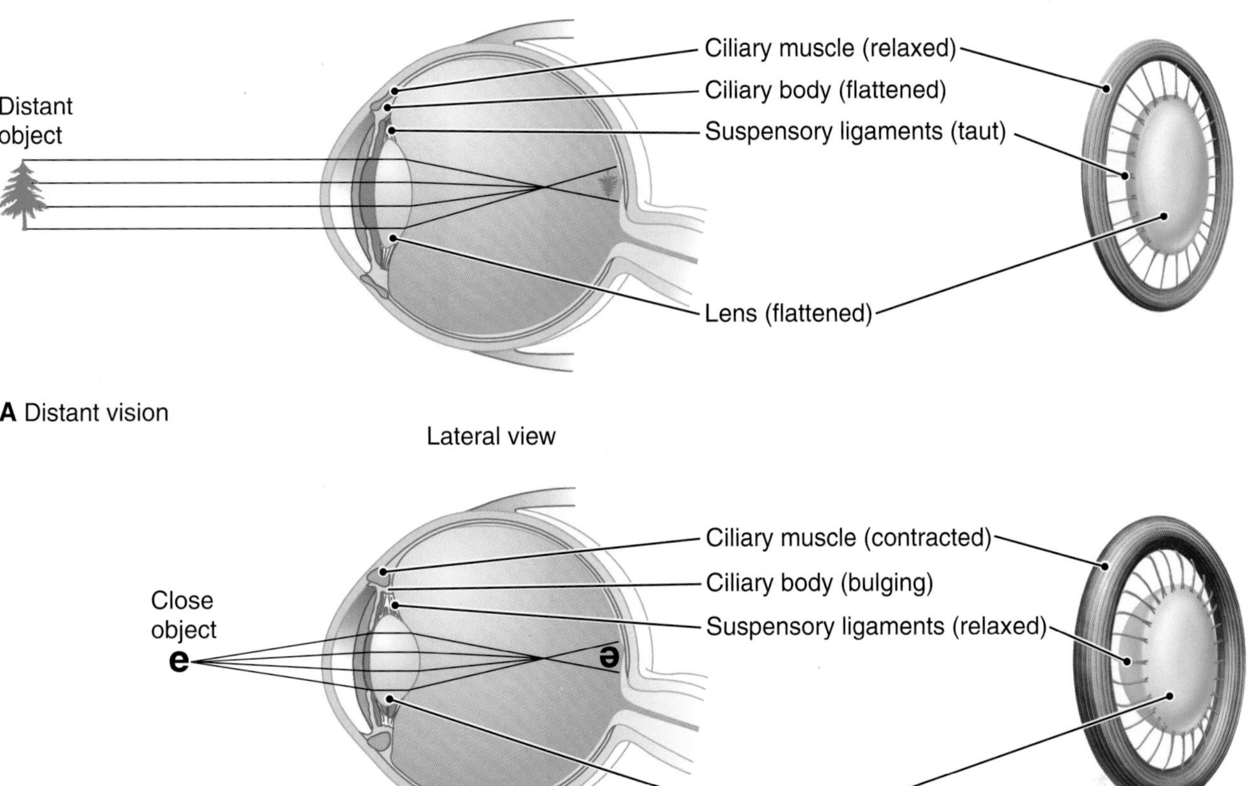

Figure 11-6 **Accommodation. A. Distant vision.** The lens is flattened to lessen refraction when viewing distant objects. **B. Close vision.** The lens is rounded to increase refraction when viewing close objects. **KEY POINT** When viewing a close object, the lens must become more rounded to focus light rays on the retina. **ZOOMING IN** When the lens is rounded, what is the position of the ciliary muscle?

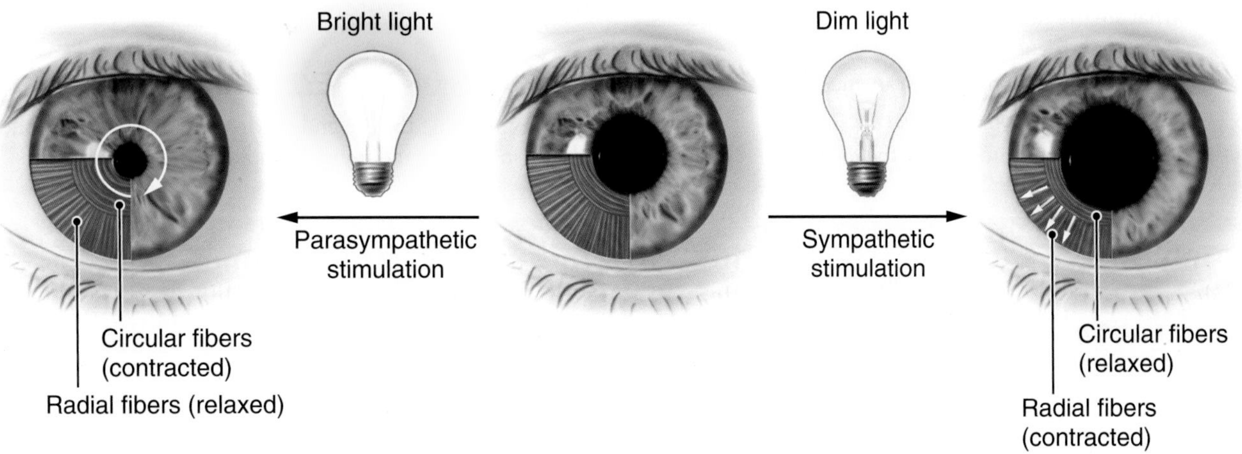

Figure 11-7 **Function of the iris.** **KEY POINT** In bright light, circular muscles contract and constrict the pupil, limiting the light that enters the eye. In dim light, the radial muscles contract and dilate the pupil, allowing more light to enter the eye. **ZOOMING IN** What muscle fibers of the iris contract to make the pupil smaller and to make the pupil larger?

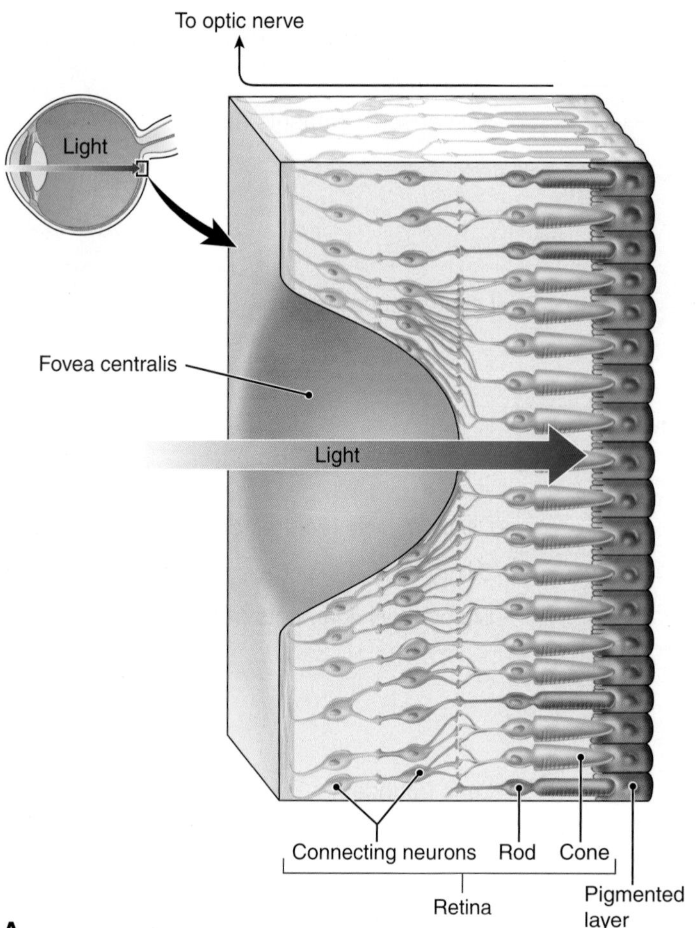

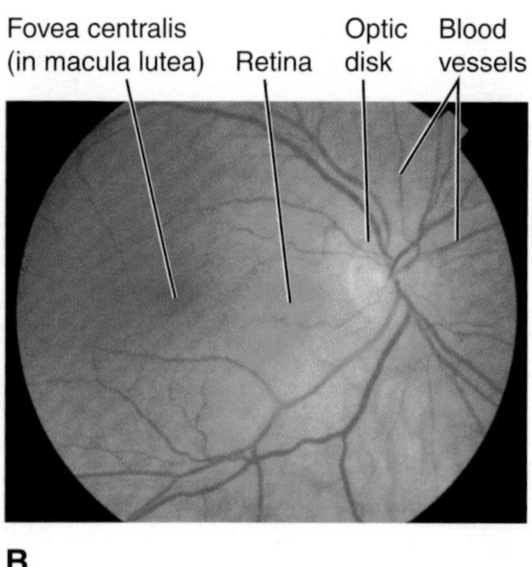

Figure 11-8 **The retina. A. Structure of the retina.** Rods and cones form a deep layer of the retina near the choroid. Connecting neurons carry visual impulses toward the optic nerve. **B. The fundus (back) of the eye as seen through an ophthalmoscope.** An abnormal appearance of the fundus can indicate disease.

CHECKPOINTS

- ☐ **11-8** What are the three tunics of the eyeball?
- ☐ **11-9** What are the structures that refract light as it passes through the eye?
- ☐ **11-10** What is the function of the ciliary muscle?
- ☐ **11-11** What is the function of the iris?
- ☐ **11-12** What are the receptor cells of the retina?

See the Student Resources on thePoint to view the animation "The Retina," which illustrates the structure and function of this receptor.

DISORDERS OF THE EYE AND VISION

Disorders affecting the eye and vision include structural abnormalities, infections and injuries, and degenerative disorders.

Table 11-1 Comparison of the Rods and Cones of the Retina

Characteristic	Rods	Cones
Shape	Cylindrical	Flask shaped
Number	About 120 million in each retina	About 6 million in each retina
Distribution	Toward the periphery (anterior) of the retina	Concentrated at the center of the retina
Stimulus	Dim light	Bright light
Visual acuity (sharpness)	Low	High
Pigments	Rhodopsin	Pigments sensitive to red, green, or blue
Color perception	None; shades of gray	Respond to color

Errors of Refraction **Hyperopia** (hi-per-O-pe-ah), or farsightedness, usually results from an abnormally short eyeball (Fig. 11-9A). In this situation, light rays focus behind the retina because the lens cannot bend them sharply enough to focus them on the retina. The lens can thicken only to a given limit to accommodate for near vision. If the need for refraction exceeds this limit, a person must move an object away from the eye to see it clearly. Glasses with convex lenses that increase light refraction can correct for hyperopia.

Myopia (mi-O-pe-ah), or nearsightedness, is another eye defect related to development. In this case, either the eyeball is too long or the cornea bends the light rays too sharply so that the focal point is in front of the retina (see Fig. 11-9B). Distant objects therefore appear blurred, and only near objects can be seen clearly. A concave lens corrects for myopia by refracting light peripherally and moving the focal point backward. Nearsightedness in a young person becomes worse each year until the person reaches his or her 20s.

Presbyopia (pres-be-O-pe-ah, literally "old eye") results from an age-related loss of elasticity in the lens. As a result of this change, the lens can no longer thicken sufficiently to bring near objects into focus. One can be both myopic and presbyopic, requiring different lenses for close vision and far vision. Bifocal glasses combine different lenses, incorporating a small convex lens for reading within a large concave lens for far vision.

Another common visual defect, **astigmatism** (ah-STIG-mah-tizm), is caused by irregularity in the curvature of the cornea or the lens. As a result, light rays are incorrectly bent, causing blurred vision. Astigmatism is often found in combination with hyperopia or myopia, so a careful eye examination is needed to obtain an accurate prescription for corrective lenses.

Surgical techniques are now available to correct some visual defects and eliminate the need for eyeglasses or contact lenses. Such refractive surgery can be used to correct nearsightedness, farsightedness, and astigmatism. In these procedures, the cornea is reshaped, often using a laser, to enable optimal light refraction.

Strabismus **Strabismus** (strah-BIZ-mus) is a disorder of the extrinsic eye muscles in which the two eyes do not move together. In convergent strabismus, one eye deviates toward the nasal side, or medially. This disorder gives an appearance of being cross-eyed. In divergent strabismus, the affected eye deviates laterally.

Persistent strabismus impacts the interpretation of visual impulses from the affected eye. Eventually, the brain no longer "sees" images from that eye. Loss of vision in a healthy eye because it cannot work properly with the other eye is termed **amblyopia** (am-ble-O-pe-ah). Care by an ophthalmologist as soon as the condition is detected may result in restoration of muscle balance. In some cases, eye exercises, eye glasses, and patching of one eye correct the defect. In other cases, surgery is required to alter muscle action.

11

Infections Inflammation of the conjunctiva is called **conjunctivitis** (kon-junk-tih-VI-tis). It may be acute or chronic and may be caused by a variety of irritants and pathogens. "Pinkeye" is a highly contagious acute conjunctivitis that is usually caused by cocci or bacilli. Sometimes, irritants such as wind and excessive glare cause an inflammation, which may in turn increase susceptibility to bacterial infection.

Inclusion conjunctivitis is an acute eye infection caused by *Chlamydia trachomatis* (klah-MID-e-ah trah-KO-mah-tis). This same organism causes a sexually transmitted reproductive tract infection. Certain strains of chlamydia cause a chronic infection known as **trachoma** (trah-KO-mah).

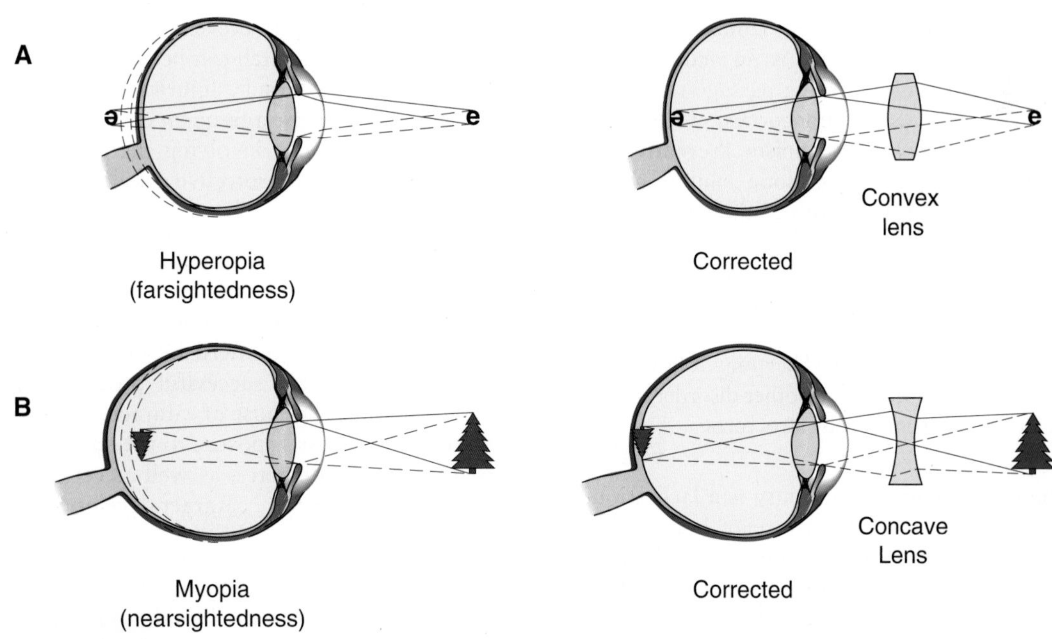

Figure 11-9 **Errors of refraction.** **KEY POINT A.** In hyperopia (farsightedness), an image falls behind the retina. **B.** In myopia (nearsightedness), an image falls in front of the retina. A convex lens corrects for hyperopia; a concave lens corrects for myopia, as shown in the right column of the figure.

Figure 11-10 Visual disorders. A scene as it might be viewed by individuals with various eye disorders. **A.** Normal vision. **B.** Cataract. **C.** Diabetic retinopathy. **D.** Macular degeneration.

Trachoma is endemic in some areas of Africa. If not treated, scarring of the conjunctiva and cornea can cause blindness. The use of antibiotics and proper hygiene to prevent reinfection has reduced the incidence of blindness from this disorder in many parts of the world.

Neonatal conjunctivitis, or ophthalmia neonatorum (of-THAL-me-ah ne-o-na-TO-rum), is an acute infection caused by organisms acquired during passage through the birth canal. The usual cause is gonococcus, chlamydia, or some other sexually transmitted organism. Preventive doses of an appropriate antiseptic or antibiotic solution are routinely administered onto a newborn's conjunctiva just after birth.

Infections of the iris, choroid coat, ciliary body, and other parts of the eyeball are rare but potentially serious. Syphilis spirochetes, tubercle bacilli, and a variety of cocci may cause these painful infections. They may follow a sinus infection, tonsillitis, conjunctivitis, or other disorder in which an infection can spread from nearby structures.

Injuries The most common eye injury is a laceration or scratch of the cornea caused by a foreign body. Injuries caused by foreign objects or by infection may result in corneal scar formation, leaving an area of opacity that light rays cannot penetrate. If such an injury involves the central area anterior to the pupil, blindness may result.

Because the cornea lacks blood vessels, a person can receive a corneal transplant without danger of rejection. Eye banks store corneas obtained from donors, and corneal transplantation is a fairly common procedure.

Severe eye injuries may not be subject to surgical repair. In such cases, a surgeon may perform an operation to remove the eyeball, a procedure called **enucleation** (e-nu-kle-A-shun).

It is important to prevent infection in cases of eye injury. Even a tiny scratch can become so seriously infected that blindness may result. Injuries by pieces of glass and other sharp objects are a frequent cause of eye damage. The incidence of accidents involving the eye has been greatly reduced by the use of protective goggles.

Cataract A **cataract** is an opacity (cloudiness) of the lens or the lens's outer covering. An early cataract causes a gradual loss of visual acuity (sharpness) (**Fig. 11-10B**). An untreated cataract leads to complete vision loss. Surgical removal of the lens followed by implantation of an artificial lens is a highly successful procedure for restoring vision. Although the cause of cataracts is not known, age is a factor, as is excess exposure to the sun's ultraviolet rays. Diseases such as diabetes, as well as certain medications, are known to accelerate cataract development.

Glaucoma **Glaucoma** is a condition characterized by excess pressure of the aqueous humor. This fluid is produced constantly from the blood, and after circulation in the eye, it is reabsorbed into the bloodstream. Interference with this fluid's normal reentry into the bloodstream leads to increased pressure inside the eyeball.

The most common type of glaucoma usually progresses rather slowly, with vague visual disturbances being the only symptom. In most cases, the aqueous humor's high pressure causes destruction of some optic nerve fibers before the person is aware of visual change. Many glaucoma cases are diagnosed by the measurement of pressure in the eye, which should be part of a routine eye examination for people older than 35 years or for those with a family history of glaucoma. Early diagnosis and continuous treatment with medications to reduce pressure usually result in the preservation of vision.

Disorders Involving the Retina Diabetes is a leading cause of blindness among adults in the United States. In **diabetic retinopathy** (ret-in-OP-ah-the), damage to the small blood vessels of the choroid results in bleeding and the death of retinal neurons. Other eye disorders directly related to diabetes include optic atrophy, in which the optic nerve fibers die; cataracts, which occur earlier and with greater frequency among diabetics than among nondiabetics; and retinal detachment.

In cases of **retinal detachment**, the retina separates from the underlying choroid layer as a result of trauma or an accumulation of fluid or tissue between the layers (Fig. 11-11). This disorder may develop slowly or may occur suddenly. If it is left untreated, complete detachment can occur, resulting in blindness. Surgical treatment includes a sort of "spot welding" with an electric current or a weak laser beam. A series of pinpoint scars (connective tissue) develops to reattach the retina.

Macular degeneration is another leading cause of blindness. This name refers to the macula lutea, the yellow area of the retina that contains the fovea centralis. Changes in this area distort the center of the visual field (see Fig. 11-10D). When associated with aging, this deterioration is described as *age-related macular degeneration* (AMD). In the "dry" form of AMD, material accumulates on the retina, causing gradual vision loss. In the "wet" form, abnormal blood vessels grow under the retina, causing it to detach. Laser surgery may stop the growth of these vessels and delay vision loss. More recently, ophthalmologists have had success in delaying the progress of wet AMD with regular intraocular injections of drugs that inhibit blood vessel formation. Factors contributing to macular degeneration are smoking, exposure to sunlight, and a high-cholesterol diet. Some forms are known to be hereditary.

Box 11-1 provides information on new methods of treating eye disorders.

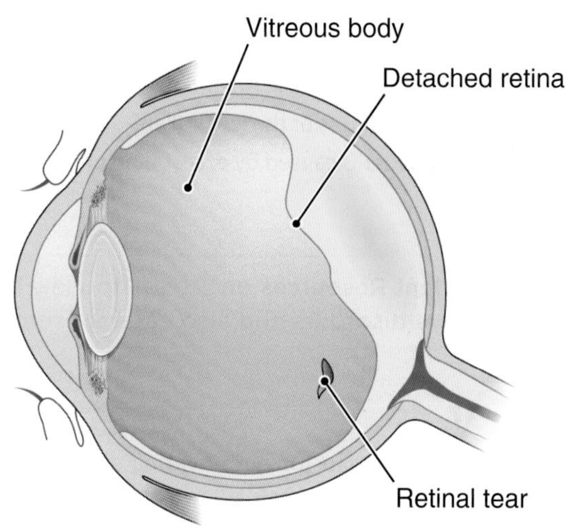

Figure 11-11 Retinal detachment. The retina separates from the underlying choroid.

11

Box 11-1

HOT TOPICS

Eye Surgery: A Glimpse of the Cutting Edge

Cataracts, glaucoma, and refractive errors are the most common eye disorders affecting Americans. In the past, cataract and glaucoma treatments concentrated on managing the diseases. Refractive errors were corrected using eyeglasses and, more recently, contact lenses. Today, laser and microsurgical techniques can remove cataracts, reduce glaucoma, and allow people with refractive errors to put their eyeglasses and contacts away. These cutting-edge procedures include the following:

- *Laser in situ keratomileusis (LASIK)* to correct refractive errors. During this procedure, a surgeon uses a laser to reshape the cornea so that it refracts light directly onto the retina, rather than in front of or behind it. A microkeratome (surgical knife) is used to cut a flap in the cornea's outer layer. A computer-controlled laser sculpts the middle layer of the cornea, and then the flap is replaced. The procedure takes only a few minutes, and patients recover their vision quickly and usually with little postoperative pain.
- *Laser trabeculoplasty* to treat glaucoma. This procedure uses a laser to help drain fluid from the eye and lower intraocular pressure. The laser is aimed at drainage canals located between the cornea and iris and makes several burns that are believed to open the canals and allow fluid to drain better. The procedure is typically painless and takes only a few minutes.
- *Phacoemulsification* to remove cataracts. During this surgical procedure, a very small incision (approximately 3 mm long) is made through the sclera near the cornea's outer edge. An ultrasonic probe is inserted through this opening and into the center of the lens. The probe uses sound waves to emulsify the lens's central core, which is then suctioned out. Then, an artificial lens is permanently implanted in the lens capsule. The procedure is typically painless, although the patient may feel some discomfort for one to two days afterward.

CHECKPOINTS

- [ ] **11-13** What are four errors of refraction?
- [ ] **11-14** What is cloudiness of the lens called?
- [ ] **11-15** What disorder is caused by excess fluid pressure in the eye?

See the Student Resources on thePoint to view illustrations of eye disorders and also cataract surgery.

The Ear

The ear is the sense organ for both hearing and equilibrium (**Fig. 11-12**). It is divided into three main sections:

- The **outer ear** includes an outer projection and a canal ending at a membrane.
- The **middle ear** is an air space containing three small bones.
- The **inner ear** is the most complex and contains the sensory receptors for hearing and equilibrium.

THE OUTER EAR

The external portion of the ear consists of a visible projecting portion, the **pinna** (PIN-nah), also called the *auricle* (AW-rih-kl), and the **external auditory canal**, or *meatus* (me-A-tus), that leads into the ear's deeper parts. The pinna directs sound waves into the ear, but it is probably of little importance in humans. The external auditory canal extends medially from the pinna for about 2.5 cm or more, depending on which wall of the canal is measured. The skin lining this tube is thin and, in the first part of the canal, contains many wax-producing **ceruminous** (seh-RU-mih-nus) **glands**. The wax, or **cerumen** (seh-RU-men), may become dried and impacted in the canal and must then be removed. The same kinds of disorders that involve the skin elsewhere—atopic dermatitis, boils, and other infections—may also affect the skin of the external auditory canal.

The **tympanic** (tim-PAN-ik) **membrane**, or *eardrum*, is at the end of the external auditory canal and separates this canal from the middle ear cavity. The tympanic membrane vibrates freely when struck with sound waves that enter the ear.

THE MIDDLE EAR AND OSSICLES

The middle ear cavity is a small, flattened space that contains three small bones, or **ossicles** (OS-ih-klz) (**see Fig. 11-12**). The three ossicles are joined in such a way that they amplify the sound waves received by the tympanic membrane as they transmit the sounds to the inner ear. The first bone is shaped like a hammer (or mallet) and is called the **malleus**

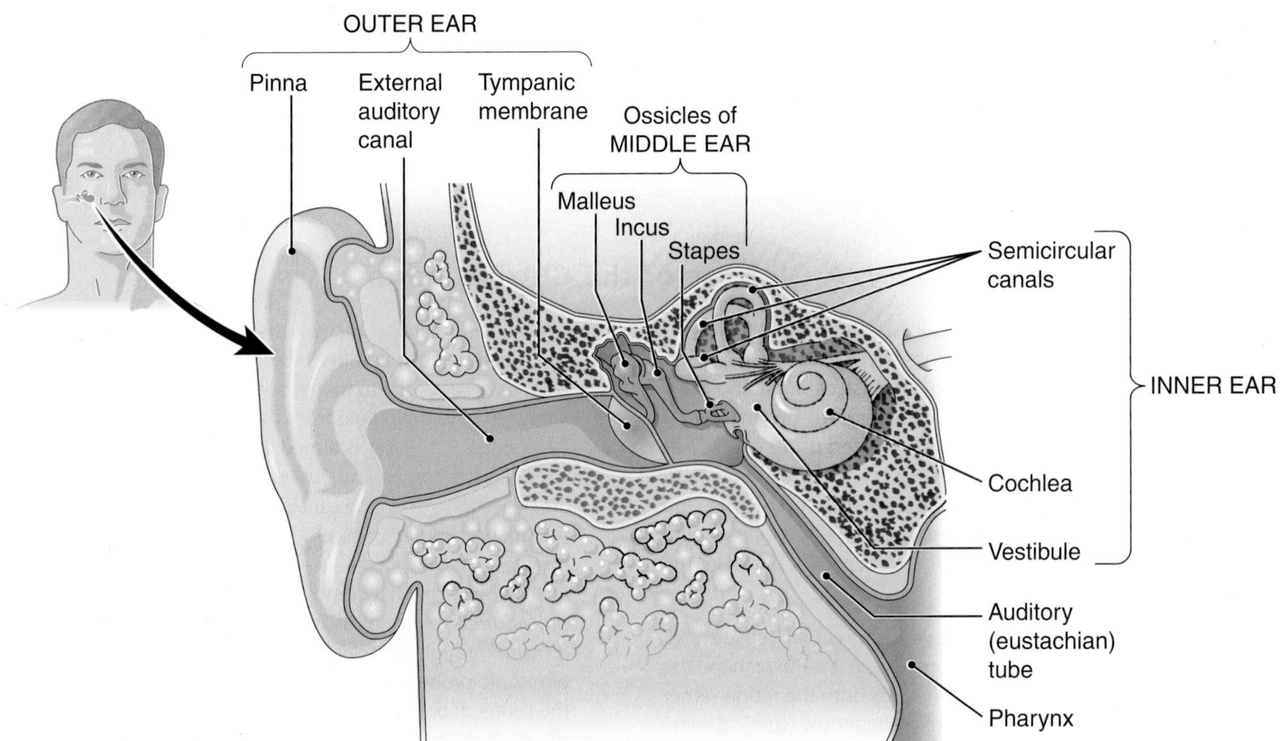

Figure 11-12 The ear. **KEY POINT** Structures in the outer, middle, and inner divisions are shown. **ZOOMING IN** What structure separates the outer ear from the middle ear?

(MAL-e-us). The handle-like part of the malleus is attached to the tympanic membrane, whereas the headlike part is connected to the second bone, the **incus** (ING-kus). The incus is shaped like an anvil, an iron block used by blacksmiths to shape metal. The innermost ossicle is shaped somewhat like the stirrup of a saddle and is called the **stapes** (STA-peze), which is Latin for *stirrup*. The base of the stapes is in contact with the inner ear.

The **auditory tube**, also called the *eustachian* (u-STA-shun) *tube*, connects the middle ear cavity with the throat, or pharynx (FAR-inks) (**see Fig. 11-12**). This tube opens to allow pressure to equalize on the two sides of the tympanic membrane. A valve that closes the tube can be forced open by swallowing hard, yawning, or blowing with the nose and mouth sealed, as a person often does when experiencing pain from pressure changes in an airplane.

The mucous membrane of the pharynx is continuous through the auditory tube into the middle ear cavity. The posterior wall of the middle ear cavity contains an opening into the mastoid air cells, which are spaces inside the temporal bone's mastoid process (**see Fig. 7-5B**).

THE INNER EAR

The ear's most complicated and important part is the internal portion, which is described as a *labyrinth* (LAB-ih-rinth) because it has a complex, mazelike construction (**Fig. 11-13**). The outer shell of the inner ear is composed of hollow bone comprising the **bony labyrinth.** This outer portion is filled with a fluid called **perilymph** (PER-e-limf).

Within the bony labyrinth is an exact replica of this bony shell made of membrane, much like an inner tube within a tire. The tubes and chambers of this **membranous labyrinth** are filled with a fluid called **endolymph** (EN-do-limf) (**see Fig. 11-13A**). Thus, the endolymph is within the membranous labyrinth, and the perilymph surrounds the membranous labyrinth. These fluids are important to the sensory functions of the inner ear. The inner ear has three divisions:

- The **vestibule** consists of two chambers (the utricle and the saccule) that contain some of the receptors for equilibrium.
- The **semicircular canals** are three projecting tubes located toward the posterior. The area at the base of each semicircular canal contains receptors for equilibrium.
- The **cochlea** (KOK-le-ah) is coiled like a snail shell (*cochlea* is Latin for "snail"). It contains the receptors for hearing.

HEARING

Within the cochlea, the membranous labyrinth is known as the **cochlear duct** (**Fig. 11-13A and B**). It bisects the bony labyrinth into a superior portion, the **vestibular duct**, and an inferior portion, the **tympanic duct. Hair cells**, the receptors for hearing, sit on the lower membrane of the cochlear duct. The long strip of hair cells is also known as the **spiral organ** (organ of Corti [KOR-te]). The tips of the hair cells are embedded in a gelatinous membrane called the **tectorial membrane.** (The membrane is named from a Latin word that means "roof.")

The numbers below match the numbers in **Figure 11-14.** The cochlea is pictured as unrolled to more easily show how sound waves progress through the inner ear chambers. The steps in hearing are as follows:

1. Sound waves first enter the external auditory canal. The pitch (high or low) and amplitude (loud or soft) depend on the characteristic of the wave.
2. The sound waves set up vibrations in the tympanic membrane. The ossicles amplify these vibrations, which can be large (loud) or small (soft), and fast (high pitch) or slow (low pitch).
3. The stapes transmits the vibrations to the **oval window** of the inner ear. This membrane then transmits the sound waves to the perilymph within the vestibular duct.
4. This fluid wave in the vestibular duct sets up vibrations in the membranes of the cochlear duct. High-pitched sounds cause the proximal portions of the cochlear membranes to vibrate, whereas low-pitched sounds initiate vibrations in the distal portion.
5. The cochlear duct vibrations initiate a second fluid wave in the tympanic duct. This wave dissipates when it hits the round window.
6. Recall that the hair cells perch upon the lower cochlear duct membrane with their tips (cilia) embedded in the tectorial membrane. Vibrations in these membranes move the cilia back and forth.
7. Cilia movement produces an electrical signal in the hair cell.
8. This motion sets up nerve impulses that travel to the brain in the **vestibulocochlear nerve**, a branch of the eighth cranial nerve (formerly called the *auditory* or *acoustic nerve*).

Hearing receptors respond to both the pitch (tone) of sound and its intensity (loudness). Loud sounds stimulate more cells and produce more vibrations, sending more nerve impulses to the brain. Exposure to loud noises, such as very loud music, jet plane noise, or industrial noises, can damage the receptors for particular pitches of sound and lead to hearing loss for those tones.

EQUILIBRIUM

The other sensory receptors in the inner ear are those related to equilibrium (balance). They are located in the vestibule and the semicircular canals. Receptors for the sense of equilibrium respond to acceleration and like the hearing receptors, are ciliated hair cells. As the head moves, a shift in the position of the cilia within a thick material around them generates a nerve impulse.

Receptors located in the vestibule's two small chambers sense the position of the head relative to the force of gravity

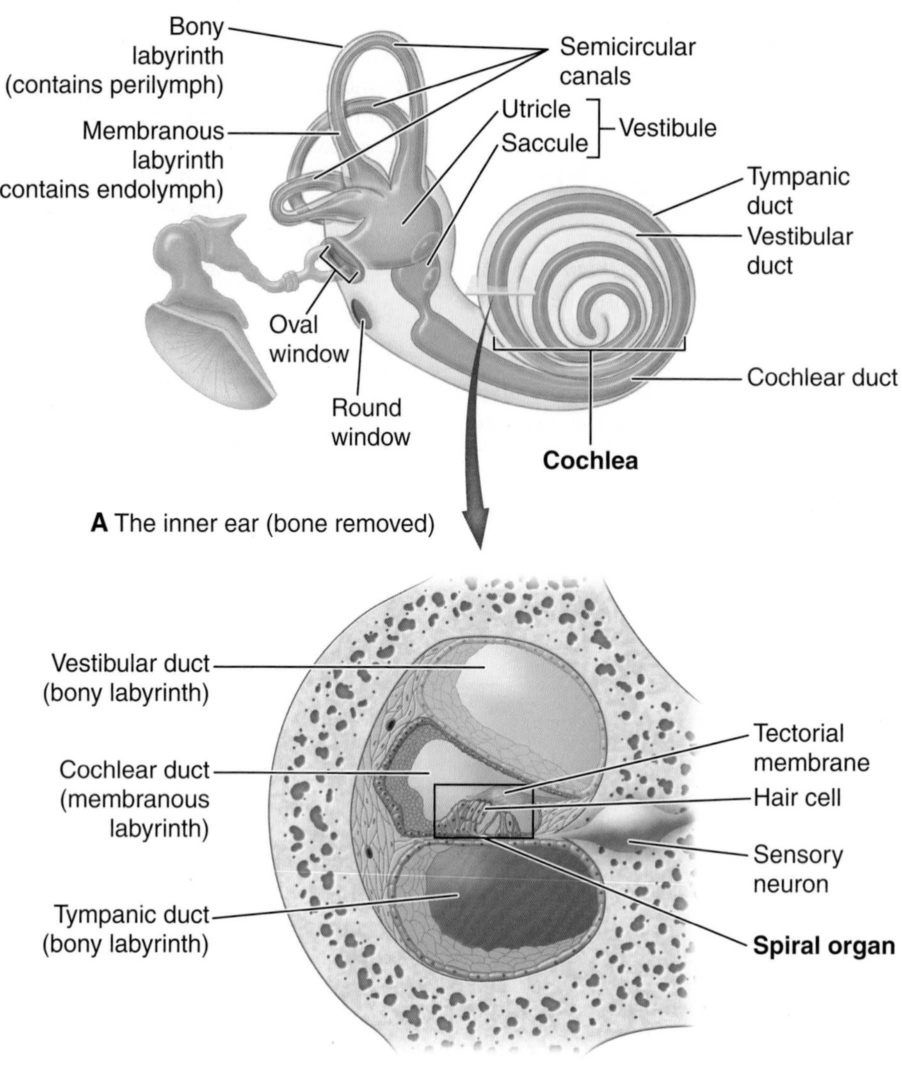

Figure 11-13 The inner ear. **KEY POINT A.** The inner ear. The vestibule, semicircular canals, and cochlea are made of a bony shell, described as a bony labyrinth, with an interior membranous labyrinth. Endolymph fills the membranous labyrinth, and perilymph surrounds it in the bony labyrinth. The cochlea is the organ of hearing. The semicircular canals and vestibule are concerned with equilibrium. **B. Cross-section of the cochlea.** **ZOOMING IN** What type of fluid is in contact with the vestibular membrane?

and also respond to acceleration in a straight line, as in a forward-moving vehicle or an elevator. Each receptor is called a **macula.** (There is also a macula in the eye, but *macula* is a general term that means "spot.") The macular hair cells are embedded in a gelatinous material, the **otolithic** (o-to-LITH-ik) **membrane**, which is weighted with small crystals of calcium carbonate, called **otoliths** (O-to-liths). The force of gravity pulls the membrane downward, which bends the cilia of the hair cells, generating a nerve impulse (**Fig. 11-15**). In linear acceleration, the otolithic membrane lags behind the forward motion, bending the cilia in a direction opposite to the direction of acceleration. Picture sweeping mud off a garden path with a broom. As you move forward, the thick mud is dragging the broom straws in the opposite direction.

The receptors for detecting rotation, such as when you shake your head or twirl in a circle, are located at the bases of the semicircular canals (**Fig. 11-16**). These receptors, called **cristae** (KRIS-te), are hair cells embedded in a gelatinous material called the cupula (KU-pu-lah). As with the maculae, when the head moves, the cupula lags behind a

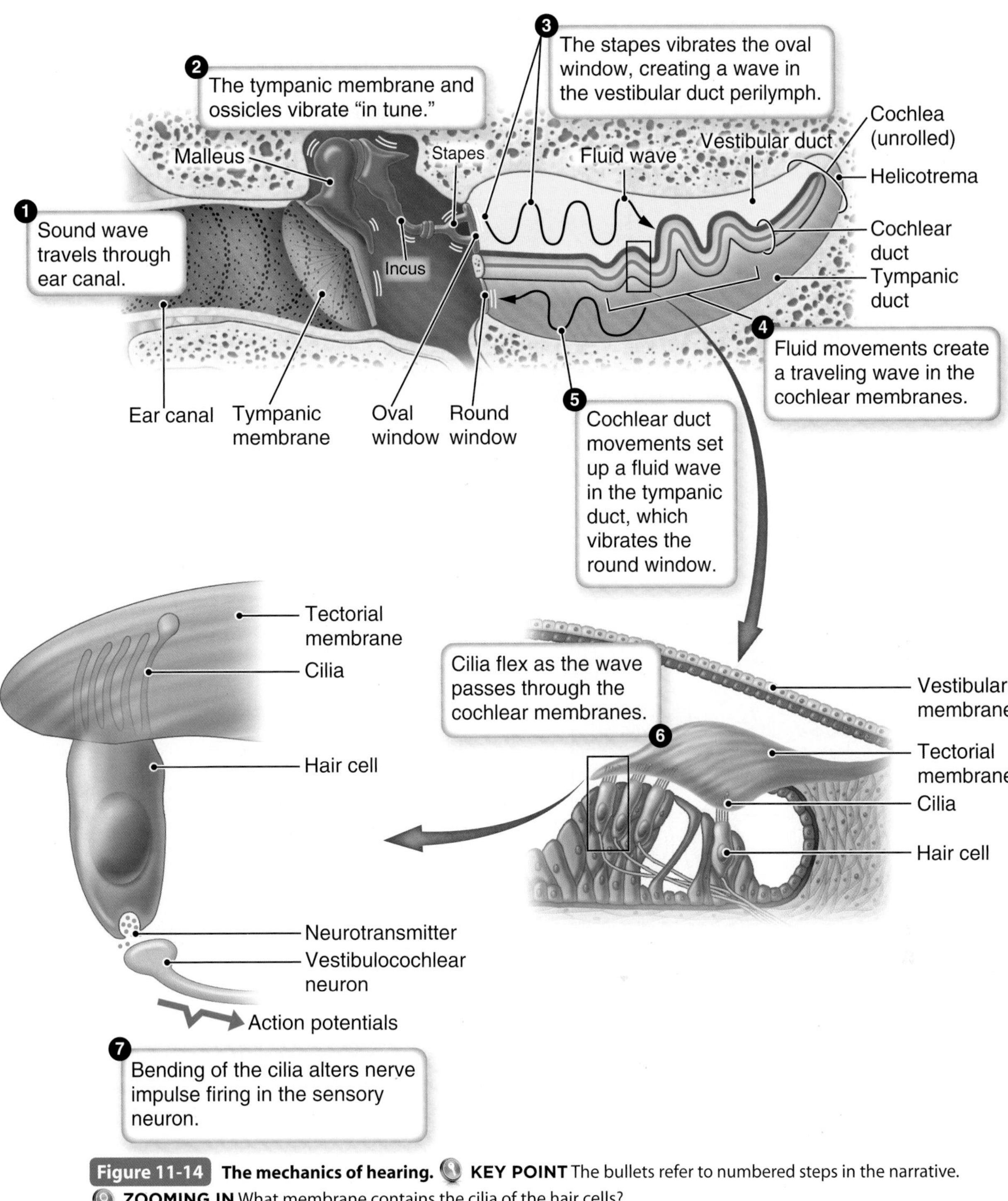

1 Sound wave travels through ear canal.
2 The tympanic membrane and ossicles vibrate "in tune."
3 The stapes vibrates the oval window, creating a wave in the vestibular duct perilymph.
Malleus
Stapes
Incus
Fluid wave
Vestibular duct
Cochlea (unrolled)
Helicotrema
Cochlear duct
Tympanic duct
4 Fluid movements create a traveling wave in the cochlear membranes.
Ear canal
Tympanic membrane
Oval window
Round window
5 Cochlear duct movements set up a fluid wave in the tympanic duct, which vibrates the round window.
Tectorial membrane
Cilia
Hair cell
Neurotransmitter
Vestibulocochlear neuron
Action potentials
Cilia flex as the wave passes through the cochlear membranes.
6
Vestibular membrane
Tectorial membrane
Cilia
Hair cell
7 Bending of the cilia alters nerve impulse firing in the sensory neuron.

Figure 11-14 **The mechanics of hearing.** **KEY POINT** The bullets refer to numbered steps in the narrative. **ZOOMING IN** What membrane contains the cilia of the hair cells?

bit, bending the cilia in the opposite direction. It's easy to remember what these receptors do, because the semicircular canals go off in three different directions. The crista in the horizontal canal responds to horizontal rotation, as in a dancer's spin; the one in the superior canal responds to forward and backward rotation, as in somersaulting; the one in the posterior canal responds to left–right rotation, as in doing a cartwheel.

Nerve fibers from the vestibule and from the semicircular canals form the **vestibular** (ves-TIB-u-lar) **nerve,** which joins the cochlear nerve to form the vestibulocochlear nerve, the eighth cranial nerve (**see Fig. 11-16**).

11

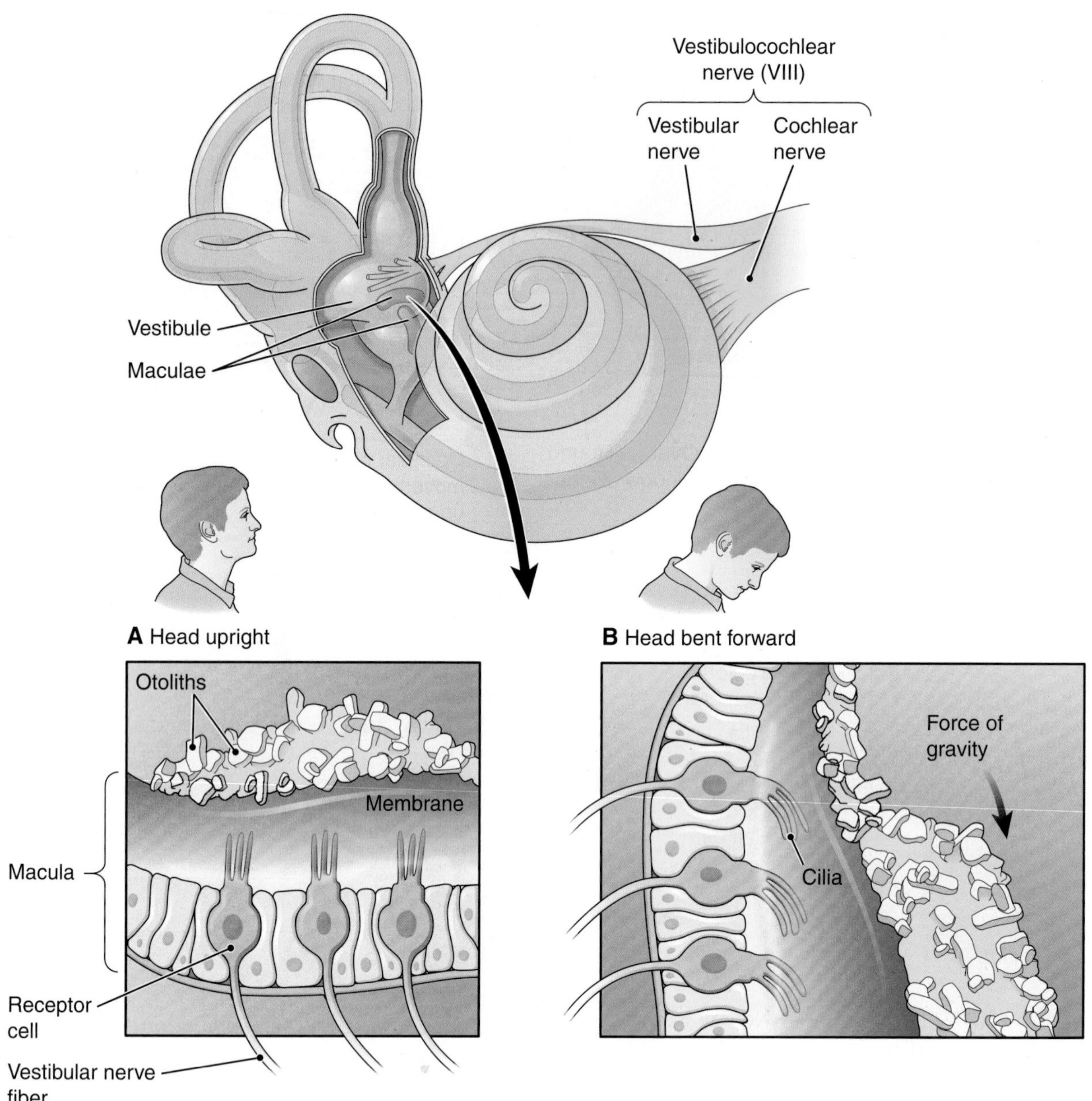

Figure 11-15 Action of the vestibular equilibrium receptors (maculae). **KEY POINT** As the head moves, the otolithic membrane, weighted with otoliths, pulls on the receptor cells' cilia, generating a nerve impulse. These receptors also function in linear acceleration. **ZOOMING IN** What happens to the cilia of the macular cells when the otolithic membrane moves?

CHECKPOINTS

- ☐ **11-16** What are the three divisions of the ear?
- ☐ **11-17** What are the names of the ear ossicles, and what do they do?
- ☐ **11-18** What are the two fluids found in the inner ear, and where are they located?
- ☐ **11-19** What is the name of the hearing organ, and where is it located?
- ☐ **11-20** Where are the receptors for equilibrium located?

DISORDERS OF THE EAR

Disorders involving the ear most commonly include infection, hearing loss, and disturbances of equilibrium.

Ear Infection Infection and inflammation of the middle ear cavity, **otitis media** (o-TI-tis ME-de-ah), is relatively common. A variety of bacteria and viruses may cause otitis media, and it is a frequent complication of the common cold, influenza, and other infections, especially those of the pharynx. Pathogens are transmitted from the pharynx

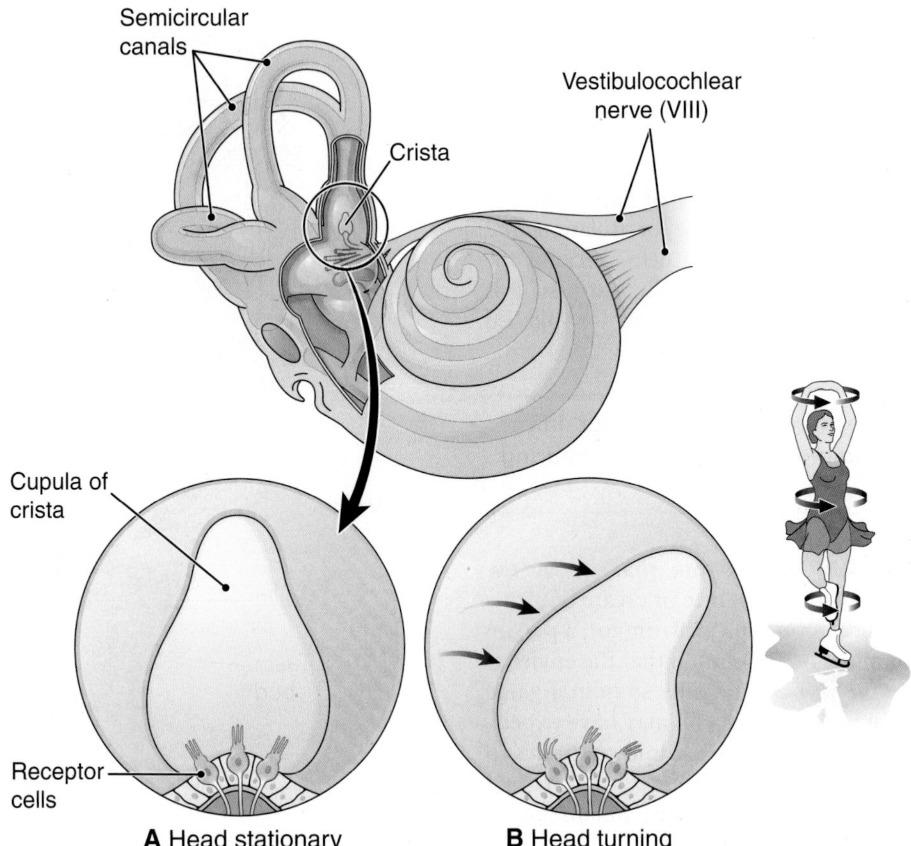

Figure 11-16 Action of the equilibrium receptors (cristae) in the semicircular canals. KEY POINT As the body spins or moves in different directions, the receptor cells' cilia bend, generating nerve impulses.

11

to the middle ear most often in children, partly because the auditory tube is relatively short and horizontal in the child; in the adult, the tube is longer and tends to slant down toward the pharynx. Antibiotics have reduced complications and have caused a marked reduction in the amount of surgery done to drain middle ear infections. In some cases, however, pressure from pus or exudate in the middle ear can be relieved only by cutting the tympanic membrane, a procedure called a **myringotomy** (mir-in-GOT-o-me). Placement of a **tympanostomy** (tim-pan-OS-to-me) **tube** in the eardrum allows pressure to equalize and prevents further damage to the eardrum and additional ear infections.

Otitis externa is inflammation of the external auditory canal. Infections in this area may be caused by a fungus or bacterium. They are most common among those living in hot climates and among swimmers, leading to the alternate name "swimmer's ear."

Hearing Loss Another disorder of the ear is hearing loss, which may be partial or complete. When the loss is complete, the condition is called **deafness**. The two main types of hearing loss are conductive hearing loss and sensorineural hearing loss.

Conductive hearing loss results from interference with the passage of sound waves from the outside to the inner ear. In this condition, wax or a foreign body may obstruct the external canal. Blockage of the auditory tube prevents the equalization of air pressure on both sides of the tympanic membrane, thereby decreasing the membrane's ability to vibrate. Another cause of conductive hearing loss is damage to the tympanic membrane or ossicles resulting from chronic otitis media or from **otosclerosis** (o-to-skle-RO-sis), a hereditary bone disorder that prevents normal vibration of the stapes. Surgical removal of the diseased stapes and its replacement with an artificial device allows sound conduction from the ossicles to the cochlea.

Sensorineural hearing loss may involve the cochlea, the vestibulocochlear nerve, or the brain areas concerned with hearing. It may result from prolonged exposure to loud noises and has become increasingly common in young people who listen to music through headphones. This is not only because these devices preserve excellent sound quality even at high volumes but also because their small size and wearability allows people to listen to them for long periods of time. Sensorineural hearing loss may also be caused by the long-term use of certain drugs or exposure to various infections and toxins. People with severe hearing loss that originates in the inner ear may benefit from a cochlear implant. This prosthetic device stimulates the cochlear nerve directly, bypassing the receptor cells, and it may restore hearing for medium to loud sounds. A cochlear implant is described in the opening case study.

Presbycusis (pres-be-KU-sis) is age-related hearing loss, usually reflecting the progressive atrophy of hair cells. The affected person may experience a sense of isolation and depression and may need psychological help. Because the ability to hear high-pitched sounds is usually lost first, it is important to address elderly people in clear, low-pitched tones.

See the Student Resources on thePoint for information on how audiologists help treat hearing disorders.

Equilibrium Disturbances Factors that affect the equilibrium receptors can affect balance or create unpleasant sensations of motion. In **vertigo** (VER-tih-go), a person has a sensation of spinning or a sensation that the environment is spinning. If you have ever rapidly spun in a game or ridden on a twirling carnival ride, you may have experienced this sensation when you stopped. Fluid continues to move in the semicircular canals for a few seconds, creating a spinning sensation. Conditions that cause inflammation of the inner ear or displacement of the otoliths may cause vertigo, and the term is loosely used to mean dizziness or light-headedness. Other causes of vertigo include alcohol and other drugs, low blood pressure, and CNS lesions.

CHECKPOINTS

- ☐ 11-21 What are the two types of hearing loss?
- ☐ 11-22 What is the term for an abnormal sensation of spinning?

Other Special Sense Organs

The sense organs of taste and smell respond to chemical stimuli.

SENSE OF TASTE

The sense of taste, or **gustation** (gus-TA-shun), involves receptors in the tongue and two different nerves that carry taste impulses to the brain (**Fig. 11-17**). The gustatory sensory organs, known as **taste buds**, are located mainly on the superior surface of the tongue. They are enclosed in raised projections called **papillae** (pah-PIL-e), which give the tongue's surface a rough texture and help manipulate food when chewing. Taste buds are stimulated only if the substance to be tasted is in solution or dissolves in the fluids of the mouth. Within each taste bud, modified epithelial cells (the gustatory cells) respond to one of five basic tastes:

- **Sweet** receptors respond to simple sugars.
- **Salty** receptors respond to sodium.

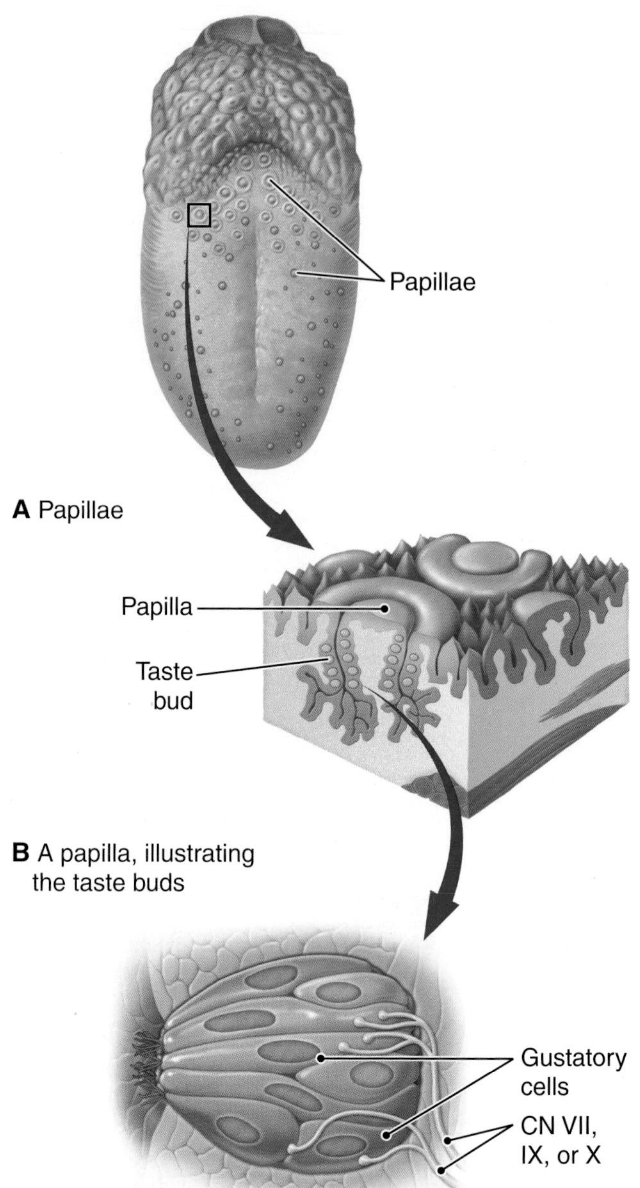

Figure 11-17 **Taste. A.** Papillae are small bumps on the tongue. **B.** A papilla containing taste buds. **C.** Anatomy of a taste bud. **KEY POINT** Gustatory cells in the taste bud respond to chemicals dissolved in saliva.

- **Sour** receptors detect hydrogen ions.
- **Bitter** receptors respond to various organic compounds.
- **Umami** (u-MOM-e) is a pungent or savory taste based on a response to the amino acids glutamate and aspartate, which add to the meaty taste of protein. Glutamate is found in MSG (monosodium glutamate), a flavor enhancer used in some processed foods and some restaurants.

Some investigators consider spiciness to be a sixth taste, but the chemicals involved (such as capsaicin) activate pain/touch receptors, namely the trigeminal nerve, rather than specialized gustatory cells. Other tastes are a combination of these five with additional smell sensations. The nerves of taste include the facial and the glossopharyngeal cranial nerves (VII and IX) (**see Fig. 11-17**). The interpretation of taste impulses is probably accomplished by the brain's lower frontal cortex, although there may be no sharply separate gustatory center.

SENSE OF SMELL

The importance of the sense of smell, or **olfaction** (ol-FAK-shun), is often underestimated. This sense helps detect gases and other harmful substances in the environment and helps warn of spoiled food. Smells can trigger memories and other psychological responses. Smell is also important in sexual behavior.

The olfactory receptor cells are neurons embedded in the epithelium of the nasal cavity's superior region (**see Fig. 11-18**). These neurons extend dendrites into the nasal cavity that interact with smell chemicals (odorants). Again, the chemicals detected must be dissolved in the mucus that lines the nose. Because these receptors are high in the nasal cavity, you must "sniff" to bring odors upward in your nose.

The axons of the olfactory receptor cells pass through the ethmoid bone to synapse with other neurons in the *olfactory bulb*, the enlarged ending of the olfactory nerve (cranial nerve I). The olfactory nerve carries smell impulses directly to the olfactory center in the brain's temporal cortex as well as to the limbic system.

The interpretation of smell is closely related to the sense of taste, but a greater variety of dissolved chemicals can be detected by smell than by taste. We have hundreds of different types of odor receptors; **Figure 11-18** illustrates two types. Different odors can also activate specific combinations of receptors so that we can detect over 10,000 different smells. The smell of foods is just as important in stimulating appetite

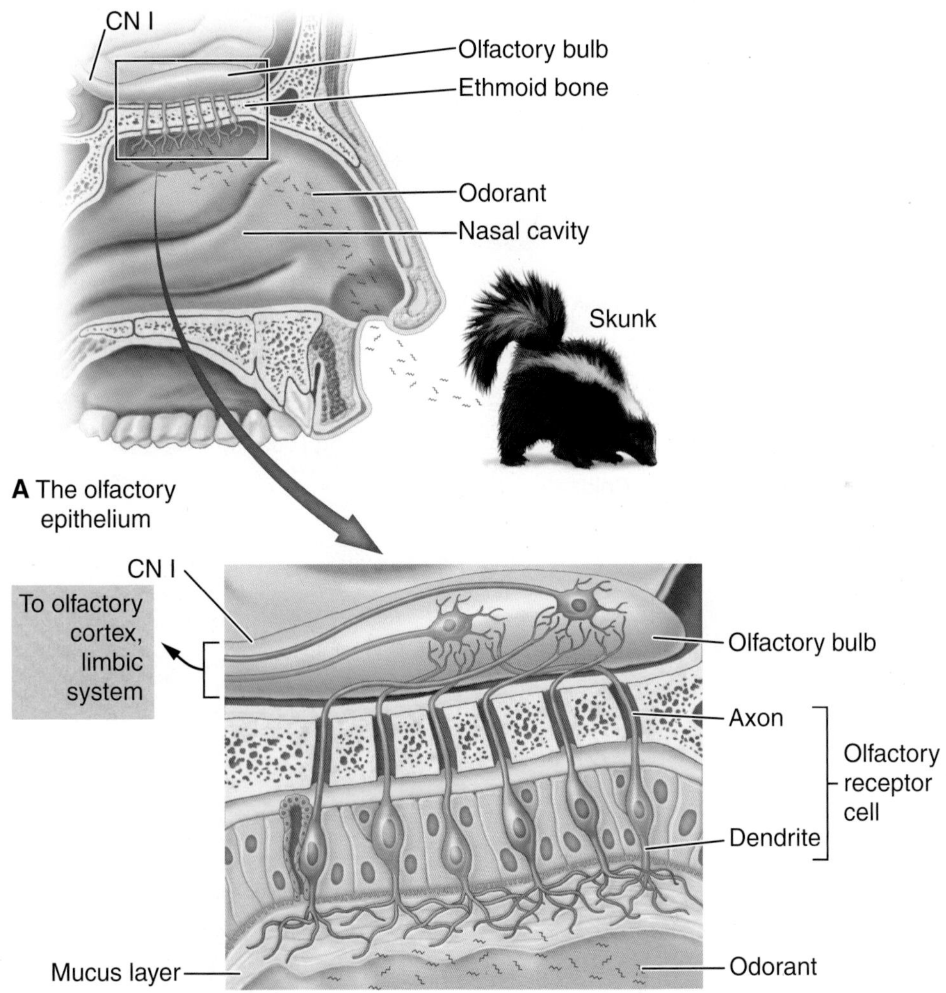

Figure 11-18 **Smell. A.** Olfactory cells in the superior portion of the nasal cavity detect smells. **B.** Olfactory receptor cells detect an odorant and convey a signal to a neuron in the olfactory bulb.

ZOOMING IN What part of an olfactory receptor cell interacts with an odorant?

and the flow of digestive juices as is the sense of taste. When you have a cold, food often seems tasteless and unappetizing because nasal congestion reduces your ability to smell the food.

The olfactory receptors deteriorate with age, and food may become less appealing. It is important when presenting food to elderly people that the food look inviting so as to stimulate their appetites.

CHECKPOINT

- [ ] **11-23** What are the special senses that respond to chemical stimuli?

The General Senses

Unlike the special sensory receptors, which are localized within specific sense organs and are limited to a relatively small area, the general sensory receptors are scattered throughout the body. These include receptors for touch, pressure, temperature, position, and pain (**Fig. 11-19**).

SENSE OF TOUCH

The touch receptors, **tactile** (TAK-til) **corpuscles**, are found mostly in the dermis of the skin and around hair follicles. Touch sensitivity varies with the number of touch receptors in different areas. They are especially numerous and close together in the tips of the fingers and the toes. The lips and the tip of the tongue also contain many of these receptors and are very sensitive to touch. Other areas, such as the back of the hand and the back of the neck, have fewer receptors and are less sensitive to touch.

The sensation of tickle is related to the sense of touch but is still something of a mystery. Tickle receptors are free nerve endings associated with the tactile mechanoreceptors. No one knows the value of tickling, but it may be a form of social interaction. Oddly, we experience tickling only when touched by someone else. Apparently, the brain inhibits these sensations when you are trying to tickle yourself and know the tickling site, eliminating the element of surprise.

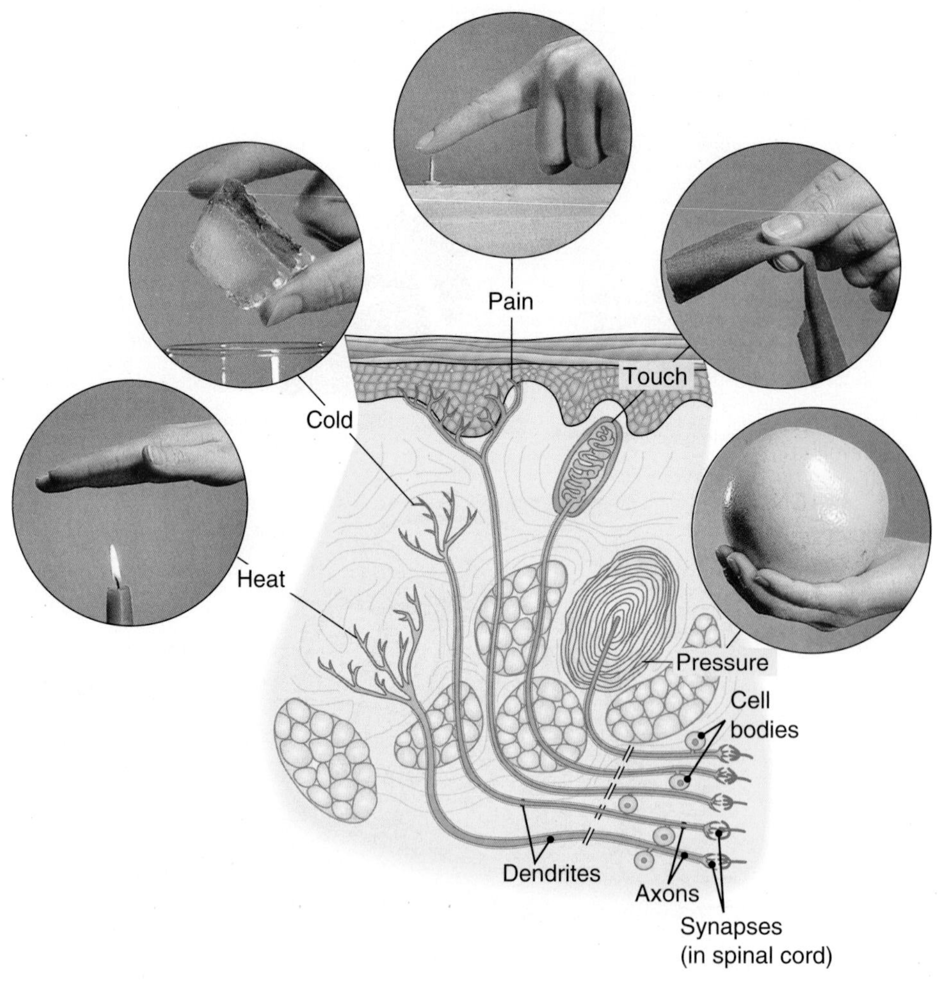

Figure 11-19 Sensory receptors in the skin. **KEY POINT** The skin has a variety of sensory receptors. Synapses with interneurons are in the spinal cord.

SENSE OF PRESSURE

Even when the skin is anesthetized, it can still respond to pressure stimuli. These sensory receptors for deep pressure are located in the subcutaneous tissues beneath the skin and also near joints, muscles, and other deep tissues. They are sometimes referred to as *receptors for deep touch*.

SENSE OF TEMPERATURE

The temperature receptors are **free nerve endings**, receptors that are not enclosed in capsules but are simply branchings of nerve fibers. Temperature receptors are widely distributed in the skin, and there are separate receptors for heat and cold. A warm object stimulates only the heat receptors, and a cool object affects only the cold receptors. Internally, there are temperature receptors in the brain's hypothalamus, which help adjust body temperature according to the temperature of the circulating blood.

SENSE OF POSITION

Receptors located in muscles, tendons, and joints relay impulses that aid in judging body position and relative changes in the locations of body parts. They also inform the brain of the amount of muscle contraction and tendon tension. These rather widespread receptors, known as **proprioceptors** (pro-pre-o-SEP-tors), are aided in this function by the internal ear's equilibrium receptors. The term **kinesthesia** (kin-es-THE-ze-ah) is sometimes used to describe dynamic, or movement-associated, aspects of proprioception.

Information received by proprioceptors is needed for muscle coordination and is important in such activities as walking, running, and many more complicated skills, such as playing a musical instrument. Proprioceptors play an important part in maintaining muscle tone and good posture. They also help assess the weight of an object to be lifted so that the right amount of muscle force is used.

The nerve fibers that carry impulses from these receptors enter the spinal cord and ascend to the brain in the posterior part of the cord. The cerebellum is a main coordinating center for these impulses.

SENSE OF PAIN

Pain is the most important protective sense. The receptors for pain are widely distributed free nerve endings. They are found in the skin, muscles, and joints and to a lesser extent in most internal organs (including the blood vessels and viscera). Two pathways transmit pain to the CNS. One is for acute, sharp pain, and the other is for slow, chronic pain. Thus, a single strong stimulus can produce an immediate sharp pain, followed in a second or so by a slow, diffuse pain that increases in severity with time.

Referred Pain Sometimes, pain that originates in an internal organ is experienced as coming from a more superficial part of the body, particularly the skin. This phenomenon is known as *referred pain*. Liver and gallbladder disease often cause referred pain in the skin over the right shoulder. Spasm of the coronary arteries that supply the heart may cause pain in the left shoulder and arm. Infection of the appendix can be felt as pain of the skin covering the lower right abdominal quadrant.

Apparently, some interneurons in the spinal cord have the twofold duty of conducting impulses from visceral pain receptors in the chest and abdomen and from somatic pain receptors in neighboring areas of the skin, resulting in referred pain. The brain cannot differentiate between these two possible sources, but because most pain sensations originate in the skin, the brain automatically assigns the pain to this more likely place of origin. Knowing where visceral pain is referred to in the body is of great value in diagnosing chest and abdominal disorders.

Itch Itch receptors are free nerve endings that may be specific for that sensation or may share pathways with other receptors, such as those for pain. There are multiple causes for itching, including skin disorders, allergies, kidney disease, infection, and a host of chemicals. Usually, itching is a mild, short-lived annoyance, but for some, it can be chronic and debilitating. No one knows why scratching helps alleviate itch. It may replace the itch sensation with pain or send signals to the brain to relieve the sensation.

Methods of Pain Relief Sometimes, the cause of pain cannot be remedied quickly, and occasionally, it cannot be remedied at all. In the latter case, it is desirable to lessen the pain as much as possible. Some pain relief methods that have been found to be effective include:

- **Analgesic drugs**. An analgesic (an-al-JE-zik) is a drug that relieves pain. There are two main categories of such agents:
  - **Nonnarcotic analgesics** are effective for mild to moderate pain. Most of these drugs reduce inflammation and are commonly known as nonsteroidal antiinflammatory drugs (NSAIDs). Examples are aspirin, ibuprofen (i-bu-PRO-fen), and naproxen (na-PROK-sen). Acetaminophen (e.g., Tylenol) and related compounds reduce pain and fever but have little effect on inflammation.
  - **Narcotics** act on the CNS to alter the perception and response to pain. Effective for severe pain, narcotics are administered by varied methods, including orally, intravenously, and by intramuscular injection. They are also effectively administered into the space surrounding the spinal cord (see Fig. 9-15). An example of a narcotic drug is morphine.
- **Anesthetics**. Although most commonly used to prevent pain during surgery, anesthetic injections are also used to relieve certain types of chronic pain.
- **Endorphins** (en-DOR-fins) are released naturally from certain brain regions and are associated with

11

pain control. Massage, acupressure, and electric stimulation are among the techniques that are thought to activate this system of natural pain relief.

- **Applications of heat or cold** can be a simple but effective means of pain relief, either alone or in combination with medications. Care must be taken to avoid injury caused by excessive heat or cold.
- **Relaxation or distraction techniques** include several methods that reduce pain perception in the CNS. Relaxation techniques counteract the fight-or-flight response to pain and complement other pain control methods.

CHECKPOINTS

- [ ] 11-24 What are five examples of general senses?
- [ ] 11-25 What are proprioceptors, and where are they located?

Disease in Context Revisited

Evan's Cochlear Implant

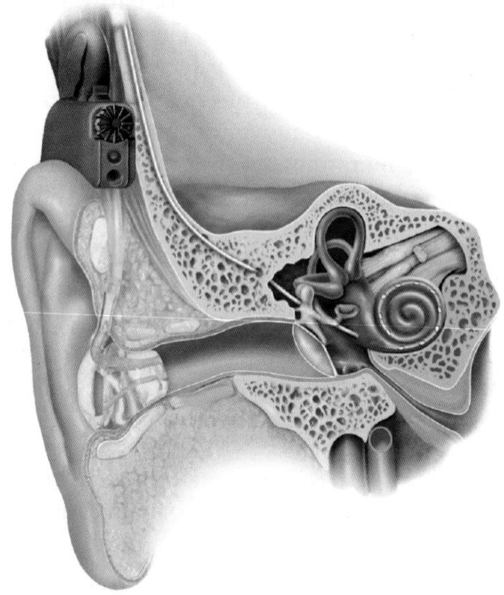

"Hi. How are we doing today?" asked Dr. Sanchez. Evan had undergone a right cochlear implant four weeks ago and was in for another follow-up visit. He appeared to be a normal, happy 20-month-old sitting on his mother's lap. A couple of months earlier, meningitis had destroyed many of the essential hair cells in Evan's cochlea. The drugs used to treat the meningitis may also have been ototoxic, contributing to further destruction of these cells.

Dr. Sanchez explained, "Evan is ready to have his external fitting. If you recall, the implant has three main components: the internal receiver that we implanted behind Evan's ear with electrodes going to the inner ear, and the external parts, a transmitter and sound processor. These components allow the brain to interpret the frequency of sound as it would if the hair cells were functioning properly. Today, we are going to activate the implant by connecting the processor to the internal device. Evan will need some help from a speech therapist and audiologist, but we're sure he will make progress. If all goes as expected, the device will complement the limited hearing in his left ear, and he will not need a second implant."

"This entire process has been very difficult for us all," said Evan's father. "But to think that Evan will be able to hear well again is truly amazing!"

In this case, we saw how the sense of hearing can be compromised by damaged hair cells in the cochlea, but new and advancing technology is helping to treat sensorineural hearing loss.

CHAPTER 11

Chapter Wrap-Up

Summary Overview

A detailed chapter outline with space for note taking is on *thePoint*. The figure below illustrates the main topics covered in this chapter.

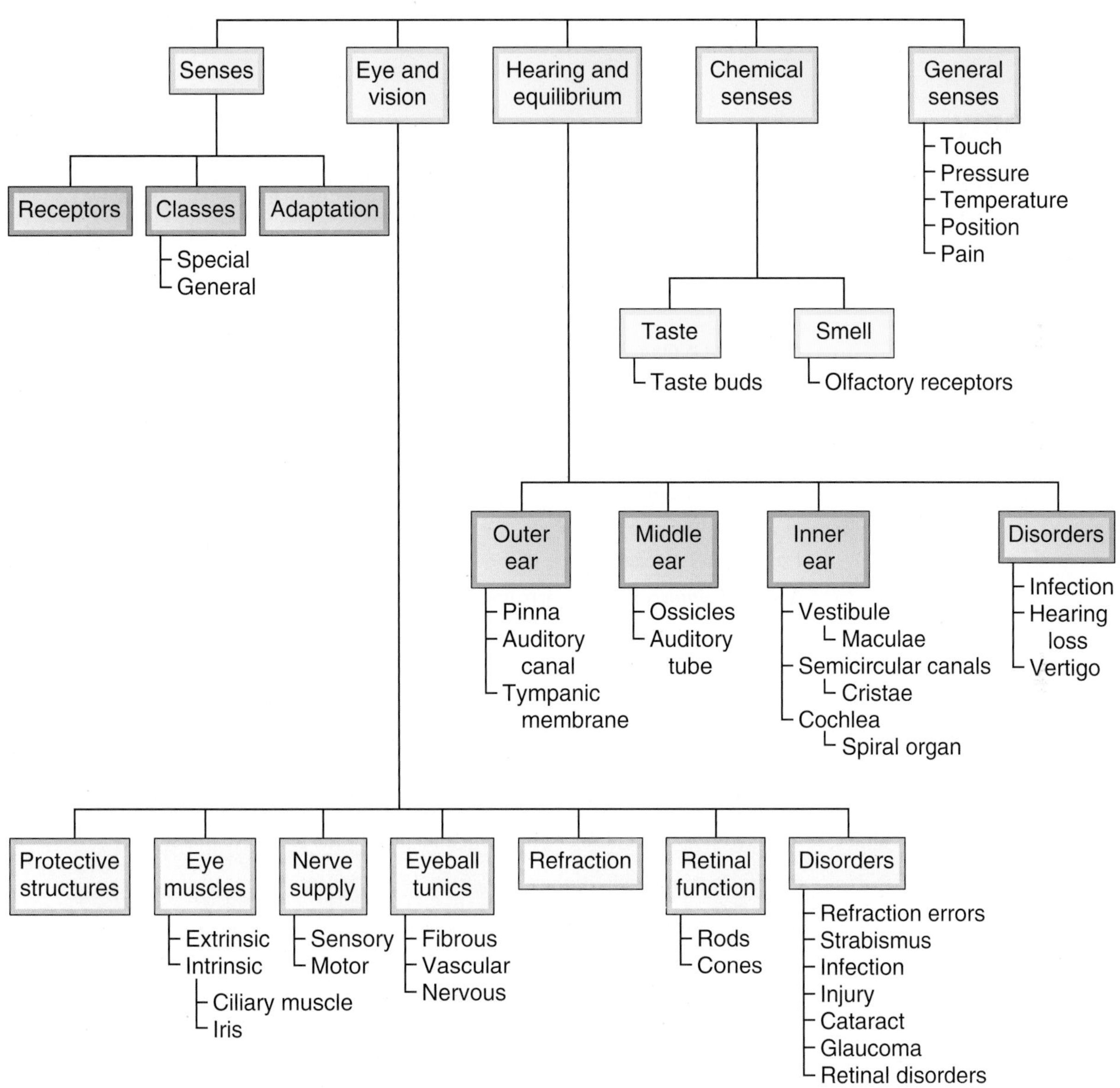

Key Terms

The terms listed below are emphasized in this chapter. Knowing them will help you organize and prioritize your learning. These and other boldface terms are defined in the Glossary with phonetic pronunciations.

accommodation
astigmatism
auditory tube
aqueous humor
cataract
choroid
cochlea
conjunctiva
convergence
cornea
glaucoma
gustation
hyperopia
iris
kinesthesia
lacrimal apparatus
lens (crystalline lens)
macula
myopia
olfaction
ossicle
proprioceptor
refraction
retina
sclera
semicircular canal
sensory adaptation
sensory receptor
spiral organ
strabismus
tympanic membrane
vestibule
vitreous body

Word Anatomy

Medical terms are built from standardized word parts (prefixes, roots, and suffixes). Learning the meanings of these parts can help you remember words and interpret unfamiliar terms.

WORD PART	MEANING	EXAMPLE
The Eye and Vision		
ambly/o	dimness	*Amblyopia* is poor vision in a healthy eye that cannot work properly with the other eye.
e-	out	*Enucleation* is removal of the eyeball.
lute/o	yellow	The macula *lutea* is a yellowish spot in the retina that contains the fovea centralis.
ophthalm/o	eye	An *ophthalmologist* is a physician who specializes in treatment of the eye.
-opia	disorder of the eye or vision	*Hyperopia* is farsightedness.
presby-	old	*Presbyopia* is farsightedness that occurs with age.
-scope	instrument for examination	An *ophthalmoscope* is an instrument used to examine the posterior of the eye.
The Ear		
-cusis	hearing	*Presbycusis* is hearing loss associated with age.
equi-	equal	*Equilibrium* is balance (*equi-* combined with the Latin word *libra* meaning "balance").
lith	stone	*Otoliths* are small crystals in the inner ear that aid in static equilibrium.
myring/o	tympanic membrane	*Myringotomy* is a cutting of the tympanic membrane to relieve pressure.
ot/o	ear	*Otitis* is inflammation of the ear.
-stomy	creation of an opening	A *tympanostomy* tube creates a passageway between two structures through the tympanic membrane.
tympan/o	drum	The *tympanic* membrane is the eardrum.
The General Senses		
alges/i	pain	An *analgesic* is a drug that relieves pain.
-esthesia	sensation	*Anesthesia* is a loss of sensation, as of pain.
kine	movement	*Kinesthesia* is a sense of body movement.
narc/o	stupor	A *narcotic* is a drug that alters the perception of pain.
propri/o-	own	*Proprioception* is perception of one's own body position.

Questions for Study and Review

BUILDING UNDERSTANDING

Fill in the Blanks

1. The part of the nervous system that detects a stimulus is a(n) ______.
2. The bending of light rays as they pass from one substance to another is called ______.
3. Nerve impulses are carried from the ear to the brain by the ______ nerve.
4. A receptor that senses body position is a(n) ______.
5. A receptor's loss of sensitivity to a continuous stimulus is called ______.

Matching > Match each numbered item with the most closely related lettered item.

___ **6.** Progressive loss of near accommodation ability with age

___ **7.** Irregularity in the curvature of the cornea or lens

___ **8.** Deviation of the eye due to lack of coordination of the eyeball muscles

___ **9.** Increased pressure inside the eyeball

___ **10.** Loss of vision in a healthy eye because it cannot work properly with the other eye

a. glaucoma
b. amblyopia
c. presbyopia
d. astigmatism
e. strabismus

Multiple Choice

___ **11.** Which of the following is a general sense?

a. taste
b. smell
c. equilibrium
d. touch

___ **12.** From superficial to deep, what is the order of the eyeball's tunics?

a. nervous, vascular, fibrous
b. fibrous, nervous, vascular
c. vascular, nervous, fibrous
d. fibrous, vascular, nervous

___ **13.** Which eye structure has the greatest effect on light refraction?

a. cornea
b. lens
c. vitreous body
d. retina

___ **14.** Which nerve carries sensory signals from the retina to the brain?

a. ophthalmic
b. optic
c. oculomotor
d. abducens

___ **15.** What do receptors in the vestibule sense?

a. muscle tension
b. sound
c. light
d. acceleration

UNDERSTANDING CONCEPTS

16. Differentiate between the terms in each of the following pairs:
  a. special sense and general sense
  b. extrinsic and intrinsic eye muscles
  c. rods and cones
  d. endolymph and perilymph
  e. maculae and cristae

17. Trace the path of a light ray from the outside of the eye to the retina.

18. Define *convergence* and *accommodation*, and describe several disorders associated with them.

19. List in order the structures that sound waves pass through in traveling through the ear to the receptors for hearing.

20. Compare and contrast conductive hearing loss and sensorineural hearing loss. Which type of hearing loss was involved in Evan's opening case study, and what was its cause?

21. Name the five basic tastes. Where are the taste receptors? Name the nerves of taste.

22. Trace the pathway of a nerve impulse from the olfactory receptors to the olfactory center in the brain.

23. Name several types of pain-relieving drugs. Describe methods for relieving pain that do not involve drugs.

CONCEPTUAL THINKING

24. Maia T., 2 years old, is taken to see the pediatrician because of a severe earache. Examination reveals that the tympanic membrane is red and bulging outward toward the external auditory canal. What disorder does Maia have? Why is the incidence of this disorder higher in children than in adults? What treatment options are available to Maia?

25. You and a friend have just finished riding the roller coaster at the amusement park. As you walk away from the ride, your friend stumbles and comments that the ride has affected her balance. How do you explain this?

26. Referring to the case study, imagine you were Dr. Sanchez speaking to Evan's parents. How would you explain the role of hair cells in hearing and how the cochlear implant can overcome their loss?

For more questions, see the Learning Activities on thePoint.

CHAPTER 12

The Endocrine System: Glands and Hormones

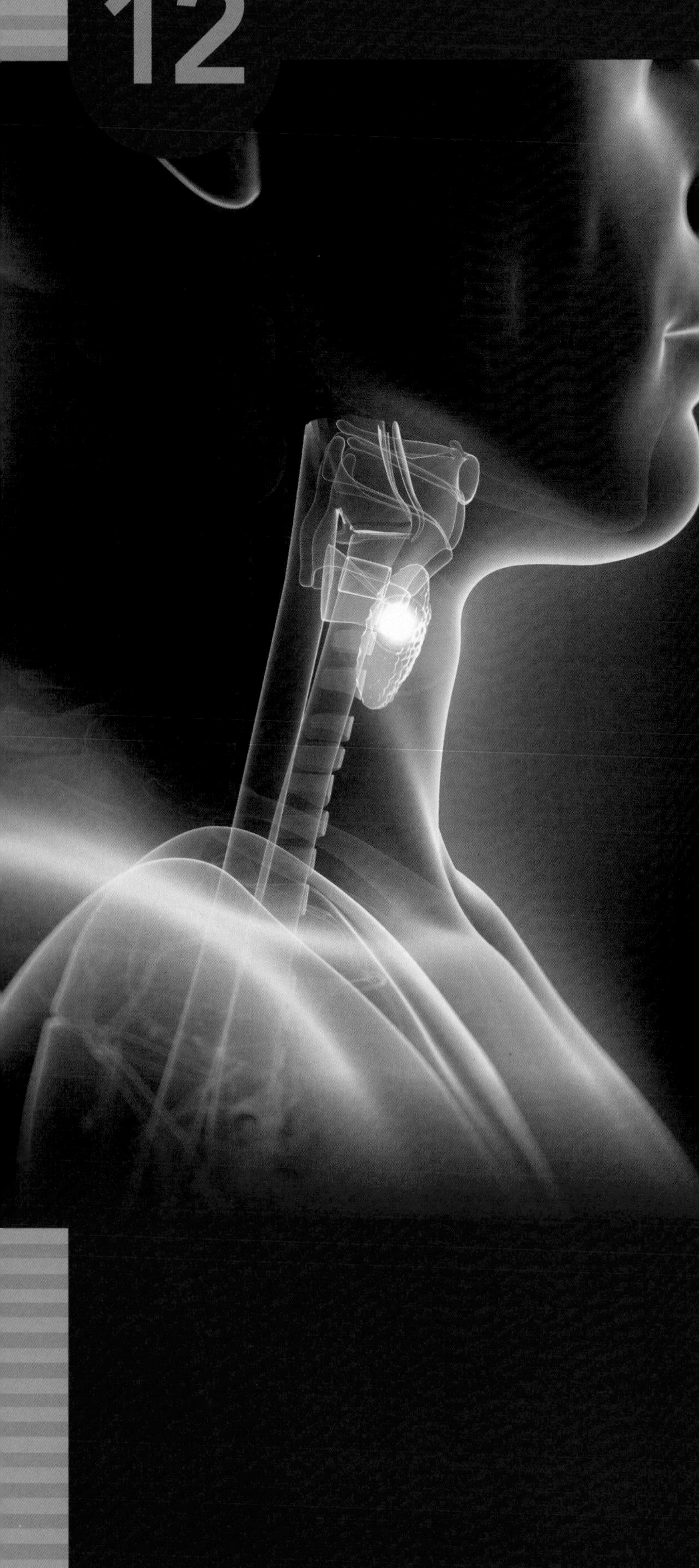

Learning Objectives

After careful study of this chapter, you should be able to:

1 ▶ Compare the effects of the nervous system and the endocrine system in controlling the body. ***p. 270***

2 ▶ Describe the functions of hormones. ***p. 270***

3 ▶ Identify the glands of the endocrine system on a diagram. ***p. 270***

4 ▶ Discuss the chemical composition of hormones. ***p. 271***

5 ▶ Explain how hormones are regulated. ***p. 271***

6 ▶ List the hormones produced by each endocrine gland, and describe the effects of each on the body. ***p. 271***

7 ▶ Describe how the hypothalamus controls the posterior and anterior pituitary. ***p. 271***

8 ▶ Describe the effects of hyposecretion and hypersecretion of the various hormones. ***p. 273***

9 ▶ List seven tissues other than the endocrine glands that produce hormones. ***p. 282***

10 ▶ Explain the origin and function of prostaglandins. ***p. 283***

11 ▶ List eight medical uses of hormones. ***p. 283***

12 ▶ Explain how the endocrine system responds to stress. ***p. 284***

13 ▶ Referring to the case study, discuss the signs, symptoms, and treatment of diabetes mellitus type 1. ***pp. 269, 285***

14 ▶ Show how word parts are used to build words related to the endocrine system (see Word Anatomy at the end of the chapter). ***p. 287***

Disease in Context *Becky's Case: When an Endocrine Organ Fails*

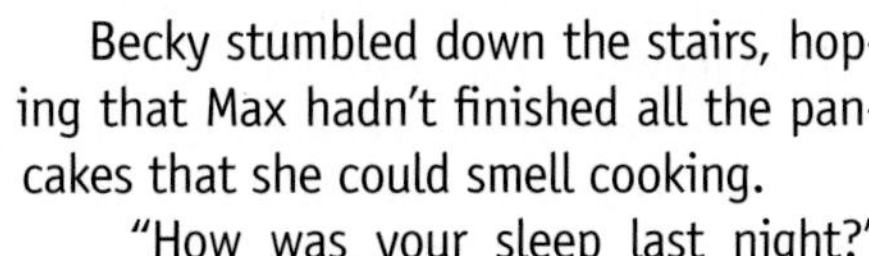

Becky stumbled down the stairs, hoping that Max hadn't finished all the pancakes that she could smell cooking.

"How was your sleep last night?" asked Becky's mother.

"Awful," sighed Becky, drowning the pancakes she was served in a lake of syrup. "I woke up a bunch of times to go to the bathroom."

"Were you actually able to make it this time?" chimed Becky's little brother. Becky wished Max hadn't brought *that* up. She hoped he wasn't blabbing to his friends that she was wetting the bed again.

"You know, if you didn't drink so much, you wouldn't have to pee so much," explained Max, as his sister gulped down her orange juice. Becky pretended that she didn't care about Max's comment. But he was right. She was so thirsty—and hungry!

It had been a long day when the bell rang and Becky boarded the bus for home. Math class had been a disaster, because she couldn't concentrate. During gym, she was tired and had a stomach ache. And she had to keep asking for permission to go to the bathroom! Now, she was exhausted and her head hurt. Fighting tears, she remembered that during breakfast, her mom had mentioned that she'd made an appointment for Becky to see her doctor. She hadn't been too keen on the idea, but now she was relieved.

Later that week, Becky's pediatrician weighed and measured her and asked her a bunch of questions.

"So, Becky," said Dr. Carter. "For the past couple of weeks, you say you've felt lethargic and sick to your stomach. You've been really thirsty and have needed to go to the bathroom a lot. You've also been really hungry. You've had headaches and some difficulty concentrating at school, and have felt tired when playing sports." Becky wasn't too sure what lethargic meant, but other than that he seemed to have gotten the facts right. So Becky nodded her head yes.

Turning to Becky's mother, Dr. Carter said, "Checking her chart, it appears that she's lost several pounds since her last appointment despite her appetite. I'm going to order urine and blood tests. I'd like to see what her glucose levels are." Becky didn't enjoy the tests one bit. Having to pee in a cup was gross, and as for the blood test, that was the worst.

The next day, Dr. Carter called Becky's mother. "The urinalysis was positive for glucose and ketones, suggesting that Becky is not metabolizing glucose correctly. Her blood test revealed that she's hyperglycemic; her blood sugar is too high. My diagnosis so far is that Becky has type 1 diabetes mellitus and needs insulin."

Dr. Carter suspects that Becky's pancreas does not produce enough insulin, a hormone needed to utilize glucose. As we will see later, diabetes has a dramatic effect on Becky's health.

ANCILLARIES *At-A-Glance*

Visit thePoint to access the following resources. For guidance in using these resources most effectively, see pp. xv–xvii.

Learning RESOURCES

- Tips for Effective Studying
- Web Figure: Primary Gigantism
- Web Figure: Child with Growth Hormone Deficiency
- Web Figure: Clinical Manifestations of Acromegaly
- Web Figure: Acromegaly
- Web Figure: Bronze Skin Pigmentation of Addison Disease
- Web Figure: Clinical Manifestations of Addison Disease
- Web Figure: Hypothyroidism and Hyperthyroidism Compared
- Web Figure: Clinical Manifestations of Cushing Syndrome
- Web Figure: Cushing Syndrome
- Web Figure: Nontoxic Goiter
- Web Figure: Clinical Manifestations of Hyperparathyroidism
- Web Figure: Metabolic Syndrome
- Animation: Hormonal Control of Glucose
- Animation: Diabetes
- Health Professions: Exercise and Fitness Specialist
- Detailed Chapter Outline
- Answers to Questions for Study and Review
- Audio Pronunciation Glossary

Learning ACTIVITIES

- Pre-Quiz
- Visual Activities
- Kinesthetic Activities
- Auditory Activities

A LOOK BACK

The past several chapters have described the nervous system and its role in regulating body responses. The endocrine system is also viewed as a controlling system, exerting its effects through hormones. The endocrine glands differ from the exocrine glands described in Chapter 4 because they secrete directly into body fluids and not through ducts. Control of the endocrine system relies mainly on negative feedback, described in Chapter 1.

The **endocrine system** consists of a group of glands that produces regulatory chemicals called **hormones**. These glands specialize in hormone secretion and are illustrated in **Figure 12-1**. The endocrine system and the nervous system work together to control and coordinate all other body systems. The nervous system controls such rapid actions as muscle movement and intestinal activity by means of electrical and chemical stimuli. The effects of the endocrine system occur more slowly and over a longer period. They involve chemical stimuli only, and these chemical messengers have widespread effects on the body.

Although the nervous and endocrine systems differ in some respects, the two systems are closely related. For example, the activity of the pituitary gland, which in turn regulates other glands, is controlled by the brain's hypothalamus. You can see both structures in **Figure 12-1**. The connections between the nervous system and the endocrine system enable endocrine function to adjust to the demands of a changing environment.

Hormones

Hormones are chemical messengers that have specific regulatory effects on certain cells or organs. Hormones from the endocrine glands are released, not through ducts, but directly into surrounding tissue fluids. Most

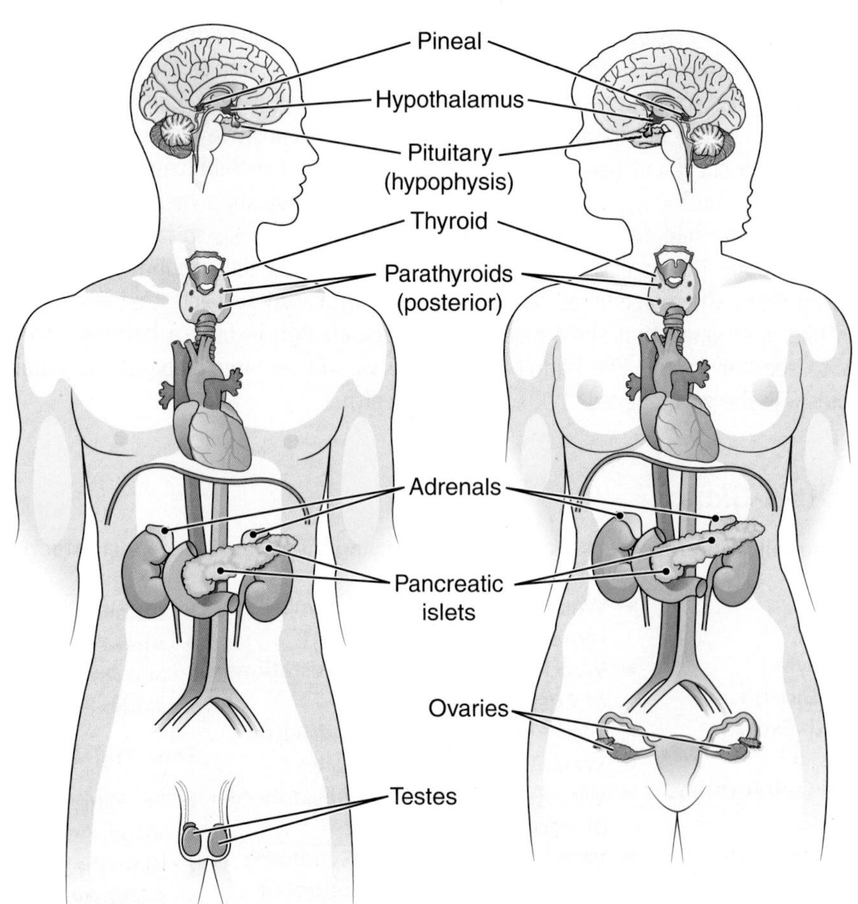

Figure 12-1 **The endocrine glands.** **KEY POINT** The endocrine system comprises glands with a primary function of hormone secretion.

then diffuse into the bloodstream, which carries them throughout the body. The specific tissue acted on by each hormone is the **target tissue.** The cells that make up these tissues have **receptors** in the plasma membrane or within the cytoplasm to which the hormone attaches. Once a hormone binds to a receptor on or in a target cell, the bound receptor affects cell activities such as regulating the manufacture of specific proteins, changing the membrane's permeability to specific substances, or affecting metabolic reactions. Since blood carries hormones throughout the body, any cell possessing receptors for a specific hormone will respond to the hormone, be it a close neighbor of the secreting cell or not.

HORMONE CHEMISTRY

Chemically, hormones fall into two main categories:

- **Amino acid compounds.** These hormones are proteins or related compounds also made of amino acids. All hormones except those of the adrenal cortex and the sex glands fall into this category.
- **Steroids.** These hormones are derived from the steroid cholesterol, a type of lipid (**see Fig. 2-9**). Steroid hormones are produced by the adrenal cortex and the sex glands. Many can be recognized by the ending *sterone*, as in progesterone and testosterone.

HORMONE REGULATION

As discussed in Chapter 1, the process of negative feedback keeps the level of a particular parameter within a specific range. **Figure 1-4** illustrates how the hormone insulin is a signal in the negative feedback loop regulating blood glucose concentrations. When blood glucose increases, insulin secretion rises. Insulin actions reduce blood glucose, reversing the initial stimulus and restoring homeostasis.

Hormone release may fall into a rhythmic pattern. Hormones of the adrenal cortex follow a 24-hour cycle related to a person's sleeping pattern, with the secretion level greatest just before arising and least at bedtime. Hormones of the female menstrual cycle follow a monthly pattern.

CHECKPOINTS 

- ☐ 12-1 What are hormones, and what are some effects of hormones?
- ☐ 12-2 What name is given to the specific tissue that responds to a hormone?
- ☐ 12-3 Hormones belong to what two chemical categories?
- ☐ 12-4 What is the most common mechanism used to regulate hormone secretion?

The Endocrine Glands and Their Hormones

The remainder of this chapter discusses hormones and the tissues that produce them. Although most of the discussion centers on the endocrine glands, which specialize in hormone production, it is important to note that many tissues—other than the endocrine glands—also secrete hormones. These tissues include the brain, digestive organs, and kidneys. Some of these other tissues are discussed later in the chapter. **Table 12-1** summarizes the information on the endocrine glands and their hormones. Each section of the chapter also includes information on the effects of a hormone's hypersecretion (oversecretion) or hyposecretion (undersecretion), summarized in **Table 12-2.**

See the Student Resources on thePoint for more details and illustrations of the disorders in Table 12-2.

THE PITUITARY

The **pituitary** (pih-TU-ih-tar-e), or *hypophysis* (hi-POF-ih-sis), is a gland about the size of a cherry. It is located in a saddle-like depression of the sphenoid bone just posterior to the point where the optic nerves cross. It is surrounded by bone except where it connects with the brain's **hypothalamus** by a stalk called the **infundibulum** (in-fun-DIB-u-lum). The gland is divided into two parts: the **anterior lobe** and the **posterior lobe** (**Fig. 12-2**). The anterior lobe is a true endocrine gland, composed of epithelial tissue. The posterior lobe, however, is not a true gland. It consists of the axons and axon terminals of neurons that originate in the hypothalamus. The two lobes are considered separately below.

Posterior Lobe The two hormones of the posterior pituitary (antidiuretic hormone, or ADH, and oxytocin) are actually produced in the hypothalamus and only stored in the posterior pituitary (**Fig. 12-2**). Their release is controlled by nerve impulses that travel over pathways (tracts) between the hypothalamus and the posterior pituitary. Their actions are as follows:

- **Antidiuretic** (an-ti-di-u-RET-ik) **hormone (ADH)** promotes the reabsorption of water from the kidney tubules and thus decreases water excretion. A large amount of this hormone causes contraction of smooth muscle in blood vessel walls and raises blood pressure. An inadequate amount of ADH causes excessive water loss and results in a disorder called **diabetes insipidus.** This type of diabetes should not be confused with diabetes

Table 12-1 The Endocrine Glands and Their Hormones

Gland	Hormone	Principal Functions
Hypothalamus	Releasing hormones	Control the release of anterior pituitary hormones
Hypothalamus and posterior pituitary	ADH (antidiuretic hormone)	Promotes water reabsorption in kidney tubules; at high concentration, stimulates constriction of blood vessels
	Oxytocin	Causes uterine muscle contraction; causes milk ejection from mammary glands
Anterior pituitary	GH (growth hormone)	Promotes growth of all body tissues
	TSH (thyroid-stimulating hormone)	Stimulates thyroid gland to produce thyroid hormones
	ACTH (adrenocorticotropic hormone)	Stimulates adrenal cortex to produce glucocorticoids (cortisol) and androgens
	PRL (prolactin)	Stimulates milk production by mammary glands
	FSH (follicle-stimulating hormone)	Stimulates growth and hormonal activity of ovarian follicles; stimulates growth of testes; promotes sperm cell development
	LH (luteinizing hormone)	Initiates ovulation, corpus luteum formation, and progesterone production in the female; stimulates testosterone secretion in male
Thyroid	Thyroxine (T_4) and triiodothyronine (T_3)	Increase metabolic rate, influencing both physical and mental activities; required for normal growth
Parathyroids	PTH (parathyroid hormone)	Regulates exchange of calcium between blood and bones; increases calcium level in blood
Adrenal medulla	Epinephrine	Increases blood pressure and heart rate; activates cells influenced by sympathetic nervous system plus many not supplied by sympathetic nerves
Adrenal cortex	Cortisol (95% of glucocorticoids)	Increases blood glucose concentration in response to stress
	Aldosterone (95% of mineralocorticoids)	Promotes salt (and thus water) retention and potassium excretion
	Weak androgens	Contribute to some secondary sex characteristics in women
Pancreatic islets	Insulin	Reduces blood glucose concentrations by promoting glucose uptake into cells and glucose storage; promotes fat and protein synthesis
	Glucagon	Stimulates the liver to release glucose, thereby increasing blood glucose levels
Testes	Testosterone	Stimulates growth and development of sexual organs (testes and penis) plus development of secondary sexual characteristics, such as hair growth on the body and face and deepening of voice; stimulates sperm cell maturation
Ovaries	Estrogens (e.g., estradiol)	Stimulates growth of primary sexual organs (uterus and tubes) and development of secondary sexual organs, such as breasts; stimulates development of ovarian follicles
	Progesterone	Stimulates development of mammary glands' secretory tissue; prepares uterine lining for implantation of fertilized ovum; aids in maintaining pregnancy
Pineal	Melatonin	Regulates mood, sexual development, and daily cycles in response to the amount of light in the environment

mellitus, which is due to inadequate amounts of insulin. In fact, the name *insipidus* refers to the dilute, pale, "insipid" urine produced in this form of diabetes, as compared to the glucose-containing urine characteristic of diabetes mellitus.

- **Oxytocin** (ok-se-TO-sin) causes uterine contractions and triggers milk ejection from the breasts. Under certain circumstances, commercial preparations of this hormone are administered during childbirth to promote uterine contraction.

Table 12-2 Disorders Associated with Endocrine Dysfunction

Hormone	Effects of Hypersecretion	Effects of Hyposecretion
Growth hormone	Gigantism (children), acromegaly (adults)	Dwarfism (children)
Antidiuretic hormone	Syndrome of inappropriate antidiuretic hormone (SIADH)	Diabetes insipidus
Aldosterone	Aldosteronism	Addison disease
Cortisol	Cushing syndrome	Addison disease
Thyroid hormone	Graves disease, thyrotoxicosis	Congenital and adult hypothyroidism
Insulin	Hypoglycemia	Diabetes mellitus; hyperglycemia
Parathyroid hormone	Bone degeneration	Tetany (muscle spasms)

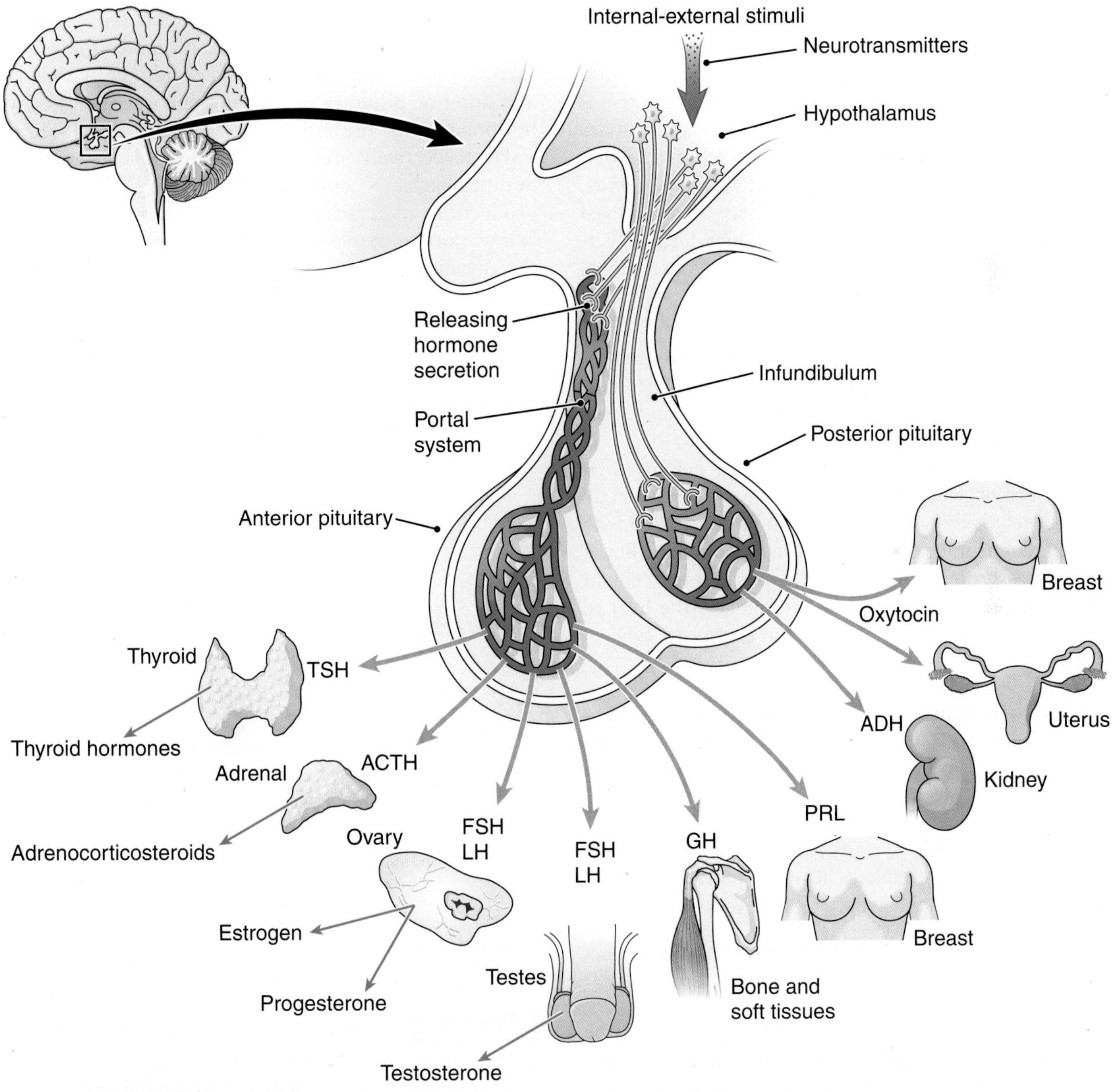

Figure 12-2 **The hypothalamus, pituitary gland, and target tissues.** **KEY POINT** The hypothalamus synthesizes hormones secreted by the posterior pituitary, and synthesizes releasing hormones that regulate the anterior pituitary. **ZOOMING IN** What two structures does the infundibulum connect?

Anterior Lobe The hormone-producing cells of the anterior pituitary are controlled by secretions called **releasing hormones** produced in the hypothalamus (**see Fig. 12-2**). These releasing hormones travel to the anterior pituitary by way of a special type of circulatory pathway called a **portal system.** By this circulatory "detour," some of the blood that leaves the hypothalamus travels to capillaries in the anterior pituitary before returning to the heart. Each pituitary cell produces a particular hormone and is stimulated by specific hypothalamic releasing hormones. Hypothalamic releasing hormones are indicated with the abbreviation *RH* added to an abbreviation for the name of the hormone stimulated. For example, the releasing hormone (RH) that controls growth hormone (GH) is GHRH. Inhibitory hormones from the hypothalamus also regulate the anterior pituitary hormones.

Anterior Lobe Hormones The anterior pituitary is often called the *master gland* because it releases hormones that affect the working of other glands, such as the thyroid, gonads (ovaries and testes), and adrenal glands. (Hormones that stimulate other glands may be recognized by the ending *tropin* as in *thyrotropin,* which means "acting on the thyroid gland.") The major hormones are as follows (**Fig. 12-2**):

- **Growth hormone** (**GH**), or *somatotropin* (so-mah-to-TRO-pin), acts directly on most body tissues, promoting protein manufacture that is essential for growth. GH causes increases in size and height to occur in youth, before the closure of long bone epiphyses. A young person with a GH deficiency will remain small, though relatively well-proportioned unless treated with adequate hormone. GH is produced throughout life. It stimulates protein synthesis and is needed for cellular maintenance and repair. It also stimulates the liver to release fatty acids and glucose for energy in time of stress.
- **Thyroid-stimulating hormone** (TSH), or *thyrotropin* (thi-ro-TRO-pin), stimulates the thyroid gland (a large gland found in the neck) to produce thyroid hormones.
- **Adrenocorticotropic** (ad-re-no-kor-tih-ko-TRO-pik) **hormone** (ACTH) stimulates hormone production in the cortex of the adrenal glands.
- **Prolactin** (pro-LAK-tin) (**PRL**) stimulates milk production in the breasts.
- **Follicle-stimulating hormone** (**FSH**) stimulates the development of ovarian follicles in which egg cells mature and the development of sperm cells in the testes.
- **Luteinizing** (LU-te-in-i-zing) **hormone** (**LH**) causes ovulation in females and promotes progesterone secretion in females and testosterone secretion in males.

FSH and LH are classified as **gonadotropins** (gon-ah-do-TRO-pinz), hormones that act on the gonads to regulate growth, development, and reproductive function in both males and females.

Negative Feedback and the Anterior Lobe Negative feedback homeostatically regulates most hormone secretions of the anterior pituitary and its target glands. That is, a hormone itself inhibits further hormone secretion. An example is the secretion of thyroid hormones (**Fig. 12-3**). Thyrotropin-releasing hormone (TRH) from the hypothalamus stimulates the production of TSH from the anterior pituitary gland. TSH promotes the release of thyroid hormones from the thyroid gland. The hypothalamus and anterior pituitary gland sense any increase in thyroid hormone levels and reduce their production of TRH and TSH, respectively. As a result, thyroid hormone levels decline back to normal. Conversely, TRH and TSH production increase if thyroid hormone levels decline below normal levels. As mentioned above, TSH signals the thyroid gland (a gland in the neck) to secrete more hormones.

Growth hormone, ACTH, and, to a certain extent, the gonadal steroids are all subject to similar negative feedback loops. These self-regulating systems keep hormone levels within a set normal range.

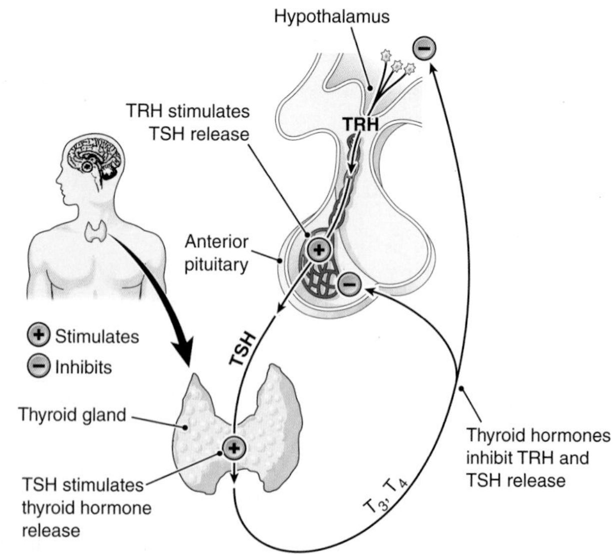

Figure 12-3 Negative feedback control of thyroid hormones. **KEYPOINT** Thyroid hormone levels are kept constant by negative feedback. **ZOOMING IN** What gland controls the thyroid gland?

Pituitary Tumors Pituitary tumors usually develop from one of the cell types in the anterior lobe and can result in excess production of a specific hormone. For instance, some tumors contain an excessive number of the cells that produce growth hormone. A person who develops such a tumor in childhood will grow to an abnormally tall stature, a condition called **gigantism** (ji-GAN-tizm) (see Table 12-2).

If the GH-producing cells become overactive in the adult, a disorder known as **acromegaly** (ak-ro-MEG-ah-le) develops. In acromegaly, the bones of the face, hands, and feet widen. The fingers thicken, and the face takes on a coarse appearance; the nose widens, the lower jaw protrudes, and the forehead bones may bulge. Multiple body systems may be affected by acromegaly, and life expectancy is considerably reduced. The most common cause of death is cardiovascular disease.

Some pituitary tumors do not secrete hormones, but they grow so large that they destroy normal pituitary tissue. The symptoms of the tumor depend on which pituitary cells are affected. For instance, some pituitary tumors interfere with the release of antidiuretic hormone (ADH), resulting in diabetes insipidus. Pituitary tumors may also involve the optic nerves and cause headaches and blindness.

Evidence of tumor formation in the pituitary gland may be obtained by radiographic examinations of the skull. The pressure of the tumor distorts the sella turcica, the saddle-like space in the sphenoid bone that holds the pituitary. Physicians also use computed tomography and magnetic resonance imaging scans to diagnose pituitary abnormalities.

CHECKPOINTS

- ☐ 12-5 What part of the brain controls the pituitary?
- ☐ 12-6 What hormones are released from the posterior pituitary?
- ☐ 12-7 What hormones does the anterior pituitary secrete?

THE THYROID GLAND

12

The **thyroid**, located in the neck, is the largest of the endocrine glands (Fig. 12-4) (see also Figure 5-5A in the Dissection Atlas). The thyroid has two roughly oval lateral lobes on either side of the larynx (voice box) connected by a narrow band called an *isthmus* (IS-mus). A connective tissue capsule encloses the entire gland.

Thyroid Hormones The thyroid produces two hormones that regulate metabolism. The principal hormone is **thyroxine** (thi-ROK-sin), which is symbolized as T_4, based on the four iodine atoms contained in each molecule. The other hormone, which contains three atoms of iodine, is **triiodothyronine** (tri-i-o-do-THI-ro-nene), or T_3. These

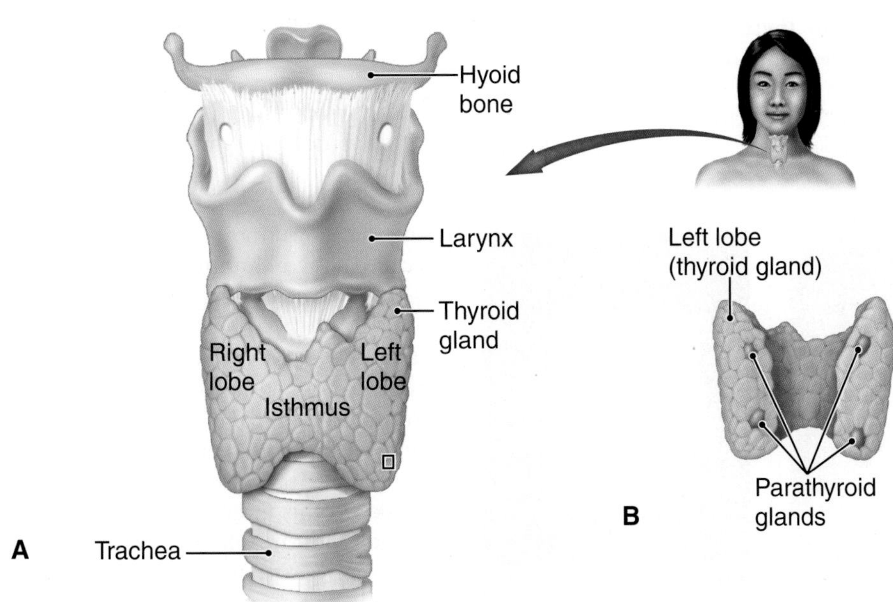

Figure 12-4 **Thyroid and parathyroid glands. A.** The thyroid has two lobes connected by an isthmus. These are shown here in relation to other structures in the throat. The epiglottis is a cartilage of the larynx. **B.** The parathyroid glands are embedded in the posterior surface of the thyroid gland. **ZOOMING IN** What structure is superior to the thyroid? Inferior to the thyroid?

hormones increase the metabolic rate in body cells. That is, they increase the rate at which cells use nutrients to generate ATP and heat. Both thyroid hormones and growth hormone are needed for normal growth. As we saw in **Figure 12-3**, thyroid hormone production is under the control of TSH from the anterior pituitary, and thyroid hormones feed back to inhibit TSH production.

Disorders of the Thyroid Gland A **goiter** (GOY-ter) is an enlargement of the thyroid gland. Usually, a goiter results from excess stimulation by TSH, because the pituitary hormones stimulate the growth (as well as the activity) of their target glands. A simple goiter is the uniform overgrowth of the thyroid gland, with a smooth surface appearance. An adenomatous (ad-eh-NO-mah-tus), or nodular, goiter is an irregularly appearing goiter accompanied by tumor formation.

Inadequate production of thyroid hormones results in **hypothyroidism** (hi-po-THI-royd-izm):

- **Congenital hypothyroidism** usually results from a failure of the thyroid gland to form during fetal development. The infant suffers lack of physical growth and lack of mental development. Early and continuous treatment with replacement hormone can alter the outlook of this disease. By state law, all newborns are tested for hypothyroidism in the United States.
- Hypothyroidism in adults usually results from thyroid autoimmunity, that is, abnormal production of antibodies to the thyroid gland. Thyroiditis (thyroid inflammation) results as the antibodies attack and destroy thyroid tissue. This form of hypothyroidism, also called **Hashimoto thyroiditis**, is most common in women between 30 and 50 years of age (**Fig. 12-5A**). Other causes of hypothyroidism in adults include destruction of thyroid tissue by surgery or radiation for treatment of cancer or thyroid hyperactivity, certain drugs, or pituitary failure to produce TSH. Hypothyroidism causes fatigue, intolerance to cold, dry skin and hair, and swelling in the legs and face. Oral administration of thyroid hormone usually restores health, although treatment must be maintained throughout life.
- **Endemic goiter** results from lack of iodine in the diet, limiting the thyroid's ability to manufacture hormones. Without normal negative feedback from thyroid hormones, the anterior pituitary increases production of TSH, and a goiter develops. **See Figure 12-3** to review this feedback pathway. Iodine deficiency is rare now in developed nations because of widespread availability of this mineral in iodized salt; however, iodine deficiency is still prevalent in developing nations. In addition to iodized salt, seafood, dairy products, and processed foods provide some iodine.

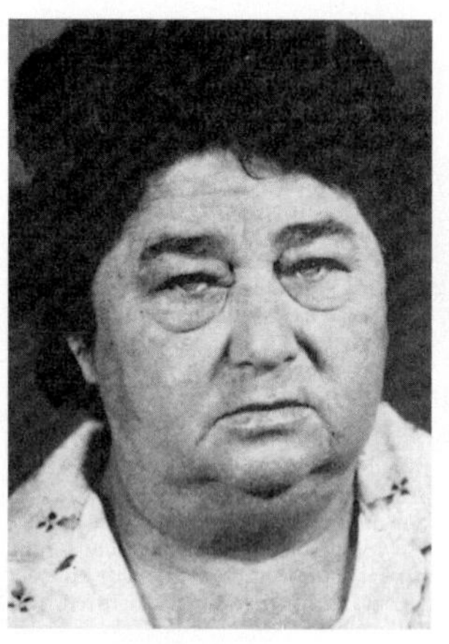

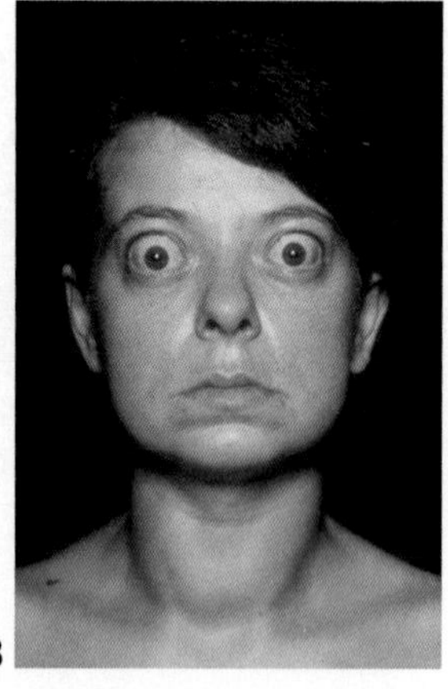

Figure 12-5 **Thyroid hormone disorders. A.** Adult hypothyroidism. **B.** Graves disease, showing goiter and exophthalmos. **KEY POINT** Goiter is enlargement of the thyroid; exophthalmos is bulging of the eyes.

Hyperthyroidism is the opposite of hypothyroidism, that is, overactivity of the thyroid gland with excessive hormone secretion. A common form of hyperthyroidism is **Graves disease**, which is characterized by a goiter, a strained appearance of the face, intense nervousness, weight loss, a rapid pulse, sweating, tremors, and an abnormally quick metabolism. Another characteristic symptom is protrusion (bulging) of the eyes, known as **exophthalmos** (ek-sof-THAL-mos), which is caused by swelling of the tissue behind the eyes and of the eyeballs themselves (**Fig. 12-5B**). Graves disease results from antibodies that mimic TSH, causing increased thyroid growth (goiter) and activity.

Treatment of hyperthyroidism may take the following forms:

- Suppression of hormone production with medication
- Destruction of thyroid tissue with radioactive iodine
- Partial surgical removal of the thyroid gland

An exaggerated form of hyperthyroidism with a sudden onset is called a **thyroid storm**. Untreated, it is usually fatal, but with appropriate care, most affected people survive.

Tests of Thyroid Function Thyroid function can be studied most easily by measuring blood levels of thyroid hormones and TSH. However, a more accurate test measures the thyroid's uptake of radioactive iodine. Because the thyroid takes up iodine to make its hormones, low uptake of this mineral indicates hypothyroidism and high uptake indicates hyperthyroidism. Physicians use these very sensitive tests not only to detect abnormal thyroid function but also to monitor response to drug therapy.

THE PARATHYROID GLANDS

The four tiny **parathyroid glands** are embedded in the thyroid's posterior capsule or in the surrounding connective tissue (**see Fig. 12-4**). The secretion of these glands, **parathyroid**

hormone (**PTH**), promotes calcium release from bone tissue, thus increasing the amount of calcium circulating in the bloodstream. PTH also causes the kidney to conserve calcium. PTH levels are controlled by negative feedback based on the amount of calcium in the blood; when calcium is low, PTH is produced.

Calcium Metabolism Calcium balance is required not only for the health of bones and teeth but also for the proper function of the nervous system and muscles. Another hormone, in addition to PTH, is needed for calcium balance. This hormone is **calcitriol** (kal-sih-TRI-ol), technically called dihydroxycholecalciferol (di-hi-drok-se-ko-le-kal-SIF-eh-rol), the active form of vitamin D. Calcitriol is produced by modification of vitamin D in the liver and then the kidney, a process stimulated by PTH. Calcitriol increases intestinal absorption of calcium to raise blood calcium levels.

PTH and calcitriol work together to regulate the amount of calcium in the blood and provide calcium for bone maintenance and other functions.

Disorders of the Parathyroid Glands Inadequate production of parathyroid hormone, as a result of removal or damage to the parathyroid glands, for example, causes a series of muscle contractions, particularly involving the hands and face. These spasms result from a low blood calcium concentration, and the condition is called **tetany** (TET-ah-ne). This low-calcium tetany should not be confused with the infection called *tetanus* (lockjaw).

In contrast, if there is excess production of PTH, as may happen in parathyroid tumors, calcium is removed from its normal storage place in the bones and released into the bloodstream. The loss of calcium from the bones leads to fragile bones that fracture easily. Because the kidneys ultimately excrete the calcium, kidney stone formation is common in such cases.

CHECKPOINTS

- 12-8 What is the effect of thyroid hormones on cells?
- 12-9 What mineral is needed to produce thyroid hormones?
- 12-10 What is the term for an enlarged thyroid gland?
- 12-11 What mineral do parathyroid hormone (PTH) and calcitriol regulate?

See the Student Resources on thePoint to view illustrations of a goiter and hyperparathyroidism.

THE ADRENAL GLANDS

The **adrenals**, also called the *suprarenal glands*, are two small glands located atop the kidneys (see **Figure A5-8** in the Dissection Atlas). Each adrenal gland has two parts that act as separate glands. The inner area is called the **medulla**, and the outer portion is called the **cortex** (**Fig. 12-6**).

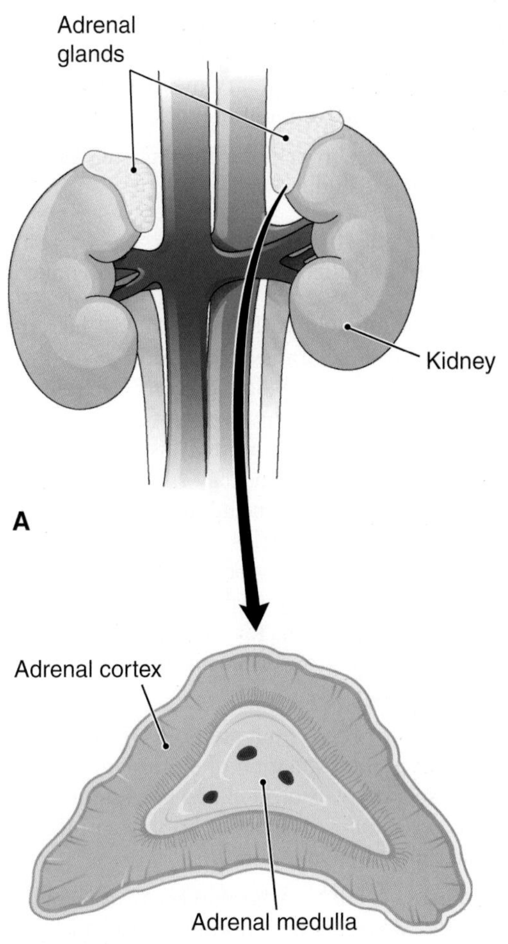

Figure 12-6 **The adrenal gland.** **KEY POINT** The medulla secretes epinephrine. The cortex secretes steroid hormones. **A.** Location of the adrenal glands on the kidneys. **B.** Structure of the adrenal gland. **ZOOMING IN** What is the outer region of the adrenal gland called? The inner region?

Hormones from the Adrenal Medulla The hormones of the adrenal medulla are released in response to stimulation by the sympathetic nervous system. The principal hormone produced by the medulla is **epinephrine,** also called *adrenaline*. Epinephrine is chemically and functionally similar to norepinephrine, the neurotransmitter active in the sympathetic nervous system, as described in Chapter 9. However, epinephrine is generally considered to be a hormone because it is released into the bloodstream instead of being released locally at synapses. Both epinephrine and norepinephrine are responsible for *fight-or-flight* responses during emergency situations. Some of their effects are as follows:

- Stimulation of smooth muscle contraction in the walls of some arterioles, causing them to constrict and blood pressure to rise accordingly
- Increase in the heart rate
- Increase in the metabolic rate of body cells

12

- Conversion of glycogen stored in the liver into glucose; the glucose enters the blood and travels throughout the body, allowing the voluntary muscles and other tissues to do an increased amount of work
- Dilation of the bronchioles through relaxation of the smooth muscle in their walls

Hormones from the Adrenal Cortex There are three main groups of hormones secreted by the adrenal cortex:

- **Glucocorticoids** (glu-ko-KOR-tih-koyds) help the body respond to unfavorable conditions such as starvation. They maintain blood glucose levels in times of stress by stimulating the liver to convert amino acids into glucose instead of protein (as indicated by *gluco* in the name). In addition, they raise the levels of other nutrients in the blood, including amino acids from tissue proteins and fatty acids from fats stored in adipose tissue. Glucocorticoids also have the ability to suppress the inflammatory response and are often administered as medication for this purpose. The major hormone of this group is **cortisol**, which is also called *hydrocortisone*.
- **Mineralocorticoids** (min-er-al-o-KOR-tih-koyds) are important in the regulation of electrolyte balance. They control sodium reabsorption and potassium secretion by the kidney tubules. The major hormone of this group is **aldosterone** (al-DOS-ter-one).
- **Androgens** ("male" sex hormones) are secreted in small amounts. Whereas these hormone quantities are insignificant in males (the testes produce large amounts of androgens), they constitute over 50% of the androgens in premenopausal women and are the only source of sex hormones in postmenopausal women. In normal amounts, they promote some bone and muscle growth and stimulate libido (sexual desire).

Disorders of the Adrenal Cortex Hypofunction of the adrenal cortex gives rise to a condition known as Addison disease, a disease characterized chiefly by muscle atrophy (loss of tissue), weakness, sometimes heightened skin pigmentation, and disturbances in salt and water balance. The disorder usually results from autoimmune destruction of the gland. The relationship between the skin pigmentation of Addison disease and melanocyte-stimulating hormone from the pituitary is explained in Box 12-1.

Hypersecretion of cortisol, for instance from a tumor, results in a condition known as **Cushing syndrome**, the symptoms of which include central body obesity with a round ("moon") face, thin skin that bruises easily, thin extremities and muscle weakness, bone loss, and elevated blood glucose (Fig. 12-7). Medical use of cortisol to treat inflammatory or autoimmune disorders can also cause these symptoms.

If the excess cortisol production results from excess pituitary stimulation from ACTH, adrenal androgens are also overproduced. In women, excess androgens cause **virilization**, the appearance of male characteristics such as facial hair and premature balding (See Fig. 12-7).

Excess aldosterone production, known as *aldosteronism*, results in abnormal salt and water retention and thus high blood pressure. Since aldosterone stimulates potassium excretion, blood potassium levels are also low, leading to neurological and muscular problems.

CHECKPOINTS 

- ☐ 12-12 What is the main hormone produced by the adrenal medulla?
- ☐ 12-13 What three categories of hormones are released by the adrenal cortex?
- ☐ 12-14 What effect does cortisol have on blood glucose levels?

Box 12-1

A CLOSER LOOK

Melanocyte-Stimulating Hormone: More Than a Tan?

In amphibians, reptiles, and certain other animals, melanocyte-stimulating hormone (MSH) darkens skin and hair by stimulating melanocytes to manufacture the pigment melanin. In humans, though, MSH levels are usually so low that its role as a primary regulator of skin pigmentation and hair color is questionable. What, then, is its function in the human body?

Recent research suggests that MSH is probably more important as a neurotransmitter in the brain than as a hormone in the rest of the body. A narrow region between the anterior and posterior pituitary, the intermediate lobe, produces MSH. When the pituitary gland secretes adrenocorticotropic hormone (ACTH), it secretes MSH as well. This is so because pituitary cells do not produce ACTH directly but produce a large precursor molecule, proopiomelanocortin (POMC), which enzymes cut into ACTH and MSH. In Addison disease, the pituitary tries to compensate for decreased glucocorticoid levels by increasing POMC production. The resulting increased levels of ACTH and MSH appear to cause the blotchy skin pigmentation that characterizes the disease.

MSH's other roles include helping the brain regulate food intake, fertility, and even the immune response. Interestingly, despite MSH's relatively small role in regulating pigmentation, women do produce more MSH during pregnancy and often have darker skin.

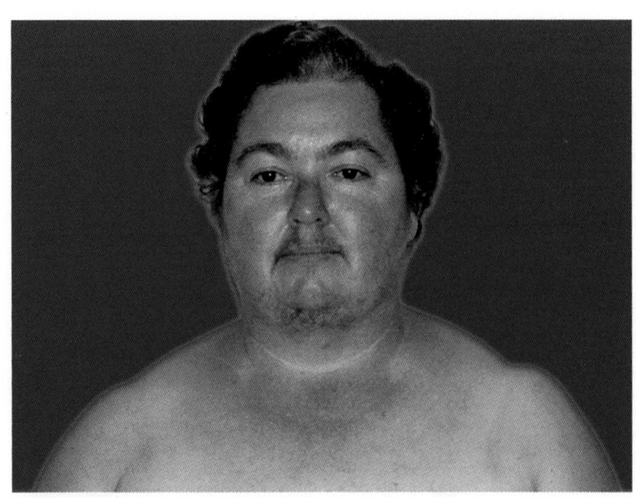

Figure 12-7 **Cushing disease.** A woman with an ACTH-producing tumor. Excess cortisol resulted in a moon face and increased truncal fat deposits, and excess DHEA caused increased facial hair and thinning of scalp hair.

THE ENDOCRINE PANCREAS

The **pancreas** is located in the left upper quadrant of the abdomen, inferior to the liver and gallbladder and lateral to the first portion of the small intestine, the duodenum (**Fig. 12-8**). It has two main types of cells that perform different functions. The most abundant type forms small clusters called *acini* (AS-ih-ni) (singular *acinus*) that resemble blackberries (*acinus* comes from a Latin word meaning "berry"). Acini secrete digestive enzymes through ducts directly into the small intestine (see Chapter 19), thus making up the exocrine portion of the pancreas. In addition, scattered throughout the pancreas are specialized cells that form "little islands" called **islets** (I-lets), originally the *islets of Langerhans* (LAHNG-er-hanz) (**see Fig. 12-8**). These cells produce hormones that diffuse into the bloodstream, thus making up the endocrine portion of the pancreas, which we discuss further here.

Pancreatic Hormones The most important hormone secreted by the islets is **insulin** (IN-su-lin), which is produced by beta (β) cells. As illustrated in **Figure 1-4**, insulin is an important signal in the regulation of glucose homeostasis by negative feedback. Increased blood glucose levels stimulate insulin production, and insulin stimulates glucose uptake and use by body cells. All cells use more glucose for energy, and liver and muscle cells convert excess glucose into glycogen, the storage form of glucose. As a result of these actions, blood glucose levels decline, and insulin secretion declines.

In addition to its role in glucose homeostasis, insulin also promotes overall tissue building. Under its effects, tissues store fatty acids as triglycerides (fats) and use amino acids to build proteins.

A second islet hormone, produced by alpha (α) cells, is **glucagon** (GLU-kah-gon), which works with insulin to regulate blood glucose levels. When blood glucose levels decrease, such as during an overnight fast, glucagon secretion increases. The liver responds to glucagon by increasing glucose production from amino acids and from glycogen. The increased hepatic glucose production increases blood glucose levels.

To summarize, insulin is known as the *hormone of feasting,* because food intake stimulates its release, and insulin acts to stimulate glucose use and storage. Glucagon, on the other hand, is the *hormone of fasting,* because starvation stimulates its release and glucagon promotes glucose production by the liver. The activities of insulin and glucagon are summarized in **Figure 12-9**.

Diabetes Mellitus When the pancreatic islet cells fail to produce enough insulin or body cells do not respond adequately to the insulin, metabolic processes throughout the body are disrupted. This condition, **diabetes mellitus** (di-ah-BE-teze mel-LI-tus), is the most common endocrine disorder. It is this form of diabetes (in contrast to diabetes insipidus) that is meant when the term *diabetes* is used alone. The metabolic problems associated with diabetes mellitus are as follows:

12

- Elevated blood glucose concentration, known as **hyperglycemia** (hi-per-gli-SE-me-ah). Insulin deficiency prevents most cells from taking up and using glucose, so the glucose remains in the blood. Also, in the absence of insulin, the liver produces glucose from glycogen and amino acids, and this newly synthesized glucose diffuses into blood. Hyperglycemia causes many of the signs, symptoms, and complications of diabetes mellitus, as discussed shortly.
- Elevated blood lipids, known as **hyperlipidemia**. Low insulin levels result in the release of more fatty acids from adipose cells. The liver converts the fatty acids into phospholipids and cholesterol, resulting in high blood levels of fats and the accelerated development of atherosclerosis (arterial degeneration).
- **Ketoacidosis** results from the presence of ketones (partially metabolized fatty acids) in blood. Ketones are an alternate energy source that the brain uses, and insulin normally inhibits their production. Ketones are acidic, so blood pH decreases (acidosis). Acidosis can cause confusion, coma, and eventually death.
- Impaired amino acid uptake and protein synthesis result in the weakness and poor tissue repair seen in people who have been diabetic for many years. It may also explain the reduced resistance to infection noted in diabetic patients.

Many of the acute and chronic problems associated with diabetes mellitus result from hyperglycemia. For instance:

- The kidneys excrete some of the excess glucose, drawing more water into the urine. The increased urine volume (polyuria) causes dehydration, which results in excess thirst (polydipsia), and potentially low blood pressure. Low blood pressure can cause coma and even death. Diabetes mellitus is named for Greek words that mean "siphon," based on the high output of urine, and "honey" because of the urine's sweetness. Frequent urination is an early sign of diabetes in Becky's case study.

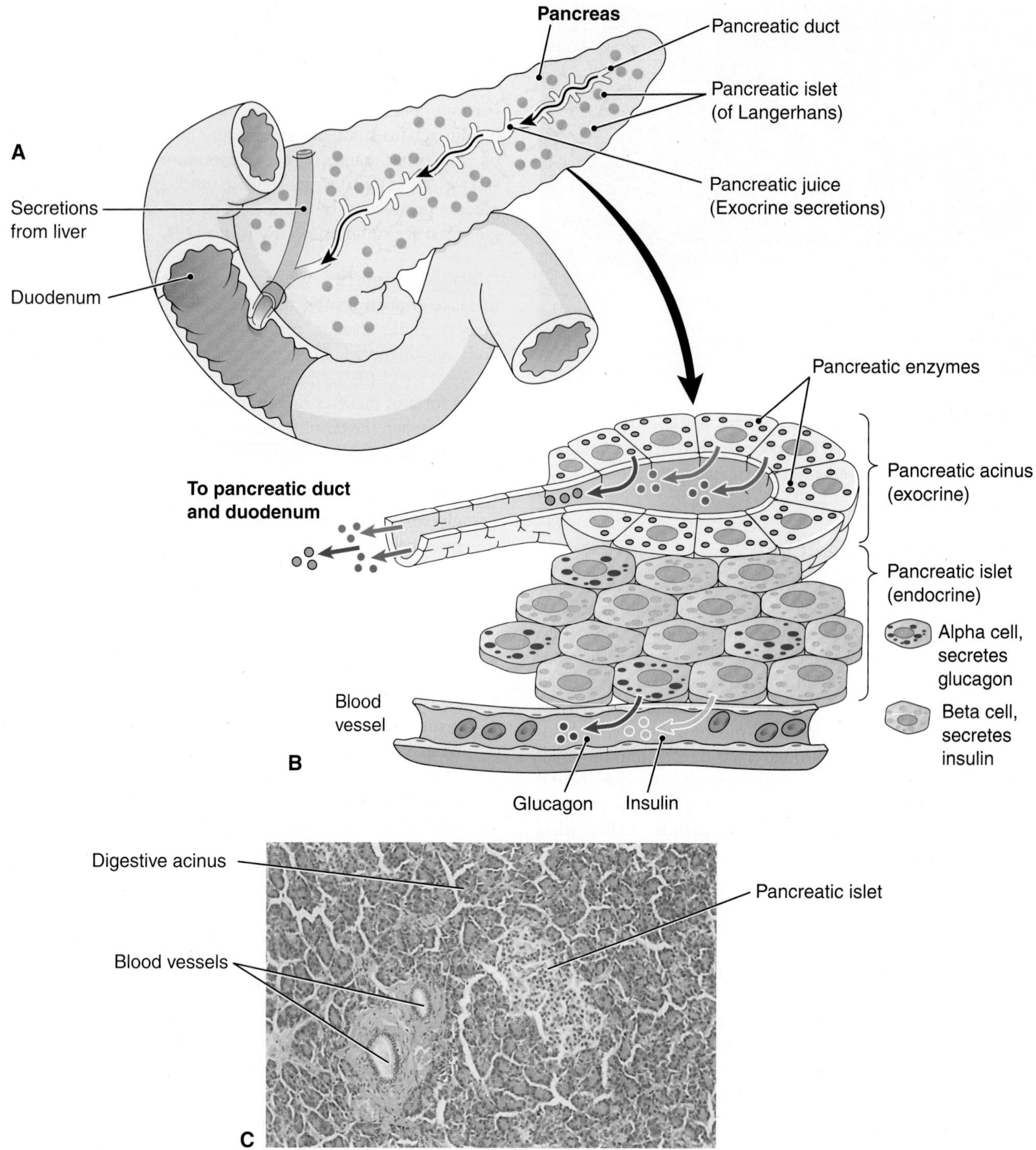

Figure 12-8 The pancreas and its functions. **KEY POINT** The pancreas has both exocrine and endocrine functions. **A.** The pancreas in relation to the duodenum (small intestine). **B.** Diagram of an acinus, which secretes digestive juices into ducts, and an islet, which secretes hormones into the bloodstream. **C.** Photomicrograph of pancreatic cells. Light-staining islets are visible among the many acini that produce digestive juices.

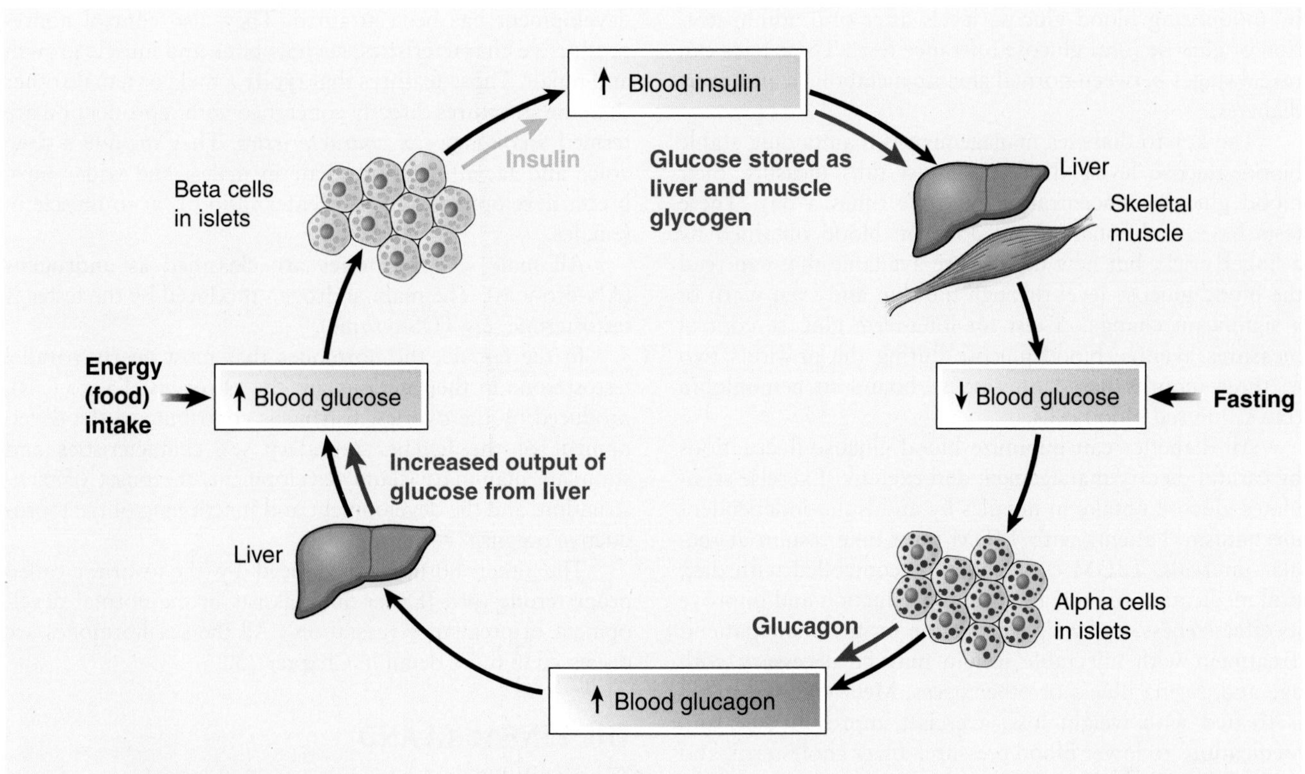

Figure 12-9 The effects of insulin and glucagon. **KEY POINT** These two hormones work together to regulate blood glucose levels. **ZOOMING IN** What organs do insulin and glucagon mainly influence?

12

- Persistent hyperglycemia and high lipid levels damage arteries, including the coronary arteries that supply blood to the heart muscle. These changes increase the risk for heart disease. Damage to the arteries that nourish the retina leads to diabetic retinopathy, a leading cause of blindness in American adults. Capillaries, such as those in the kidney, are often damaged as well.
- Persistent hyperglycemia damages peripheral nerves with accompanying pain and loss of sensation. Damage to the autonomic nervous system can result in poor stomach emptying.

Classification Diabetes is divided into two main types:

- Type 1 diabetes mellitus (T1DM) is less common but more severe. This disease usually appears before the age of 30 years and is brought on by an autoimmune destruction of the insulin-producing β cells in the islets.
- Type 2 diabetes mellitus (T2DM) characteristically occurs in adults, although the incidence has gone up considerably in the United States in recent years among younger people. It is typically associated with overweight in both adults and children. In this form of diabetes, the ability of the pancreas to secrete insulin may be mildly or severely impaired. However, in all cases of T2DM, the ability of body cells to respond to the hormone is diminished, a state referred to as *insulin resistance*. **Metabolic syndrome** (also called *syndrome X* or *insulin resistance syndrome*) is a set of clinical signs that increase the risk of both T2DM and cardiovascular disease. These signs include hyperglycemia, high levels of plasma triglycerides (fats), low levels of high-density lipoproteins (HDLs, the healthy form of cholesterol), and hypertension (high blood pressure). The underlying causes are thought to be central body obesity and insulin resistance.

Diabetes that develops during pregnancy is termed *gestational diabetes*. This form of diabetes usually disappears after childbirth, although it is a risk factor for T2DM later in life. Gestational diabetes usually affects women with a family history of diabetes, those who are obese, or those who are of older age. Diagnosis and treatment are important because of a high risk of complications for both the mother and the fetus.

Diabetes may also develop in association with other disorders, including pancreatic disease or other endocrine disorders. Viral infections, toxic chemicals in the environment, and drugs may also be involved.

Diagnosis and Treatment Diabetes mellitus is diagnosed by measuring blood glucose levels with or without fasting and by monitoring blood glucose levels after oral administration of glucose (oral glucose tolerance test). These tests can reveal stages between normal glucose metabolism and overt diabetes.

The key to diabetes management is maintaining stable blood glucose levels. Diabetics must thus measure their blood glucose concentration multiple times a day. These tests have traditionally been done on blood obtained by a finger prick, but new devices are available that can read the blood glucose level through the skin and even warn of a significant change. A test for long-term glucose control measures average blood glucose during the previous two to three months based on glucose bound to hemoglobin (HbA_{1c}) in red blood cells.

All diabetics can minimize blood glucose fluctuations by careful dietary management and exercise. Exercise stimulates glucose uptake in muscles by an insulin-independent mechanism. Patients with T1DM must take insulin at regular intervals. T2DM can usually be controlled with diet, oral medication to increase insulin production and improve its effectiveness, and weight reduction for the obese patient. Treatment with injectable insulin may be necessary with age and during illness or other stress. Metabolic syndrome is treated with weight loss, exercise, improved diet, and medications to lower blood pressure, lower cholesterol, and decrease insulin resistance.

Diabetics dependent on insulin require multiple injections each day. An alternate method for insulin administration is by means of a pump that provides an around-the-clock supply. The insulin is placed in a device—typically no larger than a small cell phone—that the patient wears close to the body. The device injects insulin through tubing into the subcutaneous tissues of the abdomen. People taking insulin injections are subject to episodes of low blood glucose and should carry notification of their disease.

Methods of administering insulin by pills or capsules, inhaler spray, or skin patches are still in the experimental stage. Researchers are also studying the possibility of transplanting the pancreas or islet cells to take over for failed cells in people with diabetes.

See the Student Resources on thePoint to view the animations "Hormonal Control of Glucose" and "Diabetes." See also an illustration on the effects of metabolic syndrome.

THE SEX GLANDS

The sex glands, the female ovaries and the male testes, not only produce the sex cells (sperm and ova) but are also important endocrine organs. The hormones produced by these organs are needed in the development of the sexual characteristics, which usually appear in the early teens, and for the maintenance of the reproductive organs once full development has been attained. They also control nonreproductive characteristics, such as bone and muscle growth and repair. Those features that typify a male or female other than the structures directly concerned with reproduction are termed *secondary sex characteristics*. They include a deep voice and facial and body hair in males, and wider hips, breast development, and a greater ratio of fat to muscle in females.

All male sex hormones are classified as **androgens** (AN-dro-jens). The main androgen produced by the testes is **testosterone** (tes-TOS-ter-one).

In the female, the hormones that most nearly parallel testosterone in their actions are the **estrogens** (ES-tro-jens), produced by the ovaries. Estrogens contribute to the development of the female secondary sex characteristics and stimulate mammary gland development, the onset of menstruation, and the development and functioning of the reproductive organs.

The other hormone produced by the ovaries, called **progesterone** (pro-JES-ter-one), assists in the normal development of pregnancy (gestation). All the sex hormones are discussed in more detail in Chapter 23.

THE PINEAL GLAND

The **pineal** (PIN-e-al) **gland** is a small, flattened, cone-shaped structure located posterior to the midbrain and connected to the roof of the third ventricle (**see Fig. 12-2**).

The pineal gland produces the hormone **melatonin** (mel-ah-TO-nin) during dark periods; little hormone is produced during daylight hours. This pattern of hormone secretion influences the regulation of sleep–wake cycles (**see Box 12-2**). Melatonin also appears to delay the onset of puberty.

CHECKPOINTS

- ☐ 12-15 What two hormones produced by the islets of the pancreas regulate blood glucose levels?
- ☐ 12-16 What hormone is low or ineffective in cases of diabetes mellitus?
- ☐ 12-17 Sex hormones confer certain features associated with male and female gender. What are these features called as a group?
- ☐ 12-18 What hormone does the pineal gland secrete?

Other Hormone-Producing Tissues

Originally, the word *hormone* applied to the secretions of the endocrine glands only. The term now includes various body substances that have regulatory actions, either locally or at a distance from where they are produced. Many body organs and tissues produce such regulatory substances.

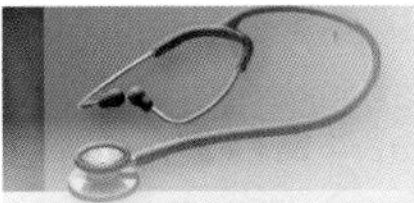

CLINICAL PERSPECTIVES

Box 12-2

Seasonal Affective Disorder: Seeing the Light

Most of us find that long, dark days make us blue and sap our motivation. Are these learned responses or is there a physical basis for them? Studies have shown that the amount of light in the environment does have a physical effect on behavior. Evidence that light alters mood comes from people who are intensely affected by the dark days of winter—people who suffer from **seasonal affective disorder**, aptly abbreviated SAD. When days shorten, these people feel sleepy, depressed, and anxious. They tend to overeat, especially carbohydrates. Research suggests that SAD has a genetic basis and may be associated with decreased levels of the neurotransmitter serotonin.

As light strikes the retina of the eye, it sends impulses that decrease the amount of melatonin produced by the pineal gland in the brain. Because melatonin depresses mood, the final effect of light is to elevate mood. Daily exposure to bright lights has been found to improve the mood of most people with SAD. Exposure for 15 minutes after rising in the morning may be enough, but some people require longer sessions both morning and evening. Other aids include aerobic exercise, stress management techniques, and antidepressant medications.

12

HORMONE-PRODUCING ORGANS

Some body organs that produce hormones include the following:

- Adipose tissue (fat) produces **leptin**, a hormone that controls appetite.
- The small intestine secretes hormones that control appetite and help regulate the digestive process.
- The kidneys produce a hormone called **erythropoietin** (e-rith-ro-POY-eh-tin) (EPO), which stimulates red blood cell production in the bone marrow. Production of this hormone increases when there is a decreased oxygen supply in the blood.
- **Osteocalcin**, produced in bone, stimulates such diverse processes as bone formation and insulin secretion.
- The atria (upper chambers) of the heart produce a substance called **atrial natriuretic** (na-tre-u-RET-ik) **peptide** (**ANP**) in response to their increased filling with blood. ANP increases sodium excretion by the kidneys and lowers blood pressure.
- The thymus is a mass of lymphoid tissue that lies in the upper part of the chest superior to the heart (see **Figure A5-7** in the Dissection Atlas). This organ is important in the development of immunity early in life, but it shrinks and becomes less important in adulthood. Its hormone, thymosin (THI-mo-sin), assists in the maturation of certain white blood cells known as T cells (T lymphocytes) after they have left the thymus gland and taken up residence in lymph nodes throughout the body. The immune system is discussed in Chapter 17.
- The **placenta** (plah-SEN-tah) produces several hormones during pregnancy. These cause changes in the uterine lining, and later in pregnancy, they help prepare the breasts for lactation. Pregnancy tests are based on the presence of placental hormones (see Chapter 24).

PROSTAGLANDINS

Prostaglandins (pros-tah-GLAN-dins) are a group of hormone-like substances derived from fatty acids. The name *prostaglandin* comes from the fact that they were first discovered in semen and thought to be derived from the male prostate gland. We now know that prostaglandins are synthesized by almost all body cells. One reason that they are not strictly classified as hormones is that they are produced, act, and are rapidly inactivated in or close to where they are produced. In addition, they are produced not at a defined location, but throughout the body.

A bewildering array of functions has been ascribed to prostaglandins. Some cause constriction of blood vessels, bronchial tubes, and the intestine, whereas others cause dilation of these same structures. Prostaglandins are active in promoting inflammation; certain antiinflammatory drugs, such as aspirin, act by blocking prostaglandin production. Some prostaglandins have been used to induce labor or abortion and have been recommended as possible contraceptive agents.

Overproduction of prostaglandins by the uterine lining (endometrium) can cause painful cramps of the uterine muscle during menstruation. Treatment with prostaglandin inhibitors has been successful in some cases. Much has been written about these substances, and extensive research on them continues.

Hormones and Treatment

Hormones used for medical treatment are obtained from several different sources. Some are extracted from animal tissues. Some hormones and hormone-like substances are available in synthetic form, meaning that they are manufactured in commercial laboratories. A few hormones are produced by the genetic engineering technique of recombinant DNA. In this method, a gene for the cellular manufacture of a given product is introduced in the laboratory into a harmless strain of the common bacterium *Escherichia coli*. The organisms are then grown in quantity, and the desired substance is harvested and purified.

A few examples of natural and synthetic hormones used in treatment are as follows:

- *Growth hormone* is used for the treatment of children and adults with a deficiency of this hormone. Adequate supplies are produced by recombinant DNA techniques.
- *Insulin* is used in the treatment of diabetes mellitus. Pharmaceutical companies now produce "human" insulin by recombinant DNA methods.
- *Adrenal steroids*, primarily the glucocorticoids, are used for the relief of inflammation in such diseases as rheumatoid arthritis, lupus erythematosus, asthma, and cerebral edema; for immunosuppression after organ transplantation; and for relief of symptoms associated with circulatory shock.
- *Epinephrine* (adrenaline) has many uses, including stimulation of the heart muscle when rapid response is required; treatment of asthmatic attacks by relaxation of the small bronchial tubes; and treatment of the acute allergic reaction called **anaphylaxis** (an-ah-fi-LAK-sis).
- *Thyroid hormones* are used in the treatment of hypothyroidism and as replacement therapy after surgical removal of the thyroid gland.
- *Oxytocin* is used to cause uterine contractions and induce labor.
- *Androgens*, including testosterone and androsterone, are used in severe chronic illness to aid tissue building and promote healing.
- *Estrogen and progesterone* are used as oral contraceptives (e.g., birth control pills, "the pill"). They are highly effective in preventing pregnancy. Occasionally, they give rise to unpleasant side effects, such as nausea. More rarely, they cause serious complications, such as thrombosis (blood clots) or hypertension (high blood pressure). These adverse side effects are more common among women who smoke. Any woman taking birth control pills should have a yearly medical examination.

In women experiencing menopause, levels of estrogen and progesterone begin to decline. Thus, preparations of synthetic estrogen and progesterone, called *hormone replacement therapy* (*HRT*) or *menopausal hormone therapy (MHT)*, have been developed to treat the symptoms associated with this decline and to protect against the adverse changes—such as decreased bone density—that typically occur in the years after menopause. Recent studies on the most popular forms of MHT have raised questions about their benefits and revealed some risks associated with their use. See Chapter 23 for more information.

Hormones and Stress

Stress in the form of physical injury, disease, emotional anxiety, and even pleasure calls forth a specific physiologic response that involves both the nervous system and the endocrine system. The nervous system response, the "fight-or-flight" response, is mediated by parts of the brain, especially the hypothalamus, and by the sympathetic nervous system, which releases norepinephrine. The hypothalamus also coordinates the endocrine response to stress, stimulating the production of some hormones and inhibiting the production of others. Some of these hormones include:

- Epinephrine released from the adrenal medulla facilitates the fight or flight response.
- Cortisol released from the adrenal gland helps deal with the stress of starvation, increasing nutrient availability in blood, and inhibiting inflammation.
- ADH released from the posterior pituitary promotes water conservation.
- Growth hormone released from the anterior pituitary also increases nutrient availability and helps repair damaged tissues. GH is only released in response to physical stress, such as exercise.

Stress inhibits the production of insulin in order to maximize nutrient availability. It also suppresses the production of thyroid hormones and sex hormones, because these hormones do not facilitate short-term survival.

These stimulatory and inhibitory changes in hormone levels help the body meet stressful situations. Unchecked, however, they are harmful and may lead to such stress-related disorders as hyperglycemia, high blood pressure, heart disease, ulcers, insomnia, back pain, and headaches. Cortisol decreases the immune response, leaving the body more susceptible to infection.

Although no one would enjoy a life totally free of stress in the form of stimulation and challenge, unmanaged stress, or "distress," has negative physical effects. For this reason, techniques such as biofeedback and meditation to control stress are useful. The simple measures of setting priorities, getting adequate periods of relaxation, and getting regular physical exercise are important in maintaining total health.

CHECKPOINTS

- ☐ **12-19** What organs produce each of the following: erythropoietin, osteocalcin, ANP?
- ☐ **12-20** What are four hormones released in times of stress?

See the Student Resources on thePoint for information on careers in exercise and fitness.

Effects of Aging on the Endocrine System

The incidence of endocrine diseases, particularly hypothyroidism and type 2 diabetes mellitus, increases with age. In addition, some of the changes associated with healthy aging, such as loss of muscle and bone tissue, can be linked to changes in the endocrine system.

GH declines, accounting for some losses in strength, immunity, skin thickness, and healing. Sex hormones decline during the middle-age years in both males and females. These changes come from decreased activity of the gonads and also decreased anterior pituitary activity, resulting in decline of gonadotropic hormone secretion. Production of adrenal androgens also declines. Decrease in bone mass leading to osteoporosis is one result of declining sex steroid production. HRT has shown some beneficial effects on mucous membranes and bone mass, but as noted above, its use is controversial. In contrast, the production of thyroid hormone, cortisol, and pancreatic hormones remains relatively constant in healthy old age.

Disease in Context Revisited

Becky's New "Normal"

Becky made her way to the kitchen, hoping she was still in time for pancakes. "Good morning, sleepyhead," greeted her mother as she handed Becky the glucose monitor and lancet. "How was your sleep last night?"

"Great," yawned Becky as she lanced the side of her finger and squeezed a tiny drop of blood onto the monitor's test strip. After a few seconds, the monitor beeped and displayed her blood glucose concentration. "I'm normal," said Becky, half-expecting a wisecrack from her little brother, but he kept on eating.

Since Dr. Carter's diagnosis, Becky had been getting used to her new "normal." It wasn't easy being diabetic. She had to be really careful about what she ate and when. She had to measure her glucose before meals and inject herself with insulin after. Monitoring her glucose wasn't too bad, but Becky didn't think she would ever get used to the needles. She was also a little worried about what the kids at school were saying about her and her disease. One unexpected benefit was that Max seemed to have a newfound respect for her and her ability to inject herself. "What a weirdo!" she thought as she carefully poured a little bit of syrup on her pancakes.

During this case, we saw that the lack of the hormone insulin had negative effects on Becky's whole body. In this and later chapters, we learn more about the endocrine system's role in regulating body functions.

CHAPTER 12

Chapter Wrap-Up

Summary Overview

A detailed chapter outline with space for note taking is on *thePoint*. The figure below illustrates the main topics covered in this chapter.

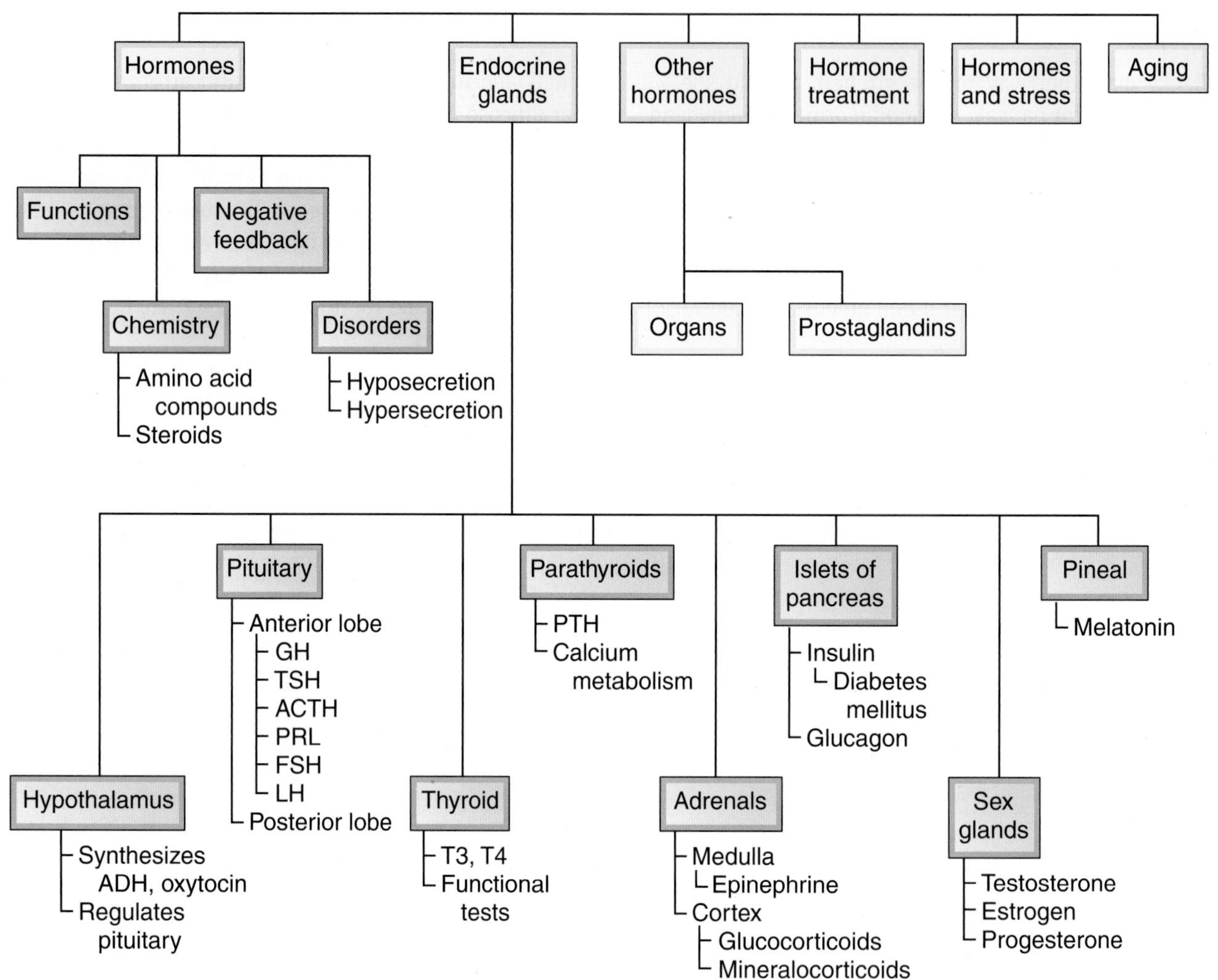

Key Terms

The terms listed below are emphasized in this chapter. Knowing them will help you organize and prioritize your learning. These and other boldface terms are defined in the Glossary with phonetic pronunciations.

endocrine system	hypothalamus	prostaglandin	steroid
hormone	pituitary (hypophysis)	receptor	target tissue

Word Anatomy

Medical terms are built from standardized word parts (prefixes, roots, and suffixes). Learning the meanings of these parts can help you remember words and interpret unfamiliar terms.

WORD PART	MEANING	EXAMPLE
The Endocrine Glands and Their Hormones		
acr/o	end, extremity	*Acromegaly* causes enlargement of the hands and feet.
andr/o	male	An *androgen* is any male sex hormone.
cortic/o	cortex	*Adrenocorticotropic* hormone acts on the adrenal cortex.
glyc/o	glucose, sugar	*Hyperglycemia* is high blood glucose.
insul/o	pancreatic islet, island	*Insulin* is a hormone produced by pancreatic islets.
lact/o	milk	*Prolactin* stimulates milk production in the breasts.
-megaly	enlargement	See "acr/o" example.
nephr/o	kidney	*Epinephrine* (adrenaline) is secreted by the adrenal gland near the kidney.
oxy	sharp, acute	*Oxytocin* stimulates uterine contractions during labor.
ren/o	kidney	The *adrenal* glands are near (ad-) the kidneys.
-sterone	steroid hormone	*Testosterone* is a steroid hormone from the testes.
toc/o	labor	See "oxy" example.
trop/o	acting on, influencing	*Testosterone* is a steroid hormone from the testes.
ur/o	urine	*Antidiuretic* hormone promotes reabsorption of water in the kidneys and decreases excretion of urine.
Other Hormone-Producing Tissues		
natri	sodium (*L. natrium*)	Atrial *natriuretic* peptide stimulates release of sodium in the urine.
-poiesis	making, forming	*Erythropoietin* is a hormone from the kidneys that stimulates production of red blood cells.

Questions for Study and Review

BUILDING UNDERSTANDING

Fill in the Blanks

1. Chemical messengers secreted by the endocrine glands are called ______.

2. The part of the brain that regulates pituitary gland activity is the ______.

3. Red blood cell production in the bone marrow is stimulated by the hormone ______.

4. The main androgen produced by the testes is ______.

5. A hormone produced by the heart is ______.

Matching > Match each numbered item with the most closely related lettered item.

___ **6.** A disorder caused by overproduction of growth hormone in the adult

___ **7.** A disorder caused by underproduction of parathyroid hormone

___ **8.** A disorder caused by overproduction of insulin

___ **9.** A disorder caused by overproduction of growth hormone in a child

___ **10.** A disorder caused by underproduction of antidiuretic hormone

a. hypoglycemia

b. gigantism

c. tetany

d. diabetes insipidus

e. acromegaly

Multiple Choice

___ **11.** To what do hormones bind?

a. lipid bilayer
b. transporters
c. ion channels
d. receptors

___ **12.** Which hormone promotes uterine contractions and milk ejection?

a. prolactin
b. oxytocin
c. estrogen
d. luteinizing hormone

___ **13.** Choose the principal hormonal regulator of metabolism.

a. thyroxine
b. triiodothyronine
c. aldosterone
d. progesterone

___ **14.** Which structure secretes epinephrine?

a. adrenal cortex
b. adrenal medulla
c. kidneys
d. pancreas

___ **15.** Which organ regulates sleep–wake cycles?

a. pituitary
b. thyroid
c. thymus
d. pineal

UNDERSTANDING CONCEPTS

16. With regard to regulation, what are the main differences between the nervous system and the endocrine system?

17. Explain how the hypothalamus and pituitary gland regulate certain endocrine glands. Use the thyroid as an example.

18. Name the two divisions of the pituitary gland. List the hormones released from each division, and describe the effects of each.

19. Compare and contrast the following hormones:

a. thyroxine and triiodothyronine
b. cortisol and aldosterone
c. insulin and glucagon
d. testosterone and estrogen

20. Describe the anatomy of the following endocrine glands:

a. thyroid
b. pancreas
c. adrenals

21. Compare and contrast the following diseases:

a. Hashimoto thyroiditis and Graves disease
b. type 1 diabetes and type 2 diabetes
c. Addison disease and Cushing syndrome

22. Name the hormone released by the kidneys and by the pineal gland. What are the effects of each?

23. List several hormones released during stress. What is the relationship between prolonged stress and disease?

24. Referring to Appendix 2, cite changes in the urine that accompany diabetes mellitus. What is the normal fasting level for glucose in the blood?

25. Referring to Appendix 5, give the figure and diagram numbers of any endocrine glands.

CONCEPTUAL THINKING

26. In the case study Dr. Carter noted that Becky presented with the three cardinal signs of type 1 diabetes mellitus. What are they? What causes them?

27. How is type 1 diabetes mellitus similar to starvation?

28. Mr. Jefferson has rheumatoid arthritis, which is being treated with glucocorticoids. During a recent checkup, his doctor notices that Mr. Jefferson's face is "puffy" and his arms are bruised. Why does the doctor decide to lower his patient's glucocorticoid dosage?

For more questions, see the Learning Activities on thePoint.

UNIT

V Circulation and Body Defense

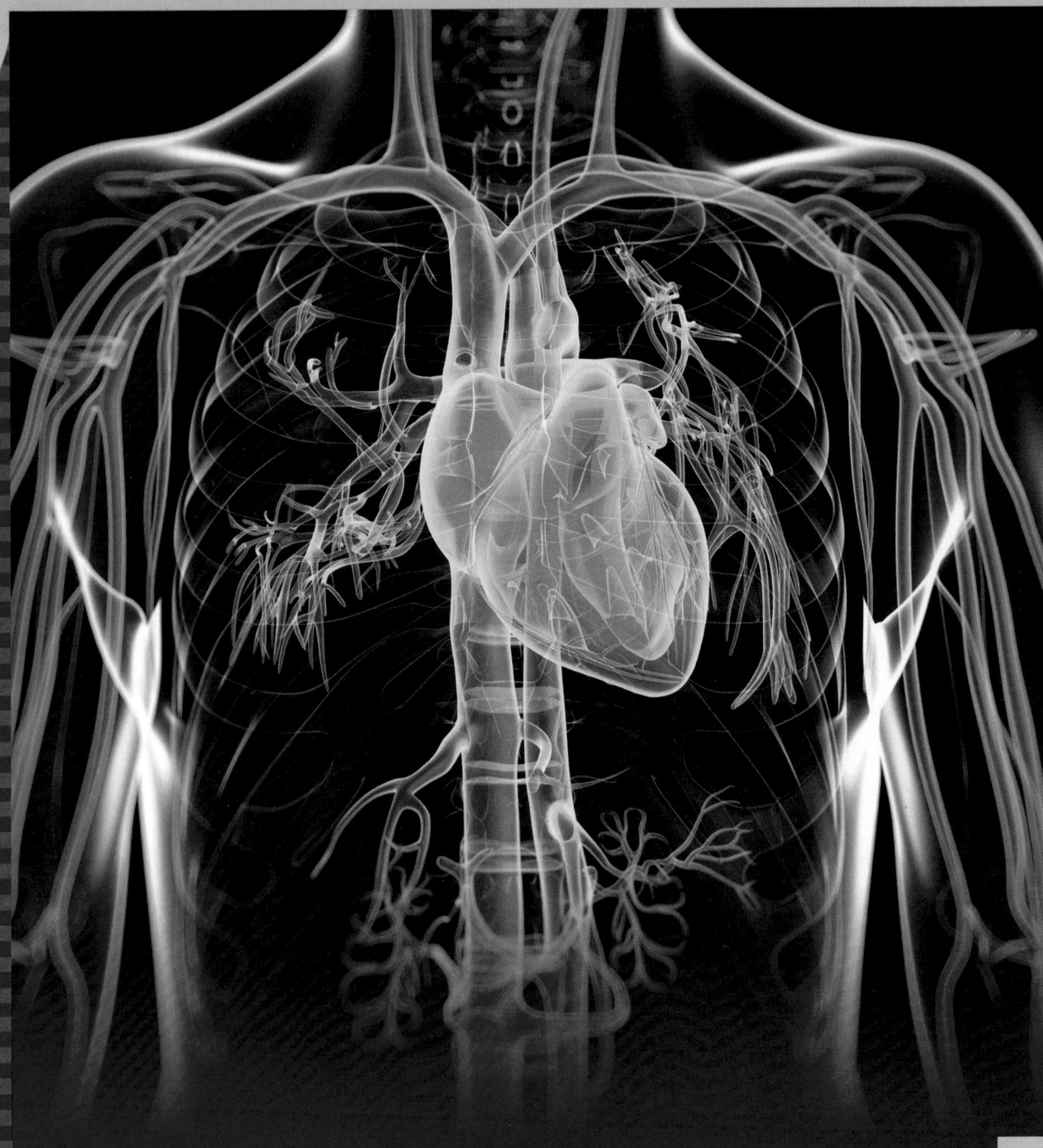

The chapters in this unit discuss the systems that move materials through the body. The blood is the main transport medium. It circulates through the cardiovascular system, consisting of the heart and the blood vessels. The lymphatic system, in addition to other functions, takes up excess tissue fluid and returns it to the cardiovascular system. Components of the blood and the lymphatic system also participate in the activities of the immune system, fighting external threats from harmful microbes and internal threats from cancerous cells.

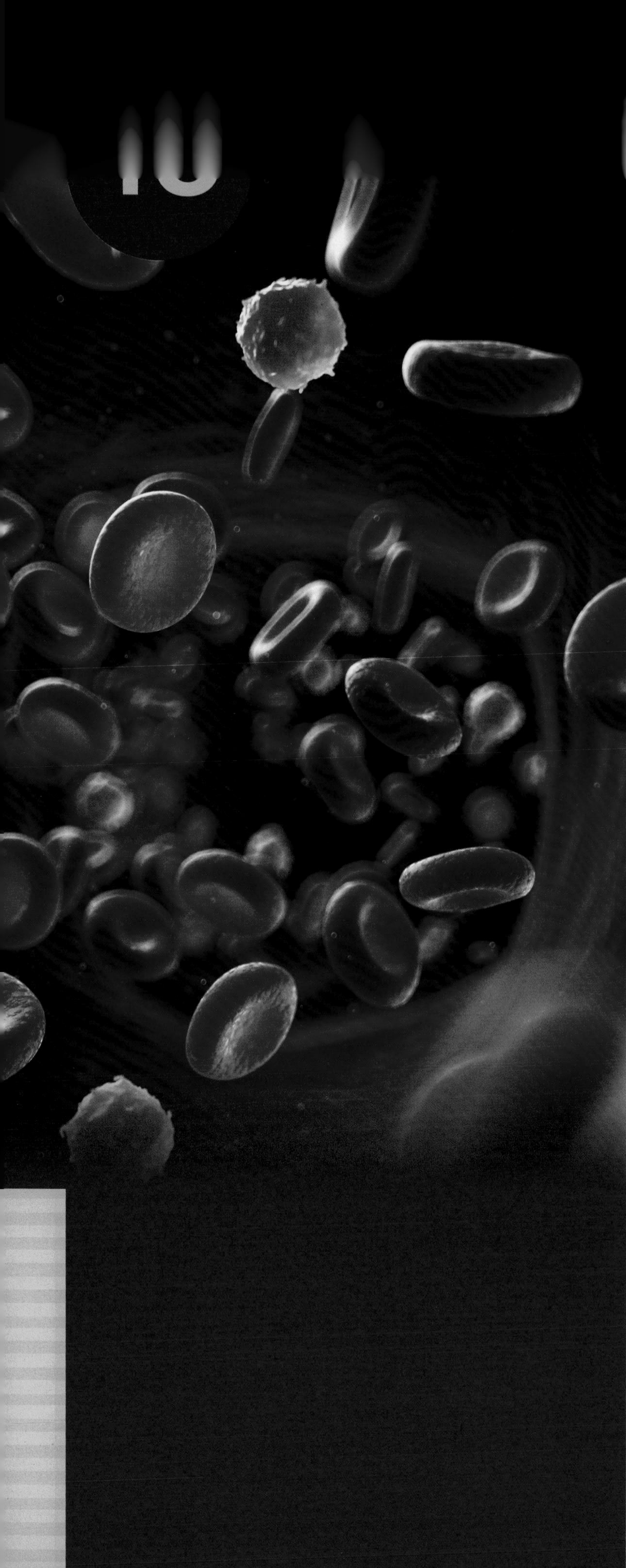

Learning Objectives

After careful study of this chapter, you should be able to:

1 ▶ List the functions of the blood. ***p. 292***

2 ▶ Identify the main components of plasma. ***p. 293***

3 ▶ Describe the formation of blood cells. ***p. 294***

4 ▶ Name and describe the three types of formed elements in the blood, and give the functions of each. ***p. 294***

5 ▶ Characterize the five types of leukocytes. ***p. 295***

6 ▶ Define *hemostasis*, and cite three steps in hemostasis. ***p. 298***

7 ▶ Briefly describe the steps in blood clotting. ***p. 298***

8 ▶ Define *blood type*, and explain the relation between blood type and transfusions. ***p. 299***

9 ▶ Explain the basis of Rh incompatibility and its possible consequences. ***p. 300***

10 ▶ List four possible reasons for transfusions of whole blood and blood components. ***p. 300***

11 ▶ Define *anemia*, and list six causes of anemia. ***p. 303***

12 ▶ Define *leukemia*, and name the two types of leukemia. ***p. 304***

13 ▶ Describe four forms of clotting disorders. ***p. 305***

14 ▶ Identify six types of tests used to study blood. ***p. 305***

15 ▶ Referring to the case study, discuss the adverse effects of bone marrow damage. ***pp. 291, 308***

16 ▶ Show how word parts are used to build words related to the blood (see Word Anatomy at the end of the chapter). ***p. 310***

Disease in Context ***Eleanor's Bone Marrow Failure***

Eleanor, a 52-year-old professor of anthropology, presented to the emergency room of a large metropolitan hospital with a severe nosebleed.

"I never have nosebleeds," she said. "This takes the cake; it just won't stop. My life is running out of my nose!"

Questioning by hospital staff revealed that eight years earlier she had surgery for breast cancer. Physicians had followed her closely until three years earlier, when she divorced and moved to her current job in a new city.

"I'm embarrassed to admit that I haven't seen a doctor in three years," she said. "I've just been too busy to have my regular checkups." Further questioning revealed nothing medically unusual. She mentioned, however, having felt extremely tired in the last few months.

"I seem to wear out at even the smallest tasks," she said. "Last week I stopped for a rest on a park bench on my way home. That's never been necessary before."

Eleanor was pale, and her skin contained numerous pinpoint hemorrhages. Otherwise, her physical examination was unremarkable. Blood analysis revealed a marked deficiency in red blood cells, white blood cells, and platelets. Her blood type was determined to be O positive.

In the emergency room, Eleanor's nose was packed with cotton strips, and she was transfused with platelets, which stopped the nosebleed. She was admitted to the hospital and transfused with red blood cells. Knowing that breast cancer has a tendency to spread to bones, the examining physician suspected that cancer cells had taken over her red bone marrow, the site of blood cell production. A bone marrow biopsy, which showed nearly complete replacement of normal bone marrow by cancer cells, confirmed his suspicion.

Eleanor was treated with additional chemotherapy but continued to need red blood cell transfusions to maintain an adequate hemoglobin level. She required antibiotic treatment on multiple occasions to treat pneumonia, skin abscesses, and recurrent diarrhea.

This case study shows the importance of the bone marrow in producing all of the cells found in blood. We'll learn more about the functions of the different blood cells in this chapter and follow up on Eleanor's case.

ANCILLARIES *At-A-Glance*

Visit thePoint to access the following resources. For guidance in using these resources most effectively, see pp. xv–xvii.

Learning RESOURCES

- Tips for Effective Studying
- Web Figure: Hematopoiesis
- Web Figure: Hemostasis
- Web Figure: Production, Circulation, and Death of Red Blood Cells
- Web Figure: The DNA Mutation in Sickle Cell Disease
- Web Figure: Clinical and Pathologic Findings in Sickle Cell Anemia
- Web Figure: Acute Myelogenous Leukemia
- Web Figure: Chronic Lymphocytic Leukemia
- Animation: Hemostasis
- Health Professions: Hematology Specialist
- Detailed Chapter Outline
- Answers to Questions for Study and Review
- Audio Pronunciation Glossary

Learning ACTIVITIES

- Pre-Quiz
- Visual Activities
- Kinesthetic Activities
- Auditory Activities

A LOOK BACK

Blood is one of the two types of circulating connective tissue introduced in Chapter 4. The other is lymph, discussed more fully in Chapter 16. Blood was mentioned in previous chapters as a transport medium for hormones (Chapter 12) and for substances needed for muscle contraction (Chapter 8). In this chapter, we fully discuss the structure and function of this all-important body fluid.

Blood is a life-giving fluid that brings nutrients and oxygen to the cells and carries away waste. The heart pumps blood continuously through a closed system of vessels. The heart and blood vessels are described in Chapters 14 and 15, respectively.

Blood is classified as a connective tissue because it consists of cells suspended in an extracellular background material, or matrix. However, blood differs from other connective tissues in that its cells are not fixed in position; instead, they move freely in the plasma, the blood's liquid matrix.

Whole blood is a viscous (thick) fluid that varies in color from bright scarlet to dark red, depending on how much oxygen it is carrying. (It is customary in drawings to color blood high in oxygen as red and blood low in oxygen as blue.) The blood volume accounts for approximately 8% of total body weight. The actual quantity of circulating blood differs with a person's size and gender; adult men have about 75 mL/kg body weight, while adult women have about 65 mL/kg body weight. So, a 150-lb (68-kg) male contains about 5.1 L (5.4 qt) of blood, and a 120-lb (54-kg) woman contains about 3.5 L (3.7 qt).

Functions of the Blood

The circulating blood serves the body in three ways: transportation, regulation, and protection.

TRANSPORTATION

- **Gases.** Oxygen from inhaled air diffuses into the blood through thin membranes in the lungs and is carried by the circulation to all body tissues. Carbon dioxide, a waste product of cellular metabolism, is carried from the tissues to the lung capillaries, where it diffuses into the air sacs (alveoli) and is breathed out.
- **Nutrients.** The blood transports nutrients, water, vitamins, and electrolytes to the cells. These materials enter the blood from the digestive system or are released into the blood from body reserves.
- **Waste.** The blood transports waste products from cells to organs that can destroy them or eliminate them from the body. For example, the kidney excretes excess water, acid, electrolytes, and urea (a nitrogen-containing waste) in urine. The liver inactivates hormones and drugs and sends blood pigments and other substances into the digestive tract for elimination. The lungs exhale carbon dioxide.
- **Hormones.** The blood carries hormones from their sites of synthesis to their target organs.

REGULATION

Water and most dissolved substances (except proteins) can move freely between the blood and the fluid surrounding cells (the interstitial fluid). So, by regulating the composition of blood, the body also regulates the interstitial fluid.

- **pH.** Buffers in the blood help keep the pH of body fluids steady at about 7.4. (The actual range of blood pH is 7.35 to 7.45.) Recall that pH is a measure of a solution's acidity or alkalinity. At an average pH of 7.4, blood is slightly alkaline (basic).
- **Fluid balance.** The blood regulates the amount of fluid in the tissues by means of substances (mainly proteins) that maintain the proper osmotic pressure. Recall from Chapter 3 that osmotic pressure is related to the concentration of dissolved and suspended materials in a solution; as their concentration increases, osmotic pressure increases. Normal blood concentrations of proteins and electrolytes are necessary to maintain normal blood volume, as described in Chapter 15.
- **Heat.** The blood transports heat that is generated in the muscles to other parts of the body, thus aiding in the regulation of body temperature.

PROTECTION

- **Disease.** The blood is important in defense against disease. It carries the substances and cells of the immune system that protect against pathogens.
- **Blood loss.** The blood contains substances called *clotting factors* that protect against blood loss from the site of an injury.

CHECKPOINTS

- ☐ 13-1 What are four types of substances transported in the blood?
- ☐ 13-2 What is the pH range of the blood?

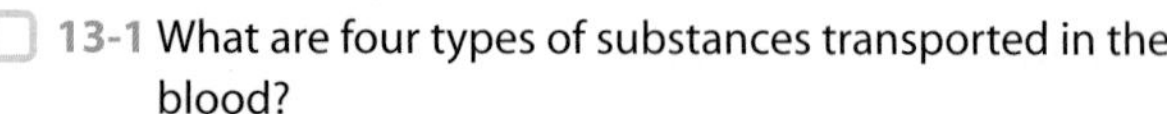

See the Student Resources on thePoint for information on careers in hematology, the study of blood.

Blood Constituents

The blood is divided into two main components (**Fig. 13-1**). The liquid portion is the **plasma**. The **formed elements**, which include cells and cell fragments, can be further divided into three categories:

- **Erythrocytes** (eh-RITH-ro-sites), from *erythro*, meaning "red," are the red blood cells, which transport oxygen.

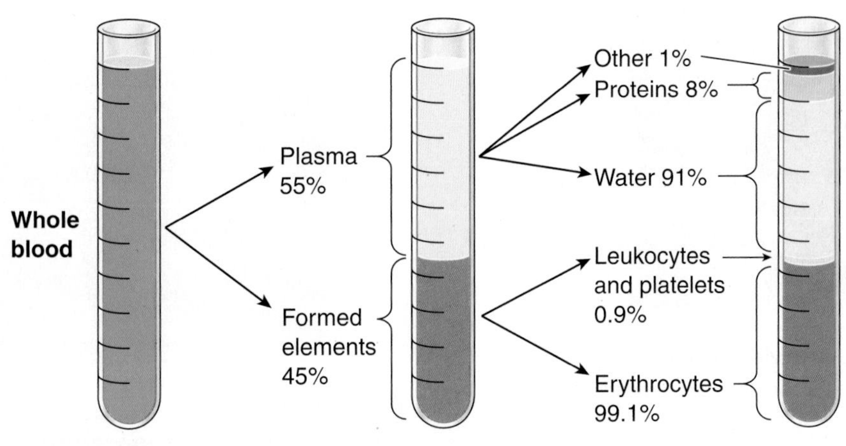

Figure 13-1 **Composition of whole blood.** **KEY POINT** Percentages show the relative proportions of the different components of plasma and formed elements.

- **Leukocytes** (LU-ko-sites), from *leuko*, meaning "white," are the several types of white blood cells (WBCs), which protect against infection.
- **Platelets**, also called **thrombocytes** (THROM-bo-sites), are cell fragments that participate in blood clotting.

Table 13-1 summarizes information on the different types of formed elements. Figure 13-2 shows all the categories of formed elements in a blood smear, that is, a blood sample spread thinly over the surface of a glass slide, as viewed under a microscope.

BLOOD PLASMA

About 55% of the total blood volume is plasma. The plasma itself is 91% water. Many different substances, dissolved or suspended in the water, make up the other 9% by weight (see Fig. 13-1). The plasma content may vary somewhat because substances are removed and added as the blood circulates to and from the tissues. However, the body tends to maintain a fairly constant level of most substances. For example, the level of glucose, a simple sugar, is maintained at a remarkably constant level of about one-tenth of 1% (0.1%) in solution.

After water, the next largest percentage (about 8%) of material in the plasma is protein. The liver synthesizes most plasma proteins. They include the following:

- **Albumin** (al-BU-min), the most abundant protein in plasma, is important for maintaining the blood's osmotic pressure. Recall from Chapter 3 that osmotic pressure reflects the ability of a solution to attract water. Albumin is thus necessary to maintain normal blood volume.
- **Clotting factors**, necessary for preventing blood loss from damaged vessels, are discussed later.
- **Antibodies**, substances that combat infection and are made by certain WBCs involved in immunity.
- **Complement** consists of a group of enzymes that participate in immunity (see Chapter 17).

The remaining 1% of the plasma consists of nutrients, electrolytes, and other materials that must be transported. With regard to the nutrients, glucose is the principal carbohydrate found in the plasma. This simple sugar is absorbed from digested foods in the intestine. It can also be released from the liver, where it is stored as glycogen. Amino acids, the products of protein digestion, also circulate in the plasma. Lipids constitute a

Table 13-1 Formed Elements of Blood

Elements	Number per mcL of Blood	Description	Function
Erythrocytes (red blood cells)	5 million	Tiny (7 mcm diameter), biconcave disks without nucleus (anuclear)	Carry oxygen bound to hemoglobin; also carry some carbon dioxide and buffer blood
Leukocytes (white blood cells)	5,000–10,000	Larger than red cells with prominent nucleus that may be segmented (granulocyte) or unsegmented (agranulocyte); vary in staining properties	Active in immunity; located in blood, tissues, and lymphatic system
Platelets	150,000–450,000	Fragments of large cells (megakaryocyte)	Hemostasis; form a platelet plug and start blood clotting (coagulation)

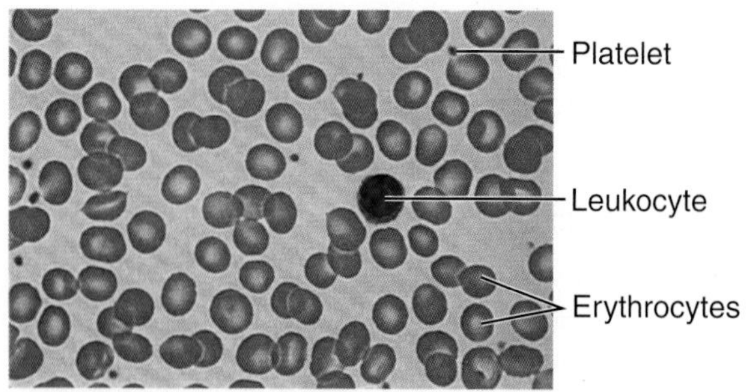

Figure 13-2 Blood cells as viewed under the microscope. **KEY POINT** All three types of formed elements are visible. **ZOOMING IN** Which cells are the most numerous in the blood?

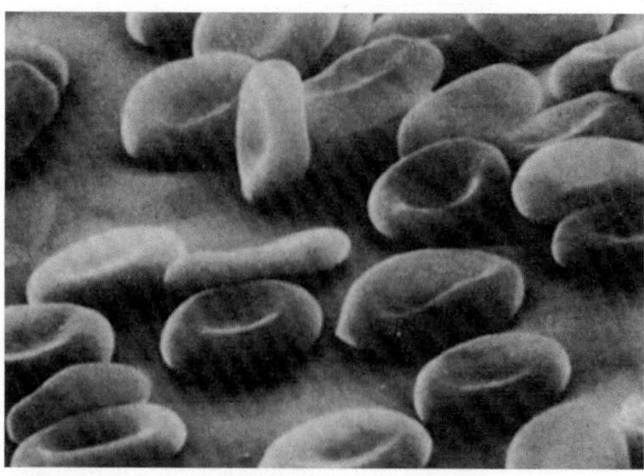

Figure 13-3 Red blood cells as seen under a scanning electron microscope. **KEY POINT** This type of microscope provides a three-dimensional view of the cells, revealing their shape. **ZOOMING IN** Why are these cells described as biconcave?

small percentage of blood plasma. Lipid components include cholesterol and fats. As lipids are not soluble in plasma, they combine with proteins to form lipoproteins. The electrolytes in the plasma include sodium, potassium, calcium, magnesium, chloride, carbonate, and phosphate. These electrolytes have a variety of functions, including the formation of bone (calcium and phosphorus), the production of certain hormones (such as iodine for the production of thyroid hormones), and maintenance of the acid–base balance (such as sodium and potassium carbonates and phosphates present in buffers).

Other materials transported in plasma include hormones, waste products, drugs, and dissolved gases, primarily oxygen and carbon dioxide.

THE FORMED ELEMENTS

All of the blood's formed elements are produced in red bone marrow, which is located in the ends of long bones and in the inner portion of all other bones. The ancestors of all the blood cells are called **hematopoietic** (blood-forming) **stem cells.** These cells reproduce frequently, and their offspring differentiate into the different blood cell types discussed below.

In comparison with other cells, most blood cells are short-lived. The need for constant blood cell replacement means that normal activity of the red bone marrow is absolutely essential to life. In the opening case study, Eleanor has lost the protective functions of the blood cells due to bone marrow damage.

See the Student Resources on thePoint for a figure on hematopoiesis detailing the development of all the formed elements and a figure on the life cycle of red cells.

Erythrocytes Erythrocytes, the red blood cells (RBCs or red cells), measure about 7 mcm (micrometers) in diameter (a micrometer is one-millionth of a meter). They are disk-shaped bodies with a depression on both sides. This biconcave shape creates a central area that is thinner than the edges (**Fig. 13-3**). Erythrocytes are different from other cells in that the mature form found in the circulating blood lacks a nucleus (is anuclear) and also lacks most of the other organelles commonly found in cells. As red cells mature, these components are lost, providing more space for the cells to carry oxygen. This vital gas is bound in the red cells to **hemoglobin** (he-mo-GLO-bin), a protein that contains iron. (see **Box 13-1** on the structure and function of hemoglobin.) Hemoglobin, combined with oxygen, gives the blood its characteristic red color. The more oxygen carried by the hemoglobin, the brighter is the blood's red color. Therefore, the blood that goes from the lungs to the tissues is a bright red because it carries a great supply of oxygen; in contrast, the blood that returns to the lungs is a much darker red because it has given up some of its oxygen to the tissues.

Hemoglobin has two lesser functions in addition to the transport of oxygen. Hemoglobin can carry hydrogen ions, so it acts as a buffer and plays an important role in acid–base balance (see Chapter 21). Hemoglobin also carries some carbon dioxide from the tissues to the lungs for elimination.

Hemoglobin's ability to carry oxygen can be blocked by carbon monoxide. This odorless and colorless but harmful gas combines with hemoglobin to form a stable compound that can severely restrict the erythrocytes' ability to carry oxygen. Carbon monoxide is a byproduct of the incomplete burning of fuels, such as gasoline and other petroleum products and coal, wood, and other carbon-containing materials. It also occurs in cigarette smoke and automobile exhaust.

Erythrocytes are by far the most numerous of the blood cells, averaging from 4.5 to 5 million per microliter (mcL) of blood. (A microliter is one-millionth of a liter.) Because mature red cells have no nucleus and cannot divide or repair themselves, they must be replaced constantly. After leaving the bone marrow, they circulate in the bloodstream for about 120 days before their membranes deteriorate, and they are destroyed by the liver and spleen. Red cell production is stimulated by the hormone **erythropoietin** (eh-rith-ro-POY-eh-tin) (**EPO**), which is released from the kidney in response to decreased oxygen. Constant red cell production requires an adequate supply of nutrients, particularly protein; the vitamins B_{12} and folic acid, required for the production of DNA; and the minerals iron and copper for the production of hemoglobin. Vitamin C is also important for the proper absorption of iron from the small intestine.

Box 13-1

A CLOSER LOOK

Hemoglobin: Door-to-Door Oxygen Delivery

The hemoglobin molecule is a protein made of four amino acid chains (the globin part of the molecule), each of which holds an iron-containing heme group. Each of the four hemes can bind one molecule of oxygen.

Hemoglobin allows the blood to carry much more oxygen than it could were the oxygen simply dissolved in the plasma. A red blood cell contains about 250 million hemoglobins, each capable of binding four molecules of oxygen. So a single red blood cell can carry about 1 billion oxygen molecules! Hemoglobin reversibly binds oxygen, picking it up in the lungs and releasing it in the body tissues. Active cells need more oxygen and also generate heat and acidity. These changing conditions promote the release of oxygen from hemoglobin into metabolically active tissues.

Immature red blood cells (erythroblasts) produce hemoglobin as they mature into erythrocytes in the red bone marrow. When the liver and spleen destroy old erythrocytes, they break down the released hemoglobin. Some of its components are recycled, and the remainder leaves the body as a brown fecal pigment called stercobilin. Despite some conservation, dietary protein and iron are still essential to maintain hemoglobin supplies.

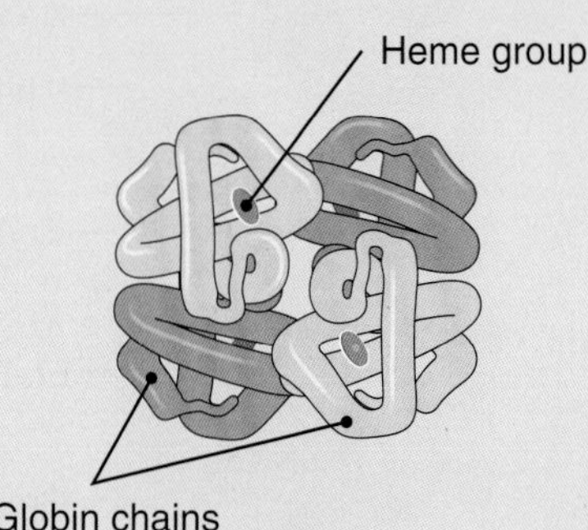

Hemoglobin. This protein in red blood cells consists of four amino acid chains (globins), each with an oxygen-binding heme group.

13

CHECKPOINTS 

- ☐ 13-3 What are the two main components of blood?
- ☐ 13-4 Next to water, what is the most abundant type of substance in plasma?
- ☐ 13-5 Where do blood cells form?
- ☐ 13-6 What type of cell gives rise to all blood cells?
- ☐ 13-7 What is the main function of hemoglobin?

Leukocytes The leukocytes, or white blood cells (WBCs or white cells), differ from the erythrocytes in appearance, quantity, and function. The cells themselves are round, but they contain prominent nuclei of varying shapes and sizes. Occurring at a concentration of 5,000 to 10,000/mcL of blood, leukocytes are outnumbered by red cells by about 700 to one. Although the red cells have a definite color, the leukocytes are colorless and must be stained if we are to study them under the microscope.

Types of Leukocytes The different types of white cells are identified by their size, the shape of the nucleus, and the appearance of granules in the cytoplasm when the cells are stained (Table 13-2). The stain commonly used for blood is the Wright stain, which is a mixture of dyes that differentiates the various blood cells. The "granules" in the white cells are actually lysosomes and secretory vesicles. They are present in all WBCs, but they are more easily stained and more visible in some cells than in others. Leukocytes are active in immunity. As we discuss later in this chapter, the relative percentage of the different types of leukocytes is a valuable clue in arriving at a medical diagnosis.

The granular leukocytes, or **granulocytes** (GRAN-u-lo-sites), are so named because they show visible granules in the cytoplasm when stained. Each has a very distinctive, highly segmented nucleus (see Table 13-2). The different types of granulocytes are named for the type of dyes they take up when stained. They include the following:

- **Neutrophils** (NU-tro-fils) stain with either acidic or basic dyes.
- **Eosinophils** (e-o-SIN-o-fils) stain with acidic dyes (eosin is one).
- **Basophils** (BA-so-fils) stain with basic dyes.

Neutrophils are the most numerous of the white cells, constituting approximately 60% of all circulating leukocytes. Because the nuclei of the neutrophils have various shapes, these cells are also called **polymorphs** (meaning "many forms") or simply *polys*. Other nicknames are *segs*, referring to the segmented nucleus, and *PMNs*, an abbreviation of polymorphonuclear neutrophils. Before reaching full maturity and becoming segmented, a neutrophil's nucleus looks like a thick, curved band (Fig. 13-4). An increase in the number of these **band cells** (also called *stab* or *staff cells*) is a sign of infection and active neutrophil production. Eosinophils and basophils make up a small percentage of the white cells but increase in number during allergic reactions.

The agranular leukocytes, or **agranulocytes**, are so named because they lack easily visible granules (see Table 13-2). Their nuclei are round or curved and are not segmented. There are two types of agranular leukocytes:

Table 13-2 Leukocytes (White Blood Cells)

Cell Type		Relative Percentage (Adult)	Description	Function
Neutrophils	Nucleus; Granules; Erythrocyte	54%–62%	Stain with either acidic or basic dyes; show lavender granules when stained	Phagocytosis
Eosinophils	Erythrocyte; Granules; Nucleus	1%–3%	Stain with acidic dyes; show beadlike, bright pink granules when stained	Allergic reactions; defense against parasites
Basophils	Nucleus; Granules	less than 1%	Stain with basic dyes; have large, dark blue granules that can obscure the nucleus	Allergic reactions; inflammatory reactions
Lymphocytes	Platelet; Nucleus; Erythrocyte	25%–38%	Mature and can multiply in lymphoid tissue	Immunity (T cells and B cells)
Monocytes	Erythrocyte; Nucleus	3%–7%	Largest of leukocytes	Phagocytosis

- **Lymphocytes** (LIM-fo-sites) are the second most numerous of the white cells. Although lymphocytes originate in the red bone marrow, they develop to maturity in lymphoid tissue and can multiply in this tissue as well. They are more abundant in the lymphatic system than in blood (see Chapter 16).
- **Monocytes** (MON-o-sites) are the largest of all white cells. They average about 5% of the leukocytes.

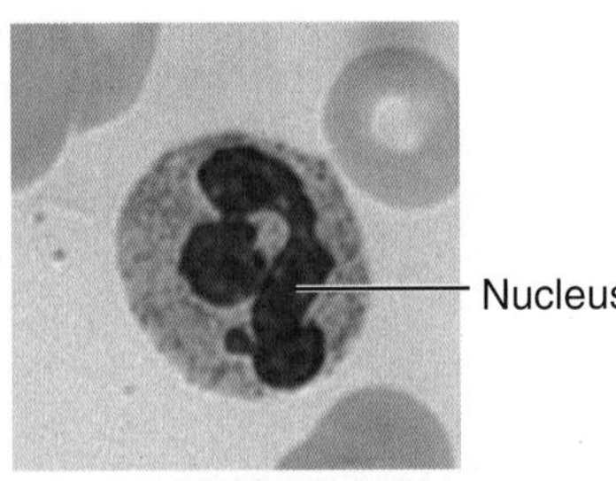

A Mature neutrophil

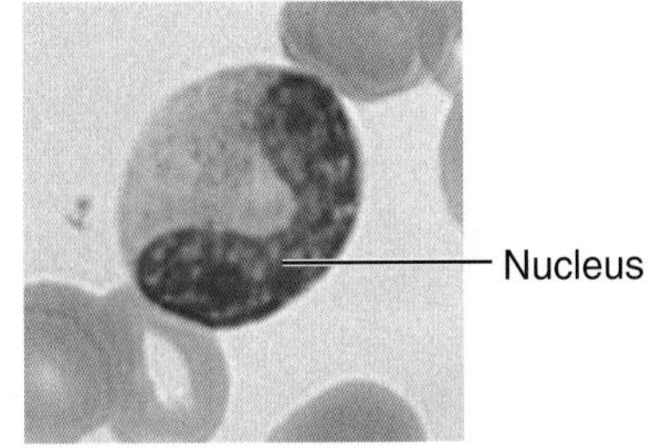

B Band cell (immature neutrophil)

Figure 13-4 **Stages in neutrophil development.** **KEY POINT** **A.** A mature neutrophil has a segmented nucleus. **B.** An immature neutrophil is called a band cell because the nucleus is shaped like a thick, curved band.

Functions of Leukocytes Leukocytes clear the body of foreign material and cellular debris. Most importantly, they destroy pathogens that may invade the body. Neutrophils and monocytes engage in **phagocytosis** (fag-o-si-TO-sis), the engulfing of foreign matter (**Fig. 13-5**). Whenever pathogens enter the tissues, as through a wound, phagocytes are attracted to the area. They squeeze between the cells of the capillary walls and proceed by ameboid (ah-ME-boyd), or ameba-like, motion to the area of infection where they engulf the invaders. Lysosomes in the cytoplasm then digest the foreign organisms, and the cells eliminate the waste products.

When foreign organisms invade, the bone marrow and lymphoid tissue go into emergency production of white cells, and their number increases enormously as a result. Detection of an abnormally large number of white cells in the blood is an indication of infection. In battling pathogens, leukocytes themselves are often destroyed. A mixture of dead and living bacteria, together with dead and living leukocytes, forms **pus**. A collection of pus localized in one area is known as an **abscess**.

When monocytes enter the tissues, they enlarge and mature into **macrophages** (MAK-ro-faj-ez). These phagocytic superstars are highly active in disposing of invaders and foreign material. Although most circulating lymphocytes live only six to eight hours, those that enter the tissues may survive for longer periods—days, months, or even years.

Some lymphocytes become **plasma cells**, active in the production of circulating antibodies needed for immunity. The activities of the various white cells are further discussed in Chapter 17.

13

Platelets **Blood platelets** (thrombocytes) are the smallest of all the formed elements (**Fig. 13-6A**). These tiny structures are not cells in themselves but rather fragments constantly released from giant bone marrow cells called **megakaryocytes** (meg-ah-KAR-e-o-sites) (**see Fig. 13-6B**). Platelets do not have nuclei or DNA, but they do contain active enzymes and mitochondria. The number of platelets in the circulating blood normally varies from 150,000 to 450,000/mcL. They have a life span of about 10 days.

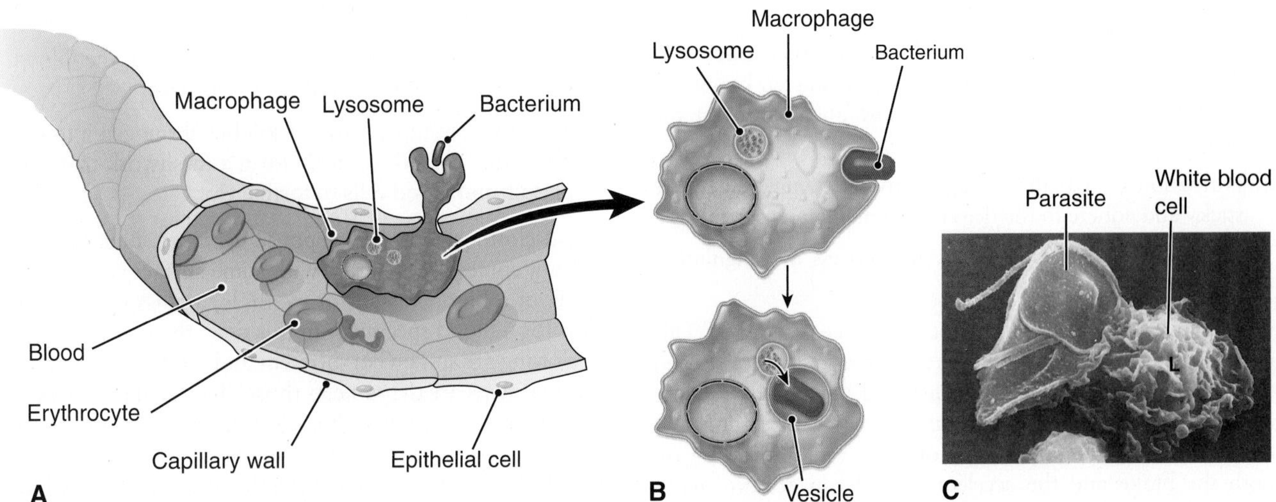

Figure 13-5 **Phagocytosis.** **KEY POINT** Phagocytosis is the engulfing of foreign matter by white cells. **A.** A phagocytic leukocyte (white blood cell) squeezes through a capillary wall in the region of an infection and engulfs a bacterium. **B.** The bacterium is enclosed in a vesicle and digested by a lysosome. **C.** A scanning electron microscope image of a phagocyte ingesting a parasite. **ZOOMING IN** What type of epithelium makes up the capillary wall?

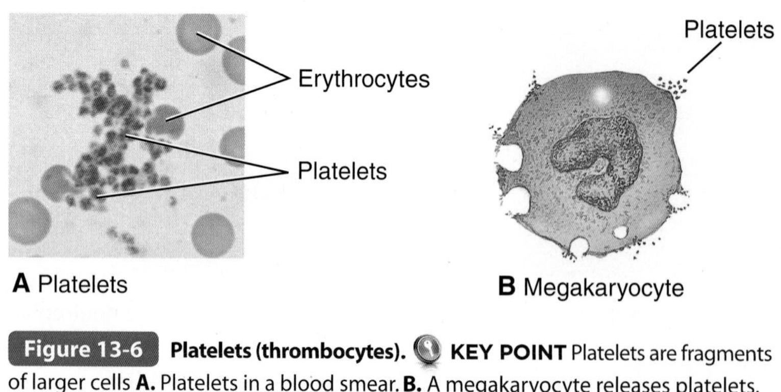

Figure 13-6 **Platelets (thrombocytes).** **KEY POINT** Platelets are fragments of larger cells **A.** Platelets in a blood smear. **B.** A megakaryocyte releases platelets.

Platelets are essential for the prevention of blood loss (hemostasis) and blood coagulation (clotting), discussed next. When blood comes in contact with any tissue other than the smooth lining of the blood vessels, as in the case of injury, the platelets stick together and form a plug that seals the wound. The platelets then release chemicals that participate in the formation of a clot to stop blood loss. More details on these reactions follow.

CHECKPOINTS

- 13-8 What are the three types of granular leukocytes? What are the two type of agranular leukocytes?
- 13-9 What is the most important function of leukocytes?
- 13-10 What is the function of blood platelets?

Hemostasis

Hemostasis (he-mo-STA-sis) is the process that prevents blood loss from the circulation when a blood vessel is ruptured by an injury. Events in hemostasis include the following:

1. **Contraction** of the smooth muscles in the blood vessel wall. The resulting reduction in the vessel's diameter, known as *vasoconstriction*, reduces blood flow and loss from the defect in the vessel.
2. Formation of a **platelet plug**. Activated platelets become sticky and adhere to the defect to form a temporary plug.
3. Formation of a **blood clot**, by the process of **coagulation** (ko-ag-u-LA-shun).

Once initiated, the clotting process proceeds rapidly. This quick response is possible because all of the substances needed for clotting are already present in blood but usually in their inactive forms. Think of a racecar driver at the beginning of the race. They maintain a state of readiness by pressing on both the brake and the accelerator, and can achieve maximum speed within a short time interval. Similarly, a balance is maintained between compounds that promote clotting, known as **procoagulants**, and those that prevent clotting, known as **anticoagulants.** Under normal conditions, the substances that prevent clotting (the anticoagulants) prevail. When an injury occurs, however, the procoagulants are activated, and a clot is formed (**Fig. 13-7**).

The clotting process is a well-controlled series of separate events involving 12 different clotting factors, each designated by a Roman numeral. Calcium ion (Ca^{2+}) is one such factor. Others are released from damaged tissue and activated platelets. Still others are enzyme precursors, made in the liver and released into the bloodstream, which can be activated in the clotting process. To manufacture these enzymes, the liver requires vitamin K. We obtain some of this vitamin in food from green vegetables and grains, but a large proportion is made by bacteria living symbiotically in the large intestine. The final step in the clotting reaction is the conversion of a plasma protein called **fibrinogen** (fi-BRIN-o-jen) into solid threads of **fibrin**, in which blood cells are trapped to form the clot. The final steps involved in blood clot formation are described below and illustrated in **Figure 13-7**.

1. Substances released from damaged tissue and sticky platelets initiate a reaction sequence that leads to the formation of an active enzyme called **prothrombinase** (pro-THROM-bih-nase).
2. Prothrombinase converts prothrombin in the blood to **thrombin.** Calcium is needed for this step.
3. Thrombin, in turn, converts soluble fibrinogen into insoluble fibrin. Threads of fibrin form a meshwork that entraps plasma and blood cells to form a clot.

Blood clotting occurs in response to injury. Blood also clots when it comes into contact with some surface other than the lining of a blood vessel, for example, a glass or plastic tube used for a blood specimen. In this case, the preliminary steps of clotting are somewhat different and require more time, but the final steps are the same as those illustrated in **Figure 13-7**. The fluid that remains after clotting has occurred is called **serum** (plural, *sera*). Serum contains all the components of blood plasma *except* the clotting factors, as expressed in the formula:

$$\text{Plasma} = \text{serum} + \text{clotting factors}$$

Several methods used to measure the body's ability to coagulate blood are described later in this chapter.

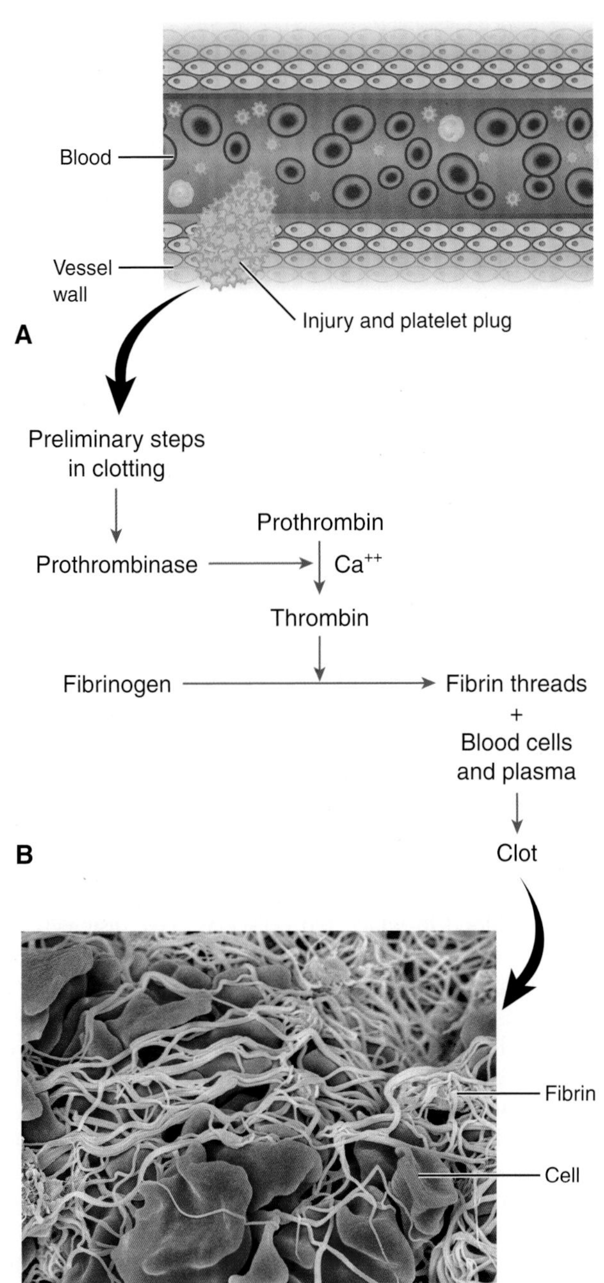

Figure 13-7 Blood clotting (coagulation). KEY POINT Blood coagulation requires a complex series of reactions that lead to the formation of fibrin, an insoluble protein. Fibrin threads trap blood cells to form a clot. **A.** Substances released from damaged tissue and sticky platelets initiate the preliminary steps in clotting. **B.** The final coagulation reactions lead to formation of a fibrin clot. **C.** Scanning electron micrograph of blood cells trapped in fibrin. ZOOMING IN What part of the word *prothrombinase* indicates that it is an enzyme? What part of the word *prothrombin* indicates that it is a precursor?

See the Student Resources on thePoint for a summary diagram and an animation on hemostasis.

CHECKPOINTS

- 13-11 What is the general term for the process that stops blood loss?
- 13-12 What substance in the blood forms a clot?
- 13-13 How does serum differ from blood plasma?

Blood Types

Like all body cells, blood cells contain substances (usually proteins) capable of activating an immune response. These substances, known as **antigens** (AN-ti-jens), trigger the immune system to make specialized proteins called *antibodies* that help destroy any cell with the offending antigen, as discussed more fully in Chapter 17.

Blood cell antigens vary among individuals and become important when blood components are donated from one individual to another in a process called a **transfusion.** Antibodies that recognize red cell antigens are known as *agglutinins*, because they cause red cells to undergo **agglutination** (ah-glu-tih-NA-shun) (clumping). The cells then rupture and release their hemoglobin by a process called **hemolysis** (he-MOL-ih-sis). The resulting condition is dangerous to a patient who has received incompatible blood. There are many types of red blood cell antigens, but only two groups are particularly likely to cause a transfusion reaction: the so-called A and B antigens and the Rh factor.

THE ABO BLOOD TYPE GROUP

There are four blood types involving the A and B antigens: A, B, AB, and O (**Table 13-3**). These letters indicate the type of antigen present on the red cells. If only the A antigen is present, the person has type A blood; if only the B antigen is present, he or she has type B blood. Type AB red cells have both antigens, and type O blood has neither. Of course, no one has antibodies to his or her own blood type antigens, or their plasma would destroy their own cells. Each person does, however, produce antibodies that react with the AB antigens he or she is lacking. (These antibodies are produced early in life from exposure to A and B antigens in the environment.) It is these antibodies in the patient's plasma that can react with antigens on the donor's red cells to cause a transfusion reaction.

Testing for Blood Type Blood type can be tested using blood sera containing antibodies to the A or B antigens. These **antisera** are prepared in animals using either the A or the B antigens to induce a response. Blood serum containing antibodies that recognize the A antigen is called **anti-A serum**; blood serum containing antibodies that recognize the B antigen is called **anti-B serum**. When combined with a blood sample in the laboratory, each antiserum causes the corresponding red cells to agglutinate. The blood's agglutination pattern when mixed with these two sera *one at a time* reveals its blood type (**Fig. 13-8**). Type A reacts with anti-A serum only; type B reacts with anti-B serum only. Type AB agglutinates with both, and type O agglutinates with neither A nor B.

13

Table 13-3 The ABO Blood Group System

Blood Type	Red Blood Cell Antigen	Reacts with Antiserum	Plasma Antibodies	Can Take From	Can Donate To
A	A	Anti-A	Anti-B	A, O	A, AB
B	B	Anti-B	Anti-A	B, O	B, AB
AB	A, B	Anti-A, Anti-B	None	AB, A, B, O	AB
O	None	None	Anti-A, Anti-B	O	O, A, B, AB

Blood Compatibility Heredity determines a person's blood type, and the percentage of people with each of the different blood types varies in different populations. For example, about 45% of the white population of the United States have type O blood, 40% have A, 11% have B, and only 4% have AB. The percentages vary within other population groups.

In an emergency, type O blood can be given to any ABO type because the cells lack both A and B antigens and will not react with either A or B antibodies (see Table 13-3). People with type O blood are called *universal donors*. Conversely, type AB blood contains no antibodies to agglutinate red cells, and people with this blood type can therefore receive blood from any ABO type donor. Those with AB blood are described as *universal recipients*. Whenever possible, it is safest to give the same blood type as the recipient's blood.

THE Rh FACTOR

More than 85% of the United States population has another red cell antigen group called the **Rh factor**, named for *Rh*esus monkeys, in which it was first found. Rh is also known as the *D antigen*. People with this antigen are said to be **Rh positive**; those who lack this protein are said to be **Rh negative**. If Rh-positive blood is given to an Rh-negative person, he or she may produce antibodies to the "foreign" Rh antigens. The blood of this "Rh-sensitized" person will then destroy any Rh-positive cells received in a later transfusion.

Rh incompatibility is a potential problem in certain pregnancies (Fig. 13-9). A mother who is Rh negative may develop antibodies to the Rh protein of an Rh-positive fetus (the fetus having inherited this factor from the father). Red cells from the fetus that enter the mother's circulation during pregnancy and childbirth evoke the response. In a subsequent pregnancy with an Rh-positive fetus, some of the anti-Rh antibodies may pass from the mother's blood into the blood of her fetus and destroy the fetus's red cells. This condition is called **hemolytic disease of the newborn** (HDN). An older name is *erythroblastosis fetalis*. HDN is now prevented by administration of immune globulin $Rh_0(D)$, trade name RhoGAM, to the mother during pregnancy and shortly after delivery. These preformed antibodies clear the mother's circulation of Rh antigens and prevent stimulation of an immune response. In many cases, a baby born with HDN could be saved by a transfusion that replaces much of the baby's blood with Rh-negative blood.

CHECKPOINTS

- ☐ 13-14 What is the term for any substance that activates an immune response?
- ☐ 13-15 What are the four ABO blood types?
- ☐ 13-16 What blood factor is associated with incompatibility during pregnancy?

Uses of Blood and Blood Components

Blood can be packaged and kept in blood banks for emergencies. To keep the blood from clotting, a solution such as citrate–phosphate–dextrose–adenine (CPDA-1) is added. The blood may then be stored for up to 35 days. The blood supplies in the bank are dated with an expiration date to prevent the use of blood in which red cells may have disintegrated. Blood banks usually have all types of blood and blood products available. It is important that there be an extra supply of type O, Rh-negative blood because in an emergency, this type can be used for any patient. It is normal procedure, however, to test the recipient and give blood of the same type. In this chapter's case study, Eleanor's blood was typed, and she was given red cells and platelets to overcome the effects of her bone marrow failure.

A person can donate his or her own blood before undergoing elective (planned) surgery to be used during surgery if needed. This practice eliminates the possibility of incompatibility and of disease transfer as well. Such **autologous** (aw-TOL-o-gus) (self-originating) blood is stored in a blood bank only until the surgery is completed.

WHOLE BLOOD TRANSFUSIONS

The transfer of whole human blood from a healthy person to a patient is often a life-saving process. Whole blood transfusions may be used for any condition in which there is loss of a large volume of blood, known as a **hemorrhage** (HEM-eh-rij). We may be familiar with the hemorrhages resulting from external wounds, but less evident are internal hemorrhages resulting from ruptures in deeper vessels. Regardless of the cause, severe hemorrhage starves body cells of oxygen and nutrients. Serious mechanical injuries, surgical operations, and internal injuries such as bleeding ulcers are all commonly treated using whole-blood transfusions.

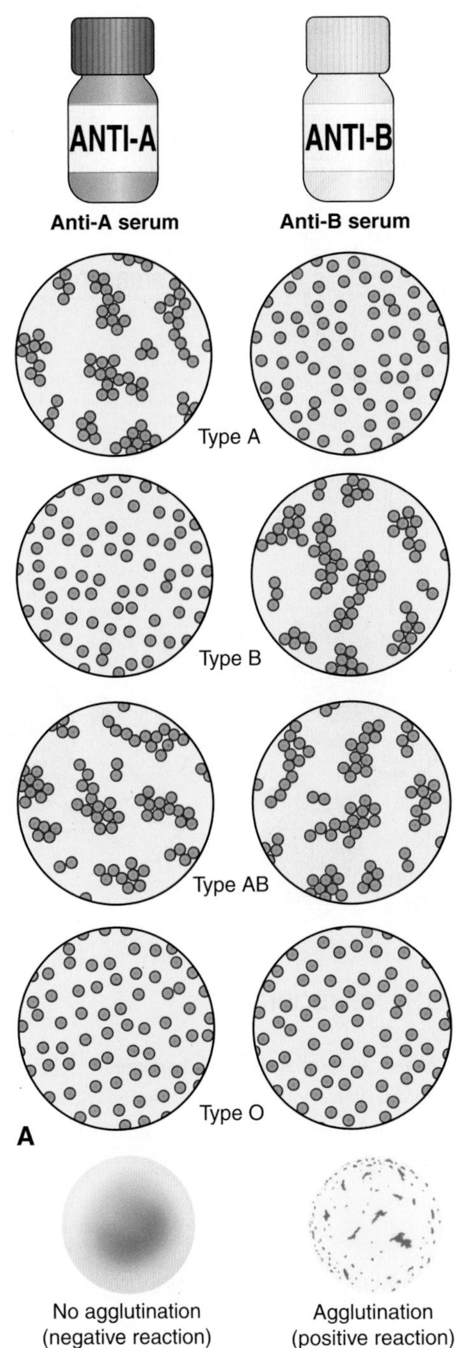

Figure 13-8 **Blood typing.** **KEY POINT** Blood type can be determined by mixing small volumes with antisera prepared against the different red cell antigens (proteins). Agglutination (clumping) with an antiserum indicates the presence of the corresponding antigen. **A.** Labels at the top of each column denote the kind of antiserum added to the blood samples. Anti-A serum agglutinates red cells in type A blood, but anti-B serum does not. Anti-B serum agglutinates red cells in type B blood, but anti-A serum does not. Both sera agglutinate type AB blood cells, and neither serum agglutinates type O blood. **B.** Photographs of blood typing reactions. **ZOOMING IN** Can you tell from these reactions whether these cells are Rh positive or Rh negative?

Caution and careful evaluation of the need for a blood transfusion is the rule, however, because of the risk for transfusion reactions and the possible transmission of viral diseases, particularly hepatitis and AIDS. (In developed countries, careful screening has virtually eliminated transmission of these viruses in donated blood.) Ideally, the compatibility of donor and recipient blood is tested prior to transfusion, a process called *cross-matching*. Both the red cells and the serum are tested separately for any possible cross-reactions with donor blood. This procedure is particularly important in individuals who have previously received transfused blood.

USE OF BLOOD COMPONENTS

Blood can be separated into its various parts, which may be used for different purposes.

Preparation of Blood Components A common method for separating the blood plasma from the formed elements is by use of a **centrifuge** (SEN-trih-fuje), a machine that spins in a circle at high speed to separate a mixture's components according to density. When a container of blood is spun rapidly, all the blood's heavier formed elements are pulled to the bottom of the container. They are thus separated from the plasma, which is less dense. The formed elements may be further separated, for example, packed red cells alone for the treatment of hemorrhage or platelets alone for the treatment of clotting disorders.

Blood losses to the donor can be minimized if the blood is removed, the desired components are separated, and the remainder is returned to the donor. The general term for this procedure is **hemapheresis** (hem-ah-fer-E-sis) (from the Greek word *apheresis* meaning "removal"). If the plasma is removed and the formed elements returned to the donor, the procedure is called **plasmapheresis** (plas-mah-fer-E-sis).

Uses of Plasma Blood plasma alone may be given in an emergency to replace blood volume and prevent circulatory failure (shock). Plasma is especially useful when blood typing and the use of whole blood are not possible, such as in natural disasters or in emergency rescues. Because the red cells have been removed from the plasma, there are no incompatibility problems; plasma can be given to anyone. Plasma separated from the cellular elements is usually further separated by chemical means into various components, such as plasma protein fraction, serum albumin, immune serum, and clotting factors.

The packaged plasma that is currently available is actually plasma protein fraction. Further separation yields serum albumin that is available in solutions of 5% or 25% concentration. In addition to its use in treatment of circulatory shock, these solutions are given when plasma proteins are deficient. They increase the blood's osmotic pressure and thus draw fluids back into circulation. The use of plasma proteins and serum albumin has increased because these blood components can be treated with heat to prevent transmission of viral diseases.

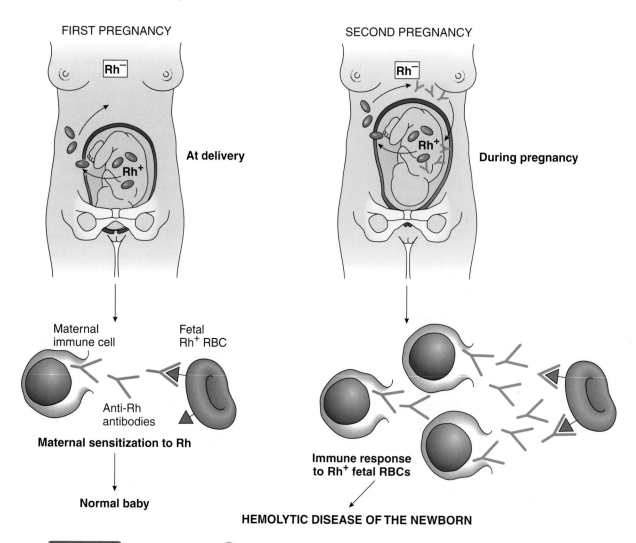

Figure 13-9 Rh incompatibility. **KEY POINT** An Rh-negative mother can form antibodies (become sensitized) to an Rh-positive fetus's red cells when exposed to the antigen during delivery. Unless she is treated with RhoGAM to prevent a response, her Rh antibodies can cross the placenta in a subsequent pregnancy and destroy fetal red cells if they are Rh positive. The result is hemolytic disease of the newborn.

In emergency situations, healthcare workers may administer fluids known as *plasma expanders*. These are cell-free isotonic solutions used to maintain blood fluid volume to prevent circulatory shock.

Fresh plasma may be frozen and saved. Plasma frozen when it is less than six hours old contains all the factors needed for clotting. When frozen plasma is thawed, a white precipitate called **cryoprecipitate** (kri-o-pre-SIP-ih-tate) forms in the bottom of the container. Cryoprecipitate is especially rich in fibrinogen and clotting factors. It may be given when there is a special need for these substances.

A portion of the plasma called the *gamma globulin fraction* contains antibodies produced by lymphocytes when they come in contact with foreign agents, such as bacteria and viruses. Antibodies play an important role in immunity (see Chapter 17). Commercially prepared immune sera are available for administration to patients in immediate need of antibodies, such as infants born to mothers with active hepatitis.

CHECKPOINT

- 13-17 How is blood commonly separated into its component parts?

Blood Disorders

Abnormalities involving the blood may be divided into three groups:

- **Anemia** (ah-NE-me-ah), a disorder in which there is an abnormally low level of hemoglobin or red cells in the blood and thus impaired delivery of oxygen to the tissues

- **Leukemia** (lu-KE-me-ah), a neoplastic blood disease characterized by an increase in the number of white cells
- **Clotting disorders**, conditions characterized by an abnormal tendency to bleed because of a breakdown in the body's clotting mechanism

ANEMIA

Anemia can result from excessive red cell loss or from inadequate red cell production.

Excessive Loss or Destruction of Red Cells Red cells may form properly but be lost through injury or destroyed after they are formed.

Hemorrhagic Anemia Hemorrhagic (hem-eh-RAJ-ik) loss of red cells may be sudden and acute or gradual and chronic. The average adult has about 5 L of blood. If a person loses as much as 2 L suddenly, death usually results. However, the body can withstand a loss equivalent to the entire blood volume over a period of weeks or months by increasing blood production. Possible causes of chronic blood loss include bleeding ulcers, excessive menstrual flow, and bleeding hemorrhoids (piles). If the cause of the blood loss can be corrected, the body is usually able to restore homeostasis. This process can take as long as six months, and until the blood returns to normal, the affected person may have **hemorrhagic anemia.**

Hemolytic Anemia Anemia caused by excessive red cell destruction is called **hemolytic** (he-mo-LIH-tik) **anemia**. The spleen, along with the liver, normally destroys old red cells. Occasionally, an enlarged, overactive spleen destroys the cells too rapidly, causing anemia. Infections may also cause red cell loss. For example, the malarial parasite multiplies in and destroys red cells, and certain bacteria, particularly streptococci, produce a toxin that causes hemolysis.

Certain inherited diseases that cause the production of abnormal hemoglobin may also result in hemolytic anemia. The hemoglobin in normal adult cells is of the A type and is designated *HbA*. In the inherited disease **sickle cell anemia**, also called *sickle cell disease*, the hemoglobin in many of the red cells is abnormal and is designated *HbS*. When these cells give up some of their oxygen to the tissues, they are transformed from the normal disk shape into a sickle shape (**Fig. 13-10**). These sickle cells are fragile and tend to break easily. Because of their odd shape, they also tend to become tangled in masses that can block smaller blood vessels, depriving tissues supplied by those vessels of oxygen. Where obstruction occurs, there may be severe joint swelling and pain, especially in the fingers and toes, as well as abdominal pain. This aspect of sickle cell anemia is referred to as *sickle cell crisis*.

People who inherit the HbS hemoglobin gene from just one parent, not both, are said to have **sickle cell trait.** They can pass this gene to their children and may experience sickle cell disease symptoms in certain circumstances, such as when severely stressed, doing intense exercise, or under conditions of low atmospheric oxygen. Carrying the sickle cell gene confers significant protection against malaria which, as noted above, hemolyzes normal red blood cells. Consequently, this characteristic occurs almost exclusively among people who live in tropical regions, where malaria is common, or among their descendants. In the United States, about 8% of African Americans have the sickle cell trait.

Figure 13-10 **A blood smear in sickle cell anemia.** **KEY POINT** Abnormal cells take on a crescent (sickle) shape when they give up oxygen. **ZOOMING IN** What kind of microscope was used to take this picture?

It is only when the sickle cell gene is transmitted from both parents that the clinical disease appears. About one in 5,000 (0.02%) of African Americans have two HbS genes and thus have sickle cell disease.

One drug has been found to reduce the frequency of painful crises in certain adults. Hydroxyurea causes the body to make some hemoglobin of an alternate form (fetal hemoglobin) so that the red cells are not as susceptible to sickling. People taking hydroxyurea require blood tests every two weeks to assess for drug-induced bone marrow suppression.

See the Student Resources on thePoint for figures explaining and illustrating sickle cell disease in greater detail.

Impaired Production of Red Cells or Hemoglobin Red cell production requires a specific set of nutrients and occurs within the red bone marrow. Any situation that reduces the body's nutrient stores or interferes with normal bone marrow function can thus cause anemia.

Nutritional Anemia Anemia that results from a deficiency of a specific nutrient is referred to as *nutritional anemia*. These conditions may arise from a dietary deficiency, from an inability to absorb the nutrient, or from drugs that interfere with the body's use of the nutrient.

The most common nutritional anemia is **iron deficiency anemia.** Iron is an essential constituent of hemoglobin. Although it is available in a wide range of animal and plant foods, it is most readily absorbed from meat, poultry, and fish. The average American diet usually provides enough iron to meet the needs of the adult male, but it can be inadequate to meet the needs of menstruating females and growing children, especially if they do not consume much of the foods that provide usable iron.

A diet deficient in proteins or vitamins can also result in anemia. Folic acid, one of the B complex vitamins, is necessary for blood cell production. Folic acid deficiency anemia occurs in people with alcoholism, in elderly people on poor diets, and in infants or others suffering from intestinal disorders that interfere with the absorption of this water-soluble vitamin.

Pernicious (per-NISH-us) **anemia** is characterized by a deficiency of vitamin B_{12}, a substance essential for proper red cell formation. The cause is a permanent deficiency of **intrinsic factor**, a gastric juice secretion that is responsible for vitamin B_{12} absorption from the intestine. Neglected pernicious anemia can bring about deterioration in the nervous system, causing difficulty in walking, weakness and stiffness in the extremities, mental changes, and permanent damage to the spinal cord. Early treatment, including the intramuscular injection of vitamin B_{12} and attention to a prescribed diet, ensures an excellent outlook. This treatment must be kept up for the rest of the patient's life to maintain good health.

Thalassemia (thal-ah-SE-me-ah) is a group of hereditary anemias resulting from defects in one of the hemoglobin genes. The defect impacts the production of normal hemoglobin molecules. To compound the problem, erythrocytes may be destroyed in the bone marrow before they mature. The red cells are small and pale, as in iron deficiency anemia. But these patients frequently have abnormally high iron reserves because of the many transfusions they receive and from excess iron absorption from the digestive tract. The two main types of thalassemia are α (alpha) and β (beta), named according to the part of the hemoglobin molecule affected (see Box 13-1). Severe beta thalassemia is also called *Cooley anemia* or *beta thalassemia major*. The beta thalassemias are found mostly in populations of Mediterranean descent (the name comes from the Greek word for "sea").

Bone Marrow Suppression Bone marrow suppression or failure also leads to decreased red cell production. One type of bone marrow failure, **aplastic** (a-PLAS-tik) **anemia**, may be caused by a variety of physical and chemical agents. Chemical substances that injure the bone marrow include certain prescribed drugs and toxic agents such as gold compounds, arsenic, and benzene. Physical agents that may injure the marrow include x-rays, atomic radiation, radium, and radioactive phosphorus.

Because damaged bone marrow also fails to produce white cells, aplastic anemia is accompanied by **leukopenia** (lu-ko-PE-ne-ah), a drop in the number of white cells. Platelet levels are also affected, as described shortly under clotting disorders. Removal of the toxic agent, followed by blood transfusions until the marrow is able to resume its activity, may result in recovery. Bone marrow transplantations have also been successful.

Bone marrow suppression also may develop in patients with certain chronic diseases, such as cancer, kidney or liver disorders, rheumatoid arthritis, and some viral infections. Some medications are now available to stimulate bone marrow production of specific types of blood cells. The hormone EPO made by recombinant methods (genetic engineering) can be given in cases of severe anemia to stimulate red cell production.

LEUKEMIA

Leukemia is a neoplastic disease of blood-forming tissue. It is characterized by an enormous increase in the number of white cells. Although the cells are high in number, they are incompetent and cannot perform their normal jobs. They also crowd out the other blood cells.

As noted earlier, white cells develop in two locations: red marrow, also called *myeloid tissue*, and lymphoid tissue. If this wild proliferation of white cells stems from cancer of the bone marrow, the condition is called **myelogenous** (mi-eh-LOJ-en-us) **leukemia** (Fig. 13-11). When the cancer arises in the lymphoid tissue, so that most of the abnormal cells are lymphocytes, the condition is called **lymphocytic** (lim-fo-SIT-ik) **leukemia.** Both types of leukemia are further classified into acute and chronic forms.

The cause of leukemia is unknown. Both inborn factors and various environmental agents have been implicated. Among the latter are chemicals (such as benzene), x-rays, radioactive substances, and viruses.

Patients with leukemia exhibit the general symptoms of anemia because the neoplastic tissue takes over the red marrow that normally produces red cells. In addition, they have a tendency to bleed easily, owing to a lack of platelets. The white cells circulating in blood are often immature and nonfunctional, so infections are both frequent and more serious. The spleen is greatly enlarged, and several other organs may increase in size because of internal accumulation of white cells.

Leukemia is usually treated with chemotherapy and sometimes radiation therapy. The prognosis depends on the type of leukemia and the patient's age. Chronic myelogenous leukemia has a five-year survival rate of 90% with current treatment methods, but treatment must continue indefinitely. Acute myelogenous leukemia has a much lower survival rate of 40%. Some childhood leukemias are curable.

A bone marrow transplant can cure leukemia by replacing the neoplastic tissue with normal tissue. First, the patient's existing bone marrow stem cells (hematopoietic cells) are destroyed by radiation or chemotherapy. Then, donor bone marrow cells (injected intravenously) travel to the recipient's bone marrow where they begin to produce new blood cells. Sometimes, the patient's own bone marrow is removed and treated and then used for the transplant, but most patients receive transplants of genetically similar bone marrow cells from a close relative. It is important that the donor marrow match the recipient's marrow as closely as possible to avoid rejection, so potential donors undergo blood tests to determine compatibility. Hematologists also use this procedure

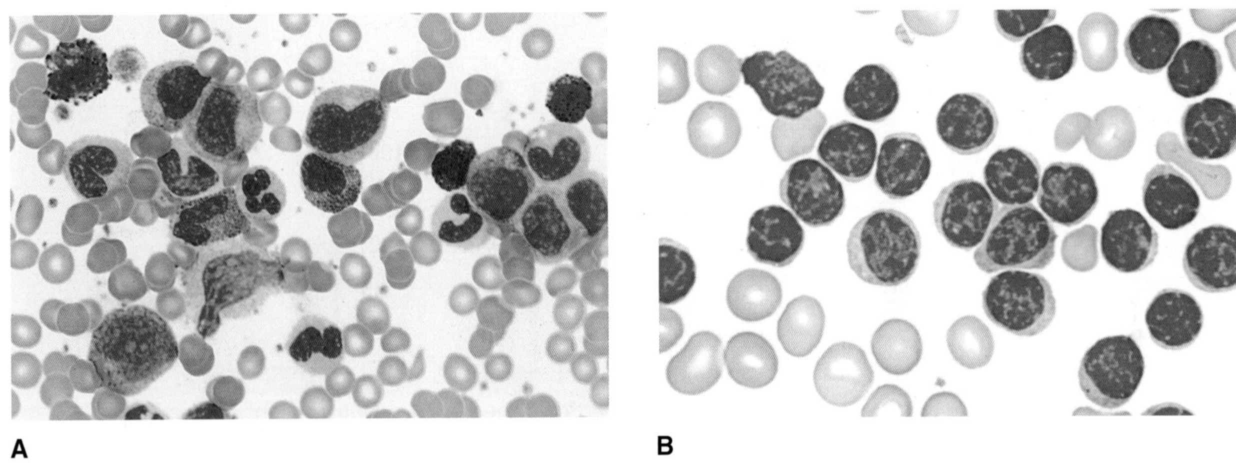

Figure 13-11 **Leukemia.** **KEY POINT** Leukemia is malignant overproduction of white cells that can originate in bone marrow or lymphatic tissue. Compare these photos with **Figure 13-2**. **A.** Chronic myelogenous leukemia showing overproduction of all categories of white cells. **B.** Chronic lymphocytic leukemia showing numerous lymphocytes.

to treat sickle cell anemia, thalassemia, aplastic anemia, and some immune disorders.

See the Student Resources on thePoint for additional photographs of leukemic cells.

CLOTTING DISORDERS

Most clotting disorders involve a disruption of platelet plug formation or the coagulation cascade, which brings about abnormal bleeding. Alternately, the disturbance may originate with excess clotting.

Hemophilia (he-mo-FIL-e-ah) is a rare hereditary bleeding disorder. It was once called "the royal disease," because Queen Victoria, who ruled Great Britain through most of the 19th century, carried this trait and passed it through her children to some royal families of Western Europe and Russia. All forms of hemophilia are characterized by a deficiency of a specific clotting factor, most commonly factor VIII. In those with hemophilia, any injury may cause excess bruising and serious abnormal bleeding. There is also spontaneous internal bleeding, especially in the digestive tract, brain, and other soft tissues. Bleeding into the joints, a common effect of hemophilia, is not only painful but can lead to serious disability if untreated. As mentioned earlier, the needed clotting factors are now available in purified concentrated form for treatment in cases of injury, preparation for surgery, or abnormal bleeding. Cryoprecipitate contains factor VIII, and clotting factors are also produced by recombinant (genetic engineering) methods.

Von Willebrand disease is another hereditary clotting disorder. It involves a shortage of von Willebrand factor, a plasma component that helps platelets adhere (stick) to damaged tissue. It also transports clotting factor VIII. This disorder is treated by administration of the appropriate clotting factor. In mild cases, a drug similar to ADH (antidiuretic hormone) may work to prevent bleeding by raising the level of von Willebrand factor in the blood.

The most common clotting disorder is a deficient number of circulating platelets (thrombocytes). The condition, called **thrombocytopenia** (throm-bo-si-to-PE-ne-ah), results in hemorrhage in the skin or mucous membranes. The decrease in platelets may result from their decreased production in red bone marrow or increased destruction. There are several possible causes of thrombocytopenia, including diseases of the bone marrow, liver disorders, and various drug toxicities. When a drug causes the disorder, its withdrawal leads to immediate recovery.

Disseminated intravascular coagulation (DIC) is a serious clotting disorder involving excessive coagulation. This disease occurs in cases of tissue damage caused by massive burns, trauma, certain acute infections, cancer, and some disorders of childbirth. Small clots develop in blood vessels throughout the body, reducing blood flow to tissues and organs. Moreover, during the progression of DIC, platelets and various clotting factors are used up faster than they can be produced; thus, serious hemorrhaging may result.

CHECKPOINTS

- ☐ 13-18 What is anemia?
- ☐ 13-19 What is leukemia?
- ☐ 13-20 What blood components are low in cases of thrombocytopenia?

Blood Studies

Many kinds of studies can be done on blood, and some of these have become standard parts of a routine physical examination. The tests included in a **complete blood count** (CBC) are shown in Appendix 2-2. Machines that are able to perform

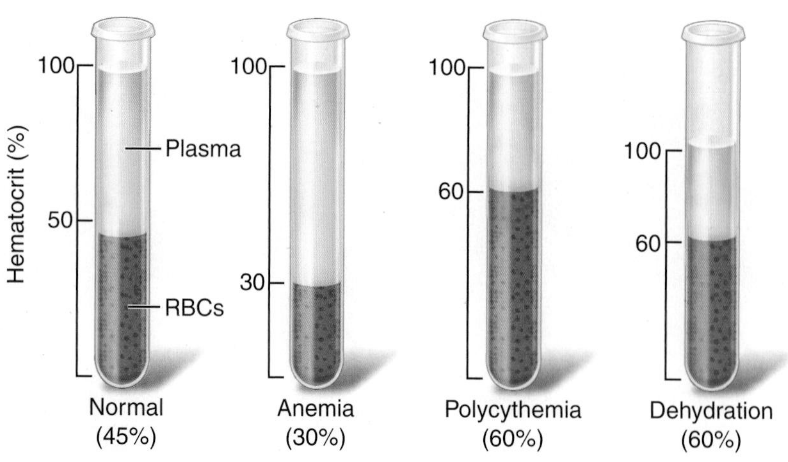

Figure 13-12 Hematocrit. **KEY POINT** The hematocrit tests the volume percentage of red cells in whole blood. The **tube on the left** shows a normal hematocrit. Abnormal hematocrit results can indicate disease (**two middle tubes**) or simply dehydration (**far right tube**).

several tests at the same time have largely replaced manual procedures, particularly in large institutions. Standard blood chemistry tests are listed in Appendix 2-3.

THE HEMATOCRIT

The **hematocrit** (he-MAT-o-krit) (Hct) measures how much of the blood volume is taken up by red cells. It reflects both the size and the number of cells and can provide an estimate of the oxygen-carrying capacity of blood.

The hematocrit is determined by spinning a blood sample in a high-speed centrifuge for three to five minutes to separate the cellular elements from the plasma (Fig. 13-12). It is expressed as the volume of packed red cells per unit volume of whole blood. For example, "hematocrit, 38%" in a laboratory report means that the patient has 38 mL red cells/dL (deciliter; 100 mL) of blood; red cells comprise 38% of the total blood volume. For adult men, the normal range is 42% to 54%, whereas for adult women, the range is slightly lower, 36% to 46%. These normal ranges, like all normal ranges for humans, may vary depending on the method used and the interpretation of the results by an individual laboratory. Usually, a decreased hematocrit is diagnostic of anemia, and an elevated hematocrit indicates *polycythemia*, a disorder of increased blood cell production discussed shortly. However, an elevated hematocrit can also reflect dehydration, because plasma volume decreases, but the RBC volume remains constant.

HEMOGLOBIN TESTS

Hemoglobin carries the oxygen in blood cells, and the best estimate of the blood's oxygen-carrying capacity is the hemoglobin concentration. New techniques can measure the concentration directly in a blood sample. Older techniques require lysing the cells to release the hemoglobin into solution, then quantifying the color intensity. Hemoglobin (Hb) is expressed in grams per deciliter of whole blood. Normal hemoglobin concentrations for adult males range from 14 to 17 g/dL blood. Values for adult women are in a somewhat lower range, at 12 to 15 g/dL blood. The hemoglobin reading can also be expressed as a percentage of a given standard, usually the average male normal of 15.6 g Hb/dL. Thus, a reading of 90% would mean 90% of 15.6 or 14 g Hb/dL. A decrease in hemoglobin to below normal levels signifies anemia.

Normal and abnormal types of hemoglobin can be separated and measured by the process of **electrophoresis** (e-lek-tro-fo-RE-sis). In this procedure, an electric current is passed through the liquid that contains the hemoglobin to separate different components based on their size. This test is useful in the diagnosis of sickle cell anemia and other disorders caused by abnormal types of hemoglobin.

BLOOD CELL COUNTS

Laboratories use automated methods for obtaining the data for blood counts. Visual counts are sometimes done using a **hemocytometer** (he-mo-si-TOM-eh-ter), a ruled slide used to count the cells in a given volume of blood under the microscope.

Red Cell Counts The normal red cell count varies from 4.5 to 5.5 million cells/mcL of blood. An increase in the red cell count is called **polycythemia** (pol-e-si-THE-me-ah). People who live at high altitudes develop polycythemia, as do patients with the disease **polycythemia vera**, a disorder of the bone marrow that causes red cell proliferation. As discussed earlier, low red cell counts are usually indicative of anemia.

White Cell Counts The leukocyte count varies from 5,000 to 10,000 cells/mcL of blood. In leukopenia, the white count is below 5,000 cells/mcL. This condition indicates depressed bone marrow or a bone marrow neoplasm. In **leukocytosis** (lu-ko-si-TO-sis), the white cell count exceeds 10,000 cells/mcL. This condition is characteristic of most bacterial infections. It may also occur after hemorrhage, in cases of gout (a type of arthritis), and in uremia, the presence of nitrogenous waste in the blood as a result of kidney disease.

Platelet Counts It is difficult to count platelets visually because they are so small. Laboratories can obtain more accurate counts with automated methods. These counts are necessary for the evaluation of platelet loss (thrombocytopenia) such as occurs after radiation therapy or cancer chemotherapy. The normal platelet count ranges from 150,000 to 450,000/mcL of blood, but counts may fall to 100,000 or less without causing serious bleeding problems. If a count is very low, a platelet transfusion may be given.

THE BLOOD SLIDE (SMEAR)

In addition to the above tests, a CBC includes the examination of a stained blood slide (see Fig. 13-2). In this procedure, a drop of blood is spread thinly and evenly over a glass slide, and a special stain (Wright) is applied to differentiate the otherwise colorless white cells. The slide is then studied under the microscope. The red cells are examined for abnormalities in size, color, or shape and for variations in the percentage of immature forms, known as reticulocytes. (see Box 13-2 to learn about reticulocytes and how their counts are used to diagnose disease.) The number of platelets is estimated. Parasites, such as the malarial organism and others, may be found. In addition, a **differential white count** is done. This is an estimation of the percentage of each white cell type in the smear. Because each type has a specific function, changes in their proportions can be a valuable diagnostic aid (see Table 13-2).

BLOOD CHEMISTRY TESTS

Batteries of tests on blood serum are often done by machine. The "Chem-7" test quantifies levels of four electrolytes (sodium, potassium, chloride, and bicarbonate), blood glucose, blood urea nitrogen, and **creatinine** (kre-AT-in-in).

Other tests check for enzymes. Increased levels of **creatine kinase** (CK), **lactic dehydrogenase** (LDH), and other enzymes indicate tissue damage, such as that resulting from heart disease. An excess of **alkaline phosphatase** (FOS-fah-tase) could indicate a liver disorder or metastatic cancer involving bone (see Table A2-3 in Appendix 2).

Blood can be tested for amounts of lipids, such as cholesterol, triglycerides (fats), and lipoproteins, or for amounts of plasma proteins. For example, the presence of more than the normal amount of glucose in the blood indicates uncontrolled diabetes mellitus. The list of blood chemistry tests is extensive and constantly increasing. We may now obtain values for various hormones, vitamins, antibodies, and toxic or therapeutic drug levels.

13

COAGULATION STUDIES

Before surgery and during treatment of certain diseases, hemophilia, for example, it is important to know that coagulation will take place within normal time limits. Because clotting is a complex process involving many reactants, a delay may result from a number of different causes, including lack of certain hormones, calcium, or vitamin K. The amounts of the various

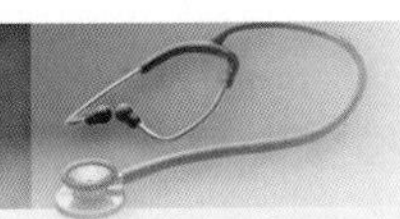

CLINICAL PERSPECTIVES

Box 13-2

Counting Reticulocytes to Diagnose Disease

As erythrocytes mature in the red bone marrow, they go through a series of stages in which they lose their nucleus and most other organelles, maximizing the space available to hold hemoglobin. In one of the last stages of development, small numbers of ribosomes and some rough endoplasmic reticulum remain in the cell. These appear as a network (or reticulum) when stained. Cells at this stage are therefore called **reticulocytes**. Reticulocytes leave the red bone marrow and enter the bloodstream where they become fully mature erythrocytes in about 24 to 48 hours. The average number of red cells maturing through the reticulocyte stage at any given time is about 1% to 2%. Changes in these numbers can be used in diagnosing certain blood disorders.

When erythrocytes are lost or destroyed, as from chronic bleeding or some form of hemolytic anemia, red blood cell production is "stepped up" to compensate for the loss. Greater numbers of reticulocytes are then released into the blood before reaching full maturity, and counts increase above normal. On the other hand, a decrease in the number of circulating reticulocytes suggests a problem with red blood cell production, as in cases of deficiency anemias or suppression of bone marrow activity.

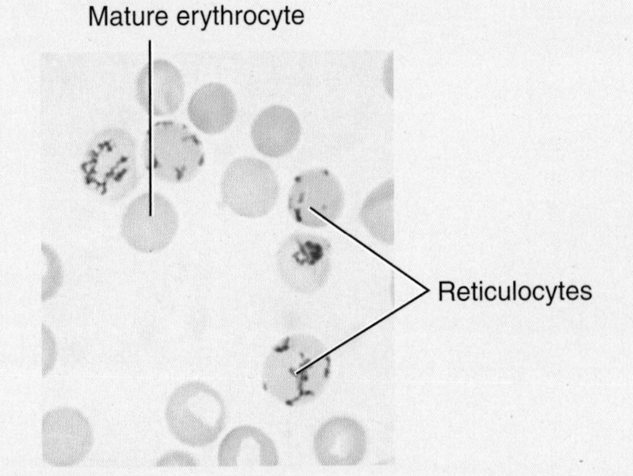

Reticulocytes. Some ribosomes and rough ER appear as a network in a late stage of erythrocyte development.

clotting factors are measured to aid in the diagnosis and treatment of bleeding disorders.

Additional tests for coagulation include tests for bleeding time, clotting time, capillary strength, and platelet function.

BONE MARROW BIOPSY

A special needle is used to obtain a small sample of red marrow from the sternum, sacrum, or iliac crest in a procedure called a **bone marrow biopsy**. If marrow is taken from the sternum, the procedure may be referred to as a **sternal puncture**. Examination of the cells gives valuable information that can aid in the diagnosis of bone marrow disorders, including leukemia and certain kinds of anemia.

CHECKPOINTS 

- [ ] 13-21 What test measures the relative volume of red cells in blood?
- [ ] 13-22 What are two ways of expressing hemoglobin level?

Disease in Context Revisited

Eleanor's Outcome

Nine months after Eleanor's initial visit, she was brought to the emergency room by ambulance for severe bloody vomiting. She was pale and confused, with a blood pressure of 60/20 mm Hg and a heart rate of 140 bpm. Despite heroic efforts to save her, Eleanor's heart stopped and could not be restarted.

Lab studies from blood collected before her death showed the counts of red and white blood cells and platelets to be very low. *Staphylococcus aureus* bacteria were cultured from her blood. At autopsy, the bone cavities normally containing red marrow were filled with tumor cells; little normal marrow remained. She was also found to have severe bacterial pneumonia and an extensive fungal infection in her esophagus. This had produced a large esophageal ulcer, which was the source of her fatal hemorrhage.

CHAPTER 13

Chapter Wrap-Up

Summary Overview

A detailed chapter outline with space for note taking is on *thePoint*. The figure below illustrates the main topics covered in this chapter.

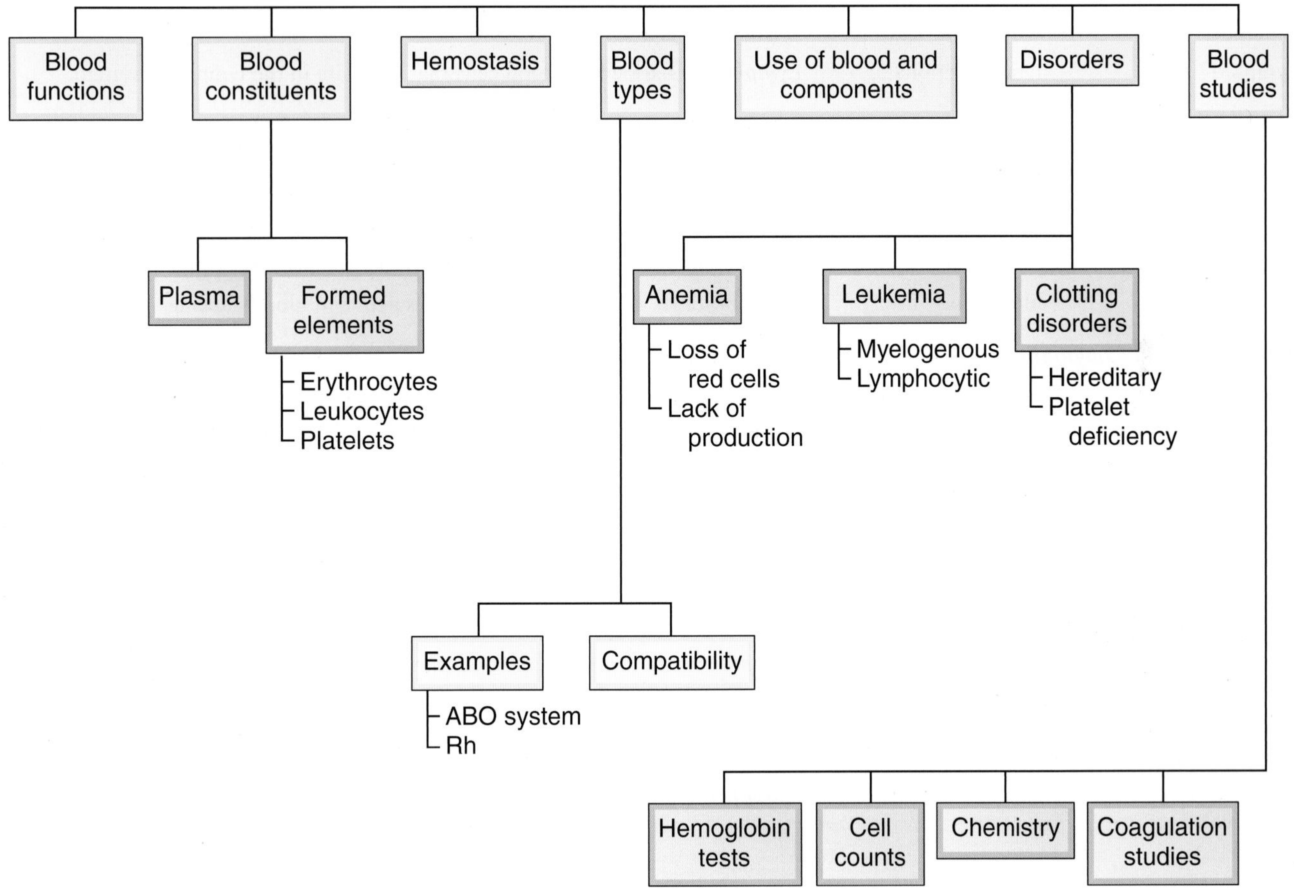

Key Terms

The terms listed below are emphasized in this chapter. Knowing them helps you organize and prioritize your learning. These and other boldface terms are defined in the Glossary with phonetic pronunciations.

agglutination
albumin
anemia
antigen
antiserum
basophil
centrifuge
coagulation
cryoprecipitate
eosinophil
erythrocyte
fibrin
hematocrit
hematopoietic
hemoglobin
hemolysis
hemorrhage
hemostasis
leukemia
leukocyte
lymphocyte
megakaryocyte
monocyte
neutrophil
plasma
platelet (thrombocyte)
serum
thrombin
thrombocytopenia
transfusion

Word Anatomy

Medical terms are built from standardized word parts (prefixes, roots, and suffixes). Learning the meanings of these parts can help you remember words and interpret unfamiliar terms.

WORD PART	MEANING	EXAMPLE
Blood Constituents		
erythr/o	red, red blood cell	An *erythrocyte* is a red blood cell.
hemat/o	blood	*Hematopoietic* stem cells form (–poiesis) all of the blood cells.
hemo	blood	*Hemoglobin* is a protein that carries oxygen in the blood.
kary/o	nucleus	A *megakaryocyte* has a very large nucleus.
leuk/o	white, colorless	A *leukocyte* is a white blood cell.
lymph/o	lymph, lymphatic system	*Lymphocytes* are white blood cells that circulate in the lymphatic system.
macr/o	large	A *macrophage* takes in large amounts of foreign matter by phagocytosis.
mon/o	single, one	A *monocyte* has a single, unsegmented nucleus.
morph/o	shape	The nuclei of *polymorphs* have many shapes.
phag/o	eat, ingest	Certain leukocytes take in foreign matter by the process of phagocytosis.
thromb/o	blood clot	A *thrombocyte* is a cell fragment that is active in blood clotting.
Hemostasis and Coagulation		
-gen	producing, originating	*Fibrinogen* converts to fibrin in the formation of a blood clot.
pro-	before, in front of	Prothrombinase is an enzyme (–ase) that converts *prothrombin* to thrombin.
Blood Types		
-lysis	loosening, dissolving, separating	A recipient's antibodies to donated red cells can cause hemolysis of the cells.
Uses of Blood and Blood Components		
cry/o	cold	*Cryoprecipitate* forms when blood plasma is frozen and then thawed.
Blood Disorders		
-emia (from -hemia)	blood	*Anemia* is a lack (an-) of red cells or hemoglobin.
-penia	lack of	*Leukopenia* is a lack of white cells.

Questions for Study and Review

BUILDING UNDERSTANDING

Fill in the Blanks

1. The liquid portion of blood is called ______.
2. The ancestors of all blood cells are called ______cells.
3. Platelets are produced by certain giant cells called ______.
4. Some monocytes enter the tissues and mature into phagocytes called ______.
5. Erythrocytes have a life span of approximately ______ days.

Matching > Match each numbered item with the most closely related lettered item:

___ **6.** an increased erythrocyte count
___ **7.** a decreased erythrocyte count
___ **8.** an increased leukocyte count
___ **9.** a decreased leukocyte count
___ **10.** a decreased platelet count

a. thrombocytopenia
b. anemia
c. leukopenia
d. leukocytosis
e. polycythemia

Multiple Choice

___ **11.** What iron-containing protein transports oxygen?

a. erythropoietin
b. complement
c. hemoglobin
d. thrombin

___ **12.** What is the correct sequence for hemostasis?

a. vessel contraction, plug formation, blood clot
b. blood clot, plug formation, vessel contraction
c. plug formation, blood clot, vessel contraction
d. vessel contraction, blood clot, plug formation

___ **13.** The hematology specialist needs to measure the number of eosinophils in a blood sample. Which test should she conduct?

a. hematocrit
b. electrophoresis
c. bone marrow biopsy
d. differential white blood cell count

___ **14.** What vitamin is needed for blood clotting?

a. vitamin A
b. vitamin K
c. biotin
d. vitamin E

UNDERSTANDING CONCEPTS

15. List the three main functions of blood. What is the average volume of circulating blood in the body?

16. Referring to Appendix 2-3, name a substance that is a breakdown product of hemoglobin.

17. Compare and contrast the following:

a. formed elements and plasma
b. erythrocyte and leukocyte
c. hemorrhage and transfusion
d. hemapheresis and plasmapheresis

18. List four main types of proteins in blood plasma, and state their functions. What are some other substances carried in blood plasma?

19. Describe the structure and function of erythrocytes. State the normal blood cell count for erythrocytes.

20. Construct a chart that compares the structure and function of the five types of leukocytes. State the normal blood cell count for leukocytes.

21. Diagram the three final steps in blood clot formation.

22. Name the four blood types in the ABO system. What antigens and antibodies (if any) are found in people with each type?

23. Is an Rh-negative fetus of an Rh-negative mother in any danger of HDN? Explain.

24. Compare and contrast the following disease conditions:

a. hemolytic anemia and aplastic anemia
b. myelogenous leukemia and lymphocytic leukemia
c. hemophilia and von Willebrand disease

CONCEPTUAL THINKING

25. J. Regan, a 40-year-old firefighter, has just had his annual physical. He is in excellent health, except for his red blood cell count, which is elevated. How might Mr. Regan's job explain his polycythemia?

26. If leukemia is associated with an elevated white blood cell count, why is it also associated with an increased risk of infection?

27. List the symptoms Eleanor experienced in the opening case study, and relate them to the problem with her bone marrow.

For more questions, see the Learning Activities on thePoint.

CHAPTER

14

The Heart and Heart Disease

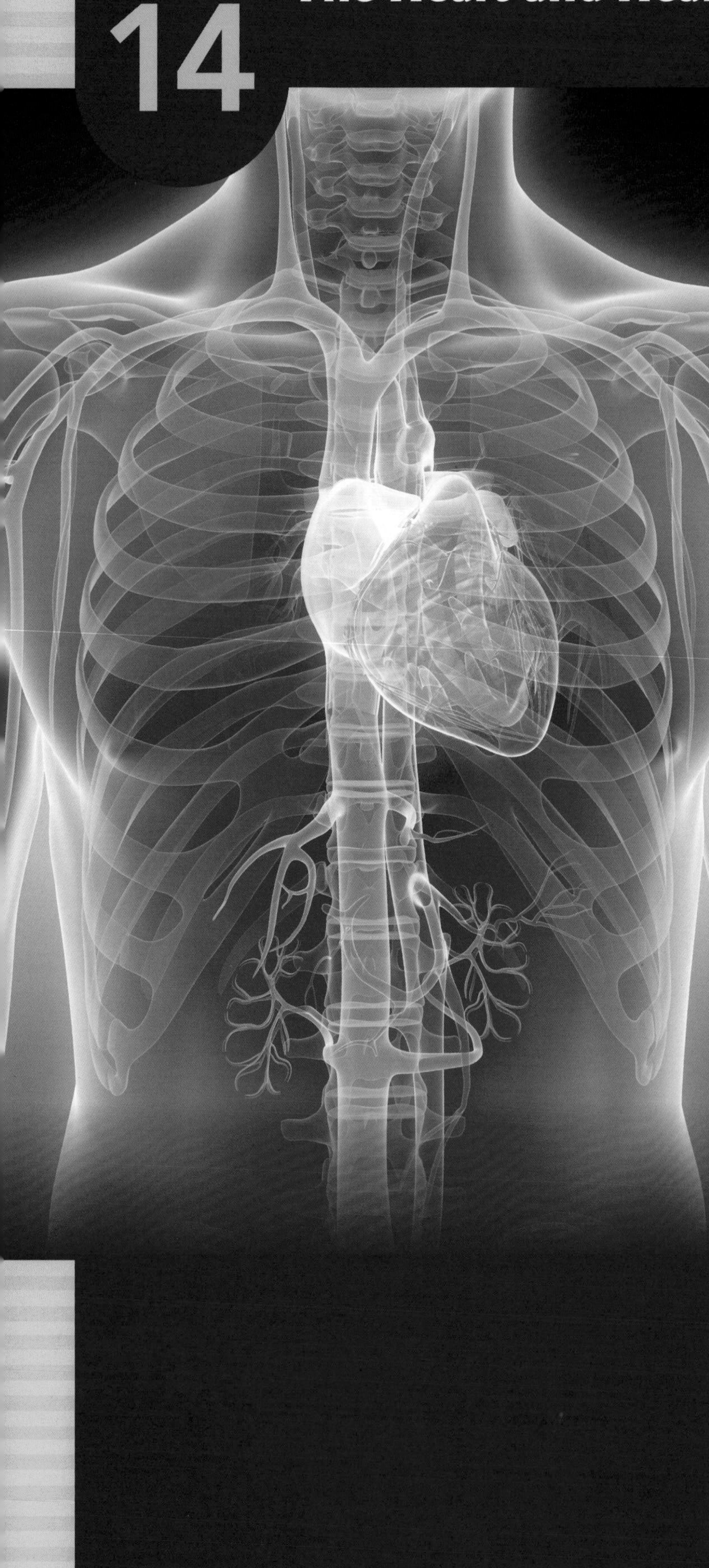

Learning Objectives

After careful study of this chapter, you should be able to:

1 ▶ Describe the three tissue layers of the heart wall. ***p. 314***

2 ▶ Describe the location and structure of the pericardium, and cite its functions. ***p. 315***

3 ▶ Compare the functions of the right and left chambers of the heart. ***p. 315***

4 ▶ Name the valves at the entrance and exit of each ventricle, and identify the function of each. ***p. 316***

5 ▶ Briefly describe blood circulation through the myocardium. ***p. 318***

6 ▶ Briefly describe the cardiac cycle. ***p. 319***

7 ▶ Name and locate the components of the heart's conduction system. ***p. 320***

8 ▶ Explain the effects of the autonomic nervous system (ANS) on the heart rate. ***p. 321***

9 ▶ List and define several terms that describe variations in heart rates. ***p. 322***

10 ▶ Explain what produces each of the two normal heart sounds, and identify the usual cause of a murmur. ***p. 322***

11 ▶ Briefly describe five methods used to study the heart. ***p. 322***

12 ▶ Describe six types of heart disease. ***p. 323***

13 ▶ List four risk factors for coronary artery disease that cannot be modified. ***p. 327***

14 ▶ List seven risk factors for coronary artery disease that can be modified. ***p. 327***

15 ▶ Describe three approaches to the treatment of heart disease. ***p. 328***

16 ▶ List four changes that may occur in the heart with age. ***p. 332***

17 ▶ Referring to the case study, list the emergency and surgical procedures commonly performed following a myocardial infarction, and explain why they are done. ***pp. 313, 332***

18 ▶ Show how word parts are used to build words related to the heart (see Word Anatomy at the end of the chapter). ***p. 334***

Disease in Context *Jim's Coronary Emergency*

The emergency room's dispatch radio echoed from the triage desk.

"This is Medic 5 en route with Jim, a 58-year-old Caucasian male. Suspected acute myocardial infarction while playing basketball. Cardiopulmonary resuscitation was initiated on scene. Patient was defibrillated in ambulance twice. Portable electrocardiography (ECG) indicates S-T interval depression and an inverted T wave. Patient is receiving oxygen through nasal cannulae. Estimated time of arrival (ETA) approximately 10 minutes."

When Jim arrived at the ER, the emergency team rushed to stabilize him. A trauma nurse measured his vital signs—he was hypertensive with tachycardia—while another inserted an IV needle into his arm and placed an oxygen mask over his nose and mouth. Meanwhile, a phlebotomist drew blood from Jim's other arm for testing in the lab. A cardiology technician attached ECG leads to his chest and began to record his cardiac muscle's electrical activity. The emergency doctor looked at the printout from the electrocardiograph and confirmed that Jim was having a heart attack. The doctor knew that one or more of the coronary arteries feeding Jim's heart muscle was blocked with a thrombus (blood clot). He administered several medications in an attempt to restore blood flow to the heart and minimize myocardial damage. Aspirin, which prevents platelets from adhering to each other, was given to inhibit the formation of any more thrombi. Nitroglycerin, a potent vasodilator, was given to widen Jim's coronary arteries and thus increase blood flow to the heart. Morphine was administered to manage his pain and lower his cardiac output in order to reduce the heart's workload. Finally, tissue plasminogen activator was administered to dissolve the thrombi present in his coronary arteries.

Thanks to the quick action of the paramedics and emergency team, Jim was resting comfortably in the intensive care unit a few hours after thrombolytic treatment—he was lucky to be alive! Later in the chapter, we will visit Jim again and learn how cardiac surgeons repair coronary arteries in cases of infarctions.

ANCILLARIES *At-A-Glance*

Visit thePoint to access the following resources. For guidance in using these resources most effectively, see pp. xv–xvii.

Learning RESOURCES

- Tips for Effective Studying
- Web Figure: Interior View of the Left Atrium and Ventricle
- Web Figure: Pathway of Blood through the Heart
- Web Figure: Tetralogy of Fallot
- Web Figure: Clinical Picture of Acute Myocardial Infarction
- Web Figure: Clinical Findings in Congestive Heart Failure
- Web Chart: Layers of the Heart Wall
- Web Chart: Layers of the Pericardium
- Web Chart: Chambers of the Heart
- Web Chart: Valves of the Heart
- Animation: Blood Circulation
- Animation: Cardiac Cycle
- Animation: Myocardial Blood Flow
- Animation: Heart Failure
- Health Professions: Surgical Technologist
- Detailed Chapter Outline
- Answers to Questions for Study and Review
- Audio Pronunciation Glossary

Learning ACTIVITIES

- Pre-Quiz
- Visual Activities
- Kinesthetic Activities
- Auditory Activities

A LOOK BACK

In Chapter 4, we learned that cardiac muscle is one of the three types of muscle in the body. Now it is time to study this tissue and the organ where it is found—the heart. We will also see that, even though the heart can work on its own, the nervous and endocrine systems influence its actions.

The next two chapters investigate how the blood delivers oxygen and nutrients to the cells and carries away the waste products of cellular metabolism. The continuous one-way circuit of blood through the blood vessels is known as **circulation**. The prime mover that propels blood throughout the body is the **heart**. This chapter examines the heart's structure and function as a foundation for the detailed discussion of blood vessels that follows.

The heart's importance has been recognized for centuries. Strokes (contractions) of this pump average about 72 per minute and continue unceasingly for a lifetime. The beating of the heart is affected by the emotions, which may explain the frequent references to it in song and poetry. However, the heart's vital functions and its disorders are of more practical concern.

Structure of the Heart

The heart is slightly bigger than a person's fist. It is located between the lungs in the center and a bit to the left of the body's midline (Fig. 14-1). It occupies most of the **mediastinum** (me-de-as-TI-num), the central region of the thorax. The heart's **apex**, the pointed, inferior portion, is directed toward the left. The broad, superior **base**, directed toward the right, is the area of attachment for the large vessels carrying blood into and out of the heart. See Dissection Atlas Figure A5-5 for a photograph of the heart in position in the thorax.

TISSUE LAYERS OF THE HEART WALL

The heart is a hollow organ, with walls formed of three different layers. Just as a warm coat might have a smooth lining, a thick interlining, and an outer covering of a third fabric, so the heart wall has three tissue layers (Fig. 14-2). Starting with the innermost layer, these are as follows:

1. The **endocardium** (en-do-KAR-de-um) is a thin, smooth layer of epithelial cells that lines the heart's interior. The endocardium provides a smooth surface for easy flow as blood travels through the heart. Extensions of this membrane cover the flaps (cusps) of the heart valves.
2. The **myocardium** (mi-o-KAR-de-um), the heart muscle, is the thickest layer and pumps blood through the vessels. The cardiac muscle's unique structure will be described in more detail shortly.
3. The **epicardium** (ep-ih-KAR-de-um) is a serous membrane that forms the thin, outermost layer of the heart wall. It is also considered the visceral layer of the pericardium, discussed next.

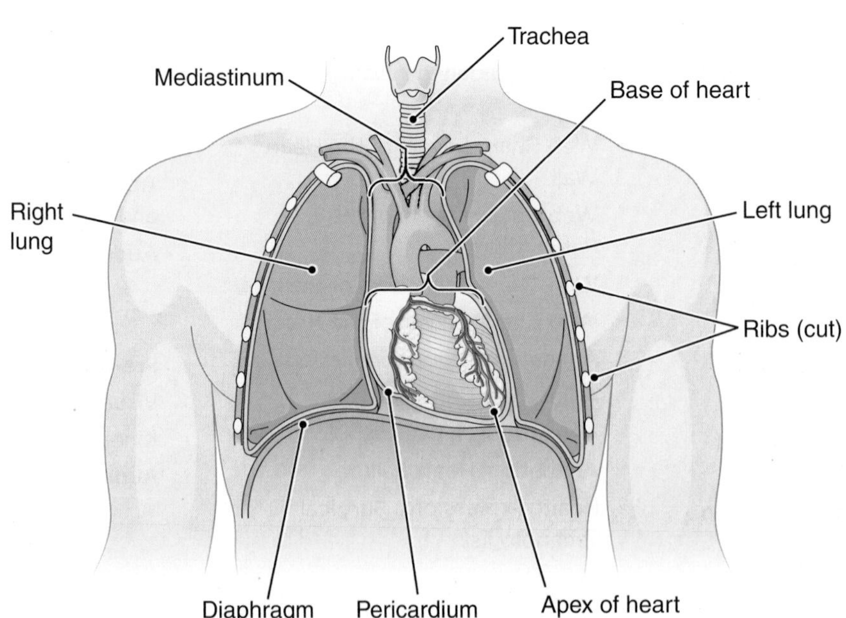

Figure 14-1 The heart in position in the thorax (anterior view). **KEY POINT** The heart is located between the lungs and just superior to the diaphragm in a region known as the mediastinum. **ZOOMING IN** Why is the left lung smaller than the right lung?

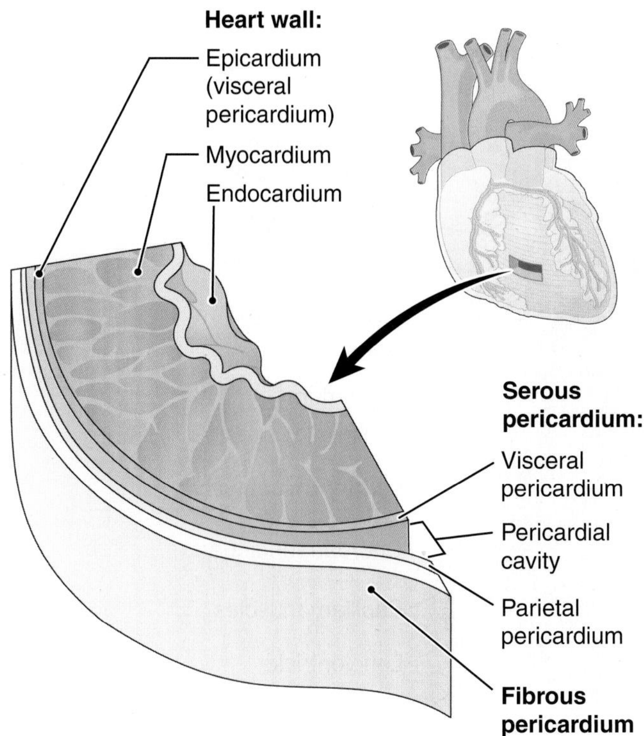

Figure 14-2 Layers of the heart wall and pericardium. **KEY POINT** The serous pericardium covers the heart and lines the fibrous pericardium. **ZOOMING IN** Which layer of the heart wall is the thickest?

THE PERICARDIUM

The **pericardium** (per-ih-KAR-de-um) is the sac that encloses the heart (see Fig. 14-2). The formation of the pericardial sac was described and illustrated in Chapter 4 under the discussion of membranes (see Fig. 4-8). This sac's outermost and heaviest layer is the fibrous pericardium, a connective tissue membrane. Additional connective tissue anchors this pericardial layer to the diaphragm, located inferiorly; to the sternum, located anteriorly; and to other structures surrounding the heart, thus holding the heart in place. A serous membrane forms the inner layer of the pericardium. This membrane, known as the serous pericardium, consists of an outer, parietal layer that lines the fibrous pericardium and an inner, visceral layer (the epicardium) that covers the myocardium. A thin film of fluid between these two layers reduces friction as the heart moves within the pericardium. Normally, the visceral and parietal layers are very close together, but fluid may accumulate in the region between them, the pericardial cavity, under certain disease conditions.

SPECIAL FEATURES OF THE MYOCARDIUM

Cardiac muscle cells are lightly striated (striped) based on alternating actin and myosin filaments, as seen in skeletal muscle cells (see Table 8-1). Unlike skeletal muscle cells, however, cardiac muscle cells have a single nucleus instead of multiple nuclei. Also, cardiac muscle tissue is involuntarily controlled; it typically contracts independently of conscious thought. There are specialized partitions between cardiac muscle cells that show faintly under a microscope (Fig. 14-3). These **intercalated** (in-TER-cah-la-ted) **disks** are actually plasma membranes of adjacent cells that are tightly joined together by specialized membrane proteins. Other membrane proteins within the disks permit electric impulses to travel between adjacent cells. Such electrical synapses, mentioned in Chapter 9, provide rapid and coordinated communication between cells.

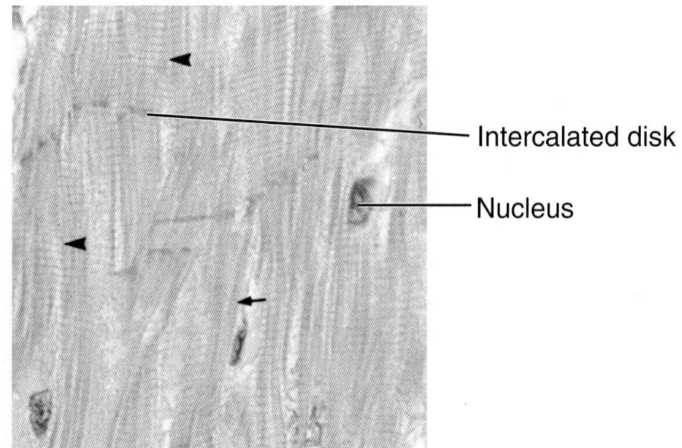

Figure 14-3 Cardiac muscle tissue viewed under the microscope (×540). **KEY POINT** The sample shows light striations (see the *arrowheads*), intercalated disks, and branching fibers (*arrow*).

Another feature of cardiac muscle tissue is the branching of the muscle fibers (cells). These branched fibers are interwoven so that the stimulation that causes the contraction of one fiber results in the contraction of a whole group. The intercalated disks between the fibers and the branching cellular networks allow cardiac muscle cells to contract in a coordinated manner for effective pumping.

DIVISIONS OF THE HEART

Healthcare professionals often refer to the *right heart* and the *left heart*, because the human heart is really a double pump (Fig. 14-4). The right side receives blood low in oxygen content that has already passed through the body and pumps it to the lungs through the pulmonary circuit. The left side receives highly oxygenated blood from the lungs and pumps it throughout the body via the systemic circuit. Each side of the heart is divided into two chambers. See Dissection Atlas **Figure A5-6** for a photograph of the human heart showing the chambers and the vessels that connect to the heart.

Four Chambers The upper chambers on the right and left sides, the **atria** (A-tre-ah), are mainly blood-receiving chambers (see Fig. 14-4). The lower chambers on the right and left sides, the **ventricles** (VEN-trih-klz), are forceful pumps. The chambers, listed in the order in which blood originating in the body tissues flows through them, are as follows:

1. The **right atrium** (A-tre-um) is a thin-walled chamber that receives the blood returning from the body tissues. This blood, which is comparatively low in oxygen, is carried

14

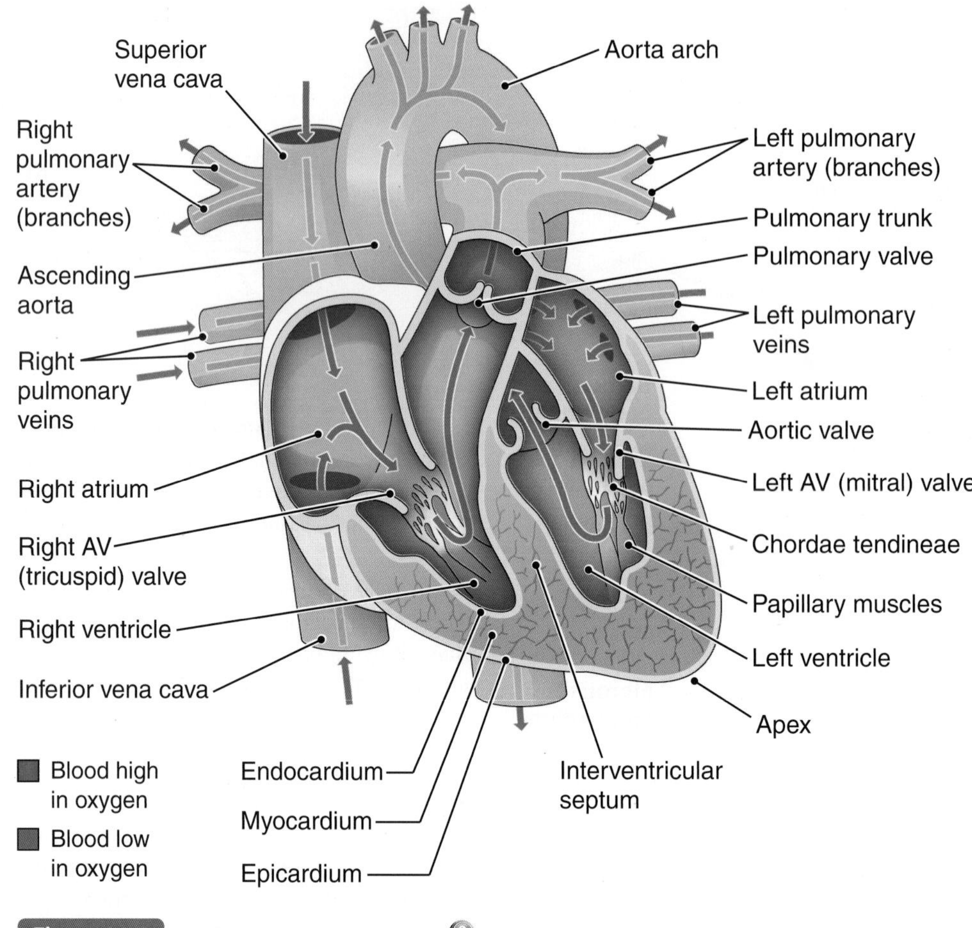

Figure 14-4 **The heart and great vessels.** **KEY POINT** The right heart has blood low in oxygen; the left heart has blood high in oxygen. The *arrows* show the direction of blood flow through the heart. The abbreviation AV means atrioventricular. **ZOOMING IN** Which heart chamber has the thickest wall?

in veins, the blood vessels leading back to the heart. The superior vena cava brings blood from the head, chest, and arms; the inferior vena cava delivers blood from the trunk and legs. A third vessel that opens into the right atrium brings blood from the heart muscle itself, as described later in this chapter.

2. The **right ventricle** receives blood from the right atrium and pumps it to the lungs. Blood passes from the right ventricle into a large pulmonary trunk, which then divides into right and left pulmonary arteries. Branches of these arteries carry blood to the lungs. An artery is a vessel that takes blood from the heart to the tissues. Note that the pulmonary arteries in **Figure 14-4** are colored blue because they are carrying blood low in oxygen, unlike other arteries, which carry blood high in oxygen.
3. The **left atrium** receives oxygen-rich blood as it returns from the lungs in pulmonary veins. Note that the pulmonary veins in **Figure 14-4** are colored red because they are carrying blood high in oxygen content, unlike other veins, which carry blood low in oxygen.
4. The **left ventricle**, which is the chamber with the thickest wall, pumps highly oxygenated blood to all parts of the body, including the lung tissues. This blood goes first into the aorta (a-OR-tah), the largest artery, and then into the branching systemic arteries that take blood to the tissues. The heart's apex, the lower pointed region, is formed by the wall of the left ventricle.

The heart's right and left chambers are completely separated from each other by partitions, each of which is called a **septum.** The **interatrial** (in-ter-A-tre-al) **septum** separates the two atria, and the **interventricular** (in-ter-ven-TRIK-u-lar) **septum** separates the two ventricles. The septa, like the heart wall, consist largely of myocardium.

Four Valves One-way valves that direct blood flow through the heart are located at the entrance and exit of each ventricle (**see Fig. 14-4**). The entrance valves are the **atrioventricular** (a-tre-o-ven-TRIK-u-lar) (AV) **valves,** so named because they are between the atria and ventricles. The exit valves are the **semilunar** (sem-e-LU-nar) **valves,**

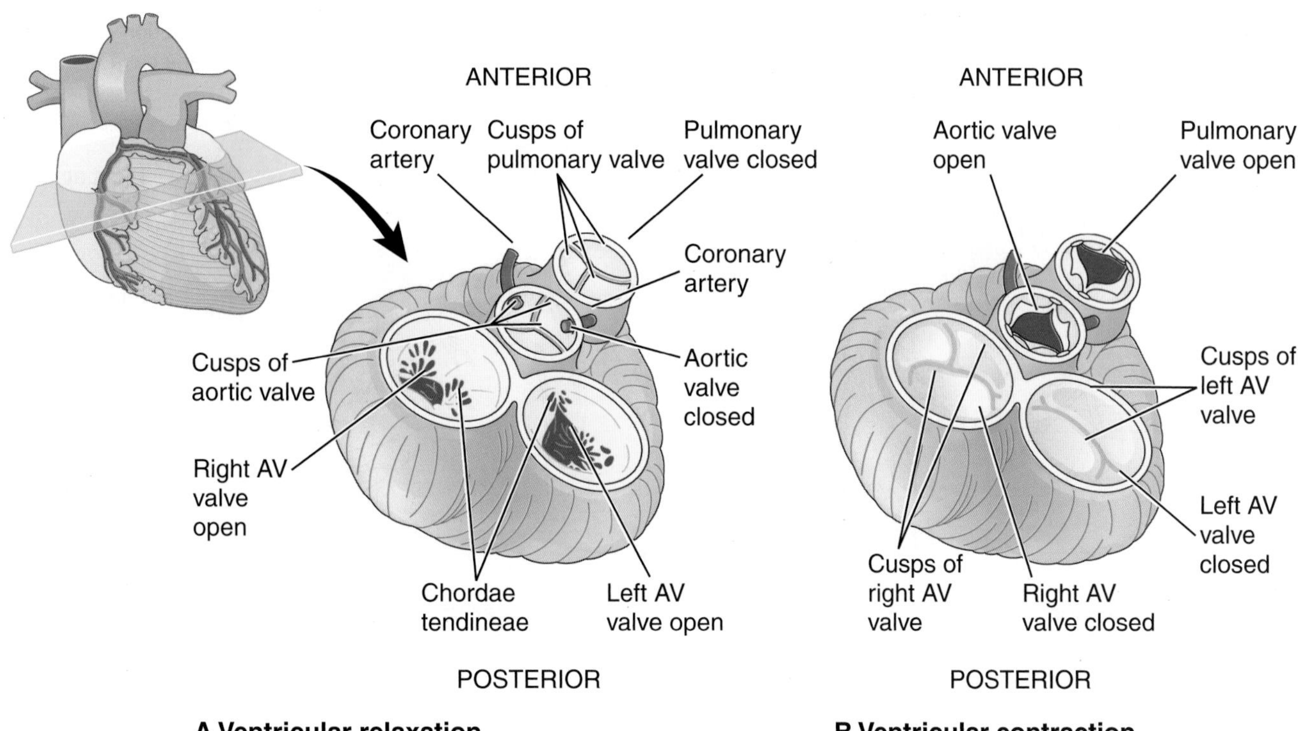

Figure 14-5 **Heart valves (superior view from posterior, atria removed).** **KEY POINT** Valves keep blood flowing in a forward direction through the heart. **A.** When the ventricles are relaxed, the AV valves are open and blood flows freely from the atria to the ventricles. The pulmonary and aortic valves are closed. **B.** When the ventricles contract, the AV valves close, and blood pumped out of the ventricles opens the pulmonary and aortic valves. **ZOOMING IN** How many cusps does the right AV valve have? The left?

so named because each flap of these valves resembles a half-moon. Each valve has a specific name, as follows:

- The **right atrioventricular** (AV) **valve** is also known as the **tricuspid** (tri-KUS-pid) **valve** because it has three cusps, or flaps, that open and close (Fig. 14-5). When this valve is open, blood flows freely from the right atrium into the right ventricle. When the right ventricle begins to contract, however, the valve is closed by blood pressing against the cusps. With the valve closed, blood cannot return to the right atrium but must flow forward into the pulmonary trunk.
- The **left atrioventricular** (AV) **valve** is the bicuspid valve, but it is commonly referred to as the **mitral** (MI-tral) **valve** (named for a miter, the pointed, two-sided hat worn by bishops). It has two heavy cusps that permit blood to flow freely from the left atrium into the left ventricle. The cusps close when the left ventricle begins to contract; this closure prevents blood from returning to the left atrium and ensures the forward flow of blood into the aorta. Both the right and left AV valves are attached by means of thin fibrous threads to **papillary** (PAP-ih-lar-e) **muscles** arising from the walls of the ventricles. The function of these threads, called the **chordae tendineae** (KOR-de ten-DIN-e-e) (see Fig. 14-4), is to stabilize the valve flaps when the ventricles contract so that the blood's force will not push the valves up into the atria. In this manner, they help prevent a backflow of blood when the heart beats.
- The **pulmonary** (PUL-mon-ar-e) **valve** is a semilunar valve located between the right ventricle and the pulmonary trunk that leads to the lungs. When the right ventricle relaxes, pressure in that chamber drops. The higher pressure in the pulmonary artery, described as *back pressure*, closes the valve and prevents blood from returning to the ventricle.
- The **aortic** (a-OR-tik) **valve** is a semilunar valve located between the left ventricle and the aorta. When the left ventricle relaxes, back pressure closes the aortic valve and prevents the backflow of blood from the aorta into the ventricle.

Note that blood passes through the heart twice in making a trip from the heart's right side through the pulmonary circuit to the lungs and back to the heart's left side to start on its way through the systemic circuit. However, it is important to bear in mind that the heart's two sides function in unison to pump identical volumes of blood through both circuits at the same time.

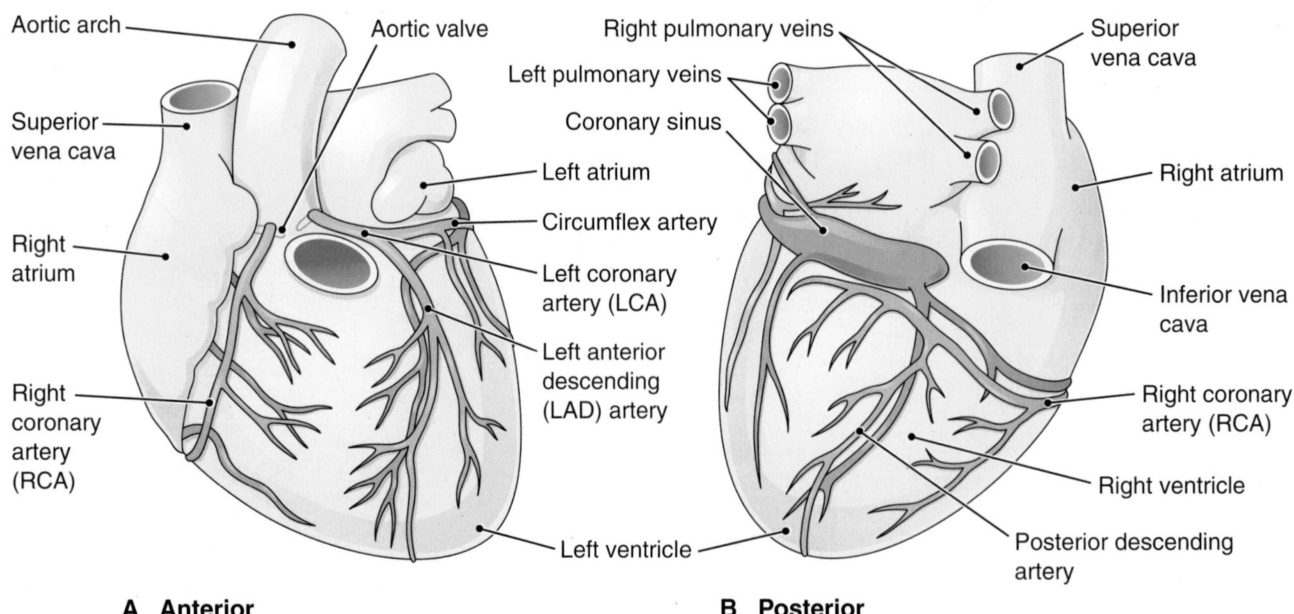

Figure 14-6 **Blood vessels that supply the myocardium.** **KEY POINT** Coronary arteries and cardiac veins constitute the heart's circulatory pathways. **A.** Anterior view. **B.** Posterior view. **ZOOMING IN** What is the largest cardiac vein, and where does it lead?

See the Student Resources on thePoint for charts summarizing the structure of the heart and pericardium and for a detailed picture of the heart's interior. See also the animation "Blood Circulation" and a numbered diagram showing blood flow through the heart.

BLOOD SUPPLY TO THE MYOCARDIUM

Only the endocardium comes into contact with the blood that flows through the heart chambers. Therefore, the myocardium must have its own blood vessels to provide oxygen and nourishment and to remove waste products. Together, these blood vessels form the **coronary** (KOR-o-na-re) **circulation.** It is the coronary circulation that is involved in Jim's case study. The main arteries that supply blood to the heart muscle are the right and left coronary arteries (Fig. 14-6), named because they encircle the heart like a crown. These arteries, which are the first to branch off the aorta, arise just above the cusps of the aortic valve and branch to all regions of the heart muscle. They receive blood only when the ventricles relax because the aortic valve must be closed to expose the entrance to these vessels (Fig. 14-7). The left coronary artery (LCA) branches into the circumflex artery and the left anterior descending (LAD) artery (also known as the anterior interventricular branch of the LCA). The right coronary artery (RCA) snakes around the heart just inferior to the right atrium, giving off a major branch called the posterior descending artery (also known as the posterior interventricular artery). After passing through the capillaries in the myocardium, blood drains into a system of cardiac veins that brings blood back toward the right atrium. Blood finally collects in the **coronary sinus,** a dilated vein that opens into the right atrium near the inferior vena cava (see Fig. 14-6).

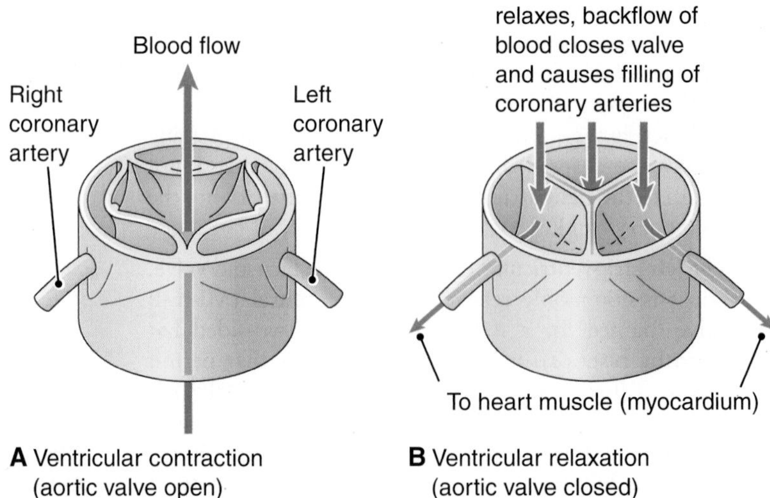

Figure 14-7 **Opening of coronary arteries in the aortic valve (anterior view).** **KEY POINT A.** When the left ventricle contracts, the aortic valve opens. The valve cusps prevent filling of the coronary arteries. **B.** When the left ventricle relaxes, backflow of blood closes the aortic valve, and the coronary arteries fill.

CHECKPOINTS

- 14-1 What are the names of the innermost, middle, and outermost layers of the heart wall?
- 14-2 What is the name of the sac that encloses the heart?
- 14-3 What is the heart's upper receiving chamber on each side called? What is the lower pumping chamber called?
- 14-4 What is the purpose of the valves in the heart?
- 14-5 What is the name of the system that supplies blood to the myocardium?

Heart Function

Although the heart's right and left sides are separated from each other, they work together. A heart muscle contraction begins in the thin-walled upper chambers, the atria, and is followed by a contraction of the thick muscle of the lower chambers, the ventricles. In each case, the active phase, called **systole** (SIS-to-le), is followed by a resting phase known as **diastole** (di-AS-to-le). One complete sequence of heart contraction and relaxation is called the **cardiac cycle** (Fig. 14-8). Each cardiac cycle represents a single heartbeat. At rest, one cycle takes an average of 0.8 seconds.

The cardiac cycle begins with contraction of both atria, which forces blood through the AV valves into the ventricles. The atrial walls are thin, and their contractions are not very powerful. However, they do improve the heart's efficiency by forcing blood into the ventricles before these lower chambers contract. Atrial contraction ends before ventricular contraction begins. Thus, atrial diastole begins at the same time ventricular systole begins. While the ventricles are contracting, forcing blood through the semilunar valves, the atria are relaxed and again are filling with blood (see Fig. 14-8).

After the ventricles have contracted, all the chambers are relaxed for a short period. During this period of complete relaxation, blood enters the atria from the great veins and passively drains into the ventricles. Then, another cycle begins with an atrial contraction followed by a ventricular contraction. Although both upper and lower chambers have a systolic and diastolic phase in each cardiac cycle, discussions of heart function usually refer to these phases as they occur in the ventricles, because these chambers contract more forcefully and drive blood into the arteries.

CARDIAC OUTPUT

A unique property of heart muscle is its ability to adjust the strength of contraction to the amount of blood received. When the heart chamber is filled and the wall stretched (within limits), the contraction is strong. As less blood enters the heart, contractions become less forceful. Thus, as more blood enters the heart, the muscle contracts with greater strength to push the larger volume of blood out into the blood vessels. The heart's ability to pump out all of the blood it receives prevents blood from pooling in the chambers.

The volume of blood pumped by each ventricle in one minute is termed the **cardiac output** (CO). It is the product of the **stroke volume** (SV)—the volume of blood ejected from the ventricle with each beat—and the **heart rate** (HR)—the number of times the heart beats per minute. To summarize:

$$CO = HR \times SV$$

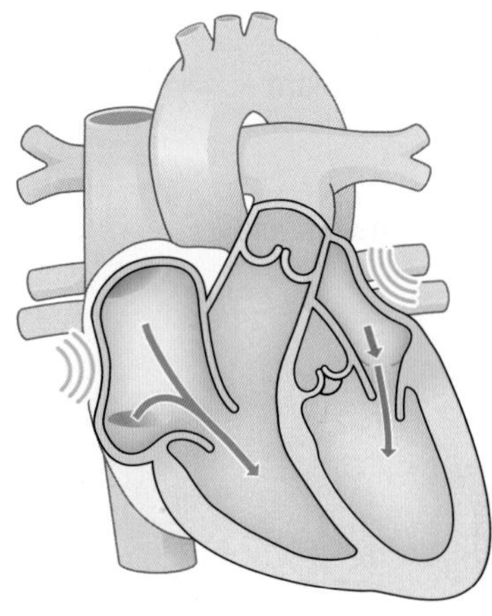

Atrial systole
Contraction of atria pumps additional blood into the ventricles.

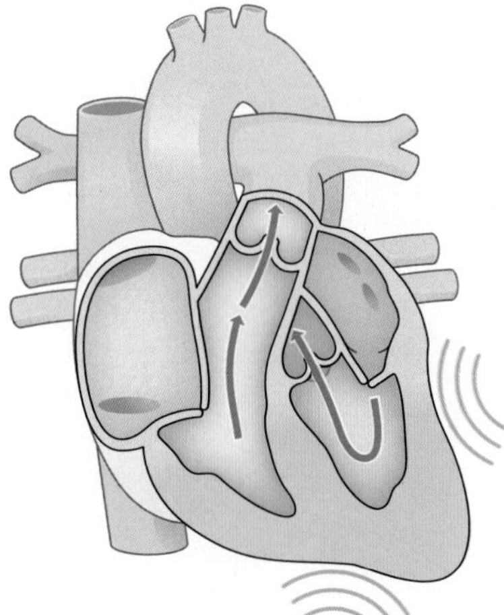

Ventricular systole
Contraction of ventricles pumps blood into aorta and pulmonary arteries.

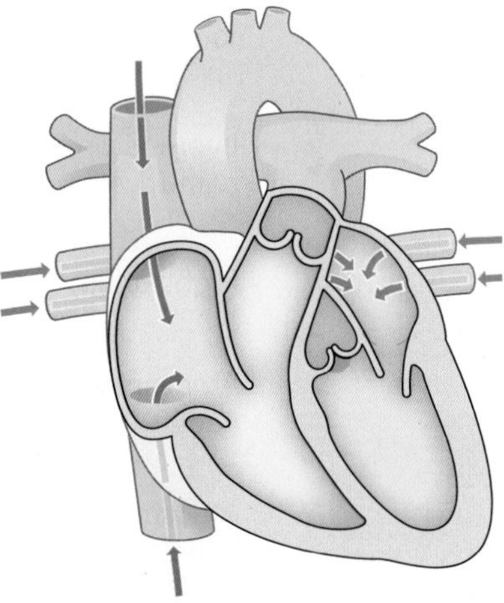

Complete diastole
Atria fill with blood, which flows directly into the relaxed ventricles.

Figure 14-8 The cardiac cycle. **KEY POINT** In one cardiac cycle, contraction of both atria is followed by contraction of both ventricles. The entire heart relaxes briefly before the next cardiac cycle begins. The green shading indicates that the chamber is contracting. **ZOOMING IN** When the ventricles contract, what valves close? What valves open?

Based on a heart rate of 75 bpm and a stroke volume of 70 mL/beat, the average cardiac output for an adult at rest is about 5 L/min. This means that at rest, the heart pumps the equivalent of the total blood volume each minute. But like many other organs, the heart has great reserves of strength. The cardiac reserve is a measure of how many times more than resting output the heart can produce when needed.

During mild exercise, cardiac output might double. During strenuous exercise, it might double again. In other words, for most people, the cardiac reserve is four to five times the resting output. This increase is achieved by an increase in stroke volume and heart rate. In athletes exercising vigorously, the ratio may reach six to seven times the resting volume. In contrast, those with heart disease may have little or no cardiac reserve. They may be fine at rest but quickly become short of breath or fatigued when exercising or even when carrying out the simple tasks of daily living.

See the Student Resources on thePoint for the animations "Myocardial Blood Flow" and "The Cardiac Cycle."

THE HEART'S CONDUCTION SYSTEM

Like other muscles, the heart muscle is stimulated to contract by a wave of electric energy that passes along the cells. This action potential is generated by specialized tissue within the heart and spreads over structures that form the heart's conduction system (**Fig. 14-9**). Two of these structures are tissue masses called **nodes**, and the remainder consists of specialized fibers that branch through the myocardium.

The **sinoatrial** (SA) **node** is located in the upper wall of the right atrium in a small depression described as a sinus. This node initiates the heartbeats by generating an action potential at regular intervals. Because the SA node sets the rate of heart contractions, it is commonly called the **pacemaker**. The second node, located in the interatrial septum at the bottom of the right atrium, is called the **atrioventricular** (AV) **node.**

The **atrioventricular** (AV) **bundle**, also known as the *bundle of His*, is located at the top of the interventricular septum. Fibers travel first down both sides of the interventricular septum in groups called the right and left bundle branches. Smaller **Purkinje** (pur-KIN-je) **fibers** then travel in a branching network throughout the myocardium of the ventricles. Intercalated disks allow the rapid flow of impulses throughout the heart muscle.

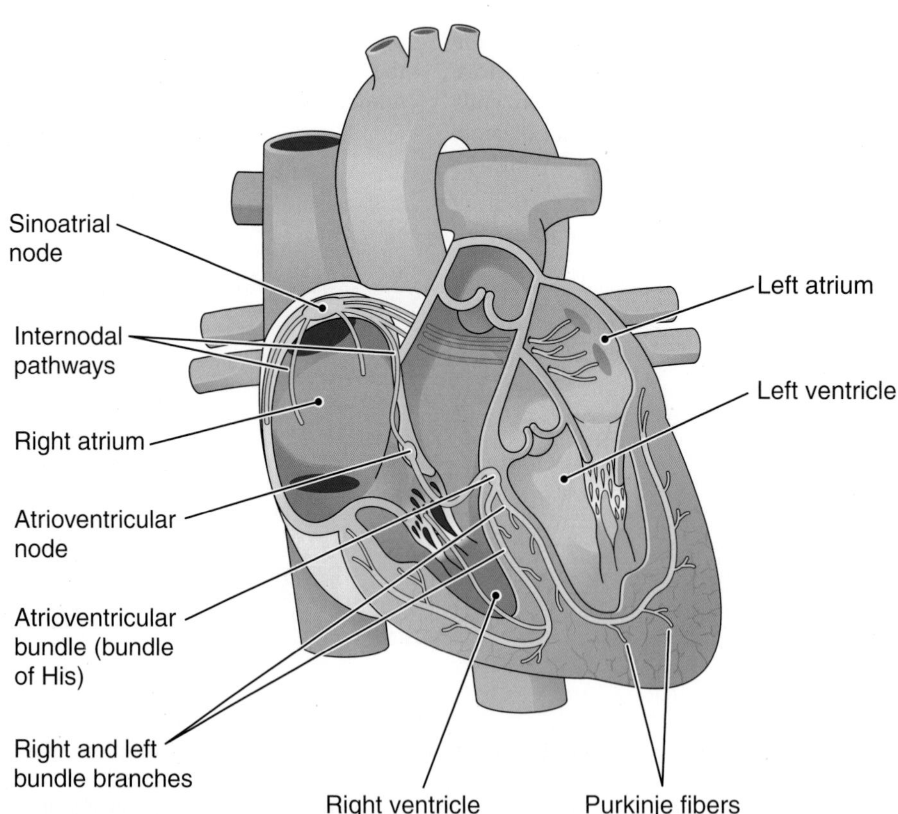

Figure 14-9 **Conduction system of the heart.** **KEY POINT** The sinoatrial (SA) node, the atrioventricular (AV) node, and specialized fibers conduct the electric signal that stimulates the heart muscle to contract. **ZOOMING IN** What parts of the conduction system do the internodal pathways connect?

The order in which impulses travel through the heart is as follows:

1. The SA generates the electric impulse that begins the heartbeat (see Fig. 14-9).
2. The excitation wave travels throughout the myocardium of each atrium, causing the atria to contract. At the same time, impulses also travel directly to the AV node by means of fibers in the wall of the atrium that make up the **internodal pathways**.
3. The atrioventricular node is stimulated. A relatively slower rate of conduction through the AV node allows time for the atria to contract and complete the filling of the ventricles before the ventricles contract.
4. The excitation wave rapidly travels through the AV bundle and then throughout the ventricular walls by means of the bundle branches and Purkinje fibers. The entire ventricular musculature contracts in a wave, beginning at the apex and squeezing the blood upward toward the aorta and pulmonary artery.

A normal heart rhythm originating at the SA node is termed a **sinus rhythm**. As a safety measure, a region of the conduction system other than the SA node can generate a heartbeat if the SA node fails, but it does so at a slower rate.

CONTROL OF THE HEART RATE

Although the heart's fundamental beat originates within the heart itself, the heart rate can be influenced by the nervous system, hormones, and other factors in the internal environment.

The ANS modifies heart rate according to changing body conditions (Fig. 14-10). Stressors, such as excitement and exercise, activate the sympathetic nervous system. Sympathetic fibers increase the contraction rate by stimulating the SA and AV nodes. They also increase the contraction force and thus the stroke volume by acting directly on the fibers of the myocardium. These actions translate into increased cardiac output. Trained athletes, for instance, can increase their cardiac output by 10 times. Parasympathetic stimulation decreases the heart rate. The parasympathetic nerve that supplies the heart is the vagus nerve (cranial nerve X). It slows the heart rate by acting on the SA and AV nodes but does not influence the stroke volume (see Fig. 14-10).

The heart rate is also affected by substances circulating in the blood, including hormones, such as epinephrine and thyroxine; ions, primarily K^+, Na^+, and Ca^{2+}; and drugs. Regular exercise strengthens the heart and increases the amount of blood ejected with each beat. Consequently, the body's circulatory needs at rest can be met with a lower heart rate. Trained athletes usually have a low resting heart rate.

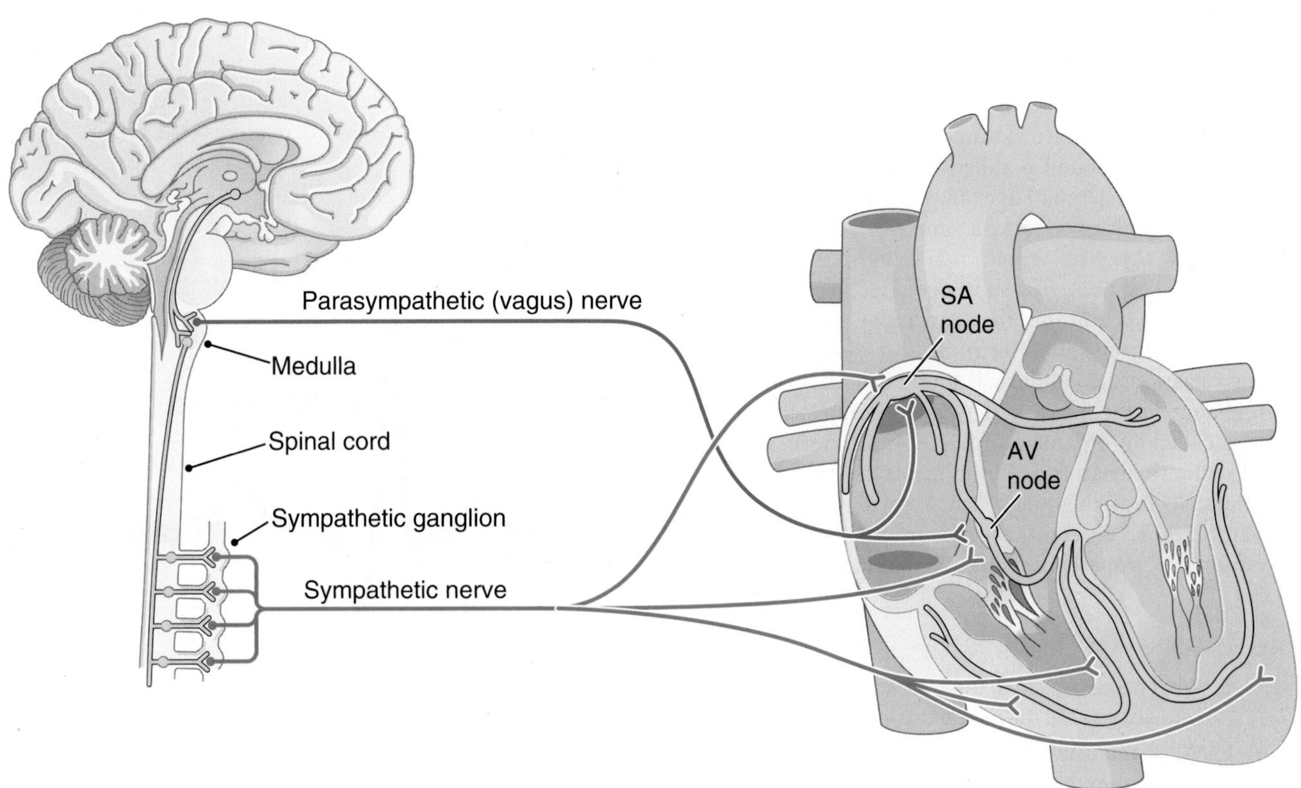

Figure 14-10 Autonomic nervous system (ANS) regulation of the heart. **KEY POINT** The ANS affects the rate and force of heart contractions by acting on the SA and AV nodes and the myocardium itself. **ZOOMING IN** Which cranial nerve carries parasympathetic impulses to the heart?

The following variations in heart rate occur commonly. Note that these variations do not necessarily indicate pathology:

- **Bradycardia** (brad-e-KAR-de-ah) is a relatively slow heart rate of less than 60 bpm. During rest and sleep, the heart may beat less than 60 bpm, but the rate usually does not fall below 50 bpm.
- **Tachycardia** (tak-e-KAR-de-ah) refers to a heart rate of more than 100 bpm. Tachycardia is normal during exercise or stress, or with excessive caffeine intake but may also occur with certain disorders.
- **Sinus arrhythmia** (ah-RITH-me-ah) is a regular variation in heart rate caused by changes in the rate and depth of breathing. It is a normal phenomenon.
- **Premature ventricular contraction** (PVC), also called *ventricular extrasystole*, is a ventricular contraction initiated by the Purkinje fibers rather than the SA node. It can be experienced as a palpitation between normal heartbeats or as a skipped beat. PVCs may be initiated by caffeine, nicotine, or psychological stresses. They are also common in people with heart disease.

NORMAL AND ABNORMAL HEART SOUNDS

The normal heart sounds are usually described by the syllables "lub" and "dup." The first heart sound (S_1), the "lub," is a longer, lower-pitched sound that occurs at the start of ventricular systole. It is caused by a combination of events, mainly closure of the AV valves. This action causes vibrations in the blood passing through the valves and in the tissue surrounding the valves. The second heart sound (S_2), the "dup," is shorter and sharper. It occurs at the beginning of ventricular relaxation and is caused largely by sudden closure of the semilunar valves.

An abnormal sound is called a **murmur** and is usually due to faulty valve action. For example, if a valve fails to close tightly and blood leaks back, a murmur is heard. Another condition giving rise to an abnormal sound is the narrowing, or **stenosis** (sten-O-sis), of a valve opening.

The many conditions that can cause abnormal heart sounds include congenital (birth) defects, disease, and physiologic variations. An abnormal sound caused by any structural change in the heart or the vessels connected with the heart is called an **organic murmur**. Certain normal sounds heard while the heart is working may also be described as murmurs, such as the sound heard during rapid filling of the ventricles. To differentiate these from abnormal sounds, they are more properly called **functional murmurs**.

CHECKPOINTS

- ☐ 14-6 What name is given to the contraction phase of the cardiac cycle? To the relaxation phase?
- ☐ 14-7 What is cardiac output? What two factors determine cardiac output?
- ☐ 14-8 What is the scientific name of the heart's pacemaker?
- ☐ 14-9 What system exerts the main influence on the rate and strength of heart contractions?
- ☐ 14-10 What is a heart murmur?

Heart Studies

Experienced listeners can gain important information about the heart using a **stethoscope** (STETH-o-skope). This relatively simple instrument is used to convey sounds from within the patient's body to an examiner's ear.

The **electrocardiograph** (**ECG** or **EKG**) is used to record the electrical activity of the heart as it functions. (The abbreviation EKG comes from the German spelling of the word.) This activity corresponds to the depolarization and repolarization that occur during an action potential, as described in Chapters 8 and 9. The ECG may reveal certain myocardial injuries. Electrodes (leads) placed on the skin surface pick up electric activity, and the ECG tracing, or electrocardiogram, represents this activity as waves (**Fig. 14-11**). These waves are identified by consecutive letters of the alphabet. The P wave corresponds to depolarization of the atria; the QRS wave corresponds to depolarization of the ventricles. The T wave shows ventricular repolarization, but atrial repolarization is hidden by the QRS wave. Cardiologists use changes in the waves and the intervals between them to diagnose heart damage and arrhythmias.

Many people with heart disease undergo **catheterization** (kath-eh-ter-i-ZA-shun). In right heart catheterization, an extremely thin tube (catheter) is passed through the veins of the right arm or right groin and then into the right side of the heart. This procedure gives diagnostic information and monitors heart function. A **fluoroscope** (flu-OR-o-skope), an instrument for examining deep structures with x-rays, is used to show the route taken by the catheter. The tube is passed all the way through the pulmonary valve into the large lung arteries. Blood samples are obtained along the way for testing, and pressure readings are taken. In left

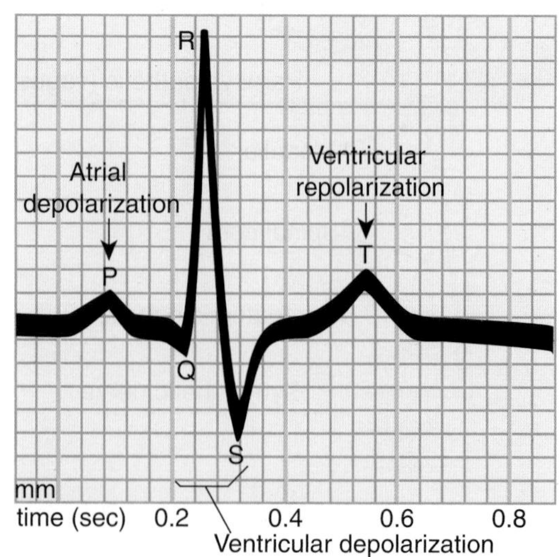

Figure 14-11 Normal electrocardiography (ECG) tracing. **KEY POINT** Electric activity in the myocardium produces ECG waves. Changes in the wave patterns indicate a disorder. The tracing shows one cardiac cycle. **ZOOMING IN** What is the length of the cardiac cycle shown in this diagram?

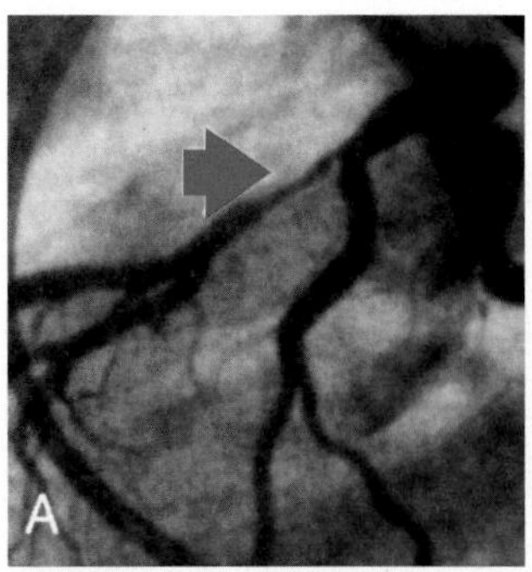

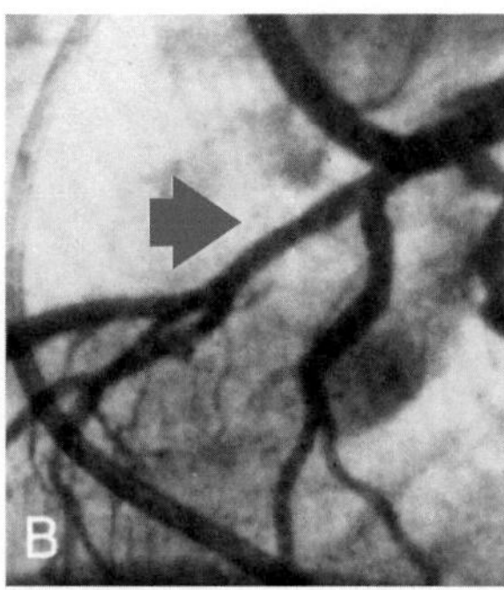

Figure 14-12 **Coronary angiography.** **KEY POINT** The coronary vessels are imaged following administration of a dye. **A.** Coronary angiography shows narrowing in the mid-left anterior descending (LAD) artery (*arrow*). **B.** The same vessel after a procedure to remove plaque. Note the improved blood flow through the artery.

heart catheterization, a catheter is passed through an artery in the left groin or arm to the heart. The tube may be passed through the aortic valve into the left ventricle for studies of pressure and volume in that chamber.

During catheterization, dye can be injected into the coronary arteries to map vascular damage, a procedure known as **coronary angiography** (an-je-OG-rah-fe) (Fig. 14-12). The root *angi/o* means "vessel." **Coronary computed tomography angiography** (coronary CTA) uses advanced radiographic techniques for visualizing the coronary arteries. The necessary dye is injected intravenously, and any abnormalities in the coronary arteries can be seen in computed tomography (CT) scans. Patients complaining of chest pain who show negative CTA results and have no other risk factors for heart attack can safely avoid cardiac catheterization and spend less time in the hospital for observation and testing.

Ultrasound consists of sound waves generated at a frequency above the human ear's range of sensitivity. In **echocardiography** (ek-o-kar-de-OG-rah-fe), also known as *ultrasound cardiography*, high-frequency sound waves are sent to the heart from a small instrument on the chest surface. The ultrasound waves bounce off the heart and are recorded as they return, showing the heart in action. Movement of the echoes is traced on an electronic instrument called an *oscilloscope* and recorded on film. (The same principle is employed by submarines to detect ships.) The method is safe and painless, and it does not use x-rays. It provides information on the size and shape of heart structures, on cardiac function, and on possible heart defects.

CHECKPOINTS

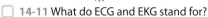

- 14-11 What do ECG and EKG stand for?
- 14-12 What is the general term for using a thin tube threaded through a vessel for diagnosis or repair?
- 14-13 What techniques use a dye and x-rays to visualize the coronary arteries?

Heart Disease

Diseases of the heart and circulatory system are the most common causes of death in industrialized countries. Few people escape having some damage to the heart and blood vessels in a lifetime. Heart diseases include inflammatory processes, rhythm abnormalities, congenital defects, valve malfunctions, coronary artery disease, and heart failure.

HEART INFLAMMATION

Any of the three heart layers can become inflamed:

- **Endocarditis** (en-do-kar-DI-tis) means "inflammation of the heart's lining." Endocarditis may involve the lining of the chambers, but the term most commonly refers to inflammation of the endocardium covering the valves.
- **Myocarditis** (mi-o-kar-DI-tis) is inflammation of heart muscle. In severe cases, the heart muscle becomes damaged and may undergo necrosis.
- **Pericarditis** (per-ih-kar-DI-tis) refers to inflammation of the serous or fibrous membrane surrounding the heart.

These inflammatory diseases are often caused by infection but may also be caused by autoimmune diseases, toxins, radiation, or electric shock. Treatments address the infectious agent (if relevant) and reduce inflammation with corticosteroids. Severe cases may require additional medications or surgery to improve heart function. These measures are discussed later.

ABNORMALITIES OF HEART RHYTHM

A dysfunction anywhere within the heart's conducting system can cause an abnormal rhythm of the heartbeat, or **arrhythmia** (ah-RITH-me-ah), also called *dysrhythmia*. Extremely rapid but coordinated contractions, numbering up to 300 per minute, are described as a flutter. An episode of rapid, wild, and uncoordinated heart muscle contractions is called fibrillation (fih-brih-LA-shun), which may involve the atria only or both the atria and the ventricles. Ventricular fibrillation is a serious disorder because there is no effective heartbeat. It must be corrected by a **defibrillator**, a device that generates a strong electric current to discharge all the cardiac muscle cells at once, allowing a normal rhythm to resume. Atrial fibrillation is not as serious as ventricular fibrillation with regard to survival. However, blood pooling in the atria promotes clot formation and the risk of stroke. Atrial fibrillation can also weaken the heart muscle over time.

An interruption of electric impulses in the heart's conduction system is called **heart block**. The seriousness of this condition depends on how completely the impulses are blocked. It may result in independent beating of the chambers if the ventricles respond to a second pacemaker.

CONGENITAL HEART DISEASE

Congenital heart diseases, that is, heart diseases present at birth, often are the result of defects in fetal development. Two of these disorders represent the abnormal persistence of structures that are part of the normal fetal circulation (**Fig. 14-13A**). Because the lungs are not used until a child is born, the fetus has some adaptations that allow blood to bypass the lungs. The fetal heart has a small hole, the **foramen ovale** (for-A-men o-VAL-e), in the septum between the right and left atria. This opening allows some blood to flow directly from the right atrium into the left atrium, thus bypassing the lungs. Failure of the foramen ovale to close is one cause of an abnormal opening known as an **atrial septal defect** (**see Fig. 14-13B**), sometimes described as a hole in the heart.

The **ductus arteriosus** (ar-te-re-O-sus) in the fetus is a small blood vessel that connects the pulmonary artery and the aorta so that some blood headed toward the lungs will enter the aorta instead. The ductus arteriosus normally closes on its own once the lungs are in use. Persistence of the vessel after birth is described as **patent** (open) **ductus arteriosus** (**see Fig. 14-13C**).

In the fetus, the foramen ovale shunts blood from the right to the left heart, and the ductus arteriosus shunts blood from the pulmonary artery to the aorta. However, if these adaptations persist after birth, increased left-side pressure during normal heart function will drive some oxygen-rich blood from the left heart into the right and from the aorta into the pulmonary artery. Both of these defects are thus categorized as left-to-right shunts because part of the heart's oxygenated left-side output goes to the right heart and then to the lungs instead of out to the body. A congenital hole in the interventricular septum, known as a **ventricular septal defect**, is another relatively common example of a left-to-right shunt (**see Fig. 14-13D**).

A small left-to-right shunt may cause no difficulty and is often not diagnosed until an adult is examined for other cardiac problems. More serious defects greatly increase the pressure in the pulmonary circulation and cause the right ventricle to work harder. If left untreated, the pulmonary arterial walls thicken, and pulmonary hypertension (high blood pressure) develops. Eventually, pulmonary pressure equals that of the systemic system, and poorly oxygenated blood passes through the arterial or septal defect from the right heart directly into the left heart and out to the body. At this point, no cure is possible, so early treatment is critical.

Other congenital defects that tax the heart involve restriction of outward blood flow. **Coarctation** (ko-ark-TA-shun) **of the aorta** is a localized narrowing of the aortic arch (**see Fig. 14-13E**). Another example is obstruction or narrowing of the pulmonary trunk that prevents blood from passing in sufficient quantity from the right ventricle to the lungs.

In many cases, several congenital heart defects occur together. The most common combination is that of four specific defects known as the **tetralogy of Fallot** (FAH-yo): pulmonary artery stenosis, interventricular septal defect, aortic displacement to the right, and right ventricular hypertrophy (size increase). This combination of defects results in a right-to-left shunt, so poorly oxygenated blood is sent directly out to the body instead of to the lungs. So-called

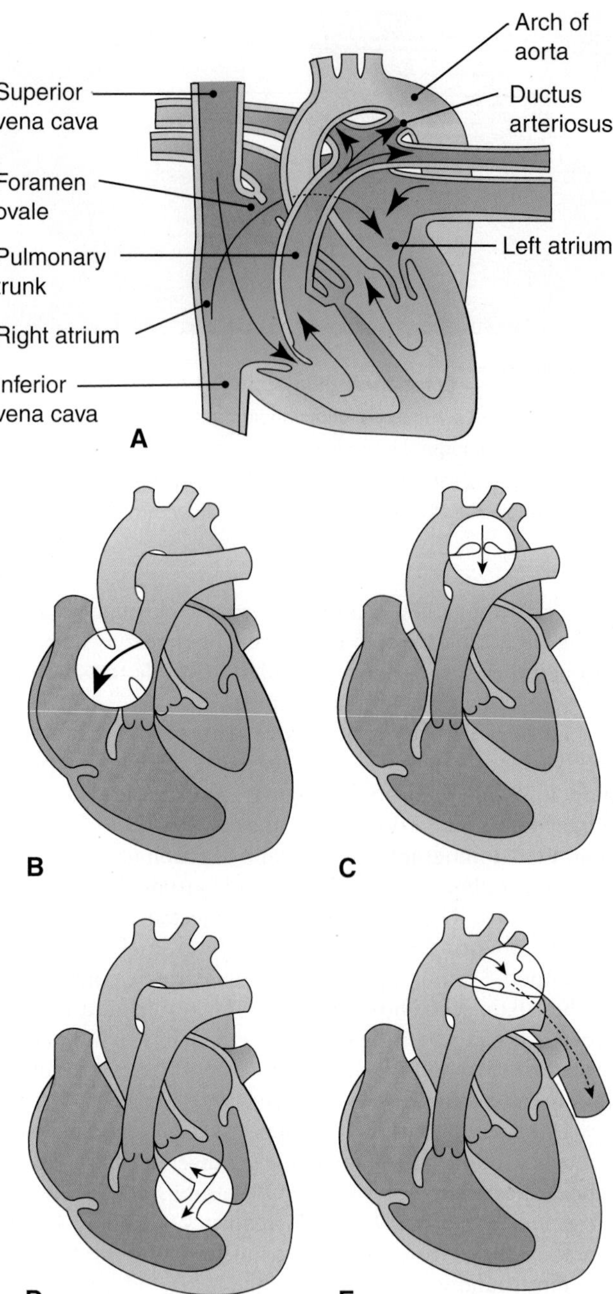

Figure 14-13 Congenital heart defects. **KEY POINT** Some congenital heart defects involve persistence of fetal structures. Others are defects in the septa or vessels associated with the heart. **A.** Normal fetal heart showing the foramen ovale and ductus arteriosus. **B.** Persistence of the foramen ovale results in an atrial septal defect. **C.** Persistence of the ductus arteriosus (patent ductus arteriosus) forces blood back into the pulmonary artery. **D.** A ventricular septal defect. **E.** Coarctation of the aorta restricts outward blood flow in the aorta.

"blue babies" commonly have this disorder. The blueness, or **cyanosis** (si-ah-NO-sis), of the skin and mucous membranes is caused by a relative lack of oxygen. (See Chapter 18 for other causes of cyanosis.)

See the Student Resources on thePoint for a tetralogy of Fallot diagram.

Most U.S. hospitals are now required to screen for congenital heart defects at birth with pulse oximetry. The device detects abnormalities that might not appear until after the baby leaves the hospital. The test is inexpensive and has a low false-positive rate.

In recent years, it has become possible to remedy many congenital defects by heart surgery, one of the more spectacular advances in modern medicine. A patent ductus arteriosus may also respond to drug treatment. During fetal life, prostaglandins (described in Chapter 12) keep the ductus arteriosus open. Drugs that inhibit prostaglandins can promote the duct's closing after birth.

VALVE DISORDERS

Recall that valves ensure one-way flow through the heart. **Valvular stenosis** describes valves that fail to open completely or have narrowed openings, reducing blood flow within and out of the heart. **Valvular insufficiency** describes valves that fail to close properly, leading to backflow. Valvular disorders can result from congenital defects, degenerative changes in the valve tissue, or infection and are usually diagnosed by auscultation and echocardiography. Damaged valves can sometimes be repaired but must be replaced if the damage is too extensive. Substitute valves made of a variety of natural and artificial materials have been used successfully.

Rheumatic (ru-MAT-ik) **heart disease** originates with an attack of rheumatic fever in childhood or in youth. A certain type of streptococcal infection, the type that causes "strep throat," is indirectly responsible for rheumatic fever and rheumatic heart disease. The toxin produced by these streptococci causes a normal immune response. However, in some cases, the initial infection may be followed some two to four weeks later by rheumatic fever, a generalized inflammatory disorder with marked swelling of the joints. The antibodies (immune proteins) formed to combat the toxin are believed to cause this disease. These antibodies may also attack the heart valves, producing a condition known as *rheumatic endocarditis*. The heart valves, particularly the mitral valve, become inflamed, and the normally flexible valve cusps thicken and harden. The mitral valve may not open sufficiently (mitral stenosis) to allow enough blood into the ventricle or may not close effectively, allowing blood to return to the left atrium (mitral regurgitation or insufficiency). Either condition interferes with blood flow from the left atrium into the left ventricle, causing pulmonary congestion, an important characteristic of mitral heart disease. Rheumatic fever most commonly occurs in children 5 to 15 years old but can appear in younger children and adults. The incidence of rheumatic heart disease has declined with antibiotic treatment of streptococcal infections, and it is now rare in industrialized countries. However, those who do not receive adequate diagnosis and treatment are subject to developing the disease.

14

CORONARY ARTERY DISEASE

Coronary artery disease involves the walls of the blood vessels that supply the heart muscle. Like vessels elsewhere in the body, the coronary arteries can undergo degenerative changes with time. The lumen (space) inside the vessel may gradually narrow because of a progressive deposit of fatty material known as **plaque** (PLAK) in the lining of the vessels, usually the arteries. This process, called **atherosclerosis** (ath-er-o-skleh-RO-sis), causes thickening and hardening of the vessels with a loss of elasticity (Fig. 14-14). The *athero* part of the name means "gruel," because of the porridge-like material that adheres to the vessel walls. The vessels' narrowing leads to **ischemia** (is-KE-me-ah), a lack of blood supply to the areas fed by those arteries. Degenerative changes in the arterial wall also may cause the inside vascular surface to become roughened,

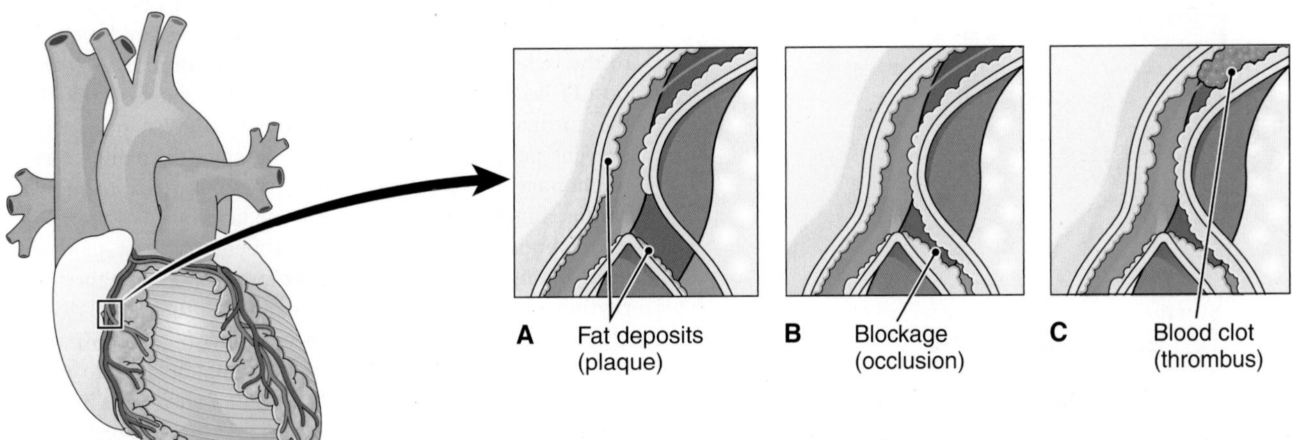

Figure 14-14 Coronary atherosclerosis. **KEY POINT** Fatty deposits in the arterial wall limit blood supply to the tissues and may close off a vessel. **A.** Fat deposits (plaque) narrow an artery, leading to ischemia (lack of blood supply). **B.** Plaque causes blockage (occlusion) of a vessel. **C.** Formation of a blood clot (thrombus) in a vessel leads to myocardial infarction (MI).

promoting blood clot (thrombus) formation (**see Fig. 14-14C**). Jim's heart attack in the opening case study most likely involved coronary artery disease.

Angina Pectoris Moderate ischemia causes a characteristic discomfort, called **angina pectoris** (an-JI-nah PEK-to-ris), felt in the region of the heart and in the left arm and shoulder. Angina pectoris may be accompanied by a feeling of suffocation and a general sensation of forthcoming doom. Women often experience angina differently than men and may report it as noncardiac chest pain. Their symptoms may include pain in the jaw, neck, and upper back or epigastric pain. They may also feel fatigued and be short of breath with exertion. The discomfort of angina can be relieved by rest (which reduces myocardial demand) or by drugs that dilate (widen) the coronary arteries and improve the heart's blood supply. Nitroglycerin, a component of explosives, is one such drug. Coronary artery disease is a common cause of angina pectoris, although the condition has other causes as well.

Myocardial Infarction Thrombus formation in a coronary artery results in a life-threatening condition known as **coronary thrombosis**, as is described in Jim's case study. Sudden occlusion (ok-LU-zhun), or closure, of a coronary vessel with complete obstruction of blood flow is commonly known as a heart attack. Because the area of tissue damaged in a heart attack is described as an infarct (IN-farkt), the medical term for a heart attack is myocardial infarction (MI) (**Fig. 14-15**). The oxygen-deprived tissue will eventually undergo necrosis (death), impairing the heart's ability to convey electrical signals and to generate force.

The symptoms of MI commonly include the abrupt onset of severe, constricting chest pain that may radiate to the left arm, back, neck, or jaw. Patients may experience shortness of breath, sweating, nausea, vomiting, or pain in the epigastric region, which can be mistaken for indigestion. They may feel weak, restless, or anxious, and the skin may be pale, cool, and moist. In women, because degenerative changes more commonly affect multiple small vessels rather than the major coronary pathways, MI symptoms are often more long-term, subtle, and diffuse than the intense chest pain that is more typical in men. For this reason, symptoms may be ignored by the patient, and a correct diagnosis may be delayed or overlooked.

MI is diagnosed by ECG and assays for specific substances in the blood. **Creatine kinase** (CK) is an enzyme normal to muscle cells. It is released in increased amounts when any muscle tissue is damaged. The form of CK specific to cardiac muscle cells is creatine kinase MB (CK-MB) (**see Box 8-1**). Troponin (Tn) is a protein that regulates muscle cell contraction (see Chapter 8). Increased plasma levels of troponin subunits specific to myocardium indicate recent MI or other types of heart damage.

See the Student Resources on thePoint for an illustration of the clinical signs of MI.

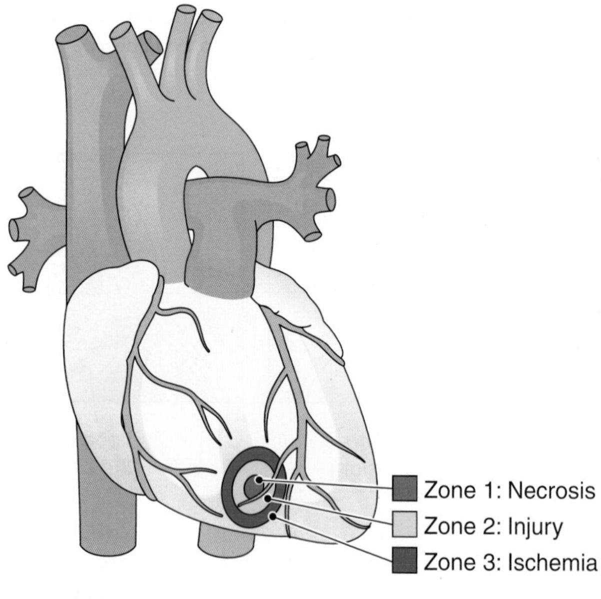

Figure 14-15 Myocardial infarction (MI). KEY POINT Occlusion (closure) of a coronary artery causes ischemia (lack of blood supply) and then tissue death (necrosis).

Treatment of Myocardial Infarctions The outcome of a myocardial infarction depends largely on the extent and location of the damage. Complete and prolonged lack of blood to any part of the myocardium results in tissue necrosis and weakening of the heart wall. Many people die within the first hour after onset of symptoms, but prompt, aggressive treatment can improve outcomes. The vast majority of patients who reach the hospital alive survive.

Initial treatment involves cardiopulmonary resuscitation (CPR) and defibrillation at the scene when needed. The American Heart Association is adding training in the use of the automated external defibrillator (AED) to the basic course in CPR. The AED detects fatal arrhythmia and automatically delivers the correct preprogrammed shock. Work is underway to place machines in shopping centers, sports venues, and other public settings.

Prompt transport by paramedics who are able to monitor the heart and give emergency drugs helps people survive and reach a hospital. The next step is to restore blood flow to the ischemic areas by administering **thrombolytic** (thrombo-LIT-ik) **drugs**, which act to dissolve the clots blocking the coronary arteries. Therapy must be given promptly to prevent permanent heart muscle damage. In many cases, a pulmonary artery catheter (tube) is put in place to monitor cardiac function and response to medication.

Supportive care includes treatment of chest pain with intravenous (IV) morphine. Healthcare workers monitor heart rhythm and give medications to maintain a functional rhythm. Oxygen is given to improve heart function. Some patients require a surgical procedure, such as angioplasty, to reopen vessels or a vascular graft to bypass damaged vessels; others may need an artificial pacemaker to maintain a normal heart rhythm. These measures are discussed in greater detail below.

Recovery from a heart attack and resumption of normal activities is often possible if the patient follows his or her prescribed drug therapy plan and takes steps to reduce cardiac risk factors.

Prevention of Coronary Artery Disease Prevention of coronary artery disease is based on identification of cardiovascular risk factors and modification of those factors that can be changed. Risk factors that cannot be modified include the following:

- Age. The risk of coronary heart disease increases with age.
- Gender. Until middle age, men have greater risk than do women. Women older than 50 years or past menopause have risk equal to that of men.
- Heredity. Those with immediate family members with heart disease are at greater risk.
- Body type. In particular, the hereditary tendency to deposit fat in the abdomen or on the chest surface increases risk.

Risk factors that can be modified include the following:

- Smoking and other forms of tobacco use, which lead to spasm and hardening of the arteries. These arterial changes result in decreased blood flow and poor supply of oxygen and nutrients to the myocardium.
- Physical inactivity. Lack of exercise weakens the heart muscle and decreases the heart's efficiency. It also decreases the efficiency of the skeletal muscles, which further taxes the heart.
- Weight over the ideal increases risk.
- Saturated fat in the diet. Elevated fat levels in the blood lead to blockage of the coronary arteries by plaque (Box 14-1).
- Hypertension damages heart muscle. Smoking cessation, regular physical activity, a healthful, low-sodium diet, and appropriate medication, if needed, are all important in reducing this risk factor.
- Diabetes causes damage to small blood vessels. Type 2 diabetes can be managed with diet, exercise, and proper medication, if needed.
- Individuals suffering from sleep apnea, that is, people who frequently stop breathing for short periods when they sleep, have a higher risk of coronary artery disease. Sleep apnea can be treated with devices to aid breathing during the night and sometimes with surgery to remove obstructions of air passageways.

Efforts to prevent heart disease should include having regular physical examinations and minimizing the controllable risk factors. Jim's physicians will undoubtedly discuss these lifestyle changes with him during the course of his treatment.

In recent years, researchers have identified several substances circulating in the blood that, at high levels, are associated with a risk of heart disease. These so-called *markers* may reflect the controllable and uncontrollable risk factors described above. They include the following:

- C-reactive protein (CRP), a protein produced in the liver in response to inflammation. Chronic inflammation appears to promote atherosclerosis. Some of its possible sources include infection, arthritis, and periodontal (gum) disease. CRP can predict a risk of MI, even when blood cholesterol is low, and testing for CRP is recommended for people with normal cholesterol levels and moderate risk of MI.
- Homocysteine (ho-mo-SIS-tene), an amino acid. Higher than average levels of this amino acid in the blood are linked to cardiovascular disease.
- Lipoprotein (a), a type of low-density lipoprotein (LDL) (see Box 14-1). A high level of lipoprotein (a) promotes atherosclerosis.

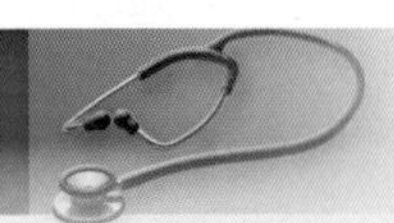

CLINICAL PERSPECTIVES

Box 14-1

Lipoproteins: What's the Big DL?

Although cholesterol has received a lot of bad press in recent years, it is a necessary substance in the body. It is found in bile salts needed for digestion of fats, in hormones, and in the cell's plasma membrane. However, high levels of cholesterol in the blood have been associated with atherosclerosis and heart disease.

It now appears that the total amount of blood cholesterol is not as important as the form in which it occurs. Cholesterol is transported in the blood in combination with other lipids and with protein, forming compounds called lipoproteins. These compounds are distinguished by their relative density. High-density lipoprotein (HDL) is composed of a high proportion of protein and relatively little cholesterol. HDLs remove cholesterol from the tissues, including the arterial walls, and carry it back to the liver for reuse or disposal. In contrast, low-density lipoprotein (LDL) contains less protein and a higher proportion of cholesterol. LDLs carry cholesterol from the liver to the tissues, making it available for membrane or hormone synthesis. However, excess LDLs can deposit cholesterol along the lining of arterial walls. Thus, high levels of HDLs (60 mg/dL and above) indicate efficient removal of arterial plaques, whereas high levels of LDLs (130 mg/dL and above) suggest that arteries will become clogged.

Diet is an important factor in regulating lipoprotein levels. Saturated fatty acids (found primarily in animal fats) raise LDL levels, while unsaturated fatty acids (found in most vegetable oils) lower LDL levels and stimulate cholesterol excretion. Thus, a diet lower in saturated fat and higher in unsaturated fat may reduce the risk of atherosclerosis and heart disease. Other factors that affect lipoprotein levels include cigarette smoking, caffeine, and stress, which raise LDL levels, and exercise, which lowers LDL levels.

14

HEART FAILURE

Heart failure is a condition in which the heart is unable to pump efficiently. It often occurs after the heart muscle is damaged by an MI but can also result from hypertension (high blood pressure) or valvular abnormalities, both of which force the heart to work harder. Infection that damages the myocardium can cause heart failure, as can lung disease, coronary artery disease, or any other condition that restricts oxygen supply to the myocardium. In cases of heart failure, the heart cannot pump out all the blood it receives, and blood accumulates in the ventricles. At first, the increased volume stretches the cardiac muscle fibers and increases their contraction strength to maintain adequate cardiac output. Over time, however, the chambers enlarge and the myofibers become permanently stretched, further reducing their pumping ability. In an attempt to increase blood flow, the nervous system increases the HR and constriction of blood vessels, increasing blood pressure. Blood backs up into the pulmonary circulation, increasing pressure and fluid in the lungs and causing shortness of breath and a persistent cough. Fluid is also retained in the legs and feet, liver, and abdomen, leading to the original name for this disorder—*congestive heart failure* (CHF).

Drugs that cardiologists use to treat heart failure include those that strengthen heart contractions, reduce blood pressure, regulate the heartbeat, or limit water and salt retention. People with heart failure can often live comfortably by practicing healthful living habits, taking appropriate medication, and monitoring themselves for sudden weight gain, which may indicate fluid retention.

See the Student Resources on thePoint for an illustration of the clinical effects of CHF and for the animation "Heart Failure."

CHECKPOINTS

- ☐ 14-14 What are the three types of heart inflammation?
- ☐ 14-15 What is an abnormal heart rhythm called?
- ☐ 14-16 What is congenital heart disease?
- ☐ 14-17 What types of organisms cause rheumatic fever?
- ☐ 14-18 What degenerative process commonly causes narrowing of the vessels in coronary artery disease?
- ☐ 14-19 What is the medical term for a "heart attack?"
- ☐ 14-20 What is the general name for a device that restores a normal heart rhythm?

Treatment of Heart Disease

Lifestyle changes are the first approach to prevention and treatment of many heart ailments. If these have failed to achieve desired results, cardiologists may use medical and surgical therapies, often in combination, to treat heart disease.

MEDICATIONS

For someone with persistent high blood cholesterol, a factor associated with atherosclerosis, or someone who has suffered a heart attack, a physician might recommend a drug categorized as a **statin** (STAT-in). These drugs lower blood cholesterol by inhibiting an enzyme the liver needs to manufacture it. Statins also lower CRP, a blood marker associated with atherosclerosis, as previously described.

Anticoagulants (an-ti-ko-AG-u-lants) are valuable drugs for some heart patients. They may be used to prevent clot formation in patients with damage to heart valves or blood vessels or in patients who have had a myocardial infarction. **Aspirin** (AS-pir-in), chemically known as acetylsalicylic (a-SE-til-sal-ih-sil-ik) acid (ASA), is an inexpensive and time-tested drug for pain and inflammation that reduces blood clotting by interfering with platelet activity. A small daily dose of aspirin is recommended for patients with angina pectoris, those who have suffered a myocardial infarction, and those who have undergone surgery to open or bypass narrowed coronary arteries. Aspirin is contraindicated for people with bleeding disorders or gastric ulcers, because it irritates the stomach lining. Newer antiplatelet medications include clopidogrel (trade name Plavix). **Warfarin** (WAR-fah-rin) is an anticoagulant that works by limiting vitamin K activity. It is used to prevent thrombosis or prevent further clot formation following thrombosis. Warfarin, widely known by its most common trade name *Coumadin* (KU-mah-din), is used in cases of atrial fibrillation, for people with artificial valves, and in cases of thrombosis in deep veins. Warfarin therapy is difficult to manage because the drug interacts with some common medications (certain antibiotics and herbal remedies) and even some foods and vitamins (foods that contain vitamin K, garlic, fish oils, vitamin E). A physician must monitor treatment carefully with regular blood tests to ensure proper dosage.

One of the oldest drugs for heart treatment, and still the most important drug for many patients, is **digitalis** (dij-ih-TAL-is). This agent, which slows and strengthens heart muscle contractions, is obtained from the leaf of the foxglove, a plant originally found growing wild in many parts of Europe. Foxglove is now cultivated to ensure a steady supply of digitalis for medical purposes.

Beta-adrenergic blocking agents ("beta-blockers") reduce sympathetic stimulation of the heart. They reduce the rate and strength of heart contractions, thus reducing the heart's oxygen demand. Propranolol is one example.

Antiarrhythmic agents (e.g., quinidine) are used to regulate the rate and rhythm of the heartbeat.

Slow calcium channel blockers aid in the treatment of coronary heart disease and hypertension by several mechanisms. They may dilate vessels, control the force of heart contractions, or regulate conduction through the AV node. Their actions are based on the fact that calcium ions must enter muscle cells before contraction can occur.

PACEMAKERS

If the SA node fails to generate a normal heartbeat or there is some failure in the cardiac conduction system, an electric,

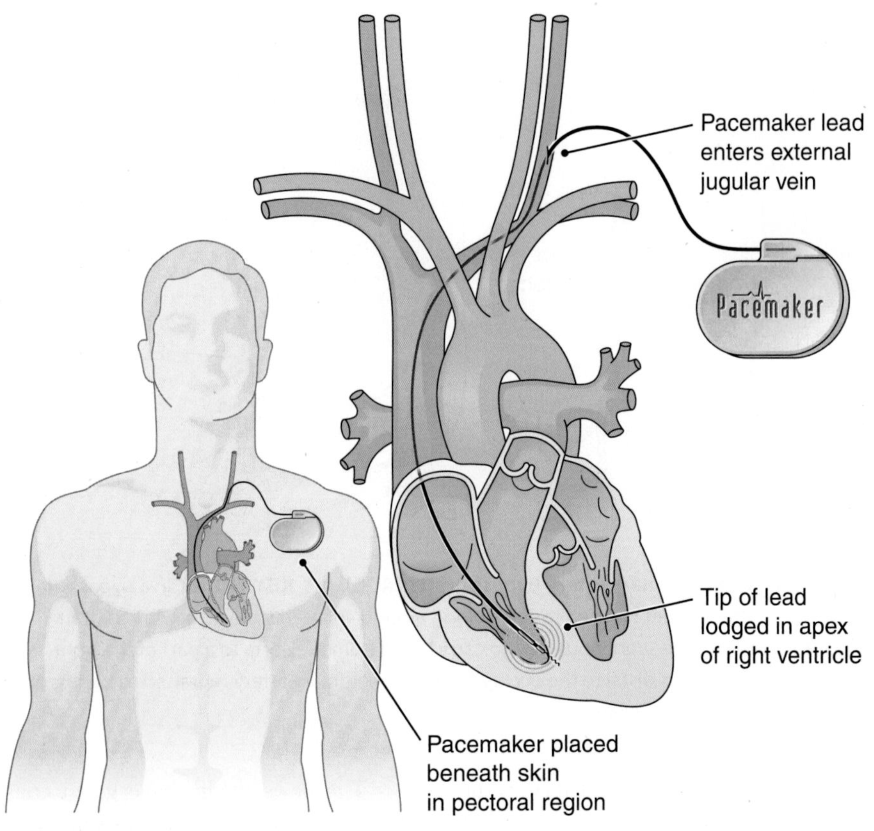

Figure 14-16 **Placement of an artificial pacemaker.** **KEY POINT** The lead is placed in an atrium or ventricle (usually on the right side). A dual-chamber pacemaker has leads in both chambers.

battery-operated **artificial pacemaker** can be employed (**Fig. 14-16**). This device, implanted under the skin, supplies regular impulses to stimulate the heartbeat. The implantation site is usually in the left chest area. A pacing wire (lead) from the pacemaker is then passed into the heart through a vessel and lodged in the heart. The lead may be fixed in an atrium or a ventricle (usually on the right side). A dual-chamber pacemaker has a lead in each chamber to coordinate beats between the atrium and ventricle. Some pacemakers operate at a set rate; others can be set to stimulate a beat only when the heart fails to do so on its own. Another type of pacemaker adjusts its pacing rate in response to changing activity, as during exercise. This rather simple device has saved many people whose hearts cannot effectively beat alone. In an emergency, a similar stimulus can be supplied to the heart muscle through electrodes placed externally on the chest wall.

In cases of chronic ventricular fibrillation, a battery-powered device can be implanted in the chest to restore a normal rhythm. The device detects a rapid abnormal rhythm and delivers a direct shock to the heart. The restoration of a normal heartbeat, either by electric shock or by drugs, is called *cardioversion*, and this device is called an implantable cardioverter-defibrillator (ICD). A lead wire from the defibrillator is placed in the right ventricle through the pulmonary artery. In cases of severe tachycardia, tissue that is causing the rhythmic disturbance can be destroyed by heart surgery or catheterization, as described shortly.

HEART SURGERY

The heart–lung machine makes many operations on the heart and other thoracic organs possible. There are several types of machines in use, all of which serve as temporary substitutes for the patient's heart and lungs. The machine siphons off the blood from the large vessels entering the heart on the right side, so that no blood passes through the heart and lungs. While in the machine, the blood is oxygenated, and carbon dioxide is removed chemically. The blood is also "defoamed," or rid of air bubbles, which could fatally obstruct blood vessels. The machine then pumps the processed blood back into the general circulation through a large artery. Modern advances have enabled cardiac surgeons to perform certain procedures without bypassing the circulation, partially immobilizing the heart while it continues to beat.

Coronary artery bypass graft (CABG) restores the heart's blood supply following an obstruction in the coronary arteries (**Fig. 14-17**). While the damaged coronary arteries remain in place, healthy segments of blood vessels from other parts of the patient's body are grafted onto the vessels to bypass any obstructions. Usually, parts of the saphenous vein

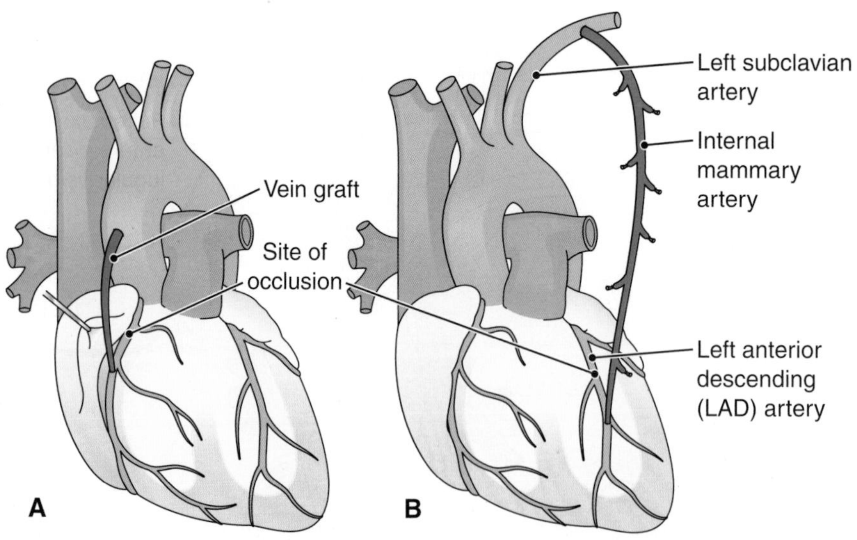

Figure 14-17 Coronary artery bypass graft (CABG). **KEY POINT** In a bypass graft, a healthy vessel segment is used to carry blood around an arterial blockage. **A.** This graft uses a segment of the saphenous vein to carry blood from the aorta to a part of the right coronary artery that is distal to the occlusion. **B.** The mammary artery is grafted to bypass an obstruction in the LAD artery.

(a superficial vein in the leg) or the mammary artery in the chest are used. Coronary bypass surgery is explained further in Jim's follow-up case study.

Sometimes, as many as six or seven segments are required to establish an adequate blood supply. The mortality associated with this operation is low, and most patients are able to return to a nearly normal lifestyle after recovery from the surgery. The effectiveness of this procedure diminishes over a period of years, however, due to blockage of the replacement vessels.

Less invasive surgical procedures include the technique of **angioplasty** (AN-je-o-plas-te), which is used to open restricted arteries in the heart and other areas of the body. In balloon angioplasty, a fluoroscope is used to guide a catheter with a balloon to the affected area (**Fig. 14-18**). There, the balloon is inflated to break up the blockage in the coronary artery, thus restoring effective circulation to the myocardium. To prevent repeated blockage, a small tube called a **stent** may be inserted in the vessel to keep it open (**Fig. 14-19**). Modern stents are coated with drugs that prevent reocclusion.

Cardiologists may also use a catheter to cut away plaque in a narrowed artery. In a process called **coronary atherectomy** (ath-eh-REK-to-me), plaque is removed by a cutting or grinding device that also captures small pieces of the disrupted plaque that might enter the blood and block smaller arteries. Angiograms showing a coronary artery before and after atherectomy are in **Figure 14-12**. Cardiac surgeons can also correct arrhythmia using catheterization. In this case, radio waves or some other form of energy, such as a laser beam, microwaves, or cryothermy (cold), are used to ablate (destroy) heart tissue that is generating abnormal signals. The procedure is called **cardiac ablation** (ab-LA-shun).

The news media have given considerable attention to the **surgical transplantation** of human hearts and sometimes of lungs and hearts together. This surgery is done in specialized centers and is available to some patients with degenerative heart disease and cardiomyopathies (diseases of the heart muscle) who are otherwise in good health. Tissues of the recently deceased donor and of the recipient must be as closely matched as possible to avoid rejection.

Efforts to replace a damaged heart with a completely artificial heart have not met with long-term success so far. There are devices available, however, to assist a damaged heart in pumping during recovery from heart attack or while a patient is awaiting a donor heart. The ventricular assist device (VAD) draws blood from a ventricle and pumps it into the aorta (on the left) or the pulmonary artery (on the right). Medical researchers are also testing a fully implantable artificial heart to use for the same purpose.

Box 14-2 discusses a new approach for repairing a damaged heart.

CHECKPOINTS

- [ ] 14-21 How do aspirin and warfarin act to prevent heart attacks?
- [ ] 14-22 What is the name for a device that is wired to the heart to regulate the heartbeat?
- [ ] 14-23 What does CABG stand for?

See the Student Resources on thePoint for career information on surgical technologists, who assist in all types of surgical operations.

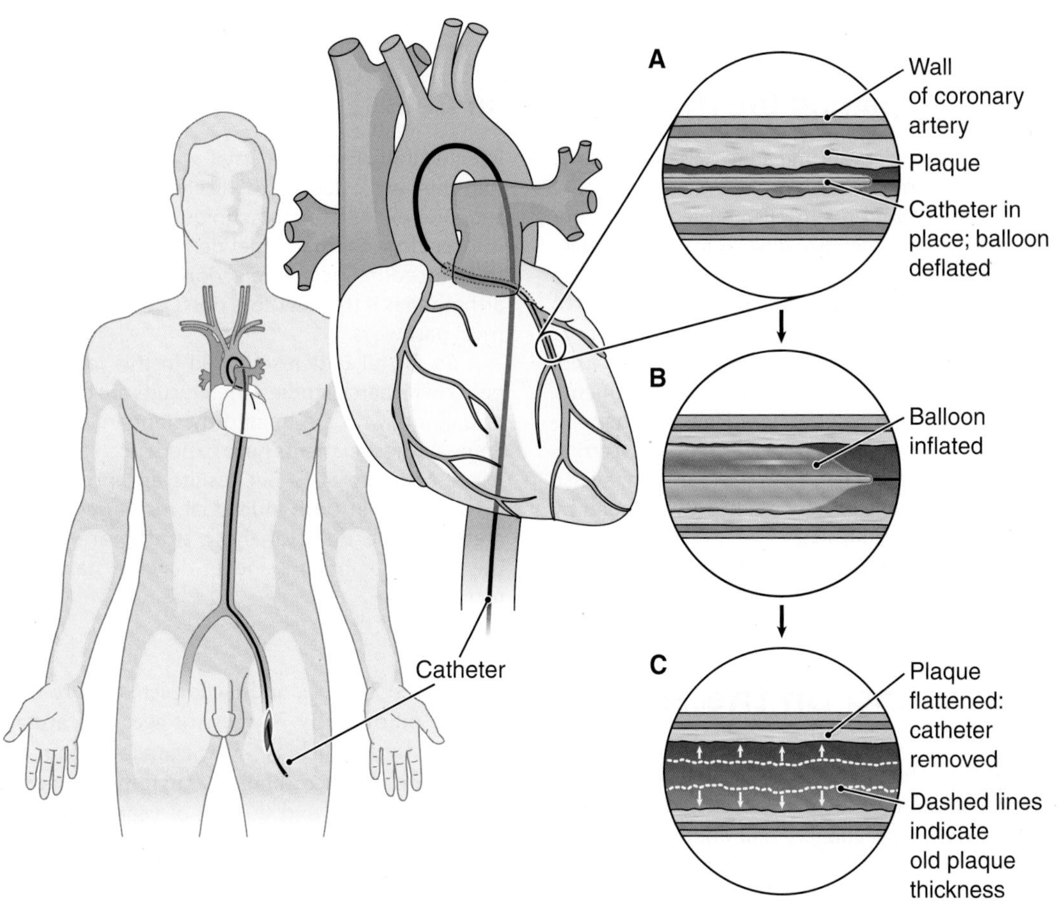

Figure 14-18 **Coronary angioplasty.** **KEY POINT** Angioplasty can be used to open an occluded vessel. **A.** A guide catheter is threaded into the coronary artery. **B.** A balloon catheter is inserted through the occlusion and inflated. **C.** The balloon is inflated and deflated until plaque is flattened and the vessel is opened.

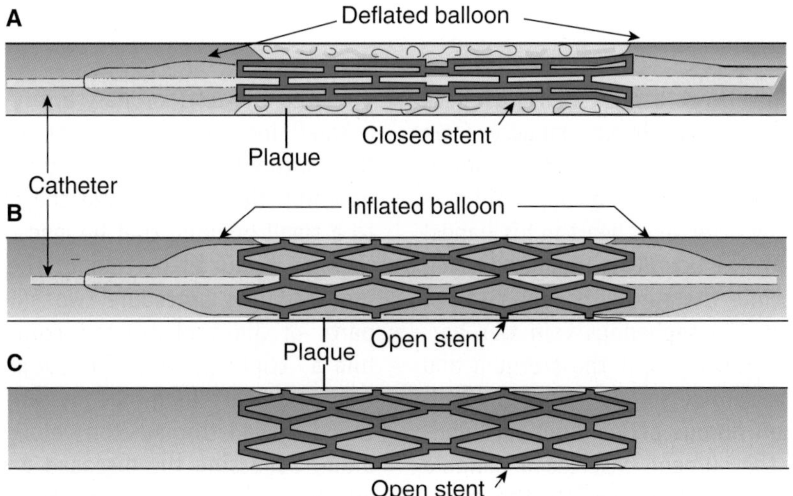

Figure 14-19 **Arterial stent.** **KEY POINT** A stent (tube) may be used to keep a vessel open after angioplasty. **A.** Stent closed, before balloon inflation. **B.** Stent open, balloon inflated. The stent will remain expanded after balloon is deflated and removed. **C.** Stent open, balloon removed.

14

HOT TOPICS

Box 14-2

Repairing the Heart with Stem Cells

Stem cells are immature cells that have the potential to develop into multiple types of tissue. Mixed in with differentiated mature cells, they can be triggered to multiply in order to regenerate or repair tissue.

In the past few years, medical researchers have tried to repair damaged heart muscle with infusions of stem cells. These cells are taken from the bone marrow of a donor or from the patient's own heart. With modern medical treatments, more patients are surviving heart attacks. However, the temporary loss of oxygen to cardiac muscle can produce scarring that eventually weakens the heart. Stem cells can help reverse this damage and also help strengthen a heart weakened by heart failure or a valvular insufficiency.

The stem cells from a donor or from a small biopsy of the patient's heart are grown in the laboratory and then injected into the heart itself with a catheter, in the hope that they will establish themselves and repair damaged tissue. One experimental technique to keep the cells in place is to enclose them in a gel derived from algae. The risks of this procedure include rejection of donor cells and possible arrhythmias if the new cells don't tie in with the heart's electrical pathways.

So far, all patients treated in this manner are participants in research studies. Their results are compared to those of patients who are treated by standard medical practices. Some of these patients have experienced a reduction in scar tissue of up to 50%. But results are inconsistent because of variations in both individual responses and in research methods. The goals of these studies are to determine if the therapy is helpful and to develop standard treatment techniques.

Effects of Aging on the Heart

There is a great deal of individual variation in the way the heart ages, depending on heredity, environmental factors, diseases, and personal habits such as diet, exercise patterns, and tobacco use. However, many changes still commonly occur with age. The heart chambers become smaller, and myocardial tissue atrophies and gets replaced with connective tissue. These changes significantly reduce cardiac output. The valves become less flexible, and incomplete closure may produce an audible murmur. By 70 years of age, the cardiac output may decrease by as much as 35%. Damage within the conduction system can produce abnormal rhythms, including extra beats, rapid atrial beats, and slowing of ventricular contraction rate. Temporary failure of the conduction system (heart block) can cause periodic loss of consciousness. Because of the decrease in the heart's reserve strength, elderly people may be less able to respond efficiently to physical or emotional stress.

Disease in Context Revisited

Jim's Heart Surgery

Several weeks after his heart attack, Jim was back in the hospital for his coronary bypass surgery. Even though his cardiologist had fully explained the procedure to him, Jim was still nervous—in a couple of hours a surgeon would literally have Jim's heart in his hands!

Jim was brought to the operating room and given general anesthesia. While the cardiac surgeon sawed through Jim's sternum, the saphenous vein was harvested from Jim's leg. Having split the sternum and retracted the ribs, the cardiac surgeon made an incision through the tough fibrous pericardium surrounding his heart. Next, the surgeon inserted a cannula into the right atrium and another one into the aorta. The doctor connected the cannulae to the heart–lung machine and stopped Jim's heart from beating. Now, venous blood from Jim's right atrium flowed through the heart–lung machine where it was oxygenated before being pumped into his aorta. Then, the surgeon prepared the left coronary artery for bypass. He made a small incision through the arterial wall and carefully sutured the cut end of the harvested vein to the opening. Next, he sutured the other end of the vein to a small opening that he made in the aorta, bypassing the occluded portion of the coronary artery. He repeated this procedure two more times in different parts of Jim's obstructed coronary arteries—giving him a "triple bypass." The surgeon disconnected Jim from the heart–lung machine and restarted his heart. Blood flowed through the vein grafts to the myocardium, bypassing the diseased parts of Jim's coronary arteries. Jim's surgery was a success.

Although this chapter concentrates on medical terms related to the heart, Jim's case also contains terminology about blood vessels. In Chapter 15, Blood Vessels and Blood Circulation, you will examine these terms in more detail.

CHAPTER 14

Chapter Wrap-Up

Summary Overview

A detailed chapter outline with space for note taking is on *thePoint*. The figure below illustrates the main topics covered in this chapter.

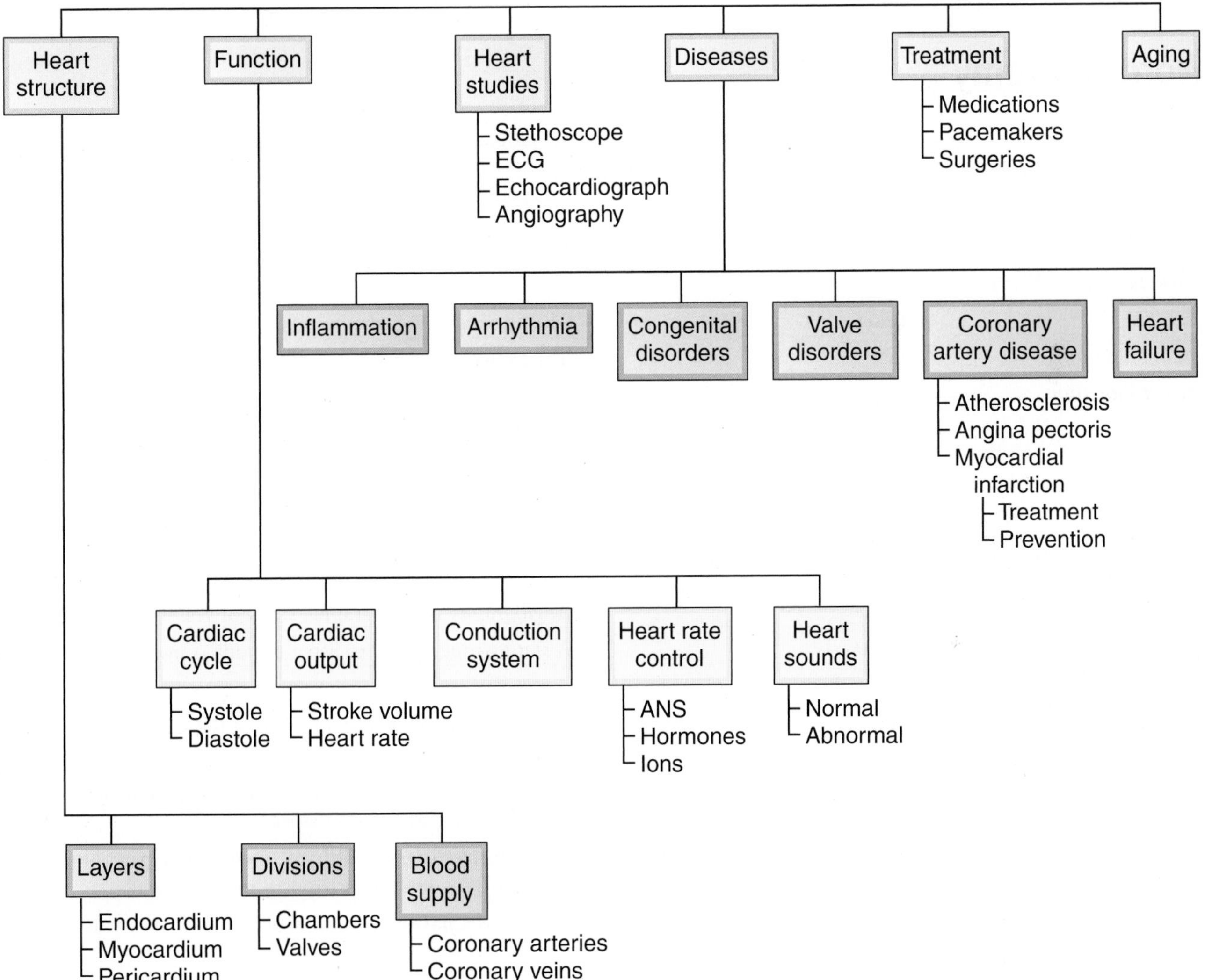

Key Terms

The terms listed below are emphasized in this chapter. Knowing them will help you organize and prioritize your learning. These and other boldface terms are defined in the Glossary with phonetic pronunciations.

angina pectoris
angiography
arrhythmia
atherosclerosis
atrium
bradycardia
cardiac output
coronary
coronary thrombosis
diastole
echocardiography
electrocardiograph
endocardium
epicardium
fibrillation
infarct
ischemia
mediastinum
murmur
myocardium
pacemaker
pericardium
plaque
septum
stenosis
systole
tachycardia
valve
ventricle

Word Anatomy

Medical terms are built from standardized word parts (prefixes, roots, and suffixes). Learning the meanings of these parts can help you remember words and interpret unfamiliar terms.

WORD PART	MEANING	EXAMPLE
Structure of the Heart		
cardi/o	heart	The *myocardium* is the heart muscle.
pulmon/o	lung	The *pulmonary* circuit carries blood to the lungs.
Heart Function		
brady-	slow	*Bradycardia* is a slow heart rate.
sin/o	sinus	The *sinoatrial* node is in a space (sinus) in the wall of the right atrium.
tachy-	rapid	*Tachycardia* is a rapid heart rate.
Heart Studies		
angi/o	vessel	*Angiography* is radiographic study of vessels.
steth/o	chest	A *stethoscope* is used to listen to body sounds.
Heart Disease		
cyan/o	blue	*Cyanosis* is a bluish discoloration of the skin caused by lack of oxygen.
isch	suppression	Narrowing of blood vessels leads to *ischemia*, a lack of blood (emia) supply to tissues.
scler/o	hard	In *atherosclerosis*, vessels harden with fatty, gruel-like (ather/o) material that deposits on vessel walls.
sten/o	narrowing, closure	*Stenosis* is a narrowing of a structure, such as a valve.
Treatment of Heart Disease		
-ectomy	surgical removal	*Atherectomy* is removal of atherosclerotic (ecto [out] + tom/o [cut]) plaque from a vessel.
-plasty	molding, surgical formation	*Angioplasty* is used to reshape vessels that are narrowed by disease.

Questions for Study and Review

BUILDING UNDERSTANDING

Fill in the Blanks

1. The heart's pointed inferior portion is the _____.
2. The layer of the heart wall that pumps blood is the _____.
3. The heartbeat is initiated by electrical impulses from the _____.
4. Lack of blood to a tissue due to blocked arteries is called _____.
5. Pain caused by lack of blood to heart muscle is called _____.

Matching > Match each numbered item with the most closely related lettered item.

___ **6.** receives blood low in oxygen from the body
___ **7.** receives blood high in oxygen from the lungs
___ **8.** sends blood low in oxygen to the lungs
___ **9.** sends blood high in oxygen to the body

a. right ventricle
b. left ventricle
c. right atrium
d. left atrium

Multiple Choice

___ **10.** Which structural characteristic promotes rapid transfer of electrical signals between cardiac muscle cells?

a. the striated nature of the cells
b. branching of the cells
c. the abundance of mitochondria within the cells
d. intercalated disks between the cells

___ **11.** What separates the upper chambers of the heart from each other?

a. intercalated disk
b. interatrial septum
c. interventricular septum
d. ductus arteriosus

___ **12.** Which term describes one complete sequence of heart contraction and relaxation?

a. systole
b. diastole
c. cardiac cycle
d. cardiac output

___ **13.** Cardiac output is the product of which factors?

a. stroke volume and heart rate
b. cardiac reserve and atrial systole
c. heart rate and ventricular diastole
d. stroke volume and dysrhythmia

___ **14.** Which medication reduces cardiac output by lowering sympathetic responses?

a. anticoagulant
b. antiarrhythmic agent
c. slow calcium channel blocker
d. beta-adrenergic blocking agent

___ **15.** Which variation in heart rate can be due to changes in the rate and depth of breathing?

a. murmur
b. cyanosis
c. sinus arrhythmia
d. stent

UNDERSTANDING CONCEPTS

16. Referring to *The Body Visible* at the beginning of this book, give the name and number of the following:

a. two structures that keep the AV valves from opening into the atria
b. the heart's pacemaker
c. the two vessels that carry blood into the coronary circulation
d. the vein that drains blood from the coronary circulation and empties into the right atrium

17. Differentiate between the terms in each of the following pairs:

a. serous pericardium and fibrous pericardium
b. atrium and ventricle
c. foramen ovale and ductus arteriosus
d. systole and diastole
e. bradycardia and tachycardia

18. Explain the purpose of the four heart valves, and describe their structures and locations.

19. Trace a drop of blood from the superior vena cava to the lungs and then from the lungs to the aorta.

20. Describe the order in which electrical impulses travel through the heart. What is an interruption of these impulses in the heart's conduction system called?

21. Compare the effects of the sympathetic and parasympathetic nervous systems on heart function.

22. Compare and contrast the following disease conditions:

a. endocarditis and pericarditis
b. functional murmur and organic murmur
c. flutter and fibrillation

23. Referring to Appendix 2-3, name four substances in blood that can indicate myocardial infarction. What types of substances are these?

24. List some age-related changes to the heart.

CONCEPTUAL THINKING

25. Using Jim's case study and the text, explain what is meant by a myocardial infarction (MI). How is it diagnosed and treated?

26. Certain risk factors for coronary artery disease may have contributed to Jim's heart attack. What can Jim do to lower his risk of having it happen again? What risk factors is he unable to change? Apply your knowledge of these factors to your own life or the life of someone you know.

27. Three-month-old Hannah R is brought to the doctor by her parents. They have noticed that when she cries, she becomes breathless and turns blue. The doctor examines Hannah and notices that she is lethargic, is small for her age, and has a loud mitral valve murmur. With this information, explain the likely cause of Hannah's symptoms.

For more questions, see the Learning Activities on thePoint.

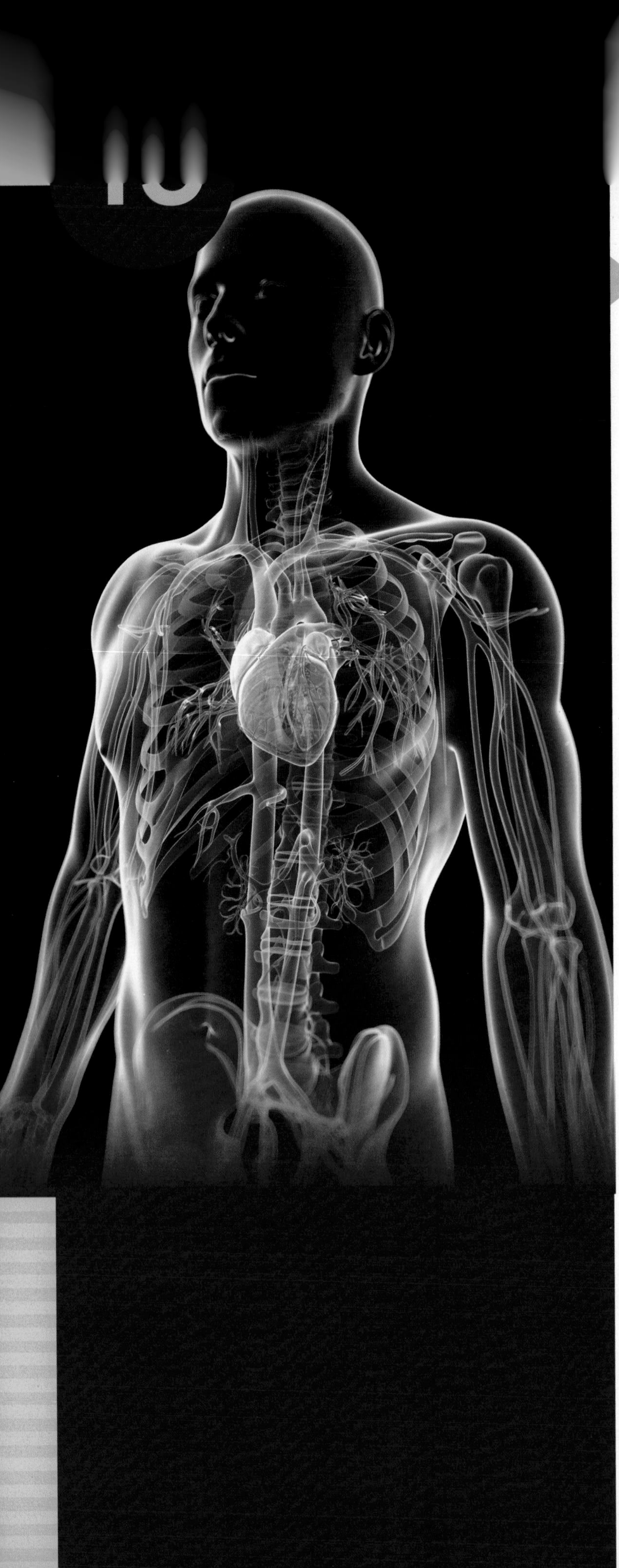

Learning Objectives

After careful study of this chapter, you should be able to:

1 ▶ Differentiate among the five types of blood vessels with regard to structure and function. ***p. 340***

2 ▶ Compare the pulmonary and systemic circuits relative to location and function. ***p. 340***

3 ▶ Name the four sections of the aorta, and list the main branches of each section. ***p. 343***

4 ▶ Trace the pathway of blood through the main arteries of the upper and lower limbs. ***p. 346***

5 ▶ Define *anastomosis*, cite its function, and give four examples of anastomoses. ***p. 346***

6 ▶ Compare superficial and deep veins, and give examples of each type. ***p. 346***

7 ▶ Name the main vessels that drain into the superior and inferior venae cavae. ***p. 349***

8 ▶ Define *venous sinus*, and give four examples of venous sinuses. ***p. 349***

9 ▶ Describe the structure and function of the hepatic portal system. ***p. 350***

10 ▶ Explain the forces that affect exchange across the capillary wall. ***p. 352***

11 ▶ Describe five factors that regulate blood flow. ***p. 352***

12 ▶ Define *pulse*, and list six factors that affect pulse rate. ***p. 353***

13 ▶ List four factors that affect blood pressure. ***p. 354***

14 ▶ Explain the role of baroreceptors in controlling blood pressure. ***p. 354***

15 ▶ Explain how blood pressure is commonly measured. ***p. 355***

16 ▶ Discuss six disorders involving the blood vessels. ***p. 358***

17 ▶ Based on the opening case study, discuss the dangers of thrombosis, and describe one approach to its treatment. ***pp. 339, 361***

18 ▶ Show how word parts are used to build words related to the blood vessels and circulation (see Word Anatomy at the end of the chapter). ***p. 363***

Disease in Context *Jocelyn's Circulation Crisis*

"I'm feel like I'm getting old," Jocelyn lamented to her husband, John, as they relaxed after dinner. "This left knee that I strained at work has been bothering me for a few weeks, and it doesn't seem to be getting much better. The elastic brace and the antiinflammatory drug the orthopedist prescribed are not helping much. My leg looks swollen, and the pain actually seems to be moving into my calf."

Jocelyn, age 52, works at a day care center for the elderly and thinks she may have injured herself in working with her clients. Taking John's advice to see the doctor again, she visited Dr. Rennard's office, and considering her continuing pain, he ordered a venous ultrasound of her leg.

"Get this," Jocelyn said, in tears, as she hung up the phone. "The report came back that I have a blood clot in my leg. The condition might be really dangerous, and the doctor wants me to come in this afternoon."

Dr. Rennard explained to Jocelyn that she had a deep vein thrombosis, or DVT, in the popliteal vein behind her knee. She would need injections and medication daily until the clot resolved. He told her to watch for any signs of phlebitis, such as pain, redness, or swelling in the affected limb, or pulmonary embolism, that is, a blot clot breaking loose and traveling to her lungs. Signs of this dangerous development include shortness of breath, chest pain, cough, or fainting.

"Your body should develop alternate blood routes to compensate for the clot, but we still have to get rid of it," he explained.

"I really appreciate your giving me these shots," Jocelyn told her husband, "but you know how I hate needles, and I might be a really bad patient."

John was to give her twice-daily injections of Lovenox into her lateral abdomen for 11 days. In addition, she took Coumadin orally once a day, adjusting the dose based on the drug's activity as measured with twice-weekly blood tests. Lovenox is a form of the anticoagulant heparin, which inhibits certain clotting factors, and Coumadin interferes with the action of vitamin K needed for clotting. To everyone's relief, she was able to stop the injections after 11 days but had to continue the Coumadin for a total of six months. She also required regular tests to be sure that her blood would clot properly if necessary.

"Your ultrasound shows no blood clot," Dr. Rennard was pleased to tell Jocelyn at the end of her treatment. "However, I highly recommend that you see a hematologist. Maybe we can find out why this clot formed and how you might be able to prevent recurrences."

Dr. Rennard recognizes the dangers of a blockage anywhere in the circulatory system. In this chapter, we'll learn about the normal blood routes, the physiology of capillary exchanges and blood pressure, and some of the disorders that can affect the circulatory system.

ANCILLARIES *At-A-Glance*

Visit thePoint to access the following resources. For guidance in using these resources most effectively, see pp. xv–xvii.

Learning RESOURCES

- Tips for Effective Studying
- Web Figure: Capillary Micrograph
- Web Figure: Evolution of Atherosclerosis
- Web Figure: Formation of an Atheroma
- Web Figure: Thrombus in an Aortic Aneurysm
- Animation: Blood Circulation
- Animation: Hypertension
- Health Professions: Vascular Technologist
- Detailed Chapter Outline
- Answers to Questions for Study and Review
- Audio Pronunciation Glossary

Learning ACTIVITIES

- Pre-Quiz
- Visual Activities
- Kinesthetic Activities
- Auditory Activities

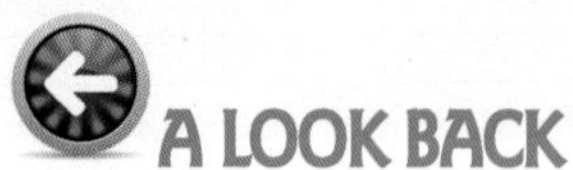

The story of circulation continues with a discussion of the vessels that carry blood away from and then back to the heart. In describing how materials flow between the tissues and the bloodstream, we return to discussions of diffusion, filtration, osmosis, and osmotic pressure introduced in Chapter 3.

The blood vessels, together with the four chambers of the heart, form a closed system in which blood is carried to and from the tissues. Although whole blood does not leave the vessels, components of the plasma and tissue fluids can be exchanged through the walls of the tiniest vessels—the capillaries (**Fig. 15-1**).

The vascular system is easier to understand if you refer to the appropriate illustrations in this chapter as the vessels are described. When this information is added to what you already know about the blood and the heart, a picture of the cardiovascular system as a whole will emerge.

Overview of Blood Vessels

Blood vessels may be divided into five groups, named according to the sequence of blood flow from the heart:

1. **Arteries** carry blood away from the heart and toward the tissues. The heart's ventricles pump blood into the arteries (**see Fig. 15-1**).
2. **Arterioles** (ar-TE-re-olz) are small subdivisions of the arteries (**see Fig. 15-2**). They divide into the capillaries.
3. **Capillaries** are tiny, thin-walled vessels that allow for exchanges between systems. These exchanges occur between the blood and the body cells and between the blood and the air in the lung tissues. The capillaries connect the arterioles and venules.
4. **Venules** (VEN-ulz) are small vessels that receive blood from the capillaries and begin its transport back toward the heart (**see Fig. 15-2**).
5. **Veins** are vessels formed by the merger of venules. They continue blood's transport until it is returned to the heart.

BLOOD CIRCUITS

The vessels together may be subdivided into two groups, or circuits: pulmonary and systemic. **Figure 15-1** diagrams blood flow through these two circuits.

The Pulmonary Circuit The **pulmonary circuit** delivers blood to the lungs, where some carbon dioxide is eliminated and oxygen is replenished. The pulmonary vessels that carry blood to and from the lungs include the following:

1. The pulmonary trunk and its arterial branches, which carry blood low in oxygen from the right ventricle to the lungs

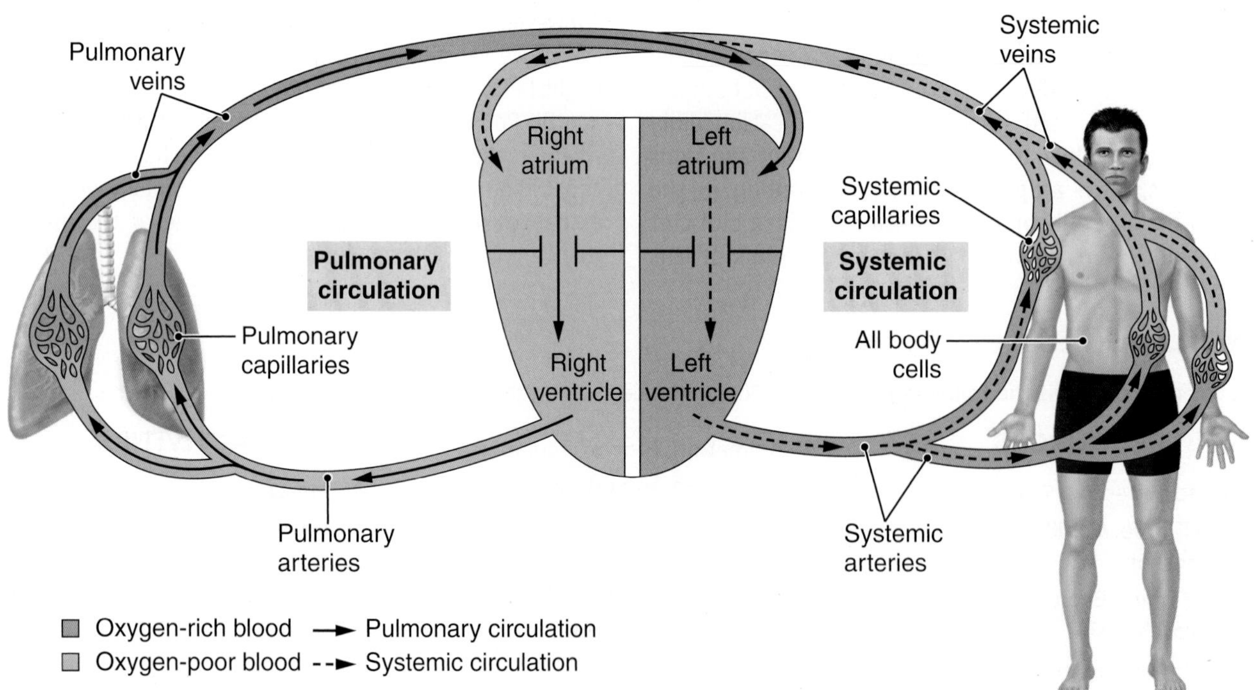

Figure 15-1 **The cardiovascular system.** **KEY POINT** Blood flows in a closed system with exchanges of material between the blood and tissues through the capillary walls. There are two circuits, pulmonary and systemic. **ZOOMING IN** Which arteries contain oxygen-poor blood? Which veins contain oxygen-rich blood?

2. The capillaries in the lungs, through which gases, nutrients, and wastes are exchanged
3. The pulmonary veins, which carry freshly oxygenated blood back to the left atrium

Note that the pulmonary vessels differ from those in the systemic circuit in that the pulmonary arteries carry blood that is *low* in oxygen content, and the pulmonary veins carry blood that is *high* in oxygen content. In contrast, the systemic arteries carry highly oxygenated blood, and the systemic veins carry blood that is low in oxygen.

The Systemic Circuit The **systemic** (sis-TEM-ik) **circuit** supplies nutrients and oxygen to all the tissues and carries waste materials away from the tissues for disposal. The systemic vessels include the following:

1. The **aorta** (a-OR-tah) receives freshly oxygenated blood from the left ventricle and then branches into the systemic arteries carrying blood to the tissues.
2. The systemic capillaries are the blood vessels through which materials are exchanged.
3. The systemic veins carry blood low in oxygen back toward the heart. The venous blood flows into the right atrium of the heart through the superior vena cava and inferior vena cava.

VESSEL STRUCTURE

The arteries have thick walls because they must be strong enough to receive blood pumped under pressure from the heart's ventricles (**Fig. 15-2**). The three tunics (coats) of the arteries resemble the heart's three tissue layers. Named from internal to external, they are as follows:

1. The inner tunic, a membrane of simple, squamous epithelial cells making up the **endothelium** (en-do-THE-le-um), forms a smooth surface over which the blood flows easily.
2. The middle tunic, the thickest layer, is made up of smooth (involuntary) muscle, which is under the control of the autonomic nervous system.
3. The outer tunic is made of supporting connective tissue.

Elastic tissue between the layers of the arterial wall allows these vessels to stretch when receiving blood and then return to their original size. This elastic force propels blood forward between heartbeats, ensuring continuous blood flow. The

15

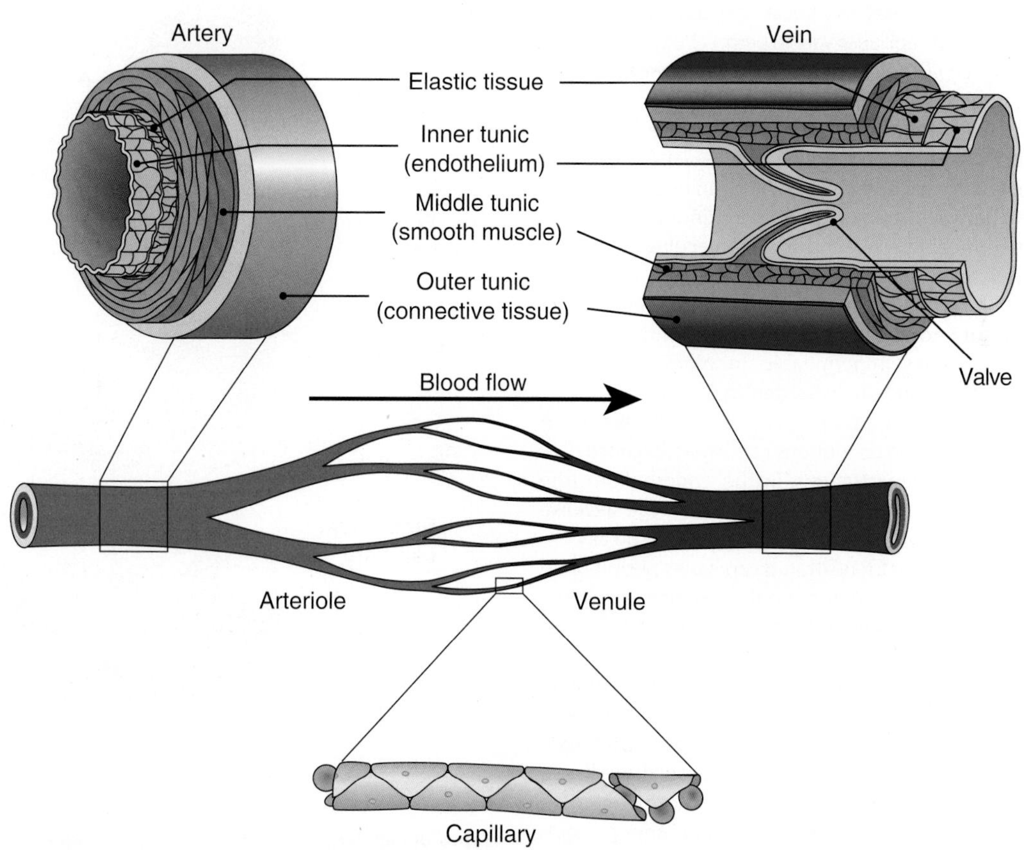

Figure 15-2 **Sections of small blood vessels.** **KEY POINT** Drawings show the thick wall of an artery, the thin wall of a vein, and the single-layered wall of a capillary. A valve is also shown. The *arrow* indicates the direction of blood flow. **ZOOMING IN** Which vessels have valves that control blood flow?

amount of elastic tissue diminishes as the arteries branch and become smaller.

The small subdivisions of the arteries, the arterioles, have thinner walls in which there is little elastic connective tissue but relatively more smooth muscle. The autonomic nervous system controls this involuntary muscle. The vessel lumens (openings) become narrower (constrict) when the muscle contracts, and they widen (dilate) when the muscle relaxes. In this manner, the arterioles regulate the amount of blood that enters the various tissues at a given time. As discussed later, change in the diameter of many arterioles at once alters blood pressure.

The microscopic capillaries that connect arterioles and venules reach a maximum diameter of 10 mcm, just about wide enough for a blood cell to pass through. They have the thinnest walls of any vessels: one cell layer (**see Fig. 15-2**). The transparent capillary walls are a continuation of the smooth endothelium that lines the arteries. The thinness of these walls allows for exchanges between the blood and the body cells and between the lung tissue and the outside air. The capillary boundaries are the most important center of activity for the entire circulatory system. Their function is explained later in this chapter. Capillary structure varies according to function, as described in **Box 15-1**.

See the Student Resources on thePoint to review the animation "Blood Circulation" and for a micrograph of a capillary in cross-section.

The smallest veins, the venules, are formed by the union of capillaries, and their walls are only slightly thicker than are those of the capillaries. As the venules merge to form veins, the smooth muscle in the vessel walls becomes thicker, and the venules begin to acquire the additional layers found in the larger vessels.

Box 15-1

A CLOSER LOOK

Capillaries: The Body's Free Trade Zones

The exchange of substances between body cells and the blood occurs along about 50,000 miles (80,000 km) of capillaries. Exchange rates vary because based on their structure, different types of capillaries vary in permeability.

Continuous capillaries (top) are the most common type and are found in muscle, connective tissue, the lungs, and the central nervous system (CNS). These capillaries are composed of a continuous layer of endothelial cells. Adjacent cells are loosely attached to each other with small openings called intercellular clefts between them. Although continuous capillaries are the least permeable, water and small molecules can diffuse easily through their walls. Large molecules, such as plasma proteins and blood cells, cannot. In certain body regions like the CNS, adjacent endothelial cells are joined tightly together, making the capillaries impermeable to many substances (see **Box 10-1**, "The Blood–Brain Barrier," in Chapter 10).

Fenestrated (FEN-es-tra-ted) *capillaries* (middle) are much more permeable than are continuous capillaries, because they have many holes, or fenestrations, in the endothelium (the word is derived from Latin meaning "window"). These sievelike capillaries are permeable to water and solutes as large as peptides. In the digestive tract, fenestrated capillaries permit rapid absorption of water and nutrients into the bloodstream. In the kidneys, they permit rapid filtration of blood plasma, the first step in urine formation.

Sinusoidal capillaries (bottom) are the most permeable. In addition to fenestrations, they have large spaces between endothelial cells that allow the exchange of water, large solutes, such as plasma proteins, and even blood cells. Sinusoidal capillaries, also called *sinusoids*, are found in the liver and red bone marrow, for example. Albumin, clotting factors, and other proteins formed in the liver enter the bloodstream through sinusoidal capillaries. In red bone marrow, newly formed blood cells travel through sinusoidal capillary walls to join the bloodstream.

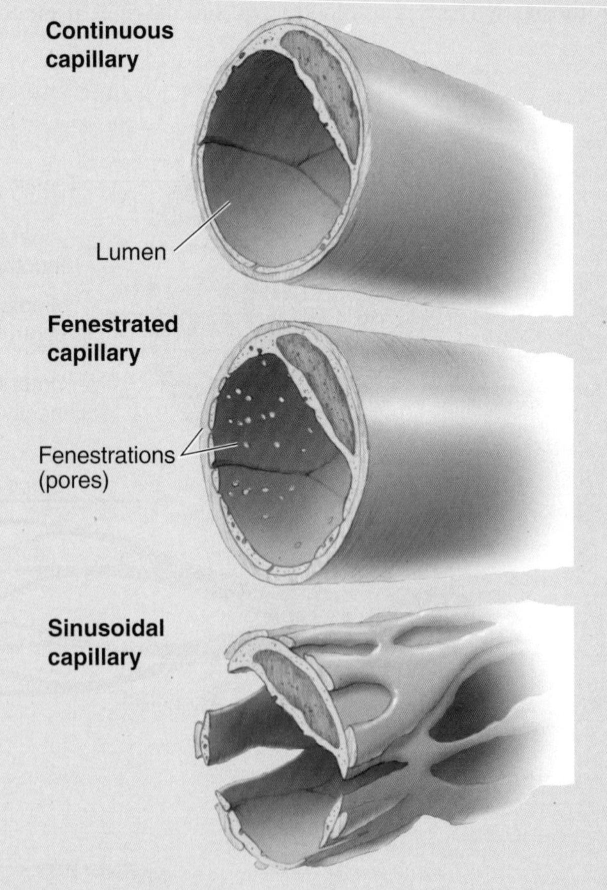

Types of capillaries. (Reprinted with permission from *The Massage Connection Anatomy and Physiology*. Philadelphia, PA: Lippincott Williams & Wilkins, 2004.)

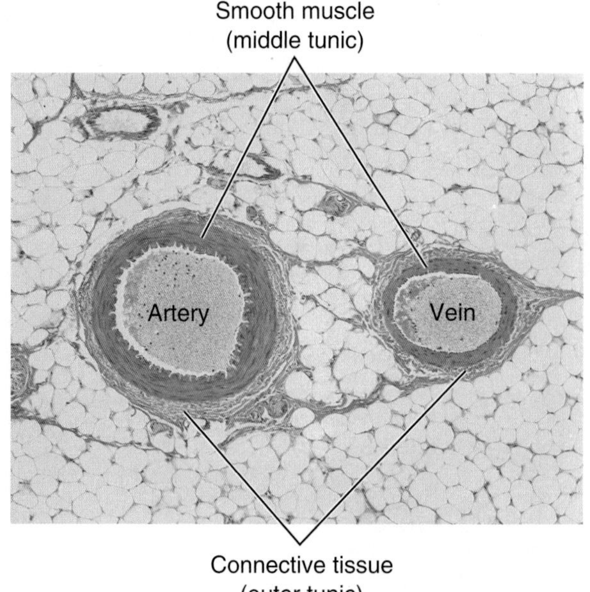

Figure 15-3 Cross section of an artery and vein. The smooth muscle and connective tissue of the vessels are visible in this photomicrograph. **ZOOMING IN** Which type of vessel shown has a thicker wall?

The walls of the veins have the same three layers as those of the arteries. However, the middle smooth muscle tunic is relatively thin in the veins. A vein wall is much thinner than is the wall of a comparably sized artery (**see Fig. 15-2**). These vessels also have less elastic tissue between the layers, so they expand easily and carry blood under much lower pressure. Because of their thinner walls, the veins collapse easily. Even a slight pressure on a vein by a tumor or other mass may interfere with blood flow.

Most veins are equipped with one-way valves that permit blood to flow in only one direction: toward the heart (**see Fig. 15-2**). Such valves are most numerous in the veins of the extremities. **Figure 15-3** is a cross-section of an artery and a vein as seen through a microscope.

CHECKPOINTS

- 15-1 What are the five types of blood vessels?
- 15-2 What are the two blood circuits, and what areas does each serve?
- 15-3 What type of tissue makes up the middle tunic of arteries and veins, and how is this tissue controlled?
- 15-4 How many cell layers make up the wall of a capillary?

Systemic Arteries

The systemic arteries begin with the aorta, the largest artery, which measures about 2.5 cm (1 in) in diameter. This vessel receives blood from the left ventricle, ascends from the heart, and then arches back to travel downward through the body, branching to all organs. (See Dissection Atlas **Figure A5-4** for a photograph showing the major vessels of the trunk.)

THE AORTA AND ITS PARTS

The aorta is a thick-walled vessel about the diameter of your thumb (**Fig. 15-4**). It is one continuous artery, but its regions are named as follows:

1. The **ascending aorta** extends upward and slightly to the right from the left ventricle. It lies within the pericardial sac.
2. The **aortic arch** curves from the right to the left and also extends posteriorly.
3. The **thoracic aorta** descends just anterior to the vertebral column posterior to the heart in the mediastinum.
4. The **abdominal aorta** is the longest section of the aorta, beginning at the diaphragm and spanning the abdominal cavity.

The thoracic and abdominal aorta together make up the descending aorta.

15

Branches of the Ascending Aorta and Aortic Arch The aorta's ascending part has two branches near the heart, called the *left* and *right coronary arteries*, which supply the heart muscle (see **Figs. 14-6 and 14-7**). As noted in Chapter 14, these arteries form a crown around the heart's base and give off branches to all parts of the myocardium.

The aortic arch, located immediately past the ascending aorta, gives rise to three large branches.

1. The first, the **brachiocephalic** (brak-e-o-seh-FAL-ik) **artery**, is a short vessel that supplies the arm and the head on the right side (**see Fig. 15-4**). After extending upward about 5 cm (2 in), it divides into the **right subclavian** (sub-KLA-ve-an) **artery**, which extends under the right clavicle (collar bone) and supplies the right upper extremity (arm) and part of the brain, and the **right common carotid** (kah-ROT-id) **artery**, which supplies the right side of the neck, head, and brain. Note that the brachiocephalic artery is unpaired.
2. The second, the **left common carotid artery**, extends upward from the highest part of the aortic arch. It supplies the left side of the neck and the head.
3. The third, the **left subclavian artery**, extends under the left clavicle and supplies the left upper extremity and part of the brain. This is the aortic arch's last branch.

Branches of the Descending Aorta The thoracic aorta supplies branches to the chest wall and esophagus (e-SOF-ah-gus), the bronchi (subdivisions of the trachea), the lungs, and the muscles of the chest wall (**Fig. 15-5**). There are usually nine to 10 pairs of **intercostal** (in-ter-KOS-tal) **arteries** that extend between the ribs, sending branches to the muscles and other structures of the chest wall.

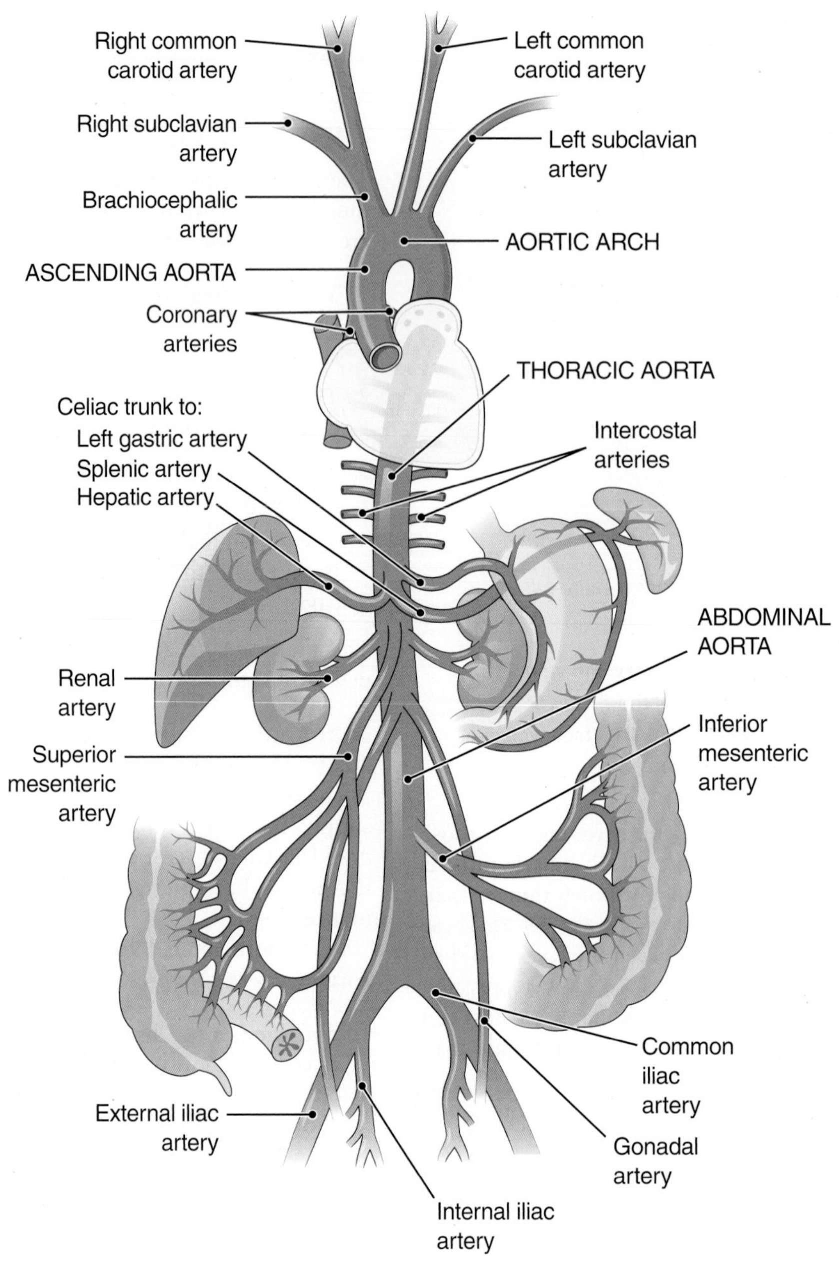

Figure 15-4 The aorta and its branches. **KEY POINT** As the aorta travels from the heart through the body, it branches to all tissues. **ZOOMING IN** How many brachiocephalic arteries are there?

The abdominal aorta has unpaired branches extending anteriorly and paired branches extending laterally. The unpaired vessels are large arteries that supply the abdominal viscera. The most important of these visceral branches are as follows:

1. The **celiac** (SE-le-ak) **trunk** is a short artery, about 1.25 cm (1/2 in) long, that subdivides into three branches: the **left gastric artery** goes to the stomach, the **splenic** (SPLEN-ik) **artery** goes to the spleen, and the **hepatic** (heh-PAT-ik) **artery** goes to the liver (**see Fig. 15-4**).
2. The large **superior mesenteric** (mes-en-TER-ik) **artery** carries blood to most of the small intestine and to the first half of the large intestine.
3. The much smaller **inferior mesenteric artery,** located below the superior mesenteric artery and near the end of the abdominal aorta, supplies the second half of the large intestine.

The abdominal aorta's paired lateral branches include the following right and left vessels:

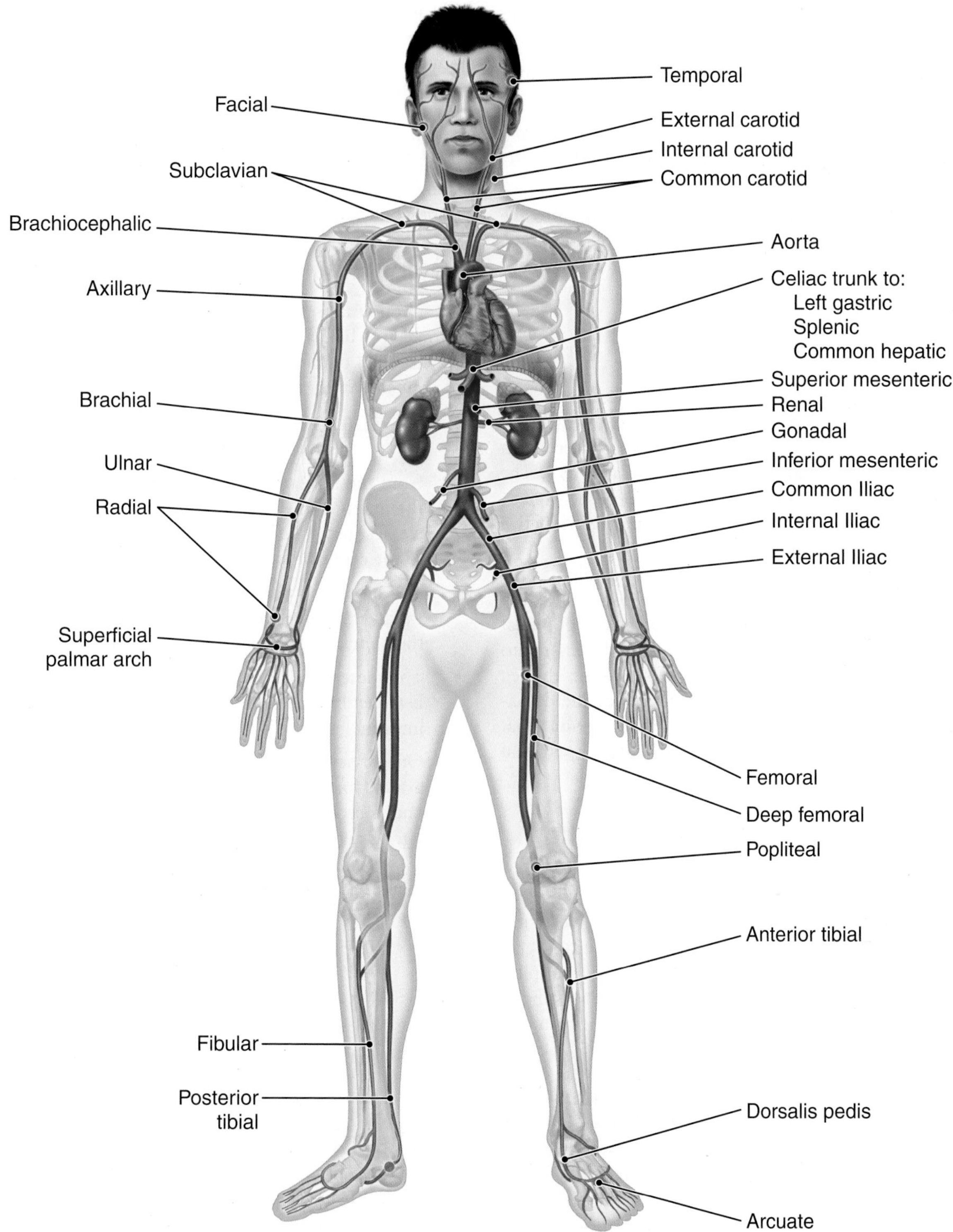

Figure 15-5 **Principal systemic arteries.** **ZOOMING IN** What large vessels branch from the terminal aorta?

1. The **superior** and **inferior phrenic** (FREN-ik) **arteries** supply the diaphragm. (These are not shown in the figures.)
2. The **renal** (RE-nal) **arteries**, the largest in this group, carry blood to the kidneys.
3. The **gonadal** (go-NAD-al) **arteries**—the **ovarian** (o-VAR-e-an) **arteries** in females and **testicular** (tes-TIK-u-lar) **arteries** in males—supply the sex glands.
4. Four pairs of **lumbar** (LUM-bar) **arteries** extend into the musculature of the abdominal wall. (These are not shown in the figures.)

THE ILIAC ARTERIES AND THEIR SUBDIVISIONS

The abdominal aorta finally divides into two **common iliac** (IL-e-ak) **arteries** (see Fig. 15-5). These vessels, which are about 5 cm (2 in) long, extend into the pelvis, where each subdivides into an **internal** and an **external iliac artery**. The internal iliac vessels then send branches to the pelvic organs, including the urinary bladder, the rectum, and reproductive organs other than the gonads.

Each external iliac artery continues into the thigh as the **femoral** (FEM-or-al) **artery**. This vessel gives rise to the **deep femoral artery** in the thigh and then becomes the **popliteal** (pop-LIT-e-al) **artery**, which subdivides below the knee into the anterior and posterior **tibial arteries**, supplying the leg and foot. The anterior tibial artery terminates as the **dorsalis pedis** (dor-SA-lis PE-dis) at the foot. The posterior tibial artery gives rise to the **fibular** (FIB-u-lar) **artery** (peroneal artery) in the leg.

ARTERIES THAT BRANCH TO THE ARM AND HEAD

The **subclavian** (sub-KLA-ve-an) **artery** supplies blood to the arm and hand. Its first branch, however, is the **vertebral** (VER-te-bral) **artery**, which passes through the transverse processes of the first six cervical vertebrae and supplies blood to the posterior brain (see Fig. 15-6). The subclavian artery changes names as it travels through the arm and branches to the forearm and hand (see Fig. 15-5). It first becomes the **axillary** (AK-sil-ar-e) **artery** in the axilla (armpit). The longest part of this vessel, the **brachial** (BRA-ke-al) **artery**, is in the arm proper. The brachial artery subdivides into two branches near the elbow: the **radial artery**, which continues down the thumb side of the forearm and wrist, and the **ulnar artery**, which extends along the medial or little finger side into the hand. These two arteries unite in the **palmar arches**, which give off smaller **digital arteries** that supply the hand and fingers.

The right and left common carotid arteries travel along either side of the trachea enclosed in a sheath with the internal jugular vein and the vagus nerve. Just anterior to the angle of the mandible (lower jaw), each branches into the **external** and **internal carotid arteries** (Fig. 15-6). You can feel the pulse of the carotid artery just anterior to the large sternocleidomastoid muscle in the neck and below the jaw. The internal carotid artery travels into the head and branches to supply the eye, the anterior portion of the brain, and other structures in the cranium. The external carotid artery branches to the thyroid gland and to other structures in the head and upper part of the neck.

Just as the larger branches of a tree divide into limbs of varying sizes, so the arterial tree has a multitude of subdivisions. Hundreds of names might be included. We have mentioned only some of them.

ANASTOMOSES

A communication between two vessels is called an **anastomosis** (ah-nas-to-MO-sis). By means of arterial anastomoses, blood reaches vital organs by more than one route. Some examples of such end-artery unions are as follows:

- The **cerebral arterial circle** (circle of Willis) (Fig. 15-6) receives blood from the two internal carotid arteries and from the **basilar** (BAS-il-ar) **artery**, which is formed by the union of the two vertebral arteries. This arterial circle lies just under the brain's center and sends branches to the cerebrum and other parts of the brain.
- The **superficial palmar arch** is formed by the union of the radial and ulnar arteries in the hand. It sends branches to the hand and the fingers (see Fig. 15-5).
- The **mesenteric** arches are made up of communications between branches of the vessels that supply blood to the intestinal tract.
- One of several arterial arches in the foot is the **arcuate artery**, formed by the union of the dorsalis pedis artery and a second branch of the anterior tibial artery (the lateral tarsal artery).

CHECKPOINTS

- ☐ 15-5 What are the subdivisions of the aorta, the largest artery?
- ☐ 15-6 What are the three branches of the aortic arch?
- ☐ 15-7 What areas are supplied by the brachiocephalic artery?
- ☐ 15-8 What is an anastomosis?

Systemic Veins

Whereas most arteries are located in protected and rather deep areas of the body, many of the principal systemic veins are found near the surface (Fig. 15-7). The most important of the **superficial veins** are in the extremities and include the following:

- The veins on the back of the hand and at the front of the elbow. Those at the elbow are often used for drawing blood for test purposes, as well as for intravenous injections. The largest of this venous group are the **cephalic** (seh-FAL-ik), the **basilic** (bah-SIL-ik), and the **median cubital** (KU-bih-tal) **veins**.

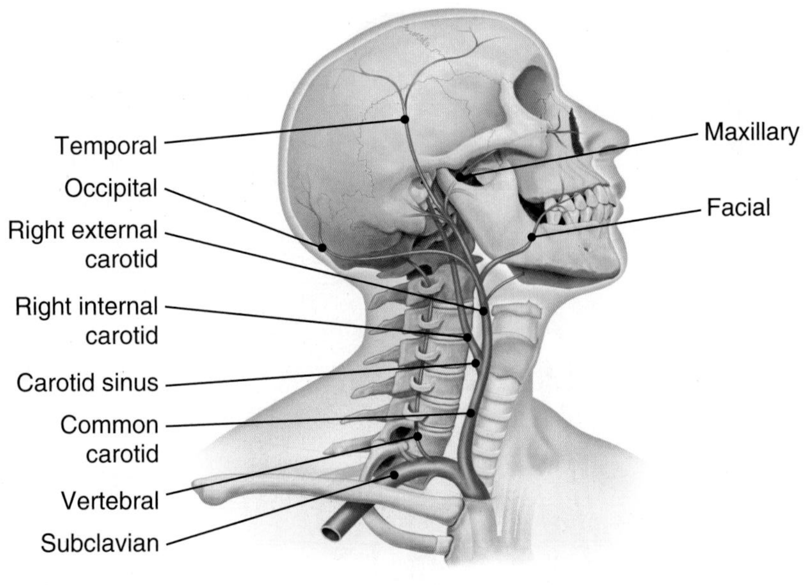

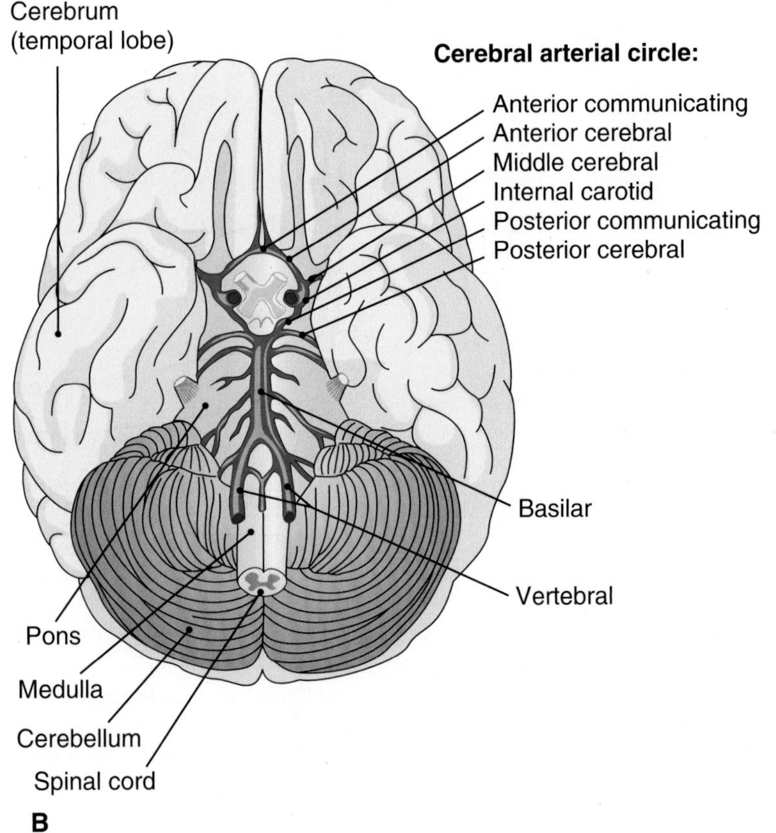

Figure 15-6 Arteries of the neck and head. A. Arteries of the head and the neck, lateral view. **B.** Arteries supplying the brain. The arteries of the cerebral arterial circle (CAC) are shown. **KEY POINT** Anastomosis of arteries help preserve blood supply to the brain.

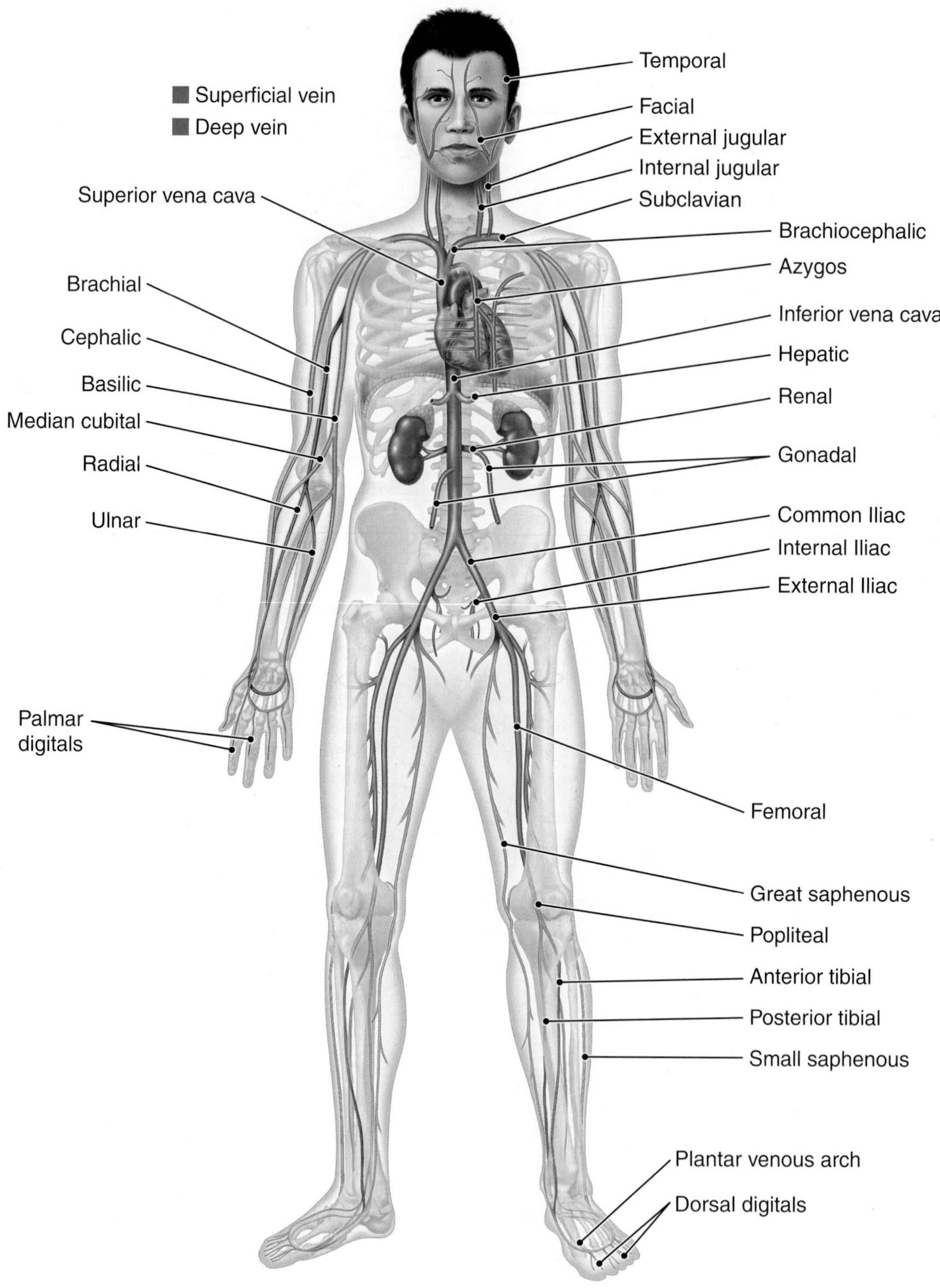

Figure 15-7 **Principal systemic veins.** Anterior view. **KEY POINT** Deep veins (*in blue*) usually parallel arteries and carry the same names. Superficial veins are shown *in purple*. **ZOOMING IN** How many brachiocephalic veins are there?

- The **saphenous** (sah-FE-nus) **veins** of the lower extremities, which are the body's longest veins. The great saphenous vein begins in the foot and extends up the medial side of the leg, the knee, and the thigh. It finally empties into the femoral vein near the groin.

The **deep veins** tend to parallel arteries and usually have the same names as the corresponding arteries (**see Fig. 15-7**). Examples of these include the **femoral** and the **external** and **internal iliac vessels** of the lower body, and the **radial, ulnar, brachial, axillary,** and **subclavian vessels** of the upper extremities. (A deep vein was involved in Jocelyn's thrombosis in the opening case study.) The veins of the head and the neck, however, have different names than the arteries. The two jugular (JUG-u-lar) veins on each side of the neck drain the areas supplied by the carotid arteries (*jugular* is from a Latin word meaning "neck"). The larger of the two veins, the **internal jugular**, receives blood from the large veins (cranial venous sinuses) that drain the head and also from regions of the face and neck (**see Fig. 15-8**). The smaller **external jugular** drains the areas supplied by the external carotid artery. Both veins empty directly into a **subclavian vein.** On each side, the subclavian, external jugular, and internal jugular veins join to form a **brachiocephalic vein.** (Remember, there is only *one* brachiocephalic artery.)

THE VENAE CAVAE AND THEIR TRIBUTARIES

Two large veins receive blood from the systemic vessels and empty directly into the heart's right atrium. The veins of the head, neck, upper extremities, and chest all drain into the **superior vena cava** (VE-nah KA-vah). This vessel is formed by the union of the right and left brachiocephalic veins. The unpaired **azygos** (AZ-ih-gos) **vein** drains the veins of the chest wall and empties into the superior vena cava just before the latter empties into the right atrium of the heart (**see Fig. 15-7**) (*azygos* is from a Greek word meaning "unpaired").

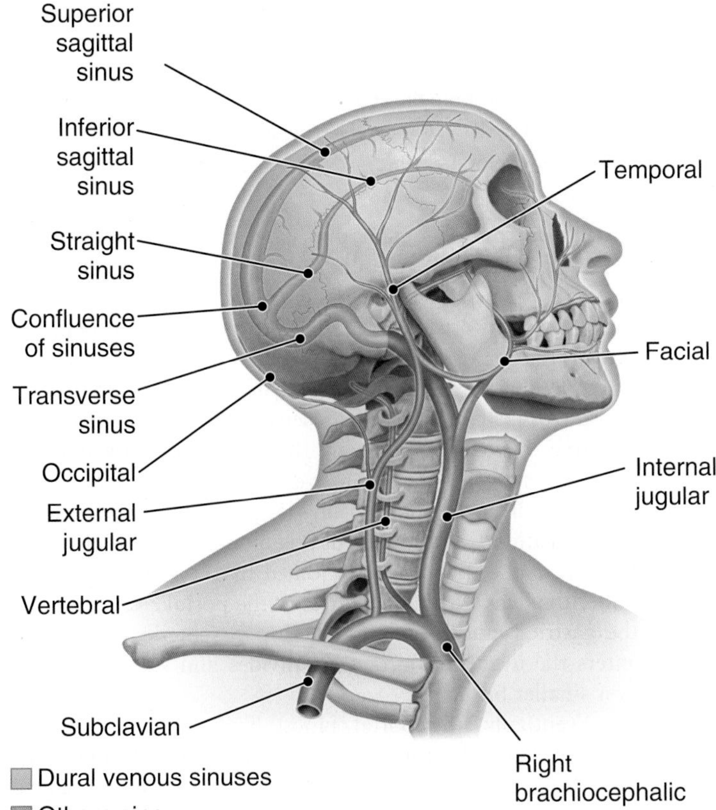

Figure 15-8 Veins of the head and the neck and the cranial sinuses, lateral view. **KEY POINT** The cranial sinuses are spaces between the two layers of the dura mater. **ZOOMING IN** Which vein receives blood from the transverse sinus?

The **inferior vena cava**, which is much longer than is the superior vena cava, returns blood from areas below the diaphragm (see Dissection Atlas **Fig. A5-4**). It begins in the lower abdomen with the union of the two common iliac veins. It then ascends along the abdomen's posterior wall, through a groove in the posterior part of the liver, through the diaphragm, and finally through the lower thorax to empty into the heart's right atrium.

The large veins below the diaphragm may be divided into two groups:

1. The right and left veins that drain paired parts and organs. They include the **external** and **internal iliac veins** from near the groin that join to form the **common iliac veins**; four pairs of **lumbar veins** from the dorsal trunk and from the spinal cord; the **gonadal veins**—the **testicular veins** from the male testes and the **ovarian veins** from the female ovaries; the **renal veins** from the kidneys; and finally the large **hepatic veins** from the liver. For the most part, these vessels empty directly into the inferior vena cava. The left testicular in the male and the left ovarian in the female empty into the left renal vein, which carries this blood to the inferior vena cava; these veins thus constitute exceptions to the rule that the paired veins empty directly into the vena cava.
2. Unpaired veins that drain the spleen and parts of the digestive tract (stomach and intestine) empty into the **hepatic portal vein**, discussed shortly and shown in **Figure 15-9**. Unlike other lower veins, which empty directly into the inferior vena cava, the hepatic portal vein is part of a special system that enables blood to circulate through the liver before returning to the heart. This system, the hepatic portal system, will be described in more detail later.

VENOUS SINUSES

The word *sinus* means "space" or "hollow." A **venous sinus** is a large channel that drains blood low in oxygen but does not have a vein's usual tubular structure. One example of a venous sinus is the **coronary sinus**, which receives most of the blood from the heart wall (**see Fig. 14-6** in Chapter 14). It lies between the left atrium and the left ventricle on the heart's posterior surface and empties directly into the right atrium, along with the two venae cavae.

Other important venous sinuses are the **cranial venous sinuses**, which are spaces between the two layers of the dura mater. Veins from throughout the brain drain into these channels (**Fig. 15-8**). They also collect cerebrospinal fluid from the CNS and return it to the bloodstream. The largest of the cranial venous sinuses are the following:

- The **superior sagittal** (SAJ-ih-tal) **sinus** is a single long space located in the midline above the brain and in the fissure between the two cerebral hemispheres. It ends in an enlargement called the **confluence** (KON-flu-ens) **of sinuses**.

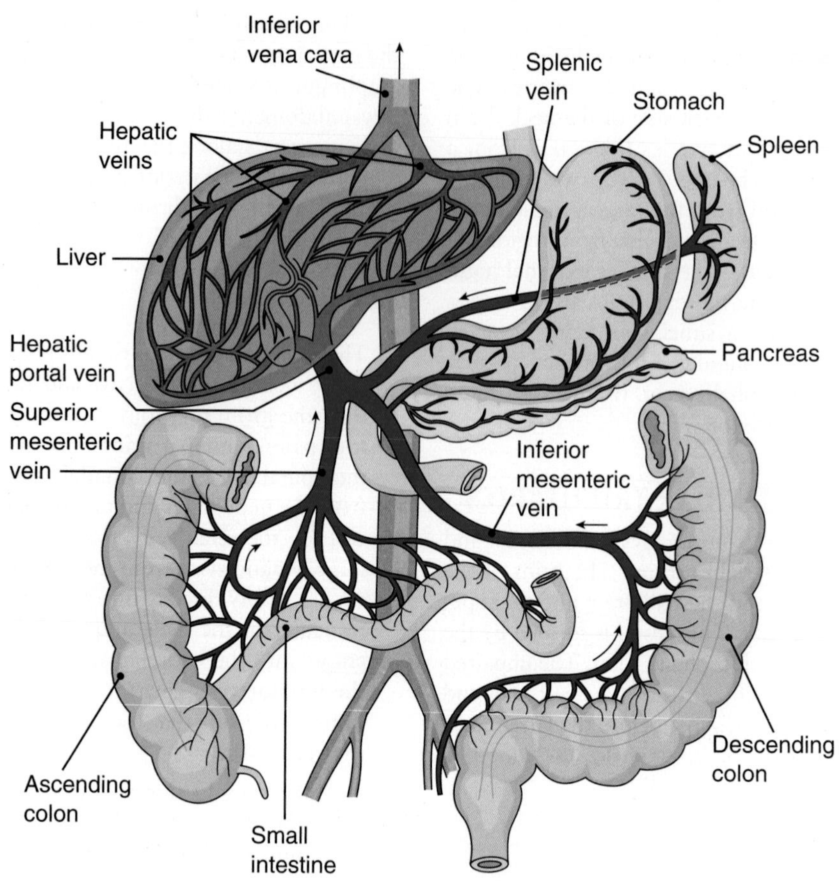

Figure 15-9 **Hepatic portal system.** **KEY POINT** Veins from the abdominal organs carry blood to the hepatic portal vein leading to the liver. *Arrows* show the direction of blood flow. **ZOOMING IN** What vessel do the hepatic veins drain into?

- The **inferior sagittal sinus** parallels the superior sagittal sinus and merges with the straight sinus.
- The **straight sinus** receives blood from the inferior sagittal sinus and a large cerebral vein and flows into the confluence of sinuses.
- The two **transverse sinuses**, also called the **lateral sinuses**, begin posteriorly from the confluence of sinuses and then extend laterally. As each sinus extends around the skull's interior, it picks up additional blood. Nearly all of the blood leaving the brain eventually empties into one of the transverse sinuses. Each of these extends anteriorly to empty into an internal jugular vein, which then passes through a channel in the skull to continue downward in the neck.

THE HEPATIC PORTAL SYSTEM

Almost always, when blood leaves a capillary bed, it flows through venules and veins directly back to the heart. In a portal system, however, blood circulates through a second capillary bed in a second organ before it returns to the heart. Portal circulations enable all of the products of one organ to pass directly to another organ. Chapter 12 described the small local portal system that carries secretions from the hypothalamus to the pituitary gland. A much larger portal system is the **hepatic portal system**, which carries blood from the abdominal organs to the liver to be processed before it returns to the heart (**Fig. 15-9**).

The hepatic portal system includes the veins that drain blood from capillaries in the spleen, stomach, pancreas, and intestine. Instead of emptying their blood directly into the inferior vena cava, they deliver it through the hepatic portal vein to the liver. The portal vein's largest tributary is the **superior mesenteric vein**, which drains blood from the proximal portion of the intestine. It is joined by the **splenic vein** just under the liver. Other tributaries of the portal circulation are the **gastric, pancreatic**, and **inferior mesenteric veins.** As it enters the liver, the portal vein divides and subdivides into ever smaller branches.

Eventually, the portal blood flows into a vast network of sinus-like capillaries called **sinusoids** (SI-nus-oyds) (see Box 15-1). These leaky vessels allow free exchange of proteins, nutrients, and dissolved substances between liver cells (hepatocytes) and blood. (Similar blood channels are found in the spleen and endocrine glands, including the thyroid and adrenals.) After leaving the sinusoids, blood is finally collected by the hepatic veins, which empty into the inferior vena cava.

The hepatic portal system ensures that most substances absorbed from the intestines can be processed by the liver before they encounter body cells. For example, the liver can inactivate some of the ingested toxins (such as alcohol and drugs) before they reach the general circulation. The liver processes nutrients and can store or release them according to body needs. The liver also receives arterial blood and breaks down alcohol, certain drugs, and various other toxins that have already reached the general circulation.

CHECKPOINTS 

- 15-9 What is the difference between superficial and deep veins?
- 15-10 What two large veins drain the systemic blood vessels and empty into the right atrium?
- 15-11 What is a venous sinus?
- 15-12 The hepatic portal system takes blood from the abdominal organs to which organ?

Circulation Physiology

Circulating blood might be compared to a bus that travels around the city, picking up and delivering passengers at each stop on its route. Take gases, for example. As blood flows through capillaries surrounding the air sacs in the lungs, it picks up oxygen and unloads carbon dioxide. Later, when this oxygen-rich blood is pumped to systemic capillaries, it unloads the oxygen and picks up carbon dioxide and other substances generated by the cells (**Fig. 15-10**). The microscopic capillaries are of fundamental importance in these activities. It is only through the cells of these thin-walled vessels that the necessary exchanges can occur.

All living cells are immersed in a slightly salty liquid, the **interstitial** (in-ter-STISH-al) **fluid**, or *tissue fluid*. Looking again at **Figure 15-10**, one can see how this fluid serves as a "middleman" between the capillary membrane and the neighboring cells. As water, oxygen, electrolytes, and other necessary cellular materials pass through the capillary walls, they enter the tissue fluid. Then, these substances make their way by diffusion, osmosis, or active transport to the cells. At the same

15

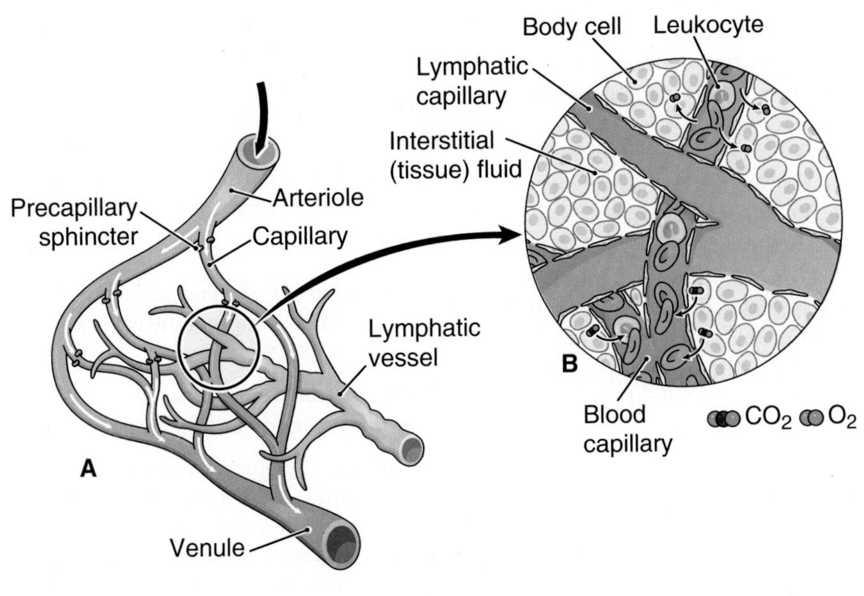

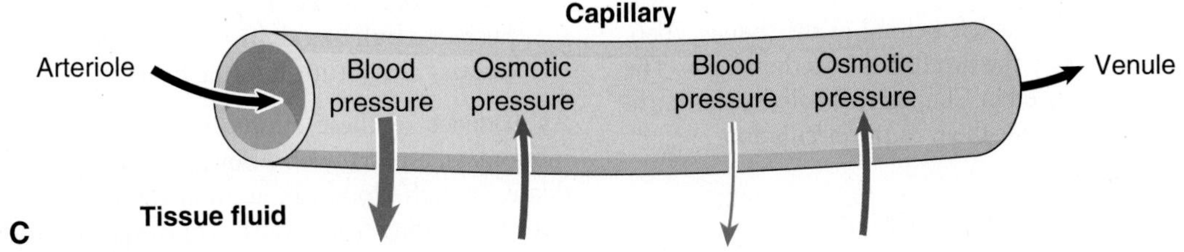

Figure 15-10 The role of capillaries. **KEY POINT** Capillaries are the point of exchanges between the bloodstream and the tissues. **A.** A capillary network. Note the lymphatic capillaries, which aid in tissue drainage. **B.** Materials, such as the gases oxygen and carbon dioxide, diffuse between the blood and the interstitial fluid. **C.** At the start of a capillary bed, blood pressure helps push materials out of the blood. At the end of the capillary bed, osmotic pressure is the greater force and draws materials into the blood. The lymphatic system picks up excess water and proteins for return to the circulation.

time, carbon dioxide and other metabolic end products leave the cells and move in the opposite direction. These substances enter the capillaries and are carried away in the bloodstream for processing in other organs or elimination from the body.

CAPILLARY EXCHANGE

Many substances move between the cells and the capillary blood by diffusion. Recall that diffusion is the movement of a substance from an area where it is in higher concentration to an area where it is in lower concentration. Diffusion does not require cellular energy, but water-soluble substances need channels or transporters to cross cell membranes. A select few substances move by active transport. For example, the sodium–potassium pump moves sodium out of cells and potassium into cells.

Recall from Chapter 3 that filtration is the movement of a fluid down a pressure gradient through a membrane. This process pushes water and dissolved materials from capillary blood into tissue fluid through the gaps between capillary cells (**see Fig. 15-10C**). Blood pressure creates the pressure gradient that drives filtration. Fluid is drawn back into the capillary by osmotic pressure, the "pulling force" of substances dissolved or suspended in the blood. Osmotic pressure is maintained by plasma proteins (mainly albumin), which are too large in molecular size to pass between the capillary cells. Filtration and osmosis result in the constant exchange of fluids across the capillary wall.

The balance between filtration pressure and osmotic pressure determines the net fluid movement. As blood enters the capillary bed, the force of its fluid pressure is greater than its opposing osmotic pressure. The tendency is for water and dissolved materials to move out of the capillaries and into the interstitial fluid. Water loss lowers blood pressure as blood flows through the capillaries. Thus, as blood leaves the capillary bed, the "pulling in" force of the blood osmotic pressure exceeds the "pushing out" force of the blood fluid pressure, and materials will tend to enter the capillaries.

The movement of blood through the capillaries is relatively slow, owing to the much larger cross-sectional area of the capillaries compared with that of the vessels from which they branch. This slow progress through the capillaries allows time for exchanges to occur.

Note that even when the capillary exchange process is most efficient, some water is left behind in the tissues. Also, some proteins escape from the capillaries into the tissues. The lymphatic system, discussed in Chapter 16, collects this extra fluid and protein and returns them to the circulation.

THE DYNAMICS OF BLOOD FLOW

Blood flow is carefully regulated to supply tissue needs without unnecessary burden on the heart. Some organs, such as the brain, liver, and kidneys, require large quantities of blood even at rest. The requirements of some tissues, such as the skeletal muscles and digestive organs, increase greatly during periods of activity. For example, the blood flow in muscle can increase up to 25 times during exercise. The volume of blood flowing to a particular organ can be regulated by changing the size of the blood vessels supplying that organ.

Vasomotor Changes An increase in a blood vessel's internal diameter is called **vasodilation**. This change allows for the delivery of more blood to an area. **Vasoconstriction** is a decrease in a blood vessel's internal diameter, causing a decrease in blood flow. These *vasomotor activities* result from the contraction or relaxation of smooth muscle in the walls of the blood vessels, mainly the arterioles. A **vasomotor** center in the medulla of the brain stem regulates changes in vessel diameter, sending its messages through the autonomic nervous system.

A **precapillary sphincter** of smooth muscle encircles the entrance to each capillary, controlling its blood supply (**see Fig. 15-10A**). This sphincter widens to allow more blood to enter when the nearby cells need more oxygen.

Blood's Return to the Heart Blood leaving the capillary networks returns in the venous system to the heart, and even picks up some speed along the way, despite factors that work against its return. Blood flows in a closed system and must continually move forward, whether the heart is contracting or relaxing. However, by the time blood arrives in the veins, little force remains from the heart's pumping action. Also, because the veins expand easily under pressure, blood tends to pool in the veins. Considerable amounts of blood are normally stored in these vessels. Finally, the force of gravity works against upward flow from regions below the heart. Several mechanisms help overcome these forces and promote blood's return to the heart in the venous system. These are as follows:

- **Contraction of skeletal muscles.** As skeletal muscles contract, they compress the veins and squeeze blood forward (**Fig. 15-11**).
- **Valves in the veins.** They prevent backflow and keep blood flowing toward the heart.
- **Breathing.** Pressure changes in the abdominal and thoracic cavities during breathing also promote blood return in the venous system. During inhalation, the diaphragm flattens and puts pressure on the large abdominal veins. At the same time, chest expansion causes pressure to drop in the thorax. Together, these actions serve to both push and pull blood through the abdominal and thoracic cavities and return it to the heart.

As evidence of these effects, if a person stands completely motionless, especially on a hot day when the superficial vessels dilate, enough blood can accumulate in the lower extremities to cause fainting from insufficient oxygen to the brain.

CHECKPOINTS 

- [ ] 15-13 What force helps push materials out of a capillary? What force helps draw materials into a capillary?
- [ ] 15-14 Name the two types of vasomotor changes.
- [ ] 15-15 Where are vasomotor activities regulated?

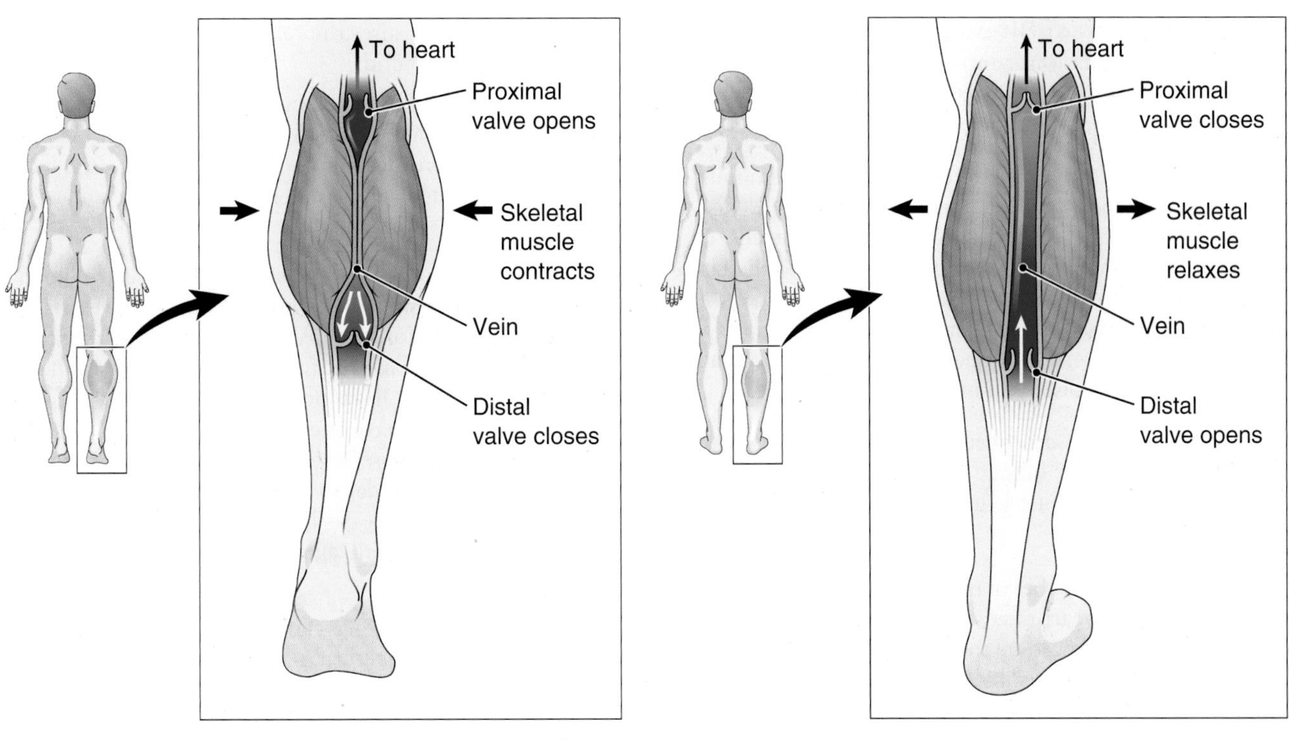

Figure 15-11 Blood return. **KEY POINT** Muscle contraction and valves keep blood flowing back toward the heart. **A.** Contracting skeletal muscle compresses the vein and drives blood forward, opening the proximal valve, while the distal valve closes to prevent backflow of blood. **B.** When the muscle relaxes again, the distal valve opens, and the proximal valve closes until blood moving in the vein forces it to open again. **ZOOMING IN** Which of the two valves shown is closer to the heart?

15

THE PULSE

The ventricles regularly pump blood into the arteries about 70 to 80 times a minute. The force of ventricular contraction starts a wave of increased pressure that begins at the heart and travels along the arteries. This wave, called the **pulse**, can be felt in any artery that is relatively close to the surface, particularly if the vessel can be pressed down against a bone. At the wrist, the radial artery passes over the bone on the forearm's thumb side, and the pulse is most commonly obtained here. Other vessels sometimes used for taking the pulse are the carotid artery in the neck and the dorsalis pedis on the top of the foot.

Normally, the pulse rate is the same as the heart rate, but if a heartbeat is abnormally weak, or if the artery is obstructed, the beat may not be detected as a pulse. In checking another person's pulse, it is important to use your second or third finger. If you use your thumb, you may be feeling your own pulse. When taking a pulse, it is important to gauge its strength as well as its regularity and rate.

The heart rate (and thus the pulse rate) can vary significantly without compromising homeostasis. Various factors may influence the pulse rate. We describe just a few here:

- The pulse is somewhat faster in small people than in large people, and usually slightly faster in women than in men.
- In a newborn infant, the rate may be from 120 to 140 bpm. As the child grows, the rate tends to become slower.
- Muscular activity influences the pulse rate. During sleep, the pulse may slow down to 60 bpm, whereas during strenuous exercise, the rate may go up to well over 100 bpm. For a person in good condition, the pulse does not go up as rapidly as it does in an inactive person, and it returns to a resting rate more quickly after exercise.
- Emotional disturbances may increase the pulse rate.
- Pulse rate increases with increased temperature, as in cases of infection.
- Excessive secretion of thyroid hormone may cause a rapid pulse.

BLOOD PRESSURE

Blood pressure is the force exerted by the blood against the walls of the vessels and is the force propelling blood to the tissues and back to the heart. This must be adjusted constantly to guarantee adequate blood flow to the tissues while preventing stress on the cardiovascular system.

Factors that Affect Blood Pressure As you will see from the next discussion, control of blood pressure is very complex and involves many systems. Some of the factors that affect blood pressure include the following:

Total Blood Volume This refers to the total amount of blood that is in the vascular system at a given time. Just as pressure will increase within a water balloon as it fills, converting a floppy sac into a ball, adding volume to the circulatory system increases blood pressure. The reverse is also true; loss of volume, as by hemorrhage, for example, will lower blood pressure. The kidneys are important regulators of blood volume, as discussed in Chapter 22.

Cardiac Output As described in Chapter 14, the output of the heart, or cardiac output (CO), is the volume of blood pumped out of each ventricle in a minute. CO is the product of two factors:

- **Heart rate**, the number of times the heart beats each minute. The basic heart rate is set internally by the SA node but can be influenced by the autonomic nervous system, hormones, and other substances circulating in the blood, such as ions.
- **Stroke volume**, the volume of blood ejected from the ventricle with each beat. The sympathetic nervous system can stimulate more forceful heart contractions to increase blood ejection. Also, if more blood returns to the heart in the venous system, the increased blood volume stretches the heart muscle and promotes more forceful contractions. This response ensures that the heart will pump out as much blood as it receives and prevents pooling of blood in the ventricles.

Resistance to Blood Flow Resistance is opposition to blood flow. Because the effects of resistance are seen mostly in small arteries and arterioles that are at a distance from the heart, this factor is often described as *peripheral resistance*. Resistance in the vessels is affected by the following factors:

- Blood vessel diameter: One form of resistance is the friction generated as blood slides along the vessel walls. A narrow vessel offers more resistance to blood flow than does a wider vessel, just as it is harder to draw fluid through a narrow straw than through a wide straw. Vasoconstriction increases resistance to flow, and vasodilation lowers resistance.
- Blood viscosity, or thickness: Just as a milkshake is harder to suck through a straw than milk is, increased blood viscosity will increase blood pressure. Under normal circumstances, blood viscosity remains within a constant range. However, loss of red cells, as in anemia, or loss of plasma proteins will decrease viscosity. Conversely, increased numbers of red blood cells, as in polycythemia, or a loss of plasma volume, as by dehydration, will increase viscosity. The hematocrit test described in Chapter 13 is one measure of blood viscosity; it measures the relative percentage of packed cells in whole blood.
- Blood vessel length: A longer vessel offers more resistance to blood flow than does a shorter vessel. As blood vessel length is ordinarily constant, this is not a significant physiologic factor in peripheral resistance.

Blood Vessel Compliance and Elasticity

- The ease with which arteries expand to receive blood is termed their **compliance** (kom-PLI-ans). If vessels lose their capacity for expansion, as occurs with atherosclerosis, for example, they offer more resistance to blood flow. You have probably experienced this phenomenon if you have tried to blow up a firm, new balloon. More pressure is generated as you blow, and the balloon is a lot harder to inflate than is a softer balloon, which expands more easily under pressure. Blood vessels lose compliance with aging, thus increasing resistance and blood pressure.
- The term **elasticity** (e-las-TIH-sih-te) describes the ability of blood vessels to return to their original size after being stretched. Think of an elasticized garment, a bathing suit for example. When new, its compliance (stretchability) is low, and it may be hard to get into, but it readily springs back to shape when removed. With time, its elasticity has decreased and its compliance has increased. It may be easier to put on, but it does not resume its original shape after wearing. Arteries stretch as they receive blood, near the heart for example, and then tend to recoil to their original size, putting pressure on the blood. This response helps prevent wide fluctuations in blood pressure as the heart contracts and relaxes. Because arteries have more elastic tissue than do veins, pressure is higher in the arterial system than in the venous system, and pressure drops continuously as blood travels away from the heart (**Fig. 15-12**).

Control of Blood Pressure A negative feedback loop enables the body to maintain relatively constant blood pressure despite changes in body position, tissue needs, and overall blood volume. The sensors for blood pressure consist of **baroreceptors** (bar-o-re-SEP-torz) in the walls of the carotid arteries and the aorta. Baroreceptors are stretch receptors; increased pressure stretches the vessel wall, increasing the signal rate from the baroreceptors, while decreased pressure decreases the signal rate. **Figure 15-13** shows how the negative feedback loop compensates for hemorrhage. The resulting decrease in blood pressure and blood vessel stretching reduces the signal rate from the baroreceptors to the cardiovascular control center in the medulla oblongata. The control center responds by increasing sympathetic activation, resulting in increased heart rate, stroke volume, and vasoconstriction. Thus, both cardiac output and peripheral resistance increase, and blood pressure rises. On the other hand, high blood pressure increases baroreceptor signaling. The control center responds by sending signals via the parasympathetic nervous system that result in a slower heart rate and dilation of peripheral vessels, and blood pressure decreases.

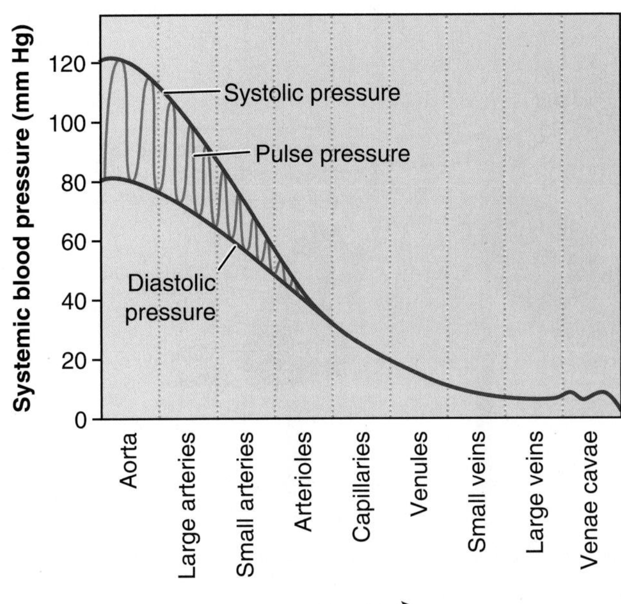

Figure 15-12 **Blood pressure.** **KEY POINT** Blood pressure declines as blood flows farther from the heart. Systolic pressure is the maximum pressure that develops in the arteries after heart muscle contraction; diastolic pressure is the lowest pressure in the arteries after relaxation of the heart. The difference between the two pressures (pulse pressure) declines as blood flows through the arteries. The small rises in the venae cavae represent the breathing pump that promotes venous return. **ZOOMING IN** In which vessels does the pulse pressure drop to zero?

Measurement of Blood Pressure The measurement and careful interpretation of blood pressure may prove a valuable guide in the care and evaluation of a person's health. Because blood pressure decreases as the blood flows from arteries into capillaries and finally into veins, healthcare providers ordinarily measure arterial pressure only, most commonly in the brachial artery of the arm. In taking blood pressure, two variables are measured:

- **Systolic pressure**, the maximum pressure that develops in the arteries after heart muscle contraction.
- **Diastolic pressure**, the lowest pressure measured in the arteries after relaxation of the heart muscle.

The wave of increased pressure that develops when the heart contracts represents the pulse, as previously described. The difference between the systolic and the diastolic pressures is called the *pulse pressure*. This value declines continuously as blood gets farther from the heart, dropping to zero by the time blood reaches the capillaries (**see Fig. 15-12**).

The instrument used to measure blood pressure is a **sphygmomanometer** (sfig-mo-mah-NOM-eh-ter) (**Fig. 15-14**), more simply called a blood pressure cuff or blood pressure apparatus. The sphygmomanometer is an inflatable cuff attached to a pressure gauge. Pressure is expressed in millimeters mercury (mm Hg), that is, the height to which the pressure can push a column of mercury in a tube. The examiner wraps the cuff around the patient's upper arm and inflates it with air until the brachial artery is compressed and the blood flow is cut off. Then, listening with a stethoscope, he or she slowly lets air out of the cuff until the first pulsations are heard. At this point, the pressure in the cuff is equal to the systolic pressure, and this pressure is read. Then, more air is let out gradually until a characteristic muffled sound indicates that the vessel is open and the diastolic pressure is read. Original-style sphygmomanometers display readings on a graduated column of mercury, but newer types display them on a dial. The newest devices measure blood pressure electronically: the examiner simply applies the cuff, which self-inflates and provides a digital reading. A typical normal systolic pressure is less than 120 mm Hg; a typical normal diastolic pressure is less than 80 mm Hg. Blood pressure is reported as systolic pressure first, then diastolic pressure, separated by a slash, such as 120/80. This reading would be reported verbally as "120 over 80."

Considerable experience is required to ensure an accurate blood pressure reading. Often, it is necessary to repeat measurements. Note also that blood pressure varies throughout the day and under different conditions, so a single reading does not give a complete picture. Some people typically have a higher reading in a doctor's office because of stress. People who experience such "white coat hypertension" may need to take their blood pressure at home while relaxed to get a more accurate reading. **Box 15-2** explains how cardiac catheterization is used to measure blood pressure with high accuracy.

15

Abnormal Blood Pressure Lower-than-normal blood pressure is called **hypotension** (hi-po-TEN-shun). Because of individual variations in normal pressure levels, however, what would be a low pressure for one person might be normal for someone else. For this reason, hypotension is best evaluated in terms of how well the body tissues are being supplied with blood. A person whose systolic blood pressure drops to below his or her normal range may experience fainting episodes because of inadequate blood flow to the brain. The sudden lowering of blood pressure to below a person's normal level is one sign of shock; it may also occur in certain chronic diseases and in heart block.

Hypertension (hi-per-TEN-shun), or high blood pressure, has received a great deal of attention in medicine. Hypertension normally occurs temporarily as a result of excitement or exertion. However, it may persist in a number of conditions, including the following:

- Kidney disease and uremia (excess nitrogenous waste in the blood) or other toxic conditions
- Endocrine disorders, such as hyperthyroidism and acromegaly
- Arterial disease, including hardening of the arteries (atherosclerosis), which reduces vascular compliance
- Tumors of the adrenal medulla (central portion), with the release of excess epinephrine

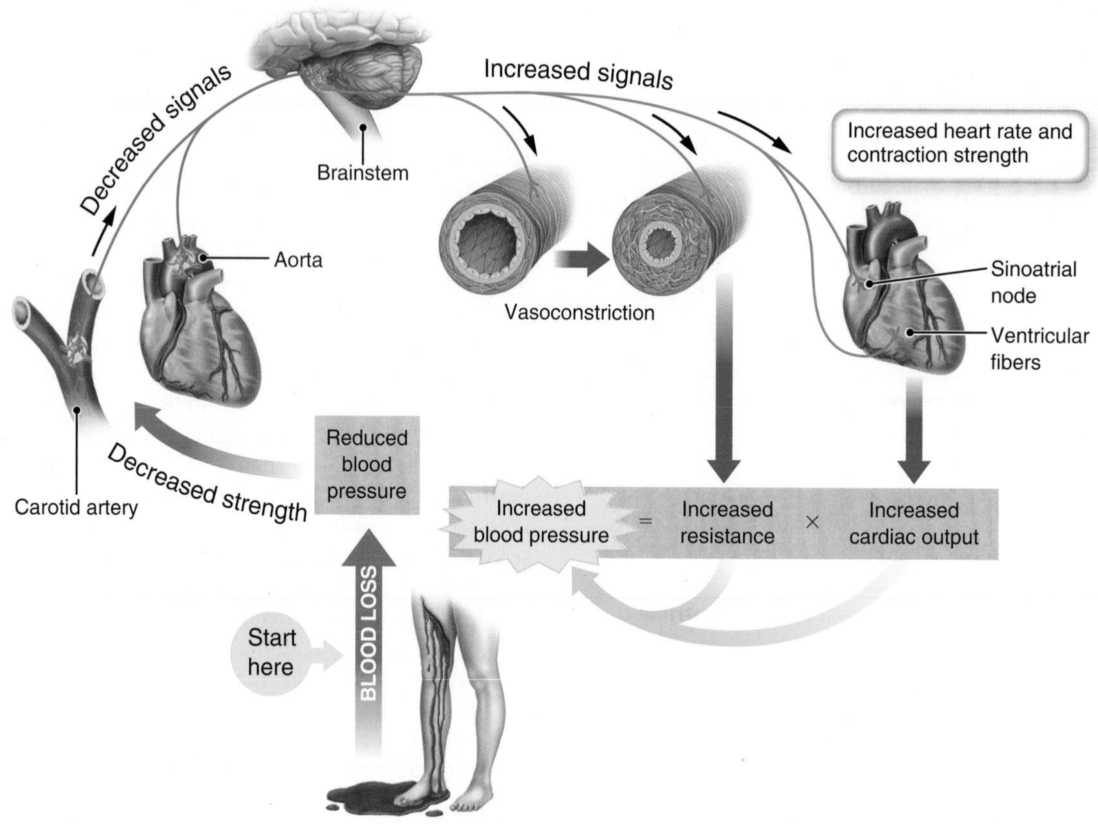

Figure 15-13 **The baroreceptor response.** **KEY POINT** The baroreceptors act as sensors in a negative feedback loop that keeps blood pressure relatively constant. **ZOOMING IN** What happens to the heart rate when blood pressure falls?

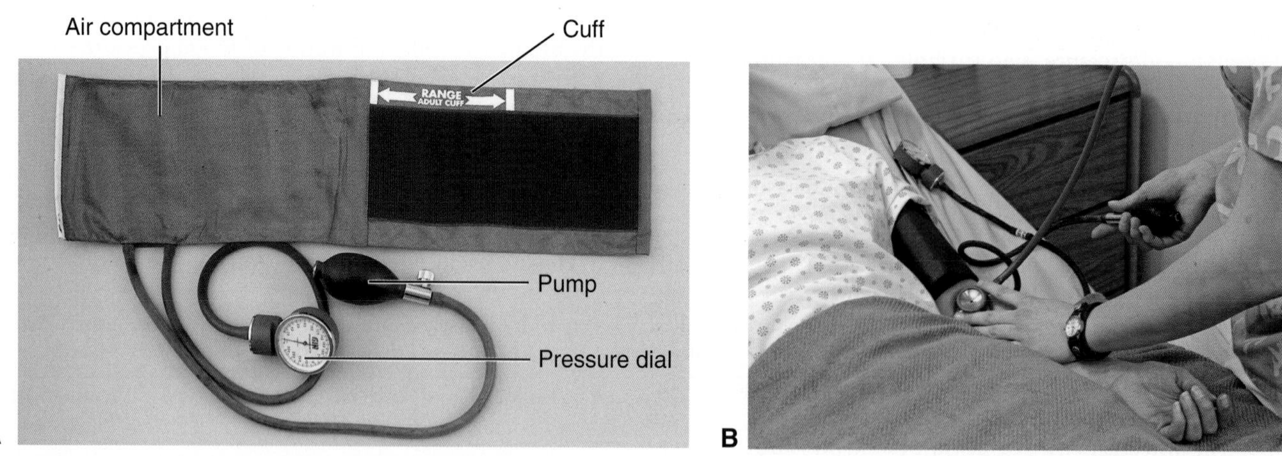

Figure 15-14 **Measurement of blood pressure. A.** A sphygmomanometer, or blood pressure cuff. **B.** Once the cuff is inflated, the examiner releases the pressure and listens for sounds in the vessels with a stethoscope.

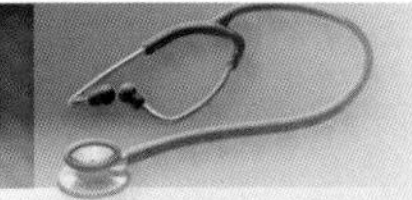

CLINICAL PERSPECTIVES

Box 15-2

Cardiac Catheterization: Measuring Blood Pressure from Within

Because arterial blood pressure decreases as blood flows farther away from the heart, measurement of blood pressure with a simple inflatable cuff around the arm is only a reflection of the pressure in the heart and pulmonary arteries. Precise measurement of pressure in these parts of the cardiovascular system is useful in diagnosing certain cardiac and pulmonary disorders.

More accurate readings can be obtained using a catheter (thin tube) inserted directly into the heart and large vessels. One type commonly used is the pulmonary artery catheter (also known as the Swan-Ganz catheter), which has an inflatable balloon at the tip. This device is threaded into the right side of the heart through a large vein. Typically, the right internal jugular vein is used because it is the shortest and most direct route to the heart, but the subclavian and femoral veins may be used instead. The catheter's position in the heart is confirmed by a chest x-ray, and, when it is appropriately positioned, the atrial and ventricular blood pressures are recorded. As the catheter continues into the pulmonary artery, one can read the pressure in this vessel. When the balloon is inflated, the catheter becomes wedged in a branch of the pulmonary artery, blocking blood flow. The reading obtained is called the **pulmonary capillary wedge pressure.** It gives information on pressure in the heart's left side and on resistance in the lungs. Combined with other tests, cardiac catheterization can be used to diagnose cardiac and pulmonary disorders such as shock, pericarditis, congenital heart disease, and heart failure.

Hypertension that has no apparent medical cause is called **essential hypertension**. Excess of an enzyme called **renin** (RE-nin), produced in the kidney, appears to play a role in the severity of essential hypertension. Renin raises blood pressure by activating pathways that promote vasoconstriction and stimulate water retention (see Chapter 22).

It is important to treat even mild hypertension because this condition can eventually

- Weaken vessels and lead to saclike bulges (aneurysms) in vessel walls that are likely to rupture. In the brain, vessel rupture is one cause of stroke. Rupture of a vessel in the eye may lead to blindness.
- Stress the heart by causing it to work harder to pump blood into the arterial system. In response to this greater effort, the heart enlarges, but eventually it weakens and becomes less efficient.
- Stress the kidneys and damage renal vessels.
- Damage the lining of vessels, predisposing one to atherosclerosis.

See the Student Resources on thePoint for an animation on hypertension.

Although medical caregivers often place emphasis on the systolic blood pressure, in many cases, the diastolic pressure is even more important. The total fluid volume in the vascular system and the condition of small arteries may have a greater effect on diastolic pressure. Table 15-1 lists degrees of hypertension as compared with normal blood pressure values. Physicians describe blood pressure in the range just above normal as prehypertension. People with prehypertension have a higher-than-normal risk of developing true hypertension and cardiovascular disease.

Treatment of Hypertension Even though there is much individual variation in blood pressure, physicians have established guidelines for the diagnosis and treatment of hypertension. Stage 1 hypertension describes readings between 140/90 mm Hg and 160/100 mm Hg. It is treated by lifestyle modifications such as increased physical

Table 15-1 Blood Pressure

Blood Pressure Classification (Adults)[a]

Category	Systolic (mm Hg)	Diastolic (mm Hg)
Normal	Under or equal to 120	Under or equal to 80
Prehypertension	120–139	80–89
Hypertension		
Stage 1	140–159	90–99
Stage 2	Over or equal to 160	Over or equal to 100

[a]When the systolic and diastolic pressures are in different categories, the higher category is used.

exercise, a low-salt, low-fat diet, and if necessary, weight loss and smoking cessation. Stage 2 hypertension, in which pressure readings are above 160/100 mm Hg, requires the addition of drug therapy to the lifestyle changes. Drugs used to treat hypertension include the following:

- Diuretic (di-u-RET-ik) drugs that promote water loss through the kidneys (see Chapter 22)
- Drugs that limit production of renin or block its action
- Drugs that relax blood vessels, including adrenergic blockers and calcium channel blockers

CHECKPOINTS

- 15-16 What is the definition of a pulse?
- 15-17 What is the definition of blood pressure?
- 15-18 What is the most significant factor in determination of peripheral resistance?
- 15-19 What are the sensors for blood pressure called?
- 15-20 What two components of blood pressure are measured?
- 15-21 What is the medical term for high blood pressure?

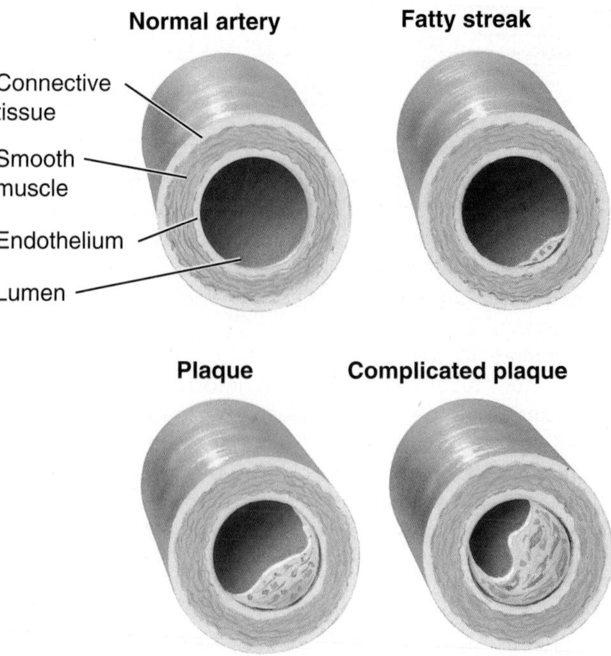

Figure 15-15 **Stages in atherosclerosis.** **KEY POINT** Accumulation of fatty material in the wall of a vessel obstructs blood flow.

Vascular Disorders

A variety of underlying disease processes, including hypertension and diabetes mellitus, can contribute to vascular disorders. In addition, injury, infection, and even chronic pressure from pregnancy or long periods of standing can damage blood vessels and threaten the health of the vascular system.

ARTERIAL DEGENERATION

As a result of age or other degenerative changes, materials may be deposited within the arterial walls. These deposits cause an irregular thickening of the wall at the expense of the lumen (space inside the vessel) as well as a loss of compliance. In some cases, calcium salts and scar tissue may cause this hardening of the arteries, technically called **arteriosclerosis** (ar-te-re-o-skle-RO-sis). The most common form of this disorder is atherosclerosis, described in regard to the coronary arteries in Chapter 14. Atherosclerosis begins with microscopic damage to the arterial endothelium caused by direct contact with LDL (the "bad" cholesterol), oxidizing chemicals such as free radicals, and some proteins. Streaky areas of yellow, fatlike material (plaque) appear in the vessels (**Fig. 15-15**). As this plaque accumulates, it separates the wall's muscle and elastic tissue layers. The arterial wall bulges into the vessel's lumen and interferes with blood flow. With time, the plaque can become complicated by calcification. Platelets can aggregate at the site of plaque formation, leading to development of a blood clot (thrombus) and partial or complete obstruction of the vessel. The serious consequences of coronary thrombosis were described in Chapter 14. Arteries in the heart, brain, kidneys, and the extremities seem to be especially vulnerable to this process.

Atherosclerosis and its complications (heart disease, stroke, and thrombosis) account for 40% of all deaths in the United States. A diet high in fats, particularly saturated fats, is known to contribute to atherosclerosis. Cigarette smoking also increases the extent and severity of this disorder.

Signs and Symptoms of Arterial Degeneration

Arterial damage may be present for years without causing any noticeable symptoms. As the thickening of the wall continues and the lumen's diameter decreases, limiting blood flow, a variety of signs and symptoms can appear. The nature of these disturbances varies with the area affected and with the extent of the arterial changes. Some examples are as follows:

- Leg cramps, pain, and sudden lameness while walking may be caused by insufficient blood supply to the lower extremities.
- Headaches, dizziness, and mental disorders can result from cerebral artery sclerosis.
- Hypertension may result from a decrease in lumen size within many arteries throughout the body. Although hypertension may be present in young people with no apparent arterial damage, and atherosclerosis may be present without causing hypertension, the two are often found together in elderly people.
- Palpitations, dyspnea (difficulty in breathing), paleness, weakness, and other symptoms may be the result of coronary artery arteriosclerosis. The severe pain of angina pectoris may follow the lack of oxygen and the myocardial damage associated with sclerosis of the vessels that supply the heart.
- The presence of albumin in urine, known as albuminuria, indicates damage to renal vessels. Albuminuria is a significant risk factor for cardiovascular disorders.

- Ulceration and tissue necrosis (death), especially in the extremities, results when damaged arteries can no longer supply the oxygen needs of peripheral tissues. If the dead tissue is invaded by bacteria, the result is **gangrene** (GANG-grene). The arterial damage that is caused by diabetes, for example, often leads to gangrene in the extremities of elderly diabetic patients.

See the Student Resources on thePoint for illustrations of atherosclerosis development and formation of an atheroma (plaque).

Treatment for Arterial Degeneration Chapter 14 explained how balloon catheterization, bypass grafts, and stents (small tubes used to keep vessels open) can be used to treat coronary artery disease. These measures can also treat degenerative changes in other arteries. An additional treatment approach is **endarterectomy** (end-ar-ter-EK-to-me), removal of a vessel's thickened, atheromatous lining. Common sites for this procedure are the carotid artery or vertebral artery leading to the brain and the common iliac or femoral arteries leading to the lower limbs. Surgeons can remove a blockage by direct incision of a vessel. More commonly, to remove plaque, they use a cutting tool inserted with a catheter through the vessel opening.

ANEURYSM

An **aneurysm** (AN-u-rizm) is a bulging sac in a blood vessel's wall caused by a localized weakness in that part of the vessel (**Fig. 15-16**). The aorta and vessels in the brain are common aneurysm sites. The damage to the wall may be congenital or a result of arteriosclerosis. Whatever the cause, the aneurysm may continue to grow in size. As it swells, it may cause some derangement of other structures, in which case definite symptoms are present. If undiagnosed, the weakened area eventually yields to the pressure, and the aneurysm bursts like a balloon—a life-threatening event. Surgical replacement of the damaged segment with a synthetic graft may save a person's life.

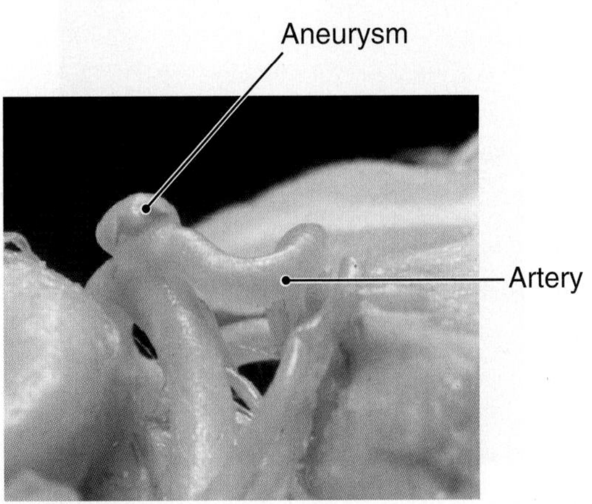

Figure 15-16 A cerebral aneurysm in the cerebral arterial circle. **KEY POINT** An aneurysm results from a localized weakness in a vessel wall that enlarges and may rupture.

See the Student Resources on thePoint for a photograph of an aortic aneurysm.

HEMORRHAGE

A profuse escape of blood from the vessels is known as **hemorrhage** (HEM-or-ij), a word that means "a bursting forth of blood." Such bleeding may be external or internal, from vessels of any size, and may involve any part of the body. Capillary oozing is usually stopped by the normal process of clot formation.

The loss of a small amount of blood will cause no problem for a healthy adult, but rapid loss of 1 L or more of blood is life threatening. In an emergency, application of pressure where a local artery can be pressed against a bone will slow bleeding. The most important of these "pressure points" are the following (**see Fig. 15-5**):

- The **facial artery** may be pressed against the lower jaw for hemorrhage around the nose, mouth, and cheek. One can feel the pulse of the facial artery in the depression about 1 in anterior to the lower jaw's angle.
- The **temporal artery** may be pressed against the side of the skull just anterior to the ear to stop hemorrhage on the side of the face and around the ear.
- The **common carotid artery** in the neck may be pressed back against the spinal column for bleeding in the neck and the head.
- The **subclavian artery** may be pressed against the first rib by a downward push with the thumb to stop bleeding from the shoulder or arm.
- The **brachial artery** may be pressed against the humerus (arm bone) by a push inward along the natural groove between the arm's two large muscles. This stops hand, wrist, and forearm hemorrhage.
- The **femoral artery** (in the groin) may be pressed to avoid serious hemorrhage of the lower extremity.

It is important not to leave the pressure on too long, as this may cause damage to tissues supplied by arteries past the pressure point.

SHOCK

The word **shock** has a number of meanings. In terms of the circulating blood, it refers to a life-threatening condition in which there is inadequate blood flow to the body tissues. A wide range of conditions that reduce effective circulation can cause shock. The exact cause is often not known. However, a widely used classification is based on causative factors, the most important of which include the following:

- **Cardiogenic** (kar-de-o-JEN-ik) **shock**, the leading cause of shock death, occurs when the ventricles fail to pump blood effectively. It is often a complication of heart muscle damage, as occurs in myocardial infarction.

- **Septic shock** is second only to cardiogenic shock as a cause of shock death. It is usually the result of an overwhelming bacterial infection.
- **Hypovolemic** (hi-po-vo-LE-mik) **shock** is caused by a decrease in the volume of circulating blood and may follow severe hemorrhage or burns.
- **Anaphylactic** (an-ah-fih-LAK-tik) **shock** is a severe allergic reaction to foreign substances to which the person has been sensitized (see Chapter 17 on Immunity).

When the cause is not known, shock is classified according to its severity.

In mild shock, regulatory mechanisms relieve the circulatory deficit. Signs include acidosis, increased heart rate, and a minor decrease in blood pressure. Constriction of small blood vessels and the detouring of blood away from certain organs increase the effective circulation. Mild shock may develop into a severe, life-threatening circulatory failure.

Severe shock occurs if the body's compensatory mechanisms fail to maintain adequate blood pressure. A profound decrease in blood pressure results in the following changes, all of which further reduce blood pressure:

1. Blood accumulates in the capillaries, increasing the amount of fluid passing from blood to the interstitial space. Blood volume decreases.
2. The reduced blood volume and impaired coronary circulation weaken cardiac contractions. This reduction in stroke volume reduces cardiac output.
3. Metabolic waste accumulates, dilating vessels and reducing peripheral resistance.

The baroreceptor reflex discussed above attempts to compensate for the decreased blood pressure, by increasing heart rate, stroke volume, and vasoconstriction, but cannot fully correct the deficit. Late shock symptoms include clammy skin, anxiety, low blood pressure, rapid pulse, and rapid, shallow breathing.

The victim of shock should first be placed in a horizontal position and covered with a blanket. If there is bleeding, it should be stopped. The patient's head should be kept turned to the side to prevent aspiration (breathing in) of vomited material, an important cause of death in shock cases. Further treatment of shock depends largely on the treatment of the causative factors. For example, shock resulting from fluid loss, as in hemorrhage or burns, is best treated with blood products or plasma expanders (intravenous fluids). Shock caused by heart failure should be treated with drugs that improve heart muscle contractions. In any case, all measures are aimed at supporting the circulation and improving the cardiac output. Oxygen is frequently administered to improve oxygen delivery to the tissues.

THROMBOSIS

As mentioned earlier, formation of a blood clot in a vessel is **thrombosis** (throm-BO-sis). A blood clot in a vein, termed *deep venous thrombosis*, most commonly develops in the deep veins of the calf muscle, although it may appear elsewhere. Thromboses typically occur in people who are recovering from surgery, injury, or childbirth or those who are bedridden, but they can also occur in healthy people immobilized for long periods (such as on airplanes). Immobilization of Jocelyn's injured knee might have contributed to the formation of her deep vein thrombosis in the opening case study. Clot formation may also be associated with some diseases, with obesity, and with certain drugs, such as hormonal medications. Symptoms are pain and swelling, often with warmth and redness below or around the clot. Thrombosis can be diagnosed with ultrasound or with magnetic resonance imaging.

A dangerous complication of thrombosis is the formation of an **embolus** (EM-bo-lus), a piece of the clot that becomes loose and floats in the blood. An embolus is carried through the circulatory system until it lodges in a vessel. If it reaches the lungs, sudden death from **pulmonary embolism** (EM-bo-lizm) may result. Prevention of infections, early activity to promote circulation after an injury or an operation, and the use of anticoagulant drugs when appropriate have greatly reduced the incidence of this condition.

Phlebitis (fleh-BI-tis), inflammation of a vein, may contribute to clot formation, in which case the condition is called **thrombophlebitis** (throm-bo-fleh-BI-tis).

VARICOSE VEINS

Varicose veins, or **varices** (VAR-ih-seze), are superficial veins that have become swollen, distorted, and ineffective. (The singular form is *varix* [VAR-iks]). These changes result from valvular damage that allows blood to pool in the vessels. The saphenous veins of the legs are commonly involved (**Fig. 15-17**). This condition frequently occurs in people who stand for long periods, such as salespeople, because gravity

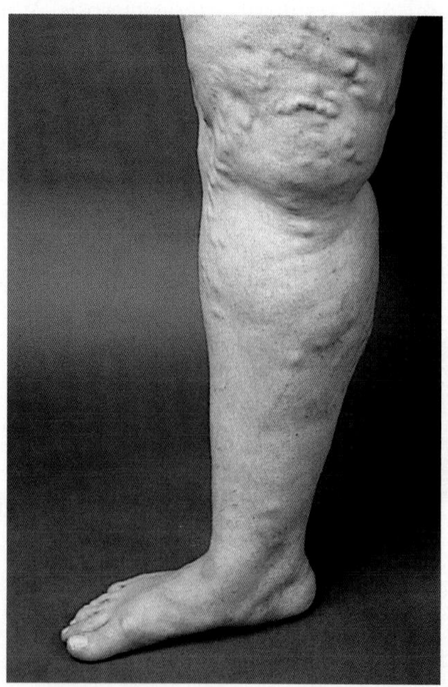

Figure 15-17 **Varicose veins.** **KEY POINT** Valvular damage allows blood to pull in superficial vessels causing swelling and distortion.

and lack of muscle contraction cause blood to accumulate in the legs. Pregnancy, with its accompanying pressure on the pelvic veins, may also be a predisposing factor. Elastic stockings help prevent varicosities by compressing the veins and decreasing their compliance. Varicose veins in the rectum are called **hemorrhoids** (HEM-o-royds), or *piles*. Varices can also appear in the lower esophagus as a result of hypertension in the portal system caused by liver disease.

See the Student Resources on thePoint for career information on vascular technology. Vascular technologists collect information on the blood vessels and circulation to aid in diagnosis.

CHECKPOINTS

- 15-22 What is the general term for any hardening of the arteries?
- 15-23 What is a bulging sac in a vessel wall called?
- 15-24 What is the term for profuse bleeding?
- 15-25 What is circulatory shock?
- 15-26 What term describes a twisted and swollen vein?

Disease in Context Revisited

Jocelyn Sees a Hematologist

In her first visit with Dr. Schuman, the hematologist, Jocelyn discussed the history of the blood clot in her left leg. In a second visit, the doctor explained the meaning of the blood test she had done for hereditary protein S deficiency.

"This protein normally helps block blood clotting," she said. "You have a mild deficiency, so your blood might tend to clot easily. I don't recommend any further treatment for you at this point, but you should report this deficiency if you are in an accident or undergoing surgery. You will probably always have poor circulation in that leg, so try not to sit or stand in one position for long. If, for example, you are traveling for a long time, wear a compression stocking on that leg and move around at least once every hour. I also recommend that you have your two young adult daughters tested for this deficiency. Definitely no menopausal hormone replacement therapy for you, and no hormonal contraceptives for the girls if they have this genetic trait. And of course, not smoking is important."

15

CHAPTER 15

Chapter Wrap-Up

Summary Overview

A detailed chapter outline with space for note taking is on *thePoint*. The figure below illustrates the main topics covered in this chapter.

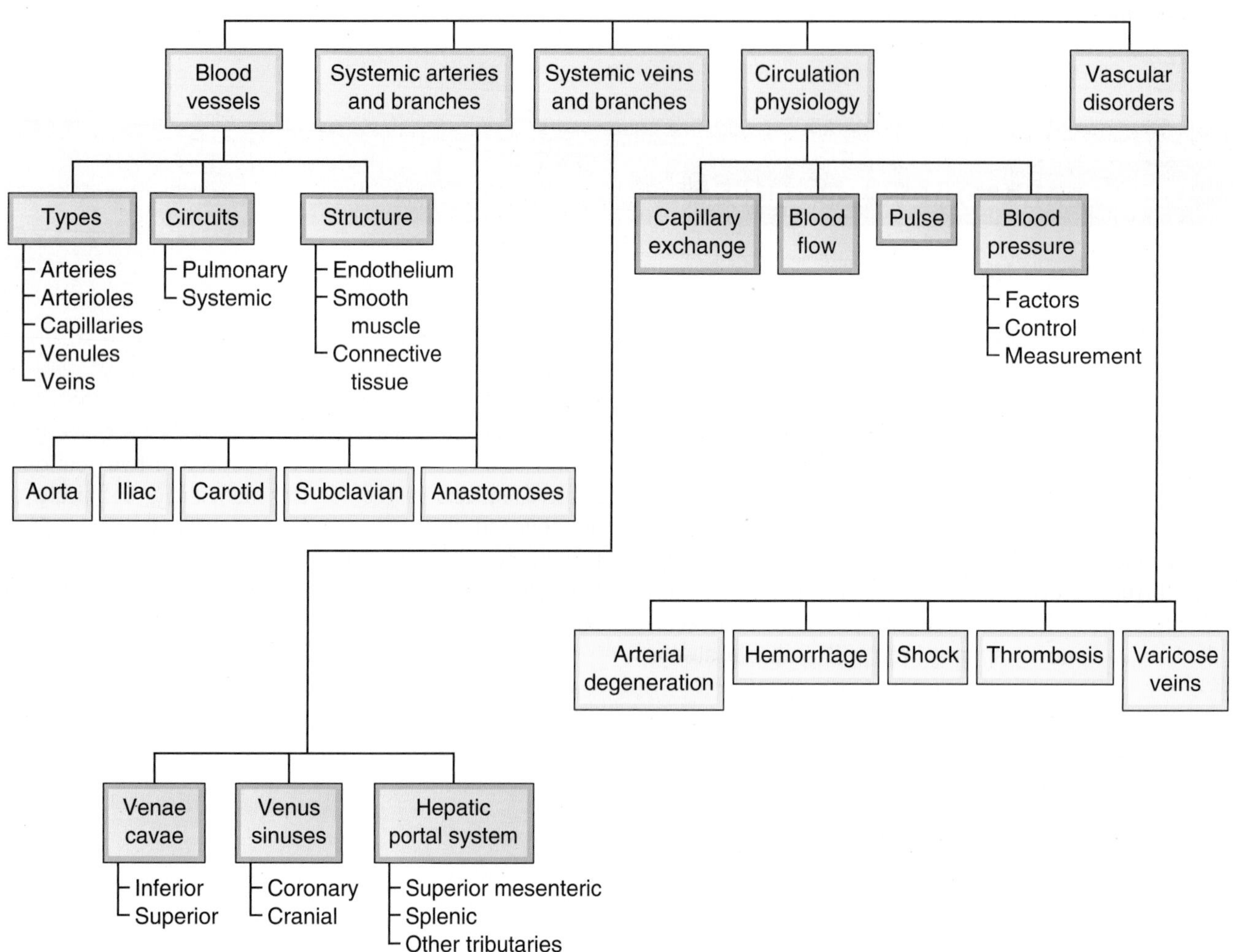

Key Terms

The terms listed below are emphasized in this chapter. Knowing them will help you organize and prioritize your learning. These and other boldface terms are defined in the Glossary with phonetic pronunciations.

aneurysm
aorta
arteriole
arteriosclerosis
artery
baroreceptor
compliance
capillary
elasticity
embolus
endarterectomy
endothelium
hemorrhage
hypertension
hypotension
phlebitis
pulse
shock
sinusoid
sphygmomanometer
thrombosis
varices
varicose vein
vasoconstriction
vasodilation
vasomotor
vein
vena cava
venous sinus
venule

Word Anatomy

Medical terms are built from standardized word parts (prefixes, roots, and suffixes). Learning the meanings of these parts can help you remember words and interpret unfamiliar terms.

WORD PART	MEANING	EXAMPLE
Systemic Arteries		
brachi/o	arm	The *brachiocephalic* artery supplies blood to the arm and head on the right side.
celi/o	abdomen	The *celiac* trunk branches to supply blood to the abdominal organs.
cephal/o	head	See "brachi/o" example.
clav/o	clavicle	The *subclavian* artery extends under the clavicle on each side.
cost/o	rib	The *intercostal* arteries are between the ribs.
enter/o	intestine	The *mesenteric* arteries supply blood to the intestines.
gastr/o	stomach	The *gastric* artery goes to the stomach.
hepat/o	liver	The *hepatic* artery supplies blood to the liver.
ped/o	foot	The dorsalis *pedis* artery supplies blood to the foot.
phren/o	diaphragm	The *phrenic* artery supplies blood to the diaphragm.
splen/o	spleen	The *splenic* artery goes to the spleen.
stoma	mouth	An *anastomosis* is a communication between two vessels.
Circulation Physiology		
bar/o	pressure	A *baroreceptor* responds to changes in pressure.
man/o	pressure	See next example.
sphygm/o	pulse	A *sphygmomanometer* is used to measure blood pressure.
Vascular Disorders		
phleb/o	vein	*Phlebitis* is inflammation of a vein.

Questions for Study and Review

BUILDING UNDERSTANDING

Fill in the Blanks

1. Capillaries receive blood from vessels called ______.
2. The specific part of the medulla oblongata that regulates blood flow is the ______.
3. The flow of blood into an individual capillary is regulated by a(n) ______.
4. Lower-than-normal blood pressure is called ______.
5. Inflammation of a vein is called______.

Matching > Match each numbered item with the most closely related lettered item.

___ **6.** Lack of blood supply to a tissue or organ
___ **7.** Bulging sac in the wall of a vessel
___ **8.** Profuse loss of blood
___ **9.** An immobile blood clot within a vessel
___ **10.** A mobile blood clot within a vessel

a. embolus
b. thrombus
c. ischemia
d. aneurysm
e. hemorrhage

Multiple Choice

___ **11.** Which tissue makes up a blood vessel's inner tunic?
a. smooth muscle
b. epithelium
c. connective tissue
d. nervous tissue

___ **12.** What is the name of either large vein that drains into the right atrium?
a. vena cava
b. jugular
c. carotid
d. iliac

___ **13.** What is the main process of capillary exchange?
a. endocytosis
b. exocytosis
c. osmosis
d. diffusion

___ **14.** Which vessel supplies oxygenated blood to the stomach, spleen, and liver?
a. hepatic portal system
b. superior mesenteric artery
c. inferior mesenteric artery
d. celiac trunk

___ **15.** What is the medical term for "hardening of the arteries?"
a. shock
b. gangrene
c. arteriosclerosis
d. stasis

UNDERSTANDING CONCEPTS

16. Differentiate between the terms in each of the following pairs:
a. artery and vein
b. arteriole and venule
c. anastomosis and venous sinus
d. vasoconstriction and vasodilation
e. systolic and diastolic pressure

17. How does the structure of the blood vessels correlate with their function?

18. Trace a drop of blood from the left ventricle to the:
a. right side of the head and the neck
b. lateral surface of the left hand
c. right foot
d. liver
e. small intestine

19. Trace a drop of blood from capillaries in the wall of the small intestine to the right atrium. What is the purpose of going through the liver on this trip?

20. Describe three mechanisms that promote the return of blood to the heart in the venous system.

21. What physiological factors influence blood pressure?

22. What are some symptoms of arteriosclerosis, and how are these produced?

23. What is shock, and why is it so dangerous? Name some symptoms of shock, and identify the types of shock based on (a) cause and (b) severity.

24. Based on **Figure A5-4** in the Dissection Atlas, give the names and numbers of the following:
a. the veins that carry blood from the lungs to the heart
b. two arteries that carry blood to the head
c. a vein that drains blood from the kidney
d. an artery that branches from the descending aorta at its termination
e. the third artery to branch off the aortic arch

CONCEPTUAL THINKING

25. Kidney disease usually results in the loss of protein from the blood into the urine. One common sign of kidney disease is edema. From your understanding of capillary exchange, explain why edema is often associated with kidney disease.

26. Cliff C., a 49-year-old self-described "couch potato," has a blood pressure of 162/100 mm Hg. What is Cliff's diagnosis? What can this disorder lead to? If you were Cliff's healthcare provider, what treatments might you discuss with him?

27. Based on Jocelyn's DVT study, cite the dangers of thrombosis, and research the methods used to treat thrombosis.

For more questions, see the Learning Activities on thePoint.

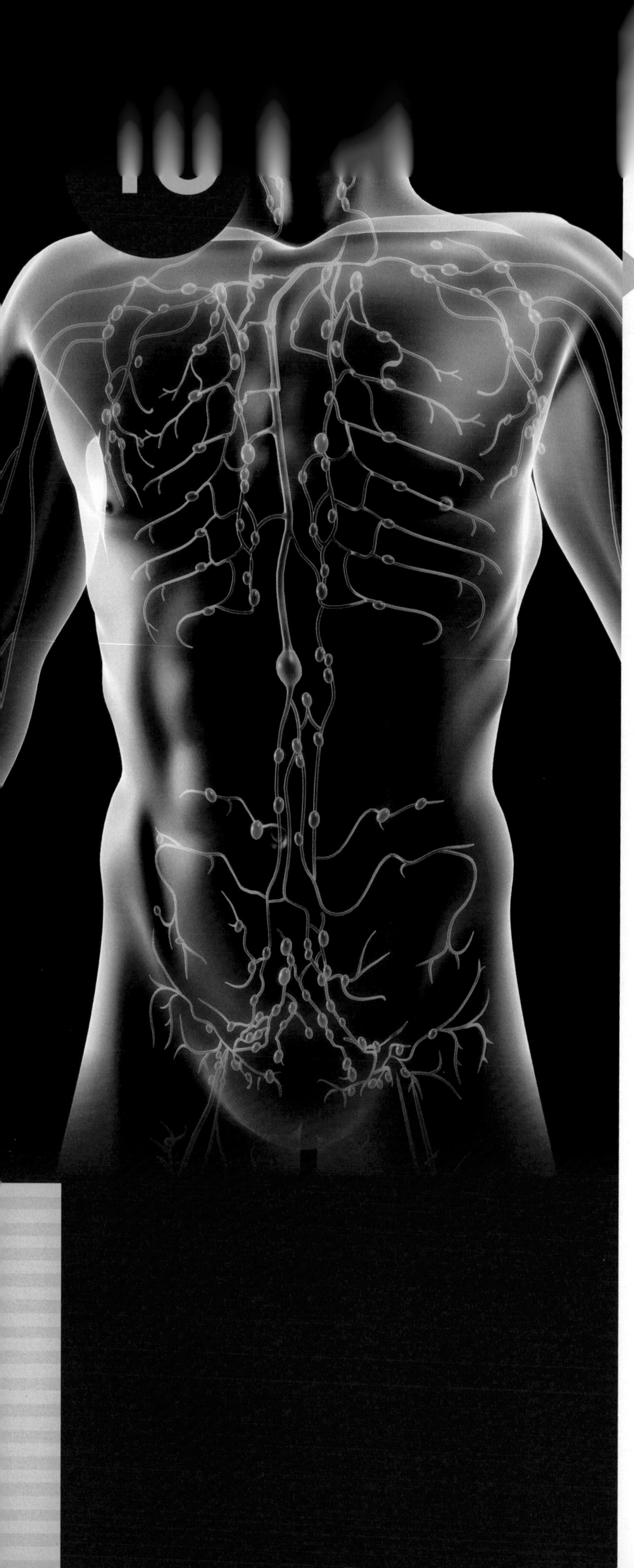

Learning Objectives

After careful study of this chapter, you should be able to:

1 ▶ List the functions of the lymphatic system. ***p. 368***

2 ▶ Explain how lymphatic capillaries differ from blood capillaries. ***p. 369***

3 ▶ Name the two main lymphatic ducts, and describe the area drained by each. ***p. 371***

4 ▶ Name and give the locations of five types of lymphatic tissue, and list the functions of each. ***p. 371***

5 ▶ Describe four lymphatic system disorders. ***p. 375***

6 ▶ Cite the causes and symptoms of infectious mononucleosis, as described in the case study that opens this chapter. ***pp. 367, 377***

7 ▶ Show how word parts are used to build words related to the lymphatic system (see Word Anatomy at the end of the chapter). ***p. 379***

Disease in Context *Lucas's Mononucleosis*

Lucas was looking forward to the upcoming high school marching band season. His band had been putting in many hours practicing new tunes and marching sequences. Lucas recently had been feeling pretty tired, which he attributed to the late hours and twice-a-day practices. His 30-lb sousaphone seemed to weigh more these days.

"Come on Lucas, you have to eat dinner before going to band practice tonight," his mom urged one evening.

"I don't feel like eating, and besides, my throat is kind of sore," Lucas replied.

Thinking about her son's answer, his mom recalled that Lucas had been sleeping more than usual the past few days. She worried about his hectic schedule, the frequent crowded bus trips, and the intense classroom sessions with the 125-plus band members. She was aware that two of the band members had come down with mono the past week and that illness was already a topic of discussion at the parent booster club meeting. She decided to call Lucas's physician the next day and schedule an appointment.

"Hi Lucas, how's the band season coming along?" Dr. Fischer asked when he saw Lucas later that week.

"Pretty good; we've got some new songs and I think we are going to score well in the band competition. We have a new tuba player, and he's got one of those new lightweight sousaphones, pretty cool" Lucas said. "Everyone was trying it out, it's really neat."

Dr. Fischer took a history and then asked Lucas to lie down on the examination table. "Let's take a look and see why you are feeling so tired lately and what might be causing the sore throat."

Dr. Fischer considered the symptoms: general malaise for seven to 10 days, fever, loss of appetite, and a sore throat. He began the physical examination by observing Lucas's throat and noted it was red and swollen. He palpated the lymph nodes in the cervical, axillary, and inguinal regions. They were all enlarged. He also palpated the left upper quadrant (LUQ) of the abdomen and noted that the spleen was enlarged.

"It looks like you might have come down with a viral infection called mononucleosis," Dr. Fischer told Lucas. "I'm going to take a throat culture and blood sample to confirm my suspicions."

Later, we will check on the results of Lucas's laboratory tests and his diagnosis. In this chapter, you will learn about the lymphatic system, its functions, how it helps protect us from infections, and how it can also be a vehicle for the spread or metastasis of disease.

ANCILLARIES *At-A-Glance*

Visit thePoint to access the following resources. For guidance in using these resources most effectively, see pp. xv–xvii.

Learning RESOURCES

- Tips for Effective Studying
- Web Figure: Cervical Lymphadenopathy
- Web Chart: Lymphoid Tissue
- Health Professions: Clinical Massage Therapist
- Detailed Chapter Outline
- Answers to Questions for Study and Review
- Audio Pronunciation Glossary

Learning ACTIVITIES

- Pre-Quiz
- Visual Activities
- Kinesthetic Activities
- Auditory Activities

A LOOK BACK

In Chapter 15, we learned that the blood leaves some fluid behind in the tissues as it travels through the capillary networks. The lymphatic system collects this fluid and returns it to the circulation. Lymph's return to the heart is governed by the same mechanisms that promote venous return of blood. The lymphatic system has other functions besides aiding in circulation, as we will see in this chapter.

The lymphatic system is a widespread system of tissues and vessels. Its organs are not grouped together but are scattered throughout the body, and it services almost all regions. Only bone marrow, cartilage, epithelium, and the central nervous system are not in direct communication with this system.

Functions of the Lymphatic System

The lymphatic system's functions are just as varied as its locations. These functions fall into three categories:

- **Fluid balance.** As blood circulates through the capillaries in the tissues, water and dissolved substances are constantly exchanged between the bloodstream and the interstitial (in-ter-STISH-al) fluids that bathe the cells. The volume of fluid that leaves the blood is not quite matched by the amount that returns to the blood, so there is always a slight excess of fluid left behind in the tissues. In addition, some proteins escape from the blood capillaries and are left behind. This fluid and protein would accumulate in the tissues if not for a second drainage pathway through lymphatic vessels (**Fig. 16-1**).

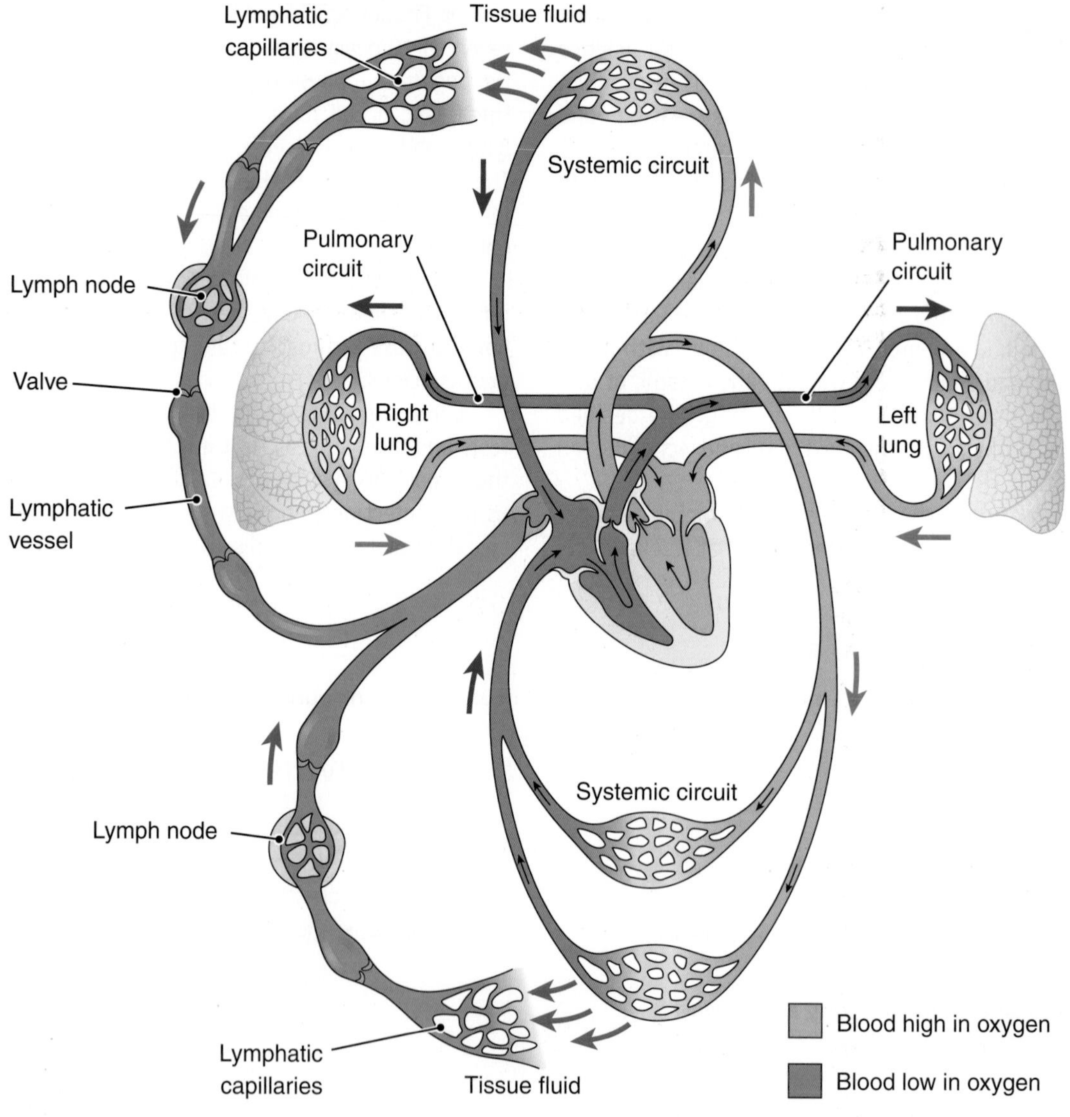

Figure 16-1 The lymphatic system in relation to the cardiovascular system. **KEY POINT** Lymphatic vessels pick up fluid in the tissues and return it to the blood in vessels near the heart. **ZOOMING IN** What type of blood vessel receives lymph collected from the body?

In addition to the blood-carrying capillaries, the tissues also contain microscopic lymphatic capillaries (**Fig. 16-2**). These small vessels pick up excess fluid and protein from the tissues. They then drain into larger vessels, which eventually return these materials to the venous system near the heart.

The fluid that circulates in the lymphatic system is called **lymph** (limf), a clear fluid similar in composition to interstitial fluid. Although lymph is formed from the components of blood plasma, it differs from the plasma in that it has much less protein.

- **Protection.** The lymphatic system is an important component of the immune system, which fights infection and helps prevent cancer. One group of white blood cells, the lymphocytes, can live and multiply in the lymphatic system, where they attack and destroy foreign organisms. Lymphoid tissue scattered throughout the body filters out pathogens, other foreign matter, tumor cells, and cellular debris found in body fluids. More will be said about the lymphocytes and immunity in Chapter 17.
- **Absorption of fats.** Following the chemical and mechanical breakdown of food in the digestive tract, most nutrients are absorbed into the blood through intestinal capillaries. Many digested fats, however, are too large to enter the blood capillaries and are instead absorbed into specialized lymphatic capillaries in the lining of the small intestine. Fats taken into these **lacteals** (LAK-te-als) are transported in lymphatic vessels until the lymph is added to the blood. More information on the lymphatic system's role in digestion is found in Chapter 19.

☐ 16-1 What are the three functions of the lymphatic system?

Lymphatic Circulation

Lymph travels through a network of small and large channels that are in some ways similar to the blood vessels. However, the system is not a complete circuit. It is a one-way system that begins in the tissues and ends when the lymph joins the blood (**see Fig. 16-1**).

LYMPHATIC CAPILLARIES

The walls of the lymphatic capillaries resemble those of the blood capillaries in that they are made of one layer of flattened (squamous) epithelial cells. This thin layer, also called *endothelium*, allows for easy passage of soluble materials and water through the capillary wall (**see Fig. 16-2B**). The gaps between the endothelial cells in the lymphatic capillaries are larger than those of the blood capillaries. The lymphatic capillaries are thus more permeable, allowing for easier entrance of relatively large protein molecules. The proteins do not move back out of the vessels because the endothelial cells overlap slightly, forming one-way valves to block their return.

Unlike the blood capillaries, the lymphatic capillaries arise blindly; that is, they are closed at one end and do not form a bridge between two larger vessels. Instead, one end simply lies within a lake of tissue fluid, and the other communicates with a larger lymphatic vessel that transports the lymph toward the heart (**see Fig. 16-2A**).

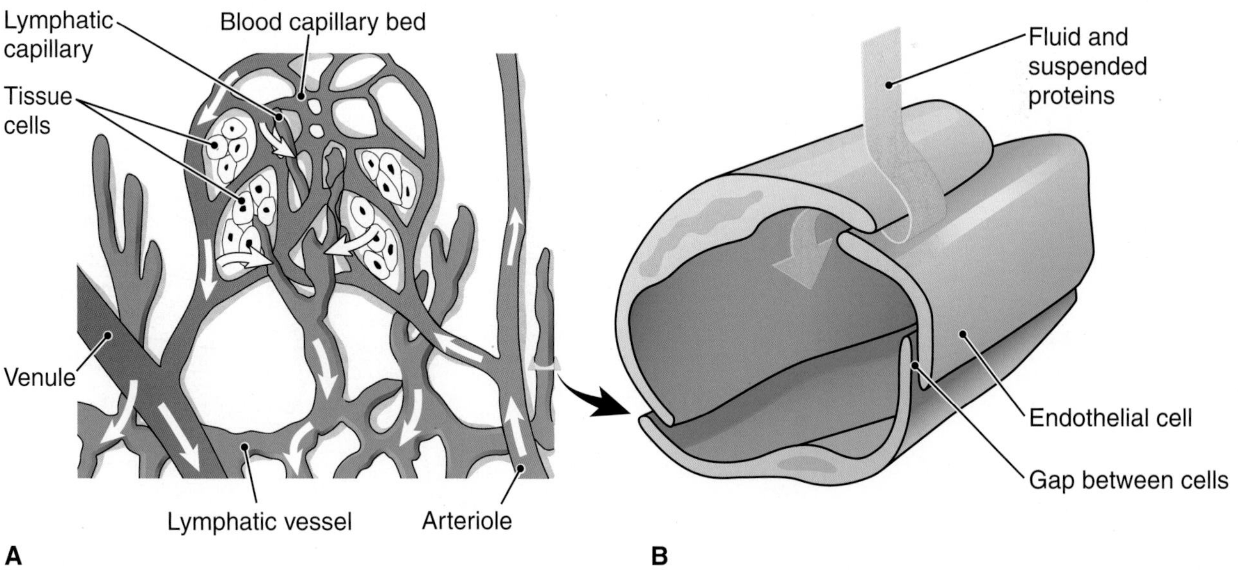

Figure 16-2 **Lymphatic drainage in the tissues.** **KEY POINT** Lymphatic capillaries pick up fluid and proteins from the tissues for return to the heart in lymphatic vessels. **A.** Blind-ended lymphatic capillaries in relation to blood capillaries. *Arrows* show the direction of flow. **B.** Structure of a lymphatic capillary. Fluid and proteins can enter the capillary with ease through gaps between the endothelial cells. Overlapping cells act as valves to prevent the material from leaving.

LYMPHATIC VESSELS

The lymphatic vessels are thin walled and delicate and have a beaded appearance because of indentations where valves are located (**see Fig. 16-1**). These valves prevent backflow in the same way as do those found in veins.

Lymphatic vessels include superficial and deep sets (**Fig. 16-3**). The surface lymphatics are immediately below the skin, often lying near the superficial veins. The deep vessels are usually larger and accompany the deep veins.

Lymphatic vessels are named according to location. For example, those in the breast are called mammary lymphatic vessels (**see Fig. 16-3D**), those in the thigh are called femoral lymphatic vessels, and those in the leg are called tibial lymphatic vessels. At certain points, the vessels drain through

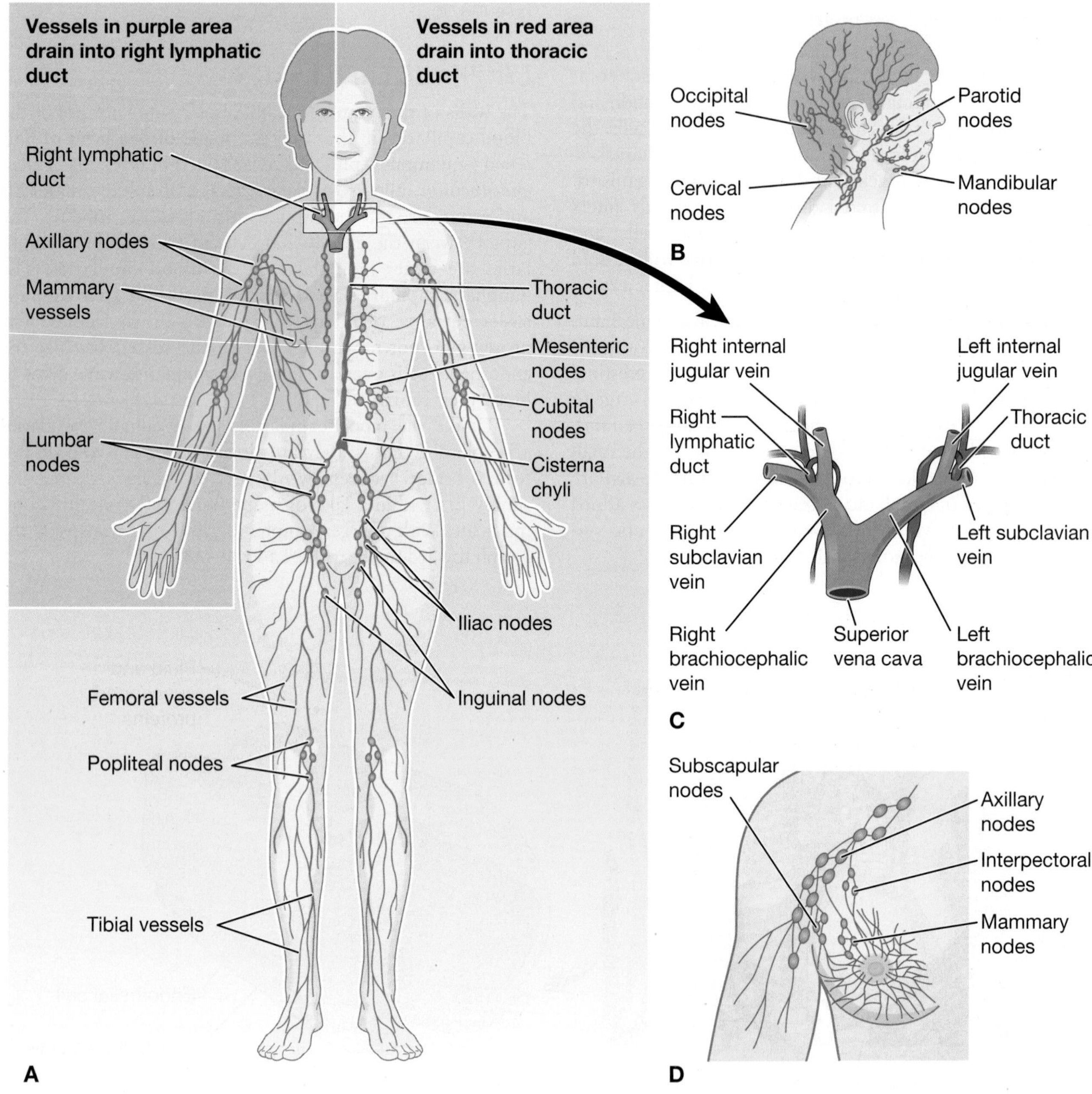

Figure 16-3 **Vessels and nodes of the lymphatic system.** **KEY POINT** Lymphatic vessels serve almost every area in the body. Lymph nodes are distributed along the path of the vessels. **A.** Lymph nodes and vessels, showing areas draining into the right lymphatic duct (*purple*) and the thoracic duct (*red*). **B.** Lymph nodes and vessels of the head. **C.** Drainage of right lymphatic duct and thoracic duct into subclavian veins. **D.** Lymph nodes and vessels of the mammary gland and surrounding areas. **ZOOMING IN** What are some nodes that receive lymph drainage from the breast?

lymph nodes, small masses of lymphatic tissue that filter the lymph. The nodes are in groups that serve a particular region, as will be described shortly. The lymph nodes and other lymphoid tissue are involved in Lucas's case study on infectious mononucleosis. Lymphatic vessels carrying lymph away from the regional nodes eventually drain into one of two terminal vessels, the right lymphatic duct or the thoracic duct, both of which empty into the bloodstream near the heart.

The Right Lymphatic Duct The **right lymphatic duct** is a short vessel, approximately 1.25 cm (1/2 in) long, that receives only the lymph that comes from the body's superior right quadrant: the right side of the head, neck, and thorax, as well as the right upper extremity. The right lymphatic duct empties into the right subclavian vein near the heart (**see Fig. 16-3C**). Its opening into this vein is guarded by two pocket-like semilunar valves to prevent blood from entering the duct. The rest of the body is drained by the thoracic duct.

The Thoracic Duct The **thoracic duct**, or left lymphatic duct, is the larger of the two terminal vessels, measuring approximately 40 cm (16 in) in length. As shown in **Figure 16-3**, the thoracic duct receives lymph from all parts of the body except those superior to the diaphragm on the right side. It then drains into the left subclavian vein. This duct begins in the posterior part of the abdominal cavity, inferior to the attachment of the diaphragm. The duct's first part is enlarged to form a cistern, or temporary storage pouch, called the **cisterna chyli** (sis-TER-nah KI-li). **Chyle** (kile) is the milky fluid that drains from the intestinal lacteals; it is formed by the combination of fat globules and lymph. Chyle passes through the intestinal lymphatic vessels and the lymph nodes of the mesentery (the membrane around the intestines), finally entering the cisterna chyli. In addition to chyle, all the lymph from below the diaphragm empties into the cisterna chyli and subsequently the thoracic duct.

The thoracic duct extends upward from the cisterna chyli through the diaphragm and along the posterior thoracic wall into the base of the neck on the left side. Here, it receives the left jugular lymphatic vessels from the head and neck and the left subclavian vessels from the left upper extremity. In addition to the valves along the duct, there are two valves at its opening into the left subclavian vein to prevent the passage of blood into the duct.

MOVEMENT OF LYMPH

The segments of lymphatic vessels located between the valves contract rhythmically, propelling the lymph forward. The contraction rate is related to the fluid volume in the vessel—the more fluid, the more rapid the contractions.

The same mechanisms that promote venous return of blood to the heart also move lymph. As skeletal muscles contract during movement, they compress the lymphatic vessels and drive lymph forward. Changes in pressures within the abdominal and thoracic cavities caused by breathing aid lymphatic movement from the abdomen to the thorax. When lymph does not flow properly, a condition called lymphedema results. This is discussed later in this chapter.

Massage therapy can improve lymphatic drainage and blood circulation as well as benefit muscles. See the Student Resources on thePoint for more information about this field.

CHECKPOINTS 

- 16-2 What are the two differences between blood capillaries and lymphatic capillaries?
- 16-3 What are the two main lymphatic vessels?

Lymphoid Tissue

Lymphoid (LIM-foyd) **tissue** is distributed throughout the body and makes up the lymphatic system's specialized organs. As previously mentioned, the lymph nodes are part of the network of lymphatic vessels. In contrast, the spleen, thymus, tonsils, and other lymphoid organs do not encounter lymph.

See the Student Resources on thePoint for a quick study chart on lymphoid tissue.

LYMPH NODES

The lymph nodes, as noted, filter the lymph as it travels through the lymphatic vessels (**Fig. 16-4**). They are also sites where lymphocytes of the immune system multiply and work to combat foreign organisms. The lymph nodes are small, rounded masses varying from pinhead size to as long as 2.5 cm (1 in). Each has a fibrous connective tissue capsule from which partitions called *trabeculae* extend into the node's substance. At various points in the node's surface, afferent lymphatic vessels pierce the capsule to carry lymph toward open channels, or sinuses, in the node. An indented area, the *hilum* (HI-lum) is the exit point for efferent lymphatic vessels carrying lymph out of the node. At this location, other structures, including blood vessels and nerves, link with the node. (*Hilum* is a general term for an indented region of an organ where vessels and nerves connect.)

A lymph node is divided into two regions: an outer cortex and an inner medulla. Each of these regions includes lymph-filled sinuses and cords of lymphatic tissue. The cortex contains pulplike *cortical nodules*, each of which has a *germinal center* where certain lymphocytes multiply. The medulla has populations of immune cells, including lymphocytes and macrophages (phagocytes) along the *medullary sinuses*, which drain into the efferent lymphatic vessels.

Lymph nodes are seldom isolated. As a rule, they are grouped together in numbers varying from two or three to well over 100. Some of these groups are placed deeply,

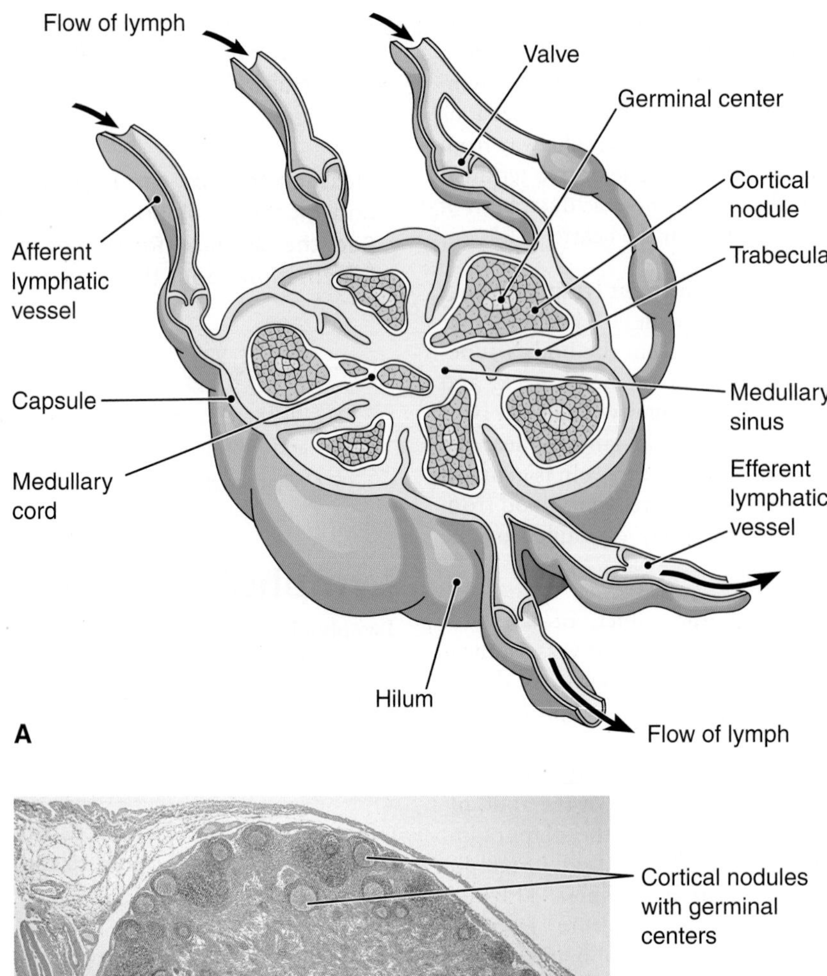

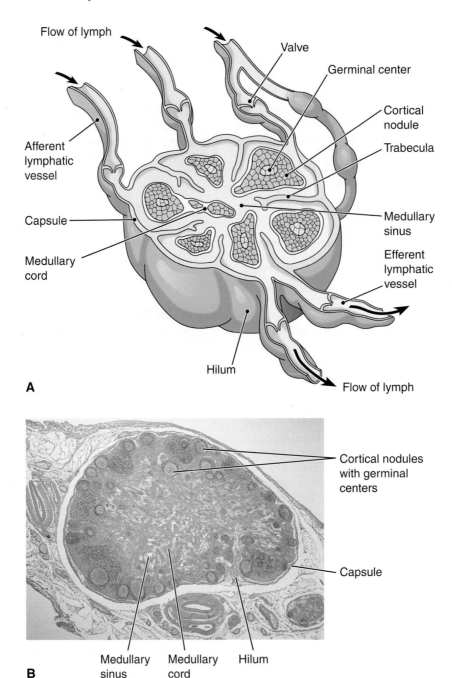

Figure 16-4 **Lymph nodes.** **KEY POINT** Lymph is filtered as it travels through a lymph node. Cells of the immune system also multiply in the nodes. **A.** Structure of a lymph node. *Arrows* indicate the flow of lymph. **B.** Section of a lymph node as seen under the microscope (low power). **ZOOMING IN** What type of lymphatic vessel carries lymph into a node? What type of lymphatic vessel carries lymph out of a node?

whereas others are superficial. The main groups include the following:

- **Cervical nodes**, located in the neck in deep and superficial groups, drain various parts of the head and neck. They often become enlarged during upper respiratory infections.
- **Axillary nodes**, located in the axillae (armpits), may become enlarged after infections of the upper extremities and the breasts. Cancer cells from the breasts often metastasize (spread) to the axillary nodes, as noted in Box 16-1, which explains the role of lymph node biopsy in the treatment of cancer.

Box 16-1

HOT TOPICS

Sentinel Node Biopsy: Finding Cancer Before It Spreads

Ordinarily, the lymphatic system is one of the body's primary defenses against disease. In cancer, though, it can be a vehicle for the spread (metastasis) of disease. When cancer cells enter the lymphatic vessels, they are often trapped and killed within the lymph nodes by immune system cells. However, if the immune system is overwhelmed by cancer cells, some may escape the lymph nodes and travel to other parts of the body, where they may establish new tumors.

In breast cancer, the degree of invasion of nearby lymph nodes helps determine what treatments are required after surgical removal of the tumor. Until recently, a mastectomy often included the removal of nearby lymphatic vessels and nodes (a procedure called axillary lymph node dissection). Biopsy of the nodes determined whether or not they contained cancerous cells. If they did, radiation treatment or chemotherapy was required. In many women with early-stage breast cancer, however, the axillary nodes do not contain cancerous cells. In addition, about 20% of the women whose lymphatic vessels and nodes have been removed suffer impaired lymph flow, resulting in lymphedema, pain, disability, and an increased risk of infection.

Sentinel node biopsy is a diagnostic procedure that may minimize the need to perform axillary lymph node dissection, while still detecting metastasis. Surgeons use radioactive tracers to identify the first nodes that receive lymph from the area of a tumor. Biopsy of only these "sentinel nodes" reveals whether tumor cells are present, providing the earliest indication of metastasis. Research shows that sentinel lymph node biopsy is associated with less pain, fewer complications, and faster recovery than is axillary lymph node dissection. However, clinical trials are ongoing to determine whether sentinel node biopsy is as successful as axillary dissection in finding cancer before it spreads.

- **Tracheobronchial** (tra-ke-o-BRONG-ke-al) **nodes** are found near the trachea and around the larger bronchial tubes. In people who smoke, are exposed to smoke, or live in highly polluted areas, these nodes become filled with airborne contaminants.
- **Mesenteric** (mes-en-TER-ik) **nodes** are found between the two layers of peritoneum that form the mesentery. There are some 100 to 150 of these nodes.
- **Inguinal nodes**, located in the groin region, receive lymph drainage from the lower extremities and from the external reproductive organs. When they become enlarged, they are often referred to as **buboes** (BU-bose), from which bubonic plague got its name.

THE SPLEEN

The **spleen** filters blood, much like the lymph nodes filter lymph. It is located in the superior left hypochondriac region of the abdomen, high up under the dome of the diaphragm, and normally is protected by the lower part of the rib cage (**Fig. 16-5**). The spleen is a soft, purplish, and somewhat flattened organ, measuring approximately 12.5 to 16 cm (5 to 6 in) long and 5 to 7.5 cm (2 to 3 in) wide. The spleen's capsule, as well as its framework, is more elastic than that of the lymph nodes. It contains an involuntary muscle, which enables the splenic capsule to contract and also to withstand some swelling.

Not surprisingly, considering its role in blood filtration, the spleen has an unusually rich blood supply. The organ is filled with a soft pulp rich in phagocytes and lymphocytes. As blood slowly percolates through this tissue, connective tissue fibers trap cellular debris and other impurities for destruction by phagocytes. The spleen is classified as part of the lymphatic system because it contains prominent masses of lymphoid tissue. However, it has wider functions than other lymphatic structures, including the following:

- Destroying old, worn-out red blood cells. The iron and other breakdown products of hemoglobin are then carried to the liver by the hepatic portal system to be reused or eliminated from the body.

16

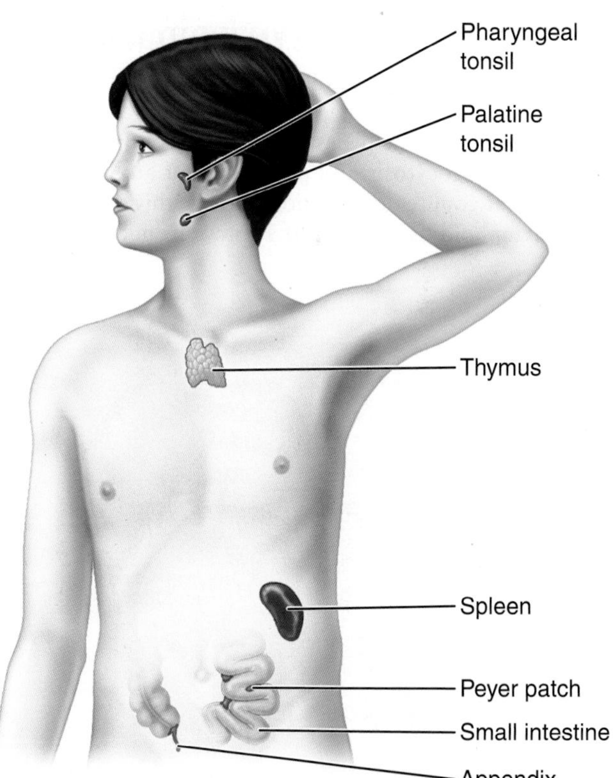

Figure 16-5 Location of lymphoid organs. **KEY POINT** In addition to lymphatic vessels and nodes, the lymphatic system includes the thymus, spleen, and mucosa-associated lymphoid tissue (MALT). MALT includes the tonsils, Peyer patches, and appendix.

- Producing red blood cells before birth
- Serving as a reservoir for blood, which can be returned to the bloodstream in case of hemorrhage or other emergency

Splenectomy (sple-NEK-to-me), or surgical removal of the spleen, is usually a well-tolerated procedure. Although the spleen is the body's largest lymphoid organ, other lymphoid tissues can take over its functions. There is evidence that splenectomy carries a risk of developing certain infections, especially in younger patients. Physicians might prescribe prophylactic antibiotics, educate patients about the risk, or perform a partial splenectomy if possible.

THE THYMUS

The **thymus** (THI-mus), located in the superior thorax deep to the sternum, plays a role in immune system development during fetal life and infancy (**see Fig. 16-5**). Certain lymphocytes, T cells, must mature in the thymus gland before they can perform their functions in the immune system (see Chapter 17). Removal of the thymus causes a generalized decrease in the production of T cells as well as a decrease in the size of the spleen and of lymph nodes throughout the body.

The thymus is most active during early life. After puberty, the tissue undergoes changes; it shrinks in size and is gradually replaced by connective tissue and fat.

MUCOSA-ASSOCIATED LYMPHOID TISSUE

In the mucous membranes lining portions of the digestive, respiratory, and urogenital tracts, there are areas of lymphatic tissue that help destroy foreign contaminants. By means of phagocytosis and production of antibodies, substances that counteract infectious agents, this mucosa-associated lymphoid tissue, or **MALT**, prevents microorganisms from invading deeper tissues. Along with the mucosal membranes themselves, MALT is now recognized as an important barrier against infection.

Some of the largest aggregations of MALT are found in the digestive tract and are known as **GALT**, or gut-associated lymphoid tissue. **Peyer** (PI-er) **patches** are clusters of lymphoid nodules located in the mucous membranes lining the distal small intestine. The **appendix** (ah-PEN-diks) is a finger-like tube of lymphatic tissue, measuring approximately 8 cm (3 in) long. It is attached, or "appended," to the first portion of the large intestine (**see Fig. 16-5**). Like the tonsils, the appendix seems to be noticed only when it becomes infected, causing appendicitis. However, it may, like the tonsils, figure in the development of immunity.

TONSILS

The **tonsils** are unencapsulated masses of GALT located in the vicinity of the pharynx (throat). They help protect against ingested or inhaled contaminants (**Fig. 16-6**). The tonsils have deep grooves lined with lymphatic nodules. Lymphocytes attack pathogens trapped in these grooves. The tonsils are located in the following three areas:

- The **palatine** (PAL-ah-tine) **tonsils** are oval bodies located at each side of the soft palate. These are generally what is meant when one refers to "the tonsils."
- The single **pharyngeal** (fah-RIN-je-al) **tonsil** is commonly referred to as the **adenoid** (AD-eh-noyd) (a general term that means "glandlike"). It is located behind the nose on the posterior wall of the upper pharynx.
- The **lingual** (LING-gwal) **tonsils** are little mounds of lymphoid tissue at the posterior of the tongue.

Any of these tonsils may become so loaded with bacteria that they become reservoirs for repeated infections, and their

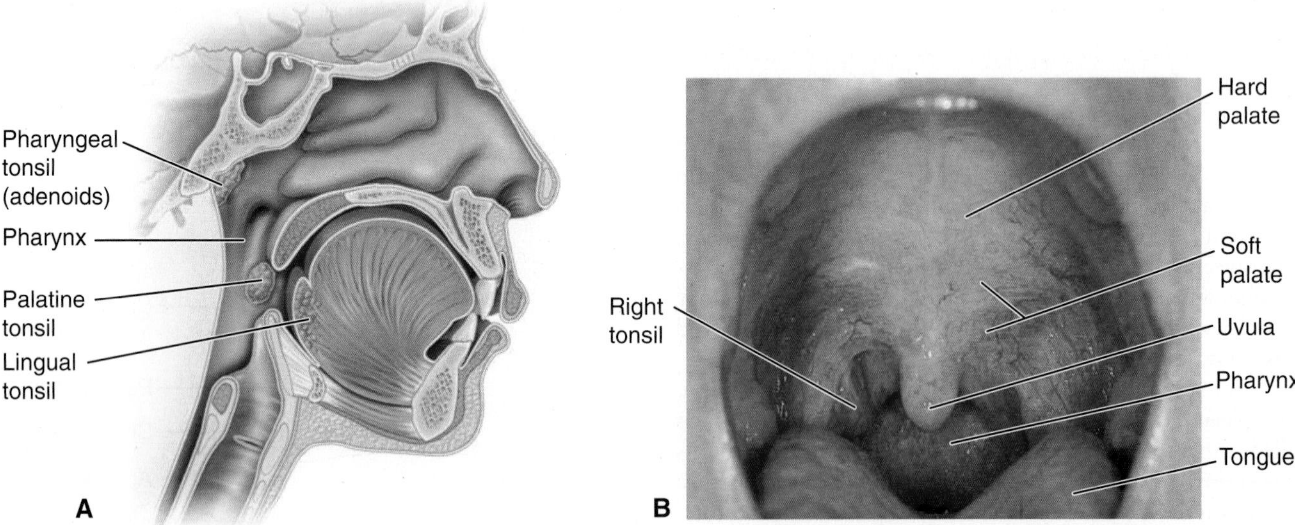

Figure 16-6 **The tonsils.** **KEY POINT** All of the tonsils are in the vicinity of the pharynx (throat) where they filter impurities. **A.** The tonsils in a sagittal section. **B.** Healthy tonsils in a healthy adult.

removal is advisable. In children, a slight enlargement of any of them is not an indication for surgery, however, because all lymphoid tissue masses tend to be larger in childhood. A physician must determine whether these masses are abnormally enlarged taking the patient's age into account, because the tonsils are important in immune function during early childhood. Surgery may be advisable in cases of recurrent infection or if the enlarged tonsils make swallowing or breathing difficult. Their removal may also help children suffering from otitis media, because bacteria infecting the tonsils may travel to the middle ear. The surgery to remove the palatine tonsils is a tonsillectomy; an adenoidectomy is removal of the adenoid. Often these two procedures are done together and abbreviated as T&A. Most tonsillectomies are performed by electrocautery, which uses an electric current to burn the tissue away. A newer technique, which allows faster recovery and fewer complications, uses radio waves to break down the tonsillar tissue.

CHECKPOINTS

- 16-4 What is the function of the lymph nodes?
- 16-5 What does the spleen filter?
- 16-6 What kind of immune system cells develop in the thymus?
- 16-7 What is the meaning of the acronyms MALT and GALT?
- 16-8 What is the general location of the tonsils?

Disorders of the Lymphatic System

Conditions commonly affecting the lymphatic system include infection, edema, and tumors.

DISORDERS RELATED TO INFECTION

Lymphadenopathy (lim-fad-en-OP-ah-the) is a term meaning "disease of the lymph nodes." Although enlarged lymph nodes can result from cancer (discussed shortly), they are more commonly caused by infectious disease. For example, generalized lymphadenopathy is an early sign of infection with HIV (human immunodeficiency virus), the virus that causes AIDS (acquired immunodeficiency syndrome).

In **lymphadenitis** (lim-fad-en-I-tis), or inflammation of the lymph nodes, the nodes become enlarged and tender. This condition reflects the body's attempt to combat an infection. Cervical lymphadenitis occurs during measles, scarlet fever, septic sore throat, diphtheria, and, frequently, the common cold. Chronic lymphadenitis may be caused by the bacillus that causes tuberculosis. Infections of the upper extremities cause enlarged axillary nodes; infections of the external genitals or the lower extremities may cause enlargement of the inguinal lymph nodes.

Infectious mononucleosis (mon-o-nu-kle-O-sis) ("mono") is an acute viral infection characterized by marked cervical lymphadenitis. Other possible symptoms are sore throat, fever, fatigue, weight loss, and enlarged spleen. The causative agent is Epstein-Barr virus (EBV), a type of herpes virus, which is frequently spread by saliva. The virus infects B lymphocytes causing overgrowth and abnormality (**Fig. 16-7**). The disease's name refers to the fact that it involves a type of white blood cell with a single nucleus (in contrast to an agranular leukocyte, which has a segmented nucleus). As noted in the case study, mononucleosis is fairly common among adolescents and young adults, who may refer to their enlarged lymph nodes as "swollen glands." However, lymph nodes do not produce secretions and are not glands.

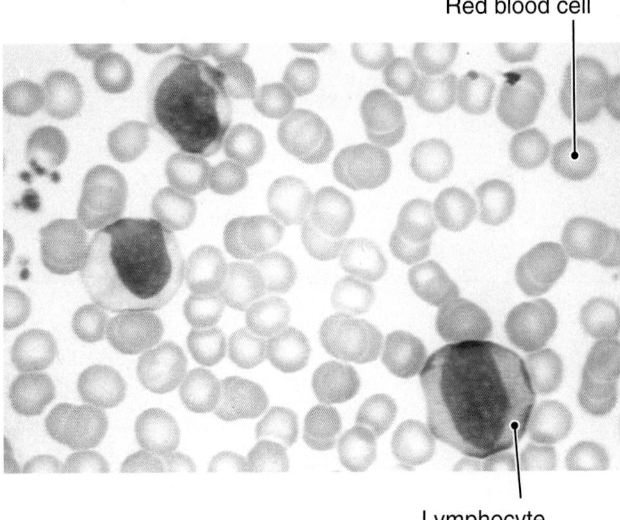

Figure 16-7 **Infectious mononucleosis.** **KEY POINT** Overgrowth of atypical lymphocytes is characteristic of infectious mononucleosis.

Lymphangitis (lim-fan-JI-tis), which is inflammation of lymphatic vessels, usually begins in the region of an infected and neglected injury and can be seen as red streaks extending along an extremity (**Fig. 16-8A**). Such inflamed vessels are a sign that bacteria have spread into the lymphatic system. If the lymph nodes are not able to stop the infection, pathogens may enter the bloodstream, causing **septicemia** (sep-tih-SE-me-ah), or blood poisoning. Streptococci are often the invading organisms in such cases.

See the Student Resources on thePoint for a picture of cervical lymphadenopathy.

LYMPHEDEMA

Edema is tissue swelling due to excess fluid. The condition has a variety of causes, but edema due to obstruction of lymph flow is called **lymphedema** (lim-feh-DE-mah) (**Fig. 16-8B**). Possible causes of lymphedema include infection of the lymphatic vessels, a malignant growth that obstructs lymph flow, or loss of lymphatic vessels and nodes as a result of injury or surgery. Areas affected by lymphedema are more prone to infection because the lymphatic system's filtering activity is diminished. Mechanical

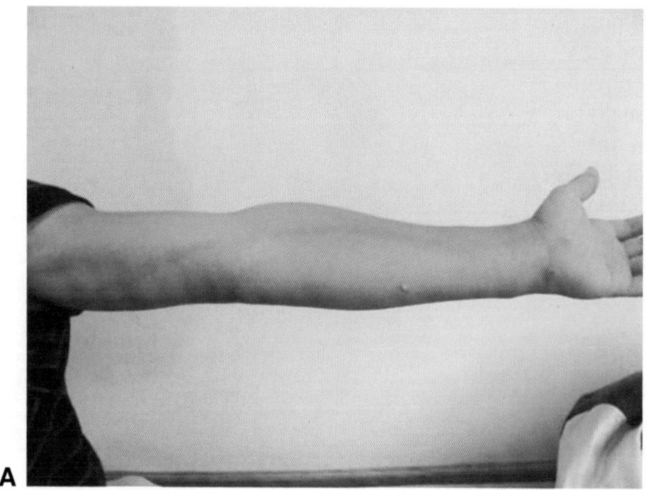

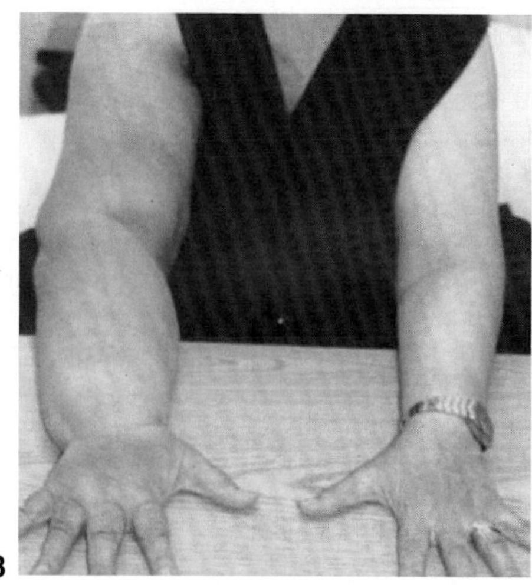

Figure 16-8 **Lymphatic disorders. A.** Lymphangitis is inflammation of lymphatic vessels. Note the *linear red streak* proximal to a skin infection. **B.** Lymphedema of the right upper extremity subsequent to the removal of axillary lymph nodes.

methods to improve drainage and drugs to promote water loss are possible treatments for lymphedema (see Box 16-2).

As illustrated in Figure 5-14, **elephantiasis** (lymphatic filariasis) is a pronounced enlargement of the lower extremities resulting from lymphatic vessel blockage by small worms called filariae (fi-LA-re-e). These tiny parasites, carried by insects such as flies and mosquitoes, invade the tissues as larvae or immature forms. They then grow in the lymph channels and obstruct lymphatic flow. The swelling of the legs or, as sometimes happens in men, the scrotum, may be so great that the victim becomes incapacitated. This disease is especially common in certain parts of Asia and in some of the Pacific islands. No cure is known.

SPLENOMEGALY

Enlargement of the spleen, known as **splenomegaly** (spleno-MEG-ah-le), accompanies certain acute infectious diseases, including infectious mononucleosis (as previously noted), scarlet fever, typhus fever, typhoid fever, and syphilis. Many tropical parasitic diseases cause splenomegaly. A certain blood fluke (flatworm) that is fairly common in Japan and other parts of Asia causes marked splenic enlargement.

Other causes of splenomegaly include liver disease, blood cancer, such as leukemia, metabolic disorders, and hypertension in the splenic or hepatic vein.

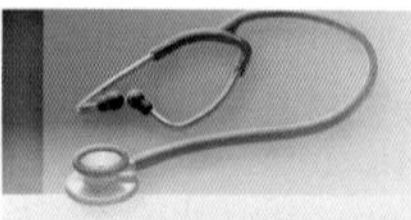

CLINICAL PERSPECTIVES — Box 16-2

Lymphedema: When Lymph Stops Flowing

The body's fluid balance requires appropriate fluid distribution among the cardiovascular system, lymphatic system, and the tissues. **Edema** occurs when the balance is tipped toward excess fluid in the tissues. Often, edema is due to heart failure. However, blockage of lymphatic vessels (and the resulting fluid accumulation in the subcutaneous tissues) can cause another form of edema called lymphedema. The clinical hallmark of lymphedema is chronic swelling of an arm or leg, whereas heart failure usually causes swelling of both legs.

Lymphedema may be either primary or secondary. Primary lymphedema is a rare congenital condition caused by abnormal development of lymphatic vessels. Secondary lymphedema, or acquired lymphedema, can develop as a result of trauma to a limb, surgery, radiation therapy, or lymphangitis. One of lymphedema's most common causes is the removal of axillary lymph nodes during mastectomy (breast removal), which disrupts lymph flow from the adjacent arm. Lymphedema may also occur following prostate surgery.

Therapies that encourage flow through the lymphatic vessels are useful in treating lymphedema. These therapies may include elevation of the affected limb, manual lymphatic drainage through massage, light exercise, and firm wrapping of the limb to apply compression. In addition, changes in daily habits can lessen lymphedema's effects. For example, further blockage of lymph drainage can be prevented by wearing loose-fitting clothing and jewelry, carrying a purse or handbag on the unaffected arm, and sitting with legs uncrossed. Lymphangitis requires the use of appropriate antibiotics. Prompt treatment is necessary because, in addition to swelling, other complications include poor wound healing, skin ulcers, and increased risk of infection.

LYMPHOMA

Lymphoma (lim-FO-mah) is any tumor, benign or malignant, that occurs in lymphoid tissue. Two examples of malignant lymphoma are described next.

Hodgkin lymphoma is a chronic malignant disease of lymphoid tissue, especially the lymph nodes. The median age at the time of diagnosis is 38 years. About 32% of cases occur between 20 and 34 years of age; 12% occur before age 20, and the remainder occur in diminishing percentages throughout the remainder of the life span. The cause of Hodgkin lymphoma is unknown, but in some cases may involve a viral infection. The disease appears as painless enlargement of a lymph node or close group of nodes, often in the neck, but also in the armpit, thorax, and groin. It may spread throughout the lymphatic system and eventually to other systems if not controlled by treatment. Early signs are weight loss, fever, night sweats, fatigue, anemia, and decline in immune defenses. A clear sign of the disease is the presence of Reed-Sternberg cells in lymph node biopsy tissue (**Fig. 16-9**). Chemotherapy and radiation therapy, either separately or in combination, have been used with good results; approximately 90% of patients whose disease is confined to the primary site or has spread only to regional lymph nodes survive for five or more years after treatment, and almost 75% of those with distant metastasis survive.

Non-Hodgkin lymphoma is more common than Hodgkin disease. It appears mostly in older adults and patients with deficient immune systems, such as those with AIDS. Enlargement of the lymph nodes (lymphadenopathy), especially in the cervical region, is an early sign in many cases. It is more widespread through the lymphatic system than Hodgkin disease and spreads more readily to other tissues, such as the liver. Like Hodgkin disease, it may be related to a viral infection. It shares many of the same symptoms as are seen in Hodgkin disease, but there are no Reed-Sternberg cells on biopsy. The current cure rate with chemotherapy and radiation ranges from approximately 80% with localized cases down to just under 60% with distant metastases.

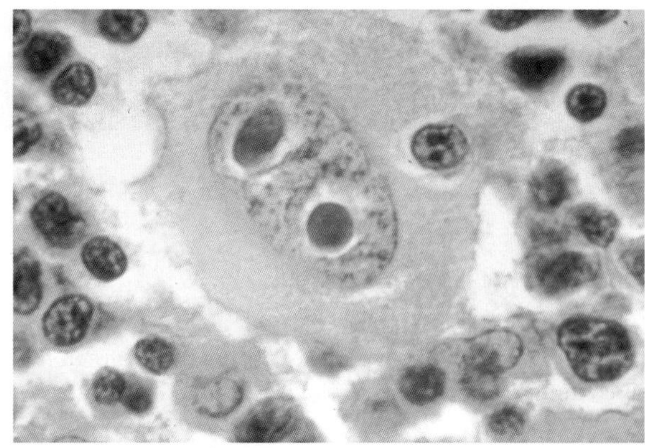

Figure 16-9 Reed-Sternberg cell. **KEY POINT** Reed-Sternberg cells are characteristic of Hodgkin lymphoma. A typical cell has two nuclei with large, dark-staining nucleoli.

16

CHECKPOINTS

- ☐ 16-9 What is lymphadenopathy?
- ☐ 16-10 What condition results from blockage of lymph flow?
- ☐ 16-11 What is the medical term for enlargement of the spleen?
- ☐ 16-12 What is lymphoma, and what are the two examples of malignant lymphoma?

Disease in Context Revisited

Lucas's Mononucleosis

Lucas's throat culture came back positive for streptococcal pharyngitis. His blood tests showed leukocytosis, with an increase in the number of lymphocytes. Lucas also tested positive in the monospot test for antibodies to the Epstein-Barr virus (EBV), the usual cause of infectious mononucleosis. That result in combination with the pharyngitis, lymphadenopathy, splenomegaly, fever, fatigue, and loss of appetite all pointed to mono. Dr. Fischer told Lucas that this illness is frequently spread through contact with oral secretions and that sharing the new band member's tuba might not have been a good idea.

"We can treat your symptoms and offer support, but there is no cure for mono. We just have to let it run its course" Dr. Fischer said. "That means you need to get lots of rest, and you will need to drink lots of fluids."

Lucas was disappointed when he learned that he could not participate in marching band practice for a while. Dr. Fischer explained that Lucas would be tired for a couple more weeks, but that he should recover completely and most likely be able to participate in the band competitions the following month.

CHAPTER 16

Chapter Wrap-Up

Summary Overview

A detailed chapter outline with space for note taking is on *thePoint*. The figure below illustrates the main topics covered in this chapter.

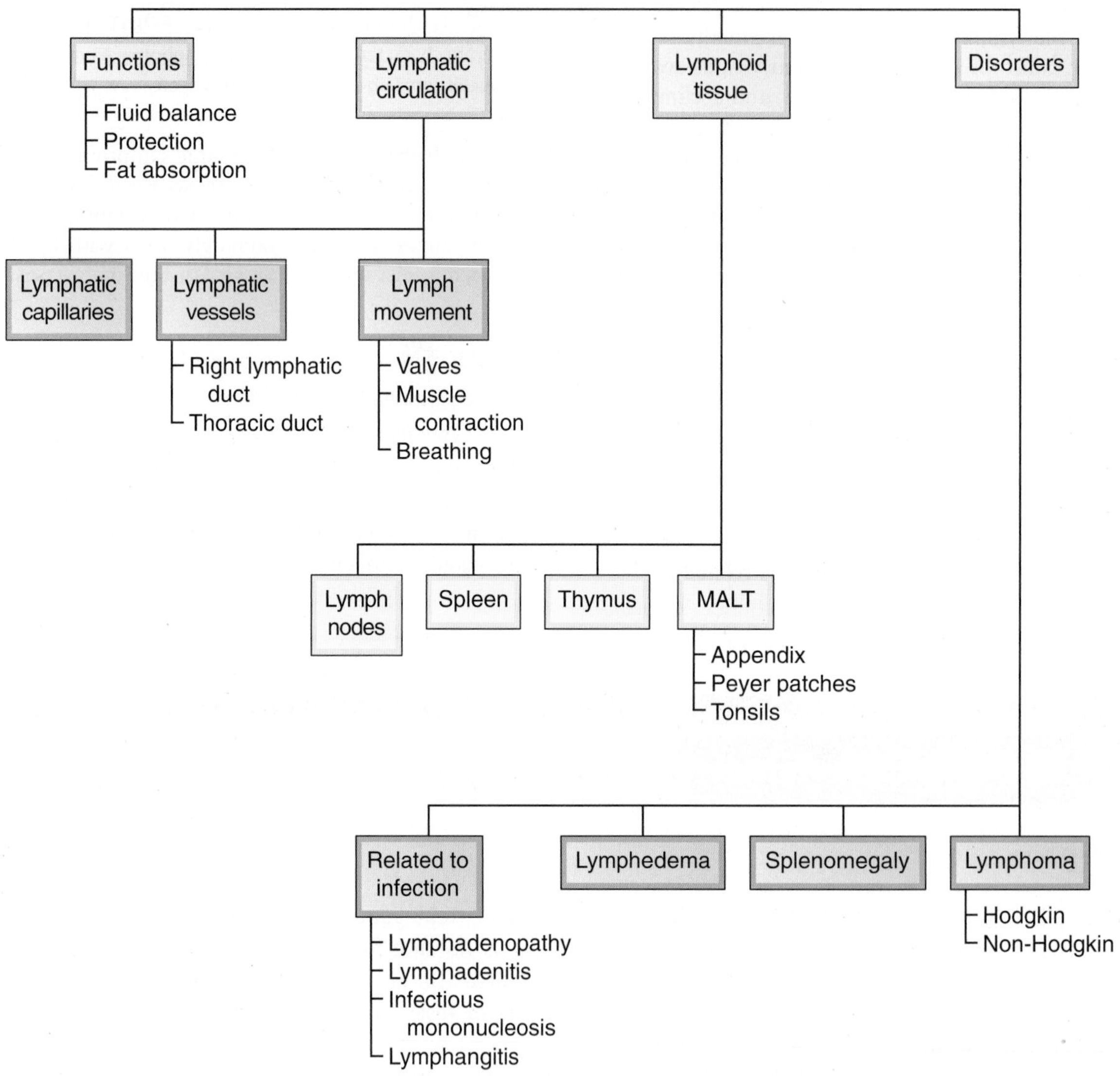

Key Terms

The terms listed below are emphasized in this chapter. Knowing them will help you organize and prioritize your learning. These and other boldface terms are defined in the Glossary with phonetic pronunciations.

adenoid
chyle
GALT
lymph
lymphadenitis
lymphadenopathy
lymphangitis
lymphatic duct
lymph node
lymphoma
MALT
spleen
thymus
tonsil

Word Anatomy

Medical terms are built from standardized word parts (prefixes, roots, and suffixes). Learning the meanings of these parts can help you remember words and interpret unfamiliar terms.

WORD PART	MEANING	EXAMPLE
Lymphoid Tissue		
aden/o	gland	The *adenoids* are glandlike tonsils.
lingu/o	tongue	The *lingual* tonsils are at the back of the tongue.
-oid	like, resembling	*Lymphoid* tissue makes up the specialized organs of the lymphatic system.
Disorders of the Lymphatic System		
-megaly	excessive enlargement	*Splenomegaly* is excessive enlargement of the spleen.
-pathy	any disease	*Lymphadenopathy* is any disease of the lymph nodes.

Questions for Study and Review

BUILDING UNDERSTANDING

Fill in the Blanks

1. The fluid that circulates in the lymphatic system is called ______.

2. Digested fats enter the lymphatic circulation through vessels called ______.

3. Digested fats and lymph combine to form a milky fluid called ______.

4. Enlargement of the spleen is termed ______.

5. When filariae block lymphatic vessels, they cause the disease called ______.

Matching > Match each numbered item with the most closely related lettered item.

___ **6.** Fluid retention due to obstruction of lymph vessels

___ **7.** Inflammation of lymph nodes

___ **8.** Inflammation of lymphatic vessels

___ **9.** Tumor that occurs in lymphoid tissue

a. lymphadenitis
b. lymphangitis
c. lymphedema
d. lymphoma

Multiple Choice

___ **10.** Compared to plasma, lymph contains much less
- **a.** fat
- **b.** carbohydrate
- **c.** protein
- **d.** water

___ **11.** Which vessel returns lymph from the lower extremities to the cardiovascular system?
- **a.** appendix
- **b.** lacteal
- **c.** right lymphatic duct
- **d.** thoracic duct

___ **12.** Which tonsil is located behind the nose on the posterior wall of the upper pharynx?
- **a.** appendix
- **b.** lingual
- **c.** palatine
- **d.** pharyngeal

___ **13.** What is the hallmark clinical sign of infectious mononucleosis?
- **a.** edema
- **b.** lymphadenopathy
- **c.** lymphangitis
- **d.** splenomegaly

UNDERSTANDING CONCEPTS

14. How does the structure of lymphatic capillaries correlate with their function? List some differences between lymphatic and blood capillaries.

15. Describe three mechanisms that propel lymph through the lymphatic vessels.

16. Trace a globule of fat from a lacteal in the small intestine to the right atrium.

17. Describe the structure of a typical lymph node.

18. Name two examples of GALT.

19. List three disorders of the lymphatic system related to infection.

20. Based on the opening case study, list the symptoms of infectious mononucleosis.

21. Describe two forms of lymphoma.

CONCEPTUAL THINKING

22. Explain the absence of arteries in the lymphatic circulatory system.

23. If the spleen is severely damaged, as in an accident, its removal might become necessary. What is the name for this type of surgery, and what possible dangers might that cause?

24. Name several locations of MALT, and explain the significance of those locations.

For more questions, see the Learning Activities on thePoint.

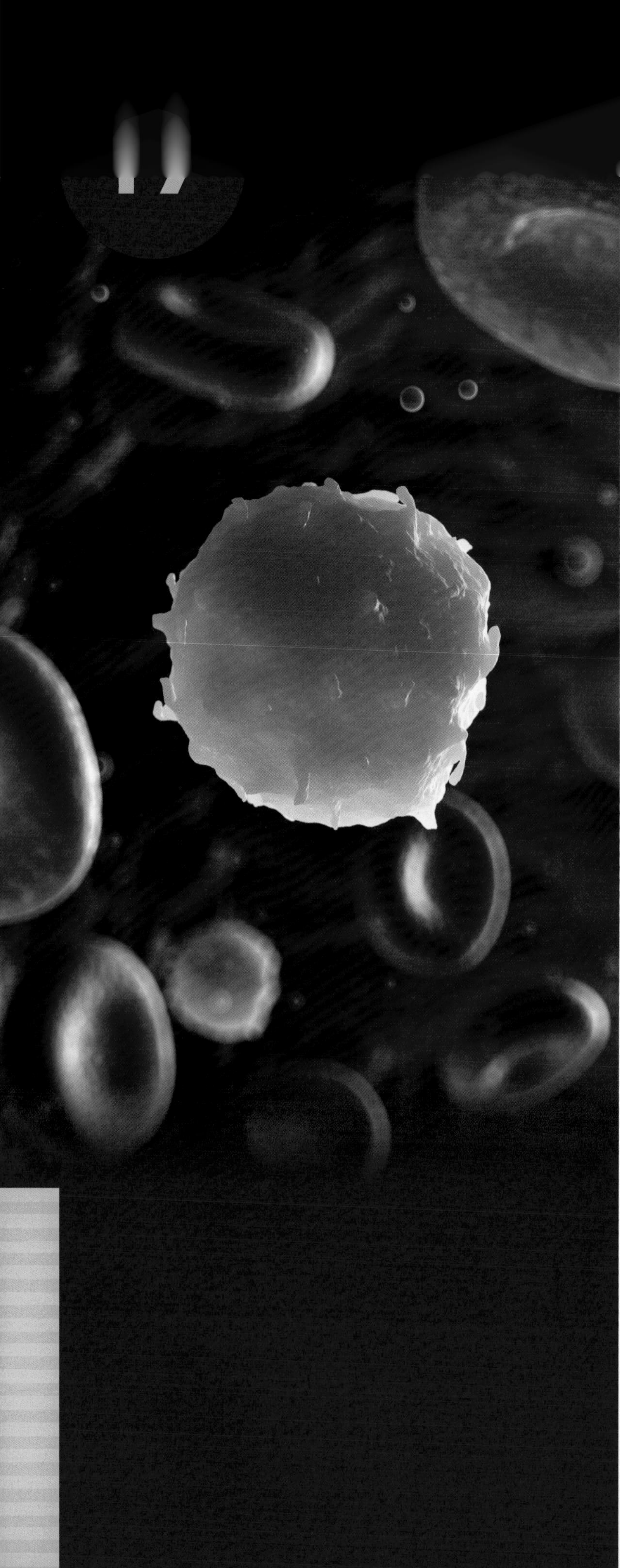

Learning Objectives

After careful study of this chapter, you should be able to:

1 ▶ List four factors that determine the occurrence of infection. ***p. 384***

2 ▶ Differentiate between innate and adaptive immunity, and give examples of each. ***p. 384***

3 ▶ Name three types of cells and three types of chemicals active in the second line of defense against disease. ***p. 385***

4 ▶ Briefly describe the inflammatory reaction. ***p. 387***

5 ▶ Define *antigen* and *antibody*. ***p. 388***

6 ▶ Compare and contrast T cells and B cells with respect to development and type of activity. ***p. 388***

7 ▶ Describe the activities of four types of T cells. ***p. 388***

8 ▶ Explain the role of antigen-presenting cells in adaptive immunity. ***p. 389***

9 ▶ Differentiate between natural and artificial adaptive immunity. ***p. 391***

10 ▶ Differentiate between active and passive immunity. ***p. 391***

11 ▶ Define the term *vaccine*, and give three examples of vaccine types. ***p. 391***

12 ▶ Define the term *antiserum*, and give five examples of antisera. ***p. 394***

13 ▶ Discuss three types of immune disorders. ***p. 395***

14 ▶ Explain the role of the immune system in preventing cancer. ***p. 397***

15 ▶ Explain the role of immunity in tissue transplantation. ***p. 397***

16 ▶ Based on the case study, describe the causes and symptoms of the autoimmune disorder rheumatoid arthritis. ***pp. 383, 399***

17 ▶ Show how word parts are used to build words related to immunity (see Word Anatomy at the end of the chapter). ***p. 401***

Disease in Context *Meredith Learns about Her Immune System*

Meredith, a 46-year-old high school mathematics teacher, read a snowboarding magazine in the waiting room at the local health center. An avid winter athlete, she was no longer able to keep up with her children on the mountain because she felt so tired and achy. At her family's urging, she had made the appointment with her primary care physician.

Dr. Bedell met her in the examination room. "What brings you in today, Meredith?"

"I ache all over," she replied. "I'm so stiff in the morning, it's difficult to get up and going, even on vacation. And my hands and wrists hurt most of the time. It's even hard to hold the chalk and write on the board at school. I've noticed that my hands and wrists have been a little swollen lately. What's going on with me? It feels like arthritis, but I think I'm too young for that. My mother didn't get arthritis until she was in her late 60s!"

The doctor conducted a physical examination and noted that areas over Meredith's wrists, elbows, and ankles were warm to touch. He checked the range of motion in her joints and found that mobility was reduced in her shoulders and knees. Meredith flinched when he manipulated her wrists and hands, saying she felt pain. Her vital signs were normal except for her temperature, which was elevated slightly. Meredith's history of joint pain was leading Dr. Bedell to think she might have rheumatoid arthritis (RA).

"Your symptoms are suspect of an autoimmune condition called rheumatoid arthritis," he said. "This disorder affects women more frequently than men and may start as early as 30 to 50 years of age. I would like to obtain x-rays of certain joints and also have some blood drawn for lab tests."

He went on to explain that RA is a systemic inflammatory disease resulting from antibodies produced against a person's own tissues. Checking her blood for these antibodies, or immunoglobulins (Ig), could help lead to a diagnosis. In addition to joint pain, the disorder can cause a low-grade fever and sometimes altered blood counts. Dr. Bedell also looked at Meredith's hemoglobin levels to check for anemia, which would account for her fatigue.

Later we'll follow up with Meredith's test results and see if the diagnosis of rheumatoid arthritis was confirmed. In this chapter, we will see how the immune system defends us against disease, but can also lead to disorders such as allergy and autoimmunity.

ANCILLARIES *At-A-Glance*

Visit thePoint to access the following resources. For guidance in using these resources most effectively, see pp. xv–xvii.

Learning RESOURCES

- Tips for Effective Studying
- Web Figure: Chain of Events in Inflammation
- Web Figure: Phases of HIV Infection and AIDS
- Web Figure: Clinical and Pathologic Features of AIDS
- Animation: Acute Inflammation
- Animation: Immune Response
- Health Professions: Nurse Practitioner
- Detailed Chapter Outline
- Answers to Questions for Study and Review
- Audio Pronunciation Glossary

Learning ACTIVITIES

- Pre-Quiz
- Visual Activities
- Kinesthetic Activities
- Auditory Activities

A LOOK BACK

The lymphatic system plays an important role in immunity, as just described in Chapter 16. In Chapter 3, we discussed phagocytosis as one of the methods by which materials enter cells. Phagocytosis is a critical part of immunity. Now is also a good time to review Chapter 5 and the relevant Student Resources as a review of disease organisms and the disease process.

Chapter 5 presents a rather frightening list of harmful organisms that surround us in our environment. Fortunately, most of us survive contact with these invaders and even become more resistant to disease in the process. All of the defenses that protect us against disease constitute **immunity**. These defenses protect us against any harmful agent that enters the body, such as an infectious organism. They also protect us against abnormal cells that arise within the body, such as tumor cells. Although immune defenses are critical to protection from cancer, this chapter concentrates on immunity as it applies to invasion by infectious organisms. Some immune defenses combat any foreign agent, while others are effective against a particular pathogen type or a specific pathogen. Scientists use the term *immune system* to describe all the cells and tissues that protect us against foreign organisms or any cells different from our own normal cells.

Why Do Infections Occur?

Although the body is constantly exposed to pathogenic invasion, many conditions determine whether an infection will actually occur. Pathogens have a decided preference for certain body tissues and must have access to these tissues to establish an infection. The polio virus, for example, may be inhaled or swallowed in large numbers and may encounter the mucous membranes lining the respiratory and digestive tracts. However, it causes no apparent disorder of these tissues but goes on to attack only nervous tissue. In contrast, the viruses that cause influenza and the common cold do attack the respiratory mucous membranes. Human immunodeficiency virus (HIV), the virus that causes acquired immunodeficiency syndrome (AIDS), attacks a certain type of T cell (T lymphocyte), which has surface receptors for the virus.

The **portal of entry** is an important condition influencing infection. The respiratory tract is a common entrance route for pathogens. Other important entry points include the digestive system and the tubes that open into the urinary and reproductive systems. Any break in the skin or in a mucous membrane allows organisms, such as staphylococci, easy access to deeper tissues and may lead to infection, whereas unbroken skin or mucous membranes are usually not affected. As discussed in Chapter 16, mucosal-associated lymphoid tissue (MALT) helps protect mucous membranes.

The **virulence** (VIR-u-lens) of an organism, or the organism's power to overcome its host's defenses, is another important factor. Virulence has two aspects: one may be thought of as "aggressiveness," or invasive power; the other is the organism's ability to produce **toxins** (poisons) that damage the body. Not only do different microbial species vary in virulence, but different strains of the same organism can vary in this regard as well. The strains of influenza virus circulating in a given year can be more or less dangerous than those prevalent in other years. Also, organisms may gain virulence as they pass from one host to another.

The **dose** (number) of pathogens that invade the body is also a determining factor in whether an infection develops. Even if the virulence of a particular organism happens to be low, infection may occur if a large number enter the body.

Finally, an individual's condition is important. Disease organisms are around us all the time. Why does one person rarely get a cold, flu, or other infection, whereas another person gets several each year? Part of the answer lies in the person's predisposition to disease, as influenced by genetic makeup, general physical and emotional health, nutrition, living habits, and age. We noted in Chapter 5 that an infection may take hold because the host has been compromised (weakened) by a disease, such as AIDS, or by medical treatment that depresses the immune system, such as cancer chemotherapy.

CHECKPOINT

☐ 17-1 What term describes the ability of an organism to overcome host defenses?

Innate Immunity

The features that protect the body against disease are usually considered successive "lines of defense" beginning with the relatively simple outer barriers and proceeding through progressively more complicated, internal responses. Our defenses can be categorized as *innate* or *adaptive*. Innate defenses are inborn, that is, they are inherited along with all of a person's other characteristics. They include barriers and certain internal cellular and metabolic responses that protect against any foreign invaders or harmful substances. Innate responses are very rapid, are relatively nonspecific, and can help prevent or slow infections. Adaptive defenses, discussed later, develop *after* exposure to a particular pathogen and are specific to that pathogen. Adaptive defenses are slower but can completely eliminate a specific pathogen from the body as well as prevent future infections.

THE FIRST LINE OF DEFENSE: INNATE BARRIERS

The first defense against invading organisms includes chemical and mechanical barriers, such as the following (**Fig. 17-1**):

- The skin serves as a mechanical barrier as long as it remains intact. A serious danger to burn victims, for example, is the risk of infection resulting from skin destruction.
- The mucous membranes that line the passageways leading into the body also act as barriers, trapping foreign

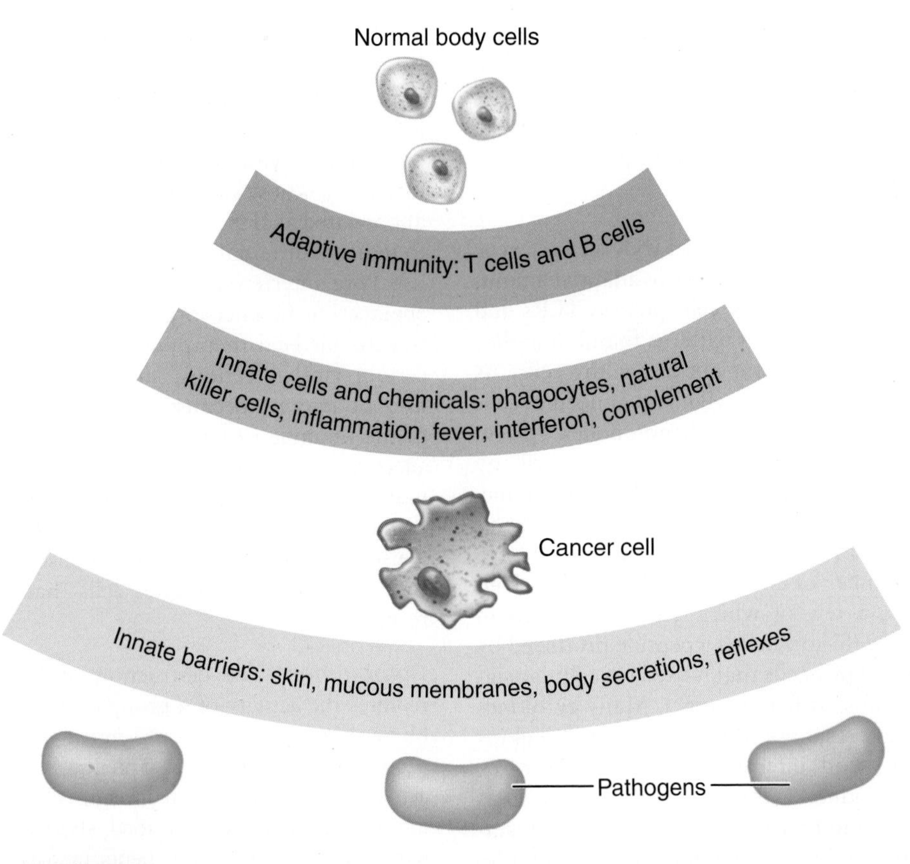

Figure 17-1 Lines of defense. **KEY POINT** Physical barriers are the first line of defense against pathogens. The innate and adaptive arms of the immune system are the second and third lines.

material in their sticky secretions. The cilia in membranes in the upper respiratory tract help sweep impurities out of the body.

- Body secretions, such as tears, perspiration, and saliva, wash away microorganisms and may contain acids, enzymes, or other chemicals that destroy invaders. Digestive juices destroy many ingested bacteria and their toxins.
- Certain reflexes aid in the removal of pathogens. Sneezing and coughing, for instance, tend to remove foreign matter including microorganisms from the upper respiratory tract. Vomiting and diarrhea are ways in which ingested toxins and bacteria may be expelled.

THE SECOND LINE OF DEFENSE: INNATE CELLS AND CHEMICALS

If an organism has overcome initial defenses, we have a number of internal activities that constitute a second line of defense (**see Fig. 17-1**). Although we present them in a separate category, you will see later in this chapter that many participate in and promote adaptive immune responses. Innate immunity was previously described as *nonspecific immunity*, but we now know that bacteria, viruses, and cancer cells induce distinct innate responses.

For many years, scientists did not know how the cells of the innate immune system were able to differentiate between body cells and foreign cells. Recently, it was discovered that innate immune cells possess a specific class of receptors called **toll-like receptors (TLRs)**. (*Toll* means "great" in German.) Different TLRs recognize different types of pathogens. For instance, one TLR recognizes bacterial cell walls, whereas other TLRs recognize characteristics of viral DNA. While TLRs cannot identify specific pathogens, they can identify specific patterns that strongly suggest that a cell is "nonself," and they can also distinguish between different pathogen types. These harmful cells can then be eliminated by various pathways, including phagocytosis.

Phagocytosis In the process of **phagocytosis**, white blood cells take in and destroy waste (including worn-out cells), cancer cells, and pathogens. (See **Fig. 13-5** in Chapter 13.) Neutrophils, a category of granular leukocytes, are important phagocytic white blood cells. (See **Table 13-2** in Chapter 13.) Another active phagocyte is the **macrophage** (MAK-ro-faj). (The name *macrophage* means "big eater.") Macrophages are large white blood cells derived from monocytes, a type of agranular leukocyte. Monocytes develop into macrophages upon entering the tissues. Some macrophages remain fixed in the tissues, for example, in the skin, liver, lungs, lymphoid tissue, bone marrow, and soft connective tissue throughout the body. In some organs, macrophages are given special names. For example, Kupffer (KOOP-fer) cells are macrophages located in the lining of the liver sinusoids (blood channels). In the lungs, where they ingest solid particles, macrophages are called *dust cells*.

Natural Killer Cells The **natural killer (NK) cell** is a type of lymphocyte different from those active in adaptive immunity, which are described later. NK cells possess TLRs and other receptors that detect antigen patterns found in pathogens and cancerous body cells but not in healthy cells. As the name indicates, NK cells kill foreign and abnormal cells on contact. NK cells are found in the lymph nodes, spleen, bone marrow, and blood. They destroy abnormal cells by secreting a protein that breaks down the plasma membrane.

Cytokines and Other Chemicals Scientists have identified a huge number of chemical mediators involved in innate immunity, a few of which are summarized in Table 17-1. A **cytokine** (SI-to-kine) is a peptide produced by immune cells or other body cells that is used for cellular signaling (the name means "cell activator"). Many cytokines modulate the immune response; these include **interleukins** (in-ter-LU-kinz) (IL) and **interferons** (in-ter-FERE-onz). Immune cells also produce many noncytokine molecules that activate innate immune responses, including histamine (HIS-tah-mene) and prostaglandins. Finally, **complement** (KOM-ple-ment) is a group of proteins that circulate in blood in inactive forms. They are activated by tissue damage or the presence of pathogens. These chemicals work together with immune cells to promote *inflammation* and *fever*, two generalized innate immune responses that fight pathogens. Before we explain these important processes, we must first discuss interferons and complement in greater detail.

Interferons Certain cells infected with a virus release a substance that prevents nearby cells from producing more viruses. This substance was first found in cells infected with influenza virus, and it was called "interferon" because it "interferes" with multiplication and spread of the virus. Interferon is now known to be a group of substances. Each is abbreviated as IFN with a Greek letter, alpha (α) or beta (β), to indicate different categories.

Pure interferons are now being produced by genetic engineering in microorganisms, making adequate quantities for medical therapy available. They are used to treat certain viral infections, such as hepatitis. Interferons are also of interest because they act nonspecifically on cells of the immune system. They have been used with varying success to boost the immune response in the treatment of malignancies, such as melanoma, leukemia, and Kaposi sarcoma, a cancer associated with AIDS. Interestingly, IFN-β is used to treat the autoimmune disorder multiple sclerosis (MS), because it stimulates cells that depress the immune response.

Complement The destruction of foreign cells sometimes requires the activity of a group of nonspecific proteins in the blood, together called complement (Fig. 17-2). Complement proteins are always present in the blood, but they must be activated by contact with foreign cell surfaces or by specific immune complexes (described shortly). Complement is so named because it complements (assists with) immune reactions. Some of complement's actions are to

- Bind to foreign cells to help phagocytes recognize and engulf them (Fig. 17-2A)
- Destroy cells by forming channels called **membrane attack complexes** (MACs) in pathogens' membranes; water enters the pathogen, resulting in cell rupture (Fig. 17-2B)

Table 17-1 Selected Chemicals of Innate Immunity[a]

Chemical	Producing Cells	Actions
Cytokines		
Interleukins	Activated macrophages and other immune cells	Various actions; some promote inflammation and fever
Interferons (alpha and beta)	Released by virus-infected cells	Block viruses from infecting other cells; activate natural killer cells
Noncytokines		
Histamine	Mast cells	Dilates arterioles and increases capillary leakiness, resulting in redness and swelling
Prostaglandins	Mast cells, neutrophils, and other immune cells	Similar effects as histamine; activate pain receptors
Complement	Synthesized in the liver; found in blood in their inactive forms	Activates mast cells; lyses pathogens; stimulates phagocytosis

[a]Note that many of these chemicals are also active in adaptive immunity.

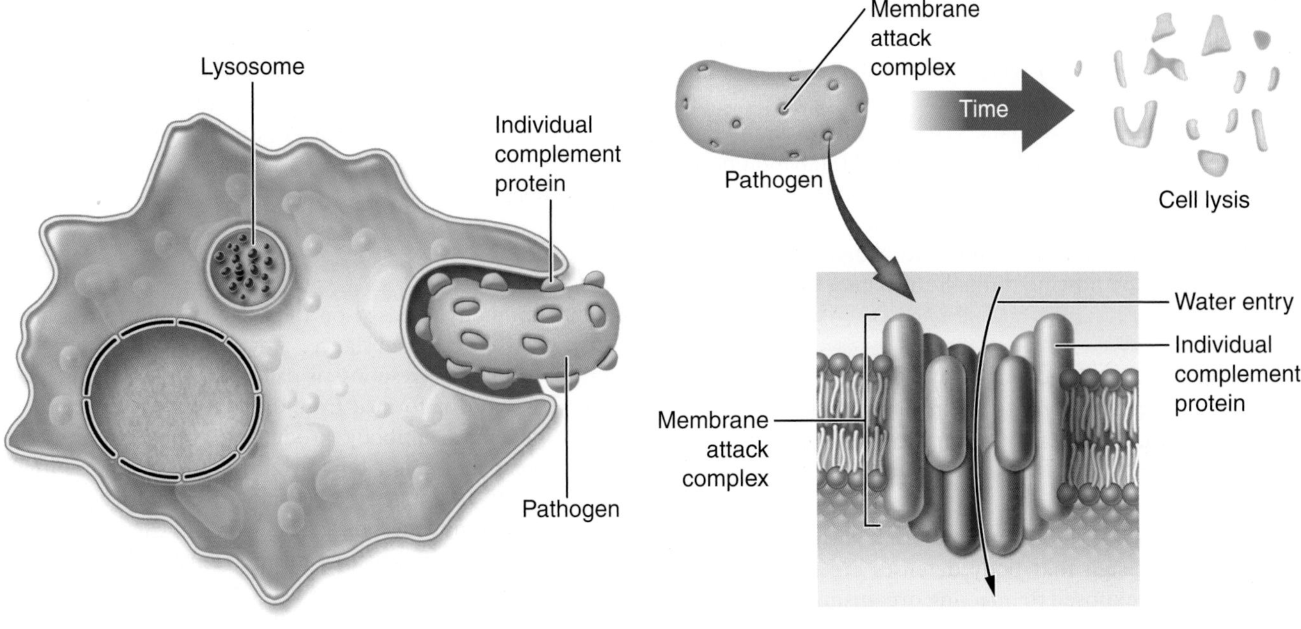

Figure 17-2 Complement. **KEY POINT** Complement participates in the destruction of pathogens. **A.** Complement proteins attract phagocytes to an area of inflammation and help them attach to and engulf a foreign organism. The organism is enclosed in a phagocytic vesicle, which then merges with a lysosome. Lysosomal enzymes digest the organism. **B.** A complex of specific complement proteins forms the membrane attack complex.

- Promote inflammation by increasing capillary permeability
- Attract phagocytes to an area of inflammation

Inflammation **Inflammation** is a nonspecific defensive response to a tissue-damaging irritant. Any irritant can cause inflammation: friction, x-rays, fire, extreme temperatures, and wounds, as well as caustic chemicals and contact with allergens. These can all be classified as irritants. Often, however, inflammation results from irritation caused by infection. With the entrance and multiplication of pathogens, a whole series of defensive processes begins (**Fig. 17-3**). This inflammatory reaction is accompanied by four classic symptoms: heat, redness, swelling, and pain, as described below.

Tissue injury or pathogens activate **mast cells**, basophil-like immune cells in tissues that produce histamine. Histamine causes local blood vessels to dilate (widen) and become leaky. Blood plasma leaks out of the vessels into the tissues and begins to clot, thus limiting the spread of infection to other areas. The increased blood flow causes heat and redness.

Pathogens also activate resident immune cells, which release cytokines and other inflammatory chemicals. Some of these chemicals attract leukocytes to the area, especially neutrophils and monocytes. These cells enter the tissue through gaps in the capillary wall, and monocytes convert into macrophages. The resident immune cells and the new arrivals work together to phagocytose pathogens and damaged cells.

The mixture of leukocytes and fluid, the **inflammatory exudate**, contributes to swelling and puts pressure on nerve endings, causing pain. Substances secreted from the activated immune cells, including prostaglandins, cytokines, and histamine, also contribute to the pain of inflammation.

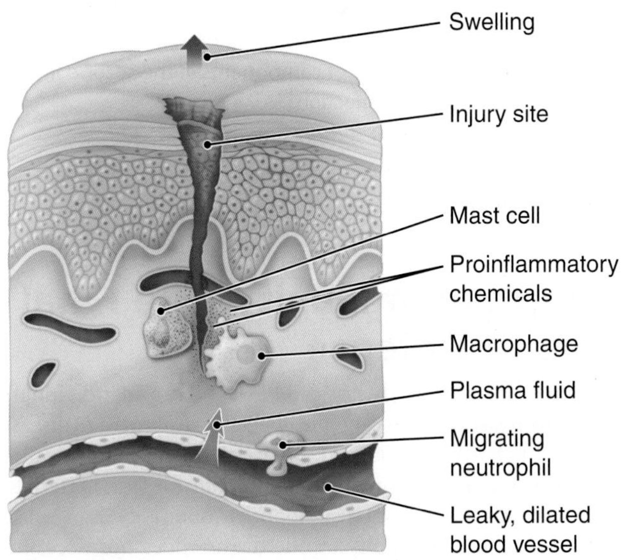

Figure 17-3 Inflammation. **KEY POINT** Acute inflammation involves components in blood and tissues. **ZOOMING IN** What causes the heat, redness, swelling, and pain characteristic of inflammation?

Phagocytes are destroyed in large numbers as they work, and dead cells gradually accumulate in the area. The mixture of exudate, living and dead white blood cells, pathogens, and destroyed tissue cells is called "pus."

Meanwhile, the lymphatic vessels begin to drain fluid from the inflamed area and carry it toward the lymph nodes for filtration. The regional lymph nodes become enlarged and tender, a sign that they are performing their protective function by working overtime to produce phagocytic cells that "clean" the lymph flowing through them (see Fig. 17-1).

See the Student Resources on thePoint for a diagram summarizing the events in inflammation and for the animation "Acute Inflammation."

Fever An increase in body temperature above the normal range can be a sign that body defenses are active. When phagocytes are exposed to infecting organisms, they release substances that raise body temperature. Fever boosts the immune system in several ways. It stimulates phagocytes, increases metabolism, and decreases certain disease organisms' ability to multiply.

A common misperception is that fever is a dangerous symptom that should always be eliminated. Control of fever in itself does little to alter the course of an illness. Healthcare workers, however, should always be alert to fever development as a possible sign of a serious disorder and should recognize that an increased metabolic rate may have adverse effects on a weak patient's heart.

CHECKPOINTS

- ☐ 17-2 What constitutes the first line of defense against the invasion of pathogens?
- ☐ 17-3 What are two types of components in the second line of defense against infection?
- ☐ 17-4 What are the four signs of inflammation?
- ☐ 17-5 What are three ways that fever boosts the immune system?

Adaptive Immunity: The Final Line of Defense

Adaptive immunity to disease can be defined as an individual's power to resist or overcome the effects of a *particular* disease agent or its harmful products. These defense mechanisms also recognize and attack potentially useful, but foreign, materials, such as transplanted organs. Adaptive immunity is a selective process (i.e., immunity to one disease does not necessarily cause immunity to another). This selective characteristic is called **specificity** (spes-ih-FIS-ih-te).

Adaptive immunity is also termed **acquired immunity**, because it develops during a person's lifetime as he or she encounters various specific harmful agents. If the following description of adaptive immunity seems complex, bear in mind that from infancy onward, your immune system is able to protect you from millions of foreign substances, even synthetic substances not found in nature. All the while, the system is kept in check so that it does not usually overreact to produce allergies or mistakenly attack and damage your own body tissues.

ANTIGENS

Specific immunity is based on the body's ability to recognize a particular foreign substance. Any foreign substance that induces an immune response is called an **antigen** (AN-te-jen) (**Ag**). (The word is formed from *anti*body + *gen*, because an antigen generates antibody production.) Most antigens are large protein molecules, but carbohydrates and some lipids also may act as antigens. Normally, only nonself antigens stimulate an immune response. Such antigens can be found on the surfaces of pathogenic organisms, transfused blood cells, transplanted tissues, cancerous cells, and also on pollens, in toxins, and in foods. The critical feature of any substance described as an antigen is that it stimulates the activity of certain lymphocytes classified as T or B cells.

T CELLS

Both T and B cells come from hematopoietic (blood-forming) stem cells in bone marrow, as do all blood cells. The T and B cells differ, however, in their development and their methods of action. Some of the immature stem cells migrate to the thymus and become T cells, which constitute about 80% of the lymphocytes in the circulating blood. While in the thymus, these T lymphocytes multiply and become capable of combining with specific foreign antigens, at which time they are described as *sensitized*. These thymus-derived cells produce an immunity that is said to be **cell-mediated immunity**.

There are several types of T cells, each with different functions. The different types and some of their functions are as follows:

- **Cytotoxic T cells** (T_c) destroy certain abnormal cells directly. They recognize cells infected with viruses or other intracellular pathogens, cancer cells, and foreign antigens present in transplanted tissue. They are able to form pores in the plasma membranes of these cells and insert enzymes that destroy the cell. They also produce substances that cause the cells to "self-destruct" by apoptosis.
- **Helper T cells** (T_h) are essential to the immune response through the release of interleukins. These substances stimulate the production of cytotoxic T cells as well as B cells and macrophages. (Interleukins are so named because they act between white blood cells.) There are several subtypes of helper T cells, one of which is infected and destroyed by the AIDS virus (HIV). The HIV-targeted T cells have a special surface receptor (CD4) to which the virus attaches, as described later.

- **Regulatory T cells** (T_{reg}) suppress the immune response in order to prevent overactivity. These T cells may inhibit or destroy active lymphocytes.
- **Memory T cells** remember an antigen and start a rapid response if that antigen is contacted again.

The T-cell portion of the immune system is generally responsible for defense against cancer cells, certain viruses, and other pathogens that grow within cells (intracellular parasites), as well as for the rejection of tissue transplanted from another person.

Antigen-Presenting Cells T cells cannot respond to foreign antigens directly. Instead, the antigen must be "presented" to them by an antigen-presenting cell (APC). The most important APCs are **dendritic** (den-DRIT-ik) **cells**, large phagocytic cells derived from monocytes or lymphocytes and named for their many fibrous processes. Macrophages and some lymphocytes can also act as APCs. An APC is a processing center for foreign antigens. First, it ingests the foreign material, such as a disease organism, enclosing it in a vesicle. As is typical in phagocytosis, this vesicle then merges with a lysosome filled with digestive enzymes that break down the organism (**Fig. 17-4**). However, the APC then inserts fragments of the foreign antigen into its plasma membrane—in a sense, advertising them to helper T cells. They display the foreign antigen in combination with antigens that a T_h cell can recognize as belonging to the "self." Self-antigens are known as MHC (major histocompatibility complex) antigens because of their importance in cross-matching for tissue transplantation. They are also known as HLAs (human leukocyte antigens), because white blood cells are used in testing tissues for compatibility.

For a T_h cell to react with a foreign antigen, that antigen must be presented to the T_h cell along with the MHC proteins. A special receptor on the T_h cell must bind with both the MHC protein and the foreign antigen fragment (**see Fig. 17-4**). The activated T_h cell then produces various cytokines that stimulate the production and activity of different lymphocytes. Because cytokines stimulate the cells active in immunity, they are used medically to boost the immune system.

CHECKPOINTS

- ☐ 17-6 What is adaptive immunity?
- ☐ 17-7 What is an antigen?
- ☐ 17-8 List four types of T cells.
- ☐ 17-9 What is the role of APCs in immunity?

B CELLS AND ANTIBODIES

B cells (B lymphocytes) are the second main class of lymphocytes active in immunity. Whereas T cells mature in the thymus, B cells mature in the red bone marrow before becoming active in the blood. B cells function in immunity by producing Y-shaped proteins called **antibodies (Ab)**, also known as *immunoglobulins* (Ig), in response to a foreign antigen. (Globulins, in general, are a class of folded proteins described in Chapter 2. Another example of these proteins is hemoglobin.)

B cells have surface receptors that bind with a specific type of antigen (**Fig. 17-5**). Exposure to the antigen stimulates the cells to rapidly multiply and produce large numbers (clones) of **plasma cells**. These mature cells produce antibodies against the original antigen and release them into the blood, providing a form of immunity described as **humoral immunity** (the term *humoral* refers to body fluids). Humoral immunity is long term and generally protects against circulating antigens and bacteria that grow outside the cells

17

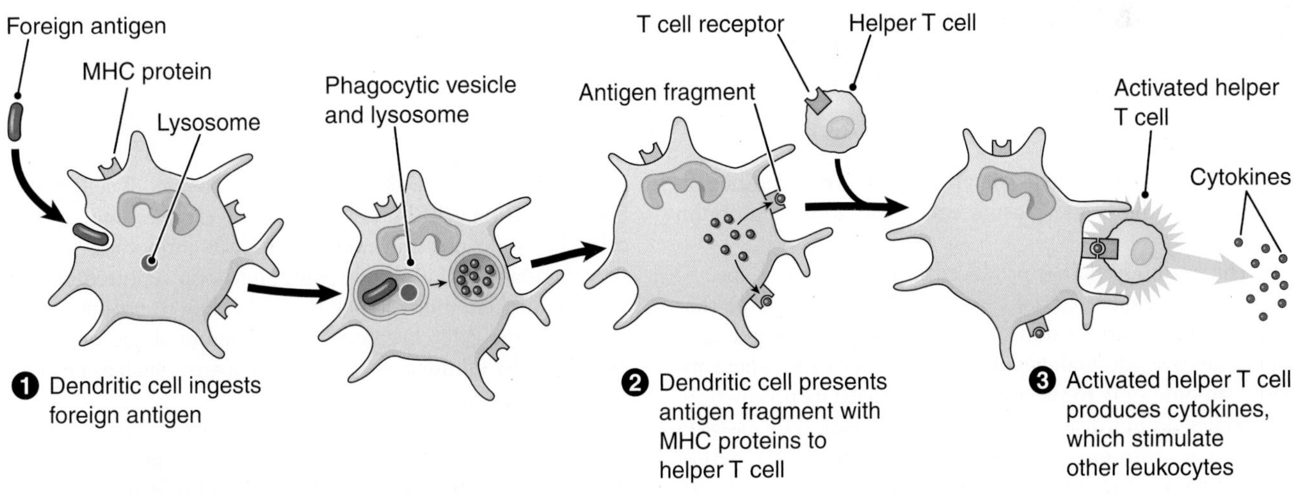
Foreign antigen
MHC protein
Lysosome
Phagocytic vesicle and lysosome
T cell receptor
Helper T cell
Antigen fragment
Activated helper T cell
Cytokines
1 Dendritic cell ingests foreign antigen
2 Dendritic cell presents antigen fragment with MHC proteins to helper T cell
3 Activated helper T cell produces cytokines, which stimulate other leukocytes

Figure 17-4 Activation of a helper T cell by a dendritic cell (antigen-presenting cell, APC). KEY POINT An antigen-presenting cell (APC) displays digested foreign antigen on its surface along with self major histocompatibility complex (MHC) antigen. A helper T cell is activated by contact with this complex and produces stimulatory interleukins. **ZOOMING IN** What is contained in the lysosome that joins the phagocytic vesicle?

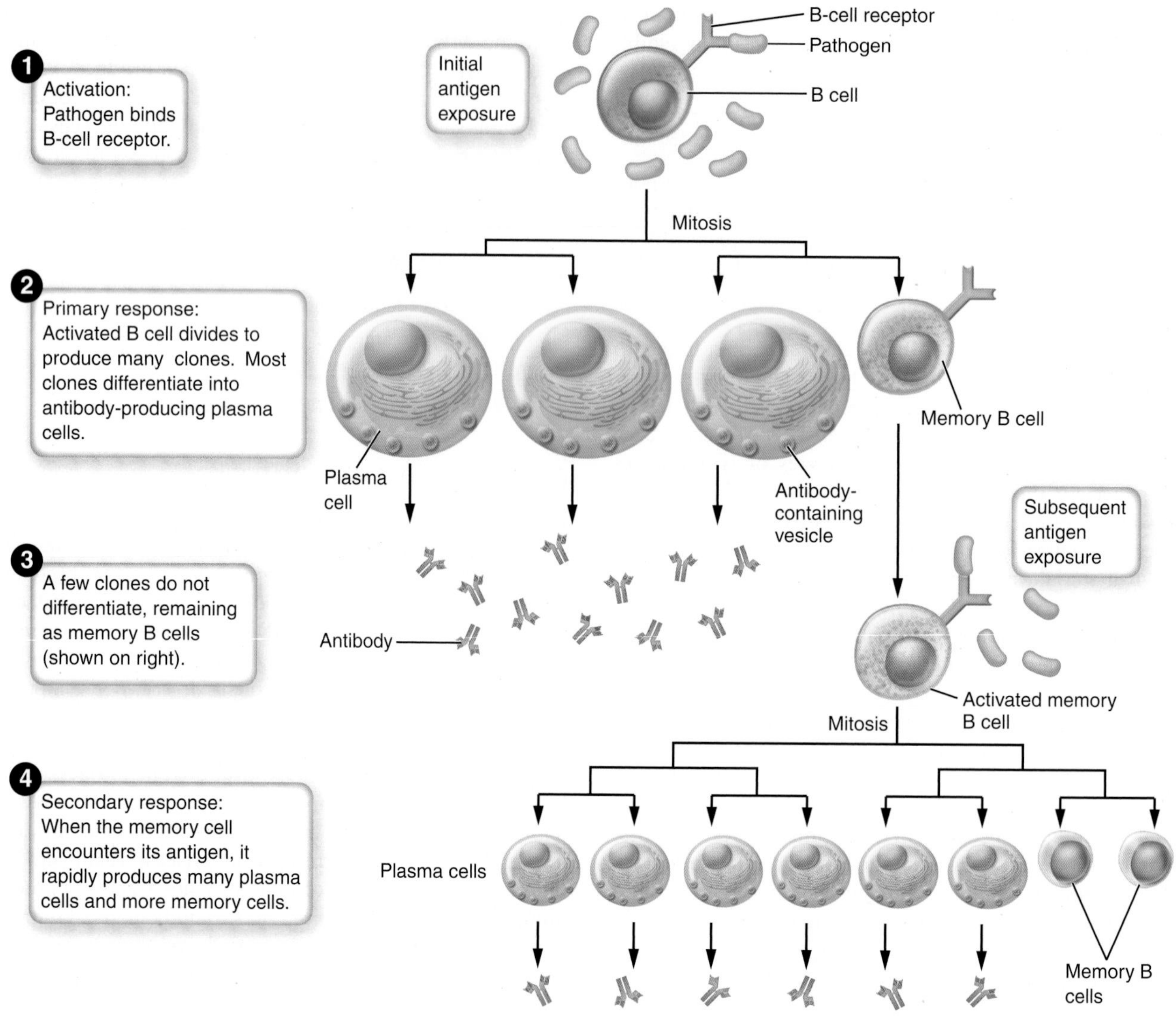
1
Activation:
Pathogen binds
B-cell receptor.
Initial
antigen
exposure
B-cell receptor
Pathogen
B cell
Mitosis
2
Primary response:
Activated B cell divides to
produce many clones. Most
clones differentiate into
antibody-producing plasma
cells.
Memory B cell
Plasma
cell
Antibody-
containing
vesicle
Subsequent
antigen
exposure
3
A few clones do not
differentiate, remaining
as memory B cells
(shown on right).
Antibody
Activated memory
B cell
Mitosis
4
Secondary response:
When the memory cell
encounters its antigen, it
rapidly produces many plasma
cells and more memory cells.
Plasma cells
Memory B
cells

Figure 17-5 **Activation of B cells.** **KEY POINT** The B cell combines with a specific antigen. The cell divides to form plasma cells, which produce antibodies. Some of the cells develop into memory cells, which protect against reinfection. **ZOOMING IN** What two types of cells develop from activated B cells?

(extracellular pathogens). All antibodies are contained in a portion of the blood plasma called the **gamma globulin** fraction.

The antibody that is produced in response to a specific antigen, such as a bacterial cell or a toxin, has a shape that matches some part of that antigen, much in the same way that a key's shape matches the shape of its lock. The antibody can bind only to the antigen that caused its production. Antibodies do not destroy cells directly; rather, they assist in the immune response. For example, they prevent a pathogen's attachment to a host cell; help with phagocytosis; activate NK cells; and neutralize toxins. The antigen–antibody complex may activate the complement system, which helps in immunity, as previously described. These antigen–antibody interactions are illustrated, and their protective effects are described in **Table 17-2**.

Notice in **Figure 17-5** that some of the activated B cells do not become plasma cells but, like certain T cells, become memory cells. These do not immediately produce antibodies. Instead, they circulate in the bloodstream and upon repeated contact with an antigen, immediately begin dividing to produce many active plasma cells. Because of this "immunologic memory," one is usually immune to a childhood disease, such as chickenpox, after having it.

Figures 17-5 and 17-6 illustrate this secondary response. There are five different classes of antibodies distinguished by their locations and functions. The antibodies in this figure are designated as IgM (immunoglobulin M) and IgG (immunoglobulin G). IgM is the first type of antibody produced in an immune response, followed shortly by IgG. A second encounter with the antigen stimulates production of both types of antibodies but has a much greater effect on IgG production.

Table 17-2 Antigen–Antibody Interactions and Their Effects

Interaction	Effects
Prevention of attachment	A pathogen coated with antibody is prevented from attaching to a cell.
Clumping of antigen	Antibodies can link antigens together, forming a cluster that phagocytes can ingest.
Neutralization of toxins	Antibodies bind to toxin molecules to prevent them from damaging cells.
Help with phagocytosis	Phagocytes can attach more easily to antigens that are coated with antibody.
Activation of complement	When complement attaches to antibody on a cell surface, a series of reactions begins that activates complement to destroy cells.
Activation of NK cells	NK cells respond to antibody adhering to a cell surface and attack the cell.

These and the other classes of immunoglobulins are described in Box 17-1.

CHECKPOINTS 

- 17-10 What is an antibody?
- 17-11 What type of cells produce antibodies?

See the Student Resources on thePoint for the animation "Immune Response."

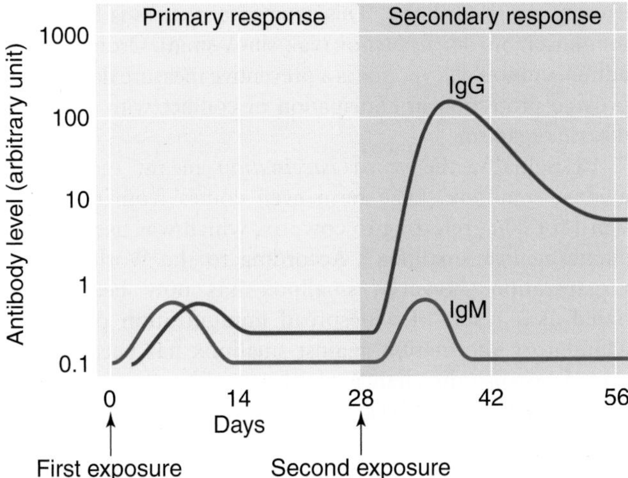

Figure 17-6 Production of antibodies (Ab). KEY POINT Antibodies or immunoglobulins (Ig) are produced in response to a first encounter with a foreign antigen. A second exposure produces a greater response. Immunoglobulin M (IgM) and immunoglobulin G (IgG) are two of the five classes of antibodies.

TYPES OF ADAPTIVE IMMUNITY

As we have just seen, adaptive immunity may develop naturally through contact with a specific disease organism. In this case, the infected person's T cells and antibodies act against the infecting agent or its toxins. The infection that triggers the immunity may be so mild as to cause no symptoms (subclinical). Nevertheless, it stimulates the host's immune response. Moreover, each time a person is invaded by the disease organism, his or her cells will respond to the infection. Such immunity may last for years, and in some cases for life. Because the host is actively involved in generating protection, this type of immunity is described as *active*. Because the immunity is formed against harmful agents encountered in the normal course of life, it is called **natural active immunity** (Fig. 17-7).

Immunity also may be acquired naturally by the passage of antibodies from a mother to her fetus through the placenta. Because these antibodies come from an outside source, this type of immunity is called **natural passive immunity**. The antibodies obtained in this way do not last as long as actively produced antibodies, but they do help protect the infant for about six months, by which time the child's own immune system begins to function. Nursing an infant can lengthen this protective period because the mother's specific antibodies are present in her breast milk and colostrum (the first breast secretion). These are the only known examples of naturally acquired passive immunity.

A person who has not been exposed to a particular pathogen has no antibodies or T cells against that organism and may be defenseless against infection. Therefore, medical personnel may use artificial measures to establish immunity. As with naturally acquired immunity, artificially acquired immunity may be active or passive (see Fig. 17-7). Artificial active immunity results from the use of a **vaccine** (vak-SENE), a prepared substance that initiates an immune response against a particular pathogen. This type of immunity is described as active because the recipient's own immune system is at work. Artificial passive immunity, on the other hand, involves the administration of antibodies obtained from an outside source, a preparation known as an antiserum. Vaccines and antisera are discussed in greater detail next.

VACCINES

The administration of virulent pathogens to stimulate immunity obviously would be dangerous. Instead, laboratory workers treat the harmful agent to reduce its virulence before it is administered. In this way, the antibodies are produced without

Box 17-1

A CLOSER LOOK

Antibodies: A Protein Army That Fights Disease

Antibodies are proteins secreted by plasma cells (activated B cells) in response to specific antigens. They are all contained in a fraction of the blood plasma known as gamma globulin. Because the plasma contains other globulins as well, antibodies have become known as immunoglobulins (Ig). Immunologic studies have shown that there are several classes of immunoglobulins that vary in molecular size and in function (see below). Studies of these antibody fractions can be helpful in making a diagnosis. For example, high levels of IgM antibodies, because they are the first to be produced in an immune response, indicate a recent infection.

Class	Abundance	Characteristics and Function
IgG	75%	Found in the blood, lymph, and intestines Enhances phagocytosis, neutralizes toxins, and activates complement Crosses the placenta and confers passive immunity from mother to fetus
IgA	15%	Found in glandular secretions such as sweat, tears, saliva, mucus, and digestive juices Provides local protection in mucous membranes against bacteria and viruses Also found in breast milk, providing passive immunity to newborn
IgM	5–10%	Found in the blood and lymph The first antibody to be secreted after infection Stimulates agglutination and activates complement
IgD	<1%	Located on the surface of B cells
IgE	<0.1%	Located on basophils and mast cells Active in allergic reactions and parasitic infections

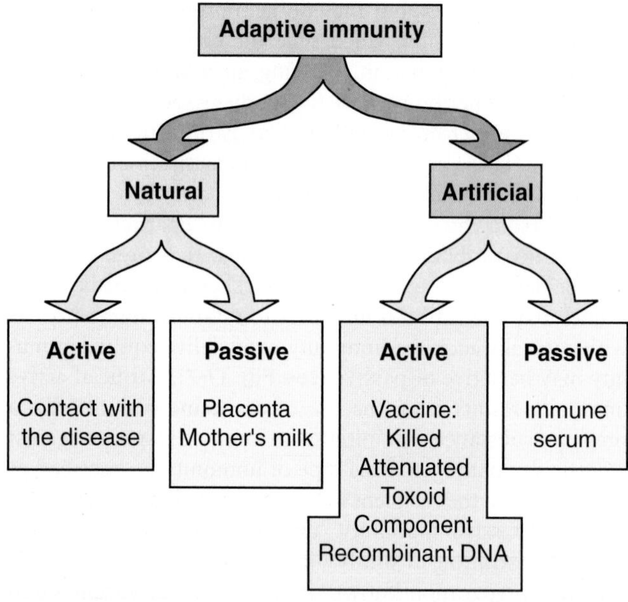

Figure 17-7 Adaptive immunity. **KEY POINT** Adaptive immunity, also called acquired or specific immunity, is a response to a particular antigen. Adaptive immunity may develop naturally (as by contact with the disease) or by artificial means (as by vaccination). Both natural and artificial adaptive immunity may be active (generated by the individual) or passive (provided from an outside source).

causing a serious illness. This protective process is known as **immunization**, or *vaccination* (vak-sin-A-shun). Ordinarily, the administration of a vaccine is a preventive measure designed to provide protection in anticipation of contact with a specific disease organism.

Originally, the word *vaccination* meant inoculation against smallpox. (The term even comes from the Latin word for *cow*, referring to cowpox, which was used to vaccinate against smallpox.) According to the World Health Organization, however, smallpox has now been eliminated as a result of widespread immunization programs. Mandatory vaccination against smallpox has been discontinued because the chance of adverse side effects from the vaccine is thought to be greater than the probability of contracting the disease.

All vaccines carry a small risk of adverse side effects and may be contraindicated in some cases. People who are immunosuppressed, for example, should not be given vaccines that contain a live virus. Also, pregnant women should not receive live virus vaccines because the virus could cross the placenta and harm the fetus. In general, however, vaccines are extensively tested for safety, and for most people, their potential benefits far outweigh their risks. Moreover, the vaccination of most people in a given area protects individuals who cannot be vaccinated for some reason because the pathogen will lack a susceptible population in which to spread. Declining levels of pertussis (whooping cough) vaccinations, for instance, have recently led to increased infection rates.

Table 17-3 Immunizations[a]

Vaccine	Disease(s)	Schedule
DTaP	Diphtheria, tetanus, pertussis (whooping cough)	2, 4, 6, and 15–18 mo; 4–6 y; pregnant women Diphtheria and tetanus toxoid (Td) at 11–12 y; booster every 10 y
Hib	*Haemophilus influenza* type b (spinal meningitis)	2 and 4 mo or 2, 4, and 6 mo depending on type used
HepA	Hepatitis A virus	12–23 mo; second dose 6 mo later
HepB	Hepatitis B	Birth, 1–2 mo, 6–18 mo
Influenza	Influenza ("flu")	Yearly from 6 mo to 6 y
MMR	Measles, mumps, rubella	15 mo and 4–6 y
PCV	Pneumococcus (pneumonia, meningitis)	2, 4, 6, and 12–15 mo; after 65 y
Polio vaccine (IPV)	Poliomyelitis	2 and 4 mo, 6–18 mo, and 4–6 y
Rotavirus (RV)	Rotavirus gastroenteritis	2 and 4 mo or 2, 4, and 6 mo, depending on the version used
Varicella	Chickenpox	12–15 mo and 4–6 y
HPV	Human papillomavirus	Three doses administered between 11 and 26 y (sooner is better)
Meningococcal	Meningococcus	High risk: 9 mo to 10 y Routine: 11–12 y, 16 y
Zoster	Shingles	After 60 y of age

[a]Recommended by the Advisory Committee on Immunization Practices (www.cdc.gov/vaccines/recs/acip), the American Academy of Pediatrics (www.aap.org), and the American Academy of Family Physicians (www.aafp.org). Information is also available through the National Immunization Program Web site (www.cdc.gov/vaccines).

17

Types of Vaccines Vaccines can be made with live organisms or with organisms killed by heat or chemicals. If live organisms are used, they must be nonvirulent for humans, such as the cowpox virus used for smallpox immunization, or they must be treated in the laboratory to weaken them as human pathogens. An organism weakened for use in vaccines is described as **attenuated.** Another type of vaccine is made from the toxin produced by a disease organism. The toxin is altered with heat or chemicals to reduce its harmfulness, but it can still function as an antigen to induce immunity. Such an altered toxin is called a **toxoid.**

The newest types of vaccines are produced from antigenic components of pathogens or by genetic engineering. By techniques of recombinant DNA, the genes for specific disease antigens are inserted into the genetic material of harmless organisms. The antigens produced by these organisms are extracted and purified and used for immunization. The hepatitis B vaccine is produced in this manner.

Boosters In many cases, an active immunity acquired by artificial (or even natural) means does not last a lifetime. Circulating antibodies can decline with time. To help maintain a high titer (level) of antibodies in the blood, repeated inoculations, called *booster shots*, are administered at intervals. The number of booster injections recommended varies with the disease and with an individual's environment or degree of exposure. On occasion, epidemics in high schools or colleges may prompt recommendations for specific boosters. **Table 17-3** lists the vaccines currently recommended in the United States. The number and timing of doses vary with the different vaccines.

> **Nurse practitioners often prescribe and administer vaccines. See the Student Resources on thePoint to read about this career and specifically about pediatric nurse practitioners.**

Examples of Bacterial Vaccines Children are routinely immunized with vaccines against bacteria or their toxins. Because of whooping cough's seriousness in young infants, early inoculation with whooping cough, or **pertussis** (per-TUS-is), vaccine is recommended. A new form of the vaccine containing pertussis toxoid causes fewer adverse reactions than did older types that contained heat-killed organisms. This acellular (aP) vaccine is usually administered in a mixture with diphtheria and tetanus toxoid. The combination, referred to as *DTaP*, may be given as early as 2 months of age. Diphtheria and tetanus toxoid (Td) is administered again at 11 to 12 years of age. A tetanus booster is administered when there is a disease risk and the last booster was administered more than 10 years prior to exposure.

Routine inoculation against *Haemophilus influenzae* type B (Hib) has nearly eliminated the life-threatening meningitis caused by this organism among preschool children.

Hib also causes pneumonia and recurrent ear infections in young children.

Pneumococcal vaccine (PCV) protects against infection with pneumococcus, an organism that can cause pneumonia and meningitis. Four doses are administered between the ages of 2 and 15 months, and boosters are administered to adults over 65 years of age.

Examples of Viral Vaccines Intensive research on viruses has resulted in the development of vaccines for an increasing number of viral diseases.

- Worldwide public health immunization programs have almost eliminated poliomyelitis, a disease that causes paralysis and sometimes death. The first available polio vaccine was an injected inactivated type (IPV) developed by Dr. Jonas Salk and made with killed poliovirus. Dr. Albert Sabin later developed a more convenient oral vaccine (OPV) made with live attenuated virus. The OPV is no longer routinely used in the United States.
- MMR, made with live attenuated viruses, protects against measles (rubeola), mumps, and rubella (German measles). Rubella is a very mild disease, but it causes birth defects in a developing fetus. (See **Table A3-2** in Appendix 3.)
- Infants are now routinely immunized against hepatitis B (hepB). A newer vaccine used in children includes DTaP, the inactivated polio vaccine, and hepatitis B. The hepB vaccine is also recommended for adults at high risk of hepatitis B infection, including healthcare workers, people on kidney dialysis, people receiving blood clotting factors, injecting drug users, and those with multiple sexual partners. A vaccine against hepatitis A virus is recommended for travelers to developing nations and others at high risk for infection.
- A vaccine against chickenpox (varicella) has been available since 1995. Children who have not had the disease by 1 year of age should be vaccinated. Although chickenpox is usually a mild disease, it can cause encephalitis, and infection in a pregnant woman can cause congenital malformation of the fetus. Because varicella is the same virus that causes shingles, a painful nerve disorder, vaccination may prevent this late-life sequel. Now, a live virus shingles vaccine is also available for people aged 60 years or older.
- A number of vaccines have been developed against influenza, which is caused by a variety of different viral strains. Laboratories produce a new vaccine each year to combat what they expect will be the most common strains in the population. A nasal spray containing live, attenuated vaccines is now available, a boon for needle-phobic children and adults (**Fig. 17-8**). The elderly, the debilitated, and children, especially those with certain risk factors, including asthma, heart disease, sickle cell disease, HIV infection, and diabetes, should be immunized yearly against influenza.

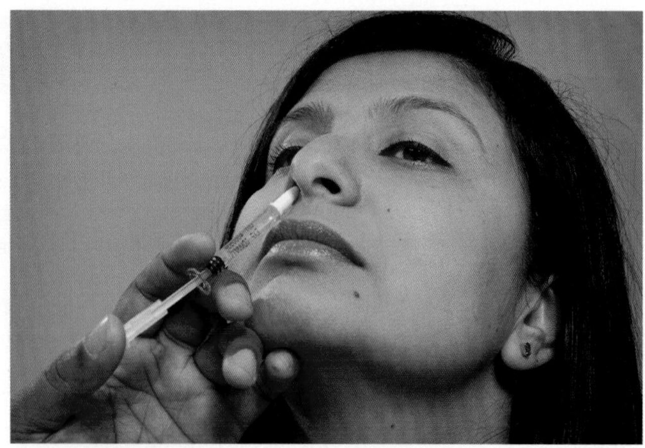

Figure 17-8 The intranasal vaccine. This recipient is receiving a live, attenuated H1N1 intranasal vaccine instead of an injected version.

- Rotavirus causes a highly contagious gastrointestinal infection among babies and toddlers worldwide. The vomiting and diarrhea that result from infection can lead rapidly to life-threatening dehydration, so infants are now routinely vaccinated.
- The human papilloma virus (HPV) causes sexually transmitted genital warts in both men and women and is associated with almost all cases of cervical cancer in women. Vaccines against the most prevalent HPV strains are now available, and immunization is recommended for both boys and girls aged 11 to 12 years through 18 years of age.
- The rabies vaccine is an exception to the rule that a vaccine should be given before invasion by a disease organism. Rabies is a viral disease transmitted by the bite of wild animals such as raccoons, bats, foxes, and skunks. Mandatory vaccination of domestic animals has practically eliminated this source of rabies in some countries, including the United States, but worldwide, a variety of wild and domestic animals are host to the virus. There is no cure for rabies; it is fatal in nearly all cases. The disease develops so slowly, however, that affected people vaccinated after transmission of the organism still have time to develop an active immunity. The vaccine may be given preventively to people who work with animals.

ANTISERUM

Active immunity, either natural or artificial, requires several weeks or longer to fully develop. Therefore, a person who receives a large dose of virulent organisms and has no established immunity to them is in great danger. To prevent illness, the person must quickly receive counteracting antibodies from an outside source. This is accomplished through the administration of an **antiserum** or *immune serum*. The "ready-made" serum gives short-lived but effective protection against the organisms in the form of an artificially acquired passive immunity.

Preparation of Antisera Immune sera often are derived from animals, mainly horses. It has been found that the horse's tissues produce large quantities of antibodies in response to the injection of organisms or their toxins. After repeated injections, the horse is bled according to careful sterile technique; because of the animal's size, it is possible to remove large quantities of blood without causing injury. The blood is allowed to clot, and the serum is removed and packaged in sterile containers.

Injecting humans with serum derived from animals is not without its problems. The foreign proteins in animal sera may cause an often serious sensitivity reaction called **serum sickness.** To avoid this problem, human antibody in the form of gamma globulin may be used.

Examples of Antisera Some immune sera contain antibodies, known as **antitoxins**, that neutralize toxins but have no effect on the toxic organisms themselves. Certain antibodies act directly on pathogens, inactivating them or increasing their phagocytic destruction. Some antisera are obtained from animal sources, others from human sources. Examples of immune sera are as follows:

- Diphtheria antitoxin is obtained from immunized horses.
- Tetanus immune globulin is effective in preventing lockjaw (tetanus), which is often a complication of neglected wounds and if left untreated can be fatal. Because tetanus immune globulin is of human origin, it carries less risk of adverse reactions than do sera obtained from horses.
- Immune globulin (human) is given to nonvaccinated people exposed to hepatitis A, measles, or polio. It is also given on a regular basis to people with congenital (present at birth) immune deficiencies.
- Hepatitis B immune globulin, used after hepatitis B exposure, is given principally to infants born to mothers who have hepatitis.
- The immune globulin Rh_0(D) (trade name RhoGAM) is a concentrated human antibody given to prevent an Rh-negative mother from forming Rh antibodies. It is given during pregnancy if maternal antibodies develop and after the birth of an Rh-positive infant (or even after a miscarriage of a presumably Rh-positive fetus) (see Chapter 13). It is also given when Rh transfusion incompatibilities occur.
- Anti–snake bite sera or **antivenins** (an-te-VEN-ins) are used to combat the effects of certain poisonous snake bites.
- Botulism antitoxin, an antiserum from horses, offers the best hope for botulism victims, although it is effective only if given early.
- Rabies antiserum, from humans or horses, is used along with the vaccine to treat victims of rabid animal bites.

CHECKPOINTS

- 17-12 What is the difference between active and passive adaptive immunity?
- 17-13 What is the difference between natural and artificial adaptive immunity?
- 17-14 What is a vaccine?
- 17-15 What is a booster?
- 17-16 What is an antiserum, and when are antisera used?

Immune Disorders

Immune disorders may result from overactivity or underactivity. Allergy and autoimmune diseases fall into the first category; hereditary, infectious, and environmental immune deficiency diseases fall into the second.

ALLERGY

An **allergy** (AL-er-je) is informally defined as an unfavorable immune response to a substance that is commonly encountered and otherwise harmless, such as pollen or shrimp. A broader term—**hypersensitivity**—refers to any deleterious immune response, which also includes autoimmune diseases and transplantation responses. Environmental substances that induce hypersensitivity reactions are called **allergens** (AL-er-jens). Examples of typical allergens are pollens, house dust, animal dander (*dander* is the term for the minute scales that are found on hairs and feathers), and certain food proteins. Many drugs can be allergens, particularly aspirin, barbiturates, and antibiotics (especially penicillin). (Chapter 6 has illustrations of allergic responses that affect the skin.)

When a susceptible person is exposed to an allergen, inhaled pollens, for example, B cells produce a specific class of antibodies called IgE against them (**Fig. 17-9**, left side; **see Table 17-2**). Normally, these antibodies are only produced in response to a parasitic infection. The antibodies insert into the membrane of mast cells and basophils. When the person is next exposed to that antigen, the allergens bind to the antibodies in the immune cells' membranes, with results that are disagreeable and sometimes dangerous. When pollen allergens encounter sensitized mast cells in the nasal mucosa, for instance, they cause the cells to release histamine and other inflammatory chemicals (**Fig. 17-9**, right side). This results in inflammation of the nasal mucosa and the characteristic runny nose of **hay fever.** In the upper respiratory tract, the histamine released by activated immune cells causes bronchial constriction leading to breathing difficulties.

Antihistamines are drugs that counteract histamine and may be effective in treating the symptoms of certain allergies. Sometimes, it is possible to desensitize an allergic person by repeated intermittent injections of the offending allergen. Unfortunately, this form of protection does not last long.

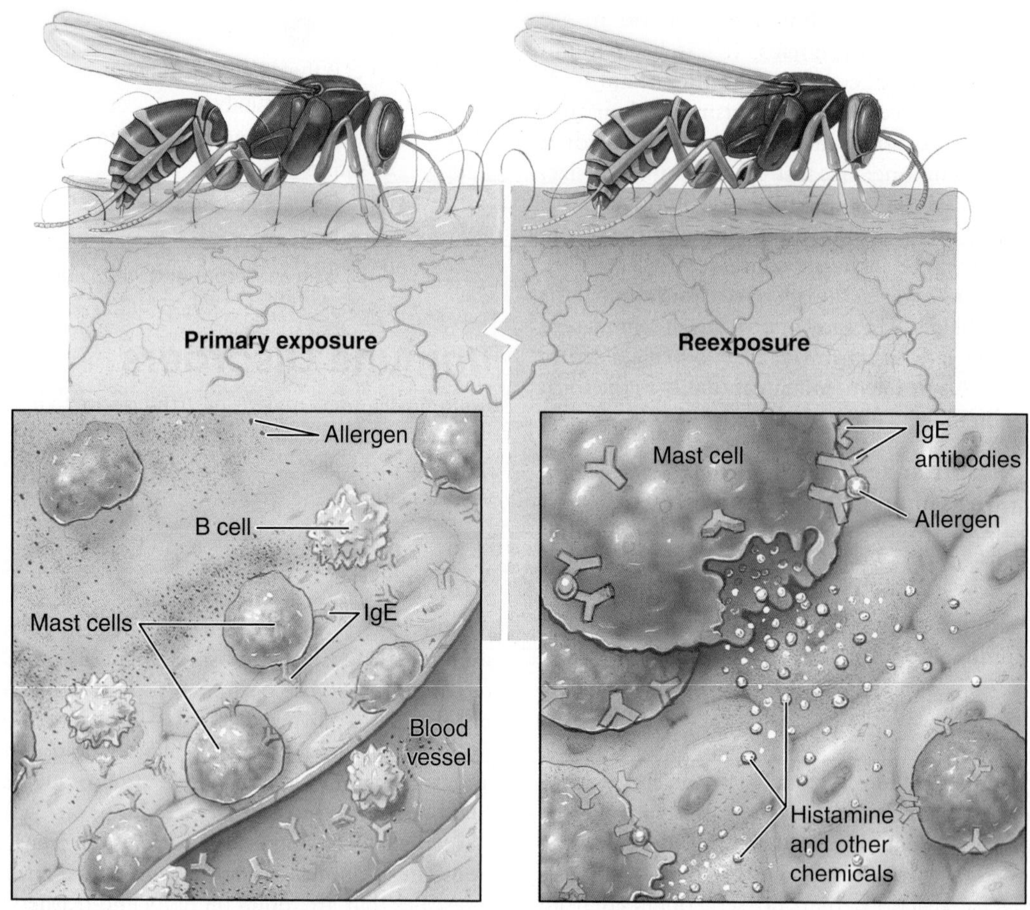

Figure 17-9 **Allergy.** **KEY POINT** The first (primary) exposure to an allergen stimulates IgE production by B cells. IgE inserts into the mast cell membranes. Upon reexposure, the allergen attaches to the IgE molecules and activates the mast cell. Activated mast cells release histamine and other proinflammatory chemicals.

Anaphylaxis (an-ah-fih-LAK-sis) is a severe, life-threatening allergic response in a sensitized individual. (The term actually means excess "guarding," in this case, immune protection, from the Greek word *phylaxis*.) Any allergen can provoke an anaphylactic response, but common causes are drugs, insect venom, and foods. Symptoms appear within seconds to minutes after contact and include breathing problems, swelling of the throat and tongue, urticaria, and edema. Peripheral blood vessels dilate, sometimes resulting in a dangerous drop in blood pressure known as anaphylactic shock (see Chapter 15). Anaphylaxis is treated with injectable epinephrine, antihistamine, administration of oxygen, and plasma expanders to increase blood volume and thus blood pressure. People subject to severe allergic reactions must avoid contact with known allergens. They should be sensitivity tested before administration of a new drug and should also carry injectable epinephrine and wear a medical bracelet identifying their allergy.

AUTOIMMUNITY

The term **autoimmunity** refers to an abnormal reactivity (hypersensitivity) to one's own tissues. In autoimmunity, the immune system reacts to the body's own antigens, described as "self," as if they were foreign antigens, or "nonself." Normally, the immune system learns before birth to ignore (tolerate) the body's own tissues by eliminating or inactivating those lymphocytes that will attack them. Some factors that might result in autoimmunity include the following:

- A change in "self" proteins, as a result of disease, for example
- Loss of immune system control, as through loss of regulatory T-cell activity, for example
- Antibodies developed against pathogens that also react with "self" antigens; this reaction occurs in rheumatic fever, for example, when antibodies to streptococci damage the heart valves

Autoimmunity is involved in a long list of diseases, including rheumatoid arthritis, multiple sclerosis, lupus erythematosus, psoriasis, inflammatory bowel diseases, Graves disease, glomerulonephritis, and type 1 diabetes. All of these diseases probably result in varying degrees from the interaction of individual genetic makeup with environmental factors, including infections. Autoimmune diseases are three times more prevalent in women than in men, perhaps related to hormonal differences.

Common treatments for autoimmune diseases include drugs that suppress the immune system (such as corticosterone) and antilymphocyte antibodies prepared in a laboratory. A newer approach uses chemotherapy to destroy immune cells followed by their replacement with healthy stem cells from bone marrow.

IMMUNE DEFICIENCY DISEASES

An immune deficiency results from failure of one or more components of the immune system, such as a specific lymphoid organ or lymphocyte type. Such disorders may be congenital (present at birth) or may be acquired as a result of malnutrition, infection, or treatment with x-rays or certain drugs. Immunodeficiencies can be transient or permanent. Children, for instance, can suffer from temporary immunodeficiency subsequent to a viral infection. Excessive stress can also temporarily impair immune function, mainly through increased production of cortisol, which suppresses inflammation and T-cell activity. You may have noticed that you often get sick when you are tired and under stress, during exam time, for example.

The disease **AIDS**, autoimmune deficiency syndrome, is a devastating example of an infection that attacks the immune system. It is caused by **HIV**, human immunodeficiency virus, which can bind to a so-called CD4 receptor on T_h cells. The CD4 receptor is normally involved in T_h-cell activation, but HIV uses it to gain access to the cell. Its first appearance in the United States in the early 1980s was among homosexual men and injecting drug users. It now occurs worldwide in heterosexual populations of all ages. AIDS is considered to be a pandemic, especially in sub-Saharan Africa and in some parts of Asia. It is spread through unprotected sexual activity and the use of contaminated injection needles. It can also be transmitted from a mother to her fetus. The testing of donated blood has virtually eliminated the spread of HIV through blood transfusions. Diagnosis of HIV infection is based on the presence of HIV antibodies, the virus, or viral components in the blood. The disease is monitored with CD4 T-cell counts and measurement of HIV RNA in the blood. Note that an individual can have an HIV infection for years or decades without developing AIDS.

T_h cells are central to our defenses against all pathogens and cancer cells. Patients with AIDS thus succumb easily to disease, including rare diseases such as a fungal (*Pneumocystis*) pneumonia and an especially malignant skin cancer, **Kaposi** (KAP-o-se) **sarcoma**. Drugs active against HIV stop viral growth at different stages of replication. Some, such as AZT, inhibit *reverse transcriptase*, an enzyme critical to HIV reproduction. These drugs, often used in combination, can slow the progress of the disease to such an extent that HIV infection is now considered to be a chronic disease (like diabetes mellitus) rather than a death sentence. Clinical trials of HIV vaccines are underway.

> **See the Student Resources on thePoint for illustrations on the course of HIV infection and the pathology of AIDS.**

MULTIPLE MYELOMA

Multiple myeloma is a bone marrow tumor resulting in the overproduction of B lymphocytes, more specifically plasma cells. The plasma cells accumulate in the bone marrow and interfere with the production of normal blood cells, causing anemia and low resistance to infection. The often nonfunctional antibodies they produce increase the protein content of blood to an extent that damages the kidneys. The cancerous plasma cells also secrete a specific factor that increases bone resorption, leading to frequent fractures and hypercalcemia. Multiple myeloma is treated with chemotherapy and medications to reduce bone loss. A new approach is high-dose chemotherapy combined with bone marrow transplants. Blood-forming stem cells in the bone marrow replace cells killed by the chemotherapy. This treatment is expensive, and stem cell transplants in themselves are dangerous, but this combined treatment has improved survival rates.

☐ 17-17 What are four types of immune disorders?

17

The Immune System and Cancer

Cancer cells differ slightly from normal body cells and therefore the immune system should recognize them as "nonself." The fact that people with AIDS and other immune deficiencies develop cancer at a higher rate than normal suggests that this is true. Cancer cells probably form continuously in the body but normally are destroyed by NK cells, cytotoxic T cells, and macrophages in a process called **immune surveillance** (sur-VAY-lans) (**Fig. 17-10**). As a person ages, cell-mediated immunity declines, and cancer is more likely to develop.

Medical scientists are attempting to treat cancer by stimulating the patient's immune responses, a practice called **immunotherapy**. In one approach, T cells are removed from the patient, activated with cytokines, and then reinjected. This method has given some positive results, especially in treatment of melanoma, a highly malignant form of skin cancer. In the future, a vaccine against cancer may become a reality. Vaccines that target specific proteins produced by cancer cells have already been tested in a few forms of cancer. Read more about using the immune system to combat cancer in **Box 17-2**.

Transplantation and Rejection Syndrome

Transplantation is the transfer of an organ or tissue from one organism (the donor) to another (the recipient), in order to replace an injured or incompetent body part. Tissues that have been transplanted include the bone marrow, lymphoid tissue, skin, corneas, parathyroid glands, ovaries, kidneys, lungs, heart, and liver.

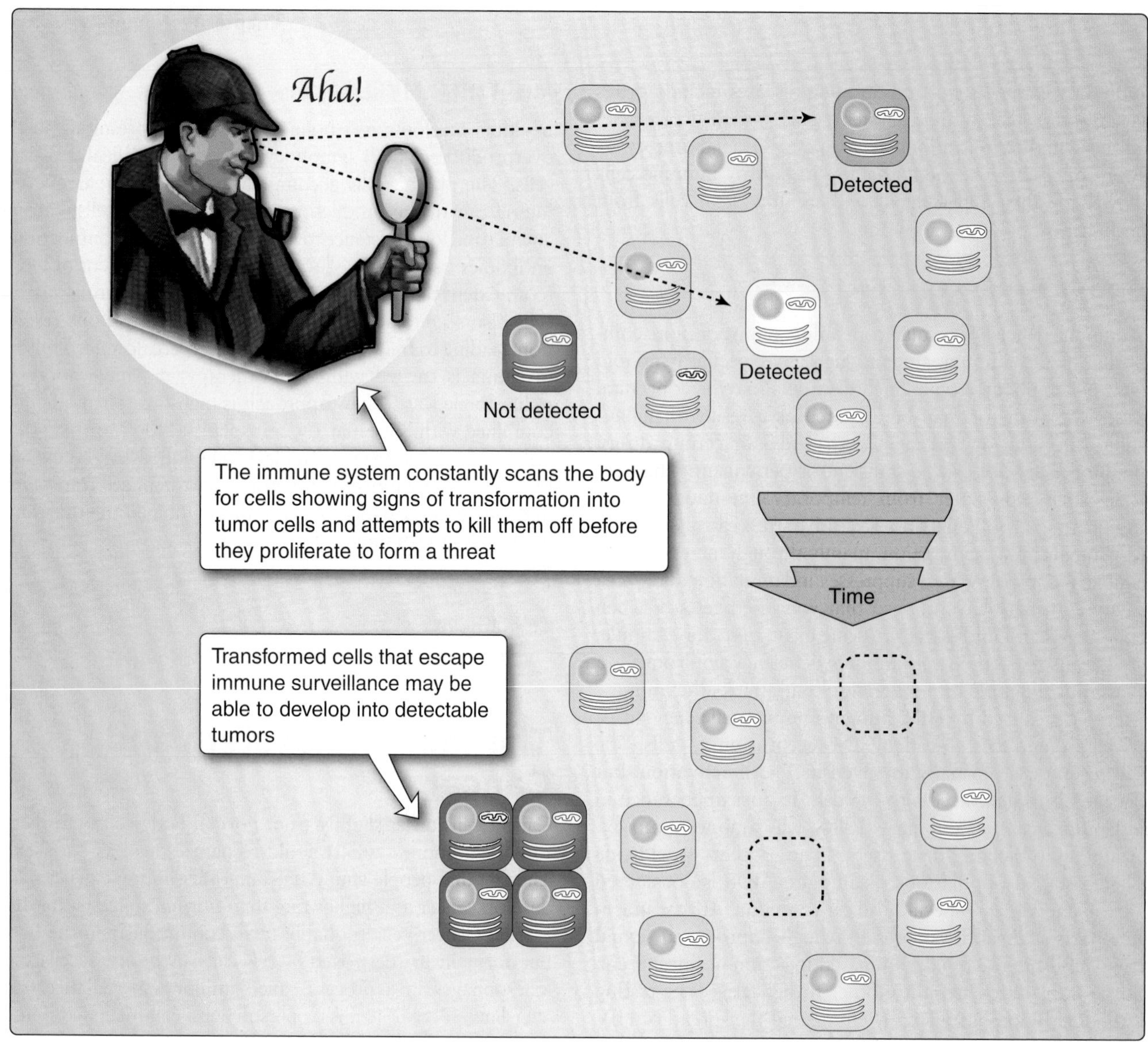

Figure 17-10 **Immune surveillance.** Tumors can occur when abnormal cells evade immune surveillance.

HOT TOPICS

Box 17-2

Harnessing the Immune System to Fight Cancer

Cancer therapy is beginning to emerge from its long history of treating the disease by harming cells, both normal and diseased, with chemotherapy and radiation. Investigators are trying to make treatment more individualized, for example, by selecting a chemotherapy drug based on the genetic makeup of a person's tumor cells. Other avenues of research involve using the immune system.

An immunity approach that has met with some success is using T cells to combat leukemia. A patient's own T cells are engineered in the laboratory to attack cancer cells bearing a specific antigen, or "marker." If successful, the transfused cells continue to multiply internally and eliminate the abnormal cancer cells.

So far, this method has been most successful in treating acute lymphoblastic leukemia (ALL), which is the most common form of leukemia in children. The trick to this method is finding an antigen that is unique to the cancer cells, which after all, arise from normal body cells. Other obstacles have been the unexplained loss of the modified T cells and the unwanted destruction of the patient's healthy B cells, the cell line involved in ALL.

Medical researchers have also used pure antibodies made in the lab to specifically target cancer cells or to deliver chemotherapeutic drugs directly to cancer cells. Future medical studies hold the promise of killing unwanted cells while leaving normal cells intact.

The immune system's ability to destroy foreign substances, including tissues from another person or any other animal, has been the most formidable obstacle to complete success. This normal antigen–antibody reaction has, in this case, been called **rejection syndrome**.

Rejection syndrome can be minimized by close matching of the donor and recipient antigens. The process of *tissue typing* is more complicated than the blood typing procedure described in Chapter 13 because tissue antigens are more varied than are blood cell antigens. Close relatives usually have many matching antigens, so they are often the best donor candidates. For instance, a person can receive a kidney, lung lobe, or liver section from a living relative. (One exception to the need for careful cross-matching is corneal transplantation in the eye. The cornea does not have blood vessels, so antibodies cannot reach the transplanted tissue.)

Because it is impossible to match all of a donor's antigens with those of the recipient, physicians give the recipient drugs that will suppress an immune response to the transplanted tissue. These include drugs that suppress synthesis of nucleic acids; drugs or antibodies that inhibit lymphocytes; and adrenal glucocorticoid hormones, such as cortisol, that suppress immunity. These drugs cause a variety of adverse side effects, such as hypertension, kidney damage, and osteoporosis (glucocorticoids). Most importantly, they reduce a patient's ability to fight infection. Because T cells cause much of the reaction against the foreign material in transplants, scientists are trying to use drugs and antibodies to suppress the action of these lymphocytes without damaging the B cells. B cells produce circulating antibodies and are most important in preventing infections. Success with transplantation will increase when methods are found to selectively suppress the immune attack on transplants without destroying the recipient's ability to combat disease.

CHECKPOINTS

- ☐ **17-18** How does the immune system guard against cancer?
- ☐ **17-19** What is the greatest obstacle to tissue transplantation from one person to another?

17

Disease in Context Revisited

Meredith's Rheumatoid Arthritis

On a follow-up visit, Dr. Bedell explained Meredith's condition to her.

"The results of your laboratory tests do show that you have elevated levels of the immunoglobulins IgA and IgM," he said. "Your complement levels, which sometimes rise in cases of autoimmunity, are slightly above normal. Your hemoglobin is below normal, and your white blood count (WBC) is slightly elevated, all of which coincide with the diagnosis of rheumatoid arthritis. The x-rays of your wrists and knees show some soft tissue swelling, again leading to an RA diagnosis. For some reason we don't understand, your body is reacting to your own antigens as if they were foreign. At present, there's no cure for RA, but there are drugs that can reduce the pain it causes and even slow the joint degeneration that occurs with time."

"I'll recommend some rheumatologists who can work with you. They will set up a treatment plan that will address your pain and stiffness and help you maintain optimal mobility. You'll receive education about the disease; explanations on medications and their effects; plans for physical and emotional rest; and therapeutic exercises. You can still live an active life, but you'll have to really take care of yourself."

Meredith was upset with the diagnosis, but she was thankful that the doctor had given her some guidance and information on what she and her family could expect.

CHAPTER 17

Chapter Wrap-Up

Summary Overview

A detailed chapter outline with space for note taking is on *thePoint*. The figure below illustrates the main topics covered in this chapter.

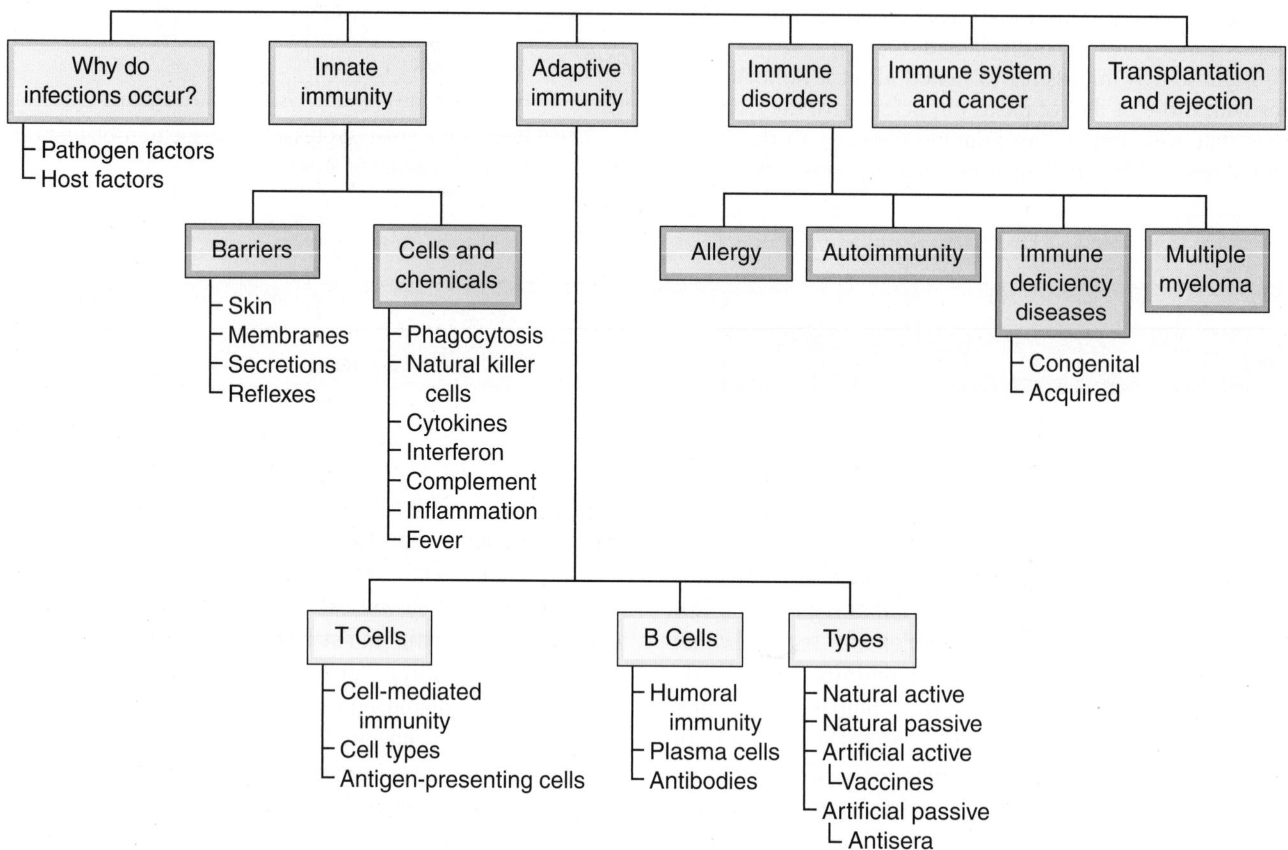

Key Terms

The terms listed below are emphasized in this chapter. Knowing them will help you organize and prioritize your learning. These and other boldface terms are defined in the Glossary with phonetic pronunciations.

allergy
anaphylaxis
antibody
antigen
antiserum
antitoxin
attenuated
autoimmunity
B cell
complement
cytokine
dendritic cells
gamma globulin
immunity
immunization
immunoglobulin
immunotherapy
inflammation
interferon
interleukin
macrophage
mast cell
natural killer (NK) cell
phagocytosis
plasma cell
T cell
toxin
toxoid
transplantation
vaccine

Word Anatomy

Medical terms are built from standardized word parts (prefixes, roots, and suffixes). Learning the meanings of these parts can help you remember words and interpret unfamiliar terms.

WORD PART	MEANING	EXAMPLE
Why Do Infections Occur?		
tox	poison	A *toxin* is a substance that is poisonous.
Immune Disorders		
ana-	back, again	*Anaphylaxis* is a life-threatening condition that results from an exaggerated immune reaction.
erg	work	In cases of *allergy*, the immune system overworks.
myel/o	marrow	Multiple *myeloma* is a cancer (–oma) of blood-forming cells in bone marrow.

Questions for Study and Review

BUILDING UNDERSTANDING

Fill in the Blanks

1. The power of an organism to overcome its host's defenses is called _____.

2. Heat, redness, swelling, and pain are classic signs of _____.

3. Any foreign substance that enters the body and induces an immune response is called a(n) _____.

4. All antibodies are contained in a portion of the blood plasma termed the _____.

5. Any substance capable of inducing an allergic reaction is called a(n) _____.

Matching > Match each numbered item with the most closely related lettered item.

___ **6.** Destroy foreign cells directly

___ **7.** Release interleukins, which stimulate other cells to join the immune response

___ **8.** Suppress the immune response in order to prevent overactivity

___ **9.** Recognize an antigen and start a rapid response if the antigen is contacted again

___ **10.** Produce antibodies when activated by antigens

a. regulatory T cells
b. memory cells
c. cytotoxic T cells
d. B cells
e. helper T cells

Multiple Choice

___ **11.** Which of the following is NOT part of barriers, the first line of defense?

a. tears
b. saliva
c. complement
d. skin

___ **12.** Which of the following is an active phagocyte?

a. NK cell
b. plasma cell
c. macrophage
d. toll-like receptor

___ **13.** Which cell matures in the thymus?

a. T cell
b. B cell
c. dendritic cell
d. mast cell

___ **14.** What is an abnormal reactivity to one's own tissues called?

a. anaphylaxis
b. autoimmunity
c. toxicity
d. virulence

___ **15.** Allergy and autoimmunity are types of

a. hypersensitivity
b. serum sickness
c. myeloma
d. immune deficiency

UNDERSTANDING CONCEPTS

16. Describe four factors that influence the occurrence of infection.

17. What causes the symptoms of inflammation?

18. Differentiate between the terms in each of the following pairs:

a. first and second lines of defense
b. innate immunity and adaptive immunity
c. cell-mediated immunity and humoral immunity
d. active immunity and passive immunity
e. toxin and toxoid

19. Define the following abbreviations

a. Ig ______________________
b. NK ______________________
c. MAC ______________________
d. APC ______________________
e. MHC ______________________

20. Describe the events that must occur for a T cell to react with a foreign antigen. Once activated, what do the T cells do?

21. Define *cytokine*. Give two examples of cytokines, and explain what they do.

22. What role do antibodies play in immunity? How are they produced? How do they work?

23. What is an immune serum? Give examples. Define antitoxin.

24. Define allergy. How does the process of allergy resemble that of immunity, and how do they differ?

25. What is meant by rejection syndrome, and what is being done to offset this syndrome?

CONCEPTUAL THINKING

26. Maria's young son wakes during the night with a low-grade fever. Give some reasons why she should take steps to limit the fever. Give some reasons why she might not want to eliminate his fever.

27. While in the garden with his father, Alek, a 4-year-old boy, was in his own words, "kicked by a bee." Shortly afterward, Alek developed hives near the affected area, which he began to scratch. About 10 minutes later, Alek's father noticed that his son was wheezing. What is happening to Alek? Describe the inflammatory events that are occurring in his body. How should Alek's father respond?

28. In the opening case study, Meredith has the autoimmune disease rheumatoid arthritis. What might be causing her disease and its symptoms?

For more questions, see the Learning Activities on thePoint.

UNIT

VI Energy: Supply and Use

The five chapters in this unit show how oxygen and nutrients are processed, taken up by the body fluids, and used by the cells to yield energy. This unit also describes the actions of the respiratory and urinary system in eliminating metabolic waste products. Some of the processes discussed in these chapters, such as food intake, drinking, and breathing, can be controlled voluntarily. Others, such as body temperature and urine output, cannot. You will learn that negative feedback mechanisms can influence these conscious and unconscious processes in order to maintain homeostasis of body temperature, blood acidity, blood composition and volume, and even—to a certain extent—body weight.

CHAPTER 18

The Respiratory System

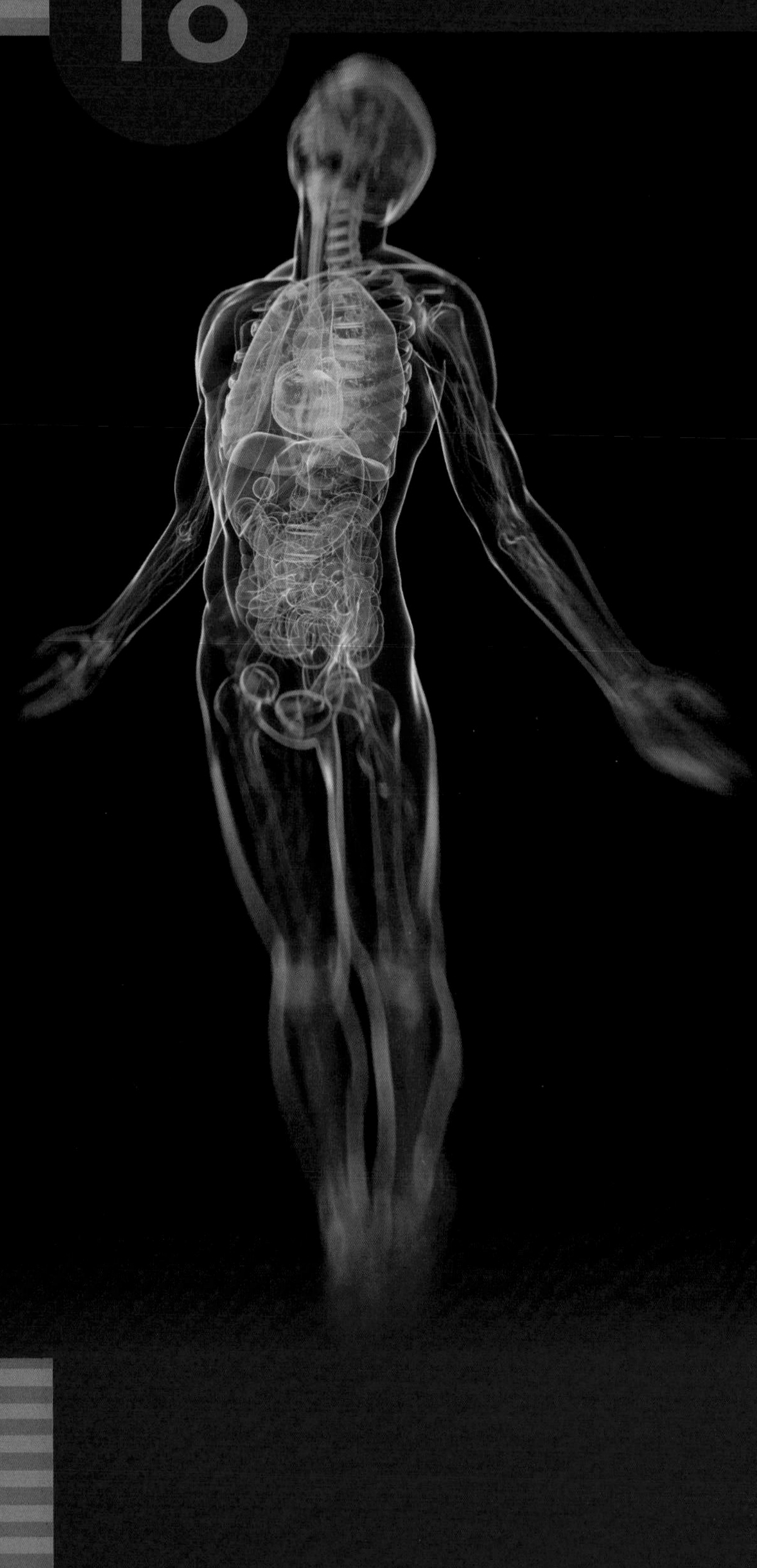

Learning Objectives

After careful study of this chapter, you should be able to:

1 ▶ Define respiration, and describe its four phases. ***p. 406***

2 ▶ Name and describe all the structures of the respiratory system. ***p. 407***

3 ▶ Explain the mechanism for pulmonary ventilation. ***p. 411***

4 ▶ Discuss the processes of internal and external gas exchange. ***p. 413***

5 ▶ List the ways in which oxygen and carbon dioxide are transported in the blood. ***p. 415***

6 ▶ Describe factors that control respiration. ***p. 416***

7 ▶ Describe normal and altered breathing patterns. ***p. 418***

8 ▶ Compare hyperventilation and hypoventilation, and cite three possible results of hypoventilation. ***p. 419***

9 ▶ List nine types of respiratory disorders. ***p. 420***

10 ▶ List six types of respiratory infections. ***p. 420***

11 ▶ Identify the diseases involved in chronic obstructive pulmonary disease (COPD). ***p. 423***

12 ▶ Describe some disorders that involve the pleura. ***p. 425***

13 ▶ Describe five types of equipment used to treat respiratory disorders. ***p. 426***

14 ▶ Referring to the case study, discuss how asthma can be diagnosed and treated. ***pp. 405, 427***

15 ▶ Show how word parts are used to build words related to respiration (see Word Anatomy at the end of the chapter). ***p. 429***

Disease in Context *Emily's Case: Advances in Asthma Therapy*

"Remind me to mention to Dr. Martinez that Emily still has that nagging cough," Nicole told her husband.

"I've been worried about that," he replied. "You know, I had asthma as a kid—I hope she doesn't. I could hardly do any sports without taking a couple puffs of my inhaler."

Later that week, Dr. Martinez listened carefully to 3-year-old Emily's lungs. He knew that the common symptoms of asthma—coughing, wheezing, and shortness of breath—were due to swelling of the airway tissues and spasm of the smooth muscle around them.

"I don't hear any wheezing, but given the family history, we can't rule out asthma," he said. "In addition to the genetic component, asthma can have several environmental triggers such as respiratory infections, allergies, cold air, and exercise."

"Well," replied Nicole, "Emily did have a cold right before the coughing began. I haven't noticed any allergies, but now that I think about it, she did have a persistent cough last winter too. And she is getting lots of exercise at preschool and dance class. If Emily does have asthma, will that limit her activities?"

"The asthma drugs we have now are much improved since your husband's youth," the doctor replied. "But first we need to figure out what is causing Emily's cough. I'm going to order a chest x-ray to rule out infection. If Emily were a little bit older, we could measure her lung function with a spirometry test. For now, I'm going to ask you to monitor her for the next few weeks and see if anything exacerbates her symptoms. If she has asthma, we'll start daily treatment with a corticosteroid inhaler to control airway inflammation. We could also go with a long-term oral medication that prevents the lungs from producing leukotrienes, substances that cause constriction of smooth muscle in the airways. Emily will also need a short-term rescue inhaler in case of an acute attack."

Asthma is the most common chronic respiratory disease of childhood. In this chapter, we'll examine the respiratory system and its involvement in this disease. Later in the chapter, we'll check in on Emily and learn about other medications used to treat asthma.

ANCILLARIES *At-A-Glance*

Visit thePoint to access the following resources. For guidance in using these resources most effectively, see pp. xv–xvii.

Learning RESOURCES

- Tips for Effective Studying
- Web Figure: Principal Muscles of Breathing and Lateral Chest
- Web Figure: Respiratory Infections
- Web Figure: Mycobacterium Tuberculosis
- Web Figure: Natural History of Tuberculosis
- Web Figure: Emphysema
- Web Figure: Effects of Smoking
- Web Figure: The Heimlich Maneuver
- Animation: Pulmonary Ventilation
- Animation: Asthma
- Animation: Oxygen Transport
- Animation: Carbon Dioxide Exchange
- Health Professions: Respiratory Therapist
- Detailed Chapter Outline
- Answers to Questions for Study and Review
- Audio Pronunciation Glossary

Learning ACTIVITIES

- Pre-Quiz
- Visual Activities
- Kinesthetic Activities
- Auditory Activities

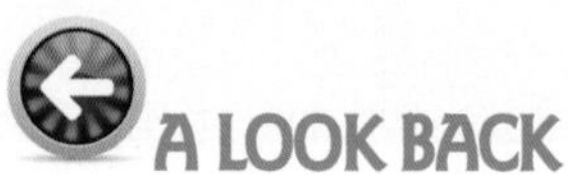

A LOOK BACK

In Chapter 2, we talked about the properties of water that make it such a unique substance. Now we expand on the discussion to understand how water functions in respiration and gas exchange. The principles of diffusion, introduced in Chapter 3, apply to these exchanges. In Chapter 4, we also introduced the types of tissues that make up the respiratory system, including pseudostratified epithelium and simple squamous epithelium. Finally, the ideas on compliance and elasticity, discussed in relation to the cardiovascular system, apply to respiratory physiology as well.

Phases of Respiration

Most people think of respiration simply as the process by which air moves into and out of the lungs—that is, *breathing*. By scientific definition, respiration is the process by which oxygen is obtained from the environment and delivered to the cells. Carbon dioxide is transported to the outside in a reverse pathway (**Fig. 18-1**).

The coordinated actions of the respiratory and cardiovascular systems accomplish the four phases of respiration:

- **Pulmonary ventilation,** which is the exchange of air between the atmosphere and the air sacs (alveoli) of the lungs. This is normally accomplished by the inhalation and exhalation of breathing.
- **External gas exchange,** which occurs in the lungs as oxygen (O_2) diffuses from the air sacs into the blood and carbon dioxide (CO_2) diffuses out of the blood to be eliminated.
- **Gas transport in the blood.** The circulating blood carries gases between the lungs and the tissues, supplying oxygen to the cells and bringing back carbon dioxide.

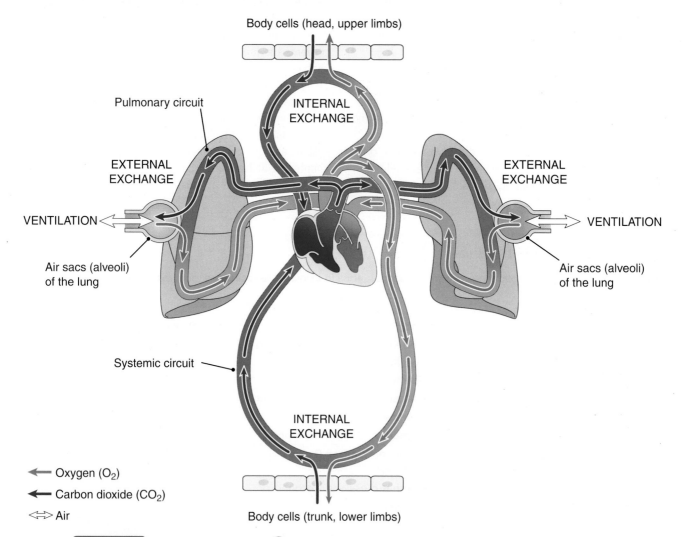

Figure 18-1 **Overview of respiration.** **KEY POINT** In ventilation, gases are moved into and out of the lungs. In external exchange, gases move between the air sacs (alveoli) of the lungs and the blood. In internal exchange, gases move between the blood and body cells. The circulation transports gases in the blood. **ZOOMING IN** From which side of the heart does blood leave to travel to the lungs? To which side does it return?

- **Internal gas exchange**, which occurs in the tissues as oxygen diffuses from the blood to the cells, whereas carbon dioxide travels from the cells into the blood.

The term *respiration* is also used to describe a related process that occurs at the cellular level. In *cellular respiration*, the cells take in oxygen and use it in the breakdown of nutrients. In this process, the cells release energy and carbon dioxide, a waste product of cellular respiration, as described in Chapter 20.

CHECKPOINT

☐ 18-1 What are the four phases of respiration?

Structure of the Respiratory System

The respiratory system is a complex series of spaces and passageways that conduct air into and through the lungs (**Fig. 18-2**). These spaces include the nasal cavities; the pharynx, which is common to the digestive and respiratory systems; the voice box, or larynx; the windpipe, or trachea; and the lungs themselves, with their conducting tubes and air sacs.

THE NASAL CAVITIES

Air enters the respiratory system through the openings in the nose called the **nostrils** or *nares* (NA-reze) (sing. naris). Immediately inside the nostrils, located between the roof of the mouth and the cranium, are the two spaces known as the **nasal cavities**. These two spaces are separated from each other by a partition, the **nasal septum**. The septum's superior portion is formed by a thin plate of the ethmoid bone, and the inferior portion is formed by the vomer (see **Fig. 7-5A** in Chapter 7). An anterior extension of the septum is made of hyaline cartilage. The septum and the walls of the nasal cavity are covered with mucous membrane, consisting of stratified squamous (flat) epithelium.

On the lateral walls of each nasal cavity are three projections called the **conchae** (KONG-ke) (see **Figs. 7-5A and C** and later **Fig. 18-12**). The shell-like conchae greatly increase the surface area of the mucous membrane over which air travels on its way through the nasal cavities. This membrane contains many blood vessels that deliver heat and moisture. The membrane's cells secrete a large amount of fluid—up to 1 qt each day. Breathing through the nose (instead of the mouth) enables the nasal mucosa to warm and moisten inhaled air. Moreover, foreign bodies, such as dust and pathogens, can be trapped in the nasal hairs or the surface mucus.

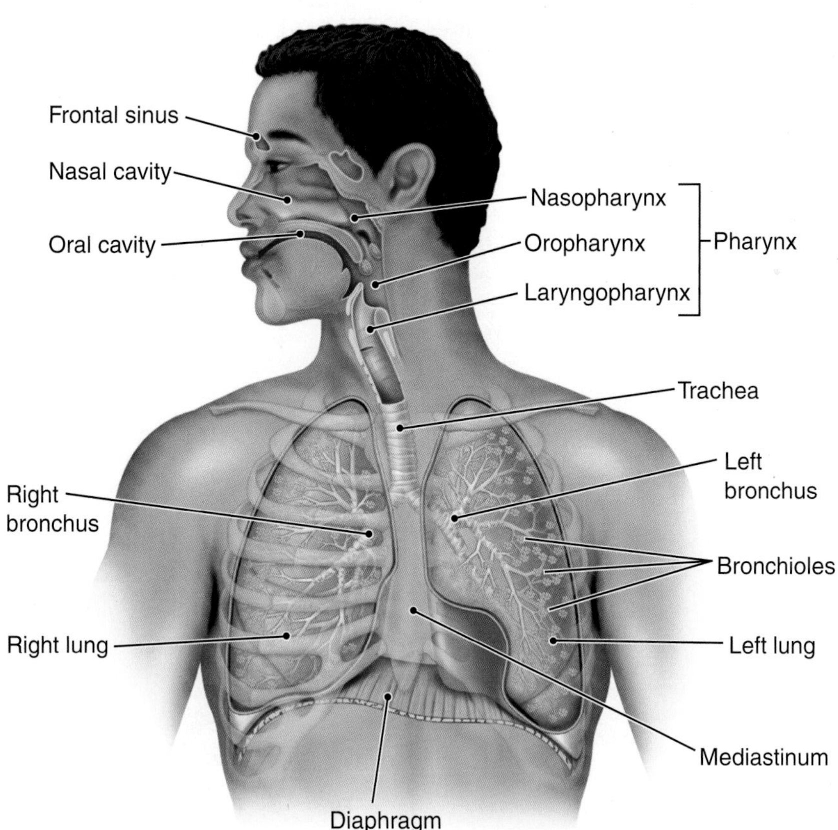

Figure 18-2 **The respiratory system.** **KEY POINT** The respiratory system consists of a series of airways that finally branch through the lungs. The esophagus, posterior to the trachea, is not shown. **ZOOMING IN** What organ is located in the medial depression of the left lung?

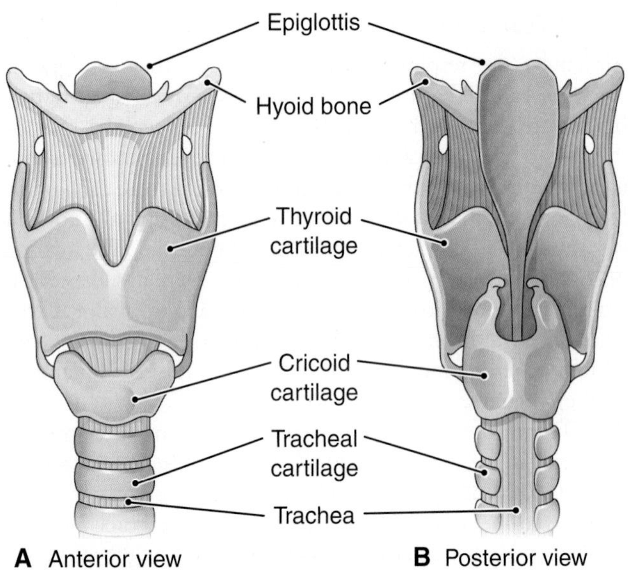

Figure 18-3 The larynx. **KEY POINT** The larynx is reinforced by hyaline cartilage, as is the trachea. The laryngeal cartilages include the epiglottis, thyroid cartilage, and cricoid cartilage.

As noted in Chapter 7, the paranasal sinuses are small cavities in the skull bones near the nose (**see Fig. 7-4**). They are resonating chambers for the voice and lessen the skull's weight. These sinuses are lined with mucous membrane and communicate with the nasal cavities. They are highly susceptible to infection traveling from the nose and throat.

THE PHARYNX

The muscular **pharynx** (FAR-inks), or throat, carries air into the respiratory tract and carries foods and liquids into the digestive system (**see Fig. 18-2**). The superior portion, located immediately behind the nasal cavity, is called the **nasopharynx** (na-zo-FAR-inks); the middle section, located posterior to the mouth, is called the **oropharynx** (o-ro-FAR-inks); and the most inferior portion is called the **laryngopharynx** (lah-rin-go-FAR-inks). This last section opens into the larynx toward the anterior and into the esophagus toward the posterior.

THE LARYNX

The **larynx** (LAR-inks), commonly called the *voice box* (**Fig. 18-3**), connects the pharynx with the trachea. Its rigid framework consists of nine portions of hyaline cartilage. The anterior part, the *thyroid cartilage*, protrudes at the anterior of the neck. This projection is commonly called the *Adam's apple* because it is considerably larger in men than in women. The inferior *cricoid* (KRI-koyd) *cartilage* forms a ring below the thyroid cartilage. It is used as a landmark for medical procedures involving the trachea.

Folds of mucous membrane used in producing speech are located centrally in the superior larynx. These are the **vocal folds**, or *vocal cords* (**Fig. 18-4**), which vibrate as

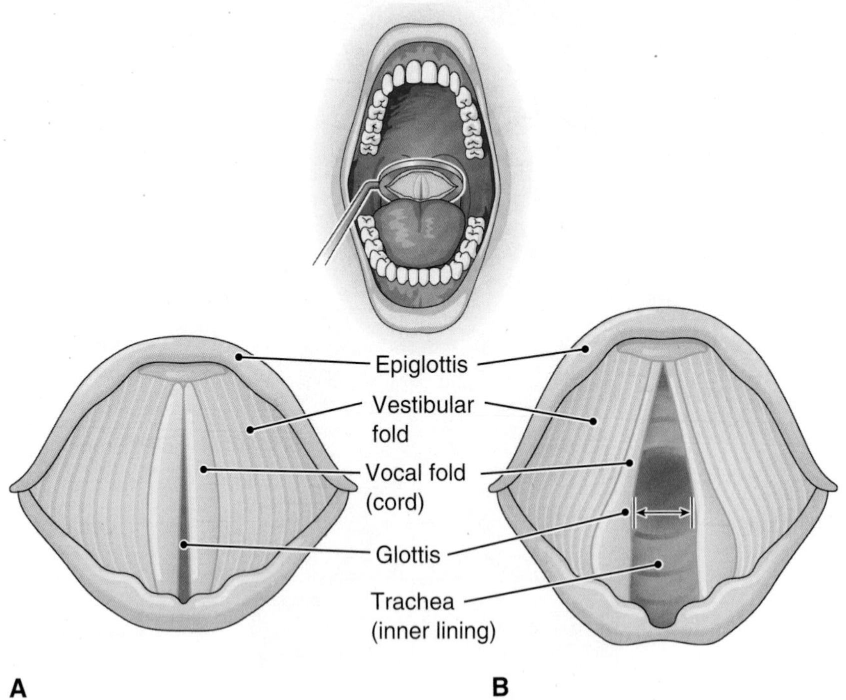

Figure 18-4 The vocal folds, superior view. **KEY POINT** The larynx contains the vocal cords, used in speech production. The glottis is the space between the vocal cords. The vestibular folds help close off the glottis when necessary. **A.** The glottis in closed position. **B.** The glottis in open position. **ZOOMING IN** What cartilage is named for its position above the glottis?

air flows over them. You can feel this vibration by placing your fingertips over the larynx at the center of your anterior throat and saying "ah." Variations in the length and tension of the vocal folds and the distance between them regulate the pitch of sound. The amount of air forced over them regulates volume. A difference in the size of the larynx and the vocal folds is what accounts for the difference between adult male and female voices. In general, a man's larynx is larger than a woman's. His vocal folds are thicker and longer, so they vibrate more slowly, resulting in a lower range of pitch. Muscles of the pharynx, tongue, lips, and face also are used to articulate words. The mouth, nasal cavities, paranasal sinuses, and the pharynx all serve as resonating chambers for speech, just as does the cabinet for an audio speaker. The space between the vocal folds is called the **glottis** (GLOT-is). This is partially open during normal breathing but widely open during forced breathing (**see Fig. 18-4**).

Superior to the vocal folds are additional folds in the laryngeal mucous membrane. These are known as the *vestibular folds* (**see Fig. 18-4**), sometimes called the "false vocal folds" because they do not contribute to speech production. Muscles in the larynx can bring these folds together to close off the glottis and help keep materials out of the respiratory tract during swallowing. They are also closed to help in holding one's breath against pressure in the thoracic cavity, as when straining to lift a heavy weight or to defecate.

The little leaf-shaped cartilage that covers the larynx during swallowing is called the **epiglottis** (ep-ih-GLOT-is). (The name means "above the glottis.") The glottis and epiglottis help keep food and liquids out of the remainder of the respiratory tract. As the larynx moves upward and forward during swallowing, the epiglottis moves downward, covering the opening into the larynx. You can feel the larynx move upward toward the epiglottis during this process by placing the pads of your fingers on your larynx as you swallow.

Despite laryngeal protections, it is possible to inhale or **aspirate** (AS-pih-rate) material into the respiratory tract. This might occur when someone is laughing or talking vigorously while eating, or an incapacitated individual inhales vomited gastric contents. Children can aspirate small objects or pieces of slippery food, such as hot dogs. Objects that enter the respiratory tract may cut off air supply completely, resulting in suffocation, or become lodged in the respiratory passageways. If the object is not removed, infection and inflammation due to irritation are likely to result.

THE TRACHEA

The **trachea** (TRA-ke-ah), commonly called the *windpipe*, is a tube that extends from the inferior edge of the larynx to the mediastinum, just superior to the heart (**see Fig. 18-2**). The trachea conducts air between the larynx and the lungs.

A framework of separate cartilages reinforces the trachea and keeps it open. These cartilages, each shaped somewhat like a tiny horseshoe or the letter C, are found along the trachea's entire length (**see Fig. 18-3**). The open sections in the cartilages are lined up at their posterior so that the esophagus can expand into this region during swallowing.

CHECKPOINTS

- ☐ 18-2 What happens to air as it passes over the nasal mucosa?
- ☐ 18-3 What are the three regions of the pharynx?
- ☐ 18-4 What are the scientific names for the throat, voice box, and windpipe?

THE BRONCHI

At its inferior end, the trachea divides into two mainstem, or primary, **bronchi** (BRONG-ki), which enter the lungs (**Fig. 18-5**). Cartilage rings stabilize the bronchi and keep them open to permit air passage. The right bronchus is considerably larger in diameter than is the left and extends downward in a more vertical direction (**see Fig. 18-19B** later in the chapter). Therefore, if a foreign body is inhaled, it is likely to enter the right lung. Each bronchus enters the lung at a depression called the hilum (HI-lum). Blood vessels and nerves also connect with the lung here and, together with the bronchus, make up a region known as the *root* of the lung.

18

THE LINING OF THE AIR PASSAGEWAYS

The trachea, bronchi, and other conducting passageways of the respiratory tract are lined with a mucous membrane (**see Fig. 4-3A**). Basically, it is simple, ciliated columnar epithelium with an underlying lamina propria, but the cells are arranged in such a way that they appear stratified. The tissue is thus described as *pseudostratified*, meaning "falsely stratified." Goblet cells within this epithelial membrane secrete mucus to trap impurities. The cilia on nearby cells sweep the mucus upward toward the throat, where it can be swallowed or eliminated by coughing, sneezing, or blowing the nose.

THE LUNGS

The **lungs** contain both air passageways and minute, thin-walled sacs called *alveoli*. They are the site of external gas exchange, that is, the exchange of gases between air and blood. The two lungs are set side by side in the thoracic (chest) cavity (**see Fig. 18-5A**). Between them are the heart, the great blood vessels, and other organs of the mediastinum (the space and organs between the lungs). (See **Fig. A5-5** in the Dissection Atlas for a photograph showing the lungs in relation to the heart and diaphragm.) The left lung has an indentation on its medial side that accommodates the heart.

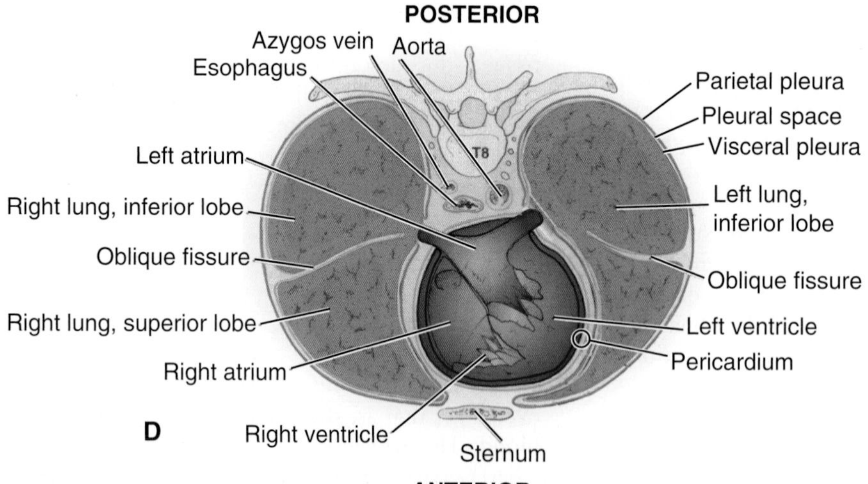

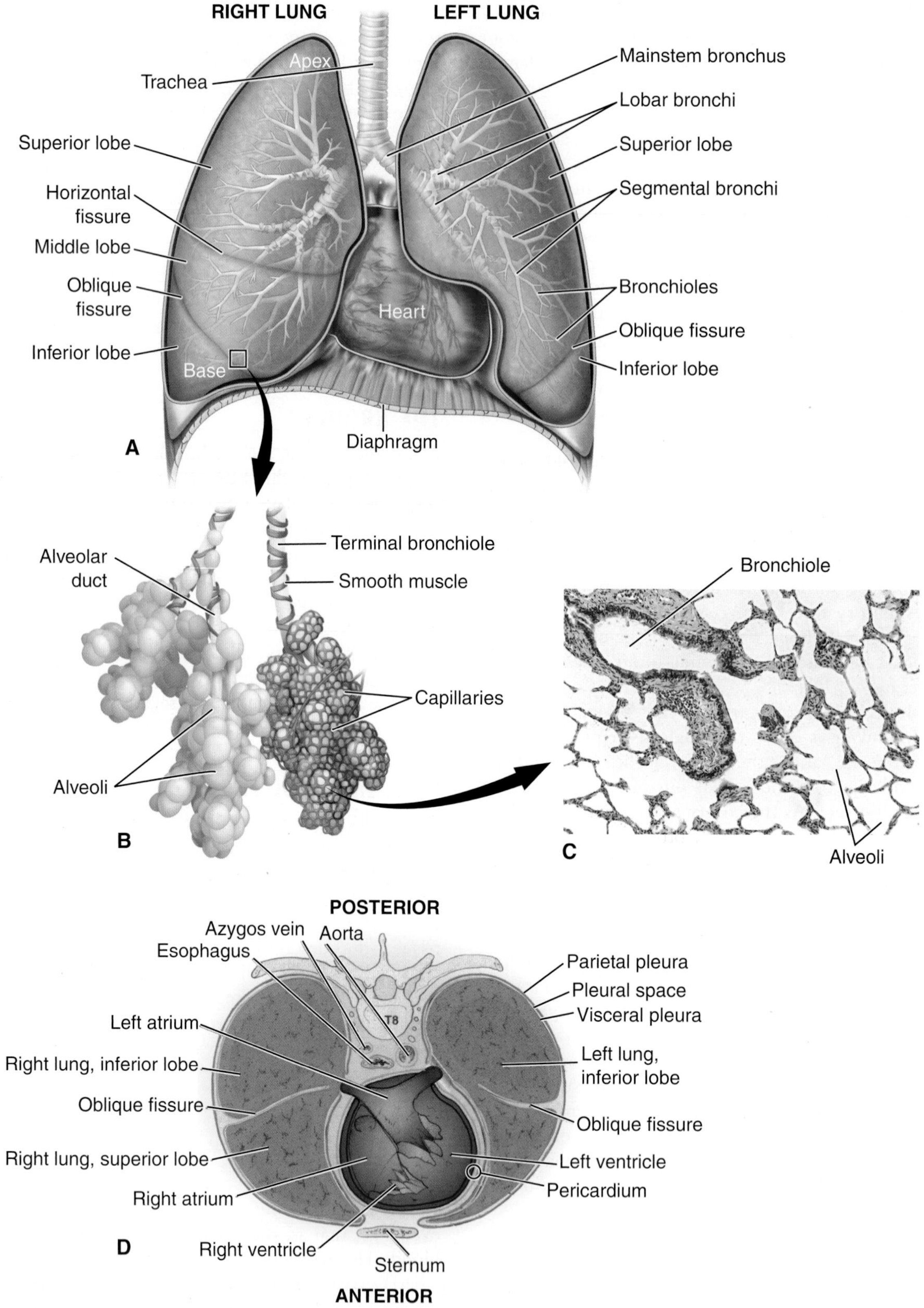

Figure 18-5 **The lungs.** **KEY POINT** The lungs are divided into lobes and segments that correspond to divisions of the bronchial tree. **A.** Position and structure of the lungs. **B.** Alveoli (air sacs). The left section of the diagram shows alveoli with capillaries removed. **C.** Histology of lung tissue. **D.** Cross-section of the thorax through the lungs (superior view), showing the visceral and parietal pleurae and the pleural space.

Divisions of the Lungs The right lung is subdivided by a horizontal and an oblique fissure into three lobes (superior, middle, and inferior); the left lung is divided by a single oblique fissure into two lobes (superior and inferior). Each lobe is then further subdivided into segments and then lobules. These subdivisions correspond to subdivisions of the bronchi as they branch throughout the lungs.

Each mainstem bronchus enters the lung at the hilum and immediately subdivides. The right bronchus divides into three lobar, or secondary bronchi, each of which enters one of the right lung's three lobes. The left bronchus gives rise to two lobar bronchi, which enter the left lung's two lobes. The bronchi subdivide again and again, becoming progressively smaller as they branch through lung tissue. The lobar bronchi become segmental bronchi as they branch into smaller segments of the lung. Because the bronchial subdivisions resemble the branches of a tree, they have been given the common name *bronchial tree*.

The Bronchioles The smallest of these conducting tubes are called **bronchioles** (BRONG-ke-oles). The histology of the tubes gradually changes as they become smaller. The amount of cartilage decreases until it is totally absent in the bronchioles; what remains is mostly smooth muscle, which is under the control of the autonomic (involuntary) nervous system. In Emily's asthma case study, spasms of this smooth muscle constrict her airways and make breathing difficult.

The Alveoli At the end of the **terminal bronchioles**, the smallest subdivisions of the bronchial tree, there are clusters of tiny air sacs in which most external gas exchange takes place. These sacs are the **alveoli** (al-VE-o-li) (sing. alveolus) (**see Fig. 18-5B**). The wall of each alveolus is made of simple squamous epithelium. This thin wall provides easy passage for the gases entering and leaving the blood as the blood circulates through the millions of tiny capillaries covering the alveoli.

There are about 300 million alveoli in the human lungs. The resulting surface area in contact with gases approximates 60 m^2 (some sources say even more). This area is equivalent, as an example, to the floor surface of a classroom that measures about 24 by 24 ft. As with many other body systems, there is great functional reserve; we have about three times as much lung tissue as is minimally necessary to sustain life. Because of the many air spaces, the lung is light in weight; normally, a piece of lung tissue dropped into a glass of water will float. **Figure 18-5C** shows a microscopic view of lung tissue.

The pulmonary circuit brings blood to and from the lungs. In the lungs, blood passes through the capillaries around the alveoli, where gas exchange takes place.

The Lung Cavities and Pleura The lungs occupy a considerable portion of the thoracic cavity, which is separated from the abdominal cavity by the muscular partition known as the **diaphragm** (DI-ah-fram). A continuous doubled sac, the **pleura** (PLU-ra), covers each lung (**see Fig. 18-5D**). As discussed in Chapter 4, the pleura is a serous membrane composed of an epithelial layer overlying areolar tissue. The two layers of the pleura are named according to location. The portion that is attached to the chest wall is the parietal pleura, and the portion that is attached to the lung surface is the visceral pleura. Each closed sac completely surrounds the lung, except at the hilum, where the bronchus and blood vessels enter the lung.

Between the two layers of the pleura is the **pleural space**, containing a thin film of fluid that lubricates the membranes. The effect of this fluid is the same as between two flat pieces of glass joined by a film of water; that is, the surfaces slide easily on each other but strongly resist separation. Thus, the lungs are able to expand and contract effortlessly during breathing, but the pleural fluid keeps them from separating from the chest wall.

CHECKPOINTS 

- 18-5 What structures do the inferior branching of the trachea form?
- 18-6 What feature of the cells lining the respiratory passageways enables them to move impurities away from the lungs?
- 18-7 In what structures does gas exchange occur in the lung?
- 18-8 What is the name of the membrane that encloses the lung?

The Process of Respiration

Respiration involves ventilation of the lungs, exchange of gases at the lungs and the body tissues, and their transport in the blood. Respiratory needs are met by central and peripheral controls of breathing.

PULMONARY VENTILATION

Ventilation is the movement of air into and out of the lungs, normally accomplished by breathing. There are two phases of ventilation: **inhalation**, which is the drawing of air into the lungs, and **exhalation**, or expiration, which is the expulsion of air from the lungs. Air moves down pressure gradients from an area of higher pressure to an area of lower pressure. Thus, inhalation occurs when lung pressure drops below atmospheric pressure, and exhalation occurs when lung pressure rises above atmospheric pressure. As discussed next, we change lung pressure by altering lung volume.

Inhalation In inhalation, the active phase of quiet breathing, respiratory muscles of the thorax and diaphragm contract to enlarge the thoracic cavity (**Fig. 18-6**). During quiet breathing, the diaphragm's movement accounts for most of the increase in thoracic volume. The diaphragm is a strong, dome-shaped muscle attached to the body wall around the base of the rib cage. The diaphragm's contraction and flattening cause a piston-like downward motion that increases the chest's vertical

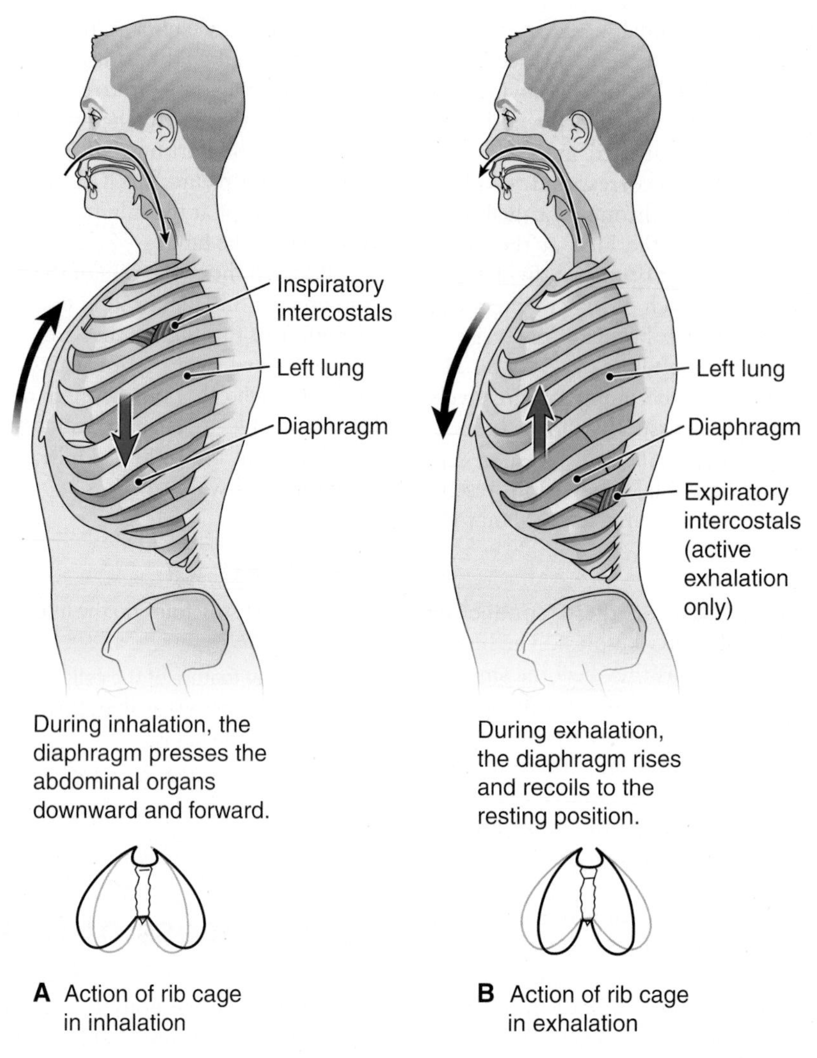
Inspiratory intercostals
Left lung
Diaphragm
During inhalation, the diaphragm presses the abdominal organs downward and forward.
A Action of rib cage in inhalation
Left lung
Diaphragm
Expiratory intercostals (active exhalation only)
During exhalation, the diaphragm rises and recoils to the resting position.
B Action of rib cage in exhalation

Figure 18-6 **Pulmonary ventilation.** **KEY POINT** The diaphragm and intercostal muscles are involved in inhalation, the active phase of quiet breathing. These muscles relax in exhalation, the passive phase of quiet breathing. **A.** Inhalation. **B.** Exhalation. **ZOOMING IN** What muscles are located between the ribs?

dimension. Other muscles that participate in breathing are the intercostal muscles located between the ribs. One subset of the intercostal muscles contracts during inhalation, lifting the rib cage upward and outward. Put the palms of your hands on either side of the rib cage to feel this action as you inhale. During forceful inhalation, the rib cage is moved further up and out by contraction of muscles in the neck and chest wall. Paralysis or weakness in the respiratory muscles thus interferes with one's ability to breathe.

As the thoracic cavity increases in size, gas pressure within the cavity decreases. This phenomenon follows a law in physics stating that when the volume of a given amount of gas increases, the pressure of the gas decreases. Conversely, when the volume decreases, the pressure increases. If you blow air into a tight balloon, the gas particles are in close contact and will hit the wall of the balloon frequently, creating greater pressure (**Fig. 18-7**). If you poke this balloon, it will spring back to its original shape. When you blow into a soft balloon that expands easily, the gas particles spread out into a larger volume and will not hit the balloon's wall as often. If you poke the balloon, your finger will make an indentation. Thus, pressure in the chest cavity drops as the thorax expands. When the pressure drops to slightly below the air pressure outside the lungs, air moves down the pressure gradient into the lungs as by suction.

Air enters the respiratory passages and flows through the ever-dividing tubes of the bronchial tree. As the air travels this route, it moves more and more slowly through the many bronchial tubes until there is virtually no forward flow as it reaches the alveoli. The incoming air mixes with the residual air remaining in the respiratory passageways so that the gases soon are evenly distributed. Each breath causes relatively little change in the gas composition of the alveoli, but normal continuous breathing ensures the presence of adequate oxygen and the removal of carbon dioxide.

As with expansion of blood vessels discussed in Chapter 15, the ease with which one can expand the lungs

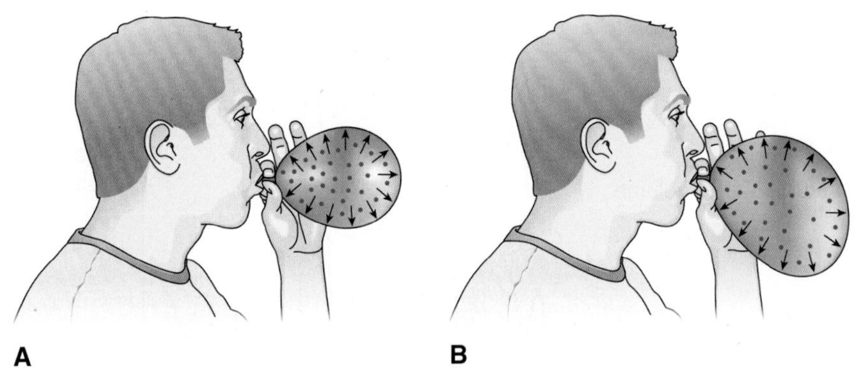

Figure 18-7 **The relationship of gas pressure to volume.** **KEY POINT A.** Inflation of a stiff balloon creates strong air pressure against the wall of the balloon. **B.** The same amount of air in a soft balloon spreads out into the available space, resulting in lower gas pressure. **ZOOMING IN** What happens to gas pressure as the volume of its container increases?

and thorax in inhalation is called **compliance**. To understand some of the conditions that can affect lung compliance, we must look again at the properties of water, first introduced in Chapter 2. A thin film of water lines the alveoli. The moisture is necessary, because gases must go into solution before they can diffuse across the membrane in traveling between the air and capillary blood. Water has the property of high *surface tension* based on water molecules' attraction for each other as a result of hydrogen bonding (see Box 2-1). You may have noticed that water will climb a short way up the inside of a narrow glass tube—the result of surface tension. A kind of "skin" forms at the surface of water that will support a light object or small insect, for example. Water's surface tension exerts an inward "pull" on the alveoli, causing them to resist expansion. To counteract this force, certain alveolar cells produce **surfactant** (sur-FAK-tant), a substance that reduces the surface tension of the fluids that line the alveoli. Surfactant is a mixture of lipoproteins that behaves much like a dish detergent, which reduces surface tension to aid in breaking down fats. Normal compliance of the lung tissue, aided by surfactant, allows the lungs to expand and fill adequately with air during inhalation. Compliance is decreased when the lungs resist expansion. Conditions that can decrease compliance include diseases that damage or scar lung tissue or deficiency of surfactant.

See the Student Resources on thePoint for illustrations of the breathing muscles and for the animation "Pulmonary Ventilation."

Exhalation In exhalation, the passive phase of quiet breathing, the respiratory muscles relax, allowing the ribs and diaphragm to return to their original positions. The lung tissues are elastic and recoil to their original size during exhalation. Surface tension within the alveoli aids in this return to resting size. During forced exhalation, a different subset of the intercostal muscles contracts, pulling the bottom of the rib cage in and down. The muscles of the abdominal wall contract, pushing the abdominal viscera upward against the relaxed diaphragm. Even with maximum exhalation, however, you cannot expel all the air from your lungs. There is always a certain residual volume left to fill the airways and keep the lungs inflated.

18

Spirometry The effectiveness of ventilation is measured by a **spirometer** (spi-ROM-eh-ter), an instrument for recording volumes of air inhaled and exhaled (Fig. 18-8). The tracing is a **spirogram** (SPI-ro-gram). Table 18-1 gives the definitions and average values for some of the breathing volumes and capacities that are important in evaluating respiratory function. A lung *capacity* is a sum of volumes. These same values are shown on a graph as they might appear on a spirogram.

Respiratory therapists evaluate and treat breathing disorders. See the Student Resources on thePoint for a description of this career.

GAS EXCHANGE

External gas exchange is the movement of gases between the alveoli and the capillary blood in the lungs (Fig. 18-9). The barrier that separates alveolar air from the blood is composed of the alveolar wall and the capillary wall, both of which are extremely thin. This respiratory membrane is not only very thin, but it is also moist. As previously noted, the moisture is important because the oxygen and carbon dioxide must go into a solution before they can diffuse across the membrane.

The principle that governs gas exchange between the outside air and the tissues is the now familiar concept of diffusion. Recall that solutes diffuse down concentration gradients. Gases diffuse instead down individual *pressure*

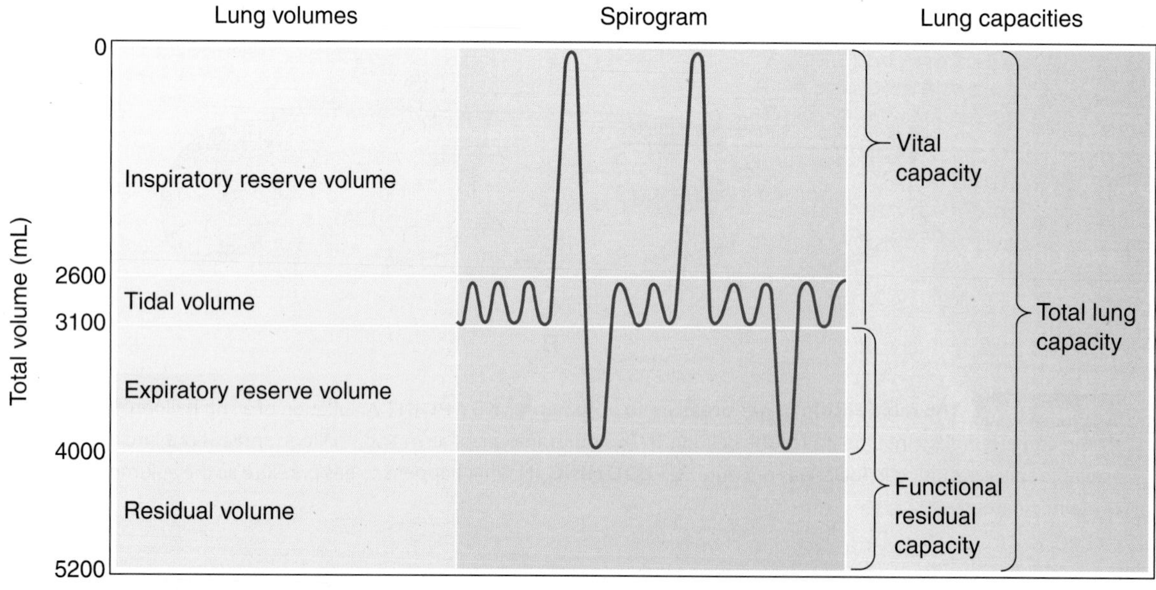

Figure 18-8 A spirogram. **KEY POINT** A spirogram is a tracing of lung volumes made with a spirometer, an instrument that measures volumes of air inhaled and exhaled. **ZOOMING IN** What lung volume cannot be measured with a spirometer? Which lung capacities cannot be measured with a spirometer?

gradients, and the pressure of a gas is expressed in millimeters of mercury (mm Hg), as is blood pressure. The pressure of a gas within a gas mixture (such as air) is called its **partial pressure**, symbolized with a P and a subscript of its formula. Thus, the partial pressures of oxygen and carbon dioxide are symbolized as Po_2 and Pco_2, respectively. Note that even though the gases are in a mixture, each diffuses independently of any other gas in that mixture.

The relative pressure of a gas on the two sides of a membrane determines the direction of diffusion (**see Fig. 18-9**). Recall that blood in the pulmonary capillaries has already encountered body cells, so its Po_2 is lower than that in inspired air. In the lungs, therefore, oxygen will diffuse across the alveolar wall and into the capillaries. Because the cells generate carbon dioxide in cellular respiration, the Pco_2 in the pulmonary capillaries is higher than the Pco_2 in inspired air, and carbon dioxide diffuses out of the blood and into the alveoli.

In contrast, internal gas exchange takes place between the blood and the tissues, again based on the relative pressures of the two gases. The blood arriving in the tissues has received additional oxygen in the lungs, and oxygen will pass into the oxygen-poor tissues. Carbon dioxide will diffuse out of the tissues and into the blood. Blood returning from the tissues and entering the lung capillaries through the pulmonary circuit is relatively low in oxygen and high in carbon

Table 18-1 Lung Volumes and Capacities

Volume	Definition	Average Value (mL)
Tidal volume	The amount of air moved into or out of the lungs in quiet, relaxed breathing	500
Residual volume	The volume of air that remains in the lungs after maximum exhalation	1,200
Inspiratory reserve volume	The additional amount that can be breathed in by force after a normal inhalation	2,600
Expiratory reserve volume	The additional amount that can be breathed out by force after a normal exhalation	900
Vital capacity	The volume of air that can be expelled from the lungs by maximum exhalation after maximum inhalation	4,000
Functional residual capacity	The amount of air remaining in the lungs after normal exhalation	2,100
Total lung capacity	The total volume of air that can be contained in the lungs after maximum inhalation	5,200

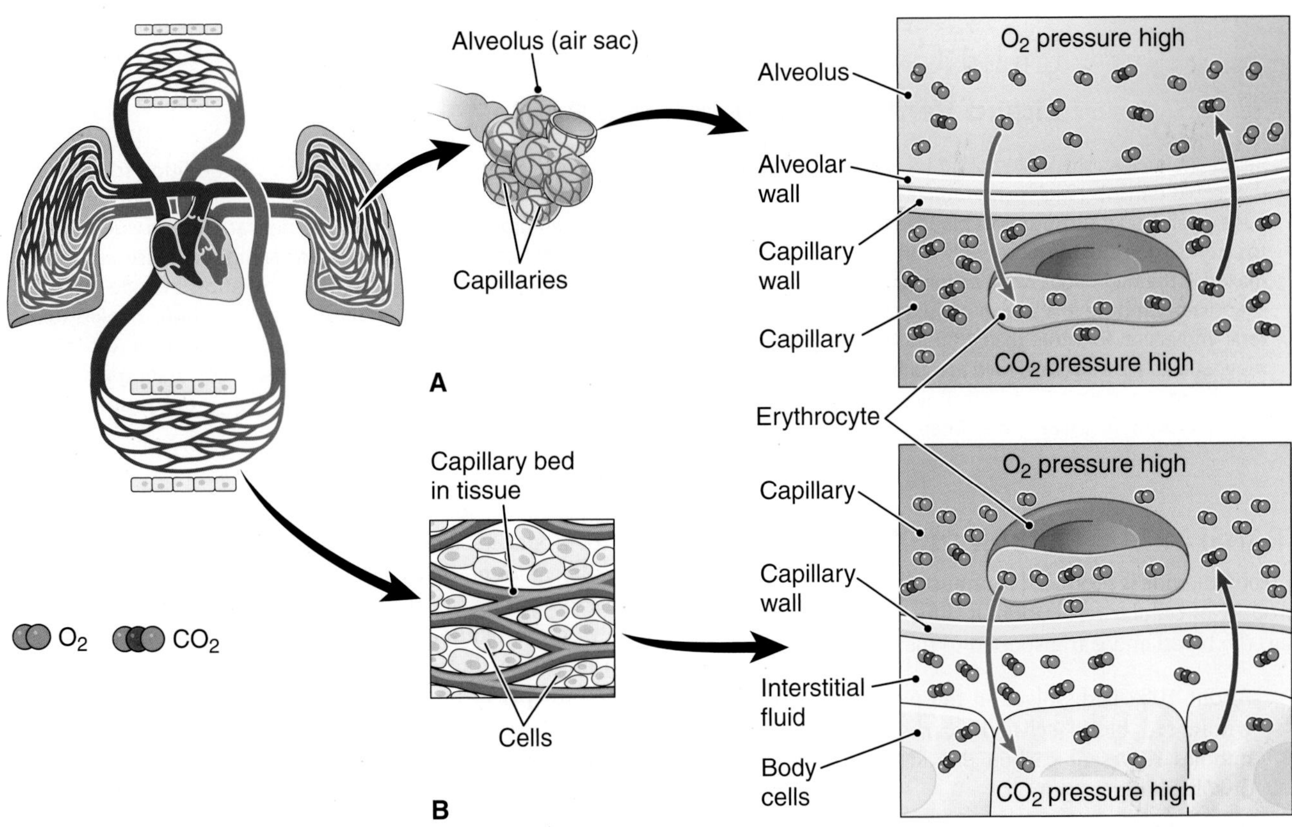

Figure 18-9 **Gas exchange.** **KEY POINT** Gas exchanges are based on relative partial pressures of oxygen and carbon dioxide on either side of a membrane. **A.** External exchange between the alveoli and the blood. Oxygen diffuses into the blood and carbon dioxide diffuses out based on pressures of the two gases in the alveoli and in the blood. **B.** Internal exchange between the blood and the cells. Oxygen diffuses out of the blood and into tissues, while carbon dioxide diffuses from the cells into the blood.

dioxide. Again, the blood will pick up oxygen and give up carbon dioxide. After a return to the left side of the heart, it starts once more on its route through the systemic circuit.

TRANSPORT OF OXYGEN

A very small amount (1.5%) of the oxygen in the blood is carried in solution in the plasma. (Oxygen does dissolve in water, as shown by the fact that aquatic animals get their oxygen from water.) However, almost all (98.5%) of the oxygen that diffuses into the capillary blood in the lungs binds to hemoglobin in the red blood cells. If not for hemoglobin and its ability to hold oxygen in the blood, it would be impossible for the heart to supply enough oxygen to the tissues. The hemoglobin molecule is a large protein with four small iron-containing "heme" regions. Each heme portion can bind one molecule of oxygen.

Highly oxygenated blood (in systemic arteries and pulmonary veins) is 97% saturated with oxygen. That is, only 3% of all the oxygen-binding sites on all of the hemoglobin molecules are unoccupied. Many hemoglobin molecules are carrying four oxygen molecules, but some carry only three. Blood low in oxygen (in systemic veins and pulmonary arteries) is usually about 70% saturated with oxygen. This 27% difference represents the oxygen that has been taken up by the cells. The terms *oxygenated* and *deoxygenated* are often used to describe blood that is high and low in oxygen, respectively. Note, however, that even blood that is described as deoxygenated still has a large reserve of oxygen. Even under conditions of high oxygen consumption, as in vigorous exercise, for example, the blood is never totally depleted of oxygen.

To enter the cells, oxygen must separate from hemoglobin. Normally, the bond between oxygen and hemoglobin is easily broken, and oxygen is released as blood travels into tissues where the oxygen pressure is relatively low. Cells are constantly using oxygen in cellular respiration and obtaining fresh supplies by diffusion from the blood. In addition, some conditions increase the rate of oxygen's release from hemoglobin. Increasing body temperature and increasing acidity, both generated by cellular activity, promote its release to the tissues.

The poisonous gas carbon monoxide (CO), at low partial pressure, binds with hemoglobin at the same molecular sites as does oxygen. However, it binds more tightly and displaces oxygen. Even a small amount of carbon monoxide causes a serious reduction in the blood's ability to carry oxygen.

For an interesting variation on normal gas transport, see Box 18-1 on liquid ventilation.

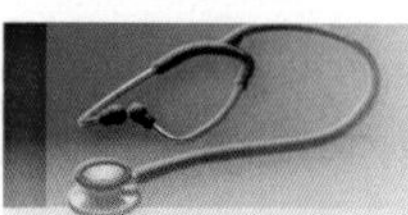

CLINICAL PERSPECTIVES

Box 18-1

Liquid Ventilation: Breath in a Bottle

Researchers have been attempting for years to develop a fluid that could transport high concentrations of oxygen in the body. Such a fluid could substitute for blood in transfusions or be used to carry oxygen into the lungs. Early work on liquid ventilation climaxed in the mid-1960s when a pioneer in this field submerged a laboratory mouse in a beaker of fluid and the animal survived total immersion for more than 10 minutes. The fluid was a synthetic substance that could hold as much oxygen as does air.

A newer version of this fluid, a fluorine-containing chemical known as PFC, has been tested to ventilate the collapsed lungs of premature babies. In addition to delivering oxygen to the lung alveoli, it also removes carbon dioxide. The fluid is less damaging to delicate lung tissue than is air, which has to be pumped in under higher pressure. Others who might benefit from liquid ventilation include people whose lungs have been damaged by infection, inhaled toxins, asthma, emphysema, and lung cancer, but more clinical research is required. Scientists are also investigating whether liquid ventilation could be used to deliver drugs directly to lung tissue.

TRANSPORT OF CARBON DIOXIDE

Carbon dioxide is produced continuously in the tissues as a byproduct of cellular respiration. It diffuses from the tissue cells into the blood and is transported to the lungs in three ways:

- About 10% is dissolved in the plasma and in the fluid within red blood cells. (In carbonated beverages, the bubbles represent carbon dioxide that was previously dissolved in the solution.)
- About 15% is combined with the protein portion of hemoglobin and with plasma proteins.
- About 75% is transported as an ion, known as a **bicarbonate** (bi-KAR-bon-ate) **ion**, which is formed when carbon dioxide undergoes a chemical change after it enters red blood cells. It first combines with water to form **carbonic** (kar-BON-ik) **acid**, which then separates (ionizes) into hydrogen and bicarbonate ions. This reaction is catalyzed by an enzyme called **carbonic anhydrase** (an-HI-drase) (CA).

The bicarbonate formed in the red blood cells moves to the plasma and then is carried to the lungs. In the lungs, the process is reversed as bicarbonate reenters the red blood cells, joins with a hydrogen ion to form carbonic acid, and again under the effects of carbonic anhydrase, releases carbon dioxide and water. The carbon dioxide diffuses into the alveoli and is exhaled. For those with a background in chemistry, the equation for these reactions follows. The arrows going in both directions signify that the reactions are reversible. The upper arrows describe what happens as CO_2 enters the blood; the lower arrows indicate what happens as CO_2 is released from the blood to be exhaled from the lungs. CA represents the enzyme carbonic anhydrase.

$$\underset{\text{Carbon dioxide}}{CO_2} + \underset{\text{Water}}{H_2O} \overset{CA}{\rightleftharpoons} \underset{\text{Carbonic acid}}{H_2CO_3} \rightleftharpoons \underset{\text{Hydrogen ion}}{H^+} + \underset{\text{Bicarbonate ion}}{HCO_3^-}$$

Carbon dioxide is important in regulating the blood's pH (acid–base balance). As a bicarbonate ion is formed from carbon dioxide in the plasma, a hydrogen ion (H^+) is also produced. Therefore, the blood becomes more acidic as the amount of carbon dioxide in the blood increases. The exhalation of carbon dioxide shifts the blood's pH more toward the alkaline (basic) range. The bicarbonate ion is also an important buffer in the blood, acting chemically to help keep the pH of body fluids within a steady range of 7.35 to 7.45. Chapters 21 and 22 have more information on body fluids and the kidneys' role in regulating the composition of body fluids.

CHECKPOINTS

- ☐ **18-9** What are the two phases of quiet breathing? Which is active and which is passive?
- ☐ **18-10** What substance produced by lung cells aids in compliance?
- ☐ **18-11** What property of a gas determines its direction of diffusion across a membrane, and in what units is this property expressed?
- ☐ **18-12** What substance in red blood cells holds almost all of the oxygen carried in the blood?
- ☐ **18-13** What is the main form in which carbon dioxide is carried in the blood?

See the Student Resources on thePoint to view the animation "Oxygen Transport" and "Carbon Dioxide Exchange."

REGULATION OF RESPIRATION

Regulation of respiration is a complex process that must keep pace with moment-to-moment changes in cellular oxygen requirements and carbon dioxide production. It must also be coordinated with complex behaviors such as laughing, swallowing, coughing, and singing. Centers in the central nervous system control the fundamental respiratory pattern.

Central Nervous Control The respiratory control center is a complex network of neurons located in the medulla and pons of the brain stem. The control center's main part, located in the medulla, sets the basic pattern of respiration (**Fig. 18-10**). At rest, these centers fire about 12 times per minute, so we breathe about every five seconds. This pattern changes in response to input from other brain regions, as discussed later.

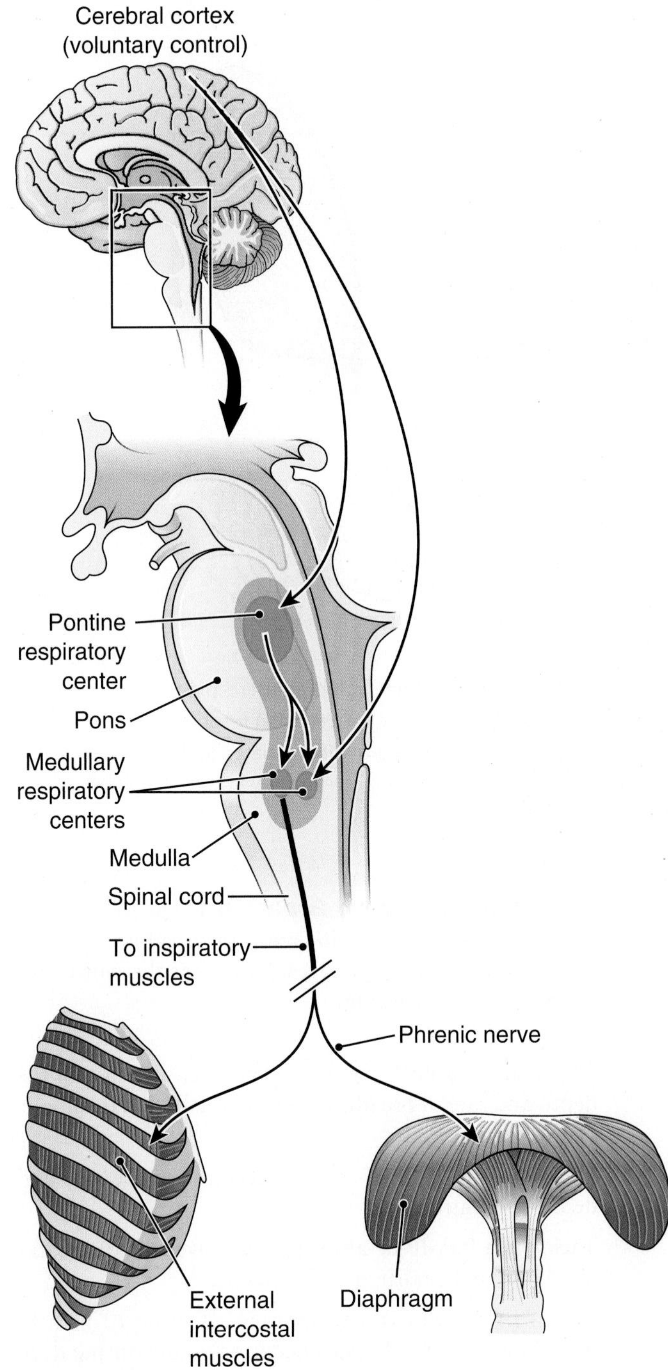

Figure 18-10 Brain centers controlling respiration. KEY POINT Regulatory centers are located in the medulla and pons. They control contractions of the diaphragm and intercostal muscles through the vagus nerve (X). Voluntary control comes from the cerebral cortex.

From the respiratory center in the medulla, motor nerve fibers extend into the spinal cord. From the cervical (neck) part of the cord, these nerve fibers continue through the **phrenic** (FREN-ik) **nerve** (a branch of the vagus nerve) to the diaphragm and also to the intercostal muscles. The diaphragm and the other respiratory muscles are voluntary in the sense that they can be regulated consciously by messages from the higher brain centers, notably the cerebral cortex (**see Fig. 18-10**). It is possible for you to deliberately breathe more rapidly or more slowly or to hold your breath and not breathe at all for a while. In a short time, however, the respiratory center in the brain stem will override the voluntary desire to hold your breath, and breathing will resume. Most of the time, we breathe without thinking about it, and the respiratory center is in control.

Chemical Control Of vital importance in the control of respiration are **chemoreceptors** (ke-mo-re-SEP-tors), which, like the receptors for taste and smell, are sensitive to chemicals dissolved in body fluids (**Fig. 18-11**). The chemoreceptors that regulate respiration are located centrally (near the brain stem) and peripherally (in arteries).

The central chemoreceptors are on either side of the brain stem near the medullary respiratory center. These receptors respond to the CO_2 level in circulating blood, but the gas acts indirectly. CO_2 is capable of diffusing through the capillary blood–brain barrier. It dissolves in medullary interstitial fluid and separates into hydrogen ion and bicarbonate ion, as explained previously. It is the presence of hydrogen ion and its effect in lowering pH that actually stimulates the central chemoreceptors. The rise in blood CO_2 level, known as **hypercapnia** (hi-per-KAP-ne-ah), thus increases the rate and depth of breathing.

The peripheral chemoreceptors that regulate respiration are found in structures called the *carotid* and *aortic bodies* (**see Fig. 18-11**). The carotid bodies are located near the bifurcation (forking) of the common carotid arteries in the neck, whereas the aortic bodies are located in the aortic arch. These bodies contain sensory neurons that respond mainly to a decrease in oxygen supply. They are not usually involved in regulating breathing because they do not act until oxygen drops to a very low level. Because there is usually an ample reserve of oxygen in the blood, carbon dioxide has the most immediate effect in regulating respiration at the level of the central chemoreceptors. When the carbon dioxide level increases, breathing must be increased to blow off the excess gas. Oxygen only becomes a controlling factor when its level falls considerably below normal, as in cases of lung disease or in high-altitude environments where oxygen partial pressure is low.

Other Factors in Respiratory Control Illustrating the complexity of respiratory control, there are still additional factors that influence breathing (**see Fig. 18-11**). Stimulation of pain receptors throughout the body can increase ventilation, acting through the brain's hypothalamus. Emotional responses can do the same. In the airways of the lungs, there are stretch receptors that stop inhalation to prevent overexpansion of the lungs. These are examples

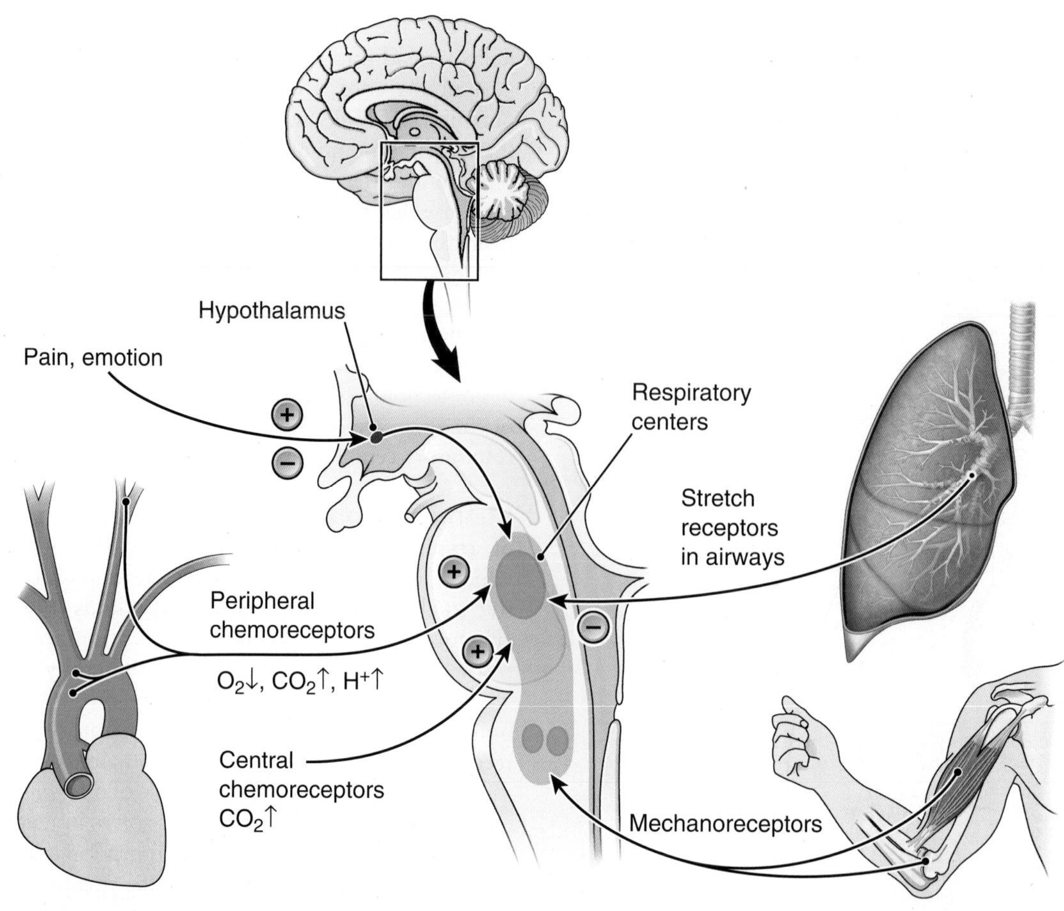

Figure 18-11 Regulation of respiration. **KEY POINT** Chemical controls and other factors adjust respiration to keep pace with need. + indicates increased respiration; – indicates decreased respiration; ± means that response may vary with circumstances. For simplicity, the specific areas of the respiratory centers are not differentiated in this diagram.

of mechanoreceptors, described in Chapter 11, that respond to changes in position. Finally, in muscles and joints, we have other mechanoreceptors that respond to movement and increase respiration as we move. All these mechanoreceptors act through the medullary respiratory center.

BREATHING PATTERNS

Normal breathing rates vary from 12 to 20 breaths per minute for adults. In children, rates may vary from 20 to 40 breaths per minute, depending on age and size. In infants, the respiratory rate may be more than 40 breaths per minute. Changes in respiratory rates are important in various disorders, and a healthcare provider should record them carefully. To determine the respiratory rate, the clinician counts the client's breathing for at least 30 seconds, usually by watching the chest rise and fall with each inhalation and exhalation. The count is then multiplied by 2 to obtain the rate in breaths per minute. It is best if the person does not realize that he or she is being observed because awareness of the measurement may cause a change in the breathing rate.

Some Terms for Altered Breathing The following is a list of terms designating various respiratory abnormalities. These are symptoms, not diseases. Note that the word ending *-pnea* refers to breathing.

- **Hyperpnea** (hi-PERP-ne-ah) refers to an increase in the depth and rate of breathing to meet the body's metabolic needs, as in exercise.
- **Hypopnea** (hi-POP-ne-ah) is a decrease in the rate and depth of breathing.
- **Tachypnea** (tak-IP-ne-ah) is an excessive rate of breathing that may be normal, as in exercise.
- **Apnea** (AP-ne-ah) is a temporary cessation of breathing. Short periods of apnea occur normally during deep sleep. More severe sleep apnea can result from obstruction of the respiratory passageways or, less commonly, by failure in the central respiratory center.
- **Dyspnea** (disp-NE-ah) is a subjective feeling of difficult or labored breathing.

- **Orthopnea** (or-THOP-ne-ah) refers to a difficulty in breathing that is relieved by sitting in an upright position, either against two pillows in bed or in a chair.
- **Kussmaul** (KOOS-mowl) **respiration** is deep, rapid respiration characteristic of acidosis as seen in uncontrolled diabetes.
- **Cheyne-Stokes** (CHANE stokes) **respiration** is a rhythmic variation in the depth of respiratory movements alternating with periods of apnea. It is caused by depression of the breathing centers and is noted in certain critically ill patients.

ABNORMAL VENTILATION

In **hyperventilation** (hi-per-ven-tih-LA-shun), the rate and depth of breathing increases above optimal levels. This condition can occur during anxiety attacks or when a person is experiencing pain or other forms of stress. Hyperventilation increases the exhalation of carbon dioxide and decreases the level of that gas in the blood, a condition called **hypocapnia** (hi-po-KAP-ne-ah). Recall the equation that links levels of carbon dioxide with blood pH. Hyperventilation shifts the equation to the left, removing acidic products from the blood and increasing its pH. This condition is referred to as *alkalosis*, excess alkalinity of body fluids, and can result in dizziness and tingling sensations.

In the absence of continued overriding inputs from other brain centers prompting hyperventilation, the respiratory center responds to decreased carbon dioxide levels by decreasing the rate and depth of respiration. Gradually, the carbon dioxide level returns to normal, and a regular breathing pattern is resumed. In extreme cases, a person may faint, and then breathing will involuntarily return to normal. In assisting a person who is hyperventilating, one should speak calmly, reassure him or her that the situation is not dangerous, and encourage even breathing using the diaphragm.

In **hypoventilation**, an amount of air insufficient to meet the body's metabolic needs enters the alveoli. The many possible causes of this condition include respiratory obstruction, lung disease, injury to the respiratory center, depression of the respiratory center, as by drugs, and chest deformity. Hypoventilation increases the blood's carbon dioxide concentration, shifting the equation cited previously to the right and decreasing the blood's pH. This condition is called *acidosis*, excess acidity of body fluids. The respiratory center responds to this condition by attempting to increase the rate and depth of respiration.

Results of Hypoventilation Conditions that may result from hypoventilation include the following:

- **Cyanosis** (si-ah-NO-sis) is a bluish color of the skin and mucous membranes caused by an insufficient amount of oxygen in the blood (**see Fig. 6-6**).
- **Hypoxia** (hi-POK-se-ah) means a lower-than-normal oxygen level in the tissues. The term **anoxia** (ah-NOK-se-ah) is sometimes used instead, but it is not as accurate because it means a total lack of oxygen.
- **Hypoxemia** (hi-pok-SE-me-ah) refers to a lower-than-normal oxygen partial pressure in arterial blood.

Box 18-2 offers information on adjusting to high altitudes and other hypoxic conditions.

18

CHECKPOINTS

- ☐ 18-14 Where in the brain stem are the centers that set the basic pattern of respiration?
- ☐ 18-15 What is the name of the motor nerve that controls the diaphragm?
- ☐ 18-16 What gas is the main chemical controller of respiration?
- ☐ 18-17 What is the meaning of the word ending *-pnea*?

Box 18-2

A CLOSER LOOK

Adaptations to High Altitude: Living with Hypoxia

Our bodies work best at low altitudes where oxygen is plentiful. However, people are able to live at high altitudes where oxygen is scarce and can even survive climbing Mount Everest, the tallest peak on our planet, showing that the human body can adapt to both short-term and long-term hypoxic conditions. This adaptation process compensates for decreased atmospheric oxygen by increasing the efficiency of the respiratory and cardiovascular systems.

The body's immediate response to high altitude is to increase the rate of ventilation (hyperventilation) and raise the heart rate to increase cardiac output. Hyperventilation makes more oxygen available to the cells and increases blood pH (alkalosis), which boosts hemoglobin's capacity to bind oxygen. Over time, the body adapts in additional ways. Hypoxia stimulates the kidneys to secrete erythropoietin, prompting red bone marrow to manufacture more erythrocytes and hemoglobin. Also, capillaries proliferate, increasing blood flow to the tissues. Some people are unable to adapt to high altitudes, and for them, hypoxia and alkalosis lead to potentially fatal **altitude sickness**.

Successful adaptation to high altitude illustrates the principle of homeostasis and also helps to explain how the body adjusts to hypoxia associated with disorders such as chronic obstructive pulmonary disease.

Respiratory Disorders

Infection is a major cause of respiratory disorders. These may involve any portion of the system. Allergies and environmental factors also affect respiration, and lung cancer is a major cause of cancer deaths in both men and women.

DISORDERS OF THE NASAL CAVITIES AND RELATED STRUCTURES

The paranasal sinuses are located in the skull bones in the vicinity of the nasal cavities. Infection may easily travel into these sinuses from the mouth, nose, and throat along their mucous membranes. The resulting inflammation is called **sinusitis**. Chronic (long-standing) sinus infection may stimulate proliferation of epithelial cells, resulting in benign tumor formation. Some of these growths have a grapelike appearance and cause airway obstruction; such tumors are called *polyps* (POL-ips).

As noted, the partition separating the two nasal cavities is the nasal septum. Because of minor structural variations, the nasal septum is rarely exactly in the midline. If it is markedly to one side, it is described as a **deviated septum** (**Fig. 18-12**). In this condition, one nasal space may be considerably smaller than the other. If an affected person has an attack of hay fever or develops a cold with accompanying swelling of the mucosa, the smaller nasal cavity may be completely closed. Sometimes, the septum is curved in such a way that both nasal cavities are occluded, forcing the person to breathe through his or her mouth. Such an occlusion may also prevent proper drainage from the sinuses and aggravate a case of sinusitis.

The most common cause of nosebleed, also called **epistaxis** (ep-e-STAK-sis) (from a Greek word meaning "to drip"), is injury to the mucous membranes in the nasal cavity. Causes of injury include infection, drying of the membranes, picking the nose, or other forms of trauma. These simple nosebleeds usually stop on their own, but some measures that help are applying pressure to the upper lip under the nose, compressing both nostrils, or inserting gauze or cotton into the bleeding nostril. Tilting the head forward can prevent blood from dripping down the throat and causing nausea. Epistaxis may signal more serious problems, such as blood clotting abnormalities, excessively high blood pressure, or tumors. Serious injuries that lead to nosebleed, or a nosebleed that will not stop, require professional medical care. Treatment may include packing the nose with gauze or other material, administering vasoconstrictors, or cauterizing the wound.

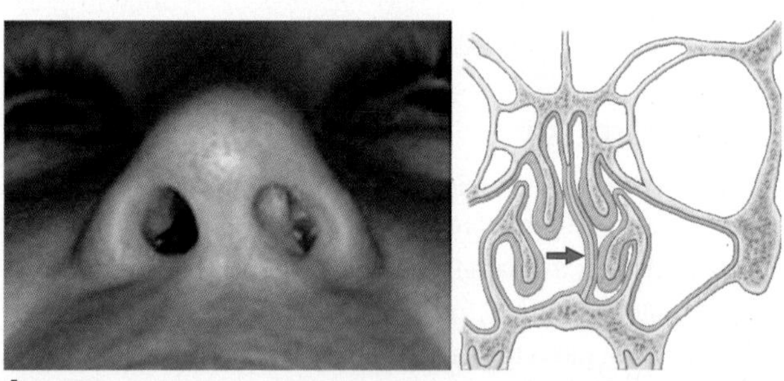

A

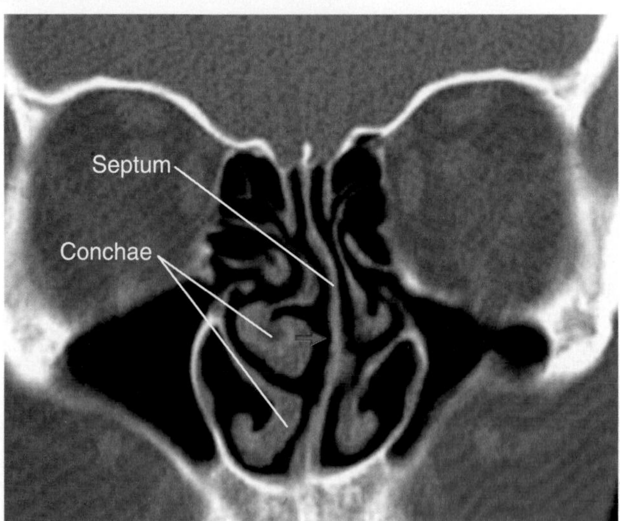

B

Figure 18-12 **Deviated nasal septum. A.** Nasal septum deviated to left side. **B.** Anterior view, CT scan.

INFECTION

The respiratory tract mucosa is one of the main portals of entry for disease-producing organisms. The transfer of disease organisms from one person's respiratory system to another's occurs most rapidly in crowded places, such as schools, theaters, and institutions. Droplets from one sneeze may be loaded with many billions of disease-producing organisms.

Mucous membranes can resist infection to some degree by producing larger quantities of mucus. The runny nose, an unpleasant symptom of the common cold, is the body's attempt to wash away pathogens and protect deeper tissues from further infection. If the membrane's resistance is reduced, however, the membrane may act as a pathway for the spread of disease. The infection may travel by that route into the nasal sinuses, or middle ear, along the respiratory passageways, or into the lung. Each infection is named according to the part involved, such as pharyngitis (commonly called a *sore throat*), laryngitis, or bronchitis.

Infections transmitted through the respiratory passageways include the common cold, chickenpox, measles, influenza, pneumonia, and tuberculosis. Any infection that is confined to the nose and throat is called an **upper respiratory infection** (URI). Often, an URI is the first evidence of infectious disease in children. Such an infection may precede the onset of a serious disease, such as rheumatic fever, which may follow a streptococcal throat infection.

See the Student Resources on thePoint for a summary diagram of respiratory infections.

The Common Cold The common cold is the most widespread of all respiratory diseases—of all communicable diseases, for that matter. More time is lost from school and work because of the common cold than any other disorder (see **Box 5-2** in Chapter 5). The causative agents are viruses that probably number more than 200 different types. Medical science has yet to establish the effectiveness and safety of any method for preventing the common cold. Because there are so many organisms involved, the production of an effective vaccine against colds seems unlikely.

The symptoms of the common cold are familiar: first the swollen and inflamed mucosa of the nose and the throat, then the copious discharge of watery fluid from the nose, and finally the thick and ropy discharge that occurs when the cold is subsiding. The scientific name for the common cold is **acute coryza** (ko-RI-zah); the word coryza can also mean simply "a nasal discharge."

Respiratory Syncytial Virus Respiratory syncytial virus (RSV) is the most common cause of lower respiratory tract infections in infants and young children worldwide. The name comes from the fact that the virus induces fusion of cultured cells (formation of a syncytium) when grown in the laboratory. Infection may result in bronchiolitis or pneumonia, but the virus may affect the upper respiratory tract as well. Most susceptible are premature infants, those with congenital heart disease, and those who are immunodeficient. Exposure to cigarette smoke is a definite risk factor.

The virus usually enters through the eyes and nose following contact with contaminated air, nasal secretions, or objects. The incubation period is three to five days, and an infected person sheds virus particles during the incubation period and up to two weeks thereafter. Infection usually resolves in five to seven days, although some cases require hospitalization and antiviral drug treatments.

Croup Croup usually affects children under 3 years of age and is associated with a number of different infections that result in upper respiratory inflammation. Airway constriction produces a loud, barking cough, wheezing, difficulty in breathing, and hoarseness. If croup is severe, the child may produce a harsh, squeaking noise (stridor) when breathing in through a narrowed trachea. Viral infections, such as those involving parainfluenza, adenovirus, RSV, influenza, or measles, are usually the cause. Although croup may be frightening to parents and children, recovery is complete in most cases within a week. However, medical treatment is warranted if the child's respiratory rate is very high and if the ribs become visible with each inhalation. Home treatments include humidifying room air or having the child breathe in steam. Also, cool air may shrink the respiratory tissues enough to bring relief. Medical interventions usually involve the administration of corticosteroids and bronchodilators.

Influenza Influenza (in-flu-EN-zah), or "flu," is an acute contagious disease characterized by an inflammatory condition of the upper respiratory tract accompanied by generalized aches and pains and high fever. It is caused by a virus and may spread throughout the upper and lower respiratory tracts. Inflammation of the trachea and the bronchi causes the characteristic cough of influenza, and the general infection causes an extremely weakened condition. The great danger of influenza is its tendency to develop into a particularly severe form of pneumonia. At intervals in history, there have been tremendous epidemics of influenza in which millions of people have died. Vaccines have been effective, although the protection is of short duration.

Pneumonia Pneumonia (nu-MO-ne-ah) is an infectious inflammation of the lungs in which the air spaces become filled with fluid. A variety of bacteria and viruses may be responsible. Many of these pathogens may be carried by a healthy person in the upper respiratory mucosa. If the person remains in good health, they may be carried for a long time with no ill effect. If the patient's resistance to infection is lowered, however, the pathogens may invade the tissues and cause disease.

Susceptibility to pneumonia is increased in patients with chronic, debilitating illness or chronic respiratory disease, in smokers, and in people with alcoholism. It is also increased in cases of exposure to toxic gases, suppression of the immune system, or viral respiratory infections.

There are two main kinds of pneumonia as determined by the extent of lung involvement and other factors:

- **Lobar pneumonia**, in which an entire lobe of the lung is infected at one time (**Fig. 18-13**, right lung). The causative organism is usually a pneumococcus, although other pathogens may also cause this disease. *Legionella*, as noted, is the agent of Legionnaires disease, or legionellosis, which is a severe lobar pneumonia that occurs mostly in localized epidemics. 18

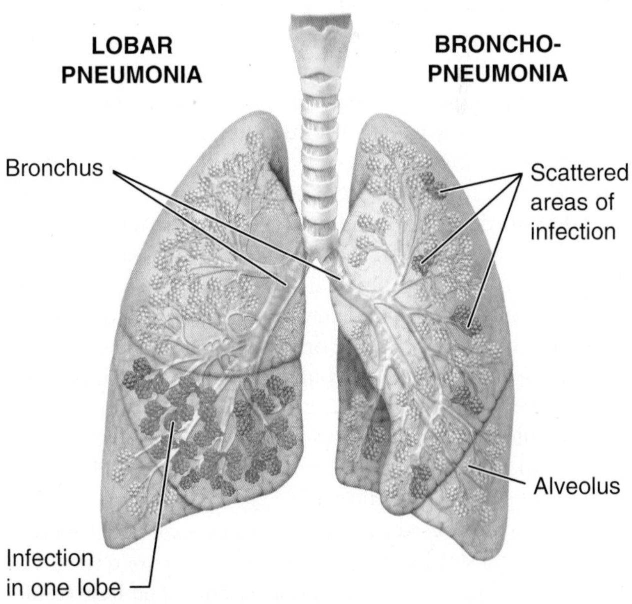

Figure 18-13 **Pneumonia.** **KEY POINT** Lobar pneumonia (right lung) involves an entire lobe of the lung. Bronchopneumonia (left lung) involves patchy areas of the entire lung.

- **Bronchopneumonia**, in which the disease process is scattered throughout the lung (**Fig. 18-13**, left lung). The cause may be infection with a staphylococcus, gram-negative *Proteus* species, colon bacillus (not normally pathogenic), or a virus. Bronchopneumonia most often is secondary to an infection or to some factor that has lowered the patient's resistance to disease. This is the most common form of pneumonia.

Most types of pneumonia are characterized by the formation of a fluid, or **exudate** (EKS-u-date), in the infected alveoli; this fluid consists chiefly of serum and pus, products of infection. Some red blood cells may be present, as indicated by red streaks in the sputum. Sometimes, so many air sacs become filled with fluid that the victim finds it hard to absorb enough oxygen to maintain life.

Pneumocystis pneumonia (PCP) occurs mainly in people with weakened immune systems, such as HIV-positive individuals or transplant recipients on immunosuppressant drugs. The infectious agent was originally called *P. carinii* and classified as a protozoon. It has now been reclassified as an atypical fungus and renamed *P. jiroveci*. The organism grows in the fluid that lines the alveoli of the lungs. Laboratories diagnose the disease by microscopic identification of the organism in a sputum sample, bronchoscopy specimen, or lung biopsy specimen. PCP is treated with antimicrobial drugs, although these medications may cause serious side effects in immunocompromised patients.

Tuberculosis **Tuberculosis** (tu-ber-ku-LO-sis) (TB) is an infectious disease caused by the bacillus *Mycobacterium tuberculosis*. Although this bacillus may invade any body tissue, it usually grows in the lung. In the United States, tuberculosis remains a leading cause of death from communicable disease, primarily because of the relatively large numbers of cases among recent immigrants, elderly people, and poor people in metropolitan areas. The spread of AIDS has been linked with a rising incidence of TB because this viral disease weakens host defenses.

The name *tuberculosis* comes from the small lesions, or *tubercles*, that form where the organisms grow. If unchecked, these lesions degenerate and may even liquefy to cause cavities within an organ. In early stages, the disease may lie dormant, only to flare up at a later time. The tuberculosis organism can readily spread into the lymph nodes or into the blood and be carried to other organs. The lymph nodes in the thorax, especially those surrounding the trachea and bronchi, are frequently involved. Infection of the pleura results in tuberculous pleurisy (inflammation of the pleura). In this case, a collection of fluid, known as an effusion (e-FU-zhun), accumulates in the pleural space.

Drugs can be used successfully in many cases of tuberculosis, although strains of the TB organism that are resistant to multiple antibiotics have appeared recently. The best results have been obtained by use of a combination of several drugs, with prompt, intensive, and uninterrupted treatment once a program is begun. Therapy is usually continued for a minimum of six to 18 months; therefore, close supervision by the healthcare practitioner is important. Adverse drug reactions are rather common, necessitating changes in the drug combinations. Drug treatment of patients whose infection has not progressed to active disease is particularly effective.

Researchers have recently developed a rapid molecular test for TB in sputum that gives results in two hours rather than the two to three days for a skin test or the four to six weeks needed for culturing the organism in the laboratory. The test also reveals whether the strain of TB in question is resistant to the drug commonly used to treat the disease (rifampin). Use of this test would lead to earlier treatment and help prevent spread of the disease.

See the Student Resources on thePoint for a micrograph of the acid-fast TB organism, *Mycobacterium tuberculosis*, and a diagram showing the course of a tuberculosis infection.

ALLERGIC RHINITIS

Hypersensitivity to an inhaled substance or substances is the cause of *allergic rhinitis* (ri-NI-tis) (*rhin/o* is the word root for "nose"). Common causative allergens include plant pollens, dust, mold, and animal dander. When seasonal pollens are the cause, the disorder is commonly called *hay fever*. Allergic rhinitis is characterized by a watery discharge from the eyes and nose, sneezing, and coughing. Stuffy nose, sore throat, and headache may also develop, leading to fatigue and irritability.

While avoiding the offending substance prevents allergic rhinitis, identifying the particular allergen involved can be difficult. Allergists usually give a number of skin tests, but in most cases, the results are inconclusive. Antihistamines, corticosteroids, and decongestants can help treat the symptoms, but these have significant side effects. People with these allergies may benefit from a series of injections to reduce their sensitivity to specific substances.

ASTHMA

Another disorder often related to allergy is **asthma** (AZ-mah). The causes of asthma and individual responses may vary, but respiratory infection, noxious fumes, tobacco smoke, cold air, and allergies to inhaled substances or drugs can initiate episodes. Most asthma cases involve multiple causes. One of the more common triggers for asthma attacks is exercise: the rapid air movement in sensitive airways causes smooth muscle spasm in the breathing passages.

The symptoms of asthma are caused by reversible changes in the airways, including inflammation and excessive mucus production and spasm of the involuntary muscle in bronchial tubes. These changes narrow the bronchial lumen, causing resistance to air flow. The person experiences a sense of suffocation and has labored breathing (dyspnea), often with wheezing.

Asthma treatment may include inhaled steroids to prevent inflammation and inhaled bronchodilators to open airways during acute episodes. Emily's physician in the case study recommends both of these measures.

CHRONIC OBSTRUCTIVE PULMONARY DISEASE

Chronic obstructive pulmonary disease (COPD) is the term used to describe several associated lung disorders, including **chronic bronchitis** and **emphysema** (em-fih-SE-mah). Most affected patients have symptoms and lung damage characteristic of both diseases. In chronic bronchitis, the airway linings are chronically inflamed and produce excessive secretions. Emphysema is characterized by dilation and finally destruction of the alveoli (**Fig. 18-14**).

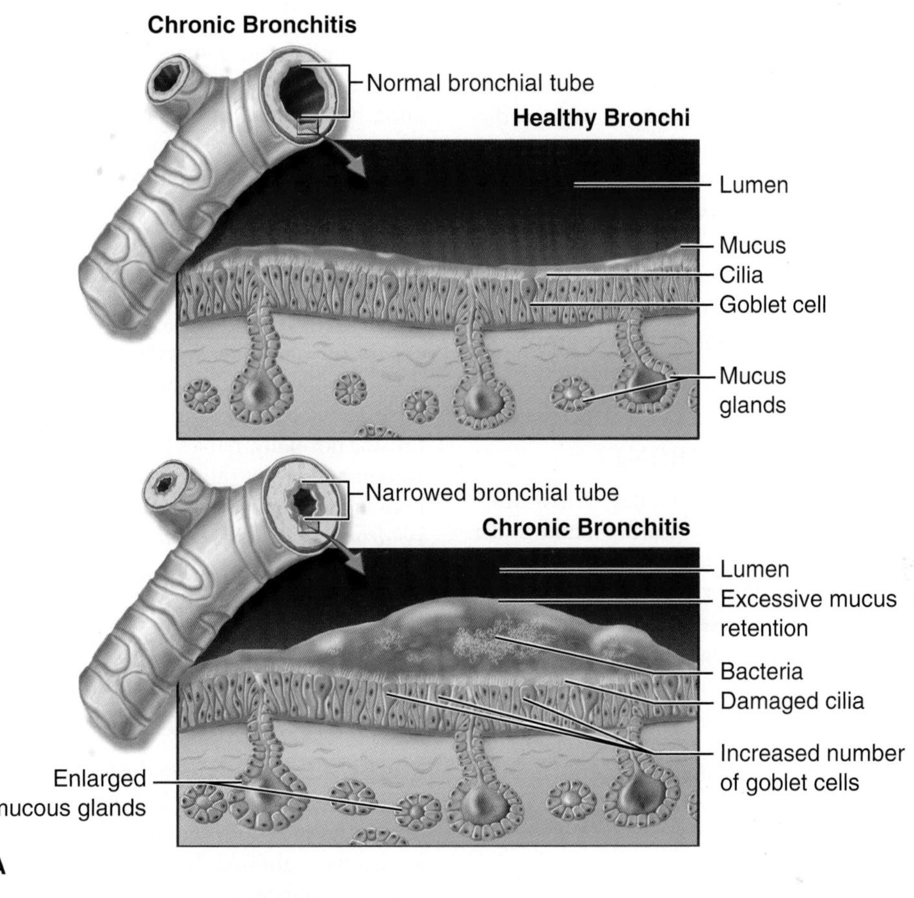

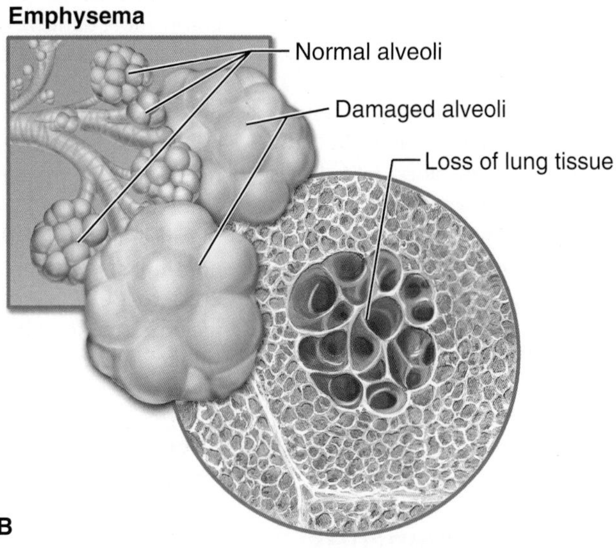

Figure 18-14 Chronic obstructive pulmonary disease (COPD). **KEY POINT** The main disorders involved in COPD are chronic bronchitis and emphysema. **A.** Changes seen in chronic bronchitis, involving airway inflammation, damage to cilia, and excess mucus secretion. **B.** Emphysema results in dilation and destruction of alveoli.

In COPD, the airway and lung changes impede airflow, particularly during exhalation. Exhalation becomes more difficult as the disease progresses, until the person cannot exhale all of the air inhaled. Increasing amounts of air become trapped in the lungs, and the resulting lung overexpansion makes inhalation difficult as well. (You can experience this sensation by performing deep inhalations but shallow exhalations.) To make matters worse, destruction of the alveoli decreases the lung area available for gas diffusion. Carbon dioxide accumulates in the blood, and eventually oxygen transport is also reduced.

In certain smokers, COPD may result in serious disability and death within 10 years of the onset of symptoms. Giving up smoking slows and may reverse progression of the disorder, especially when it is diagnosed early and other respiratory irritants also are eliminated.

See the Student Resources on thePoint for an animation on asthma and a photograph showing emphysema in lung tissue.

SUDDEN INFANT DEATH SYNDROME

Sudden infant death syndrome (SIDS), also called "crib death," is the unexplained death of a seemingly healthy infant under 1 year of age. Death usually occurs during sleep, leaving no signs of its cause. Autopsy, family history, and circumstances of death provide few clues. However, exposure to cigarette smoke, infant prematurity, and low birth weight are the highest risk factors for SIDS. Certain maternal conditions during pregnancy are also associated with an increased risk of SIDS, though none is a sure predictor. These include cigarette smoking, age under 20, low weight gain, anemia, and illegal drug use.

Some guidelines that have reduced the SIDS incidence are as follows:

- Place the baby supine (on his or her back) for sleep, unless there is some medical reason not to do so. A reminder for parents and other caregivers is the slogan "back to sleep." A side position is not as effective, as the baby can roll over. The baby should never be put to sleep prone (face down). In the prone position, the baby can rebreathe its own CO_2, and high blood CO_2 levels can depress the respiratory center. The position may also cause obstruction in the upper airways or lead to overheating by reducing loss of body heat.
- Keep the baby in a smoke-free environment.
- Use a firm, flat baby mattress, not soft foam, fur, or fiber-filled bedding.
- Avoid overheating the baby with room air, clothing, or bedding, especially when the baby has a cold or other infection.

ACUTE RESPIRATORY DISTRESS SYNDROME

Acute respiratory distress syndrome (ARDS) is a sudden, potentially reversible inflammatory condition of the lung resulting from injury or infection. It is frequently fatal. Some causes of ARDS are:

- Airway obstruction as from mucus, foreign bodies, emboli, or tumors
- Sepsis (systemic infection)
- Aspiration (inhalation) of stomach contents
- Allergy
- Lung trauma

Inflammation and damage to the alveoli result in pulmonary edema, dyspnea (difficulty in breathing), decreased compliance, hypoxemia, and formation of fibrous scar tissue in the lungs. The incomplete expansion of a lung or portion of a lung, such as results from ARDS, is called **atelectasis** (at-eh-LEK-tah-sis), or collapsed lung.

SURFACTANT DEFICIENCY DISORDER

Recall that surfactant is a soapy substance that facilitates lung expansion and thus inhalation. Specialized fetal lung cells begin producing surfactant at about 26 weeks, so babies born after about 35 weeks in the womb usually have enough to breathe normally. Babies born too early, however, can lack surfactant and find it difficult to inflate their lungs. This disorder, known as **surfactant deficiency disorder** or *respiratory distress syndrome (RDS) of the newborn*, is now treated by administration of surfactant that is produced by genetic engineering.

CANCER

Tumors can arise in all portions of the respiratory tract. Two common sites are described next.

Lung Cancer Cancer of the lungs is the most common cause of cancer-related deaths in both men and women. The incidence rate in women continues to increase, whereas the rate in men has stabilized. By far, the most important cause of lung cancer is cigarette smoking, and modern cigarettes are considerably more dangerous than are older versions. Smokers suffer from lung cancer 10 times as often as do nonsmokers. The risk of getting lung cancer is increased in people who started smoking early in life, smoke large numbers of cigarettes daily, and inhale deeply. Smokers who are exposed to toxic chemicals or particles in the air have an even higher lung cancer rate. Smoking has also been linked with an increase in other types of cancer and in COPD. Studies have also shown that lifetime nonsmokers who are exposed to secondhand smoke have a 20% to 30% increase in cancer risk. Secondhand smoke includes smoke exhaled by a smoker as well as "sidestream smoke" which comes out of a cigarette, cigar, or pipe.

See the Student Resources on thePoint for an illustration of the effects of smoking.

A common form of lung cancer is **bronchogenic** (brong-ko-JEN-ik) **carcinoma**, so-called because the malignancy originates in a bronchus. **Figure 18-15** shows a particularly large tumor in the right lung. The tumor may grow until the bronchus is blocked, cutting off the air supply to that lung.

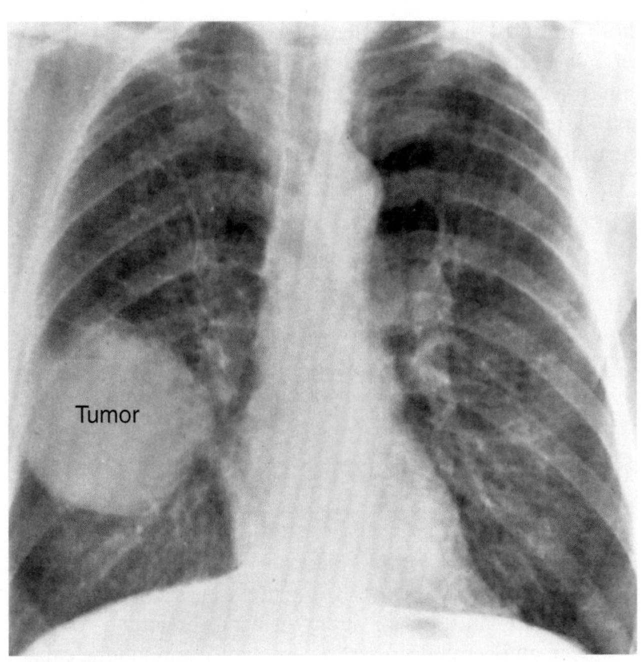

Figure 18-15 **Bronchogenic carcinoma.** **KEY POINT** A well-defined mass is seen in one lung.

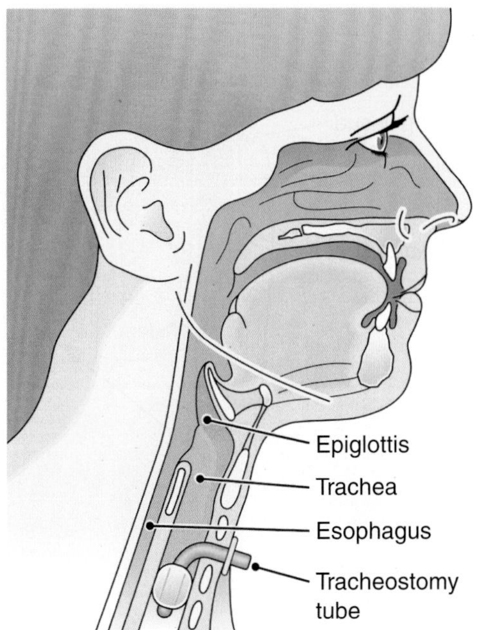

Figure 18-16 **A tracheostomy tube in place.** **KEY POINT** A tracheostomy tube creates an artificial airway. **ZOOMING IN** What structure is posterior to the trachea?

The lung then collapses, and the secretions trapped in the lung spaces become infected with a resulting pneumonia or lung abscess formation. This type of lung cancer can spread, causing secondary growths in the lymph nodes of the chest and neck and in the brain and other parts of the body. A treatment that offers a possibility of cure before secondary growths have had time to form is complete removal of the lung. This operation is called a **pneumonectomy** (nu-mo-NEK-to-me).

Malignant tumors of the stomach, breast, and other organs may spread to the lungs as secondary growths (metastases).

Cancer of the Larynx Cancer of the larynx usually involves squamous cell carcinoma, and its incidence is definitely linked to cigarette smoking and alcohol consumption. Symptoms include sore throat, hoarseness, ear pain, and enlarged cervical lymph nodes. The cure rate is high for small tumors that have not spread when treated with radiation and (sometimes) surgery. Total or partial removal of the larynx is necessary in cases of advanced cancer that do not respond to radiation or chemotherapy. During and after laryngectomy a patient may need a surgically created opening, or *stoma*, in the trachea with a tube (tracheostomy tube) for breathing. Patients have to care for the stoma and the "trach" (trake) tube (**Fig. 18-16**). If the vocal cords are removed, a patient can learn to speak by using air forced out of the esophagus or by using a mechanical device to generate sound.

DISORDERS INVOLVING THE PLEURA

Pleurisy (PLUR-ih-se), or inflammation of the pleura, usually accompanies a lung infection—particularly pneumonia or tuberculosis. This condition can be quite painful because the inflammation produces a sticky exudate that roughens the pleura of both the lung and the chest wall; when the two surfaces rub together during ventilation, the roughness causes acute irritation. The sticking together of two surfaces is called an adhesion (ad-HE-zhun). Infection of the pleura also causes an increase in the amount of pleural fluid. This fluid may accumulate in the pleural space in quantities large enough to compress the lung, resulting in an inability to obtain enough air.

Pneumothorax (nu-mo-THOR-aks) is an accumulation of air in the pleural space (**Fig. 18-17**). The lung on the affected side collapses partially or completely causing the patient to have great difficulty breathing. Pneumothorax may be caused by a wound in the chest wall or by rupture of the lung's air

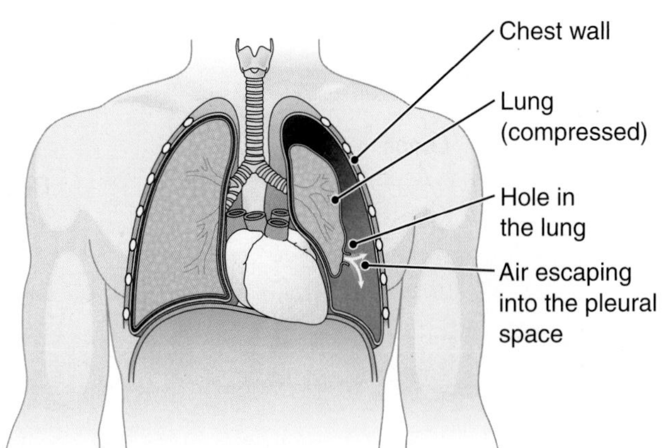

Figure 18-17 **Pneumothorax.** **KEY POINT** Injury to lung tissue allows air to leak into the pleural space and put pressure on the lung.

18

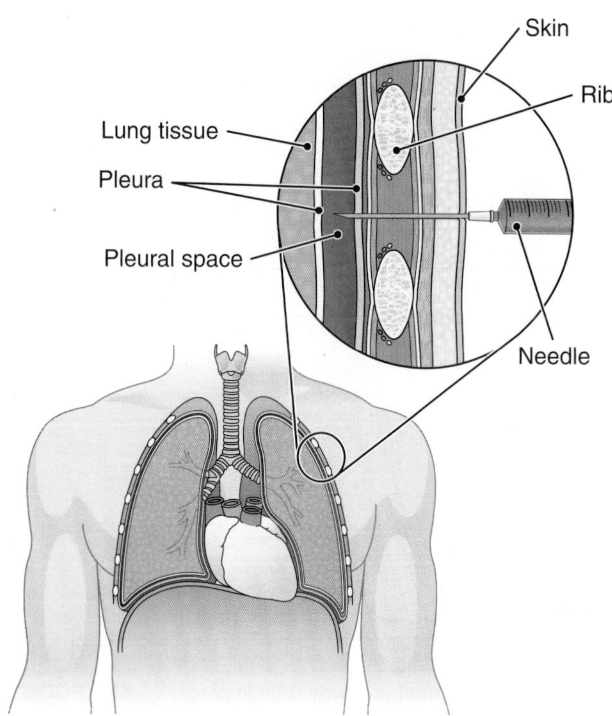

Figure 18-18 **Thoracentesis.** **KEY POINT** A needle is inserted into the pleural space to withdraw fluid.

spaces. In pneumothorax caused by a penetrating wound in the chest wall, an airtight cover over the opening prevents further air from entering. The remaining lung can then function to provide adequate ventilation until the wound heals.

Blood in the pleural space, a condition called **hemothorax,** is also caused by penetrating chest wounds. In such cases, the first priority is to stop the bleeding.

The abnormal accumulation of fluid or air in the pleural space from any of the above conditions may call for procedures to promote lung expansion. In **thoracentesis** (thor-ah-sen-TE-sis), a large-bore needle is inserted between ribs into the pleural space to remove fluid (**Fig. 18-18**). The presence of both air and fluid in the pleural space may require insertion of a chest tube, a large tube with several openings along the internal end. The tube is securely connected to a chest drainage system. These procedures restore negative pressure in the pleural space and allow re-expansion of the lung.

CHECKPOINTS

- [ ] **18-18** What are the names of six respiratory disorders caused by infection?
- [ ] **18-19** What are some causes of allergic rhinitis? Asthma?
- [ ] **18-20** What does COPD mean, and what two diseases are commonly involved in COPD?

Effects of Aging on the Respiratory Tract

With age, the tissues of the respiratory tract lose elasticity and become more rigid. Similar rigidity in the chest wall, combined with arthritis and loss of strength in the breathing muscles, results in an overall decrease in compliance and in lung capacity. However, there is a great deal of variation among individuals, and regular aerobic activity throughout adulthood (walking, running, swimming, etc.) can contribute significantly to maintaining respiratory function.

Reduction in protective mechanisms in the lungs, such as phagocytosis, leads to increased susceptibility to infection. The incidence of lung disease increases with age but is hastened by cigarette smoking and by exposure to other environmental irritants.

Special Equipment for Respiratory Treatment

The **bronchoscope** (BRONG-ko-skope) is a rigid or flexible fiberoptic tubular instrument used for inspection of the primary bronchi and the larger bronchial tubes (**Fig. 18-19**). Most bronchoscopes are now attached to video recording

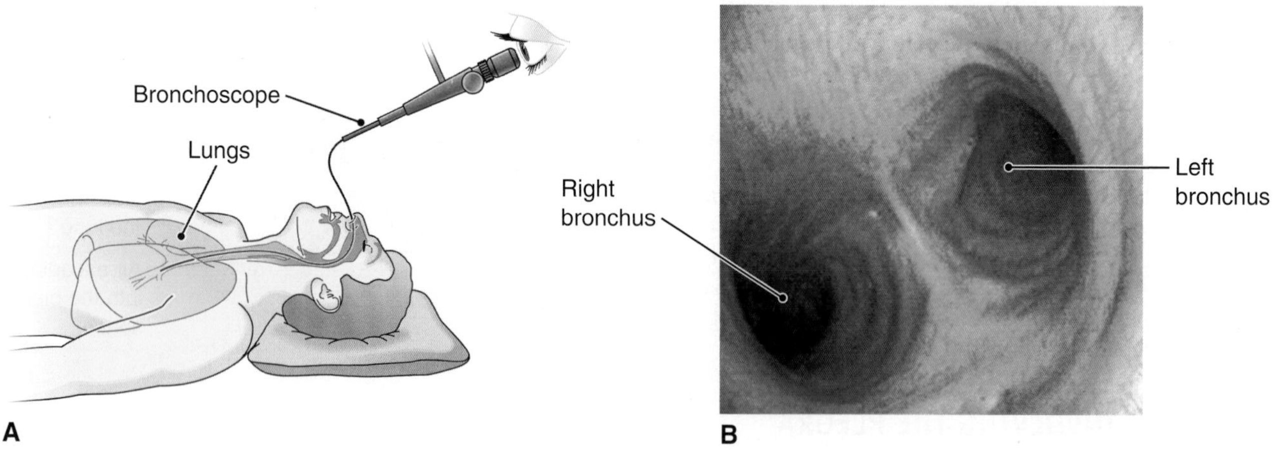

Figure 18-19 **Use of a bronchoscope. A.** A bronchoscope is a lighted tube used to inspect the bronchi, remove specimens, and remove foreign objects. **B.** View of the bronchial openings through a bronchoscope. Note the larger right bronchus.

equipment. The bronchoscope is passed into the respiratory tract by way of the nose or mouth and the pharynx. Physicians may use a bronchoscope to remove foreign bodies, inspect and take tissue samples (biopsies) from tumors, or collect other specimens. Children may aspirate (inhale) a variety of objects, such as pins, beans, pieces of nuts, and small coins, all of which a physician may remove with the aid of a bronchoscope. As previously noted, if such items are left in the lung, they may cause an infection or other serious complications and may even lead to death.

Oxygen therapy is used to sustain life when a condition interferes with adequate oxygen supply to the tissues. The oxygen must first have moisture added by bubbling it through water that is at room temperature or heated. Therapists may deliver oxygen to the patient by mask, catheter, or nasal prongs. Because there is danger of fire when oxygen is being administered, smoking in the room is absolutely prohibited.

A **suction apparatus** is used for removing mucus or other substances from the respiratory tract by means of negative pressure. A container to trap secretions is located between the patient and the machine. The tube leading to the patient has an opening to control the suction. When suction is applied, the drainage flows from the patient's respiratory tract into the collection container.

A **tracheostomy** (tra-ke-OS-to-me) **tube** is used if the pharynx or the larynx is obstructed. It is a small metal or plastic tube that is inserted through a cut (stoma) made in the trachea. As discussed earlier, it acts as an artificial airway for ventilation (**see Fig. 18-16**). The procedure for the insertion of such a tube is called a *tracheostomy*. The word **tracheotomy** (tra-ke-OT-o-me) refers to the incision in the trachea.

Artificial respiration is used when a patient has temporarily lost the capacity to breathe independently. Such emergencies include cases of gas or smoke inhalation, electric shock, drowning, poisoning, or paralysis of the breathing muscles. A number of different apparatuses are used for artificial respiration in a clinical setting. There are also some techniques that can be used in an emergency.

Many public agencies offer classes in techniques to revive people experiencing respiratory or cardiac arrest. The original technique of **cardiopulmonary resuscitation**, or **CPR**, involved cardiac massage and mouth-to-mouth respiration. New studies have found that manual chest compression alone, or "hands-only CPR," is more effective than is the older method in most cases. It is faster, focuses attention on restoring the heartbeat, and is not as likely to discourage potential helpers. Public service classes also include instruction in emergency airway clearance using abdominal thrusts (Heimlich maneuver) or back blows to open obstructed airways.

See the Student Resources on thePoint for diagrams showing the Heimlich maneuver, a procedure for ejecting an object someone has aspirated.

18

Disease in Context Revisited

Emily's Asthma

It had been about a month since Emily's appointment with Dr. Martinez. During that time, her parents had monitored her breathing carefully and had noticed patterns in her coughing bouts.

"It does seem that Emily gets out of breath sooner than the other kids when she exercises," Emily's mother reported to Dr. Martinez during her follow-up appointment. "And now that the weather has cooled off, I do notice that she coughs more."

"Let's take a listen to Emily's lungs," replied the doctor as he placed his stethoscope on the little girl's chest. "Yes, I hear wheezing today, which suggests that Emily's airways are narrowed. Emily's chest x-ray came back negative for infection, but the radiologist did detect thickening of the bronchial walls. The wheezing, your observations, and the family history all lead to the diagnosis that Emily has asthma."

Although Nicole had expected the doctor's diagnosis, it was still a shock to hear him say it. Seeing her look of alarm, Dr. Martinez continued.

"Most kids with asthma lead very normal, active lives. The medications available today target asthma right at its source—inflammation. In fact, with proper treatment, many kids maintain near-normal pulmonary function. Let's start with an antiinflammatory in the form of a low-dose corticosteroid inhaler. If, after a few weeks, we don't see any improvement, I'll start Emily on an oral leukotriene modifier that she will take daily to control airway inflammation. I'm also going to prescribe another type of inhaler to use if Emily has a severe asthma episode. The inhaler contains a medication that relaxes the smooth muscle of her airways, giving her short-term relief of symptoms."

In this case, we learned that Emily's asthma was caused by airway inflammation. Medications that limit the inflammatory response in the respiratory passages can prevent the symptoms of asthma. For a review of the role of inflammation in normal body defense mechanisms, see Chapter 17.

CHAPTER 18

Chapter Wrap-Up

Summary Overview

A detailed chapter outline with space for note taking is on *thePoint*. The figure below illustrates the main topics covered in this chapter.

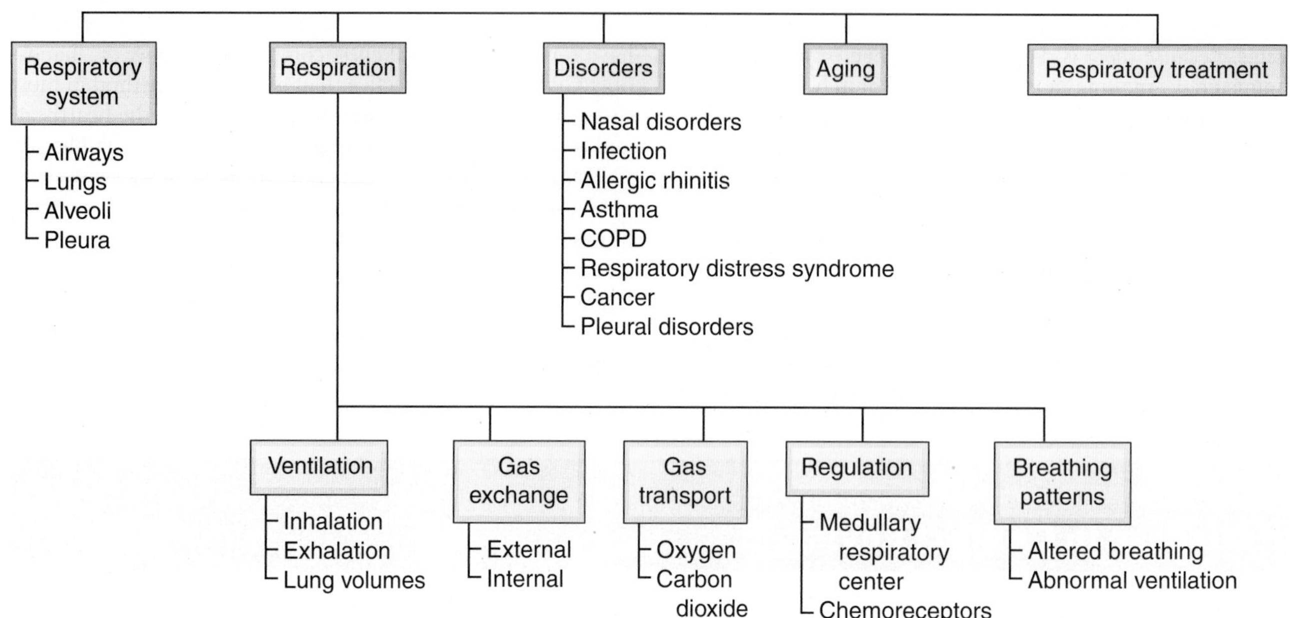

Key Terms

The terms listed below are emphasized in this chapter. Knowing them will help you organize and prioritize your learning. These and other boldface terms are defined in the Glossary with phonetic pronunciations.

alveoli (sing., alveolus)
asthma
bicarbonate ion
bronchiole
bronchus (pl., bronchi)
carbonic acid
carbonic anhydrase
chemoreceptor
compliance
diaphragm
dyspnea
emphysema
epiglottis
epistaxis
hypercapnia
hypoxia
larynx
lung
pharynx
phrenic nerve
pleura
pneumothorax
pulmonary ventilation
respiration
spirometer
surfactant
trachea

Word Anatomy

Medical terms are built from standardized word parts (prefixes, roots, and suffixes). Learning the meanings of these parts can help you remember words and interpret unfamiliar terms.

WORD PART	MEANING	EXAMPLE
Structure of the Respiratory System		
laryng/o	larynx	The *laryngeal* pharynx opens into the larynx.
nas/o	nose	The *nasopharynx* is behind the nasal cavity.
or/o	mouth	The *oropharynx* is behind the mouth.
pleur/o	side, rib	The *pleura* covers the lung and lines the chest wall (rib cage).
The Process of Respiration		
capn/o	carbon dioxide	*Hypercapnia* is a rise in the blood level of carbon dioxide.
orth/o	straight	*Orthopnea* can be relieved by sitting in an upright position.
-pnea	breathing	*Hypopnea* is a decrease in the rate and depth of breathing.
spir/o	breathing	A *spirometer* is an instrument used to record breathing volumes.
Respiratory Disorders		
atel/o	incomplete	*Atelectasis* is incomplete expansion of the lung.
-centesis	tapping, perforation	In *thoracentesis*, a needle is inserted into the pleural space to remove fluid.
pneum/o	air, gas	*Pneumothorax* is accumulation of air in the pleural space.
pneumon/o	lung	*Pneumonia* is inflammation of the lung.
rhin/o	nose	*Rhinitis* is inflammation of the nose.

Questions for Study and Review

BUILDING UNDERSTANDING

Fill in the Blanks

1. The exchange of air between the atmosphere and the lungs is called ______.
2. The space between the vocal cords is the ______.
3. The ease with which the lungs and thorax can be expanded is termed ______.
4. A lower-than-normal level of oxygen in the tissues is called ______.
5. The serous membrane around the lung is the ______.

Matching > Match each numbered item with the most closely related lettered item.

___ **6.** A decrease in the rate and depth of breathing
___ **7.** A normal increase in breathing rate, as during exercise
___ **8.** A temporary cessation of breathing
___ **9.** Difficult or labored breathing
___ **10.** Difficult breathing that is relieved by sitting up

a. dyspnea
b. tachypnea
c. orthopnea
d. hypopnea
e. apnea

Multiple Choice

___ **11.** Which of the following are bony projections in the nasal cavity?

a. nares
b. septa
c. conchae
d. sinuses

___ **12.** Which structure contains the vocal folds?

a. pharynx
b. larynx
c. trachea
d. lungs

___ **13.** What covers the larynx during swallowing?

a. epiglottis
b. glottis
c. conchae
d. sinus

___ **14.** Which structure has the centers that regulate respiration?

a. cerebral cortex
b. diencephalon
c. brain stem
d. cerebellum

___ **15.** Which term describes incomplete lung expansion?

a. effusion
b. adhesion
c. epistaxis
d. atelectasis

UNDERSTANDING CONCEPTS

16. Differentiate between the terms in each pair:

a. internal and external gas exchange
b. carbonic acid and carbonic anhydrase
c. pneumonia and pleurisy
d. pneumothorax and hemothorax

17. Referring to Appendix 2-3, give the normal value for carbon dioxide in the blood, and list conditions that can increase or decrease this level.

18. Referring to Appendix 3-1, give the name of:

a. the bacillus that causes whooping cough
b. one coccus that can cause pneumonia
c. one bacillus that can cause pneumonia

19. Describe the characteristics of the tissue that lines the respiratory passageways, and explain how this tissue protects us.

20. Trace the path of air from the nostrils to the lung capillaries.

21. Compare and contrast oxygen and carbon dioxide transport in the blood.

22. Define hyperventilation and hypoventilation. What is the effect of each on blood CO_2 levels and blood pH?

23. What are chemoreceptors, and how do they function to regulate breathing?

24. Describe the structural and functional changes that occur in the respiratory system in chronic obstructive pulmonary disease (COPD).

25. Give the meaning of the following abbreviations:

a. CPR ______________________________
b. RSV ______________________________
c. ARDS ______________________________
d. SIDS ______________________________

CONCEPTUAL THINKING

26. Jake, a sometimes exasperating 4-year-old, threatens his mother that he will hold his breath until "he dies." Should his mother be concerned that he might succeed?

27. Why is it important that airplane interiors are pressurized? If the cabin lost pressure, what physiological adaptations to respiration might occur in the passengers?

28. In Emily's case, an antiinflammatory medication was used to control her asthma symptoms. Explain how this drug works in the respiratory system.

For more questions, see the Learning Activities on thePoint.

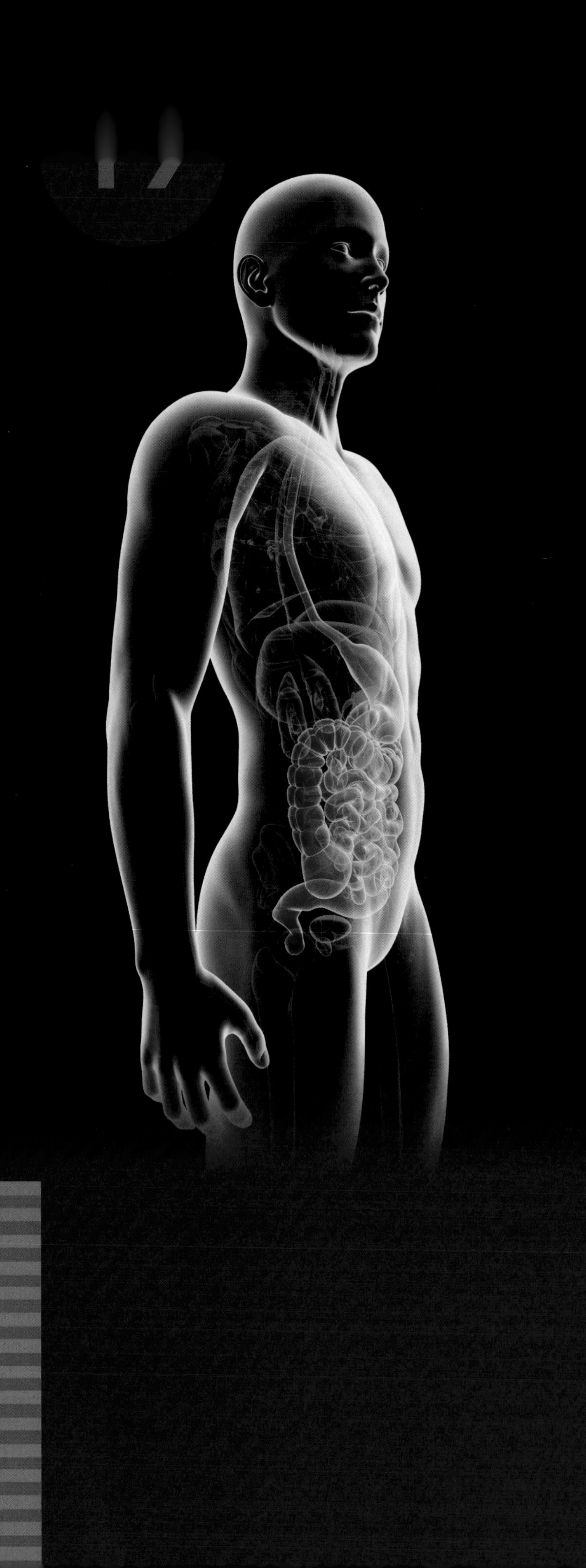

Learning Objectives

After careful study of this chapter, you should be able to:

1 ▶ Name the three main functions of the digestive system. ***p. 434***

2 ▶ Name and locate the two main layers and the subdivisions of the peritoneum. ***p. 434***

3 ▶ Describe the four layers of the digestive tract wall. ***p. 435***

4 ▶ Describe the two types of muscular contractions important in the digestive process. ***p. 436***

5 ▶ Name and describe the functions of the digestive tract organs. ***p. 436***

6 ▶ Name and locate the different types of teeth. ***p. 437***

7 ▶ Name and describe the functions of the accessory organs of digestion. ***p. 442***

8 ▶ Describe how bile travels into the digestive tract and functions in digestion. ***p. 443***

9 ▶ Explain the role of enzymes in digestion, and list enzymes involved in the digestion of fats, carbohydrates, proteins, and nucleic acids. ***p. 444***

10 ▶ Define the term *hydrolysis*, and explain its role in digestion. ***p. 444***

11 ▶ Name the digestion products of fats, proteins, and carbohydrates. ***p. 444***

12 ▶ Define *absorption*, and state how villi function in absorption. ***p. 445***

13 ▶ Explain nervous control of digestion and the role of the enteric nervous system. ***p. 446***

14 ▶ List four hormones involved in regulating digestion, and explain the function of each. ***p. 447***

15 ▶ Describe common disorders of the digestive tract and its accessory organs. ***p. 447***

16 ▶ Using the case study, describe the colonoscopy procedure and its role in diagnosing certain colon disorders. ***pp. 433, 454***

17 ▶ Show how word parts are used to build words related to digestion (see Word Anatomy at the end of the chapter). ***p. 456***

Disease in Context *Adam's Case: The Picture of Health*

Adam was OK with everything his family doctor had described about his routine physical examination.

"We'll draw some blood for testing," Dr. Michaels explained. "You didn't have anything to eat since last night, did you? We'll send your specimen to the lab to get information on your blood cells and blood chemistry, hemoglobin, lipoproteins, and such. You need to leave a urine sample to check for sugar, and then Annette will take your blood pressure and run an ECG. I'll be in to ask you some general questions and do some hands-on examination including, of course, a check on your prostate. And Adam, since you are well past your 50th birthday, you need to stop stalling on getting that colonoscopy."

Sending in the stool sample later to test for signs of blood was not a problem for Adam, and he could live with the prostate exam, but he was not so happy about the colonoscopy! He had heard that the prep to clean out the colon was unpleasant.

"You know," Adam protested, "I'm healthy and have no family history of colon cancer or polyps, so why do I need to do this?"

"Actually," replied Dr. Michaels, "most colorectal cancers appear in people with no symptoms and with no family history or genetic predisposition. In high-risk populations, we recommend testing even earlier and more frequently. There is a new virtual colonoscopy procedure that uses computerized x-rays instead of an endoscope to generate detailed images of the colon, but for a baseline study, your proctologist might prefer the routine method. Besides, if we have to remove any polyps or other abnormal tissue, we'd have to resort to that anyway. When you're ready to leave the office, ask Jean at the front desk to set up an appointment for you."

As part of Adam's physical, the doctor examined many of Adam's organ systems, including his digestive system. Because of his age, the doctor recommended a closer inspection of Adam's colon. In this chapter, we will learn about the digestive tract and the accessory organs that contribute to digestion. We will also visit Adam again and find out how the colonoscopy went.

ANCILLARIES *At-A-Glance*

Visit thePoint to access the following resources. For guidance in using these resources most effectively, see pp. xv–xvii.

Learning RESOURCES

- Tips for Effective Studying
- Web Figure: Ulcers
- Web Figure: Pyloric Stenosis
- Web Figure: Complications of Ulcerative Colitis
- Web Figure: Colonic Diverticulosis
- Web Figure: Acute and Chronic Diverticulitis
- Web Figure: Cirrhosis
- Web Figure: Clinical Features of Cirrhosis
- Web Figure: Portal Hypertension
- Web Figure: Formation of Gallstones
- Web Figure: Cholecystitis
- Animation: General Digestion
- Animation: Digestion of Carbohydrates
- Animation: Enzymes
- Animation: The Liver in Health and Disease
- Health Professions: Dental Hygienist
- Detailed Chapter Outline
- Answers to Questions for Study and Review
- Audio Pronunciation Glossary

Learning ACTIVITIES

- Pre-Quiz
- Visual Activities
- Kinesthetic Activities
- Auditory Activities

A LOOK BACK

This chapter discusses some of the membranes and other tissue types introduced in Chapter 4. You should also review the structure of organic compounds and the activity of enzymes described in Chapter 2.

General Structure and Function of the Digestive System

Every body cell needs a constant supply of nutrients. Cells use the energy contained in nutrients to do their work. In addition, they rearrange the nutrients' chemical building blocks to manufacture materials the body needs for metabolism, growth, and repair.

Most ingested nutrients are too large to enter cells. They must first be broken down into smaller components, a process known as **digestion**. The transfer of nutrients from the digestive tract to the circulation is called **absorption**. The circulatory system carries nutrients to body cells for use and storage. Finally, undigested waste material must be eliminated. Digestion, absorption, and elimination are the three chief functions of the digestive system.

For our purposes, the digestive system may be divided into two groups of organs (**Fig. 19-1**):

- The **digestive tract**, a continuous passageway beginning at the mouth, where food is taken in, and terminating at the anus, where the solid waste products of digestion are expelled. The remainder of the digestive tract consists of the pharynx, esophagus, stomach, and small and large intestines, illustrated shortly.
- The **accessory organs**, which are necessary for the digestive process but are not a direct part of the digestive tract. The teeth and the tongue physically manipulate food prior to digestion. The other accessory organs release substances into the digestive tract through ducts. These secretory organs include the salivary glands, liver, gallbladder, and pancreas.

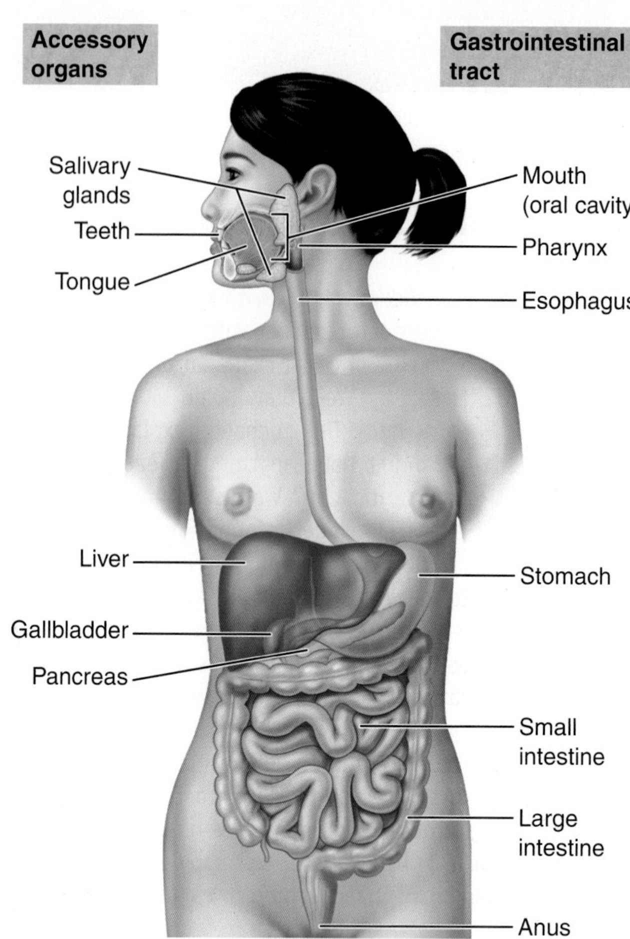

Figure 19-1 The digestive system. **KEY POINT** The digestive system extends from the mouth to the anus. Accessory organs secrete into the digestive tract. **ZOOMING IN** Which accessory organs of digestion secrete into the mouth?

THE PERITONEUM

The **peritoneum** (per-ih-to-NE-um) is a thin, shiny serous membrane that lines the abdominopelvic cavity and also folds back to cover most of the organs contained within the cavity (**Fig. 19-2**). As noted in Chapter 4, the outer portion of this membrane, the layer in contact with the body wall, is called the *parietal* (pah-RI-eh-tal) **peritoneum**; the layer that covers the organs is called the *visceral* (VIS-eh-ral) *peritoneum*. This slippery membrane allows the organs to slide over each other as they function. The peritoneum also carries blood vessels, lymphatic vessels, and nerves. In some places, it supports the organs and binds them to each other. The peritoneal cavity is the potential space between the membrane's two layers and contains serous fluid (peritoneal fluid).

Subdivisions of the peritoneum around the various organs have special names. The **mesentery** (MES-en-ter-e) is a double-layered portion of the peritoneum shaped somewhat like a fan (**see Fig. 19-2**). The handle portion is attached to the posterior abdominal wall, and the expanded long edge is attached to the small intestine. Between the two membranous layers of the mesentery are the vessels and nerves that supply the intestine. The section of the peritoneum that extends from the colon to the posterior abdominal wall is the mesocolon (mes-o-KO-lon).

A large double layer of the peritoneum containing much fat hangs like an apron over the front of the intestine. This **greater omentum** (o-MEN-tum) extends from the lower border of the stomach into the pelvic cavity and then

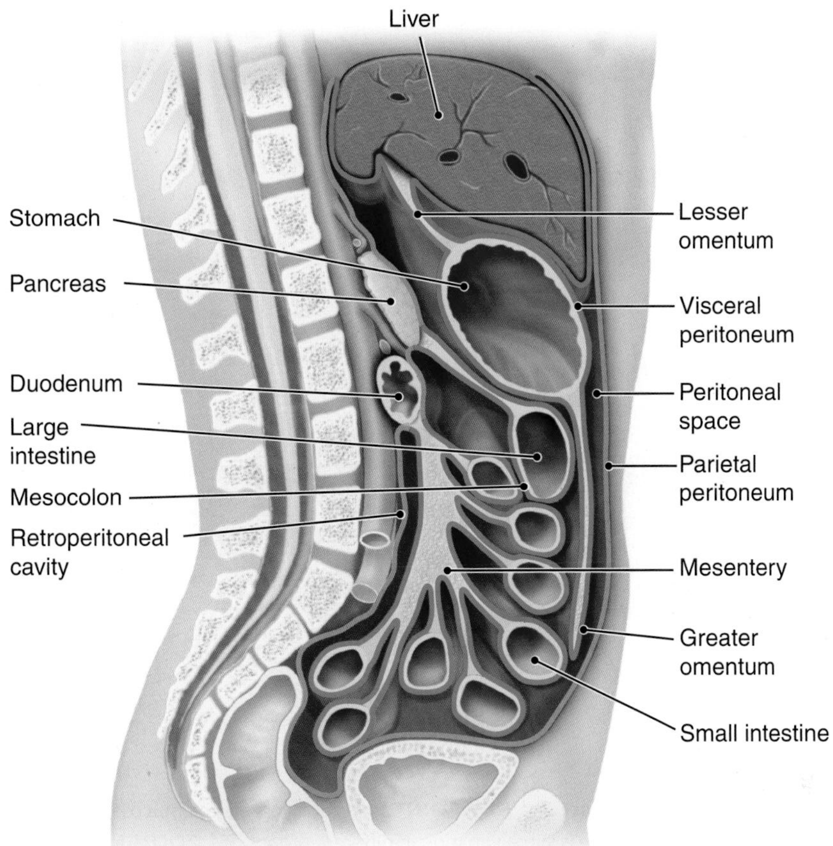

A The peritoneum

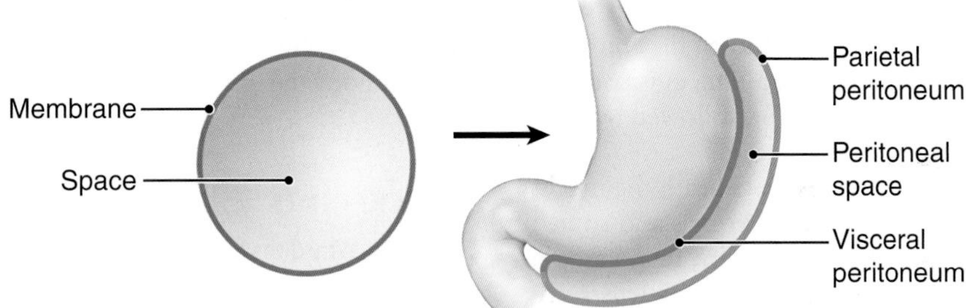

B Formation of the peritoneum

Figure 19-2 The abdominopelvic cavity and peritoneum. **KEY POINT** The parietal peritoneum lines the abdominopelvic cavity. Subdivisions of the visceral peritoneum fold over, supporting and separating individual organs. **ZOOMING IN** Which part of the peritoneum is around the small intestine?

loops back up to the transverse colon. A smaller membrane, called the **lesser omentum**, extends between the stomach and the liver.

THE WALL OF THE DIGESTIVE TRACT

Although modified for specific tasks in different organs, the wall of the digestive tract, from the esophagus to the anus, is similar in structure throughout (Fig. 19-3). The general pattern consists of four layers, which are, from innermost to outermost:

1. Mucous membrane, or mucosa
2. Submucosa
3. Smooth muscle, the muscularis externa
4. Serous membrane, or serosa

First is the mucous membrane, or **mucosa**. From the mouth through the esophagus, and also in the anus, the mucosal epithelium consists of stratified squamous cells, which help protect deeper tissues. Throughout the remainder of the digestive tract, the mucosa contains simple columnar

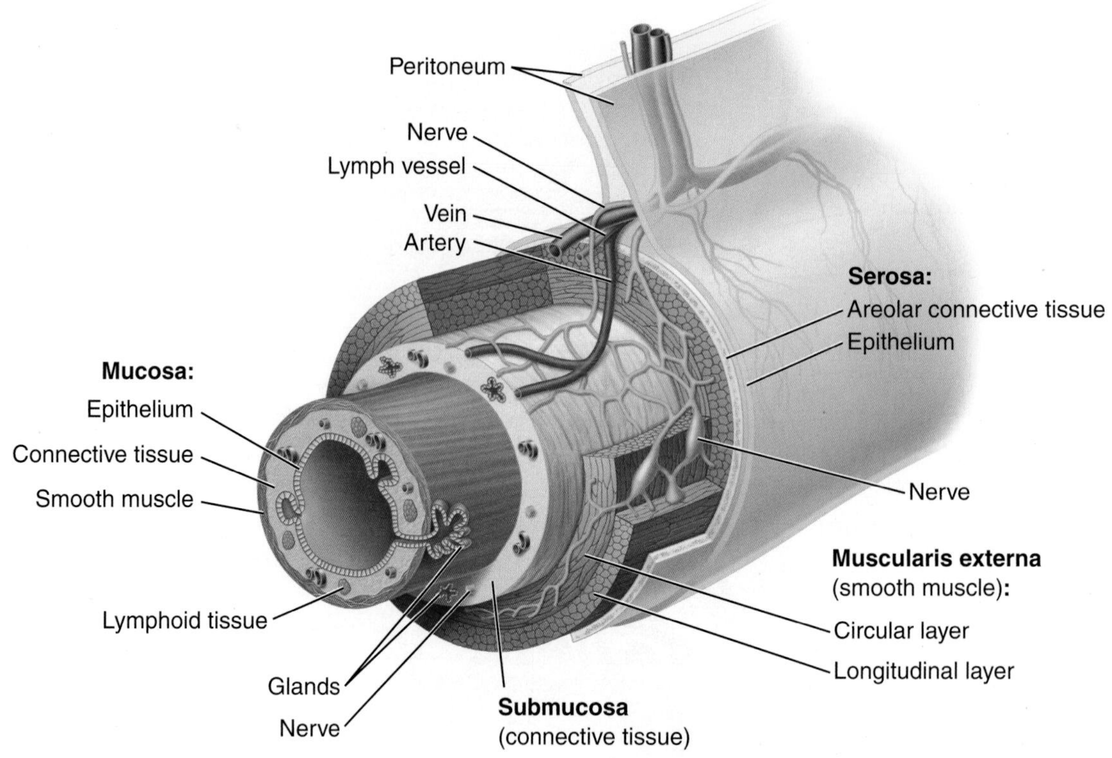

Figure 19-3 Wall of the digestive tract. **KEY POINT** The four layers are the mucosa, submucosa, muscularis externa (smooth muscle), and serosa. There is some variation in different organs. **ZOOMING IN** What type of tissue is between the submucosa and the serous membrane in the digestive tract wall?

epithelium. Goblet cells within this epithelium secrete mucus to protect the system's lining. Beneath the epithelium is a thin layer of connective tissue containing gut-associated lymphoid tissue (GALT; see Chapter 16), followed by a very thin layer of smooth muscle.

The thick layer of connective tissue beneath the mucosa is the **submucosa**, which contains blood vessels and some of the nerves that help regulate digestive activity. In the esophagus and small intestine, this layer also contains mucus-secreting glands.

The next layer, the **muscularis externa**, is composed of smooth muscle. Most of the digestive organs have two layers of smooth muscle: an inner layer of circular fibers, and an outer layer of longitudinal fibers. When a section of the circular muscle contracts, the organ's lumen narrows; when the longitudinal muscle contracts, a section of the wall shortens. A wave of circular muscle contractions propels food through parts of the digestive tract, a movement known as **peristalsis** (per-ih-STAL-sis) (**Fig. 19-4**). Alternatively, rhythmic contractions of the circular muscle mix food with digestive juices, a process known as **segmentation**. Different regions of the digestive tract use these different forms of motility, as discussed later. Nerves that control motility (movement) of the digestive organs are located in this smooth muscle layer.

The digestive organs in the abdominopelvic cavity have an outermost layer of serous membrane, or **serosa**, a thin, moist tissue composed of simple squamous epithelium and areolar (loose) connective tissue. This membrane forms the inner layer of the large serous membrane that extends throughout the abdominopelvic cavity, discussed next.

CHECKPOINTS

- [ ] 19-1 Why does food have to be digested before cells can use it?
- [ ] 19-2 What is the name of the large serous membrane that lines the abdominopelvic cavity and covers the organs it contains?
- [ ] 19-3 What are the four layers of the digestive tract wall?

Organs of the Digestive Tract

As we study the organs of the digestive system, locate each in **Figure 19-1**. The digestive tract is a muscular tube extending through the body. It is composed of several parts: the **mouth**, **pharynx**, **esophagus**, **stomach**, **small intestine**, and **large intestine**. It is more commonly referred to as the **gastrointestinal (GI) tract** because of the major importance of the stomach and intestine in the digestive process.

The next section describes the structure and function of each digestive organ. These descriptions are followed by an overview of how the organs work together in digestion. See Dissection Atlas **Figure A5-7** for a photograph showing the abdominal digestive organs in place.

THE MOUTH

The mouth, also called the *oral cavity*, is where a substance begins its travels through the digestive tract (**Fig. 19-5A**). The mouth has the following digestive functions:

- It receives food, a process called **ingestion**.
- It breaks food into small portions. This is done mainly by the teeth in the process of chewing or **mastication** (mas-tih-KA-shun), but the tongue, cheeks, and lips are also used.
- It mixes the food with **saliva** (sah-LI-vah), which is produced by the salivary glands and secreted into the mouth. The salivary glands will be described with the other accessory organs.
- It moves controlled amounts of food toward the throat to be swallowed, a process called **deglutition** (deg-lu-TISH-un).

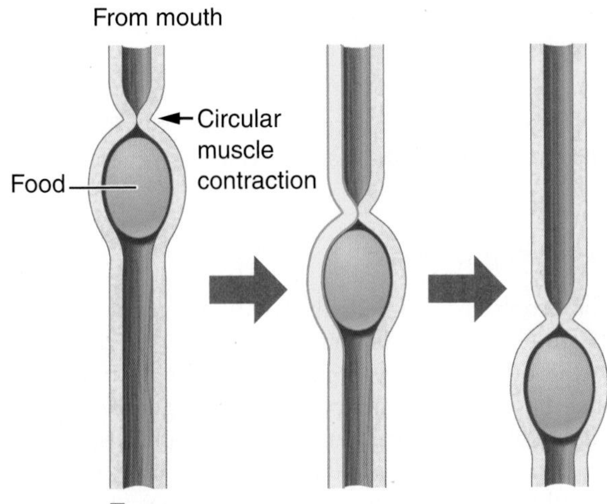

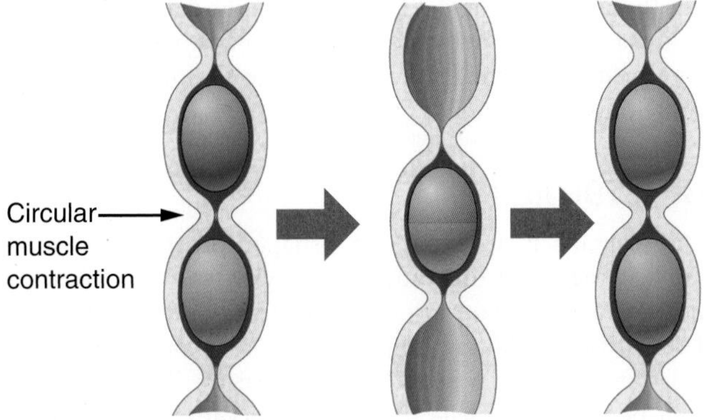

Figure 19-4 Gastrointestinal motility. A. Peristalsis moves the tube's contents ahead of the wave of contraction. **B.** Segmentation squishes the contents back and forth to mix them. **ZOOMING IN** Which type of motility would be most useful in the esophagus where the contents should move quickly?

The **tongue** (TUNG), a muscular organ that projects into the mouth, aids in chewing and swallowing and is one of the principal organs of speech. The tongue has a number of special surface receptors, called *taste buds*, which can differentiate taste sensations (e.g., bitter, sweet, sour, or salty) (see Chapter 11).

THE TEETH

The oral cavity also contains the teeth (**see Fig. 19-5A**). Young children have 20 teeth, known as the baby teeth or **deciduous** (de-SID-u-us) teeth. (The word deciduous means "falling off at a certain time," such as the leaves that fall off the trees in autumn.) A complete set of adult permanent teeth numbers 32. The cutting teeth, or **incisors** (in-SI-sors), occupy the anterior part of the oral cavity. The **cuspids** (KUS-pids), commonly called the *canines* (KA-nines) or *eyeteeth*, are lateral to the incisors. They are pointed teeth with deep roots that are used for more forceful gripping and tearing of food. The posterior **molars** (MO-lars) are the larger grinding teeth. There are two premolars and three molars. In an adult, each quadrant (quarter) of the mouth, moving from anterior to posterior, has two incisors, one cuspid, and five molars.

The first eight deciduous (baby) teeth to appear through the gums are usually the incisors. Later, the cuspids and molars appear. Usually, the 20 baby teeth all have appeared by the time a child has reached the age of 2 to 3 years. During the first 2 years, the permanent teeth develop within the upper jaw (maxilla) and lower jaw (mandible) from buds that are present at birth. The first permanent teeth to appear are the four 6-year molars, which usually come in before any baby teeth are lost. Because decay and infection of deciduous molars may spread to new, permanent teeth, deciduous teeth need proper care.

19

As a child grows, the jawbones grow, making space for additional teeth. After the 6-year molars have appeared, the baby incisors loosen and are replaced by permanent incisors. Next, the baby canines (cuspids) are replaced by permanent canines, and finally, the baby molars are replaced by the permanent bicuspids (premolars).

At this point, the larger jawbones are ready for the appearance of the 12-year, or second, permanent molar teeth. During or after the late teens, the third molars, or so-called *wisdom teeth*, may appear. In some cases, the jaw is not large enough for these teeth, or there are other abnormalities, so that the third molars may not erupt or may have to be removed. **Figure 19-5B** shows the parts of a molar.

The main substance of the tooth is **dentin**, a calcified substance harder than bone. Within the tooth is a soft pulp containing blood vessels and nerves. The tooth's *crown* projects above the gum, the **gingiva** (JIN-jih-vah), and is covered with **enamel**, the hardest substance in the body. The *root* or roots of the tooth, located below the gum line in a bony socket, are covered with a rigid connective tissue (cementum) that helps hold the tooth in place. A connective tissue sheet called the *periodontal ligament* joins the cementum to the tooth socket. Each root has a canal containing extensions of the pulp.

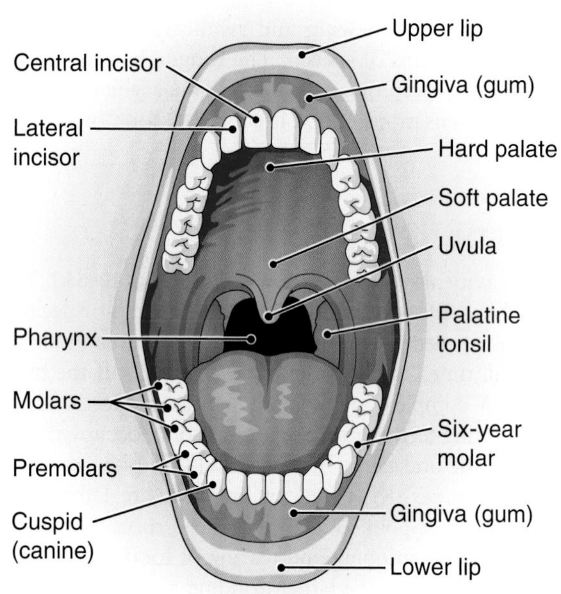

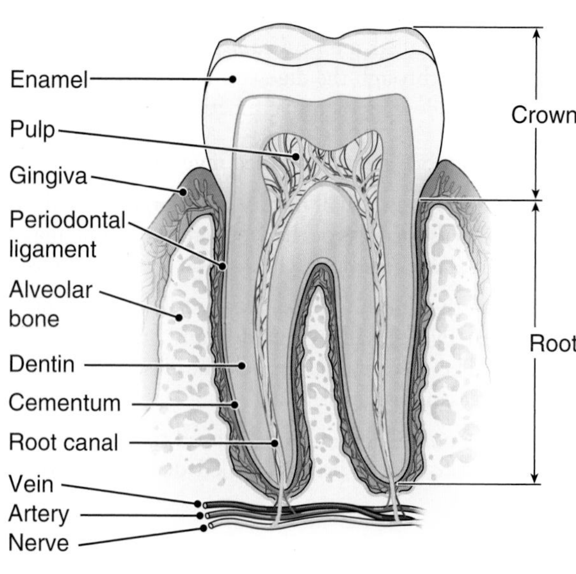

Figure 19-5 **The mouth and teeth.** **KEY POINT** The mouth breaks food into smaller pieces and mixes it with saliva. **A.** The mouth. The teeth and tonsils are visible in this view. **B.** A molar tooth. **ZOOMING IN** What is the common name for the gingiva?

Dental hygienists are concerned with oral health. See the Student Resources on thePoint for information on this career.

THE PHARYNX

The pharynx (FAR-inks) is commonly referred to as the throat. It is a combined pathway for the respiratory and digestive systems and was described in Chapter 18 (see **Fig. 18-2**). The oral part of the pharynx, the oropharynx, is visible when you look into an open mouth and depress the tongue. The palatine tonsils may be seen at either side of the oropharynx. The pharynx also extends upward to the nasal cavity, where it is referred to as the nasopharynx, and downward to the larynx, where it is called the laryngopharynx. The **soft palate** forms the posterior roof of the oral cavity. From it hangs a soft, fleshy, V-shaped mass called the **uvula** (U-vu-lah).

In swallowing, the tongue pushes a **bolus** (BO-lus) of food, a small portion of chewed food mixed with saliva, into the pharynx. Once the food reaches the pharynx, swallowing occurs rapidly by an involuntary reflex action. At the same time, the soft palate and uvula are raised to prevent food and liquid from entering the nasal cavity, and the tongue is raised to seal the back of the oral cavity. As described in Chapter 18, the entrance of the trachea is guarded during swallowing by the leaf-shaped cartilage, the epiglottis, which covers the opening of the larynx (**see Fig. 18-3**). The swallowed food is then moved into the esophagus.

THE ESOPHAGUS

The esophagus (eh-SOF-ah-gus) is a muscular tube about 25 cm (10 in) long. Its musculature differs slightly from that of the other digestive organs because it has voluntary striated muscle in its upper portion, which gradually changes to smooth muscle along its length. In the esophagus, food is lubricated with mucus, and peristalsis moves it into the stomach. No additional digestion occurs in the esophagus.

Before joining the stomach, the esophagus must pass through the diaphragm. It travels through an opening in the diaphragm called the **esophageal hiatus** (eh-sof-ah-JE-al hi-A-tus) (**Fig. 19-6**). If there is a weakness in the diaphragm at this point, a portion of the stomach or other abdominal organ may protrude through the space, a condition called *hiatal hernia,* discussed later.

THE STOMACH

The stomach is an expanded J-shaped organ in the superior left region of the abdominal cavity (**see Fig. 19-6**). In addition to the two muscle layers already described, it has a third, inner oblique (angled) layer that aids in grinding food and mixing it with digestive juices. The left-facing arch of the stomach is the **greater curvature**, whereas the smaller right surface forms the **lesser curvature.** The superior rounded portion under the left side of the diaphragm is the stomach's **fundus**. The region of the stomach leading into the small intestine is the **pylorus** (pi-LOR-us). The large region between the fundus and the pylorus is known as the stomach's body.

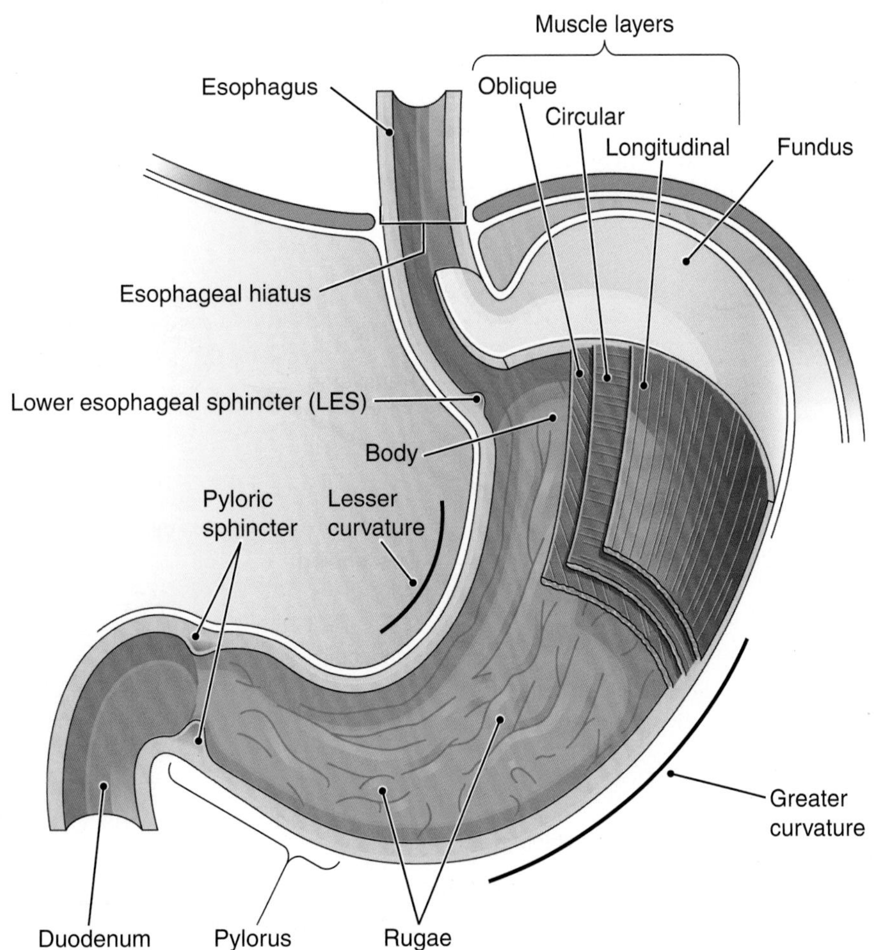

Figure 19-6 Longitudinal section of the stomach. KEY POINT The stomach's interior is shown along with a portion of the esophagus and the duodenum. Sphincters regulate the organ's entrance and exit openings. **ZOOMING IN** Which additional muscle layer is in the wall of the stomach that is not found in the rest of the digestive tract?

Sphincters A **sphincter** (SFINK-ter) is a muscular ring that regulates the size of an opening. There are two sphincters that separate the stomach from the organs above and below.

The **lower esophageal sphincter (LES)** controls the passage of food from the esophagus to the stomach. This muscle has also been called the *cardiac sphincter* because of its proximity to the heart. We are sometimes aware of the existence of this sphincter when it does not relax as it should, producing a feeling of being unable to swallow past that point.

Between the distal, or far, end of the stomach and the small intestine is the **pyloric** (pi-LOR-ik) **sphincter.** This sphincter and the stomach's pylorus, which leads to it, are important in regulating how rapidly food moves into the small intestine.

Functions of the Stomach The stomach serves as a storage pouch, digestive organ, and churn. When the stomach is empty, the lining forms many folds called **rugae** (RU-je) (**see Fig. 19-6**). These folds disappear as the stomach expands. (The stomach can stretch to hold one-half of a gallon of food and liquid.) Special cells in the stomach's lining secrete substances that mix together to form gastric juice (*gastr/o* is the word root for "stomach.") Some of the cells secrete a great amount of mucus to protect the organ's lining from digestive secretions. Other cells produce the active components of the gastric juice, which are

- Hydrochloric acid (HCl), a strong acid that denatures (unwinds) proteins to prepare them for digestion and also destroys foreign organisms. HCl is produced in anticipation of eating and is produced in greater amounts when food enters the stomach.
- Pepsin, a protein-digesting enzyme. Pepsin is produced in an inactive form called *pepsinogen*, which is activated only when it contacts HCl or previously activated pepsin molecules.

Chyme (*kime*), from a Greek word meaning "juice," is the highly acidic, semiliquid mixture of gastric juice and food that leaves the stomach to enter the small intestine.

19

THE SMALL INTESTINE

The small intestine is the longest part of the digestive tract (**Fig. 19-7**). It is known as the small intestine because, although it is longer than is the large intestine, it is smaller in diameter, with an average width of approximately 2.5 cm (1 in). After death, when relaxed to its full length, the small intestine is approximately 6 m (20 ft) long. In life, contraction of the longitudinal muscle shortens the small intestine to about 3 m (10 ft) in length. The first 25 cm (10 in) or so of the small intestine make up the **duodenum** (du-o-DE-num) (named for the Latin word for "twelve," based on its length of 12 finger widths). Beyond the duodenum are two more divisions: the

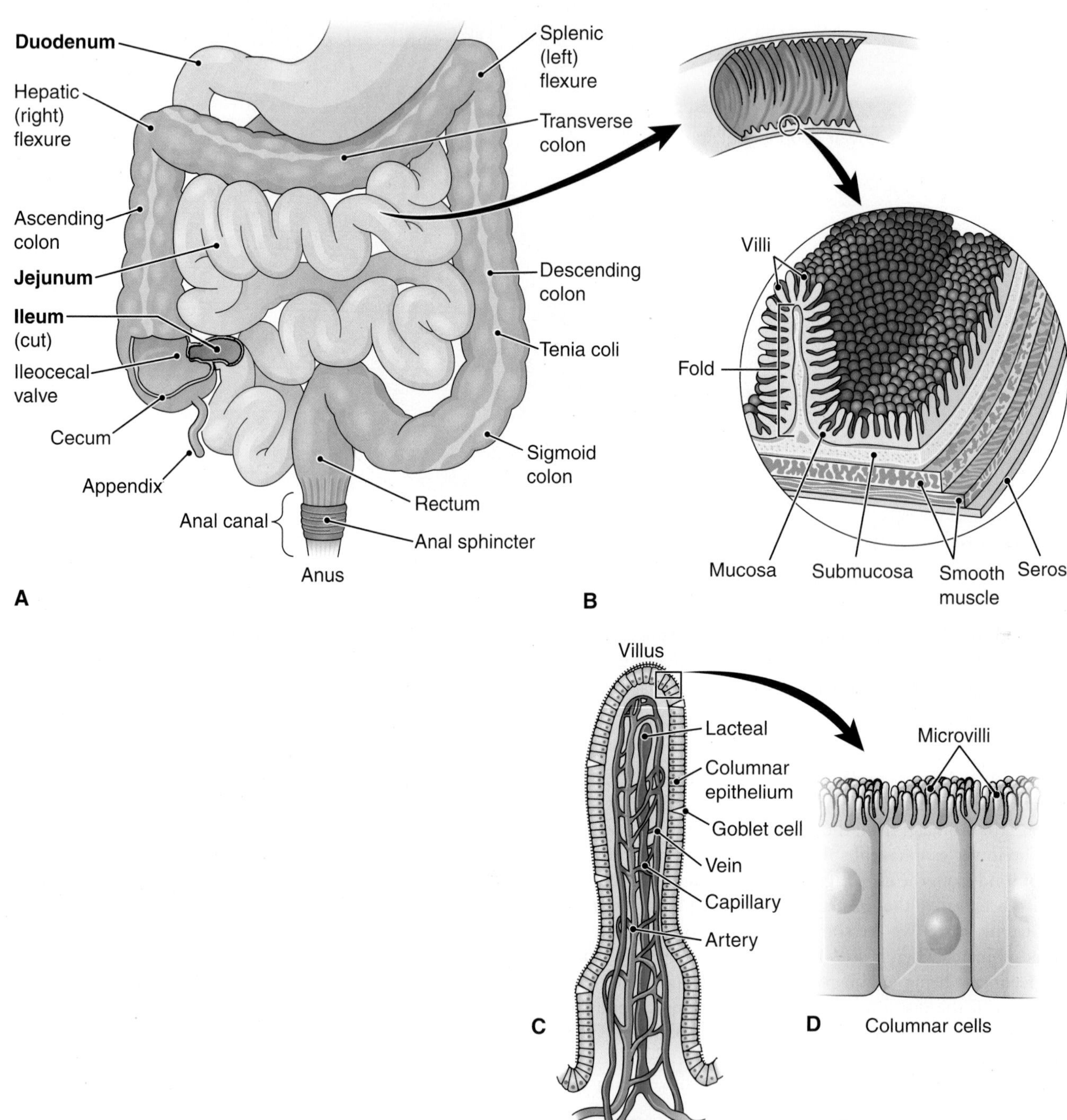

Figure 19-7 **The intestines.** **KEY POINT** The intestines are greatly folded to increase surface area. **A.** The small and large intestines. The three sections of the small intestine are labeled in bold type. The colon is the main portion of the large intestine. **B.** The wall of the small intestine. **C.** Drawing of a villus. **D.** The plasma membrane of the columnar epithelial cells is folded into microvilli. **ZOOMING IN** Which portions of the small and large intestines join at the ileocecal valve?

jejunum (je-JU-num), which forms the next two-fifths of the small intestine, and the **ileum** (IL-e-um), which constitutes the remaining portion.

Functions of the Small Intestine The small intestine participates in all aspects of digestive function: secretion, motility, digestion, and absorption.

Secretion Glands in the duodenal mucosa and submucosa secrete large amounts of mucus to protect the small intestine from the strongly acidic chyme entering from the stomach. Mucosal cells of the small intestine also produce enzymes that participate in the final stages of carbohydrate and protein digestion (discussed later under the digestive process). These enzymes are inserted into the cells' plasma membrane and act on nutrients that come in contact with the intestinal lining.

Motility Minimal peristalsis occurs in the small intestine. This form of motility is too rapid to allow for effective digestion and absorption in the small intestine. Instead, the process of segmentation (**see Fig. 19-4**) ensures that the food is thoroughly mixed with digestive juices and is placed in contact with the enzymes at the mucosal surface. Segmentation is regulated so that proximal segments contract before distal segments, slowly propelling the intestinal contents from the duodenum to the end of the ileum.

Digestion Most digestion takes place in the small intestine's lumen under the influence of secretions from the liver and pancreas as well as enzymes in the membranes of intestinal cells. These digestive processes are discussed in detail later.

Absorption Most digested nutrients, water, and electrolytes are absorbed in the small intestine. **Figure 19-7** shows three modifications of the small intestine lining that increase the surface area and thus maximize nutrient absorption.

1. The mucosa and submucosa are formed into large circular folds (**see Fig. 19-7B**).
2. The mucosa of each fold is formed into millions of tiny, finger-like projections, or **villi** (VIL-li), which give the inner surface a velvety appearance. Each villus contains a capillary and a specialized lymphatic capillary called a **lacteal** (LAK-tele) (**see Fig. 19-7C**).
3. The epithelial cells of the villi have small projecting folds of the plasma membrane known as **microvilli** (**see Fig. 19-7D**).

Box 19-1 provides more information on the relationship of surface area to absorption.

THE LARGE INTESTINE

The large intestine is approximately 6.5 cm (2.5 in) in diameter and approximately 1.5 m (5 ft) long (**see Fig. 19-7A**). It is named for its wide diameter, rather than its length. The outer longitudinal muscle fibers in its wall gather into three separate surface bands. These bands, known as **teniae** (TEN-e-e) **coli**, draw up the organ's wall to give it its distinctive puckered appearance. (The name is also spelled *taeniae*; the singular is *tenia* or *taenia*).

Subdivisions of the Large Intestine The large intestine begins in the lower right region of the abdomen. The first part is a small pouch called the **cecum** (SE-kum). The **ileocecal** (il-e-o-SE-kal) **valve** permits food passage from the ileum of the small intestine into the cecum, but not vice versa. Attached to the cecum is a small, blind tube containing lymphoid tissue; its full name is **vermiform** (VER-mih-form) **appendix** (*vermiform* means "worm-like") but usually just "appendix" is used.

The second portion, the **ascending colon**, extends superiorly along the right side of the abdomen toward the liver. It bends near the liver at the hepatic (right) flexure and extends across the abdomen as the **transverse colon.** It bends again sharply at the splenic (left) flexure and extends inferiorly on the left side of the abdomen into the pelvis, forming the

Box 19-1

A CLOSER LOOK

The Folded Intestine: More Absorption with Less Length

Whenever materials pass from one system to another, they must travel through a cellular membrane. A major factor in how much transport can occur per unit of time is the total surface area of the membrane; the greater the surface area, the higher the rate of transport. The problem of packing a large amount of surface into a small space is solved in the body by folding the membranes. We do the same thing in everyday life. Imagine trying to store a bed sheet in the closet without folding it!

In the small intestine, where digested food must absorb into the bloodstream, there is folding of membranes down to the level of single cells.

- The 6-m long organ is coiled to fit into the abdominal cavity.
- The inner wall of the organ is thrown into circular folds called plicae circulares, which not only increase surface area, but aid in mixing.
- The mucosal villi project into the lumen, providing more surface area than a flat membrane would.
- The individual cells that line the small intestine have microvilli, tiny finger-like folds of the plasma membrane that increase surface area tremendously.

Together, these structural features of the small intestine result in an absorptive surface area estimated to be about 250 m^2! Folding is present in other parts of the digestive system and in other areas of the body as well. Can you name other systems that show this folding pattern?

descending colon. The distal colon bends backward into an S shape forming the **sigmoid colon** (named for the Greek letter *sigma*), which continues downward to empty into the **rectum,** a temporary storage area for indigestible or nonabsorbable food residue (see Fig. 19-7A). The narrow terminal portion of the large intestine is the **anal canal,** which leads to the outside of the body through an opening called the **anus** (A-nus). Visual examination of the colon is a part of Adam's physical examination in the case study.

Functions of the Large Intestine The colonic epithelium produces a great quantity of mucus, but no enzymes. Minimal digestion occurs in this organ, but some water is reabsorbed, and undigested food is stored, formed into solid waste material, called **feces** (FE-seze) or stool, and then eliminated.

Food waste moves through the large intestine by intermittent **mass movements,** waves of peristalsis that propel the contents several feet at a time toward the rectum. These movements often occur just after meals. Stretching of the rectum stimulates smooth muscle contraction in the rectal wall. Aided by voluntary contractions of the diaphragm and the abdominal muscles, the feces are eliminated from the body in a process called **defecation** (def-e-KA-shun). An anal sphincter provides voluntary control over defecation (see Fig. 19-7A).

Food waste remains quite stationary between mass movements. During these periods, bacteria that normally live in the colon act on it to produce vitamin K and some of the B-complex vitamins. Systemic antibiotic therapy may destroy these symbiotic (helpful) bacteria living in the large intestine, causing undesirable side effects.

CHECKPOINTS

- 19-4 Which form of motility occurs in the esophagus? In the small intestine?
- 19-5 What type of food is digested in the stomach?
- 19-6 What are the three divisions of the small intestine?
- 19-7 How does the small intestine function in the digestive process?
- 19-8 What are the functions of the large intestine?

The Accessory Organs

The accessory organs release secretions through ducts into the digestive tract. Specifically, the salivary glands deliver their secretions into the mouth. The liver, gallbladder, and pancreas release secretions into the duodenum.

THE SALIVARY GLANDS

Saliva (sah-LI-vah) is a watery solution that moistens food and facilitates mastication (chewing) and deglutition (swallowing). Saliva helps keep the teeth and mouth clean. It also contains some antibodies and an enzyme (lysozyme) that help reduce bacterial growth.

This watery mixture contains mucus and an enzyme called **salivary amylase** (AM-ih-laze), which initiates carbohydrate digestion. Saliva is manufactured by three pairs of glands (Fig. 19-8):

- The **parotid** (pah-ROT-id) **glands,** the largest of the group, are located inferior and anterior to the ear.
- The **submandibular** (sub-man-DIB-u-lar) **glands,** also called *submaxillary* (sub-MAK-sih-ler-e) *glands,* are located near the body of the lower jaw.
- The **sublingual** (sub-LING-gwal) **glands** are under the tongue.

All these glands empty through ducts into the oral cavity.

THE LIVER

The **liver** (LIV-er), often referred to by the word root *hepat*, is the largest accessory organ (Fig. 19-9). It is located in the superior right portion of the abdominal cavity under the dome of the diaphragm. The lower edge of a normal-sized liver is level with the ribs' lower margin. The human liver is the same reddish brown color as animal liver seen in the supermarket. It has a large right lobe and a smaller left lobe; the right lobe includes two inferior smaller lobes that are not illustrated. The liver is supplied with blood through two vessels: the portal vein and the hepatic artery (the portal system and blood supply to the liver were described in Chapter 15). These vessels deliver about 1.5 qt (1.6 L) of blood to the liver every minute. The hepatic artery carries blood high in oxygen, whereas the venous portal system carries blood that is lower in oxygen and rich in digestive end products.

Functions of the Liver This most remarkable organ has many functions relating to digestion, metabolism, blood composition, and elimination of waste. Some of its major activities are:

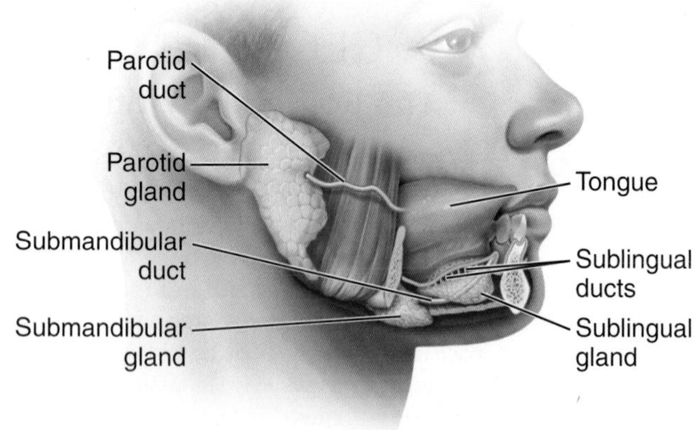

Figure 19-8 Salivary glands. KEY POINT Three groups of glands secrete into the mouth through ducts. **ZOOMING IN** Which salivary glands are directly below the tongue?

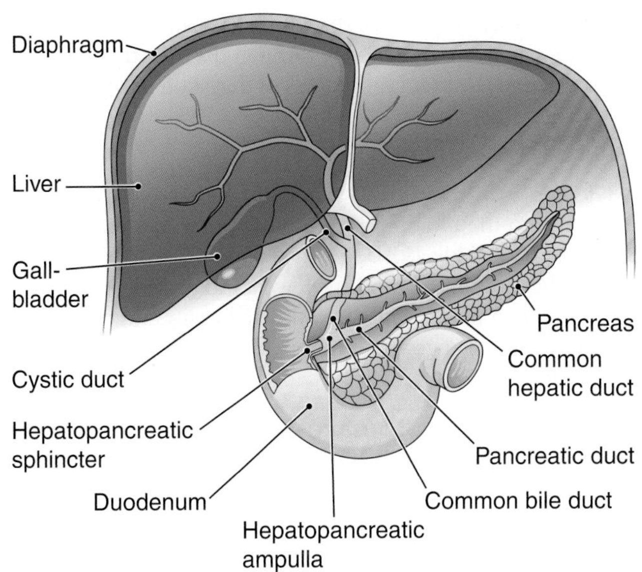

Figure 19-9 **Accessory organs of digestion.** **KEY POINT** The accessory organs secrete digestive substances into the small intestine. **ZOOMING IN** Into which part of the small intestine do these accessory organs secrete?

- The manufacture of **bile**, a substance needed for the digestion of fats, discussed shortly.
- The control of blood glucose levels under the influence of hormones. The liver stores glucose in the form of glycogen, the animal equivalent of the starch found in plants. When the blood glucose level falls below normal, liver cells convert glycogen to glucose and synthesize new glucose molecules from amino acids. The glucose enters blood to restore a normal concentration.
- The modification of fats so that they can be used more efficiently by cells all over the body. The liver is also an important site for fat storage.
- The storage of iron and some vitamins.
- The formation of blood plasma proteins, such as albumin, globulins, and clotting factors.
- The destruction of old red blood cells and the recycling or elimination of their breakdown products. One by-product, a pigment called **bilirubin** (BIL-ih-ru-bin), is eliminated in bile and gives the stool its characteristic dark color.
- The synthesis of **urea** (u-RE-ah), a waste product of protein metabolism. Urea is released into the blood and transported to the kidneys for elimination.
- The **detoxification** (de-tok-sih-fih-KA-shun) (removal of toxicity) of harmful substances, such as alcohol and certain drugs.

Bile The liver's main digestive function is the production of bile, a substance needed for the processing of fats. The salts contained in bile act like a detergent to **emulsify** fat; that is, to break up fat into small droplets that can be acted on more effectively by digestive enzymes. Bile also aids in absorption of digested fats from the small intestine lumen into the epithelial cells.

Bile leaves the lobes of the liver by two ducts that merge to form the **common hepatic duct** (**Fig. 19-9**). This duct joins with the **cystic** (SIS-tik) **duct** (from the gallbladder) to form the **common bile duct**. The common bile duct joins with the pancreatic duct to form a very short, wide channel called the **hepatopancreatic ampulla.** The **hepatopancreatic sphincter** controls the emptying of pancreatic fluids and bile from the ampulla into the duodenum, as discussed next.

THE GALLBLADDER

Although the liver may manufacture bile continuously, the body needs it only a few times a day, so the hepatopancreatic sphincter is usually closed. Consequently, bile from the liver flows up through the cystic duct into the **gallbladder** (GAWL-blad-er) a muscular sac on the inferior surface of the liver that stores bile (**see Fig. 19-9**). When chyme enters the duodenum, the gallbladder contracts and the hepatopancreatic sphincter opens. Bile flows out of the gallbladder and liver into the duodenum.

THE PANCREAS

The **pancreas** (PAN-kre-as) is a long gland that extends from the duodenum to the spleen (**see Fig. 19-9**). The pancreas produces enzymes that digest fats, proteins, carbohydrates, and nucleic acids. The protein-digesting enzymes are produced in inactive forms which must be converted to active forms in the small intestine by other enzymes.

The pancreas also releases large amounts of sodium bicarbonate ($NaHCO_3$), an alkaline (basic) fluid that neutralizes the acidic chyme in the small intestine, thus protecting the digestive tract's lining. These juices collect in the pancreatic duct and along with bile, enter the duodenum via the hepatopancreatic sphincter. Most people have an additional smaller pancreatic duct that opens directly into the duodenum and bypasses the hepatopancreatic ampulla.

As described in Chapter 12, the pancreas also functions as an endocrine gland, producing the hormones insulin and glucagon that regulate glucose metabolism. These islet cell secretions are released into the blood.

CHECKPOINTS

- ☐ **19-9** What are the names and locations of the salivary glands?
- ☐ **19-10** Which accessory organ secretes bile, and what is the function of bile in digestion?
- ☐ **19-11** What is the role of the gallbladder?
- ☐ **19-12** What accessory organ secretes sodium bicarbonate, and what is the function of this substance in digestion?

Enzymes and the Digestive Process

Although the individual organs of the digestive tract are specialized for digesting different types of food, the fundamental chemical process of digestion is the same for fats, proteins, and carbohydrates. In every case, this process requires enzymes. Recall from Chapter 2 (**see Fig. 2-11**) that enzymes are catalysts, substances that speed the rate of chemical reactions, but are not themselves changed or used up in the reaction.

Almost all enzymes are proteins, and they are highly specific in their actions. In digestion, an enzyme acts only in a certain type of reaction involving a certain type of nutrient molecule. For example, the carbohydrate-digesting enzyme amylase only splits starch into the disaccharide (double sugar) maltose. Another enzyme is required to split maltose into two molecules of the monosaccharide (simple sugar) glucose. Other enzymes split fats, or triglycerides, into their building blocks, glycerol, and fatty acids (**see Fig. 2-9A**). Still others split proteins into smaller units called *peptides* and into their building blocks, amino acids (**see Fig. 2-10**).

See the Student Resources on thePoint to review the animations "Enzymes," "General Digestion," and "Digestion of Carbohydrates."

THE ROLE OF WATER

Because water is added to nutrient molecules as they are split by enzymes, the process of digestion is referred to chemically as **hydrolysis** (hi-DROL-ih-sis), which means "splitting (lysis) by means of water (hydr/o)." In this chemical process, water's hydroxyl group (OH^-) is added to one fragment and the hydrogen ion (H^+) is added to the other, splitting the molecule. **Figure 19-10** shows the hydrolysis of a disaccharide into two monosaccharides. The building blocks of fats and proteins are separated in the same manner. Each hydrolysis reaction requires a specific enzyme and uses one molecule of water. About 7 L of water are secreted into the digestive tract each day, in addition to the nearly 2 L taken in with food and drink. You can now understand why so large an amount of water is needed. Water is not only used to produce digestive juices and to dilute food so that it can move more easily through the digestive tract but is also used in the chemical process of digestion itself.

DIGESTION, STEP-BY-STEP

Let us see what happens to a mass of food from the time it is taken into the mouth to the moment that it is ready to be absorbed (**see Table 19-1**).

In the mouth, the food is chewed and mixed with saliva, softening it so that it can be swallowed easily. Salivary amylase initiates the digestive process by digesting some starch into maltose.

Digestion in the Stomach When the food reaches the stomach, it is acted on by gastric juice, with its hydrochloric acid (HCl) and enzymes. The hydrochloric acid has the important function of denaturing proteins, that is, unfolding them to prepare them for digestion. In addition, HCl activates the enzyme pepsin, which is secreted by gastric cells in an inactive form, as previously noted. Once activated by hydrochloric acid, pepsin works to cleave long protein chains into shorter chains of amino acids.

The food, gastric juice, and mucus (which is also secreted by cells of the gastric lining) are mixed to form chyme. This semiliquid substance is released gradually from the stomach through the pyloric sphincter into the small intestine for further digestion.

Digestion in the Small Intestine Most digestion occurs in the duodenum, the first part of the small intestine, with the assistance of pancreatic and hepatic secretions. In brief:

- **Fats.** Most ingested fat is in the form of **triglycerides**, which are composed of glycerol and three fatty acids. After bile emulsifies fats into tiny droplets, the pancreatic enzyme **lipase** digests the triglycerides into free fatty acids (two from each triglyceride) and monoglycerides (glycerol combined with one fatty acid). If pancreatic lipase is absent, fats are expelled with the feces in an undigested form.
- **Carbohydrates.** Like salivary amylase, pancreatic amylase changes starch to maltose. Enzymes in the intestinal cell membranes digest maltose and the other

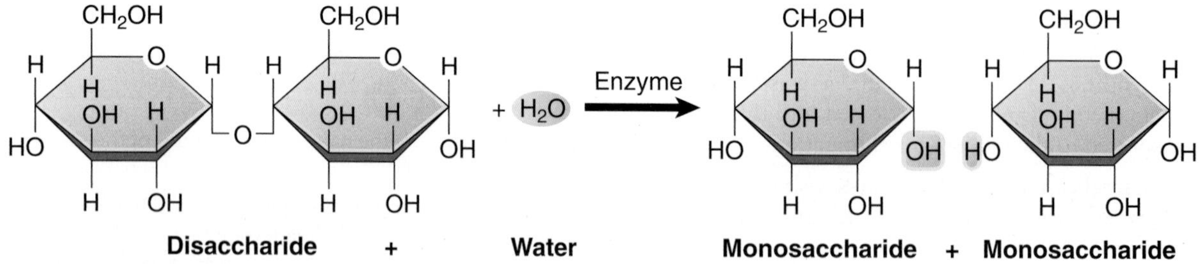

CH₂OH O H H HO H OH H H OH O CH₂OH O H H H OH H OH H OH + H_2O Enzyme

Disaccharide + **Water** → **Monosaccharide** + **Monosaccharide**

Figure 19-10 Hydrolysis. **KEY POINT** In digestion, water is added to compounds to split them into simpler building blocks. The figure shows the splitting of a disaccharide (double sugar) into two monosaccharides (simple sugars) by the addition of H^+ to one and OH^- to the other. The chemical bonds between the building blocks of fats and proteins are split in the same way. A specific enzyme is needed for every hydrolysis reaction.

Table 19-1 Summary of Digestion

Organ	Activity	Nutrients Digested	Active Substances
Mouth	Chews food and mixes it with saliva; forms it into bolus for swallowing	Starch	Salivary amylase
Esophagus	Moves food by peristalsis into stomach	—	—
Stomach	Stores food, churns food, and mixes it with digestive juices	Proteins	Hydrochloric acid, pepsin
Small intestine	Mixes food with secretions from pancreas and liver, digests food, absorbs nutrients and water into the blood or lymph	Fats, proteins, carbohydrates, nucleic acids	Intestinal enzymes (in cell membranes), pancreatic enzymes, bile from liver
Large intestine	Reabsorbs some water; forms, stores, and eliminates stool	—	—

disaccharides into monosaccharides. These enzymes are **maltase**, **sucrase**, and **lactase**, which act on the disaccharides maltose, sucrose, and lactose, respectively (**Fig. 19-11**).

- **Proteins. Trypsin** (TRIP-sin) and other pancreatic enzymes split proteins into small peptides. Other pancreatic enzymes and enzymes in the intestinal cell membranes digest these short peptides into individual amino acids.
- **Nucleic acids. Nucleases** (NU-kle-ases) in the pancreatic juice digest the nucleic acids DNA and RNA.

It is important to note that most digestive enzymes are produced by the pancreas. When pancreatic juice is absent, serious digestive disturbances always occur.

Table 19-2 summarizes the main substances used in digestion. Note that, except for HCl, sodium bicarbonate, and bile salts, all the substances listed are enzymes.

CHECKPOINTS

- ☐ **19-13** What is an enzyme?
- ☐ **19-14** What process means "splitting by means of water," as in digestion?
- ☐ **19-15** Which organ produces the most complete digestive secretions?

Absorption

The means by which digested nutrients reach the blood is known as absorption. Most absorption takes place through the villi in the mucosa of the small intestine (**see Fig. 19-7D**). Within each villus is an arteriole and a venule bridged with capillaries. Simple sugars (monosaccharides), small peptides, amino acids, a few simple fatty acids, nucleic acids, and most of the water in the digestive tract are absorbed into the blood through these capillaries. From here, they pass by way of the portal system to the liver, to be processed, stored, or released as needed.

ABSORPTION OF FATS

Most fats have an alternative method of reaching the blood. Instead of entering the blood capillaries, they are absorbed by the villi's more permeable lymphatic capillaries, the lacteals. The absorbed fat droplets give the lymph a milky appearance. The mixture of lymph and fat globules that drains from the small intestine after fat has been digested is called **chyle** (kile). Chyle merges with the lymphatic circulation and eventually enters the blood when the lymph drains into veins near the heart. The absorbed fats then move to the liver for further processing.

ABSORPTION OF VITAMINS AND MINERALS

Minerals and vitamins ingested with food are also absorbed from the small intestine. The minerals and some of the vitamins mix with water and are absorbed directly into the blood.

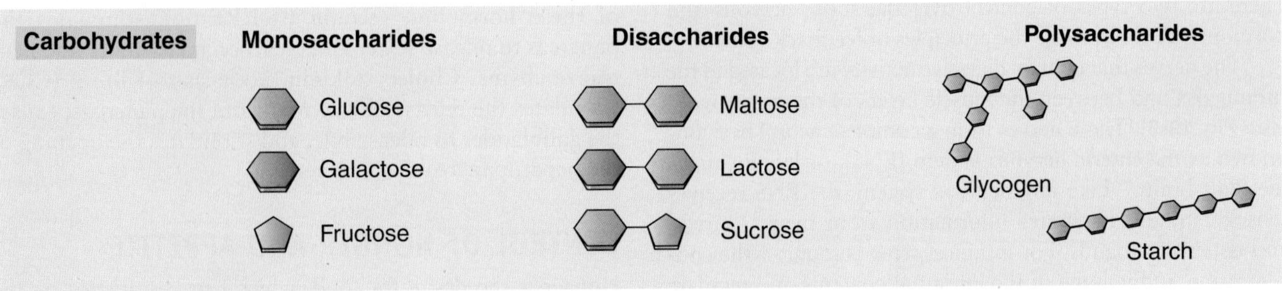

Figure 19-11 Carbohydrates. **KEY POINT** We can consume carbohydrates in many forms, but they are all eventually converted to glucose in the body. **ZOOMING IN** Name the two monosaccharide components of sucrose.

Table 19-2 Digestive Agents Produced by Digestive Tract Organs and Accessory Organs

Organ	Substance	Action
Salivary glands	Salivary amylase	Converts starch to maltose
Stomach	Hydrochloric acid (HCl)[a]	Denatures proteins; activates pepsin
	Pepsin	Digests proteins to peptides
Liver	Bile salts[a]	Emulsify fats
Pancreas	Sodium bicarbonate[a]	Neutralizes HCl
	Amylase	Digests starch to maltose
	Trypsin	Digests protein to peptides
	Lipase	Digests fats to fatty acids and monoglycerides
	Nucleases	Digest nucleic acids
Small intestine (enzymes embedded in cell membranes)	Peptidases	Digest peptides to amino acids
	Lactase, maltase, sucrase	Digest disaccharides to monosaccharides

[a]Not enzymes.

Other vitamins are incorporated in fats and are absorbed along with the fats. Vitamin K and some B vitamins are produced by bacterial action in the colon and are absorbed from the large intestine.

CHECKPOINT

☐ 19-16 What is absorption?

Control of Digestion and Eating

As food moves through the digestive tract, its rate of movement and the activity of each organ it passes through must be carefully regulated. If food moves too slowly or digestive secretions are inadequate, the body will not get enough nourishment. If food moves too rapidly or excess secretions are produced, digestion and absorption may be incomplete, or the digestive tract's lining may be damaged.

CONTROL OF DIGESTION

There are two types of control over digestion: nervous and hormonal. Both illustrate the principles of feedback control.

The nerves that control digestive activity are located in the submucosa and between the muscle layers of the organ walls (see Fig. 19-3). These nerves form a complete neural network known as the **enteric nervous system (ENS)**, or less formally, the "gut brain." Like any nervous system, the ENS receives sensory stimuli, integrates information from many sources, and issues commands. For instance, sensors monitor the concentration and volume of the intestinal contents and regulate both glands and smooth muscle. The activity of the ENS can be modified by the autonomic (visceral) nervous system. In general, parasympathetic stimulation increases activity, and sympathetic stimulation decreases activity. Excess sympathetic stimulation, as can be caused by stress, can slow food's movement through the digestive tract and inhibit mucus secretion, which is crucial to protecting the digestive tract's lining.

The digestive organs themselves produce the hormones involved in regulating digestion. The following is a discussion of some of these controls (**Table 19-3**).

The sight, smell, thought, taste, or feel of food in the mouth stimulates, through the nervous system, the secretion of saliva and the release of gastric juice. Once in the stomach, food stimulates the release into the blood of the hormone **gastrin**, which promotes stomach secretions and motility.

When chyme enters the duodenum, nerve impulses inhibit stomach motility, so that food will not move too rapidly into the small intestine. This action is a good example of negative feedback. At the same time, hormones released from the duodenum not only stimulate intestinal activity but also feed back to the stomach to reduce its activity. **Gastric-inhibitory peptide (GIP)** is one such hormone. It acts on the stomach to inhibit the release of gastric juice. Its more important action is to stimulate insulin release from the pancreas when glucose enters the duodenum. Another of these hormones, **secretin** (se-KRE-tin), stimulates the pancreas to release water and bicarbonate to dilute and neutralize chyme. **Cholecystokinin** (ko-le-sis-to-KI-nin) (**CCK**) stimulates the release of enzymes from the pancreas, causes the gallbladder to release bile, and stimulates the opening of the hepatopancreatic sphincter.

CONTROL OF HUNGER AND APPETITE

Hunger is the desire for food, which can be satisfied by the ingestion of a filling meal. Appetite differs from hunger in that although it is basically a desire for food, it often has no

Table 19-3 Hormones Active in Digestion

Hormone	Source	Action
Gastrin	Stomach	Stimulates release of gastric juice
Gastric-inhibitory peptide (GIP)	Duodenum	Stimulates insulin release from pancreas when glucose enters duodenum; inhibits release of gastric juice
Secretin	Duodenum	Stimulates release of water and bicarbonate from pancreas; inhibits the stomach
Cholecystokinin (CCK)	Duodenum	Stimulates release of digestive enzymes from pancreas; stimulates release of bile from gallbladder; inhibits the stomach

relationship to the need for food. Even after an adequate meal that has relieved hunger, a person may still have an appetite for additional food. A variety of factors, such as emotional state, cultural influences, habit, and memories of past food intake can affect appetite.

Hunger is regulated by hypothalamic centers that respond to nutrient levels, neural input from the digestive tract, and hormones.

Short-Term Regulation of Hunger The hypothalamus induces hunger sensations when we haven't eaten in a while and fullness sensations when we should stop eating. Between meals, the empty stomach releases a hormone called **ghrelin** (GREL-in) that stimulates hunger (**Fig. 19-12A**). Low nutrient levels, particularly hypoglycemia, also activate the hypothalamic hunger center. On the other hand, food consumption distends the digestive tract, sending neural signals that decrease hunger. Also, remember that food arrival in the duodenum stimulates the release of the hormones CCK and GIP, and GIP stimulates insulin production. CCK and insulin further suppress hunger, as does the increased blood glucose level resulting from nutrient digestion (**Fig. 19-12B**). These neural and chemical signals prevent the consumption of more food than the intestine can process.

Long-Term Regulation of Body Weight Despite day-to-day variations in food intake and physical activity, a healthy individual maintains a constant body weight and energy reserves of fat over long periods. With the discovery of the hormone leptin (from the Greek word *leptos*, meaning "thin"), researchers have been able to piece together one long-term mechanism for regulating weight. Leptin is produced by adipocytes (fat cells). When fat stores increase because of excess food intake or inadequate activity, the cells release more leptin (**Fig. 19-12C**). Centers in the hypothalamus respond to the hormone by decreasing food intake and increasing energy expenditure, resulting in weight loss. If this feedback mechanism is disrupted, obesity will result. Early hopes of using leptin to treat human obesity have dimmed, however, because obese people do not have a leptin deficiency. This system's failure in humans appears to be caused by the hypothalamus's inability to respond to leptin rather than our inability to make the hormone.

CHECKPOINTS

- [ ] 19-17 What are the two types of control over the digestive process?
- [ ] 19-18 What is the difference between hunger and appetite?

Disorders of the Digestive System

Infections, ulcers, cancer, and structural abnormalities all affect the digestive system at almost any level. Stones may form in the accessory organs or their ducts. Mechanical, nervous, chemical, genetic, and hormonal factors may be at the source of digestive problems.

19

PERITONITIS

Inflammation of the peritoneum, termed **peritonitis** (per-ih-to-NI-tis), is a serious complication that may follow infection of an organ covered by the peritoneum—often, the appendix. The frequency and severity of peritonitis have been greatly reduced by the use of antibiotics. The disorder still occurs, however, and can be dangerous. If the infection is kept in one area, it is said to be *localized peritonitis*. A *generalized peritonitis*, as may be caused by a ruptured appendix, a perforated ulcer, or a penetrating wound, may lead to the growth of so many disease organisms and the release of so much bacterial toxin as to be fatal. Immediate surgery to repair the rupture and medical care are needed.

DISEASES OF THE MOUTH AND TEETH

Tooth decay is also termed dental **caries** (KA-reze) (from Latin meaning "rottenness"). It has a number of causes, including diet, heredity, mechanical problems, and endocrine disorders. People who ingest a lot of sugar are particularly prone to this disease. Because a baby's teeth begin to develop before birth, a mother's diet during pregnancy is important in ensuring the formation of healthy teeth in her baby.

Any infection of the gum is called **gingivitis** (jin-jih-VI-tis). If such an infection continues untreated, it may lead to a more serious condition, **periodontitis** (per-e-o-don-TI-tis), which involves not only the gum tissue but also the teeth's supporting bone. Tooth loosening and bone destruction

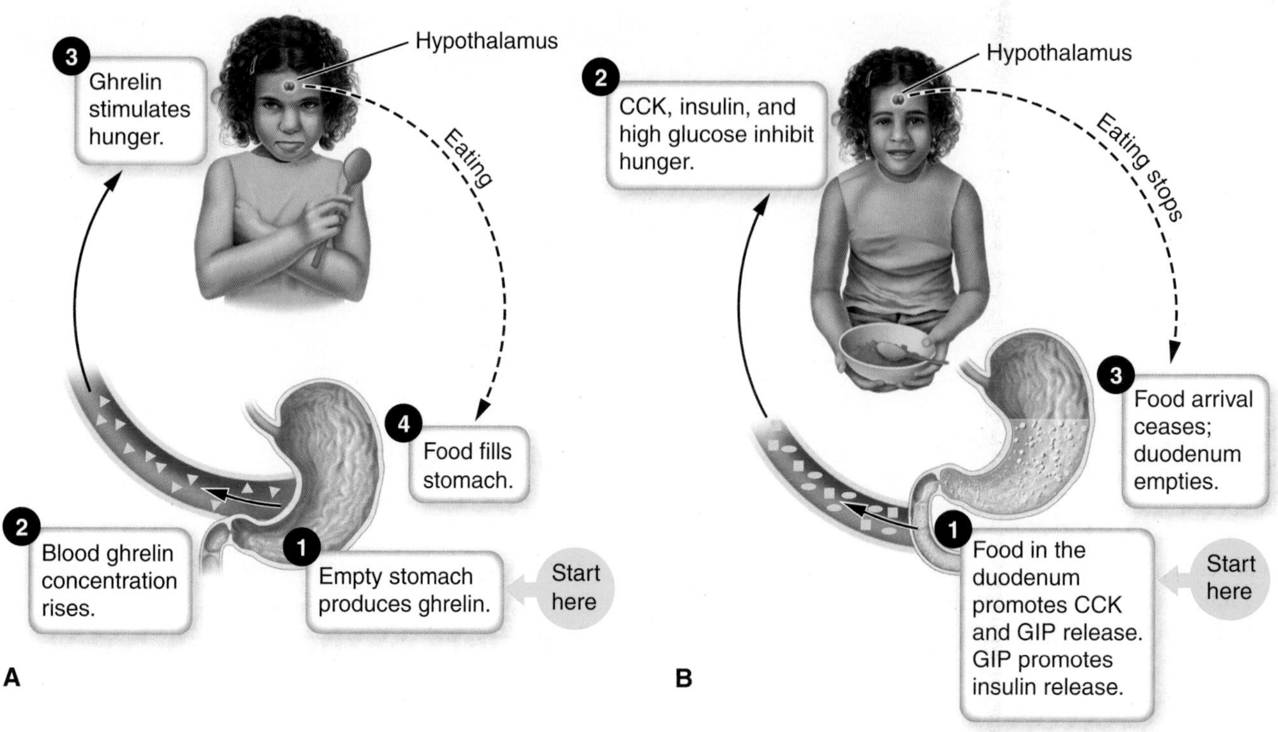

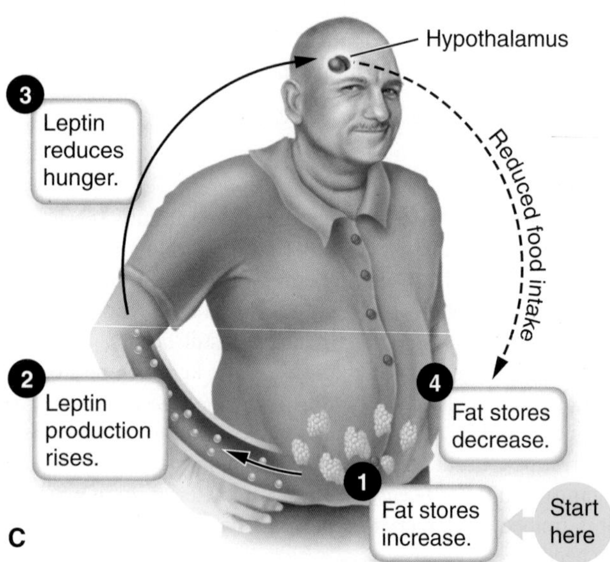

Figure 19-12 **Regulation of food intake. A.** Ghrelin initiates hunger. **B.** CCK (among other factors) inhibits hunger. **C.** Leptin governs food intake over the long term.

follow unless proper treatment and improved dental hygiene halt periodontitis. This condition is responsible for nearly 80% of tooth loss in people older than 45 years of age.

Vincent disease, a kind of gingivitis caused by a spirochete or a bacillus, is most prevalent in teenagers and young adults. Characterized by inflammation, ulceration, and infection of the oral mucous membranes, this disorder is highly contagious, particularly by oral contact.

Fungal infection of the mouth caused by a species of yeast called *Candida albicans* is commonly called *oral thrush*. It causes creamy white lesions on the tongue and oral mucosa that can bleed. Patients on antibiotic therapy are prone to develop oral fungal infections because these drugs may destroy the normal bacterial flora and allow fungi to grow. They are also more common in patients with compromised immunity.

Parotitis (par-o-TI-tis) is inflammation of the parotid glands. The contagious parotitis, commonly called *mumps,* is caused by a viral infection. Mumps may lead to inflammation of the testicles by the same virus. Males affected after puberty are at risk for permanent damage to these sex organs, resulting in sterility. Another complication that occurs in about 10% of cases is meningitis. Mumps can be prevented by childhood immunization with a vaccine (MMR).

DISORDERS OF THE ESOPHAGUS AND STOMACH

We discuss the esophagus and stomach together, as some disorders involve both, especially the juncture of these two organs.

Hiatal Hernia and Gastric Reflux A weakness in the diaphragm at the point where the esophagus joins the stomach may allow the stomach to protrude upward as a **hiatal** (hi-A-tal) **hernia** (**Fig. 19-13A**). Minor irregularities in this area are common and may cause no problem, but the incidence and severity of hiatal hernia increase with age. A hiatal hernia may cause discomfort after meals, gastritis, or ulceration, and serious cases may need surgical repair.

Weakness in the lower esophageal sphincter (LES) may allow the acidic stomach contents to flow back into the distal esophagus (**Fig. 19-13B**). The result is a burning sensation below or behind the sternum that is described as *heartburn.* The symptom does not involve the heart in any way, but is felt in the vicinity of the heart. Often it is mistaken for a heart attack. More dangerously, a heart attack can be mistaken for heartburn, and people may fail to seek medical attention thinking they have a minor disturbance. Chronic reflux is referred to as **gastroesophageal reflux disease** (**GERD**). Overfilling of the stomach and meals high in fat contribute to reflux and GERD by initiating nervous responses that relax the LES. Acid reflux irritates the esophageal mucous membrane, leading to esophagitis. Eventually there may be edema and scar tissue formation that narrows the esophagus and interferes with digestion. GERD increases the risk of esophageal cancer. It also increases the risk of lung cancer, as a result of acid reflux into the lungs.

Acid reflux and GERD are treated with antacids and drugs that inhibit the production of HCl. People with this problem should avoid certain foods and beverages, including fats, caffeine, chocolate, and alcohol, and should not smoke. Other measures that may help are eating in an upright position, not bending down for long periods, not lying down for several hours after eating, and sleeping with the head elevated. Weight loss may also help in cases of obesity.

Nausea and Vomiting **Vomiting,** also called **emesis** (EM-eh-sis), is the expulsion of gastric (and sometimes intestinal) contents through the mouth. The contraction of muscles in the abdominal wall forcibly empties the stomach. Peristaltic waves sometimes pass from the intestine toward the esophagus to aid in the propulsion. **Nausea** (NAW-ze-ah) is an unpleasant sensation of queasiness that suggests that vomiting may soon occur.

19

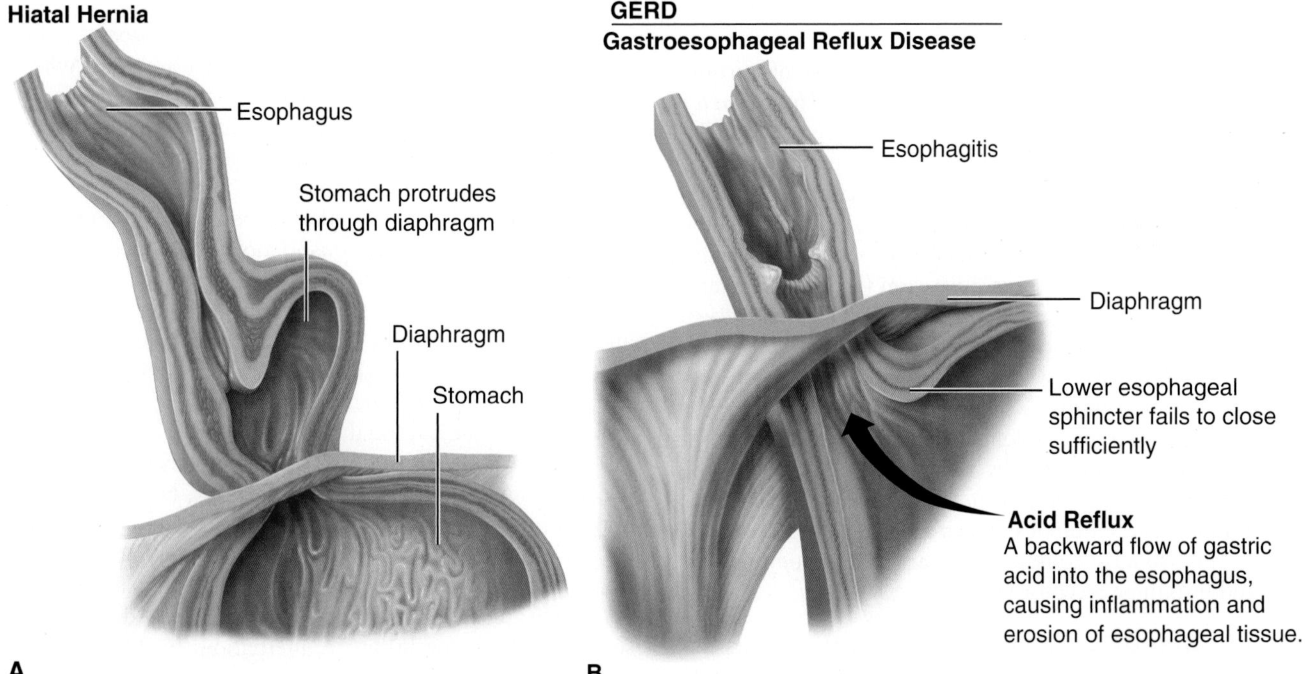

Figure 19-13 **Disorders involving the esophagus and stomach.** **KEY POINT** Anatomic problems allow acidic stomach contents to flow up into the esophagus. **A.** Hiatal hernia. A weakness in the diaphragm allows a portion of the stomach to protrude upward through that muscle. **B.** Gastroesophageal reflux disease (GERD). Weakness of the lower esophageal sphincter allows acid reflux into the esophagus with damaging results.

Nausea and vomiting can reflect multiple causes, including

- Inflammation or excess distention of the esophagus, stomach, intestine, or peritoneum
- Problems with the vestibular system, such as inner ear infection or motion sickness
- Drugs (such as chemotherapy drugs) that directly activate the brain centers responsible for vomiting
- Pregnancy, via an unknown mechanism

Stomach Cancer Although stomach cancer has become rare in the United States, it is common in many parts of the world, and it is a serious disorder because of its high death rate. Men are more susceptible to it than are women. The tumor nearly always develops from the stomach's epithelial or mucosal lining and is often of the type called **adenocarcinoma** (ad-en-o-kar-sih-NO-mah). Sometimes, the victim has suffered from long-standing indigestion (discomfort after meals) but has failed to consult a physician until the cancer has metastasized to other organs, such as the liver or lymph nodes. Persistent indigestion is one of the important warning signs of stomach cancer.

Gastritis and Peptic Ulcers **Gastritis** (gas-TRI-tis) is inflammation of the stomach lining. Some agents, such as strong chemicals, nicotine, radiation therapy, and chemotherapy, directly damage the gastric mucosal cells. Other agents, particularly aspirin and antiinflammatory drugs, decrease the production of mucous and/or increase the production of acid, thereby enabling the stomach juices themselves to damage the cells. An important cause of chronic gastritis is infection with the bacterium *Helicobacter pylori*. This pathogen lives in the mucous coating of the stomach's pylorus and stimulates acid production.

Severe gastritis can erode the mucosa, causing an ulcer. **Peptic ulcers** (named for the enzyme pepsin) occur in the mucous membrane of the esophagus, stomach, or duodenum and are most common in people between the ages of 30 and 45 years. Peptic ulcers in the stomach are termed *gastric ulcers*; those in the duodenum are *duodenal* (du-o-DE-nal) *ulcers*.

Factors promoting gastritis favor ulcer formation, particularly infection with *H. pylori*, excess antiinflammatory drug use, and cigarette smoking. Most ulcers are cured when the *H. pylori* is eliminated by antibiotics, emphasizing the importance of this pathogen in the disease process. Drugs that inhibit the secretion of stomach acid can help treat peptic ulcers. The two most common types are proton pump inhibitors and histamine receptor blockers.

Pyloric Stenosis Normally, the stomach contents are moved through the pyloric sphincter within approximately two to six hours after eating. Some infants, however, most often boys, are born with an obstruction of the pyloric sphincter, a condition called **pyloric stenosis** (steh-NO-sis). Usually, surgery is required in these cases to modify the muscle so that food can pass from the stomach into the duodenum.

See the Student Resources on thePoint for an illustration of common ulcer types and sites and for an illustration of pyloric stenosis.

INTESTINAL DISORDERS

A wide variety of disorders can affect the intestines, impairing digestion, absorption, and/or elimination.

Inflammatory Disorders Many intestinal disorders involve inflammation. **Appendicitis** (ah-pen-dih-SI-tis) is inflammation of the appendix and may result from infection or obstruction. The cause of obstruction is usually a **fecalith** (FE-cah-lith), a hardened piece of fecal material. The first sign of acute appendicitis is usually abdominal pain with loss of appetite and sometimes nausea or vomiting. Pain eventually localizes in the right lower quadrant of the abdomen. Laboratory blood tests show elevated leukocytes. Surgery (appendectomy) is required to remove an inflamed appendix. Untreated, it can rupture to spread infection into the peritoneal cavity.

Two similar diseases are included under the heading of **inflammatory bowel disease** (IBD): **Crohn** (krone) **disease** and **ulcerative colitis.** Both occur mainly in adolescents and young adults and cause similar symptoms of pain, diarrhea, weight loss, and rectal bleeding. Crohn disease usually involves inflammation of the distal small intestine. It is an autoimmune disease, which may in part be hereditary. Ulcerative colitis involves inflammation and ulceration of the colon, and usually the rectum.

Celiac disease (SE-le-ak) is a digestive disease caused by an inability to tolerate gluten, a protein found in wheat, rye, and barley. When people with celiac disease eat foods containing gluten, their immune system sets up an inflammatory response that damages or destroys the intestinal villi. They thus have a reduced ability to absorb nutrients from digested food and become malnourished. Other symptoms of celiac disease are diarrhea, constipation, abdominal pain, and bloating after consuming foods containing gluten. Continued inflammation of the villi in untreated celiac disease is associated with a rare form of intestinal cancer. Celiac disease is inherited, and it is now estimated to affect about one in 200 people. The only treatment is total avoidance of gluten, which enables the intestinal villi to regenerate over time. The disorder is often misdiagnosed as IBD or other bowel disorder, sometimes with serious consequences, but both genetic and antibody screening tests are now available.

Irritable bowel syndrome (IBS) is a common gastrointestinal disorder seen typically in young to middle-aged adults. Symptoms include pain and constipation or diarrhea, or sometimes both conditions in alternation. In IBS, the intestine is overly sensitive to stimulation, often brought on by stress. Although the condition is chronic and causes much pain, frustration, and anxiety, it is not life-threatening and does not develop into more serious bowel diseases.

Difficulties with digestion or absorption may be due to **enteritis** (en-ter-I-tis), an intestinal inflammation. When both the stomach and the small intestine are involved, the disorder is called **gastroenteritis** (gas-tro-en-ter-I-tis). The symptoms of gastroenteritis include nausea, vomiting, and diarrhea as well as acute abdominal pain (colic). The disorder may be caused by a variety of pathogenic organisms, including viruses, bacteria, and protozoa, and it is sometimes described as "stomach flu." Important viral agents include rotavirus, which can cause life-threatening gastroenteritis in babies, and norovirus, which frequently causes epidemics throughout cruise ships, schools, and nursing homes. Chemical irritants, such as alcohol, certain drugs (e.g., aspirin), and other toxins, have been known to cause this disorder as well. The condition often resolves spontaneously in a few hours or days.

Diverticula (di-ver-TIK-u-lah) are small pouches in the wall of the intestine, most commonly in the colon. A diet low in fiber contributes to formation of large numbers of diverticula, a condition called **diverticulosis** (di-ver-tik-u-LO-sis). Collection of waste and bacteria in these sacs leads to **diverticulitis** (di-ver-tik-u-LI-tis), which is accompanied by pain and sometimes bleeding. There is no cure for diverticulitis; it is treated with diet, stool softeners, and drugs to reduce intestinal spasms and pain if needed.

See the Student Resources on thePoint for information on the complications of ulcerative colitis and illustrations of diverticulitis.

Diarrhea Diarrhea is a symptom characterized by abnormally frequent watery bowel movements. The danger of diarrhea is dehydration and loss of salts, especially in infants. Diarrhea may result from excess activity of the colon, faulty absorption, or infection. Infections resulting in diarrhea include cholera, dysentery, and food poisoning. **Tables A3-1 and A3-5** in Appendix 3 list some of the organisms causing these diseases. Such infections are often spread by poor sanitation and contaminated food, milk, or water. A stool examination may be required to establish the cause of diarrhea; examination may reveal the presence of pathogenic organisms, worm eggs, or blood.

Constipation Millions of dollars are spent each year in an effort to remedy a condition called **constipation**. What is constipation? Many people erroneously think they are constipated if they go a day or more without having a bowel movement. Actually, what is normal varies greatly; one person may normally have a bowel movement only once every two or three days, whereas another may normally have more than one movement daily. The term *constipation* is also used to refer to hard stools or difficulty with defecation.

On the basis of its onset, constipation may be classified as acute or chronic. Acute constipation occurs suddenly and may be due to an intestinal obstruction, such as a tumor or diverticulitis. Extreme constipation is termed **obstipation** (ob-stih-PA-shun). Chronic constipation, in contrast, has a more gradual onset and may be divided into two groups:

- **Spastic constipation**, in which the intestinal musculature is overstimulated so that the canal becomes narrowed and the lumen (space) inside the intestine is not large enough to permit the passage of fecal material.
- **Flaccid** (FLAK-sid) **constipation**, which is characterized by a lazy, or *atonic* (ah-TON-ik), intestinal muscle. Elderly people and those on bed rest are particularly susceptible to this condition. Often, it results from repeated denial of the urge to defecate. Regular bowel habits, moderate exercise, eating more vegetables and other bulky foods, and an increase in fluid intake may help people who have sluggish intestinal muscles.

People should avoid the chronic use of laxatives and enemas, which interfere with the natural defecation reflex. They may also alter electrolyte balance and result in fluid loss. The streams of fluid used in enemas may injure the intestinal lining by removing the normal protective mucus. In addition, enemas aggravate hemorrhoids. Enemas should be done sparingly and only on a physician's order.

Intestinal Obstruction **Intussusception** (in-tuh-suh-SEP-shun) is the slipping of a part of the intestine into an adjacent part (**Fig. 19-14A**). It occurs mainly in male infants in the ileocecal region. **Volvulus** (VOL-vu-lus) is a twisting of the intestine, usually the sigmoid colon (**Fig. 19-14B**). It may be a congenital malformation or the result of a foreign body or tumor. Both intussusception and volvulus can be fatal if not treated quickly. **Ileus** (IL-e-us) is an intestinal obstruction caused by lack of peristalsis or by muscle contraction. A physician can insert a tube to release intestinal material. **Hemorrhoids** (HEM-o-roydz) are varicose veins in the rectum. These enlarged veins may cause pain and bleeding and may eventually extend out through the rectum.

Cancer of the Colon and Rectum Tumors of the colon and rectum are among the six most common types of cancer in the United States. These tumors are usually adenocarcinomas that arise from benign polyps (small growths) in the mucosal lining (**see Fig. 19-15C**). The occurrence of colon cancer is evenly divided between the

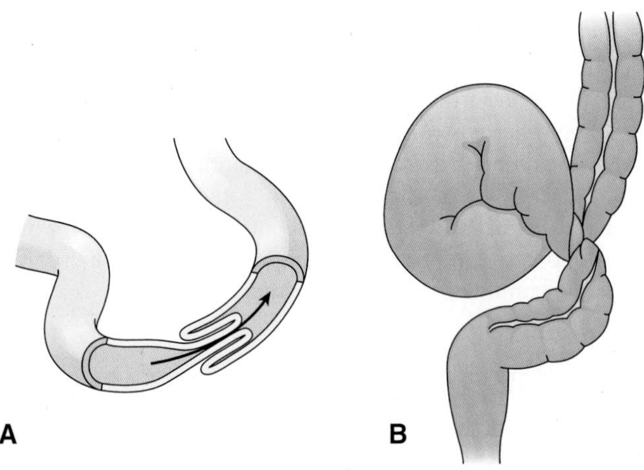

Figure 19-14 **Intestinal obstructions. A.** Intussusception. A part of the intestine slips into an adjacent part. **B.** Volvulus. A twisting of the intestine.

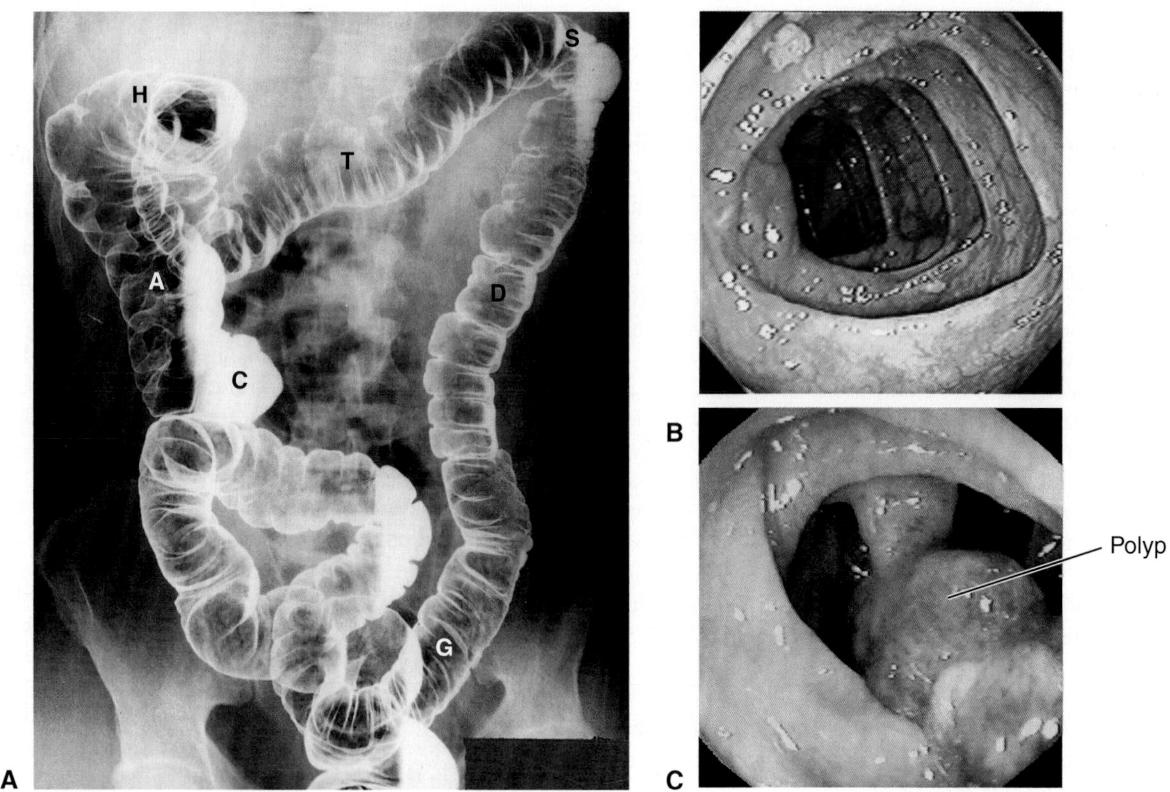

Figure 19-15 **Imaging of the colon. A.** X-ray study of the colon with barium contrast. Labels indicate the cecum (C) as well as the ascending (A), transverse (T), descending (D), and sigmoid (G) colon. The hepatic (H) and splenic (S) flexures are also labeled. **B.** Endoscopic image of the cecum, the first portion of the large intestine. **C.** Endoscopic image of a colonic polyp.

sexes, but malignant tumors of the rectum are more common in men than in women.

In the past, most colon tumors were identified by x-ray using a contrast medium, usually barium sulfate, to enhance the image (**Fig. 19-15A**). This technique is still used to identify structural disorders of the colon but has been largely replaced by direct visual examination of the rectum and lower colon with an instrument called a **sigmoidoscope** (sig-MOY-do-skope) (named for the sigmoid colon). A **colonoscope** (ko-LON-o-skope) is used to examine deeper regions of the colon (**Fig. 19-15B and C**) (**see Box 19-2**). The presence of blood in the stool may indicate cancer of the bowel or some other gastrointestinal disturbance. A simple chemical test can detect extremely small quantities of blood in the stool, referred to as *occult* ("hidden") *blood*. Early detection and treatment are the keys to increasing survival rates. A colonoscopy and test for occult blood are part of Adam's physical examination in the case study.

CIRRHOSIS AND OTHER LIVER DISEASES

Cirrhosis (sih-RO-sis) of the liver is a chronic disease in which active liver cells are replaced by inactive connective (scar) tissue. The most common type of cirrhosis is alcoholic (portal) cirrhosis. Alcohol has a direct damaging effect on liver cells that is compounded by malnutrition. Cirrhosis is one possible cause of *portal hypertension*, in which damaged hepatic tissue obstructs blood flow and leads to increased pressure in the portal vein. As a result of the increased pressure, blood accumulates in the gastrointestinal tract, spleen, and peritoneal cavity. Varices (varicose veins) may appear in organs of the digestive tract, commonly in the esophagus and stomach. Many people with advanced cirrhosis die from sudden bleeding of such varices.

See the Student Resources on thePoint for a diagram on the clinical features of cirrhosis, a photograph of a cirrhotic liver, an illustration on the effects of portal hypertension, and the animation "The Liver in Health and Disease."

Jaundice Recall that the liver takes up bilirubin, a by-product of blood cell destruction, and excretes it in bile. Damage to liver cells or blockage in any of the bile ducts interferes with this activity, leading to the buildup of bilirubin in blood. As a result, the stool may become pale in color and the skin and sclera of the eyes may become yellowish; this symptom is called **jaundice** (JAWN-dis) (from French *jaune* for "yellow") (**see Fig. 6-6C**). Jaundice may also be caused by excess destruction of red blood cells. In addition, it is often seen in newborns, in whom the liver is immature and not yet functioning efficiently.

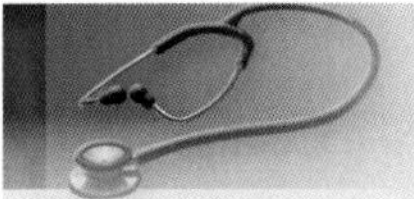

CLINICAL PERSPECTIVES

Box 19-2

Endoscopy: A View from Within

Modern medicine has made great strides toward looking into the body without resorting to invasive surgery. An instrument that has made this possible in many cases is the **endoscope**, which is inserted into the body through an orifice or small incision and used to examine passageways, hollow organs, and body cavities. The first endoscopes were rigid lighted telescopes that could be inserted only a short distance into the body. Today, physicians are able to navigate the twists and turns of the digestive tract using long **fiberoptic endoscopes** composed of flexible bundles of glass or plastic that transmit light.

In the gastrointestinal tract, endoscopy can detect structural abnormalities, bleeding ulcers, inflammation, and tumors. In addition, endoscopes can be used to remove fluid samples or tissue specimens. Some surgery can even be done with an endoscope, such as polyp removal from the colon or expansion of a sphincter. Endoscopy can also be used to examine and operate on joints (arthroscopy), the bladder (cystoscopy), respiratory passages (bronchoscopy), and the abdominal cavity (laparoscopy).

Capsular endoscopy, a recent technological advance, has made examination of the gastrointestinal tract even easier. It uses a pill-sized camera that can be swallowed. As the camera moves through the digestive tract, it transmits video images to a data recorder worn on the patient's belt. Each camera is used only once; after about 24 hours, it is eliminated with the stool and flushed away.

Hepatitis Inflammation of the liver, called **hepatitis** (hep-ah-TI-tis), may be caused by drugs, alcohol, or infection (see **Table A3-2** in Appendix 3). The known viruses that cause hepatitis are named A through E. These vary in route (pathway) of infection, severity, and complications. All types of hepatitis are marked by hepatic cell destruction and such symptoms as loss of appetite, jaundice, and liver enlargement. In most patients, the liver cells regenerate with little residual damage. The types of hepatitis virus and their primary routes of transmission are as follows:

- **Hepatitis A** (HAV): commonly transmitted in fecal matter and contaminated food and water. Children are now routinely vaccinated against hepatitis A.
- **Hepatitis B** (HBV): transmitted by exposure to the virus in blood or body fluids, although it also can be spread by fecal contamination. This is the most prevalent form of hepatitis. HBV and other blood-borne types of hepatitis viruses (C and D) have been linked to long-term development of liver cancer. Also, people with these forms of hepatitis may develop into carriers, able to transmit the disease but not showing any symptoms. HBV is usually transmitted by use of improperly sterilized needles. A vaccine is available that is now recommended for childhood immunization and for people working in healthcare and child care.
- **Hepatitis C** (HCV): transmitted primarily by exposure to infected blood. There is some evidence of limited sexual transmission. Clinical symptoms of hepatitis C may develop many years after exposure to the virus.
- **Hepatitis D** (HDV): transmitted by direct exchange of blood. It occurs only in those with hepatitis B infection.
- **Hepatitis E** (HEV): transmitted by fecal contamination of water. Most cases have been linked to epidemics in Asia, Africa, and Central America.

There is no specific treatment for hepatitis.

Liver Cancer The metastasis of cancer to the liver is common in cases that begin as cancer in one of the abdominal organs; the tumor cells are carried in the blood through the portal system to the liver.

GALLSTONES

The most common disease of the gallbladder is the formation of stones, or **cholelithiasis** (ko-le-lih-THI-ah-sis) (**Fig. 19-16**). Stones are formed from the substances contained in bile, mainly cholesterol and sometimes calcium salts. They may remain in the gallbladder or may lodge in the bile ducts, causing extreme pain. Cholelithiasis is usually associated with inflammation of the gallbladder, or **cholecystitis** (ko-le-sis-TI-tis).

See the Student Resources on thePoint for an illustration on cholecystitis and the formation of gallstones.

PANCREATIC DISORDERS

Because they are usually activated only when they reach the duodenum, pancreatic enzymes do not damage pancreatic tissues. However, if the bile ducts become blocked, as from gallstones, some activated enzymes can flow back from the duodenum into the pancreas and activate enzymes there. The pancreas suffers destruction by its own secretions, a condition called *pancreatitis*. The result is digestive enzyme and electrolyte imbalances and disruption in production of glucose-regulating hormones (insulin and glucagon). Other possible causes of pancreatitis include trauma, alcohol abuse, drug treatment, infections, tumors, or genetic abnormalities.

Acute pancreatitis comes on suddenly and is usually short-lived if treated. Treatment usually involves removal of gallstones or the gallbladder or surgical widening of the bile ducts. Chronic pancreatitis worsens over time, leading to permanent damage and malnutrition. Chronic pancreatitis is more difficult to diagnose, as it may be painless. People with chronic pancreatitis may need to supplement their diets with digestive enzymes and may require surgery to remove damaged areas of the pancreas.

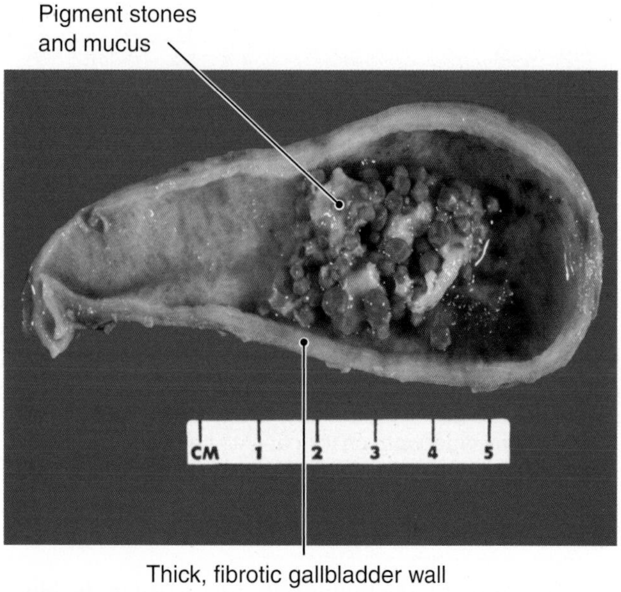

Pigment stones and mucus

Thick, fibrotic gallbladder wall

Figure 19-16 Gallstones. **KEY POINT** Formation of gallstones (cholelithiasis) causes obstruction of bile and gallbladder inflammation (cholecystitis). Numerous gallstones with mucus and a thickened gallbladder wall caused by chronic inflammation are apparent in this figure.

Chronic pancreatitis is a risk factor in pancreatic cancer. This type of cancer is hard to predict and diagnose; by the time it is found, it has usually metastasized, and the prognosis is poor. There is no treatment except early surgery.

CHECKPOINTS

- 19-19 What are six common diseases of the mouth and teeth?
- 19-20 What does GERD stand for?
- 19-21 Which two diseases fall into the category of inflammatory bowel disease?
- 19-22 What are the two forms of constipation?
- 19-23 What is hepatitis?
- 19-24 What is the common term for cholelithiasis?

Effects of Aging on the Digestive System

With age, receptors for taste and smell deteriorate, leading to a loss of appetite and decreased enjoyment of food. A decrease in saliva and poor gag reflex make swallowing more difficult. Tooth loss or poorly fitting dentures may make chewing food more difficult.

Activity of the digestive organs decreases. These changes can be seen in poor absorption of certain vitamins and poor protein digestion. Slowing of peristalsis in the large intestine and increased consumption of easily chewed, refined foods contribute to the common occurrence of constipation.

The tissues of the digestive system require constant replacement. Slowing of this process contributes to a variety of digestive disorders, including gastritis, ulcers, and diverticulosis. As with many body systems, tumors and cancer occur more frequently with age.

Disease in Context Revisited

Adam's Colonoscopy

At his scheduled time, Adam reported to the hospital as an outpatient for his colonoscopy. He had stayed on a clear liquid diet for a day and done the required laxative prep to clear his colon. He met with Dr. Clarkson, a gastroenterologist (a physician who specializes in disorders of the gastrointestinal tract). Dr. Clarkson described the procedure.

"We'll give you light sedation, and then use a flexible lighted endoscope with a camera to examine the entire colon. The procedure should take only about half an hour and has a very low risk. You have made arrangements for someone to drive you home, haven't you?"

Adam said his brother was coming.

After the test, Dr. Clarkson reported that everything looked fine and that he would send the results to Dr. Michaels.

"The good news, Adam, is that you have 10 years before you have to do this again. Maybe next time we'll be able to get our pictures with a small camera in a pill that you can swallow, that is, we'll do a virtual colonoscopy. Although it hasn't replaced the standard method, the virtual colonoscopy is used in some cases. Scientists are also working on developing a screening test done by genetic study of cells sloughed off in the stool—and that won't require a prep."

Adam's case shows the importance of anatomic studies in the diagnosis and treatment of disease. Box 1-2 has general information on medical imaging, and various methods are mentioned in chapters and cases throughout this book.

Chapter Wrap-Up

Summary Overview

A detailed chapter outline with space for note taking is on *thePoint*. The figure below illustrates the main topics covered in this chapter.

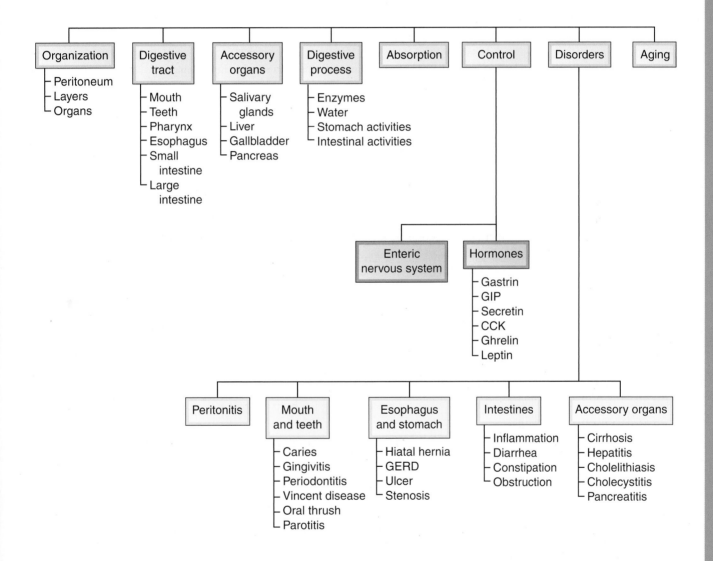

- **Organization**
  - Peritoneum
  - Layers
  - Organs
- **Digestive tract**
  - Mouth
  - Teeth
  - Pharynx
  - Esophagus
  - Small intestine
  - Large intestine
- **Accessory organs**
  - Salivary glands
  - Liver
  - Gallbladder
  - Pancreas
- **Digestive process**
  - Enzymes
  - Water
  - Stomach activities
  - Intestinal activities
- **Absorption**
- **Control**
  - Enteric nervous system
  - Hormones
    - Gastrin
    - GIP
    - Secretin
    - CCK
    - Ghrelin
    - Leptin
- **Disorders**
  - Peritonitis
  - Mouth and teeth
    - Caries
    - Gingivitis
    - Periodontitis
    - Vincent disease
    - Oral thrush
    - Parotitis
  - Esophagus and stomach
    - Hiatal hernia
    - GERD
    - Ulcer
    - Stenosis
  - Intestines
    - Inflammation
    - Diarrhea
    - Constipation
    - Obstruction
  - Accessory organs
    - Cirrhosis
    - Hepatitis
    - Cholelithiasis
    - Cholecystitis
    - Pancreatitis
- **Aging**

Key Terms

The terms listed below are emphasized in this chapter. Knowing them will help you organize and prioritize your learning. These and other bold face terms are defined in the Glossary with phonetic pronunciations.

absorption
bile
chyle
chyme
defecation
deglutition
digestion
duodenum
emulsify
esophagus
gallbladder
hydrolysis
ingestion
intestine
lacteal
liver
mastication
mesentery
pancreas
peptic ulcer
peristalsis
peritoneum
saliva
segmentation
sphincter
stomach
villi

Word Anatomy

Medical terms are built from standardized word parts (prefixes, roots, and suffixes). Learning the meanings of these parts can help you remember words and interpret unfamiliar terms.

WORD PART	MEANING	EXAMPLE
General Structure and Function of the Digestive System		
ab-	away from	In absorption, digested materials are taken from the digestive tract into the circulation.
enter/o	intestine	The mesentery is the portion of the peritoneum around the intestine.
mes/o-	middle	The mesocolon, like the mesentery, comes from the middle layer of cells in the embryo, the mesoderm.
Organs of the Digestive Tract		
gastr/o	stomach	The gastrointestinal tract consists mainly of the stomach and intestine.
The Accessory Organs		
amyl/o	starch	The starch-digesting enzyme in saliva is salivary amylase.
bil/i	bile	Bilirubin is a pigment found in bile.
cyst/o	bladder, sac	The cystic duct carries bile into and out of the gallbladder.
hepat/o	liver	The hepatic portal system carries blood to the liver.
lingu/o	tongue	The sublingual salivary glands are under the tongue.
Control of Digestion		
chole	bile, gall	Cholecystokinin is a hormone that activates the gallbladder (cholecyst/o).
Disorders of the Digestive System		
-lith	stone	A fecalith is a hardened piece of fecal material.
odont/o	tooth	Periodontitis is a disease of the gums and the tissue around a tooth.
-rhea (the "r" is doubled when added to a word)	flow, discharge	Diarrhea is a flow of watery bowel movements through (dia-) the digestive tract.

Questions for Study and Review

BUILDING UNDERSTANDING

Fill in the Blanks

1. The wavelike movement of the digestive tract wall is called _____.
2. The small intestine is connected to the posterior abdominal wall by _____.
3. The liver can store glucose in the form of _____.
4. The parotid glands secrete _____.
5. Inflammation of the gallbladder is termed _____.

Matching > **Match each numbered item with the most closely related lettered item.**

___ **6.** Digests starch
___ **7.** Begins protein digestion
___ **8.** Digests fats
___ **9.** Splits protein into amino acids
___ **10.** Emulsifies fats

a. lipase
b. amylase
c. trypsin
d. pepsin
e. bile salt

Multiple Choice

___ **11.** The teeth break up food into small parts by a process called

a. absorption
b. deglutition
c. ingestion
d. mastication

___ **12.** Which organ secretes hydrochloric acid and pepsin?

a. salivary glands
b. stomach
c. pancreas
d. liver

___ **13.** What is the main substance of a tooth?

a. gingiva
b. cuspid
c. dentin
d. cementum

___ **14.** What is the soft, fleshy V-shaped mass of tissue that hangs from the soft palate called?

a. epiglottis
b. esophageal hiatus
c. uvula
d. gingiva

___ **15.** What is the scientific name for tooth decay?

a. periodontitis
b. leukoplakia
c. chyme
d. caries

UNDERSTANDING CONCEPTS

16. Referring to **Figure A5-7** in Appendix 5, give the name and number of the peritoneal layer that extends from the stomach over the intestines.

17. Referring to the digestive system in "The Body Visible" at the beginning of this book, give the name and number of the following:

a. the first portion of the large intestine
b. the muscle that controls movement of food from the stomach to the small intestine.
c. the largest salivary gland
d. folds in the lining of the stomach
e. the middle portion of the small intestine

18. Differentiate between the terms in each of the following pairs:

a. deciduous and permanent teeth
b. digestion and absorption
c. parietal and visceral peritoneum
d. gastrin and gastric-inhibitory peptide
e. secretin and cholecystokinin

19. Name the four layers of the digestive tract. Which tissue makes up each layer? What is the function of each layer?

20. Trace the path of a bolus of food through the digestive system.

21. Describe the structure and function of the liver.

22. Describe the system that delivers secretions from the liver, pancreas, and gallbladder to the duodenum.

23. What is hydrolysis? Give three examples of hydrolysis.

24. Where does absorption occur in the digestive tract, and what structures are needed for absorption? Which types of nutrients are absorbed into the blood? Into the lymph?

25. Name two hormones that regulate eating, and give the function of each.

26. Compare and contrast the following disorders:

a. inflammatory bowel disease and irritable bowel syndrome
b. intussusception and volvulus

27. What are the causes and effects of hepatitis?

CONCEPTUAL THINKING

28. Cholelithiasis can cause pancreatitis. Why?

29. In Adam's case, the doctor described a future test in which pathologists would look for genetic changes in sloughed off colon cells. What kind of changes might they look for?

30. Colorectal cancer often develops from polyps in the intestinal lining. What is a polyp, and why might a polyp become cancerous?

For more questions, see the Learning Activities on thePoint.

CHAPTER 20

Metabolism, Nutrition, and Body Temperature

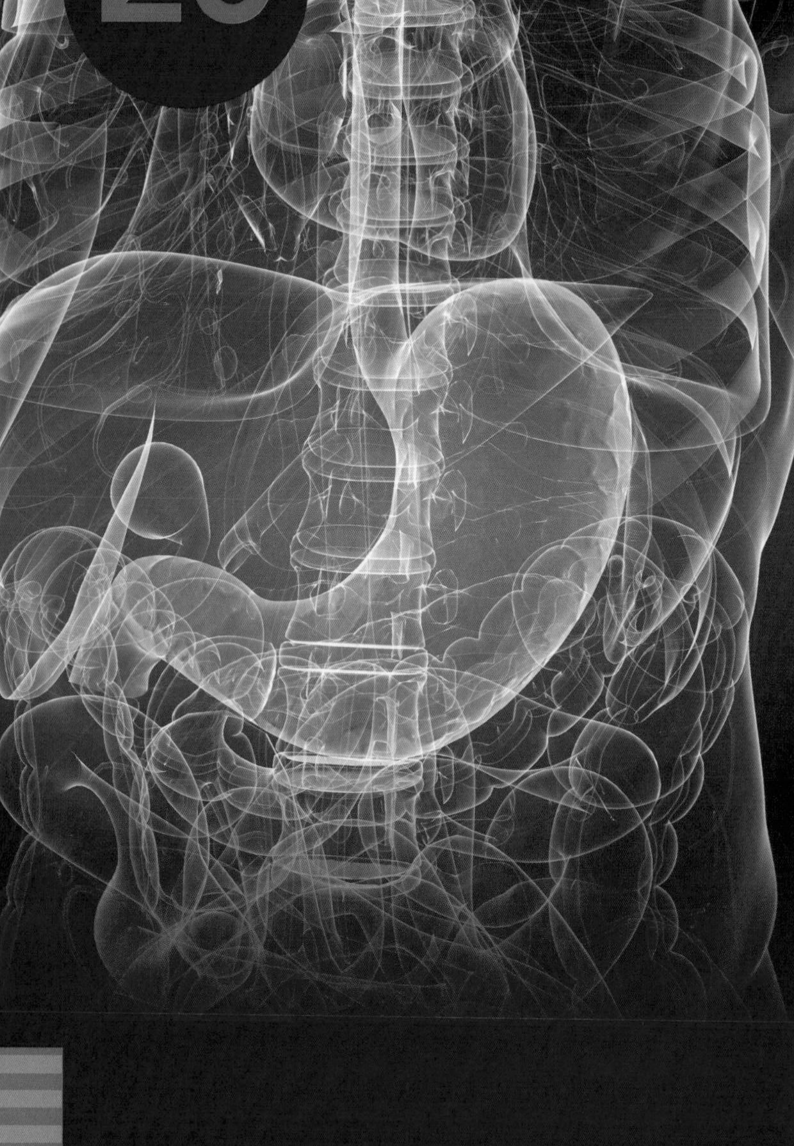

Learning Objectives

After careful study of this chapter, you should be able to:

1 ▶ Differentiate between catabolism and anabolism. ***p. 462***

2 ▶ Differentiate between the anaerobic and aerobic phases of glucose catabolism and give the end products and the relative amount of energy released by each. ***p. 462***

3 ▶ Define metabolic rate, and name six factors that affect it. ***p. 463***

4 ▶ Explain how carbohydrates, fats, and proteins are metabolized for energy. ***p. 463***

5 ▶ Compare the energy contents of carbohydrates, fats, and proteins. ***p. 464***

6 ▶ List the recommended percentages of carbohydrate, fat, and protein in the diet. ***p. 465***

7 ▶ Distinguish between simple and complex carbohydrates, giving examples of each. ***p. 465***

8 ▶ Compare saturated and unsaturated fats. ***p. 466***

9 ▶ Define *essential amino acid*. ***p. 466***

10 ▶ Explain the roles of minerals and vitamins in nutrition, and give examples of each. ***p. 467***

11 ▶ List six adverse effects of alcohol consumption. ***p. 470***

12 ▶ Describe four nutritional disorders. ***p. 470***

13 ▶ Explain how heat is produced and lost in the body. ***p. 473***

14 ▶ Describe the role of the hypothalamus in regulating body temperature. ***p. 474***

15 ▶ Explain the role of fever in disease. ***p. 475***

16 ▶ Describe responses to excessive heat and cold. ***p. 475***

17 ▶ Using the case study and the text, define anorexia nervosa, and list some of its adverse effects. ***pp. 461, 476***

18 ▶ Show how word parts are used to build words related to metabolism, nutrition, and body temperature (see Word Anatomy at the end of the chapter). ***p. 478***

Disease in Context *Claudia's Friends Are Concerned*

Dr. Wade, program advisor of Health Sciences at the university, was wrapping things up in her office for the day when Josie, a nursing student, stopped by.

"Hi Dr. Wade, do you have a minute to talk? I am worried about Claudia. She and I are roommates and close friends. After your lecture on eating disorders this week, well, I think she might have anorexia."

Dr. Wade told Josie to come in and sit down, and then closed the door.

"I appreciate your concern for your classmate," she said. "What have you noticed?"

"We are in a couple of study groups together, and when we break for lunch, she always has an excuse for why she doesn't eat," Josie explained. "Either says she's not hungry, didn't bring a lunch, or doesn't have any money. We always offer to share our food but she won't eat. And yesterday when we all insisted she eat something, she had a few bites then went to the restroom. I think she may have made herself throw up. Also, after class, a bunch of us take the shuttle back to the dorm for dinner but Claudia has been jogging back and stops at the gym to work out. That might not sound like much but you should see her workout schedule; it's intense. She's always weighing herself, and I think she's overdoing the exercise. And recently when we were practicing measuring our BMIs for class, I noticed she didn't really participate in the exercise. She's pretty thin, and I know she thinks about her weight since she joined the cheerleading squad. She gets cold really easily, and I don't think she's had her period for months. I hate being a tattletale, but I'm really getting worried!"

"Josie, over half of teenage girls engage in unhealthy eating behaviors, and it can be hard to ask for help," Dr. Wade said. "You've made some good observations about your friend, and your timing is good in coming to me. I'll try to speak with Claudia tomorrow after class, but see if you can get her to seek help at the Student Health Center this evening. Also, let me give you some Web sites with tips on supporting friends with eating disorders."

Dr. Wade thought about Claudia after Josie left. She herself had notice Claudia's lack of concentration in class, and although a perfectionist, her academic performance was slipping. This student was showing clear signs of having an eating disorder.

Later we will revisit Claudia and check on her condition. Nutrition and weight control will be discussed in this chapter, and we will see how malnutrition and being underweight can negatively affect one's health.

ANCILLARIES *At-A-Glance*

Visit thePoint to access the following resources. For guidance in using these resources most effectively, see pp. xv–xvii.

Learning RESOURCES

- Tips for Effective Studying
- Web Figure: Effects of Chronic Alcohol Abuse
- Web Figure: Fatty Liver of Chronic Alcoholism
- Web Figure: Cerebellar Atrophy of Chronic Alcoholism
- Web Figure: Long-Term Complications of Diabetes
- Health Professions: Dietitian and Nutritionist
- Detailed Chapter Outline
- Answers to Questions for Study and Review
- Audio Pronunciation Glossary

Learning ACTIVITIES

- Pre-Quiz
- Visual Activities
- Kinesthetic Activities
- Auditory Activities

A LOOK BACK

The concept of metabolism was introduced in Chapter 1 and applied to discussion of muscle function in Chapter 8. Nutritional studies also require knowledge of organic chemistry, introduced in Chapter 2 and further discussed in the previous chapter on digestion. Some metabolic reactions are regulated by hormones, as discussed in Chapter 12. Finally, reviewing homeostasis and negative feedback will help you understand the maintenance of normal body temperature.

Metabolism

Chapter 19 discussed how nutrients absorbed from the digestive tract are chemically transformed into other substances for the body's use. These reactions, along with those involving stored body substances, are described as the body's **metabolism.** In Chapter 1, we said that there are two types of metabolic activities:

- **Catabolism,** the breakdown of complex compounds into simpler components. Catabolism includes the digestion of food into small molecules and the release of energy from these molecules within the cell.
- **Anabolism,** the building of simple compounds into nutrient storage compounds, structural materials, and functional molecules such as enzymes and transporters.

CELLULAR RESPIRATION

Energy is released from nutrients in a series of catabolic reactions called **cellular respiration** (Fig. 20-1 and Table 20-1). These reactions can begin with glucose, fatty acids, or, more rarely, amino acids. We begin our explanation with glucose, a simple sugar that is a principal energy source for cells.

Glucose Catabolism: Anaerobic Phase The first steps in the breakdown of glucose do not require oxygen; that is, they are **anaerobic.** This phase of catabolism, known as **glycolysis** (gli-KOL-ih-sis), occurs in the cell's cytoplasm. It yields a small amount of energy, which is used to make ATP, the cell's energy compound. Each glucose molecule yields enough energy by this process to produce two molecules of ATP.

Glycolysis ends with formation of an organic product called **pyruvic** (pi-RU-vik) **acid** (pyruvate). This organic acid is further metabolized in the next phase of cellular respiration, which requires oxygen. In muscle cells that need to produce large amounts of ATP rapidly, for example, at the start of exercise or during very intense exercise, pyruvic acid cannot be metabolized fast enough and accumulates in the cells. In this case, pyruvic acid is converted to a related compound called lactic acid. As lactic acid accumulates in the cells, it can spill over into the blood. Most physiologists do not believe that this lactic acid harms or inhibits the muscle cells in any way. It is simply converted back to pyruvic acid and fully metabolized either by other less active cells or by the cells that produced it when the later steps in metabolism catch up with the anaerobic yield.

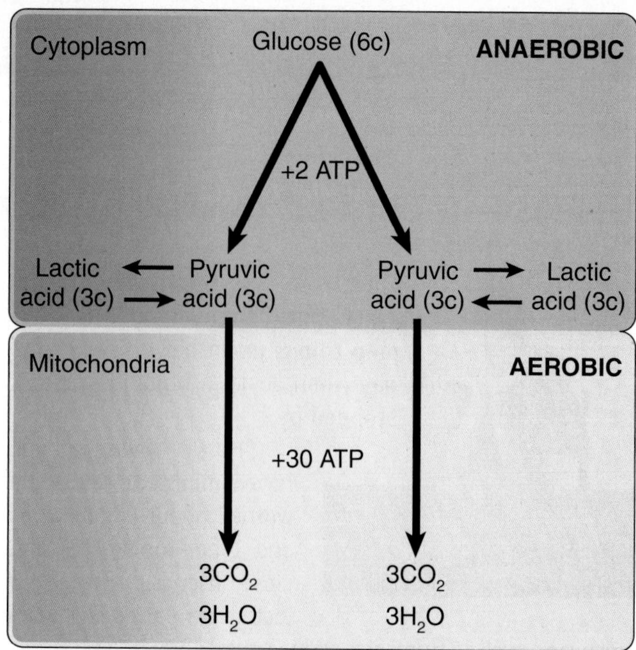

Figure 20-1 Cellular respiration. **KEY POINT** There are two stages of glucose catabolism, the first occurring without oxygen (anaerobic), followed by steps occurring with oxygen (aerobic). (The letter *c* represents a carbon atom, and the numbers show the number of carbon atoms in one molecule of the named substance.) In cellular respiration, glucose first yields two molecules of pyruvic acid. These initial steps occur in the cell's cytosol and do not require oxygen. The pyruvic acid is then fully catabolized in the mitochondria using oxygen. The final products of this phase are carbon dioxide and water and a large amount of adenosine triphosphate (ATP). Under anaerobic conditions, as during intense exercise, pyruvic acid is temporarily converted to lactic acid. When oxygen becomes available, the lactic acid is converted back to pyruvic acid for complete oxidation. **ZOOMING IN** What does pyruvic acid produce when it is metabolized anaerobically? What does it produce when metabolized completely using oxygen?

Glucose Catabolism: Aerobic Phase To generate enough energy for survival, the body's cells must break pyruvic acid down more completely. These next **aerobic** (oxygen-requiring) reactions occur within the cell's mitochondria. They result in transfer of most of the energy remaining in the nutrients to ATP. On average, cells are able to form about 30 molecules of ATP aerobically per glucose molecule. Statements on energy yields may differ slightly, because cells in different tissues vary in their metabolic pathways and in the amount of energy they use to power cellular respiration. In any case, this additional yield is quite an increase over glycolysis alone, generally resulting in a total of 32 molecules of ATP per glucose as compared to two.

Table 20-1 Summary of Cellular Respiration of Glucose

Phase	Location in Cell	End Product(s)	Energy Yield/Glucose
Anaerobic (glycolysis)	Cytoplasm	Pyruvic acid	Two adenosine triphosphate (ATP)
Aerobic	Mitochondria	Carbon dioxide and water	30 ATP

During the aerobic steps of cellular respiration, the mitochondria produce carbon dioxide and water. Because of the type of chemical reactions involved and because oxygen is used in the final steps, cellular respiration is described as nutrient **oxidation**. Note that enzymes are required as catalysts in all these metabolic reactions. Many of the vitamins and minerals described later in this chapter are parts of these enzymes.

Although the oxidation of food is often compared to the burning of fuel, this comparison is inaccurate. Burning fuel results in a sudden and often wasteful release of energy in the form of heat and light. In contrast, metabolic oxidation occurs in regulated steps, and much of the energy released is stored as ATP for later use by the cells; some of the energy is released as heat, which is used to maintain body temperature, as discussed later in this chapter.

For those who know how to read chemical equations, the net balanced equation for cellular respiration, starting with glucose, is as follows:

$$\underset{\text{glucose}}{C_6H_{12}O_6} + \underset{\text{oxygen}}{6O_2} \rightarrow \underset{\text{carbon dioxide}}{6CO_2} + \underset{\text{water}}{6H_2O}$$

Cellular Respiration of Fatty Acids and Amino Acids Fatty acids and, to a lesser extent, amino acids can also be oxidized to generate energy. Unlike glucose, these nutrients can generate ATP only by aerobic mechanisms, so they are of minimal use in high-intensity exercise and in cells with few mitochondria. As with pyruvate, fatty acids and amino acids are broken down completely to yield ATP, water, and carbon dioxide. The ATP yield and the precise sequence of chemical reactions vary among different fatty acids and amino acids.

Before amino acids are oxidized for energy, they must have their nitrogen (amine) groups removed. This removal, called **deamination** (de-am-ih-NA-shun), occurs in the liver, where the nitrogen groups are then formed into urea by combination with carbon dioxide. The blood transports urea to the kidneys to be eliminated.

METABOLIC RATE

Metabolic rate refers to the rate at which cellular respiration converts nutrients into ATP. Since the body produces ATP on demand, the metabolic rate relates to overall energy requirements. It is affected by a person's size, body fat, sex, age, activity, and hormones, especially thyroid hormone (thyroxine). Metabolic rate is high in children and adolescents and decreases with age. **Basal metabolism** is the amount of energy needed to maintain life functions while the body is at rest. Thus, your *basal metabolic rate* (*BMR*) is the energy you expend each day simply to stay alive. Any activity you perform, even tapping a toe, increases your energy expenditure above your BMR.

The unit used to measure energy is the **kilocalorie** (kcal), which is the amount of heat needed to raise 1 kg of water 1°C. Nutrition information for the general public typically replaces the word **kilocalorie** with *calorie (C)*. To estimate how many calories you need each day taking into account your activity level, see **Box 20-1**.

NUTRIENT METABOLISM

Instead of being processed for energy, nutrients may have other fates. They can be converted into storage forms, functional molecules, or even other nutrient types. Enzymes catalyze all of these chemical reactions.

20

Carbohydrates We ingest carbohydrates in many different forms, as shown in **Figure 19-11**. As discussed in Chapter 19, dietary carbohydrates are digested into monosaccharides (simple sugars) before absorption into blood. The liver then converts the monosaccharides into glucose, which can be used for energy.

When nutrients are abundant and the body's energy needs are low, liver and muscle cells convert glucose into **glycogen** (GLI-ko-jen), the storage form of carbohydrates. Recall that the hormone insulin promotes this reaction. When glucose is needed for energy, glycogen is broken down under the influence of the hormone glucagon to yield glucose. Muscle glycogen is used specifically in the muscle cell that produced it. The glucose from liver glycogen can be released into the bloodstream to power other cells. If we have more glucose in the blood than is needed for energy and glycogen storage, it is converted to fat and stored in adipose tissue and the liver.

Fats Most tissues can use fatty acids for energy. Some organs, such as the liver, rely exclusively on fatty acids, and other tissues, such as muscle, use fatty acids during rest and low-intensity exercise. Brain tissue is one exception; it relies on glucose or if glucose is not available, partially catabolized fatty acids called **ketone bodies**. Large amounts of ketone bodies are produced by the liver in cases of starvation or low carbohydrate intake, and the ketones can provide energy for the brain and other tissues. They are also produced in uncontrolled diabetes mellitus, in which insulin deficiency "tricks" the liver into thinking that glucose is unavailable. However, ketone bodies are acidic, and excesses disrupt the body's acid–base balance. We will say more about these changes in Chapters 21 and 22.

Box 20-1

A CLOSER LOOK

Calorie Counting: Estimating Daily Energy Needs

Have you ever wondered how many calories you need to eat each day in order to avoid gaining or losing weight? To answer that question, you first need to calculate your basal metabolic rate (BMR) and then estimate the calories you burn each day in physical activity. Adding those two numbers together should give you your daily calorie needs.

You can estimate your BMR with a simple formula. An average woman requires 0.9 kcal/kg/h, and a man, 1.0 kcal/kg/h. If you need to convert your body weight from pounds to kilograms (kg), divide your weight in pounds by 2.2. Next, multiply your body weight in kilogram by 0.9 if you are female, and by 1.0 if you are male. This gives you kcal burned per hour. Finally, multiply by 24 to find your BMR (the number of kcal you expend at rest per day).

For example, if you are female and weigh 132 lb, your equation would be as follows:

$$132\ \text{lb} \div 2.2\ \text{lb/kg} = 60\ \text{kg}$$

$$0.9\ \text{kcal/kg/h} \times 60\ \text{kg} = 54\ \text{kcal/h}$$

$$54\ \text{kcal/h} \times 24\ \text{h/d} = 1{,}296\ \text{kcal/d}$$

Notice that, if you are male, you can skip step 2, since you'd simply be multiplying by 1.

To estimate your total energy needs for a day, you need to add to your BMR a percentage based on your current activity level ("couch potato" to serious athlete). These percentages are shown in the table that follows.

The equation to calculate total energy needs for a day is

$$\text{BMR} + (\text{BMR} \times \text{activity level})$$

Using the BMR from our previous example with different activity levels, the following equations apply:

At 25% activity:

$$1{,}296\ \text{kcal/d} + (1{,}296\ \text{kcal/d} \times 25\%) = 1{,}620\ \text{kcal/d}$$

At 60% activity:

$$1{,}296\ \text{kcal/d} + (1{,}296\ \text{kcal/d} \times 60\%) = 2{,}073.6\ \text{kcal/d}$$

As you can see, physical activity can help you maintain a healthful body weight. In this case, increased activity increased the individual's energy needs by more than 450 kcal/d!

Activity Level	Male	Female
Little activity ("couch potato")	25%–40%	25%–35%
Light activity (e.g., walking to and from class, but little or no intentional exercise)	50%–75%	40%–60%
Moderate activity (e.g., aerobics several times a week)	65%–80%	50%–70%
Heavy activity (serious athlete)	90%–120%	80%–100%

Fatty acids are a highly concentrated energy source and generate more ATP per molecule than does glucose. In fact, fat in the diet yields more than twice as much energy as do protein and carbohydrate; fat yields 9 kcal of energy per gram, whereas protein and carbohydrate each yields 4 kcal/g.

Because fats are such a concentrated energy source, they are the most efficient way to store excess calories. Excess caloric intake, be it carbohydrate, protein, or fat, can be converted into triglycerides and stored in adipose tissue. Then, in times of nutrient deficiency, the triglycerides can be converted to fatty acids for use throughout the body. Adipose tissue growth is virtually unlimited. According to the Guinness Book of World Records in 2014, the world's heaviest man, Jon Brower Minnoch, lost 924 lb from his peak weight of 1,400 lb, most of which would have been adipose tissue.

Proteins There are no specialized storage forms of proteins, as there are for carbohydrates (glycogen) and fats (triglycerides), because specific proteins are synthesized to meet specific body needs. For instance, trying to lift a heavy weight places a load on muscle tissue, stimulating the production of actin and myosin proteins. Protein consumed in excess of daily needs is not stored as protein, but is catabolized for energy or converted to triglycerides. Conversely, when dietary proteins do not meet a person's needs, they must be obtained from structural proteins such as those constituting muscle tissue. Moreover, the synthesis of needed functional proteins slows down. Fats and carbohydrates are described as "protein sparing," because they are used for energy before proteins are and thus spare proteins for the synthesis of necessary body components.

CHECKPOINTS

- ☐ 20-1 What are the two types of activities that make up metabolism?
- ☐ 20-2 What name is given to the series of cellular reactions that releases energy from nutrients?
- ☐ 20-3 What is the organic end product of glycolysis?
- ☐ 20-4 What element is required for aerobic cellular respiration but not for glycolysis?
- ☐ 20-5 What is removed from amino acids before they are metabolized for energy?

Nutritional Guidelines

The relative amounts of carbohydrates, fats, and proteins that should be in the daily diet vary somewhat with the individual. Typical recommendations for the number of calories derived each day from the three types of food are as follows:

- Carbohydrate: 55% to 60%
- Fat: 30% or less
- Protein: 15% to 20%

It is important to realize that the type as well as the amount of each nutrient is a factor in good health. A weight loss diet should follow the same proportions as given above but with a reduction in portion sizes.

CARBOHYDRATES

A healthful diet provides abundant complex carbohydrates, whereas simple sugars are kept to a minimum. Simple sugars are monosaccharides, such as glucose and fructose (fruit sugar), and disaccharides, such as sucrose (table sugar) and lactose (milk sugar). Simple sugars are a source of fast energy because they are metabolized rapidly. However, they boost pancreatic insulin output, and as a result, they cause blood glucose levels to rise and fall rapidly. It is healthier to maintain steady glucose levels, which normally range from approximately 85 to 125 mg/dL throughout the day.

The **glycemic effect** is a measure of how rapidly a particular food raises the blood glucose level and stimulates the release of insulin. The effect is generally low for whole grains, fruit, vegetables, legumes, and dairy products, and high for refined sugars and refined ("white") grains. Note, however, that the glycemic effect of a food also depends on when it is eaten during the day and if or how it is combined with other foods.

Complex carbohydrates are polysaccharides. Examples are:

- Starches, found in grains, legumes, and potatoes and other starchy vegetables
- Fibers, such as cellulose, pectins, and gums, which are the structural materials of plants

Fiber cannot be used for energy, but it adds bulk to the stool and promotes elimination of toxins and waste. It also slows the digestion and absorption of other carbohydrates, thus regulating the release of glucose. It helps in weight control by providing a sense of fullness and limiting caloric intake. Adequate fiber in the diet lowers cholesterol and helps to prevent diabetes, colon cancer, hemorrhoids, appendicitis, and diverticulitis. Foods high in fiber, such as whole grains, fruits, and vegetables, are also rich in vitamins and minerals (see Box 20-2).

FATS

Moderate amounts of fats are necessary for health, providing energy and cell components, and adding taste to other critical nutrients. However, not all fats are the same. The healthiness of a triglyceride is determined by the characteristics of its three fatty acids.

Essential Fatty Acids While any fatty acid can provide energy, some cell components require specific fatty acids. Most fatty acid varieties can be synthesized by body cells. However, there are two essential fatty acids, **linoleic** (lin-o-LE-ik) **acid** and **alpha-linolenic** (lin-o-LEN-ik) **acid**, which must be taken in as food. Linoleic acid is easily obtained through a healthful, balanced diet that includes plenty of vegetables and vegetable oils. It is used to make prostaglandins, among other functions. In contrast, alpha-linolenic acid is found primarily in fatty fish and shellfish, dark green, leafy vegetables, and flaxseeds, soybeans, walnuts, and their oils. Thus, it is somewhat more difficult to obtain.

20

HEALTH MAINTENANCE **Box 20-2**

Dietary Fiber: Bulking Up

Dietary fiber is best known for its ability to improve bowel habits and promote weight loss. But fiber may also help to prevent diabetes, heart disease, and certain digestive disorders such as diverticulitis and gallstones.

Dietary fiber is an indigestible type of carbohydrate found in fruit, vegetables, and whole grains. The amount of fiber recommended for a 2,000-cal diet is 25 g/d, but most people in the United States tend to get only half this amount. One should eat fiber-rich foods throughout the day to meet the requirement. It is best to increase fiber in the diet gradually to avoid unpleasant symptoms, such as intestinal bloating and flatulence. If your diet lacks fiber, try adding the following foods over a period of several weeks:

- Whole grain breads, cereals, pasta, and brown rice. These add 1 to 3 more grams of fiber per serving than the "white" product.
- Legumes, which include beans, peas, and lentils. These add 4 to 12 g of fiber per serving.
- Fruits and vegetables. Whole, raw, unpeeled versions contain the most fiber and juices the least. Apple juice has no fiber, whereas a whole apple has 3 g.
- Unprocessed bran. This can be sprinkled over almost any food: cereal, soups, and casseroles. One tablespoon adds 2 g of fiber. Be sure to take adequate fluids with bran.

Saturated and Unsaturated Fats Fats are subdivided into saturated and unsaturated forms based on their chemical structures. The fatty acids in **saturated fats** have more hydrogen atoms in their molecules because they have no double bonds between carbons atoms. In other words, their carbon atoms are fully "saturated" with hydrogen (**Fig. 20-2**). Most saturated fats are from animal sources and are solid at room temperature, such as butter and lard.

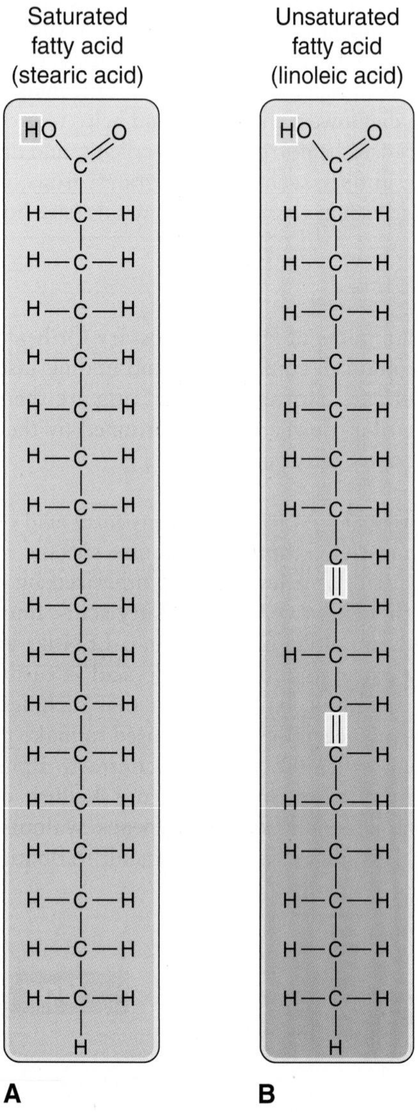

Saturated fatty acid (stearic acid)

Unsaturated fatty acid (linoleic acid)

A B

Figure 20-2 Saturated and unsaturated fats. **KEY POINT** Saturated and unsaturated fats differ in their bonding structures. **A.** Saturated fatty acids contain the maximum numbers of hydrogen atoms attached to carbons and no double bonds between carbon atoms. **B.** Unsaturated fatty acids have less than the maximum number of hydrogen atoms attached to carbons and one or more double bonds between carbon atoms (highlighted). The first carbon in the chain (**top**) is the alpha carbon; the last (**bottom**) is the omega carbon. The "omega" fatty acids are described according to how many carbons there are from the last double-bonded carbon in the chain to the omega carbon, such as omega-3, omega-6. Linoleic acid is an omega-6 fatty acid.

Also included in this group are the so-called tropical oils: coconut oil and palm oil.

Unsaturated fats are derived from plants. They are liquid at room temperature and are generally referred to as oils, such as corn, peanut, olive, and canola oils. Their fatty acids have one or more double carbon bonds, which exclude hydrogen (**see Fig. 20-2**). An unsaturated fatty acid is *monounsaturated* if it has a single double bond and polyunsaturated if it has more than one double bond. You may have heard or read about fatty acids described as "omega" acids with a number. This terminology refers to the position of the last double-bonded carbon in the chain with regard to the last carbon, named the omega carbon for the last letter of the Greek alphabet. In the essential fatty acid linoleic acid, for example, the last double-bonded carbon is the sixth carbon from the omega carbon, so linoleic acid is an omega-6 fatty acid (**see Fig. 20-2B**). The essential fatty acid alpha-linolenic acid is an omega-3 fatty acid. As to fat intake, a predominance of unsaturated fats in general, in addition to the previously mentioned essential fatty acids, contributes to a healthful diet.

Saturated fats should make up less than one-third of the fat in the diet (less than 10% of total calories). Diets high in saturated fats are associated with a higher than normal incidence of cancer, heart disease, and cardiovascular problems, although the relation between these factors is not fully understood.

Because most unsaturated fats are liquid and easily become rancid (spoil), commercial food manufacturers have used fats that are artificially saturated to extend shelf life and improve consistency of foods. These fats are listed on food labels as partially hydrogenated (hi-DRO-jen-a-ted) vegetable oils and are found commonly in baked goods, processed peanut butter, vegetable shortening, and some solid margarines. Evidence shows that components of hydrogenated fats, known as *trans fatty acids*, are more harmful than even natural saturated fats. They raise blood levels of LDL (the less healthful lipoproteins) and lower levels of HDLs (the healthful lipoproteins). Food manufacturers are responding to evidence that trans fats are harmful and are making efforts to remove them from processed foods. Many countries now require labeling of trans fat content on food labels. In the United States, trans fat contents of 0.5 g or more per serving must be listed. Bear in mind, though, that restaurants still may be using trans fats in cooking, especially deep-frying.

PROTEINS

The synthesis of each protein requires a specific set of amino acids. Of the 20 amino acids needed to build proteins, metabolic reactions can synthesize 11 internally. These 11 amino acids are described as *nonessential* because they need not be taken in as food (**Table 20-2**). The remaining nine amino acids cannot be made metabolically and therefore must be taken in as part of the diet; these are the **essential amino acids**. Note that some nonessential amino acids may become essential under certain conditions, as during extreme physical stress, or in certain hereditary metabolic diseases.

Because proteins, unlike carbohydrates and fats, are not stored in special reserves, protein foods should be consumed

Table 20-2 Amino Acids

Nonessential Amino Acids[a]		Essential Amino Acids[b]	
Name	**Pronunciation**	**Name[c]**	**Pronunciation**
Alanine	AL-ah-nene	Histidine	HIS-tih-dene
Arginine	AR-jih-nene	Isoleucine	i-so-LU-sene
Asparagine	ah-SPAR-ah-jene	Leucine	LU-sene
Aspartic acid	ah-SPAR-tik AH-sid	Lysine	LI-sene
Cysteine	SIS-teh-ene	Methionine	meh-THI-o-nene
Glutamic acid	glu-TAM-ik AH-sid	Phenylalanine	fen-il-AL-ah-nene
Glutamine	GLU-tah-mene	Threonine	THRE-o-nene
Glycine	GLY-sene	Tryptophan	TRIP-to-fan
Proline	PRO-lene	Valine	VA-lene
Serine	SERE-ene		
Tyrosine	TI-ro-sene		

[a]Nonessential amino acids can be synthesized by the body.

[b]Essential amino acids cannot be synthesized by the body; they must be taken in as part of the diet.

[c]If you are ever called upon to memorize the essential amino acids, the mnemonic (memory device) Pvt. T. M. Hill gives the first letter of each name.

regularly with attention to obtaining the essential amino acids. Most animal proteins supply all of the essential amino acids and are described as complete proteins. Most plant proteins are lacking in one or more of the essential amino acids. People on strict vegetarian diets must learn to combine foods, such as legumes (e.g., beans and peas) with grains (e.g., rice, corn, or wheat), to obtain all the essential amino acids each day. **Figure 20-3** demonstrates the principles of combining two foods, legumes and grains, to supply essential amino acids that might be missing in one food or the other. Legumes are rich in isoleucine and lysine but poor in methionine and tryptophan, while grains are just the opposite. For illustration purposes, **Table 20-2** includes only the four missing essential amino acids (there are nine total). Traditional ethnic diets reflect these healthful combinations, for example, beans with corn or rice in Mexican dishes or chickpeas and lentils with wheat in Middle Eastern fare. Nuts and vegetables also provide plant proteins; thus, a peanut butter sandwich on whole grain bread or vegetables over rice are other complementary combinations.

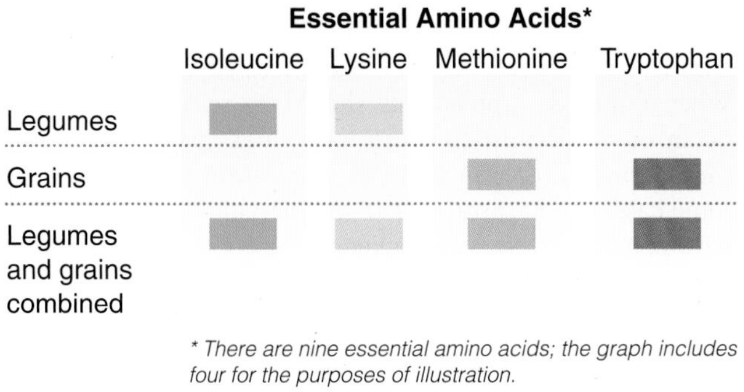

Figure 20-3 Combining foods to obtain the essential amino acids. KEY POINT There are nine essential amino acids. If someone does not eat animal proteins, most of which supply all of the essential amino acids, foods must be combined to supply all the needed amino acids within the day. In this illustration, legumes and grains are combined to provide four of the nine essential amino acids.

MINERALS AND VITAMINS

In addition to needing carbohydrates, fats, and proteins, the body requires minerals and vitamins.

Minerals are chemical elements needed for body structure, fluid balance, and such activities as muscle contraction, nerve impulse conduction, and blood clotting. Some minerals are components of vitamins. A list of the main minerals needed in a proper diet is given in **Table 20-3**. Some additional minerals not listed are also required for good health. Minerals needed in extremely small amounts are referred to as **trace elements.**

Vitamins are complex organic substances needed in small quantities. Vitamins are parts of enzymes or other substances essential for metabolism, and vitamin deficiencies lead to a variety of nutritional diseases.

The water-soluble vitamins are the B vitamins and vitamin C. These are not stored and must be taken in regularly with food. The fat-soluble vitamins are A, D, E, and K. These vitamins are kept in reserve in fatty tissue. Excess intake of the fat-soluble vitamins can lead to toxicity. A list of vitamins is given in **Table 20-4**.

Certain substances are valuable in the diet as **antioxidants.** They defend against the harmful effects of reactive oxygen species (ROS), also described as *free radicals*, highly reactive and unstable molecules produced from oxygen in the normal course of metabolism (and also resulting from UV

Table 20-3 Minerals

Mineral	Functions	Sources	Results of Deficiency
Calcium (Ca)	Formation of bones and teeth; blood clotting; nerve conduction; muscle contraction	Dairy products, eggs, green vegetables, legumes (peas and beans)	Rickets, tetany, osteoporosis
Phosphorus (P)	Formation of bones and teeth; found in ATP, nucleic acids	Meat, fish, poultry, egg yolk, dairy products	Osteoporosis, abnormal metabolism
Sodium (Na)	Fluid balance; nerve impulse conduction; muscle contraction	Most foods, especially processed foods, table salt	Weakness, cramps, diarrhea, dehydration
Potassium (K)	Fluid balance; nerve and muscle activity	Fruits, meats, seafood, milk, vegetables, grains	Muscular and neurologic disorders
Chloride (Cl)	Fluid balance; hydrochloric acid in stomach	Meat, milk, eggs, processed food, table salt	Rarely occurs
Iron (Fe)	Oxygen carrier (hemoglobin, myoglobin)	Meat, eggs, fortified cereals, legumes, dried fruit	Anemia, dry skin, indigestion
Iodine (I)	Thyroid hormones	Seafood, iodized salt	Hypothyroidism, goiter
Magnesium (Mg)	Catalyst for enzyme reactions; carbohydrate metabolism	Green vegetables, grains, nuts, legumes	Spasticity, arrhythmia, vasodilation
Manganese (Mn)	Catalyst in actions of calcium and phosphorus; facilitator of many cellular processes	Many foods	Possible reproductive disorders
Copper (Cu)	Necessary for absorption and use of iron in formation of hemoglobin; part of some enzymes	Meat, water	Anemia
Chromium (Cr)	Works with insulin to regulate blood glucose levels	Meat, unrefined food, fats and oils	Inability to use glucose
Cobalt (Co)	Part of vitamin B_{12}	Animal products	Pernicious anemia
Zinc (Zn)	Promotes carbon dioxide transport and energy metabolism; found in enzymes	Meat, fish, poultry, grains, vegetables	Alopecia (baldness); possibly related to diabetes
Fluoride (F)	Prevents tooth decay	Fluoridated water, tea, seafood	Dental caries

radiation, air pollution, and tobacco smoke). Free radicals contribute to aging and disease. Antioxidants react with ROS to stabilize them and minimize their harmful effects on cells. Vitamins C and E and beta-carotene, an orange pigment found in plants that is converted to vitamin A, are antioxidants. There are also many compounds found in plants (e.g., soybeans and tomatoes) that are antioxidants.

Mineral and Vitamin Supplements The need for adding mineral and vitamin supplements to the diet of healthy individuals is a subject of controversy. Some researchers maintain that adequate amounts of these substances can be obtained from a varied, healthful diet. Many commercial foods, including milk, cereal, and bread, are already fortified with minerals and vitamins. Others hold that pollution, depletion of the soils, and the storage, refining, and processing of foods make additional supplementation beneficial. At present, research does not support claims that multivitamins increase longevity, improve brain function, prevent cancer, or convey infection resistance in healthy adults, including in the elderly. Earlier studies noted that individuals with chronic illnesses often have vitamin deficiencies, but the deficiencies may be the result rather than the cause of the disease.

Nevertheless, specific populations benefit from specific supplements. Pregnant women and their babies, for instance, benefit from supplemental iron and folic acid, and individuals with low bone mass improve when treated with vitamin D. When required, supplements should be selected by a physician or nutritionist to fit an individual's particular needs. Megavitamin dosages may cause unpleasant reactions and in some cases are hazardous. Fat-soluble vitamins have caused serious toxic effects when taken in excess of the established safe upper limit (UL).

USDA DIETARY GUIDELINES

The United States Department of Agriculture (USDA) has published dietary guidelines at regular intervals since 1916. The newest guidelines, MyPlate (**Fig. 20-4**), were updated and simplified in 2011 to replace the earlier Food Guide Pyramid. The plate in the new graphic is divided into four sections for fruits, grains, vegetables, and proteins in the relative proportion that each should make up in the diet. A fifth category,

Table 20-4 Vitamins

Vitamins	Functions	Sources	Results of Deficiency
A (retinol)	Required for healthy epithelial tissue and for eye pigments; involved in reproduction and immunity	Orange fruits and vegetables, liver, eggs, dairy products, dark green vegetables	Night blindness, dry, scaly skin, decreased immunity
B_1 (thiamin)	Required for enzymes involved in oxidation of nutrients; nerve function	Pork, cereal, grains, meats, legumes, nuts	Beriberi, a disease of nerves
B_2 (riboflavin)	In enzymes required for oxidation of nutrients	Milk, eggs, liver, green leafy vegetables, grains	Skin and tongue disorders
B_3 (niacin, nicotinic acid)	Involved in oxidation of nutrients	Yeast, meat, liver, grains, legumes, nuts	Pellagra with dermatitis, diarrhea, mental disorders
B_6 (pyridoxine)	Amino acid and fatty acid metabolism; formation of niacin; manufacture of red blood cells	Meat, fish, poultry, fruit grains, legumes, vegetables	Anemia, irritability, convulsions, muscle twitching, skin disorders
Pantothenic acid	Essential for normal growth; energy metabolism	Yeast, liver, eggs, and many other foods	Sleep disturbances, digestive upset
B_{12} (cyanocobalamin)	Production of cells; maintenance of nerve cells; fatty acid and amino acid metabolism	Animal products	Pernicious anemia
Biotin (a B vitamin)	Involved in fat and glycogen formation, amino acid metabolism	Peanuts, liver, tomatoes, eggs, oatmeal, soy, and many other foods	Lack of coordination, dermatitis, fatigue
Folate (folic acid, a B vitamin)	Required for amino acid metabolism; DNA synthesis; maturation of red blood cells	Vegetables, liver, legumes, seeds	Anemia, digestive disorders, neural tube defects in the embryo
C (ascorbic acid)	Maintains healthy skin and mucous membranes, involved in synthesis of collagen; antioxidant	Citrus fruits, green vegetables, potatoes, orange fruits	Scurvy, poor wound healing, anemia, weak bones
D (calciferol)	Aids in absorption of calcium and phosphorus from intestinal tract	Fatty fish, liver, eggs, fortified milk, cereal	Rickets, bone deformities, osteoporosis
E (tocopherol)	Protects cell membranes; antioxidant	Seeds, green vegetables, nuts, grains, oils	Anemia, muscle and liver degeneration, pain
K	Synthesis of blood clotting factors; bone formation	Liver, cabbage, and leafy green vegetables. Bacteria in digestive tract.	Hemorrhage

dairy, is represented by a cup, suggesting milk and milk products, such as cheeses and yogurt. Low-fat varieties are recommended, and for those who can't consume milk because of lactose intolerance or who choose not to consume milk, intakes of lactose-free products or other good sources of calcium are recommended. This mineral is also found in green vegetables and legumes.

A full description of the guidelines and much supplementary information can be found online at ChooseMyPlate.gov. Here you can click on each section of the plate to see what foods are in each group. The protein section specifies a nutrient category rather than a type of food. Foods high in proteins are meat, poultry, seafood, and eggs, but proteins are also found in plant products, such as seeds, nuts, legumes, and grains. The guidelines accommodate individual variation. On the website, you can get a personalized estimate of what and how much you should eat based on your height, weight, age, sex, and level of physical activity. You can also assess your diet and plan healthy menus.

The guidelines stress the following:

- Moderation. A single serving or portion is smaller than most people think.
- Variety in the diet. Foods in all the groups are needed each day for good health.
- Eating fruits and vegetables. Most people need more of these in their diets.
- Choice of unrefined foods, such as whole grains, and unprocessed foods. At least one-half the grains in the diet should be whole grains.
- Drinking water instead of sugary beverages.
- The importance of exercise for good health.

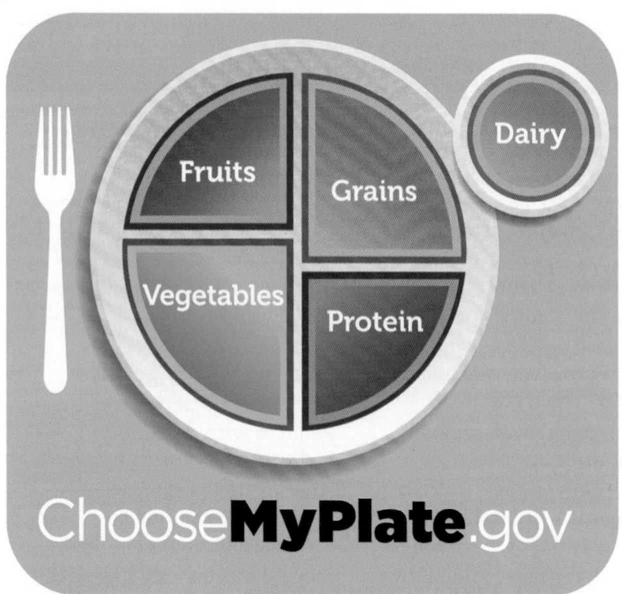

Figure 20-4 **USDA dietary guidelines.** The plate shows the recommended proportion of each food category in the diet.

The MyPlate graphic does not include oils, solid fats, or sugars. Although oils in moderation are important for health, solid fats and added sugars (SoFAS) are described as "empty calories" that provide calories but little nutrition. These are "extras" that you can eat within your recommended daily energy limit after you meet your nutritional needs. Of course, you could also select additional foods from among the recommended nutrient-rich groups to satisfy your energy needs. Limiting salt intake, especially in processed foods and restaurant fare, is also important because excess salt intake may contribute to high blood pressure.

The dietary guidelines are updated regularly. Issues always under consideration include weight control and giving attention to special groups, such as children, older adults, and pregnant or breast-feeding women. Any guidelines also should accommodate people with special needs or dietary preferences, such as people with gluten allergies, those who are lactose intolerant, strict vegetarians (vegans) who eat no animal products, and vegetarians who eat dairy products and eggs (lacto-ovo vegetarians).

See the Student Resources on thePoint for information on dietitians and nutritionists, who study nutrition and metabolism and help people plan healthful diets.

ALCOHOL

Alcohol yields energy in the amount of 7 kcal/g, but it is not considered a nutrient because it does not yield useful end products. In fact, alcohol interferes with metabolism and contributes to a variety of disorders.

The body can metabolize about 1/2 oz of pure alcohol (ethanol) per hour. This amount translates into one glass of wine, one can of beer, or one shot of hard liquor. Consumed at a more rapid rate, alcohol accumulates in the bloodstream and affects many cells, notably in the brain.

Alcohol is rapidly absorbed through the stomach and small intestine and is detoxified by the liver. When delivered in excess to the liver, alcohol can lead to the accumulation of fat as well as inflammation and scarring of liver tissue. It can eventually cause cirrhosis (*sih-RO-sis*), which involves irreversible changes in liver structure. Alcohol metabolism ties up enzymes needed for oxidation of nutrients and also results in byproducts that acidify body fluids. Other effects of alcoholism include obesity, malnutrition, cancer, ulcers, and fetal alcohol syndrome. Health professionals advise pregnant women not to drink any alcohol. In addition, alcohol impairs judgment and leads to increased involvement in motor vehicle accidents, drownings, falls, and other accidental injuries.

Moderate alcohol consumption is defined as no more than one drink per day for women and two drinks per day for men. Current research suggests that moderate alcohol consumption is compatible with good health and may even have a beneficial effect on the cardiovascular system.

See the Student Resources on thePoint for a summary of the effects of alcohol abuse and photographs of its effects on liver and brain tissue.

CHECKPOINTS 

- ☐ **20-6** What is the term for how rapidly a food raises the blood glucose level?
- ☐ **20-7** What is meant when an amino acid or a fatty acid is described as essential?
- ☐ **20-8** What is the difference between saturated and unsaturated fats?
- ☐ **20-9** What is the difference between vitamins and minerals?
- ☐ **20-10** How much pure alcohol can the average person metabolize per hour?

Nutritional Disorders

Diet-related problems may originate from an excess or shortage of necessary nutrients. Another issue in the news today is weight control. Food allergies may also affect some people.

FOOD ALLERGIES

Some people develop clear allergic symptoms if they eat certain foods. Common food allergens are wheat, nuts, milk, shellfish, and eggs, but almost any food might cause an allergic reaction in a given individual. People may also have allergic reactions to food additives, such as flavorings, colorings, or preservatives. Signs of allergic reactions usually involve the skin, respiratory tract, or gastrointestinal tract. Food allergies may provoke potentially fatal anaphylactic shock in extremely sensitive individuals.

Whereas food allergies are actually quite rare, afflicting about 5% of children and 4% of adults, adverse reactions to food are relatively common. Some of these adverse reactions reflect a specific enzyme deficiency. As discussed in Chapter 19, examples include celiac disease, resulting from a deficiency of the enzyme that digests wheat protein (gluten), and lactose intolerance. Much more common is nonallergic food sensitivity, which describes unpleasant gastrointestinal symptoms (acid reflux, nausea, and cramping) following ingestion of a specific food. The underlying mechanisms of these responses are not well understood, but the immune system is not involved.

MALNUTRITION

If any vital nutrient is missing from the diet, the body will suffer from **malnutrition.** One commonly thinks of a malnourished person as someone who does not have enough to eat, but malnutrition can also occur from eating too much of the wrong foods. Factors that contribute to malnutrition are poverty, old age, chronic illness, anorexia, poor dental health, and drug or alcohol addiction.

In poor and underdeveloped countries, many children suffer from deficiencies of specific nutrients, including vitamin A, iodine, and iron. A diet deficient in protein and total energy is called protein-energy malnutrition (PEM). **Marasmus** (mah-RAZ-mus) is a term used for severe PEM in infancy and childhood (from Greek meaning "dying away.") Affected children lose weight and have muscle wasting and wrinkled skin from loss of underlying tissue.

Kwashiorkor (kwash-e-OR-kor) typically affects toddlers when they are weaned because another child is born (and the name means just that) (**Fig. 20-5**). A low albumin level in the blood plasma interferes with fluid return to the capillaries, resulting in edema. Often, excess fluid accumulates in the abdomen as ascites (ah-SI-teze) fluid, causing the stomach to bulge. Loss of intestinal villi results in poor nutrient absorption and diarrhea. Protein insufficiency causes anemia and loss of pigmentation in skin and hair. Finally, abnormal metabolism affects the liver, causing it to accumulate fatty substances.

Malnutrition greatly increases a child's vulnerability to infection, a primary cause of death before age 5 in the developing world. It also causes stunted growth, wasting, reduced mental capacity, and reduced work capacity, contributing to a family's and community's inability to improve their living conditions.

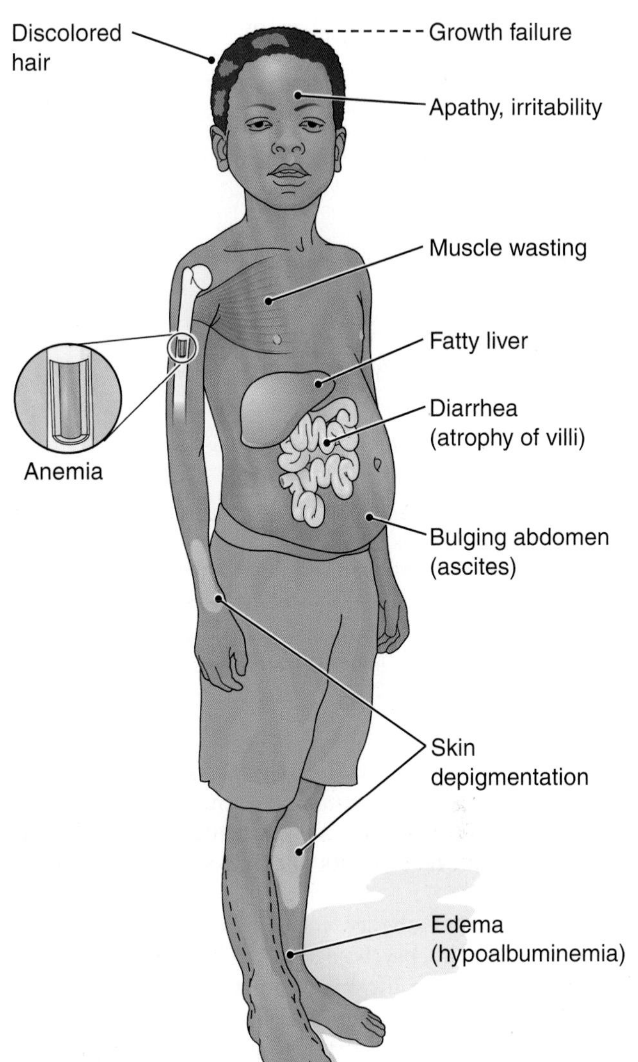

Figure 20-5 Effects of kwashiorkor. KEY POINT Kwashiorkor is a nutritional deficiency that typically affects older children in poor countries when they are weaned. Lack of protein causes wasting, anemia, edema, and pigment loss in hair and skin among other changes.

WEIGHT CONTROL

Body mass index (BMI) is a measurement used to evaluate body size. It is based on the ratio of weight to height (**Fig. 20-6**). BMI is calculated by dividing weight in kilograms by height in meters squared. (For those not accustomed to using the metric system, an alternate method is to divide weight in pounds by the square of height in inches and multiply by 703.) A healthy range for this measurement is 18.5 to 24.9.

BMI does not take into account the relative amount of muscle and fat in the body. For example, a bodybuilder might be healthy with a higher than typical BMI because muscle has a higher density than fat. Researchers have also noted that people of different ethnicities commonly have different levels of body fat at the same BMI. Finally, BMI does not give a fair indication of overweight or obesity in people over age 65 or in children. Alternative measures of adiposity include the body adiposity index (BAI), which compares hip circumference and height, and the waist to height ratio (WtH). However, none of these measures have received widespread acceptance.

Overweight and Obesity Overweight is defined as a BMI of 25 to 30 and obesity as a BMI greater than 30. Although the prevalence of overweight among Americans

Calculation of body mass index (BMI)

Formula:

$$\text{BMI} = \frac{\text{Weight (kg)}}{\text{Height (m)}^2}$$

Conversion:

Kilograms = pounds ÷ 2.2

Meters = inches ÷ 39.4

Example:

A woman who is 5′4″ tall and weighs 134 pounds has a BMI of 23.5.

Weight: 134 pounds ÷ 2.2 = 61 kg

Height: 64 inches ÷ 39.4 = 1.6 m; $(1.6)^2 = 2.6$

$$\text{BMI} = \frac{61 \text{ kg}}{2.6 \text{ m}} = 23.5$$

Figure 20-6 Calculation of body mass index (BMI). **KEY POINT** BMI is a measurement used to evaluate body size based on the ratio of weight to height. **ZOOMING IN** What is the BMI of a male 5 ft 10 in in height who weighs 170 lb? (Round off to one decimal place.)

has fluctuated between 31% and 34% since national surveys began in 1960, it is common knowledge that obesity has increased in these decades. In 1962, the U.S. obesity rate was below 14%. In 2009–2010, 35.7% of Americans were obese.

The causes of obesity are complex, involving social, economic, genetic, psychological, and metabolic factors. Obesity shortens the life span and is associated with cardiovascular disease, diabetes, some cancers, and other diseases. Unfortunately, obesity rates have also increased among American children. Not surprisingly, the incidence of type 2 diabetes, once considered to have an adult onset, has also increased greatly among children. Some researchers hold that obesity has a closer correlation to chronic disease than poverty, smoking, or drinking alcohol.

See the Student Resources on thePoint for a summary of the long-term complications of diabetes.

Scientists are studying the nervous and hormonal controls over weight, but so far they have not found any effective and safe drugs for weight control. For most people, a varied diet eaten in moderation and regular exercise are the surest ways to avoid obesity. Recent studies have questioned the belief that weight loss is best accomplished with large volume, low-intensity exercise. A number of investigators have found that high-intensity intermittent exercise (HIIE) results in much more fat loss (both subcutaneous and visceral) than low-intensity exercise. HIIE protocols are quite short, involving perhaps 20 minutes of exercise alternating between eight-second sprints and 12-second recovery periods. Researchers are still determining the mechanisms underlying this unexpected result, but they suspect decreased appetite, good adherence to the exercise program, and decreased insulin resistance are involved.

Underweight A BMI of less than 18.5 is defined as underweight. People who are underweight can have as much difficulty gaining weight as others have losing it. The problem of underweight may result from eating disorders, rapid growth, allergies, illness, or psychological factors. It is associated with low reserves of energy, reproductive disturbances, and nutritional deficiencies. In women, inadequate fat stores lead to a lack of estrogen production which may cause menstrual periods to cease. To gain weight, people have to increase their intake of calories, but they should also engage in regular exercise to add muscle tissue and not just fat.

A chronic loss of appetite, called **anorexia** (an-o-REK-se-ah), may be caused by a great variety of physical and mental disorders. In addition, patients receiving certain medications, including the chemotherapy used in cancer, can experience anorexia. Because the hypothalamus and the higher brain centers are involved in the regulation of hunger, it is also possible that emotional and social factors contribute to the development of anorexia.

Anorexia nervosa is a psychological disorder that predominantly afflicts young women, as illustrated in Claudia's opening case study. In a desire to be excessively thin, affected people literally starve themselves, sometimes to the point of death. As in all cases of starvation, the body resorts to breaking down body protein to generate amino acids, which can then be used for energy or to make glucose. The loss of bone protein weakens bone, the loss of cardiac muscle protein can lead to heart failure, and the loss of skeletal muscle protein results in muscle weakness. A related disorder, **bulimia** (bu-LIM-e-ah), is also called the *binge–purge syndrome*. Affected individuals eat huge quantities of food at one time and then induce vomiting or take large doses of laxatives to prevent absorption of the food. The reflux of acidic substances in bulimia causes erosion of the esophagus and destruction of tooth enamel.

CHECKPOINTS 

- ☐ **20-11** What condition results from lack of any specific nutrient in the diet?
- ☐ **20-12** What does the abbreviation BMI stand for?
- ☐ **20-13** What term applies to any chronic loss of appetite?

Nutrition and Aging

With age, a person may find it difficult to maintain a balanced diet. Often, the elderly lose interest in buying and preparing food or are unable to do so. Because metabolism generally slows and less food is required to meet energy needs, nutritional deficiencies may develop. With age, the ability to synthesize vitamin D declines, and the kidneys are less able to convert it to its active form. Older adults may need supplements of vitamin D as well as calcium to prevent loss of bone density. The senses of smell and taste become less acute with age, diminishing appetite. Medications may interfere with appetite and with the absorption and use of specific nutrients.

It is important for older people to seek out foods that are "nutrient dense," that is, foods that have a high proportion of nutrients in comparison with the number of calories they provide. Exercise helps boost appetite and maintain muscle tissue, which is more metabolically active.

Body Temperature

Heat is an important byproduct of the many chemical activities constantly occurring in body tissues. At the same time, heat is always being lost through a variety of outlets. Under normal conditions, negative feedback keeps body temperature constant within quite narrow limits. Maintenance of a constant temperature despite both internal and external influences is one phase of homeostasis, the tendency of all body processes to maintain a normal state despite forces that tend to alter them.

HEAT PRODUCTION

A mitochondrion produces a certain amount of heat every time it makes a molecule of ATP. Therefore, heat production directly correlates with tissue activity. The activity of most tissues, including the brain and liver, remains constant. (An increase in nervous tissue activity is not associated with increased ATP production, so studying hard doesn't burn more calories or generate additional heat.)

Muscle cell activity and heat production, though, increase enormously with exercise. Food intake also augments heat production, because of the high energy cost of processing and storing nutrients. This change is greatest with protein-rich meals and lowest with fat-rich meals.

High heat-generating tissues do not become much warmer than others because the circulating blood distributes the heat fairly evenly. For instance, strenuous exercise can make your head feel hot even though the heat was generated in the leg muscles.

In addition to exercise and digestion, hormones, such as thyroxine from the thyroid gland, epinephrine (adrenaline) from the adrenal medulla, and progesterone from the ovary increase heat production.

HEAT LOSS

Although 15% to 20% of heat loss occurs through the respiratory system and with urine and feces, more than 80% of heat loss occurs through the skin. Networks of blood vessels in the skin's dermis (deeper part) can bring considerable quantities of blood near the surface, so that heat can be dissipated to the outside.

Mechanisms of Heat Loss Heat loss from the skin can occur in several ways (**Fig. 20-7**):

- In **radiation**, heat travels from its source as heat waves or rays. Radiation does not require contact between the heat source and the heat receiver.
- In **convection**, heat transfer is promoted by movement of a cooler contacting medium. For example, if the air around the skin is put in motion, as by an electric fan, the layer of heated air next to the body is constantly carried away and replaced with cooler air. Another example of convection is the transfer of heat from muscle tissue to circulating blood.
- In **evaporation**, heat is lost in the process of changing a liquid to the vapor state. Heat from the skin provides the energy to evaporate sweat from the skin's surface, just as heat applied to a pot of water converts the water to steam.
- In **conduction**, a warm object transfers heat energy to a cooler object. For example, heat can be transferred from the skin directly to an ice pack.

To illustrate evaporation, rub some alcohol on your skin; it evaporates rapidly, using so much heat that your skin feels cold. Perspiration does the same thing, although not as quickly.

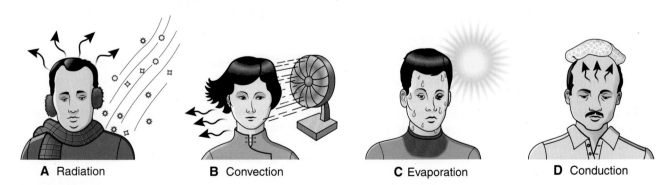

20

Figure 20-7 **Mechanisms of heat loss.** **KEY POINT** There are four processes that promote heat loss of body heat to the environment. **A.** Radiation—Heat travels away as heat waves or rays. **B.** Convection—Heat transfer is promoted by movement of a cooler medium. **C.** Evaporation—Heat is used to change a liquid (such as sweat) to a vapor. **D.** Conduction—Heat is transferred to a cooler object. **ZOOMING IN** What will happen in **(B)** if the fan speed is increased? What will happen in **(C)** if environmental humidity increases?

Prevention of Heat Loss Factors that play a part in heat loss include the volume of tissue compared with the amount of skin surface. A child loses heat more rapidly than does an adult. Such parts as fingers and toes are affected most by exposure to cold because they have a great amount of skin compared with total tissue volume.

An effective natural insulation against heat loss is the layer of fat under the skin. The degree of insulation depends on the thickness of the subcutaneous layer. Even when skin temperature is low, this fatty tissue prevents the deeper tissues from losing much heat. On average, this layer is slightly thicker in females than in males.

TEMPERATURE REGULATION

Body temperature remains within narrow limits despite wide variations in the rate of heat production or loss because of internal temperature-regulating mechanisms.

Recall from Chapter 1 that a negative feedback loop includes sensors, an integrating center, and effectors. The integrating center in heat regulation is the **hypothalamus**, the area of the brain located just above the pituitary gland. The hypothalamus receives information from temperature receptors in the skin and within the hypothalamus itself and sends signals to numerous effectors, including muscles and blood vessels.

Responses to Cold Conditions If the sensors indicate that body temperature is lower than the set point, hypothalamic signals stimulate constriction of blood vessels in skin to reduce heat loss (**Fig. 20-8A**). In addition, impulses sent to the skeletal muscles cause shivering, rhythmic contractions that result in increased metabolic heat production. Equally important are our conscious responses to cold conditions. The hypothalamus notifies the cerebral cortex that body temperature is unpleasantly low, inspiring behavioral responses such as finding a warmer environment or adding extra clothing. Clothes reduce heat loss by convection and radiation.

Responses to Hot Conditions Hot environmental conditions can provide a particular challenge to thermoregulation, because the body must disperse heat gained from solar radiation (**Fig. 20-8B**). Moreover, the smaller difference between body temperature and air temperature decreases the body's ability to lose heat by convection and conduction.

If body temperature increases above the set point, signals from the hypothalamus cause cutaneous blood vessels to dilate, so that increased blood flow will promote heat loss. The hypothalamus also stimulates the sweat glands to increase their activity, thus maximizing evaporation. However, the rate of heat loss through evaporation depends on the humidity of the surrounding air. When humidity exceeds 60% or so, perspiration does not evaporate as readily, making one feel uncomfortable unless some other means of heat loss is available, such as convection caused by a fan. It's important to remember that sweat only cools the body if it evaporates. Wiping away sweat eliminates its usefulness.

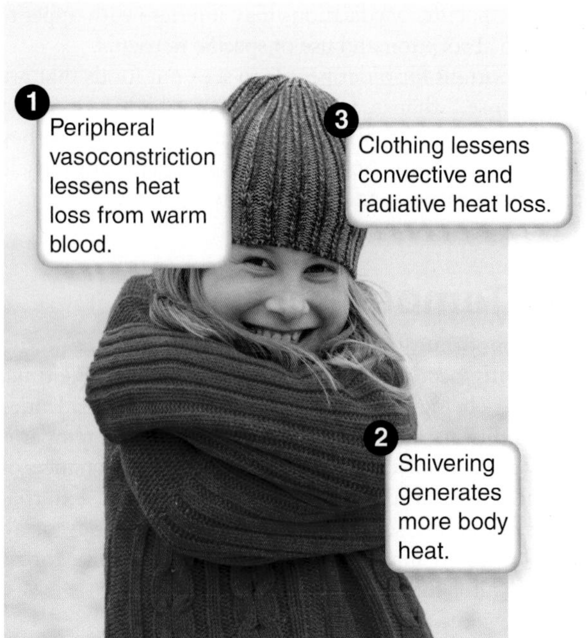
1
Peripheral vasoconstriction lessens heat loss from warm blood.
3
Clothing lessens convective and radiative heat loss.
2
Shivering generates more body heat.

A Cold conditions

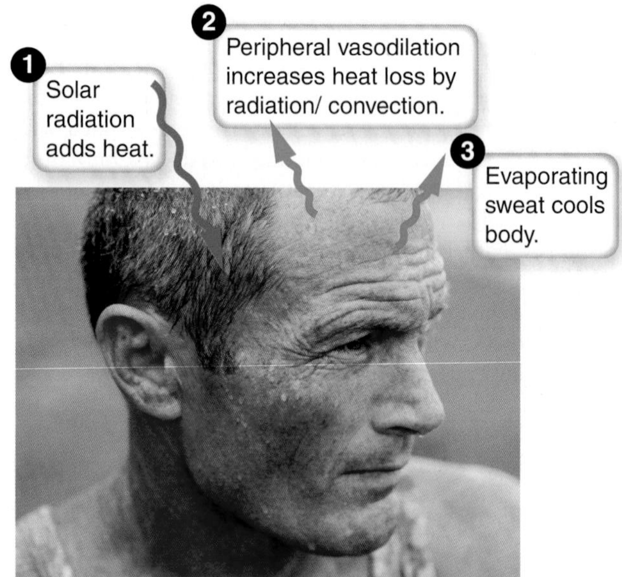
1
Solar radiation adds heat.
2
Peripheral vasodilation increases heat loss by radiation/ convection.
3
Evaporating sweat cools body.

B Hot conditions

Figure 20-8 **Thermoregulation.** **KEY POINT** Negative feedback loops help maintain body temperature within normal limits. **A.** Under cold conditions, physiological and behavioral responses limit heat loss and generate additional metabolic heat. **B.** Under hot conditions, sweating and peripheral vasodilation help the body shed metabolic heat and heat absorbed from solar radiation. **ZOOMING IN** Is blood flow in the skin higher under cold conditions or hot conditions?

Muscles are especially important in temperature regulation because variations in the activity of these large tissue masses can readily increase or decrease heat generation. Because muscles form roughly one-third of the body, either an involuntary or an intentional increase in their activity can produce enough heat to offset a considerable decrease in the environmental temperature.

Age Factors Very young and very old people are limited in their ability to regulate body temperature when exposed to environmental extremes. A newborn infant's body temperature decreases if the infant is exposed to a cool environment for a long period. Elderly people are also not able to produce enough heat to maintain body temperature in a cool environment.

With regard to overheating in these age groups, heat loss mechanisms are not fully developed in the newborn. The elderly do not lose as much heat from their skin as do younger people. Both groups should be protected from extreme temperatures.

Normal Body Temperature The normal temperature range obtained by either a mercury or an electronic thermometer may vary from 36.2°C to 37.6°C (97°F to 100°F). Body temperature varies with the time of day. Usually, it is lowest in the early morning because the muscles have been relaxed and no food has been taken in for several hours. Body temperature tends to be higher in the late afternoon and evening because of physical activity and food consumption.

Normal temperature also varies in different parts of the body. Skin temperature obtained in the axilla (armpit) is lower than mouth temperature, and mouth temperature is a degree or so lower than rectal temperature. It is believed that, if it were possible to place a thermometer inside the liver, it would register a degree or more higher than rectal temperature. The temperature within a muscle might be even higher during activity.

Although the Fahrenheit scale is used in the United States, in most parts of the world, temperature is measured with the Celsius (SEL-se-us) thermometer. On this scale, the ice point is at 0° and the normal boiling point of water is at 100°, the interval between these two points being divided into 100 equal units. The Celsius scale is also called the centigrade scale (think of 100 cents in a dollar).

FEVER

Fever (FE-ver) is a condition in which the body temperature is higher than normal. An individual with a fever is described as **febrile** (FEB-ril). Usually, the presence of fever is due to an infection, but there can be many other causes, such as malignancies, brain injuries, toxic reactions, reactions to vaccines, and diseases involving the central nervous system (CNS). Sometimes, emotional upsets can bring on a fever. Whatever the cause, the effect is to alter the hypothalamic temperature set point.

Negative feedback mechanisms change and then maintain body temperature at the new set point. For this reason, fever is usually preceded by a chill—that is, a violent attack of shivering and a sensation of cold that blankets and heating pads seem unable to relieve. These responses generate heat; body temperature increases to the new set point, and the chills subside.

The old belief that a fever should be starved is completely wrong. During a fever, there is an increase in metabolism that is usually proportional to the fever's intensity. The body uses available sugars and fats, and there is an increase in the use of protein. During the first week or so of a fever, there is definite evidence of protein destruction, so a high-calorie diet with plenty of protein is recommended.

When a fever ends, sometimes the return of temperature to normal occurs very rapidly. This sudden drop in temperature is called the **crisis**, and it is usually accompanied by symptoms indicating rapid heat loss: profuse perspiration, muscular relaxation, and dilation of blood vessels in the skin. A gradual drop in temperature, in contrast, is known as **lysis**. A drug that reduces fever is described as **antipyretic** (an-ti-pi-RET-ik).

The mechanism of fever production is not completely understood, but we might think of the hypothalamus as a thermostat that during fever is set higher than normal. This change in the heat-regulating mechanism often follows the introduction of a foreign protein or the entrance into the bloodstream of bacteria or their toxins. Substances that produce fever are called **pyrogens** (PI-ro-jens).

Up to a point, fever may be beneficial because it steps up phagocytosis (the process by which white blood cells destroy bacteria and other foreign material), inhibits the growth of certain organisms, and increases cellular metabolism, which may help recovery from disease.

RESPONSES TO EXCESSIVE HEAT

The body's heat-regulating devices are efficient, but there is a limit to what they can accomplish. High outside temperature may overcome the body's heat loss mechanisms, in which case body temperature rises and cellular metabolism with accompanying heat production increases. When body temperature rises, the affected person is apt to suffer from a series of disorders: heat cramps are followed by heat exhaustion, which, if untreated, is followed by heat stroke.

In **heat cramps**, there is localized muscle cramping of the extremities and occasionally of the abdomen. The condition abates with rest in a cool environment with intake of adequate fluids.

With further heat retention and more fluid loss, **heat exhaustion** occurs. Symptoms of this disorder include headache, tiredness, vomiting, and a rapid pulse. The victim feels hot, but the skin is cool due to evaporation of sweat. There may be a decrease in circulating blood volume and lowered blood pressure. Heat exhaustion is also treated by rest and fluid replacement.

Heat stroke (also called *sunstroke*) is a medical emergency. Heat stroke can be recognized by a body temperature of up to 41°C (105°F); hot, dry skin (because the sweating mechanism has failed); and CNS symptoms, including confusion, dizziness, and loss of consciousness. The body has responded to the loss of circulating fluid by reducing blood flow to the skin and sweat glands.

It is important to lower the heat stroke victim's body temperature immediately by removing the individual's clothing, placing him or her in a cool environment, and cooling the body with cold water or ice. The patient should be treated with appropriate fluids containing necessary electrolytes, including sodium, potassium, calcium, and chloride. Emergency medical care is necessary to avoid fatal complications.

RESPONSES TO EXCESSIVE COLD

The body is no more capable of coping with prolonged exposure to cold than with prolonged exposure to heat. If, for example, the body is immersed in cold water for a time, the water (a better heat conductor than air) removes more heat from the body than can be replaced, and temperature falls. Cold air can produce the same result, particularly when clothing is inadequate. The main effects of an excessively low body temperature, termed **hypothermia** (hi-po-THER-me-ah), are uncontrolled shivering, lack of coordination, and decreased heart and respiratory rates. Speech becomes slurred, and there is overpowering sleepiness, which may lead to coma and death. Outdoor activities in cool, not necessarily cold, weather cause many unrecognized cases of hypothermia. Wind, fatigue, and depletion of water and energy stores all play a part.

When the body is cooled below a certain point, cellular metabolism slows, and heat production is inadequate to maintain a normal temperature. The person must then be warmed gradually by heat from an outside source. The best first aid measure is to remove the person's clothing and put him or her in a warmed sleeping bag with an unclothed companion until shivering stops. Administration of warm, sweetened fluids also helps.

Exposure to cold, particularly to moist cold, may result in **frostbite**, which can cause permanent local tissue damage. The areas most likely to be affected by frostbite are the face, ears, and extremities. Formation of ice crystals and reduction of blood supply to an area leads to necrosis (death) of tissue and possible gangrene. The very young, the very old, and those who suffer from circulatory disorders are particularly susceptible to cold injuries.

A frostbitten area should never be rubbed; rather, it should be thawed by application of warm towels or immersion in circulating lukewarm (not hot) water for 20 to 30 minutes. The affected area should be treated gently; a person with frostbitten feet should not be permitted to walk. People with cold-damaged extremities frequently have some lowering of body temperature. The whole body should be warmed at the same time that the affected part is warmed.

CHECKPOINTS 

- ☐ **20-14** What part of the brain is responsible for regulating body temperature?
- ☐ **20-15** What does an antipyretic drug do?
- ☐ **20-16** What change occurs in cutaneous blood vessels under cold conditions? Under hot conditions?
- ☐ **20-17** What is the term for excessively low body temperature?

Disease in Context Revisited

Claudia's Anorexia Nervosa

Claudia approached Dr. Wade at the end of the next class, waiting until all of the other students had left.

"Do you have a minute to talk?" she asked. "I was hoping to speak with you today."

Dr. Wade replied, "Let's walk down toward the river where we can have some privacy. What's been happening with you?"

Claudia started to talk. "My friend Josie took me to the student health center last night. They think I have anorexia nervosa! At first I didn't believe it, but they did some tests, and my results were similar to that case study we did in class about starving kids in war zones. My BMI is 16, and there were ketones in my urine. They did a heart test, and my ECG showed an arrhythmia. I don't know how this happened; I'm enjoying college life and I thought I was dealing with the stress, but I'm really in trouble. The psychologist wants me to keep a food diary, try to eat 2,000 calories a day, and limit my daily exercise to an hour. I have to meet with her every day, and if I can't turn things around in the next few weeks they might refer me to an inpatient treatment center. I don't know if I can do it!"

Dr. Wade took a moment to think, and said carefully, "Claudia, I'm so relieved that you are dealing with your illness. You have a support network of health professionals and friends, and I'm very hopeful for you. Please check in with me every week or so to let me know how you are doing, and let me know if you need to get some assignments or tests rescheduled."

CHAPTER 20

Chapter Wrap-Up

Summary Overview

A detailed chapter outline with space for note taking is on *thePoint*. The figure below illustrates the main topics covered in this chapter.

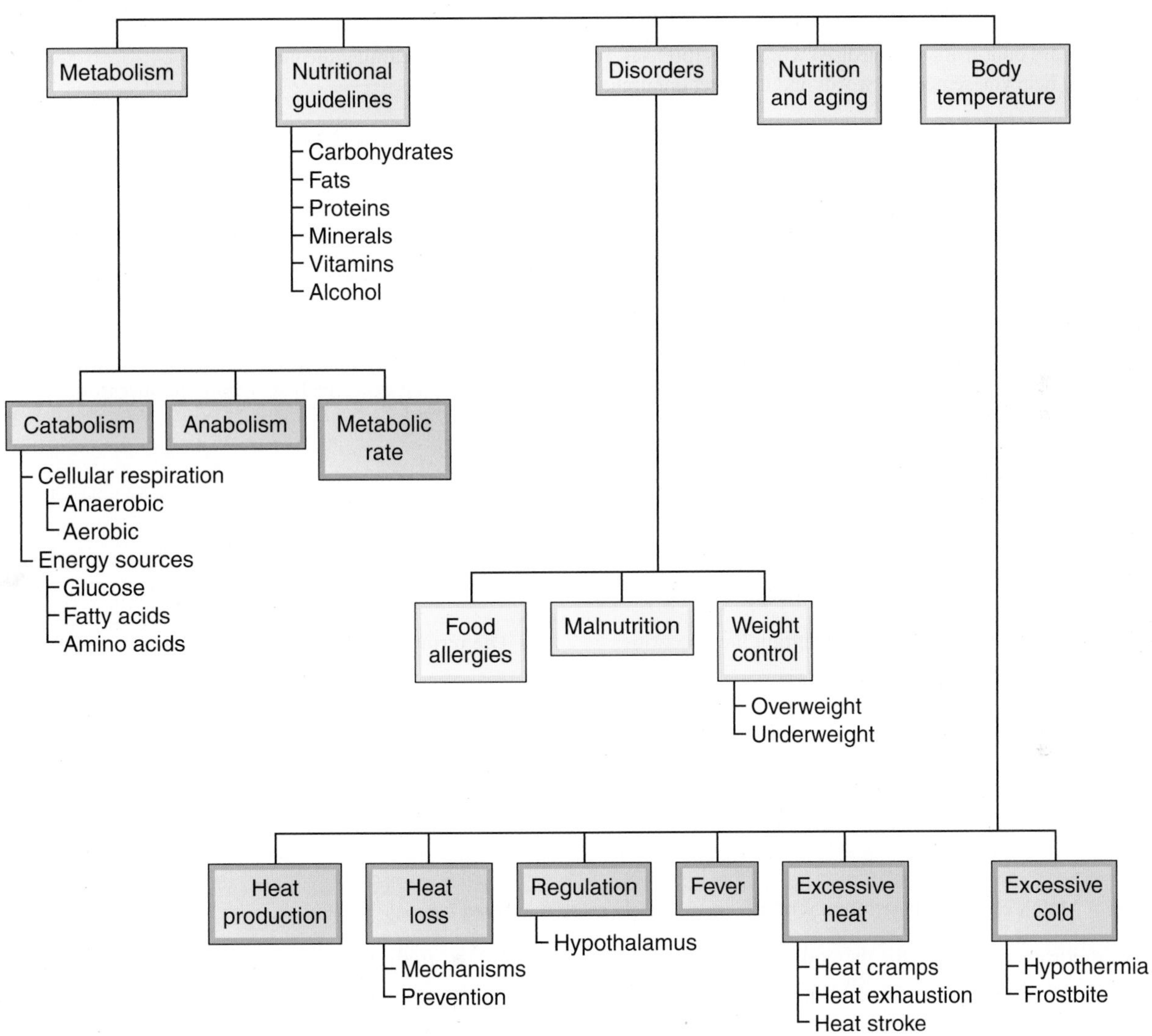

Key Terms

The terms listed below are emphasized in this chapter. Knowing them will help you organize and prioritize your learning. These and other boldface terms are defined in the Glossary with phonetic pronunciations.

anabolism
basal metabolism
catabolism
cellular respiration
fever
glucose
glycogen
glycolysis
hypothalamus
hypothermia
kilocalorie
malnutrition
metabolic rate
mineral
oxidation
pyrogen
vitamin

Word Anatomy

Medical terms are built from standardized word parts (prefixes, roots, and suffixes). Learning the meanings of these parts can help you remember words and interpret unfamiliar terms.

WORD PART	MEANING	EXAMPLE
Metabolism		
glyc/o	sugar, sweet	*Glycogen* yields glucose molecules when it breaks down.
-lysis	separating, dissolving	*Glycolysis* is the breakdown of glucose for energy.
Body Temperature		
pyr/o	fire, fever	An *antipyretic* drug reduces fever.
therm/o	heat	*Hypothermia* is an excessively low body temperature.

Questions for Study and Review

BUILDING UNDERSTANDING

Fill in the Blanks

1. The amount of energy needed to maintain life functions while at rest is _______.
2. Reserves of glucose are stored in liver and muscle as _______.
3. The area of the brain most important for temperature regulation is the _______.
4. The term *febrile* describes a person who has a(n) _______.
5. The removal of nitrogen groups from amino acids in metabolism is called _______.

Matching > Match each numbered item with the most closely related lettered item:

___ **6.** Major energy source for the body
___ **7.** Chemical element required for normal body function
___ **8.** Complex organic substance required for normal body function
___ **9.** A byproduct of fatty acid metabolism
___ **10.** A simple fat

a. triglyceride
b. vitamin
c. mineral
d. ketone bodies
e. glucose

Multiple Choice

___ **11.** Which of the following is NOT an example of catabolism?

a. glycolysis
b. deamination
c. cellular respiration
d. glycogen formation

___ **12.** Which of the following would have the lowest glycemic effect?

a. glucose
b. sucrose
c. lactose
d. starch

___ **13.** Which organ catabolizes alcohol?

a. small intestine
b. liver
c. pancreas
d. spleen

___ **14.** What is the term for a gradual decrease in a fever?

a. lysis
b. pyrogenesis
c. crisis
d. anorexia

UNDERSTANDING CONCEPTS

15. In what part of the cell does anaerobic respiration occur, and what are its end products? In what part of the cell does aerobic respiration occur? What are its end products?

16. About how many kilocalories are released from a tablespoon of butter (14 g)? A tablespoon of sugar (12 g)? A tablespoon of egg white (15 g)?

17. If you eat 2,000 kcal a day, how many kilocalories should come from carbohydrates? From fats? From protein?

18. How is heat produced in the body? What structures produce the most heat during increased activity?

19. During a severe cold, Emily's body temperature increased from 36.4°C to 37.1°C and then decreased to 36.4°C again. Describe the feedback mechanism regulating Emily's body temperature.

20. What is hypothermia? Under what circumstances does it usually occur? List some of its effects.

21. Differentiate between the terms in the following pairs:

a. essential and nonessential amino acids
b. saturated and unsaturated fats
c. marasmus and kwashiorkor
d. radiation and conduction
e. heat exhaustion and heat stroke

CONCEPTUAL THINKING

22. The oxidation of glucose to form ATP is often compared to the burning of fuel. Why is this analogy inaccurate?

23. Richard M, a self-described couch potato, is 6 ft tall and weighs 240 lb. Calculate Richard's body mass index. Is Richard overweight or obese? List some diseases associated with obesity.

24. Referring to Claudia's opening case study, explain how her body is compensating metabolically for her anorexia nervosa and what effects her disorder might have.

25. Jennifer's 85-year-old grandmother is a widow and lives alone. Should Jennifer be concerned about her nutritional well-being? What can she do to prevent possible malnutrition?

For more questions, see the Learning Activities on thePoint.

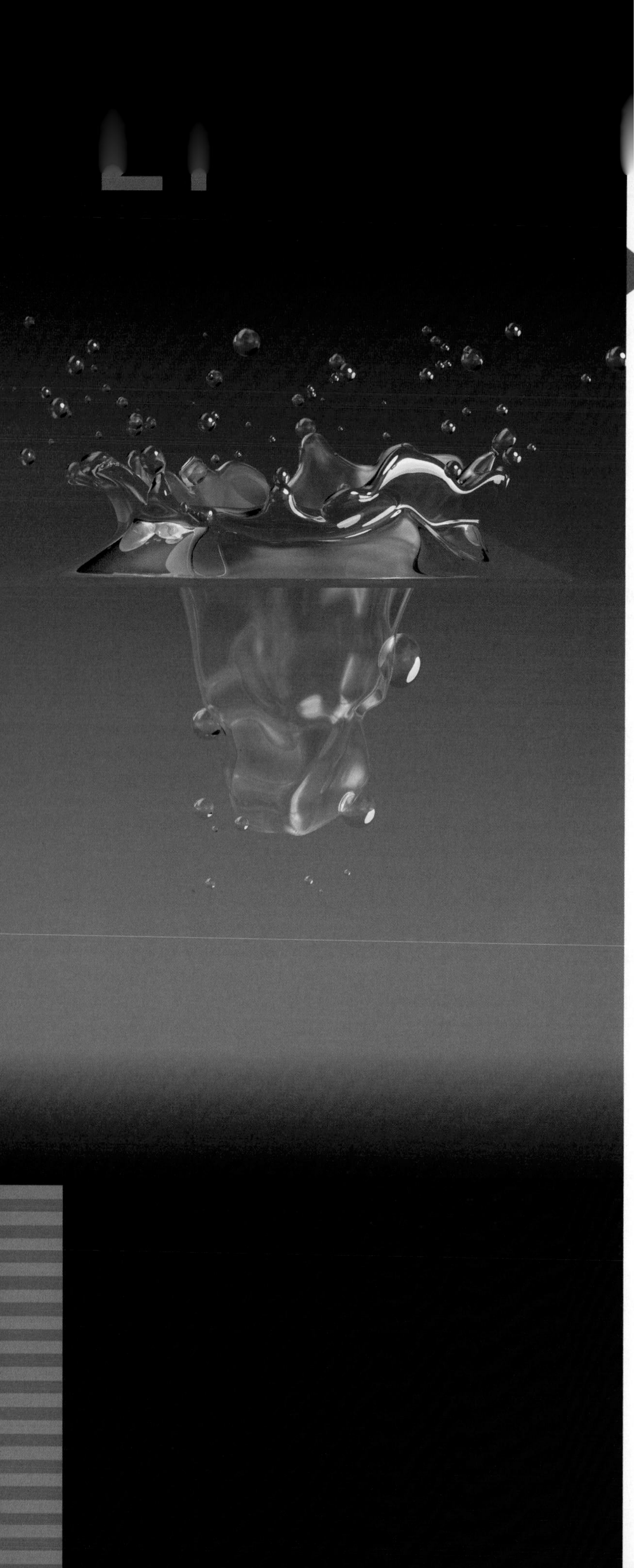

Learning Objectives

After careful study of this chapter, you should be able to:

1. ▶ Compare intracellular and extracellular fluids. ***p. 482***
2. ▶ List four types of extracellular fluids. ***p. 482***
3. ▶ Define *electrolytes*, and describe some of their functions. ***p. 483***
4. ▶ Name four systems that are involved in water balance. ***p. 484***
5. ▶ Explain how thirst is regulated. ***p. 484***
6. ▶ Describe the role of hormones in electrolyte balance. ***p. 486***
7. ▶ Describe three methods for regulating the pH of body fluids. ***p. 487***
8. ▶ Compare acidosis and alkalosis, including possible causes. ***p. 488***
9. ▶ Describe four disorders involving body fluids. ***p. 488***
10. ▶ Cite instances in which fluids are administered therapeutically, and cite four fluids that are used. ***p. 489***
11. ▶ Referring to the opening case study, explain the dangers of overhydrating. ***pp. 481, 490***
12. ▶ Show how word parts are used to build words related to bodily fluids (see Word Anatomy at the end of the chapter). ***p. 492***

Disease in Context *Ethan's Triathlon Emergency*

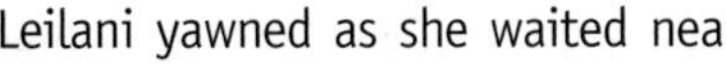

Leilani yawned as she waited near the finish line of the Ironman Triathlon in Kona, Hawai'i. She had volunteered as a finish line attendant, which meant that she and her partner "caught" the athletes as they crossed the finish line and escorted them to get their medals, find their families, or, increasingly, to get a bed in the medical tent. The race was over in three minutes, and there were still two athletes out on the course. They had already swum 2.4 miles in choppy surf, biked over 180 miles through lava fields, and had nearly finished the 26.2-mile run portion of the punishing event. Despite all of their efforts, they had to complete their run before midnight in order to qualify as Ironman finishers. The crowd noise increased as a tall man stumbled over to the finish line, to the announcer's cry, "Ethan Landry, you are an Ironman!" Leilani and her partner raced over to catch him.

"We better get this one straight to the medical tent," her partner gasped as they tried to keep the man from sinking to the pavement. They were met by Darren, a local physician assistant, who helped them lay the man down on a stretcher.

"What's your name? Do you know where you are?" Darren asked the man as he strapped on a blood pressure monitor.

"Um, Ethan? I'm in Hawai'i? I'm really tired and I don't feel so good," the athlete replied. Darren pinched Ethan's arm and noted that the pinched skin returned immediately to its original state. His face seemed swollen, and his wedding ring was sunk deep into his engorged fingers.

"Should I get him some water?" Leilani asked.

"I don't think so," Darren answered, as he tilted Ethan's bed backward. "His blood pressure is low, which is why I'm lowering his head below his body, but he has some evidence of edema. I think that Ethan is actually overhydrated and suffering from exercise-associated hyponatremia, or EAH. This condition used to be called water intoxication, and it's pretty common in endurance events. Basically, consuming too much water can dilute your blood and other body fluids and lead to swelling. Look at his face and fingers; it's as if they are waterlogged. Let's get him to the hospital for some tests."

In Chapter 3, we discussed the effect of osmotic pressure on cells. We continue in this chapter with a more detailed study of body fluids and the need to keep the total volume and composition of these fluids within normal limits. Later, we'll find out if Darren's suspicions were correct.

ANCILLARIES *At-A-Glance*

Visit thePoint to access the following resources. For guidance in using these resources most effectively, see pp. xv–xvii.

Learning RESOURCES

- Tips for Effective Studying
- Health Professions: Emergency Medical Technician
- Detailed Chapter Outline
- Answers to Questions for Study and Review
- Audio Pronunciation Glossary

Learning ACTIVITIES

- Pre-Quiz
- Visual Activities
- Kinesthetic Activities
- Auditory Activities

A LOOK BACK

In Chapter 2, we introduced the chemistry of water. In that chapter, and also in Chapters 18 and 19, we discussed its importance in the body as a transport medium and in nutrient hydrolysis (digestion). Now we discuss its distribution into different compartments and list the substances that are dissolved and suspended in body fluids. Chapter 2 also discussed acids, bases, and the pH scale. Chapter 18 pointed out the importance of the respiratory system in maintaining a normal acid–base balance. Now we introduce other mechanisms for regulating pH. The integumentary (Chapter 6), respiratory, and digestive systems are avenues for water loss, as is the urinary system, to be described in Chapter 22. Finally, in Chapter 20, we listed some of the elements important in nutrition. Now we look at how they are distributed in body fluids.

Body Fluids

Despite the body's apparent solidity, fluids make up over half of its mass. Composed primarily of water, body fluids also contain electrolytes (salts), nutrients, gases, waste, and special substances, such as enzymes and hormones. This chapter begins with a discussion of the location and composition of body fluids and then discusses some important aspects of body fluid regulation.

FLUID COMPARTMENTS

Although body fluids have much in common no matter where they are located, there are some important differences between fluid inside and outside cells. Accordingly, fluids are grouped into two main compartments (**Fig. 21-1**):

- **Intracellular fluid** (ICF) is contained within the cells. About two-thirds to three-fourths of all body fluids are in this category.
- **Extracellular fluid** (ECF) includes all body fluids outside of cells. In this group are the following:
  - **Interstitial** (in-ter-STISH-al) **fluid**, or more simply, tissue fluid. This fluid is located in the spaces between the cells in tissues throughout the body. It is estimated that tissue fluid constitutes about 15% of body weight.
  - **Blood plasma**, which constitutes about 4% of body weight.

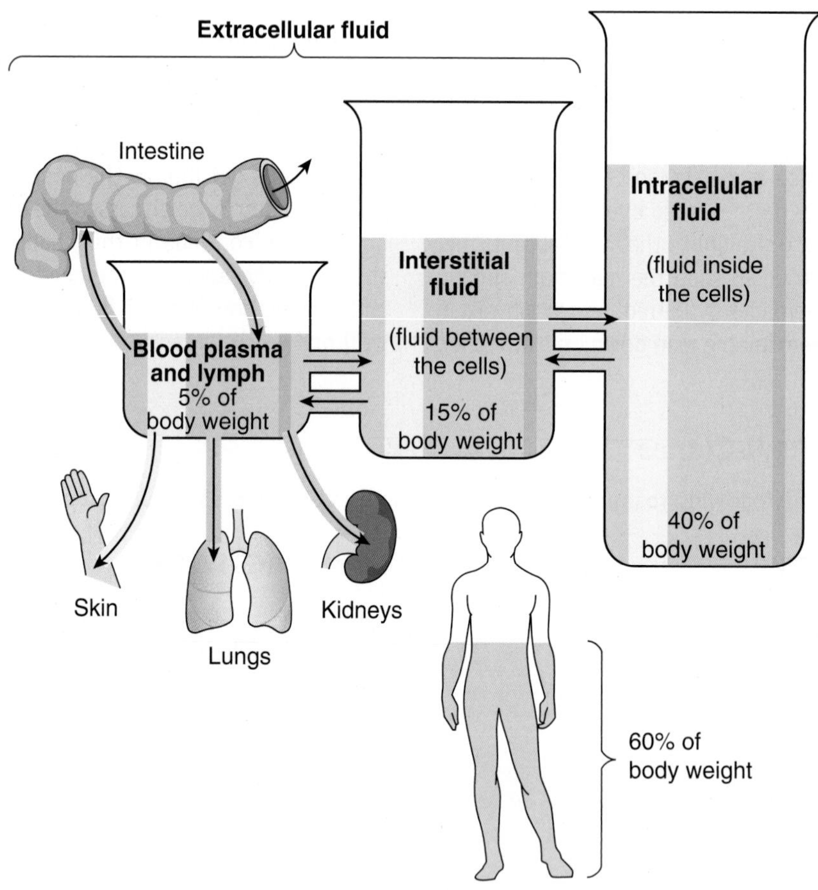

Figure 21-1 Main fluid compartments showing relative percentage by weight of body fluid. **KEY POINT** Fluid percentages vary but total about 60% of body weight. Fluids are constantly exchanged among compartments, and each day fluids are lost and replaced. **ZOOMING IN** What are some avenues through which water is lost?

- **Lymph,** the fluid that drains from the tissues into the lymphatic system. This is about 1% of body weight.
- **Fluid in special compartments,** such as cerebrospinal fluid, the aqueous and vitreous humors of the eye, serous fluid, and synovial fluid. Together, these make up about 1% to 3% of total body fluids.

Fluids are not locked into one compartment. There is a constant interchange between compartments as fluids move between blood and interstitial fluid and between interstitial fluid and intracellular fluid (**see Fig. 21-1**).

WATER AND ITS FUNCTIONS

Water is important to living cells as a solvent, a transport medium, and a participant in metabolic reactions. The force of water in the vessels helps maintain blood pressure, which is important for maintaining blood flow and capillary exchange. The normal proportion of body water varies from 50% to 70% of a person's weight. It is highest in the young and in thin, muscular individuals. As the amount of fat increases, the percentage of water in the body decreases, because adipose tissue holds little water compared with muscle tissue. In infants, water makes up 75% of the total body mass. That's why infants are in greater danger from dehydration than are adults.

ELECTROLYTES AND THEIR FUNCTIONS

Electrolytes are important constituents of body fluids. These compounds separate into positively and negatively charged ions in solution. Positively charged ions are called **cations;** negatively charged ions are called **anions.** Electrolytes are so named because they conduct an electric current in solution.

Major Ions A few of the most important cations and anions are reviewed next. **Figure 21-2** shows the distribution of common electrolytes in the different fluid compartments.

- Cations (positive ions):
  - **Sodium** is chiefly responsible for maintaining osmotic balance and body fluid volume. It is the main positive ion in extracellular fluids. Sodium is required for nerve impulse conduction and is important in maintaining acid–base balance.
  - **Potassium** is also important in the transmission of nerve impulses and is the major positive ion in ICF. Potassium is involved in enzymatic activities, and it helps regulate the chemical reactions by which carbohydrate is converted to energy and amino acids are converted to protein.
  - **Calcium** is required for bone formation, muscle contraction, nerve impulse transmission, and blood clotting.
  - **Magnesium** is necessary for muscle contraction and for the action of some enzymes.
- Anions (negative ions):
  - **Bicarbonate** is an important buffer in body fluids.
  - **Chloride** is essential for the formation of hydrochloric acid in the stomach. It also helps regulate fluid balance and pH. It is the most abundant anion in extracellular fluids.
  - **Phosphate** is essential in carbohydrate metabolism, bone formation, and acid–base balance. Phosphates are found in plasma membranes, nucleic acids (DNA and RNA), and ATP.

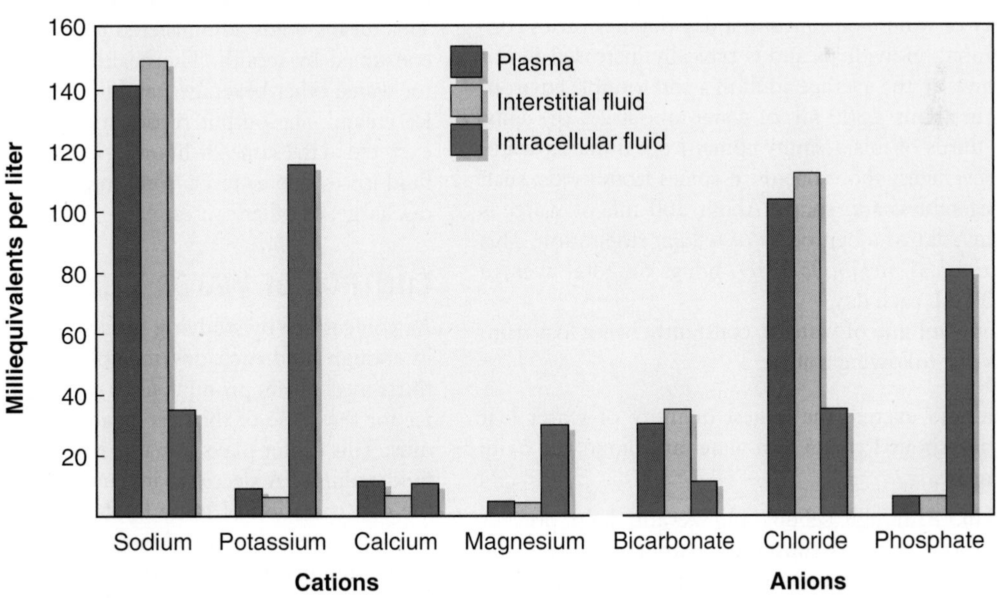

Figure 21-2 Distribution of some major ions in intracellular and extracellular fluids. **KEY POINT** Ion distribution varies in the different fluid compartments. Plasma and interstitial fluid are extracellular fluids. Ions are measured in milliequivalents (mEq), a unit based on atomic weight and valence. **ZOOMING IN** What ions are the most concentrated in extracellular fluids? In intracellular fluids?

CHECKPOINTS

- ☐ 21-1 Where are intracellular fluids located? Extracellular fluids?
- ☐ 21-2 What is interstitial fluid?
- ☐ 21-3 What are cations and anions?

Regulation of Body Fluids

Negative feedback mechanisms maintain the composition and volume of body fluids within narrow limits. Whenever the volume or chemical makeup of these fluids deviates even slightly from normal, disease results. (See Appendix 2, **Table A2-3**, for normal values.) The constancy of body fluids is maintained in the following ways:

- The thirst mechanism helps maintain a constant fluid volume by adjusting water intake.
- Kidney activity helps keep water volume and electrolytes constant by adjusting the loss of water and dissolved substances in the urine (see Chapter 22).
- Buffers, respiration, and kidney function regulate body fluid pH (acidity and alkalinity).

The maintenance of proper fluid balance involves many of the principles discussed in earlier chapters, such as pH and buffers, the effects of respiration on pH, tonicity of solutions, and forces influencing capillary exchange. Some of these chapters are referenced in the following sections. Additional information follows in Chapter 22 on the urinary system.

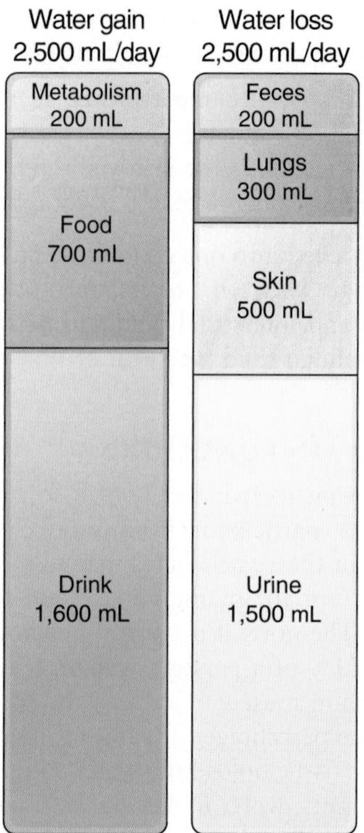

Figure 21-3 **Daily gain and loss of water.** **KEY POINT** In a healthy person, water gained in a day is approximately equal to the quantity lost. **ZOOMING IN** In what way is the most water lost in a day?

WATER BALANCE

In a healthy person, the quantity of water gained in a day is approximately equal to the quantity lost (output) (**Fig. 21-3**). The quantity of water consumed in a day (intake) varies considerably among individuals and is typically increased in hot weather; however, the average adult in a comfortable environment takes in about 2,300 mL of water (about 2½ qt) daily. About two-thirds of this quantity comes from drinking water and other beverages; about one-third comes from foods, such as fruits, vegetables, and soups. About 200 mL of water is produced each day as a byproduct of cellular respiration. This water, described as *metabolic water*, brings the total average gain to 2,500 mL each day.

The same volume of water is constantly being lost from the body by the following routes:

- The **kidneys** excrete the largest quantity of water lost each day. About 1 to 1.5 L of water are eliminated daily in the urine.
- The **skin**. Although sebum and keratin help prevent dehydration, water is constantly evaporating from the skin's surface. Larger amounts of water are lost from the skin as sweat when it is necessary to cool the body.
- The **lungs** expel water along with exhaled air.
- The **intestinal tract** eliminates water along with the feces.

In many disorders, it is important for the healthcare team to know whether a patient's intake and output are equal; in such a case, a 24-hour intake–output record is kept. The intake record includes *all* the liquid the patient has taken in. This means fluids administered intravenously as well as those consumed by mouth. The healthcare provider must account for water, other beverages, and liquid foods, such as soup and ice cream. The output record includes the quantity of urine excreted in the same 24-hour period as well as an estimation of fluid losses due to fever, vomiting, diarrhea, bleeding, wound discharge, or other causes.

CONTROL OF WATER INTAKE

As you can see by studying **Figure 21-3**, it is essential to take in enough fluid each day to replace physiologic losses. Our thirst mechanism prompts us to drink water. The control center for the sense of thirst is located in the brain's hypothalamus. This center plays a major role in the regulation of total fluid volume. A decrease in fluid volume or an increase in the concentration of body fluids stimulates the thirst center, causing a person to drink water or other fluids containing large amounts of water. Dryness of the mouth also causes a sensation of thirst. Excessive thirst, such as that caused by excessive urine loss in cases of diabetes, is called **polydipsia** (pol-e-DIP-se-ah). (See **Box 21-1** on receptors involved in water balance.)

Box 21-1

A CLOSER LOOK

Osmoreceptors: Thinking about Thirst

Osmoreceptors are specialized neurons that help maintain water balance by detecting changes in the concentration of extracellular fluid (ECF). They are located in the hypothalamus of the brain in an area adjacent to the third ventricle, where they monitor the osmotic pressure (concentration) of the circulating blood plasma.

Osmoreceptors respond primarily to small increases in sodium, the most common cation in ECF. As the blood becomes more concentrated, sodium draws water out of the cells, initiating nerve impulses. Traveling to different regions of the hypothalamus, these impulses may have two different but related effects:

- They stimulate the hypothalamus to produce antidiuretic hormone (ADH), which is then released from the posterior pituitary. ADH travels to the kidneys and causes these organs to conserve water.
- They stimulate the thirst center of the hypothalamus, causing increased consumption of water. Almost as soon as water consumption begins, however, the sensation of thirst disappears. Receptors in the throat and stomach send inhibitory signals to the thirst center, preventing overconsumption of water and allowing time for ADH to affect the kidneys.

Both of these mechanisms serve to dilute the blood and other body fluids. Both are needed to maintain optimal water balance. If either fails, a person soon becomes dehydrated.

Thirst and drinking are our only defenses against decreased fluid volume, because the kidneys can only conserve existing body fluids. Ideally, the thirst mechanism should stimulate enough drinking to balance fluids, but this is not always the case. During vigorous exercise, especially in hot weather, the body can dehydrate rapidly, and people may not drink enough to replace needed fluids. While plain water is usually the best choice, prolonged dehydration should be treated with beverages containing sodium and other electrolytes in order to maximize water absorption from the intestine.

While the thirst center effectively stimulates thirst, it does not prevent excess beverage consumption, as illustrated in Ethan's opening case study. As discussed later in this chapter, the kidneys compensate for excess beverage consumption by excreting more urine.

CONTROL OF URINE OUTPUT

The kidneys play a critical role in the regulation of blood volume and electrolyte concentrations in extracellular fluids by modifying the volume and composition of urine (see Box 21-2). The maintenance of water and electrolyte balance is one of the most difficult problems for health providers in caring for patients.

As mentioned earlier, sodium plays a critical role in water balance. An increase in body sodium stimulates water consumption and decreases urine volume in order to restore

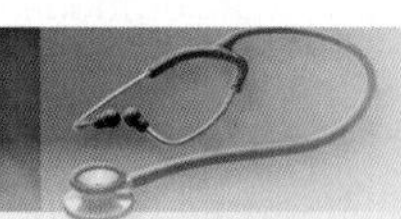

Box 21-2

CLINICAL PERSPECTIVES

Potassium Balance

Regulation of potassium ion (K^+) concentrations in body fluids is critical to normal body function. In Chapter 9, we discussed the key role of potassium in action potentials, which are necessary for neural signaling and muscle contractions. Increased extracellular potassium makes it harder for cells to fire action potentials, so electric signals might not be sent. Conversely, decreased extracellular potassium makes it too easy to generate action potentials, and cells may fire them at the wrong time. Potassium is also important as a part of many enzymes as well as ATP.

The term **hypokalemia** is taken from the Latin name for potassium, *kalium*. It refers to low levels of potassium in body fluids. Symptoms include muscle fatigue, paralysis, confusion, hypoventilation, and cardiac arrhythmias. Hypokalemia can result from chronic diarrhea, prolonged vomiting, pH imbalance, and secretion of excess aldosterone from the adrenal cortex. Medical treatment with certain diuretics is another possible cause of hypokalemia. These drugs used to treat hypertension and edema can cause loss of potassium along with increased urination. Diuretics that work indirectly by blocking aldosterone can avoid this adverse side effect and are thus described as "potassium-sparing." Hypokalemia can also result from intense physical exertion during athletic events or low potassium in the diet. Foods that contain high levels of potassium include oranges, bananas, squashes, melons, and dried fruit.

Hyperkalemia refers to excess potassium in body fluids, which can be caused by kidney disease, acidosis, hemolysis, or severe bleeding. Its symptoms include nausea, muscle weakness, a weak or irregular pulse, and heart symptoms such as slowing or sudden stopping of the heart and cardiac fibrillation. This condition can be treated with sodium bicarbonate to correct acidosis, calcium for muscle and heart function, and diuretics that promote potassium loss.

21

Table 21-1 Hormonal Regulation of Fluid Homeostasis

Hormone	Stimulus for Secretion	Action
Aldosterone	Low blood pressure or elevated blood potassium concentration	Increased sodium and water retention; increased blood pressure; increased potassium loss
Antidiuretic hormone (ADH)	Elevated blood sodium concentration	Increased water retention
Angiotensin II	Low blood pressure	Thirst, increased aldosterone and ADH secretion
Atrial natriuretic peptide (ANP)	High blood pressure in the atrium	Increased excretion of sodium and water; decreased blood pressure

sodium homeostasis. On the other hand, a decrease in body fluid volume stimulates water intake and also reduces sodium loss to prevent excessive dilution of body fluids.

Thirst and urine output are controlled by several different hormones. Chapter 12 introduced these hormones, and Chapter 22 discusses the kidney processes these hormones regulate. In brief (**Table 21-1**):

- High blood potassium and low blood pressure stimulate the production of aldosterone from the adrenal cortex (**Fig. 21-4**). Aldosterone promotes the reabsorption of sodium (and thus, water) and the elimination of potassium in the kidney. In Addison disease, in which the adrenal cortex does not produce enough aldosterone, there is a loss of sodium and water and excess potassium retention (hyperkalemia). Blood pressure can fall dangerously low, and the excess potassium can disrupt the heart rhythm.
- When the blood concentration of sodium rises above the normal range, the pituitary secretes more antidiuretic hormone (ADH). This hormone increases water reabsorption in the kidney to dilute the excess sodium. ADH deficiency is called diabetes insipidus because of the large amount of dilute urine produced.
- Low blood pressure also stimulates the production of angiotensin II (ATII), a hormone discussed in greater detail in Chapter 22. ATII stimulates thirst as well as the release of both aldosterone and ADH.
- Atrial natriuretic peptide (ANP) is secreted by specialized atrial myocardial cells when blood pressure rises too high. ANP causes the kidneys to excrete sodium and water, thus decreasing blood volume and lowering blood pressure. The name comes from *natrium*, the Latin name for sodium, and the adjective for *uresis*, which refers to urination.

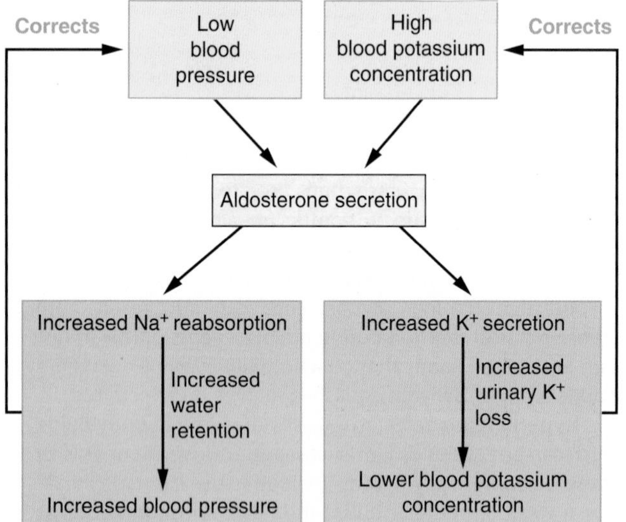

Figure 21-4 Aldosterone. **KEY POINT** Aldosterone increases water retention and potassium loss. **ZOOMING IN** How would the overconsumption of potassium-rich foods like bananas alter aldosterone secretion?

CHECKPOINTS

- ☐ 21-4 What are four routes for water loss from the body?
- ☐ 21-5 What hormone from the adrenal cortex promotes sodium reabsorption in the kidney?
- ☐ 21-6 What pituitary hormone increases water reabsorption in the kidney?

Acid-Base Balance

The **pH scale** is a measure of how acidic or basic (alkaline) a solution is. As described in Chapter 2, the pH scale measures the hydrogen ion (H^+) concentration in a solution. Body fluids are slightly alkaline, with a pH range of 7.35 to 7.45. These fluids must be kept within a narrow pH range, or damage, even death, will result. A shift in either direction by three-tenths of a point on the pH scale, to 7.0 or 7.7, is fatal.

REGULATION OF pH

Recall from Chapter 18 that carbon dioxide, generated by cellular respiration, dissolves in the blood and yields carbonic acid. Carbonic acid lowers blood pH. This change usually appears exclusively in venous blood, because healthy lungs exhale as much CO_2 as the body produces.

However, any impairment of lung function can cause CO_2 to accumulate in arterial blood and thus lower blood pH.

The body constantly produces acids in the course of metabolism that cannot be eliminated in exhaled air. Catabolism of fats yields fatty acids and ketones, and anaerobic metabolism during intense exercise generates hydrogen ions. Moreover, significant amounts of base are lost in the feces in the form of bicarbonate ions. Normal metabolic activities thus tend to reduce blood pH. Conversely, a few abnormal conditions (discussed later) may cause alkaline shifts in pH. Several systems act together to counteract these changes and maintain acid–base balance (**Fig. 21-5**):

- **Buffer systems.** Buffers are substances that prevent sharp changes in hydrogen ion (H^+) concentration and thus maintain a relatively constant pH. Buffers work by accepting or releasing these ions as needed to keep the pH steady. The main buffer systems in the body are bicarbonate buffers, phosphate buffers, and proteins, such as hemoglobin in red blood cells and plasma proteins.
- **Respiration.** The respiratory system can compensate for increased metabolic acid production, because faster, deeper breathing eliminates more carbon dioxide and makes the blood more alkaline. Conversely, abnormal loss of metabolic acids (say, from excess vomiting) reduces ventilation, so carbon dioxide is retained, and blood becomes a bit more acidic. However, the respiratory system uses up buffers as it compensates for pH disturbances, so its actions are temporary.
- **Kidney function.** The kidneys regulate pH by reabsorbing or eliminating hydrogen ions as needed and creating new buffer molecules. The kidneys are responsible for long-term pH regulation. The activity of the kidneys is described in Chapter 22.

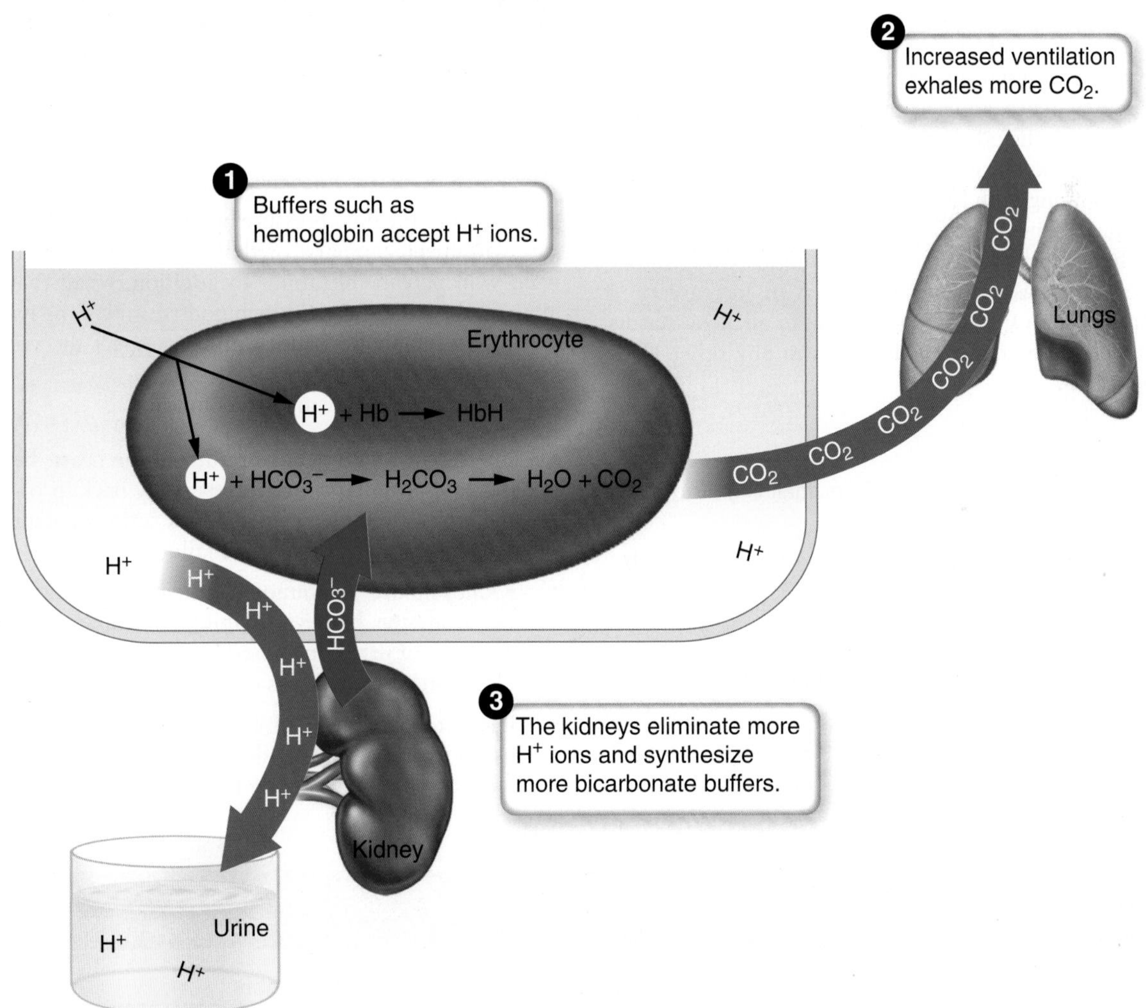

2 Increased ventilation exhales more CO_2.
1 Buffers such as hemoglobin accept H^+ ions.
H^+
Erythrocyte
H^+ + Hb → HbH
H^+ + HCO_3^- → H_2CO_3 → H_2O + CO_2
CO_2
Lungs
HCO_3^-
3 The kidneys eliminate more H^+ ions and synthesize more bicarbonate buffers.
Kidney
Urine

Figure 21-5 **pH Regulation.** This figure illustrates the response to acids that are ingested or generated by metabolic processes. **KEY POINT** The body compensates for these acids using buffers and increased ventilation, but renal actions are required to eliminate the excess acids. **ZOOMING IN** Which organ generates bicarbonate ions?

ABNORMAL pH

If shifts in pH cannot be controlled, either acidosis or alkalosis results (Table 21-2). **Acidosis** (as-ih-DO-sis) is a condition produced by a drop in the pH of body fluids to a pH of less than 7.35. Acidosis depresses the nervous system, leading to mental confusion and ultimately coma. It may result from a respiratory obstruction or any lung disease that prevents the release of CO_2. Acidosis may also arise from kidney failure or prolonged diarrhea, which drains the alkaline contents of the intestine. Intense, long-term exercise can result in exercise-induced acidosis.

Acidosis may also result from inadequate carbohydrate metabolism, as occurs in uncontrolled diabetes mellitus, ingestion of a low-carbohydrate diet, or starvation. In these cases, the body metabolizes too much fat and protein from food or body materials, leading to the production of excess acid. When acidosis results from the accumulation of ketone bodies, the condition is more accurately described as **ketoacidosis**.

Alkalosis (al-kah-LO-sis) is a condition that results from an increase in pH to greater than 7.45. This abnormality excites the nervous system to produce tingling sensations, muscle twitches, and eventually paralysis. The possible causes of alkalosis include hyperventilation (the release of too much carbon dioxide), ingestion of too much antacid, and prolonged vomiting with loss of stomach acids.

It is convenient to categorize acidosis and alkalosis as having either respiratory or metabolic origins. Respiratory acidosis or alkalosis results from either an increase or a decrease in blood CO_2. Metabolic acidosis or alkalosis results from unregulated increases or decreases in any other acids (see Table 21-2).

CHECKPOINTS

- 21-7 What are three mechanisms for maintaining the acid–base balance of body fluids?
- 21-8 What condition arises from abnormally low pH of body fluids? From high pH of body fluids?

Disorders of Body Fluids

Edema is the accumulation of excessive interstitial fluid (Fig. 21-6). Recall from Chapter 15 (Fig. 15-10) that fluid filters out of capillaries into the interstitial spaces. Anything that increases the fluid pressure or decreases the osmotic pressure within the capillary, or that compromises the capillary wall, will increase filtration. In addition, recall that excess interstitial fluid returns to the bloodstream via the lymphatic system, so lymphatic issues can also impact the volume of interstitial fluid.

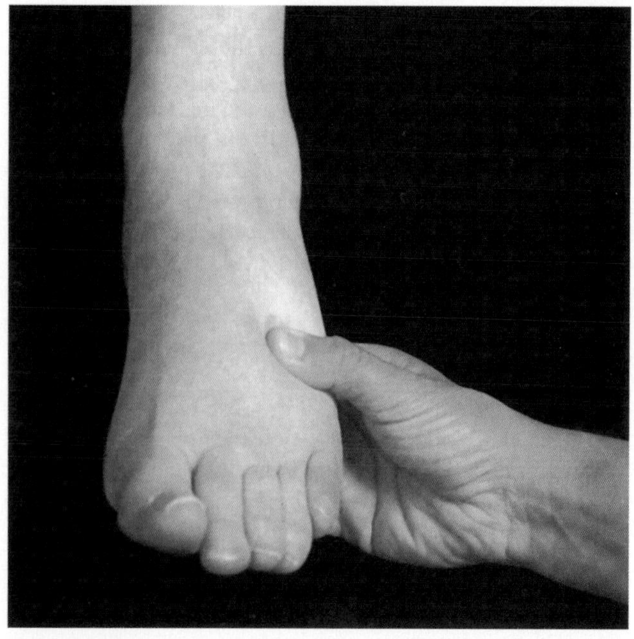

Figure 21-6 Edema of the foot. KEY POINT Finger pressure creates an indentation in edematous tissue.

Some causes of edema are as follows:

- Interference with venous return to the heart, because it increases capillary fluid pressure. A backup of fluid in the lungs, **pulmonary edema**, is a serious potential consequence of congestive heart failure.
- Kidney failure, a common clinical cause of edema, resulting from the kidneys' inability to eliminate adequate amounts of urine. Increased body fluid volume increases blood pressure and thus capillary fluid pressure.
- Lack of plasma protein, because it decreases the plasma osmotic pressure. This deficiency may result from protein loss or ingestion of too little dietary protein for

Table 21-2 Causes of Acidosis and Alkalosis

	Acidosis	Alkalosis
Metabolic	Kidney failure; intense, prolonged exercise; lack of carbohydrate metabolism, as in uncontrolled diabetes mellitus, starvation; prolonged diarrhea	Overuse of antacids; prolonged vomiting
Respiratory	Respiratory obstruction; lung disease such as asthma or emphysema; apnea or decreased ventilation	Hyperventilation (overbreathing due to anxiety or oxygen deficiency)

an extended period. It may also result from failure of the liver to manufacture adequate amounts of the protein albumin, as frequently occurs in liver disease. The decrease in protein lowers the blood's osmotic pressure and reduces fluid return to the circulation. Diminished fluid return results in accumulation of fluid in the tissues.

- Increased permeability of the capillary wall, as caused by inflammation, injury, or allergic reactions.
- Blockage in the lymphatic system, which reduces the return of excess interstitial fluid to the circulation.

Hyponatremia, sometimes known as water intoxication, describes a reduced blood sodium concentration. It is usually caused by excess dilution of extracellular body fluids, rendering them hypotonic. (While all electrolytes get more diluted, sodium is the most important solute in water balance.) Recall from Chapter 3 that water passes freely across cell membranes by osmosis, so water moves from hypotonic solutions into body cells, resulting in swelling. In the brain, cellular swelling may lead to convulsions, coma, and finally death. Causes of hyponatremia include an excess of ADH and intake of excess dilute fluids by intravenous injection or by mouth, as illustrated in Ethan's opening case study.

Effusion (e-FU-zhun) is the escape of fluid into a cavity or a space. Pleural effusion is the accumulation of fluid within the pleural space, resulting from tuberculosis, cancer, or infection. A pleural effusion makes it difficult to expand the lung, so it hampers inhalation. Effusion into the pericardial sac, which encloses the heart, may occur in certain infections and in autoimmune disorders, such as lupus erythematosus and rheumatoid arthritis. The pericardial effusion may interfere with normal heart contractions and can cause death. Effusion into the abdominal cavity causes **ascites** (ah-SI-teze), which may occur in disorders of the liver, kidneys, and heart, as well as in cancers, infection, or malnutrition. Ascites in cases of kwashiorkor, a form of malnutrition, was illustrated in Chapter 20 (**see Fig. 20-5**).

Dehydration (de-hi-DRA-shun), a severe deficit of body fluids, will result in death if it is prolonged. The causes include vomiting, diarrhea, drainage from burns or wounds, excessive perspiration, and inadequate fluid intake, as may be caused by interference with the thirst mechanism. In cases of dehydration, it may be necessary to administer intravenous fluids to correct fluid and electrolyte imbalances. **Figure 21-7** shows a clinician testing for dehydration in a child. When there is a lack of tissue fluid, a pinch of skin will remain elevated instead of returning immediately to normal position.

CHECKPOINTS

- ☐ 21-9 What is the term for accumulation of excess interstitial fluid?
- ☐ 21-10 What term describes low blood sodium concentration?
- ☐ 21-11 What is the term for effusion of fluid into the abdominal cavity?

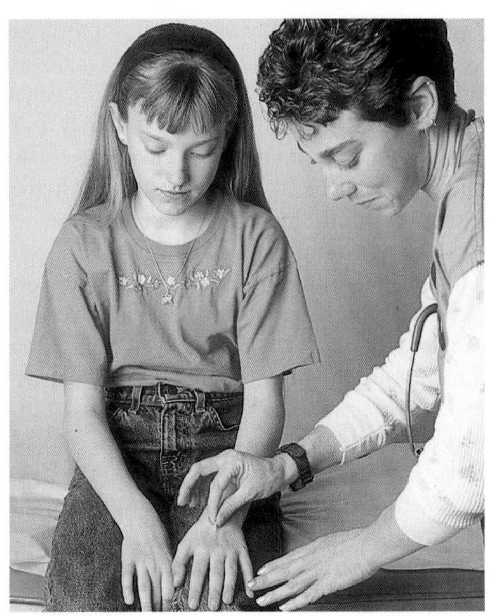

Figure 21-7 **Simple test for dehydration.** **KEY POINT** A pinch of skin does not return immediately to place if the tissue lacks fluid.

Fluid Therapy

Chapter 3 discussed the principles concerning movement of water into and out of cells when they are placed in different solutions. Recall that an isotonic solution has the same solute concentration as the cellular fluids and will not cause a net loss or gain of water. A hypertonic solution is more concentrated than cellular fluid and will draw water out of the cells. A hypotonic solution is less concentrated than the cellular fluids; a cell will take in water when placed in this type of solution. These rules must be considered when fluid is administered.

Fluids are administered into a vein under a wide variety of conditions to help maintain blood volume and normal body functions when natural intake is not possible. Fluids are also administered to provide nourishment, adjust pH of body fluids, or correct specific fluid and electrolyte imbalances caused by surgery, injury, or disease.

See the Student Resources on thePoint for information on emergency medical technicians who often must administer fluids in providing health care.

The first fluid administered intravenously in emergencies involving significant blood loss is normal saline, which contains 0.9% sodium chloride, an amount equal to the solute concentration of plasma. Because it is isotonic, this type of solution does not change the ion distribution in the body fluid compartments.

Some intravenous fluids contain dextrose (glucose) to provide nourishment. One of these contains 5% dextrose in 0.45% (1/2 normal) saline. This solution is hypertonic when infused but becomes hypotonic after the sugar is metabolized.

21

Another is 5% dextrose in water. This solution is slightly hypotonic when infused. The amount of sugar contained in a liter of this fluid is equal to 170 calories. The sugar is soon used up, resulting in a fluid that is effectively pure water. Use of these hypotonic fluids is not advisable for long-term therapy because of the potential for hyponatremia. Small amounts of potassium chloride are often added to replace electrolytes lost by vomiting or diarrhea.

Lactated Ringer solution contains sodium, potassium, calcium, chloride, and lactate. Its electrolyte concentrations are approximately equal to those of normal plasma. The lactate is metabolized to bicarbonate, which acts as a buffer and helps counteract acidosis. This fluid is used in cases of burns, to correct fluid loss due to diarrhea, and to administer medications.

In 25% serum albumin, the concentration of the plasma protein albumin is five times normal. This hypertonic solution draws fluid from the interstitial spaces into the circulation. It may be administered when plasma proteins are low, as occurs when the liver fails to produce adequate albumin. This solution is also used to correct edema because it draws fluids from the tissues into the blood.

Fluids containing varied concentrations of dextrose, sodium chloride, potassium, and other electrolytes and substances are manufactured. These fluids are used to correct specific imbalances. Nutritional solutions containing concentrated sugar, protein, and fat are available for administration when oral intake is not possible for an extended period.

CHECKPOINTS 

- [ ] **21-12** What is the percentage of sodium chloride in normal saline?
- [ ] **21-13** What is another name for dextrose?

Disease in Context Revisited

Ethan Overdid the Hydration

The next day, Leilani was heading down to the bay to go paddle boarding when she ran into Darren.

"Hey Darren, what ever happened to the late finisher, Ethan?" she asked.

"Just as I thought, he had exercise-associated hyponatremia," Darren replied. "Blood tests showed that his blood sodium level was only 129 mmol/L, and it should be between 135 and 145. We treated him with sodium tablets and a diuretic, and he was OK after about eight hours."

"Do you know what caused this?" Leilani asked. "I know athletes have to guard against dehydration, but I never knew that overhydration could be a problem."

Darren smiled, "Ethan's coach back home told him to drink as much and as often as he could, and he followed the advice too enthusiastically. When he started to feel bad, he assumed that he was dehydrated, and so he drank more water. Most people respond to the increased water intake by urinating more, and they don't get sick. But for reasons we don't yet understand, some people who overdrink retain too much water and dangerously dilute their body fluids. We estimate that Ethan must have gained about 8 lb of water weight during the race, showing that he is susceptible to EAH and will have to really watch his hydration in future races. I'm not saying that dehydration isn't dangerous—it is—but people have to trust their sense of thirst more and let their body signals regulate their fluid intake. We can't know in advance who will develop EAH, so we advise all of our triathletes to drink no more than two cups of liquid per hour. We weigh them before they start the run portion, and if they haven't lost some weight we ask them to stop drinking for a while."

CHAPTER 21

Chapter Wrap-Up

Summary Overview

A detailed chapter outline with space for note taking is on *thePoint*. The figure below illustrates the main topics covered in this chapter.

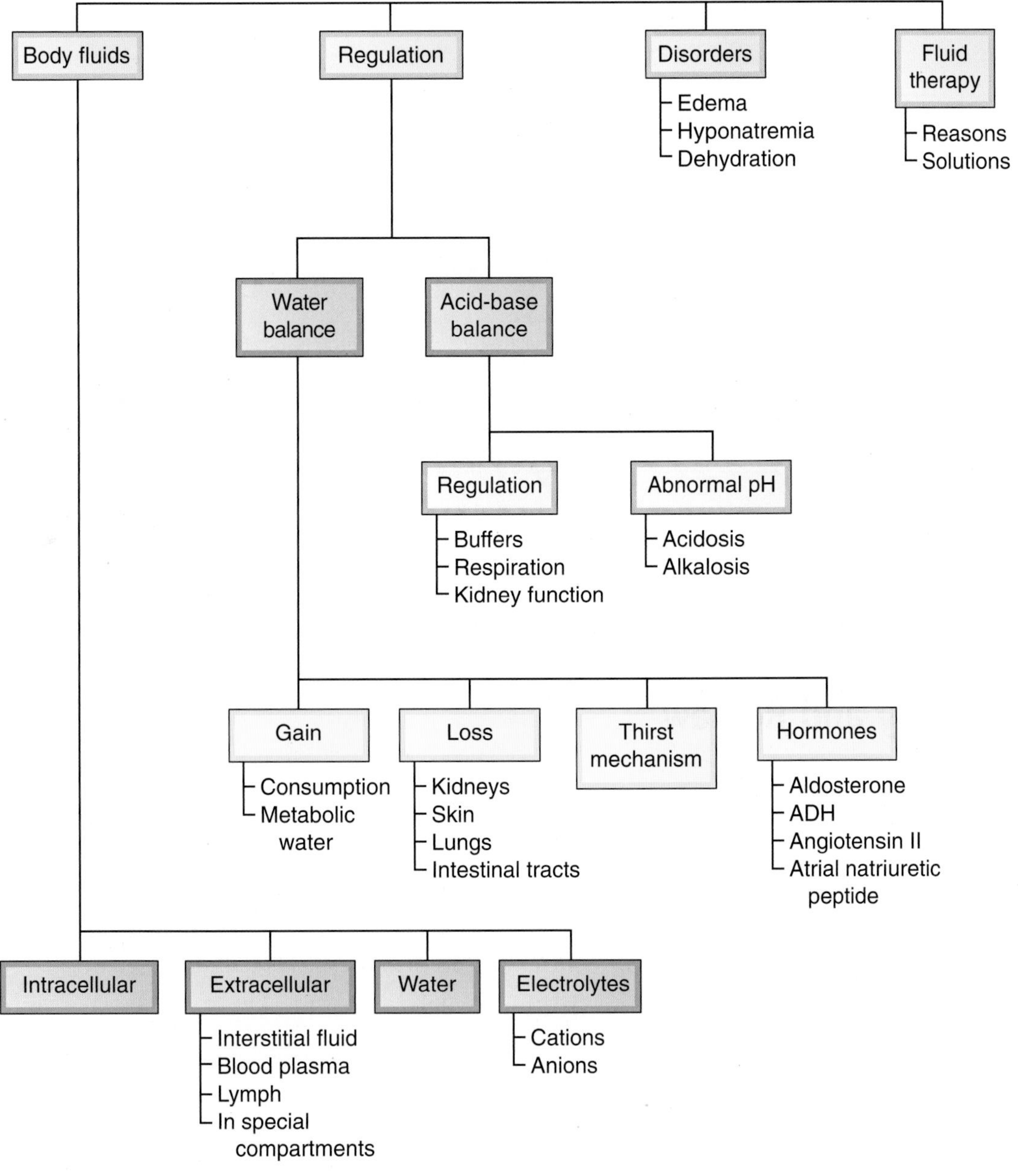

Key Terms

The terms listed below are emphasized in this chapter. Knowing them helps you organize and prioritize your learning. These and other boldface terms are defined in the Glossary with phonetic pronunciations.

acidosis
alkalosis
anion
ascites
cation
dehydration
edema
effusion
extracellular
interstitial
intracellular
ketoacidosis
pH scale
polydipsia

Word Anatomy

Medical terms are built from standardized word parts (prefixes, roots, and suffixes). Learning the meanings of these parts can help you remember words and interpret unfamiliar terms.

WORD PART	MEANING	EXAMPLE
Fluid Compartments		
extra-	outside of, beyond	*Extracellular* fluid is outside the cells.
intra-	within	*Intracellular* fluid is within a cell.
semi-	partial, half	A *semipermeable* membrane is partially permeable.
Water Balance		
osmo-	osmosis	*Osmoreceptors* detect changes in osmotic concentration of fluids.
poly-	many	*Polydipsia* is excessive thirst.
Acid–Base Balance		
o/sis	condition, process	*Acidosis* is a condition produced by a drop in the pH of body fluids.
Disorders of Body Fluids		
hydr/o	water	*Dehydration* is a severe deficit of body fluids.
tox/o	poison	Water *intoxication* is dilution of body fluids by excess water.

Questions for Study and Review

BUILDING UNDERSTANDING

Fill in the Blanks

1. Fluid located between the cells in tissues is called ______.
2. The body's hydrogen ion concentration is regulated by substances in the blood called ______.
3. Effusion with accumulation of fluid within the abdominal cavity is termed ______.
4. A severe deficit in body fluid is called ______.
5. Excessive thirst is termed ______.

Matching > Match each numbered item with the most closely related lettered item:

___ **6.** Essential for maintaining osmotic balance and body fluid volume; this cation is abundant in extracellular fluid

___ **7.** Important in the transmission of nerve impulses and enzyme activities; this cation is abundant in intracellular fluid

___ **8.** Required for bone formation, muscle contraction, and blood clotting

___ **9.** Essential in bone formation and acid–base balance; this anion is found in plasma membranes, ATP, and nucleic acids

___ **10.** Important for gastric acid formation; this anion is abundant in extracellular fluid

a. calcium

b. chloride

c. phosphate

d. potassium

e. sodium

Multiple Choice

___ **11.** Who has the greatest body water content?

a. infants
b. children
c. young adults
d. elderly adults

___ **12.** Which structure is responsible for evaporative water loss?

a. intestinal tract
b. kidney
c. lung
d. skin

___ **13.** Which of the following is an important buffer in body fluids?

a. chloride
b. bicarbonate
c. metabolic water
d. aldosterone

___ **14.** Which system is responsible for long-term pH regulation?

a. buffer system
b. digestive system
c. respiratory system
d. urinary system

___ **15.** What does increased arterial blood CO_2 cause?

a. metabolic acidosis
b. metabolic alkalosis
c. respiratory acidosis
d. respiratory alkalosis

UNDERSTANDING CONCEPTS

16. Compare the terms in each of the following pairs:

a. anions and cations
b. intracellular and extracellular fluid
c. aldosterone and antidiuretic hormone
d. angiotensin II (ATII) and atrial natriuretic peptide (ANP)

17. In a healthy person, what is the ratio of fluid intake to output?

18. Explain the role of the hypothalamus in water balance.

19. What are some metabolic sources of acids in body fluids?

20. How do the respiratory and urinary systems regulate pH?

21. Compare and contrast the following disorders:

a. acidosis and alkalosis
b. edema and effusion
c. hyponatremia and dehydration

22. List some causes of edema.

23. List three purposes for administering intravenous fluids.

24. Compare and contrast the following types of intravenous fluids:

a. normal saline and 5% dextrose saline
b. Ringer solution and 25% serum albumin

CONCEPTUAL THINKING

25. Patty Grant, 55 years old, reports severe headaches and excessive thirst and urination. Her blood glucose is normal. What is the probable cause of Patty's symptoms?

26. Why is emphysema associated with decreased urine pH?

27. In the opening case study, Ethan replaced body fluids lost in the Ironman competition with pure water. What physiologic changes resulted, and how did they relate to his symptoms?

For more questions, see the Learning Activities on thePoint.

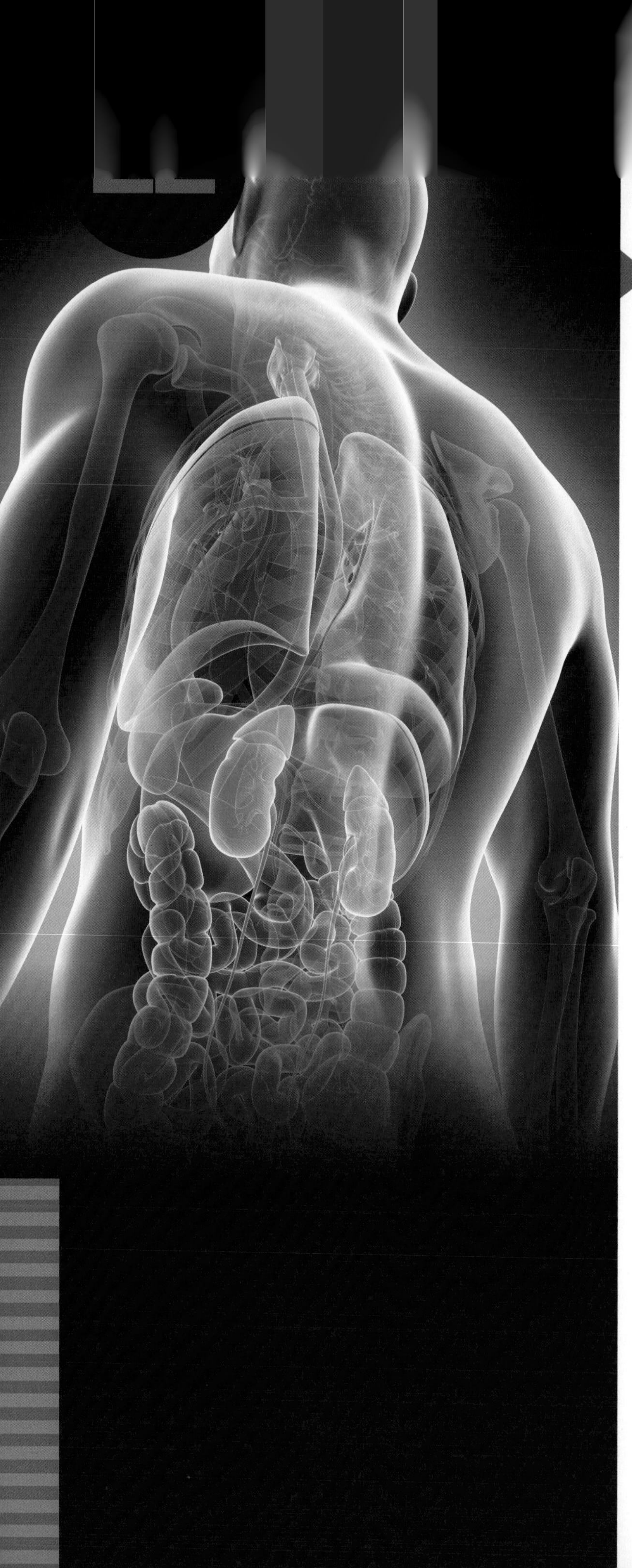

Learning Objectives

After careful study of this chapter, you should be able to:

1 ▶ Describe the organs of the urinary system, and give the functions of each. ***p. 498***

2 ▶ List four systems that eliminate waste, and name the substances each eliminates. ***p. 498***

3 ▶ List four activities of the kidneys in maintaining homeostasis. ***p. 499***

4 ▶ Describe the location and internal organization of the kidneys. ***p. 499***

5 ▶ Describe a nephron. ***p. 500***

6 ▶ Trace the path of a drop of blood as it flows through the kidney. ***p. 500***

7 ▶ Name the four processes involved in urine formation, and describe what happens during each. ***p. 500***

8 ▶ Explain the roles of juxtamedullary nephrons and antidiuretic hormone (ADH) in urine formation. ***p. 503***

9 ▶ Describe the components and functions of the juxtaglomerular (JG) apparatus. ***p. 505***

10 ▶ Describe the process of micturition. ***p. 507***

11 ▶ List normal and abnormal constituents of urine. ***p. 507***

12 ▶ Discuss six types of urinary system disorders. ***p. 508***

13 ▶ List six signs of chronic renal failure. ***p. 512***

14 ▶ Explain the principle and purpose of renal dialysis. ***p. 512***

15 ▶ Referring to the case study, describe how urethral blockage can affect kidney function. ***pp. 497, 515***

16 ▶ Show how word parts are used to build words related to the urinary system (see Word Anatomy at the end of the chapter). ***p. 517***

Disease in Context *Eric's Urinary Blockade*

Eric rolled out of bed and glanced at the clock as he rushed to the bathroom. *It's only been a couple of hours since the last time I had to pee, he thought. Hopefully this time it won't take as long.* Sure enough, Eric had difficulty voiding. Even once he was able to get started, he only managed to produce a small volume of urine. Lately, this was happening more and more frequently. At first, he chalked it up to getting older, but now, he wondered if he had some sort of bladder or kidney infection. As he climbed back into bed, Eric decided that he would make an appointment with his family doctor.

"So let me get this straight," said Dr. Michaels. "Over the last several weeks, you've experienced increased frequency and urgency of urination, even at night. While urinating, you've had hesitation in starting, decreased volume, and diminished force of the stream. And even after urination, you still feel that your urinary bladder is not completely empty."

Based on the symptoms, Dr. Michaels suspected that Eric's prostate gland (a male reproductive organ) was causing problems for his urinary system. The prostate gland lies immediately inferior to the urinary bladder, where it surrounds the first part of the urethra. If the gland becomes enlarged, it can obstruct the urethra and prevent the urinary bladder from emptying completely.

Dr. Michaels's suspicions were confirmed by the digital rectal exam. Eric's prostate was large and rubbery. The doctor carefully palpated the prostate's surface with his finger—he could not detect any nodules on its smooth surface.

"Eric, my initial diagnosis is that you have benign prostatic hypertrophy—your prostate has grown larger and is preventing urine from passing out of your urinary bladder," the doctor explained. "Your prostate's surface is smooth though, which suggests that the growth is not cancerous, but we'll have to get that ruled out. I'm going to order some blood tests and urinalyses, as well as refer you to a urologist."

Eric's prostate is causing problems for his urinary system. In this chapter, we will examine the anatomy and physiology of the urinary system. Later, we will learn how Eric's disorder is resolved.

ANCILLARIES *At-A-Glance*

Visit thePoint to access the following resources. For guidance in using these resources most effectively, see pp. xv–xvii.

Learning RESOURCES

- Tips for Effective Studying
- Web Figure: Acute Pyelonephritis
- Web Figure: Hydronephrosis
- Web Figure: Adult Polycystic Disease
- Web Figure: Urinary Obstruction, Reflux, and Infection
- Web Chart: Kidney Regulation of Blood Pressure
- Animation: Renal Function
- Health Professions: Hemodialysis Technician
- Detailed Chapter Outline
- Answers to Questions for Study and Review
- Audio Pronunciation Glossary

Learning ACTIVITIES

- Pre-Quiz
- Visual Activities
- Kinesthetic Activities
- Auditory Activities

A LOOK BACK

Study of body fluid volume and composition continues with this chapter on the urinary system. Most water lost in a day is through the urine produced by the kidneys. These organs not only help regulate total fluid volume but also the electrolyte and acid–base balance of body fluids, as noted in Chapter 21. Fluid volume is an important factor in circulation and blood pressure, first discussed in Chapter 15. This chapter also discusses the process of filtration in blood capillaries. This same process initiates urine formation in the kidney.

Systems Involved in Excretion

The urinary system is also called the *excretory system* because one of its main functions is **excretion**, removal and elimination of unwanted substances from the blood. The urinary system also regulates the volume, acid–base balance (pH), and electrolyte composition of body fluids.

The main parts of the urinary system, shown in **Figure 22-1**, are as follows:

- Two **kidneys**. These organs extract wastes from the blood, balance body fluids, and form urine.
- Two **ureters** (U-re-ters). These tubes conduct urine from the kidneys to the urinary bladder.
- A single **urinary bladder**. This reservoir receives and stores the urine brought to it by the two ureters.
- A single **urethra** (u-RE-thrah). This tube conducts urine from the bladder to the outside of the body for elimination.

Although the focus of this chapter is the urinary system, some other systems are mentioned here as well, because they also function in excretion. These systems and some of the substances they eliminate are the following:

- The *digestive system* eliminates water, some salts, and bile, in addition to digestive residue, all of which are contained in the feces. The liver is important in eliminating the products of red blood cell destruction and in breaking down certain drugs and toxins.
- The *respiratory system* eliminates carbon dioxide and small amounts of water. The latter appears as vapor, as can be demonstrated by breathing on a windowpane or a mirror, where the water condenses.
- The skin, or *integumentary system*, excretes water, salts, and very small quantities of nitrogenous wastes. These all appear in perspiration, although water also evaporates continuously from the skin without our being conscious of it.

CHECKPOINTS

- ☐ 22-1 What are the organs of the urinary system?
- ☐ 22-2 What are three systems other than the urinary system that eliminate waste?

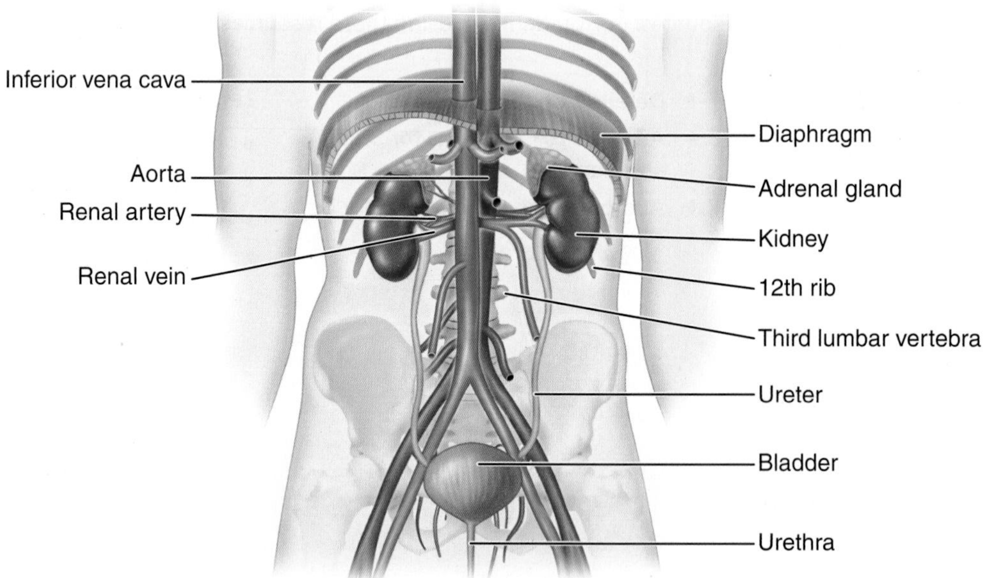

Figure 22-1 **Urinary system.** The urinary system consists of the kidneys, ureters, urinary bladder, and urethra.
ZOOMING IN Identify the structures that carry urine to and from the bladder.

The Kidneys

The kidneys interact with other systems, primarily the cardiovascular system, as prime regulators of homeostasis. After a brief overview of the kidney's many activities, we will describe its structure and specific roles in urine formation and regulation of blood pressure.

KIDNEY ACTIVITIES

The kidneys are involved in the following processes:

- Excretion of unwanted substances, such as cellular metabolic waste and toxins. One product of amino acid metabolism is nitrogen-containing waste material, chiefly **urea** (u-RE-ah). After synthesis in the liver, urea is transported in the blood to the kidneys for elimination. The kidneys have a specialized mechanism for the elimination of urea and other nitrogenous (ni-TROJ-en-us) wastes.
- Homeostasis of body fluids. As discussed in Chapter 21, the kidneys control the sodium and water content of body fluids by altering the composition and volume of **urine**. The kidneys also control blood pH by excreting variable amounts of acids and bases and by synthesizing buffers.
- Blood pressure regulation. Chapter 21 explained how the kidneys control blood pressure by altering fluid balance. They also act more directly via their participation in the renin–angiotensin system, as discussed later in this chapter.
- Hormone production. The kidneys produce erythropoietin (EPO), which stimulates red cell production in the bone marrow (see Chapter 13). They also activate vitamin D, which is necessary for bone health (see Chapter 12). Individuals with chronic kidney failure may not make enough of these two substances and thus require supplementation to prevent anemia (inadequate red blood cells) and osteomalacia (weak bones).

KIDNEY STRUCTURE

The kidneys lie against the back muscles in the upper abdomen at about the level of the last thoracic and first three lumbar vertebrae. The right kidney is slightly lower than the left to accommodate the liver (**see Fig. 22-1**). Each kidney is firmly enclosed in a membranous **renal capsule** made of fibrous connective tissue (**Fig. 22-2**). In addition, there is a protective layer

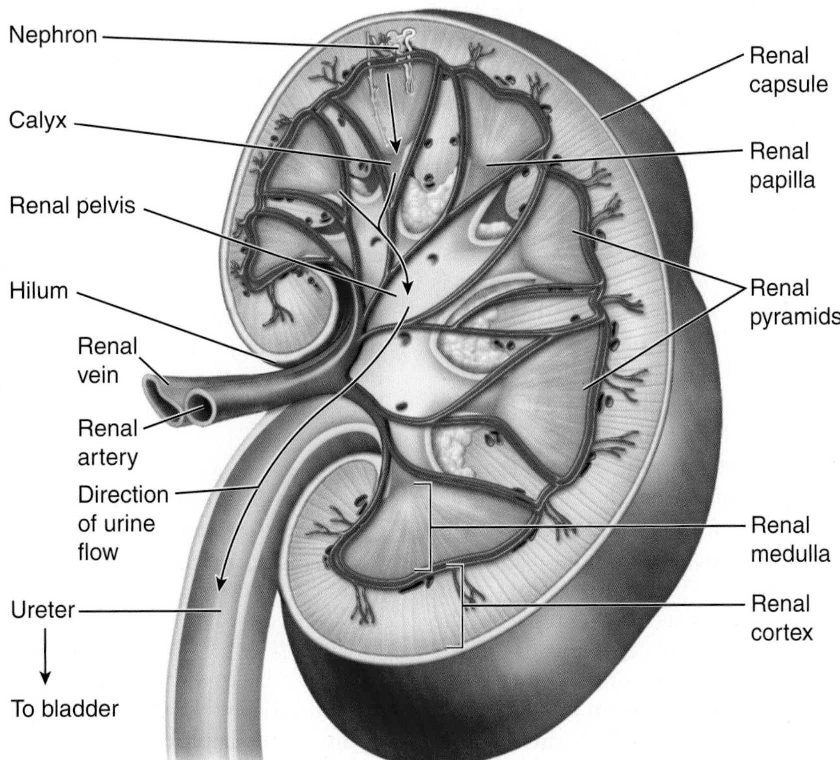

Figure 22-2 Kidney structure and the renal blood supply. A longitudinal section through the kidney shows its internal structure. **KEY POINT** Urine made in the nephrons drains into the renal pelvis and then the ureter. The renal artery branches to all regions of the kidney. Venous drainage combines to flow into the renal vein. **ZOOMING IN** What is the outer region of the kidney called? What is the inner region of the kidney called?

of fat called the *adipose capsule* around the organ that is not illustrated. An outermost layer of fascia (connective tissue) anchors the kidney to the peritoneum and abdominal wall. The kidneys, as well as the ureters, lie posterior to the peritoneum (see Chapter 19). Thus, they are not in the peritoneal cavity but rather in an area known as the **retroperitoneal** (ret-ro-per-ih-to-NE-al) **space**.

Blood is brought to the kidney by a short branch of the abdominal aorta called the **renal artery**. Blood leaves the kidney via the **renal vein**, which carries blood into the inferior vena cava for return to the heart. See **Figure A5-8** in Appendix 5 for a dissection photograph showing the kidneys and renal vessels in place.

Organization The kidney is a somewhat flattened organ approximately 10 cm (4 in) long, 5 cm (2 in) wide, and 2.5 cm (1 in) thick (**see Fig. 22-2**). On the medial border, there is a notch called the **hilum**, where the renal artery, the renal vein, and the ureter connect with the kidney. The lateral border is convex (curved outward), giving the entire organ a bean-shaped appearance.

The kidney is divided into two regions: the renal cortex and the renal medulla (**see Fig. 22-2**). The **renal cortex** is the kidney's outer portion. The internal **renal medulla** contains tubes in which urine is formed and collected. These tubes form a number of cone-shaped structures called **renal pyramids**. The tips of the pyramids point toward the **renal pelvis**, a funnel-shaped basin that forms the upper end of the ureter. Cuplike extensions of the renal pelvis surround the tips of the pyramids and collect urine; these extensions are called **calyces** (KA-lih-seze; singular, **calyx**, KA-liks). The urine that collects in the pelvis then passes down the ureters to the bladder.

The Nephron and its Blood Supply As is the case with most organs, the kidney's most fascinating aspect is too small to be seen with the naked eye. This basic unit, which actually creates urine, is the **nephron** (NEF-ron) (**Fig. 22-3A**). Each nephron begins with a hollow, cup-shaped bulb known as the **glomerular** (glo-MER-u-lar) **capsule** or *Bowman capsule*. This structure gets its name from the cluster of capillaries it surrounds, which is called the **glomerulus** (a word that comes from the Latin meaning "ball of yarn"). This combined unit of the glomerulus with its glomerular capsule is the nephron's filtering device. The remainder of the nephron is essentially a tiny coiled tube, the **renal tubule**, which consists of several parts. The portion leading from the glomerular capsule is called the **proximal tubule**. The renal tubule then untwists to form a hairpin-shaped segment called the **nephron loop**, or *loop of Henle*. The first part of the loop, which carries fluid toward the medulla, is the **descending limb**. The part that continues from the loop's turn and carries fluid away from the medulla is the **ascending limb**. Continuing from the ascending limb, the tubule folds into the **distal tubule**, so called because it is farther along the tubule from the glomerular capsule than is the proximal tubule. Many renal tubules empty into each collecting duct, which then continues through the medulla toward the renal pelvis.

In most cases, the glomerulus, glomerular capsule, and the proximal and distal tubules of the nephron are within the renal cortex. The nephron loop and collecting duct extend into the medulla (**see Fig. 22-3**, top left).

A small blood vessel, the **afferent arteriole**, supplies the glomerulus with blood; another small vessel, called the **efferent arteriole**, carries blood from the glomerulus (**Fig. 22-3B**). When blood leaves the glomerulus, it does not head immediately back toward the heart. Instead, it flows into a capillary network that surrounds the renal tubule. These **peritubular capillaries** are named for their location.

Each kidney contains about 1 million nephrons; if all these coiled tubes were separated, straightened out, and laid end to end, they would span some 120 km (75 miles)! **Figure 22-4** is a microscopic view of kidney tissue showing several glomeruli, each surrounded by a glomerular capsule. This figure also shows sections through renal tubules.

CHECKPOINTS 

- ☐ 22-3 Where is the retroperitoneal space?
- ☐ 22-4 What vessel supplies blood to the kidney, and what vessel drains blood from the kidney?
- ☐ 22-5 What is the name of the funnel-shaped collecting area that forms the upper end of the ureter?
- ☐ 22-6 What is the functional unit of the kidney called?
- ☐ 22-7 What name is given to the coil of capillaries in the glomerular capsule?

FORMATION OF URINE

The following explanation of urine formation describes a complex process, involving many back-and-forth exchanges between the bloodstream and the kidney tubules. As fluid filtered from the blood travels slowly through the nephron's twists and turns, there is ample time for exchanges to take place. These processes together allow the kidney to "fine-tune" body fluids as they adjust the urine's composition.

See the Student Resources on thePoint for the animation "Renal Function," which shows the process of urine formation in action.

Glomerular Filtration Chapter 15 explained how blood fluids cross the capillary wall and enter the interstitial fluid (IF) by the process of filtration. The same process occurs in the nephron, but the fluid enters the glomerular capsule instead of the IF. In the nephron, fluid must pass through the walls of both the capillary and the glomerular capsule. Gaps between the cells making up these walls permit the passage of water and dissolved materials; however, the capillary walls retain blood cells and large proteins, and these components remain in the blood (**Fig. 22-5**).

Because the diameter of the afferent arteriole is slightly larger than that of the efferent arteriole (**see Fig. 22-5**),

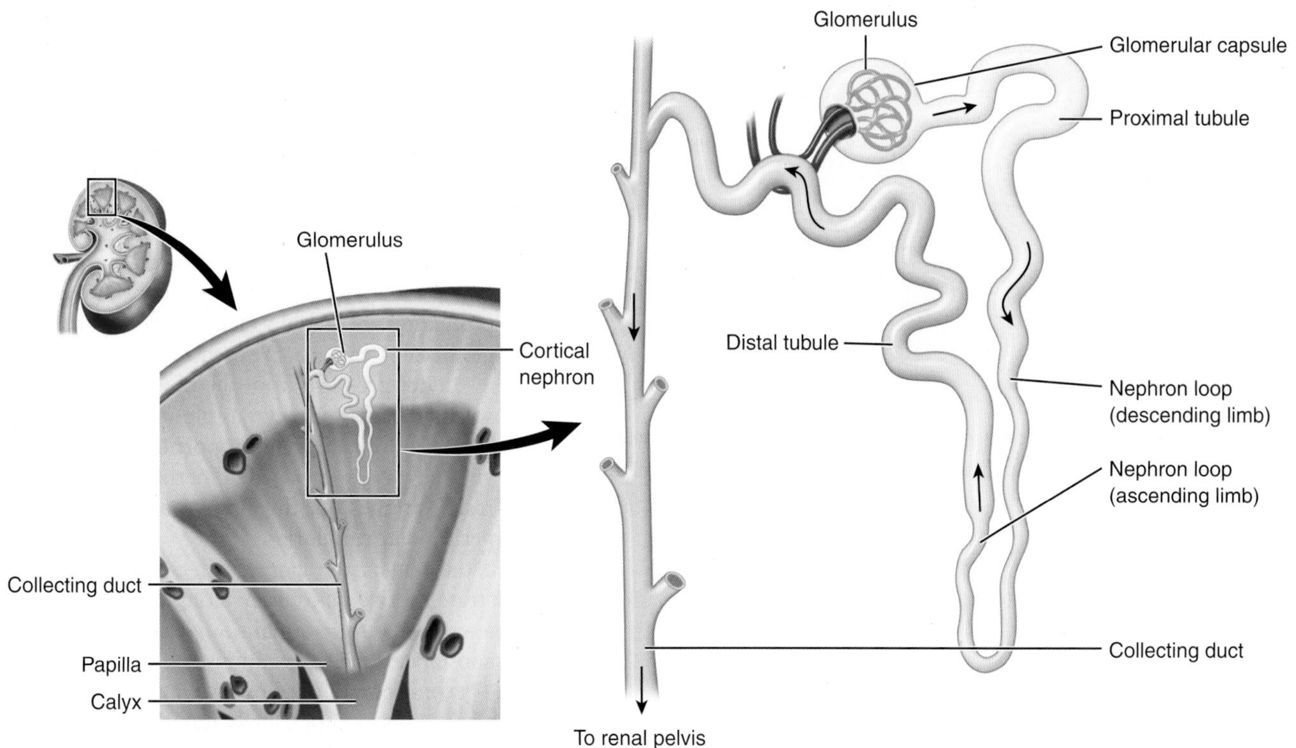

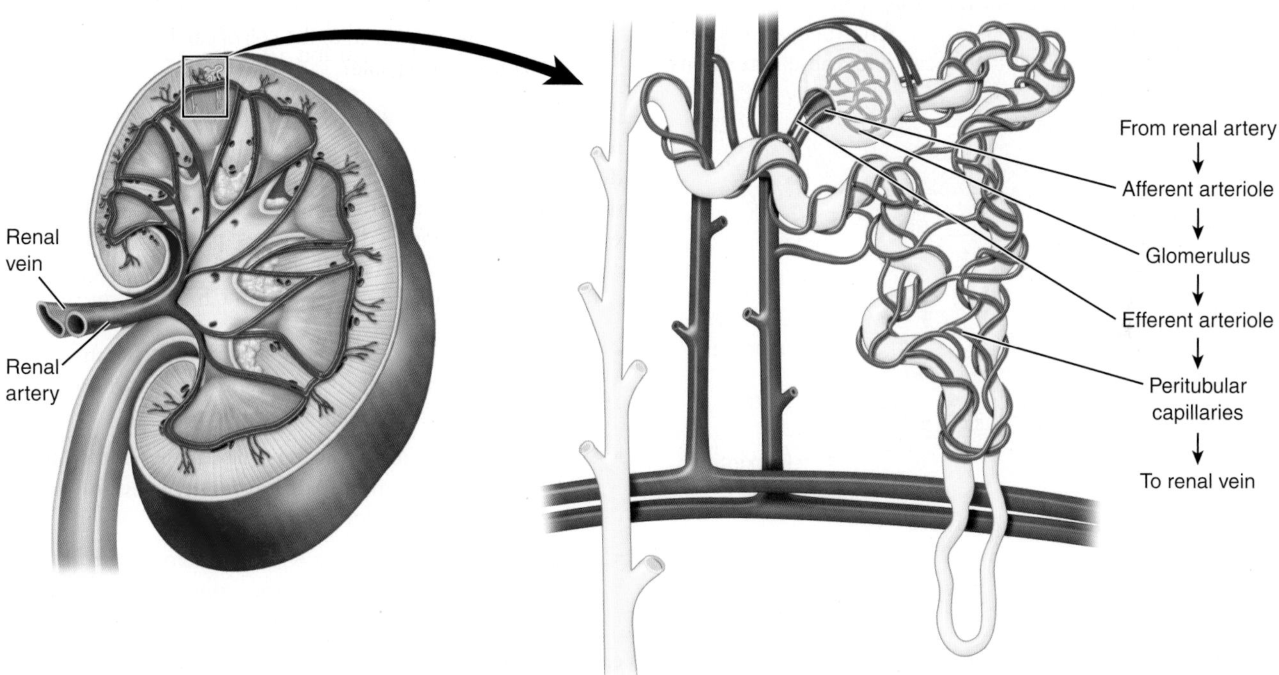

Figure 22-3 **A nephron and its blood supply.** **KEY POINT** The nephron regulates the proportions of urinary water, waste, and other materials according to the body's constantly changing needs. Materials that enter the nephron can be returned to the blood through the surrounding capillaries. **A.** A nephron with its blood vessels removed. **B.** A nephron and its associated blood vessels. **ZOOMING IN** The nephron is associated with two capillary beds. Which capillary bed receives blood first?

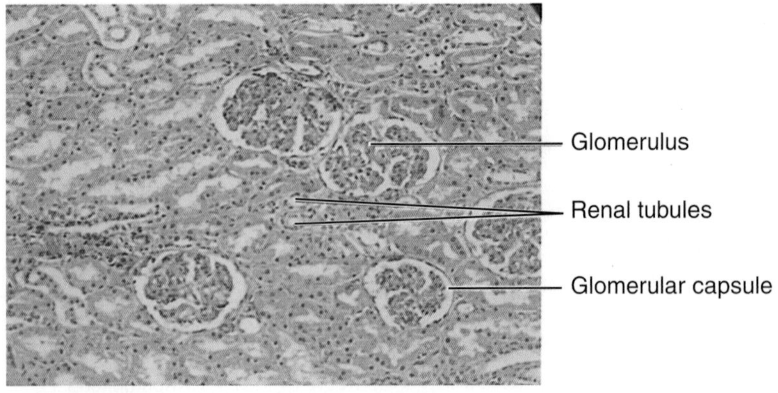

Figure 22-4 Microscopic view of the kidney. **KEY POINT** The glomeruli and glomerular capsules are visible along with cross-sections of the renal tubules.

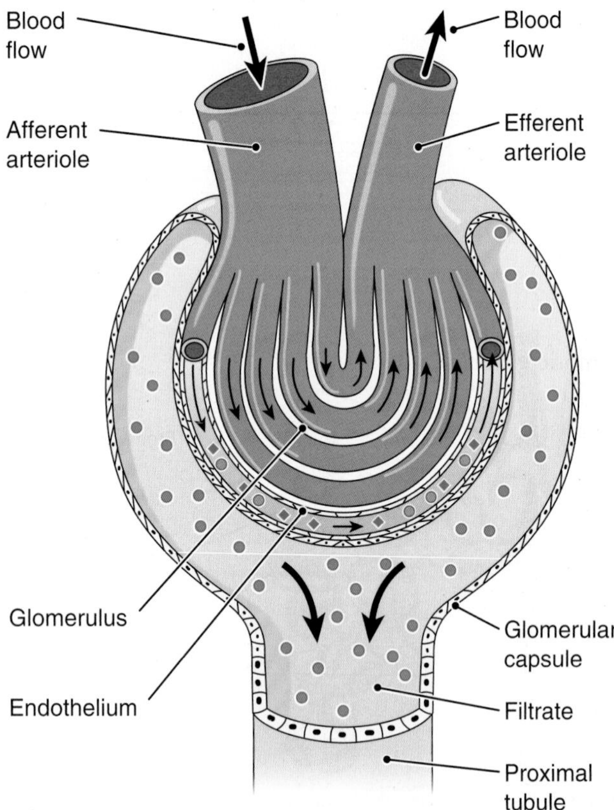

Figure 22-5 Glomerular filtration: the first step in urine formation. **KEY POINT** Blood pressure inside the glomerulus forces water and dissolved substances into the glomerular capsule. Blood cells and proteins remain behind in the blood. The smaller diameter of the efferent arteriole as compared with that of the afferent arteriole maintains high hydrostatic (fluid) pressure. In this simple illustration, only one capillary is cut away, but filtration occurs through all the vessels, and the glomerular capillaries are actually in a cluster, as shown in other figures.

blood can enter the glomerulus more easily than it can leave. Thus, blood pressure in the glomerulus is about three to four times higher than that in other capillaries. To understand this effect, think of placing your thumb over the end of a garden hose as water comes through. As you make the diameter of the opening smaller, water is forced out under higher pressure. As a result of increased fluid (hydrostatic) pressure in the glomerulus, materials are constantly being pushed out of the blood and into the nephron's glomerular capsule. This movement of materials under pressure from the blood into the capsule is therefore known as **glomerular filtration.** Note that only a small proportion of the blood's volume passes into the glomerular capsule. The rest continues out the efferent arteriole into the peritubular capillaries.

The fluid that enters the glomerular capsule, called the **glomerular filtrate**, begins its journey along the renal tubule. In addition to water and the normal soluble substances in the blood, other substances, such as vitamins and drugs, also may be filtered and become part of the glomerular filtrate. As noted, blood cells and proteins are not usually found in the filtrate.

Tubular Reabsorption The kidneys form about 160 to 180 L of filtrate each day. However, only 1 to 1.5 L of urine is eliminated daily. Clearly, most of the water that enters the nephron is not excreted with the urine but rather is returned to the circulation. In addition to water, many other substances the body needs, such as nutrients and ions, pass into the nephron as part of the filtrate, and these also must be returned. Therefore, the process of filtration that occurs in the glomerular capsule is followed by a process of **tubular reabsorption.** As the filtrate travels through the renal tubule, water and other needed substances pass through tubular cells and enter the interstitial fluid (IF). They move by several processes previously described in Chapter 3, including:

- Diffusion. The movement of substances from an area of higher concentration to an area of lower concentration (following the concentration gradient).
- Osmosis. Diffusion of water through a semipermeable membrane, from an area of low solute concentration to an area of higher solute concentration.
- Active transport. Movement of materials through the plasma membrane against the concentration gradient using energy and transporters.

Several hormones noted in Chapter 21, including aldosterone and atrial natriuretic peptide (ANP), affect tubular reabsorption in the distal portions of the tubule (Table 22-1).

Reabsorbed substances move from the nephron to the IF and then enter the peritubular capillaries and return to the circulation. Substances that are filtered but not reabsorbed remain in the filtrate and are excreted in the urine. For example, about 50% of the urea and uric acid (products of amino acid and nucleic acid breakdown,

Table 22-1 Substances that Affect Renal Function

Substance	Source	Action
Aldosterone (al-DOS-ter-one)	Hormone released from the adrenal cortex when blood pressure falls or blood potassium levels are too high	Promotes reabsorption of sodium and water in the kidney to conserve water and increase blood pressure
Atrial natriuretic peptide (na-tre-u-RET-ik) (ANP)	Atrial myocardial cells; released when blood pressure is too high	Causes kidney to excrete sodium and water to decrease blood volume and blood pressure
Antidiuretic hormone (an-te-di-u-RET-ik) (ADH)	Made in the hypothalamus and released from the posterior pituitary; released when blood becomes too concentrated or blood pressure falls	Promotes water reabsorption from the distal tubule and collecting duct to concentrate the urine and conserve water
Renin (RE-nin)	Enzyme produced by renal cells when blood pressure falls	Activates angiotensin in the blood
Angiotensin (an-je-o-TEN-sin)	Protein in the blood that is activated by renin	Causes constriction of blood vessels to raise blood pressure; increases fluid volume by stimulating thirst and promoting release of aldosterone and ADH

respectively) and all of the creatinine (a product of muscle cell breakdown) are kept within the tubule to be eliminated. Box 22-1 presents additional information on tubular reabsorption in the nephron.

Tubular Secretion The process of **tubular secretion** actively transports specific substances from the peritubular capillary into the nephron tubule. These substances were not filtered or were reabsorbed in a more proximal portion of the tubule. Potassium ions are moved into the urine in this manner. Importantly, the kidneys regulate the acid–base (pH) balance of body fluids by the active secretion of hydrogen ions. Some drugs, such as penicillin, also are actively secreted into the nephron tubule for elimination.

Concentration of the Urine The proximal tubule reabsorbs about 65% of filtered water by osmosis, as water follows reabsorbed solutes. The net result is a tubular fluid with the same **osmolarity** (solute concentration) as the IF and blood. However, the mammalian kidney has the remarkable ability to produce urine that is more concentrated than other body fluids. This ability relies on the actions of specialized **juxtamedullary nephrons** with exceptionally long nephron loops dipping deep into the renal medulla (the name means "near the medulla") (Fig. 22-6A). These long loops establish the **medullary osmotic gradient**, which is a difference in solute concentration (and thus osmotic pressure) in different regions of the medulla. The IF in deeper medullary regions has more solutes and thus greater osmotic pressure than does the IF in regions closer to the cortex.

The collecting duct of all nephrons can use the medullary gradient to produce concentrated urine. As the dilute filtrate passes through the collecting duct toward the renal pelvis, it encounters progressively more concentrated IF. Water moves by osmosis from the dilute filtrate into the more concentrated IF, and then into the blood (Fig. 22-6B). In this manner, the urine becomes more concentrated as it leaves the nephron and its volume is reduced.

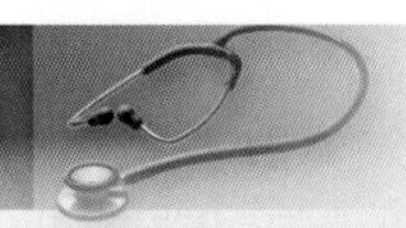

CLINICAL PERSPECTIVES **Box 22-1**

Transport Maximum

The kidneys work efficiently to return valuable substances to the blood after glomerular filtration. However, the carriers that are needed for active transport of these substances can become overloaded. Thus, there is a limit to the amount of each substance that can be reabsorbed in a given time period. The limit of this reabsorption rate is called the **transport maximum (Tm)**, or tubular maximum, and it is measured in milligrams (mg) per minute. For example, the Tm for glucose is approximately 375 mg/min.

If a substance is present in excess in the blood, it may exceed its transport maximum and then, because it cannot be totally reabsorbed, some will be excreted in the urine. Thus, the transport maximum determines the **renal threshold**—the plasma concentration at which a substance will begin to be excreted in the urine, which is measured in mg per deciliter (dL). For example, if the concentration of glucose in the blood exceeds its renal threshold (180 mg/dL), glucose will begin to appear in the urine, a condition called glycosuria. The most common cause of glycosuria is uncontrolled diabetes mellitus.

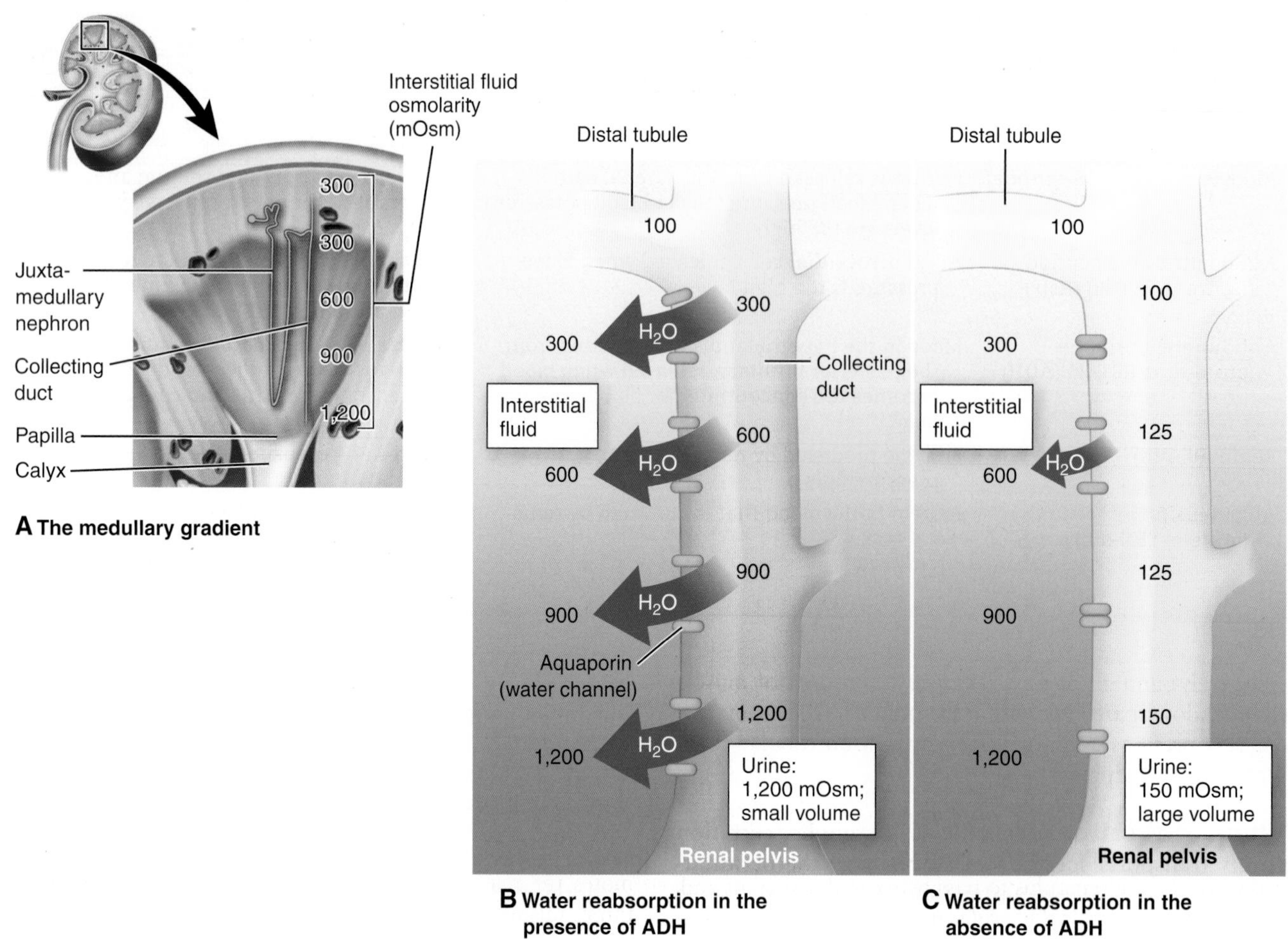

Figure 22-6 **Urine concentration.** **KEY POINT** The renal tubule establishes the medullary gradient, as measured in milliosmoles (mOsm). The collecting duct, under the influence of antidiuretic hormone (ADH), concentrates urine. **A.** The nephron loop establishes an osmotic gradient in the renal medulla. **B.** If ADH is present, water leaves the collecting duct by osmosis, resulting in concentrated urine. **C.** If ADH is absent, water cannot leave the collecting duct and the urine remains dilute. **ZOOMING IN** Where is the osmotic gradient stronger—by the distal tubule or by the renal pelvis?

Water reabsorption from the collecting duct is controlled by **antidiuretic hormone (ADH)**, a hormone released from the posterior pituitary gland (**see Table 22-1**). Recall from Chapter 3 that water crosses plasma membranes through channels called aquaporins. ADH makes the walls of the collecting duct more permeable to water by stimulating the insertion of aquaporins. As a result, more water will be reabsorbed and less water will be excreted with the urine (**see Fig. 22-6B**). In the absence of ADH, collecting duct plasma membranes contain very few aquaporins, so very little water moves down the osmotic gradient from tubular fluid into medullary IF (**Fig. 22-6C**).

The release of ADH from the posterior pituitary is regulated by a feedback system. As the blood becomes more concentrated, the hypothalamus triggers more ADH release from the posterior pituitary; as the blood becomes more dilute, less ADH is released. In the disease diabetes insipidus, there is inadequate secretion of ADH from the hypothalamus. The collecting duct is not very permeable to water in the absence of ADH, so large amounts of dilute urine are produced.

Summary of Urine Formation The processes involved in urine formation are summarized below and illustrated in **Figure 22-7**.

1. Glomerular filtration moves water and solutes from the blood into the nephron tubule.
2. Tubular reabsorption moves water and other useful substances back into the blood.
3. Tubular secretion moves unfiltered or reabsorbed substances from the blood into the nephron for elimination.
4. In the presence of ADH, the collecting duct concentrates the urine and reduces the volume excreted.

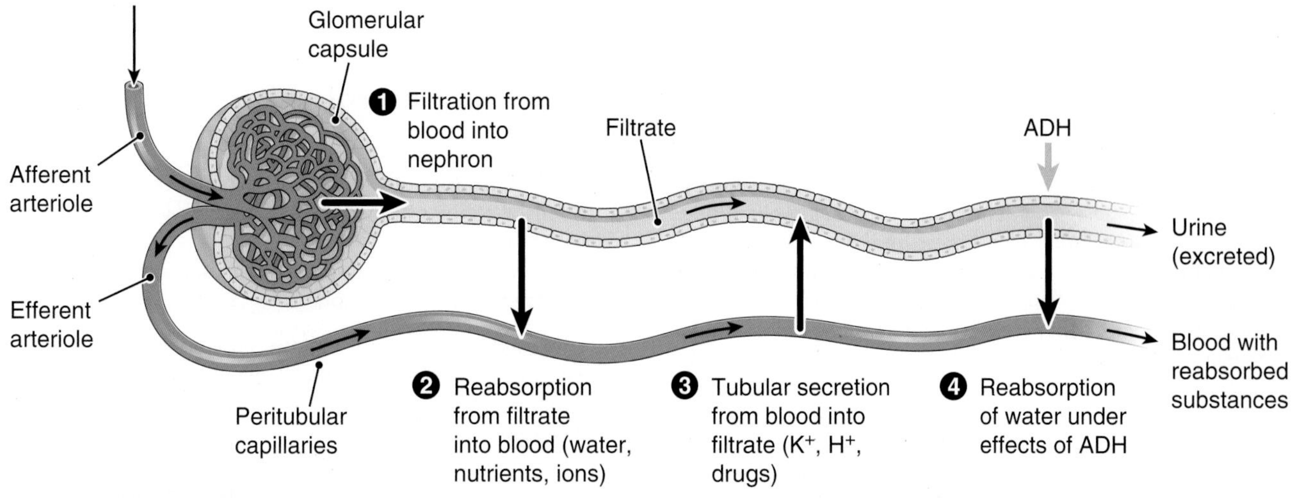

Figure 22-7 **Summary of urine formation in a nephron.** **KEY POINT** Four processes involved in urine formation are shown. **ZOOMING IN** What vessels absorb materials that leave the nephron?

The urine is not further modified once it enters the renal pelvis. Therefore, urine contains substances that were secreted as well as substances that were filtered but not reabsorbed.

THE JUXTAGLOMERULAR APPARATUS

The **juxtaglomerular (JG) apparatus** is a specialized region of the kidney involved in blood pressure regulation. As seen in **Figure 22-8**, the first portion of the distal tubule curves backward toward the glomerulus to pass between the afferent and efferent arterioles (*juxtaglomerular* means "near the glomerulus"). At the point where the distal tubule makes contact with the afferent arteriole, there are specialized cells in each that together make up the JG apparatus.

Receptors in the distal tubule respond to low sodium content in the filtrate leaving the nephron. Note that low sodium in the filtrate correlates with low volume. When stimulated, these receptors trigger cells in the afferent arteriole to secrete the enzyme **renin** (RE-nin) (**see Table 22-1**). This enzyme participates in the production of **angiotensin II** from inactive precursors. Recall from Chapter 21 that angiotensin II elevates blood pressure by several mechanisms. It promotes the release of aldosterone and ADH and stimulates thirst, raising blood pressure by increasing blood volume. It also increases blood pressure by stimulating vasoconstriction (narrowing of the arterioles). **Box 22-2** has more details on these events and their clinical applications.

22

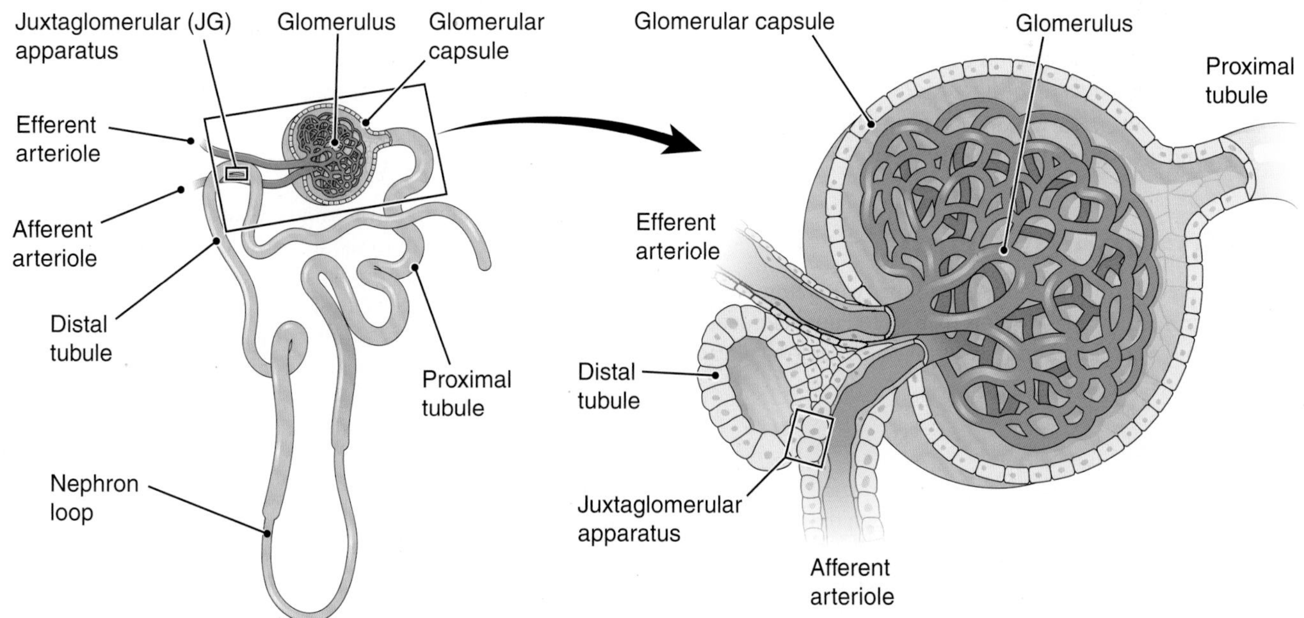

Figure 22-8 **The juxtaglomerular (JG) apparatus.** **KEY POINT** Note how the distal tubule contacts the afferent arteriole (**right**). Cells in these two structures make up the JG apparatus, which releases renin to raise blood pressure.

Box 22-2

A CLOSER LOOK

The Renin–Angiotensin Pathway: Renal Control of Blood Pressure

In addition to forming urine, the kidneys play an integral role in regulating blood pressure. When blood pressure drops, cells of the JG apparatus secrete the enzyme renin into the blood. Renin acts on another blood protein, **angiotensinogen**, which is manufactured by the liver. Renin converts angiotensinogen into **angiotensin I** by cleaving off some amino acids from the end of the protein. Angiotensin I is then converted into **angiotensin II** by yet another enzyme called angiotensin-converting enzyme (ACE), which is manufactured by capillary endothelium, especially in the lungs. Angiotensin II increases blood pressure in four ways:

1. It increases cardiac output and stimulates vasoconstriction.
2. It stimulates the release of aldosterone, a hormone that acts on the nephron's distal tubule to increase sodium reabsorption and, secondarily, water reabsorption.
3. It stimulates the release of ADH, which acts directly on the distal tubules and collecting ducts to increase water reabsorption.
4. It stimulates thirst centers in the hypothalamus, resulting in increased fluid consumption.

The combined effects of angiotensin II produce a dramatic increase in blood pressure. In fact, angiotensin II is estimated to be four to eight times more powerful than norepinephrine, another potent stimulator of hypertension, and thus, it is a good target for blood pressure–controlling drugs. One class of drugs used to treat hypertension is the **ACE inhibitors**, which control blood pressure by blocking the production of angiotensin II.

See the Student Resources on thePoint for a flow chart summarizing kidney regulation of blood pressure.

CHECKPOINTS 

- 22-8 What process drives materials out of the glomerulus and into the glomerular capsule?
- 22-9 What is the name of the process that returns materials from the nephron back to the circulation?
- 22-10 What component of the filtrate is moved by tubular secretion to balance pH?
- 22-11 What hormone controls water reabsorption from the collecting duct of the nephron?
- 22-12 What substance is produced by the JG apparatus, and under what conditions is it produced?

Elimination of Urine

Urine is excreted from the kidneys into the two ureters, which transport urine to the bladder. It is then stored until eliminated from the body via the urethra. Let us take a closer look at each of these organs.

THE URETERS

Each of the two ureters is a long, slender, muscular tube that extends from the kidney down to and through the inferior portion of the urinary bladder (see Fig. 22-1). Like the other parts of the urinary system, the ureters are entirely extraperitoneal (outside the peritoneum). Their length naturally varies with the size of the individual; they may be anywhere from 25 to 32 cm (10 to 13 in) long. Nearly 2.5 cm (1 in) of the terminal distal ureter passes obliquely (at an angle) through the inferior bladder wall. The two ureteral openings are located just superior and lateral to the urethral opening on the inferior surface of the bladder and are protected by one-way valves (Fig. 22-9). A full bladder compresses the distal ureters and closes these valves, preventing the backflow of urine.

The ureteral wall includes a lining of epithelial cells, a relatively thick layer of involuntary muscle, and finally, an outer coat of fibrous connective tissue. The epithelium is the transitional type, which flattens from a cuboidal shape as the tube stretches. This same type of epithelium lines the renal pelvis, the bladder, and the proximal portion of the urethra. The ureteral muscles are capable of the same rhythmic contraction (peristalsis) that occurs in the digestive system. Urine is moved along the ureter from the kidneys to the bladder by gravity and by peristalsis at frequent intervals.

THE URINARY BLADDER

When it is empty, the urinary bladder is located posterior to the pubic symphysis, as shown in Figure 23-1. The urinary bladder is a temporary reservoir for urine, just as the gallbladder is a storage sac for bile. A full (distended) bladder lies in an unprotected position in the lower abdomen, and a blow may rupture it, necessitating immediate surgical repair.

The bladder wall has many layers. It is lined with mucous membrane containing transitional epithelium. The bladder's lining, like that of the stomach, is thrown into folds called *rugae* (see Fig. 22-9) when the organ is empty. Beneath the mucosa is a layer of connective tissue, followed by a layer of involuntary smooth muscle that can stretch considerably. Finally, the parietal peritoneum covers the bladder's superior portion.

When the bladder is empty, the muscular wall thickens, and the entire organ feels firm. As the bladder fills, the muscular wall becomes thinner, and the organ may increase from

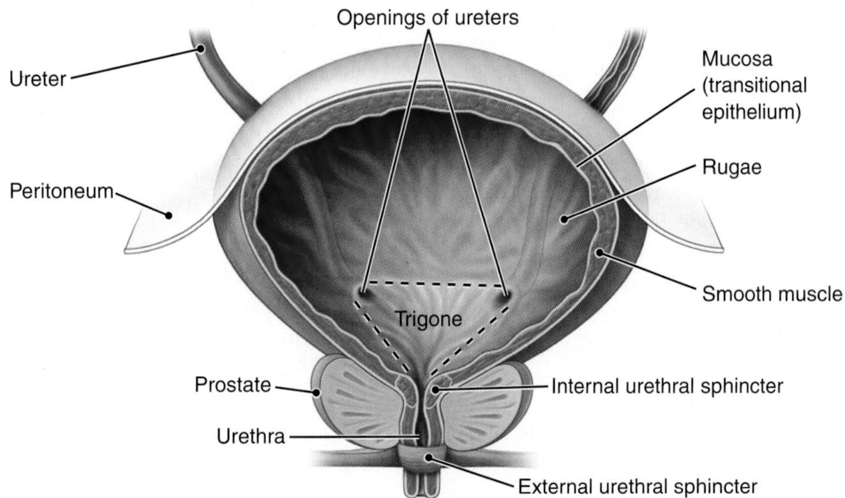

Figure 22-9 **The male urinary bladder.** **KEY POINT** The interior of the bladder is folded into rugae. Internal and external sphincters regulate urination. The trigone is a triangular region in the floor of the bladder marked by the openings of the ureters and the urethra. **ZOOMING IN** What gland does the urethra pass through in the male?

a length of 5 cm (2 in) up to as much as 12.5 cm (5 in) or even more. A moderately full bladder holds about 470 mL (1 pint) of urine.

The **trigone** (TRI-gone) is a triangular-shaped region in the floor of the bladder. It is marked by the openings of the two ureters and the urethra (see Fig. 22-9). As the bladder fills with urine, it expands upward, leaving the trigone at the base stationary. This stability prevents stretching of the ureteral openings and the possible backflow of urine into the ureters.

THE URETHRA

The urethra carries urine from the bladder to the outside (see Fig. 22-1). The urethra differs in males and females; in the male, it is part of both the reproductive system and the urinary system, and it is much longer than is the female urethra.

The male urethra is approximately 20 cm (8 in) in length. Proximally, it passes through the prostate gland, where it is joined by two ducts carrying male reproductive cells (spermatozoa) from the testes and glandular secretions (see Fig. 23-1). From here, it leads to the outside through the **penis** (PE-nis), the male organ of copulation. The male urethra serves the dual purpose of draining the bladder and conveying semen with spermatozoa.

"Straddle" injuries to the urethra are common in males. This type of injury occurs when a man or child slips and lands with a narrow rod or beam between his legs, as when riding a bike or doing construction work. Such an accident may catch the urethra between the hard surface and the pubic arch and rupture the urethra. In accidents in which the bones of the pelvis are fractured, urethral rupture is fairly common.

The urethra in the female is a thin-walled tube about 4 cm (1.5 in) long. It is posterior to the pubic symphysis and is embedded in the muscle of the anterior vaginal wall (see Fig. 23-8). The external opening, the urethral orifice, is located just anterior to the vaginal opening between the labia minora. The female urethra drains only the bladder and is entirely separate from the reproductive system.

URINATION

The process of expelling (voiding) urine from the bladder is called **urination** or **micturition** (mik-tu-RISH-un). This process is controlled both voluntarily and involuntarily with the aid of two muscular rings (sphincters) that surround the urethra (see Fig. 22-9). Near the bladder's outlet is an involuntary **internal urethral sphincter** formed by a continuation of the bladder's smooth muscle. Below this muscle is a voluntary **external urethral sphincter** formed by the muscles of the pelvic floor.

As the bladder fills with urine, stretch receptors in its wall send impulses to a center in the lower part of the spinal cord. Motor impulses from this center stimulate contraction of the bladder wall and relaxation of the urethral sphincters, forcing urine outward. In the infant, this emptying occurs automatically as a simple reflex. Impulses from higher brain centers can overcome this reflexive relaxation of the external urethral sphincter, blocking involuntary urination. Urination can also be voluntarily initiated by higher motor centers, even when the bladder is virtually empty. Early in life, a child learns to control urination, a process known as *toilet training*. The impulse to urinate will override conscious controls if the bladder becomes too full.

THE URINE

Urine is a yellowish liquid that is approximately 95% water and 5% dissolved solids and gases. The pH of freshly collected urine averages 6.0, with a range of 4.5 to 8.0. Diet may cause considerable variation in pH as well as color and odor. For example, some people's urine takes on an unusual odor when they eat asparagus, and beet consumption can give urine a worrying reddish hue.

The amount of dissolved substances in urine is indicated by its **specific gravity**. The specific gravity of pure water, used as a standard, is 1.000. Because of the dissolved materials it contains, urine has a specific gravity that normally varies from 1.002 (very dilute urine) to 1.040 (very concentrated urine). When the kidneys are diseased, they lose the ability to concentrate urine, and the specific gravity no longer varies as it does when the kidneys function normally.

Normal Constituents of Urine Some of the dissolved substances normally found in the urine are the following:

- **Nitrogenous waste products**, including the following:
  - Urea, formed from amine groups released in protein catabolism
  - Uric acid from the breakdown of purines, which are found in some foods and nucleic acids
  - Creatinine (kre-AT-ih-nin), a breakdown product of muscle creatine
- **Electrolytes**, including sodium ions, chloride, and different kinds of sulfates and phosphates. Electrolytes are excreted in appropriate amounts to keep their blood concentration constant.
- **Pigments,** like urochrome, a yellow substance derived from the breakdown of hemoglobin, is the main pigment in urine. Small amount of bilirubin and other bile pigments are also found in normal urine. Beets and other dark foods can add color, as can B vitamins, vitamin C, and food dyes. Also certain drugs can alter the color of the urine.

Abnormal Constituents of Urine Examination of urine, called a **urinalysis** (u-rin-AL-ih-sis) (**UA**), is one of the most important parts of a medical evaluation. A routine UA includes observation of color and turbidity (cloudiness) as well as the measurement of pH and specific gravity. Laboratories also test for a variety of abnormal components, including the following:

- **Glucose** is usually an important indicator of diabetes mellitus, in which the cells do not adequately metabolize blood glucose (see Chapter 12). The excess glucose, which cannot be reabsorbed from the tubule, is excreted in the urine. The presence of glucose in the urine is known as **glycosuria** (gli-ko-SU-re-ah) or glucosuria.
- **Albumin.** Damage to the wall of the glomerulus or glomerular capsule permits passage of albumin into the tubule, and since it cannot be reabsorbed, it appears in urine. The presence of this protein, which is normally retained in the blood, may thus indicate damage to the parts of the nephron involved in filtration. Albumin in the urine is known as **albuminuria** (al-bu-mih-NU-re-ah).
- **Blood** in the urine, which is known as **hematuria** (hem-ah-TU-re-ah), is usually an important indicator of urinary system disease. Injury to the urinary tract is another possible source of hematuria.
- **Ketones** (KE-tones) are produced when fats are incompletely oxidized; ketones in the urine are seen in diabetes mellitus and starvation.
- **White blood cells** (pus) are evidence of infection of the bladder or, less commonly, the kidney itself; they can be seen by microscopic examination of a centrifuged specimen. Pus in the urine is known as **pyuria** (pi-U-re-ah).
- **Casts** are solid materials molded within the microscopic kidney tubules. They consist of cells or proteins, and when present in large number, they usually indicate disease of the nephrons.

More extensive tests on urine may include analysis for drugs, enzymes, hormones, and other metabolites, as well as cultures for microorganisms. Normal values for common urine tests are given in Appendix 2, **Table A2-1**.

CHECKPOINTS

- 22-13 What is the name of the tube that carries urine from the kidney to the bladder?
- 22-14 What openings form the bladder's trigone?
- 22-15 What is the name of the tube that carries urine from the bladder to the outside?
- 22-16 What are some normal constituents of urine? Abnormal constituents?

Disorders of the Urinary System

The urinary system is prone to a wide variety of structural, inflammatory, and genetic disorders. Since all of its components are anatomically linked, a disorder in one structure frequently affects the others.

OBSTRUCTIONS AND STRUCTURAL DISORDERS

Normally, the urinary tract functions well as a one-way conduit of urine. Obstructions in the tract can cause urine to accumulate in the tubes and eventually in the kidney itself. **Hydronephrosis** (hi-dro-nef-RO-sis) is the distention of the renal pelvis and calyces with accumulated fluid. If the obstruction is not removed, the kidney will be permanently damaged. Common causes of obstructions include tumors, scar tissue from injury or inflammation, kidney stones, and structural abnormalities.

Kidney Stones Kidney stones, or **renal calculi** (KAL-ku-li), are made of substances, such as calcium salts or uric acid, that precipitate out of the urine instead of remaining in solution. They usually form in the renal pelvis (**Fig. 22-10A**) but may also form in the bladder.

The causes of stone formation include dehydration, urinary stasis (stagnation), and urinary tract infection (UTI). The stones may vary in size from tiny grains resembling bits of gravel up to large masses that fill the renal pelvis and extend into the calyces. The latter are described as **staghorn calculi** (**see Fig. 22-10A and B**). The passage of a small stone along the ureter causes excruciating pain, called *renal colic*. Relief of this pain usually requires morphine or an equally powerful drug.

There is no way of dissolving these stones because substances that could do so would also destroy renal tissue. Sometimes, instruments can be used to crush small stones and thus allow them to be expelled with the urine, but surgical removal is often required. A **lithotriptor** (LITH-o-trip-tor), literally a "stone-cracker," is a device that employs external shock waves to shatter kidney stones. The procedure is called **lithotripsy** (LITH-o-trip-se), or more accurately *extracorporeal shock wave lithotripsy* (**Fig. 22-10C**).

The first "barber surgeons," operating without benefit of anesthesia, were permitted by their patients to cut through the skin and the muscles of the back to remove stones from the ureters. "Cutting for stone" in this way was relatively successful, despite the lack of sterile technique, because the approach through the back avoided the peritoneal cavity and the serious risk of peritonitis.

Modern surgery employs a variety of instruments for the removal of stones from the ureter, including endoscopes similar to those described in Chapter 19. The transurethral route through the urethra and urinary bladder and then into the ureter, as well as entrance through the skin and muscles

22

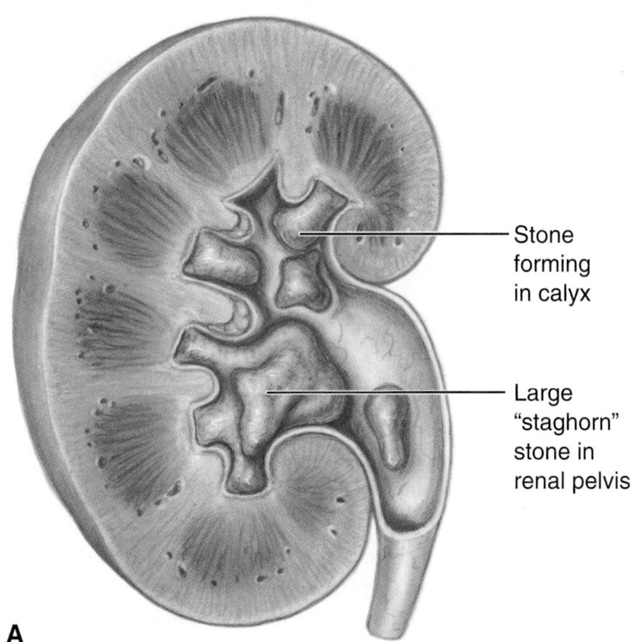

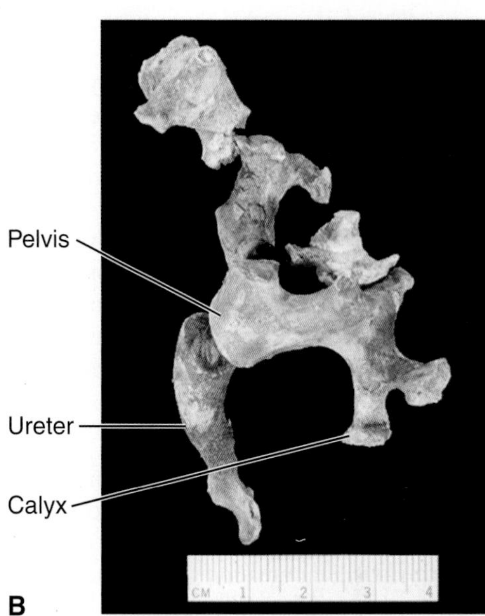

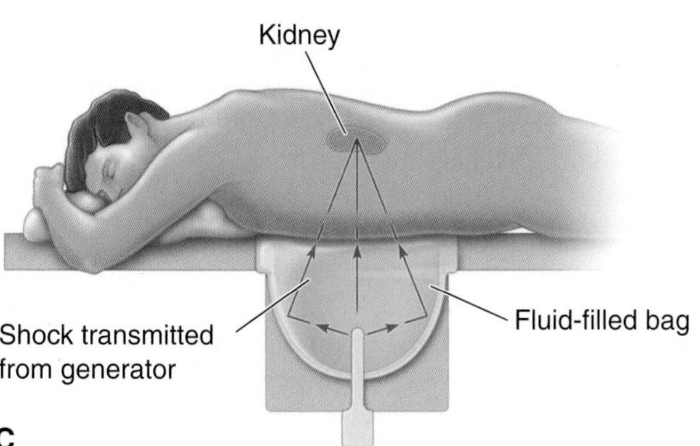

Figure 22-10 **Kidney stones.** **KEY POINT** Stones (calculi) may form in the kidney and other portions of the urinary system. **A.** Large stones that extend into the calyces are called *staghorn calculi*. **B.** A photo of a staghorn calculus. **C.** Lithotripsy. Shock waves can be used to break kidney stones and allow for their passage. The procedure is called *extracorporeal shock wave lithotripsy*. **ZOOMING IN** What does the word *extracorporeal* mean?

of the back, may be used to remove calculi from the kidney pelvis or from the ureter.

Structural Abnormalities All parts of the urinary system are prone to congenital (present at birth) abnormalities. Kidney abnormalities are the most common, afflicting about 10% of all babies. Renal hypoplasia describes kidneys that are abnormally small but normally formed, and renal dysplasia describes kidneys with malformed nephrons or connective tissue. Either of these conditions can lead to renal failure.

Abnormalities in ureteral anatomy include the doubling of each ureter at the renal pelvis, an obstruction at the junction between the ureter and the renal pelvis, and constricted or abnormally narrow parts, called **strictures** (STRICK-tures). In cases of **ureterocele** (u-RE-ter-o-sele), the end of the ureter bulges into the bladder. The usual cause of ureterocele is a congenital narrowing of the ureteral opening. These abnormalities can interfere with urine flow, causing it to distend the ureter (hydroureter) and renal pelvis (hydronephrosis).

Urine flow can also be obstructed by problems with the urethra. The opening of the urethra to the outside may be too small, or the urethra itself may be narrowed. Occasionally, an abnormal valve-like structure is found at the point where the urethra enters the bladder. Surgical correction of these conditions may prevent urinary backflow and the resultant increase in infections and distention of the urinary tract. There is also a condition in the male in which the urethra opens on the undersurface of the penis instead of at the end. This is called **hypospadias** (hi-po-SPA-de-as) (**Fig. 22-11**).

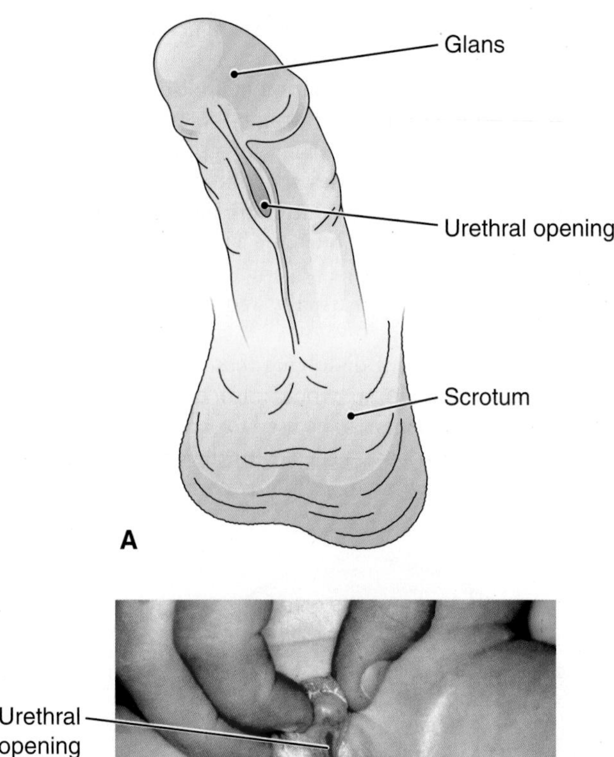

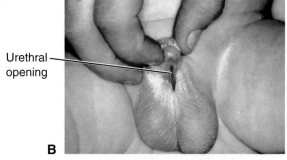

Figure 22-11 Hypospadias. A. A ventral view shows the urethra opening on the underside of the penis. **B.** In this photo of a baby with hypospadias, the urethral opening is on the scrotum.

INFLAMMATORY DISORDERS

The elements of the urinary system are connected to the body's exterior via the urethra, so they are prone to infection. A **urinary tract infection (UTI)** usually begins in the urethra. **Urethritis** is characterized by inflammation of the mucous membrane and the glands of the urethra. Most commonly, sexual activity or poor hygiene transfers *E. coli* or other fecal bacteria from the anus to the urethra, but viral or fungal agents can also cause infections in immunocompromised patients.

Cystitis Bacteria or other infectious agents frequently ascend the urethra to induce inflammation of the bladder, called **cystitis** (sis-TI-tis). Cystitis is 10 times as common in women as in men due to the short urethra of the female compared to that of the male. Catheterization to remove urine from the bladder is another possible source of infection. Pain, urgency to urinate, and urinary frequency are common symptoms of cystitis.

Retention of urine within the bladder (urinary stasis) increases the chances that bacteria will be able to establish an infection. The most common cause of stasis in men is enlargement of the prostate gland, which surrounds the first portion of the urethra. In the opening case study, Eric faced the problem of an enlarged prostate gland. The structural and hormonal changes of pregnancy also interfere with urine flow and predispose women to cystitis. Less frequent causes include neurogenic bladder, which is bladder dysfunction resulting from neurologic lesions, as seen in diabetes mellitus, and structural defects in the area where the ureters enter the bladder. Like all infectious disorders, UTIs are also more common in individuals with suppressed immune systems.

Another type of cystitis called **interstitial cystitis** causes pelvic pain with discomfort before and after urination. The tissues below the mucosa are involved. The disorder appears almost exclusively in women. The cause of interstitial cystitis in not known, but because no bacteria are involved, antibiotics are not effective treatment and may even be harmful.

Pyelonephritis Relatively infrequently, bladder infections that are not entirely cured can spread up the ureters to the renal pelvis and kidney tissue, resulting in **acute pyelonephritis** (pi-el-o-nef-RI-tis). Urinary stasis or backflow (the passage of urine from the bladder to the kidney) increases the ability of organisms to establish a kidney infection. More rarely, the blood carries bacteria to the kidney. The signs and symptoms include fever and persistent pain in the kidney region. Pyelonephritis usually

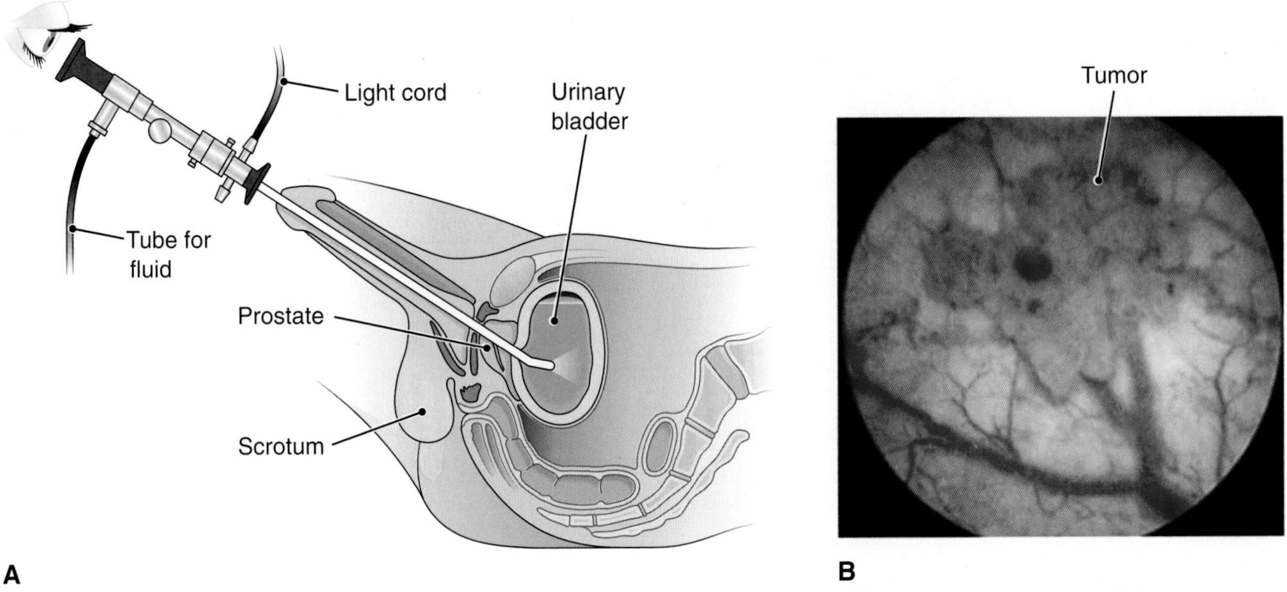

Figure 22-12 Cystoscopy. A. A lighted cystoscope is introduced through the urethra into the bladder of a male subject. Sterile fluid is used to inflate the bladder. The cystoscope is used to examine the bladder, remove specimens for biopsy, and remove tumors. **B.** A cancer of the bladder, as viewed through a cystoscope.

responds to antibiotic treatment, fluid replacement, rest, and fever control.

Chronic pyelonephritis, a more serious disease, is frequently seen in patients with persistent urinary stasis or backflow. It may be caused by persistent or repeated bacterial infections. Progressive kidney damage is evidenced by high blood pressure, continual loss of protein in the urine, and dilute urine.

Glomerulonephritis **Acute glomerulonephritis** (glo-mer-u-lo-nef-RI-tis), also known as *acute poststreptococcal glomerulonephritis*, is the most common disease of the kidney. Unlike pyelonephritis, it is not related to a UTI. This condition usually occurs in children about one to four weeks after a streptococcal throat infection. Antibodies formed in response to the streptococci attach to the glomerular membrane and cause injury. These damaged glomeruli are abnormally leaky, allowing protein, especially albumin, to filter into the glomerular capsule and ultimately to appear in the urine (albuminuria). They also allow red blood cells to filter into the urine (hematuria). Usually, the patient recovers without permanent kidney damage. In adult patients, the disease is more likely to become chronic, with a gradual decrease in the number of functioning nephrons leading to chronic renal failure.

NEOPLASMS

Tumors of the kidneys usually grow rather slowly, but rapidly invading types are occasionally found. Blood in the urine and dull pain in the kidney region are warnings that should be heeded at once. Surgical removal of the kidney offers the best chance of a cure because most renal cancers do not respond to chemotherapy or radiation.

Bladder tumors, which are most prevalent in men older than 50 years of age, include benign papillomas and various kinds of cancer. About 90% of these tumors arise from the bladder's epithelial lining. Possible causes include toxins (particularly certain aniline dyes), chronic infestations (e.g., schistosomiasis), heavy cigarette smoking, and the presence of urinary stones, which may develop and increase in size within the bladder.

Hematuria and frequent urination, in the absence of pain or fever, are early signs and symptoms of a bladder tumor. The tumor can often be visualized with the use of a **cystoscope**, a type of endoscope used to examine the bladder (Fig. 22-12). Treatment includes removal of the tumor, which may be done cystoscopically, and localized chemotherapy. More serious cases may require irradiation. Removal of the tumor before it invades the muscle wall gives the best prognosis.

If it is necessary to remove the bladder in a **cystectomy** (sis-TEK-to-me), the ureters must be vented elsewhere. They may be diverted to the body surface through a segment of the ileum, a procedure known as an **ileal conduit** (Fig. 22-13), or diverted to some other portion of the intestine. Alternatively, surgeons may create a bladder from a section of the colon.

POLYCYSTIC KIDNEY DISEASE

Polycystic (pol-e-SIS-tik) **kidney disease** is a genetic disorder in which many fluid-containing sacs develop within the kidney. These sacs can grow extremely large, and kidneys weighing up to 30 lb have been reported. This disorder is usually asymptomatic in children, but gradually the pressure from the growing cysts destroys the nephrons. Eventually, the kidney may

22

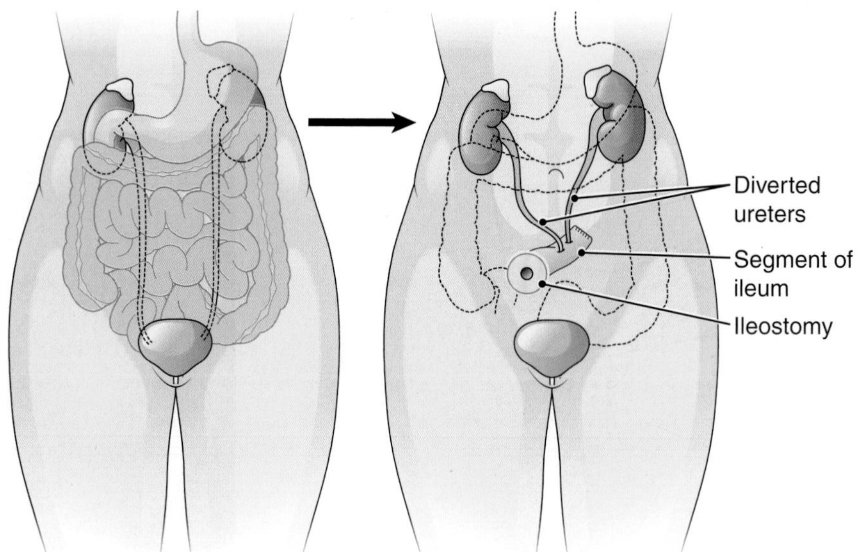

Figure 22-13 **Ileal conduit.** **KEY POINT** The ureters are vented through a segment of the ileum to open at the body surface through an ileostomy.

not have enough functioning nephrons to filter blood effectively, resulting in chronic renal failure (discussed below). This disorder runs in families, and treatment has not proved very satisfactory except for the use of dialysis machines or kidney transplantation. For some reason, cysts do not develop in the transplanted kidneys.

RENAL FAILURE

Acute renal failure may result from a medical or surgical emergency or from toxins that damage the tubules. This condition is characterized by a sudden, serious decrease in kidney function accompanied by electrolyte and acid–base imbalances. Acute renal failure occurs as a serious complication of other severe illness and may be fatal.

Chronic renal failure results from a gradual loss of nephrons. As more and more nephrons are destroyed, the kidneys slowly lose the ability to perform their normal functions. As the disease progresses, nitrogenous waste products accumulate to high levels in the blood, causing a condition known as **uremia** (u-RE-me-ah). In many cases, there is a lesser decrease in renal function, known as **renal insufficiency**, that produces fewer symptoms.

A few of the characteristic signs and symptoms of chronic renal failure are the following:

- **Dehydration** (de-hi-DRA-shun). Excessive loss of body fluid may occur early in renal failure, when the kidneys cannot concentrate the urine and large amounts of water are eliminated.
- **Edema** (eh-DE-mah). Accumulation of fluid in the tissue spaces may occur late in chronic renal disease, when the kidneys cannot eliminate water in adequate amounts.
- **Electrolyte imbalance**, including retention of sodium and potassium.
- **Hypertension** may occur as the result of fluid overload and increased renin production (see Box 22-2).
- **Anemia** occurs when the kidneys cannot produce the hormone EPO to activate red blood cell production in bone marrow.
- **Uremia** (u-RE-me-ah), an excess of nitrogenous waste products in the blood. When these levels are high, urea can be changed into ammonia in the stomach and intestine and cause ulcerations and bleeding.

Renal Dialysis and Kidney Transplantation

Dialysis (di-AL-ih-sis) means "the separation of dissolved molecules based on their ability to pass through a semipermeable membrane." Molecules that can pass through the membrane move from an area of greater concentration to one of lesser concentration. In patients who have impaired renal function, urea and other nitrogenous waste products accumulated in the blood are most commonly reduced by passage of the patient's blood through a dialysis machine, in a process called **hemodialysis** (blood dialysis). The principle of "molecules leaving the area of greater concentration" operates to remove wastes from the blood. The dialysis membrane is made of cellophane or other synthetic material (Fig. 22-14A). The fluid in the dialysis machine, the dialysate, can be adjusted to regulate the flow of substances out of the blood.

In another method, peritoneal dialysis (dialysis in the abdominal cavity), the surface area of the peritoneum acts as the dialysis membrane (see Fig. 22-14B). Dialysis fluid is introduced into the peritoneal cavity and then periodically removed along with waste products. This procedure may be done at intervals through the day or during the night. Figure 22-14C illustrates the movement of waste products from the blood to the dialysate.

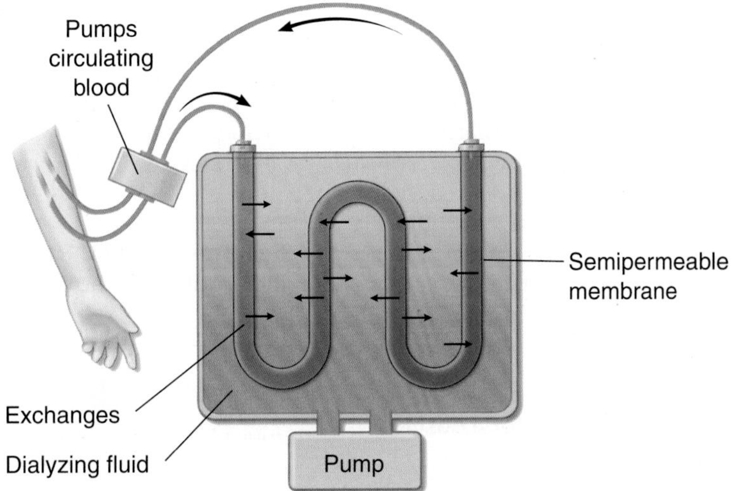

A Hemodialysis

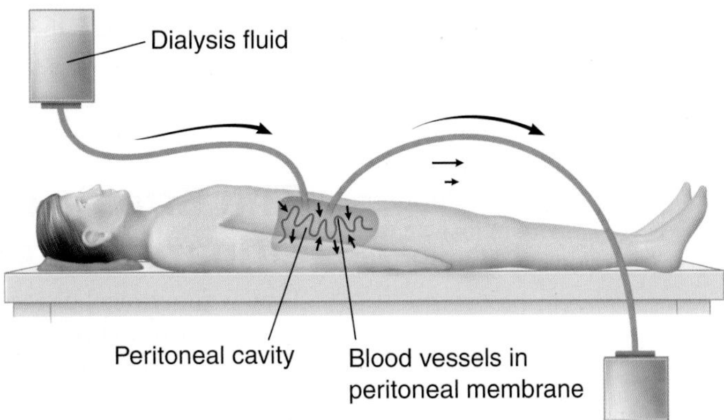

B Peritoneal dialysis

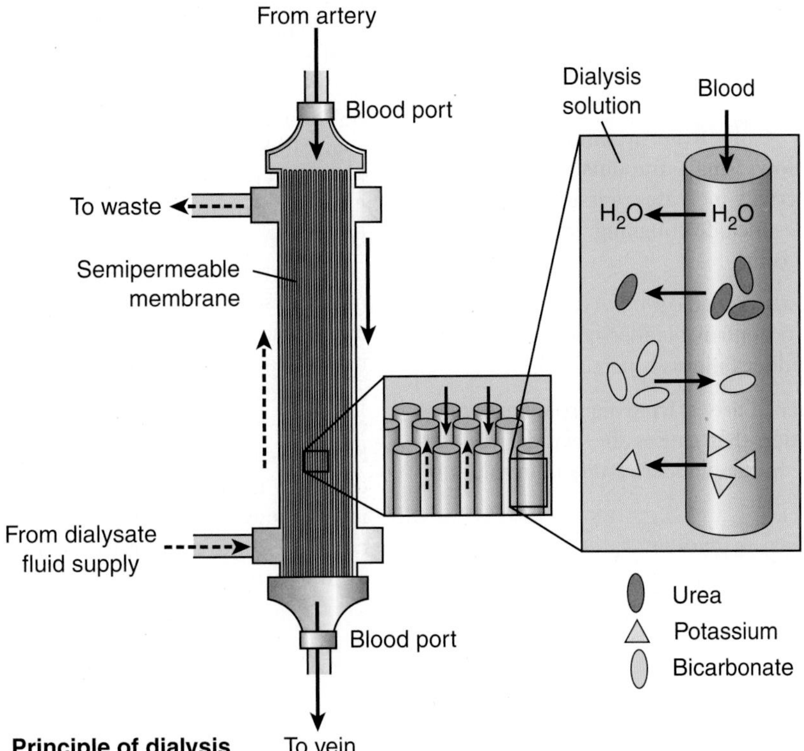

C Principle of dialysis

Figure 22-14 Dialysis. **KEY POINT** The dialysate draws fluid and waste out of the blood as a substitute for defective kidneys. **A.** Hemodialysis: The patient's blood flows through a dialysis machine. **B.** Peritoneal dialysis: The dialysate is introduced into the peritoneal cavity; the peritoneum acts as the dialysis membrane. **C.** The principle of dialysis: Materials flow through a semipermeable membrane based on their concentration on either side of the membrane.

A 1973 amendment to the Social Security Act provides federal financial assistance for people who have chronic renal disease and require dialysis. Most hemodialysis is performed in freestanding clinics. Treatment time has been reduced; a typical schedule involves two to three hours, three times a week. Access to the bloodstream has been made safer and easier through surgical establishment of a permanent exchange site (shunt). Peritoneal dialysis also has been improved and simplified, enabling patients to manage treatment at home.

See the Student Resources on thePoint for a description of the career of hemodialysis technician.

The final option for the treatment of renal failure is kidney transplantation. Surgeons have successfully performed many of these procedures. Kidneys have so much extra functioning tissue that the loss of one kidney normally poses no problem to the donor. Records show that transplantation success is greatest when surgeons use a kidney from a living donor who is closely related to the patient. Organs from deceased donors have also proved satisfactory in many cases. The problem of tissue rejection (rejection syndrome) is discussed in Chapter 17.

URINARY INCONTINENCE

Urinary incontinence (in-KON-tin-ens) refers to an involuntary loss of urine. The condition may originate with a neurologic disorder, trauma to the spinal cord, weakness of the pelvic muscles, impaired bladder function, or medications. Different forms of urinary incontinence have specific names:

- Stress incontinence results from urethral incompetence (weakness) that allows small amounts of urine to be released when an activity increases pressure in the abdomen. These activities include coughing, sneezing, laughing, lifting, or exercising.
- Urge incontinence, also called *overactive bladder*, results from an inability to control bladder contractions once the sensation of bladder fullness is perceived.
- Overflow incontinence arises from neurologic damage or urinary obstruction that causes the bladder to overfill. Excess pressure in the bladder results in involuntary urine loss.
- **Enuresis** (en-u-RE-sis) is involuntary urination, usually during the night (bed-wetting).

Some treatment approaches to incontinence include muscle exercises, dietary changes, biofeedback, medication, surgery, or in serious cases, self-catheterization.

See the Student Resources on thePoint for some illustrations of the kidney disorders discussed.

CHECKPOINTS

- ☐ 22-17 What is the scientific name for stones, as may occur in the urinary tract?
- ☐ 22-18 What is the term for inflammation of the bladder?
- ☐ 22-19 What is the term for an excess of nitrogenous waste products in the blood?
- ☐ 22-20 What process can be used to eliminate waste products from the blood in cases of renal failure?

The Effects of Aging on the Urinary System

Even without renal disease, aging causes the kidneys to lose some of their ability to concentrate urine. With aging, progressively more water is needed to excrete the same amount of waste. Older people find it necessary to drink more water than do young people, and they eliminate larger amounts of urine (polyuria), even at night (nocturia).

Beginning at about 40 years of age, there is a decrease in the number and size of the nephrons. Often, more than half of them are lost before the age of 80 years. There may be an increase in blood urea nitrogen without serious symptoms. Elderly people are more susceptible than young people to UTIs. Childbearing may cause damage to the pelvic floor musculature, resulting in urinary tract problems in later years.

Enlargement of the prostate, common in older men, may cause obstruction and back pressure in the ureters and kidneys (**see Fig. 22-1**). If this condition is untreated, it will cause permanent damage to the kidneys. In the case study, Eric is confronted with the problem of an enlarged prostate. Changes with age, including decreased bladder capacity and decreased muscle tone in the bladder and urinary sphincters, may predispose people to incontinence. However, most elderly people (60% in nursing homes and up to 85% living independently) have no incontinence.

Disease in Context Revisited

Eric Has Surgery to Prevent Kidney Damage

The urologist inserted the cystoscope into Eric's urethra, carefully guiding it toward the urinary bladder. When he examined the bladder's mucous membrane lining, he did not see any urinary stones or tumors of the rugae. He did note that the neck of Eric's urinary bladder was almost completely occluded by his enlarged prostate. This observation fit with the results of the pyelogram (a special radiograph of the urinary system) he had ordered for Eric a week earlier. The x-ray images indicated a blockage in the neck of the urinary bladder, which prevented urine from exiting. The back pressure of the urine was causing distention of the ureters (hydroureter) and kidneys (hydronephrosis). The doctor removed the cystoscope and reported his findings to his patient.

"Eric, Dr. Michaels's diagnosis was correct," the urologist said. "The blockage in your urinary system is due to enlargement of your prostate gland. If we don't treat this now, your kidneys are at risk of severe damage and renal failure. There are a couple of treatment options, but my recommendation is to do a procedure called a *transurethral prostatectomy* to remove the prostatic overgrowth and reestablish urine flow."

A few days later, Eric was back at the hospital for his surgery. The urologist inserted an instrument called a resectoscope into Eric's urethra. With the electrical loop at the end of the instrument, he removed pieces of the prostate and cauterized blood vessels to control bleeding. By the end of the surgery, the urologist had resected enough of the prostate to restore normal urine flow. It would take Eric a few weeks to recover, but soon he would be back to normal.

During this case, we saw that enlargement of the prostate gland can seriously affect urinary system function. To learn more about the prostate gland and other organs of the male reproductive system, see Chapter 23.

Chapter Wrap-Up

Summary Overview

A detailed chapter outline with space for note taking is on *thePoint*. The figure below illustrates the main topics covered in this chapter.

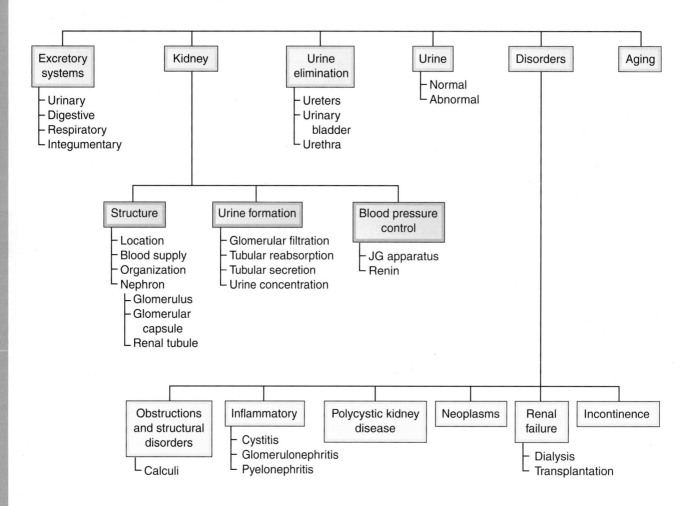

- Excretory systems
  - Urinary
  - Digestive
  - Respiratory
  - Integumentary
- Kidney
  - Structure
    - Location
    - Blood supply
    - Organization
    - Nephron
      - Glomerulus
      - Glomerular capsule
      - Renal tubule
  - Urine formation
    - Glomerular filtration
    - Tubular reabsorption
    - Tubular secretion
    - Urine concentration
  - Blood pressure control
    - JG apparatus
    - Renin
- Urine elimination
  - Ureters
  - Urinary bladder
  - Urethra
- Urine
  - Normal
  - Abnormal
- Disorders
  - Obstructions and structural disorders
    - Calculi
  - Inflammatory
    - Cystitis
    - Glomerulonephritis
    - Pyelonephritis
  - Polycystic kidney disease
  - Neoplasms
  - Renal failure
    - Dialysis
    - Transplantation
  - Incontinence
- Aging

Key Terms

The terms listed below are emphasized in this chapter. Knowing them will help you organize and prioritize your learning. These and other boldface terms are defined in the Glossary with phonetic pronunciations.

angiotensin
antidiuretic hormone (ADH)
calculi
cystitis
dialysis
excretion
glomerular filtrate
glomerulonephritis
glomerulus
hemodialysis
kidney
micturition
nephron
osmolarity
pyelonephritis
renin
tubular reabsorption
urea
ureter
urethra
urinalysis
urinary bladder
urine

Word Anatomy

Medical terms are built from standardized word parts (prefixes, roots, and suffixes). Learning the meanings of these parts can help you remember words and interpret unfamiliar terms.

WORD PART	MEANING	EXAMPLE
The Kidneys		
juxta	next to	The *juxtaglomerular* apparatus is next to the glomerulus.
nephr/o	kidney	The *nephron* is the functional unit of the kidney.
ren/o	kidney	The *renal* artery carries blood to the kidney.
retro	backward, behind	The *retroperitoneal* space is posterior to the peritoneal cavity.
The Ureters		
extra	beyond, outside of	The ureters are *extraperitoneal.*
Disorders of the Urinary System		
cele	swelling, enlarged space	A *ureterocele* is formed as the end of the ureter bulges into the bladder.
cyst/o	sac, bladder	A *polycystic* kidney develops many fluid-containing sacs.
dia	through	*Dialysis* is the separation (lysis) of molecules based on their ability to pass through a semipermeable membrane.
pyel/o	renal pelvis	*Pyelonephritis* is inflammation of the nephrons and renal pelvis.
trans	across, through	A *transurethral* route is through the urethra.
The Effects of Aging		
noct/i	night	*Nocturia* is excessive urination at night.

Questions for Study and Review

BUILDING UNDERSTANDING

Fill in the Blanks

1. Each kidney is located outside the abdominal cavity in the _____ space.
2. The renal artery, renal vein, and ureter connect to the kidney at the _____.
3. The functional unit of the kidney is the _____.
4. The amount of dissolved substances in urine is indicated by its _____.
5. The presence of glucose in the urine is known as _____.

Matching > Match each numbered item with the most closely related lettered item.

___ **6.** Produced by the kidney in response to low blood pressure
___ **7.** Stimulates vasoconstriction
___ **8.** Produced by the kidney in response to hypoxia
___ **9.** Stimulates kidneys to produce concentrated urine
___ **10.** Produced by the liver during protein catabolism

a. urea
b. erythropoietin
c. antidiuretic hormone
d. renin
e. angiotensin

Multiple Choice

___ **11.** What is the cluster of capillaries through which materials enter the nephron tubule?
a. renal capsule
b. juxtaglomerular apparatus
c. glomerulus
d. glomerular capsule

___ **12.** Which of the following is NOT a nitrogenous waste product?
a. urochrome
b. creatinine
c. urea
d. uric acid

___ **13.** Reabsorbed or unfiltered materials can be moved into the nephron by what process?
a. filtration
b. tubular secretion
c. diffusion
d. osmosis

___ **14.** Which structure is responsible for voluntary control of urination?
a. trigone
b. internal urethral sphincter
c. external urethral sphincter
d. urinary meatus

___ **15.** Which term means "pus in the urine?"
a. pyuria
b. uremia
c. anemia
d. enuresis

UNDERSTANDING CONCEPTS

16. List four organ systems active in excretion. What are the products eliminated by each?

17. Referring to The Body Visible in the front of the book, give the name and number of:
a. the triangle at the base of the bladder
b. the vessel that collects blood from the renal vein
c. a vessel that travels along the base of a renal pyramid
d. vessels that surround the ascending and descending limbs of the nephron
e. the tube that empties into a minor calyx

18. Compare and contrast the following terms:
a. glomerular filtration and tubular reabsorption
b. afferent arteriole and efferent arteriole
c. proximal tubule and distal tubule
d. ureter and urethra

19. Trace the pathway of a urea molecule from the afferent arteriole to the urinary meatus.

20. Describe the four processes involved in the formation of urine.

21. Compare the male urethra and female urethra in structure and function. Why is cystitis more common in women than in men?

22. Write the meaning of the following abbreviations:
a. UTI ______________________________
b. EPO ______________________________
c. ADH ______________________________
d. IF ______________________________
e. JG______________________________

23. Differentiate between the following disorders:
a. albuminuria and hematuria
b. glomerulonephritis and pyelonephritis
c. hydronephrosis and polycystic kidney
d. renal ptosis and ureterocele

24. What is meant by the word *dialysis*, and how is this principle used for patients with kidney failure? What kinds of membranes are used for hemodialysis? For peritoneal dialysis?

CONCEPTUAL THINKING

25. Referring to Appendix 2-1, discuss three causes of decreased specific gravity of urine.

26. A class of antihypertensive drugs called loop diuretics prevents sodium reabsorption in the nephron loop. How could a drug like this lower blood pressure?

27. In Eric's case study, enlargement of the prostate gland led to hydronephrosis. What effect might this have on glomerular filtration?

For more questions, see the Learning Activities on thePoint.

UNIT VII

Perpetuation of Life

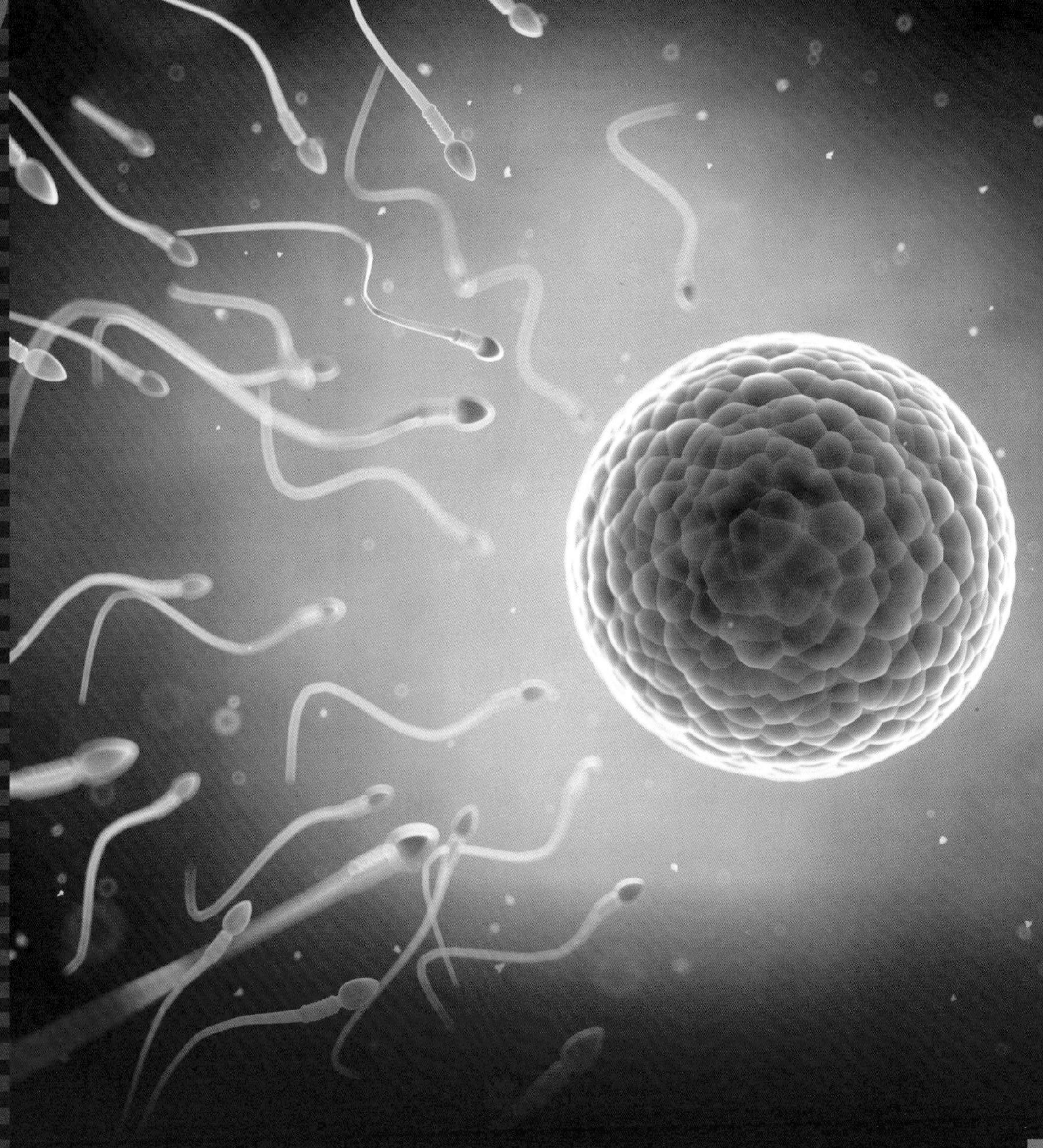

The final unit includes three chapters on the structures and functions related to reproduction and heredity. The reproductive system is not necessary for the continuation of the life of the individual but rather is needed for the continuation of the human species. The reproductive cells and their genes have been studied intensively during recent years as part of the rapidly advancing science of genetics.

CHAPTER 23

The Male and Female Reproductive Systems

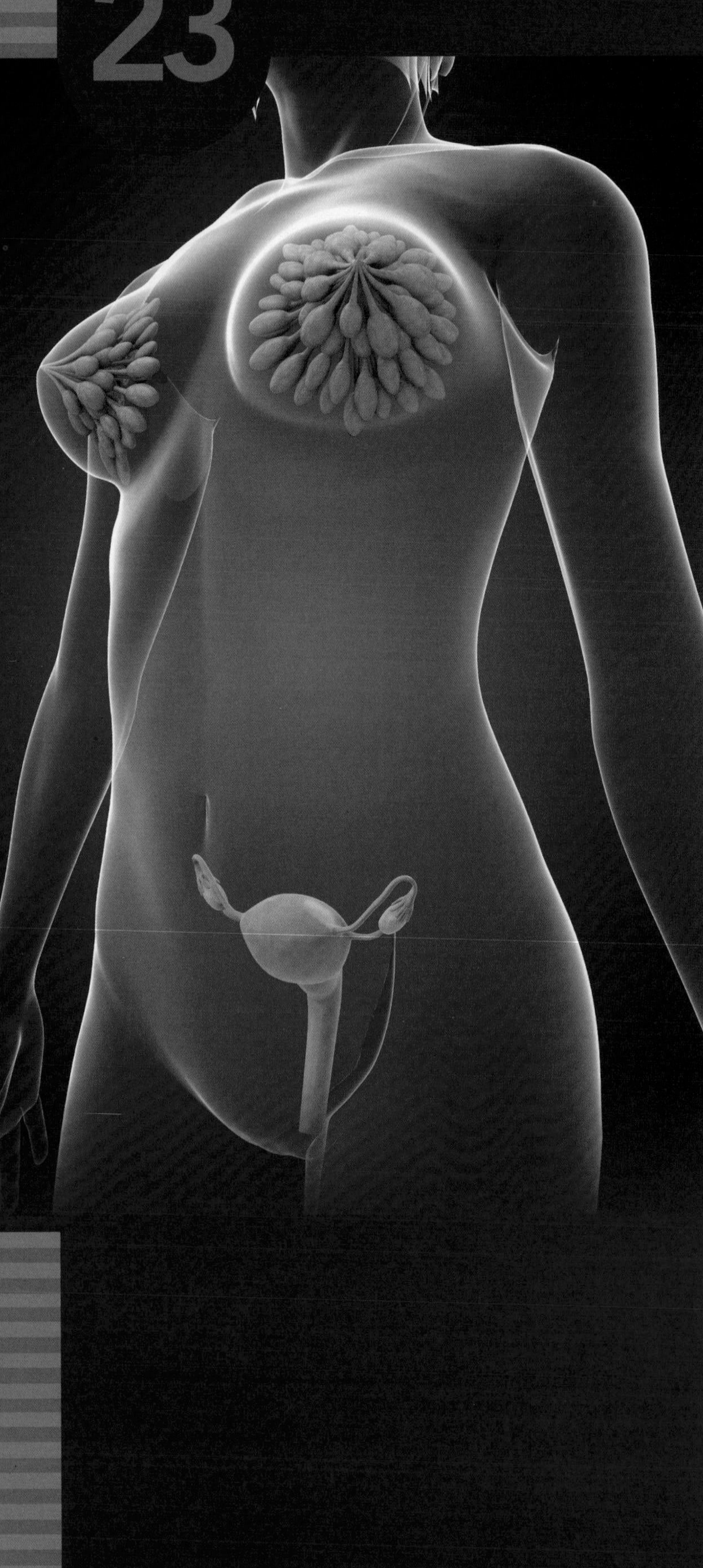

Learning Objectives

After careful study of this chapter, you should be able to:

1 ▸ Identify the male and female gametes, and state the purpose of meiosis. ***p. 524***

2 ▸ Name the accessory organs and gonads of the male reproductive system, and cite the function of each. ***p. 524***

3 ▸ Describe the composition and function of semen. ***p. 525***

4 ▸ Draw and label a spermatozoon. ***p. 527***

5 ▸ Identify the two hormones that regulate the production and development of the male gametes. ***p. 527***

6 ▸ Discuss three types of male reproductive system disorders, and give examples of each. ***p. 529***

7 ▸ Name the accessory organs and gonads of the female reproductive system, and cite the function of each. ***p. 531***

8 ▸ In the correct order, list the hormones produced during the menstrual cycle, citing the source and function of each. ***p. 534***

9 ▸ Describe the changes that occur during and after menopause. ***p. 536***

10 ▸ Cite the main methods of birth control in use. ***p. 536***

11 ▸ Discuss three types of female reproductive system disorders, and give examples of each. ***p. 536***

12 ▸ Using the text and information in the case study, discuss possible causes of infertility in men and women. ***p. 523, 542***

13 ▸ Show how word parts are used to build words related to the reproductive systems (see Word Anatomy at the end of the chapter). ***p. 544***

Disease in Context *Jessica and Brett's Infertility Problems*

"Do you think my history of endometriosis is the reason I am unable to conceive? In my early 20s, I started taking the pill to help with my heavy and painful periods. My gynecologist said that I had stage III endometriosis."

Jessica, a 34-year-old computer systems analyst, and her husband were conferring with Dr. Christensen, an infertility specialist recommended by her family physician.

"I stopped taking the pill about two years ago when Brett and I decided to start a family," Jessica said. "I know it can take a while for the effect of contraceptive pills to wear off, but two years seems really long."

"Your medical history is definitely relevant," the specialist replied and went on to explain. "As you are aware, endometriosis is a condition in which the lining of the uterus grows in locations outside of the uterus. These deposits break down and bleed every month, just like the endometrium within your uterus. This might cause inflammation and changes in the pelvic organs that would affect your ability to conceive. Your physician has already done the preliminary tests for both you and Brett. Your thyroid hormones and FSH levels are normal. Also I see that you monitored your LH levels over several months with urine tests, and the results showed that you are ovulating at about Day 12 of your cycle. Brett, your sperm count was well within normal limits, and the cells were normal in shape and active. This means it appears that you are both able to make healthy gametes. The next step is to look for structural problems in Jessica's reproductive tract that might interfere with fertilization or nourishing the fertilized egg. I'll schedule a laparoscopic exam of your uterus and uterine tubes."

Brett chimed in, "What if we find out that there's something wrong? Do we have any other options?"

"Well, let's take this one step at a time," the doctor responded. "There's always the possibility of in vitro fertilization, or IVF, but let's first see what might be preventing you from becoming pregnant."

Later we will see the results of Jessica's laparoscopy and the couple's options for starting a family. This chapter discusses the male and female reproductive systems and some of the clinical problems associated with each.

ANCILLARIES *At-A-Glance*

Visit thePoint to access the following resources. For guidance in using these resources most effectively, see pp. xv–xvii.

Learning RESOURCES

- Tips for Effective Studying
- Web Figure: Descent of the Testes
- Web Figure: Prostate Surgery Procedures
- Web Figure: Uterine Fibroids
- Web Figure: Laparoscopic Sterilization
- Web Chart: Reproductive Hormones
- Animation: Ovulation and Fertilization
- Health Professions: Physician Assistant
- Detailed Chapter Outline
- Answers to Questions for Study and Review
- Audio Pronunciation Glossary

Learning ACTIVITIES

- Pre-Quiz
- Visual Activities
- Kinesthetic Activities
- Auditory Activities

A LOOK BACK

In this chapter, we return once again to the concept of negative feedback, which regulates some reproductive activities in males and females. We also provide more details on the production and action of the sex hormones, first introduced in Chapter 12. The reproductive system shares some structures with the urinary tract, discussed in Chapter 22.

The chapters in this unit deal with what is certainly one of the most interesting and mysterious attributes of life: the ability to reproduce. The simplest forms of life, one-celled organisms, usually need no partner to reproduce; they simply divide by themselves. This form of reproduction is known as **asexual** (nonsexual) reproduction.

In most animals, however, reproduction is **sexual**, meaning that there are two kinds of individuals, males and females, each of which has specialized cells designed specifically for the perpetuation of the species. These specialized sex cells are known as **gametes** (GAM-etes), or *germ cells*. In the male, they are called **spermatozoa** (sper-mah-to-ZO-ah; sing., spermatozoon) or simply sperm cells; in the female, they are called **ova** (O-vah; sing., ovum) or eggs.

Gametes are characterized by having half as many chromosomes as are found in any other body cell. During their formation, they go through a special process of cell division, called **meiosis** (mi-O-sis), which halves the number of chromosomes (**see Fig. 25-2**). In humans, meiosis reduces the chromosome number in a cell from 46 to 23. The role of meiosis in reproduction is explained in more detail in Chapter 25.

The male and female reproductive systems each include two groups of organs, primary and accessory:

- The primary organs are the **gonads** (GO-nads), or sex glands; they produce the gametes and manufacture hormones. The male gonad is the **testis** (pl., testes), and the female gonad is the **ovary**.
- The **accessory organs** include a series of ducts that transport the gametes as well as various exocrine glands.

The Male Reproductive System

The male reproductive system functions to manufacture spermatozoa (sperm cells) and to deliver them to the female. As illustrated in **Figure 23-1**, this system includes two testes, which are the sites of spermatozoa production, and the accessory organs. We begin our discussion with the ductal system that stores sperm cells and delivers them outside the body. (See **Figure A5-9** in the Dissection Atlas for a photograph of the male reproductive system.)

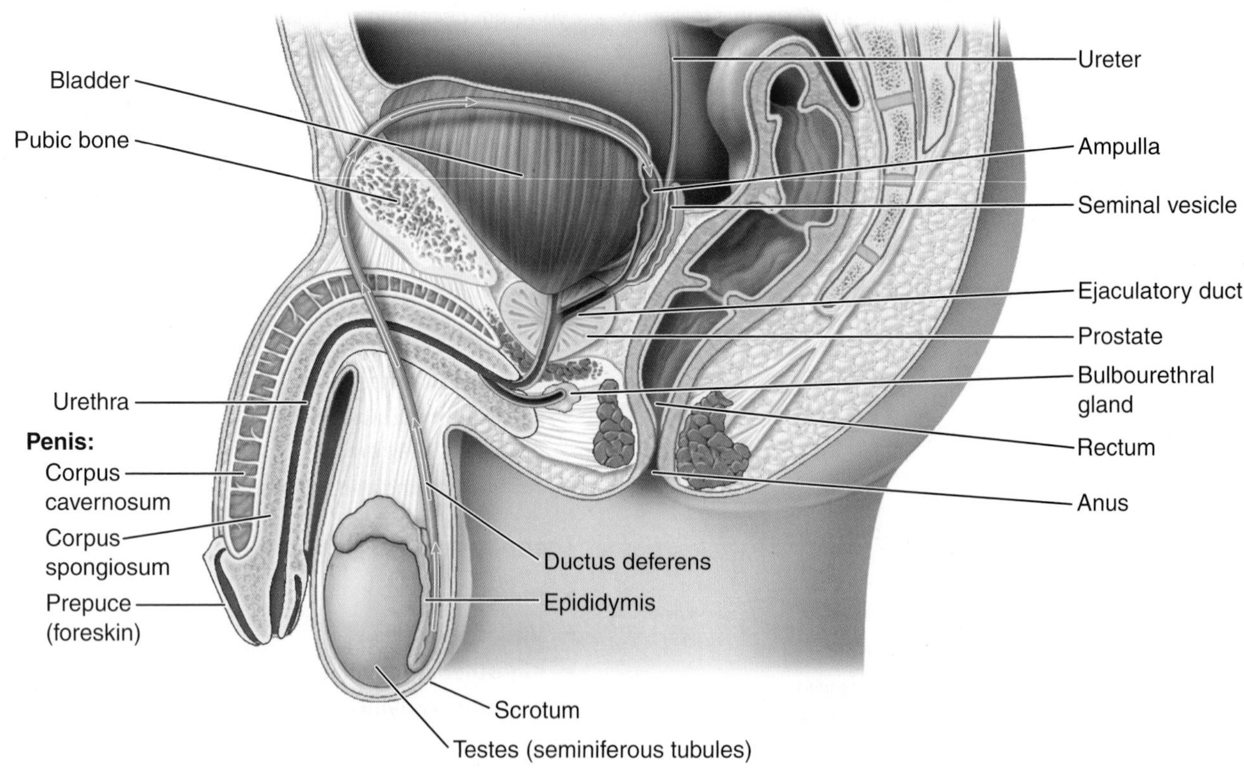

Figure 23-1 **Male reproductive system.** This sagittal section illustrates the organs of the male reproductive system. **ZOOMING IN** What four glands empty secretions into the urethra? What duct receives secretions from the epididymis?

ACCESSORY ORGANS

The testes deliver spermatozoa into a greatly coiled tube called the **epididymis** (ep-ih-DID-ih-mis), which is 6 m (20 ft) long and is located on the surface of the testis (**see Fig. 23-1**). While they are temporarily stored in the epididymis, the sperm cells mature and become motile, able to move or "swim" by themselves.

The epididymis extends upward as the **ductus deferens** (DEF-er-enz), also called the *vas deferens*. This tube loops over the pubic bone and curves behind the urinary bladder. The ductus deferens widens to form an **ampulla** (dilation) just before it joins with the duct of the **seminal vesicle** (VES-ih-kl) on the same side to form the **ejaculatory** (e-JAK-u-lah-to-re) **duct**. The right and left ejaculatory ducts travel through the body of the prostate gland and then empty into the urethra.

SEMEN

Semen (SE-men) (meaning "seed") is the mixture of sperm cells and various secretions that is expelled from the body. It is a sticky fluid with a milky appearance. The pH is in the alkaline range of 7.2 to 7.8. The secretions in semen serve several functions:

- Nourish the spermatozoa
- Transport the spermatozoa
- Neutralize the acidity of the male urethra and the female vaginal tract
- Lubricate the reproductive tract during sexual intercourse
- Prevent infection by means of antibacterial enzymes and antibodies

The glands discussed next contribute secretions to the semen (**see Fig. 23-1**).

The Seminal Vesicles The seminal vesicles are twisted muscular tubes with many small outpouchings. They are approximately 7.5 cm (3 in) long and are attached to the connective tissue at the posterior of the urinary bladder. The glandular lining produces a thick, yellow, alkaline secretion containing large quantities of simple sugar and other substances that provide nourishment for the spermatozoa. The seminal fluid makes up a large part of the semen's volume.

The Prostate Gland The **prostate gland** lies immediately inferior to the urinary bladder, where it surrounds the first part of the urethra. Ducts from the prostate carry its secretions into the urethra. The thin, alkaline prostatic secretion helps neutralize vaginal acidity and enhances the spermatozoa's motility. The prostate gland is also supplied with muscular tissue, which, upon signals from the nervous system, contracts to aid in the expulsion of the semen from the body.

Bulbourethral Glands The **bulbourethral** (bul-bo-u-RE-thral) **glands**, also called *Cowper glands*, are a pair of pea-sized organs located in the pelvic floor just inferior to the prostate gland. They secrete mucus to lubricate the urethra and tip of the penis during sexual stimulation. The ducts of these glands extend approximately 2.5 cm (1 in) from each side and empty into the urethra before it extends into the penis.

Other very small glands secrete mucus into the urethra as it passes through the penis.

THE URETHRA AND PENIS

The male urethra, as discussed in Chapter 22, serves the dual purpose of conveying urine from the bladder and semen from the ejaculatory duct to the outside. Semen ejection is made possible by **erection**, the stiffening and enlargement of the penis, through which the major portion of the urethra extends. The penis is made of spongy tissue containing many blood spaces that are relatively empty when the organ is flaccid but that fill with blood and distend when the penis is erect. This tissue is subdivided into three segments, each called a **corpus** (body) (**see Figs. 23-1 and 23-2**). A single, ventrally located **corpus spongiosum** contains the urethra. On either side is a larger **corpus cavernosum** (pl., corpora cavernosa). At the distal end of the penis, the corpus spongiosum enlarges to form the **glans penis**, which is covered with a loose fold of skin, the **prepuce** (PRE-puse), commonly called the *foreskin*. It is the end of the foreskin that is removed in a **circumcision** (sir-kum-SIZH-un), a surgery frequently performed on male infants for religious or cultural reasons. Experts vary in their opinions on the medical value of circumcision with regard to improved cleanliness and disease prevention.

Ejaculation (e-jak-u-LA-shun) is the forceful expulsion of semen through the urethra to the outside. The process is initiated by reflex centers in the spinal cord that stimulate smooth muscle contraction in the prostate. This is followed by contraction of skeletal muscle in the pelvic floor, which provides the force needed for expulsion. During ejaculation, the involuntary sphincter at the base of the bladder closes to prevent the release of urine.

A male typically ejaculates 2 to 5 mL of semen containing 50 to 150 million sperm cells per mL. Out of the millions of spermatozoa in an ejaculation, only one, if any, can fertilize an ovum. The remainder of the cells live from only a few hours up to a maximum of five to seven days.

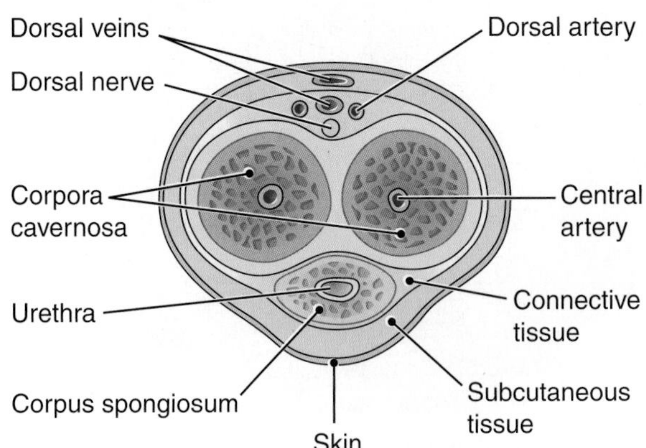

Figure 23-2 **Cross-section of the penis.** The subdivisions of the penis are shown along with associated vessels and a nerve.

ZOOMING IN What subdivision of the penis contains the urethra?

23

THE TESTES

The testes are located outside of the body proper, suspended between the thighs in a sac called the **scrotum** (SKRO-tum) (**Fig. 23-3**). The penis and scrotum together make up the male **external genitalia** (jen-ih-TA-le-ah). The scrotum also contains the epididymis and the proximal portion of the ductus deferens. The testes are oval organs measuring approximately 4.0 cm (1.5 in) in length and approximately 2.5 cm (1 in) in each of the other two dimensions. During embryonic life, each testis develops from tissue near the kidney.

A month or two before birth, the testis normally descends (moves downward) through the **inguinal** (ING-gwih-nal) **canal** in the abdominal wall into the scrotum. There, the testis is suspended by a **spermatic cord** that extends through the inguinal canal (**see Fig. 23-7A** later in this chapter). This cord contains blood vessels, lymphatic vessels, nerves, and the ductus deferens. The gland must descend completely if it is to function normally; to produce spermatozoa, the testis must be kept at the temperature of the scrotum, which is several degrees lower than that of the abdominal cavity.

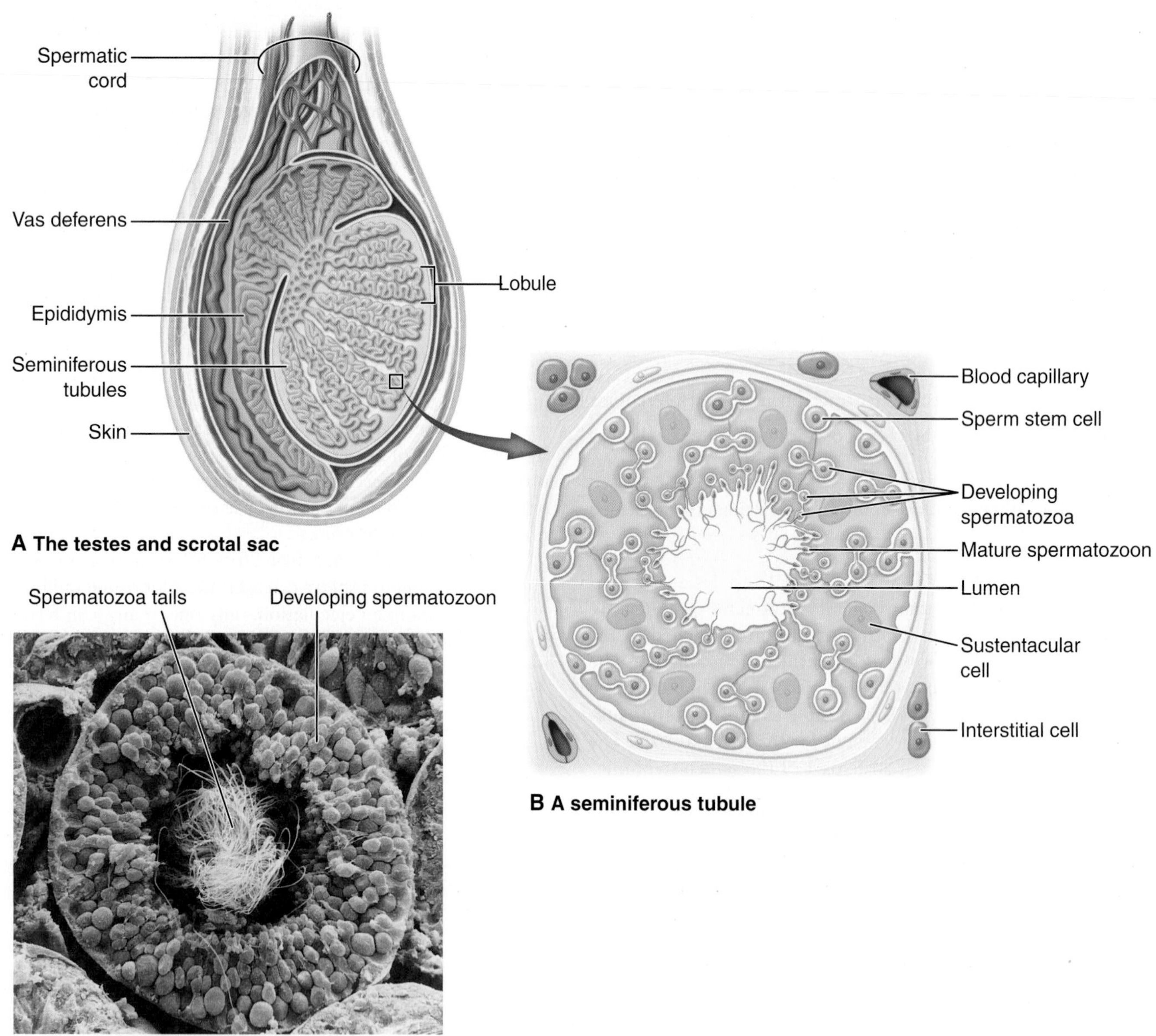

Figure 23-3 **The testis.** **KEY POINT** Spermatozoa develop within the testis's seminiferous tubules. **A.** The testes and scrotal sac. **B.** A seminiferous tubule. **C.** A micrograph illustrating a cross-section of a seminiferous tubule. **ZOOMING IN** Where are the interstitial cells located?

See the Student Resources on thePoint for a figure showing descent of the testes.

Internal Structure Most of the specialized tissue of the testis consists of tiny, coiled **seminiferous** (seh-mih-NIF-er-us) **tubules** (see Fig. 23-3A). Primitive cells in the walls of these tubules develop into spermatozoa, aided by neighboring cells called **sustentacular** (sus-ten-TAK-u-lar) (Sertoli) **cells** (Fig. 23-3B). These so-called "nurse cells" nourish and protect the developing gametes. They also secrete a protein that binds testosterone in the seminiferous tubules. Once produced, spermatozoa pass into the epididymis for storage and final maturation.

Specialized **interstitial** (in-ter-STISH-al) **cells** that secrete the male sex hormone **testosterone** (tes-TOS-teh-rone) are located between the seminiferous tubules. An older name for these cells is *Leydig* (LI-dig) *cells*. Figure 23-3C is a microscopic view of a seminiferous tubule in cross-section, showing the developing spermatozoa.

The Spermatozoa Spermatozoa are tiny individual cells illustrated in Figure 23-4. They are so small that at least 200 million can be contained in the average ejaculation. Beginning at puberty, sperm cells are manufactured continuously in the seminiferous tubules.

The spermatozoon has an oval head that is mostly a nucleus containing chromosomes. The **acrosome** (AK-ro-some), which covers the head like a cap, contains enzymes that help the sperm cell penetrate the ovum.

Whiplike movements of the tail (flagellum) propel the sperm through the female reproductive tract to the ovum. The cell's middle region (midpiece) contains many mitochondria that provide energy for movement.

CHECKPOINTS 

- 23-1 What is the process of cell division that halves the chromosome number in a cell to produce a gamete?
- 23-2 What is the male gamete called?
- 23-3 What is the male gonad?
- 23-4 What is the structure on the surface of the testis that stores sperm?
- 23-5 What glands, aside from the testis, contribute secretions to semen?
- 23-6 What are the main subdivisions of a spermatozoon?

Hormonal Control of Male Reproduction

The activities of the testes are under the control of two hormones produced by the anterior pituitary as well as testicular hormones themselves (Fig. 23-5). These hormones are named for their activity in female reproduction (described later), although they are chemically the same in both males and females. They are:

- **Follicle-stimulating hormone** (**FSH**) stimulates the sustentacular cells to produce growth factors that promote the formation of spermatozoa.
- **Luteinizing hormone** (**LH**) stimulates the interstitial cells between the seminiferous tubules to produce testosterone, which is also needed for sperm cell development.

Starting at puberty, the hypothalamus begins to secrete a hormone (gonadotropin-releasing hormone, known as GnRH) that triggers the release of FSH and LH. These hormones are continuously secreted in the male.

The activity of the hypothalamus is in turn regulated by a negative feedback mechanism involving testosterone. As the blood level of testosterone increases, the hypothalamus secretes less GnRH; as the level of testosterone decreases, the hypothalamus secretes more GnRH (Fig. 23-5).

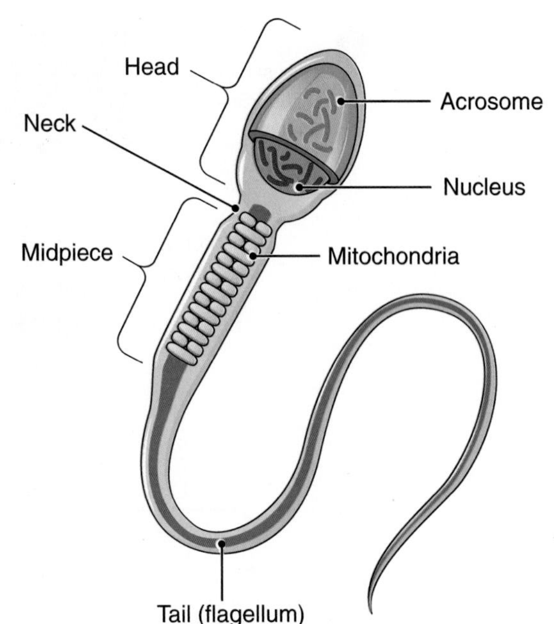

Figure 23-4 Human spermatozoon. The diagram shows major structural features of the male gamete. **ZOOMING IN** What organelles provide energy for sperm cell motility?

TESTOSTERONE

From the testis, testosterone diffuses into surrounding fluids and is then absorbed into the bloodstream. This hormone has many functions, including (see Fig. 23-5)

- Development and maintenance of the male reproductive accessory organs
- Development of spermatozoa
- Development of **secondary sex characteristics**, traits that characterize males and females but are not directly concerned with reproduction. In males, these traits include a deeper voice, broader shoulders, narrower hips, a greater percentage of muscle tissue, and more body hair than found in females.

23

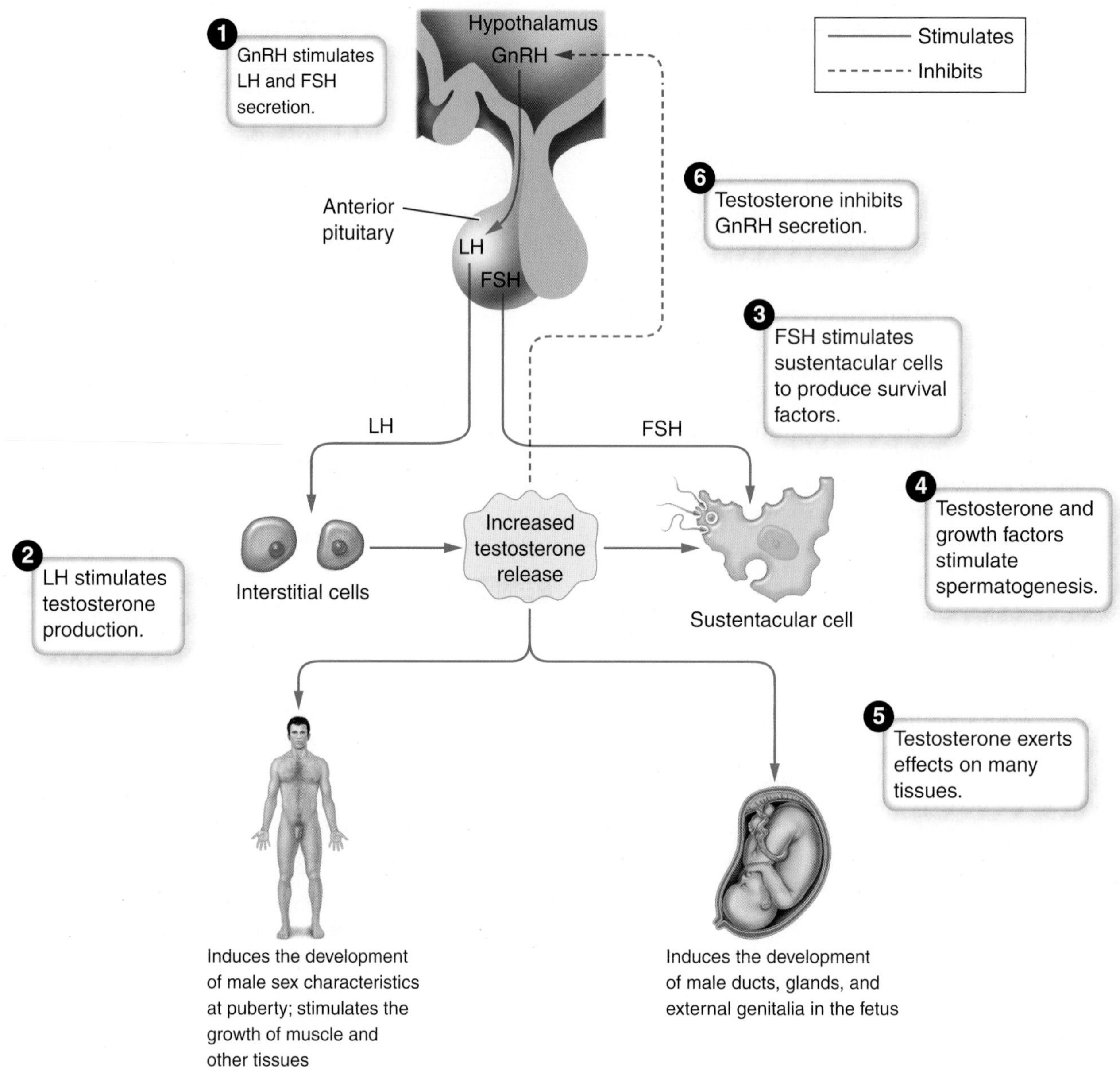
Hypothalamus
GnRH
Stimulates
Inhibits
1 GnRH stimulates LH and FSH secretion.
Anterior pituitary
LH
FSH
6 Testosterone inhibits GnRH secretion.
3 FSH stimulates sustentacular cells to produce survival factors.
LH
FSH
Increased testosterone release
Interstitial cells
Sustentacular cell
2 LH stimulates testosterone production.
4 Testosterone and growth factors stimulate spermatogenesis.
5 Testosterone exerts effects on many tissues.
Induces the development of male sex characteristics at puberty; stimulates the growth of muscle and other tissues
Induces the development of male ducts, glands, and external genitalia in the fetus

Figure 23-5 **Hormonal control of male reproduction.** The hypothalamus and anterior pituitary gland control testosterone synthesis by the testes. Testosterone regulates sperm production and other functions relating to male reproduction and also feeds back to inhibit its own secretion. **ZOOMING IN** Which hormone stimulates testosterone secretion—LH or FSH?

CHECKPOINTS

- 23-7 What two pituitary hormones regulate both male and female reproduction?
- 23-8 Which cell type in the testis produces the main male sex hormone?

The Effects of Aging on Male Reproduction

A gradual decrease in the production of testosterone and spermatozoa begins at about the age of 40 and continues throughout life. Sperm motility and quality also decline in middle and later life, but some men retain fertility into their 90s. Secretions from the prostate and seminal vesicles decrease in amount and become less viscous.

Erectile dysfunction may affect men at any age, but its prevalence increases with age. It is defined as an inability to attain or maintain an erection adequate to engage in sexual intercourse. Box 23-1 has more information on the causes and treatments of this disorder.

Nonmalignant enlargement of the prostate, known as benign prostatic hyperplasia (BPH), exists in nearly all men by the age of 80 and causes symptoms in about half of them. As noted in Chapter 22, an enlarged prostate can put pressure on the urethra and interfere with urination. In severe cases, urine

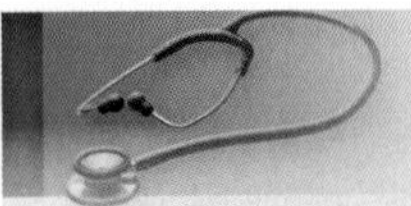

Box 23-1

CLINICAL PERSPECTIVES

Treating Erectile Dysfunction

Approximately 25 million American men and their partners are affected by **erectile dysfunction** (ED), the inability to achieve or sustain an erection long enough to have satisfying sexual intercourse. Although ED is more common in men over the age of 65, it can occur at any age and can have many causes. Until recently, ED was believed to be caused by psychological factors, such as stress or depression. It is now known that many cases of ED are caused by physical factors, including cardiovascular disease, diabetes, spinal cord injury, and damage to penile nerves during prostate surgery. Antidepressant and antihypertensive medications also can produce ED.

Erection results from interaction between the autonomic nervous system and penile blood vessels. Sexual arousal stimulates parasympathetic nerves in the penis to release a compound called nitric oxide (NO), which activates the vascular smooth muscle enzyme guanylyl cyclase. This enzyme catalyzes production of cyclic GMP (cGMP), a potent vasodilator that increases blood flow into the penis to cause erection. Physical factors that cause ED prevent these physiologic occurrences.

Until recently, treatment options for ED, such as penile injections, vacuum pumps, and insertion of medications into the penile urethra, were inadequate, inconvenient, and painful. Today, drugs that target the physiologic mechanisms that underlie erection are giving men who suffer from ED new hope. The best known of these is sildenafil (Viagra), which works by inhibiting the enzyme that breaks down cGMP, thus prolonging the effects of NO. Because of its short duration of action, Viagra must be taken shortly before sexual intercourse. Other drugs, such as tadalafil (Cialis) can be taken once daily, removing the need to plan the timing of sexual activity.

Although effective in about 80% of all ED cases, Viagra can cause some relatively minor side effects, including headache, nasal congestion, stomach upset, and blue-tinged vision. Viagra should never be used by men who are taking nitrate drugs to treat angina. Because nitrate drugs elevate NO levels, taking them with Viagra, a drug that prolongs the effects of NO, can cause life-threatening hypotension.

stagnates in the bladder, increasing susceptibility to infection. BPH may respond to medication to shrink the prostate. If urinary function is threatened, however, surgery is needed to reduce the obstruction. Traditionally, this surgery has been performed through the urethra, in a transurethral resection of the prostate (TURP). Newer methods include use of laser and ultrasound to destroy excess tissue or placement of a stent to widen the urethra.

See the Student Resources on thePoint for illustrations of prostate surgery procedures.

Disorders of the Male Reproductive System

The most common disorders of the male reproductive system are structural problems, infections, and tumors. Some of these can reduce a man's fertility.

STRUCTURAL DISORDERS

Cryptorchidism (kript-OR-kid-izm) is a failure of the testis to descend into the scrotum. Unless corrected in childhood, this condition results in infertility. Undescended testes are also particularly subject to tumor formation. Most testes that are undescended at birth descend spontaneously by 1 year of age. Surgical correction is the usual remedy in the remaining cases.

Testicular torsion occurs when the testis rotates within the scrotum, twisting the spermatic cord and interfering with the testis's blood supply (**Fig. 23-6**). Normally, the testis is firmly anchored to the scrotum and cannot twist; however, this attachment is lacking in some boys, predisposing them to testicular torsion. Torsion may occur during descent of the testis or during postnatal testicular growth, most commonly between the ages of 12 to 18 years. The condition is accompanied by acute pain and swelling and must be treated within six hours in order to prevent death of testicular tissue.

Hernia (HER-ne-ah), or *rupture*, refers to the abnormal protrusion of an organ or organ part through the wall of the cavity in which it is normally contained (**Fig. 23-7**). This condition was described in Chapter 19 with regard to hiatal hernia. An **inguinal hernia** occurs when a congenital weakness in the abdominal wall allows a portion of the small intestine to pass through the inguinal canal into the scrotum. This condition can also result from excessive pressure in the abdominal

23

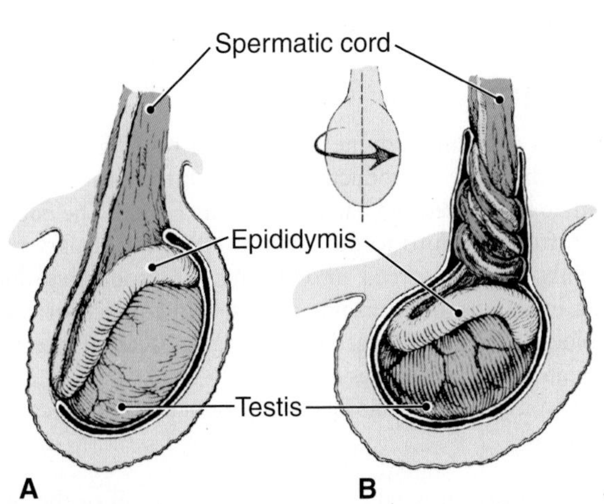

Figure 23-6 Torsion of the testis. **KEY POINT** The testis rotates, twisting the spermatic cord. **A.** Normal. **B.** Torsion.

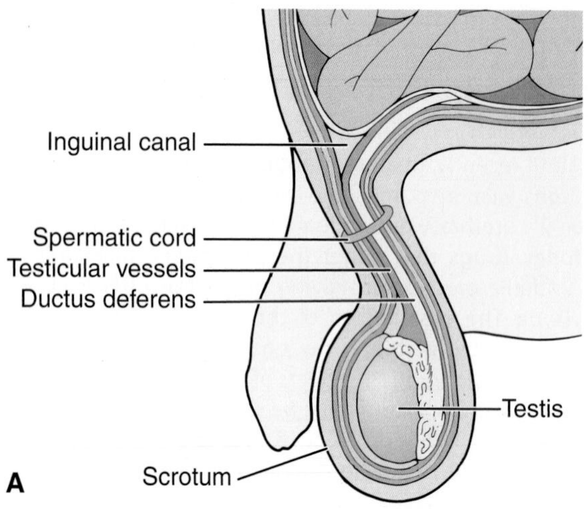

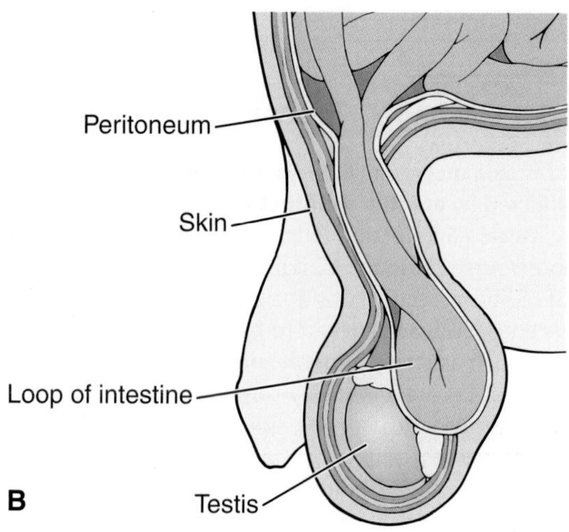

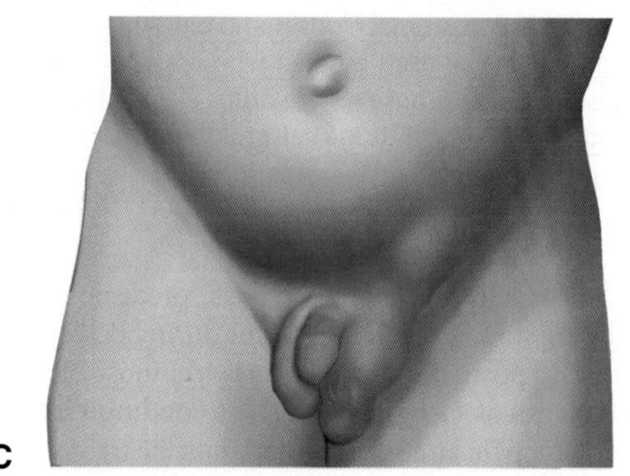

Figure 23-7 Inguinal hernia. **KEY POINT** The intestine protrudes through a weakness in the abdominal wall at the inguinal canal. **A.** Normal. **B.** Inguinal hernia. **C.** An inguinal hernia can cause a visible bulge in the scrotum and inguinal area.

cavity, during heavy lifting, for example. Corrective surgery may be necessary in order to prevent death of intestinal tissue.

Phimosis (fi-MO-sis) is a tightness of the foreskin (prepuce) so that it cannot be drawn back. Phimosis may be remedied by circumcision, in which part or all of the foreskin is surgically removed.

INFECTIONS

Infections and autoimmune diseases can result in inflammation of the male reproductive organs. Infections frequently originate from intestinal bacteria, such as *E. coli*. The bacteria spread from the anus to the urethra, infect the urinary tract, and subsequently spread into the male reproductive ducts. Other infections are transmitted by sexual contact, or rarely, via blood or lymph. Regardless of origin, bacterial infections are treated with antibiotics and rest.

Any region of the male reproductive tract can be inflamed:

- **Prostatitis** (pros-tah-TI-tis) is inflammation of the prostate gland. In addition to infection, other possible causes of prostatitis are autoimmune disorders and injury.
- A congenital malformation in the urinary tract may favor the development of **epididymitis** (ep-ih-did-ih-MI-tis), or inflammation of the epididymis.
- **Orchitis** (or-KI-tis), inflammation of the testis, may result from the spread of infectious agents from the epididymis or prostate to the seminiferous tubules. Certain infections, particularly mumps, may also spread through the blood to infect and inflame the testes. Orchitis can result in infertility if it occurs during or after puberty.

Sexually Transmitted Infections

Sexually transmitted infections (STIs), formerly known as *sexually transmitted diseases* (STDs) or *venereal diseases* (VDs), are spread through sexual contact in both males and females. They most commonly involve chlamydial infections and gonococcal infections (gonorrhea). In males, these diseases are manifested by a discharge from the urethra, which may be accompanied by a burning sensation and pain, especially during urination. The infection may travel along the mucous membrane to cause prostatitis and epididymitis. If untreated, the infection can spread to the seminiferous tubules and cause orchitis. If both sides of the scrotum are affected and enough scar tissue is formed to destroy the tubules, infertility is likely.

Another common STI is a persistent infection called **genital herpes.** Caused by a virus, this disorder is characterized by fluid-filled vesicles (blisters) on and around the genitalia. This disease cannot be cured, but medication can reduce the number and severity of outbreaks.

The STI **syphilis** is caused by a spirochete (*Treponema pallidum*). Because syphilis spreads quickly in the bloodstream, it is regarded as a systemic disorder, but it can be successfully treated with antibiotics (see Appendix 3, **Table 1**). The genital ulcers caused by syphilis increase the chances of infection with the AIDS virus. HIV itself is considered an STI because sexual activity is its most common route of transmission. (See **Box 23-2** on avoiding STIs.)

CANCER

Prostatic cancer is the most common cancer of males in the United States, especially among men older than 50 years of age. Other risk factors, in addition to age, are race, family history, and certain environmental agents. A high-fat diet may increase risk by promoting production of male sex hormones. Prostatic cancer is frequently detected as a nodule during rectal examination. Blood tests for prostate-specific antigen (PSA) can help in early detection. This protein increases in cases of prostate cancer, although it may increase in other prostatic disorders as well. Depending on the age of the patient and the nature of the cancer, the course of treatment may include surveillance, radiation therapy, surgery, or hormone treatments.

Testicular cancer affects young to middle-aged adults. Almost all testicular cancers arise in the germ cells, and a tumor can metastasize through the lymphatic system at an early stage of development. Early detection improves the chances for effective treatment, and a testicular examination should be part of any regular physical checkup. Men should discuss with their doctor the advisability of doing a regular testicular self-examination. According to the American Cancer Society, the five-year survival rate for localized testicular cancer is now greater than 95%. Often fertility can be preserved, although sperm banking is an option for men about to undergo treatment for testicular cancer.

CHECKPOINTS 

- ☐ **23-9** What is the term for failure of the testis to descend?
- ☐ **23-10** What does the abbreviation STI mean?

The Female Reproductive System

The female gonads are the paired ovaries (O-vah-reze), where the female gametes, or ova, are formed. The remainder of the female reproductive tract consists of an organ (uterus) to hold and nourish a developing infant, various passageways, and the external genital organs (**Fig. 23-8**). As with our discussion of the male, we introduce the accessory organs before focusing on the structure and function of the ovary. (See **Figure A5-10** in the Dissection Atlas for a photograph of the female reproductive system.)

ACCESSORY ORGANS

The accessory organs in the female are the uterus, the uterine tubes, the vagina, the greater vestibular glands (not illustrated), and the vulva (**see Fig. 23-8**). The breasts are also considered to be accessory reproductive organs in the female; because of their role in child nourishment, they are discussed in Chapter 24.

The Uterus The **uterus** (U-ter-us) is a pear-shaped, muscular organ approximately 7.5 cm (3 in) long, 5 cm (2 in) wide, and 2.5 cm (1 in) deep, in which the fetus develops to maturity (**see Fig. 23-8**). (The organ is typically larger in women who have borne children and smaller in postmenopausal women.) The uterus's superior portion rests on the upper surface of the urinary bladder; the inferior

23

Box 23-2

HEALTH MAINTENANCE

Sexually Transmitted Infections: Lowering Your Risks

Sexually transmitted infections (STIs) such as chlamydia, gonorrhea, genital herpes, HIV, and syphilis are some of the most common infectious diseases in the United States, affecting more than 13 million men and women each year. These diseases are associated with complications such as pelvic inflammatory disease, epididymitis, infertility, liver failure, neurological disorders, cancer, and AIDS. Women are more likely to contract STIs than are men. The same mechanisms that transport sperm cells through the female reproductive tract also move infectious organisms. The surest way to prevent STIs is to avoid sexual contact with others. If you are sexually active, the following techniques can lower your risks:

- Maintain a monogamous sexual relationship with an uninfected partner.
- Correctly and consistently use a male or female condom. Although not 100% effective, condoms greatly reduce the risk of contracting an STI.
- Avoid contact with body fluids such as blood, semen, and vaginal fluids, all of which may harbor infectious organisms.
- Urinate and wash the genitals after sex. This may help remove infectious organisms before they cause disease.
- Have regular checkups for STIs. Most of the time STIs cause no symptoms, particularly in women.

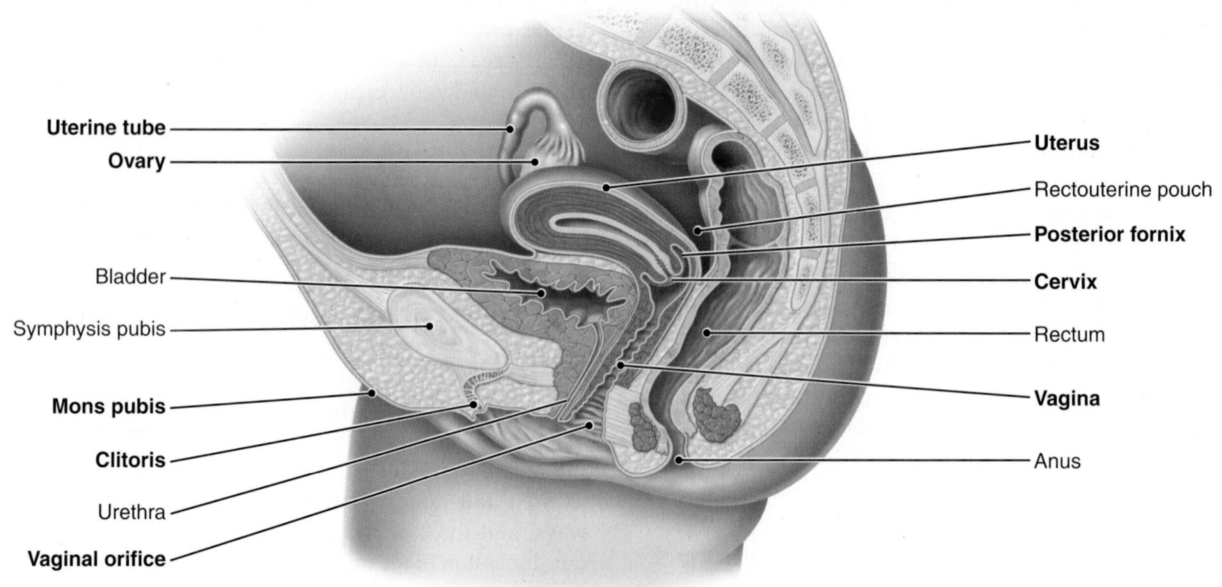

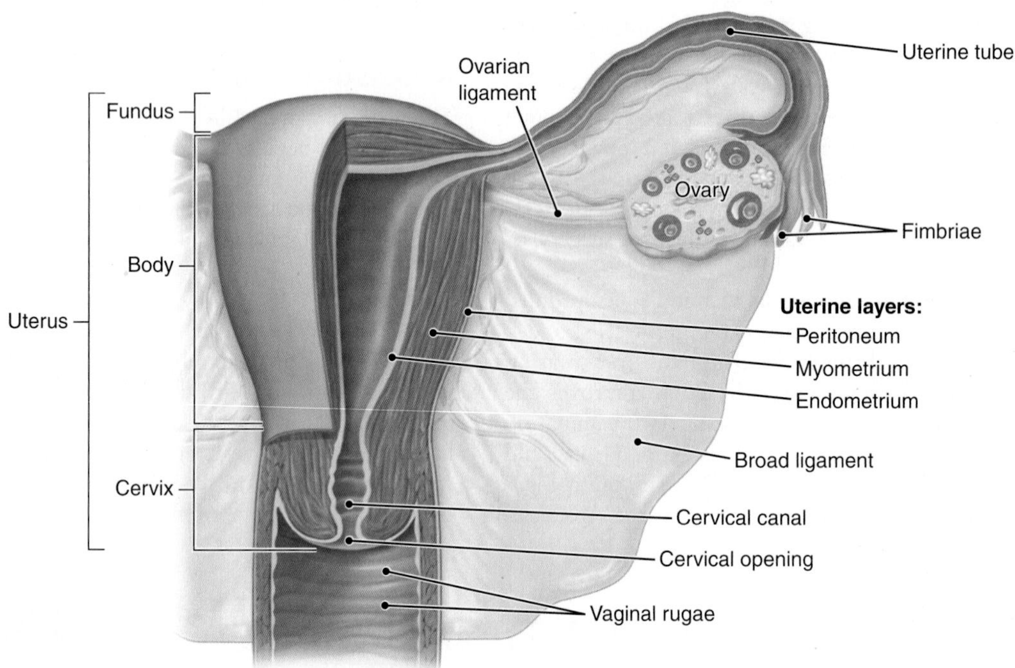

Figure 23-8 **Female reproductive system. A.** As shown in this sagittal view, the female internal genitalia are sandwiched between structures of the urinary and gastrointestinal systems, which are also shown. Structures of the reproductive system are labeled in bold type. **B.** Ligaments hold the uterus and ovaries in place. **ZOOMING IN** What is the deepest part of the uterus called? The most inferior portion?

portion is embedded in the pelvic floor between the bladder and the rectum. Its wider upper region is called the **body**, or corpus; the lower, narrower region is the **cervix** (SER-viks), or neck. The small, rounded region above the level of the tubal entrances is known as the **fundus** (FUN-dus).

Folds of the peritoneum called the *broad ligaments* support the uterus, extending from each side of the organ to the lateral body wall. Along with the uterus, these two membranes form a partition dividing the female pelvis into anterior and posterior areas. The ovaries are suspended from the broad ligaments, and the uterine tubes lie within the upper borders. Blood vessels that supply these organs are found between the layers of the broad ligaments.

The muscular wall of the uterus is called the **myometrium** (mi-o-ME-tre-um) (**Fig. 23-8B**). The lining of the uterus is a specialized epithelium known as **endometrium** (en-do-ME-tre-um). This inner layer undergoes monthly changes during a woman's reproductive years, first building up to nourish a fertilized egg, then breaking down if no fertilization has occurred. This degenerated endometrial tissue is released as the menstrual flow. The cavity inside the uterus is shaped somewhat like a capital T, but it is capable of changing shape and dilating as a fetus develops.

The Uterine Tubes Each **uterine** (U-ter-in) **tube** is a small, muscular structure, nearly 12.5 cm (5 in) long, extending from the uterus to a point near the ovary (**see Fig. 23-8**). They are also known as *oviducts* (O-vih-dukts) or *fallopian* (fah-LO-pe-an) *tubes*. There is no direct connection between the ovary and uterine tube. Instead, the ova are released from the ovary into the abdominal cavity. **Fimbriae** (FIM-bre-e), small, fringelike extensions of the tube's opening, create currents in the peritoneal fluid that sweep the ovum into the uterine tube.

Unlike the sperm cell, the ovum cannot move by itself. Its progress through the uterine tube toward the uterus depends on the sweeping action of cilia in the tube's lining and on peristalsis of the tube. It takes about five days for an ovum to reach the uterus from the ovary.

The Vagina The **vagina** is a muscular tube approximately 7.5 cm (3 in) long (**Fig. 23-8A**). The cervix dips into the vagina's superior portion forming a circular recess known as the **fornix** (FOR-niks). The deepest area of the fornix, located behind the cervix is the **posterior fornix**. This recess in the posterior vagina lies adjacent to the most inferior portion of the peritoneal cavity, a narrow passage between the uterus and the rectum named the **rectouterine pouch**. A rather thin layer of tissue separates the posterior fornix from this region so that abscesses or tumors in the peritoneal cavity can sometimes be detected by vaginal examination.

The lining of the vagina is a wrinkled mucous membrane similar to that found in the stomach (**Fig. 23-8B**). The rugae (folds) permit enlargement so that childbirth usually does not tear the lining. In addition to being a part of the birth canal, the vagina is the organ that receives the penis during sexual intercourse. A fold of membrane called the **hymen** (HI-men) may sometimes be found at or near the vaginal (VAJ-ih-nal) canal opening.

The Greater Vestibular Glands Just superior and lateral to the vaginal opening are the two mucus-producing **greater vestibular** (ves-TIB-u-lar) (Bartholin) **glands** (not illustrated). These glands secrete into an area near the vaginal opening known as the *vestibule*. Like the bulbourethral glands in males, these glands provide lubrication during intercourse. If a gland becomes infected, a surgical incision may be needed to reduce swelling and promote drainage.

The Vulva and the Perineum The external female genitalia make up the **vulva** (VUL-vah) (**Fig. 23-9**). This includes two pairs of **labia** (LA-be-ah), the larger *labia majora* (sing., labium majus) and smaller *labia minora* (sing., labium minus). It also includes the **clitoris** (KLIT-o-ris), a small organ of great sensitivity, as well as the openings of the urethra and vagina, and the *mons pubis*, a pad of fatty tissue over the pubic symphysis (joint) (**see Fig. 23-8A**). Although the entire pelvic floor in both the male and female is properly called the **perineum** (per-ih-NE-um), those who care for pregnant women usually refer to the limited area between the vaginal opening and the anus as the *perineum* or *obstetric perineum*.

THE OVARIES AND OVA

The ovary is a small, somewhat flattened oval body measuring approximately 4 cm (1.6 in) in length, 2 cm (0.8 in) in width, and 1 cm (0.4 in) in depth (**Fig. 23-10**). Like the testes, the ovaries descend, but only as far as the pelvic cavity. Here, they are held in place by ligaments, including the broad ligaments, the ovarian ligaments, and others, that attach them to the uterus and the body wall.

The outer layer of each ovary is made of a single layer of epithelium. The ova are produced beneath this layer. Each ovum is contained within a small cluster of cells called an **ovarian follicle** (o-VA-re-an FOL-ih-kl) (**Fig. 23-10**). The follicular cells protect the ovum and produce the ovarian hormones.

Unlike males, females are born with all of the gametes they will ever produce. The ovaries of a newborn female contain more than a million immature follicles, but only a few hundred thousand remain viable at puberty. As puberty approaches, a few follicles begin to develop. Their oocytes enlarge, and the follicular cells multiply. During each month of a woman's reproductive years, one developing follicle

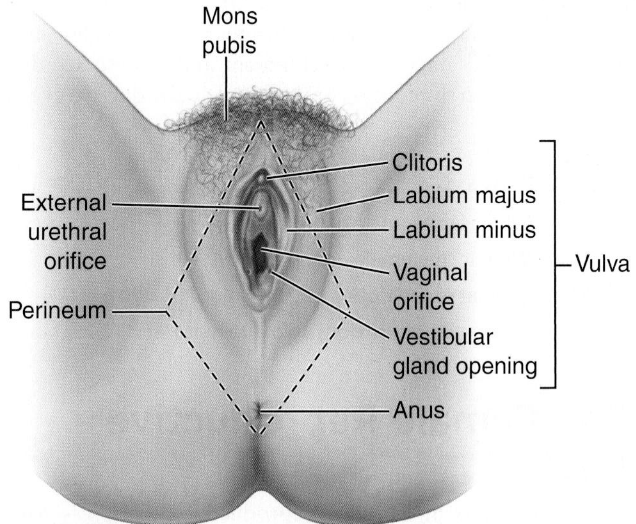

Figure 23-9 External parts of the female reproductive system. **KEY POINT** The external female genitalia, or vulva, includes the labia, clitoris, and mons pubis. Nearby structures are also shown.

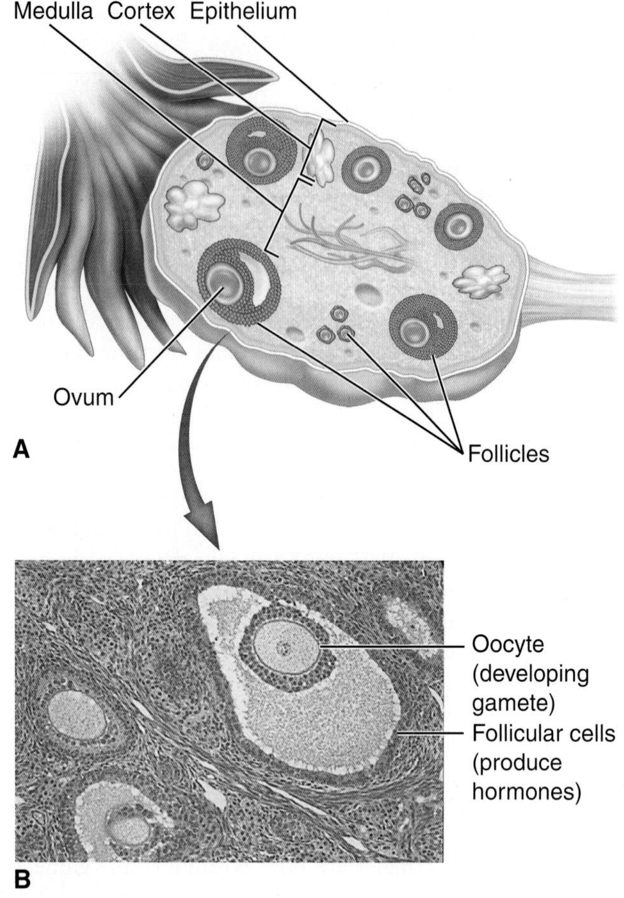

Figure 23-10 The ovary. **KEY POINT** The gametes and hormone-producing cells of the ovary are organized in follicles. **A.** The ovary, containing follicles in different stages of maturation. **B.** This micrograph shows an oocyte within a mature follicle.

(usually) is chosen to fully mature and release its ovum. **Figure 23-10** shows follicles at different stages of development. The final stages of development and release are discussed shortly in the context of the female reproductive cycle.

CHECKPOINTS

- 23-11 What is the female gamete called?
- 23-12 What is the female gonad called?
- 23-13 In what organ does a fetus develop?
- 23-14 In what structure does an ovum mature?

The Female Reproductive Cycle

In the female, as in the male, reproductive function is controlled by pituitary hormones that are regulated by the hypothalamus. Female activity differs, however, in that it is cyclic; it shows regular patterns of increases and decreases in hormone levels. The most obvious sign of these changes is periodic vaginal bleeding, or **menstruation** (men-stru-A-shun), so this cycle is known as the **menstrual cycle.** The typical length of the menstrual cycle varies between 22 and 45 days, but 28 days is taken as an average, with the first day of menstrual flow being considered the first day of the cycle (**Fig. 23-11**).

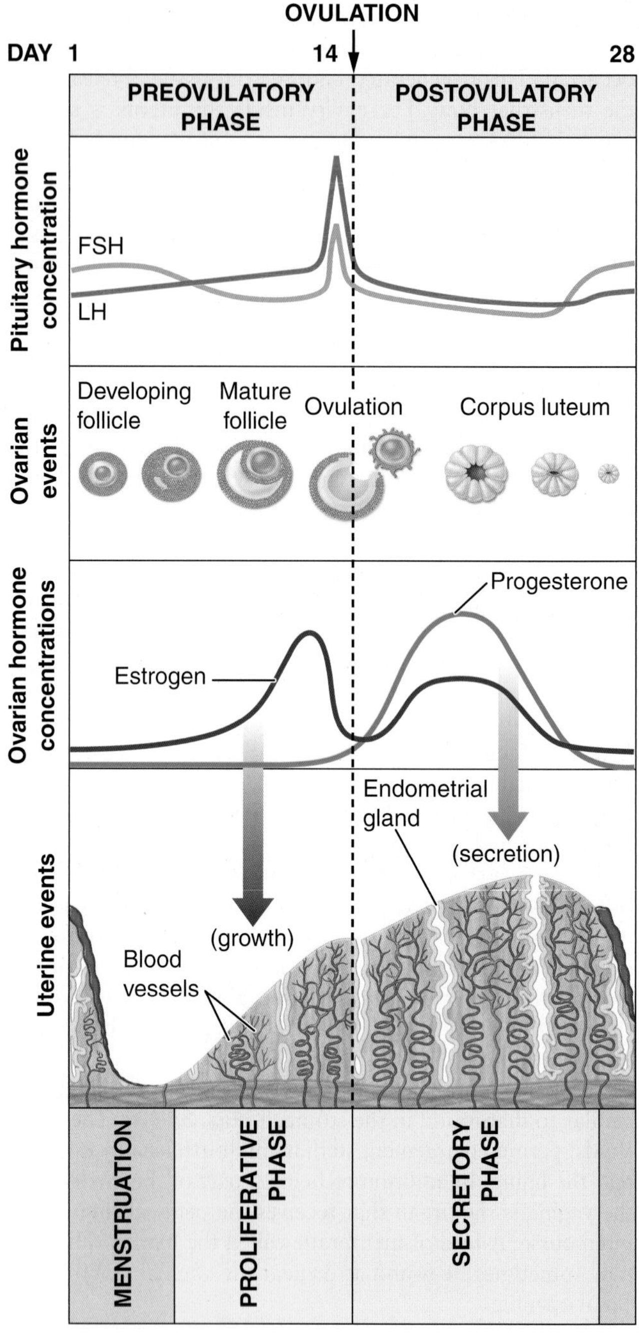

Figure 23-11 The female reproductive cycle. **KEY POINT** The anterior pituitary gland secretes FSH and LH, which control the follicle and corpus luteum. The follicle and corpus luteum secrete ovarian hormones, which control changes in the uterine endometrium. This figure illustrates an average 28-day menstrual cycle with ovulation on Day 14. **ZOOMING IN** What ovarian hormone peaks closest to ovulation? What ovarian hormone peaks after ovulation?

Before we delve into the details of the cycle, take a moment to scan **Figure 23-11.** **Ovulation** (ov-u-LA-shun), the release of the gamete from the ovary, separates the reproductive cycle into the preovulatory (follicular) and postovulatory (luteal) phases. The first row illustrates changes in the production of pituitary hormones. The second row illustrates how pituitary hormones alter follicular development in the ovary. As the follicle grows and transforms, it produces varying amounts of hormones, as shown in the third row. These hormones cause cyclic changes in the uterus, as shown at the bottom of the diagram.

PREOVULATORY PHASE

At the start of each cycle, under the influence of pituitary FSH, several follicles, each containing an ovum, enter the final stages of maturation. As each follicle matures, it enlarges, and fluid accumulates in its central cavity. While it grows, the follicle secretes increasing amounts of the ovarian hormone **estrogen** (ES-tro-jen), which stimulates further growth of the follicle (**see Fig. 23-11**, second row). (*Estrogen* is the term used for a group of related hormones, the most active of which is estradiol.) Eventually, most of the developing follicles die off, and only a single follicle survives to release its ovum, as discussed shortly. Any maturing follicles that do not release their ova simply degenerate.

The estrogen produced by the follicles travels through the blood to the uterus (as well as to other tissues), where it starts preparing the endometrium for a possible pregnancy. This preparation includes thickening of the endometrium and elongation of the glands that produce uterine secretions. The time period between menstruation and ovulation is known as the **proliferative phase** of uterine development (**see Fig. 23-11**, bottom).

Estrogen in the blood also acts as a feedback messenger to inhibit the release of FSH. When estrogen reaches high levels, it stimulates the **LH surge**, a sharp rise of LH in the blood (**see Fig. 23-11**, first row). (Note that there is also a small rise in FSH at this time caused by the increase in hypothalamic-releasing hormone that promotes the rise in LH.) The LH surge triggers the next stage of the reproductive cycle—ovulation.

OVULATION AND THE POSTOVULATORY PHASE

Approximately one day after the LH surge, ovulation occurs. The wall of the mature follicle and the adjacent ovarian tissue both rupture, releasing the ovum into the peritoneum. Currents in the peritoneal fluid created by the fimbriae of the uterine tubes usually propel the ovum into the tube. In some women, ovulation causes a stabbing abdominal pain that can persist minutes or hours.

In an average 28-day cycle, ovulation occurs on Day 14 and is followed two weeks later by the start of the menstrual flow (the beginning of a new cycle). However, an ovum can be released any time from Days 7 to 21, thus accounting for the variation in the length of normal cycles.

In addition to causing ovulation, LH transforms the ruptured follicle into a solid mass called the **corpus luteum** (LU-te-um). This structure secretes estrogen and also **progesterone** (pro-JES-ter-one), a hormone that promotes the survival of a fertilized ovum and eventually an embryo. Under the influence of estrogen and progesterone, the endometrium continues to thicken, and the glands and blood vessels increase in size. The glands also secrete increasing amounts of fluid and nutrients, so the time period after ovulation and before menstruation is known as the **secretory phase** of the uterine cycle. The rising levels of estrogen and progesterone feed back to inhibit the release of FSH and LH from the pituitary.

During the postovulatory phase, the ovum makes its journey to the uterus by way of the uterine tube. If the ovum is fertilized, it begins to secrete hormones that maintain the corpus luteum for a few months. However, if the ovum is not fertilized, it dies within two to three days, and the corpus luteum degenerates by about 11 days postovulation. Sometimes, as a result of normal ovulation, the corpus luteum persists and forms a small ovarian cyst (fluid-filled sac). This condition usually resolves without treatment, but large cysts may rupture or leak, causing severe abdominal pain. These larger cysts may require surgery.

MENSTRUATION

If fertilization does not occur, the corpus luteum degenerates, and the levels of estrogen and progesterone decrease. Without the hormones to support growth, the endometrium degenerates. Small hemorrhages appear in this tissue, producing the bloody discharge known as the **menses** (MEN-seze) or *menstrual flow*. Bits of endometrium break away and accompany the blood flow during this period of menstruation. The average duration of menstruation is two to six days.

Even before the menstrual flow ceases, the endometrium begins to repair itself through the growth of new cells. The low levels of estrogen and progesterone allow the release of FSH from the anterior pituitary. FSH causes new follicles to begin to ripen within the ovaries, and the cycle begins anew.

The activity of ovarian hormones as negative feedback messengers is the basis of hormonal methods of contraception (birth control). The estrogen and progesterone in birth control pills inhibit the release of FSH and LH from the pituitary, preventing follicular development and ovulation. The menstrual period that follows withdrawal of this pharmaceutical estrogen and progesterone is anovulatory (an-OV-u-lah-tor-e); that is, it is not preceded by ovulation.

See the Student Resources on thePoint for a summary chart on reproductive hormones and the animation "Ovulation and Fertilization."

CHECKPOINTS

- ☐ 23-15 What are the two hormones produced in the ovaries?
- ☐ 23-16 What process releases an ovum from the ovary?
- ☐ 23-17 What does the follicle become after its ovum is released?

23

Menopause

Menopause (MEN-o-pawz) is the time period during which menstruation ceases altogether. It ordinarily occurs gradually between the ages of 45 and 55 years and is caused by a normal decline in ovarian function. The ovary becomes chiefly scar tissue and no longer produces mature follicles or appreciable amounts of estrogen and progesterone. Eventually, the uterus, uterine tubes, vagina, and vulva all become somewhat atrophied, and the vaginal mucosa becomes thinner, dryer, and more sensitive.

Menopause is an entirely normal condition, but its onset sometimes brings about effects that are temporarily disturbing. The decrease in estrogen levels can cause nervous symptoms, such as anxiety and insomnia. Because estrogen also helps maintain the vascular dilation that promotes heat loss, low levels may result in "hot flashes."

Physicians may prescribe hormone replacement therapy (HRT) to relieve the discomforts associated with menopause. This medication is usually a combination of estrogen with a synthetic progesterone (progestin), which is included to prevent overgrowth of the endometrium and the risk of endometrial cancer. Studies with the most commonly prescribed form of HRT have shown it lowers the incidence of colorectal cancer. It also lowers the incidence of hip fractures, a sign of osteoporosis. However, in addition to an increased risk of breast cancer, HRT also carries a risk of thrombosis and embolism, which is highest among women who smoke. All HRT risks increase with the duration of therapy. Therefore, treatment should be given for a short time and at the lowest effective dose. Women with a history or family history of breast cancer or circulatory problems should not take HRT.

Because of its beneficial effects, studies are continuing with estrogen alone, generally prescribed for women who have undergone a hysterectomy and do not have a uterus.

- ☐ 23-18 What is the term describing the complete cessation of menstrual cycles?

Birth Control

Birth control is most commonly achieved by **contraception**, which is the use of artificial methods to prevent fertilization of the ovum. Birth control measures that prevent implantation of the fertilized ovum are also considered contraceptives, although technically they do not prevent conception and are more accurately called **abortifacients** (ah-bor-tih-FA-shents) (agents that cause abortion). Some birth control methods act by both mechanisms. **Table 23-1** presents a brief description of the main contraceptive methods currently in use along with some advantages and disadvantages of each. The list is given in a rough order of decreasing effectiveness.

The surest contraceptive method is surgical sterilization. Tubal ligation and vasectomy work by severing and tying off (or cauterizing) the ducts that carry the gametes: the uterine tubes in women and the ductus deferens in men. A man who has had a vasectomy retains the ability to produce hormones and semen as well as the ability to engage in sexual intercourse, but no fertilization can occur.

The various hormonal methods of birth control basically differ in how the hormones are administered. In addition to delivery by pills, an injection, a skin patch, or a vaginal ring, birth control hormones can be implanted as a thin rod under the skin of the upper arm. This method is highly effective and lasts for about three years, but it must be implanted and removed by a health professional. The emergency contraceptive pill is a synthetic progesterone (progestin) taken within 72 hours after intercourse, usually in two doses 12 hours apart. It reduces the risk of pregnancy following unprotected intercourse. This so-called "morning-after pill" is intended for emergency use and not as a regular birth control method. Researchers have done trials with a male contraceptive pill, but none is on the market as yet. The male version of "the pill" also works by suppressing GnRH to inhibit release of FSH and LH, which are important in spermatogenesis. Use of testosterone as a negative feedback messenger requires regular injections and has some undesirable side effects at the doses needed. Administration of the female hormone progesterone prevents spermatogenesis, but also inhibits normal testosterone production. Studies are ongoing to find the best way to deliver the right male contraceptive hormones at safe and effective doses.

Mifepristone (RU-486) is a drug taken after conception to terminate an early pregnancy. It blocks the action of progesterone, causing the uterus to shed its lining and release the fertilized egg. It must be combined with administration of prostaglandins to expel the uterine tissue.

Note in **Table 23-1** that only male and female condoms protect against the spread of STIs (**Fig. 23-12**). The female condom is a sheath that fits into the vagina. A flexible inner ring inserted into the top of the vagina holds the condom in place; an outer ring remains outside the vaginal opening. While more expensive than the male condom, it gives a woman control over last-minute contraception and does not require any drugs or medical visits.

See the Student Resources on thePoint for an illustration of laparoscopic sterilization.

- ☐ 23-19 What term describes the use of artificial means to prevent fertilization of an ovum?

Disorders of the Female Reproductive System

Female reproductive disorders include menstrual disturbances, various forms of tumors, and infections, any of which can contribute to infertility.

Table 23-1 Main Methods of Birth Control Currently in Use

Method	Description	Advantages	Disadvantages
Surgical			
Vasectomy/tubal ligation	Cutting and tying of tubes carrying gametes	Nearly 100% effective; involves no chemical or mechanical devices	Not usually reversible; rare surgical complications
Hormonal			
Birth control pills	Estrogen and progestin, or progestin alone, taken orally to prevent ovulation	Highly effective; requires no last-minute preparation	Alters physiology; return to fertility may be delayed; risk of cardiovascular disease in older women who smoke or have hypertension
Birth control shot	Injection of synthetic progesterone every 3 mo to prevent ovulation	Highly effective; lasts for 3–4 mo	Alters physiology; same side effects as birth control pill; possible menstrual irregularity, amenorrhea
Birth control patch	Adhesive patch placed on body that administers estrogen and progestin through the skin; left on for 3 wk and removed for a 4th wk	Protects long-term; less chance of incorrect use; no last-minute preparation	Alters physiology; same possible side effects as birth control pill
Birth control ring	Flexible ring inserted into vagina that releases hormones internally; left in place for 3 wk and removed for a 4th wk	Long-lasting; highly effective; no last-minute preparation	Possible infections, irritation; same possible side effects as birth control pill
Barrier			
Condom	Sheath that prevents semen from contacting the female reproductive tract	Readily available; does not affect physiology; does not require medical consultation; protects against STIs	
Male	Sheath that fits over erect penis and prevents release of semen	Inexpensive	Must be applied just before intercourse; may slip or tear
Female	Sheath that fits into vagina and covers cervix	Gives women control over last-minute contraception	Relatively expensive; may be difficult or inconvenient to insert
Diaphragm (with spermicide)	Rubber cap that fits over cervix and prevents entrance of sperm	Does not affect physiology; no side effects	Must be inserted before intercourse and left in place for 6 h; requires fitting by physician
Contraceptive sponge (with spermicide)	Soft, disposable foam disk containing spermicide which is moistened with water and inserted into the vagina	Protects against pregnancy for 24 h; nonhormonal; available without prescription; inexpensive	85%–90% effective depending on proper use; possible skin irritation
Intrauterine device	Metal or plastic device inserted into uterus through vagina; prevents fertilization and implantation by release of copper or birth control hormones	Highly effective for 5–10 y depending on type; reversible; no last-minute preparation	Must be introduced by health professional; heavy menstrual bleeding
Other			
Spermicide	Chemicals used to kill sperm; best when used in combination with a barrier method	Available without prescription; inexpensive; does not affect physiology	May cause local irritation; must be used just before intercourse
Fertility awareness	Abstinence during fertile part of cycle as determined by menstrual history, basal body temperature, or quality of cervical mucus	Does not affect physiology; accepted by certain religions	High failure rate; requires careful record keeping

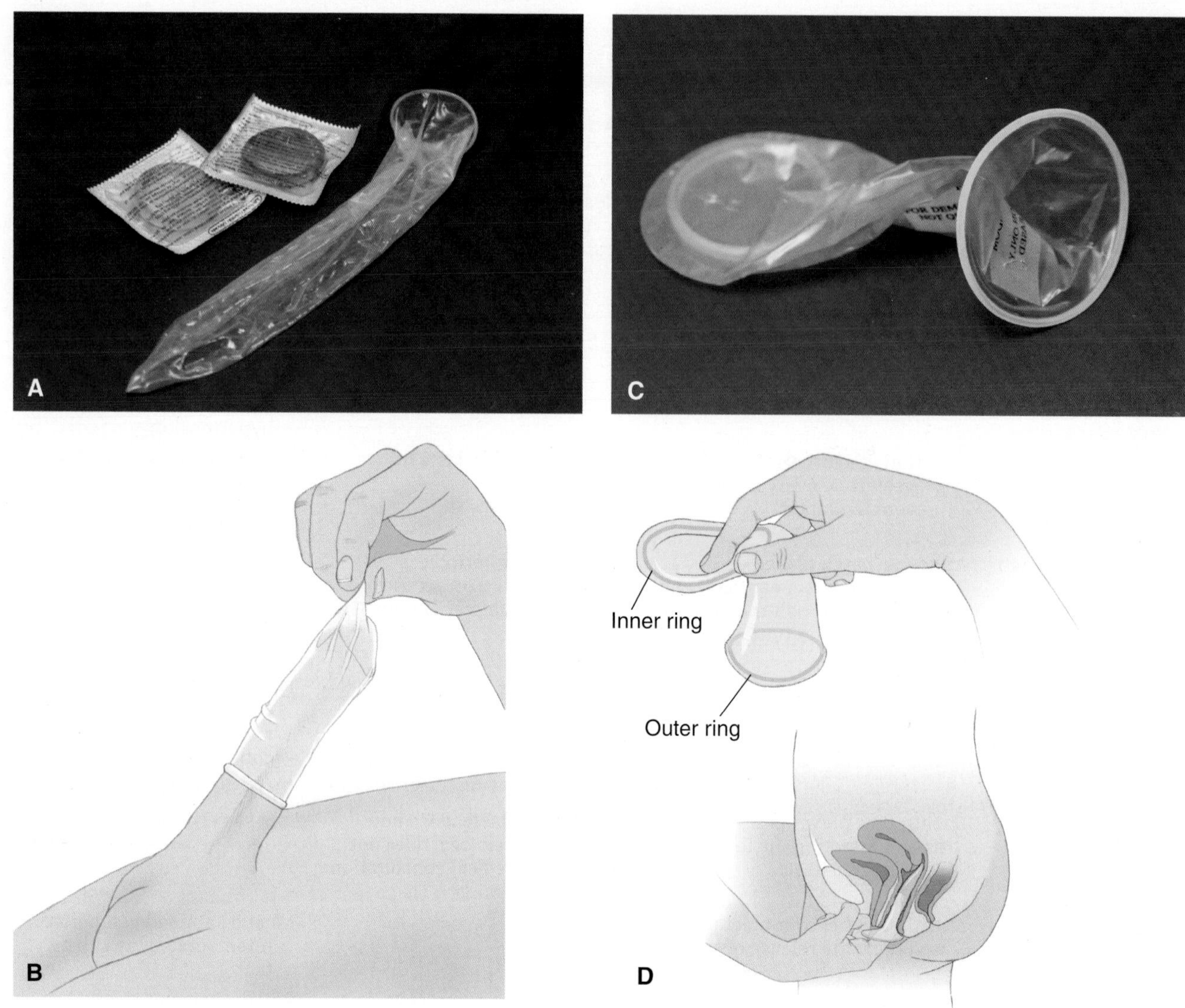

Figure 23-12 **Condoms. A.** The male condom consists of a flexible sheath. **B.** When applying a male condom, leaving a space at the tip helps prevent breakage. **C.** The female condom has a flexible inner ring and an outer ring supporting a sheath. **D.** The flexible inner ring is compressed between the thumb and middle finger and inserted with the index finger deep into the vagina.

MENSTRUAL DISORDERS

Absence of the menses is known as **amenorrhea** (ah-men-o-RE-ah). This condition can be symptomatic of insufficient hormone secretion or a congenital (inborn) abnormality of the reproductive organs. Stress and other psychological factors often play a part in cessation of the menstrual flow. For example, any significant change in a woman's general state of health or change in her living habits, such as a shift in working hours, can interfere with menstruation. Very low body weight with a low percentage of body fat can lead to amenorrhea by reducing estrogen synthesis, as may occur in athletes who overtrain without eating enough and in women who are starving or have eating disorders.

Dysmenorrhea (dis-men-o-RE-ah) means painful or difficult menstruation. Primary dysmenorrhea has no known cause and often declines with age or following childbirth. Prostaglandins are thought to be involved in the uterine contractions that cause the pain, so prostaglandin inhibitors are often used to treat dysmenorrhea. The application of heat over the abdomen may relieve the pain, just as it may ease other types of muscular cramps. Often, such health measures as sufficient rest, a well-balanced diet, and appropriate exercise reduce the severity of the disorder. Secondary dysmenorrhea results from an underlying condition, usually endometriosis or uterine fibroids (discussed below). Common pain relief measures are often less effective with this form of dysmenorrhea, and treatment of the underlying condition may be necessary.

Abnormal uterine bleeding includes excessive menstrual flow (menorrhagia), too-frequent menstruation, and nonmenstrual bleeding. Any of these may cause serious anemias and deserve careful medical attention. Nonmenstrual bleeding may be an indication of a tumor, possibly cancer.

Endometriosis (en-do-me-tre-O-sis) is growth of endometrial tissue outside the uterus, commonly on the ovaries, uterine tubes, peritoneum, or other pelvic organs (**Fig. 23-13**). Endometriosis results in inflammation and other complications and may require surgical removal. This disorder figures in Jessica's infertility problems in the opening case study.

Premenstrual syndrome (PMS) is a condition characterized by physical and emotional symptoms occurring one to two weeks before the menstrual period. Among the most common symptoms are fluid retention (bloating), swollen or tender breasts, headache, fatigue, irritability, anxiety, and depression. Despite many years of study, the cause of PMS has not been determined. It may be caused by the fluctuating hormone levels that accompany the menstrual cycle and may involve the interaction of sex hormones with brain neurotransmitters. Sometimes, a healthful, balanced, low-salt diet and appropriate medication for two weeks before the menses prevent this disorder. This treatment may also avert dysmenorrhea.

BENIGN AND MALIGNANT TUMORS

Fibroids, which are more correctly called *myomas*, are common tumors of the uterus. Studies indicate that about 50% of women who reach the age of 50 have one or more of these growths in the uterine wall. Often, these tumors are small; they usually remain benign and produce no symptoms. They develop between puberty and menopause and ordinarily stop growing after a woman has reached the age of 50. In some cases, the growths interfere with pregnancy. In a patient younger than 40 years of age, a surgeon may simply remove the tumor and leave the uterus fairly intact.

Fibroids may become so large that pressure on adjacent structures causes problems. Sometimes, invasion of blood vessels near the uterine cavity causes serious hemorrhages. Treatment may be suppression of hormones that stimulate fibroid development, blocking blood supply to the fibroid, or surgical removal of the growth. In some cases, surgeons may need to remove the entire uterus or a large part of it, a procedure called a **hysterectomy** (his-ter-EK-to-me).

See the Student Resources on thePoint for an illustration of possible fibroid formation sites.

Endometrial Cancer The most common cancer of the female reproductive tract is cancer of the endometrium (the lining of the uterus). This type of cancer usually affects women during or after menopause. It is seen most frequently in women who have been exposed to high levels of estrogen, which causes endometrial overgrowth. This group includes those who have received estrogen therapy unopposed by progesterone, those who have had few or no pregnancies, and the obese. Symptoms include an abnormal discharge or irregular bleeding; later, there is cramping and pelvic pain. This type of cancer is diagnosed by endometrial biopsy. The usual methods of treatment include surgery and irradiation. Endometrial cancer grows slowly in its beginning stages, so early aggressive treatment usually saves a patient's life.

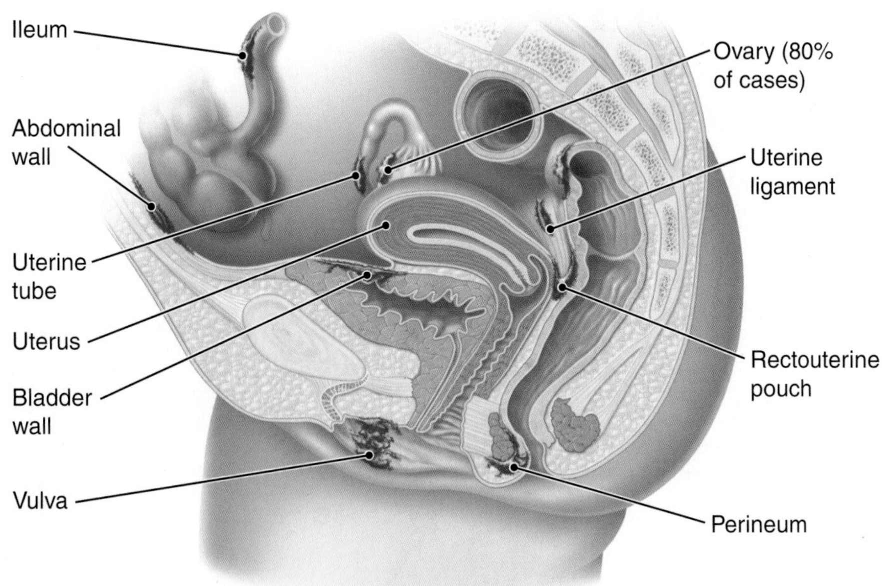

Figure 23-13 Possible sites of endometriosis. **KEY POINT** Endometrial tissue can grow outside the uterus almost anywhere in the peritoneal cavity, causing inflammation and other complications.

Ovarian Cancer Ovarian cancer is the second most common reproductive tract cancer in women, usually occurring between the ages of 40 and 65 years. It is a leading cause of cancer deaths in women. Although most ovarian cysts are not malignant, they should always be investigated for possible malignant change. Ovarian cancer is highly curable if treated before it has spread to other organs. However, these malignancies tend to progress rapidly, and they are difficult to detect because symptoms are vague, there are few recognized risk factors, and at present, there is no reliable screening test.

Cervical Cancer Cancer of the cervix is linked to infection with human papilloma virus (HPV), which causes genital warts and is spread through sexual contact. Thus, cervical cancer can be considered a STI. Risk factors for the disease are related to HPV exposure, such as early age of sexual activity and multiple sex partners. Certain strains of the virus are found in cervical carcinomas and precancerous cervical cells. A vaccine against the most prevalent HPV strains is recommended for females from 11 to 12 years of age.

Early detection is often possible because the cancer develops slowly from atypical cervical cells. The decline in the death rate from cervical cancer is directly related to use of the **Papanicolaou** (pap-ah-nik-o-LAH-o) **test**, also known as the *Pap test* or *Pap smear* (**Fig. 23-14**). The Pap smear is a microscopic examination of cells obtained from cervical scrapings and swabs of the cervical canal. All women should be encouraged to have this test every year. Even girls younger than 18 years of age should be tested if they are sexually active. Guidelines now recommend less frequent testing after normal results are obtained in three annual tests. A newer procedure that can supplement or replace the Pap test involves testing a cervical cell sample for the DNA of cancer-causing HPV strains.

INFECTIONS

The same STIs that occur in men also occur in women; however, their signs and symptoms are often less apparent in women. Also, infections in women can spread more easily throughout the urinary and reproductive systems, resulting in complications such as kidney infection and infertility (**Fig. 23-15**).

The most common STIs in women are chlamydial infections, gonorrhea, HIV, and genital herpes, caused by herpes simplex virus. Syphilis also occurs in women and can be passed through the placenta from mother to fetus, causing stillbirth or birth of an infected infant.

The incidence of **genital warts**, caused by HPV, has increased in recent years. HPV infections have been linked to cancer of the reproductive tract, especially, as noted earlier, cancer of the uterine cervix.

Salpingitis (sal-pin-JI-tis) means inflammation of any tube, but usually refers to disease of the uterine tubes.

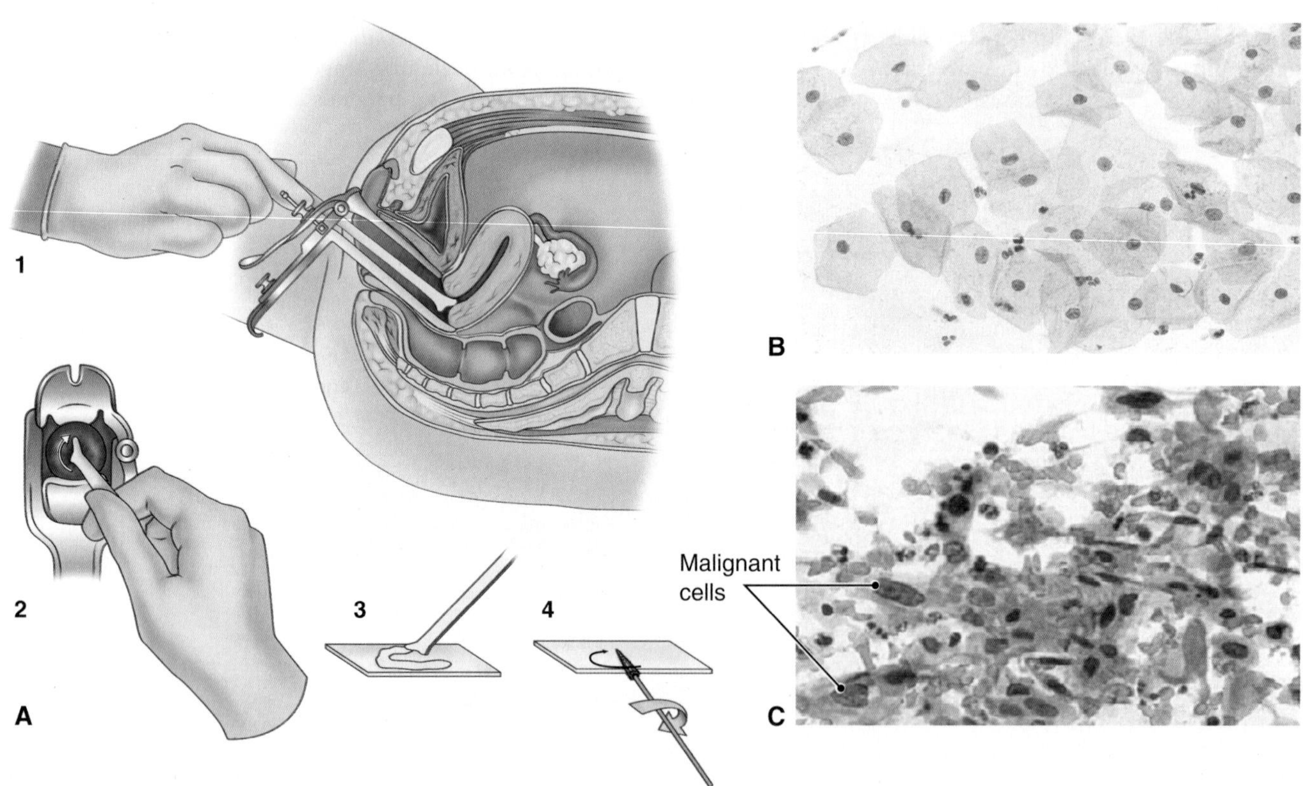

Figure 23-14 **Pap smear. A.** With the speculum in place to dilate the vagina, the examiner uses a spatula to obtain cervical secretions (steps 1 and 2). The secretions are smeared onto a glass slide for microscopic examination (steps 2 and 3). **B.** A normal Pap smear, showing healthy cervical cells. **C.** An abnormal Pap smear, showing a carcinoma.

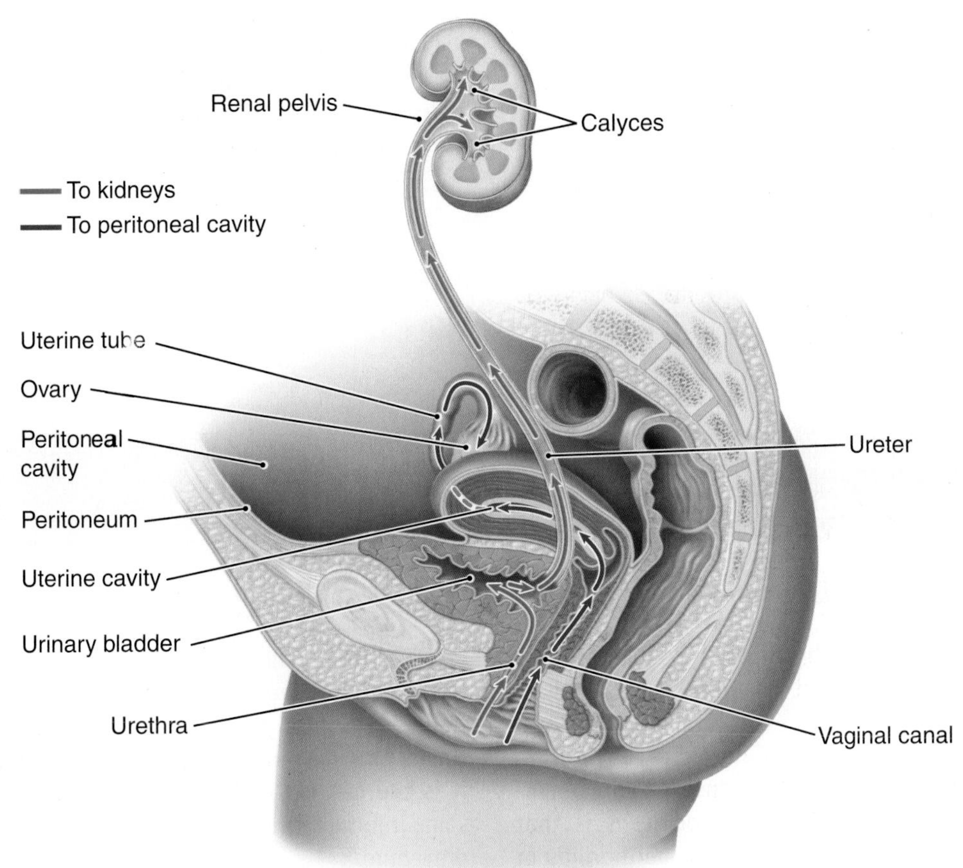

Figure 23-15 **Pathway of infection.** **KEY POINT** Disease organisms can travel from outside to the peritoneal cavity and into the urinary system. **ZOOMING IN** Do the pathways show an ascending or descending infection?

Most uterine tube infections are caused by gonococci or by the bacterium *Chlamydia trachomatis*, but other bacteria may be the cause. Salpingitis may lead to infertility by obstructing the tubes, thus preventing the passage of ova.

Pelvic inflammatory disease (PID) results from extension of infections from the reproductive organs into the pelvic cavity, and it often involves the peritoneum. (See the purple arrow pathway in **Figure 23-15**.) Gonococcus or chlamydia is usually the initial cause of infection, but most cases of PID involve multiple organisms. PID is also associated with an increased risk of infertility.

Infertility

We have mentioned a few of the variety of disorders that can contribute to **infertility**, an inability of a couple to achieve pregnancy after having regular, unprotected sexual intercourse for at least one year. About a third of cases of infertility are caused by factors in only the male, another third are caused by factors in only the female, and a third of cases involve factors in both partners.

In all cases of apparent infertility, the male partner should be investigated first because the procedures for determining lack of fertility in males are much simpler and less costly than those in females. Adequate numbers of sperm are required to degrade the coating around the ovum so that one sperm can fertilize it. Thus, a low sperm count, a condition called **oligospermia** (ol-ih-go-SPER-me-ah), is a significant cause of male infertility. The seminiferous tubules can be damaged by x-rays, infections, toxins, and malnutrition, resulting in decreased sperm production. Smoking, alcohol abuse, and even excessive environmental heat—such as from regular use of a hot tub—can also reduce sperm counts. Other causes of male infertility include impaired sperm motility; blockage within the duct system that impedes the flow of sperm; testosterone deficiency; and drug abuse.

As just noted, infertility is much more difficult to diagnose and evaluate in women than in men. Whereas a microscopic examination of properly collected semen may be enough to determine the presence of abnormal or too few sperm cells in a male, no such simple study can be made in a female. Causes of female infertility include infections, endocrine disorders, and abnormalities in the structure and function of the reproductive organs themselves. One of the most common causes is the presence of inflammation or scar tissue within the uterine tubes. This condition, which can develop as a result of a STI, blocks the ovum's transit to the uterus and blocks sperm cells' access to the ovum. Other causes include endometriosis, early menopause, and malnutrition (sufficient calorie intake is essential to maintain the production of estrogen).

Clinics and medical offices, like those of gynecologists and other doctors, may employ physician assistants. See the Student Resources on thePoint for a description of this career.

CHECKPOINTS

- ☐ 23-20 What is amenorrhea?
- ☐ 23-21 What is the common term for a myoma, a benign uterine tumor?
- ☐ 23-22 What is oligospermia?

Disease in Context Revisited

Jessica and Brett Discuss IVF

Dr. Christensen greeted the anxious couple as they waited in his office to get the results of Jessica's laparoscopy.

"It appears that your uterine tubes have been scarred and blocked, possibly because of your endometriosis," he told Jessica. "The good news is that your uterus looks healthy. I didn't see any fibroids or malformations that would interfere with your ability to carry a fetus to term. As I mentioned previously, you might want to consider IVF to become pregnant."

"Just how is that done?" Brett asked.

"Well first, Jessica, you would go on a schedule of hormone supplements to induce the ripening of multiple eggs," the doctor explained. "These eggs are then removed from your ovaries prior to ovulation. They are fertilized with Brett's sperm in a laboratory and then transferred into your uterus, bypassing the uterine tubes. This may have to be done more than once, but it is the method of choice in your case. I'll give you some information on costs and reimbursements for this procedure."

The couple thanked Dr. Christensen for his guidance and said they would think about their options.

Jessica's case illustrates a clinical condition associated with the female reproductive system and some procedures available to sidestep infertility. Box 24-1 in the next chapter has more information on assisted reproductive techniques currently in use.

CHAPTER 23

Chapter Wrap-Up

Summary Overview

A detailed chapter outline with space for note taking is on *thePoint*. The figure below illustrates the main topics covered in this chapter.

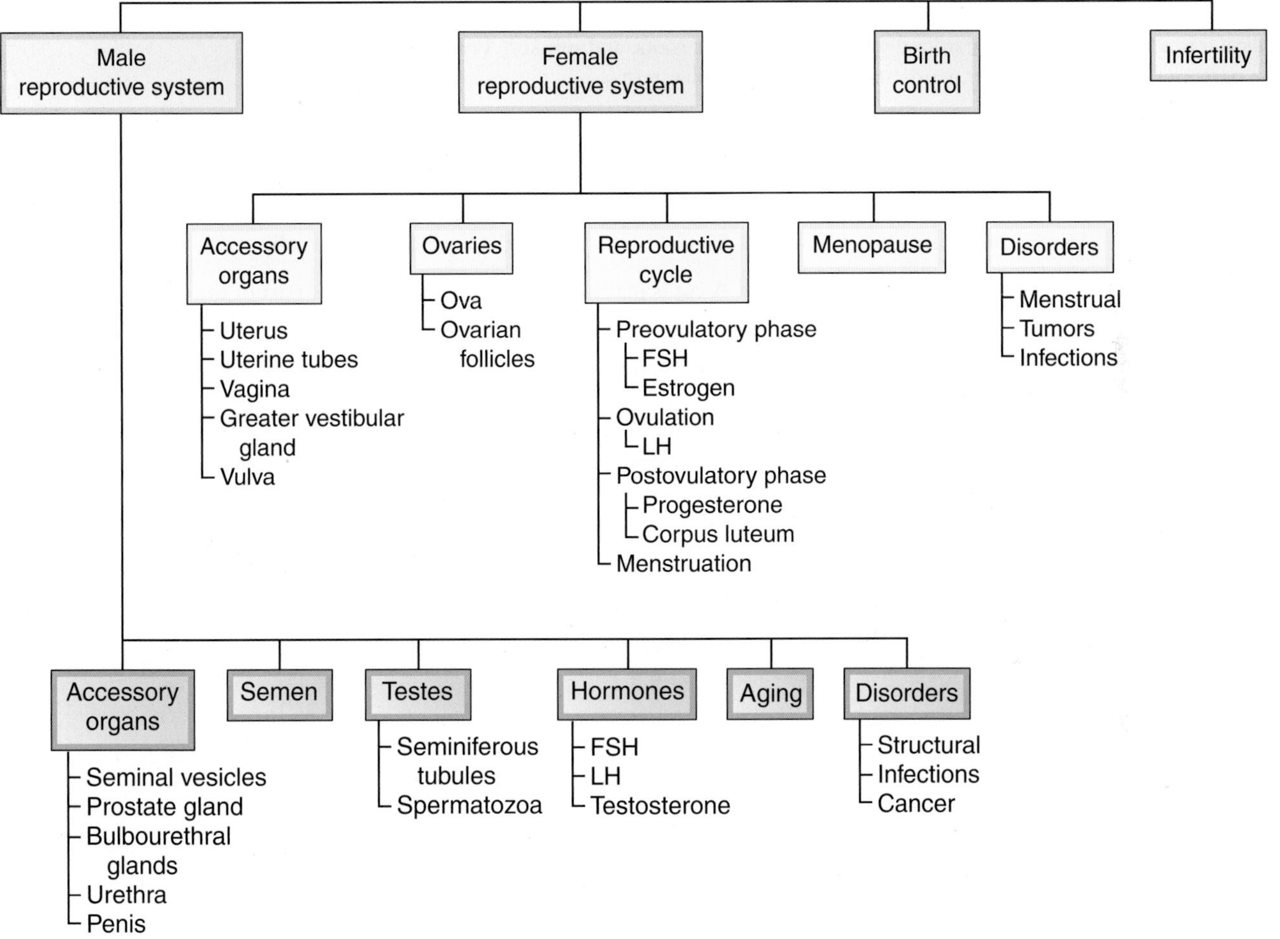

Key Terms

The terms listed below are emphasized in this chapter. Knowing them will help you organize and prioritize your learning. These and other boldface terms are defined in the Glossary with phonetic pronunciations.

corpus luteum
endometrium
estrogen
follicle-stimulating hormone (FSH)
gamete
infertility
luteinizing hormone (LH)
menopause
menses
menstruation
ovarian follicle
ovary
ovulation
ovum (pl., ova)
progesterone
semen
spermatozoon (pl., spermatozoa)
testis (pl., testes)
testosterone
uterus

Word Anatomy

Medical terms are built from standardized word parts (prefixes, roots, and suffixes). Learning the meanings of these parts can help you remember words and interpret unfamiliar terms.

WORD PART	MEANING	EXAMPLE
The Male Reproductive System		
acr/o	extremity, end	The *acrosome* covers the head of a sperm cell.
circum-	around	A cut is made around the glans to remove part of the foreskin in a *circumcision*.
fer	to carry	The ductus *deferens* carries spermatozoa away from (de-) the testis.
semin/o	semen, seed	Sperm cells are produced in the *seminiferous* tubules.
test/o	testis	The hormone *testosterone* is produced in the testis.
Disorders of the Male Reproductive System		
crypt/o-	hidden	*Cryptorchidism* refers to an undescended testis (orchid/o).
olig/o-	few, deficiency	*Oligospermia* is a deficiency in the numbers of spermatozoa produced.
orchid/o, orchi/o	testis	*Orchiectomy* is removal of the testis.
The Female Reproductive System		
metr/o	uterus	The *myometrium* is the muscular (my/o) layer of the uterus.
ovar, ovari/o	ovary	An *ovarian* follicle encloses an ovum.
ov/o, ov/i	egg	An *ovum* is an egg cell.
rect/o	rectum	The *rectouterine* pouch is between the uterus and the rectum.
Disorders to the Female Reproductive System		
hyster/o	uterus	*Hysterectomy* is surgical removal of the uterus.
men/o	uterine bleeding, menses	*Amenorrhea* is absence of menstrual flow.
salping/o	tube	*Salpingitis* is inflammation of a tube, such as the uterine tube.

Questions for Study and Review

BUILDING UNDERSTANDING

Fill in the Blanks

1. Gametes go through a special process of cell division called _____.
2. Spermatozoa begin their development in tiny coiled _____.
3. An ovum matures in a small fluid-filled cluster of cells called the _____.
4. Failure of the testis to descend into the scrotum results in the disorder _____.
5. Surgical removal of the uterus is called a(n) _____.

Matching > Match each numbered item with the most closely related lettered item.

___ **6.** A hormone released by the pituitary that promotes follicular development in the ovary

___ **7.** A hormone released by developing follicles that promotes thickening of the endometrium

___ **8.** A hormone released by the pituitary that stimulates ovulation

___ **9.** A hormone released by the corpus luteum that peaks after ovulation

___ **10.** The main male sex hormone

a. follicle-stimulating hormone
b. testosterone
c. estrogen
d. luteinizing hormone
e. progesterone

Multiple Choice

___ **11.** What structure does the fetal testis travel through to descend into the scrotum?

a. spermatic cord
b. inguinal canal
c. seminiferous tubule
d. vas deferens

___ **12.** Where are the enzymes that help the sperm cell penetrate the ovum located?

a. acrosome
b. head
c. midpiece
d. flagellum

___ **13.** What is testicular inflammation called?

a. phimosis
b. epididymitis
c. prostatitis
d. orchitis

___ **14.** What structure(s) suspend(s) the uterus and ovaries in the pelvic cavity?

a. uterine tubes
b. broad ligaments
c. fimbriae
d. fornix

___ **15.** What is the term for the area between the vaginal opening and the anus?

a. vestibule
b. vulva
c. hymen
d. perineum

___ **16.** Where is the most common site of cancer in the female reproductive tract?

a. endometrium
b. myometrium
c. ovaries
d. cervix

___ **17.** Give the meanings of the following acronyms.

a. PID ______________________
b. STI ______________________
c. HPV ______________________
d. LH ______________________
e. BPH ______________________

UNDERSTANDING CONCEPTS

18. Referring to the male and female reproductive systems in The Body Visible at the beginning of the book, give the name and number of

- **a.** the duct that delivers semen to the urethra
- **b.** the section of the penis that surrounds the urethra
- **c.** a gland that secretes mucus into the urethra in males
- **d.** structure formed by the follicle after ovulation
- **e.** extensions of the uterine tube that sweep an ovum into the tube
- **f.** organ of sexual stimulation in females
- **g.** portion of the recess around the cervix

19. Compare and contrast the following terms:

- **a.** asexual reproduction and sexual reproduction
- **b.** spermatozoa and ova
- **c.** sustentacular cell and interstitial cell
- **d.** ovarian follicle and corpus luteum
- **e.** myometrium and endometrium

20. Trace the pathway of sperm from the site of production to the urethra.

21. Describe the components of semen, their sites of production, and their functions.

22. List the hormones that control male reproduction, and state their functions.

23. Trace the pathway of an ovum from the site of production to the site of implantation.

24. Beginning with the first day of the menstrual flow, describe the events of one complete cycle, including the role of the hormones involved.

25. Define *contraception*. Describe methods of contraception that involve (1) barriers, (2) chemicals, (3) hormones, and (4) prevention of implantation.

26. Compare and contrast the following disorders:

- **a.** epididymitis and prostatitis
- **b.** benign prostatic hyperplasia and prostatic cancer
- **c.** amenorrhea and dysmenorrhea
- **d.** fibroids and endometrial cancer

CONCEPTUAL THINKING

27. Jodie, a middle-aged mother of three, is considering a tubal ligation, a contraceptive procedure that involves cutting the uterine tubes. Jodie is worried that this might cause her to enter early menopause. Should she be worried?

28. In the opening case study, Jessica has fertility problems, possibly related to endometriosis. Explain what endometriosis is and give possible locations for endometriosis.

For more questions, see the Learning Activities on thePoint.

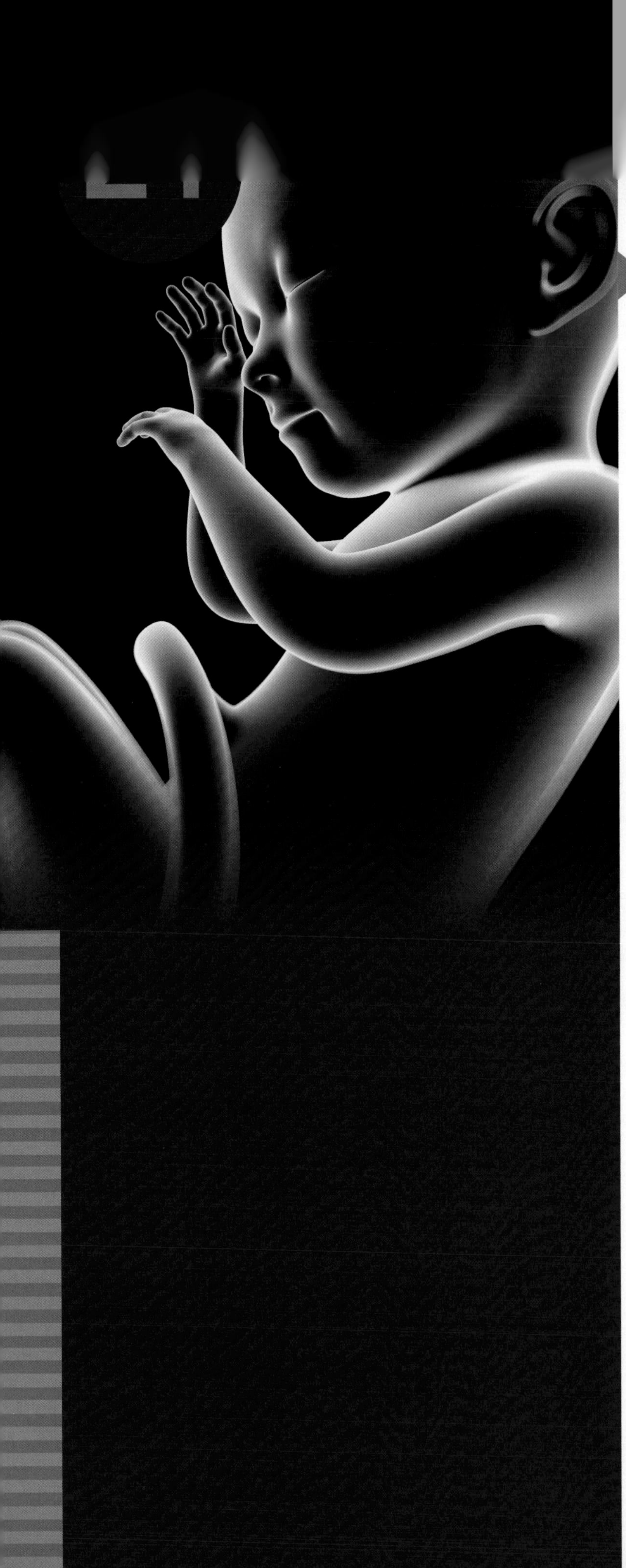

Learning Objectives

After careful study of this chapter, you should be able to:

1 ▶ Describe fertilization and the early development of the fertilized egg. ***p. 550***

2 ▶ Describe the structure and function of the placenta. ***p. 550***

3 ▶ Describe how fetal circulation differs from adult circulation. ***p. 550***

4 ▶ Name five hormones active during pregnancy, and describe the function of each. ***p. 551***

5 ▶ Briefly describe changes that occur in the embryo, fetus, and mother during pregnancy. ***p. 552***

6 ▶ Briefly describe the four stages of labor. ***p.556***

7 ▶ Name five hormones active in lactation, and describe the action of each. ***p. 560***

8 ▶ Cite the advantages of breast-feeding. ***p. 560***

9 ▶ Describe four disorders associated with the placenta. ***p. 561***

10 ▶ Explain how breast cancer is diagnosed and treated. ***p. 562***

11 ▶ Cite four possible causes of lactation disturbances. ***p. 563***

12 ▶ Referring to the case study and the text, discuss possible causes of high-risk pregnancies. ***pp. 549, 563***

13 ▶ Show how word parts are used to build words related to development and birth (see Word Anatomy at the end of the chapter). ***p. 565***

Disease in Context *Amy's High-Risk Pregnancy*

Rick was writing his last exam of the semester when he noticed a security guard entering the exam room. His heart sank when the proctor walked over to his desk and told him that he had to leave the exam. His wife Amy was 33 weeks' pregnant and had been feeling tired and achy for a few days.

"Is it my wife?" he asked the guard.

"I'm afraid so" he replied. "She's at the emergency room, and she wants you to come right away."

Rick drove to the local hospital and raced in to find his wife. "Amy, are you OK? Is the baby all right?"

"The baby's fine so far," she said. "But I'm not! You know how I've had those pains in my back and chest and horrible heartburn? I finally went to see Dr. Borin today, and the first thing she did was check my blood pressure. As soon as she saw the result she called an ambulance, and here I am. They've done all sorts of tests, but I haven't talked to a doctor yet. I was worried that I had the flu, but I'm scared that it's something even worse!"

A doctor entered Amy's cubicle five minutes after Rick's arrival.

"Hi Amy, I'm the obstetrician for high-risk pregnancies," said Dr. Renaud. "I've read Dr. Borin's report and reviewed your test results. I want to reassure you that your baby is OK, but your test results are troubling. Your blood pressure is too high, and there is protein in your urine. Your platelet count is 80,000 per microliter of blood, about half of normal levels, and your blood also contains increased concentrations of liver enzymes. There's also evidence that your blood cells are undergoing hemolysis, or destruction. When I listen to your lungs, I hear congestion. This finding suggests that your heart is struggling to pump enough blood. The bottom line is that you have a variant of preeclampsia known as HELLP. If we deliver your baby right away, there's a very strong chance that both of you will be fine. Unfortunately, your platelet count is too low for us to do a caesarean section, so we will transfer you to the obstetrical wing and induce labor."

Amy was scared, and she protested, "But he is premature and not ready!"

"Yes he is premature, but your condition is serious," Dr. Renaud explained. "There's no time to waste, so we can talk more after your baby is born."

Later in this chapter, we will see how Amy fares in delivery and learn more about HELLP. In addition to learning about normal pregnancy and childbirth, this chapter will also look at some disorders associated with pregnancy.

ANCILLARIES *At-A-Glance*

Visit thePoint to access the following resources. For guidance in using these resources most effectively, see pp. xv–xvii.

Learning RESOURCES

- Tips for Effective Studying
- Web Figure: Apgar Scoring System
- Web Chart: Placental Hormones
- Animation: Fetal Circulation
- Health Professions: Midwives and Doulas
- Detailed Chapter Outline
- Answers to Questions for Study and Review
- Audio Pronunciation Glossary

Learning ACTIVITIES

- Pre-Quiz
- Visual Activities
- Kinesthetic Activities
- Auditory Activities

This chapter concludes our discussion of the reproductive system by discussing the production of offspring. We elaborate on the action of some hormones introduced in Chapter 12 as they apply to pregnancy, childbirth, and lactation. We also provide more details on fetal circulation, first discussed in Chapter 14 in connection with possible heart defects. Negative feedback mechanisms control many aspects of reproduction. However, childbirth and lactation represent a different type of feedback process—positive feedback—a system that accelerates rather than reverses an action. A thorough understanding of the female and male reproductive systems will help you understand the topics in this chapter.

Pregnancy

Pregnancy begins with fertilization of an ovum and ends with childbirth. During this approximately 38-week period of development, known as **gestation** (jes-TA-shun), a single fertilized egg divides repeatedly, and the new cells differentiate into all the tissues of the developing offspring. Along the way, many changes occur in the mother as well as her baby. **Obstetrics** (ob-STET-riks) (OB) is the branch of medicine that is concerned with the care of women during pregnancy, childbirth, and the six weeks after childbirth. A physician who specializes in obstetrics is an *obstetrician.* Some obstetricians specialize in high-risk pregnancies, as does Amy's physician in the case study.

FERTILIZATION AND THE START OF PREGNANCY

When semen is deposited in the vagina, the millions of spermatozoa immediately wriggle about in all directions, some traveling into the uterus and uterine tubes. If an oocyte (egg cell) is present in the uterine tube, many spermatozoa cluster around it (**Fig. 24-1**). Enzymes released from acrosomes on the heads of the sperm cells dissolve the coating around the ovum so that the plasma membranes of the ovum and the sperm can fuse. The nuclei of the sperm and egg then combine in **fertilization** (see Box 24-1 for information on assisted reproduction).

The result of this union is a single cell called a **zygote** (ZI-gote), which has the full human chromosome number of 46. The zygote divides rapidly into two cells and then four cells and soon forms a ball of identical cells called a **morula** (MOR-u-lah). During this time, the cell cluster is traveling toward the uterine cavity, pushed along by cilia lining the oviduct and by peristalsis (contractions) of the tube. Before it reaches the uterus, the morula develops into a partially hollow structure called a **blastocyst** (BLAS-to-sist). The blastocyst then burrows into the greatly thickened uterine lining and is soon implanted and completely covered. After **implantation** in the uterus, a group of cells within the blastocyst, called the **inner cell mass**, becomes an **embryo** (EM-bre-o), the term used for the growing offspring in the early stage of gestation. The rest of the blastocyst cells, known as **trophoblasts**, will differentiate into tissue that will support the developing **fetus** (FE-tus), the growing offspring from the beginning of the third month of gestation until birth.

THE PLACENTA

For a few days after implantation, the embryo gets nourishment from its surrounding fluids. As it grows, the embryo's increasing needs are met by the **placenta** (plah-SEN-tah), a flat, circular organ that consists of a spongy network of blood-filled sinuses and capillary-containing villi (**Fig. 24-2A**). (*Placenta* is from a Latin word meaning "pancake.") The placenta consists of both maternal and embryonic tissue. The maternal portion is simply a well-vascularized internal portion of the endometrium named the **decidua** (de-SID-u-ah). The embryonic portion (derived from trophoblasts) is called the **chorion** (KO-re-on), which forms projections called *chorionic villi.* The chorionic villi break down the endometrial tissue, creating a network of venous sinuses filled with maternal blood. The placenta is the organ of nutrition, respiration, and excretion for the developing offspring throughout gestation. Although the blood of the mother and her offspring do not mix—each has its own blood and cardiovascular system—exchanges take place between maternal blood in the sinuses and embryonic blood in the capillaries of the chorionic villi. In this manner, gases (CO_2 and O_2) are exchanged, nutrients are provided to the developing offspring, and waste products are released into the maternal blood to be eliminated.

The Umbilical Cord The embryo is connected to the developing placenta by a stalk of tissue that eventually becomes the **umbilical** (um-BIL-ih-kal) **cord.** This structure carries blood to and from the embryo, later called the fetus. The cord encloses two arteries that carry blood low in oxygen from the fetus to the placenta and one vein that carries blood high in oxygen from the placenta to the fetus (**see Fig. 24-2**). (Note that, like the pulmonary vessels, these arteries carry blood low in oxygen, and this vein carries blood high in oxygen.)

Fetal Circulation The fetus has special circulatory adaptations to carry blood to and from the umbilical cord and to bypass its nonfunctional lungs (**Fig. 24-2B**). A small amount of the oxygen-rich blood traveling toward the fetus in the **umbilical vein** is delivered directly to the liver. However, most of the blood bypasses the liver and is added to the oxygen-poor blood in the inferior vena cava through a small vessel, the **ductus venosus**. Although mixed, this blood still contains enough oxygen to nourish fetal tissues. Once in the right atrium, most of the blood flows directly into the left atrium through a small hole in the atrial septum, the **foramen ovale** (o-VA-le). This blood has bypassed the right ventricle and the pulmonary circuit. Blood that does enter the right ventricle is pumped into the pulmonary artery. Although a small amount of this blood goes to the lungs, most of it shunts directly into the systemic circuit through a small vessel, the **ductus arteriosus**, which connects the pulmonary artery to the descending aorta. After traveling

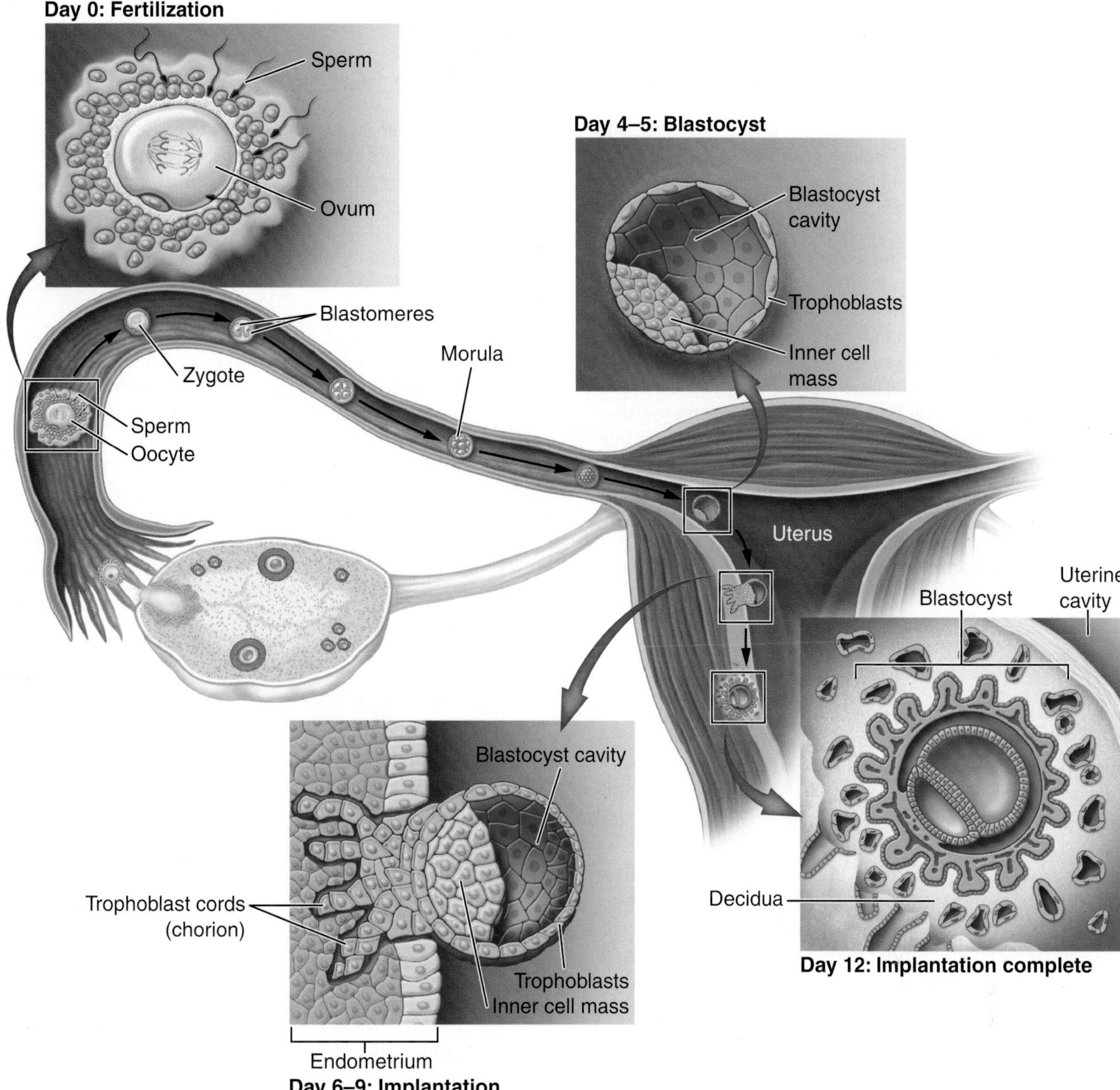

Figure 24-1 **Fertilization and implantation.** **KEY POINT** The union of oocyte and sperm produces a zygote, which develops into a blastocyst and implants into the endometrium. **ZOOMING IN** Where is the ovum fertilized?

throughout fetal tissue, blood returns to the placenta to pick up oxygen through the two **umbilical arteries**.

After birth, when the baby's lungs are functioning, these adaptations begin to close. The foramen ovale gradually seals, and the various vessels constrict into fibrous cords, usually within minutes after birth (only the proximal parts of the umbilical arteries persist as arteries to the urinary bladder). Failure to close results in congenital heart defects, as shown in **Figure 14-13B and C**.

See the Student Resources on thePoint to view the animation "Fetal Circulation."

HORMONES AND PREGNANCY

Beginning soon after implantation, the blastocyst produces the hormone **human chorionic gonadotropin** (ko-re-ON-ik gon-ah-do-TRO-pin) (**hCG**). This hormone is very similar in structure to luteinizing hormone (LH), and like LH, it stimulates growth of the ovarian corpus luteum. The corpus luteum continues to grow and produce increasing amounts of progesterone and estrogen for about 11 or 12 weeks postfertilization, at which point it degenerates. It is hCG that is used in tests as an indicator of pregnancy.

Progesterone is essential for the maintenance of pregnancy. It promotes endometrial secretions to nourish the

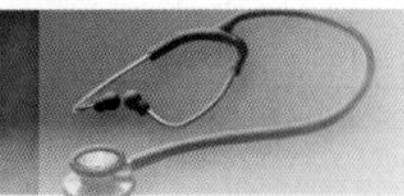

CLINICAL PERSPECTIVES

Box 24-1

Assisted Reproductive Technology: The "Art" of Conception

At least one in 10 American couples is affected by infertility. Assisted reproductive technologies such as in vitro fertilization (IVF), gamete intrafallopian transfer (GIFT), and zygote intrafallopian transfer (ZIFT) can help these couples become pregnant.

In vitro fertilization refers to fertilization of an ovum outside the mother's body in a laboratory dish (in vitro means "in glass"). It is often used when a woman's uterine tubes are blocked or when a man has a low sperm count. The woman participating in IVF is given hormones to cause ovulation of several ova. These are then withdrawn with a needle and fertilized with the father's sperm. After a few divisions, some of the fertilized ova are placed in the uterus, thus bypassing the uterine tubes. Additional fertilized ova can be frozen to repeat the procedure in case of failure or for later pregnancies.

GIFT can be used when the woman has at least one normal uterine tube and the man has an adequate sperm count. As in IVF, the woman is given hormones to cause ovulation of several ova, which are collected. Then, the ova and the father's sperm are placed into the uterine tube using a catheter. Thus, in GIFT, fertilization occurs inside the woman, not in a laboratory dish.

ZIFT is a combination of both IVF and GIFT. Fertilization takes place in a laboratory dish, and then the zygote is placed into the uterine tube.

Because of a lack of guidelines or restrictions in the United States in the field of assisted reproductive technology, some problems have arisen. These issues concern the use of stored embryos and gametes, use of embryos without consent, and improper screening for disease among donors. In addition, the implantation of more than one fertilized ovum has resulted in a high incidence of multiple births, even up to seven or eight offspring in a single pregnancy, a situation that imperils the survival and health of the babies.

embryo, maintains the endometrium, and decreases the uterine muscle's ability to contract, thus preventing the embryo from being expelled from the body. During pregnancy, progesterone also helps prepare the breasts for milk secretion. Estrogen promotes enlargement of the uterus and breasts.

The placenta also secretes hormones. By the 11th or 12th week of pregnancy, the corpus luteum is no longer needed; by this time, the placenta has developed the capacity to secrete adequate amounts of progesterone and estrogen, and the corpus luteum disintegrates. Miscarriages (loss of an embryo or fetus) are most likely to occur during this critical time when hormone secretion is shifting from the corpus luteum to the placenta.

Human placental lactogen (**hPL**), also known as human chorionic somatomammotropin, is another hormone secreted by the placenta during pregnancy, reaching a peak at term, the normal conclusion of pregnancy. This hormone stimulates growth of the breasts to prepare the mother for milk production, or **lactation** (lak-TA-shun). More importantly, it increases glucose availability for the fetus by reducing glucose use by the mother's tissues.

Relaxin is a placental hormone that softens the cervix and relaxes the sacral joints and the pubic symphysis. These changes help widen the birth canal and aid in birth.

See the Student Resources on thePoint for a summary chart on placental hormones.

CHECKPOINTS

- [ ] 24-1 What structure is formed by the union of an ovum and a spermatozoon?
- [ ] 24-2 What structure nourishes the developing fetus?
- [ ] 24-3 What is the function of the umbilical cord?
- [ ] 24-4 Fetal circulation is adapted to bypass what organs?
- [ ] 24-5 What embryonic hormone maintains the corpus luteum early in pregnancy?

DEVELOPMENT OF THE EMBRYO

The developing offspring is referred to as an embryo for the first eight weeks of life (**Fig. 24-3**), and the study of growth during this period is called **embryology** (em-bre-OL-o-je). The beginnings of all body systems are established during this time. The heart and the brain are among the first organs to develop. A primitive nervous system begins to form in the third week. The heart and blood vessels originate during the second week, and the first heartbeat appears during week 4 at the same time other muscles begin to develop.

By the end of the first month, the embryo is approximately 0.62 cm (0.25 in) long with four small swellings at the sides called **limb buds**, which will develop into the four extremities. At this time, the heart produces a prominent bulge at the anterior of the embryo.

By the end of the second month, the embryo takes on an appearance that is recognizably human. In male embryos, the primitive testes have formed and have begun to secrete testosterone, which will direct formation of the male reproductive organs as gestation continues. **Figure 24-4** shows photographs of embryonic and early fetal development.

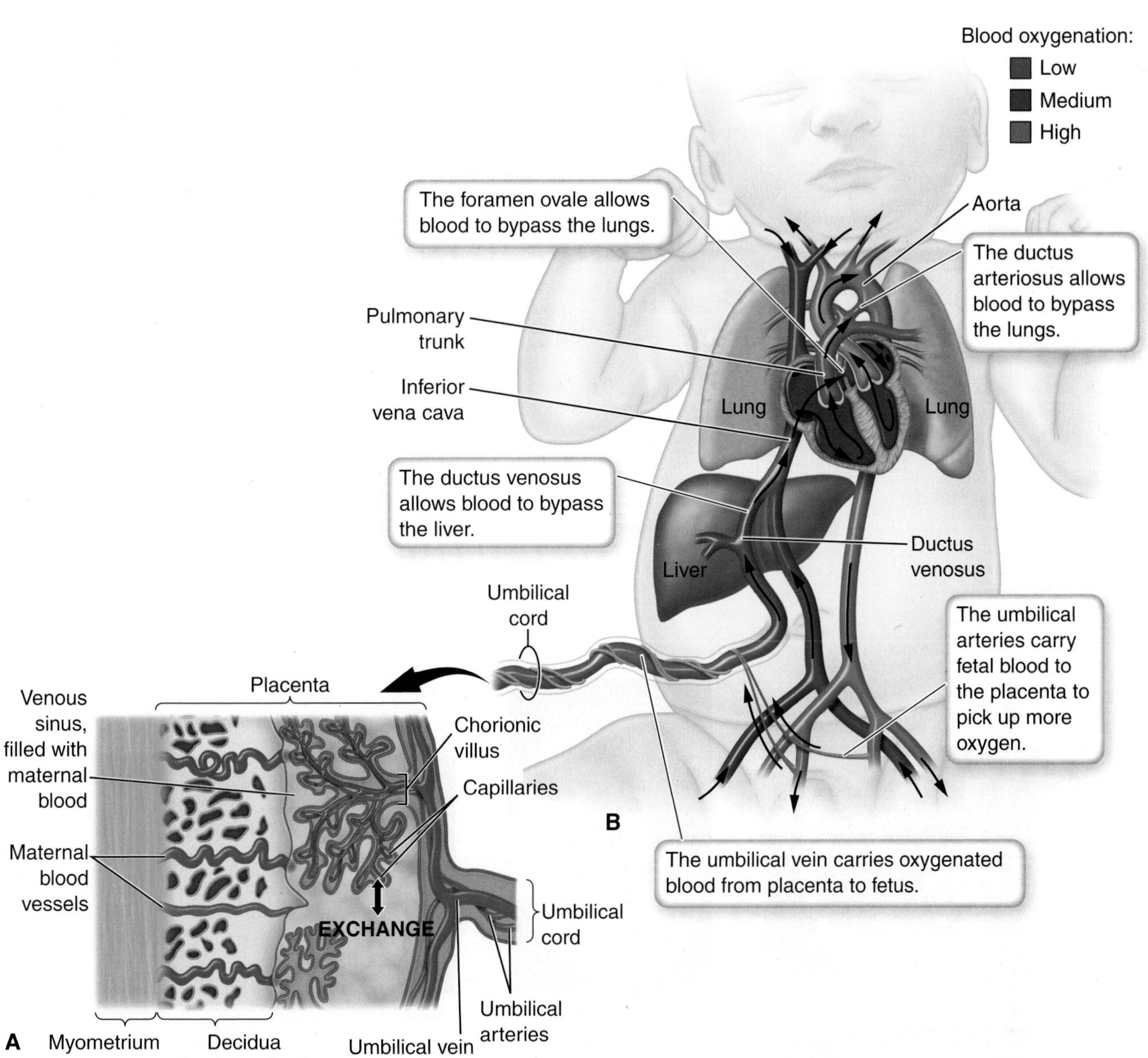
Blood oxygenation:
Low
Medium
High
The foramen ovale allows blood to bypass the lungs.
Aorta
The ductus arteriosus allows blood to bypass the lungs.
Pulmonary trunk
Inferior vena cava
Lung
Lung
The ductus venosus allows blood to bypass the liver.
Ductus venosus
Liver
Umbilical cord
The umbilical arteries carry fetal blood to the placenta to pick up more oxygen.
Placenta
Venous sinus, filled with maternal blood
Chorionic villus
Capillaries
Maternal blood vessels
B
The umbilical vein carries oxygenated blood from placenta to fetus.
EXCHANGE
Umbilical cord
Umbilical arteries
A
Myometrium
Decidua (endometrium)
Umbilical vein

Figure 24-2 The placenta and the fetal circulation. **KEY POINT** The placenta acts as the organ of gas exchange and waste removal for the fetus. **A.** The placenta. The placenta is formed by the maternal endometrium and fetal chorion. Exchanges between fetal and maternal blood occur through the capillaries of the chorionic villi. **B.** Fetal circulation. During fetal life, specialized vessels and openings shunt blood away from nonfunctioning lungs and toward the placenta. Colors show relative oxygen content of blood. **ZOOMING IN** What is signified by the purple color in this illustration?

A developing embryo is especially sensitive to harmful substances and poor maternal nutrition during this early stage of pregnancy when so many important developmental events are occurring. In Chapter 25, we discuss developmental disorders that begin during embryologic development.

DEVELOPMENT OF THE FETUS

During the period of fetal development, from the beginning of the third month until birth, the organ systems continue to grow and mature. The ovaries form in the female early in this fetal period, and at this stage, they contain all the primitive cells (oocytes) that can later develop into mature ova (eggs).

For study, the entire gestation period may be divided into three equal segments or **trimesters**. The fetus's most rapid growth occurs during the second trimester (months 4 to 6). By the end of the fourth month, the fetus is almost 15 cm (6 in) long, and its external genitalia are sufficiently developed to reveal its sex. By the seventh month, the fetus is usually approximately 35 cm (14 in) long and weighs approximately 1.1 kg (2.4 lb). At the end of pregnancy, the normal length of

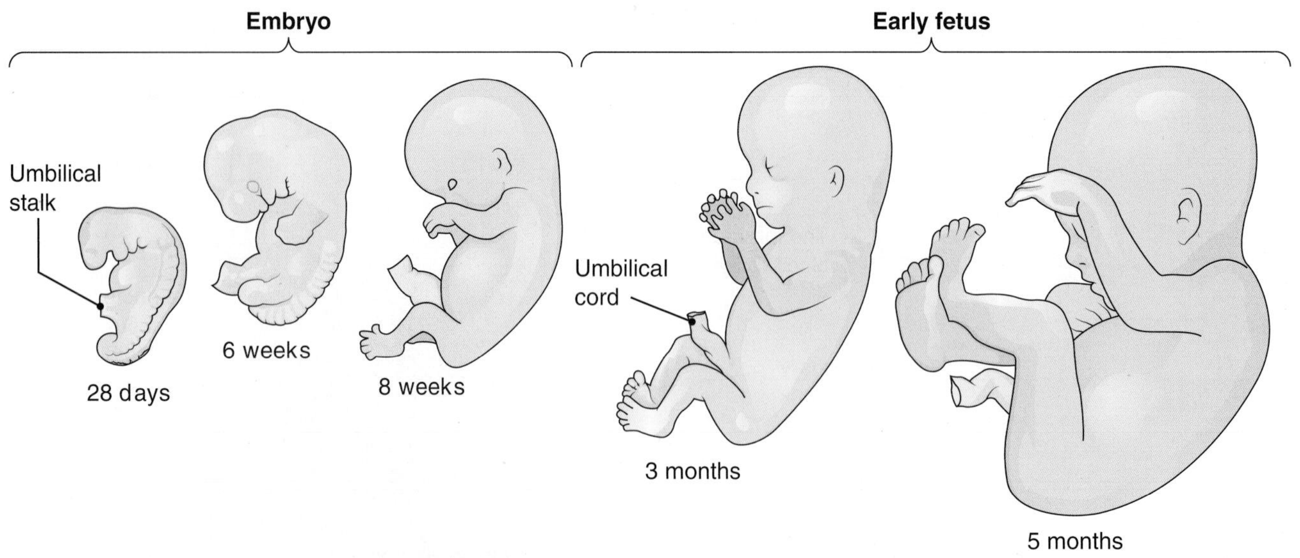

Figure 24-3 **Development of an embryo and early fetus.**

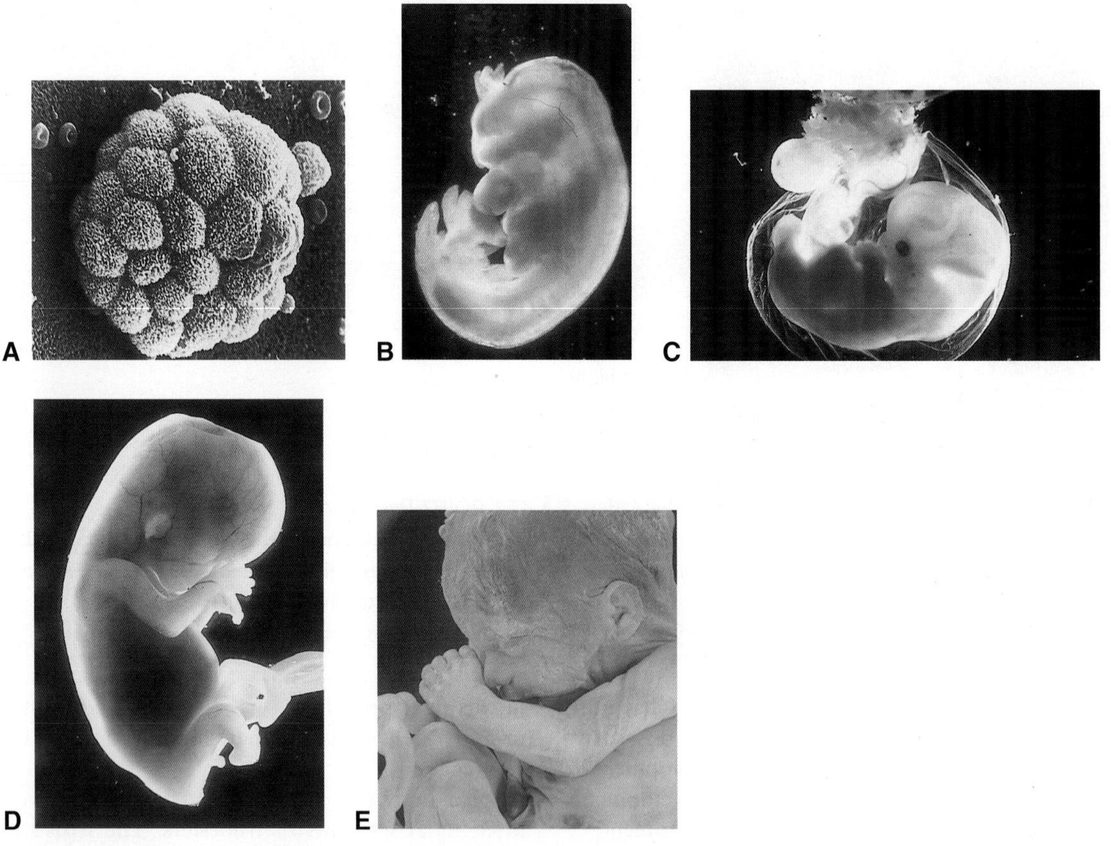

Figure 24-4 **Human embryos at different stages and early fetus.** **KEY POINT** The embryonic stage lasts for the first two months of pregnancy. Thereafter, the developing offspring is called a fetus. **A.** Implantation in uterus seven to eight days after conception. **B.** Embryo at 32 days. **C.** At 37 days. **D.** At 41 days. **E.** Fetus between 12 and 15 weeks.

the fetus is 45 to 56 cm (18 to 22.5 in), and the weight varies from 2.7 to 4.5 kg (6 to 10 lb).

The **amniotic** (am-ne-OT-ik) **sac**, which is filled with a clear liquid known as **amniotic fluid**, surrounds the fetus and serves as a protective cushion for it (**Fig. 24-5**). The amniotic sac ruptures at birth, an event marked by the common expression that the mother's "water broke."

During development, the fetal skin is protected by a layer of cheeselike material called the **vernix caseosa** (VER-niks ka-se-O-sah) (literally, "cheesy varnish").

THE MOTHER

The total period of pregnancy, from fertilization of the ovum to birth, is approximately 266 days, also given as 280 days or 40 weeks from the last menstrual period (LMP). During this time, the mother must supply all the food and oxygen for the fetus and eliminate its waste materials. To support the additional demands of the growing fetus, the mother's metabolism changes markedly, and several organ systems increase their output:

- The heart pumps more blood to supply the needs of the uterus and the fetus.
- The lungs provide more oxygen and eliminate more carbon dioxide by increasing the rate and depth of respiration.
- The kidneys excrete nitrogenous wastes from both the fetus and the mother.
- The digestive system supplies additional nutrients for the growth of maternal organs (uterus and breasts) and fetal growth, as well as for subsequent labor and milk secretion.

Nausea and vomiting are common discomforts in early pregnancy. These most often occur upon arising or during periods of fatigue and are more common in women who smoke. The specific cause of these symptoms is not known, but they may be due to the great changes in hormone levels that occur at this time. The nausea and vomiting usually last for only a few weeks to several months but may persist for the entire gestational period.

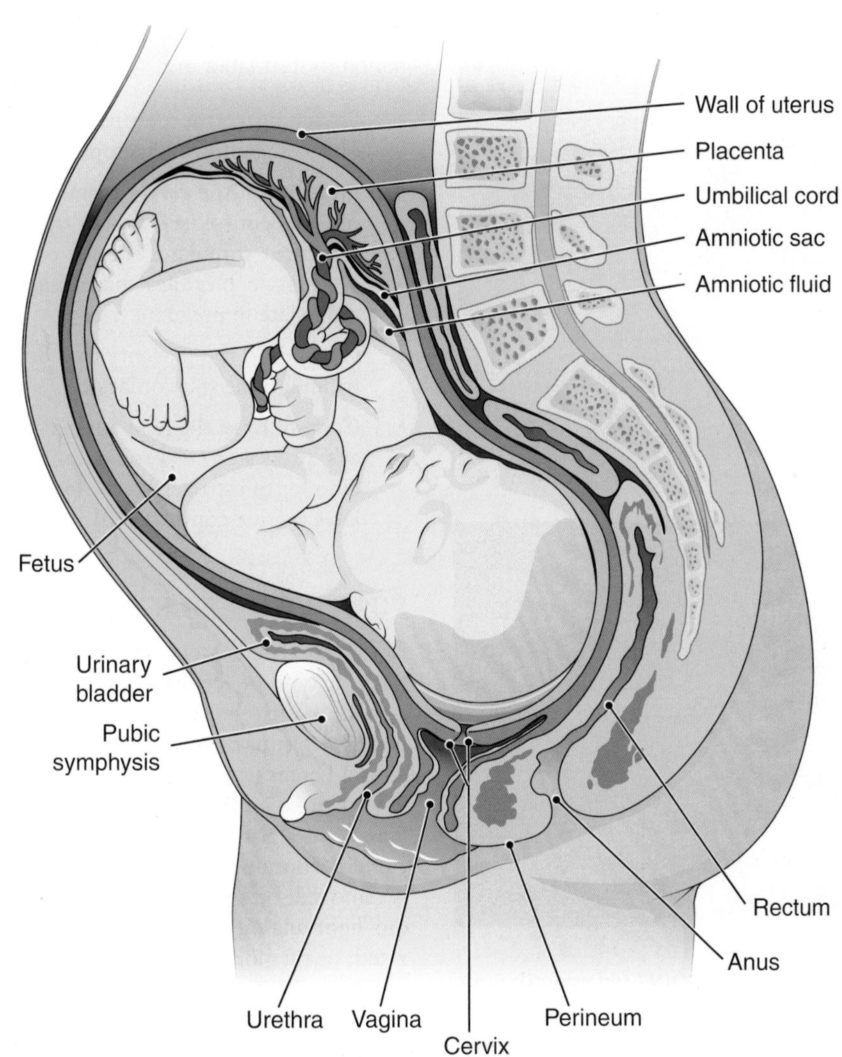

Figure 24-5 **Midsagittal section of a pregnant uterus with intact fetus.** **KEY POINT** The fetus is contained in an amniotic sac filled with amniotic fluid. **ZOOMING IN** What structure connects the fetus to the placenta?

Urinary frequency and constipation are often present during the early stages of pregnancy and then usually disappear. They may reappear late in pregnancy as the fetus's head drops from the abdominal region down into the pelvis, pressing on the urinary bladder and the rectum.

THE USE OF ULTRASOUND IN OBSTETRICS

Ultrasonography (ul-trah-son-OG-rah-fe) is a safe, painless, and noninvasive method for studying soft tissue. It has proved extremely valuable for monitoring pregnancies and childbirth.

An ultrasound image, called a *sonogram*, is made by sending high-frequency sound waves into the body (**Fig. 24-6**). Each time a wave meets an interface between two tissues of different densities, an echo is produced. An instrument called a *transducer* converts the reflected sound waves into electrical energy, and a computer is used to generate an image on a viewing screen.

A

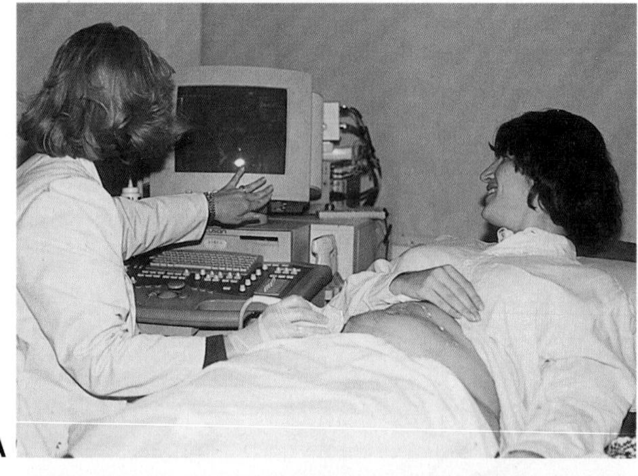

B

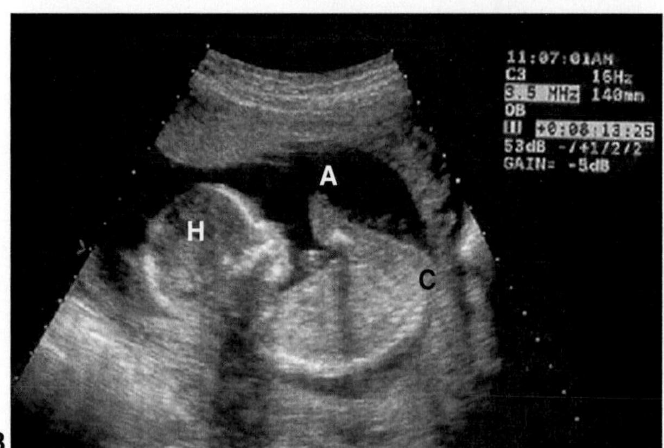

Figure 24-6 **Sonography.** **KEY POINT** Ultrasound is used to monitor pregnancy and childbirth. **A.** A sonogram is recorded as a mother views her baby on the monitor. **B.** Sonogram of a pregnant uterus at 10 to 11 weeks showing the amniotic cavity (A) filled with amniotic fluid. The fetus is seen in longitudinal section showing the head (H) and coccyx (C).

Ultrasound scans can be used in obstetrics to diagnose pregnancy, judge fetal age, and determine the location of the placenta. The technique can also show the presence of excess amniotic fluid and fetal abnormalities (**see Fig. 24-6B**).

Obstetricians use additional methods to test fetal health and identify possible genetic abnormalities in a fetus during pregnancy. These tests are described in Chapter 25.

CHECKPOINTS

- ☐ 24-6 At about what time in gestation does the heartbeat first appear?
- ☐ 24-7 What is the name of the fluid-filled sac that holds the fetus?
- ☐ 24-8 What is the approximate duration of pregnancy in days from the time of fertilization?

Childbirth

The exact mechanisms that trigger the beginning of uterine contractions for childbirth are still not completely known. Some fetal and maternal factors that probably work in combination to start labor are:

- Stretching of the uterine muscle stimulates production of prostaglandins which promote uterine contractions.
- Pressure on the cervix from the baby stimulates release of **oxytocin** (ok-se-TO-sin) from the posterior pituitary. Oxytocin stimulates uterine contractions, and the uterine muscle becomes increasingly sensitive to this hormone late in pregnancy.
- Changes in the placenta that occur with time may contribute to the start of labor.
- Cortisol from the fetal adrenal cortex inhibits the mother's progesterone production. Increase in the relative amount of estrogen as compared to progesterone stimulates uterine contractions.

POSITIVE FEEDBACK AND OXYTOCIN

The events that occur in childbirth involve **positive feedback**, a type of feedback mechanism we have not yet discussed. Positive feedback is much less common than the negative feedback illustrated many times in previous chapters. Whereas negative feedback reverses a condition to bring it back to a norm, positive feedback intensifies a response. (Think of a fist fight escalating among a group of people.) Activity continues until resources are exhausted, the stimulus is removed, or some outside force interrupts the activity. (In our fight analogy, people tire, the original fighters are separated, or the police arrive to break things up!) The diagrams in **Figure 24-7** compare negative and positive feedback.

During childbirth, for example, cervical stretching stimulates the pituitary to release oxytocin. This hormone stimulates further uterine contractions. As contractions increase in force, the cervix is stretched even more, causing further release of oxytocin. The escalating contractions

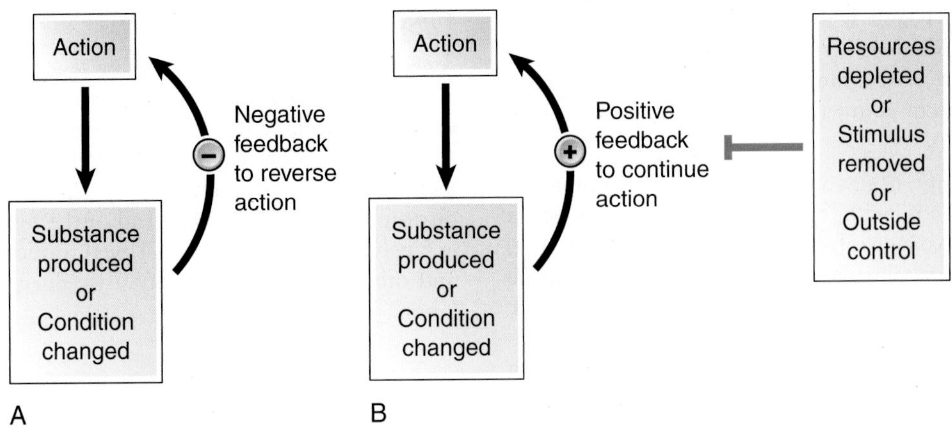

Figure 24-7 Comparison of positive and negative feedback. **KEY POINT A.** Negative feedback maintains homeostasis, bringing changes back to a norm. **B.** Positive feedback intensifies responses until the feedback loop is interrupted. **ZOOMING IN** What possible forces can stop a positive feedback system?

and hormone release continue until the baby is born and the stimulus is removed. Oxytocin in the form of the trade name drug Pitocin can be used to hasten labor, as was about to happen in Amy's case study during childbirth.

THE FOUR STAGES OF LABOR

The process by which the fetus is expelled from the uterus is known as **parturition** (par-tu-RISH-un), or *labor* (**Fig. 24-8**). It is divided into four stages:

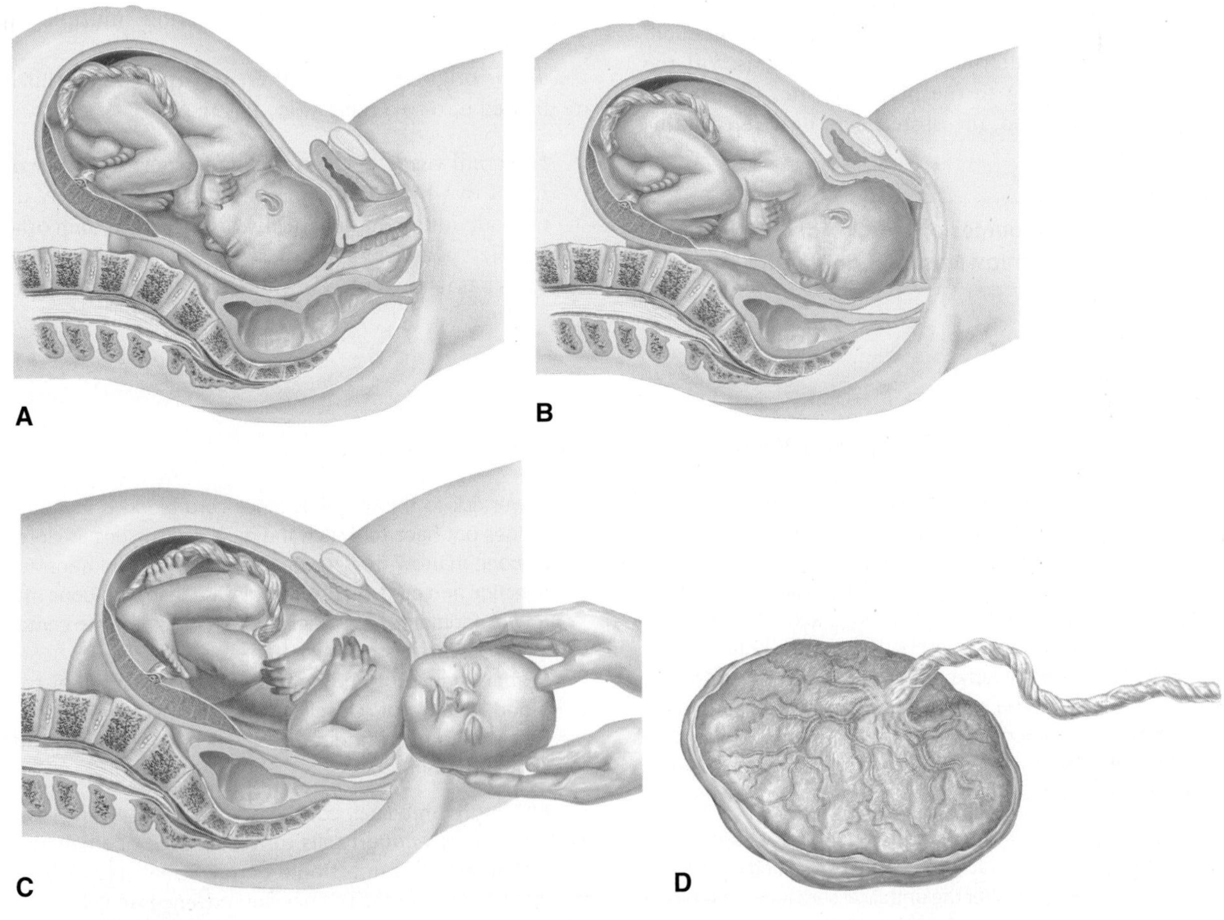

A B C D

Figure 24-8 Stages of labor. The first stage **(A)** begins with the onset of uterine contractions; the second stage **(B, C)** begins when the cervix is completely dilated and ends with birth of the baby; the third stage **(D)** includes expulsion of the afterbirth. A fourth stage (not shown) involves control of bleeding and uterine contractions.

1. The *first stage* begins with the onset of regular uterine contractions. With each contraction, the cervix becomes thinner and the cervical canal larger. Rupture of the amniotic sac may occur at any time, with a gush of fluid from the vagina.
2. The *second stage* begins when the cervix is completely dilated and ends with the baby's birth. This stage involves the passage of the fetus, usually head first, through the cervical canal and the vagina to the outside (**see Fig. 24-8B and C**). To prevent tissues of the pelvic floor from being torn during childbirth, as often happens, the obstetrician may cut the mother's perineum just before her infant is born; such an operation is called an **episiotomy** (eh-piz-e-OT-o-me). The area between the vagina and the anus that is cut in an episiotomy is referred to as the surgical or obstetrical perineum. (See **Fig. 23-9** in Chapter 23.)
3. The *third stage* begins after the child is born and ends with the expulsion of the *afterbirth*; that is, the placenta, the membranes of the amniotic sac, and the umbilical cord, except for a small portion remaining attached to the baby's umbilicus (um-BIL-ih-kus), or navel (**see Fig. 24-8D**). (See Box 24-2 on medical uses of umbilical cord blood.)
4. The *fourth stage* begins after expulsion of the afterbirth and constitutes a period in which bleeding is controlled. Contraction of the uterine muscle acts to close off the blood vessels leading to the placental site. If an episiotomy was performed, the obstetrician now repairs this clean cut.

See the Student Resources on thePoint for information on birth assistants, midwives, and doulas.

CESAREAN SECTION

A **cesarean** (se-ZAR-re-an) **section** (C section) is an incision made in the abdominal wall and uterine wall through which the fetus is manually removed from the mother's body. A cesarean section may be required for a variety of reasons, including placental abnormalities; abnormal fetal position; disproportion between the head of the fetus and the mother's pelvis that makes vaginal birth difficult or dangerous; and other problems that may arise during pregnancy and labor.

MULTIPLE BIRTHS

Until recently, statistics indicated that twins occurred in about one of every 80 to 90 births, varying somewhat in different countries. Triplets occurred much less frequently, usually once in several thousand births, whereas quadruplets occurred very rarely. The birth of quintuplets represented a historic event unless the mother had taken fertility drugs. Now these fertility drugs, usually gonadotropins, are given more commonly, and the number of multiple births has increased significantly. Multiple fetuses tend to be born prematurely and therefore have a high death rate. However, better care of infants and newer treatments have resulted in more living multiple births than ever.

Twins originate in two different ways, and on this basis are divided into two types (**Fig. 24-9**):

- **Fraternal twins** result from the fertilization of two different ova by two spermatozoa. Two completely different individuals, as genetically distinct from each other as brothers and sisters of different ages, are produced. Each fetus has its own placenta and surrounding amniotic sac.

HOT TOPICS

Box 24-2

Umbilical Cord Blood: Giving Life after Birth

Following childbirth, the umbilical cord and placenta are usually discarded. However, research suggests that blood harvested from these structures could save lives. Like bone marrow, umbilical cord blood contains stem cells capable of differentiating into all blood cell types. Cancer patients whose bone marrow is destroyed by chemotherapy often require stem cell transplants, as do those with leukemia, anemia, or certain immune disorders.

Stem cells obtained from umbilical cord blood offer some important advantages over those acquired from bone marrow. These advantages include

- Greater ease of collection and storage. Whereas bone marrow collection is a surgical procedure, cord blood can be collected immediately after the umbilical cord is cut. The blood can then be stored frozen in a blood bank.
- No risk to the donor. Because cord blood is collected after the cord is cut, the procedure is not dangerous to the donor.
- Lower risk to the recipient. Since cord blood is immature, it does not have to match the recipient's tissues as closely as bone marrow does, so there is less chance of transplant rejection and graft-versus-host disease than with bone marrow. In addition, umbilical cord blood is less likely to contain infectious organisms than is bone marrow.
- Higher chance of finding a donor. Since umbilical cord blood need not closely match a recipient's tissues, the likelihood of finding a match between donor and recipient is higher.

Although umbilical cord blood is a promising stem cell source, only enough cells to treat a child or small adult can be harvested from a single donor. Scientists hope that improved collection techniques and advances in cell culture will increase the number of stem cells available from a single donor, enabling all patients in need to benefit from this treatment.

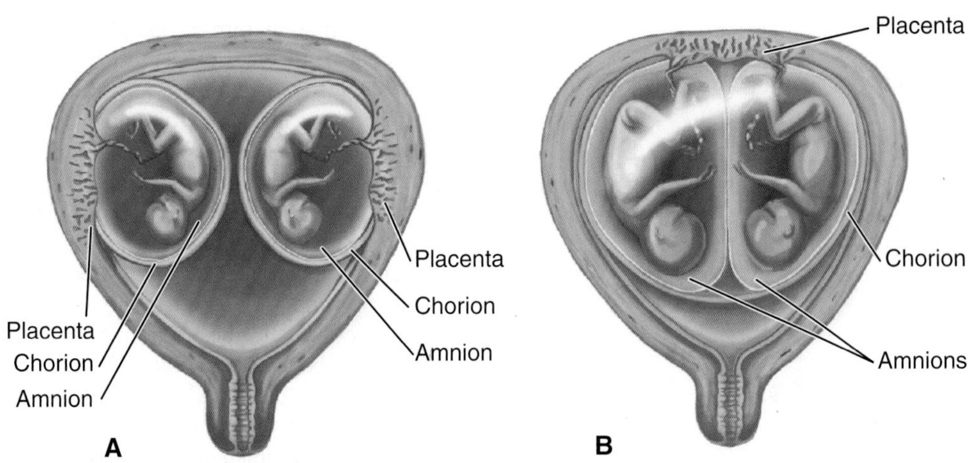

Figure 24-9 **Twins. A.** Fraternal twins. **B.** Identical twins. **ZOOMING IN** Which type of twins shares a single placenta?

- **Identical twins** develop from a single zygote formed from a single ovum fertilized by a single spermatozoon. Sometime during the early stages of development, the embryonic cells separate into two units. Usually, there is a single placenta, although there must be a separate umbilical cord for each fetus. Identical twins are always the same sex and carry the same inherited traits.

Other multiple births may be fraternal, identical, or combinations of these. The tendency for multiple births seems to be hereditary.

PREGNANCY OUTCOMES

An infant born between 37 and 42 weeks of gestation is described as a **term infant.** However, a pregnancy may end before its full term has been completed. The term **live birth** is used if the baby breathes or shows any evidence of life such as heartbeat, pulsation of the umbilical cord, or movement of voluntary muscles. Infants born before the 37th week of gestation are considered **preterm.** Often these babies have low birth weights of less than 2,500 g (5.5 lb) and are immature (premature) in development. They are subject to a number of medical conditions, including anemia, jaundice, respiratory problems, and feeding difficulties. The underlying causes of preterm labor are not well understood, but risk factors include trauma, infections, poor maternal health, and chronic health conditions.

Loss of the fetus is classified according to the duration of the pregnancy:

- The term **abortion** refers to loss of the embryo or fetus before the 20th week or weight of about 500 g (1.1 lb). This loss can be either spontaneous or induced.
  - **Spontaneous abortion,** commonly known as a **miscarriage,** occurs naturally with no interference. It is estimated that 22% of pregnancies end in spontaneous abortion. About 50% of these cases reflect a chromosomal abnormality in the embryo or fetus. Other causes include luteal phase insufficiency, in which the corpus luteum does not secrete enough progesterone, or a structural abnormality of the mother's reproductive organs, such as uterine fibroids. Acute maternal infections (e.g., listeria or toxoplasma) can spread from the blood or from the urinary tract to infect the uterine cavity and cause miscarriage. Finally, chronic disorders, such as uncontrolled diabetes mellitus, kidney disease, hypertension, or clotting disorders, are associated with increased risk of spontaneous abortion.
  - **Induced abortion** occurs as a result of deliberate interruption of pregnancy. A **therapeutic abortion** is an abortion performed by a physician as a treatment for a variety of reasons. More liberal access to this type of abortion has dramatically reduced the incidence of death related to illegal abortion.
- The term **fetal death** refers to loss of the fetus after the 20th week of pregnancy. **Stillbirth** refers to the birth of an infant who is lifeless. The loss of a fetus often indicates an infection or interference with the fetal blood supply by cord compression or placental dysfunction.

Immaturity is a leading cause of death in the newborn. After the 20th week of pregnancy, the fetus is considered **viable,** that is, able to live outside the uterus. A fetus expelled before the 24th week or before reaching a weight of 1,000 g (2.2 lb) has little more than a 50% chance of survival; one born at a point closer to the full 38 weeks of gestation stands a much better chance of living. However, increasing numbers of immature infants are being saved because of advances in neonatal intensive care.

Hospitals use the Apgar score to assess a newborn's health and predict survival. Five features, such as respiration, pulse, etc., are rated as 0, 1, or 2 at one minute and five minutes after birth. The maximum possible score on each test is 10. Infants with low scores require medical attention and have lower survival rates.

See the Student Resources on thePoint for the Apgar scoring system.

24

CHECKPOINTS

- ☐ 24-9 What pituitary hormone stimulates uterine contractions?
- ☐ 24-10 What is parturition?
- ☐ 24-11 During what stage of labor is the afterbirth expelled?
- ☐ 24-12 What surgical procedure is used to remove a fetus manually from the uterus?
- ☐ 24-13 What term describes a fetus able to live outside the uterus?

The Mammary Glands and Lactation

The **mammary glands**, contained in the female breasts, are associated organs of the reproductive system. They provide nourishment for the baby after its birth. The mammary glands are similar in construction to the sweat glands. Each gland is divided into a number of lobes composed of glandular tissue and fat, and each lobe is further subdivided. Secretions from the lobes are conveyed through **lactiferous** (lak-TIF-er-us) **ducts**, all of which converge at the papilla (nipple) (**Fig. 24-10**).

The mammary glands begin developing during puberty, but they do not become functional until the end of a pregnancy. As already noted, estrogen and progesterone during pregnancy help prepare the breasts for lactation, as do prolactin (PRL) from the anterior pituitary and placental lactogen (hPL). The first mammary gland secretion is a thin liquid called **colostrum** (ko-LOS-trum). It is nutritious but has a somewhat different composition from milk. Milk secretion begins within a few days following birth and can continue for several years as long as milk is frequently removed by the suckling baby or by pumping. In another example of positive feedback, the nursing infant's suckling at the breast promotes release of PRL, which activates the mammary glands' secretory cells. Nursing also stimulates oxytocin release from the posterior pituitary. This hormone causes the milk ducts to contract, resulting in the ejection, or *letdown*, of milk.

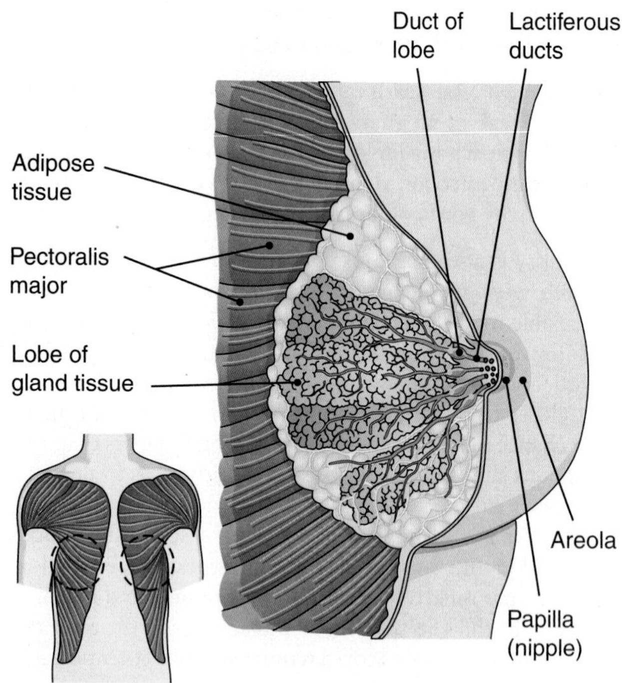

Figure 24-10 Section of the breast (mammary gland). **KEY POINT** The glands are divided into lobes containing milk ducts that converge at the nipple. **ZOOMING IN** What muscle underlies the breast?

The newborn baby's digestive tract is not ready for the usual adult mixed diet. Mother's milk is more desirable for the young infant than milk from other animals for several reasons, some of which are listed below:

- Infections that may be transmitted by foods exposed to the outside air are avoided by nursing.
- Both breast milk and colostrum contain maternal antibodies that help protect the baby against pathogens.
- The proportions of various nutrients and other substances in human milk are perfectly suited to the human infant, changing within a single feeding from a watery fluid that quenches the infant's thirst to a fluid rich in fat that satisfies the infant's hunger. The composition of breast milk also changes over time as the infant grows. Substitutes are not exact imitations of human milk. Nutrients are present in more desirable amounts if the mother's diet is well balanced.
- The psychological and emotional benefits of nursing are of infinite value to both the mother and the infant.

CHECKPOINTS

- ☐ 24-14 What are the ducts that carry milk out of the breast?
- ☐ 24-15 What substance is secreted by the breast before milk production begins?

Disorders of Pregnancy, Childbirth, and Lactation

Many disorders can affect the pregnant woman and her developing fetus. Only the most common are discussed here.

PREGNANCY-RELATED DISORDERS

A pregnancy that develops in a location outside the uterine cavity is said to be an **ectopic** (ek-TOP-ik) **pregnancy** (**Fig. 24-11**). The most common type is the tubal ectopic pregnancy, in which the embryo begins to grow in the uterine tube. This structure cannot expand to contain the growing embryo and may rupture, causing severe pain. Without prompt surgical treatment, tubal ectopic pregnancy may threaten the mother's life. Tubal pregnancies are more common in women with scarred uterine tubes, resulting from pelvic inflammatory disease (PID) or endometriosis, or in women using an intrauterine device (IUD) for birth control.

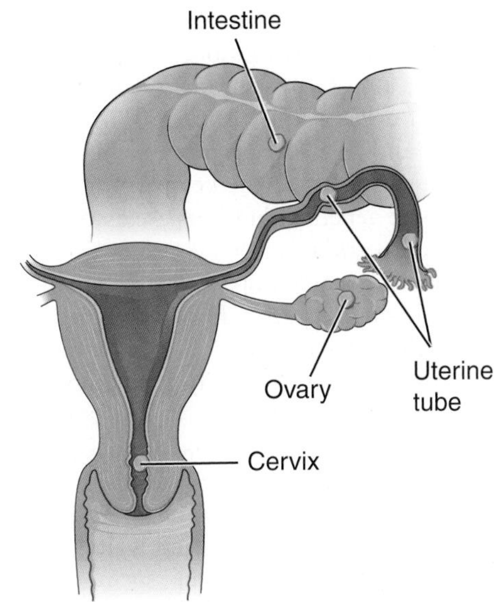

Figure 24-11 **Ectopic pregnancy.** **KEY POINT** Various possible sites are shown. The most common site is the uterine tube, in which case it is a tubal ectopic pregnancy.

Placental Disorders In **placenta previa** (PRE-ve-ah), the placenta, which is usually attached to the superior part of the uterus, instead becomes attached at or near the internal opening of the cervix (**Fig. 24-12A**). Dilation of the cervix in late pregnancy causes part of the placenta to separate from the rest. The result is painless bleeding and interference with the fetal oxygen supply.

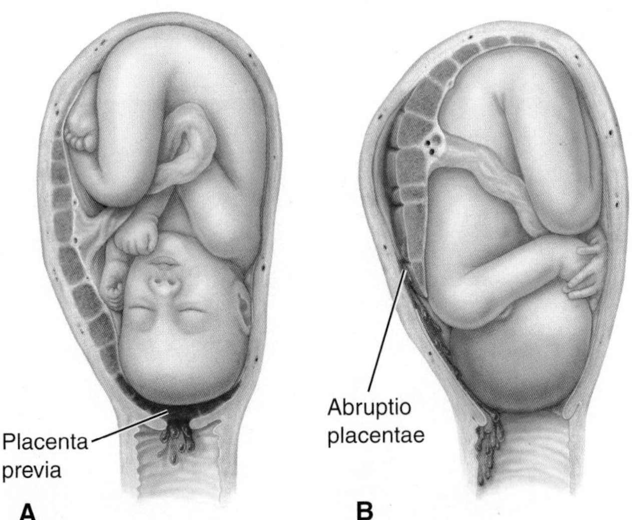

Figure 24-12 **Placental abnormalities.** **KEY POINT** If the placenta develops out of position or separates prematurely, bleeding and possible termination of pregnancy may occur. **A.** In placenta previa, the placenta attaches near the cervix instead of at the superior part of the uterus. **B.** In abruptio placentae, the placenta separates prematurely from the uterine wall.

Sometimes the placenta separates from the uterine wall prematurely, often after the 20th week of pregnancy, causing hemorrhage (**see Fig. 24-12B**). This disorder, known as **abruptio placentae** (ab-RUP-she-o plah-SEN-te), or *placental abruption*, occurs most often in multigravidas (mul-te-GRAV-ih-dahz), women who have been pregnant more than once, and in women older than 35 years of age. Placental abruption is a common cause of bleeding during the second half of pregnancy and may require termination of pregnancy to save the mother's life.

If the endometrium does not develop properly, the placenta can attach directly to the myometrium, resulting in **placenta accreta** (ah-KRE-tah). The abnormal placenta can adequately nourish the growing fetus, but it does not detach properly during the fourth stage of childbirth, potentially causing fatal postnatal bleeding. Women with placenta previa or those who have previously given birth by cesarean section are at greater risk of placenta accreta.

A **hydatidiform** (hi-dah-TID-ih-form) **mole**, or *hydatid mole*, arises from the union of abnormal gametes or fertilization of an ovum with two sperm cells. The result is a benign overgrowth of placental tissue in which the placenta dilates and resembles grapelike cysts. The growth may invade the uterine wall, causing it to rupture. Normal fetal development is not possible in such cases, and the conception products must be evacuated surgically.

Hydatidiform moles must be accurately diagnosed because they can lead to **choriocarcinoma** (ko-re-o-kar-sih-NO-mah), a very malignant tumor composed of placental tissue. Although rare, this tumor spreads rapidly, and if the mother is not treated, it may be fatal within three to 12 months. With the use of modern chemotherapy, the outlook for cure is very good. If metastases have developed, irradiation and other forms of treatment may be necessary.

Gestational Diabetes Diabetes that is first diagnosed during pregnancy is termed *gestational diabetes mellitus (GDM)*. This form of diabetes often causes no symptoms, but it may be associated with increased thirst, increased urination, frequent infections, and weight loss in spite of increased appetite. Blood glucose levels usually return to normal after childbirth. However, many women with this disorder during pregnancy go on to develop type 2 diabetes later in life. In the fetus, high glucose levels resulting from GDM cause overproduction of insulin and weight gain. These alterations can complicate childbirth and contribute to obesity starting in childhood and type 2 diabetes later in life. Treatment includes a strict diet, exercise, glucose monitoring, and sometimes medication.

GDM shares many risk factors with type II diabetes mellitus, such as obesity, impaired glucose tolerance, and a family history of diabetes. The hormone hPL appears to play a role in GDM since it induces insulin resistance in the mother in order to spare glucose for the fetus. Obstetricians should screen all pregnant women for this condition at 24 to 28 weeks' gestation, earlier if there is a high risk or suspicion of GDM. Risk factors include pregnancy over age 25, obesity, smoking, and family history of diabetes.

Pregnancy-Induced Hypertension A serious disorder that can develop in the latter part of pregnancy is **pregnancy-induced hypertension (PIH)**, also called *preeclampsia* (pre-eh-KLAMP-se-ah) or *toxemia of pregnancy*. Signs include hypertension, protein in the urine (proteinuria), general edema, and sudden weight gain, as noted in Amy's opening case study. The cause of PIH is not completely understood. Researchers currently believe that it results from inadequate formation of blood vessels in the placenta. The resulting hypoxia induces the production of various substances by placental cells, which travel through the maternal circulation to cause vasoconstriction and increased permeability of the maternal blood vessels. As soon as the placenta is delivered, the syndrome resolves. Maternal health is also involved, for example, the presence of chronic hypertension, GDM, autoimmune disease, blood disorders, or renal disease. PIH is most often seen in first pregnancies and in women whose nutritional state is poor, who are obese, and who have received little or no health care during pregnancy. If PIH remains untreated, it may lead to **eclampsia** (eh-KLAMP-se-ah) with the onset of kidney failure, convulsions, and coma during pregnancy or after birth. The result may be the death of both the mother and the infant.

POSTPARTUM DISORDERS

Childbirth-related deaths are often due to infections. **Puerperal** (pu-ER-per-al) **infections**, those related to childbirth, were once the cause of death in as many as 10% to 12% of childbearing women. Cleanliness and sterile techniques have greatly reduced maternal mortality. Nevertheless, in the United States, puerperal infection still develops in about 6% of maternity patients. Antibiotics have dramatically improved the chances of recovery for both the mother and the child.

Postpartum depression (PPD) is a major psychological depression that follows within a year of childbirth. The symptoms may be the same as any found in clinical depression but may include excessive concern with the baby's health or thoughts of harming the baby. Contributing factors are hormonal changes and general family or life stresses. PPD occurs in about 10% to 20% of new mothers.

Much more common but less serious are "postpartum blues" or "baby blues." This state involves mood swings that appear within a week of childbirth and may last several weeks as hormone levels stabilize. Persistence for longer times may indicate a more serious psychological disorder.

DISORDERS OF THE BREAST AND LACTATION

Disorders of the breast include neoplasms and lactational disturbances. While breast neoplasms are not restricted to women of childbearing age, they are discussed along with breast anatomy and physiology in this chapter.

Breast Cancer Cancer of the breast is the most commonly occurring malignant disease in women. The risk factors for breast cancer include age past 40, family history of breast cancer, and factors that increase exposure to estrogen, such as early onset of menstruation, late menopause, late or no pregnancies, long-term menopausal hormone therapy, and obesity (fat cells produce estrogen). Mutations in two genes (BRCA1 and BRCA2) are responsible for hereditary forms of breast cancer, which make up only about 8% of all cases. These same genetic mutations are associated with an increased risk of ovarian cancer.

A malignant breast tumor is usually a painless, hard, oddly shaped lump that feels firmly attached within the breast. Although the majority of breast lumps are benign cysts, a woman should report any lump, no matter how small, to her physician immediately. The **mammogram**, a radiographic study of the breast, has improved the detection of early breast cancer (**Fig. 24-13**). Some health organizations recommend regular mammograms after the age of 40 years. Other health professionals recommend waiting until age 50 unless a woman is in a high-risk group, such as having a family history of breast cancer. Some physicians now recommend adding MRI scans for high-risk patients. Suspicious areas require further study by ultrasonography or biopsy (a needle aspiration, removal of a core of tissue, or excision of the lump). In a stereotactic biopsy, a physician uses a computer-guided imaging system to locate suspicious tissue and remove samples with a needle.

A precancerous condition that originates in the milk ducts of the breast is termed *ductal carcinoma in situ* (DCIS). Comprising about 15% to 20% of breast cancer diagnoses in the United States, DCIS consists of cancer cells that have not penetrated the duct walls or spread to other locations (in situ means "in place"). It is noninvasive, but may unpredictably develop into metastatic cancer. The lesions usually cannot be felt by palpation but are diagnosed by mammography. Treatment decisions depend on tumor pathology as well as a patient's age and family history.

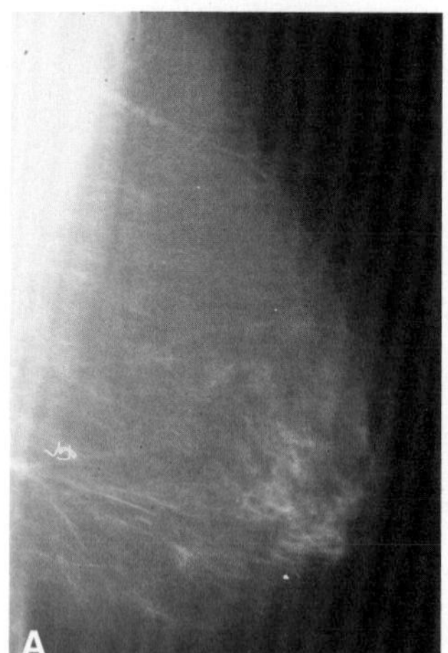

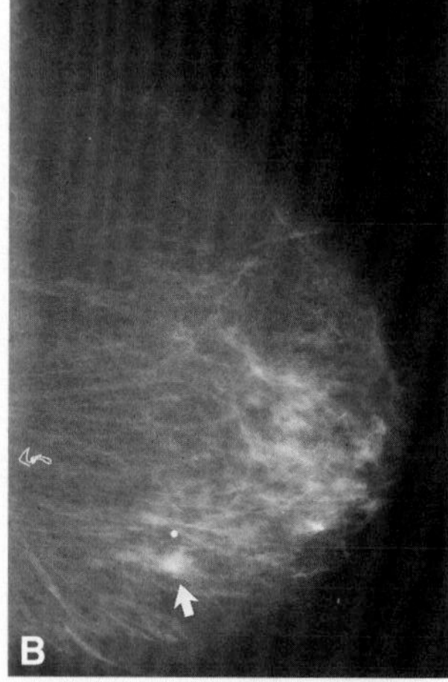

Figure 24-13 **Breast cancer.** **KEY POINT** Mammography is radiographic study of the breast for detection of tumors. Two mammograms taken 12 months apart are shown. **A.** Initial mammogram. **B.** Second mammogram shows a developing mass (*arrow*). Biopsy revealed an invasive carcinoma.

Breast cancer treatment consists of surgery with follow-up therapy of radiation, chemotherapy, or both. Surgical treatment by removal of the lump ("lumpectomy") or a segment of the breast is most common. Removal of the entire breast and dissection of the lymph nodes in the axilla (armpit) is called **modified radical mastectomy** (mas-TEK-to-me). The extent of tumor spread through the lymph nodes is an important factor in prognosis. In a sentinel lymph node biopsy, the first (sentinel) lymph nodes to receive lymph from the tumor are identified and tested for cancerous cells. Treatment is based on how much spread has occurred (see Box 16-1). Treatment of breast cancer is often followed by administration of drugs that inhibit estrogen production or block estrogen receptors in breast tissue (if the tumor responds to that hormone), or drugs that inhibit tumor growth factors. Note that the incidence of the various types of cancer should not be confused with the death rates for each type. Owing to public education and increasingly better methods of diagnosis and treatment, some forms of cancer have a higher cure rate than others. For example, breast cancer appears much more often in women than does lung cancer, but more women now die each year from lung cancer than from breast cancer.

Lactation Disturbances Disturbances in lactation may have a variety of causes, including the following:

- Malnutrition, dehydration, or anemia, which may prevent lactation entirely
- Emotional disturbances, which may affect lactation (as they may affect other glandular activities)
- Abnormalities of the mammary glands or injuries to these organs, which may interfere with their functioning
- **Mastitis** (mas-TI-tis), or "inflammation of the breast," which is caused by infection. Antibiotic treatment usually allows for the continuation of nursing.

CHECKPOINTS

- ☐ **24-16** What is the term for a pregnancy that develops outside the uterine cavity?
- ☐ **24-17** What is puerperal infection?
- ☐ **24-18** What is a radiographic study of the breast called?

Disease in Context Revisited

Disease in Context Amy's Pregnancy and HELLP Syndrome

A nurse was waiting in the room in the obstetrics wing and hooked Amy's intravenous line to a Pitocin (oxytocin) drip. Two hours later, Amy delivered a 4 lb 11 oz baby. The baby was healthy, but Amy was still in critical condition and was transferred to intensive care. A few days later, she was well enough to talk with Dr. Renaud.

"What is HELLP anyway, and why did I get so sick?" she asked.

He replied, "You may have heard during your prenatal classes of preeclampsia, also known as pregnancy-induced hypertension. Well, HELLP is a serious variant of preeclampsia. The acronym stands for three common findings—**h**emolysis, **e**levated **l**iver enzymes, and **l**ow **p**latelet count. It appears to begin with a problem in blood vessels, and in your case, it progressed to high blood pressure, liver damage, and blood cell destruction. The pains and heartburn you were having probably originated in your liver. We've given you two blood transfusions to replace the destroyed blood cells, but you're going to be anemic for a while. We don't know what causes HELLP, but we do know that you are at extreme risk of developing the same issues in subsequent pregnancies."

Amy and Rick looked at each other.

"I don't think we can go through this again," Amy said. "We have our daughter at home, and now a healthy boy. Two children will have to be enough!"

Rick took the baby home after a one-week stay in the infant nursery, and Amy was discharged after a second week.

CHAPTER 24

Chapter Wrap-Up

Summary Overview

A detailed chapter outline with space for note taking is on *thePoint*. The figure below illustrates the main topics covered in this chapter.

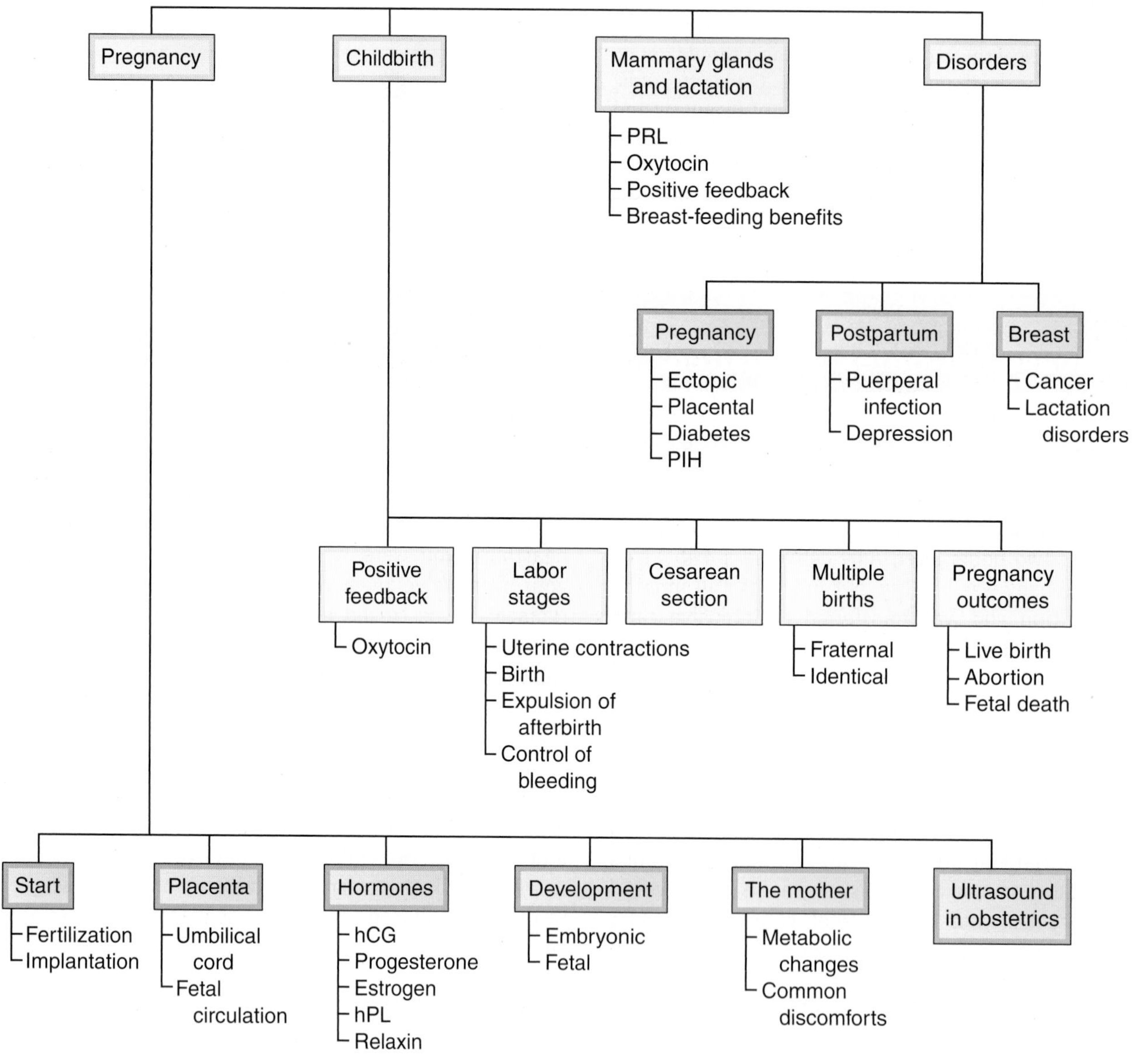

Key Terms

The terms listed below are emphasized in this chapter. Knowing them will help you organize and prioritize your learning. These and other boldface terms are defined in the Glossary with phonetic pronunciations.

abortion
amniotic sac
chorion
colostrum
decidua
embryo
fertilization
fetus
gestation
human chorionic gonadotropin (hCG)
implantation
lactation
obstetrics
oxytocin
parturition
placenta
umbilical cord
zygote

Word Anatomy

Medical terms are built from standardized word parts (prefixes, roots, and suffixes). Learning the meanings of these parts can help you remember words and interpret unfamiliar terms.

WORD PART	MEANING	EXAMPLE
Pregnancy		
chori/o	membrane, chorion	Human *chorionic* gonadotropin is produced by the chorion (outermost cells) of the embryo.
somat/o	body	Human chorionic *somatomammotropin* controls nutrients for the body and acts on the mammary glands (mamm/o).
zyg/o	joined	An ovum and spermatozoon join to form a *zygote*.
Childbirth		
ox/y	sharp, acute	*Oxytocin* is the hormone that stimulates labor.
toc/o	labor	See preceding example.
The Mammary Glands and Lactation		
lact/o	milk	The *lactiferous* ducts carry milk from the mammary glands.
mamm/o	breast, mammary gland	A *mammogram* is radiographic study of the breast.
Disorders of Pregnancy, Childbirth, and Lactation		
ecto-	outside, external	An *ectopic* pregnancy occurs outside of the uterine cavity.
mast/o	breast	A *mastectomy* is surgical removal of the breast.

Questions for Study and Review

BUILDING UNDERSTANDING

Fill in the Blanks

1. Following ovulation, the mature follicle becomes the ______.
2. The first mammary gland secretion is called ______.
3. The medical field concerned with care of a woman during pregnancy and childbirth is ______.
4. A pregnancy that develops in a location outside the uterine cavity is said to be a(n) ______ pregnancy.
5. Loss of an embryo or fetus for any reason before the 20th week of development is termed a(n) ______.

Matching > Match each numbered item with the most closely related lettered item.

___ **6.** An embryonic hormone that stimulates the ovaries to secrete progesterone
___ **7.** A placental hormone that regulates maternal blood nutrient levels
___ **8.** A placental hormone that softens the cervix to widen the birth canal
___ **9.** A pituitary hormone that stimulates uterine contractions
___ **10.** A pituitary hormone that stimulates maternal milk production

a. human placental lactogen
b. prolactin
c. oxytocin
d. relaxin
e. human chorionic gonadotropin

Multiple Choice

___ **11.** What is a morula?
a. a ball of identical cells formed from a zygote
b. the fetal portion of the placenta
c. the vessel that carries blood from the embryo to the placenta
d. an abnormality of the placenta

___ **12.** What is an acrosome?
a. the cheeselike material that protects fetal skin
b. the hollow structure that implants in the uterine lining
c. the enzyme-releasing cap over the head of a spermatozoon
d. a small vessel that bypasses the liver in fetal circulation

___ **13.** How long is the average total period of gestation from fertilization to birth?
a. 30 weeks
b. 38 weeks
c. 40 weeks
d. 48 weeks

___ **14.** With regard to identical twins, which statement is incorrect?
a. They develop from a single zygote.
b. They share one umbilical cord.
c. They are always the same sex.
d. They carry the same inherited traits.

___ **15.** What is the purpose of the Apgar score?
a. to assess a newborn's health
b. to estimate the length of pregnancy
c. to measure placental hormones
d. to measure fetal size

UNDERSTANDING CONCEPTS

16. Describe the function and structure of the placenta, including the terms for its maternal and fetal portions.

17. What is ultrasound, and how is it used during pregnancy and childbirth?

18. Is blood in the umbilical arteries relatively high or low in oxygen? In the umbilical vein? Explain your answer.

19. Describe some of the events thought to initiate labor.

20. What is the major event of each of the four stages of parturition?

21. List several reasons why breast milk is beneficial for the baby.

22. What is a cesarean section? List several reasons why it may be required.

23. Based on Amy's case study, describe the symptoms and diagnosis of the HELLP syndrome. What are some related pregnancy disorders?

24. Compare and contrast the following terms:
a. embryo and fetus
b. foramen ovale and ductus arteriosus
c. fertilization and implantation
d. placenta previa and abruptio placentae
e. puerperal infection and mastitis

CONCEPTUAL THINKING

25. Why is the risk of miscarriage highest at week 12 of pregnancy?

26. Although it is strongly suggested that a woman avoid harmful substances (like alcohol) during her entire pregnancy, why is this advice particularly important during the first trimester?

27. In the process of blood clotting, platelets adhere to damaged tissue and release chemicals that attract more platelets. This continues until a clot is formed and other blood factors stop the process. What type of control mechanism does this represent? Give two other examples of this type of action.

For more questions, see the Learning Activities on thePoint.

CHAPTER 25

Heredity and Hereditary Diseases

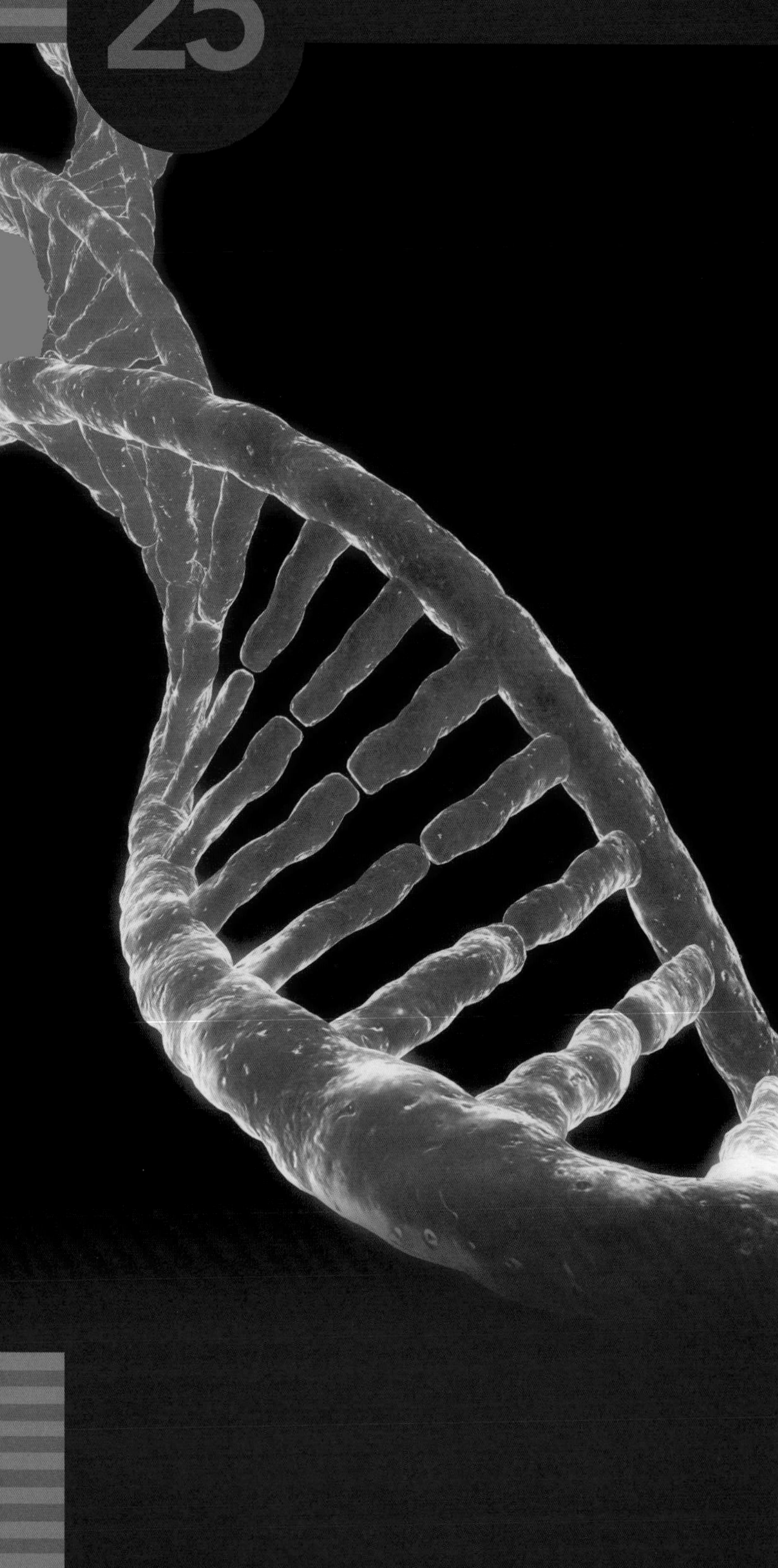

Learning Objectives

After careful study of this chapter, you should be able to:

1 ▶ Define a gene, and briefly describe how genes function. ***p. 570***

2 ▶ Explain the relationship between genotype and phenotype for dominant and recessive alleles. ***p. 570***

3 ▶ Explain how chromosomes are distributed in meiosis. ***p. 571***

4 ▶ Perform a genetic cross using a Punnett square. ***p. 571***

5 ▶ Explain how sex is determined in humans and how sex-linked traits are inherited. ***p. 572***

6 ▶ List three factors that may influence the expression of a gene. ***p. 574***

7 ▶ Define mutation. ***p. 574***

8 ▶ Explain the pattern of mitochondrial inheritance. ***p. 574***

9 ▶ Differentiate among congenital, genetic, and hereditary disorders, and give several examples of each. ***p. 575***

10 ▶ Give several examples of teratogens, and describe their effects. ***p. 575***

11 ▶ Describe the symptoms and inheritance patterns of some common genetic diseases. ***p. 576***

12 ▶ Describe four methods for diagnosing fetal disorders. ***p. 578***

13 ▶ Give examples of methods currently used to treat certain genetic disorders. ***p. 578***

14 ▶ Using the case study and information in the text, describe the role of a genetic counselor. ***pp. 569, 579***

15 ▶ Show how word parts are used to build words related to heredity (see Word Anatomy at the end of the chapter). ***p. 584***

Disease in Context *Baby Cole's Hereditary Blood Disorder*

Both baby Cole and his mother Jada were napping in the maternity ward. Although Cole was only a few hours old, he had already practiced nursing with his mom and been bathed, fed, and given a thorough examination by his pediatrician, including a hearing test and Apgar score. As part of the hospital's routine neonatal procedures, a heel stick was done to obtain a sample of Cole's blood for testing in the hematology lab.

In the lab, Andrea, the technologist, ran a panel of tests to screen for congenital hypothyroidism, PKU, and a number of other hereditary metabolic diseases, as required by the state. When Cole's blood screen showed positive for abnormal hemoglobin S, Andrea confirmed her suspicions by examining the blood cells directly under the microscope.

"Looks like your patient Cole has sickle cell anemia," she reported to Dr. Hoffman, Cole's pediatrician. In response, the physician tracked down the hospital's hematologist, and together they went to the Armstrongs' room to tell them about the lab results. Jada and her husband Carl were surprised to see the two physicians enter.

Dr. Hoffman sat down with them for a moment and then explained. "We don't mean to frighten you, but Cole's lab tests have shown some abnormalities. It appears that he has a genetic disorder called sickle cell anemia. The hematologist is here with me to tell you more about it."

Dr. Evans introduced herself and continued, "Cole's disease is caused by a mutation in the gene that directs the manufacture of the blood's hemoglobin. You both carry two versions of the gene, one normal and one abnormal. Neither of you shows the disease because your normal gene masks the effect of the abnormal gene. But Cole inherited the abnormal copy of the gene from each of you and thus shows sickle cell anemia. His red cells take on a crescent shape, blocking blood flow and the delivery of oxygen to his tissues."

Seeing the parents' look of alarm she tried to reassure them. "Starting immediately, a medical team will work with you to manage Cole's disease and minimize tissue damage and pain."

At first, Jada and Carl's attention was focused only on their child and the care he would need. Then they began to wonder how they had passed on the disease and whether or not other children they might have would be at risk.

In this chapter, we'll learn about genetic inheritance and hereditary diseases. We'll also go with the Armstrongs to visit a genetic counselor and find out the chances of their passing a disease gene to another child.

ANCILLARIES *At-A-Glance*

Visit thePoint to access the following resources. For guidance in using these resources most effectively, see pp. xv–xvii.

Learning RESOURCES

- Tips for Effective Studying
- Web Figure: Meiosis in Male and Female Gamete Formation
- Web Figure: Clinical Features of Down Syndrome
- Web Figure: Facies of Down Syndrome
- Web Figure: Marfan Syndrome
- Web Figure: Systemic Changes in Cystic Fibrosis
- Web Figure: Clinical Features of Klinefelter Syndrome
- Web Chart: Genetic Diseases
- Health Professions: Genetic Counselor
- Detailed Chapter Outline
- Answers to Questions for Study and Review
- Audio Pronunciation Glossary

Learning ACTIVITIES

- Pre-Quiz
- Visual Activities
- Kinesthetic Activities
- Auditory Activities

A LOOK BACK

In Chapter 3, we discussed the structure and function of DNA, genes, and chromosomes. Now we look further into how they function to determine hereditary traits. Chapter 3 also explained how cells divide by mitosis, the process that underlies the growth and development of offspring, which was later described in Chapter 24. Now we discuss meiosis, a different form of cell division responsible for the production of gametes, first introduced in Chapter 23. Disruption of these division processes at any time during development leads to abnormalities in the offspring.

We are often struck by the resemblance of a baby to one or both of its parents, yet rarely do we stop to consider *how* various traits are transmitted from parents to offspring. This subject—**heredity**—has fascinated humans for thousands of years. The Old Testament contains numerous references to heredity (although the word itself was unknown in ancient times). It was not until the 19th century, however, that methodical investigation into heredity was begun. At that time, an Austrian monk Gregor Mendel used garden peas to reveal a precise pattern in the appearance of characteristics among parents and their **progeny** (PROJ-eh-ne), their offspring or descendants. Mendel's most important contribution to the understanding of heredity was the demonstration that there are independent units of heredity in the cells. Later, these independent units were given the name **genes**.

Genes and Chromosomes

Genes are actually segments of DNA contained in the thread-like **chromosomes** within the nucleus of each cell. (Only the mature red blood cell, which has lost its nucleus, lacks DNA.) Genes govern cells by controlling the manufacture of proteins, especially enzymes, which are necessary for all the chemical reactions that occur within the cell. They also contain the information to make the proteins that constitute structural materials, hormones, and growth factors. This is a good time to look back at Chapter 3, which has details on chromosomes, genes, and DNA (**see Fig. 3-12**).

When body cells divide by the process of mitosis, the DNA that makes up the chromosomes is replicated and distributed to the daughter cells, so that each daughter cell gets exactly the same kind and number of chromosomes as were in the original cell (**see Fig. 3-15**). Each chromosome (aside from the Y chromosome, which determines male sex) may carry thousands of genes, and each gene carries the code for a specific **trait** (characteristic) (**Fig. 25-1**). These traits constitute the physical, biochemical, and physiologic makeup of every cell in the body (see Box 25-1 on modern studies of the human genetic makeup).

In humans, every nucleated cell except the gametes (reproductive cells) contains 46 chromosomes. The chromosomes exist in pairs. One member of each pair was received

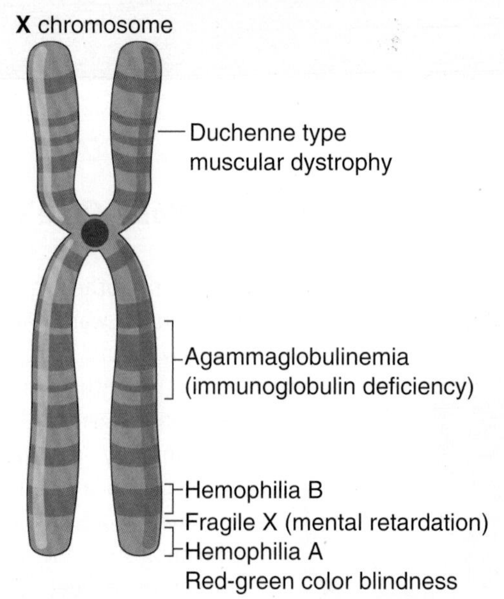

Figure 25-1 Genes and chromosomes. **KEY POINT** Genes are segments of DNA located at specific sites on a chromosome. This figure shows some genes on the X chromosome relevant to different diseases.

at the time of fertilization from the offspring's father, and one was received from the mother. The paired chromosomes, except for the pair that determines sex, are alike in size and appearance. Thus, each body cell has one pair of sex chromosomes and 22 pairs (44 chromosomes) that are not involved in sex determination and are known as **autosomes** (AW-to-somes).

The paired autosomes carry the genes for a specific trait at the same physical locations, but they might have different versions of that gene (**Fig. 25-1**). The alternative forms of a given gene have slightly different DNA sequences and encode variants of a trait (e.g., curly vs. straight hair). Each version of a specific gene is called an **allele** (al-LEEL).

DOMINANT AND RECESSIVE ALLELES

Because our chromosomes are paired, we have two alleles for every gene. When both the alleles for a trait are the same, the alleles are said to be **homozygous** (ho-mo-ZI-gus); if they are different, they are described as **heterozygous** (het-er-o-ZI-gus). The concept of dominance explains how the two alleles interact to determine a particular trait. A **dominant** allele is one that expresses its effect in the cell regardless of the identity of the other allele. The allele needs to be received from only one parent for the trait to be expressed in the offspring.

The effect of a **recessive** allele is not evident unless its paired allele on the matching chromosome is also recessive. Thus, a recessive trait appears only if the recessive gene versions for that trait are received from both parents. A simple test, often done in biology labs to study genetic inheritance, is a test for the ability to taste phenylthiocarbamide (PTC), a harmless organic chemical not found in food. The ability to detect PTC's bitter taste is inherited as a dominant allele. Inability to

Box 25-1

HOT TOPICS

The Human Genome Project: Reading the Book of Life

Packed tightly in nearly every one of your body cells (except the red blood cells) is a complete copy of your genome—the genetic instructions that direct all of your cellular activities. Written in the language of DNA, these instructions consist of genes parceled into 46 chromosomes that code for proteins. In 1990, a consortium of scientists from around the world set out to crack the genetic code and read the human genome, our "book of life." This monumental task, called the Human Genome Project, was completed in 2003 and succeeded in mapping the entire human genome—3 billion DNA base pairs arranged into about 20,000 to 25,000 genes. Now, scientists can pinpoint the exact location and chemical code of every gene in the body.

The human genome was decoded using a technique called sequencing. Samples of human DNA were fragmented into smaller pieces and then inserted into bacteria. As the bacteria multiplied, they produced more and more copies of the human DNA fragments, which the scientists extracted. The DNA copies were loaded into a sequencing machine capable of "reading" the string of DNA nucleotides that composed each fragment. Then, using computers, the scientists put all of the sequences from the fragments back together to get the entire human genome.

Now, scientists hope to use all these pages of the book of life to revolutionize the treatment of human disease. The information obtained from the Human Genome Project may lead to improved disease diagnosis, new drug treatments, and even gene therapy.

taste it is carried by a recessive allele. Thus, nontasting appears in an offspring only if alleles for nontasting are received from both parents. A recessive trait only appears if a person's alleles are homozygous for that trait. In contrast, a dominant trait will appear whether the alleles are homozygous (both dominant) or heterozygous (one of each) because a dominant allele is always expressed if it is present.

Any characteristic that can be observed or can be tested for is part of a person's **phenotype** (FE-no-tipe). Eye color, for example, can be seen when looking at a person. Blood type is not visible but can be determined by testing and is also a part of a person's phenotype. When someone has the recessive phenotype, his or her genetic makeup, or **genotype** (JEN-o-tipe), is obviously homozygous recessive. When a dominant phenotype appears, the person's genotype can be either homozygous dominant or heterozygous, as noted above. Only genetic studies or family studies can reveal which it is.

A recessive allele is not expressed if it is present in a cell together with a dominant allele. However, the recessive allele can be passed onto offspring and may thus appear in future generations. An individual who shows no evidence of a trait but has a recessive allele for that trait is described as a **carrier** of the gene. Using genetic terminology, that person shows the dominant phenotype but has a heterozygous genotype for that trait. Using our PTC tasting example, a taster might be a carrier for the nontasting gene and could pass it on to his or her children.

DISTRIBUTION OF CHROMOSOMES TO OFFSPRING

The reproductive cells (ova and spermatozoa) are produced by a special process of cell division called **meiosis** (mi-O-sis) (**Fig. 25-2**). This process divides the chromosome number in half, so that each reproductive cell has 23 chromosomes instead of the 46 in other body cells. The process of meiosis begins, as does mitosis, with replication of the chromosomes (see **Fig. 3-15** in Chapter 3). Then, these duplicated chromosomes line up across the center of the cell. However, instead of a random distribution of the 46 chromosomes, as occurs in mitosis, the chromosome pairs line up at the center of the cell (**see Fig. 25-2**). The first meiotic division (meiosis I) distributes the two members of each chromosome pair into separate cells. Then, a second meiotic division (meiosis II) separates the strands of the duplicated chromosomes and distributes each strand to an individual gamete. In the end, each gamete receives one copy of 23 different chromosomes, each representing one member of a chromosome pair.

Remember that for each chromosome pair, one chromosome was originally inherited from the father and one from the mother. It is very important to note that the separation of the chromosome pairs occurs at random, meaning that each gamete receives a mixture of maternal and paternal chromosomes. This reassortment of chromosomes in the gametes leads to increased variety within the population. (In a human cell with 46 chromosomes, the chances for varying combinations are much greater than in the example shown in **Fig. 25-2**.) Thus, children in a family resemble each other, but no two look exactly alike (unless they are identical twins), because each gamete contained a unique assortment of chromosomes from all four of their grandparents.

See the Student Resources on thePoint for a figure on meiosis in male and female gamete formation.

PUNNETT SQUARES

Geneticists use a grid called a **Punnett square** to show all the combinations of alleles that can result from a given parental cross, that is, a mating that produces offspring (**Fig. 25-3**). In these calculations, a capital letter is used for the dominant allele, and the recessive allele is represented by the lower case of the same letter. For example, if T represents the allele for the dominant trait PTC taster, then t would be the recessive allele for nontaster. In the offspring, the genotype TT is homozygous dominant and the genotype Tt is heterozygous, both of which will show the dominant phenotype taster.

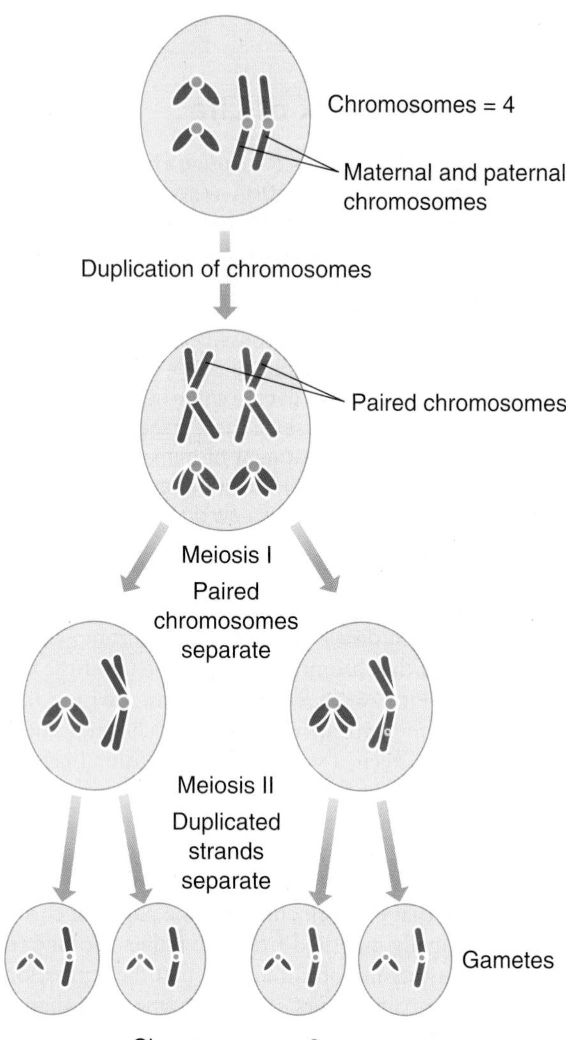

Figure 25-2 **Meiosis.** **KEY POINT** Meiosis halves the chromosome number in formation of gametes. Because the process begins with chromosome replication, two meiotic divisions are needed, first to reduce the chromosome number and then to separate the duplicated strands. The example shows a cell with a chromosome number of four, in contrast to a human cell, which has 46 chromosomes. **ZOOMING IN** How many cells are produced in one complete meiosis?

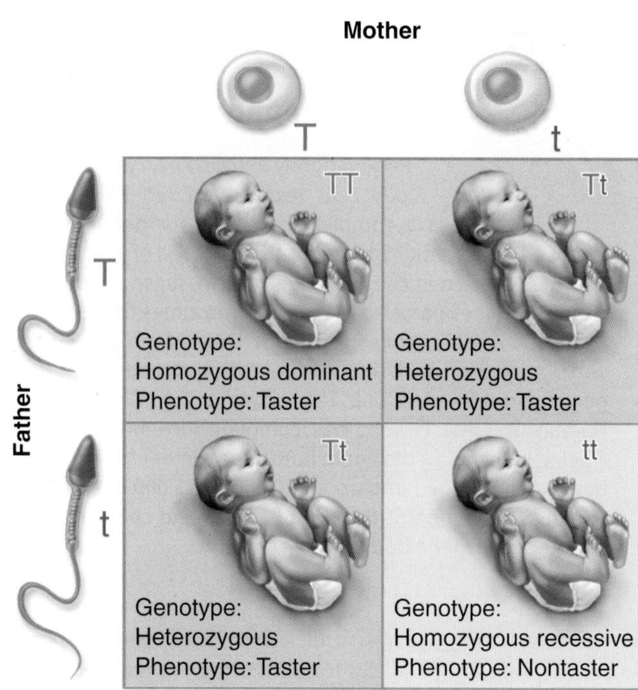

Figure 25-3 **A Punnett square.** **KEY POINT** Geneticists use this grid to show all the possible combinations of a given cross, in this case, Tt × Tt. **ZOOMING IN** What percentage of children from this cross will show the recessive phenotype? What percentage will be heterozygous?

The homozygous recessive genotype tt will show the recessive phenotype nontaster.

A Punnett square shows all the possible gene combinations of a given cross and the theoretical ratios of all the genotypes produced. For example, in the cross shown in **Figure 25-3** between two heterozygous parents, the theoretical chances of producing a baby with the genotype TT are 25% (one in four). The chances of producing a baby with the genotype of tt and the recessive phenotype nontaster are also 25%. The chances of producing heterozygous Tt offspring for this trait are 50% (two in four). In all, 75% of the offspring will have the dominant phenotype taster, because they have at least one dominant gene for the trait. In Cole's case, because sickle cell anemia is determined by a recessive allele, there was a 25% chance that he would have that disorder, because both his parents are carriers of the allele. If his parents have other children, the risk is the same with each birth.

In real life, genetic ratios may differ from the theoretical predictions, especially if the number of offspring is small. For example, the chances of having a male or female baby are 50/50 with each birth for reasons explained shortly, but a family might have several girls before having a boy, and vice versa. The chances of seeing the predicted ratios improve as the number of offspring increases.

SEX DETERMINATION

The two chromosomes that determine the offspring's sex, unlike the autosomes (the other 22 pairs of chromosomes), are not matched in size and appearance. The female X chromosome is larger than most other chromosomes and carries genes for other characteristics in addition to that for sex. The male Y chromosome is smaller than other chromosomes and mainly determines sex. A female has two X chromosomes in each body cell; a male has one X from his mother and one Y from his father.

By the process of meiosis, each sperm cell receives either an X or a Y chromosome, whereas every ovum receives only an X chromosome (**Fig. 25-4**). If a sperm cell with an X chromosome fertilizes an ovum, the resulting infant will be female; if a sperm with a Y chromosome fertilizes an ovum, the resulting infant will be male.

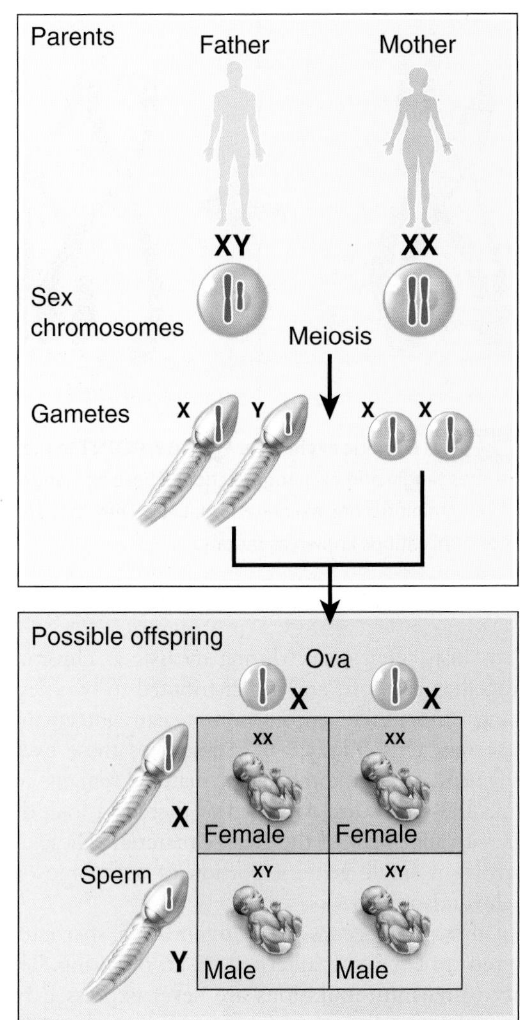

Figure 25-4 **Sex determination.** **KEY POINT** If an X chromosome from a male unites with an X chromosome from a female, the child is female (XX); if a Y chromosome from a male unites with an X chromosome from a female, the child is male (XY). **ZOOMING IN** What is the expected ratio of male to female offspring in a family?

SEX-LINKED TRAITS

Any trait that is carried on a sex chromosome is said to be **sex linked.** Because the Y chromosome carries few traits aside from sex determination, most sex-linked traits are carried on the X chromosome and are best described as *X-linked*. Examples are hemophilia, certain forms of baldness, and red–green color blindness.

Sex-linked traits appear almost exclusively in males. The reason for this is that most of these traits are recessive, and if a recessive allele is located on the X chromosome in a male, it cannot be masked by a matching dominant allele. (Remember that the Y chromosome with which the X chromosome pairs is very small and carries few genes.) Thus, a male who has only one recessive allele for a trait will exhibit that characteristic, whereas a female must have two recessive alleles to show the trait. The female must inherit a recessive allele for that trait from each parent and be homozygous recessive in order for the trait to appear. The daughter of a man with an X-linked disorder is always at least a carrier of the allele for that disorder, because a daughter always receives her father's single X chromosome. So, for example, if you are female and your father is color blind, then each of your sons will have a 50/50 chance of being color blind. See **Figure 25-5** for an example of sex-linked inheritance, illustrating a match between a normal male and a carrier female.

CHECKPOINTS

- ☐ 25-1 What is a gene? What is a gene made of?
- ☐ 25-2 What term describes a gene that always expresses its effect?
- ☐ 25-3 What is the difference between a genotype and a phenotype?
- ☐ 25-4 What is the process of cell division that forms the gametes?
- ☐ 25-5 Human body cells have 46 chromosomes. How many chromosomes are in each gamete?
- ☐ 25-6 What sex chromosome combination determines a female? A male?
- ☐ 25-7 What term describes a trait carried on a sex chromosome?

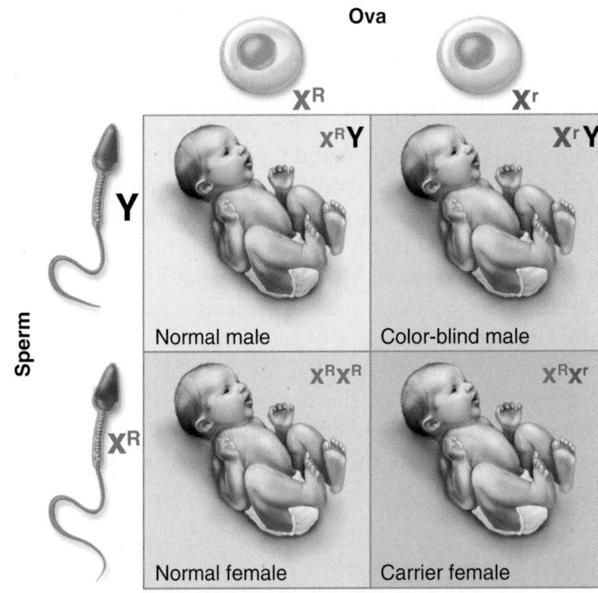

Figure 25-5 **Inheritance of sex-linked traits.** **KEY POINT** Sex-linked traits, such as color blindness, are encoded by genes on the X chromosome. This Punnett square illustrates the mating between a normal male and a female carrier. The "R" superscript is the dominant (normal) allele, and the "r" superscript represents the recessive allele for color blindness. **ZOOMING IN** What is the genotype of a carrier female?

25

Hereditary Traits

Some observable hereditary traits are skin, eye, and hair color and facial features. Also influenced by genetics are less clearly defined traits, such as weight, body build, life span, and susceptibility to disease.

Some human traits, including the traits involved in many genetic diseases, are determined by a single pair of genes; most, however, are the result of two or more gene pairs acting together in what is termed **multifactorial inheritance.** This type of inheritance accounts for the wide range of variations within populations in such characteristics as coloration, height, and weight, all of which are determined by more than one pair of genes.

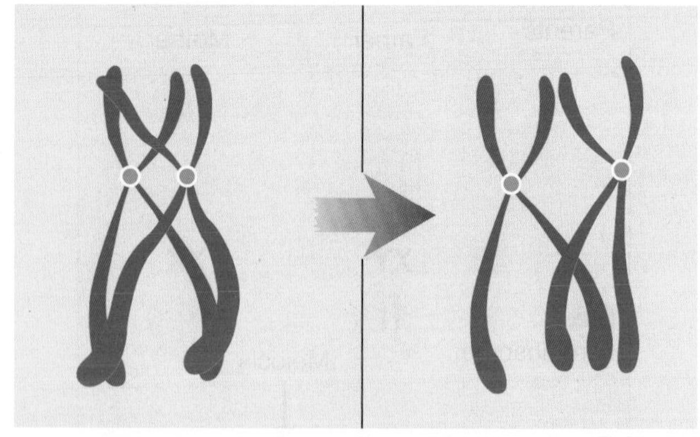

Figure 25-6 **Genetic exchange.** **KEY POINT** In meiosis, chromosomes tangle and exchange sections. These exchanges may be equal, maintaining a normal genetic makeup. They may also result in losses or duplications known as mutations.

GENE EXPRESSION

The effect of a gene on a person's phenotype may be influenced by a variety of factors, including the individual's sex or the presence of other genes. For example, the genes for certain types of baldness may be inherited by either males or females, but the traits appear mostly in males under the effects of male sex hormone.

Environment also plays a part in gene expression. For example, a person might inherit the dominant gene for freckles, but the freckles appear only when the skin is exposed to sunlight. In the case of multifactorial inheritance, gene expression might be quite complex. You inherit a potential for a given size, for example, but your actual size is additionally influenced by such factors as nutrition, development, and general state of health. The same is true of life span and susceptibility to diseases.

We should also note here that dominance is not always clear cut. Sometimes, dominant alleles are expressed together as *codominance*. Take the inheritance of blood type, for example. Alleles for type A and type B are dominant over O. A person with the genotype AO is type A; a person with the genotype BO is type B. However, neither A nor B is dominant over the other. If a person inherits an allele for both A and B, his or her blood type is AB. Also, a person heterozygous for a certain trait may show some intermediate expression of the trait. The term for this type of gene interaction is *incomplete dominance*. Sickle cell anemia, for example, is determined by a recessive allele of the hemoglobin gene. However, a heterozygous carrier may have symptoms of sickle cell anemia under conditions of respiratory stress, as described in Chapter 13.

GENETIC MUTATION

Any change in a gene or a chromosome is termed a genetic **mutation** (mu-TA-shun). Mutations may occur spontaneously or may be induced by some agent, such as ionizing radiation or a chemical, described as a **mutagen** (MU-tah-jen) or mutagenic agent. Often, small changes in gene sequence occur as copying errors during DNA synthesis. Larger mutations, in which portions of a chromosome or entire chromosomes are gained or lost, often occur during meiosis as chromosomes come together, reassort, and get distributed to two new cells. The long, threadlike chromosomes can intertwine and exchange segments (**Fig. 25-6**). Sometimes these exchanges are equivalent, and the total genetic material remains normal, but sometimes genes are damaged or there is a loss, duplication, or rearrangement of the genetic material. These changes may involve a single gene, a portion of a chromosome, or whole chromosomes.

If a mutation occurs in an ovum or a spermatozoon, the altered trait may be inherited by an offspring. The vast majority of harmful mutations are never expressed because the affected fetus dies and is spontaneously aborted. Most remaining mutations are so inconsequential that they have no detectable effect. Beneficial mutations, on the other hand, tend to survive and become increasingly common as a population evolves.

MITOCHONDRIAL INHERITANCE

Although we have emphasized the nucleus as the repository of genetic material in the cell, the mitochondria also contain some hereditary material. Mitochondria, you will recall, are the cells' powerhouse organelles, converting the energy in nutrients into ATP. Because mitochondria were separate organisms early in the evolution of life and later merged with other cells, they have their own DNA and multiply independently. Mitochondrial DNA can mutate, as does nuclear DNA, and disease can be the result. Such diseases can cause serious damage to metabolically active cells in the brain, liver, muscles, kidneys, and endocrine organs. These mitochondrial genes are passed only from a mother to her sons and daughters because almost all the cytoplasm in the zygote is contributed by the ovum. Because these cells must be smaller and lighter, the head of the spermatozoon carries very little cytoplasm and no mitochondria.

CHECKPOINTS

- [ ] 25-8 What is a mutagen?
- [ ] 25-9 From which parent does a child get mitochondrial genes?

Genetic Diseases

Any disorders that involve the genes may be said to be **genetic** but genetic diseases are not always hereditary, that is, passed from parent to offspring in the reproductive cells. Noninherited genetic disorders may begin during maturation of the sex cells or even during development of the embryo.

Advances in genetic research have made it possible to identify the causes of many hereditary disorders and to develop genetic screening methods. In people who are "at risk" for having a child with a genetic disorder, as well as fetuses and newborns with a suspected abnormality, tests can often confirm or rule out the presence of a genetic defect.

CONGENITAL VERSUS HEREDITARY DISEASES

Before we discuss hereditary diseases, we need to distinguish them from other congenital diseases. To illustrate, let us assume that two infants are born on the same day in the same hospital. It is noted that one infant has a clubfoot, a condition called **talipes** (TAL-ih-pes), and the second infant has six fingers on each hand, a condition called **polydactyly** (pol-e-DAK-til-e) (Fig. 25-7). Are both conditions hereditary? Both congenital? Is either hereditary? We can answer these questions by defining the key terms, *congenital* and *hereditary*. **Congenital** means present at the time of birth; **hereditary** means genetically transmitted or transmissible. Thus, one condition may be both congenital and hereditary and another congenital yet not hereditary.

Hereditary conditions are usually evident at birth or soon thereafter. However, certain inherited disorders, such as adult polycystic kidney disease and Huntington disease, a nervous disorder, do not manifest themselves until about midlife (40 to 50 years of age). People with these genetic defects may pass them on to their children before they are aware of them unless genetic testing reveals their presence. In the case of our earlier examples, the clubfoot is congenital, but not hereditary, having resulted from severe distortion of the developing extremities during intrauterine growth. The extra fingers are both congenital and hereditary; polydactyly is a familial trait that appears in another relative, a grandparent perhaps, or a parent, and that is evident at the time of birth.

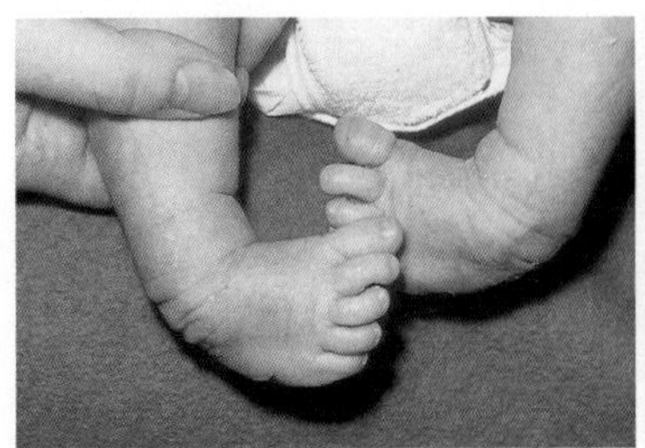
A

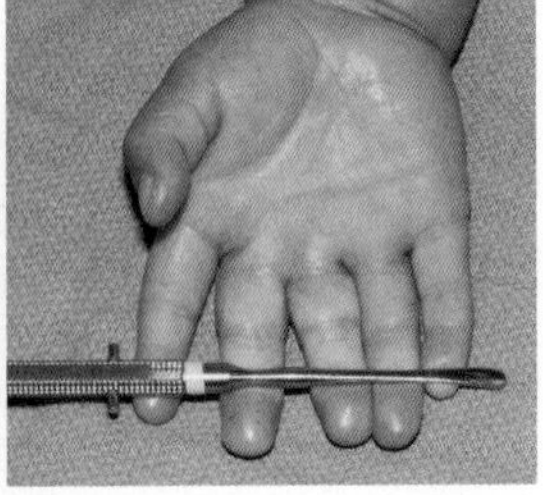
B

Figure 25-7 Congenital and hereditary diseases. A. Talipes, or clubfoot, is congenital but not hereditary. **B.** Polydactyly is both hereditary and congenital.

CAUSES OF CONGENITAL DISORDERS

Although causes of congenital deformities and birth defects often are not known, in some cases they are known and can be avoided. For example, certain infections and toxins may be transmitted from the mother's blood through the placenta to the fetal circulation. Some of these cause serious developmental disorders in affected babies. Any factor or agent, such as a drug, that causes abnormal prenatal development is termed a **teratogen** (TER-ah-to-jen) (the root *terat/o* means "malformed fetus"). As noted in Chapter 24, a developing offspring is most sensitive to teratogens during the first trimester of pregnancy when all major organ systems develop.

German measles (rubella) is a contagious viral infection that is ordinarily a mild disease, but if maternal infection occurs during the first three or four months of pregnancy, the fetus has a 40% chance of developing defects of the eye (cataracts), ear (hearing loss), brain, and heart. Appropriate immunizations can prevent maternal infections.

Ionizing radiation and various toxins may damage the genes, and the disorders they produce are sometimes transmissible. Environmental agents, such as mercury and some chemicals used in industry (e.g., certain phenols and PCB, as well as some drugs, notably LSD), are known to disrupt genetic organization (see Box 25-2).

Intake of alcohol and cigarette smoking by a pregnant woman often cause growth retardation and low birth weight in her infant. Smaller than normal infants do not do as well as average-weight babies. Healthcare professionals strongly recommend total abstinence from alcohol and cigarettes during pregnancy. Alcohol abuse during pregnancy is associated with a condition called **fetal alcohol syndrome** (FAS) (Fig. 25-8). Affected babies have malformed facial features, stunted growth, cardiac and nervous system disorders, and behavioral problems, such as hyperactivity and poor attention.

Spina bifida (SPI-nah BIF-ih-dah) is incomplete closure of the spine through which the spinal cord and its membranes may project (Fig. 25-9). The defect usually occurs in the lumbar region. If the meninges protrude (herniate), the defect is termed a meningocele (meh-NIN-go-sele) (Fig. 25-9B); if both the spinal cord and meninges protrude, it is a myelomeningocele (mi-eh-lo-meh-NIN-go-sele) (Fig. 25-9C). In the latter case, the spinal cord ends at the point of the defect, which affects function below that point. Adequate intake of folic acid, a B vitamin, reduces

Box 25-2

HEALTH MAINTENANCE

Preventing Genetic Damage

It is sometimes difficult to determine whether a genetic defect is the result of a spontaneous random mutation or is related to exposure of a germ cell to an environmental toxin. Male and female sex cells are present at birth but do not become active until the reproductive years. This leaves a long time span during which toxic exposure can occur. Most of what we know about chemicals that cause mutations has been learned from environmental accidents. For example, mercury compounds have found their way into the food chain, causing neural damage to the children of parents who consumed foods containing mercury. Lead, which is toxic when ingested, as from air or water pollution, has been implicated in sperm cell abnormalities and in reduced sperm counts. Radiation accidents have been linked to increased susceptibility to certain cancers in children born after these accidents.

Meiosis is the step most prone to genetic errors. This process can be studied in males with specimens obtained from testicular biopsy. Also, samples of sperm cells, which are produced continuously in males, can be obtained easily and studied for visible defects. Advancing age in both the mother and the father is known to increase errors in meiosis. Males between 20 and 45 years of age carry the least risk of transmitting a genetic error. For females, the least risk is between ages 15 to 35 years. Because new sperm cells are produced on a 64-day cycle, men are advised to avoid conception for a few months after exposure to x-rays, cancer chemotherapy, or other chemicals capable of causing mutations.

the risk of spina bifida and other CNS defects in the fetus. Women of childbearing age should eat foods high in folic acid (green leafy vegetables, fruits, and legumes) and should take folic acid supplements even before pregnancy, as the nervous system develops early. In the United States, folic acid is added to grain products, such as bread, pasta, rice, and cereals.

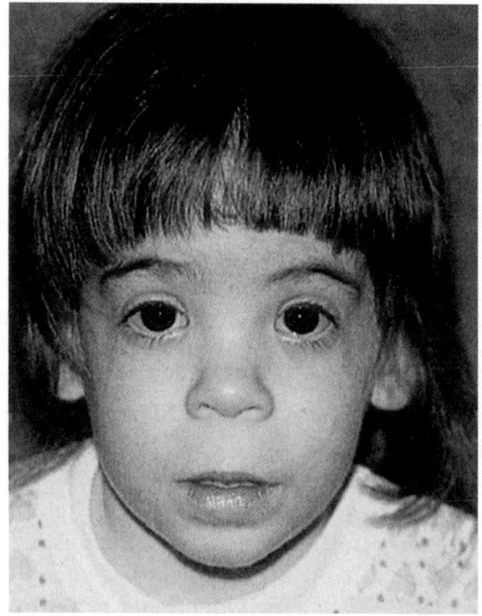

Figure 25-8 Fetal alcohol syndrome. KEY POINT Alcohol is a teratogen that can damage a fetus. This girl is mildly retarded and growth deficient with facial features typical of fetal alcohol syndrome: small eyes, short nose with flat nasal bridge, absence of a groove above the upper lip.

EXAMPLES OF GENETIC DISEASES

The best known example of a nonhereditary genetic disorder is the most common form of **Down syndrome**, also known as **trisomy 21** because it results from an extra chromosome number 21 in each cell. This abnormality arises during formation of a gamete, when both members of the chromosome pair 21 land in the same gamete. Fertilization then results in three copies of chromosome 21. The disorder is usually recognizable at birth by the child's distinctive facial features (**Fig. 25-10A**). Children with Down syndrome have poor muscle tone. They have lowered immunity and are also prone to heart disease, leukemia, and Alzheimer disease. Their intellectual function is impaired. However, the amount of skill they can gain depends on the severity of the disease and their family and school environments. Down syndrome is usually not inherited, although there is a hereditary form of the disorder. In most cases, both parents are normal, as are the child's siblings. The likelihood of having a baby with Down syndrome increases dramatically in women past the age of 35 years and may result from defects in either the male or female gametes owing to age.

Abnormal distribution of the sex chromosomes during meiosis can also result in genetic disease. For example, **Klinefelter** (KLINE-fel-ter) **syndrome**, which occurs in about one in 600 males, usually results from the union of a normal gamete with a gamete containing an extra X chromosome. All of the cells of the resulting fetus have the genotype XXY. These men are usually infertile, extremely tall, and may have cognitive and motor impairments (**Fig. 25-10B**). **Turner syndrome** results from the union of an X gamete with a gamete lacking a sex chromosome, resulting in an XO combination. This inheritance pattern is a frequent cause of spontaneous abortion (miscarriage), but about 1% of these conceptions survive and are born. As shown in **Figure 25-10C**, Turner syndrome girls have a characteristic

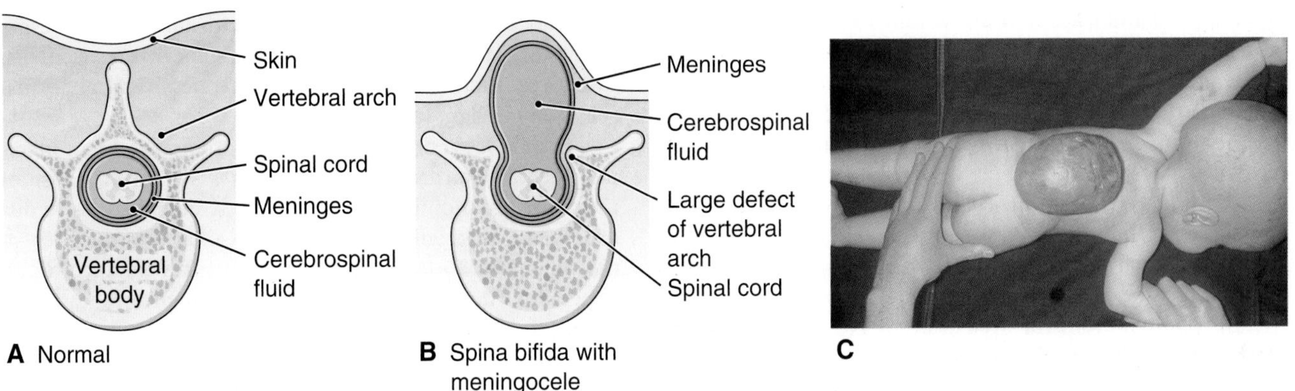

Figure 25-9 Spina bifida. **KEY POINT** Spina bifida is incomplete closure of the spinal cord. **A.** Normal spinal cord. **B.** Spina bifida with protrusion of the meninges (meningocele). **C.** Protrusion of the spinal cord and meninges (myelomeningocele).

web-necked appearance and frequently have cardiovascular and other developmental disorders.

Most genetic diseases are **familial** or hereditary; that is, they are passed on from parent to child in the ovum or sperm. In the case of a disorder carried by a dominant gene, only one parent needs to carry the abnormal gene to give rise to the disease. Any child who receives the defective gene will have the disease. Any children who do not show the disease do not have the defective gene nor are they carriers. One example is **Huntington disease**, a progressive degenerative disorder associated with rapid, involuntary muscle activity and mental deterioration. The disease does not appear until

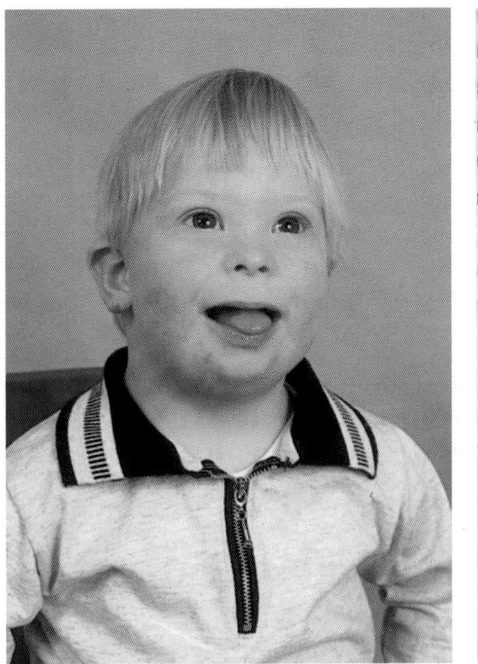
A

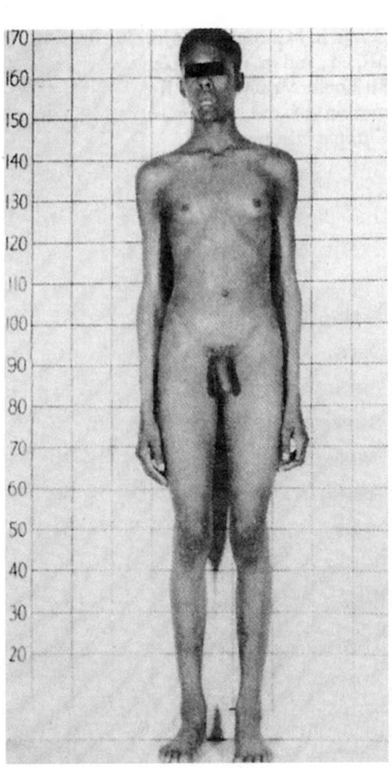

B

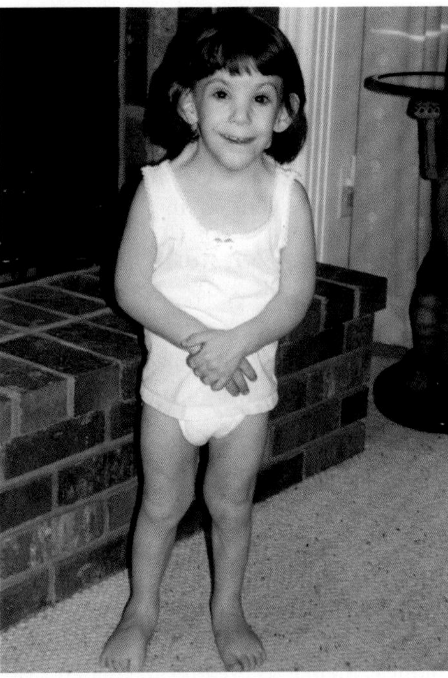
C

Figure 25-10 Chromosomal disorders. **KEY POINT** Errors in chromosomal distribution during meiosis can result in an abnormal number of chromosomes. **A.** This child with Down syndrome has three copies of chromosome 21 for a total of 47. The typical facial features are visible in this photo: almond-shaped eyes, small flat nose, small mouth with protruding tongue. **B.** Males with Klinefelter syndrome usually have an extra X chromosome for a total of 47. **C.** Females with Turner syndrome only have a single X chromosome for a total of 45.

about age 40 and leads to death within 15 years. There is no cure for Huntington disease, but people with the defective gene can be identified by genetic testing. Another example of a dominant disorder is **Marfan syndrome**, a connective tissue disease. People with Marfan syndrome are tall, are thin, and have heart defects.

If a disease trait is carried by a recessive gene, as is the case in most inheritable disorders, a defective gene must come from each parent. Some recessively inherited diseases are described next.

In **phenylketonuria** (fen-il-ke-to-NU-re-ah) (**PKU**), the lack of a certain enzyme prevents the proper metabolism of **phenylalanine** (fen-il-AL-ah-nin), one of the common amino acids. As a result, phenylalanine accumulates in the infant's blood and appears in the urine. If the condition remains untreated, it leads to mental retardation before the age of 2 years. Newborns are routinely screened for PKU.

Sickle cell anemia was first described in Chapter 13 as a blood disease found almost exclusively in black populations. Cole's opening case study involves sickle cell anemia. By contrast, **cystic fibrosis** (**CF**) is most common in white populations; in fact, it is the most frequently inherited disease within this group. CF is characterized by excessively thickened secretions in the bronchi, the intestine, and the pancreatic ducts, resulting in obstruction of these vital organs. It is associated with frequent respiratory infections, intestinal losses, particularly of fats and fat-soluble vitamins, as well as massive salt loss. Treatment includes oral administration of pancreatic enzymes and special pulmonary exercises. CF was once fatal by the time of adolescence, but now, with appropriate care, life expectancies are extending into the 30s. The gene responsible for the disease has been identified, raising hope of more widespread diagnosis, earlier treatment, and perhaps even correction of the defective gene that causes the disorder. All regions of the United States currently screen newborns for CF.

In **Tay-Sachs disease**, a certain type of fat is deposited in neurons of the CNS and retina because cellular lysosomes lack an enzyme that breaks it down. The disease occurs mainly in eastern European (Ashkenazi) Jews, French Canadians, and Cajuns and generally leads to death by age 4. There is no cure for Tay-Sachs, but a blood test can identify genetic carriers of the disease.

Another group of heritable muscle disorders is known collectively as the **progressive muscular atrophies**. Atrophy (AT-ro-fe) means wasting due to decrease in the size of a normally developed part. The absence of normal muscle movement in the infant proceeds within a few months to extreme weakness of the respiratory muscles, until ultimately the infant is unable to breathe adequately. Most afflicted babies die within several months. The name *floppy baby syndrome*, as the disease is commonly called, provides a vivid description of its effects.

Albinism is another recessively inherited disorder that affects the cells (melanocytes) that produce the pigment melanin. The skin and hair color are strikingly white and do not darken with age. The skin is abnormally sensitive to sunlight and may appear wrinkled. People with albinism are especially susceptible to skin cancer and to some severe visual disturbances, such as myopia (nearsightedness) and abnormal sensitivity to light (photophobia).

Fragile X syndrome is the most common cause of inherited mental retardation in males and a significant cause of mental retardation in females. It is an X-linked recessive disorder associated with a fragile site on an arm of the X chromosome (**Fig. 25-8**). In males, it also causes enlarged testes, hyperactivity, mitral valve malfunction, high forehead, and enlarged jaw and ears.

Other inherited disorders include **osteogenesis** (os-te-o-JEN-eh-sis) **imperfecta**, or *brittle bone disease*, in which multiple fractures may occur during and shortly after fetal life, and a disorder of skin, muscles, and bones called **neurofibromatosis** (nu-ro-fi-bro-mah-TO-sis). In the latter condition, multiple masses, often on stalks (pedunculated), grow along nerves all over the body.

More than 20 different cancers have been linked to specific gene mutations. These include cancers of the breast, ovary, and colon as well as some forms of leukemia. Note, however, that hereditary forms of cancer account for only about 1% of all cancers and that the development of cancer is complex, involving not only specific genes but also gene interactions and environmental factors.

Genetic components have been suggested in some other diseases as well, including certain forms of heart disease, diabetes mellitus, types of cleft lip and cleft palate, and perhaps Parkinson and Alzheimer disease.

CHECKPOINTS

- ☐ **25-10** What does the term congenital mean?
- ☐ **25-11** What is a teratogen?
- ☐ **25-12** What is the scientific name for Down syndrome?
- ☐ **25-13** Is fragile X syndrome more common in males or females?

See the Student Resources on thePoint for illustrations of some genetic diseases and a chart summarizing selected genetic disorders.

Treatment and Prevention of Genetic Diseases

The number of genetic diseases is so great (more than 4,000) that many pages of this book would be needed simply to list them all. Moreover, the list continues to grow as sophisticated research techniques and advances in biology make it clear that various diseases of previously unknown origin are genetic—some hereditary, others not. Can we identify which are inherited genetic disorders and which are due to environmental factors? Can we prevent the occurrence of any of them?

GENETIC COUNSELING

It is possible to prevent genetic disorders in many cases and even to treat some of them. The most effective method of preventing genetic disease is through genetic counseling, a specialized field of healthcare. Genetic counseling centers use a team approach of medical, nursing, laboratory, and social service professionals to advise and care for clients. People who might consider genetic counseling include prospective parents over the age of 35, those who have a family history of genetic disorders, and those who are considering some form of fertility treatment.

See the Student Resources on thePoint for a description of careers in genetic counseling.

The Family History An accurate and complete family history of both prospective parents is necessary for genetic counseling. This history should include information about relatives with respect to age, onset of a specific disease, health status, and cause of death. The families' ethnic origins may be relevant because some genetic diseases predominate in certain ethnic groups. Hospital and physician records are studied, as are photographs of family members. The ages of the prospective parents are considered as factors, as is parental or ancestral relationship (e.g., marriage between first cousins). The appearance of one or more particular traits in a family can be illustrated using a **pedigree chart** (**Fig. 25-11**). Pedigrees may also indicate whether a given family member is a carrier of the disease. Each successive generation in a pedigree is designated with F and a number, such as F_1, F_2. (The F stands for *filial*, referring to the sons and daughters of a given set of parents.) Note that in some cases, there is not enough family data to determine the genotype of all members with regard to a given trait. Also, small numbers of offspring may not show the strict genetic ratios expected from a given cross (see the third generation, F_3, shown in **Fig. 25-11**).

Laboratory Studies Prenatal tests can help determine if a fetus is at risk for having certain birth defects. Three tests done in the first trimester can screen for Down syndrome or other chromosomal abnormalities. These are

- Nuchal (NU-kal) transparency (NT) test—a fetal sonogram to search for excess fluid at the back of the neck (nucha)
- Pregnancy-associated plasma protein test (PAPP-A)—a maternal blood test to identify high levels of a protein produced by the placenta in early pregnancy
- Human chorionic gonadotropin (hCG) test—a maternal blood test for high levels of hCG, a hormone produced by the placenta in early pregnancy

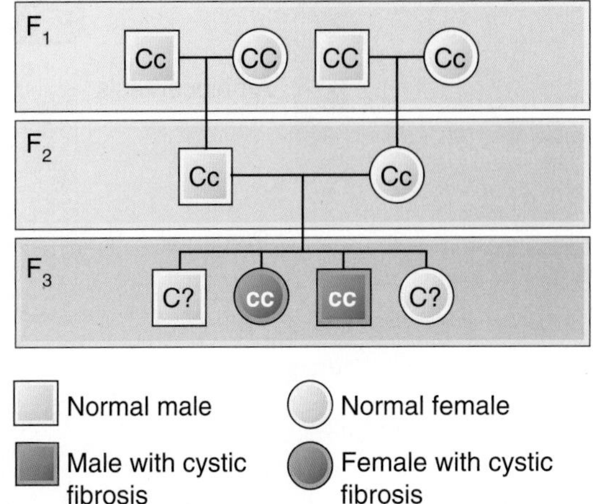

Figure 25-11 A pedigree (family history) showing three generations (F_1–F_3). **KEY POINT** The pedigree is a tool used in genetic studies and genetic counseling. In this example, one parent in each F_1 cross and both parents in the F_2 cross are normal but carry the recessive gene (c) for cystic fibrosis. For the normal children in F_3, only one gene (the dominant normal gene) is known. Note that the F_3 generation does not show strict predicted genetic ratios. Theoretically, only one in four children should be homozygous recessive (have two recessive genes). **ZOOMING IN** What are the possible genotypes of the two normal children in the F_3 generation?

Second-trimester screening for risk of fetal defects involves several maternal blood tests, including hCG levels and the following additional markers:

- Alpha-fetoprotein (AFP) screening—test for abnormal levels of a protein present in the amniotic fluid that crosses the placenta into the mother's blood
- Estriol test—measurement of a hormone produced by the placenta
- Inhibin test—measurement of a hormone produced by the placenta

If the results of these tests show abnormalities, a physician would recommend genetic counseling and further testing.

A more complex technique that enables the geneticist to study an unborn fetus is **amniocentesis** (am-ne-o-sen-TE-sis) ("amnio"). During this procedure, a small amount of the amniotic fluid that surrounds the fetus is withdrawn (**Fig. 25-12**). Fetal skin cells in the amniotic fluid are removed, grown (cultured), and separated for study. The chromosomes are examined, and the amniotic fluid is analyzed for biochemical abnormalities. With these methods, almost 200 genetic diseases can be detected before birth. Note that amniocentesis carries a small risk of spontaneous abortion (less than 0.1%) and may also cause cramping, bleeding, or infection in some cases.

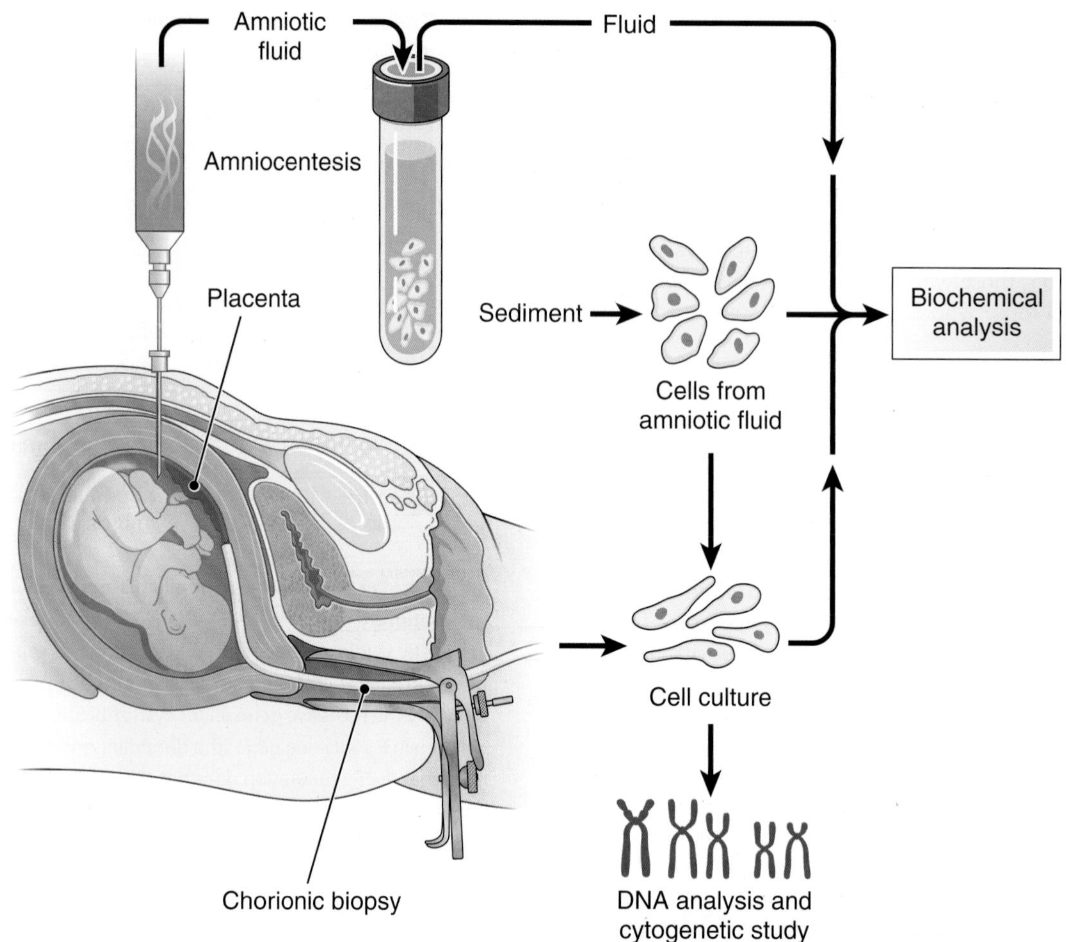

Figure 25-12 **Prenatal testing.** **KEY POINT** Amniocentesis collects cells and fluid from the amniotic fluid, while chorionic villus sampling obtains cells from the fetal portion of the placenta. Biochemical analysis of the cells and fluids can reveal some fetal abnormalities. Once the cells are cultured (increased in number), the fetal chromosomes can be analyzed.

Another method for obtaining fetal cells for study involves sampling of the chorionic villi through the cervix. The chorionic villi are hairlike projections of the membrane that surrounds the embryo in early pregnancy. The method is called **chorionic villus sampling** (CVS) (**Fig. 25-12**). Samples may be taken between eight and 10 weeks of pregnancy, and the cells obtained may be analyzed immediately. In contrast, amniocentesis cannot be done before the 14th to 16th weeks of pregnancy, and test results are not available for about two weeks. CVS carries a higher risk of spontaneous abortion (about 1%) than does amniocentesis.

Abnormalities in the chromosome number and some abnormalities within the chromosomes can be detected by **karyotype** (KAR-e-o-tipe) analysis. A karyotype is produced by growing cells obtained by amniocentesis or CVS in a special medium and arresting cell division at the metaphase stage. A technician uses special stains to reveal certain changes in fine structure within the chromosomes. The chromosomes, visible under the microscope, are then photographed and arranged in groups according to their size and form (**Fig. 25-13**). Abnormalities in the number or structure of the chromosomes can thereby be detected. **Figure 25-13** shows the karyotype of a baby with trisomy 21.

Counseling Prospective Parents Armed with all the available pertinent facts and with knowledge of genetic inheritance patterns, the counselor is equipped to inform the prospective parents of the possibility of their having genetically abnormal offspring. The couple may then use this information to make decisions about family planning. Depending on the individuals and their situation, a couple might elect to have no children, have an adoptive family, use a donor gamete, terminate a pregnancy, or accept the risk. Cole's parents in the case study will consult with a genetic counselor who will advise

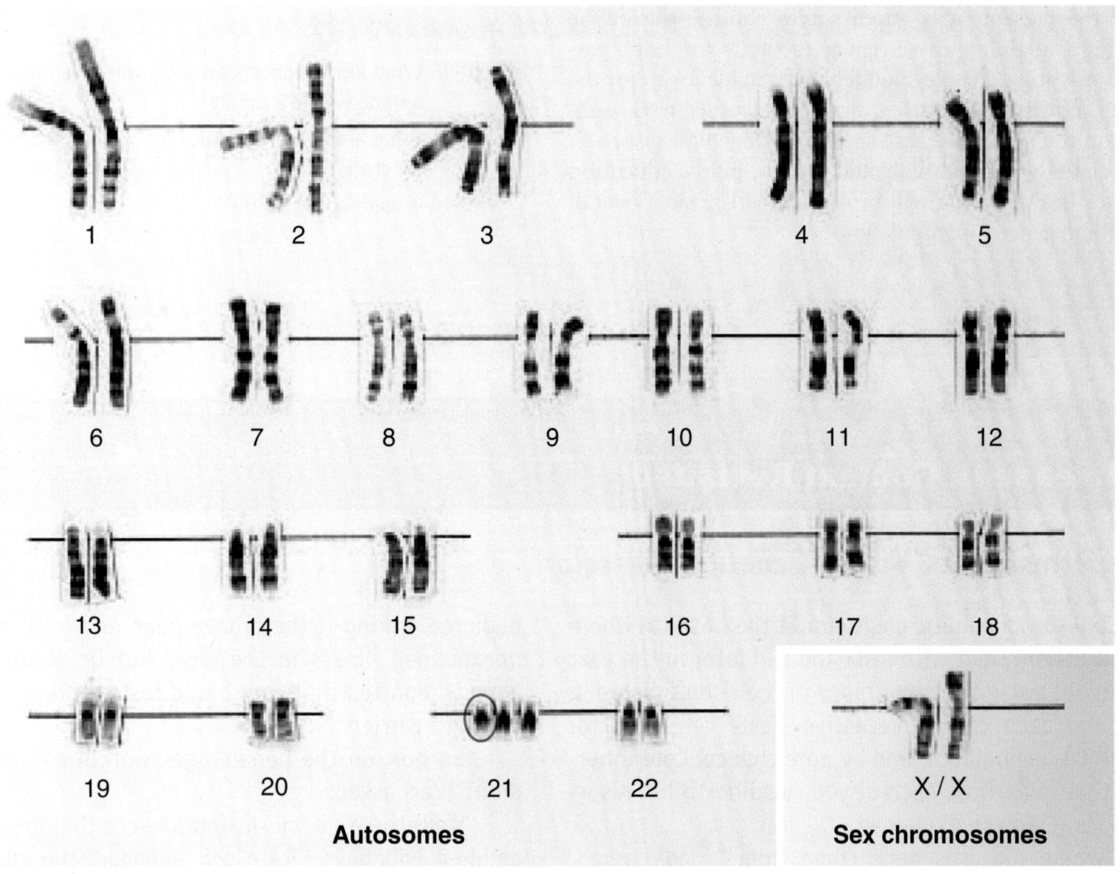

Figure 25-13 **Karyotype.** **KEY POINT** A karyotype can reveal abnormalities in the chromosomes. **ZOOMING IN** Look closely at this karyotype. How many chromosomes are present? What is the gender of the baby, and what genetic disorder is represented?

them on the risk of having other children with sickle cell anemia and points out that there is a prenatal test for the disease.

PROGRESS IN MEDICAL TREATMENT

The mental and physical ravages of many genetic diseases are largely preventable, provided the diseases are diagnosed and treated early in the individual's life. Some of these diseases respond well to dietary control. One such disease, called **maple syrup urine disease**, responds to very large doses of thiamin along with control of the intake of certain amino acids. The disastrous effects of **Wilson disease**, in which abnormal accumulations of copper in the tissue cause tremor, rigidity, uncontrollable stagger, and finally extensive liver damage, can be prevented by a combination of dietary and drug therapies.

Phenylketonuria is perhaps the best known example of dietary management of inherited disease. Babies diagnosed with PKU must be put on a phenylalanine-free formula immediately and continue with reduced protein in the diet throughout life. If the disorder is undiagnosed and untreated, 98% of affected patients will be severely mentally retarded by 10 years of age. In contrast, if the condition is diagnosed and treated before the baby reaches the age of 6 months and treatment is maintained until the age of 10 years, mental deficiency will be prevented or at least minimized. A simple blood test for PKU is now done routinely in hospitals throughout the United States. Infants are usually screened immediately after birth, but they should be retested 24 to 48 hours after taking in protein.

In the future, we can anticipate greatly improved methods of screening, diagnosis, and treatment of genetic diseases. There have been reports of fetuses being treated with vitamins or hormones after prenatal diagnosis of a genetic disorder. Ahead lies the possibility of treating or correcting genetic disorders through genetic engineering—introducing genetically altered cells to produce missing factors, such as enzymes or hormones, or even correcting faulty genes in the patient's cells. Researchers have already made some attempts to supplement failed genes with healthy ones.

For the present, it is important to educate the public about the availability of screening methods for both parents and offspring. People should also be made aware of the damaging effects of radiation, drugs, and other toxic substances on the genes. Because so many congenital disorders are associated with environmental factors, public education and better health habits will probably yield greater overall benefits than genetic manipulations.

CHECKPOINTS

- ☐ **25-14** What kind of chart can show the inheritance pattern of a gene within a family?
- ☐ **25-15** What are two procedures for removing fetal samples for study?
- ☐ **25-16** What is a karyotype?

Disease in Context Revisited

The Armstrongs Visit a Genetic Counselor

Ms. Clarkson, a genetic counselor at the hospital where Cole was born, met with Jada and Carl later in the year to explain again the inheritance of their son's disorder.

"You each carry a recessive sickle cell gene for abnormal hemoglobin, and by pure chance, Cole inherited that gene from each of you, leading to his disorder," Ms. Clarkson explained.

"Where did this gene come from?" Jada asked. "And will all of our children have this disease? Should we stop here with our family?"

"Not necessarily," said Ms. Clarkson. "A mutation might have arisen in either of your reproductive cells, but more likely this trait has been carried somewhere in your families. We can try to develop a family pedigree, asking if there have been any relatives with unexplained illness in the past, but Dr. Hoffman has already reported that your blood tests confirm that you are both carriers."

"So how do the percentages work for future children?" Carl asked.

"Well, there's a one-in-four chance that any of your children will have sickle cell anemia," the counselor said. "But a quarter will be totally free of the mutation, and half will be carriers, having what is called sickle cell trait. Those percentages apply for each birth. Fortunately, there's a prenatal test for sickle cell disease, and you can choose to make family decisions based on those results."

CHAPTER 25

Chapter Wrap-Up

Summary Overview

A detailed chapter outline with space for note taking is on *thePoint*. The figure below illustrates the main topics covered in this chapter.

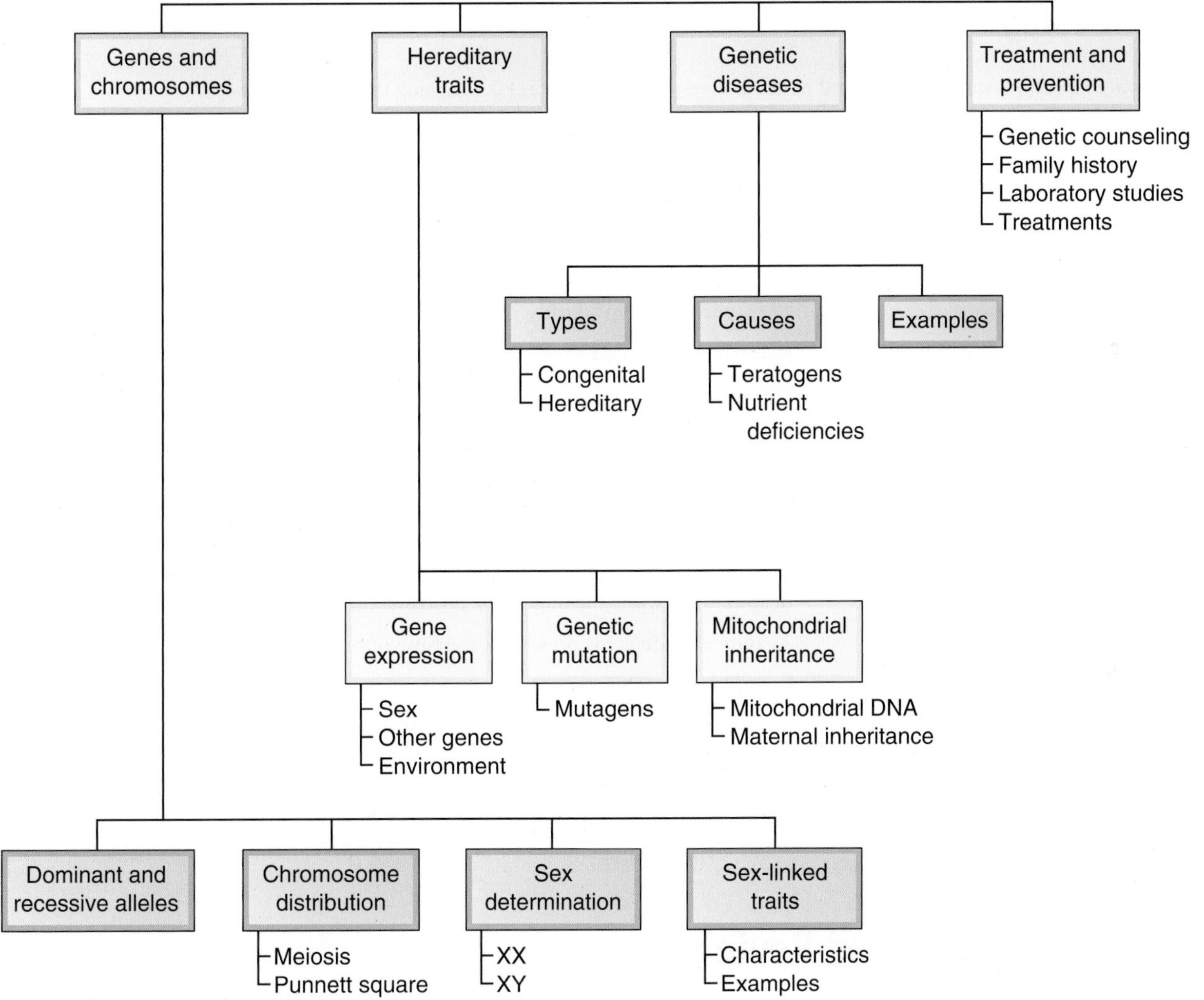

Key Terms

The terms listed below are emphasized in this chapter. Knowing them will help you organize and prioritize your learning. These and other boldface terms are defined in the Glossary with phonetic pronunciations.

allele
amniocentesis
autosome
carrier
chromosome
congenital
dominant
familial
gene
genetic
genotype
heredity
heterozygous
homozygous
karyotype
meiosis
mutagen
mutation
pedigree chart
phenotype
progeny
recessive
sex-linked
teratogen
trait

Word Anatomy

Medical terms are built from standardized word parts (prefixes, roots, and suffixes). Learning the meanings of these parts can help you remember words and interpret unfamiliar terms.

WORD PART	MEANING	EXAMPLE
Genes and Chromosomes		
aut/o-	self	*Autosomes* are all the chromosomes aside from the two that determine sex.
chrom/o	color	*Chromosomes* color darkly with stains.
heter/o	other, different	*Heterozygous* paired genes (alleles) are different from each other.
homo-	same	*Homozygous* paired genes (alleles) are the same.
phen/o	to show	Traits that can be observed or tested for make up a person's *phenotype*.
Hereditary Traits		
multi-	many	*Multifactorial* traits are determined by multiple pairs of genes.
Genetic Diseases		
-cele	swelling	In spina bifida, the meninges can protrude through the spine as a *meningocele*.
con-	with	A *congenital* defect is present at the time of birth.
dactyl/o	digit (finger or toe)	In *polydactyly*, there is an extra finger on the hand.
terat/o	malformed fetus	A *teratogen* is an agent that causes birth defects.
Treatment and Prevention of Genetic Diseases		
-centesis	tapping, perforation	*Amniocentesis* is a tap of the amniotic fluid.
kary/o	nucleus	A *karyotype* is an analysis of the chromosomes contained in the nucleus of a cell.

Questions for Study and Review

BUILDING UNDERSTANDING

Fill in the Blanks

1. The basic unit of heredity is a(n) ______.
2. Chromosomes not involved in sex determination are known as ______.
3. The number of chromosomes in each human body cell is ______.
4. Any change in a gene or chromosome is called ______.
5. Incomplete closure of the spine results in a disorder called ______.

Matching > Match each numbered item with the most closely related lettered item.

___ **6.** An allele that is always expressed if present
___ **7.** An allele that is not always expressed if present
___ **8.** Term for paired alleles for a trait that are the same
___ **9.** Term for paired alleles for a trait that are different
___ **10.** A gene carried on the X chromosome

a. sex-linked
b. recessive
c. heterozygous
d. dominant
e. homozygous

Multiple Choice

___ **11.** What is a gene?
a. a segment of DNA
b. ribonucleic acid
c. a segment of a protein
d. the nuclear membrane

___ **12.** What do genes code for?
a. carbohydrates
b. lipids
c. proteins
d. electrolytes

___ **13.** What are paired genes for a given trait known as?
a. chromosomes
b. ribosomes
c. nucleotides
d. alleles

___ **14.** What is the term for the genetic trait characterized by extra fingers or toes?
a. talipes
b. polydactyly
c. osteogenesis imperfecta
d. neurofibromatosis

___ **15.** Which technique is used to determine the genotypes of family members?
a. chorionic villus sampling
b. amniocentesis
c. a karyotype
d. a pedigree

UNDERSTANDING CONCEPTS

16. Describe the process of meiosis, and explain how it results in genetic variation.

17. Dana has one dominant allele for freckles (*F*) and one recessive allele for no freckles (f). What is Dana's genotype? What is her phenotype?

18. Explain the great variation in the color of skin and hair in humans.

19. Describe how a mutagenic agent can produce genetic variation. Give several examples of mutagenic agents.

20. Describe the cause of PKU. What are the dietary restrictions of individuals with PKU?

21. Compare and contrast amniocentesis and chorionic villus sampling. List some diseases that can be diagnosed using these techniques.

22. What is the most common inheritable disease among black people? Among white people? What are the symptoms of these two diseases?

23. Describe the causes and effects of three disorders related to abnormal numbers of chromosomes.

CONCEPTUAL THINKING

24. Can a disorder be congenital but not hereditary? Explain your answer.

25. Ms. Clarkson, the genetic counselor, is explaining to Cole's parents the chances that any future children of theirs will have sickle cell anemia. Show how she explains these percentages using a Punnett square. Use S for normal hemoglobin and s for the abnormal hemoglobin gene.

26. Richard and Tamara are expecting their first child. Tamara has normal color vision. Richard is color blind and wonders if their baby might be. A family history reveals that Tamara's father is color blind too. What are Richard's and Tamara's genotypes and phenotypes? Using a Punnett square, calculate the possible genotypic and phenotypic combinations for their offspring.

For more questions, see the Learning Activities on thePoint.

Glossary

A

abdominopelvic (ab-dom-ih-no-PEL-vik) Pertaining to the abdomen and pelvis.

abduction (ab-DUK-shun) Movement away from the midline.

abortifacient (ah-bor-tih-FA-shent) Agent that induces an abortion.

abortion (ah-BOR-shun) Loss of an embryo or fetus before the 20th week of pregnancy.

abscess (AB-ses) Area of tissue breakdown; a localized space in the body containing pus and liquefied tissue.

absorption (ab-SORP-shun) The process of taking up or assimilating a substance.

accommodation (ah-kom-o-DA-shun) Coordinated changes in the lens of the eye that enable one to focus on near and far objects.

acetylcholine (as-e-til-KO-lene) **(ACh)** Neurotransmitter; released at synapses within the nervous system and at the neuromuscular junction.

acid (AH-sid) Substance that can release a hydrogen ion when dissolved in water.

acid-fast stain Procedure used to color cells for viewing under the microscope.

acidosis (as-ih-DO-sis) Condition that results from a decrease in the pH of body fluids.

acne (AK-ne) Skin disorder resulting from overactivity of the sebaceous glands, reduced keratinocyte shedding, and infection of the hair follicles.

acquired immunodeficiency syndrome (AIDS) Viral disease that attacks the immune system, specifically the T helper lymphocytes with CD4 receptors.

acromegaly (ak-ro-MEG-ah-le) Condition caused by oversecretion of growth hormone in adults; there is overgrowth of some bones and involvement of multiple body systems.

acrosome (AK-ro-some) Caplike structure over the head of the sperm cell that helps the sperm to penetrate the ovum.

ACTH See adrenocorticotropic hormone.

actin (AK-tin) One of the two contractile proteins in muscle cells, the other being myosin.

action potential Rapid depolarization and repolarization of the plasma membrane capable of spreading down the neuron or muscle membrane; nerve impulse.

active transport Movement of molecules into or out of a cell from an area where they are in lower concentration to an area where they are in higher concentration. Such movement, which is opposite to the direction of normal flow by diffusion, requires energy and transporters.

acupuncture (AK-u-punk-chur) Ancient Chinese method of inserting thin needles into the body at specific points to relieve pain or promote healing.

acute (ah-KUTE) Referring to a severe but short-lived disease or condition.

adaptation (ad-ap-TA-shun) Change in an organ or tissue to meet new conditions.

Addison disease Condition caused by hypofunction of the adrenal cortex.

adduction (ad-DUK-shun) Movement toward the midline.

adenoids (AD-eh-noyds) Popular name for pharyngeal tonsil located in the nasopharynx.

adenosine triphosphate (ah-DEN-o-sene tri-FOS-fate) **(ATP)** Energy-storing compound found in all cells.

ADH See antidiuretic hormone.

adhesion (ad-HE-zhun) Holding together of two surfaces or parts; band of connective tissue between parts that are normally separate.

adipose (AD-ih-pose) Referring to fats or a type of connective tissue that stores fat.

adrenal (ah-DRE-nal) **gland** Endocrine gland located above the kidney; suprarenal gland.

adrenaline (ah-DREN-ah-lin) See epinephrine.

adrenergic (ad-ren-ER-jik) An activity or structure that responds to epinephrine (adrenaline).

adrenocorticotropic (ah-dre-no-kor-tih-ko-TRO-pik) **hormone (ACTH)** Hormone produced by the anterior pituitary that stimulates the adrenal cortex.

aerobic (air-O-bik) Requiring oxygen.

afferent (AF-fer-ent) Carrying toward a given point, such as a sensory neuron that carries nerve impulses toward the central nervous system.

agglutination (ah-glu-tih-NA-shun) Clumping of cells due to an antigen–antibody reaction.

agglutinin (a-GLU-tih-nin) An antibody that causes clumping or agglutination of the agent that stimulated its formation.

agonist (AG-on-ist) Muscle that contracts to perform a given movement.

agranulocyte (a-GRAN-u-lo-site) Leukocyte without visible granules in the cytoplasm when stained; lymphocyte or monocyte.

AIDS See acquired immunodeficiency syndrome.

albinism (AL-bih-nizm) A hereditary disorder that affects melanin production.

albumin (al-BU-min) Protein in blood plasma and other body fluids; helps maintain the osmotic pressure of the blood.

albuminuria (al-bu-mih-NU-re-ah) Presence of albumin in the urine, usually as a result of a kidney disorder.

aldosterone (al-DOS-ter-one) Hormone released by the adrenal cortex that promotes sodium and indirectly, water reabsorption in the kidneys.

alkali (AL-kah-li) Substance that can accept a hydrogen ion (H^+); substance that releases a hydroxide ion (OH^-) when dissolved in water; a base.

alkalosis (al-kah-LO-sis) Condition that results from an increase in the pH of body fluids.

allele (al-LEEL) A version of a gene that controls a given trait.

allergen (AL-er-jen) Environmental substance that causes hypersensitivity; substance that induces allergy.

allergy (AL-er-je) Tendency to react unfavorably to a certain environmental substance that is normally harmless to most people; hypersensitivity to environmental substances.

alopecia (al-o-PE-she-ah) Baldness.

alveolus (al-VE-o-lus) Small sac or pouch; usually a tiny air sac in the lungs through which gases are exchanged between the outside air and the blood; tooth socket; pl., alveoli.

Alzheimer (ALZ-hi-mer) **disease** Brain disorder resulting from unexplained degeneration of the cerebral cortex and hippocampus with intellectual impairment, mood changes, and confusion.

amblyopia (am-ble-O-pe-ah) Vision loss in a healthy eye because it cannot work properly with the other eye.

amino (ah-ME-no) **acid** Building block of protein.

amenorrhea (ah-men-o-RE-ah) Absence of menstrual flow.

amniocentesis (am-ne-o-sen-TE-sis) Removal of fluid and cells from the amniotic sac for prenatal diagnostic tests.

amniotic (am-ne-OT-ik) Pertaining to the sac that surrounds and cushions the developing fetus or to the fluid that fills that sac.

amniotic (am-ne-OT-ik) **sac** Fluid-filled sac that surrounds and cushions the developing fetus.

amphiarthrosis (am-fe-ar-THRO-sis) Slightly movable joint.

ampulla (am-PUL-ah) A saccular dilation of a canal or duct.

amygdala (ah-MIG-dah-lah) Two clusters of nuclei deep in the temporal lobes that coordinate emotional responses to stimuli.

amyotrophic (ah-mi-o-TROF-ik) **lateral sclerosis (ALS)** Disorder of the nervous system in which motor neurons are destroyed.

anabolism (ah-NAB-o-lizm) Metabolic building of simple compounds into more complex substances needed by the body.

anaerobic (an-air-O-bik) Not requiring oxygen.

analgesic (an-al-JE-zik) Relieving pain; a pain-relieving agent that does not cause loss of consciousness.
anaphase (AN-ah-faze) The third stage of mitosis in which chromosomes separate to opposite sides of the cell.
anaphylaxis (an-ah-fih-LAK-sis) Severe, life-threatening allergic response; anaphylactic shock.
anastomosis (ah-nas-to-MO-sis) Communication between two structures, such as blood vessels.
anatomic position Standard position of the body for anatomic studies or designations, upright, face front, arms at side with palms forward, and feet parallel.
anatomy (ah-NAT-o-me) Study of body structure.
anemia (ah-NE-me-ah) Abnormally low level of hemoglobin or red cells in the blood, resulting in inadequate delivery of oxygen to the tissues.
androgen (AN-dro-jen) Any male sex hormone.
anesthesia (an-es-THE-ze-ah) Loss of sensation, particularly of pain; drug with this effect is an anesthetic.
aneurysm (AN-u-rizm) Bulging sac in the wall of a vessel.
angina (an-JI-nah) Severe choking pain; disease or condition producing such pain.
angina pectoris (an-JI-nah PEK-to-ris) Suffocating pain in the chest usually caused by lack of oxygen supply to the heart muscle.
angiography (an-je-OG-rah-fe) Radiography of vessels after injection of contrast material.
angioplasty (AN-je-o-plas-te) Use of a balloon inserted with a catheter to open a blocked vessel.
angiotensin II (an-je-o-TEN-sin) Substance produced from inactive precursors by the action of the renal enzyme renin and other enzymes; increases blood pressure by causing vascular constriction, stimulating the release of aldosterone from the adrenal cortex and ADH from the posterior pituitary, and increasing thirst.
anion (AN-i-on) Negatively charged particle (ion).
anorexia (an-o-REK-se-ah) Chronic loss of appetite. Anorexia nervosa is a psychological condition in which a person may become seriously and even fatally weakened from lack of food.
anoxia (ah-NOK-se-ah) See hypoxia.
ANP See atrial natriuretic peptide.
ANS See autonomic nervous system.
antagonist (an-TAG-o-nist) Muscle that has an action opposite that of a given movement or muscle; substance that opposes the action of another substance.
anterior (an-TE-re-or) Toward the front or belly surface; ventral.
anthelmintic (ant-hel-MIN-tik) Agent that acts against worms; vermicide; vermifuge.
antibiotic (an-te-bi-OT-ik) Substance produced by living cells that kills or arrests the growth of bacteria.
antibody (AN-te-bod-e) **(Ab)** Substance produced in response to a specific antigen; immunoglobulin.
anticoagulant (an-te-co-AG-u-lant) Substance that prevents blood clotting.
antidiuretic (an-ti-di-u-RET-ik) **hormone (ADH)** Hormone released from the posterior pituitary gland that increases water reabsorption in the kidneys, thus decreasing the urinary output.
antigen (AN-te-jen) **(Ag)** Substance that induces an immune response.
antineoplastic (an-ti-ne-o-PLAS-tik) Acting against a neoplasm (tumor).
antioxidant (an-te-OX-ih-dant) Substances in the diet that protect against harmful free radicals.
antipyretic (an-ti-pi-RET-ik) Drug that reduces fever.
antiseptic (an-tih-SEP-tik) Disinfectant applied to skin and other living surfaces.
antiserum (an-te-SE-rum) Serum containing antibodies that may be given to provide passive immunity; immune serum.
antitoxin (an-te-TOKS-in) Antibody that neutralizes a toxin.
antivenin (an-te-VEN-in) Antibody that neutralizes snake venom.
anus (A-nus) Inferior opening of the digestive tract.
aorta (a-OR-tah) The largest artery; carries blood out of the heart's left ventricle.
apex (A-peks) Pointed region of a cone-shaped structure.
aphasia (ah-FA-ze-ah) Loss or defect in language communication; loss of the ability to speak or write is expressive aphasia; loss of understanding of written or spoken language is receptive aphasia.
apicomplexans (ap-i-kom-PELK-sans) Nonmotile, obligate parasitic protozoon; members of the group cause malaria and other diseases; sporozoon.
aplastic (a-PLAS-tik) **anemia** Anemia caused by bone marrow failure; may result from damage caused by physical or chemical agents.
apnea (AP-ne-ah) Temporary cessation of breathing.
apocrine (AP-o-krin) Referring to a gland that releases some cellular material along with its secretions.
aponeurosis (ap-o-nu-RO-sis) Broad sheet of fibrous connective tissue that attaches muscle to bone or to other muscle.
appendicular (ap-en-DIK-u-lar) **skeleton** Part of the skeleton that includes the bones of the upper extremities, lower extremities, shoulder girdle, and hips.
appendix (ah-PEN-diks) Finger-like tube of lymphatic tissue attached to the first portion of the large intestine; vermiform (wormlike) appendix.
aquaporin (ak-kwah-POR-in) A transmembrane channel protein that enables water movement across the plasma membrane.
aqueous (A-kwe-us) Pertaining to water; an aqueous solution is one in which water is the solvent.
aqueous (A-kwe-us) **humor** Watery fluid that fills much of the eyeball anterior to the lens.
arachnoid (ah-RAK-noyd) Middle layer of the meninges.
areolar (ah-RE-o-lar) Referring to loose connective tissue, any small space, or to an areola, a circular area of marked color.
arrector pili (ah-REK-tor PI-li) Muscle attached to a hair follicle that raises the hair.
arrhythmia (ah-RITH-me-ah) Abnormal rhythm of the heartbeat; dysrhythmia.
arteriole (ar-TE-re-ole) Vessel between a small artery and a capillary.
arteriosclerosis (ar-te-re-o-skle-RO-sis) Hardening of the arteries.
artery (AR-ter-e) Vessel that carries blood away from the heart.
arthritis (arth-RI-tis) Inflammation of the joints.
arthrocentesis (ar-thro-sen-TE-sis) Puncture of a joint to withdraw fluid.
arthroplasty (AR-thro-plas-te) Surgical replacement or repair of a damaged joint.
arthropod (AR-thro-pod) One of a diverse group of animals that includes spiders and mites. Arthropods may cause disease or be vectors in the spread of disease.
arthroscope (AR-thro-skope) Instrument for examining the interior of the knee and surgically repairing the knee.
articular (ar-TIK-u-lar) Pertaining to a joint.
articulation (ar-tik-u-LA-shun) A joint; an area of junction or union between two or more structures.
ascaris (AS-kah-ris) A common type of roundworm.
ascites (ah-SI-teze) Abnormal collection of fluid in the abdominal cavity.
asepsis (a-SEP-sis) Condition in which no pathogens are present; adj., aseptic.
asexual (a-SEX-u-al) Nonsexual; referring to reproduction that does not require fusion of two different gametes.
aspirate (AS-pih-rate) To inhale fluid or solid material into the respiratory tract; to withdraw by suction, as from a body cavity.
asthma (AZ-mah) Inflammation and constriction of the air passageways.
astigmatism (ah-STIG-mah-tizm) Visual defect due to an irregularity in the curvature of the cornea or the lens.
astrocyte (AS-tro-site) A star-shaped type of neuroglial cell located in the CNS.
ataxia (ah-TAK-se-ah) Lack of muscular coordination; irregular muscular action.
atelectasis (at-e-LEK-tah-sis) Incomplete lung expansion; collapsed lung.
atherosclerosis (ath-er-o-skleh-RO-sis) Hardening of the arteries due to the deposit of yellowish, fatlike material in the lining of these vessels.
atom (AT-om) Smallest subunit of a chemical element.
atomic number The number of protons in the nucleus of an element's atoms; a number characteristic of each element.
atopic dermatitis (ah-TOP-ik der-mah-TI-tis) Skin condition that may involve redness, blisters, pimples, scaling, and crusting; eczema.
ATP See adenosine triphosphate.
atrial natriuretic (na-tre-u-RET-ik) **peptide (ANP)** Hormone produced by the atria of the heart that lowers blood pressure by promoting excretion of sodium and water.

atrioventricular (a-tre-o-ven-TRIK-u-lar) (**AV**) **node** Part of the heart's conduction system.
atrium (A-tre-um) One of the heart's two upper chambers; adj., atrial.
atrophy (AT-ro-fe) Wasting or decrease in size of a part.
attenuated (ah-TEN-u-a-ted) Weakened.
auditory (AW-dih-tor-e) **tube** Tube that connects the middle ear cavity to the throat; eustachian tube.
autoclave (AW-to-clave) Instrument used to sterilize material with steam under pressure.
autoimmunity (aw-to-ih-MU-nih-te) Abnormal immune response to one's own tissues.
autologous (aw-TOL-o-gus) Related to self, such as blood or tissue taken from one's own body.
autolysis (aw-TOL-ih-sis) Lysosomal self-digestion of a cell.
autonomic (aw-to-NOM-ik) **nervous system** (**ANS**) The part of the nervous system that controls smooth muscle, cardiac muscle, and glands; the visceral or involuntary nervous system.
autophagy (aw-TOF-ah-je) Lysosomal self-digestion of cellular structures.
autosome (AW-to-some) Any chromosome not involved in sex determination. There are 44 autosomes (22 pairs) in humans.
AV node See atrioventricular node.
axial (AK-se-al) **skeleton** The part of the skeleton that includes the skull, spinal column, ribs, and sternum.
axilla (ak-SIL-ah) Hollow beneath the arm where it joins the body; armpit.
axis (AK-sis) A central pole or line around which a structure may revolve. The second cervical vertebra.
axon (AK-son) Fiber of a neuron that conducts impulses away from the cell body.

B

bacillus (bah-SIL-us) Rod-shaped bacterium; pl., bacilli (bah-SIL-i).
bacteriology (bak-te-re-OL-o-je) The study of bacteria.
bacterium (bak-TE-re-um) Type of microorganism; pl., bacteria (bak-TE-re-ah).
band cell Immature neutrophil.
baroreceptor (bar-o-re-SEP-ter) Receptor that responds to pressure, such as those in vessel walls that respond to stretching and help regulate blood pressure; a type of mechanoreceptor.
basal ganglia (BA-sal GANG-le-ah) Gray masses in the lower part of the forebrain that aid in motor planning; basal nuclei.
basal (BA-sal) **metabolism** The amount of energy needed to maintain life functions while the body is at rest.
basal nuclei (BA-sal NU-kle-i) Interconnected masses of gray matter spread throughout the brain that modulate motor inputs and facilitate routine motor tasks; basal ganglia.
base A lower portion or foundation. Substance that can accept a hydrogen ion (H^+); substance that releases a hydroxide ion (OH^-) when dissolved in water; an alkali. The broad, superior portion of the heart.
basophil (BA-so-fil) Granular white blood cell that shows large, dark blue cytoplasmic granules when stained with basic stain.
B cell Agranular white blood cell that gives rise to antibody-producing plasma cells in response to an antigen; B lymphocyte.
Bell palsy Facial paralysis caused by damage to the facial nerve (VII), usually on one side of the face.
benign (be-NINE) Denoting a mild condition, favorable for recovery, or a tumor that does not metastasize (spread).
bicarbonate ion (bi-KAR-bon-ate I-on) Molecule that combines with hydrogen ion to form carbonic acid, which separates into carbon dioxide and water; an important buffer in body fluids.
bile Substance produced in the liver that emulsifies fats.
bilirubin (BIL-ih-ru-bin) Pigment derived from the breakdown of hemoglobin and found in bile.
biofeedback (bi-o-FEED-bak) A method for controlling involuntary responses by means of electronic devices that monitor changes and feed information back to a person.
biological therapy Stimulation of the immune system to combat disease, such as cancer; immunotherapy.
biopsy (BI-op-se) Removal of tissue or other material from the living body for examination, usually under the microscope.
blastocyst (BLAS-to-sist) An early stage in embryonic development consisting of a hollow structure formed from a morula that implants in the uterine lining.
blood urea nitrogen (**BUN**) Amount of nitrogen from urea in the blood; test to evaluate kidney function.
body (BOD-e) The principal mass of any structure. The main portion of a vertebra. A corpus or subdivision of the penis.
bolus (BO-lus) A concentrated mass; the portion of food that is moved to the back of the mouth and swallowed.
bone Hard connective tissue that makes up most of the skeleton, or any structure composed of this type of tissue.
bone marrow Substance that fills the central cavity of a long bone (yellow marrow) and the spaces in spongy bone (red marrow).
bony labyrinth (LAB-ih-rinth) The outer, hollow bony shell of the inner ear that is filled with perilymph.
Bowman capsule See glomerular capsule.
bradycardia (brad-e-KAR-de-ah) Heart rate of less than 60 beats per minute.
brain The central controlling area of the central nervous system (CNS).
brain stem Portion of the brain that connects the cerebrum with the spinal cord; contains the midbrain, pons, and medulla oblongata.
Broca (bro-KAH) **area** Area of the cerebral cortex concerned with motor control of speech; motor speech area.
bronchiole (BRONG-ke-ole) Microscopic branch of a bronchus.
bronchoscope (BRONG-ko-skope) Endoscope for examination of the bronchi and removal of small objects from the bronchi.
bronchus (BRONG-kus) Large air passageway in the lung; pl., bronchi (BRONG-ki).
buboes (BU-bose) Enlarged inguinal lymph nodes, as from infection.
buffer (BUF-er) Substance that prevents sharp changes in a solution's pH.
bulbourethral (bul-bo-u-RE-thral) **gland** Gland that secretes mucus to lubricate the urethra and tip of penis during sexual stimulation; Cowper gland.
bulimia (bu-LIM-e-ah) Eating disorder also known as binge–purge syndrome.
bulk transport Movement of large amounts of material through a cell's plasma membrane.
bulla (BUL-ah) Fluid-filled sac in skin; vesicle.
BUN See blood urea nitrogen.
bunion (BUN-yun) A bone enlargement commonly found at the base and medial side of the great toe.
bursa (BER-sah) Small, fluid-filled sac found in an area subject to stress around bones and joints; pl., bursae (BER-se).
bursitis (ber-SI-tis) Inflammation of a bursa.

C

calcitriol (kal-sih-TRI-ol) The active form of vitamin D; dihydroxycholecalciferol (di-hi-drok-se-ko-le-kal-SIF-eh-rol).
calcium (KAL-se-um) An element required for bone formation, muscle contraction, nerve impulse conduction, and blood clotting.
calculus (KAL-ku-lus) Stone, such as a urinary stone; pl., calculi (KAL-ku-li).
calyx (KA-liks) Cuplike extension of the renal pelvis that collects urine; pl., calyces (KA-lih-seze).
cancellous (KAN-sel-us) Referring to spongy bone tissue.
cancer (KAN-ser) Tumor that spreads to other tissues; a malignant neoplasm.
capillary (CAP-ih-lar-e) Microscopic vessel through which exchanges take place between the blood and the tissues.
carbohydrate (kar-bo-HI-drate) Simple sugar or compound made from simple sugars linked together, such as starch or glycogen.
carbon Element that is the basis of organic chemistry.
carbon dioxide (di-OX-ide) (CO_2) Gaseous waste product of cellular metabolism.
carbonic acid (kar-BON-ik) Acid formed when carbon dioxide dissolves in water; carbonic acid then separates into hydrogen ion and bicarbonate ion.
carbonic anhydrase (an-HI-drase) Enzyme that catalyzes the interconversion of carbon dioxide with bicarbonate ion and hydrogen ion.
carcinogen (kar-SIN-o-jen) Cancer-causing substance.
carcinoma (kar-sih-NO-mah) Malignant growth of epithelial cells; a form of cancer.
cardiac (KAR-de-ak) Pertaining to the heart.
cardiac (KAR-de-ak) **output** (**CO**) Volume of blood pumped by each ventricle in one minute. CO is the product of stroke volume and heart rate.
cardiopulmonary resuscitation (**CPR**) Method to restore heartbeat and breathing by closed chest cardiac massage.

cardiovascular system (kar-de-o-VAS-ku-lar) System consisting of the heart and blood vessels that transports blood throughout the body.
caries (KA-reze) Tooth decay.
carotenemia (kar-o-te-NE-me-ah) Yellowish skin color caused by eating excessive amounts of carrots and other deeply colored vegetables.
carrier Individual who has a recessive allele of a gene that is not expressed in the phenotype but that can be passed to offspring.
cartilage (KAR-tih-lij) Type of hard connective tissue found at the ends of bones, the tip of the nose, larynx, trachea, and the embryonic skeleton.
cast Mold of a kidney tubule formed of solid material, such as cells, salts, or proteins. Rigid encasement for immobilization of a part, as during healing of a bone fracture.
CAT See computed tomography.
catabolism (kah-TAB-o-lizm) Metabolic breakdown of substances into simpler substances; includes the digestion of food and the oxidation of nutrient molecules for energy.
catalyst (KAT-ah-list) Substance that speeds the rate of a chemical reaction.
cataract (KAT-ah-rakt) Opacity of the eye's lens or lens capsule.
catheter (KATH-eh-ter) Tube that can be inserted into a vessel or cavity; may be used to remove fluid, such as urine or blood; v., catheterize.
cation (KAT-i-on) Positively charged particle (ion).
cauda equina (KAW-dah eh-KWI-nah) Bundle composed of the spinal nerves that arise from the terminal region of the spinal cord; the individual nerves gradually exit from their appropriate segments of the spinal column.
caudal (KAWD-al) Toward or nearer to the sacral region of the spinal column.
cavity (KAV-ih-te) Hollow space; hole.
CCK See cholecystokinin.
cecum (SE-kum) Small pouch at the beginning of the large intestine.
cell Basic unit of life.
cell membrane See plasma membrane.
cellular respiration Series of reactions by which nutrients are oxidized for energy within the mitochondria of body cells.
central canal Channel in the center of an osteon of compact bone; channel in the center of the spinal cord that contains CSF.
central nervous system (CNS) Part of the nervous system that includes the brain and spinal cord.
centrifuge (SEN-trih-fuje) An instrument that separates materials in a mixture based on density.
centriole (SEN-tre-ole) Rod-shaped body near the nucleus of a cell; functions in cell division.
cerebellum (ser-eh-BEL-um) Small section of the brain inferior to the cerebral hemispheres; functions in coordination, balance, and muscle tone.
cerebral (SER-e-bral) **cortex** Very thin outer layer of gray matter on the surface of the cerebral hemispheres.
cerebral palsy (PAWL-ze) Disorder caused by brain damage occurring before or during the birth process.
cerebrospinal (ser-e-bro-SPI-nal) **fluid (CSF)** Fluid that circulates in and around the brain and spinal cord.
cerebrovascular (ser-e-bro-VAS-ku-lar) **accident (CVA)** Condition involving obstruction of blood flow to brain tissue or bleeding into brain tissue, usually as a result of hypertension or atherosclerosis; stroke.
cerebrum (SER-e-brum) Largest part of the brain; composed of two cerebral hemispheres.
cerumen (seh-RU-men) Earwax; adj., ceruminous (seh-RU-min-us).
cervical (SER-vih-kal) Pertaining to the neck or the uterine cervix.
cervix (SER-vix) Constricted portion of an organ or part, such as the lower portion of the uterus; neck; adj., cervical.
cesarean (se-ZAR-re-en) **section** Incision in the abdominal and uterine walls for delivery of a fetus; C-section.
chemistry (KEM-is-tre) Study of the composition and properties of matter.
chemoreceptor (ke-mo-re-SEP-tor) Receptor that responds to chemicals in body fluids.
chemotherapy (ke-mo-THER-ah-pe) Treatment of a disease by administration of a chemical agent.
Cheyne-Stokes (CHANE-stokes) **respiration** Rhythmic variation in the depth of respiratory movements alternating with periods of apnea due to depression of the breathing centers.
chiropractic (ki-ro-PRAK-tic) Field that stresses manipulation to correct misalignment for treatment of musculoskeletal disorders.
chlamydia (klah-MID-e-ah) Type of very small bacterium that can exist only within a living cell; members of this group cause inclusion conjunctivitis, trachoma, sexually transmitted infections, and respiratory diseases.
chloride (KLOR-ide) Ion of chlorine.
cholecystokinin (ko-le-sis-to-KI-nin) **(CCK)** Duodenal hormone that stimulates release of enzymes from the pancreas and bile from the gallbladder.
cholelithiasis (ko-le-lih-THI-ah-sis) Condition of having gallstones.
cholesterol (ko-LES-ter-ol) Lipid synthesized by the liver that is found in all plasma membranes, in bile, myelin, steroid hormones, and elsewhere; circulates in the blood and is stored in liver and adipose tissue; implicated in the development of atherosclerosis.
cholinergic (ko-lin-ER-jik) Activity or structure that responds to acetylcholine.
chondrocyte (KON-dro-site) Cell that produces and maintains cartilage.
chondroma (kon-DRO-mah) Tumor of cartilage.
chordae tendineae (KOR-de ten-DIN-e-e) Fibrous threads that stabilize the heart's AV valve flaps.
choriocarcinoma (ko-re-o-kar-sih-NO-mah) Very malignant tumor made of placental tissue.
chorion (KO-re-on) Outer embryonic layer that, together with a layer of the endometrium, forms the placenta.
choroid (KO-royd) Pigmented middle layer of the eye.
choroid plexus (KO-royd PLEKS-us) Vascular network in the brain's ventricles that forms cerebrospinal fluid.
chromosome (KRO-mo-some) Dark-staining, threadlike body in a cell's nucleus; contains genes that determine hereditary traits.
chronic (KRON-ik) Referring to a disease that develops slowly, persists over a long time, or is recurring.
chyle (kile) Milky-appearing fluid absorbed into the lymphatic system from the small intestine. It consists of lymph and droplets of digested fat.
chyme (kime) Mixture of partially digested food, water, and digestive juices that forms in the stomach.
cicatrix (SIK-ah-trix) Scar.
cilia (SIL-e-ah) Hairs or hairlike processes, such as eyelashes or microscopic extensions from a cell's surface; sing., cilium.
ciliary (SIL-e-ar-e) **muscle** Eye muscle that controls the shape of the lens.
ciliate (SIL-e-ate) Type of protozoon covered with cilia for propulsion.
cingulated gyrus (SIN-gu-late JI-rus) Portion of the cerebral cortex that loops over the corpus callosum; region that associates emotions with memories.
circulation (ser-ku-LA-shun) Movement or flow within a closed system, as of blood or CSF.
circumcision (sir-kum-SIJ-un) Surgery to remove the foreskin of the penis.
circumduction (ser-kum-DUK-shun) Circular movement at a joint.
cirrhosis (sih-RO-sis) Chronic disease, usually of the liver, in which active cells are replaced by inactive scar tissue.
cisterna chyli (sis-TER-nah KI-li) Initial portion of the thoracic lymph duct, which is enlarged to form a temporary storage area.
CK See creatine kinase.
clitoris (KLIT-o-ris) Small organ of great sensitivity in the external genitalia of the female.
CNS See central nervous system.
coagulation (ko-ag-u-LA-shun) Clotting, as of blood.
coccus (KOK-us) Round bacterium; pl., cocci (KOK-si).
coccyx (KOK-siks) Terminal portion of the vertebral column; formed of four or five small fused bones.
cochlea (KOK-le-ah) Coiled portion of the inner ear that contains the organ of hearing.
colic (KOL-ik) Spasm of visceral muscle.
collagen (KOL-ah-jen) Flexible white protein that gives strength and resilience to connective tissue, such as bone and cartilage.
colloid (kol-OYD) Mixture in which suspended particles do not dissolve but remain distributed in the solvent because of their small size (e.g., cytoplasm); colloidal suspension.

colon (KO-lon) Main portion of the large intestine.
colostrum (ko-LOS-trum) Secretion of the mammary glands released prior to secretion of milk.
comedo (KOM-e-do) Plugged hair follicle.
communicable (kom-MU-nih-kabl) Capable of being transmitted from one person to another, such as disease; contagious.
complement (KOM-ple-ment) Group of blood proteins that helps antibodies and phagocytes to destroy foreign cells.
compliance (kom-PLI-ans) Ease with which a hollow structure, such as the thorax or alveoli of the lungs, can be expanded under pressure.
compound Substance composed of two or more chemical elements.
computed tomography (to-MOG-rah-fe) **(CT)** Imaging method in which multiple radiographic views taken from different angles are analyzed by computer to show an area cross-section; used to detect tumors and other abnormalities; also called computed axial tomography (CAT).
concha (KON-ka) Shell-like bone in the nasal cavity; pl., conchae (KON-ke).
concussion (kon-CUSH-on) Injury resulting from a violent blow or shock.
conduction (kon-DUK-shun) Transmission or conveyance of energy from one point to another as in transmission of a nerve impulse or transfer of heat from a warm object to a cooler one.
condyle (KON-dile) Rounded projection, as on a bone.
cone Receptor cell in the eye's retina; used for vision in bright light.
congenital (con-JEN-ih-tal) Present at birth.
conjunctiva (kon-junk-TI-vah) Membrane that lines the eyelid and covers the anterior part of the sclera (white of the eye).
connective tissue The supporting tissue of the body ranging in consistency from liquid (e.g., blood) to semisolid (e.g., loose and adipose tissue) to firm (e.g., ligaments, tendons, cartilage) to hard (e.g., bone).
constipation (kon-stih-PA-shun) Infrequency of or difficulty with defecation.
contraception (con-trah-SEP-shun) Prevention of an ovum's fertilization or a fertilized ovum's implantation; birth control.
contractility (kon-trak-TIL-ih-te) Capacity to undergo shortening, as in muscle tissue.
convection (kon-VEK-shun) Heat transfer promoted by movement of a cooler contacting medium, as in cooling by use of a fan.
convergence (kon-VER-jens) Centering of both eyes on the same visual field.
convulsion (kon-VUL-shun) Series of muscle spasms; seizure.
Cooley anemia Severe form of β-thalassemia, a hereditary blood disorder that impairs hemoglobin formation.
cornea (KOR-ne-ah) Clear portion of the sclera that covers the anterior of the eye.
coronary (KOR-on-ar-e) Referring to the heart or to the arteries supplying blood to the heart.
coronary thrombosis (throm-BO-sis) Formation of a blood clot in a vessel of the heart.
corpus (KOR-pus) Body; main part of an organ or other structure.
corpus callosum (kal-O-sum) Thick bundle of myelinated nerve cell fibers deep within the brain that carries nerve impulses from one cerebral hemisphere to the other.
corpus luteum (LU-te-um) Yellow body formed from ovarian follicle after ovulation; produces estrogen and progesterone.
corpus spongiosum (KOR-pus spon-je-O-sum) Central subdivision of the penis that contains the urethra.
cortex (KOR-tex) Outer layer of an organ, such as the brain, kidney, or adrenal gland.
coryza (ko-RI-zah) Nasal discharge; acute coryza is the common cold.
covalent (KO-va-lent) **bond** Chemical bond formed by the sharing of electrons between atoms.
CPR See cardiopulmonary resuscitation.
cranial (KRA-ne-al) Pertaining to the cranium, the part of the skull that encloses the brain; toward the head or nearer to the head.
craniosacral (kra-ne-o-SA-kral) Pertaining to the cranial and sacral region of the spinal cord, as describes the parasympathetic nervous system.
creatine (KRE-ah-tin) **phosphate** Compound in muscle tissue that stores energy in high-energy bonds.
creatine kinase (CK) (KRE-ah-tin KI-nase) Enzyme in muscle cells that is needed to synthesize creatine phosphate and is released in increased amounts when muscle tissue is damaged; the form specific to cardiac muscle cells is creatine kinase MB (CK-MB).
creatinine (kre-AT-ih-nin) Nitrogenous waste product eliminated in urine.
crenation (kre-NA-shun) Shrinking of a cell, as when placed in a hypertonic solution.
crisis (KRI-sis) Sudden change in the course of disease. Sudden drop in temperature.
crista (KRIS-tah) Receptor for the sense of rotational equilibrium; pl., cristae.
croup (krupe) Loud barking cough associated with upper respiratory infection in children.
cryoprecipitate (kri-o-pre-SIP-ih-tate) Precipitate formed when plasma is frozen and then thawed.
cryptorchidism (kript-OR-kid-izm) Failure of the testis to descend into the scrotum; undescended testicle.
CSF See cerebrospinal fluid.
CT See computed tomography.
Cushing syndrome Condition caused by excess glucocorticoids, as by overactivity of the adrenal cortex or administration of corticosteroids.
cutaneous (ku-TA-ne-us) Referring to the skin.
cuticle (KU-tih-kl) Extension of the stratum corneum that seals the space between the nail plate and the skin above the nail root.
cyanosis (si-ah-NO-sis) Bluish discoloration of the skin and mucous membranes resulting from insufficient oxygen in the blood.
cystic (SIS-tik) **duct** Duct that carries bile into and out of the gallbladder.
cystic fibrosis (SIS-tik fi-BRO-sis) Hereditary disease involving thickened secretions and electrolyte imbalances.
cystitis (sis-TI-tis) Inflammation of the urinary bladder.
cystoscope (SIS-to-skope) Endoscopic instrument for examining the interior of the urinary bladder.
cytokine (SI-to-kine) Peptide produced by immune cells or other cells that acts as a cellular signal; (e.g., interleukin).
cytology (si-TOL-o-je) Study of cells.
cytoplasm (SI-to-plazm) Substance that fills the cell, consisting of a liquid cytosol and organelles.
cytosol (SI-to-sol) Liquid portion of the cytoplasm, consisting of nutrients, minerals, enzymes, and other materials in water.

D

deamination (de-am-ih-NA-shun) Removal of amino groups from proteins in metabolism.
decidua (de-SID-u-ah) The vascularized internal portion of the endometrium in a pregnant uterus; the maternal portion of the placenta.
decubitus (de-KU-bih-tus) Lying down.
defecation (def-eh-KA-shun) Act of eliminating undigested waste from the digestive tract.
defibrillator (de-FIB-rih-la-tor) Device that generates a strong electric current to discharge all cardiac muscle cells at once to restore a normal rhythm.
degeneration (de-jen-er-A-shun) Breakdown, as from age, injury, or disease.
deglutition (deg-lu-TISH-un) Act of swallowing.
dehydration (de-hi-DRA-shun) Excessive loss of body fluid.
dementia (de-MEN-she-ah) Gradual and usually irreversible loss of intellectual function.
denaturation (de-nah-tu-RA-shun) Change in structure of a protein, such as an enzyme, so that it can no longer function.
dendrite (DEN-drite) Neuron fiber that conducts impulses toward the cell body.
dendritic (den-DRIT-ik) **cell** Phagocytic white blood cell active in the immune system as an antigen-presenting cell (APC).
dentin (DEN-tin) The main substance of a tooth.
deoxyribonucleic (de-OK-se-ri-bo-nu-kle-ik) **acid (DNA)** Genetic material of the cell; makes up the chromosomes in the cell's nucleus.
depolarization (de-po-lar-ih-ZA-shun) Reduction of the membrane potential (charge) resulting from the entry of cations or the exit of anions.

dermal papillae (pah-PIL-le) Extensions of the dermis that project up into the epidermis; they contain blood vessels that supply the epidermis.
dermatitis (der-mah-TI-tis) Inflammation of the skin.
dermatome (DER-mah-tome) Region of the skin supplied by a single spinal nerve.
dermatosis (der-mah-to-sis) Any skin disease.
dermis (DER-mis) True skin; deeper part of the skin.
detoxification (de-tok-sih-fih-KA-shun) Removal of the toxicity of a harmful substance, such as alcohol and certain drugs.
dextrose (DEK-strose) Glucose, as found in nature.
diabetes insipidus (in-SIP-ih-dus) Condition due to insufficient ADH secretion by the posterior pituitary, causing excessive water loss.
diabetes mellitus (di-ah-BE-teze mel-LI-tus) Disease of insufficient insulin or insufficient response to insulin in which excess glucose is found in the blood and the urine; characterized by abnormal metabolism of glucose, protein, and fat.
diagnosis (di-ag-NO-sis) Identification of an illness.
dialysis (di-AL-ih-sis) Method for separating molecules in solution based on differences in their ability to pass through a semipermeable membrane; method for removing nitrogenous waste products from the body, as by hemodialysis or peritoneal dialysis.
diaphragm (DI-ah-fram) Dome-shaped muscle under the lungs that flattens during inhalation; a separating membrane or structure.
diaphysis (di-AF-ih-sis) Shaft of a long bone.
diarrhea (di-ah-RE-ah) Abnormally frequent watery bowel movements.
diarthrosis (di-ar-THRO-sis) Freely movable joint; synovial joint.
diastole (di-AS-to-le) Relaxation; adj., diastolic (di-as-TOL-ik).
diencephalon (di-en-SEF-ah-lon) Region of the brain between the cerebral hemispheres and the midbrain; contains the thalamus, hypothalamus, and pituitary gland.
diffusion (dih-FU-zhun) Movement of solutes from a region where they are in higher concentration to a region where they are in lower concentration.
digestion (di-JEST-yun) Process of breaking down food into absorbable particles.
digestive system (di-JES-tiv) The system involved in taking in nutrients, converting them to a form the body can use, and absorbing them into the circulation.
dihydroxycholecalciferol (di-hi-drok-se-ko-le-kal-SIF-eh-rol) The active form of vitamin D.
dilation (di-LA-shun) Widening of a part, such as the pupil of the eye, a blood vessel, or the uterine cervix; dilatation.
disaccharide (di-SAK-ah-ride) Compound formed of two simple sugars linked together, such as sucrose and lactose.
disease Illness; abnormal state in which part or all of the body does not function properly.
disinfectant (dis-in-FEK-tant) Agent that kills most microorganisms in a given environment; antimicrobial agent.
disinfection (dis-in-FEK-shun) Killing of most microbes.
dissect (dis-sekt) To cut apart or separate tissues for study.
distal (DIS-tal) Farther from a structure's origin or from a given reference point.
diverticulitis (di-ver-tik-u-LI-tis) Inflammation of small pouches (diverticuli) in the wall of the intestine, usually in the colon.
diverticulosis (di-ver-tik-u-LO-sis) Presence of diverticuli (small pouches) in the intestinal wall.
DNA See deoxyribonucleic acid.
dominant (DOM-ih-nant) Referring to an allele of a gene that is always expressed in the phenotype if present.
dopamine (DO-pah-mene) Neurotransmitter.
dorsal (DOR-sal) At or toward the back; posterior.
dorsiflexion (dor-sih-FLEK-shun) Bending the foot upward at the ankle.
Down syndrome A congenital disorder usually due to an extra chromosome 21; trisomy 21.
duct Tube or vessel.
ductus arteriosus (DUK-tus ar-te-re-O-sus) Small vessel in the fetus that carries blood from the pulmonary artery to the descending aorta.
ductus deferens (DEF-er-enz) Tube that carries sperm cells from the testis to the urethra; vas deferens.
ductus venosus (ve-NO-sus) Small vessel in the fetus that carries blood from the umbilical vein to the inferior vena cava.
duodenum (du-o-DE-num) First portion of the small intestine.
dura mater (DU-rah MA-ter) Outermost layer of the meninges.
dysmenorrhea (dis-men-o-RE-ah) Painful or difficult menstruation.
dyspnea (disp-NE-ah) Difficult or labored breathing.

E

eccrine (EK-rin) Referring to sweat glands that regulate body temperature and vent sweat directly to the surface of the skin through a pore.
ECG See electrocardiograph.
echocardiograph (ek-o-KAR-de-o-graf) Instrument to study the heart by means of ultrasound; the record produced is an echocardiogram.
eclampsia (eh-KLAMP-se-ah) Serious and sometimes fatal condition involving convulsions, liver damage, and kidney failure that can develop from pregnancy-induced hypertension.
ectopic (ek-TOP-ik) Apart from a normal location, as a pregnancy or heartbeat.
eczema (EK-ze-mah) Skin disorder characterized by intense itching and inflammation; acute eczematous dermatitis.
edema (eh-DE-mah) Accumulation of fluid in the tissue spaces.
EEG See electroencephalograph.
effector (ef-FEK-tor) Muscle or gland that responds to a signal; effector organ.
efferent (EF-fer-ent) Carrying away from a given point, such as a motor neuron that carries nerve impulses away from the central nervous system.
effusion (eh-FU-zhun) Escape of fluid into a cavity or space; the fluid itself.
ejaculation (e-jak-u-LA-shun) Expulsion of semen through the urethra.
EKG See electrocardiograph.
elasticity (e-las-TIS-ih-te) Capacity of a structure to return to its original shape after being stretched.
electrocardiograph (e-lek-tro-KAR-de-o-graf) **(ECG, EKG)** Instrument to study the heart's electric activity; record made is an electrocardiogram.
electroencephalograph (e-lek-tro-en-SEF-ah-lo-graf) **(EEG)** Instrument used to study the brain's electric activity; record made is an electroencephalogram.
electrolyte (e-LEK-tro-lite) Compound that separates into ions in solution; substance that conducts an electric current in solution.
electron (e-LEK-tron) Negatively charged particle located in an energy level outside an atom's nucleus.
electrophoresis (e-lek-tro-fo-RE-sis) Separation of components in a mixture by passing an electric current through it; components separate on the basis of their charge.
element (EL-eh-ment) One of the substances from which all matter is made; substance that cannot be decomposed into a simpler substance.
elephantiasis (el-eh-fan-TI-ah-sis) Enlargement of the extremities due to blockage of lymph flow by small filariae (fi-LA-re-e) worms.
embolism (EM-bo-lizm) The condition of having an embolus (obstruction in the circulation).
embolus (EM-bo-lus) Blood clot or other obstruction in the circulation.
embryo (EM-bre-o) Developing offspring during the first eight weeks of gestation.
emesis (EM-eh-sis) Vomiting.
emphysema (em-fih-SE-mah) Pulmonary disease characterized by dilation and destruction of the alveoli.
emulsify (e-MUL-sih-fi) To break up fats into small particles; n., emulsification.
enamel (e-NAM-el) Material that covers the crown of a tooth; hardest substance in the body.
encephalitis (en-sef-ah-LI-tis) Inflammation of the brain.
endarterectomy (end-ar-ter-EK-to-me) Procedure to remove plaque associated with atherosclerosis from a vessel's lining.
endemic (en-DEM-ik) Describing a disease that is found at low incidence but constantly within the population of a given region.
endocarditis (en-do-kar-DI-tis) Inflammation of the endocardium, the inner lining of the heart.
endocardium (en-do-KAR-de-um) Membrane that lines the heart's chambers and covers the valves.
endocrine (EN-do-krin) Referring to a gland that secretes into the bloodstream.

endocrine system System composed of glands that secrete hormones.
endocytosis (en-do-si-TO-sis) Movement of large amounts of material into a cell using vesicles (e.g., phagocytosis and pinocytosis).
endolymph (EN-do-limf) Fluid that fills the membranous labyrinth of the inner ear.
endometriosis (en-do-me-tre-O-sis) Growth of endometrial tissue outside the uterus.
endometrium (en-do-ME-tre-um) Inner layer of the uterus.
endomysium (en-do-MIS-e-um) Connective tissue around an individual muscle fiber.
endoplasmic reticulum (en-do-PLAS-mik re-TIK-u-lum) **(ER)** Network of membranes in the cellular cytoplasm; may be smooth or rough based on absence or presence of ribosomes.
endorphin (en-DOR-fin) Pain-relieving substance released naturally from the brain.
endoscope (EN-do-skope) Lighted flexible instrument used to examine the interior of body cavities or organs.
endospore (EN-do-spore) Resistant forms of bacteria that can withstand harsh conditions and germinate into active forms when conditions are appropriate.
endosteum (en-DOS-te-um) Thin membrane that lines a bone marrow cavity.
endothelium (en-do-THE-le-um) Epithelium that lines the heart, blood vessels, and lymphatic vessels.
enucleation (e-nu-kle-A-shun) Removal of the eyeball.
enuresis (en-u-RE-sis) Involuntary urination, usually during the night.
enzyme (EN-zime) A protein that accelerates a specific chemical reaction.
eosinophil (e-o-SIN-o-fil) Granular white blood cell that shows bead-like, bright pink cytoplasmic granules when stained with acid stain.
ependymal (eh-PEN-dih-mal) **cell** Neuroglial cell in the CNS used to form the choroid plexus, a network in a ventricle that forms CSF.
epicardium (ep-ih-KAR-de-um) Membrane that forms the heart wall's outermost layer and is continuous with the lining of the fibrous pericardium; visceral pericardium.
epicondyle (ep-ih-KON-dile) Small projection on a bone above a condyle.
epidemic (ep-ih-DEM-ik) Occurrence of a disease among many people in a given region at the same time.
epidemiology (ep-ih-de-me-OL-o-je) The study of disease in populations.
epidermis (ep-ih-DER-mis) Outermost layer of the skin.
epididymis (ep-ih-DID-ih-mis) Coiled tube on the surface of the testis in which sperm cells are stored and in which they mature.
epigastric (ep-ih-GAS-trik) Pertaining to the region just inferior to the sternum (breastbone).
epiglottis (ep-e-GLOT-is) Leaf-shaped cartilage that covers the larynx during swallowing.
epilepsy (EP-ih-lep-se) Neurologic disorder involving abnormal electric activity of the brain; characterized by seizures of varying severity.
epimysium (ep-ih-MIS-e-um) Sheath of fibrous connective tissue that encloses a muscle.
epinephrine (ep-ih-NEF-rin) Hormone released from the adrenal medulla; adrenaline.
epiphysis (eh-PIF-ih-sis) End of a long bone; adj., epiphyseal (ep-ih-FIZ-e-al).
episiotomy (eh-piz-e-OT-o-me) Cutting of the perineum between the vaginal opening and the anus to reduce tissue tearing in childbirth.
epistaxis (ep-e-STAK-sis) Nosebleed.
epithelium (ep-ih-THE-le-um) One of the four main types of tissues; forms glands, covers surfaces, and lines cavities; adj., epithelial.
EPO See erythropoietin.
equilibrium (e-kwih-LIB-re-um) Sense of balance.
ER See endoplasmic reticulum.
erection (e-REK-shun) Stiffening and enlargement of the penis.
eruption (e-RUP-shun) Raised skin lesion; rash.
erythema (er-eh-THE-mah) Redness of the skin.
erythrocyte (eh-RITH-ro-site) Red blood cell.
erythropoietin **(EPO)** (eh-rith-ro-POY-eh-tin) Hormone released from the kidney that stimulates red blood cell production in the red bone marrow.
esophagus (eh-SOF-ah-gus) Muscular tube that carries food from the throat to the stomach.
estrogen (ES-tro-jen) Group of female sex hormones that promotes development of the ovarian follicle and the uterine lining and maintains secondary sex characteristics; the main estrogen is estradiol.
etiology (e-te-OL-o-je) Study of a disease's cause or the theory of its origin.
eustachian (u-STA-shun) **tube** See auditory tube.
evaporation (e-vap-o-RA-shun) Process of changing a liquid to a vapor.
eversion (e-VER-zhun) Turning outward, with reference to movement of the foot.
excitability In cells, the ability to transmit an electric current along the plasma membrane.
excoriation (eks-ko-re-A-shun) Scratch into the skin.
excretion (eks-KRE-shun) Removal and elimination of metabolic waste products from the blood.
exfoliation (eks-fo-le-A-shun) Loss of cells from the surface of tissue, such as the skin.
exhalation (eks-hah-LA-shun) Expulsion of air from the lungs; expiration.
exocrine (EK-so-krin) Referring to a gland that secretes through a duct.
exocytosis (eks-o-si-TO-sis) Movement of large amounts of material out of the cell using vesicles.
exophthalmos (ek-sof-THAL-mos) Protrusion (bulging) of the eyes, commonly seen in Graves disease.
extension (eks-TEN-shun) Motion that increases the angle at a joint, returning a body part to the anatomic position.
extracellular (EK-strah-sel-u-lar) Outside the cell.
extremity (ek-STREM-ih-te) Limb.
exudate (EKS-u-date) Any fluid or semisolid that oozes out of tissue, particularly as a result of injury or inflammation.

F

fallopian (fah-LO-pe-an) **tube** See uterine tube.
familial (fah-MIL-e-al) Hereditary; passed from parents to children in the genes.
fascia (FASH-e-ah) Band or sheet of fibrous connective tissue.
fascicle (FAS-ih-kl) Small bundle, as of muscle cells or nerve cell fibers.
fat Type of lipid composed of glycerol and fatty acids; triglyceride.
febrile (FEB-ril) Pertaining to fever.
fecalith (FE-kah-lith) Hardened piece of fecal material that may cause obstruction.
feces (FE-seze) Waste material discharged from the large intestine; excrement; stool.
feedback Return of information into a system, so that it can be used to regulate that system.
fertilization (fer-til-ih-ZA-shun) Union of an ovum and a spermatozoon.
fetus (FE-tus) Developing offspring from the start of the ninth week of gestation until birth.
fever (FE-ver) Abnormally high body temperature involving an increased temperature set point.
fibrillation (fih-brih-LA-shun) Very rapid, uncoordinated beating of the heart.
fibrin (FI-brin) Blood protein that forms a blood clot.
fibrinogen (fi-BRIN-o-jen) Plasma protein that is converted to fibrin in blood clotting.
fibroblast (FI-bro-blast) Cell that produces protein fibers and other components of the matrix in connective tissue.
fibroid (FI-broyd) Benign uterine tumor; myoma.
fibromyositis (fi-bro-mi-o-SI-tis) Muscular disorder combining myositis (muscle inflammation) and fibrositis (inflammation of connective tissue.
fight-or-flight response Physiologic response to a threatening or stressful situation as activated by the sympathetic nervous system.
filtration (fil-TRA-shun) Movement of material through a semipermeable membrane down a pressure gradient.
fimbriae (FIM-bre-e) Fringelike extensions of the uterine tube that sweep a released ovum into the tube.
fissure (FISH-ure) Deep groove.
flaccid (FLAK-sid) Flabby, limp, soft.
flagellate (FLAJ-eh-late) Protozoon that uses flagella for locomotion.

flagellum (flah-JEL-lum) Long whiplike extension from a cell used for locomotion; pl., flagella.
flexion (FLEK-shun) Bending motion that decreases the angle between bones at a joint, moving a body part away from the anatomic position.
fluoroscope (flu-OR-o-skope) Instrument for examining deep structures with x-rays.
follicle (FOL-lih-kl) Sac or cavity, such as the ovarian follicle or hair follicle.
follicle-stimulating hormone (FSH) Hormone produced by the anterior pituitary that stimulates development of ova in the ovary and spermatozoa in the testes.
fontanel (fon-tah-NEL) Membranous area in the infant skull where bone has not yet formed; also spelled fontanelle; "soft spot."
foramen (fo-RA-men) Opening or passageway, as into or through a bone; pl., foramina (fo-RAM-in-ah).
foramen magnum Large opening in the skull's occipital bone through which the spinal cord passes to join the brain.
foramen ovale (o-VA-le) Small hole in the fetal atrial septum that allows blood to pass directly from the right atrium to the left atrium.
formed elements Cells and cell fragments in the blood.
fornix (FOR-niks) Recess or archlike structure.
fossa (FOS-sah) Hollow or depression, as in a bone; pl., fossae (FOS-se).
fovea (FO-ve-ah) Small pit or cup-shaped depression in a surface; fovea centralis.
fovea centralis (FO-ve-ah sen-TRA-lis) Tiny depressed area near the optic nerve that is the point of sharpest vision.
fracture (FRAK-sure) To break or a break, as in a bone.
fragile (FRAH-jil) **X syndrome** Hereditary form of mental retardation; x-linked recessive disorder that appears in both males and females.
frontal (FRONT-al) Describing a plane that divides a structure into anterior and posterior parts. Pertaining to the anterior bone of the cranium.
frostbite Tissue damage caused by exposure to cold.
FSH See follicle-stimulating hormone.
fulcrum (FUL-krum) Pivot point in a lever system; joint in the skeletal system.
fundus (FUN-dus) The deepest portion of an organ, such as the eye, the stomach, or the uterus.
fungus (FUN-gus) Type of plantlike microorganism; yeast or mold; pl., fungi (FUN-ji).

G

gallbladder (GAWL-blad-er) Muscular sac on the inferior surface of the liver that stores bile.
GALT Gut-associated lymphoid tissue.
gamete (GAM-ete) Reproductive cell; ovum or spermatozoon.
gamma globulin (GLOB-u-lin) Protein fraction in the blood plasma that contains antibodies.
ganglion (GANG-le-on) Collection of nerve cell bodies located outside the central nervous system.
gangrene (GANG-grene) Death of tissue accompanied by bacterial invasion and putrefaction.
gastric-inhibitory peptide (GIP) Duodenal hormone that inhibits release of gastric juice and stimulates insulin release.
gastrin (GAS-trin) Hormone released from the stomach that stimulates stomach activity.
gastroenteritis (gas-tro-en-ter-I-tis) Inflammation of the stomach and intestine.
gastroesophageal (gas-tro-eh-sof-ah-JE-al) **reflux disease (GERD)** Chronic reflux from the stomach into the distal esophagus caused by weakness in the lower esophageal sphincter (LES).
gastrointestinal (gas-tro-in-TES-tih-nal) **(GI)** Pertaining to the stomach and intestine or the digestive tract as a whole.
gene Hereditary factor; portion of the DNA on a chromosome encoding a specific protein.
genetic (jeh-NET-ik) Pertaining to the genes or heredity.
genital (JEN-ih-tal) Pertaining to reproduction or the reproductive organs (genitalia).
genitalia (jen-ih-TA-le-ah) Reproductive organs, both external and internal.
genotype (JEN-o-tipe) Genetic makeup of an organism.
gestation (jes-TA-shun) Period of development from conception to birth.
GH See growth hormone.
GI See gastrointestinal.
gigantism (ji-GAN-tizm) Excessive growth due to oversecretion of growth hormone in childhood.
gingiva (JIN-jih-vah) Tissue around the teeth; gum.
gland Organ or cell specialized to produce a substance that is sent to other parts of the body.
glans Enlarged distal portion of the penis.
glaucoma (glaw-KO-mah) Disorder involving increased fluid pressure within the eye.
glial cells (GLI-al) Cells that support and protect the nervous system; neuroglia.
glioma (gli-O-mah) Tumor of neuroglial tissue.
glomerular (glo-MER-u-lar) **capsule** Enlarged portion of the nephron that surrounds the glomerulus; Bowman capsule.
glomerular (glo-MER-u-lar) **filtrate** Fluid and dissolved materials that leave the blood and enter the kidney tubule.
glomerulonephritis (glo-mer-u-lo-nef-RI-tis) Kidney disease often resulting from antibodies to a streptococcal infection.
glomerulus (glo-MER-u-lus) Cluster of capillaries surrounded by the kidney tubule's glomerular capsule.
glottis (GLOT-is) Space between the vocal cords.
glucagon (GLU-kah-gon) Hormone from the pancreatic islets that raises blood glucose level.
glucocorticoid (glu-ko-KOR-tih-koyd) Steroid hormone from the adrenal cortex that increases the concentration of nutrients in the blood during times of stress (e.g., cortisol).
glucose (GLU-kose) Simple sugar; main energy source for the cells; dextrose.
glycemic (gli-SE-mik) **effect** Measure of how rapidly a food raises the blood glucose level and stimulates release of insulin.
glycogen (GLI-ko-jen) Compound built from glucose molecules that is stored for energy in the liver and muscles.
glycolysis (gli-KOL-ih-sis) First, anaerobic phase of glucose's metabolic breakdown for energy; converts glucose into pyruvate.
glycosuria (gli-ko-SU-re-ah) Presence of glucose in the urine.
goblet cell Single-celled gland that secretes mucus.
goiter (GOY-ter) Enlargement of the thyroid gland.
Golgi (GOL-je) **apparatus** System of cellular membranes that modifies and sorts proteins; also called Golgi complex.
gonad (GO-nad) Organ producing gametes and sex steroids; ovary or testis; adj., gonadal.
gonadotropin (gon-ah-do-TRO-pin) Hormone that acts on a reproductive gland (ovary or testis) (e.g., FSH, LH).
gout Type of arthritis caused by a metabolic disturbance.
gradient (GRA-de-ent) Gradual change in a particular quality between two regions, such as gradients of position, concentration, or pressure within a system.
gram (g) Basic unit of weight in the metric system.
Gram stain Procedure used to color microorganisms for viewing under the microscope.
granulocyte (GRAN-u-lo-site) Leukocyte with visible granules in the cytoplasm when stained.
Graves disease Common form of hyperthyroidism.
gray commissure (KOM-ih-shure) Bridge of gray matter that connects the right and left horns of the spinal cord.
gray matter Nervous tissue composed of unmyelinated fibers and cell bodies.
greater vestibular (ves-TIB-u-lar) **gland** Gland that secretes mucus into the vagina; Bartholin gland.
growth hormone (GH) Hormone produced by anterior pituitary that promotes tissue growth; somatotropin.
gustation (gus-TA-shun) Sense of taste; adj., gustatory.
gyrus (JI-rus) Raised area of the cerebral cortex; pl., gyri (JI-ri).

H

hair cells Receptor cells for hearing and equilibrium.
hair follicle (FOL-lih-kl) Sheath that encloses a hair.
haversian (ha-VER-shan) **canal** See central canal.
haversian system See osteon.

hay fever Seasonal allergy often due to pollen.
hearing (HERE-ing) Ability to perceive sound; audition.
heart (hart) Organ that pumps blood through the cardiovascular system.
heart block Interruption of electrical impulses in the heart's conduction system.
helminth (HEL-minth) Worm.
hemapheresis (hem-ah-fer-E-sis) Return of blood components to a donor following separation and removal of desired components.
hematocrit (he-MAT-o-krit) **(Hct)** Volume percentage of red blood cells in whole blood; packed cell volume.
hematoma (he-mah-TO-mah) Tumor or swelling filled with blood.
hematopoiesis (hem-mah-to-poy-E-sis) Formation of blood cells.
hematopoietic (he-mah-to-poy-ET-ik) Pertaining to blood cell formation.
hematuria (hem-ah-TU-re-ah) Blood in the urine.
hemocytometer (he-mo-si-TOM-eh-ter) Device used to count blood cells under the microscope.
hemodialysis (he-mo-di-AL-ih-sis) Removal of impurities from the blood by their passage through a semipermeable membrane in a fluid bath.
hemoglobin (he-mo-GLO-bin) **(Hb)** Iron-containing protein in red blood cells that binds oxygen.
hemolysis (he-MOL-ih-sis) Rupture of red blood cells; v., hemolyze (HE-mo-lize).
hemolytic (he-mo-LIT-ik) **disease of the newborn (HDN)** Condition that results from Rh incompatibility between a mother and her fetus; erythroblastosis fetalis.
hemophilia (he-mo-FIL-e-ah) Hereditary bleeding disorder associated with a lack of clotting factors in the blood.
hemopoiesis (he-mo-poy-E-sis) Production of blood cells; hematopoiesis.
hemorrhage (HEM-eh-rij) Loss of blood.
hemorrhoids (HEM-o-royds) Varicose veins in the rectum.
hemostasis (he-mo-STA-sis) Stoppage of bleeding.
hemothorax (he-mo-THOR-aks) Accumulation of blood in the pleural space.
heparin (HEP-ah-rin) Substance that prevents blood clotting; anticoagulant.
hepatic (heh-PAT-ik) Pertaining to the liver.
hepatitis (hep-ah-TI-tis) Inflammation of the liver.
heredity (he-RED-ih-te) Transmission of characteristics from parent to offspring by means of the genes.
hereditary (he-RED-ih-tar-e) Transmitted or transmissible through the genes; familial.
hernia (HER-ne-ah) Protrusion of an organ or tissue through the wall of the cavity in which it is normally enclosed.
heterozygous (het-er-o-ZI-gus) Having unmatched alleles for a given trait; hybrid.
hilum (HI-lum) Indented region of an organ where vessels and nerves enter or leave.
hippocampus (hip-o-KAM-pus) Sea horse–shaped region of the limbic system that functions in learning and formation of long-term memory.
histamine (HIS-tah-mene) Substance released from tissues during an inflammatory reaction; promotes redness, pain, and swelling.
histology (his-TOL-o-je) Study of tissues.
HIV See human immunodeficiency virus.
Hodgkin disease Chronic malignant disease of lymphoid tissue.
homeostasis (ho-me-o-STA-sis) State of balance within the body; maintenance of body conditions within set limits.
homozygous (ho-mo-ZI-gus) Having identical alleles in a given gene pair.
hormone Chemical messenger secreted by a tissue that has specific regulatory effects on certain other cells.
host An organism in or on which a parasite lives.
human chorionic gonadotropin (ko-re-ON-ik gon-ah-do-TRO-pin) **(hCG)** Hormone produced by embryonic cells soon after implantation that maintains the corpus luteum and is diagnostic of pregnancy.
human immunodeficiency virus (HIV) Virus that causes AIDS.
human placental lactogen (hPL) Hormone produced by the placenta that prepares the breasts for lactation and increases nutrient levels in maternal blood.
humoral (HU-mor-al) Pertaining to body fluids, such as immunity based on antibodies circulating in the blood.
Huntington disease Progressive degenerative disorder carried by a dominant allele.
hyaline (HI-ah-lin) Clear, glasslike; referring to a type of cartilage.
hydatidiform (hi-dah-TID-ih-form) **mole** Benign overgrowth of placental tissue.
hydrocephalus (hi-dro-SEF-ah-lus) Abnormal accumulation of CSF within the brain.
hydrolysis (hi-DROL-ih-sis) Splitting of large molecules by the addition of water, as in digestion.
hydrophilic (hi-dro-FIL-ik) Mixing with or dissolving in water, such as salts; literally "water loving."
hydrophobic (hi-dro-FO-bik) Repelling and not dissolving in water, such as fats; literally "water fearing."
hymen Fold of membrane near the opening of the vaginal canal.
hypercapnia (hi-per-KAP-ne-ah) Increased level of carbon dioxide in the blood.
hyperglycemia (hi-per-gli-SE-me-ah) Abnormal increase in the amount of glucose in the blood.
hyperkalemia (hi-per-kah-LE-me-ah) Excess potassium in body fluids.
hyperopia (hi-per-O-pe-ah) Farsightedness.
hyperpnea (hi-PERP-ne-ah) Increase in the depth and rate of breathing, as during exercise.
hypersensitivity (hi-per-SEN-sih-tiv-ih-te) Any deleterious immune response, including allergy, autoimmune disease, or transplantation rejection syndrome.
hypertension (hi-per-TEN-shun) High blood pressure.
hyperthyroidism (hi-per-THI-royd-ism) Overactivity of the thyroid gland with excessive hormone secretion.
hypertonic (hi-per-TON-ik) Describing a solution that is more concentrated than the fluids within a cell.
hypertrophy (hy-PER-tro-fe) Enlargement or overgrowth of an organ or part.
hyperventilation (hi-per-ven-tih-LA-shun) Increase in the rate and depth of breathing above optimum levels.
hypocapnia (hi-po-KAP-ne-ah) Decreased level of carbon dioxide in the blood.
hypochondriac (hi-po-KON-dre-ak) Pertaining to a region on either side of the abdomen just inferior to the ribs.
hypogastric (hi-po-GAS-trik) Pertaining to an area inferior to the stomach; most inferior midline region of the abdomen.
hypoglycemia (hi-po-gli-SE-me-ah) Abnormal decrease in the concentration of glucose in the blood.
hyponatremia (hi-po-nah-TRE-me-ah) Deficiency of sodium in body fluids.
hypophysis (hi-POF-ih-sis) Pituitary gland.
hypopnea (hi-POP-ne-ah) Decrease in the rate and depth of breathing.
hypospadias (hi-po-SPA-de-as) Congenital disorder in which the urethral opening is found on the undersurface of the penis or scrotum.
hypotension (hi-po-TEN-shun) Low blood pressure.
hypothalamus (hi-po-THAL-ah-mus) Region of the brain that controls the pituitary; control center for numerous homeostatic negative feedback loops and for the autonomic nervous system.
hypothermia (hi-po-THER-me-ah) Abnormally low body temperature.
hypotonic (hi-po-TON-ik) Describing a solution that is less concentrated than are the fluids within a cell.
hypoventilation (hi-po-ven-tih-LA-shun) Insufficient amount of air entering the alveoli.
hypoxemia (hi-pok-SE-me-ah) Lower than normal concentration of oxygen in arterial blood.
hypoxia (hi-POK-se-ah) Lower than normal level of oxygen in the tissues.
hysterectomy (his-ter-EK-to-me) Surgical removal of the uterus.

I

iatrogenic (i-at-ro-JEN-ik) Resulting from the adverse effects of treatment.
idiopathic (id-e-o-PATH-ik) Describing a disease without known cause.
ileum (IL-e-um) Most distal portion of the small intestine.
ileus (IL-e-us) Intestinal obstruction caused by lack of peristalsis or by muscle contraction.
iliac (IL-e-ak) Pertaining to the ilium, the upper portion of the hipbone; pertaining to the most inferior, lateral regions of the abdomen.

immune (ih-MUNE) **system** Complex of cellular and molecular components that provides defense against foreign cells and substances as well as abnormal body cells.
immunity (ih-MU-nih-te) Power of an individual to resist or overcome the effects of a disease or other harmful agent.
immunization (ih-mu-nih-ZA-shun) Use of a vaccine to produce immunity; vaccination.
immunodeficiency (im-u-no-de-FISH-en-se) Any failure of the immune system.
immunoglobulin (im-mu-no-GLOB-u-lin) **(Ig)** See antibody.
immunotherapy (im-mu-no-THER-a-pe) Stimulation of the immune system to fight disease, such as cancer; biological therapy.
impetigo (im-peh-TI-go) Acute, contagious staphylococcal or streptococcal skin infection.
implantation (im-plan-TA-shun) Embedding of a fertilized ovum into the uterine lining.
incidence (IN-sih-dense) Occurrence; in epidemiology, the number of new disease cases appearing in a particular population during a specific time period.
infarct (IN-farkt) Area of tissue damaged from lack of blood supply due to a vessel blockage.
infection (in-FEK-shun) Invasion by pathogens.
infectious mononucleosis (mon-o-nu-kle-O-sis) Acute viral infection associated with enlargement of the lymph nodes.
inferior (in-FE-re-or) Below or lower.
inferior vena cava (VE-nah KA-vah) Large vein that drains the lower body and empties into the heart's right atrium.
infertility (in-fer-TIL-ih-te) Decreased ability to reproduce.
inflammation (in-flah-MA-shun) Response of tissues to injury or infection; characterized by heat, redness, swelling, and pain.
influenza (in-flu-EN-zah) Acute contagious viral disease of the upper respiratory tract.
infundibulum (in-fun-DIB-u-lum) Stalk that connects the pituitary gland to the brain's hypothalamus.
ingestion (in-JES-chun) Intake of food.
inguinal (IN-gwih-nal) Pertaining to the groin region or the region of the inguinal canal.
inhalation (in-hah-LA-shun) Drawing of air into the lungs; inspiration.
insertion (in-SER-shun) Muscle attachment connected to a movable part.
insula (IN-su-lah) Lobe of the cerebral hemisphere located interiorly.
insulin (IN-su-lin) Hormone from the pancreatic islets that lowers blood glucose level and promotes tissue building.
integument (in-TEG-u-ment) Skin; adj., integumentary.
integumentary system The skin and all its associated structures.
intercalated (in-TER-cah-la-ted) **disk** A modified plasma membrane in cardiac tissue that allows rapid transfer of electric impulses between cells.
intercellular (in-ter-SEL-u-lar) Between cells.
intercostal (in-ter-KOS-tal) Between the ribs.
interferon (in-ter-FERE-on) **(IFN)** Group of substances released from virus-infected cells that prevent spread of infection to other cells; also nonspecifically boosts the immune system.
interleukin (in-ter-LU-kin) Substance released by a T cell, macrophage, or endothelial cell that regulates other immune system cells.
interneuron (in-ter-NU-ron) Nerve cell that transmits impulses within the central nervous system or enteric (gastrointestinal) nervous system.
interphase (IN-ter-faze) Stage in a cell's life between one mitosis and the next when the cell is not dividing.
interstitial (in-ter-STISH-al) Between; pertaining to an organ's spaces or structures between active tissues; also refers to the fluid between cells.
intestine (in-TES-tin) Organ of the digestive tract between the stomach and the anus, consisting of the small and large intestine.
intracellular (in-trah-SEL-u-lar) Within a cell.
intrinsic (in-TRIN-sik) **factor** Gastric secretion needed for absorption of vitamin B_{12} in the intestine.
intussusception (in-tuh-suh-SEP-shun) Slipping of a part of the intestine into a part below it.
inversion (in-VER-zhun) Turning inward, with reference to movement of the foot.
ion (I-on) Atom or molecule with an electrical charge; anion or cation.
ionic bond Chemical bond formed by the exchange of electrons between atoms.
iontophoresis (i-on-to-for-E-sis) Method of drug administration that uses a mild electric current to move ionic drugs through the skin.
iris (I-ris) Circular colored region of the eye around the pupil.
ischemia (is-KE-me-ah) Lack of blood supply to an area.
islets (I-lets) Groups of cells in the pancreas that produce hormones; islets of Langerhans (LAHNG-er-hanz).
isometric (i-so-MET-rik) **contraction** Muscle contraction in which there is no change in muscle length but an increase in muscle tension, as in pushing against an immovable force.
isotonic (i-so-TON-ik) Describing a solution that has the same concentration as the fluid within a cell.
isotonic contraction Muscle contraction in which the tone within the muscle remains the same but the muscle shortens to produce movement.
isotope (I-so-tope) Form of an element that has the same atomic number as another form of that element but a different atomic weight; isotopes differ in their numbers of neutrons.
isthmus (IS-mus) Narrow band, such as the band that connects the two lobes of the thyroid gland.

J

jaundice (JAWN-dis) Yellowish skin discoloration that is usually due to the presence of excess bilirubin in the blood.
jejunum (je-JU-num) Second portion of the small intestine.
joint Area of junction between two or more bones; articulation.
juxtaglomerular (juks-tah-glo-MER-u-lar) **(JG) apparatus** Structure in the kidney composed of cells of the afferent arteriole and distal tubule that increases the secretion of the enzyme renin in response to decreased blood pressure.
juxtamedullary nephron (juks-tah-MED-u-lar-e NEF-ron) Nephron with exceptionally long loop that dips deep into the renal medulla.

K

karyotype (KAR-e-o-tipe) Picture of the chromosomes arranged according to size and form.
keloid (KE-loyd) Mass or raised area that results from excess production of scar tissue.
keratin (KER-ah-tin) Protein that thickens and protects the skin; makes up hair and nails.
ketoacidosis (ke-to-as-ih-DO-sis) Acidosis that results from accumulation of ketone bodies in the blood.
ketone (KE-tone) Acidic organic compound formed metabolically from the incomplete oxidation of fats.
kidney (KID-ne) Organ of excretion, hormone synthesis, and blood pressure regulation.
kilocalorie (kil-o-KAL-o-re) **(kcal)** Measure of the energy content of food; technically, the amount of heat needed to raise 1 kg of water 1°C; calorie (C).
kinesthesia (kin-es-THE-ze-ah) Sense of body movement.
Klinefelter (KLINE-fel-ter) **syndrome** Genetic disorder involving extra sex chromosomes, usually an extra X chromosome.
Kupffer (KOOP-fer) **cells** Macrophages in the liver that help to fight infection.
Kussmaul (KOOS-mowl) **respiration** Deep, rapid respiration characteristic of acidosis (overly acidic body fluids) as seen in uncontrolled diabetes.
kwashiorkor (kwash-e-OR-kor) Severe protein and energy malnutrition seen in children after weaning.
kyphosis (ki-FO-sis) Exaggerated lumbar curve of the spine.

L

labium (LA-be-um) Lip; pl., labia (LA-be-ah).
labor (LA-bor) Parturition; childbirth.
labyrinth (LAB-ih-rinth) Inner ear, named for its complex shape; maze.
laceration (las-er-A-shun) Rough, jagged skin wound.
lacrimal (LAK-rih-mal) Referring to tears or the tear glands.
lacrimal (LAK-rih-mal) **apparatus** Lacrimal (tear) gland and its associated ducts.

lacrimal gland Gland above the eye that secretes tears.
lactase (LAK-tase) Enzyme that aids in the digestion of lactose.
lactation (lak-TA-shun) Secretion of milk.
lacteal (LAK-te-al) Lymphatic capillary that drains digested fats from the villi of the small intestine.
lactic (LAK-tik) **acid** Organic acid produced from pyruvate during rapid carbohydrate metabolism.
laryngopharynx (lah-rin-go-FAR-inks) Lowest portion of the pharynx, opening into the larynx and esophagus.
larynx (LAR-inks) Structure between the pharynx and trachea that contains the vocal cords; voice box.
laser (LA-zer) Device that produces a very intense light beam.
lateral (LAT-er-al) Farther from the midline; toward the side.
lens Biconvex structure of the eye that changes in thickness to accommodate near and far vision; crystalline lens.
leptin (LEP-tin) Hormone produced by adipocytes that aids in weight control by decreasing food intake and increasing energy expenditure.
lesion (LE-zhun) Wound or local injury.
leukemia (lu-KE-me-ah) Malignant blood disease characterized by abnormal development of white blood cells.
leukocyte (LU-ko-site) White blood cell.
leukocytosis (lu-ko-si-TO-sis) Increase in the number of white cells in the blood, as occurs during infection.
leukopenia (lu-ko-PE-ne-ah) Deficiency of leukocytes in the blood.
leukoplakia (lu-ko-PLA-ke-ah) Thickened white patches on the oral mucous membranes, often due to smoking.
LH See luteinizing hormone.
ligament (LIG-ah-ment) Band of connective tissue that connects a bone to another bone; thickened portion or fold of the peritoneum that supports an organ or attaches it to another organ.
ligand (LIG-and) Substance that binds to a receptor in the plasma membrane or within the cell.
limbic (LIM-bik) **system** Area between the brain's cerebrum and diencephalon that is involved in emotional states, memory, and behavior.
lingual (LING-gwal) Pertaining to the tongue.
lipase (LI-pase) Enzyme that aids in fat digestion.
lipid (LIP-id) Type of organic compound, one example of which is a fat.
liter (LE-ter) (**L**) Basic unit of volume in the metric system; 1,000 mL; 1.06 qt.
lithotripsy (LITH-o-trip-se) Use of external shock waves to shatter stones (calculi).
liver (LIV-er) Large organ inferior to the diaphragm in the superior right abdomen; has many functions, including bile secretion, detoxification, storage, and interconversion of nutrients.
lobe Subdivision of an organ, as of the cerebrum, liver, or lung.
loop of Henle (HEN-le) See nephron loop.
lordosis (lor-DO-sis) Exaggerated lumbar curve of the spine.
lumbar (LUM-bar) Pertaining to the region of the spine between the thoracic vertebrae and the sacrum.
lumen (LU-men) Central opening of an organ or vessel.
lung Organ of respiration.
lunula (LU-nu-la) Pale half-moon–shaped area at the proximal end of the nail.
lupus erythematosus (LU-pus er-ih-the-mah-TO-sis) Chronic inflammatory autoimmune disease that involves the skin and sometimes other organs.
luteinizing (LU-te-in-i-zing) **hormone (LH)** Hormone produced by the anterior pituitary that induces ovulation and formation of the corpus luteum in females; in males, it stimulates cells in the testes to produce testosterone.
lymph (limf) Fluid in the lymphatic system.
lymph node Mass of lymphoid tissue along the path of a lymphatic vessel that filters lymph and harbors white blood cells active in immunity.
lymphadenitis (lim-fad-en-I-tis) Inflammation of lymph nodes.
lymphadenopathy (lim-fad-en-OP-ah-the) Any disorder of lymph nodes.
lymphangitis (lim-fan-JI-tis) Inflammation of lymphatic vessels.
lymphatic duct (lim-FAH-tic) One of two large vessels draining lymph from the lymphatic system into the venous system.
lymphatic system System consisting of the lymphatic vessels and lymphoid tissue; involved in immunity, digestion, and fluid balance.
lymphedema (lim-feh-DE-mah) Edema due to obstruction of lymph flow.
lymphocyte (LIM-fo-site) Agranular white blood cell that functions in acquired immunity.
lymphoma (lim-FO-mah) Any tumor, benign or malignant, that occurs in lymphoid tissue.
lysis (LI-sis) Loosening, dissolving, or separating; a gradual decline, as of a fever.
lysosome (LI-so-some) Cell organelle that contains digestive enzymes.

M

macrophage (MAK-ro-faj) Large phagocytic cell that develops from a monocyte; presents antigen to other leukocytes in immune response.
macula (MAK-u-lah) Spot; flat, discolored spot on the skin, such as a freckle or measles. Area of the retina that contains the point of sharpest vision; equilibrium receptor in the vestibule of the inner ear. Also macule.
macula lutea (MAK-u-lah LU-te-ah) Area of the retina that contains the fovea centralis, the point of sharpest vision.
magnetic resonance imaging (MRI) Method for studying tissue based on nuclear movement after exposure to radio waves in a powerful magnetic field.
major histocompatibility complex (MHC) Group of genes that codes for specific proteins (antigens) on cellular surfaces; these antigens are important in cross-matching for tissue transplantation; they are also important in immune reactions.
malignant (mah-LIG-nant) Describing a tumor that spreads; describing a disorder that tends to become worse and cause death.
malnutrition (mal-nu-TRISH-un) State resulting from lack of food, lack of an essential dietary component, or faulty use of food in the diet.
MALT Mucosal-associated lymphoid tissue; tissue in the mucous membranes that helps fight infection.
maltase (MAL-tase) Enzyme that aids in the digestion of maltose.
mammary (MAM-er-e) **gland** Milk-secreting portion of the breast.
mammogram (MAM-o-gram) Radiographic study of the breast.
maple syrup urine disease Recessive hereditary disease that affects amino acid metabolism and among other symptoms, produces the smell of maple syrup in urine and on the body.
marasmus (mah-RAZ-mus) Severe malnutrition in infants.
Marfan syndrome Dominant hereditary disorder that affects connective tissue.
mast cell White blood cell related to a basophil that is present in tissues; active in inflammatory and allergic reactions.
mastectomy (mas-TEK-to-me) Removal of the breast; mammectomy.
mastication (mas-tih-KA-shun) Act of chewing.
mastitis (mas-TI-tis) Inflammation of the breast.
matrix (MA-triks) The acellular background material in a tissue; the intercellular material.
meatus (me-A-tus) Short channel or passageway, such as the external opening of a canal or a channel in bone.
medial (ME-de-al) Nearer the midline of the body.
mediastinum (me-de-as-TI-num) Region between the lungs and the organs and vessels it contains.
medulla (meh-DUL-lah) Inner region of an organ; marrow.
medullary (MED-u-lar-e) **cavity** Channel at the center of a long bone that contains yellow bone marrow.
medulla oblongata (ob-long-GAH-tah) Part of the brain stem that connects the brain to the spinal cord.
megakaryocyte (meg-ah-KAR-e-o-site) Very large bone marrow cell that gives rise to blood platelets.
meibomian (mi-BO-me-an) **gland** Gland that produces a secretion that lubricates the eyelashes.
meiosis (mi-O-sis) Process of cell division that halves the chromosome number in the formation of the gametes.
melanin (MEL-ah-nin) Dark pigment found in the skin, hair, parts of the eye, and certain parts of the brain.
melanocyte (MEL-ah-no-site) Cell that produces melanin.
melanoma (mel-ah-NO-mah) Malignant tumor of melanocytes.
melatonin (mel-ah-TO-nin) Hormone produced by the pineal gland.
membrane (MEM-brane) Thin sheet of tissue; lipid bilayer surrounding a cell or an organelle.

membrane attack complex (MAC) Channel in a pathogen's membrane caused by the action of complement and aiding in destruction of the cell.
membrane potential (po-TEN-shal) Difference in electric charge on either side of a plasma membrane; transmembrane potential.
membranous labyrinth (LAB-ih-rinth) The inner membranous portion of the inner ear that is filled with endolymph.
Mendelian (men-DE-le-en) **laws** Principles of heredity discovered by an Austrian monk named Gregor Mendel.
meninges (men-IN-jeze) Three layers of fibrous membranes that cover the brain and spinal cord.
meningitis (men-in-JI-tis) Inflammation of the meninges.
menopause (MEN-o-pawz) Time during which menstruation ceases.
menses (MEN-seze) Monthly flow of blood from the female reproductive tract.
menstruation (men-stru-A-shun) Period of menstrual flow.
mesentery (MES-en-ter-e) Connective tissue membrane that attaches the small intestine to the dorsal abdominal wall.
mesocolon (mes-o-KO-lon) Connective tissue membrane that attaches the colon to the dorsal abdominal wall.
mesothelium (mes-o-THE-le-um) Epithelial tissue found in serous membranes.
metabolic rate Rate at which energy is released from nutrients in the cells.
metabolic syndrome Condition related to type 2 diabetes mellitus with insulin resistance, obesity, hyperglycemia, high blood pressure, and metabolic disturbances; also called syndrome X.
metabolism (meh-TAB-o-lizm) All the physical and chemical processes by which an organism is maintained.
metaphase (MET-ah-faze) Second stage of mitosis, during which the chromosomes line up across the equator of the cell.
metastasis (meh-TAS-tah-sis) Spread of tumor cells; pl., metastases (meh-TAS-tah-seze).
meter (ME-ter) **(m)** Basic unit of length in the metric system; 1.1 yards.
MHC See major histocompatibility complex.
microbiology (mi-kro-bi-OL-o-je) Study of microscopic organisms.
microglia (mi-KROG-le-ah) Glial cells that act as phagocytes in the CNS.
micrometer (MI-kro-me-ter) **(mcm)** 1/1,000th of a millimeter; an instrument for measuring through a microscope (pronounced mi-KROM-eh-ter).
microorganism (mi-kro-OR-gan-izm) Microscopic organism.
microscope (MI-kro-skope) Magnifying instrument used to examine cells and other structures not visible with the naked eye; examples are the compound light microscope, transmission electron microscope (TEM), and scanning electron microscope (SEM).
microvilli (mi-kro-VIL-li) Small projections of the plasma membrane that increase surface area; sing., microvillus.
micturition (mik-tu-RISH-un) Act of urination; voiding of the urinary bladder.
midbrain Upper portion of the brain stem.
milliosmole (mil-e-OZ-mole) One thousandth of an osmole, a measure of osmotic concentration.
mineral (MIN-er-al) Inorganic substance; in the diet, an element needed in small amounts for health.
mineralocorticoid (min-er-al-o-KOR-tih-koyd) Steroid hormone from the adrenal cortex that regulates electrolyte balance, e.g., aldosterone.
miscarriage Loss of an embryo or fetus; spontaneous abortion.
mitochondria (mi-to-KON-dre-ah) Cellular organelles that manufacture ATP with the energy released from the oxidation of nutrients; sing., mitochondrion.
mitosis (mi-TO-sis) Type of cell division that produces two daughter cells exactly like the parent cell.
mitral (MI-tral) **valve** Valve between the heart's left atrium and left ventricle; left AV valve; bicuspid valve.
mixture Blend of two or more substances.
mold Filamentous, multicellular fungus.
molecule (MOL-eh-kule) Particle formed by covalent bonding of two or more atoms; smallest subunit of a compound.
monocyte (MON-o-site) Phagocytic agranular white blood cell that differentiates into a macrophage.
monomer (MON-o-mer) Building block or single unit of a larger molecule.
monosaccharide (mon-o-SAK-ah-ride) Simple sugar; basic unit of carbohydrates.
morbidity (mor-BID-ih-te) **rate** Incidence rate or prevalence rate of a disease within a population.
mortality (mor-TAL-ih-te) **rate** Percentage of a population that dies from a given disease within a given time period.
morula (MOR-u-lah) An early stage in embryonic development; a ball of identical cells formed by cellular division from a zygote.
motor (MO-tor) Describing structures or activities involved in transmitting impulses away from the central nervous system; efferent.
motor end plate Region of a muscle cell membrane that receives nervous stimulation.
motor unit Group consisting of a single neuron and all the muscle fibers it stimulates.
mouth Proximal opening of the digestive tract where food is ingested, chewed, mixed with saliva, and swallowed.
MRI See magnetic resonance imaging.
mucosa (mu-KO-sah) Epithelial membrane that produces mucus; mucous membrane.
mucus (MU-kus) Thick protective fluid secreted by mucous membranes and glands; adj., mucous.
multiple sclerosis (skle-RO-sis) Disease that affects the myelin sheath around axons leading to neuron degeneration.
murmur Abnormal heart sound.
muscle (MUS-l) Tissue that contracts to produce movement or tension; includes skeletal, smooth, and cardiac types; adj., muscular.
muscular dystrophy (DIS-tro-fe) Any of a group of disorders in which there is deterioration of muscles that still have intact nerve function.
muscular (MUS-ku-lar) **system** The system of skeletal muscles that moves the skeleton, supports and protects the organs, and maintains posture.
mutagen (MU-tah-jen) Agent that causes mutation; adj., mutagenic (mu-tah-JEN-ik).
mutation (mu-TA-shun) Change in a gene or a chromosome.
myalgia (mi-AL-je-ah) Muscular pain.
mycology (mi-KOL-o-je) Study of fungi (yeasts and molds).
myelin (MI-el-in) Fatty material that covers and insulates the axons of some neurons.
myocardium (mi-o-KAR-de-um) Middle layer of the heart wall; heart muscle.
myoglobin (MI-o-glo-bin) Compound that stores oxygen in muscle cells.
myoma (mi-O-mah) Usually benign tumor of the uterus; fibroma; fibroid.
myometrium (mi-o-ME-tre-um) Muscular layer of the uterus.
myopia (mi-O-pe-ah) Nearsightedness.
myosin (MI-o-sin) One of the two contractile proteins in muscle cells, the other being actin.
myositis (mi-o-SI-tis) Inflammation of muscle tissue.
myringotomy (mir-in-GOT-o-me) Incision into the tympanic membrane.

N

narcotic (nar-KOT-ik) Drug that acts on the CNS to alter perception of and response to pain.
nasopharynx (na-zo-FAR-inks) Upper portion of the pharynx located posterior to the nasal cavity.
natural killer (NK) cell Type of lymphocyte that can nonspecifically destroy abnormal cells.
naturopathy (na-chur-OP-a-the) Philosophy of helping people to heal themselves by developing healthy lifestyles.
nausea (NAW-ze-ah) Unpleasant sensation in the upper GI tract that may precede vomiting.
necrosis (neh-KRO-sis) Tissue death.
negative feedback Self-regulating system in which the result of an action reverses that action; a method for keeping body conditions within a normal range and maintaining homeostasis.
neoplasm (NE-o-plazm) Abnormal growth of cells; tumor; adj., neoplastic.
nephron (NEF-ron) Microscopic functional unit of the kidney; consists of the glomerulus and the renal tubule.
nephron loop Hairpin-shaped segment of the renal tubule between the proximal and distal tubules; loop of Henle.

nerve Bundle of neuron fibers outside the central nervous system.
nerve impulse Electric charge that spreads along the membrane of a neuron; action potential.
nervous system (NER-vus) The system that transports information in the body by means of electric impulses and neurotransmitters.
neuralgia (nu-RAL-je-ah) Pain in a nerve.
neurilemma (nu-rih-LEM-mah) Thin sheath that covers certain peripheral axons; aids in axon regeneration.
neuritis (nu-RI-tis) Inflammation of a nerve or nerves.
neuroglia (nu-ROG-le-ah) Supporting and protective cells of the nervous system; glial cells.
neuroma (nu-RO-mah) Tumor that arises from a nerve.
neuromuscular junction Point at which a neuron's axon contacts a muscle cell.
neuron (NU-ron) Conducting cell of the nervous system.
neuropathy (nu-ROP-ah-the) Any disease of a nerve or nerves.
neurotransmitter (nu-ro-TRANS-mit-er) Chemical released from the ending of an axon that enables a nerve impulse to cross a chemical synapse.
neutron (NU-tron) Noncharged particle in an atom's nucleus.
neutrophil (NU-tro-fil) Phagocytic granular white blood cell; polymorph; poly; PMN; seg.
nevus (NE-vus) Mole or birthmark.
nitrogen (NI-tro-jen) Chemical element found in all proteins.
node Small mass of tissue, such as a lymph node; space between cells in the myelin sheath.
nodule (NOD-ule) Large firm papule or raised area on the surface of the skin.
norepinephrine (nor-epi-ih-NEF-rin) Neurotransmitter similar in composition and action to the hormone epinephrine; noradrenaline.
normal flora (FLOR-ah) Population of microorganisms that normally grows on body surfaces or cavities open to the environment.
normal saline Isotonic or physiologic salt solution.
nosocomial (nos-o-KO-me-al) Acquired in a hospital, such as an infection.
nucleic acid (nu-KLE-ik) Complex organic substance composed of nucleotides; DNA and RNA.
nucleolus (nu-KLE-o-lus) Small unit within the nucleus that assembles ribosomes.
nucleotide (NU-kle-o-tide) Building block of DNA and RNA; one is also a component of ATP.
nucleus (NU-kle-us) Largest cellular organelle, containing the DNA, which directs all cell activities; group of neurons in the central nervous system; in chemistry, the central part of an atom.

O

obstetrics (ob-STET-riks) Branch of medicine that is concerned with the care of women during pregnancy, childbirth, and the six weeks after childbirth.
obstipation (ob-stih-PA-shun) Extreme constipation.
occlusion (ok-LU-zhun) Closing, as of a vessel.
olfaction (ol-FAK-shun) Sense of smell; adj., olfactory.
oligodendrocyte (ol-ih-go-DEN-dro-site) Type of neuroglial cell that forms the myelin sheath in the CNS.
oligospermia (ol-ih-go-SPER-me-ah) Low sperm count.
omentum (o-MEN-tum) Portion of the peritoneum; greater omentum extends over the anterior abdomen; lesser omentum extends between the stomach and liver.
oncology (on-KOL-o-je) Study of tumors.
ophthalmic (of-THAL-mik) Pertaining to the eye.
ophthalmology (of-thal-MOL-o-je) Study of the eye and diseases of the eye.
ophthalmoscope (of-THAL-mo-skope) Instrument for examining the posterior (fundus) of the eye.
opportunistic (op-por-tu-NIS-tik) Describing an infection that takes hold because a host has been compromised (weakened) by disease.
orchitis (or-KI-tis) Inflammation of the testis.
organ (OR-gan) Body part containing two or more tissues functioning together for specific purposes.
organ of Corti (KOR-te) See spiral organ.
organelle (or-gan-EL) Specialized subdivision within a cell.
organic (or-GAN-ik) Referring to the typically large and complex carbon compounds found in living things; contain hydrogen and usually oxygen as well as other elements.
organism (OR-gan-izm) Any organized living thing, such as a plant, animal, or microorganism.
origin (OR-ih-jin) Source; beginning; muscle attachment connected to a nonmoving part.
oropharynx (o-ro-FAR-inks) Middle portion of the pharynx, located behind the mouth.
orthopnea (or-THOP-ne-ah) Difficulty in breathing that is relieved by sitting in an upright position.
osmolarity (os-mo-LAR-ih-te) Term that refers to the solute concentration of a solution; osmotic concentration.
osmosis (os-MO-sis) Passage of water through a semipermeable membrane from the region of lower solute concentration to the region of higher solute concentration.
osmotic (os-MOT-ik) **pressure** Tendency of a solution to draw water into it; directly related to a solution's concentration.
osseus (OS-e-us) Pertaining to bone tissue.
ossicle (OS-ih-kl) One of three small bones of the middle ear: malleus, incus, or stapes.
ossification (os-ih-fih-KA-shun) Process of bone formation.
osteoarthritis (os-te-o-arth-RI-tis) (OA) Degenerative joint disease (DJD). Arthritis that usually results from normal wear and tear.
osteoblast (OS-te-o-blast) Bone-forming cell.
osteoclast (OS-te-o-clast) Cell that breaks down bone.
osteocyte (OS-te-o-site) Mature bone cell; maintains bone but does not produce new bone tissue.
osteon (OS-te-on) Subunit of compact bone, consisting of concentric rings of bone tissue around a central channel; haversian system.
osteopenia (os-te-o-PE-ne-ah) Reduction in bone density to below average levels.
osteoporosis (os-te-o-po-RO-sis) Abnormal loss of bone tissue with tendency to fracture.
otolithic (o-to-LITH-ik) **membrane** Gelatinous material covering the tips of equilibrium receptor cells in the vestibule of the inner ear.
otoliths (O-to-liths) Crystals that add weight to the otolithic membrane of equilibrium receptors in the vestibule of the inner ear.
oval window Area of the inner ear where sound waves are transmitted from the footplate of the stapes to the fluids in the spiral organ.
ovarian follicle (o-VA-re-an FOL-ih-kl) Cluster of cells containing an ovum. A follicle can mature during a menstrual cycle and release its ovum.
ovary (O-vah-re) Female reproductive organ; produces ova and female sex steroids.
oviduct (O-vih-dukt) See uterine tube.
ovulation (ov-u-LA-shun) Release of an ovum from a mature ovarian follicle (graafian follicle).
ovum (O-vum) Female reproductive cell or gamete; pl., ova.
oxidation (ok-sih-DA-shun) Chemical breakdown of nutrients for energy usually using oxygen.
oxygen (OK-sih-jen) (O_2) Gas needed to break down nutrients completely for energy within the cell.
oxytocin (ok-se-TO-sin) Hormone from the posterior pituitary that causes uterine contraction and milk ejection ("letdown") from the breasts.

P

pacemaker Group of cells or artificial device that sets activity rate; in the heart, the sinoatrial (SA) node that normally initiates contractions.
palate (PAL-at) Roof of the oral cavity; anterior portion is hard palate, posterior portion is soft palate.
pallor (PAL-or) Paleness of the skin.
pancreas (PAN-kre-as) Large, elongated gland behind the stomach; produces digestive enzymes and hormones (e.g., insulin, glucagon).
pandemic (pan-DEM-ik) Disease that is prevalent throughout an entire country, continent, or the world.
Papanicolaou (pap-ah-nik-o-LAH-o) **test** Histologic test for cervical cancer; Pap test or smear.
papilla (pah-PIL-ah) Small nipple-like projection or elevation.
papillary muscles (PAP-ih-lar-e) Columnar muscles in the heart's ventricular walls that anchor and pull on the chordae tendineae to prevent the valve flaps from everting when the ventricles contract.
papule (PAP-ule) Firm, raised lesion of the skin.

paracentesis (par-eh-sen-TE-sis) Puncture of the abdominal cavity, usually to remove a fluid accumulation, such as ascites; abdominocentesis.

parasite (PAR-ah-site) Organism that lives on or within another (the host) at the other's expense.

parasympathetic nervous system Craniosacral division of the autonomic nervous system; generally reverses the fight or flight (stress) response.

parathyroid (par-ah-THI-royd) **gland** Any of four to six small glands embedded in the capsule enclosing the thyroid gland; produces parathyroid hormone, which raises blood calcium level by causing calcium release from bones and calcium retention in the kidney.

parietal (pah-RI-eh-tal) Pertaining to the wall of a space or cavity.

Parkinson disease Progressive neurologic condition characterized by tremors, rigidity of limbs and joints, slow movement, and impaired balance.

parotid (pah-ROT-id) **gland** Salivary gland located inferior and anterior to the ear.

partial pressure Pressure of an individual gas within a mixture.

parturition (par-tu-RISH-un) Childbirth; labor.

pathogen (PATH-o-jen) Disease-causing organism; adj., pathogenic (path-o-JEN-ik).

pathology (pah-THOL-o-je) Study of disease.

pathophysiology (path-o-fiz-e-OL-o-je) Study of the physiologic basis of disease.

pedigree (PED-ih-gre) **chart** Family history; used in the study of heredity; family tree.

pelvic inflammatory disease (PID) Ascending infection that involves the pelvic organs; common causes are gonorrhea and chlamydia.

pelvis (PEL-vis) Basin-like structure, such as the lower portion of the abdomen or the upper flared portion of the ureter (renal pelvis).

pemphigus (PEM-fih-gus) An autoimmune skin disease with skin blistering.

penis (PE-nis) Male organ of urination and sexual intercourse.

peptic ulcer (PEP-tik UL-ser) Ulcer in the esophagus, stomach or duodenum.

perforating canal Channel across a long bone that contains blood vessels and nerves; Volkmann canal.

pericarditis (per-i-kar-DI-tis) Inflammation of the pericardium.

pericardium (per-ih-KAR-de-um) Fibrous sac lined with serous membrane that encloses the heart.

perichondrium (per-ih-KON-dre-um) Layer of connective tissue that covers cartilage.

perilymph (PER-e-limf) Fluid that fills the inner ear's bony labyrinth.

perimysium (per-ih-MIS-e-um) Connective tissue around a fascicle of muscle tissue.

perineum (per-ih-NE-um) Pelvic floor; external region between the anus and genital organs.

periosteum (per-e-OS-te-um) Connective tissue membrane covering a bone.

peripheral (peh-RIF-er-al) Located away from a center or central structure.

peripheral nervous system (PNS) All the nerves and nervous tissue outside the central nervous system.

peripheral neuropathy (peh-RIF-er-al nu-ROP-ah-the) Disorder involving damage to the nerves of the peripheral nervous system.

peristalsis (per-ih-STAL-sis) Wavelike movements in the wall of an organ or duct that propel its contents forward.

peritoneum (per-ih-to-NE-um) Serous membrane that lines the abdominal cavity and forms the outer layer of the abdominal organs; forms supporting ligaments for some organs.

peritonitis (per-ih-to-NI-tis) Inflammation of the peritoneum.

pernicious (per-NISH-us) **anemia** Anemia caused by lack of intrinsic factor causing inability to absorb vitamin B_{12} from the intestine.

peroxisome (per-OK-sih-some) Cell organelle that enzymatically destroys harmful substances produced in metabolism.

Peyer (PI-er) **patches** Clusters of lymphatic nodules in the mucous membranes lining the distal small intestine.

pH Symbol indicating hydrogen ion (H^+) concentration; lower numbers indicate a higher H^+ concentration and higher acidity.

phagocyte (FAG-o-site) Cell capable of engulfing large particles, such as foreign matter or cellular debris, through the plasma membrane.

phagocytosis (fag-o-si-TO-sis) Engulfing of large particles through the plasma membrane.

pharynx (FAR-inks) Throat; passageway between the mouth and esophagus.

phenotype (FE-no-tipe) All the characteristics of an organism that can be seen or tested for.

phenylketonuria (fen-il-ke-to-NU-re-ah) **(PKU)** Hereditary metabolic disorder involving inability to metabolize the amino acid phenylalanine.

phimosis (fi-MO-sis) Tightness of the foreskin.

phlebitis (fleh-BI-tis) Inflammation of a vein.

phospholipid (fos-fo-LIP-id) Complex lipid containing phosphorus; major component of the plasma membrane.

phrenic (FREN-ik) Pertaining to the diaphragm.

phrenic nerve Nerve that activates the diaphragm.

physiology (fiz-e-OL-o-je) Study of the function of living organisms.

pia mater (PI-ah MA-ter) Innermost layer of the meninges.

PID See pelvic inflammatory disease.

pineal (PIN-e-al) **gland** Gland in the brain that is regulated by light; involved in sleep–wake cycles.

pinna (PIN-nah) Outer projecting portion of the ear; auricle.

pinocytosis (pi-no-si-TO-sis) Intake of small particles and droplets by a cell's plasma membrane.

pituitary (pih-TU-ih-tar-e) **gland** Endocrine gland located under and controlled by the hypothalamus; releases hormones that control other glands; hypophysis.

placenta (plah-SEN-tah) Structure that nourishes and maintains the developing fetus during pregnancy.

plantar flexion (PLAN-tar FLEK-shun) Bending the foot so that the toes point downward.

plaque (PLAK) Patch or flat area; fatty material that deposits in vessel linings in atherosclerosis.

plasma (PLAZ-mah) Liquid portion of the blood.

plasma cell Cell derived from a B cell that produces antibodies.

plasma membrane Outer covering of a cell; regulates what enters and leaves the cell; cell membrane.

plasmapheresis (plas-mah-fer-E-sis) Separation and removal of plasma from donated blood and return of the formed elements to the donor.

platelet (PLATE-let) Cell fragment that forms a plug to stop bleeding and acts in blood clotting; thrombocyte.

pleura (PLU-rah) Serous membrane that lines the chest cavity and covers the lungs.

pleurisy (PLUR-ih-se) Inflammation of the pleura; pleuritis.

plexus (PLEK-sus) Network of vessels or nerves.

pneumonia (nu-MO-ne-ah) Inflammation of the lungs, commonly due to infection.

pneumothorax (nu-mo-THO-raks) Accumulation of air in the pleural space.

PNS See peripheral nervous system.

poliomyelitis (po-le-o-mi-eh-LI-tis) Viral disease of the nervous system that occurs most commonly in children but has been largely eliminated by widespread vaccination; polio.

polycythemia (pol-e-si-THE-me-ah) Increase in the proportion of red cells in the blood.

polydipsia (pol-e-DIP-se-ah) Excessive thirst.

polymorph (POL-e-morf) Term for a neutrophil; polymorphonuclear neutrophil.

polyneuropathy (pol-e-nu-ROP-ah-the) Form of peripheral neuropathy that involves multiple nerves and mainly affects the feet and legs.

polyp (POL-ip) Protruding growth, often grapelike, from a mucous membrane.

polysaccharide (pol-e-SAK-ah-ride) Compound formed from many simple sugars linked together (e.g., starch, glycogen).

pons (ponz) Area of the brain between the midbrain and medulla; connects the cerebellum with the rest of the central nervous system.

portal system Venous system that carries blood to a second capillary bed through which it circulates before returning to the heart.

positive feedback Control system in which an action or the product of an action maintains or intensifies that action. The action stops when materials are depleted, the stimulus is removed, or an outside force interrupts the action.

positron emission tomography (to-MOG-rah-fe) (PET) Imaging method that uses a radioactive substance to show activity in an organ.

posterior (pos-TE-re-or) Toward the back; dorsal.
postsynaptic (post-sin-AP-tik) Distal to the synaptic cleft.
potential (po-TEN-shal) Electric charge, as on the plasma membrane of a neuron or other cell; potential difference, membrane potential, or transmembrane potential.
precipitation (pre-sip-ih-TA-shun) Settling out of a solid previously held in solution or suspension in a liquid; in immunity, clumping of small particles as a result of an antigen–antibody reaction; seen as a cloudiness.
preeclampsia (pre-eh-KLAMP-se-ah) See pregnancy-induced hypertension.
prefix A word part that comes before a root and modifies its meaning.
pregnancy (PREG-nan-se) Period during which an embryo or fetus is developing in the body.
pregnancy-induced hypertension (PIH) Hypertension, proteinuria, and edema, which may be associated with a mother's immune response to fetal antigens in the placenta; if untreated, may lead to eclampsia; preeclampsia, toxemia of pregnancy.
prepuce (PRE-puse) Loose fold of skin that covers the glans penis; foreskin.
presbycusis (pres-be-KU-sis) Slowly progressive hearing loss that often accompanies aging.
presbyopia (pres-be-O-pe-ah) Loss of visual accommodation that occurs with age, leading to farsightedness.
presynaptic (pre-sin-AP-tik) Proximal to the synaptic cleft.
preterm (PRE-term) Referring to an infant born before the 37th week of gestation.
prevalence (PREV-ah-lens) **rate** In epidemiology, the overall frequency of a disease in a given group.
prime mover The main muscle that produces a given movement.
prion (PRI-on) Infectious protein particle that causes progressive neurodegenerative disease.
PRL See prolactin.
progeny (PROJ-eh-ne) Offspring, descendent.
progesterone (pro-JES-ter-one) Hormone produced by the corpus luteum and placenta; maintains the uterine lining for pregnancy.
prognosis (prog-NO-sis) Prediction of the probable outcome of a disease based on the condition of the patient and knowledge about the disease.
prolactin (pro-LAK-tin) Hormone from the anterior pituitary that stimulates milk production in the mammary glands; PRL.
pronation (pro-NA-shun) Turning the palm down or backward.
prone Face down or palm down.
prophase (PRO-faze) First stage of mitosis, during which the chromosomes become visible and the organelles disappear.
prophylaxis (pro-fih-LAK-sis) Prevention of disease.
proprioceptor (pro-pre-o-SEP-tor) Sensory receptor that aids in judging body position and changes in position; located in muscles, tendons, and joints.
prostaglandin (pros-tah-GLAN-din) Any of a group of hormones produced by many cells that usually act on neighboring cells; these hormones are involved in pain and inflammation, as well as many other functions.
prostate (PROS-tate) **gland** Gland that surrounds the urethra below the bladder and contributes secretions to the semen.
protein (PRO-tene) Organic compound made of amino acids; found as structural materials and metabolically active compounds, such as enzymes, some hormones, pigments, antibodies, and others.
proteome (PRO-te-ome) All the proteins that can be expressed in a cell.
prothrombin (pro-THROM-bin) Clotting factor; converted to thrombin during blood clotting.
prothrombinase (pro-THROM-bih-nase) Blood clotting factor that converts prothrombin to thrombin.
proton (PRO-ton) Positively charged particle in an atom's nucleus.
protozoology (pro-to-zo-OL-o-je) Study of protozoa, single-celled animals.
protozoon (pro-to-ZO-on) Animal-like microorganism; pl., protozoa.
proximal (PROK-sih-mal) Nearer to point of origin or to a reference point.
pruritus (pru-RI-tis) Intense skin itching.
psoriasis (so-RI-ah-sis) Chronic skin disease with red, flat areas covered with silvery scales.
ptosis (TO-sis) Dropping down of a part.
puerperal (pu-ER-per-al) Related to childbirth.
pulmonary circuit Pathway that carries blood from the heart to the lungs to pick up oxygen and release carbon dioxide and then returns the blood to the heart.
pulse Wave of increased pressure in the vessels produced by heart contraction.
pulse pressure Difference between systolic and diastolic pressures.
pupil (PU-pil) Opening in the center of the eye through which light enters.
Purkinje (pur-KIN-je) **fibers** Part of the heart's conduction system located in the ventricles.
pus Mixture of bacteria and leukocytes formed in response to infection.
pustule (PUS-tule) Vesicle filled with pus.
pyelonephritis (pi-eh-lo-neh-FRI-tis) Inflammation of the kidney, calyces, and renal pelvis, often due to bacterial infection.
pyloric (pi-LOR-ik) Pertaining to the pylorus, the distal region of the stomach.
pylorus (pi-LOR-us) Distal region of the stomach that leads to the pyloric sphincter.
pyrogen (PI-ro-jen) Substance that produces fever.
pyruvic (pi-RU-vik) **acid** Intermediate product in the breakdown of glucose for energy.
pyuria (pi-U-re-ah) Presence of pus in the urine.

R

radiation (ra-de-A-shun) Emission of rays, as of light, radio, or heat waves, ultraviolet, or x-rays. Method of heat loss by heat waves traveling from their source.
radioactive (ra-de-o-AK-tive) Pertaining to isotopes that fall apart easily, giving off radiation.
radioactivity (ra-de-o-ak-TIV-ih-te) Emission of atomic particles from an element.
radiography (ra-de-og-rah-fe) Production of an image by passage of x-rays through the body onto sensitized film; record produced is a radiograph.
radiotracer Radioactive substance administered internally for purpose of diagnosis.
rash Surface skin lesion.
reabsorption (re-ab-SORP-shun) Absorbing or taking up again.
receptor (re-SEP-tor) Specialized cell or ending of a sensory neuron that can be excited by a stimulus. Protein in the plasma membrane or other part of a cell that binds a chemical signal (e.g., hormone, neurotransmitter) resulting in a change in cellular activity.
recessive (re-SES-iv) Referring to an allele that is not expressed in the phenotype if a dominant allele for the same trait is present.
rectum (REK-tum) Distal region of the large intestine between the sigmoid colon and the anal canal.
reflex (RE-fleks) Simple, rapid, automatic response to a specific stimulus.
reflex arc (ark) Pathway through the nervous system from stimulus to response; commonly involves a sensory receptor, sensory neuron, central neuron(s), motor neuron, and effector.
refraction (re-FRAK-shun) Bending of light rays as they pass from one medium to another of a different density.
relaxin (re-LAKS-in) Placental hormone that softens the cervix and relaxes the pelvic joints.
renal tubule Coiled and looped portion of a nephron between the glomerular capsule and the collecting duct.
renin (RE-nin) Enzyme released from the kidney's juxtaglomerular apparatus that indirectly increases blood pressure by activating angiotensin.
repolarization (re-po-lar-ih-ZA-shun) A change in the membrane potential that brings it closer to the resting value.
resorption (re-SORP-shun) Loss of substance from a solid tissue, such as bone or a tooth, and return of the components to the blood.
respiration (res-pih-RA-shun) Process by which oxygen is obtained from the environment and delivered to the cells as carbon dioxide is removed from the tissues and released to the environment.
respiratory system System consisting of the lungs and breathing passages involved in exchange of oxygen and carbon dioxide between the outside air and the blood.
reticular (reh-TIK-u-lar) **formation** Network in the limbic system that governs wakefulness and sleep.
reticulocyte (reh-TIK-u-lo-site) Immature form of erythrocyte.
retina (RET-ih-nah) Innermost layer of the eye; contains light-sensitive cells (rods and cones).

retroperitoneal (ret-ro-per-ih-to-NE-al) Behind the peritoneum, as are the kidneys, pancreas, and abdominal aorta.
Rh factor Red cell antigen; D antigen.
rheumatoid (RU-mah-toyd) **arthritis** Disease of connective tissue that affects the joints.
rhodopsin (ro-DOP-sin) Light-sensitive pigment in the rods of the eye.
rib One of the slender curved bones that make up most of the thorax; costa; adj., costal.
ribonucleic (RI-bo-nu-kle-ik) **acid (RNA)** Substance needed for protein manufacture in the cell.
ribosome (RI-bo-some) Small body in the cell's cytoplasm that is a site of protein manufacture.
rickets (RIK-ets) Softening of bone (osteomalacia) in children, usually caused by a deficiency of vitamin D.
rickettsia (rih-KET-se-ah) Extremely small oval- to rod-shaped bacterium that can grow only within a living cell.
RNA See ribonucleic acid.
rod Receptor cell in the retina of the eye; used for vision in dim light.
roentgenogram (rent-GEN-o-gram) Image produced by means of x-rays; radiograph.
root (rute) a basic part of a word; an attached or embedded part, such as a nail or hair root. The branched portion of a spinal nerve that joins with the spinal cord. See also word root.
rotation (ro-TA-shun) Twisting or turning of a bone on its own axis.
rugae (RU-je) Folds in the lining of an organ, such as the stomach or urinary bladder; sing., ruga (RU-gah).
rule of nines Method for estimating the extent of a burn based on multiples of nine.

S

sacrum (SA-krum) Portion of the vertebral column between the lumbar vertebrae and coccyx. It is formed of five fused bones and completes the posterior part of the bony pelvis.
sagittal (SAJ-ih-tal) Referring to a plane that divides the body into left and right portions.
saliva (sah-LI-vah) Secretion of the salivary glands; moistens food and contains an enzyme that digests starch.
salt Compound formed by reaction between an acid and a base (e.g., NaCl, table salt).
saltatory (SAL-tah-to-re) **conduction** Transmission of an electric impulse from node to node along a myelinated fiber; faster than continuous conduction along the entire membrane.
SA node See sinoatrial node.
sarcoma (sar-KO-mah) Malignant tumor of connective tissue; a form of cancer.
sarcomere (SAR-ko-mere) Contracting subunit of skeletal muscle.
sarcoplasmic reticulum (sar-ko-PLAS-mik re-TIK-u-lum) **(SR)** Intracellular membranous organelle in muscle cells that is equivalent to the endoplasmic reticulum (ER) in other cells; stores calcium needed for muscle contraction.
saturated fat Fat that has more hydrogen atoms and fewer double bonds between carbons than do unsaturated fats.
scar Fibrous connective tissue that replaces normal tissues destroyed by injury or disease; cicatrix (SIK-ah-triks).
Schwann (shvahn) **cell** Cell in the nervous system that produces the myelin sheath around peripheral axons.
sclera (SKLE-rah) Outermost layer of the eye; made of tough connective tissue; "white" of the eye.
scleroderma (skle-ro-DER-mah) An autoimmune disease associated with overproduction of collagen; systemic sclerosis.
scoliosis (sko-le-O-sis) Lateral curvature of the spine.
scrotum (SKRO-tum) Sac in which testes are suspended.
sebaceous (seh-BA-chus) Pertaining to sebum, an oily substance secreted by skin glands.
sebum (SE-bum) Oily secretion that lubricates the skin; adj., sebaceous (se-BA-shus).
secretin (se-KRE-tin) Hormone from the duodenum that stimulates pancreatic release of water and bicarbonate.
seizure (SE-zhur) Period of uncontrolled electrical activity in the brain that may result in a series of muscle spasms and/or loss of awareness; convulsion.
segmentation (seg-men-TA-shun) Alternating contraction and relaxation of the circular muscle in the small intestine's wall that mix its contents with digestive juices and move them through the organ.
selectively permeable Describing a membrane that regulates what can pass through (e.g., a cell's plasma membrane).
sella turcica (SEL-ah TUR-sih-ka) Saddle-like depression in the floor of the skull that holds the pituitary gland.
semen (SE-men) Mixture of sperm cells and secretions from several glands of the male reproductive tract.
semicircular canal One of three curved channels of the inner ear where receptors for rotational equilibrium are located.
semilunar (sem-e-LU-nar) Shaped like a half-moon, such as the flaps of the pulmonary and aortic valves.
seminal vesicle (VES-ih-kl) Gland that contributes secretions to the semen.
seminiferous (seh-mih-NIF-er-us) **tubules** Tubules in which sperm cells develop in the testis.
semipermeable (sem-e-PER-me-ah-bl) Permeable (passable) to some substances but not to others, as describes the cellular plasma membrane.
sensory (SEN-so-re) Describing cells or activities involved in transmitting impulses toward the central nervous system; afferent.
sensory adaptation Gradual loss of sensation when sensory receptors are exposed to continuous stimulation.
sensory receptor Part of the nervous system that detects a stimulus.
sepsis (SEP-sis) Presence of pathogenic microorganisms or their toxins in the bloodstream or other tissues; adj., septic.
septicemia (sep-tih-SE-me-ah) Presence of pathogenic organisms or their toxins in the bloodstream; blood poisoning.
septum (SEP-tum) Dividing wall, as between the chambers of the heart or the nasal cavities.
serosa (se-RO-sah) Serous membrane; epithelial membrane that secretes a thin, watery fluid.
serotonin (ser-o-TO-nin) Neurotransmitter involved in mood and other cognitive functions.
Sertoli cells See sustentacular cells.
serum (SE-rum) Liquid portion of blood without clotting factors; thin, watery fluid; adj., serous (SE-rus).
sex-linked Referring to a gene carried on a sex chromosome, usually the X chromosome.
sexually transmitted infection (STI) Communicable disease acquired through sexual relations; sexually transmitted disease (STD); venereal disease (VD).
shingles Viral infection that follows the nerve pathways; caused by the same virus that causes chickenpox; herpes zoster.
shock Pertaining to the circulation, a life-threatening condition in which there is inadequate blood flow to the tissues.
sickle cell anemia Hereditary disease in which abnormal hemoglobin causes red blood cells to change shape (sickle) when they release oxygen; sickle cell disease.
sigmoidoscope (sig-MOY-do-skope) Endoscope used to examine the rectum and lower colon.
sign (sine) Manifestation of a disease as noted by an observer.
sinoatrial (si-no-A-tre-al) **(SA) node** Tissue in the right atrium's upper wall that sets the rate of heart contractions; the heart's pacemaker.
sinus (SI-nus) Cavity or channel, such as the paranasal sinuses in the skull bones.
sinus rhythm Normal heart rhythm originating at the SA node.
sinusoid (SI-nus-oyd) Enlarged capillary that serves as a blood channel.
skeletal (SKEL-eh-tal) **system** The body system that includes the bones and joints.
skeleton (SKEL-eh-ton) Bony framework of the body; adj., skeletal.
skull Bony framework of the head.
sodium (SO-de-um) Positive ion found in body fluids and important in fluid balance, nerve impulse conduction, and acid–base balance.
solute (SOL-ute) Substance that is dissolved in another substance (the solvent).
solution (so-LU-shun) Homogeneous mixture of one substance dissolved in another; the components in a mixture are evenly distributed and cannot be distinguished from each other.
solvent (SOL-vent) Substance in which another substance (the solute) is dissolved.
somatic (so-MAT-ik) **nervous system** Division of the nervous system that controls voluntary activities and stimulates skeletal muscle.
somatotropin (so-mah-to-TRO-pin) Growth hormone.
spasm Sudden and involuntary muscular contraction.

specific gravity Weight of a substance as compared to the weight of an equal volume of pure water.
spermatic (sper-MAT-ik) **cord** Cord that extends through the inguinal canal and suspends the testis; contains blood vessels, nerves, and ductus deferens.
spermatozoon (sper-mah-to-ZO-on) Male reproductive cell or gamete; pl., spermatozoa; sperm cell.
sphincter (SFINK-ter) Muscular ring that regulates the size of an opening.
sphygmomanometer (sfig-mo-mah-NOM-eh-ter) Device used to measure blood pressure; blood pressure apparatus or cuff.
spina bifida (SPI-nah BIF-ih-dah) Incomplete closure of the spine.
spinal cord Nervous tissue contained in the spinal column; major relay area between the brain and the peripheral nervous system.
spine Vertebral column. Sharp projection from the surface of a bone.
spiral (SPI-ral) **organ** Receptor for hearing located in the cochlea of the internal ear; organ of Corti.
spirillum (spi-RIL-um) Corkscrew or spiral-shaped bacterium; pl., spirilla.
spirochete (SPI-ro-kete) Spiral-shaped microorganism that moves in a waving and twisting motion.
spirometer (spi-ROM-eh-ter) Instrument for recording lung volumes; tracing obtained is a spirogram.
spleen Lymphoid organ in the upper left region of the abdomen.
splenomegaly (sple-no-MEG-ah-le) Enlargement of the spleen.
spore Endospore; resistant form of bacterium; reproductive cell in lower plants.
squamous (SKWA-mus) Flat and irregular, as in squamous epithelium.
SR See sarcoplasmic reticulum.
staging A procedure for evaluating the extent of tumor spread.
stain (stane) Dye that aids in viewing structures under the microscope.
stapes (STA-peze) Innermost ossicle of the middle ear; transmits sound waves to the oval window of the inner ear.
staphylococcus (staf-ih-lo-KOK-us) Round bacterium found in a cluster resembling a bunch of grapes; pl., staphylococci (staf-ih-lo-KOK-si).
stasis (STA-sis) Stoppage in the normal flow of fluids, such as blood, lymph, urine, or contents of the digestive tract.
STD Sexually transmitted disease. See sexually transmitted infection.
stem cell Cell that has the potential to divide and produce different types of cells.
stenosis (sten-O-sis) Narrowing of a duct or canal.
stent Small tube inserted into a vessel to keep it open.
sterilization (ster-ih-li-ZA-shun) Process of killing every living microorganism on or in an object; procedure that makes an individual incapable of reproduction.
steroid (STE-royd) Category of lipids that includes the hormones of the sex glands and the adrenal cortex.
stethoscope (STETH-o-skope) Instrument for conveying sounds from the patient's body to the examiner's ears.
STI See sexually transmitted infection.
stimulus (STIM-u-lus) Change in the external or internal environment that produces a response.
stomach (STUM-ak) Organ of the digestive tract that stores food, mixes it with digestive juices, and moves it into the small intestine.
strabismus (strah-BIZ-mus) Deviation of the eye resulting from lack of eyeball muscle coordination.
strain Muscle injury caused by overuse or overstretching.
stratified In multiple layers (strata).
stratum (STRA-tum) A layer; pl., strata.
stratum basale (bas-A-le) Deepest layer of the epidermis; layer that produces new epidermal cells; stratum germinativum.
stratum corneum (KOR-ne-um) The thick uppermost layer of the epidermis.
striations (stri-A-shuns) Stripes or bands, as seen in skeletal muscle and cardiac muscle.
stricture (STRICK-ture) Narrowing of a part.
stroke Damage to the brain due to lack of oxygen; usually caused by a blood clot in a vessel (thrombus) or rupture of a vessel; cerebrovascular accident (CVA).
stroke volume Amount of blood ejected from a ventricle with each beat.
subacute (sub-a-KUTE) Not as severe as an acute infection nor as long lasting as a chronic disorder.
subcutaneous (sub-ku-TA-ne-us) Under the skin.
submucosa (sub-mu-KO-sah) Layer of connective tissue beneath the mucosa.
substrate Substance on which an enzyme works.
sudoriferous (su-do-RIF-er-us) Producing sweat; referring to the sweat glands.
suffix A word part that follows a root and modifies its meaning.
sulcus (SUL-kus) Shallow groove as between convolutions of the cerebral cortex; pl., sulci (SUL-si).
superior (su-PE-re-or) Above; in a higher position.
superior vena cava (VE-nah KA-vah) Large vein that drains the upper part of the body and empties into the heart's right atrium.
supination (su-pin-A-shun) Turning the palm up or forward.
supine (SU-pine) Face up or palm up.
surfactant (sur-FAK-tant) Substance in the alveoli that prevents their collapse by reducing surface tension of the fluid lining them.
suspension (sus-PEN-shun) Heterogeneous mixture that will separate unless shaken.
suspensory ligaments Filaments attached to the ciliary muscle of the eye that hold the lens in place and can be used to change its shape.
sustentacular (sus-ten-TAK-u-lar) **cells** Cells in the seminiferous tubules that aid in development of spermatozoa; Sertoli cells.
suture (SU-chur) Type of joint in which bone surfaces are closely united, as in the skull; stitch used in surgery to bring parts together; to stitch parts together in surgery.
sympathetic nervous system Thoracolumbar division of the autonomic nervous system; stimulates a fight or flight (stress) response.
symptom (SIMP-tom) Evidence of disease noted by the patient; such evidence noted by an examiner is called a sign or an objective symptom.
synapse (SIN-aps) Junction between two neurons or between a neuron and an effector.
synarthrosis (sin-ar-THRO-sis) Immovable joint.
syndrome (SIN-drome) Group of symptoms characteristic of a disorder.
synergist (SIN-er-jist) Substance or structure that enhances the work of another. A muscle that works with a prime mover to produce a given movement.
synovial (sin-O-ve-al) Pertaining to a thick, lubricating fluid found in joints, bursae, and tendon sheaths; pertaining to a freely movable (diarthrotic) joint.
system (SIS-tem) Group of organs functioning together for the same general purposes.
systemic (sis-TEM-ik) Referring to a generalized infection or condition.
systemic circuit Pathway that carries blood to all tissues of the body to deliver oxygen and pick up.
systemic sclerosis (skle-RO-sis) Autoimmune disease involving overproduction of collagen; scleroderma.
systole (SIS-to-le) Contraction; adj., systolic (sis-TOL-ik).

T

tachycardia (tak-e-KAR-de-ah) Heart rate more than 100 beats per minute in an adult.
tachypnea (tak-IP-ne-ah) Increased rate of respiration.
tactile (TAK-til) Pertaining to the sense of touch.
target tissue Tissue that is capable of responding to a specific hormone.
Tay-Sachs (tay-saks) **disease** Hereditary disease affecting fat metabolism.
T cell Lymphocyte active in immunity that matures in the thymus gland; may destroy foreign cells directly or help in or regulate the immune response; T lymphocyte.
tectorial (tek-TO-re-al) **membrane** Membrane overlying the sensory (hair) cells in the spiral organ of hearing.
telophase (TEL-o-faze) Final stage of mitosis, during which new nuclei form and the cell contents usually divide.
tendinitis (ten-din-I-tis) Inflammation of a tendon.
tendon (TEN-don) Cord of regular dense connective tissue that attaches a muscle to a bone.
teniae (TEN-e-e) **coli** Bands of smooth muscle in the wall of the large intestine.
teratogen (TER-ah-to-jen) Any factor or agent (e.g., alcohol, drug) that causes abnormal prenatal development.
testis (TES-tis) Male reproductive gland; pl., testes (TES-teze).
testosterone (tes-TOS-ter-one) Male sex hormone produced in the testes; promotes sperm cell development and maintains secondary sex characteristics.

tetanus (TET-an-us) Constant contraction of a muscle; infectious disease caused by a bacterium (*Clostridium tetani*); lockjaw.
tetany (TET-an-e) Muscle spasms due to low blood calcium, as in parathyroid deficiency.
thalamus (THAL-ah-mus) Region of the brain located in the diencephalon; chief relay center for sensory impulses traveling to the cerebral cortex.
thalassemia (thal-ah-SE-me-ah) Hereditary blood disorder that impairs hemoglobin production; the two forms are alpha (α) and beta (β).
therapy (THER-ah-pe) Treatment.
thoracentesis (thor-a-sen-TE-sis) Puncture of the chest for aspiration of fluid in the pleural space.
thorax (THO-raks) Chest; adj., thoracic (tho-RAS-ik).
thrombin (THROM-bin) Clotting factor in the blood needed to convert fibrinogen to fibrin.
thrombocyte (THROM-bo-site) Blood platelet; cell fragment that participates in clotting.
thrombocytopenia (throm-bo-si-to-PE-ne-ah) Deficiency of platelets in the blood.
thrombolytic (throm-bo-LIT-ik) Dissolving blood clots.
thrombophlebitis (throm-bo-fleh-BI-tis) Inflammation of a vein associated with blood clot formation.
thrombosis (throm-BO-sis) Condition of having a thrombus (blood clot in a vessel).
thrombus (THROM-bus) Blood clot within a vessel.
thymus (THI-mus) Lymphoid organ in the upper portion of the chest; site of T-cell development.
thyroid (THI-royd) Endocrine gland in the neck.
thyroiditis (thi-royd-I-tis) Inflammation of the thyroid gland.
thyroid-stimulating hormone (**TSH**) Hormone produced by the anterior pituitary that stimulates the thyroid gland; thyrotropin.
thyroxine (thi-ROK-sin) Hormone produced by the thyroid gland; increases metabolic rate and needed for normal growth; T_4.
tinea (TIN-e-ah) Common term for fungal skin infection.
tissue Group of similar cells that performs a specialized function.
toll-like receptor (TRL) Receptor on a cell of the innate immune system that can recognize components of a pathogen as being foreign.
tongue (tung) Muscular organ in the mouth that contains the taste buds and functions in chewing, swallowing, and speech production.
tonsil (TON-sil) Mass of lymphoid tissue in the region of the pharynx.
tonus (TO-nus) Partially contracted state of muscle; also, tone.
torticollis (tor-tih-KOL-is) Wryneck or stiff neck caused by contraction of neck muscles.
toxemia (tok-SE-me-ah) General toxic condition in which poisonous bacterial substances are absorbed into the bloodstream; presence of harmful substances in the blood as a result of abnormal metabolism.
toxin (TOK-sin) Poison.
toxoid (TOK-soyd) Altered toxin used to produce active immunity.
trachea (TRA-ke-ah) Tube that extends from the larynx to the bronchi; windpipe.
tracheostomy (tra-ke-OS-to-me) Surgical opening into the trachea for the introduction of a tube through which a person may breathe.
tracheotomy (tra-ke-OT-o-me) Incision into the trachea.
trachoma (trah-KO-mah) Acute eye infection caused by chlamydia.
tract Bundle of neuron fibers within the central nervous system.
trait Characteristic.
transfusion (trans-FU-zhun) Introduction of blood or blood components directly into the bloodstream of a recipient.
transplantation (trans-plan-TA-shun) The grafting to a recipient of an organ or tissue from an animal or other human to replace an injured or incompetent body part.
transport maximum Maximum amount of a particular substance that can be reabsorbed from the nephron tubule in milligrams per minute; tubular maximum.
transverse Describing a plane that divides a structure into superior and inferior parts.
trauma (TRAW-mah) Injury or wound.
tricuspid (tri-KUS-pid) **valve** Valve between the heart's right atrium and right ventricle.
trigeminal neuralgia (tri-JEM-ih-nal nu-RAL-je-ah) Severe spasmodic pain affecting the fifth cranial nerve; tic douloureux (tik du-lu-RU).
triglyceride (tri-GLIS-er-ide) Simple fat composed of glycerol and three fatty acids.
trigone (TRI-gone) Triangular-shaped region in the floor of the bladder that remains stable as the bladder fills.
triiodothyronine (tri-i-o-do-THI-ro-nin) Thyroid hormone that acts to raise cellular metabolism; T_3.
trisomy 21 (TRI-so-me) Down syndrome; genetic disorder caused by an extra chromosome 21.
trophoblast (TRO-fo-blast) Region of the blastocyst in embryonic development that forms the fetal portion of the placenta.
tropomyosin (tro-po-MI-o-sin) Protein that works with troponin to regulate contraction in skeletal muscle.
troponin (tro-PO-nin) Protein that works with tropomyosin to regulate contraction in skeletal muscle.
trypsin (TRIP-sin) Pancreatic enzyme active in the digestion of proteins.
TSH See thyroid-stimulating hormone.
tuberculosis (tu-ber-ku-LO-sis) (**TB**) Infectious disease, often of the lung, caused by the bacillus *Mycobacterium tuberculosis*.
tumor (TU-mor) Abnormal growth or neoplasm.
Turner syndrome Hereditary disorder in which cells have only a single X chromosome and no Y chromosome.
tympanic (tim-PAN-ik) **membrane** Membrane between the external and middle ear that transmits sound waves to the bones of the middle ear; eardrum.

U

ulcer (UL-ser) Sore or lesion associated with death and disintegration of tissue.
ultrasound (UL-trah-sound) Very high-frequency sound waves; used in medical imaging to visualize soft structures.
umbilical (um-BIL-ih-kal) **cord** Structure that connects the fetus with the placenta; contains vessels that carry blood between the fetus and placenta.
umbilicus (um-BIL-ih-kus) Small scar on the abdomen that marks the former attachment of the umbilical cord to the fetus; navel.
universal solvent Term used for water because it dissolves more substances than any other solvent.
unsaturated fat Fat that has fewer hydrogen atoms and more double bonds between carbons than do saturated fats.
urea (u-RE-ah) Nitrogenous waste product excreted in the urine; end product of protein metabolism.
uremia (u-RE-me-ah) Accumulation of nitrogenous waste products in the blood.
ureter (U-re-ter) Tube that carries urine from the kidney to the urinary bladder.
urethra (u-RE-thrah) Tube that carries urine from the urinary bladder to the outside of the body.
urinalysis (u-rin-AL-ih-sis) Laboratory examination of urine's physical and chemical properties.
urinary (U-rin-ar-e) **bladder** Hollow organ that stores urine until it is eliminated.
urinary system The system involved in elimination of soluble waste, water balance, and regulation of body fluids.
urination (u-rin-A-shun) Voiding of urine; micturition.
urine (U-rin) Liquid waste excreted by the kidneys.
urticaria (ur-tih-KA-re-ah) Hives; allergic skin reaction with elevated red patches (wheals).
uterine (U-ter-in) **tube** Tube that carries ova from the ovary to the uterus; oviduct; fallopian tube.
uterus (U-ter-us) Muscular, pear-shaped organ in the female pelvis within which the fetus develops during pregnancy; adj., uterine.
uvula (U-vu-lah) Soft, fleshy, V-shaped mass that hangs from the soft palate.

V

vaccination (vak-sin-A-shun) Administration of a vaccine to protect against a specific disease; immunization.
vaccine (vak-SENE) Substance used to produce active immunity; usually, a suspension of attenuated or killed pathogens or some component of a pathogen given by inoculation to prevent a specific disease.
vagina (vah-JI-nah) Distal part of the birth canal that opens to the outside of the body; female organ of sexual intercourse.
vagus (VA-gus) **nerve** Tenth cranial nerve.
valence (VA-lens) Combining power of an atom; number of electrons lost, gained, or shared by atoms of an element in chemical reactions.
valve Structure that prevents fluid from flowing backward, as in the heart, veins, and lymphatic vessels.

varicose (VAR-ih-kose) Pertaining to an enlarged and twisted vessel, as in varicose vein.
varix (VAR-iks) Varicose vein; pl., varices (VAR-ih-seze).
vas deferens (DEF-er-enz) Tube that carries sperm cells from the testis to the urethra; ductus deferens.
vascular (VAS-ku-lar) Pertaining to blood vessels.
vasectomy (vah-SEK-to-me) Surgical removal of part or all of the ductus (vas) deferens; usually done on both sides to produce sterility.
vasoconstriction (vas-o-kon-STRIK-shun) Decrease in a blood vessel's lumen diameter.
vasodilation (vas-o-di-LA-shun) Increase in a blood vessel's lumen diameter.
vasomotor (va-so-MO-tor) Pertaining to dilation or constriction of blood vessels.
vector (VEK-tor) An insect or other animal that transmits a disease-causing organism from one host to another.
vein (vane) Vessel that carries blood toward the heart.
vena cava (VE-nah KA-vah) Large vein that carries blood into the heart's right atrium; superior vena cava or inferior vena cava.
venereal (ve-NE-re-al) **disease (VD)** See sexually transmitted infection (STI).
venous sinus (VE-nus SI-nus) Large channel that drains blood low in oxygen.
ventilation (ven-tih-LA-shun) Movement of air into and out of the lungs.
ventral (VEN-tral) At or toward the front or belly surface; anterior.
ventricle (VEN-trih-kl) Cavity or chamber. One of the heart's two lower chambers. One of the brain's four chambers in which cerebrospinal fluid is produced; adj., ventricular (ven-TRIK-u-lar).
venule (VEN-ule) Vessel between a capillary and a vein.
vernix caseosa (VER-niks ka-se-O-sah) Cheeselike sebaceous secretion that covers a newborn.
vertebra (VER-teh-brah) Bone of the spinal column; pl., vertebrae (VER-teh-bre).
verruca (veh-RU-kah) Wart.
vertigo (VER-tih-go) Sensation of spinning or sensation that the environment is spinning.
vesicle (VES-ih-kl) Small sac or blister filled with fluid.
vesicular transport Use of vesicles to move large amounts of material through a cell's plasma membrane.
vestibular apparatus (ves-TIB-u-lar) Part of the inner ear concerned with equilibrium; consists of the semicircular canals and vestibule.
vestibular (ves-TIB-u-lar) **folds** Folds of mucous membrane superior to the vocal cords in the larynx. They close off the glottis during swallowing and straining down; false vocal cords.
vestibule (VES-tih-bule) Any space at the entrance to a canal or organ; in the inner ear, area that contains some receptors for the sense of equilibrium.
vestibulocochlear (ves-tib-u-lo-KOK-le-ar) **nerve** Eighth cranial nerve concerned with hearing and equilibrium.
vibrio (VIB-re-o) Slight curved or comma-shaped bacterium; pl., vibrios.
villi (VIL-li) Small finger-like projections from the surface of a membrane; projections in the lining of the small intestine through which digested food is absorbed; sing., villus.
virulence (VIR-u-lens) Power of an organism to overcome a host's defenses.
virus (VI-rus) Extremely small infectious agent that can reproduce only within a living cell.
viscera (VIS-er-ah) Organs in the ventral body cavities, especially the abdominal organs; adj., visceral.
viscosity (vis-KOS-ih-te) Thickness, as of the blood or other fluid.
vitamin (VI-tah-min) Organic compound needed in small amounts for health.
vitreous (VIT-re-us) **body** Soft, jelly-like substance that fills the eyeball and holds the shape of the eye; vitreous humor.
vocal folds Bands of mucous membrane in the larynx used in producing speech; vocal cords.
Volkmann canal See perforating canal.
volvulus (VOL-vu-lus) Twisting of the intestine.
von Willebrand disease Hereditary blood clotting disorder in which there is a shortage of von Willebrand factor.
vulva (VUL-va) External female genitalia.

W

wart Small tumor caused by a virus of the human papilloma group; verruca (veh-RU-kah).
Wernicke (VER-nih-ke) **area** Portion of the cerebral cortex concerned with speech recognition and the meaning of words.
white matter Nervous tissue composed of myelinated fibers.
Wilson disease Recessive hereditary disease associated with a defect in copper metabolism and accumulation of copper in the liver, brain, or other tissues.
word root The main part of a word to which prefixes and suffixes may be attached.

X

x-ray Ray or radiation of extremely short wavelength that can penetrate opaque substances and affect photographic plates and fluorescent screens.

Y

yeast Round, single-celled fungus that may reproduce by budding.

Z

zygote (ZI-gote) Fertilized ovum; cell formed by the union of a sperm and an egg.

Glossary of Word Parts

Use of Word Parts in Medical Terminology

Medical terminology, the special language of the health occupations, is based on an understanding of a few relatively basic elements. These elements—roots, prefixes, and suffixes—form the foundation of almost all medical terms. A useful way to familiarize yourself with each term is to learn to pronounce it correctly and say it aloud several times. Soon it will become an integral part of your vocabulary.

The foundation of a word is the word root. Examples of word roots are *abdomin*, referring to the belly region, and *aden*, pertaining to a gland. A word root is often followed by a vowel to facilitate pronunciation when an ending is added, as in *abdomino* and *adeno*. We then refer to it as a "combining form," and it usually appears in texts with a slash before the vowel, as in *abdomin/o* and *aden/o*.

A prefix is a part of a word that precedes the word root and changes its meaning. For example, the prefix *mal-* in malnutrition means "abnormal." A suffix, or word ending, is a part that follows the word root and adds to or changes its meaning. The suffix *-rhea* means "profuse flow" or "discharge," as in diarrhea, a condition characterized by excessive discharge of liquid stools.

Many medical words are compound words; that is, they are made up of more than one root or combining form. Examples of such compound words are *erythrocyte* (red blood cell) and *hydrocele* (fluid-containing sac), and many more complex words, such as *sternoclavicular* (indicating relations to both the sternum and the clavicle).

A general knowledge of language structure and spelling rules is also helpful in mastering medical terminology. For example, adjectives include words that end in *-al*, as in sternal (the noun is sternum), and words that end in *-ous*, as in mucous (the noun is mucus).

The following list includes some of the most commonly used word roots, prefixes, and suffixes, as well as examples of their use. Prefixes are followed by a hyphen; suffixes are preceded by a hyphen; and word roots have no hyphen. Commonly used combining vowels are added following a slash.

Word Parts

a-, an- not, without: *aphasia, atrophy, anemia, anuria*
ab- away from: *abduction, aboral*
abdomin/o belly or abdominal area: *abdominocentesis, abdominoscopy*
acous, acus hearing, sound: *acoustic, presbyacusis*
acr/o- end, extremity: *acromegaly, acromion*
actin/o, actin/i relation to raylike structures or, more commonly, to light or roentgen (x-) rays, or some other type of radiation: *actiniform, actinodermatitis*
ad- (sometimes converted to *ac-, af-, ag-, ap-, as-, at-*) toward, added to, near: *adrenal, accretion, agglomerated, afferent*
aden/o gland: *adenectomy, adenitis, adenocarcinoma*
aer/o air, gas, oxygen: *aerobic, aerate*
-agogue inducing, leading, stimulating: *cholagogue, galactagogue*
-al pertaining to, resembling: *skeletal, surgical, ileal*
alb/i- white: *albinism, albinuria*
alge, alg/o, alges/i pain: *algetic, algophobia, analgesic*
-algia pain, painful condition: *myalgia, neuralgia*
amb/i- both, on two sides: *ambidexterity, ambivalent*
ambly- dimness, dullness: *amblyopia*
amphi on both sides, around, double: *amphiarthrosis, amphibian*
amyl/o starch: *amylase, amyloid*
an- not, without: *anaerobic, anoxia, anemic*
ana- upward, back, again, excessive: *anatomy, anastomosis, anabolism*
andr/o male: *androgen, androgenous*
angi/o vessel: *angiogram, angiotensin*
ant/i- against; to prevent, suppress, or destroy: *antarthritic, antibiotic, anticoagulant*
ante- before, ahead of: *antenatal, antepartum*
anter/o- position ahead of or in front of (i.e., anterior to) another part: *anterolateral, anteroventral*
-apheresis take away, withdraw: *hemapheresis, plasmapheresis*
ap/o- separation, derivation from: *apocrine, apoptosis, apophysis*
aqu/e water: *aqueous, aquatic, aqueduct*
-ar pertaining to, resembling: *muscular, nuclear*
arthr/o joint or articulation: *arthrolysis, arthrostomy, arthritis*
-ary pertaining to, resembling: *salivary, dietary, urinary*
-ase enzyme: *lipase, protease*
-asis See –sis
atel/o- imperfect: *atelectasis*
ather/o gruel: *atherosclerosis, atheroma*
audi/o sound, hearing: *audiogenic, audiometry, audiovisual*
aut/o- self: *autistic, autodigestion, autoimmune*

bar/o pressure: *baroreceptor, barometer*
bas/o- alkaline: *basic, basophilic*
bi- two, twice: *bifurcate, bisexual*
bil/i bile: *biliary, bilirubin*
bio- life, living organism: *biopsy, antibiotic*
blast/o, -blast early stage of a cell, immature cell: *blastula, blastophore, erythroblast*
bleph, blephar/o eyelid, eyelash: *blepharism, blepharitis, blepharospasm*
brachi, brachi/o arm: *brachial, brachiocephalic, brachiotomy*
brachy- short: *brachydactylia, brachyesophagus*
brady- slow: *bradycardia*
bronch/o-, bronch/i bronchus: *bronchiectasis, bronchoscope*
bucc cheek: *buccal*

capn/o carbon dioxide: *hypocapnia, hypercapnia*
carcin/o cancer: *carcinogenic, carcinoma*
cardi/o, cardi/a heart: *carditis, cardiac, cardiologist*
cata- down: *catabolism, catalyst*
-cele swelling; enlarged space or cavity: *cystocele, meningocele, rectocele*
celi/o abdomen: *celiac, celiocentesis*
centi- relating to 100 (used in naming units of measurements): *centigrade, centimeter*
-centesis perforation, tapping: *amniocentesis, paracentesis*
cephal/o head: *cephalalgia, cephalopelvic*
cerebr/o brain: *cerobrospinal, cerebrum*
cervi neck: *cervical, cervix*
cheil/o lips; brim or edge: *cheilitis, cheilosis*
chem/o, chem/i chemistry, chemical: *chemotherapy, chemocautery, chemoreceptor*
chir/o, cheir/o hand: *cheiralgia, cheiromegaly, chiropractic*
chol/e, chol/o bile, gall: *chologogue, cholecyst, cholelith*
cholecyst/o gallbladder: *cholecystitis, cholecystokinin*
chondr/o, chondri/o cartilage: *chondric, chondrocyte, chondroma*
chori/o membrane: *chorion, choroid, choriocarcinoma*
chrom/o, chromat/o color: *chromosome, chromatin, chromophilic*
-cid, -cide to cut, kill, destroy: *bactericidal, germicide, suicide*
circum- around, surrounding: *circumorbital, circumrenal, circumduction*
-clast break: *osteoclast*
clav/o, cleid/o clavicle: *cleidomastoid, subclavian*
co- with, together: *cofactor, cohesion, coinfection*
colp/o vagina: *colpectasia, colposcope, colpotomy*
con- with: *concentric, concentrate, conduct*
contra- opposed, against: *contraindication, contralateral*
corne/o horny: *corneum, cornified, cornea*
cortic/o cortex: *cortical, corticotropic, cortisone*
cost/a, cost/o- ribs: *intercostal, costosternal*
counter- against, opposite to: *counteract, counterirritation, countertraction*

crani/o skull: *cranium, craniotomy*
cry/o- cold: *cryalgesia, cryogenic, cryotherapy*
crypt/o- hidden, concealed: *cryptic, cryptogenic, cryptorchidism*
-cusis hearing: *acusis, presbyacusis*
cut- skin: *subcutaneous*
cyan/o- blue: *cyanosis, cyanogen*
cyst/i, cyst/o sac, bladder: *cystitis, cystoscope*
cyt/o, -cyte cell: *cytology, cytoplasm, osteocyte*

dactyl/o digits (usually fingers, but sometimes toes): *dactylitis, polydactyly*
de- remove: *detoxify, dehydration*
dendr tree: *dendrite*
dent/o, dent/i tooth: *dentition, dentin, dentifrice*
derm/o, dermat/o skin: *dermatitis, dermatology, dermatosis*
di- twice, double: *dimorphism, dibasic, dihybrid*
dipl/o- double: *diplopia, diplococcus*
dia- through, between, across, apart: *diaphragm, diaphysis*
dis- apart, away from: *disarticulation, distal*
dors/i, dors/o- back (in the human, this combining form is the same as poster/o-): *dorsal, dorsiflexion, dorso-nuchal*
dys- disordered, difficult, painful: *dysentery, dysphagia, dyspnea*

e- out: *enucleation, evisceration, ejection*
-ectasis expansion, dilation, stretching: *angiectasis, bronchiectasis*
ecto- outside, external: *ectoderm, ectogenous*
-ectomy surgical removal or destruction by other means: *appendectomy, thyroidectomy*
edem swelling: *edema*
-emia condition of blood: *glycemia, hyperemia*
encephal/o brain: *encephalitis, encephalogram*
end/o- in, within, innermost: *endarterial, endocardium, endothelium*
enter/o intestine: *enteritis, enterocolitis*
epi- on, upon: *epicardium, epidermis*
equi- equal: *equidistant, equivalent, equilibrium*
erg/o work: *ergonomic, energy, synergy*
eryth-, erythr/o- red: *erythema, erythrocyte*
-esthesia sensation: *anesthesia, paresthesia*
eu- well, normal, good: *euphoria, eupnea*
ex/o- outside, out of, away from: *excretion, exocrine, exophthalmic*
extra- beyond, outside of, in addition to: *extracellular, extrasystole, extravasation*

fasci fibrous connective tissue layers: *fascia, fascitis, fascicle*
fer, -ferent to bear, to carry: *afferent, efferent, transfer*
fibr/o threadlike structures, fibers: *fibrillation, fibroblast, fibrositis*

gastr/o stomach: *gastritis, gastroenterostomy*
-gen an agent that produces or originates: *allergen, pathogen, fibrinogen*
-genic produced from, producing: *neurogenic, pyogenic, psychogenic*
genit/o- organs of reproduction: *genitoplasty, genitourinary*
gen/o- a relationship to reproduction or sex: *genealogy, generate, genetic, genotype*
-geny manner of origin, development or production: *ontogeny, progeny*
gest/o gestation, pregnancy: *progesterone, gestagen*
glio, -glia gluey material; specifically, the support tissue of the central nervous system: *glioma, neuroglia*
gloss/o tongue: *glossitis, glossopharyngeal*
glyc/o- relating to sugar, glucose, sweet: *glycemia, glycosuria*
gnath/o related to the jaw: *prognathic, gnathoplasty*
gnos to perceive, recognize: *agnostic, prognosis, diagnosis*
gon seed, knee: *gonad, gonarthritis*
-gram record, that which is recorded: *electrocardiogram, electroencephalogram*
graph/o, -graph instrument for recording, writing: *electrocardiograph, electroencephalograph, micrograph*
-graphy process of recording data: *photography, radiography*
gyn/o, gyne, gynec/o female, woman: *gynecology, gynecomastia, gynoplasty*
gyr/o circle: *gyroscope, gyrus, gyration*

hema, hemo, hemat/o blood: *hematoma, hematuria, hemorrhage*
hemi- one half: *hemisphere, heminephrectomy, hemiplegia*
hepat/o- liver: *hepatitis, hepatogenous*
heter/o- other, different: *heterogenous, heterosexual, heterochromia*
hist/o, histi/o tissue: *histology, histiocyte*
homeo-, homo- unchanging, the same: *homeostasis, homosexual*
hydr/o- water: *hydrolysis, hydrocephalus*
hyper- above, over, excessive: *hyperesthesia, hyperglycemia, hypertrophy*
hypo- deficient, below, beneath: *hypochondrium, hypodermic, hypogastrium*
hyster/o uterus: *hysterectomy*

-ia state of, condition of: *myopia, hypochondria, ischemia*
-iatrics, -trics medical specialty: *pediatrics, obstetrics*
iatr/o physician, medicine: *iatrogenic*
-ic pertaining to, resembling: *metric, psychiatric, geriatric*
idio- self, one's own, separate, distinct: *idiopathic, idiosyncrasy*
-ile pertaining to, resembling: *febrile, virile*
im-, in- in, into, lacking: *implantation, infiltration, inanimate*
infra- below, inferior: *infraspinous, infracortical*
insul/o pancreatic islet, island: *insulin, insulation, insulinoma*
inter- between: *intercostal, interstitial*
intra- within a part or structure: *intracranial, intracellular, intraocular*
isch suppression: *ischemia*
-ism state of: *alcoholism, hyperthyroidism*
iso- same, equal: *isotonic, isometric*
-ist one who specializes in a field of study: *cardiologist, gastroenterologist*
-itis inflammation: *dermatitis, keratitis, neuritis*

juxta- next to: *juxtaglomerular, juxtaposition*

kary/o nucleus: *karyotype, karyoplasm*
kerat/o cornea of the eye, certain cornified tissues: *keratin, keratitis, keratoplasty*
kine movement: *kinetic, kinesiology, kinesthesia*

lacri- tear: *lacrimal*
lact/o milk: *lactation, lactogenic*
laryng/o larynx: *laryngeal, laryngectomy, laryngitis*
later/o- side: *lateral*
-lemma sheath: *neurilemma, sarcolemma*
leuk/o- (also written as *leuc-, leuco-*) white, colorless: *leukocyte, leukoplakia*
lip/o lipid, fat: *lipase, lipoma*
lig- bind: *ligament, ligature*
lingu/o tongue: *lingual, linguadental*
lith/o stone (calculus): *lithiasis, lithotripsy*
-logy study of: *physiology, gynecology*
lute/o yellow: *macula lutea, corpus luteum*
lymph/o lymph, lymphatic system, lymphocyte: *lymphoid, lymphedema*
lyso-, -lysis, -lytic loosening, dissolving, separating: *hemolysis, paralysis, lysosome*

macr/o- large, abnormal length: *macrophage, macroblast.* See also mega-, megal/o-
mal- bad, diseased, disordered, abnormal: *malnutrition, malocclusion, malunion*
malac/o, -malacia softening: *malacoma, osteomalacia*
mamm/o- breast, mammary gland: *mammogram, mammoplasty, mammal*
man/o pressure: *manometer, sphygmomanometer*
mast/o breast: *mastectomy, mastitis*
meg/a-, megal/o, -megaly unusually or excessively large: *megacolon, megaloblast, splenomegaly, megakaryocyte*
melan/o dark, black: *melanin, melanocyte, melanoma*
men/o physiologic uterine bleeding, menses: *menstrual, menorrhagia, menopause*
mening/o membranes covering the brain and spinal cord: *meningitis, meningocele*
mes/a, mes/o- middle, midline: *mesencephalon, mesoderm*
meta- change, beyond, after, over, near: *metabolism, metacarpal, metaplasia*
-meter, metr/o measure: *hemocytometer, sphygmomanometer, spirometer, isometric*
metr/o uterus: *endometrium, metroptosis, metrorrhagia*
micro- very small: *microscope, microbiology, microsurgery, micrometer*

mon/o- single, one: *monocyte, mononucleosis*
morph/o- shape, form: *morphogenesis, morphology*
multi- many: *multiple, multifactorial, multipara*
my/o- muscle: *myenteron, myocardium, myometrium*
myc/o, mycet fungus: *mycid, mycete, mycology, mycosis, mycelium*
myel/o marrow (often used in reference to the spinal cord): *myeloid, myeloblast, osteomyelitis, poliomyelitis*
myring/o tympanic membrane: *myringotomy, myringitis*
myx/o mucus: *myxoma, myxovirus*

narc/o- stupor: *narcosis, narcolepsy, narcotic*
nas/o nose: *nasopharynx, paranasal*
natri sodium: *hyponatremia, natriuretic*
necr/o death, corpse: *necrosis*
neo- new: *neoplasm, neonatal*
neph, nephr/o kidney: *nephrectomy, nephron*
neur/o, neur/i nerve, nervous tissue: *neuron, neuralgia, neuroma*
neutr/o neutral: *neutrophil, neutropenia*
noct/i night: *noctambulation, nocturia, noctiphobia*

ocul/o- eye: *oculist, oculomotor, oculomycosis*
odont/o- tooth, teeth: *odontalgia, orthodontics*
-odynia pain, tenderness: *myodynia, neurodynia*
-oid like, resembling: *lymphoid, myeloid*
olig/o- few, a deficiency: *oligospermia, oliguria*
-oma tumor, swelling: *hematoma, sarcoma*
-one ending for steroid hormone: *testosterone, progesterone*
onych/o nails: *paronychia, onychoma*
oo ovum, egg: *oocyte, oogenesis* (do not confuse with oophor-)
oophor/o ovary: *oophorectomy, oophoritis, oophorocystectomy.* See also ovar-
ophthalm/o- eye: *ophthalmia, ophthalmologist, ophthalmoscope*
-opia disorder of the eye or vision: *heterotropia, myopia, hyperopia*
or/o mouth: *oropharynx, oral*
orchi/o, orchid/o testis: *orchitis, cryptorchidism*
orth/o- straight, normal: *orthopedics, orthopnea, orthosis*
-ory pertaining to, resembling: *respiratory, circulatory*
oscill/o to swing to and fro: *oscilloscope*
osmo- osmosis: *osmoreceptor, osmotic*
oss/i, osse/o, oste/o bone, bone tissue: *osseous, ossicle, osteocyte, osteomyelitis*
ot/o ear: *otalgia, otitis, otomycosis*
-ous pertaining to, resembling: *fibrous, venous, androgynous*
ov/o egg, ovum: *oviduct, ovulation*
ovar, ovari/o ovary: *ovariectomy.* See also oophor
ox-, -oxia pertaining to oxygen: *hypoxemia, hypoxia, anoxia*
oxy sharp, acute: *oxygen, oxytocia*

pan- all: *pandemic, panacea*
papill/o nipple: *papilloma, papillary*
para- near, beyond, apart from, beside: *paramedical, parametrium, parathyroid, parasagittal*
pariet/o wall: *parietal*
path/o, -pathy disease, abnormal condition: *pathogen, pathology, neuropathy*
ped/o-, -pedia child, foot: *pedophobia, pediatrician, pedialgia*
-penia lack of: *leukopenia, thrombocytopenia*
per- through, excessively: *percutaneous, perfusion*
peri- around: *pericardium, perichondrium*
-pexy fixation: *nephropexy, proctopexy*
phag/o to eat, to ingest: *phagocyte, phagosome*
-phagia, -phagy eating, swallowing: *aphagia, dysphagia*
-phasia speech, ability to talk: *aphasia, dysphasia*
phen/o to show: *phenotype*
-phil, -philic to like, have an affinity for: *eosinophilia, hemophilia, hydrophilic*
phleb/o vein: *phlebitis, phlebotomy*
-phobia fear, dread, abnormal aversion: *phobic, acrophobia, hydrophobia*
phot/o- light: *photoreceptor, photophobia*
phren/o diaphragm: *phrenic, phrenicotomy*
physi/o- natural, physical: *physiology, physician*
pil/e-, pil/i-, pil/o- hair, resembling hair: *pileous, piliation, pilonidal*
pin/o to drink: *pinocytosis*
-plasty molding, surgical formation: *cystoplasty, gastroplasty, kineplasty*
-plegia stroke, paralysis: *paraplegia, hemiplegia*
pleur/o side, rib, pleura: *pleurisy, pleurotomy*
-pnea air, breathing: *dyspnea, eupnea*
pneum/o-, pneumat/o- air, gas, respiration: *pneumothorax, pneumograph, pneumatocele*
pneumon/o lung: *pneumonia, pneumonectomy*
pod/o foot: *podiatry, pododynia*
-poiesis making, forming: *erythropoiesis, hematopoiesis*
polio- gray: *polioencephalitis, poliomyelitis*
poly- many: *polyarthritis, polycystic, polycythemia*
post- behind, after, following: *postnatal, postocular, postpartum*
pre- before, ahead of: *precancerous, preclinical, prenatal*
presby- old age: *presbycusis, presbyopia*
pro- before, in front of, in favor of: *prodromal, prosencephalon, prolapse, prothrombin*
proct/o rectum: *proctitis, proctocele, proctologist*
propri/o own: *proprioception*
pseud/o false: *pseudoarthrosis, pseudostratified, pseudopod*
psych/o mind: *psychosomatic, psychotherapy*
-ptosis downward displacement, falling, prolapse: *blepharoptosis, enteroptosis, nephroptosis*
pulm/o-, pulmon/o- lung: *pulmonic, pulmonology*
py/o pus: *pyuria, pyogenic, pyorrhea*
pyel/o renal pelvis: *pyelitis, pyelogram, pyelonephrosis*
pyr/o fire, fever: *pyrogen, antipyretic, pyromania*

quadr/i- four: *quadriceps, quadriplegic*

rachi/o spine: *rachicentesis, rachischisis*
radio- emission of rays or radiation: *radioactive, radiography, radiology*
re- again, back: *reabsorption, reaction, regenerate*
rect/o rectum: *rectal, rectouterine*
ren/o kidney: *renal, renopathy*
reticul/o network: *reticulum, reticular*
retro- backward, located behind: *retrocecal, retroperitoneal*
rhin/o nose: *rhinitis, rhinoplasty*
-rhage, -rhagia* bursting forth, excessive flow: *hemorrhage, menorrhagia*
-rhaphy* suturing or sewing up a gap or defect in a part: *herniorrhaphy, gastrorrhaphy, cystorrhaphy*
-rhea* flow, discharge: *diarrhea, gonorrhea, seborrhea*

sacchar/o sugar: *monosaccharide, polysaccharide*
salping/o tube: *salpingitis, salpingoscopy*
sarc/o flesh: *sarcolemma, sarcoplasm, sarcomere*
scler/o- hard, hardness; *scleroderma, sclerosis*
scoli/o- twisted, crooked: *scoliosis, scoliosometer*
-scope instrument used to look into or examine a part: *bronchoscope, endoscope, arthroscope*
semi- partial, half: *semipermeable, semicoma*
semin/o semen, seed: *seminiferous, seminal*
sep, septic poison, rot, decay: *sepsis, septicemia*
sin/o sinus: *sinusitis, sinusoid, sinoatrial*
-sis condition or process, usually abnormal: *dermatosis, osteoporosis*
soma-, somat/o, -some body: *somatic, somatotype, somatotropin*
son/o- sound: *sonogram, sonography*
sphygm/o- pulse: *sphygmomanometer*
spir/o breathing: *spirometer, inspiration, expiration*
splanchn/o- internal organs: *splanchnic, splanchnoptosis*
splen/o spleen: *splenectomy, splenic*
staphyl/o- grapelike cluster: *staphylococcus*
stat, -stasis stand, stoppage, remain at rest: *hemostasis, static, homeostasis*
sten/o- contracted, narrowed: *stenosis*
sthen/o, -sthenia, -sthenic strength: *asthenic, calisthenics, neurasthenia*
steth/o- chest: *stethoscope*
stoma, stomat/o mouth: *stomatitis*
-stomy surgical creation of an opening into a hollow organ or an opening between two organs: *colostomy, tracheostomy, gastroenterostomy*
strept/o chain: *streptococcus, streptobacillus*

sub- under, below, near, almost: *subclavian, subcutaneous, subluxation*
super- over, above, excessive: *superego, supernatant, superficial*
supra- above, over, superior: *supranasal, suprarenal*
sym-, syn- with, together: *symphysis, synapse*
syring/o fistula, tube, cavity: *syringectomy, syringomyelia*

tach/o-, tachy- rapid: *tachycardia, tachypnea*
tars/o eyelid, foot: *tarsitis, tarsoplasty, tarsoptosis*
-taxia, -taxis order, arrangement: *ataxia, chemotaxis, thermotaxis*
tel/o- end: *telophase, telomere*
tens- stretch, pull: *extension, tensor*
terat/o malformed fetus: *teratogen, teratogenic*
test/o testis: *testosterone, testicular*
tetr/a four: *tetralogy, tetraplegia*
therm/o-, -thermy heat: *thermalgesia, thermocautery, diathermy, thermometer*
thromb/o- blood clot: *thrombosis, thrombocyte*
toc/o labor: *eutocia, dystocia, oxytocin*
tom/o, -tomy incision of, cutting: *anatomy, phlebotomy, laparotomy*
ton/o tone, tension: *tonicity, tonic*
tox-, toxic/o- poison: *toxin, cytotoxic, toxemia, toxicology*
trache/o trachea, windpipe: *tracheal, tracheitis, tracheotomy*
trans- across, through, beyond: *transorbital, transpiration, transplant, transport*
tri- three: *triad, triceps*
trich/o hair: *trichiasis, trichosis, trichology*
troph/o, -trophic, -trophy nutrition, nurture: *atrophic, hypertrophy*
trop/o, -tropin, -tropic turning toward, acting on, influencing, changing: *thyrotropin, adrenocorticotropic, gonadotropic*
tympan/o drum: *tympanic, tympanum*

ultra- beyond or excessive: *ultrasound, ultraviolent, ultrastructure*
uni- one: *unilateral, uniovular, unicellular*
-uria urine: *glycosuria, hematuria, pyuria*
ur/o urine, urinary tract: *urology, urogenital*

vas/o vessel, duct: *vascular, vasectomy, vasodilation*
viscer/o internal organs, viscera: *visceral, visceroptosis*
vitre/o- glasslike: *vitreous*

xer/o- dryness: *xeroderma, xerophthalmia, xerosis*

-y condition of: *tetany, atony, dysentery*

zyg/o joined: *zygote, heterozygous, monozygotic*

---

*When a suffix beginning with *rh* is added to a word root, the *r* is doubled.

▶ Appendix 1

Periodic Table of the Elements

The periodic table lists the chemical elements according to their atomic numbers. The boxes in the table have information about the elements, as shown by the example at the top of the chart. The upper number in each box is the atomic number, which represents the number of protons in the nucleus of the atom. Under the name of the element is its chemical symbol, an abbreviation of its modern or Latin name. The Latin names of four common elements are shown below the chart. The bottom number in each box gives the atomic weight (mass) of that element's atoms compared with the weight of carbon atoms. Atomic weight is the sum of the weights of the protons and neutrons in the nucleus.

All the elements in a column share similar chemical properties based on the number of electrons in their outermost energy levels. Those in column VIII are nonreactive (inert) and are referred to as noble gases. The 26 elements found in the body are color coded according to quantity (see totals above the chart). Carbon, hydrogen, oxygen, and nitrogen make up 96% of body weight. The first three of these are present in all carbohydrates, lipids, proteins, and nucleic acids. Nitrogen is an additional component of all proteins. Nine other elements make up almost all the rest of body weight. The remaining 13 elements are present in very small amounts and are referred to as trace elements. Although needed in very small quantities, they are essential for good health, as they are parts of enzymes and other compounds used in metabolism.

PERIODIC TABLE OF THE ELEMENTS

- 96% of body weight
- 3.9% of body weight
- 0.1% of body weight

Notation:

6	Atomic number
Carbon	Name
C	Symbol
12.01	Atomic weight

I	II											III	IV	V	VI	VII	VIII
1 Hydrogen H 1.01																	2 Helium He 4.00
3 Lithium Li 6.94	4 Beryllium Be 9.01											5 Boron B 10.81	6 Carbon C 12.01	7 Nitrogen N 14.01	8 Oxygen O 16.00	9 Fluorine F 19.00	10 Neon Ne 20.18
11 Sodium Na 22.99	12 Magnesium Mg 24.31											13 Aluminum Al 26.98	14 Silicon Si 28.09	15 Phosphorus P 30.97	16 Sulfur S 32.07	17 Chlorine Cl 35.45	18 Argon Ar 39.95
19 Potassium K 39.10	20 Calcium Ca 40.08	21 Scandium Sc 44.96	22 Titanium Ti 47.88	23 Vanadium V 50.94	24 Chromium Cr 52.00	25 Manganese Mn 54.94	26 Iron Fe 55.85	27 Cobalt Co 58.93	28 Nickel Ni 58.69	29 Copper Cu 63.55	30 Zinc Zn 65.39	31 Gallium Ga 69.72	32 Germanium Ge 72.59	33 Arsenic As 74.92	34 Selenium Se 78.96	35 Bromine Br 79.90	36 Krypton Kr 83.80
37 Rubidium Rb 85.47	38 Strontium Sr 87.62	39 Yttrium Y 88.91	40 Zirconium Zr 91.22	41 Niobium Nb 92.91	42 Molybdenum Mo 95.94	43 Technetium Tc (98)	44 Ruthenium Ru 101.1	45 Rhodium Rh 102.9	46 Palladium Pd 106.4	47 Silver Ag 107.9	48 Cadmium Cd 112.4	49 Indium In 114.8	50 Tin Sn 118.7	51 Antimony Sb 121.8	52 Tellurium Te 127.6	53 Iodine I 126.9	54 Xenon Xe 131.3
55 Cesium Cs 132.91	56 Barium Ba 137.34		72 Hafnium Hf 178.5	73 Tantalum Ta 180.9	74 Tungsten W 183.9	75 Rhenium Re 186.2	76 Osmium Os 190.2	77 Iridium Ir 192.2	78 Platinum Pt 195.1	79 Gold Au 196.9	80 Mercury Hg 200.6	81 Thallium Tl 204.4	82 Lead Pb 207.2	83 Bismuth Bi 209.0	84 Polonium Po (210)	85 Astatine At (210)	86 Radon Rn (222)
87 Francium Fr (223)	88 Radium Ra (226)		104 Rutherfordium Rf (257)	105 Dubnium Db (260)	106 Seaborgium Sg (263)	107 Bohrium Bh (262)	108 Hassium Hs (265)	109 Meitnerium Mt (267)	110 Darmstadtium Ds (271)	111 Unnamed (272)	112 Unnamed (277)						

57–71 Lanthanides	57 Lanthanum La 138.9	58 Cerium Ce 140.1	59 Praseodymium Pr 140.9	60 Neodymium Nd 144.2	61 Promethium Pm (145)	62 Samarium Sm (150.4)	63 Europium Eu 152.0	64 Gadolinium Gd 157.3	65 Terbium Tb 158.9	66 Dysprosium Dy 162.5	67 Holmium Ho 164.9	68 Erbium Er 167.3	69 Thulium Tm 168.9	70 Ytterbium Yb 173.0	71 Lutetium Lu 175.0
89–103 Actinides	89 Actinium Ac (227)	90 Thorium Th 232.0	91 Protactinium Pa (231)	92 Uranium U (238)	93 Neptunium Np (237)	94 Plutonium Pu (244)	95 Americium Am (243)	96 Curium Cm (247)	97 Berkelium Bk (247)	98 Californium Cf (251)	99 Einsteinium Es (254)	100 Fermium Fm (257)	101 Mendelevium Md (256)	102 Nobelium No (259)	103 Lawrencium Lr (257)

Name	Latin name	Symbol
Copper	*cuprium*	Cu
Iron	*ferrum*	Fe
Potassium	*kalium*	K
Sodium	*natrium*	Na

Appendix 2

Tests

Appendix 2-1 Routine Urinalysis

Test	Normal Value	Clinical Significance
General characteristics and measurements		
Color	Pale yellow to amber	Color change can be due to concentration or dilution, drugs, or metabolic or inflammatory disorders
Odor	Slightly aromatic	Foul odor typical of urinary tract infection, fruity odor in uncontrolled diabetes mellitus
Appearance (clarity)	Clear to slightly hazy	Cloudy urine occurs with infection or after refrigeration; may indicate presence of bacteria, cells, mucus, or crystals
Specific gravity	1.003–1.030 (first morning catch; routine is random)	Decreased in diabetes insipidus, renal failure, hyponatremia; increased in liver disorders, heart failure, dehydration
pH	4.5–8.0	Acid urine accompanies acidosis, fever, high-protein diet; alkaline urine in urinary tract infection, metabolic alkalosis, vegetarian diet
Chemical determinations		
Glucose	Negative	Glucose present in uncontrolled diabetes mellitus, steroid excess
Ketones	Negative	Present in diabetes mellitus and in starvation
Protein	Negative	Present in kidney disorders, such as glomerulonephritis, acute kidney failure
Bilirubin	Negative	Breakdown product of hemoglobin; present in liver disease or in bile blockage
Urobilinogen	0.2–1.0 Ehrlich units/dL	Breakdown product of bilirubin; increased in hemolytic anemias and in liver disease; remains negative in bile obstruction
Blood (occult)	Negative	Detects small amounts of blood cells, hemoglobin, or myoglobin; present in severe trauma, metabolic disorders, bladder infections
Nitrite	Negative	Product of bacterial breakdown of urine; positive result suggests urinary tract infection and needs to be followed up with a culture of the urine
Microscopic		
Red blood cells	0–3 per high-power field	Increased because of bleeding within the urinary tract from trauma, tumors, inflammation, or damage within the kidney
White blood cells	0–4 per high-power field	Increased by kidney or bladder infection
Renal epithelial cells	Occasional	Increased number indicates damage to kidney tubules
Casts	None	Hyaline casts normal; a large number of abnormal casts indicate inflammation or a systemic disorder
Crystals	Present	Most are normal; may be acid or alkaline
Bacteria	Few	Increased in urinary tract infection or contamination from infected genitalia
Others		Any yeasts, parasites, mucus, spermatozoa, or other microscopic findings would be reported here

Appendix 2-2 Complete Blood Count

Test	Normal Value*	Clinical Significance
Red blood cell (RBC) count	Men: 4.2–5.4 million/mcL Women: 3.6–5.0 million/mcL	Decreased in anemia; increased in dehydration, polycythemia
Hemoglobin (Hb)	Men: 13.5–17.5 g/dL Women: 12–16 g/dL	Decreased in anemia, hemorrhage, and hemolytic reactions; increased in dehydration, heart disease, and lung disease
Hematocrit (Hct) or packed cell volume (PCV)	Men: 40%–50% Women: 37%–47%	Decreased in anemia; increased in polycythemia, dehydration
Red blood cell (RBC) indices (examples)		These values, calculated from the RBC count, Hb, and Hct, give information valuable in the diagnosis and classification of anemia
Mean corpuscular volume (MCV)	87–103 mcL/red cell	Measures the average size or volume of each RBC: small size (microcytic) in iron deficiency anemia; large size (macrocytic) typical of pernicious anemia
Mean corpuscular hemoglobin (MCH)	26–34 pg/red cell	Measures the weight of hemoglobin per RBC; useful in differentiating types of anemia in a severely anemic patient
Mean corpuscular hemoglobin concentration (MCHC)	31–37 g/dL	Defines the volume of hemoglobin per RBC; used to determine the color or concentration of hemoglobin per RBC
White blood cell (WBC) count	5,000–10,000/mcL	Increased in leukemia and in response to infection, inflammation, and dehydration; decreased in bone marrow suppression
Platelets	150,000–350,000/mcL	Increased in many malignant disorders; decreased in disseminated intravascular coagulation (DIC) or toxic drug effects; spontaneous bleeding may occur at platelet counts below 20,000 mcL
Differential (peripheral blood smear)		A stained slide of the blood is needed to perform the differential. The percentages of the different WBCs are estimated, and the slide is microscopically checked for abnormal characteristics in WBCs, RBCs, and platelets
WBCs		
Segmented neutrophils (SEGs, POLYs)	40%–74%	Increased in bacterial infections; low numbers leave person very susceptible to infection
Immature neutrophils (BANDs)	0%–3%	Increased when neutrophil count increases
Lymphocytes (LYMPHs)	20%–40%	Increased in viral infections; low numbers leave person dangerously susceptible to infection
Monocytes (MONOs)	2%–6%	Increased in specific infections
Eosinophils (EOs)	1%–4%	Increased in allergic disorders
Basophils (BASOs)	0.5%–1%	Increased in allergic disorders

*Values vary depending on instrumentation and type of test.

Appendix 2-3 Blood Chemistry Tests

Test	Normal Value	Clinical Significance
Basic panel: An overview of electrolytes, waste product management, and metabolism		
Blood urea nitrogen (BUN)	7–18 mg/dL	Increased in renal disease and dehydration; decreased in liver damage and malnutrition
Carbon dioxide (CO_2) (includes bicarbonate)	23–30 mmol/L	Useful to evaluate acid–base balance by measuring total carbon dioxide in the blood: elevated in vomiting and pulmonary disease; decreased in diabetic acidosis, acute renal failure, and hyperventilation

Appendix 2-3 Blood Chemistry Tests (*continued*)

Test	Normal Value	Clinical Significance
Chloride (Cl)	98–106 mEq/L	Increased in dehydration, hyperventilation, and congestive heart failure; decreased in vomiting, diarrhea, and fever
Creatinine	0.6–1.2 mg/dL	Produced at a constant rate and excreted by the kidney; increased in kidney disease
Glucose	Fasting: 70–110 mg/dL Random: 85–125 mg/dL	Increased in diabetes and severe illness; decreased in insulin overdose or hypoglycemia
Potassium (K)	3.5–5 mEq/L	Increased in renal failure, extensive cell damage, and acidosis; decreased in vomiting, diarrhea, and excess administration of diuretics or IV fluids
Sodium (Na)	101–111 mEq/L or 135–148 mEq/L (depending on test)	Increased in dehydration and diabetes insipidus; decreased in overload of IV fluids, burns, diarrhea, or vomiting
Additional blood chemistry tests		
Alanine aminotransferase (ALT)	10–40 U/L	Used to diagnose and monitor treatment of liver disease and to monitor the effects of drugs on the liver; increased in myocardial infarction
Albumin	3.8–5.0 g/dL	Albumin holds water in blood; decreased in liver disease and kidney disease
Albumin–globulin ratio (A/G ratio)	More than 1	Low A/G ratio signifies a tendency for edema because globulin is less effective than albumin at holding water in the blood
Alkaline phosphatase (ALP)	20–70 U/L (varies by method)	Enzyme of bone metabolism; increased in liver disease and metastatic bone disease
Amylase	21–160 U/L	Used to diagnose and monitor treatment of acute pancreatitis and to detect salivary gland inflammation
Aspartate aminotransferase (AST)	0–41 U/L (varies)	Enzyme present in tissues with high metabolic activity; increased in myocardial infarction and liver disease
Bilirubin, total	0.2–1.0 mg/dL	Breakdown product of hemoglobin from red blood cells; increased when excessive red blood cells are being destroyed or in liver disease
Calcium (Ca)	8.8–10.0 mg/dL	Increased in excess parathyroid hormone production and in cancer; decreased in alkalosis, elevated phosphate in renal failure, and excess IV fluids
Cholesterol	120–220 mg/dL desirable range	Screening test used to evaluate risk of heart disease; levels of 200 mg/dL or above indicate increased risk of heart disease and warrant further investigation
Creatine kinase (CK)	Men: 38–174 U/L Women: 96–140 U/L	Elevated enzyme level indicates myocardial infarction or damage to skeletal muscle. When elevated levels are present, tests are done for specific fractions (isoenzymes)
Gamma-glutamyl transferase (GGT)	Men: 6–26 U/L Women: 4–18 U/L	Used to diagnose liver disease and to test for chronic alcoholism
Globulins	2.3–3.5 g/dL	Proteins active in immunity; help albumin keep water in blood
Iron, serum (Fe)	Men: 75–175 mcg/dL Women: 65–165 mcg/dL	Decreased in iron deficiency and anemia; increased in hemolytic conditions
High-density lipoproteins (HDLs)	Men: 30–70 mg/dL Women: 30–85 mg/dL	Used to evaluate the risk of heart disease
Lactic dehydrogenase (LDH or LD)	95–200 U/L (normal ranges vary greatly)	Enzyme released in many kinds of tissue damage, including myocardial infarction, pulmonary infarction, and liver disease
Lipase	4–24 U/L (varies with test)	Enzyme used to diagnose pancreatitis
Low-density lipoproteins (LDLs)	80–140 mg/dL	Used to evaluate the risk of heart disease
Magnesium (Mg)	1.3–2.1 mEq/L	Vital in neuromuscular function; decreased levels may occur in malnutrition, alcoholism, pancreatitis, diarrhea
Phosphorus (P) (inorganic)	2.7–4.5 mg/dL	Evaluated in response to calcium; main store is in bone: elevated in kidney disease; decreased in excess parathyroid hormone

(*continued*)

Appendix 2-3 Blood Chemistry Tests (*continued*)

Test	Normal Value	Clinical Significance
Protein, total	6–8 g/dL	Increased in dehydration, multiple myeloma; decreased in kidney disease, liver disease, poor nutrition, severe burns, excessive bleeding
Serum glutamic oxalacetic transaminase (SGOT)		See aspartate aminotransferase (AST)
Serum glutamic pyruvic transaminase (SGPT)		See alanine aminotransferase (ALT)
Thyroxine (T_4)	4.5–11.2 mcg/dL (varies)	Screening test of thyroid function; increased in hyperthyroidism; decreased in myxedema and hypothyroidism
Thyroid-stimulating hormone (TSH)	0.45–5.0 mIU/L	Produced by pituitary to promote thyroid gland function; elevated when thyroid gland is not functioning
Triiodothyronine (T_3)	120–195 mg/dL	Elevated in specific types of hyperthyroidism
Triglycerides	Men: 40–160 mg/dL Women: 35–135 mg/dL	An indication of ability to metabolize fats; increased triglycerides and cholesterol indicate high risk of atherosclerosis
Uric acid	Men: 3.5–7.2 mg/dL Women: 2.6–6.0 mg/dL	Produced by breakdown of ingested purines in food and nucleic acids; elevated in kidney disease, gout, and leukemia

Appendix 3

Diseases

Appendix 3-1 Bacterial Diseases

Organism	Disease and Description
Cocci	
Neisseria gonorrhoeae (gonococcus)	Gonorrhea. Acute inflammation of mucous membranes of the reproductive and urinary tracts (with possible spread to the peritoneum in the female). Systemic infection may cause gonococcal arthritis and endocarditis. The organism also causes ophthalmia neonatorum, an eye inflammation of the newborn.
Neisseria meningitidis (meningococcus)	Epidemic meningitis. Inflammation of the membranes covering the brain and the spinal cord. A vaccine is available for use in high-risk populations.
Staphylococcus aureus and other staphylococci	Boils, carbuncles, impetigo, osteomyelitis, staphylococcal pneumonia, cystitis, pyelonephritis, empyema, septicemia, toxic shock, and food poisoning. Strains resistant to antibiotics are a cause of infections originating in hospitals, such as wound infections.
Streptococcus pneumoniae	Pneumonia; inflammation of the alveoli, bronchioles, and bronchi; middle ear infections; meningitis. May be prevented by use of polyvalent pneumococcal vaccine.
Streptococcus pyogenes, *Streptococcus hemolyticus*, and other streptococci	Septicemia, septic sore throat, scarlet fever, puerperal sepsis, erysipelas, streptococcal pneumonia, rheumatic fever, subacute bacterial endocarditis, acute glomerulonephritis.
Bacilli	
Bordetella pertussis	Pertussis (whooping cough). Severe infection of the trachea and bronchi. The "whoop" is caused by the effort to recover breath after coughing. All children should be immunized against pertussis.
Brucella abortus (and others)	Brucellosis, or undulant fever. Disease of animals such as cattle and goats transmitted to humans through unpasteurized dairy products or undercooked meat. Acute phase of fever and weight loss; chronic disease with abscess formation and depression.
Clostridium botulinum	Botulism. Very severe poisoning caused by eating food in which the organism has been allowed to grow and excrete its toxin. Causes muscle paralysis and may result in death from asphyxiation. Infant botulism results from ingestion of spores. It causes respiratory problems and flaccid paralysis, which usually respond to treatment.
Clostridium perfringens	Gas gangrene. Acute wound infection. Organisms cause death of tissues accompanied by the generation of gas within them.
Clostridium tetani	Tetanus. Acute, often fatal poisoning caused by introduction of the organism into deep wounds. Characterized by severe muscular spasms. Also called *lockjaw*.
Corynebacterium diphtheriae	Diphtheria. Acute throat inflammation with the formation of a leathery membrane-like growth (pseudomembrane) that can obstruct air passages and cause death by asphyxiation. Toxin produced by this organism can damage the heart, nerves, kidneys, and other organs. Disease preventable by appropriate vaccination.
Escherichia coli, *Proteus* spp., and other colon bacilli	Normal inhabitants of the colon, and usually harmless there. Cause of local and systemic infections, food poisoning, diarrhea (especially in children), septicemia, and septic shock. *Escherichia coli* is a common hospital-acquired infection.
Francisella tularensis	Tularemia, or deer fly fever. Transmitted by contact with an infected animal or by a tick or fly bite. Symptoms are fever, ulceration of the skin, and enlarged lymph nodes.

(continued)

Appendix 3-1 Bacterial Diseases (*continued*)

Organism	Disease and Description
Haemophilus influenzae type b (Hib)	Severe infections in children under 3 y of age. Causes meningitis, also epiglottitis, septicemia, pneumonia, pericarditis, and septic arthritis. Preschool vaccinations are routine.
Helicobacter pylori	Acute inflammation of the stomach (gastritis), ulcers of the stomach's pyloric area and the duodenum.
Legionella pneumophila	Legionnaires disease (pneumonia). Seen in localized epidemics; may be transmitted by air conditioning towers and by contaminated soil at excavation sites. Does not spread person to person. Characterized by high fever, vomiting, diarrhea, cough, and bradycardia. Mild form of the disease called Pontiac fever.
Mycobacterium leprae (Hansen bacillus)	Leprosy. Chronic illness in which hard swellings occur under the skin, particularly of the face, causing a distorted appearance. In one form of leprosy, the nerves are affected, resulting in loss of sensation in the extremities.
Mycobacterium tuberculosis (tubercle bacillus)	Tuberculosis. Infectious disease in which the organism causes primary lesions called *tubercles*. These break down into cheeselike masses of tissue, a process known as *caseation*. Any body organ can be infected, but in adults, the usual site is the lungs. Still one of the most widespread diseases in the world, tuberculosis is treated with chemotherapy; strains of the bacillus have developed resistance to drugs.
Pseudomonas aeruginosa	Common organism is a frequent cause of wound and urinary infections in debilitated hospitalized patients. Often found in solutions that have been standing for long periods.
Salmonella typhi (and others)	Salmonellosis occurs as enterocolitis, bacteremia, localized infection, or typhoid. Depending on type, presenting symptoms may be fever, diarrhea, or abscesses; complications include intestinal perforation and endocarditis. Carried in water, milk, meat, and other food.
Shigella dysenteriae (and others)	Serious bacillary dysentery. Acute intestinal infection with diarrhea (sometimes bloody); may cause dehydration with electrolyte imbalance or septicemia. Transmitted through fecal–oral route or other poor sanitation.
Yersinia pestis	Plague, the "black death" of the Middle Ages. Transmitted to humans by fleas from infected rodents. Symptoms of the most common form are swollen, infected lymph nodes, or *buboes*. Another form may cause pneumonia. All forms may lead to a rapidly fatal septicemia.
Curved rods	
Vibrios	
Vibrio cholerae	Cholera. Acute intestinal infection characterized by prolonged vomiting and diarrhea, leading to severe dehydration, electrolyte imbalance, and in some cases, death.
Spirochetes	
Borrelia burgdorferi	Lyme disease, transmitted by the extremely small deer tick. Usually starts with a bull's eye rash followed by flulike symptoms, at which time antibiotics are effective. May progress to neurologic problems and joint inflammation.
Borrelia recurrentis (and others)	Relapsing fever. Generalized infection in which attacks of fever alternate with periods of apparent recovery. Organisms spread by lice, ticks, and other insects.
Treponema pallidum	Syphilis. Infectious disease transmitted mainly by sexual intercourse. Untreated syphilis is seen in the following three stages: primary—formation of primary lesion (chancre); secondary—skin eruptions and infectious patches on mucous membranes; tertiary—development of generalized lesions (gummas) and destruction of tissues resulting in aneurysm, heart disease, and degenerative changes in the brain, spinal cord, ganglia, and meninges. Also a cause of intrauterine fetal death or stillbirth.
Treponema vincentii	Vincent disease (trench mouth). Infection of the mouth and throat accompanied by formation of a pseudomembrane, with ulceration.

Appendix 3-1 Bacterial Diseases (*continued*)

Organism	Disease and Description
Other bacteria	
(*Note*: The following organisms are smaller than other bacteria and vary in shape. Like viruses, they grow within cells, but they differ from viruses in that they are affected by antibiotics.)	
Chlamydia oculogenitalis	Inclusion conjunctivitis, acute eye infection. Carried in genital organs; transmitted during birth or through water in inadequately disinfected swimming pools.
Chlamydia psittaci	Psittacosis, also called ornithosis. Disease transmitted by various birds, including parrots, ducks, geese, and turkeys. Primary symptoms are chills, headache, and fever, more severe in older people. The duration may be from two to three wk, often with a long convalescence. Antibiotic drugs are effective remedies.
Chlamydia trachomatis	Sexually transmitted infection causing pelvic inflammatory disease and other reproductive tract infections. Also causes inclusion conjunctivitis, an acute eye infection, and trachoma, a chronic infection that is a common cause of blindness in underdeveloped areas of the world. Infection of the conjunctiva and cornea characterized by redness, pain, and lacrimation. Antibiotic therapy is effective if begun before there is scarring. The same organism causes lymphogranuloma venereum (LGV), a sexually transmitted infection characterized by swelling of inguinal lymph nodes and accompanied by signs of general infection
Coxiella burnetii	Q fever. Infection transmitted from cattle, sheep, and goats to humans by contaminated dust and also carried by arthropods. Symptoms are fever, headache, chills, and pneumonitis. This disorder is almost never fatal.
Rickettsia prowazekii	Epidemic typhus. Transmitted to humans by lice; associated with poor hygiene and war. Main symptoms are headache, hypotension, delirium, and a red rash. Frequently fatal in older people.
Rickettsia rickettsii	Rocky Mountain spotted fever. Tick-borne disease occurring throughout the United States. Symptoms are fever, muscle aches, and a rash that may progress to gangrene over bony prominences. The disease is rarely fatal.
Rickettsia typhi	Endemic or murine typhus. A milder rickettsial disease transmitted to humans from rats by fleas. Symptoms are fever, rash, headache, and cough. The disease is rarely fatal.

Appendix 3-2 Viral Diseases

Organism	Disease and Description
Common cold viruses	Common cold (coryza), viral infection of the upper respiratory tract. A wide variety of organisms may be involved. May lead to complications, such as pneumonia and influenza.
Cytomegalovirus (CMV)	Common mild salivary gland infection. In an immunosuppressed person, may cause infection of the retina, lung, and liver, GI tract ulceration, and brain inflammation. Causes severe fetal or neonatal damage.
Encephalitis viruses	Encephalitis, which usually refers to any brain inflammation accompanied by degenerative tissue changes. Encephalitis has many causes besides viruses. Viral forms of encephalitis include Western and Eastern epidemic, equine, St. Louis, Japanese B, and others. Some are known to be transmitted from birds and other animals to humans by insects, principally mosquitoes.
Epstein-Barr virus (EBV)	Mononucleosis, a highly infectious disease spread by saliva. Common among teenagers and young adults. There is fever, sore throat, marked fatigue, and enlargement of the spleen and lymph nodes. Infects B lymphocytes (mononuclear leukocytes) causing them to multiply. Virus remains latent for life after infection. EBV also causes Burkitt lymphoma, a malignant B-lymphocyte tumor common in parts of Africa.
Hantavirus	Hantavirus pulmonary syndrome (HPS) with high mortality rate. Spread through inhalation of virus from dried urine of infected rodents.

(*continued*)

Appendix 3-2 Viral Diseases (*continued*)

Organism	Disease and Description
Hepatitis viruses	Liver inflammation. Varieties A through E are recognized.
Hepatitis A virus (HAV)	Transmitted by fecal contamination. Does not become chronic or produce carrier state. Infection provides lifelong immunity. Vaccine is available.
Hepatitis B virus (HBV)	Transmitted by direct exchange of blood and body fluids. Can cause rapidly fatal disease or develop into chronic disease and carrier state. Risk of progression to liver cancer. Vaccine is available.
Hepatitis C virus (HCV)	Spread through blood exchange (usually transfusions before 1992 when screening began) or shared needles. May become chronic and lead to cirrhosis, liver failure, liver cancer. Antiviral drugs may limit infection.
Hepatitis D virus (HDV)	Spread by blood exchange and occurs as coinfection with hepatitis B. Responsible for half of rapidly fatal liver failure cases and also a high rate of chronic disease that progresses to death.
Hepatitis E virus (HEV)	Transmitted by fecal contamination and occurs in epidemics in Middle East and Asia. Resembles hepatitis A. Can be fatal in pregnant women.
Herpes simplex virus type 1	Cold sores or fever blisters that appear around the mouth and nose of people with colds or other illnesses accompanied by fever.
Herpes simplex virus type 2	Genital herpes. Acute inflammatory disease of the genitalia, often recurring. A very common sexually transmitted infection.
Human immunodeficiency virus (HIV)	Acquired immunodeficiency syndrome (AIDS). Disease that infects T cells of the immune system and is usually fatal if untreated. Diagnosed by antibody tests, decline in specific (CD4) cells, and presenting disease, including *Candida albicans* infection, *Pneumocystis jiroveci* pneumonia (PCP), Kaposi sarcoma, persistent swelling of lymph nodes (lymphadenopathy), chronic diarrhea, and wasting. Spread by contact with contaminated body fluids and by transplacental route.
Human papillomavirus (HPV)	Genital warts (condylomata acuminata). Sexually transmitted warts of the genital and perianal area in men and women. Associated with cervical dysplasia and cancer. A vaccine against the most prevalent strains is available.
Influenza virus	Epidemic viral infection, marked by chills, fever, muscular pains, and prostration. The most serious complication is bronchopneumonia caused by *Haemophilus influenzae* (a bacillus) or streptococci.
Mumps virus	Epidemic parotitis. Acute inflammation with swelling of the parotid salivary glands. Mumps can have many complications, such as orchitis (inflammation of the testes) in young men and meningitis in young children.
Pneumonia viruses	Lung infections caused by a number of different viruses, such as the influenza and parainfluenza viruses, adenoviruses, and varicella viruses.
Poliovirus	Poliomyelitis (polio). Acute viral infection that may attack the anterior horns of the spinal cord, resulting in paralysis of certain voluntary muscles. Most countries have eliminated polio through vaccination programs.
Rhabdovirus	Rabies. An acute, fatal disease transmitted to humans through the saliva of an infected animal. Rabies is characterized by violent muscular spasms induced by the slightest sensations. Because the swallowing of water causes throat spasms, the disease is also called hydrophobia ("fear of water"). The final stage of paralysis ends in death. Rabies vaccines are available for humans and animals.
Rotavirus	Attacks lining of small intestine causing severe diarrhea in children. A vaccine is available.
Rubella virus	Rubella or German measles. A less severe form of measles, but especially dangerous during the first three mo of pregnancy because the disease organism can cause heart defects, hearing loss, mental deficiency, and other permanent damage in the fetus.
Rubeola virus	Measles. An acute respiratory inflammation followed by fever and a generalized skin rash. Patients are prone to the development of dangerous complications, such as bronchopneumonia and other secondary infections caused by staphylococci and streptococci.
SARS virus	Highly infectious respiratory disease called severe acute respiratory syndrome (SARS). Emerged in China early in 2003 and spread to other countries before it was isolated and identified as a viral infection. Believed to have spread from small mammals to humans.
Varicella zoster	Chickenpox (varicella). A usually mild infection, almost completely confined to children, characterized by blister-like skin eruptions. Vaccine is now available.
	Shingles (herpes zoster). A very painful eruption of skin blisters that follows the course of certain peripheral nerves. These blisters eventually dry up and form scabs. Vaccine is now available for people 60 y and older.

Appendix 3-3 Prion Diseases

Agent	Disease
Chronic wasting disease agent	Chronic wasting disease in deer and elk.
Creutzfeldt-Jakob agent	Creutzfeldt-Jakob disease (CJD), a spongiform encephalopathy in humans.
Kuru agent	Kuru spongiform encephalopathy in humans.
Mad cow agent	Mad cow spongiform encephalopathy, or bovine spongiform encephalopathy (BSE) in cows and humans.
Scrapie agent	Scrapie spongiform encephalopathy in sheep.

Note: Prions are infectious agents that contain protein but no nucleic acid. They cause slow, spongy degeneration of the brain tissue, known as spongiform encephalitis, in humans and animals.

Appendix 3-4 Fungal Diseases

Disease/Organism	Description
Actinomycosis	"Lumpy jaw" in cattle and humans. The organisms cause the formation of large tissue masses, which are often accompanied by abscesses. The lungs and liver may be involved.
Blastomycosis (*Blastomyces dermatitidis*)	General term for any infection caused by a yeastlike organism. There may be skin tumors and lesions in the lungs, bones, liver, spleen, and kidneys.
Candidiasis (*Candida albicans*)	Infection that can involve the skin and mucous membranes. May cause diaper rash, infection of the nail beds, and infection of the mucous membranes of the mouth (oral thrush), throat, and vagina.
Coccidioidomycosis (*Coccidioides immitis*)	Systemic fungal disease also called *San Joaquin Valley fever*. Because it often attacks the lungs, it may be mistaken for tuberculosis.
Histoplasmosis (*Histoplasma capsulatum*)	Variety of disorders, ranging from mild respiratory symptoms or enlargement of the liver, spleen, and lymph nodes to cavities in the lungs with symptoms similar to those of tuberculosis.
Pneumocystis jiroveci (formerly *carinii*)	Pneumonia (PCP). Opportunistic infection in people with a depressed immune system. Invades lungs and causes a foamy exudate to collect in alveoli.
Ringworm Tinea capitis Tinea corporis Tinea pedis	Common fungal skin infections, many of which cause blisters and scaling with tinea capitis discoloration of the affected areas. All are caused by similar organisms from a group of fungi called dermatophytes. They are easily transmitted by person-to-person contact or by contaminated objects.

Appendix 3-5 Protozoal Diseases

Organism	Disease and Description
Amebae	
Entamoeba histolytica	Amebic dysentery. Severe ulceration of the wall of the large intestine caused by amebae. Acute diarrhea may be an important symptom. This organism also may cause liver abscesses.
Ciliates	
Balantidium coli	Gastrointestinal disturbances and ulcers of the colon.
Flagellates	
Giardia lamblia	Gastrointestinal disturbances.
Leishmania donovani (and others)	Kala azar. In this disease, there is enlargement of the liver and spleen as well as skin lesions.
Trichomonas vaginalis	Inflammation and discharge from the vagina. In males, it involves the urethra and causes painful urination.
Trypanosoma spp.	African sleeping sickness (trypanosomiasis). Disease begins with a high fever, followed by invasion of the brain and spinal cord by the organisms. Usually, the disease ends with continued drowsiness, coma, and death.
Apicomplexans (sporozoa)	
Cryptosporidium	Cramps and diarrhea that can be long term and severe in people with a weakened immune system, such as those with AIDS. Spread in water and by personal contact in close quarters.
Plasmodium; varieties include *vivax, falciparum, malariae*	Malaria. Characterized by recurrent attacks of chills followed by high fever. Severe malarial attacks can be fatal because of kidney failure, cerebral disorders, and other complications.
Toxoplasma gondii	Toxoplasmosis. Common infectious disease transmitted by cats and raw meat. Mild forms cause fever and lymph node enlargement. May cause fatal encephalitis in immunosuppressed patients. Infection of a pregnant woman is a cause of fetal stillbirth or congenital damage.

Appendix 4

Answers to Chapter Checkpoint and Zooming In Questions

Chapter 1

Answers to Checkpoint Questions

1-1 Study of body structure is anatomy; study of body function is physiology.

1-2 Organs working together combine to form systems.

1-3 Intracellular fluids are located inside cells; extracellular fluids are located outside of cells.

1-4 Homeostasis is a state of internal balance.

1-5 The components of a negative feedback loop are sensor, control center, and effector.

1-6 Catabolism and anabolism are the two types of metabolic activities. In catabolism, complex substances are broken down into simpler compounds. In anabolism, simple compounds are used to manufacture materials needed for growth, function, and tissue repair.

1-7 Distal describes a location farther from an origin. The wrist is distal to the elbow.

1-8 The three planes in which the body can be cut are sagittal, frontal (coronal), and transverse (horizontal).

1-9 The two main body cavities are the dorsal and ventral cavities.

1-10 The three central regions of the abdomen are the epigastric, umbilical, and hypogastric regions. The three left and right regions are the hypochondriac, lumbar, and iliac (inguinal) regions.

Answers to Zooming In Questions

1-6 The figures are standing in the anatomic position.

1-7 The transverse (horizontal) plane divides the body into superior and inferior parts. The frontal (coronal) plane divides the body into anterior and posterior parts.

1-10 The ventral cavity contains the diaphragm.

1-13 The umbilical, hypogastric, left lumbar, and left iliac (inguinal) regions are represented in the left lower quadrant.

Chapter 2

Answers to Checkpoint Questions

2-1 Atoms are subunits of elements.

2-2 Three types of particles found in atoms are protons, neutrons, and electrons.

2-3 The atom with six electrons in its outermost energy level is more likely to participate in a chemical reaction, as the outermost energy level of the atom with eight electrons is complete.

2-4 Ionic bonds are formed by exchange of electrons between atoms. Covalent bonds are formed by the sharing of electrons between atoms.

2-5 When an electrolyte goes into solution, it separates into charged particles called ions (cations and anions).

2-6 Molecules are units composed of two or more covalently bonded atoms. Compounds are substances composed of two or more different elements.

2-7 In a solution, the solute dissolves and remains evenly distributed (the mixture is homogeneous); in a suspension, the material in suspension does not dissolve and settles out unless the mixture is shaken (the mixture is heterogeneous).

2-8 Water is the most abundant compound in the body.

2-9 A value of 7.0 is neutral on the pH scale. An acid measures lower than 7.0; a base measures higher than 7.0

2-10 A buffer is a substance that maintains a steady pH of a solution.

2-11 Isotopes that give off radiation are termed radioactive and are called radioisotopes.

2-12 Carbon is the basis of organic chemistry.

2-13 The three main categories of organic compounds are carbohydrates, lipids, and proteins.

2-14 An enzyme is a catalyst that speeds the rate of chemical reactions in the body.

2-15 A nucleotide contains a sugar, a phosphate group, and a nitrogenous base. DNA, RNA, and ATP are examples of compounds composed of nucleotides.

Answers to Zooming In Questions

2-2 The number of protons is equal to the number of electrons. There are eight protons and eight electrons in the oxygen atom.

2-3 Oxygen needs two electrons to complete its outermost energy level. Magnesium needs to lose two electrons to achieve a stable outermost energy level.

2-4 A sodium atom has one electron in its outermost energy level. A sodium ion has eight electrons in its outermost energy level.

2-5 Two electrons are needed to complete the energy level of each hydrogen atom.

2-6 Two hydrogen atoms bond with an oxygen atom to form water.

2-7 The amount of hydroxide ion (OH^-) in a solution decreases when the amount of hydrogen ion (H^+) increases.

2-8 The building blocks (monomers) of disaccharides and polysaccharides are monosaccharides.

2-9 There are three carbon atoms in glycerol.

2-10 The amino group of an amino acid contains nitrogen.

2-11 The shape of an enzyme after the reaction is the same as it was before the reaction.

2-12 The prefix *tri-* means three.

Chapter 3

Answers to Checkpoint Questions

3-1 The cell shows organization, metabolism, responsiveness, homeostasis, growth, and reproduction.

3-2 Three types of microscopes are the compound light microscope, transmission electron microscope (TEM), and scanning electron microscope (SEM).

3-3 The four substances found within the plasma membrane are phospholipids, cholesterol, carbohydrates (glycoproteins and glycolipids), and proteins.

3-4 The cell organelles are specialized structures that perform different tasks.

3-5 The nucleus is called the cell's control center because it contains the chromosomes, hereditary units that control all cellular activities.

3-6 The two types of organelles used for movement are cilia, which are small and hairlike, and the flagellum, which is long and whiplike.

3-7 Diffusion, osmosis, and filtration do not directly require cellular energy. Active transport and bulk transport require cellular energy. Bulk transport includes endocytosis (phagocytosis, pinocytosis, receptor-mediated endocytosis) and exocytosis.

3-8 An isotonic solution is the same concentration as the cytoplasm; a hypotonic solution is less concentrated; a hypertonic solution is more concentrated.

3-9 Nucleotides are the building blocks of nucleic acids.

3-10 DNA codes for proteins in the cell.

3-11 The three types of RNA active in protein synthesis are messenger RNA (rRNA), ribosomal RNA (rRNA), and transfer RNA (tRNA).

3-12 Before mitosis can occur, the DNA must replicate (double). Replication occurs during interphase.

3-13 The four stages of mitosis are prophase, metaphase, anaphase, and telophase.

Answers to Zooming In Questions

3-1 The transmission electron microscope (TEM) in B shows the most internal structure. The scanning electron microscope (SEM) in C shows the cilia in three dimensions.

3-2 Ribosomes attached to the ER make it look rough. Cytosol is the liquid part of the cytoplasm.

3-3 The plasma membrane is described as a bilayer because it is constructed of two layers of phospholipids.

3-4 Epithelial cells (B) would best cover a large surface area because they are flat.

3-6 A decrease in the number of transporters would decrease solute transport.

3-7 If the solute could pass through the membrane in this system, the solute and solvent molecules would equalize on the two sides of the membrane, and the fluid level would be the same on both sides.

3-8 If the concentration of solute was increased on side B of this system, the osmotic pressure would increase.

3-9 If lost blood were replaced with pure water, red blood cells would swell because the blood would become hypotonic to the cells.

3-10 A lysosome would likely help destroy a particle taken in by phagocytosis.

3-12 The nucleotides pair up so that there is one larger nucleotide and one smaller nucleotide in each pair. Specifically, A pairs with T, and G pairs with C.

3-15 If the original cell has 46 chromosomes, each daughter cell will have 46 chromosomes after mitosis.

Chapter 4

Answers to Checkpoint Questions

4-1 The three basic shapes of epithelium are squamous (flat and irregular), cuboidal (square), and columnar (long and narrow).

4-2 Exocrine glands secrete into a nearby organ, cavity, or to the surface of the skin and generally secrete through ducts. Endocrine glands secrete directly into surrounding tissue fluids and into the blood.

4-3 The intercellular material in connective tissue is the matrix.

4-4 Collagen makes up the most abundant fibers in connective tissue.

4-5 Fibroblasts characterize dense connective tissue. Chondrocytes characterize cartilage. Osteoblasts and osteocytes characterize bone tissue.

4-6 The three types of muscle tissue are skeletal (voluntary), cardiac, and smooth (visceral) muscle.

4-7 The basic cell of the nervous system is the neuron, and it carries nerve impulses.

4-8 The nonconducting support cells of the nervous system are neuroglia (glial cells).

4-9 The three types of epithelial membranes are serous membranes, mucous membranes, and the cutaneous membrane (skin).

4-10 A parietal serous membrane is attached to the wall of a cavity or sac. A visceral serous membrane is attached to the surface of an organ.

4-11 Fascia is a fibrous band or sheet that supports and holds organs. Superficial fascia underlies the skin, and deep fascia covers muscles, nerves, and vessels.

4-12 A benign tumor does not spread; a malignant tumor spreads (metastasizes) to other tissues.

4-13 A biopsy is removal of living tissue for microscopic examination.

4-14 The three standard approaches to treatment of cancer are surgery, radiation, and chemotherapy.

Answers to Zooming In Questions

4-1 The epithelial cells are in a single layer.

4-2 Stratified epithelium protects underlying tissue from wear and tear.

4-4 Of the tissues shown, areolar connective tissue has the most fibers; adipose tissue is modified for storage.

Chapter 5

Answers to Checkpoint Questions

5-1 Disease is an abnormality of the structure or function of a part, organ, or system.

5-2 Categories of disease are infectious diseases, degenerative disease, nutritional disorders, metabolic disorders, immune disorders, neoplasms, and psychiatric disorders.

5-3 A predisposing cause of disease is a factor that may not in itself give rise to a disease but that increases the probability of a person's becoming ill.

5-4 Predisposing causes of disease include age, gender, heredity, living conditions and lifestyle, emotional disturbances, physical and chemical damage, and preexisting illness.

5-5 Epidemiologists collect statistics on disease incidence rate, prevalence rate, and mortality rate.

5-6 Identification of an illness is a diagnosis.

5-7 A symptom is a disease condition experienced by a patient. A sign is an objective manifestation of disease than can be observed.

5-8 A parasite is an organism that lives on or within a host and at the host's expense.

5-9 A communicable disease is one that can be transmitted from one person to another.

5-10 Any disease-causing organism is a pathogen.

5-11 The skin and the respiratory, digestive, urinary, and reproductive systems are portals of entry and exit for microorganisms.

5-12 Bacteria, viruses, fungi, protozoa, and algae are studied in microbiology.

5-13 Microorganisms that normally live in or on the body are referred to as the normal flora.

5-14 Endospores are resistant forms of bacteria.

5-15 The three basic shapes of bacteria are cocci (round), bacilli (rod shaped), and curved rods, including vibrios, spirilla, and spirochetes.

5-16 Viruses are smaller than bacteria, are not cellular, and have no organelles. They contain only DNA or RNA, not both.

5-17 A fungus causes a mycotic infection.

5-18 Protozoa are the microorganisms that are most animal-like.

5-19 Helminthology is the study of worms.

5-20 Insects and arachnids (spiders and mites) are two types of arthropods.

5-21 Two levels of asepsis are sterilization and disinfection.

5-22 Handwashing is the single most important measure for preventing the spread of infection.

5-23 An antibiotic is a substance produced by living cells that has the power to kill or arrest the growth of bacteria.

5-24 Stains are the dyes used to color microorganisms.

Answers to Zooming In Questions

5-8 Flagella indicate that the cells in (A) are capable of movement.

5-12 The parasites in (E) are intracellular because they are inside cells. Vectors transmit disease organisms from one host to another.

Chapter 6

Answers to Checkpoint Questions

6-1 The integumentary system comprises the skin and all its associated structures.

6-2 The epidermis is the superficial layer of the skin; the dermis is the deeper layer.

6-3 The subcutaneous layer is composed of loose connective tissue and adipose (fat) tissue.

6-4 The sebaceous glands produce an oily secretion called sebum.

6-5 The sweat glands are the sudoriferous glands.

6-6 A hair follicle is the sheath in which a hair develops.

6-7 The active cells that produce a nail are located in the nail root at the proximal end of the nail.

6-8 Keratin and sebum help prevent dehydration.

6-9 Temperature is regulated through the skin by dilation (widening) and constriction (narrowing) of blood vessels and by evaporation of perspiration from the body surface.

6-10 Melanin, hemoglobin, and carotene are pigments that give color to the skin.

6-11 Cyanosis is a bluish skin discoloration caused by insufficient oxygen.

6-12 Epithelial and connective tissue repair themselves most easily.

6-13 Four factors that affect healing are nutrition, blood supply, infection, and age.

6-14 A lesion is any wound or local damage to tissue.

6-15 Burns are assessed by the depth of the burn and the extent of body surface involved.

6-16 Bacteria, viruses, and fungi are microorganisms that affect the skin.

6-17 Dermatitis is the general term for inflammation of the skin.

6-18 Autoimmunity is the probable cause for chronic inflammatory skin diseases.

6-19 Acne results from overactivity of the sebaceous glands.

6-20 Alopecia is the technical term for baldness.

Answers to Zooming In Questions

6-1 The epidermis is supplied with oxygen and nutrients from blood vessels in the dermis. The subcutaneous layer is located beneath the skin.

6-4 The sebaceous glands and apocrine sweat glands secrete to the outside through the hair follicles. The sweat glands are made of simple cuboidal epithelium.

6-6 Blue color is associated with cyanosis. Yellow color is associated with jaundice.

Chapter 7

Answers to Checkpoint Questions

7-1 The diaphysis is the shaft of a long bone; the epiphysis is the end of a long bone.

7-2 Calcium compounds are deposited in the intercellular matrix of the embryonic skeleton to harden it.

7-3 The cells found in bone are osteoblasts, which build bone tissue; osteocytes, which maintain bone; and osteoclasts, which break down (resorb) bone.

7-4 Compact bone makes up the main shaft of a long bone and the outer layer of other bones; spongy (cancellous) bone makes up the ends of the long bones and the center of other bones. It also lines the medullary cavity of the long bones.

7-5 The epiphyseal plates are the secondary growth centers of a long bone.

7-6 The markings on bones help form joints, are points for muscle attachments, and allow passage of nerves and blood vessels.

7-7 The skeleton of the trunk consists of the vertebral column and the bones of the thorax, which are the ribs and sternum.

7-8 The five regions of the vertebral column are the cervical vertebrae, thoracic vertebrae, lumbar vertebrae, sacrum, and coccyx.

7-9 The four regions of the appendicular skeleton are the shoulder girdle, upper extremity, pelvic bones, and lower extremity.

7-10 Five categories of bone disorders are metabolic disorders, tumors, infections, structural disorders, and fractures.

7-11 The three types of joints based on the degree of movement allowed are synarthrosis, amphiarthrosis, and diarthrosis.

7-12 A synovial joint or diarthrosis is the most freely moveable type of joint.

7-13 Arthritis is the most common type of joint disorder.

7-14 An arthroscope is used to examine and repair joints.

Answers to Zooming In Questions

7-2 The periosteum is the membrane on the outside of a long bone; the endosteum is the membrane on the inside of a long bone.

7-3 Osteocytes are located in the spaces (lacunae) of compact bone.

7-5 The maxilla and palatine bones make up each side of the hard palate. A foramen is a hole. The ethmoid makes up the superior and middle conchae.

7-6 The anterior fontanel is the largest fontanel.

7-7 The cervical and lumbar vertebrae form a convex curve. The lumbar vertebrae are the largest and heaviest because they bear the most weight.

7-8 The costal cartilages attach to the ribs.

7-9 The prefix *supra-* means above; the prefix *infra-* means below.

7-10 The ulna is the medial bone of the forearm.

7-12 The olecranon of the ulna forms the bony prominence of the elbow.

7-13 There are 14 phalanges on each hand.

7-14 The ischium is nicknamed the "sit bone."

7-16 The fibula is the lateral bone of the leg; the tibia is weight bearing.

7-17 The calcaneus is the heel bone; the talus forms a joint with the tibia.

7-20 Kyphosis is an exaggerated concave curve; lordosis is an exaggerated convex curve.

7-22 The greater trochanter is a point for ligament attachment; articular (hyaline) cartilage covers the ends of a bone.

Chapter 8

Answers to Checkpoint Questions

8-1 The three types of muscle are smooth muscle, cardiac muscle, and skeletal muscle.

8-2 The three main functions of skeletal muscle are movement of the skeleton, maintenance of posture, and generation of heat.

8-3 Bundles of muscle fibers are called fascicles.

8-4 The membrane or transmembrane potential is the term for the difference in electrical charge on the two sides of a membrane.

8-5 The neuromuscular junction (NMJ) is the special synapse where a nerve cell makes contact with a muscle cell.

8-6 Acetylcholine (ACh) is the neurotransmitter involved in the stimulation of skeletal muscle cells.

8-7 Actin and myosin are the filaments that interact to produce a muscle contraction.

8-8 Calcium is needed to allow actin and myosin to interact.

8-9 ATP is the compound produced by the oxidation of nutrients that supplies the energy for muscle contraction.

8-10 Myoglobin stores reserves of oxygen in muscle cells.

8-11 The two main types of muscle contraction are isotonic and isometric contractions.

8-12 The origin is the muscle's attachment to a less movable part of the skeleton; the insertion is the attachment to a movable part of the skeleton. When a muscle contracts, the insertion is brought closer to the origin.

8-13 The muscle that produces a movement is the prime mover; the muscle that produces the opposite movement is the antagonist.

8-14 A third-class lever represents the action of most muscles. In this system, the fulcrum is behind the point of effort and the resistance (weight).

8-15 The diaphragm is the muscle most important in breathing.

8-16 The muscles of the abdominal wall are strengthened by having the muscle fibers run in different directions.

Answers to Zooming In Questions

8-1 The endomysium is the innermost layer of connective tissue in a skeletal muscle. Perimysium surrounds a fascicle of muscle fibers.

8-4 The actin and myosin filaments do not change in length as muscle contracts; they simply overlap more.

8-6 Contraction of the brachialis produces flexion at the elbow.

8-7 In a third-class lever system, the fulcrum is behind the point of effort and the resistance (weight).

8-10 The frontalis, temporalis, nasalis, and zygomaticus are named for the bones they are near.

8-11 Carpi means wrist; digitorum means fingers.

8-13 Rectus means straight; oblique means at an angle.

8-15 Four muscles make up the quadriceps femoris.

8-16 The Achilles tendon inserts on the calcaneus (heel bone).

Chapter 9

Answers to Checkpoint Questions

9-1 The two structural divisions of the nervous system are the central nervous system and peripheral nervous system.

9-2 The somatic nervous system is voluntary and controls skeletal muscle; the autonomic (visceral) nervous system is involuntary and controls involuntary muscles (cardiac and smooth muscle) and glands.

9-3 A dendrite is a neuron fiber that carries impulses toward the cell body; an axon is a neuron fiber that carries impulses away from the cell body.

9-4 Myelinated fibers are white; unmyelinated tissue is gray.

9-5 Sensory (afferent) nerves convey impulses toward the CNS; motor (efferent) nerves convey impulses away from the CNS.

9-6 A nerve is a bundle of neuron fibers in the peripheral nervous system; a tract is a bundle of neuron fibers in the central nervous system.

9-7 Neuroglia (glial cells) are the nonconducting cells of the nervous system that protect, nourish, and support the neurons.

9-8 The two stages of an action potential are the rising phase or depolarization, when the charge on the membrane reverses, and the falling phase or repolarization, when the charge returns to the resting state.

9-9 Sodium ion (Na^+) and potassium ion (K^+) are the two ions involved in generating an action potential.

9-10 The myelin sheath speeds conduction along an axon.

9-11 A synapse is the junction between two neurons.

9-12 Neurotransmitters are the chemicals that carry information across the synaptic cleft.

9-13 The gray matter forms an H-shaped section in the center of the cord and extends in two pairs of columns called the dorsal and ventral horns. The white matter is located around the gray matter.

9-14 Tracts in the white matter of the spinal cord carry impulses to and from the brain. Ascending tracts conduct toward the brain; descending tracts conduct away from the brain.

9-15 There are 31 pairs of spinal nerves.

9-16 The dorsal root of a spinal nerve contains sensory fibers; the ventral root contains motor fibers.

9-17 A plexus is a network of spinal nerves.

9-18 A reflex arc is a pathway through the nervous system from a stimulus to an effector.

9-19 There are two motor neurons in each motor pathway of the autonomic nervous system (ANS).

9-20 The sympathetic system stimulates a stress response; the parasympathetic system reverses a stress response.

9-21 Cerebrospinal fluid (CSF) is removed in a lumbar puncture.

9-22 The word root plegia means paralysis.

9-23 A neuropathy is any disorder of the nerves.

Answers to Zooming In Questions

9-1 The brain and spinal cord make up the central nervous system; the cranial nerves and spinal nerves make up the peripheral nervous system.

9-2 The neuron shown is a motor neuron because it is carrying information away from the CNS toward an effector organ. It is part of the somatic nervous system, because the effector is skeletal muscle.

9-5 The endoneurium is the deepest layer of connective tissue in a nerve; the epineurium is the outermost layer.

9-8 The charge on the membrane reverses at the point of an action potential.

9-11 The spinal cord is not as long as is the spinal column. There are seven cervical vertebrae and eight cervical spinal nerves.

9-12 The sacral spinal nerves (S1) carry impulses from the skin of the toes. The cervical spinal nerves (C6, 7, 8) carry impulses from the skin of the anterior hand and fingers.

9-13 The reflex arc shown is a somatic reflex arc, as it involves a skeletal muscle effector. An interneuron is located between the sensory and motor neurons in the CNS.

9-14 The parasympathetic system has ganglia closer to the effector organs.

Chapter 10

Answers to Checkpoint Questions

10-1 The main divisions of the brain are the cerebrum, diencephalon, brain stem, and cerebellum.

10-2 The three layers of the meninges are the dura mater, arachnoid, and pia mater.

10-3 CSF is produced in the ventricles of the brain. The two lateral ventricles are in the cerebral hemispheres, the third ventricle is in the diencephalon, and the fourth is between the brain stem and the cerebellum.

10-4 The frontal, parietal, temporal, and occipital lobes are the four surface lobes of each cerebral hemisphere.

10-5 The cerebral cortex is the outer layer of gray matter of the cerebral hemispheres where higher functions occur.

10-6 The thalamus and hypothalamus are the two main portions of the diencephalon. The thalamus directs sensory input to the cerebral cortex; the hypothalamus helps to maintain homeostasis.

10-7 The midbrain, pons, and medulla oblongata are the three subdivisions of the brain stem.

10-8 The cerebellum aids in coordination of voluntary muscles, maintenance of balance, and maintenance of muscle tone.

10-9 The cingulate gyrus, hippocampus, amygdala, and portions of the hypothalamus are four structures in the limbic system.

10-10 The basal nuclei modulate motor inputs and facilitate routine motor tasks.

10-11 The reticular activating system keeps one awake and attentive and screens out unnecessary sensory input.

10-12 Stroke is the common term for cerebrovascular accident.

10-13 Neuroglia are commonly involved in brain tumors.

10-14 There are 12 pairs of cranial nerves.

10-15 Cranial nerves may be sensory, motor, and mixed. A mixed nerve has both sensory and motor fibers.

Answers to Zooming In Questions

10-1 The cerebrum is the largest part of the brain. The brain stem, specifically the medulla oblongata, connects with the spinal cord.

10-2 The dural (venous) sinuses are formed where the dura mater divides into two layers. There are three layers of meninges.

10-3 The fourth ventricle is continuous with the central canal of the spinal cord.

10-4 The central sulcus separates the frontal from the parietal lobe; the lateral sulcus separates the temporal from the frontal and parietal lobes.

10-5 The primary somatosensory area is posterior to the central sulcus. The primary motor area is anterior to the central sulcus.

10-6 The pituitary gland is attached to the hypothalamus.

10-8 The cingulate gyrus is the part of the cerebral cortex that contributes to the limbic system.

10-13 An epidural hematoma forms outside of the dura mater. A subdural hematoma forms below the dura mater.

Chapter 11

Answers to Checkpoint Questions

11-1 A sensory receptor is a part of the nervous system that detects a stimulus.

11-2 Based on types of stimulus, there are chemoreceptors that respond to chemicals in solution; photoreceptors that respond to light; thermoreceptors that respond to temperature; and mechanoreceptors that respond to movement.

11-3 The general senses are widely distributed; the special senses are in specialized sense organs.

11-4 When a sensory receptor adapts to a stimulus, it fails to respond.

11-5 Structures that protect the eye include the skull bones, eyelids, eyelashes and eyebrow, lacrimal gland, and conjunctiva.

11-6 The extrinsic eye muscles pull on the eyeball so that both eyes center on one visual field, a process known as convergence.

11-7 The optic nerve (cranial nerve II) carries visual impulses from the retina to the brain.

11-8 The fibrous tunic is the outermost tunic; the vascular tunic is the middle tunic; the nervous tunic is the innermost tunic.

11-9 The structures that refract light as it passes through the eye are the cornea, aqueous humor, lens, and vitreous body.

11-10 The ciliary muscle contracts and relaxes to regulate the thickness of the lens and allow accommodation for near and far vision.

11-11 The iris adjusts the size of the pupil to regulate the amount of light that enters the eye.

11-12 The rods and cones are the receptor cells of the retina.

11-13 Hyperopia, myopia, astigmatism, and presbyopia are four errors of refraction.

11-14 Cataract is cloudiness of the lens.

11-15 Glaucoma is a disorder caused by excess fluid pressure in the eye.

11-16 The three divisions of the ear are the outer, middle, and inner ear.

11-17 The ear ossicles are the malleus, incus, and stapes. They transmit sound waves from the tympanic membrane to the inner ear.

11-18 The fluids in the inner ear are the endolymph and the perilymph. Endolymph is found in the membranous labyrinth; perilymph is found around the membranous labyrinth in the bony labyrinth.

11-19 The hearing organ is the spiral organ (organ of Corti) located in the cochlear duct within the cochlea.

11-20 The equilibrium receptors are located in the vestibule and semicircular canals of the inner ear.

11-21 Conductive hearing loss and sensorineural hearing loss are the two types of hearing loss.

11-22 Vertigo is an abnormal sensation of spinning.

11-23 The senses of taste and smell are the special senses that respond to chemical stimuli.

11-24 Touch, pressure, temperature, position (proprioception), and pain are the general senses.

11-25 Proprioceptors are the receptors that respond to change in position. They are located in muscles, tendons, and joints.

Answers to Zooming In Questions

11-3 Location and direction of fibers are characteristics used in naming the extrinsic eye muscles.

11-4 The cornea is the anterior structure continuous with the sclera.

11-5 The suspensory ligaments hold the lens in place.

11-6 The ciliary muscle is contracted when the lens is rounded.

11-7 The circular muscle fibers of the iris contract to make the pupil smaller. The radial muscles contract to make the pupil larger.

11-12 The tympanic membrane separates the outer ear from the middle ear.

11-13 Perilymph is in contact with the vestibular membrane.

11-14 The tectorial membrane contains the cilia of the hair cells.

11-15 The cilia of the macular cells bend when the otolithic membrane moves.

11-18 The dendrite of an olfactory receptor cell interacts with an odorant.

Chapter 12

Answers to Checkpoint Questions

12-1 Hormones are chemicals that have specific regulatory effects on certain cells or organs in the body. Some of their effects are to regulate growth, metabolism, reproduction, and behavior.

12-2 The target tissue is the specific tissue that responds to a hormone.

12-3 Hormones chemically are amino acid compounds and steroids.

12-4 Negative feedback is the main method used to regulate hormone secretion.

12-5 The hypothalamus controls the pituitary.

12-6 Antidiuretic hormone (ADH) and oxytocin are released from the posterior pituitary.

12-7 Growth hormone (GH), thyroid-stimulating hormone (TSH), adrenocorticotropic hormone (ACTH), prolactin (PRL), follicle-stimulating hormone (FSH), and luteinizing hormone (LH) are the hormones released from the anterior pituitary.

12-8 Thyroid hormones increase the metabolic rate in cells.

12-9 Iodine is needed to produce thyroid hormones.

12-10 A goiter is an enlarged thyroid gland.

12-11 Calcium is regulated by parathyroid hormone and calcitriol.

12-12 Epinephrine (adrenaline) is the main hormone produced by the adrenal medulla.

12-13 Glucocorticoids, mineralocorticoids, and sex hormones are released by the adrenal cortex.

12-14 Cortisol raises blood glucose levels.

12-15 Insulin and glucagon produced by the pancreatic islets regulate glucose levels.

12-16 Insulin is low or ineffective in cases of diabetes mellitus.

12-17 Secondary sex characteristics are features associated with gender other than reproductive activity.

12-18 The pineal gland secretes melatonin.

12-19 The kidneys produce erythropoietin. Bone tissue produces osteocalcin. The atria of the heart produce ANP.

12-20 Epinephrine, cortisol, ADH, and growth hormone are four hormones released in times of stress.

Answers to Zooming In Questions

12-2 The infundibulum connects the hypothalamus and the pituitary gland.

12-3 The anterior pituitary controls the thyroid gland.

12-4 The larynx is superior to the thyroid; the trachea is inferior to the thyroid.

12-6 The cortex is the outer region of the adrenal gland; the medulla is the inner region.

12-9 Insulin and glucagon mainly influence the liver and skeletal muscles.

Chapter 13

Answers to Checkpoint Questions

13-1 Four types of substances transported in the blood are gases, nutrients, waste, and hormones.

13-2 7.35 to 7.45 is the pH range of the blood.

13-3 The two main components of blood are the liquid portion, or plasma, and the formed elements, which include the cells and cell fragments.

13-4 Protein is the most abundant type of substance in plasma aside from water.

13-5 All blood cells form in red bone marrow.

13-6 Hematopoietic stem cells give rise to all blood cells.

13-7 The main function of hemoglobin is to carry oxygen in the blood.

13-8 The granular leukocytes are neutrophils, eosinophils, and basophils. The agranular leukocytes are lymphocytes and monocytes.

13-9 The main function of leukocytes is to destroy pathogens. They also remove other foreign material and cellular debris.

13-10 Platelets are essential to blood clotting (coagulation).

13-11 Hemostasis is the process that stops blood loss.

13-12 Fibrin forms a clot.

13-13 Serum does not contain clotting factors; plasma does.

13-14 An antigen is any substance that activates an immune response.

13-15 The four ABO blood type groups are A, B, AB, and O.

13-16 The Rh factor is associated with incompatibility during pregnancy.

13-17 Blood is commonly separated into its component parts by a centrifuge.

13-18 Anemia is an abnormally low level of red cells or hemoglobin in the blood.

13-19 Leukemia is a cancer of the tissues that produce white cells, resulting in an excess number of incompetent leukocytes in the blood.

13-20 Platelets are low in cases of thrombocytopenia.

13-21 A hematocrit measures the relative volume of red cells in blood.

13-22 Hemoglobin is expressed as grams per deciliter (dL) of whole blood or as a percentage of a given standard, usually the average male normal of 15.6 grams hemoglobin per dL.

Answers to Zooming In Questions

13-2 Erythrocytes (red blood cells) are the most numerous cells in the blood.

13-3 Erythrocytes are described as biconcave because they have an inward depression on both sides.

13-5 Simple squamous epithelium makes up the capillary wall.

13-7 The suffix *-ase* indicates that prothrombinase is an enzyme. The prefix *pro-* indicates that prothrombin is a precursor.

13-8 No. To test for Rh antigen, you have to use anti-Rh serum. The two types of antigens are independent.

13-10 A scanning electron microscope (SEM) was used to take this picture. This type of microscope gives a three-dimensional view.

Chapter 14

Answers to Checkpoint Questions

14-1 The endocardium is the innermost layer of the heart wall; the myocardium is the middle layer; the epicardium is the outermost layer.

14-2 The pericardium is the sac that encloses the heart.

14-3 The atrium is the heart's upper receiving chamber on each side. The ventricle is the lower pumping chamber on each side.

14-4 The valves direct the forward flow of blood through the heart.

14-5 The coronary circulation is the system that supplies blood to the myocardium.

14-6 Systole is the contraction phase of the cardiac cycle. Diastole is the relaxation phase.

14-7 Cardiac output is the amount of blood pumped by each ventricle in one minute. It is determined by the stroke volume, the amount of blood ejected from the ventricle with each beat, and by the heart rate, the number of times the heart beats per minute.

14-8 The scientific name for the pacemaker is the sinoatrial (SA) node.

14-9 The autonomic nervous system is the main influence on the rate and strength of heart contractions.
14-10 A heart murmur is an abnormal heart sound.
14-11 ECG (EKG) stands for electrocardiogram or electrocardiography.
14-12 Catheterization is the use of a thin tube threaded through a vessel for diagnosis or repair.
14-13 Coronary angiography and coronary computed tomography angiography (CTA) are techniques that use a dye and x-rays to visualize the coronary arteries.
14-14 Endocarditis, myocarditis, and pericarditis are three types of heart inflammation.
14-15 Arrhythmia and dysrhythmia are terms for an abnormal heart rhythm.
14-16 Congenital heart disease is a heart disorder present at birth.
14-17 Certain streptococci can cause rheumatic fever.
14-18 Atherosclerosis commonly causes narrowing of the vessels in coronary artery disease.
14-19 Myocardial infarction (MI) is the medical term for a heart attack.
14-20 A defibrillator is a device that restores a normal heart rhythm.
14-21 Aspirin and warfarin act as anticoagulants to prevent heart attacks.
14-22 An artificial pacemaker is a device wired to the heart to regulate the heartbeat.
14-23 CABG stands for coronary artery bypass graft.

Answers to Zooming In Questions

14-1 The left lung is smaller than the right because the heart is located more toward the left of the thorax.
14-2 The myocardium is the thickest layer of the heart wall.
14-4 The left ventricle of the heart has the thickest wall.
14-5 The right AV valve has three cusps; the left AV valve has two cusps.
14-6 The coronary sinus is the largest cardiac vein. It leads into the right atrium.
14-8 The AV (tricuspid and mitral) valves close when the ventricles contract. The semilunar (pulmonary and aortic valves) open when the ventricles contract.
14-9 The internodal pathways connect the sinoatrial (SA) and atrioventricular (AV) nodes.
14-10 The vagus nerve (cranial nerve X) carries parasympathetic impulses to the heart.
14-11 The length of the cardiac cycle in this diagram is 0.8 seconds.

Chapter 15

Answers to Checkpoint Questions

15-1 The five types of blood vessels are arteries, arterioles, capillaries, venules, and veins.
15-2 The pulmonary circuit carries blood from the heart to the lungs and back to the heart; the systemic circuit carries blood to and from all remaining tissues in the body.
15-3 Smooth muscle makes up the middle layer of arteries and veins. Smooth muscle is involuntary muscle controlled by the autonomic nervous system.
15-4 A single layer of squamous epithelium makes up the wall of a capillary.
15-5 The aorta is divided into the ascending aorta, aortic arch, thoracic aorta, and abdominal aorta. The thoracic aorta and abdominal aorta make up the descending aorta.
15-6 The brachiocephalic, left common carotid, and left subclavian arteries are the three branches of the aortic arch.
15-7 The brachiocephalic artery supplies the arm and head on the right side.
15-8 An anastomosis is a communication between two vessels.
15-9 Superficial veins are near the surface; deep veins are closer to the interior and generally parallel the arteries.
15-10 The superior vena cava and inferior vena cava drain the systemic circuit and empty into the right atrium.
15-11 A venous sinus is a large channel that drains blood low in oxygen.
15-12 The hepatic portal system takes blood from the abdominal organs to the liver.
15-13 Blood pressure helps push material out of a capillary; blood osmotic pressure helps draw materials into a capillary.
15-14 Vasodilation (widening) and vasoconstriction (narrowing) are the two types of vasomotor changes.
15-15 Vasomotor activities are regulated in the medulla of the brain stem.
15-16 Pulse is the wave of pressure that begins at the heart and travels along the arteries.
15-17 Blood pressure is the force exerted by blood against the walls of the vessels.
15-18 Blood vessel diameter, as regulated by vasomotor activities, is the most important factor in determination of peripheral resistance. Other factors are blood viscosity and blood vessel length.
15-19 Baroreceptors are the sensors for blood pressure.
15-20 Systolic and diastolic pressures are measured when taking blood pressure.
15-21 Hypertension is the medical term for high blood pressure.
15-22 Arteriosclerosis is the general term for hardening of the arteries. The most common form of arteriosclerosis is atherosclerosis.
15-23 An aneurysm is a bulging sac in a vessel.
15-24 Hemorrhage is the term for profuse bleeding.
15-25 Circulatory shock is inadequate blood flow to the tissues.
15-26 A twisted and swollen vein is described as varicose. The abnormal vein is a varix.

Answers to Zooming In Questions

15-1 Pulmonary arteries contain oxygen-poor blood. Pulmonary veins contain oxygen-rich blood.
15-2 Veins have valves that control blood flow.
15-3 An artery has a thicker wall than a vein of comparable size.
15-4 There is one brachiocephalic artery.
15-5 The left and right common iliac arteries branch from the terminal aorta.
15-7 There are two brachiocephalic veins.
15-8 The internal jugular vein receives blood from the transverse sinus.
15-9 The hepatic veins drain into the inferior vena cava.
15-11 The proximal valve is closer to the heart.
15-12 Pulse pressure drops to 0 in the arterioles.
15-13 Heart rate increases when blood pressure falls.

Chapter 16

Answers to Checkpoint Questions

16-1 The lymphatic system helps maintain fluid balance by bringing excess fluid and proteins from the tissues to the blood, protects against foreign material and foreign or abnormal cells, and absorbs fats from the small intestine.
16-2 The lymphatic capillaries are more permeable than are blood capillaries and begin blindly. They are closed at one end and do not bridge two vessels.

16-3 The right lymphatic duct and the thoracic duct are the two main lymphatic vessels.

16-4 Lymph nodes filter lymph. They also have lymphocytes to fight infection.

16-5 The spleen filters blood.

16-6 T cells of the immune system develop in the thymus.

16-7 MALT stand for mucosa-associated lymphoid tissue; GALT stands for gut-associated lymphoid tissue.

16-8 Tonsils are located in the vicinity of the pharynx (throat).

16-9 Lymphadenopathy is any disease of the lymph nodes.

16-10 Lymphedema results from blockage of lymph flow.

16-11 Splenomegaly is enlargement of the spleen.

16-12 Lymphoma is any tumor of lymphoid tissue. Two examples of malignant lymphoma are Hodgkin lymphoma and non-Hodgkin lymphoma.

Answers to Zooming In Questions

16-1 A vein receives lymph collected from the body.

16-3 Mammary, axillary, subscapular, and interpectoral nodes are some nodes that receive lymph drainage from the breast.

16-4 An afferent vessel carries lymph into a node; an efferent vessel carries lymph out of a node.

Chapter 17

Answers to Checkpoint Questions

17-1 Virulence is the ability of an organism to overcome host defenses.

17-2 Innate barriers such as the skin, mucous membranes, body secretions, and reflexes are the first line of defense against the invasion of pathogens.

17-3 Innate cells and chemicals are two components in the second line of defense against infection.

17-4 Heat, redness, swelling, and pain are the four signs of inflammation.

17-5 Fever boosts the immune system by stimulating phagocytes, increasing metabolism, and decreasing the ability of certain microorganisms to multiply.

17-6 Adaptive immunity is an individual's power to resist or overcome any particular disease or its products. Adaptive immunity is acquired during a person's lifetime, usually from contact with a disease organism or a vaccine.

17-7 An antigen is any foreign substance, usually a protein, that induces an immune response.

17-8 The four types of T cells are cytotoxic, helper, regulatory, and memory T cells.

17-9 Antigen-presenting cells (APC) take in and digest a foreign antigen. They then display fragments of the antigen in their plasma membrane along with self (MHC) antigens that a T cell can recognize.

17-10 An antibody is a substance produced in response to an antigen.

17-11 Plasma cells, derived from B cells, produce antibodies.

17-12 Active adaptive immunity involves a person's own immune system. Passive adaptive immunity depends on antibodies from an outside source.

17-13 Natural adaptive immunity results from contact with a disease organism or obtaining antibodies from an outside source without medical intervention. Artificial adaptive immunity results from administration of a vaccine or antiserum.

17-14 A vaccine is a prepared substance that induces an immune response.

17-15 A booster is a repeated inoculation given to increase antibodies to a disease.

17-16 An antiserum is an antibody prepared in an outside source, usually an animal. Antisera are used in emergencies to provide quick passive immunization against a disease organism, toxin, or venom.

17-17 Four types of immune disorders are allergy autoimmunity, immune deficiency diseases, and multiple myeloma.

17-18 The immune system guards against cancer by destroying abnormal cells in the process of immune surveillance.

17-19 The tendency of every organism to destroy foreign substances is the greatest obstacle to transplantation of tissues from one person to another.

Answers to Zooming In Questions

17-3 Chemicals released by damaged tissues dilate blood vessels and increase blood flow, causing heat, redness, swelling, and pain.

17-4 Digestive enzymes are contained in the lysosome that joins the phagocytic vesicle.

17-5 Plasma cells and memory cells develop from activated B cells.

Chapter 18

Answers to Checkpoint Questions

18-1 The four phases of respiration are pulmonary ventilation, external gas exchange, gas transport in the blood, and internal gas exchange.

18-2 As air passes over the nasal mucosa, it is filtered, warmed, and moistened.

18-3 The three regions of the pharynx are the nasopharynx, oropharynx, and laryngopharynx.

18-4 The scientific name for the throat is the pharynx; the voice box is the larynx; the windpipe is the trachea.

18-5 The two mainstem (primary) bronchi are formed by the inferior branching of the trachea.

18-6 Cilia on the cells that line the respiratory passageways move fluid with entrapped impurities upward to be eliminated from the respiratory system.

18-7 Gas exchange in the lungs occurs in the alveoli.

18-8 The pleura is the membrane that encloses the lung.

18-9 Inhalation and exhalation are the two phases of quiet breathing. Inhalation is active, and exhalation is passive.

18-10 Surfactant produced by lung cells aids in compliance.

18-11 A gas's partial pressure on either side of a membrane determines its direction of diffusion. Partial pressure is expressed in millimeter mercury (mm Hg).

18-12 Hemoglobin in red blood cells holds almost all of the oxygen carried in the blood.

18-13 The main form in which carbon dioxide is carried in the blood is as bicarbonate ion.

18-14 The medulla of the brain stem has the centers that set the basic pattern of respiration.

18-15 The phrenic nerve is the motor nerve that controls the diaphragm.

18-16 Carbon dioxide is the main chemical controller of respiration.

18-17 The word ending *-pnea* means breathing.

18-18 Six respiratory disorders caused by infections are the common cold, respiratory syncytial virus (RSV), croup, influenza, pneumonia, and tuberculosis.

18-19 Allergic rhinitis is commonly caused by pollen, dust, mold, and animal dander. Asthma may be caused by infection, fumes, smoke, cold air, allergies, and exercise.

18-20 COPD is chronic obstructive pulmonary disease; chronic bronchitis and emphysema are commonly involved.

Answers to Zooming In Questions

18-1 Blood travels to the lungs from the right side of the heart; blood returns to the left side of the heart.

18-2 The heart is located in the medial depression of the left lung.

18-4 The epiglottis is named for its position above the glottis.

18-6 The intercostal muscles are located between the ribs.

18-7 Gas pressure decreases as the volume of its container increases.

18-8 Residual volume cannot be measured with a spirometer; total lung capacity and functional residual capacity cannot be measured with a spirometer.

18-16 The esophagus is posterior to the trachea.

Chapter 19

Answers to Checkpoint Questions

19-1 Food must be broken down by digestion into particles small enough to pass through the plasma membrane of cells.

19-2 The peritoneum is the large serous membrane that lines the abdominopelvic cavity and covers the organs it contains.

19-3 The digestive tract typically has a wall composed of a mucous membrane (mucosa), a submucosa, smooth muscle (muscularis externa), and a serous membrane (serosa).

19-4 Peristalsis occurs in the esophagus. Segmentation occurs in the small intestine.

19-5 Proteins are digested in the stomach.

19-6 The duodenum, jejunum, and ileum are the three divisions of the small intestine.

19-7 Most digestion takes place in the small intestine under the effects of digestive juices from the small intestine and the accessory organs. Most absorption of digested food and water also occurs in the small intestine.

19-8 The large intestine reabsorbs some water and stores, forms, and eliminates the stool. It also houses bacteria that provide some vitamins.

19-9 The salivary glands are the parotid, anterior and inferior to the ear; submandibular (submaxillary), near the body of the lower jaw; and sublingual, under the tongue.

19-10 The liver secretes bile, which emulsifies fats, that is, breaks it down into small particles.

19-11 The gallbladder stores bile and contracts to release it into the duodenum.

19-12 The pancreas secretes sodium bicarbonate, which neutralizes the acidic chyme from the stomach.

19-13 An enzyme is a catalyst that speeds the rate of a chemical reaction but is not used up in the reaction.

19-14 Hydrolysis means "splitting by means of water" as in digestion.

19-15 The pancreas produces the most complete digestive secretions.

19-16 Absorption is the movement of digested nutrients into the circulation.

19-17 The two types of control over the digestive process are nervous control and hormonal control.

19-18 Hunger is the desire for food that can be satisfied by the ingestion of a filling meal. Appetite is a desire for food that is unrelated to a need for food.

19-19 Caries, gingivitis, periodontitis, Vincent disease, oral thrush, and parotitis are common diseases of the mouth and teeth.

19-20 GERD stands for gastroesophageal reflux disease.

19-21 Crohn disease and ulcerative colitis are inflammatory bowel diseases.

19-22 The two forms of constipation are flaccid and spastic.

19-23 Hepatitis is inflammation of the liver.

19-24 Gallstones is the common term for cholelithiasis.

Answers to Zooming In Questions

19-1 The salivary glands are the accessory organs that secrete into the mouth.

19-2 The mesentery is the part of the peritoneum around the small intestine.

19-3 Smooth muscle (circular and longitudinal) is between the submucosa and the serous membrane in the digestive tract wall.

19-4 Peristalsis to move food rapidly is most useful in the esophagus.

19-5 The common name for the gingiva is the gum.

19-6 The oblique muscle layer is an additional muscle layer in the stomach that is not found in the rest of the digestive tract.

19-7 The ileum of the small intestine joins the cecum of the large intestine at the ileocecal valve.

19-8 The sublingual salivary glands are directly below the tongue.

19-9 The accessory organs shown secrete into the duodenum.

19-11 Glucose and fructose are the two monosaccharide components of sucrose.

Chapter 20

Answers to Checkpoint Questions

20-1 The two type of activities that make up metabolism are catabolism, the breakdown of compounds into simpler components, and anabolism, the building of simple compounds into needed substances.

20-2 Cellular respiration is the series of cellular reactions that releases energy from nutrients.

20-3 Pyruvic acid (pyruvate) is the end product of glycolysis.

20-4 Oxygen is the element required for aerobic cellular respiration but not for glycolysis.

20-5 Nitrogen (amine groups) are removed from amino acids before they are metabolized for energy.

20-6 The glycemic index is the term for how rapidly a food raises the blood glucose level.

20-7 An essential amino acid or fatty acid cannot be made metabolically and must be taken in as part of the diet.

20-8 Saturated fats have no double-bonded carbons and have the maximum number of hydrogen atoms possible. Unsaturated fats have one or more double-bonded carbons and less than the maximum number of hydrogen atoms possible.

20-9 Minerals are chemical elements, and vitamins are complex organic substances.

20-10 The average person can metabolize 1/2 oz of pure alcohol per hour.

20-11 Malnutrition is the condition that results from lack of any specific nutrient in the diet.

20-12 BMI stands for body mass index.

20-13 Any chronic loss of appetite is called anorexia.

20-14 The hypothalamus is responsible for regulating body temperature.

20-15 An antipyretic drug reduces fever.

20-16 Cutaneous blood vessels constrict under cold conditions. Under hot conditions, they dilate.

20-17 Hypothermia is excessively low body temperature.

Answers to Zooming In Questions

20-1 Pyruvic acid produces lactic acid under anaerobic conditions; it produces CO_2 and H_2O when metabolized completely using oxygen.

20-6 The BMI is 24 (77 ÷ 3.2 = 24).

20-7 If the fan speed is increased in (B), convection, heat loss will increase. If environmental humidity increases in (C), evaporation, heat loss will decrease.

20-8 Blood flow in the skin is higher under hot conditions.

Chapter 21

Answers to Checkpoint Questions

21-1 Intracellular fluids are located within cells. Extracellular fluids are located outside of cells.

21-2 Interstitial fluid is fluid located between cells in tissues.

21-3 Cations are positively charged ions. Anions are negatively charged ions.

21-4 Water is lost from the body through the kidneys, the skin, the lungs, and the intestinal tract.

21-5 Aldosterone is the hormone from the adrenal cortex that promotes sodium reabsorption in the kidneys.

21-6 Antidiuretic hormone (ADH) is the pituitary hormone that increases water reabsorption in the kidney.

21-7 The acid–base balance of body fluids is maintained by buffer systems, respiration, and kidney function.

21-8 Acidosis is the condition that arises from abnormally low pH of body fluids. Alkalosis is the condition that arises from abnormally high pH of body fluids.

21-9 Edema is the term for accumulation of excess interstitial fluid.

21-10 Hyponatremia is the term for low blood sodium concentration.

21-11 Ascites is the term for effusion of fluid into the abdominal cavity.

21-12 The percentage of sodium chloride in normal saline is 0.9%.

21-13 Glucose is another name for dextrose.

Answers to Zooming In Questions

21-1 Water is lost in feces, through the lungs, through the skin, and in urine.

21-2 Sodium and chloride are highest in extracellular fluids; potassium and phosphate are highest in intracellular fluids.

21-3 Most water is lost in a day in urine.

21-4 High blood potassium increases aldosterone secretion and promotes loss of potassium.

21-5 The kidneys generate bicarbonate ions (buffers).

Chapter 22

Answers to Checkpoint Questions

22-1 The urinary system consists of two kidneys, two ureters, the bladder, and the urethra.

22-2 Systems other than the urinary system that eliminate waste include the digestive, respiratory, and integumentary systems.

22-3 The retroperitoneal space is posterior to the peritoneum.

22-4 The renal artery supplies blood to the kidney, and the renal vein drains blood from the kidney.

22-5 The renal pelvis is the funnel-shaped collecting area that forms the upper end of the ureter.

22-6 The nephron is the functional unit of the kidney.

22-7 The glomerulus is the coil of capillaries in the glomerular capsule.

22-8 Filtration is the process that drives materials out of the glomerulus and into the glomerular capsule.

22-9 Tubular reabsorption is the process that returns materials from the nephron back to the circulation.

22-10 Hydrogen ions (H^+) are moved by tubular secretion to balance pH.

22-11 Antidiuretic hormone (ADH) controls water reabsorption from the collecting duct of the nephron.

22-12 The juxtaglomerular apparatus produces renin when blood pressure drops. The signal for renin production is low sodium in the filtrate leaving the nephron.

22-13 The ureter carries urine from the kidney to the bladder.

22-14 The openings of the ureters and the urethra form the bladder's trigone.

22-15 The urethra carries urine from the bladder to the outside.

22-16 Nitrogen waste, electrolytes, and pigments are normal constituents of urine; glucose, albumin, blood, ketones, white blood cells, and multiple casts are some abnormal constituents of urine.

22-17 The scientific name for stones is calculi.

22-18 Cystitis is inflammation of the bladder.

22-19 Uremia is the term for an excess of nitrogenous waste products in the blood.

22-20 Renal dialysis can be used to eliminate waste products from the blood in cases of renal failure.

Answers to Zooming In Questions

22-1 The ureters carry urine to the bladder, and the urethra carries urine from the bladder.

22-2 The outer region of the kidney is the renal cortex. The inner region of the kidney is the renal medulla.

22-3 The glomerulus is the first capillary bed to receive blood.

22-6 The osmotic gradient is stronger by the renal pelvis than by the distal tubule.

22-7 The peritubular capillaries absorb materials that leave the nephron.

22-9 The urethra passes through the prostate gland in the male.

22-10 Extracorporeal means outside the body.

Chapter 23

Answers to Checkpoint Questions

23-1 Meiosis is the process of cell division that halves the chromosome number in a cell to produce a gamete.

23-2 The spermatozoon, or sperm cell, is the male gamete (sex cell).

23-3 The testis is the male gonad.

23-4 The epididymis is the structure on the surface of the testis that stores sperm.

23-5 Glands that contribute secretions to the semen, aside from the testes, are the seminal vesicles, prostate, and bulbourethral glands.

23-6 The main subdivisions of a spermatozoon are the head, midpiece, and tail (flagellum).

23-7 Follicle-stimulating hormone (FSH) and luteinizing hormone (LH) are the pituitary hormones that regulate male and female reproduction.

23-8 Specialized interstitial cells (Leydig cells) in the testis secrete the main male sex hormone (testosterone).

23-9 Cryptorchidism is the term for failure of the testis to descend.

23-10 The abbreviation STI means sexually transmitted infection.

23-11 Ovum is the female gamete.

23-12 The ovary is the female gonad.

23-13 A fetus develops in the uterus.

23-14 An ovum matures in an ovarian follicle.

23-15 Estrogen and progesterone are the two hormones produced in the ovaries.

23-16 Ovulation releases an ovum from the ovary.

23-17 The follicle becomes a corpus luteum after its ovum is released.

23-18 Menopause is the term for complete cessation of menstrual cycles.

23-19 Contraception is the use of artificial methods to prevent fertilization of an ovum.

23-20 Amenorrhea is the absence of menstrual flow.

23-21 Fibroid is the common term for a benign uterine tumor.

23-22 Oligospermia is a low sperm count.

Answers to Zooming In Questions

23-1 The four glands that empty secretions into the urethra are the testes, seminal vesicles, prostate, and bulbourethral glands. The ductus (vas) deferens receives secretions from the epididymis.

23-2 The corpus spongiosum of the penis contains the urethra.

23-3 The interstitial cells are located between the seminiferous tubules.

23-4 Mitochondria are the organelles that provide energy for sperm cell motility.

23-5 LH stimulates testosterone secretion.

23-8 The fundus of the uterus is the deepest part. The cervix is the most inferior portion.

23-11 Estrogen peaks closest to ovulation. Progesterone peaks after ovulation.

23-15 These pathways show an ascending infection.

Chapter 24

Answers to Checkpoint Questions

24-1 A zygote is formed by the union of an ovum and a spermatozoon.

24-2 The placenta nourishes the developing fetus.

24-3 The umbilical cord carries blood between the fetus and the placenta.

24-4 Fetal circulation is adapted to bypass the lungs.

24-5 The embryonic hormone hCG (human chorionic gonadotropin) maintains the corpus luteum early in pregnancy.

24-6 The heartbeat first appears during the fourth week of gestation.

24-7 The amniotic sac is the fluid-filled sac that holds the fetus.

24-8 The approximate length of pregnancy in days from the time of fertilization is 266.

24-9 Oxytocin is the pituitary hormone that stimulates uterine contractions.

24-10 Parturition is the process of labor and childbirth.

24-11 The afterbirth is expelled during the third stage of labor.

24-12 A cesarean section is used to remove a fetus manually from the uterus.

24-13 The term *viable* describes a fetus able to live outside the uterus.

24-14 The lactiferous ducts carry milk out of the breast.

24-15 The breast secretes colostrum before milk production begins.

24-16 An ectopic pregnancy is one that develops outside the uterine cavity.

24-17 Puerperal infection is an infection that is related to childbirth.

24-18 A mammogram is a radiographic study of the breast.

Answers to Zooming In Questions

24-1 The ovum is fertilized in the uterine tube.

24-2 The purple color signifies a mixture of blood that is low in oxygen with blood that is high in oxygen.

24-5 The umbilical cord connects the fetus to the placenta.

24-7 Depletion of materials, removal of the stimulus, or an outside force can stop positive feedback.

24-9 Identical twins share a single placenta.

24-10 The pectoralis major muscle underlies the breast.

Chapter 25

Answers to Checkpoint Questions

25-1 A gene is an independent unit of heredity. Each gene is a segment of DNA contained in a chromosome.

25-2 A dominant gene is always expressed, regardless of the gene on the matching chromosome.

25-3 A genotype is the genetic makeup of an individual; a phenotype is all the traits that can be observed or tested for.

25-4 Meiosis is the process of cell division that forms the gametes.

25-5 There are 23 chromosomes in each human gamete.

25-6 The sex chromosome combination that determines a female is XX; a male is XY.

25-7 A trait carried on a sex chromosome is described as sex-linked.

25-8 A mutagen is any agent that causes a mutation, a change in a gene or a chromosome.

25-9 A child gets mitochondrial genes from the mother.

25-10 Congenital refers to anything that is present at birth.

25-11 A teratogen is any agent that causes abnormal prenatal development.

25-12 Trisomy 21 is the scientific name for Down syndrome.

25-13 Fragile X syndrome is more common in males.

25-14 A pedigree chart can show the inheritance pattern of a gene within a family.

25-15 Amniocentesis and chorionic villus sampling are two procedures for removing fetal samples for study.

25-16 A karyotype is a picture of the chromosomes arranged in groups according to size and form.

Answers to Zooming In Questions

25-2 Four cells are produced in one complete meiosis.

25-3 25% of children will show the recessive phenotype nontaster. 50% of children will be heterozygous.

25-4 The expected ratio of male to female offspring in a family is 50/50.

25-5 The genotype of a female carrier is X^RX^r.

25-11 The possible genotypes of the two normal children in the F_3 generation are CC or Cc.

25-13 There are 47 chromosomes present. The baby is a girl. Trisomy 21 or Down syndrome is the genetic disorder represented.

▶ Appendix 5

Dissection Atlas

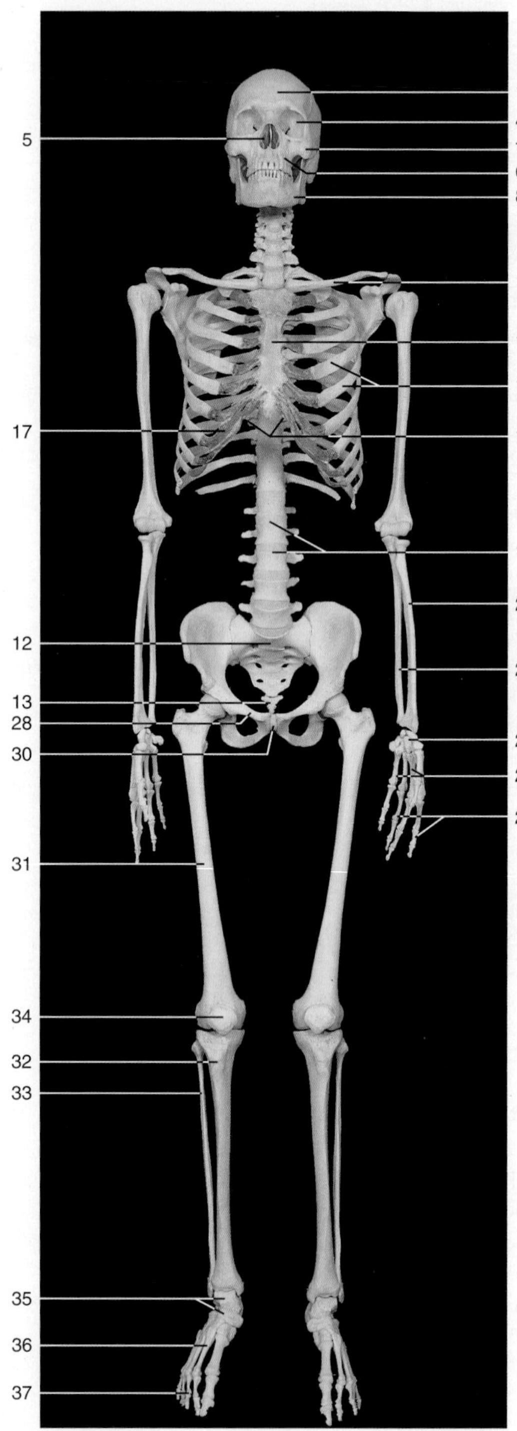

Figure A5-1 Skeleton of a female adult (anterior aspect). (Reprinted with permission from Rohen JW, Yokochi C, Lütjen-Drecoll E. *Color Atlas of Anatomy*, 6th ed. Stuttgart, New York, NY: Schattauer, 2006.)

Axial skeleton

Head

1 Frontal bone
4 Orbit
5 Nasal cavity
6 Maxilla
7 Zygomatic bone
8 Mandible

Trunk and thorax

Vertebral column

12 Sacrum
13 Coccyx
14 Intervertebral disks

Thorax

15 Sternum
16 Ribs
17 Costal cartilage
18 Infrasternal angle

Appendicular skeleton

Upper limb and shoulder girdle

19 Clavicle
22 Radius
23 Ulna
24 Carpal bones
25 Metacarpal bones
26 Phalanges of the hand

Lower limb and pelvis

28 Pubis
30 Symphysis pubis
31 Femur
32 Tibia
33 Fibula
34 Patella
35 Tarsal bones
36 Metatarsal bones
37 Phalanges of the foot

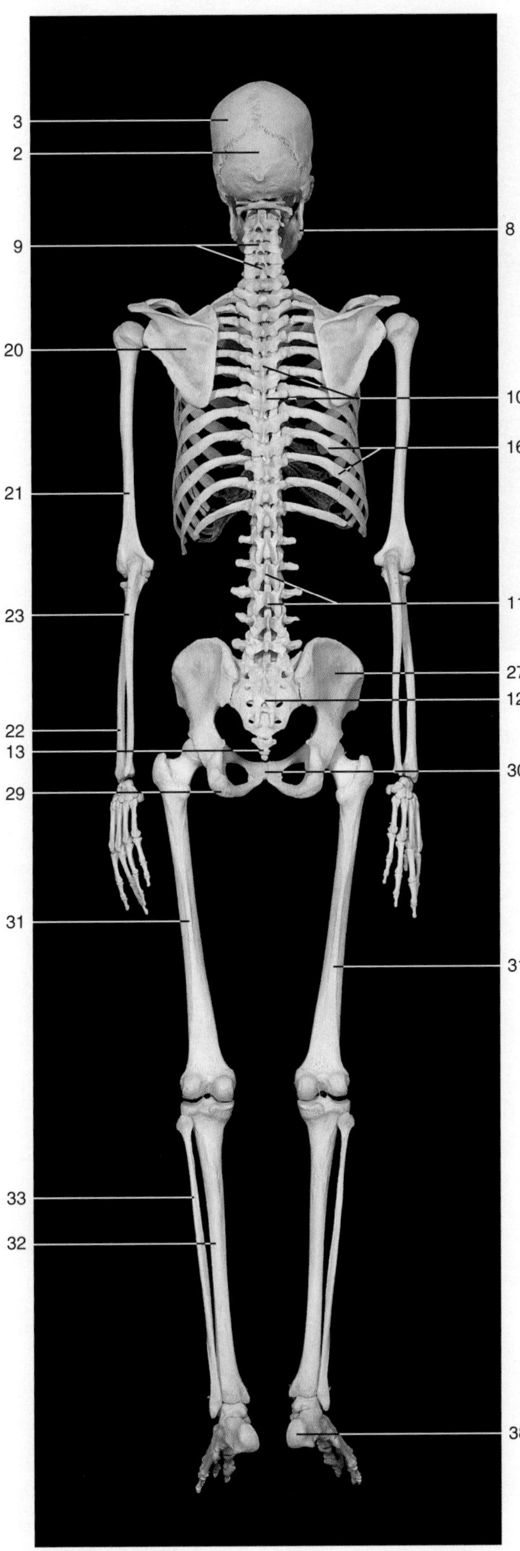

Figure A5-2 Skeleton of a female adult (posterior aspect). (Reprinted with permission from Rohen JW, Yokochi C, Lütjen-Drecoll E. *Color Atlas of Anatomy*, 6th ed. Stuttgart, New York, NY: Schattauer, 2006.)

Axial skeleton

Head

2 Occipital bone
3 Parietal bone
8 Mandible

Trunk and thorax

Vertebral column

9 Cervical vertebrae
10 Thoracic vertebrae
11 Lumbar vertebrae
12 Sacrum
13 Coccyx

Thorax

16 Ribs

Appendicular skeleton

Upper limb and shoulder girdle

20 Scapula
21 Humerus
22 Radius
23 Ulna

Lower limb and pelvis

27 Ilium
29 Ischium
30 Symphysis pubis
31 Femur
32 Tibia
33 Fibula
38 Calcaneus

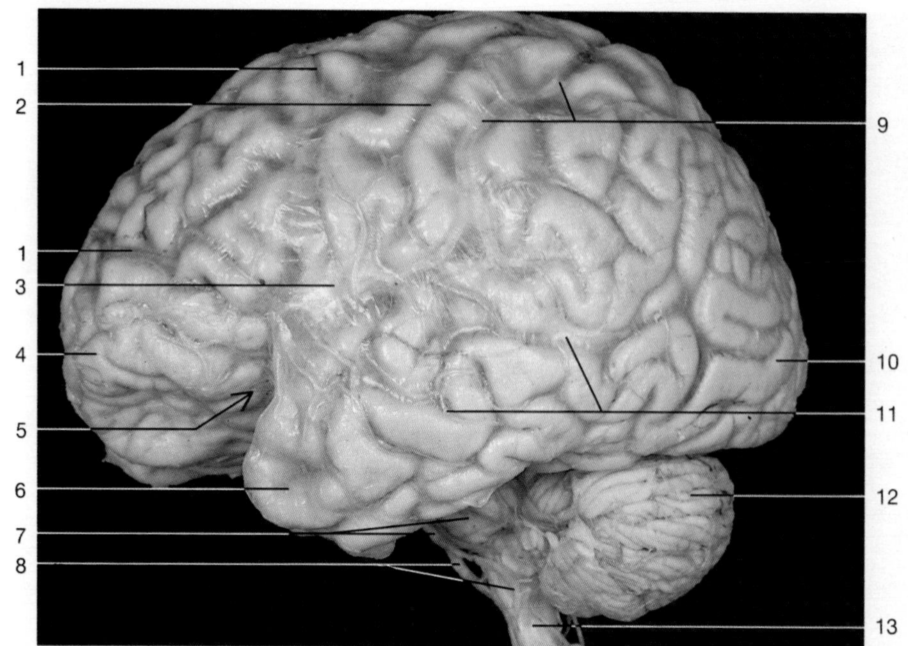

1 Superior cerebral veins
2 Position of central sulcus
3 Position of lateral sulcus
4 Frontal lobe
5 Lateral sulcus (*arrow*)
6 Temporal lobe
7 Pons and basilar artery
8 Vertebral arteries
9 Superior anastomotic vein
10 Occipital lobe
11 Inferior cerebral veins
12 Hemisphere of cerebellum
13 Medulla oblongata

Figure A5-3 Brain with pia mater and arachnoid. Frontal pole to the left (lateral aspect). (Reprinted with permission from Rohen JW, Yokochi C, Lütjen-Drecoll E. *Color Atlas of Anatomy*, 6th ed. Stuttgart, New York, NY: Schattauer, 2006.)

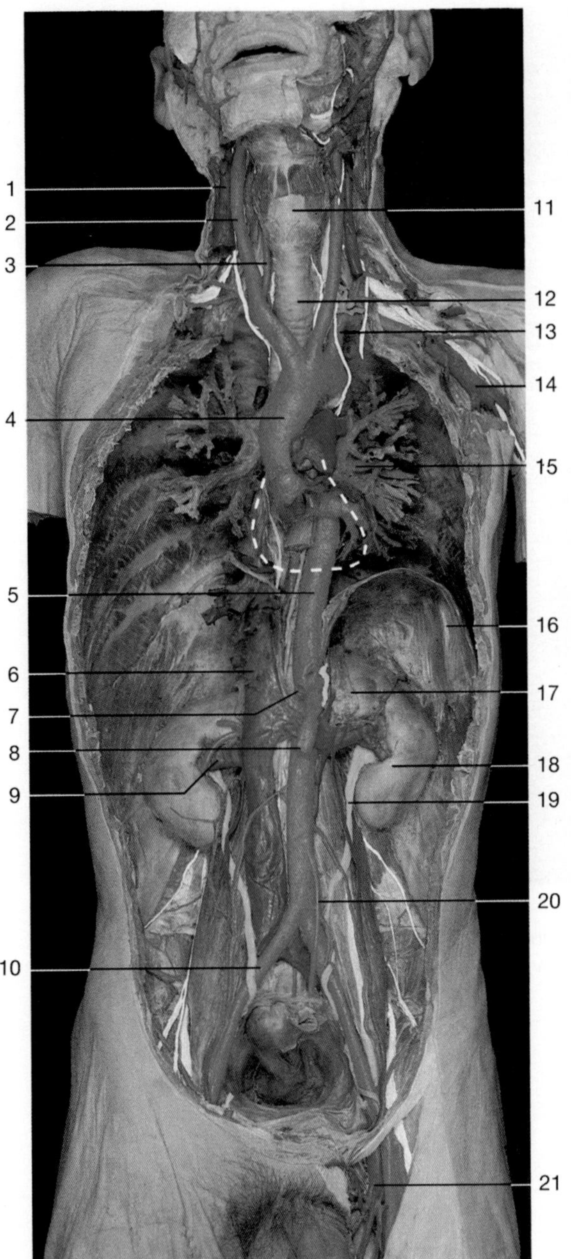

1 Internal jugular vein
2 Common carotid artery
3 Vertebral artery
4 Ascending aorta
5 Descending aorta
6 Inferior vena cava
7 Celiac trunk
8 Superior mesenteric artery
9 Renal vein
10 Common iliac artery
11 Larynx
12 Trachea
13 Left subclavian artery
14 Left axillary vein
15 Pulmonary veins
16 Diaphragm
17 Suprarenal gland
18 Kidney
19 Ureter
20 Inferior mesenteric artery
21 Femoral vein

Figure A5-4 Major vessels of the trunk. The position of the heart is indicated by the *dotted line*. (Reprinted with permission from Rohen JW, Yokochi C, Lütjen-Drecoll E. *Color Atlas of Anatomy*, 6th ed. Stuttgart, New York, NY: Schattauer, 2006.)

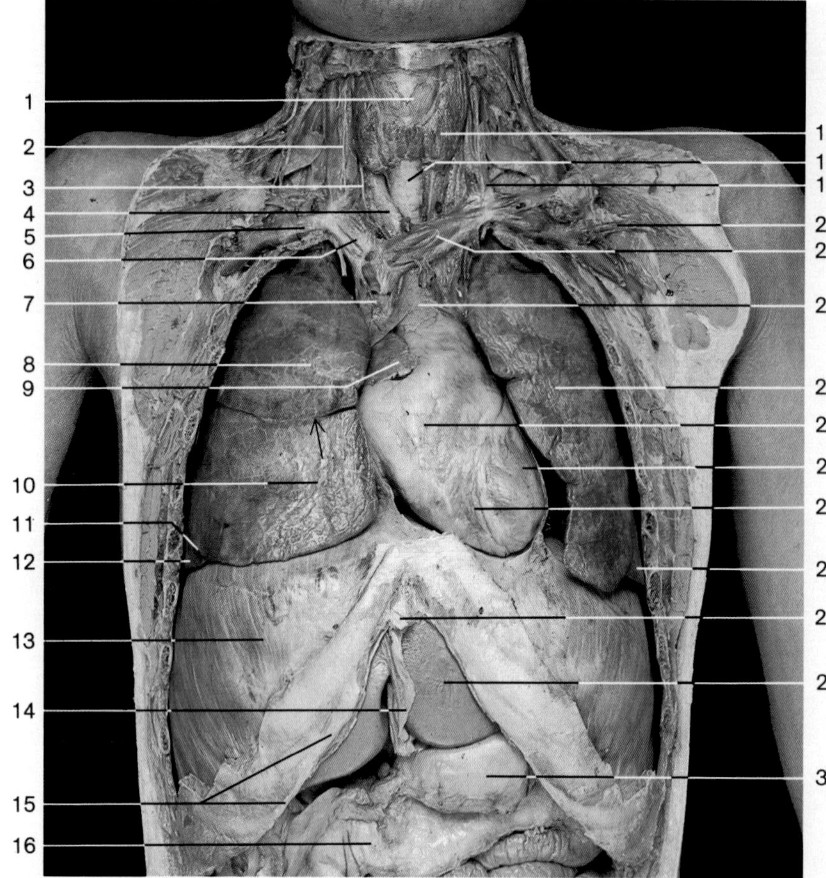

1 Cricothyroid muscle
2 Right internal jugular vein
3 Vagus nerve
4 Right common carotid artery
5 Right subclavian vein
6 Right brachiocephalic vein
7 Superior vena cava
8 Upper lobe of right lung
9 Right auricle
10 Middle lobe of right lung
11 Oblique fissure of right lung
12 Lower lobe of right lung
13 Diaphragm
14 Falciform ligament
15 Costal margin
16 Transverse colon
17 Thyroid gland
18 Trachea
19 Left internal jugular vein
20 Left cephalic vein
21 Left brachiocephalic vein
22 Pericardium (cut edge)
23 Upper lobe of left lung
24 Right ventricle
25 Left ventricle
26 Anterior interventricular sulcus
27 Lower lobe of left lung
28 Xiphoid process
29 Liver
30 Stomach

Figure A5-5 Positions of thoracic organs. The anterior thoracic wall has been removed. *Arrow*: horizontal fissure of the right lung. (Reprinted with permission from Rohen JW, Yokochi C, Lütjen-Drecoll E. *Color Atlas of Anatomy*, 6th ed. Stuttgart, New York, NY: Schattauer, 2006.)

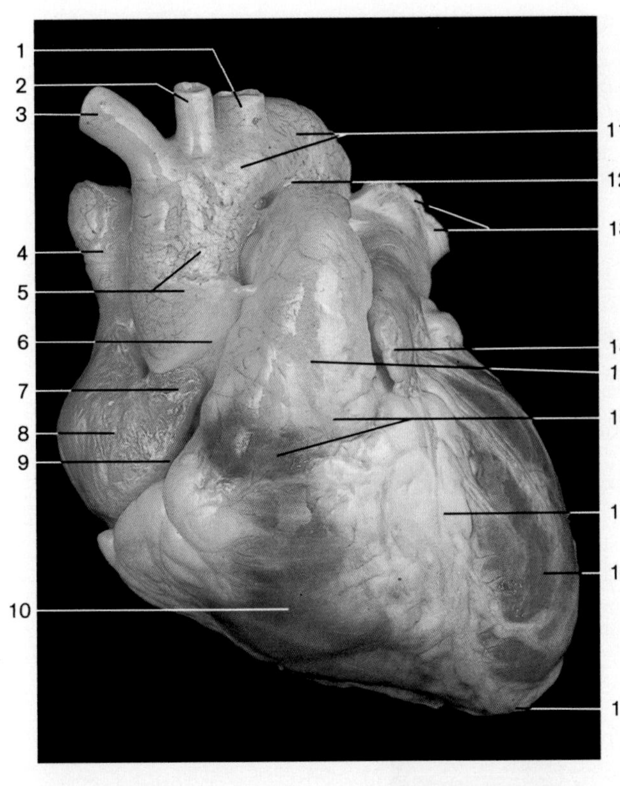

1 Left subclavian artery
2 Left common carotid artery
3 Brachiocephalic trunk
4 Superior vena cava
5 Ascending aorta
6 Bulb of the aorta
7 Right auricle
8 Right atrium
9 Coronary sulcus
10 Right ventricle
11 Aortic arch
12 Ligamentum arteriosum
13 Left pulmonary veins
14 Left auricle
15 Pulmonary trunk
16 Sinus of pulmonary trunk
17 Anterior interventricular sulcus
18 Left ventricle
19 Apex of the heart

Figure A5-6 Heart of a 30-year-old woman (anterior aspect). (Reprinted with permission from Rohen JW, Yokochi C, Lütjen-Drecoll E. *Color Atlas of Anatomy*, 6th ed. Stuttgart, New York, NY: Schattauer, 2006.)

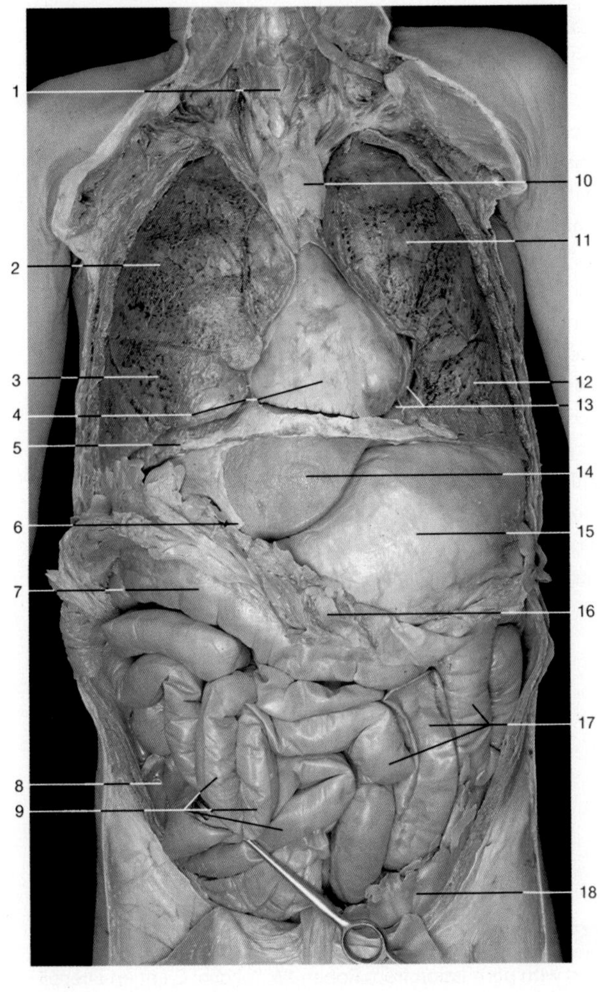

1 Thyroid gland
2 Upper lobe of right lung
3 Middle lobe of right lung
4 Heart
5 Diaphragm
6 Round ligament of liver (ligamentum teres)
7 Transverse colon
8 Cecum
9 Small intestine (ileum)
10 Thymus
11 Upper lobe of left lung
12 Lower lobe of left lung
13 Pericardium (cut edge)
14 Liver (left lobe)
15 Stomach
16 Greater omentum
17 Small intestine (jejunum)
18 Sigmoid colon

Figure A5-7 Abdominal organs in situ. The greater omentum has been partly removed or reflected. (Reprinted with permission from Rohen JW, Yokochi C, Lütjen-Drecoll E. *Color Atlas of Anatomy*, 6th ed. Stuttgart, New York, NY: Schattauer, 2006.)

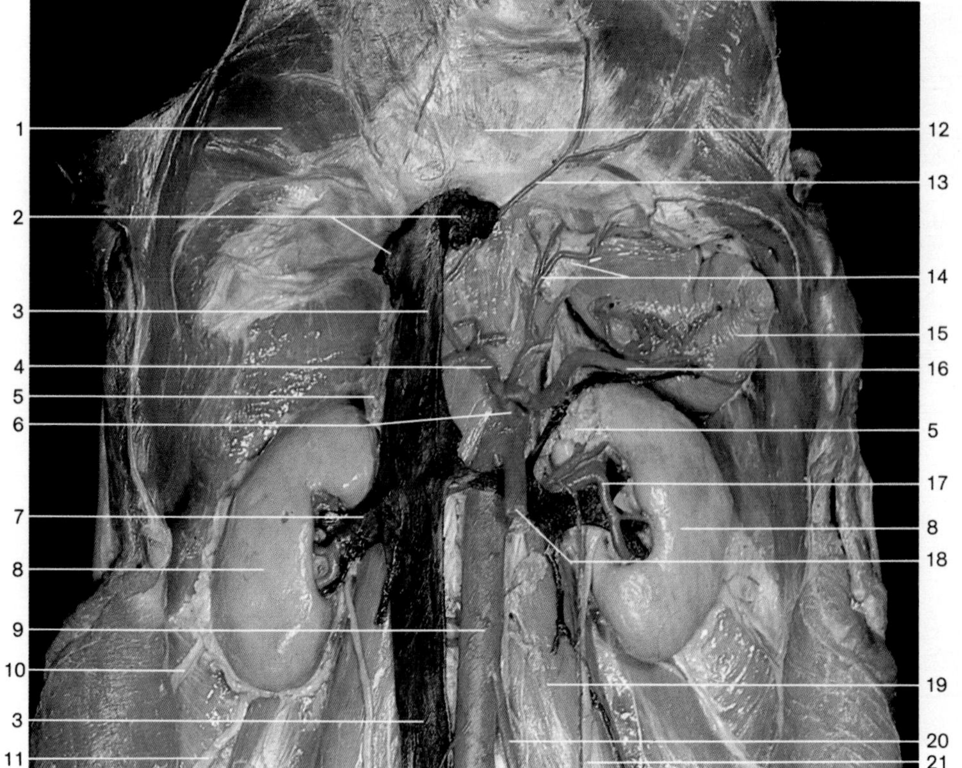

1 Diaphragm
2 Hepatic veins
3 Inferior vena cava
4 Common hepatic artery
5 Suprarenal gland
6 Celiac trunk
7 Right renal vein
8 Kidney
9 Abdominal aorta
10 Subcostal nerve
11 Iliohypogastric nerve
12 Central tendon of diaphragm
13 Inferior phrenic artery
14 Cardiac part of stomach
15 Spleen
16 Splenic artery
17 Superior renal artery
18 Superior mesenteric artery
19 Psoas major muscle
20 Inferior mesenteric artery
21 Ureter

Figure A5-8 Retroperitoneal organs, kidneys, and suprarenal glands in situ (anterior aspect). *Red*, arteries; *blue*, veins. (Reprinted with permission from Rohen JW, Yokochi C, Lütjen-Drecoll E. *Color Atlas of Anatomy*, 6th ed. Stuttgart, New York, NY: Schattauer, 2006.)

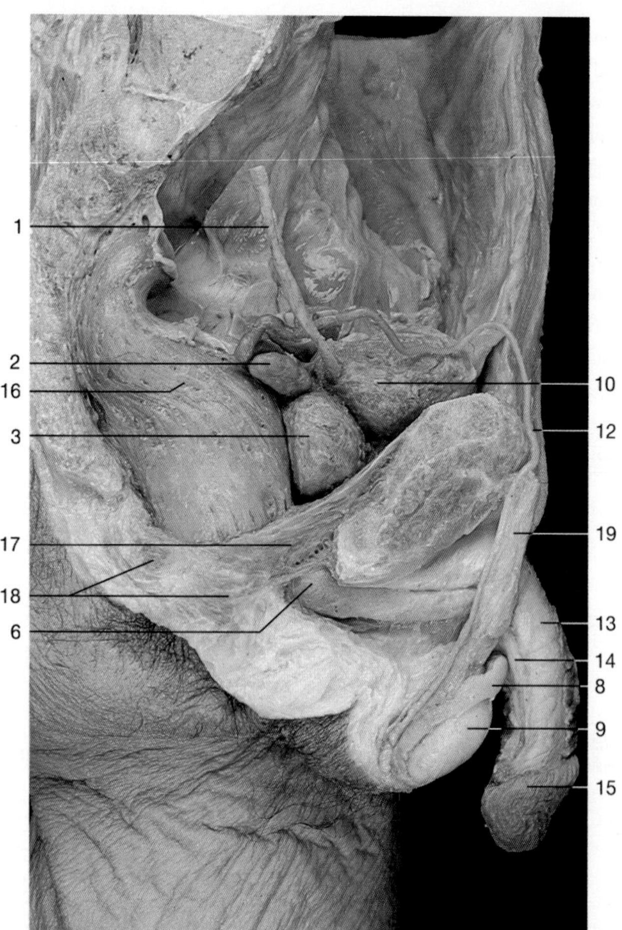

1 Ureter
2 Seminal vesicle
3 Prostate gland
6 Bulb of penis
8 Epididymis
9 Testis
10 Urinary bladder
12 Ductus deferens
13 Corpus cavernosum of penis
14 Corpus spongiosum of penis
15 Glans penis
16 Ampulla of rectum
17 Levator ani muscle
18 Anal canal and external anal sphincter muscle
19 Spermatic cord (cut)

Figure A5-9 Male genital organs in situ (right lateral aspect). (Reprinted with permission from Rohen JW, Yokochi C, Lütjen-Drecoll E. *Color Atlas of Anatomy*, 6th ed. Stuttgart, New York, NY: Schattauer, 2006.)

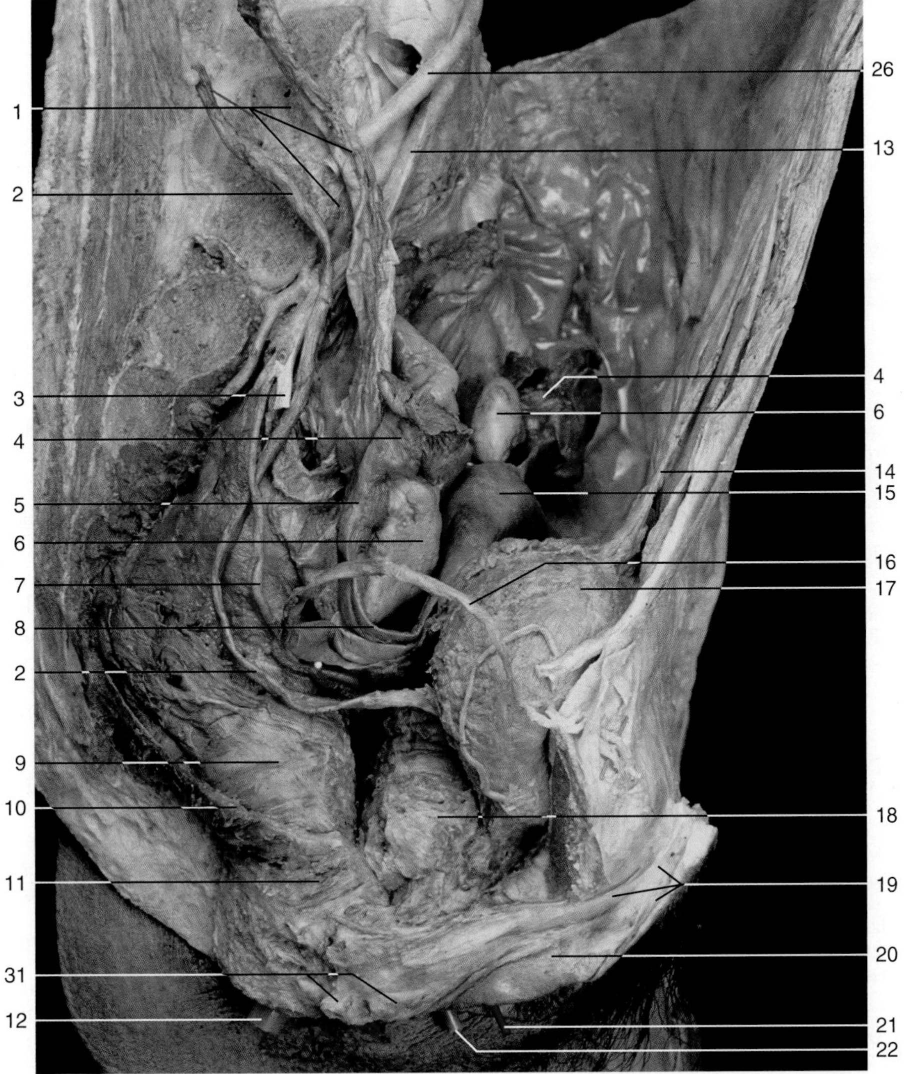

1 Body of fifth lumbar vertebra, suspensory ligament of ovary, and sacral promontory
2 Ureter
3 Medial umbilical ligament (remnant of umbilical artery) (cut)
4 Infundibulum of uterine tube
5 Ampulla of uterine tube
6 Ovary
7 Uterine artery
8 Uterine tube
9 Rectum
10 Levator ani muscle (pelvic diaphragm—cut edge)
11 External anal sphincter muscle
12 Anus (probe)
13 Internal iliac artery
14 Remnant of urachus (median umbilical ligament)
15 Uterus
16 Round ligament of uterus
17 Urinary bladder
18 Vagina
19 Clitoris
20 Labium minus
21 External orifice of urethra (red probe)
22 Vaginal orifice (green probe)
26 External iliac artery
31 Greater vestibular gland and bulb of the vestibule

Figure A5-10 Female internal genital organs in situ. Right half of the pelvis and sacrum have been removed. (Reprinted with permission from Rohen JW, Yokochi C, Lütjen-Drecoll E. *Color Atlas of Anatomy*, 6th ed. Stuttgart, New York, NY: Schattauer, 2006.)

Figure Credits

Chapter 1

1-9. Reprinted with permission from Erkonen WE, Smith WL. *Radiology 101: Basics and Fundamentals of Imaging*, 3rd ed. Philadelphia, PA: Lippincott Williams & Wilkins, 2010.

Chapter 3

3-1. A, Reprinted with permission from Cormack DH. *Essential Histology*, 2nd ed. Philadelphia, PA: Lippincott Williams & Wilkins, 2001; **B,** Reprinted with permission from Quinton P, Martinez R, eds. *Fluid and Electrolyte Transport in Exocrine Glands in Cystic Fibrosis*. San Francisco, CA: San Francisco Press, 1982; **C,** Reprinted with permission from Hafez ESE. *Scanning Electron Microscopic Atlas of Mammalian Reproduction*. Tokyo, Japan: Igaku Shoin, 1975.

3-15. Photomicrographs reprinted with permission from Cormack DH. *Essential Histology*, 2nd ed. Philadelphia, PA: Lippincott Williams & Wilkins, 2001.

Chapter 4

4-1. Reprinted with permission from Cormack DH. *Essential Histology*, 2nd ed. Philadelphia, PA: Lippincott Williams & Wilkins, 2001.

Figure 4-4. A, Reprinted with permission from McClatchey KD. *Clinical Laboratory Medicine*, 2nd ed. Baltimore, MD: Lippincott Williams & Wilkins, 2001; **B,** Reprinted with permission from Cormack DH. *Essential Histology*, 2nd ed. Philadelphia, PA: Lippincott Williams & Wilkins, 2001; **C,** Reprinted with permission from Mills SE. *Histology for Pathologists*, 3rd ed. Philadelphia, PA: Lippincott Williams & Wilkins, 2006.

4-5. Reprinted with permission from Cormack DH. *Essential Histology*, 2nd ed. Philadelphia, PA: Lippincott Williams & Wilkins, 2001; **A, B, and C,** Reprinted with permission from Mills SE. *Histology for Pathologists*, 3rd ed. Philadelphia, PA: Lippincott Williams & Wilkins, 2006; **D,** Reprinted with permission from Gartner LP, Hiatt JL. *Color Atlas of Histology*, 4th ed. Baltimore, MD: Lippincott Williams & Wilkins, 2005.

4-6. A and C, Reprinted with permission from Cormack DH. *Essential Histology*, 2nd ed. Philadelphia, PA: Lippincott Williams & Wilkins, 2001; **B,** Reprinted with permission from Gartner LP, Hiatt JL. *Color Atlas of Histology*, 4th ed. Baltimore, MD: Lippincott Williams & Wilkins, 2005.

4-7. Reprinted with permission from Cormack DH. *Essential Histology*, 2nd ed. Philadelphia, PA: Lippincott Williams & Wilkins, 2001.

4-9. A and B, Reprinted with permission from Rubin E, Farber JL. *Pathology*, 3rd ed. Philadelphia, PA: Lippincott Williams & Wilkins, 1999; **C,** Reprinted from Bullough PG, Vigorita VJ. *Atlas of Orthopaedic Pathology*. New York, NY: Gower Medical Publishing, 1984. With permission from Elsevier.

4-10. Reprinted with permission from Erkonen WE, Smith WL. *Radiology 101*, 3rd ed. Philadelphia, PA: Lippincott Williams & Wilkins, 2010.

4-11. Reprinted with permission from Okazaki H, Scheithauer BW. *Atlas of Neuropathology*. New York, NY: Gower Medical Publishing, 1988.

Chapter 5

5-3. Reprinted with permission from Smeltzer CS, Bare B. *Medical Surgical Nursing*, 10th ed. Philadelphia, PA: Lippincott Williams & Wilkins, 2004.

5-4. Reprinted with permission from Koneman EW, Alien SD, Janda WM, et al. *Color Atlas and Textbook of Diagnostic Microbiology*, 6th ed. Philadelphia, PA: Lippincott Williams & Wilkins, 2008.

5-5. Reprinted with permission from Koneman EW, Alien SD, Janda WM, et al. *Color Atlas and Textbook of Diagnostic Microbiology*, 6th ed. Philadelphia, PA: Lippincott Williams & Wilkins, 2008.

5-6. D, Reprinted with permission from Koneman EW, Alien SD, Janda WM, et al. *Color Atlas and Textbook of Diagnostic Microbiology*, 6th ed. Philadelphia, PA: Lippincott Williams & Wilkins, 2008.

5-7. Reprinted with permission from Koneman EW, Alien SD, Janda WM, et al. *Color Atlas and Textbook of Diagnostic Microbiology*, 6th ed. Philadelphia, PA: Lippincott Williams & Wilkins, 2008.

5-8. A, Image courtesy of the Centers for Disease Control (CDC); **C,** Reprinted with permission from Volk WA, et al. *Essentials of Medical Microbiology*, 5th ed. Philadelphia, PA: Lippincott-Raven, 1996.

5-10. A, Reprinted with permission from Porth C. *Pathophysiology*, 7th ed. Philadelphia, PA: Lippincott Williams & Wilkins, 2005; **B,** Courtesy of the CDC/Erskine L. Palmer, Ph.D. and M.L. Martin.

5-11. A, Reprinted with permission from *LWW's Organism Central [CD-ROM for Windows]*. Philadelphia, PA: Lippincott Williams & Wilkins, 2001; **B,** Reprinted with permission from Koneman EW, Alien SD, Janda WM, et al. *Color Atlas and Textbook of Diagnostic Microbiology*, 6th ed. Philadelphia, PA: Lippincott Williams & Wilkins, 2008; **C,** Reprinted with permission from Neville BW, et al. *Color Atlas of Oral Pathology*, 2nd ed. Philadelphia, PA: Lippincott Williams & Wilkins, 1998.

5-12. B, D, and F, Reprinted with permission from Koneman EW, Alien SD, Janda WM, et al. *Color Atlas and Textbook of Diagnostic Microbiology*, 6th ed. Philadelphia, PA: Lippincott Williams & Wilkins, 2006.

5-13. A and C, Reprinted with permission from Stedman's Medical Dictionary, 28th ed. Baltimore, MD: Lippincott Williams & Wilkins, 2006. B, Photo courtesy of the Centers for Disease Control and Dr. Melvin. **D,** Reprinted with permission from Rubin E, Farber JL. Pathology, 3rd ed. Philadelphia, PA: Lippincott Williams & Wilkins, 1999.

5-14. Reprinted with permission from Rubin E, Farber JL. *Pathology*, 3rd ed. Philadelphia, PA: Lippincott Williams & Wilkins, 1999.

5-15. Reprinted with permission from Koneman EW, Alien SD, Janda WM, et al. *Color Atlas and Textbook of Diagnostic Microbiology*, 5th ed. Philadelphia, PA: Lippincott Williams & Wilkins, 1997.

5-16. A and B, Courtesy of the Centers for Disease Control and Dr. Dennis D. Juranek. C. Reprinted with permission from Koneman EW, Alien SD, Janda WM, et al. *Color Atlas and Textbook of Diagnostic Microbiology*, 5th ed. Philadelphia, PA: Lippincott Williams & Wilkins, 2006.

5-17. A, Photograph courtesy of Dr. Janet Duben-Engelkirk and Scott & White Hospital, Temple, TX. 90621; **B,** Photo courtesy of Tech. Sgt. Dave Ahlschwede, U.S. Air Force; **C,** Courtesy of CDC/ Minnesota Department of Health, R.N. Barr Library; Librarians Melissa Rethlefsen and Marie Jones.

5-18. Reprinted with permission from Koneman EW, Alien SD, Janda WM, et al. *Color Atlas and Textbook of Diagnostic Microbiology*, 5th ed. Philadelphia, PA: Lippincott Williams & Wilkins, 2006.

5-19. Reprinted with permission from Koneman EW, Alien SD, Janda WM, et al. *Color Atlas and Textbook of Diagnostic Microbiology*, 5th ed. Philadelphia, PA: Lippincott Williams & Wilkins, 2006.

Chapter 6

6-2. Reprinted with permission from Cormack DH. *Essential Histology*, 2nd ed. Philadelphia, PA: Lippincott Williams & Wilkins, 2001.

6-3. Reprinted with permission from Cormack DH. *Essential Histology*, 2nd ed. Philadelphia, PA: Lippincott Williams & Wilkins, 2001.

6-4. A and B, Reprinted with permission from Cormack DH. *Essential Histology*, 2nd ed. Philadelphia, PA: Lippincott Williams & Wilkins, 2001.

6-5. A, Reprinted with permission from Bickley LS. *Bates' Guide to Physical Examination and History Taking*, 8th ed. Philadelphia, PA: Lippincott Williams & Wilkins, 2003.

6-6. A and B, Reprinted with permission from Bickley LS. *Bates' Guide to Physical Examination and History Taking*, 8th ed. Philadelphia, PA: Lippincott Williams & Wilkins, 2003; **C**, Reprinted with permission from Goodheart HP. *Goodheart's Photoguide of Common Skin Disorders*, 2nd ed. Philadelphia, PA: Lippincott Williams & Wilkins, 2003.

6-7. A through D, Photographs reprinted with permission from Bickley LS. *Bates' Guide to Physical Examination and History Taking*, 8th ed. Philadelphia, PA: Lippincott Williams & Wilkins, 2003; **E**, Berg D, Worzala K. *Atlas of Adult Physical Diagnosis*. Philadelphia, PA: Lippincott Williams & Wilkins, 2006; **F**, Goodheart HP. *Photoguide of Common Skin Disorders*, 2nd ed. Philadelphia, PA: Lippincott Williams & Wilkins, 2003.

6-8. A, Goodheart HP. *Goodheart's Photoguide of Common Skin Disorders*, 2nd ed. Philadelphia, PA: Lippincott Williams & Wilkins, 2003; **B and C**, Fleisher GR, Ludwig S, Baskin MN. *Atlas of Pediatric Emergency Medicine*. Philadelphia, PA: Lippincott Williams & Wilkins, 2004; **D**, Rubin E, Farber JL. *Pathology*, 3rd ed. Philadelphia, PA: Lippincott Williams & Wilkins, 1999.

6-10. A, Reprinted with permission from Goodheart HP. *Goodheart's Photoguide of Common Skin Disorders; Diagnosis and Management*, 2nd ed. Philadelphia, PA: Lippincott Williams & Wilkins, 2003; **B**, Reprinted with permission from Bickley LS. *Bates' Guide to Physical Examination and History Taking*, 8th ed. Philadelphia, PA: Lippincott Williams & Wilkins, 2003; **C**, Reprinted with permission from Rubin E, Farber JL. *Pathology*, 3rd ed. Philadelphia, PA: Lippincott Williams & Wilkins, 1999.

6-11. B, Reprinted with permission from Hall JC. *Sauer's Manual of Skin Diseases*, 8th ed. Philadelphia, PA: Lippincott Williams & Wilkins, 1999.

6-12. A, From Goodheart HP. *Goodheart's Photoguide of Common Skin Disorders*, 2nd ed. Philadelphia, PA: Lippincott Williams & Wilkins, 2003; **B**, Reprinted with permission from Bickley LS. *Bates' Guide to Physical Examination and History Taking*, 8th ed. Philadelphia, PA: Lippincott Williams & Wilkins, 2003.

6-13. A, Berg D, Worzala K. *Atlas of Adult Physical Diagnosis*. Philadelphia, PA: Lippincott Williams & Wilkins, 2006; **B and C**, From Goodheart HP. *Goodheart's Photoguide of Common Skin Disorders*, 2nd ed. Philadelphia, PA: Lippincott Williams & Wilkins, 2003; **D**, Bickley LS. *Bates' Guide to Physical Examination and History Taking*, 8th ed. Philadelphia, PA: Lippincott Williams & Wilkins, 2003.

6-14. A, McConnell TH. *The Nature of Disease Pathology for the Health Professions*. Philadelphia, PA: Lippincott Williams & Wilkins, 2007.

Chapter 7

7-1. Reprinted with permission from Erkonen WE, Smith WL. *Radiology 101*, 3rd ed. Philadelphia, PA: Lippincott Williams & Wilkins, 2010.

7-3. B, Reprinted with permission from Gartner LP, Hiatt JL. *Color Atlas of Histology*, 4th ed. Baltimore, MD: Lippincott Williams & Wilkins, 2005; **C**, Reprinted with permission from Rubin R, Strayer DS. *Rubin's Pathology: Clinicopathologic Foundations of Medicine*, 5th ed. Baltimore, MD: Lippincott Williams & Wilkins, 2008.

7-18. Reprinted with permission from Rubin E, Farber JL. *Pathology*, 5th ed. Philadelphia, PA: Lippincott Williams & Wilkins, 2008.

7-19. A and C, Reprinted with permission from Rubin E, Farber JL. *Pathology*, 3rd ed. Philadelphia, PA: Lippincott Williams & Wilkins, 2008; **B**, Becker KL, Bilezikian JP, Brenner WJ, et al. *Principles and Practice of Endocrinology and Metabolism*, 3rd ed. Philadelphia, PA: Lippincott Williams & Wilkins, 2001.

7-25. Reprinted with permission from Cohen BJ. *Medical Terminology*, 6th ed. Philadelphia, PA: Lippincott Williams & Wilkins, 2011.

7-27. Reprinted with permission from Cohen BJ. *Medical Terminology*, 6th ed. Philadelphia, PA: Lippincott Williams & Wilkins, 2011.

Chapter 8

8-1. B, Reprinted with permission from Gartner LP, Hiatt JL. *Color Atlas of Histology*, 3rd ed. Philadelphia, PA: Lippincott Williams & Wilkins, 2000.

8-2. A, Reprinted with permission from Cormack DH. *Essential Histology*, 2nd ed. Philadelphia, PA: Lippincott Williams & Wilkins, 2001; **E**, Courtesy of A. Sima.

8-3. A, Reprinted with permission from Mills SE. *Histology for Pathologists*, 3rd ed. Philadelphia, PA: Lippincott Williams & Wilkins, 2006.

8-17. A, Reprinted with permission from Bucci C. *Condition-specific Massage Therapy*. Philadelphia, PA: Lippincott Williams & Wilkins, 2012; **B**, Reprinted with permission from McConnell TH. *The Nature of Disease Pathology for the Health Professions*. Philadelphia, PA: Lippincott Williams & Wilkins, 2007.

Chapter 9

9-5. B, Reprinted with permission from Gartner LP, Hiatt JL. *Color Atlas of Histology*, 3rd ed. Philadelphia, PA: Lippincott Williams & Wilkins, 2000.

9-6. Reprinted with permission from McConnell T, Hull K. *Human Form, Human Function*. Philadelphia, PA: Lippincott Williams & Wilkins, 2011.

9-9. Reprinted with permission from Bear MF, et al. *Neuroscience, Exploring the Brain*, 3rd ed. Philadelphia, PA: Lippincott Williams & Wilkins, 2007.

9-12. B, Reprinted with permission from Mills SE. *Histology for Pathologists*, 3rd ed. Philadelphia, PA: Lippincott Williams & Wilkins, 2006.

9-15. A, From Taylor C, Lillis CA, LeMone P . *Fundamentals of Nursing*, 2nd ed. Philadelphia, PA: JB Lippincott, 1993, with permission; **B**, Photo from Dr. Vallejo showing epidural placement.

9-16. Reprinted with permission from *Stedman's Medical Dictionary*, 28th ed. Baltimore, MD: Lippincott Williams & Wilkins, 2006.

Chapter 10

10-9. A and B, Reprinted with permission from Erkonen WE. *Radiology 101*. Philadelphia, PA: Lippincott Williams & Wilkins, 1998; **C**, Courtesy of Newport Diagnostic Center, Newport Beach, CA.

10-11. Reprinted with permission from Rubin E, Farber JL. *Pathology*, 3rd ed. Philadelphia, PA: Lippincott Williams & Wilkins, 1999.

10-12. Reprinted with permission from Erkonen WE. *Radiology 101*. Philadelphia, PA: Lippincott Williams & Wilkins, 1998.

10-13. Reprinted with permission from Cohen BJ. *Medical Terminology*, 6th ed. Philadelphia, PA: Lippincott Williams & Wilkins, 2011.

10-14. B, Images courtesy of the Alzheimer's Disease Education and Referral Center, a service of the National Institute on Aging.

Chapter 11

11-1. Reprinted with permission from McConnell T, Hull K. *Human Form, Human Function*, Philadelphia, PA: Lippincott Williams & Wilkins, 2011.

11-2. Reprinted with permission from McConnell T, Hull K. *Human Form, Human Function*, Philadelphia, PA: Lippincott Williams & Wilkins, 2011.

11-6. Reprinted with permission from McConnell T, Hull K. *Human Form, Human Function*, Philadelphia, PA: Lippincott Williams & Wilkins, 2011.

11-7. Reprinted with permission from McConnell T, Hull K. *Human Form, Human Function*, Philadelphia, PA: Lippincott Williams & Wilkins, 2011.

11-8. A, Reprinted with permission from McConnell T, Hull K. *Human Form, Human Function*, Philadelphia, PA: Lippincott Williams & Wilkins, 2011.

11-10. Reprinted with permission from National Eye Institute.

11-13. Reprinted with permission from McConnell T, Hull K. *Human Form, Human Function*, Philadelphia, PA: Lippincott Williams & Wilkins, 2011.

11-14. Reprinted with permission from McConnell T, Hull K. *Human Form, Human Function*, Philadelphia, PA: Lippincott Williams & Wilkins, 2011.

11-17. Reprinted with permission from McConnell T, Hull K. *Human Form, Human Function*, Philadelphia, PA: Lippincott Williams & Wilkins, 2011.

11-18. Reprinted with permission from McConnell T, Hull K. *Human Form, Human Function*, 1st ed. Philadelphia, PA: Lippincott Williams & Wilkins, 2011.

Unnumbered Figure p. 262. Reprinted with permission from McConnell T, Hull K. *Human Form, Human Function*, Philadelphia, PA: Lippincott Williams & Wilkins, 2011.

Chapter 12

12-4. Reprinted with permission from McConnell T, Hull K. *Human Form, Human Function*, 1st ed. Philadelphia, PA: Lippincott Williams & Wilkins, 2010.

12-5. A, Reprinted with permission from Guyton A. *Medical Physiology*, 6th ed. Philadelphia, PA: WB Saunders, 1981. **B**, Sandoz Pharmaceutical Corporation.

12-6. Reprinted with permission from Cohen BJ. *Medical Terminology*, 6th ed. Philadelphia, PA: Lippincott Williams & Wilkins, 2011.

12-7. Reprinted with permission from Rubin R, Strayer DS. Rubin's Pathology, 6th ed. Philadelphia, PA: Lippincott Williams & Wilkins, 2012.

12-8. A and B, Reprinted with permission from McConnell TH. *The Nature of Disease*. Philadelphia, PA: Lippincott Williams & Wilkins, 2007; **C**, Courtesy of Dana Morse Bittus and B. J. Cohen.

Chapter 13

13-4. Reprinted with permission from Gartner LP, Hiatt JL. *Color Atlas of Histology*, 5th ed. Philadelphia, PA: Lippincott Williams & Wilkins, 2009.

13-5. B, Reprinted with permission from McConnell T, Hull K. *Human Form, Human Function*, Philadelphia, PA: Lippincott Williams & Wilkins, 2010; **C**, SEM courtesy of Erlandsen S, Engelkirk P, Kengelkirk PG, et al. *Burton's Microbiology for the Health Sciences*, 8th ed. Philadelphia, PA: Lippincott Williams & Wilkins, 2007.

13-6. B, Reprinted with permission from Gartner LP, Hiatt JL. *Color Atlas of Histology*, 5th ed. Philadelphia, PA: Lippincott Williams & Wilkins, 2009.

13-7. C, Reprinted with permission from McConnell TH, Hull KH. *Human Form Human Function*. Philadelphia, PA: Lippincott Williams & Wilkins, 2011.

13-10. Reprinted with permission from McKenzie SB. *Textbook of Hematology*, 2nd ed. Baltimore, MD: Williams & Wilkins, 1996.

13-11. A, Reprinted with permission from McClatchey KD. *Clinical Laboratory Medicine*, 2nd ed. Philadelphia, PA: Lippincott Williams & Wilkins, 2002; **B**, Reprinted with permission from Rubin R, Strayer DS. *Rubin's Pathology: Clinicopathologic Foundations of Medicine*, 5th ed. Philadelphia, PA: Lippincott Williams & Wilkins, 2008.

13-12. Reprinted with permission from McConnell T, Hull K. *Human Form, Human Function*, Philadelphia, PA: Lippincott Williams & Wilkins, 2011.

Box 13-2 Figure. Reprinted with permission from Cormack DH. *Essential Histology*, 2nd ed. Philadelphia, PA: Lippincott Williams & Wilkins, 2001.

Chapter 14

14-3. Reprinted with permission from Gartner LP, Hiatt JL. *Color Atlas of Histology*, 5th ed. Philadelphia, PA: Lippincott Williams & Wilkins, 2009; modified with permission from Moore KL, Dalley AF. *Clinically Oriented Anatomy*, 7th ed. Baltimore, MD: Lippincott Williams & Wilkins, 2013.

14-11. Reprinted with permission from Porth CM. *Pathophysiology*, 7th ed. Philadelphia, PA: Lippincott Williams & Wilkins, 2004.

14-13. Reprinted with permission from Cohen BJ. *Medical Terminology*, 6th ed. Philadelphia, PA: Lippincott Williams & Wilkins, 2011.

14-15. Reprinted with permission from Baim DS. *Grossman's Cardiac Catheterization, Angiography, and Intervention*, 7th ed. Philadelphia, PA: Lippincott Williams & Wilkins, 2006.

14-16. Reprinted with permission from Cohen BJ. *Medical Terminology*, 6th ed. Philadelphia, PA: Lippincott Williams & Wilkins, 2011.

14-17. Reprinted with permission from Cohen BJ. *Medical Terminology*, 6th ed. Philadelphia, PA: Lippincott Williams & Wilkins, 2011.

14-18. Reprinted with permission from Cohen BJ. *Medical Terminology*, 6th ed. Philadelphia, PA: Lippincott Williams & Wilkins, 2011.

14-19. Reprinted with permission from Cohen BJ. *Medical Terminology*, 6th ed. Philadelphia, PA: Lippincott Williams & Wilkins, 2011.

Chapter 15

15-3. Reprinted with permission from Cormack DH. *Essential Histology*, 2nd ed. Philadelphia, PA: Lippincott Williams & Wilkins, 2001.

15-5. Reprinted with permission from McConnell T, Hull K. *Human Form, Human Function*, Philadelphia, PA: Lippincott Williams & Wilkins, 2011.

15-6. Reprinted with permission from McConnell T, Hull K. *Human Form, Human Function*, Philadelphia, PA: Lippincott Williams & Wilkins, 2011.

15-7. Reprinted with permission from McConnell T, Hull K. *Human Form, Human Function*, Philadelphia, PA: Lippincott Williams & Wilkins, 2011.

15-8. Reprinted with permission from McConnell T, Hull K. *Human Form, Human Function*, Philadelphia, PA: Lippincott Williams & Wilkins, 2011.

15-13. Reprinted with permission from McConnell T, Hull K. *Human Form, Human Function*, Philadelphia, PA: Lippincott Williams & Wilkins, 2011.

15-14. A, Reprinted with permission from Bickley LS. *Bates' Guide to Physical Examination and History Taking*, 10th ed. Philadelphia, PA: Lippincott Williams & Wilkins, 2008; **B**, Reprinted with permission from Taylor C, Lillis C, LeMone P. *Fundamentals of Nursing*, 5th ed. Philadelphia, PA: Lippincott Williams & Wilkins, 2004.

15-15. Reprinted with permission from *Atlas of Pathophysiology*, 3rd ed. Philadelphia, PA: Lippincott Williams & Wilkins, 2009.

15-16. Reprinted with permission from Rubin E, Farber JL. *Pathology*, 4th ed. Philadelphia, PA: Lippincott Williams & Wilkins, 2005.

15-17. Reprinted with permission from Bickley LS. *Bates' Guide to Physical Examination and History Taking*, 10th ed. Philadelphia, PA: Lippincott Williams & Wilkins, 2008.

Chapter 16

16-4. B, Reprinted with permission from Cormack DH. *Essential Histology*, 2nd ed. Philadelphia, PA: Lippincott Williams & Wilkins, 2001.

16-5. Reprinted with permission from McConnell T, Hull K. *Human Form, Human Function*, Philadelphia, PA: Lippincott Williams & Wilkins, 2011.

16-6. A, Reprinted with permission from Archer P, Nelson LA. *Applied Anatomy and Physiology for Manual Therapists*. Philadelphia, PA: Lippincott Williams & Wilkins; **B**, Reprinted with permission from Bickley LS. *Bates' Guide to Physical Examination and History Taking*, 8th ed. Philadelphia, PA: Lippincott Williams & Wilkins, 2003.

16-8. A, Photo courtesy of Paul S. Matz, MD in Chung EK, et al. *Visual Diagnosis and Treatment in Pediatrics*, 2nd ed. Philadelphia, PA: Lippincott Williams & Wilkins, 2010; **B**, Reprinted with permission from Rajagopalan S, Mukherjee D, Mohler ER. *Manual of Vascular Diseases*. Philadelphia, PA: Wolters Kluwer Health, 2012.

16-9. Reprinted with permission from Rubin E, Farber JL. *Pathology*, 5th ed. Philadelphia, PA: Lippincott Williams & Wilkins, 2007.

Chapter 17

17-1. Reprinted with permission from McConnell T, Hull K. *Human Form, Human Function*, Philadelphia, PA: Lippincott Williams & Wilkins, 2011.

17-2. Reprinted with permission from McConnell T, Hull K. *Human Form, Human Function*, Philadelphia, PA: Lippincott Williams & Wilkins, 2011.

17-3. Reprinted with permission from Braun C, Anderson CM. *Pathophysiology*, 2nd ed. Philadelphia, PA: Lippincott Williams & Wilkins, 2012.

17-5. Reprinted with permission from McConnell T, Hull K. *Human Form, Human Function*, Philadelphia, PA: Lippincott Williams & Wilkins, 2011.

17-8. Photo courtesy of the CDC/Dr. Bill Atkinson.

17-9. Reprinted with permission from American Chart Company.

17-10. Reprinted with permission from Doan T, Melvold R, Viselli S, et al. *Immunology*, 2nd ed. Philadelphia, PA: Lippincott Williams & Wilkins, 2012.

Chapter 18

18-5. A and B, Reprinted with permission from McConnell TH, Hull KL. *Human Form Human Function*. Philadelphia, PA: Lippincott Williams & Wilkins, 2010; **C**. Reprinted with permission from Cui D, et al. *Atlas of Histology with Functional and Clinical Correlations*. Philadelphia, PA: Lippincott Williams & Wilkins, 2010; **D**, Reprinted with permission from Snell RS. *Clinical Anatomy*, 7th ed. Philadelphia, PA: Lippincott Williams & Wilkins, 2003.

18-12. Reprinted with permission from Moore KL, Dalley AF, Agur AMR. *Clinically Oriented Anatomy*, 7th ed. Philadelphia, PA: Lippincott Williams & Wilkins, 2013.

18-13. Reprinted with permission from *Lippincott's Visual Nursing*, 2nd ed. Philadelphia, PA: Lippincott Williams & Wilkins, 2011.

18-14. Reprinted with permission from *Anatomical Chart Company. Diseases and Disorders*. Philadelphia, PA: Lippincott Williams & Wilkins, 2008.

18-15. Reprinted with permission from Eisenberg RL. *Clinical Imaging*, 4th ed. Philadelphia, PA: Lippincott Williams & Wilkins, 2002.

18-16. Reprinted with permission from Cohen BJ. *Medical Terminology*, 6th ed. Philadelphia, PA: Lippincott Williams & Wilkins, 2011.

18-17. Reprinted with permission from Cohen BJ. *Medical Terminology*, 6th ed. Philadelphia, PA: Lippincott Williams & Wilkins, 2011.

18-18. Reprinted with permission from Cohen BJ. *Medical Terminology*, 6th ed. Philadelphia, PA: Lippincott Williams & Wilkins, 2011.

18-19. A, Reprinted with permission from Cohen BJ. *Medical Terminology*, 6th ed. Philadelphia, PA: Lippincott Williams & Wilkins, 2011; **B**, Reprinted courtesy of E.D. Andersen.

Chapter 19

19-1. Reprinted with permission from McConnell T, Hull K. *Human Form, Human Function*, Philadelphia, PA: Lippincott Williams & Wilkins, 2011.

19-2. Reprinted with permission from McConnell T, Hull K. *Human Form, Human Function*, Philadelphia, PA: Lippincott Williams & Wilkins, 2011.

19-3. Reprinted with permission from McConnell T, Hull K. *Human Form, Human Function*, Philadelphia, PA: Lippincott Williams & Wilkins, 2011.

19-4. Reprinted with permission from McConnell T, Hull K. *Human Form, Human Function*, Philadelphia, PA: Lippincott Williams & Wilkins, 2011.

19-7. D, Reprinted with permission from McConnell T, Hull K. *Human Form, Human Function*, Philadelphia, PA: Lippincott Williams & Wilkins, 2011.

19-11. Reprinted with permission from McConnell T, Hull K. *Human Form, Human Function*, Philadelphia, PA: Lippincott Williams & Wilkins, 2011.

19-12. Reprinted with permission from McConnell T, Hull K. *Human Form, Human Function*, Philadelphia, PA: Lippincott Williams & Wilkins, 2011.

19-13. Reprinted with permission from Anatomical Chart Company. *Atlas of Pathophysiology*, 3rd ed. Philadelphia, PA: Lippincott Williams & Wilkins, 2009.

19-14. Modified with permission from Cohen BJ. *Medical Terminology*, 6th ed. Philadelphia, PA: Lippincott Williams & Wilkins, 2011.

19-15. **A**, Image courtesy of Dr. E. L. Lansdown, Professor of Medical Imaging, University of Toronto, Toronto, ON, Canada; **B**, Reprinted with permission from Cohen BJ. *Medical Terminology*, 6th ed. Philadelphia, PA: Lippincott Williams & Wilkins, 2011; **C**, Credit: Mulholland MW, Maier RV, et al. *Greenfields Surgery Scientific Principles and Practice*, 4th ed. Philadelphia, PA: Lippincott Williams & Wilkins, 2006.

19-16. Reprinted with permission from McConnell TH, Hull KL. *Human Form, Human Function*. Philadelphia, PA: Lippincott Williams & Wilkins, 2011.

Chapter 20

20-4. MyPlate, U.S. Department of Agriculture/Center for Nutrition Policy and Promotion, www.choosemyplate.gov.

20-7. Reprinted with permission from Taylor C, et al. *Fundamentals of Nursing*, 5th ed. Philadelphia, PA: Lippincott Williams & Wilkins, 2005.

20-8. Reprinted with permission from McConnell T, Hull K. *Human Form, Human Function*, Philadelphia, PA: Lippincott Williams & Wilkins, 2011.

Chapter 21

21-2. Reprinted with permission from Taylor CR, Lillis C. *Fundamentals of Nursing*, 6th ed. Philadelphia, PA: Lippincott Williams & Wilkins, 2008.

21-4. Reprinted with permission from McConnell T, Hull K. *Human Form, Human Function*, Philadelphia, PA: Lippincott Williams & Wilkins, 2011.

21-6. Reprinted with permission from Bickley LS. *Bates' Guide to Physical Examination and History Taking*, 10th ed. Philadelphia, PA: Lippincott Williams & Wilkins, 2008.

21-7. Reprinted with permission from *Lippincott Manual of Nursing Practice*, 7th ed. Philadelphia, PA: Lippincott Williams & Wilkins, 2001.

Chapter 22

22-1. Reprinted with permission from McConnell T, Hull K. *Human Form, Human Function*, Philadelphia, PA: Lippincott Williams & Wilkins, 2011.

22-2. Modified with permission from McConnell T, Hull K. *Human Form, Human Function*, Philadelphia, PA: Lippincott Williams & Wilkins, 2011.

22-3. Reprinted with permission from McConnell T, Hull K. *Human Form, Human Function*, Philadelphia, PA: Lippincott Williams & Wilkins, 2011.

22-4. Courtesy of Dana Morse Bittus and BJ Cohen.

22-6. Modified with permission from McConnell T, Hull K. *Human Form, Human Function*, Philadelphia, PA: Lippincott Williams & Wilkins, 2011.

22-9. Modified with permission from McConnell T, Hull K. *Human Form, Human Function*, Philadelphia, PA: Lippincott Williams & Wilkins, 2011.

22-10. **A**, Reprinted with permission from Anatomical Chart Company. *Atlas of Pathophysiology*, 2nd ed. Philadelphia, PA: Lippincott Williams & Wilkins, 2006; **B**, Reprinted with permission from McConnell T. *The Nature of Disease*. Baltimore, MD: Lippincott Williams & Wilkins, 2007; **C**, Reprinted with permission from Nath JL. *Using Medical Terminology*. Philadelphia, PA: Lippincott Williams & Wilkins, 2006.

22-11. **A**, Reprinted with permission from Cohen BJ. *Medical Terminology*, 6th ed. Philadelphia, PA: Lippincott Williams & Wilkins, 2011; **B**, Reprinted with permission from Schaaf CP, Zachocke J, Potocki L. *Human Genetics*. Philadelphia, PA: Lippincott Williams & Wilkins, 2011.

22-12. **A**, Reprinted with permission from Cohen BJ. *Medical Terminology*, 6th ed. Philadelphia, PA: Lippincott Williams & Wilkins, 2011; **B**, Reprinted with permission from Dunnick R, Sandler C, Newhouse J. *Textbook of Uroradiology*, 5th ed. Philadelphia, PA: Lippincott Williams & Wilkins, 2012.

22-13. Reprinted with permission from Cohen BJ. *Medical Terminology*, 6th ed. Philadelphia, PA: Lippincott Williams & Wilkins, 2011.

22-14. **A and B**, Reprinted with permission from Premkumar K. *The Massage Connection*, 2nd ed. Baltimore, MD: Lippincott Williams & Wilkins, 2004; **C**, Reprinted with permission from Porth C. *Essentials of Pathophysiology*, 2nd ed. Philadelphia, PA: Lippincott Williams & Wilkins, 2006.

Chapter 23

23-1. Modified with permission from McConnell T, Hull K. *Human Form, Human Function*, Philadelphia, PA: Lippincott Williams & Wilkins, 2011.

23-3. **A and B**, Reprinted with permission from McConnell T, Hull K. *Human Form, Human Function*, Philadelphia, PA: Lippincott Williams & Wilkins, 2011; **C**, Reprinted with permission from Eroschenko VP. *Di Fiore's Atlas of Histology with Functional Correlations*, 8th ed. Philadelphia, PA: Lippincott Williams & Wilkins, 1995.

23-5. Modified with permission from McConnell T, Hull K. *Human Form, Human Function*, Philadelphia, PA: Lippincott Williams & Wilkins, 2011.

23-6. Reprinted with permission from *LifeART Pediatrics 1 [CD-ROM]*. Baltimore, MD: Lippincott Williams & Wilkins, 2000.

23-7. **A and B**, Adapted from Moore KL, Agur AM. *Essentials of Clinical Anatomy*, 2nd ed. Philadelphia, PA: Lippincott Williams & Wilkins, 2002; **C**, Reprinted with permission from Kyle T, Carman S. *Essentials of Pediatric Nursing*, 2nd ed. Philadelphia, PA: Lippincott Williams & Wilkins, 2012.

23-8. Modified with permission from McConnell T, Hull K. *Human Form, Human Function*, Philadelphia, PA: Lippincott Williams & Wilkins, 2011.

23-9. Reprinted with permission from McConnell T, Hull K. *Human Form, Human Function*, Philadelphia, PA: Lippincott Williams & Wilkins, 2011.

23-10. **A**, Reprinted with permission from McConnell T, Hull K. *Human Form, Human Function*, Philadelphia, PA: Lippincott Williams & Wilkins, 2011; **C**, Courtesy of Dana Morse Bittus and Barbara Cohen.

23-11. Modified with permission from McConnell T, Hull K. *Human Form, Human Function*, Philadelphia, PA: Lippincott Williams & Wilkins, 2011.

23-12. Reprinted with permission from Ricci SR. *Essentials of Maternity, Newborn, and Women's Health Nursing*, 2nd ed. Philadelphia, PA: Lippincott Williams & Wilkins, 2008.

23-13. Modified with permission from McConnell TH. *The Nature of Disease*, 2nd ed. Philadelphia, PA: Lippincott Williams & Wilkins, 2013.

23-14. **A**, Reprinted with permission from Timby BK, Smith NE. *Introductory Medical-Surgical Nursing*, 10th ed. Philadelphia, PA: Lippincott Williams & Wilkins, 2009; **B and C**, Reprinted with permission from McConnell TH. *The Nature of Disease*, 2nd ed. Philadelphia, PA: Lippincott Williams & Wilkins, 2013.

Chapter 24

24-1. Reprinted with permission from McConnell T, Hull K. *Human Form, Human Function*, Philadelphia, PA: Lippincott Williams & Wilkins, 2011.

24-2. Modified with permission from McConnell T, Hull K. *Human Form, Human Function*, Philadelphia, PA: Lippincott Williams & Wilkins, 2011.

24-4. Reprinted with permission from Pillitteri A. *Maternal and Child Health Nursing*, 6th ed. Philadelphia, PA: Lippincott Williams & Wilkins, 2009.

24-6. **A**, Reprinted with permission from Pillitteri A. *Maternal & Child Health Nursing*, 6th ed. Philadelphia, PA: Lippincott Williams & Wilkins, 2009; **B**, Reprinted with permission from Bushberg JT, Seibert JA, Leidholdt EM, et al. *Essential Physics of Medical Imaging*, 3rd ed. Philadelphia, PA: Lippincott Williams & Wilkins, 2011.

24-8. Reprinted with permission from Anatomical Chart Company. *OB/GYN Disorders*. Philadelphia, PA: Lippincott Williams & Wilkins, 2008.

24-9. Reprinted with permission from Ricci SS. *Essentials of Maternity, Newborn, and Women's Health Nursing*, 3rd ed. Philadelphia, PA: Lippincott Williams & Wilkins, 2012.

24-11. Reprinted with permission from Cohen BJ. *Medical Terminology*, 6th ed. Philadelphia, PA: Lippincott Williams & Wilkins, 2011.

24-12. Reprinted with permission from Anatomical Chart Company. *OB/GYN Disorders*. Philadelphia, PA: Lippincott Williams & Wilkins, 2008.

24-13. Reprinted with permission from Harris JR, et al. *Diseases of the Breast*, 4th ed. Baltimore, MD: Lippincott Williams & Wilkins, 2009.

Chapter 25

25-1. Modified with permission from McConnell TH. *The Nature of Disease*, 2nd ed. Philadelphia, PA: Lippincott Williams & Wilkins, 2013.

25-3. Modified with permission from McConnell T, Hull K. *Human Form, Human Function*, Philadelphia, PA: Lippincott Williams & Wilkins, 2011.

25-5. Modified with permission from McConnell T, Hull K. *Human Form, Human Function*, Philadelphia, PA: Lippincott Williams & Wilkins, 2011.

25-7. **A**, Reprinted with permission from Rosdahl CB, Kowalski MT. *Textbook of Basic Nursing*, 10th ed. Philadelphia, PA: Lippincott Williams & Wilkins, 2011; **B**, Reprinted with permission from Waters PM, Bae DS. *Pediatric Hand and Upper Limb Surgery*. Philadelphia, PA: Lippincott Williams & Wilkins, 2012.

25-8. Reprinted with permission from Goldman B, Schnell J, Spencer P. *Pathophysiology Concepts of Altered Health States*, 7th ed. Philadelphia, PA: Current Medicine Inc., 2008.

25-9. **A and B**, Reprinted with permission from McConnell TH. *The Nature of Disease*, 2nd ed. Philadelphia, PA: Lippincott Williams & Wilkins, 2013; **C**, Reprinted with permission from Pillitteri A. *Maternal and Child Health Nursing*, 5th ed. Philadelphia, PA: Lippincott Williams & Wilkins, 2008.

25-10. **A**, Reprinted with permission from Pillitteri A. *Maternal and Child Health Nursing*, 5th ed. Philadelphia, PA: Lippincott Williams & Wilkins, 2008; **B**, Reprinted with permission from Kyle T, Carman S. *Essentials of Pediatric Nursing*. Philadelphia, PA: Lippincott Williams & Wilkins; **C**, Reprinted with permission from Nettina SM. *Lippincott Manual of Nursing Practice*, 10th ed. Philadelphia, PA: Lippincott Williams & Wilkins, 2013.

25-12. Reprinted with permission from McConnell TH. *The Nature of Disease*, 2nd ed. Philadelphia, PA: Lippincott Williams & Wilkins, 2013.

25-13. Reprinted with permission from McConnell TH. *The Nature of Disease*, 2nd ed. Philadelphia, PA: Lippincott Williams & Wilkins, 2013.

Index of Boxes

Health Maintenance

Clinical Perspectives

A Closer Look

Hot Topics

Index

Note: Page numbers in italics denote figures; page numbers followed by *t* denote tables; and page numbers followed by *B* denote boxes.

A

B

D

E

I

J

K

L

M

N

O

P

Q

R

S